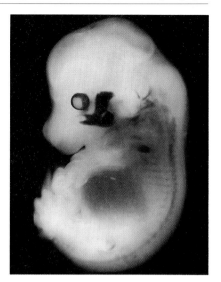

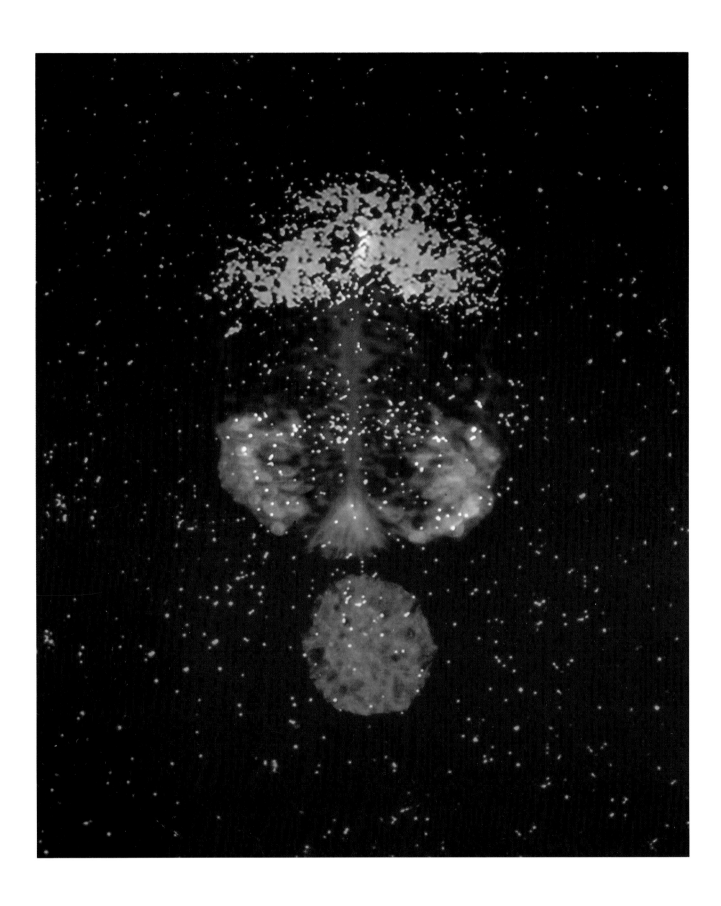

SCOTT F. GILBERT *Swarthmore College*

Developmental FOURTH
Biology EDITION

Sinauer Associates, Inc.
Publishers
Sunderland, Massachusetts

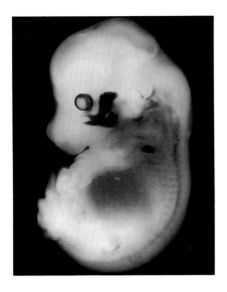

THE COVER

The Myf-5 protein is expressed in muscle cell precursors and plays a role in their becoming skeletal muscle cells. The genetic elements regulating the temporal and spatial expression of the *Myf-5* gene can be discerned by fusing a β-galactosidase gene to the sequences surrounding the *Myf-5* locus. This newly constructed gene is inserted into a pronucleus of a newly fertilized mouse egg, and the β-galactosidase protein can be stained when it appears in the embryo. Here a particular sequence upstream from the *Myf-5* gene causes the expression of the gene (black stain) in the neck muscles, pharyngeal arches, ocular muscles, forelimb muscles, and segmented myotomes of a 13.5-day mouse embryo. The gene family to which *Myf-5* belongs is discussed in Chapters 9 and 10. (Photograph courtesy of A. Patapoutian, G. Lyons, J. Miner, and B. Wold.)

THE TITLE PAGES

Left page: Diffusible signals from the notochord (green tube at bottom) induce the formation of the floor plate at the ventral side of the neural tube (green). The floor plate cells induce the formation of the two regions of motor neurons (gold) on the ventrolateral sides. The notochord also restricts the expression of dorsalin protein (needed for neural crest cell development) to the most dorsal region of the neural tube (blue). See Chapter 16. (Photograph courtesy of T. M. Jessell.)

Right page: Tadpoles of the reticulated poison-dart frog are carried on their parents' backs to small pools of water at the base of bromeliad leaves in the rain forest canopy. The female of this Peruvian Amazon species will then supply unfertilized eggs as food for the developing tadpoles. See Chapter 20. (Photograph by M. Fogden/DRK Photo.)

Developmental Biology,
Fourth Edition

Copyright © 1994 by Sinauer Associates, Inc.
All rights reserved.
This book may not be reproduced in whole or in part without permission from the publisher. For information or to order address:
Sinauer Associates Inc., Sunderland, Massachusetts 01375-0407 U.S.A.
Fax: 413-549-1118
E-mail: publish@sinauer.com

**Library of Congress
Cataloging-in-Publication Data**

Gilbert, Scott F.
    Developmental biology / Scott F. Gilbert. —
4th ed.
        p.   cm.
    Includes bibliographical references and index.
    ISBN 0-87893-249-6
    1. Embryology   I. Title.
QL955.G48   1994   93-42036
591.3—dc20   CIP

Printed in U.S.A.
6 5 4 3

Contents

PART FOUR
SPECIFICATION OF CELL FATE AND THE EMBRYONIC AXES

Preface

In 1894, three publications changed the rationale for studying development and charted new paths for embryological investigations. In that year, Wilhelm Roux founded the journal *Archiv fur Entwicklungsmechanik der Organismen* (*Archives for the Developmental Mechanics of Organisms*) and announced a new program for embryological research that would be based on experimentation. Roux exhorted young biologists to intervene experimentally in the developmental process in order to find the laws that underlie the differentiation and organization of bodily cells, tissues, and organs. Also in 1894, Hans Driesch published his *Analytische Theorie der Organischen Entwicklung* (*Analytical Theory of the Development of Organisms*) wherein, among other things, he proposed a relationship between chromosomes and cytoplasm in which the chromosomes directed the formation of new materials in the cytoplasm and the cytoplasm reciprocated by directing the chromosomes to synthesize different materials than they had before. Also that year, Oscar Hertwig's *Präformation oder Epigenese* (*Preformation and Epigenesis*) directed the attention of embryologists to the interactions between cells that cause each cell to become defined in relationship to its neighbors within the embryo. These three visionary treatises shifted the emphasis of embryology from comparative anatomy to experimentation.

A hundred years later, developmental biology is in the middle of another major revolution. The ability to introduce new genes into embryos, to activate specific genes in particular cells, and to suppress specific gene expression in cells has given us insights unimaginable even a decade ago. The structures that once were useful for presenting animal development no longer do it justice. The ability to discuss developmental anatomy as something separate from developmental genetics is gone. Similarly, the usefulness of separating vertebrate from invertebrate development has vanished with the uncovering of developmental pathways that are fundamentally the same in nematodes, flies, and humans.

This revolution of content has caused a dramatic restructuring of this book. First, there are now three introductory chapters that bring material on genetic analysis and cell membranes to the front of the book. This enables us to discuss the cellular and molecular bases of the anatomical changes seen in subsequent chapters. Second, the chapters on developmental genetics have been moved to precede the chapters that discuss the specification of the body axes and the determination of the first cell types. This was done because our understanding of the genetic bases for cell specification has expanded enormously over the past three years, and we have a much richer account when these incredible events can be approached from both embryological and genetic studies. These changes seem to reflect the way developmental biology is now being taught.

Elements present in the first three editions of this book have been strengthened. Modern developmental biology demands that both the organismal and molecular approaches to biology be respected (indeed, celebrated). The most interesting areas of the field are probably those where these approaches intersect, and I have tried to integrate them. I have also attempted to blur some of the artificial boundaries that have divided developmental biology from its sibling disciplines of evolutionary biology and genetics. However, I extend apologies to my colleagues and friends (I think I still have some) in plant developmental biology for writing a book that is all meat and no vegetables. This book would be far too long were it to do justice to the plant kingdom, and I feel strongly that plant development deserves its own course and textbook.

As the field continues to expand and deepen, a word of warning is called for: Developmental biology cannot be taught—or learned—in a single semester. This text is an attempt to provide each person with sufficient material for their course, but a teacher need not feel guilty for not assigning every chapter, and students need not feel deprived if they have not read every chapter. This is the beginning of the path, not its conclusion.

I tell my students that studying developmental biology is their reward for doing well in cell biology and genetics. Now they can have fun asking the important questions and, perhaps, finding a problem to study that is worthy of their talents. The problems are among the most important in all science: How do the egg and sperm interact to create a new organism? How does a single cell give rise to the diverse cell types of the adult body? How do neurons make their specific connections with other neurons and with peripheral tissues? How do organs form?

It was Roux's vision that developmental biology would "sometime constitute the common basis of all other biological disciplines and, in continued symbiosis with these disciplines, play a prominent part in the solutions of the problems of life." These were bold, even arrogant words one hundred years ago; today, they express a widely held assumption. Development integrates all areas of biology and plays the crucial role of relating genotype to phenotype. Development can be studied using any organism and at any level of organization, from molecules to phyla.

A final caveat: At the Marine Biology Laboratory at Woods Hole, Massachusetts (scene of some of the most important embryological investigations in the United States), there hangs a handwritten banner that reads, "Study Nature, Not Books." This sign hangs over the main entrance to the *library* as a reminder that we come to study organisms, not the history of how they are studied. This book is successful to the point that it stimulates one to relinquish it and to study the organism.

Acknowledgments

This book has benefitted enormously from the comments and criticisms of several investigators who took the time to read drafts of these chapters. These stalwarts include: R. and L. Angerer, M. Bronner-Fraser, D. D. Brown, S. Carroll, E. DeRobertis, B. A. Edgar, J. Fallon, A. Fausto-Sterling, G. Guild, R. Grainger, G. Grunwald, Z. Hall, V. Henrich, L. A. Jaffe, R. E. Keller, K. J. Kemphues, J. Kimble, W. Loomis, W. McGinnis, J. Opitz, D. Page, C. Phillips, R. Raff, L. Saxén, I. Thesleff, R. Tuan, H. Weintraub, and M. Wormington. Moreover, this book could not have been completed without the considerable effort from the scores of researchers who sent me photographs that appear herein and preprints for the Fourth Edition. This edition was substantially improved by those scientists who sent me critiques of the Third Edition, including M. Benecke, J. Koevenig, J. B. Morrill, J. E. Rebers, and F. Schwalm. Thanks also to R. Dratman for computer help when it was of crucial importance.

This edition, like its earlier incarnations, owes a great deal to the suggestions and criticisms of the students in my developmental biology and developmental genetics classes. The extremely supportive staff and faculty of Swarthmore College have also played major roles in producing this book, and science librarians E. Horikawa and M. Spencer are due special thanks for keeping recent volumes from being sent to the bindery while I was writing the book.

Andy Sinauer has yet again managed to gather the same remarkable people around this project, and it has been a privilege to work with them. My thanks to him and to editor Carol Wigg, production coordinator Chris Small, artist John Woolsey, designer Rodelinde Albrecht, copy editor Gretchen Becker, layout artist Janice Holabird, and Joe Vesely, art editor emeritus. Because publishing deadlines must be met and other work gets put aside, I have to thank my family for once again allowing me to get away with this. Especially, this book could never have been completed were it not for the encouragement of my wife, Anne Raunio, who, as an obstetrician, enjoys the more practical side of developmental biology. My thanks to you all.

Scott F. Gilbert

I

AN INTRODUCTION
TO DEVELOPMENTAL BIOLOGY

An introduction to animal development

According to Aristotle, the first major embryologist known to history, science begins with wonder. "It is owing to wonder that people began to philosophize, and wonder remains the beginning of knowledge." The development of an animal from an egg has been a source of wonder throughout human history. The simple procedure of cracking open a chick egg on each successive day of its three-week incubation provides a remarkable experience as a thin band of cells is seen to give rise to an entire bird. Aristotle performed this experiment and noted the formation of the major organs. Anyone can wonder at this remarkable—yet commonplace—phenomenon, but it is the scientist who seeks to discover how development actually occurs. And rather than dissipating wonder, new understanding increases it.

Multicellular organisms do not spring forth fully formed. Rather, they arise by a relatively slow process of progressive change that we call **development.** In nearly all cases, the development of a multicellular organism begins with a single cell—the fertilized egg, or **zygote**—which divides mitotically to produce all the cells of the body. The study of animal development has traditionally been called **embryology,** referring to the fact that between fertilization and birth the developing organism is known as an **embryo.** But development does not stop at birth, or even at adulthood. Most organisms never stop developing. Each day we replace over a gram of skin cells (the older cells being sloughed off as we move), and our bone marrow sustains the development of millions of new erythrocytes every minute of our lives. Therefore, in recent years it has become customary to speak of **developmental biology** as the discipline that involves studies of embryonic and other developmental processes.

Developmental biology is one of the most exciting and fast-growing fields of biology. Part of the excitement comes from its subject matter, for we are just beginning to understand the molecular mechanisms of animal development. Another part of the excitement comes from the unifying role that developmental biology is assuming in the biological sciences. Developmental biology is creating a framework that integrates molecular biology, physiology, cell biology, anatomy, cancer research, immunology, and even evolutionary and ecological studies. The study of development has become essential for understanding any other area of biology.

I remember perfectly well the intense satisfaction and delight with which I had listened, by the hour, to Bach's fugues . . . and it has often occurred to me that the pleasure derived from musical compositions of this kind . . . is exactly the same as in most of my problems of morphology—that you have the theme in one of the old masters' works followed out in all its endless variations, always reappearing and always reminding you of the unity in variety.
THOMAS HUXLEY (1882)

Happy is the person who is able to discern the causes of things.
VIRGIL (37 B.C.E.)

Principal features of development

Development accomplishes two major functions. It generates cellular diversity and order within each generation, and it ensures the continuity of life from one generation to the next. The first function involves the production and organization of all the diverse types of cells in the body. A single cell, the fertilized egg, gives rise to muscle cells, skin cells, neurons, lymphocytes, blood cells, and all the other cell types. This generation of cellular diversity is called **differentiation;** the processes that organize the different cells into tissues and organs are called **morphogenesis** (creation of form and structure) and **growth** (increase in size). The second major function of development is **reproduction:** the continued generation of new individuals of the species.

The major features of animal development are illustrated in Figure 1.1. The life of a new individual is initiated by the fusion of genetic material from the two **gametes**—the sperm and the egg. This fusion, called **fertilization,** stimulates the egg to begin development. The subsequent sequence of stages is collectively called **embryogenesis.** Throughout the animal kingdom an incredible variety of embryonic types exists, but most patterns of embryogenesis comprise variations on four themes.

1. Immediately following fertilization, cleavage occurs. **Cleavage** is a series of extremely rapid mitotic divisions wherein the enormous volume of zygote cytoplasm is divided into numerous smaller cells. These cells are called **blastomeres,** and by the end of cleavage they generally form a sphere known as a **blastula.**

2. After the rate of mitotic division has slowed down, the blastomeres undergo dramatic movements wherein they change their positions relative to one another. This series of extensive cell rearrangements is called **gastrulation.** As a result of gastrulation, the typical embryo contains three cell regions, called **germ layers.*** The outer layer—the **ectoderm**—produces the cells of the epidermis and the nervous system; the inner layer—the **endoderm**—produces the lining of the digestive tube and its associated organs (pancreas, liver, and so on); and the middle layer—the **mesoderm**—gives rise to several organs (heart, kidney, gonads), connective tissues (bone, muscles, tendons), and the blood cells.

3. Once the three germ layers are established, the cells interact with one another and rearrange themselves to produce the bodily organs. This process is called **organogenesis.** (In vertebrates, organogenesis is initiated when a series of cellular interactions causes the mid-dorsal ectodermal cells to form the neural tube. This tube will become the brain and spinal cord.) Many organs contain cells from more than one germ layer, and it is not unusual for the outside of an organ to be derived from one layer and the inside from another. Also during organogenesis certain cells undergo long migrations from their place of origin to their final location. These migrating cells include the precursors of blood cells, lymph cells, pigment cells, and gametes.

*From the Latin *germen,* meaning "bud" or "sprout"; the same root as germination. The names of the three germ layers are from the Greek: ectoderm from *ektos* ("outside") plus *derma* ("skin"); mesoderm from *mesos* ("middle"), and endoderm from *endon* ("within").

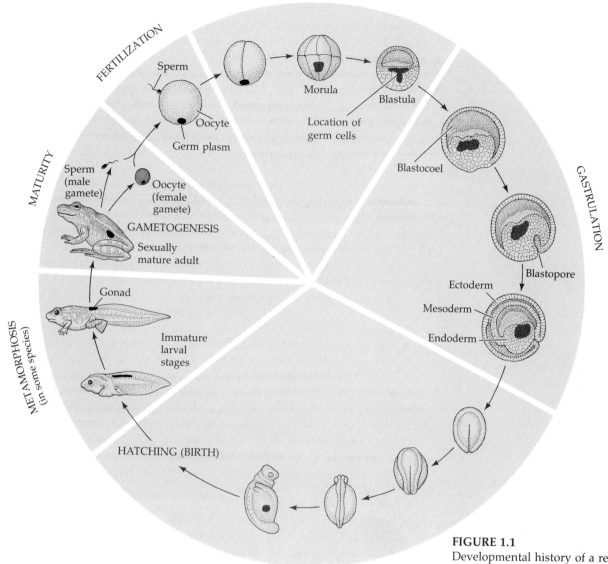

CLEAVAGE

FERTILIZATION

Sperm

Oocyte

Germ plasm

Morula

Location of
germ cells

Blastula

Blastocoel

GASTRULATION

MATURITY

Sperm
(male
gamete)

Oocyte
(female
gamete)

GAMETOGENESIS

Sexually
mature adult

Blastopore

Ectoderm

Mesoderm

Endoderm

Gonad

METAMORPHOSIS
(in some species)

Immature
larval
stages

HATCHING (BIRTH)

ORGANOGENESIS

FIGURE 1.1
Developmental history of a representative organism, the frog. The stages from fertilization through hatching (or birth) are known collectively as embryogenesis. The region set aside for producing germ cells is shown in color. Gametogenesis, which is complete in the sexually mature adult, begins at different times during development, depending upon the species.

4. As seen in Figure 1.1, a portion of egg cytoplasm gives rise to cells that are the precursors of the gametes. These cells are called **germ cells,** and they are set aside for their reproductive function. All the other cells of the body are called **somatic cells.** This separation of somatic cells (which give rise to the individual body) and germ cells (which contribute to the formation of a new generation) is typically one of the first differentiations to occur during animal development. The germ cells eventually migrate to the gonads, where they differentiate into gametes. The development of gametes, called **gametogenesis,** is usually not completed until the organism has become physically mature. At maturity, the gametes may be released and undergo fertilization to begin a new life. The adult organism eventually undergoes senescence and dies.

Our eukaryotic heritage

Organisms are divided into two major groups, their classification depending on whether their cells possess a nuclear envelope. On one side of this classification line are the **prokaryotes** (Gk. *karyon*, "nucleus"), which include the bacteria and the cyanobacteria, or blue-green algae. These cells lack a true nucleus. On the other side are the **eukaryotes,** which include protists, animals, plants, and fungi. Eukaryotic cells have a well-formed nuclear envelope surrounding their chromosomes. This fundamental difference between eukaryotes and prokaryotes influences how they arrange and utilize their genetic information. In both groups, the inherited information needed for development and metabolism is encoded in the DNA sequences of the chromosomes. The prokaryotic chromosome is generally a small, circular, double helix of DNA, consisting of approximately a million base pairs. The eukaryotic cell usually has several chromosomes, and the simplest eukaryotic protists have over 10 times the amount of DNA found in the most complex prokaryotes. Moreover, the structure of a eukaryotic gene is more complex than that of a prokaryotic gene. The amino acid sequence of a prokaryotic protein is a direct reflection of the DNA sequence in the chromosome. The protein-coding DNA of a eukaryotic gene, however, is usually divided up such that the complete amino acid sequence of a protein is derived from discontinuous segments of DNA (Figure 1.2). The intervening DNA often contains sequences that are involved with regulating the time and place that the gene is activated.

Eukaryotic chromosomes also are very different from prokaryotic chromosomes. Eukaryotic DNA is wrapped around specific proteins called **histones,** which can organize the DNA into compact structures and which are important in regulating which genes become expressed in which cells. In bacteria, there are no histones. Moreover, eukaryotic cells undergo mitosis, wherein the nuclear envelope breaks down and the replicated

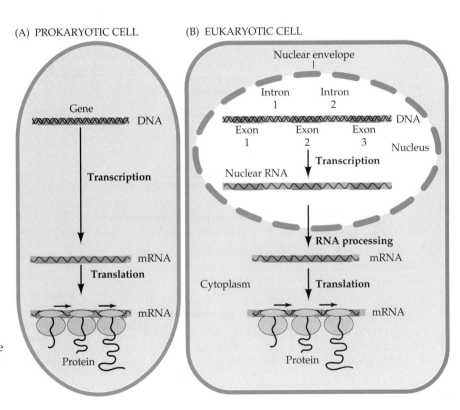

FIGURE 1.2
Summary of steps by which proteins are synthesized from DNA. (A) Prokaryotic (bacterial) gene expression: the coding regions of DNA are co-linear with the protein product. (B) Eukaryotic gene expression: the genes are discontinuous, and a nuclear envelope separates the DNA from the cytoplasm.

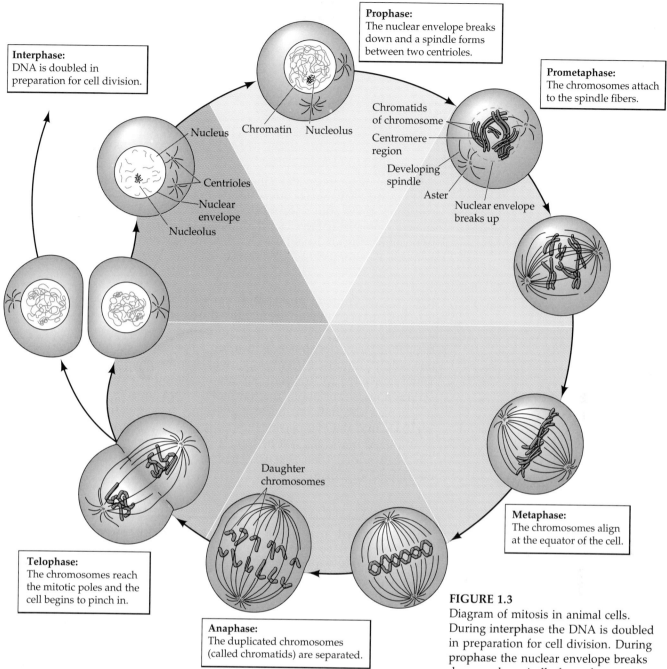

Interphase:
DNA is doubled in preparation for cell division.

Prophase:
The nuclear envelope breaks down and a spindle forms between two centrioles.

Prometaphase:
The chromosomes attach to the spindle fibers.

Nucleus

Chromatin   Nucleolus

Chromatids of chromosome

Centromere region

Developing spindle

Centrioles

Aster

Nuclear envelope breaks up

Nuclear envelope

Nucleolus

Daughter chromosomes

Metaphase:
The chromosomes align at the equator of the cell.

Telophase:
The chromosomes reach the mitotic poles and the cell begins to pinch in.

Anaphase:
The duplicated chromosomes (called chromatids) are separated.

FIGURE 1.3
Diagram of mitosis in animal cells. During interphase the DNA is doubled in preparation for cell division. During prophase the nuclear envelope breaks down and a spindle forms between the two centrioles. At metaphase the chromosomes align at the equator of the cell, and as anaphase begins the duplicated chromosomes (each duplicate called a chromatid) are separated. At telophase the chromosomes reach the mitotic poles and the cell begins to pinch in. At each pole are the same number and type of chromosomes as were present in the cell before it divided.

chromosomes are equally divided between the daughter cells (Figure 1.3). In prokaryotes, cell division is not mitotic; no mitotic spindle develops, and there is no nuclear envelope to break down. Rather, the daughter chromosomes remain attached to adjacent points on the cell membrane. These attachment points are separated by the growth of the cell membrane between them, eventually placing the chromosomes into different daughter cells.

Prokaryotes and eukaryotes have different mechanisms of gene regulation. In both prokaryotes and eukaryotes, DNA is transcribed by enzymes called RNA polymerases to make RNA. When messenger RNA—mRNA—is produced in prokaryotes, it is immediately translated into a protein while the other end of it is still being transcribed from the DNA

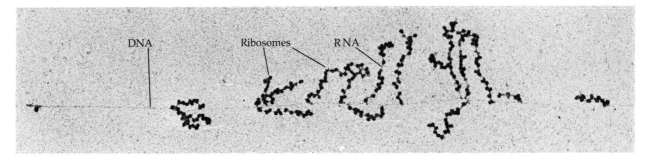

FIGURE 1.4

Concurrent transcription and translation in prokaryotes. A portion of *Escherichia coli* DNA runs horizontally across this electron micrograph. Messenger RNA transcripts can be seen on both sides. Ribosomes have attached to the mRNA and are synthesizing proteins (which cannot be seen). The size of the mRNA can be seen increasing in length from left to right, indicating the direction of transcription. (Courtesy of O. L. Miller, Jr.)

(Figure 1.4). Thus, in prokaryotes transcription and translation are simultaneous and coordinated events. But the existence of the nuclear envelope in eukaryotes provides the opportunity for an entirely new type of cell regulation. The ribosomes, which are responsible for translation, are on one side of the nuclear envelope, whereas the DNA and the RNA polymerases needed for transcription are on the other side. In between transcription and translation, the transcribed RNA must be processed so that it can pass through the nuclear envelope. By regulating which mRNAs can pass into the cytoplasm, the cell is able to select which of the newly synthesized messages will be translated. Thus, a new level of complexity has been added, one that we shall see is extremely important for the developing organism.

Development among the unicellular eukaryotes

All multicellular eukaryotic organisms have evolved from unicellular protists. It is in these protists that the basic features of development first appeared. Simple eukaryotes give us our first examples of the nucleus directing morphogenesis, the use of the cell surface to mediate cooperation between individual cells, and the first occurrences of sexual reproduction.

Control of developmental morphogenesis in *Acetabularia*

A century ago, it had not yet been proved that the nucleus contained hereditary or developmental information. Some of the best evidence for this theory came from studies in which unicellular organisms were fragmented into nucleate and anucleate pieces (reviewed in Wilson, 1896). When various protozoans were cut into fragments, nearly all the pieces died. However, the fragments containing nuclei were able to live and to regenerate entire complex cellular structures (Figure 1.5).

Nuclear control of cell morphogenesis and the interaction of nucleus and cytoplasm are beautifully demonstrated by studies of *Acetabularia*. This enormous single cell (2–4 centimeters long) consists of three parts: a cap, a stalk, and a rhizoid (Figure 1.6A). The rhizoid is located at the base of the cell and holds it to the substrate. The single nucleus of the cell resides within the rhizoid. The size of *Acetabularia* and the location of its nucleus

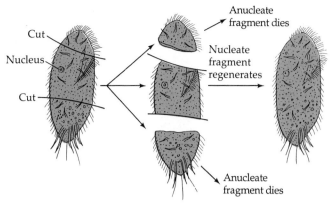

FIGURE 1.5
Regeneration of the nucleate fragment of the unicellular protist *Stylonychia*. The anucleate fragments survive for a time but finally die.

allow investigators to remove the nucleus from one cell and replace it with a nucleus from another cell. In the 1930s, J. Hämmerling took advantage of these unique features and exchanged nuclei between two morphologically distinct species, *A. mediterranea* and *A. crenulata*. As the photographs show, these two species have very different cap structures. Hämmerling found that when the nucleus from one species was transplanted into the stalk of another species, the newly formed cap eventually assumed the form associated with the donor nucleus (Figure 1.6B). Thus, the nucleus was seen to control *Acetabularia* development.

The formation of a cap is a complex morphogenic event involving the synthesis of numerous proteins, the products of which must be placed in a certain portion of the cell. The transplanted nucleus does indeed direct the synthesis of its species-specific cap, but it takes several weeks to do so. Moreover, if the nucleus is removed from an *Acetabularia* cell early in development, before it first forms a cap, a normal cap is formed weeks later, even though the organism will eventually die. These studies suggest that (1) the nucleus contains information specifying the type of cap produced (i.e., it contains the genetic information that specifies the proteins responsible for the production of a certain type of cap), and (2) the information enters the cytoplasm long before cap production occurs. The information in the cytoplasm is not used for several weeks.

One current hypothesis proposed to explain these observations is that the nucleus synthesizes a stable mRNA that lies dormant in the cytoplasm until the time of cap formation. This hypothesis is supported by an observation that Hämmerling published in 1934. Hämmerling fractionated young *Acetabularia* into several parts (Figure 1.7). The portion with the nucleus formed a cap as expected; so did the apical tip of the stalk. The remaining portion of the stalk did not form a cap. Thus, Hämmerling postulated (nearly 30 years before the existence of mRNA was known) that the instructions for the cap formation were somehow stored at the tip. Kloppstech and Schweiger (1975) have shown that nucleus-derived

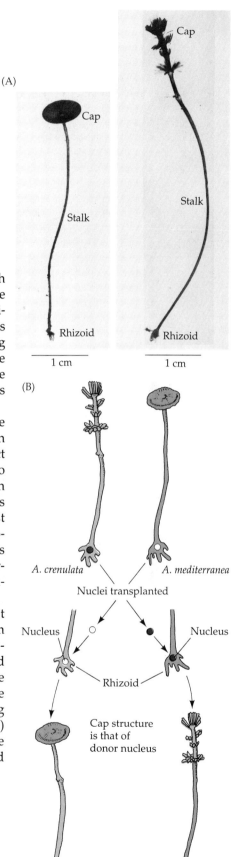

(A)

(B)

FIGURE 1.6
(A) *Acetabularia mediterranea* (left) and *A. crenulata* (right). Each individual is a single cell. The rhizoid contains the nucleus. (B) Effect of exchanging nuclei between two species of *Acetabularia;* nuclei were transplanted into enucleated rhizoid fragments. *A. crenulata* structures are shaded; *A. mediterranea* structures are unshaded. (Photographs courtesy of H. Harris.)

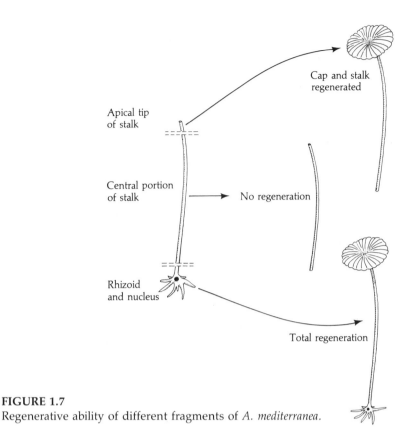

FIGURE 1.7
Regenerative ability of different fragments of *A. mediterranea.*

mRNA does accumulate at this region. Ribonuclease, an enzyme that cleaves RNA, completely inhibits cap formation when added to the seawater in which *Acetabularia* is growing. In enucleated cells, this effect is permanent; once the RNA is destroyed, no cap formation can occur. In nucleated cells, however, the cap can form after the ribonuclease is washed away—presumably because new mRNA is being made by the nucleus. Garcia and Dazy (1986) have also shown that protein synthesis is especially active in the apex of *Acetabularia*.

It is obvious from the preceding discussion that nuclear transcription plays an important role in the formation of the *Acetabularia* cap. But note that the cytoplasm also plays an essential role in cap formation. The mRNAs are not translated for weeks, even though they are in the cytoplasm. Something in the cytoplasm controls when the message is utilized. Hence, the expression of the cap is controlled not only by nuclear transcription but also by the **translational control** of the cytoplasmic RNA. In this unicellular organism, "development" is controlled at both the transcriptional and the translational levels.

Differentiation in the amoeboflagellate *Naegleria*

One of the most remarkable cases of protist "differentiation" is that of *Naegleria gruberi*. This organism occupies a special place in protist taxonomy because it can change its form from that of an amoeba to that of a flagellate (Figure 1.8). During most of its life cycle, *N. gruberi* is a typical amoeba, feeding on soil bacteria and dividing by fission. However, when the bacteria are diluted (either by rainwater or by water added by an experimenter), each *N. gruberi* rapidly develops a streamlined body shape and two long anterior flagella. Thus, instead of having several differentiated cell types in one organism, this one cell has different cell structures and biochemistry at different times of its life.

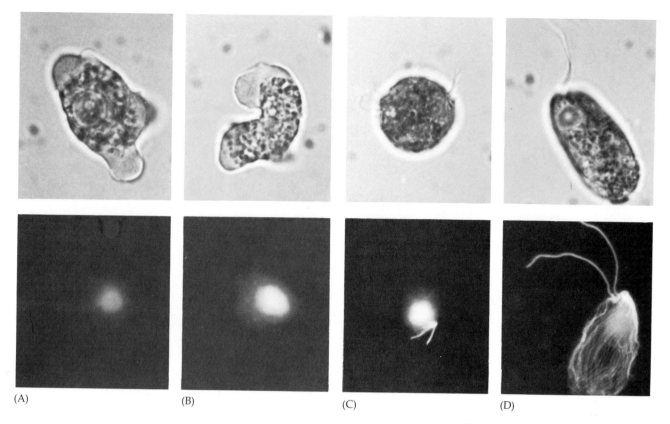

(A) (B) (C) (D)

Differentiation into the flagellate form occurs within an hour (Figure 1.9). During this time the amoeba has to create centrioles to serve as the basal bodies (microtubule organizing centers) of the flagella, as well as to create the flagella themselves. Flagella are made from many proteins, the most abundant of which is **tubulin.** The tubulin molecules are organized into microtubules, and the microtubules are further organized into an arrangement that permits flagellar movement. Fulton and Walsh (1980) showed that the tubulin for the *Naegleria* flagella does not exist in the amoeba stage. It is made de novo ("from scratch"), starting with new transcription from the nucleus. To show this, the investigators manipulated transcription at various stages with actinomycin D, an antibiotic drug that selectively inhibits RNA synthesis. When added before the dilution of the food supply, this antibiotic prevents tubulin synthesis. However, if the actinomycin D is added 20 minutes after dilution, tubulin is still made at the normal time (about 30 minutes later). Therefore, it appears that the mRNA for tubulin is made during the first 20 minutes after dilution and is used shortly thereafter. This interpretation was confirmed when it was shown that mRNA extracted from amoebae does not contain any detectable messages for flagellar tubulin, whereas mRNA extracted from differentiating cells contains a great many such messages (Walsh, 1984).

Here, then, is an excellent example of the **transcriptional control** of a development process: the *Naegleria* nucleus responds to environmental changes by synthesizing the mRNA for flagellar tubulin. We also see another process that remains extremely important in the development of all other animals and plants, namely, the assembly of tubulin molecules to produce a flagellum. This arrangement, whereby tubulin is polymerized into microtubules and the microtubules assembled into an ordered array, is seen throughout nature. In mammals it is evident in the sperm flagellum and in the cilia of the spinal cord and respiratory tract. This **posttranslational control,** whereby a protein is not functional until it is linked with

FIGURE 1.8

Transformation from amoeboid to flagellated state in *Naegleria gruberi*. Top row stained with Lugol's iodine. Bottom row stained with fluorescent antibody to the tubulin protein of the microtubules. (A) 0 minutes. (B) 25 minutes, showing new tubulin synthesis. (C) 70 minutes, showing emergence of visible flagella. (D) 120 minutes, showing mature flagella and streamlined body shape. (From Walsh, 1984, courtesy of C. Walsh.)

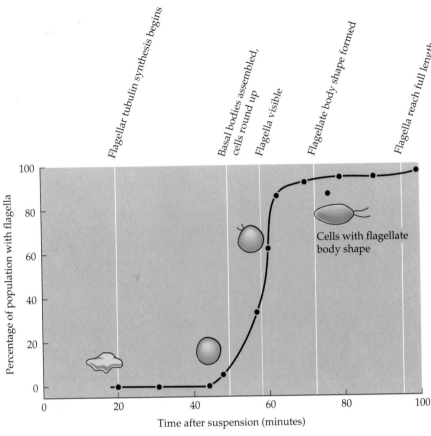

FIGURE 1.9
Differentiation of the flagellate phenotype in *Naegleria*. Amoebae that had been growing in the bacteria-enriched medium are washed free of bacteria at time zero. By 80 minutes, nearly the entire population has developed flagella. (After Fulton, 1977.)

other molecules, will be discussed more fully later. We see, however, that development in unicellular eukaryotes can be controlled at the transcriptional, translational, and posttranslational levels.

The origins of sexual reproduction

Sexual reproduction is another invention of the protists that has had a profound effect on more complex organisms. It should be noted that sex and reproduction are two distinct and separable processes. *Reproduction* involves the creation of new individuals. *Sex* involves the combining of genes from two different individuals into new arrangements. Reproduction in the absence of sex is characteristic of those organisms that reproduce by fission; there is no sorting of genes when an amoeba divides or when a hydra buds off cells to form a new colony. Sex without reproduction is also common among unicellular organisms. Bacteria are able to transmit genes from one individual to another by means of sex pili (Figure 1.10). This transmission is separate from reproduction. Protists are also able to reassort genes without reproduction. Paramecia, for instance, reproduce by fission, but sex is accomplished by **conjugation.** When two paramecia join together, they link their oral apparatuses and form a cytoplasmic connection through which they can exchange genetic material (Figure 1.11). Each macronucleus (which controls the metabolism of the

FIGURE 1.10
Sex in bacteria. Some bacterial cells are covered with numerous appendages (pili) and are capable of transmitting genes to a recipient cell (lacking pili) through a sex pilus. In this figure, the sex pilus is highlighted by viral particles that bind specifically to that structure. (Courtesy of C. C. Brinton, Jr. and J. Carnahan.)

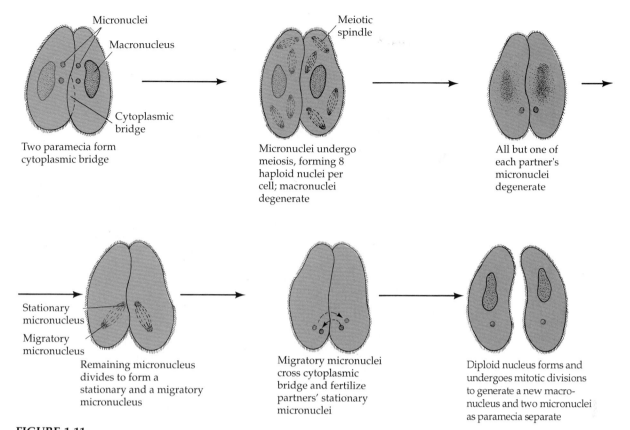

FIGURE 1.11
Conjugation across a cytoplasmic bridge in paramecia. Two paramecia can exchange genetic material, leaving each with genes that differ from those with which they started. (After Strickberger, 1985.)

organism) degenerates while each micronucleus undergoes meiosis to produce eight haploid micronuclei, of which all but one degenerate. The remaining micronucleus divides once more to form a stationary micronucleus and a migratory micronucleus. Each migratory micronucleus crosses the cytoplasmic bridge and fuses with ("fertilizes") the stationary micronucleus, thereby creating a new diploid nucleus in each cell. This diploid nucleus then divides mitotically to give rise to a new micronucleus and a new macronucleus as the two partners disengage. Therefore, no reproduction has occurred—only sex.

The union of these two distinct processes, sex and reproduction, into **sexual reproduction** is seen in unicellular eukaryotes. Figure 1.12 shows the life cycle of *Chlamydomonas*. This organism is usually haploid, having just one copy of each chromosome (like a mammalian gamete). The individuals of each species, however, are divided into two mating groups: *plus* and *minus*. When these meet, they join their cytoplasms, and their nuclei fuse to form a diploid zygote. This zygote is the only diploid cell in the life cycle, and it eventually undergoes meiosis to form four new *Chlamydomonas* cells. Here is sexual reproduction, for chromosomes are reassorted during the meiotic divisions and more individuals are formed. Note that in this protist type of sexual reproduction, the gametes are morphologically identical—the distinction between sperm and egg has not yet been made.

In evolving sexual reproduction, two important advances had to be achieved. The first is the mechanism of **meiosis** (Figure 1.13), whereby the diploid complement of chromosomes is reduced to the haploid state

FIGURE 1.12

Sexual reproduction in *Chlamydomonas*. Two strains, both haploid, can reproduce asexually when separate. Under certain conditions, the two strains can unite to produce a diploid cell that can undergo meiosis to form four new haploid organisms. (After Strickberger, 1985.)

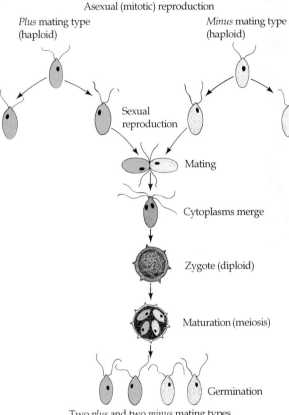

Asexual (mitotic) reproduction

Plus mating type (haploid)

Minus mating type (haploid)

Sexual reproduction

Mating

Cytoplasms merge

Zygote (diploid)

Maturation (meiosis)

Germination

Two *plus* and two *minus* mating types

FIGURE 1.13

Summary of meiosis. The DNA and associated proteins replicate during interphase. During prophase the nuclear envelope breaks down and homologous chromosomes (each chromosome being double, with the chromatids joined at the centromere) align in pairs. Chromosomal rearrangements can occur between the four homologous chromatids at this time. After the first metaphase, the two original homologous chromosomes are segregated into different cells. During the second division the centromere splits, thereby leaving each new cell with one copy of each chromosome.

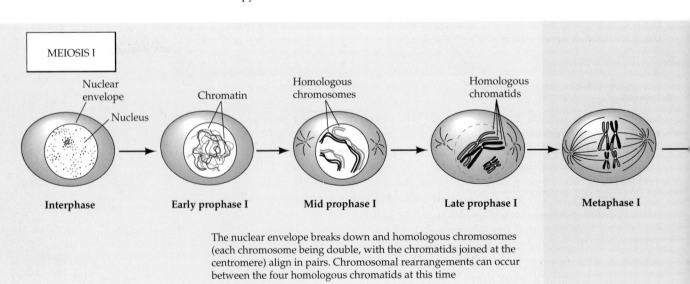

MEIOSIS I

Nuclear envelope

Nucleus

Chromatin

Homologous chromosomes

Homologous chromatids

Interphase   **Early prophase I**   **Mid prophase I**   **Late prophase I**   **Metaphase I**

The nuclear envelope breaks down and homologous chromosomes (each chromosome being double, with the chromatids joined at the centromere) align in pairs. Chromosomal rearrangements can occur between the four homologous chromatids at this time

 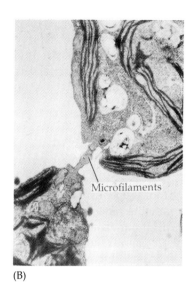

(A) (B)

FIGURE 1.14
Two-step recognition in mating *Chlamydomonas*. (A) Scanning electron micrograph (×7000) of mating pair. The interacting flagella twist about each other, adhering at the tips (arrows). (B) Transmission electron micrograph (×20,000) of a cytoplasmic bridge connecting the two organisms. The microfilaments extend from the donor (lower) cell to the recipient (upper) cell. (From Goodenough and Weiss, 1975 and Bergman et al., 1975, by permission of U. Goodenough.)

Microfilaments

(discussed in detail in Chapter 22.) The other advance is the mechanism whereby the two different mating types recognize each other. In *Chlamydomonas*, recognition occurs first on the flagellar membranes (Figure 1.14; Goodenough and Weiss, 1975; Bergman et al., 1975). The agglutination of flagella enables specific regions of the cell membranes to come together. These specialized sectors contain mating type-specific components that enable the cytoplasms to fuse. Following agglutination, the *plus* individuals initiate the fusion by extending a **fertilization tube.** This tube contacts and fuses with a specific site on the *minus* individual. Interestingly, the mechanism used to extend this tube—the polymerization of the protein **actin**—is also used to extend processes of sea urchin eggs and sperm. In Chapter 4, we will see that the recognition and fusion of sperm and egg occur in a manner amazingly similar to that of these protists.

Unicellular eukaryotes appear to have the basic elements of the developmental processes that characterize the more complex organisms of other phyla: cellular synthesis is controlled by transcriptional, transla-

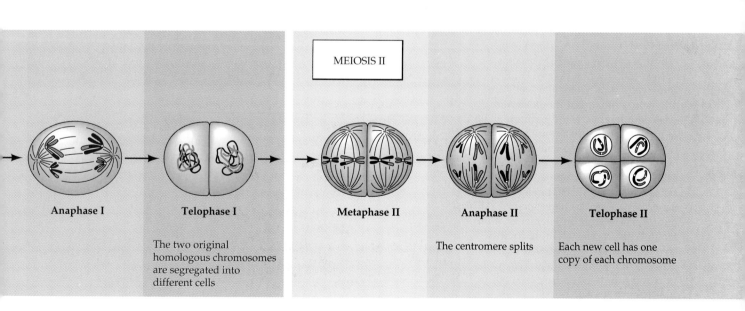

MEIOSIS II

Anaphase I

Telophase I

The two original homologous chromosomes are segregated into different cells

Metaphase II

Anaphase II

The centromere splits

Telophase II

Each new cell has one copy of each chromosome

tional, and posttranslational regulation; a mechanism for processing RNA through the nuclear membrane exists; the structures of individual genes and chromosomes are as they will be throughout eukaryotic evolution; mitosis and meiosis are perfected; and sexual reproduction exists, involving cooperation between individual cells. Such intercellular cooperation becomes even more important with the evolution of multicellular organisms.

Colonial eukaryotes: The evolution of differentiation

One of evolution's most important experiments was the creation of multicellular organisms. There appear to be several paths by which single cells evolved multicellular arrangements; we will discuss only two of them here (see Chapter 23 for a fuller discussion). The first path involves the orderly division of the reproductive cell and the subsequent differentiation of its progeny into different cell types. This path to multicellularity can be seen in a remarkable series of multicellular organisms collectively referred to as the family Volvocaceae, or the volvocaceans.

The volvocaceans

The simpler organisms among the volvocaceans are collections of numerous *Chlamydomonas*-like cells; but the more advanced members of this group have developed a second, very different, cell type.

A single organism of the volvocacean genus *Oltmannsiella* contains four *Chlamydomonas*-like cells in a row, embedded in a gelatinous matrix. In the genus *Gonium* (Figure 1.15), a single cell divides to produce a flat

FIGURE 1.15
Representatives of the order Volvocales. (A) The unicellular protist *Chlamydomonas reinhardtii*. (B) *Gonium pectorale* with eight *Chlamydomonas*-like cells in a convex disc. (C) *Pandorina morum*. (D) *Eudorina elegans*. (E) *Pleodorina californica*. Here all 64 cells are originally similar, but the posterior ones dedifferentiate and redifferentiate as gonidia, while the anterior cells remain small and biflagellate, like *Chlamydomonas*. (F) *Volvox carteri*. Here cells destined to become gonidia are set aside early in development and never have somatic characteristics. All but *Chlamydomonas* are members of the family Volvocaceae. Complexity increases from the single-celled *Chlamydomonas* to the multicellular *Volvox*. Bar in A is 5 μm; B–D, 25 μm; E–F, 50 μm. (Courtesy of D. Kirk.)

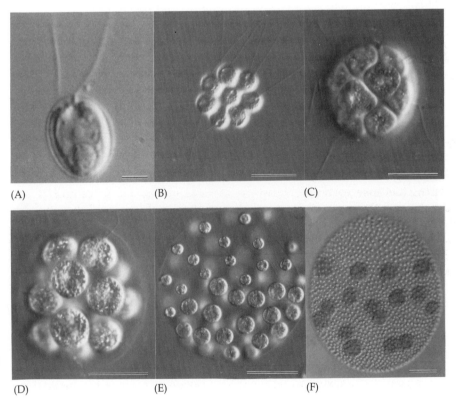

(A)   (B)   (C)

(D)   (E)   (F)

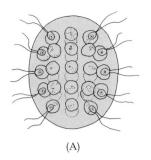
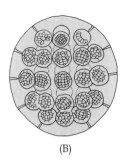
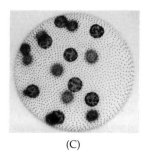

(A) (B) (C)

FIGURE 1.16
Asexual reproduction in volvocaceans. (A) Mature colony of *Eudorina elegans*. (B) Each of the *E. elegans* cells divides and produces a new colony. (C) Mature *Volvox carteri*. Most of the cells are incapable of reproduction. Germ cells (gonidia) have begun dividing into new organisms. (A and B after Hartmann, 1921; C from Kirk et al., 1982, courtesy of D. Kirk.)

plate of 4 to 16 cells, each with its own flagellum. In a related genus—*Pandorina*—the 16 cells form a sphere; and in *Eudorina*, the sphere contains 32 or 64 cells arranged in a regular pattern. In these organisms, then, a very important developmental principle has been worked out: *the ordered division of one cell to generate a number of cells that are organized in a predictable fashion.* As occurs in most animal embryos, the cell divisions by which a single volvocacean cell produces an organism of 4 to 64 cells occur in very rapid sequence and in the absence of cell growth.

The next two genera of the volvocacean series exhibit another important principle of development: *the differentiation of cell types within an individual organism.* The reproductive cells become differentiated from the somatic cells. In all the genera mentioned earler, every cell can, and normally does, produce a complete new organism by mitosis (Figure 1.16A,B). In the genera *Pleodorina* and *Volvox*, however, relatively few cells can reproduce. In *Pleodorina californica*, the cells in the anterior region are restricted to a somatic function; only those cells on the posterior side can reproduce. In *P. californica*, a colony usually has 128 or 64 cells, and the ratio of the number of somatic cells to the number of reproductive cells is usually 3:5. Thus, a 128-cell colony has 48 somatic cells and a 64-cell colony has 24.

In *Volvox*, almost all the cells are somatic and very few of the cells are able to produce new individuals. In some species of *Volvox*, reproductive cells, as in *Pleodorina*, are derived from cells that originally look and function like somatic cells before they enlarge and divide to form new progeny. However, in other members of the genus, such as *V. carteri*, there is a complete division of labor: the reproductive cells that will create the next generation are set aside during the division of the reproductive cells that are forming a new individual. The reproductive cells never develop functional flagella and never contribute to motility or other somatic functions of the individual; they are entirely specialized for reproduction. Thus, although the simpler volvocaceans may be thought of as colonial organisms (because each cell is capable of independent existence and of perpetuating the species), in *V. carteri* we have a truly multicellular organism with two distinct and interdependent cell types (somatic and reproductive), both of which are required for perpetuation of the species (Figure 1.16C). Although not all animals set aside the reproductive cells from the somatic cells (and plants hardly ever do), this separation of germ cells from somatic cells early in development is characteristic of many animal phyla and will be discussed in more detail in Chapter 22.

Although all the volvocaceans, and their unicellular relative *Chlamydomonas*, reproduce predominantly by asexual means, they are also capable of sexual reproduction. This involves the production and fusion of haploid gametes. In many species of *Chlamydomonas*, including the one illustrated in Figure 1.12, sexual reproduction is **isogamous,** since the haploid gametes that meet are similar in size, structure, and motility. However, in other species of *Chlamydomonas*—as well as many species of colonial volvocaceans—swimming gametes of very different sizes are produced by the different mating types. This is called **heterogamy.** But the larger volvocaceans have evolved a specialized form of heterogamy, called **oogamy,** which involves the production of large, relatively immotile eggs by one mating type and small, motile sperm by the other (see Sidelights & Speculations). Thus, the volvocaceans include the simplest organisms that have distinguishable male and female members of the species and that have distinct developmental pathways for the production of eggs or sperm. In all the volvocaceans, the fertilization reaction resembles that of *Chlamydomonas* in that it results in the production of a dormant diploid zygote that is capable of surviving harsh environmental conditions. When conditions allow the zygote to germinate, it first undergoes meiosis to produce haploid offspring of the two different mating types in equal numbers.

SIDELIGHTS & SPECULATIONS

Sex and individuality in Volvox

Simple as it is, *Volvox* shares many features that characterize the life cycles and developmental histories of much more complex organisms, including ourselves. As already mentioned, *Volvox* is among the simplest organisms to exhibit a division of labor between two completely different

FIGURE 1.17
Asexual reproduction in V. carteri. *When reproductive cells (gonidia) are mature, they enter a cleavage-like stage of embryonic development to produce juveniles within the adult. Through a series of cell movements resembling gastrulation, the embryonic volvox inverts and is eventually released from the parent. The somatic cells of the parent, lacking the gonidia, undergo senescence and die, while the juvenile colonies mature. The entire sexual cycle takes two days. (After Kirk, 1988.)*

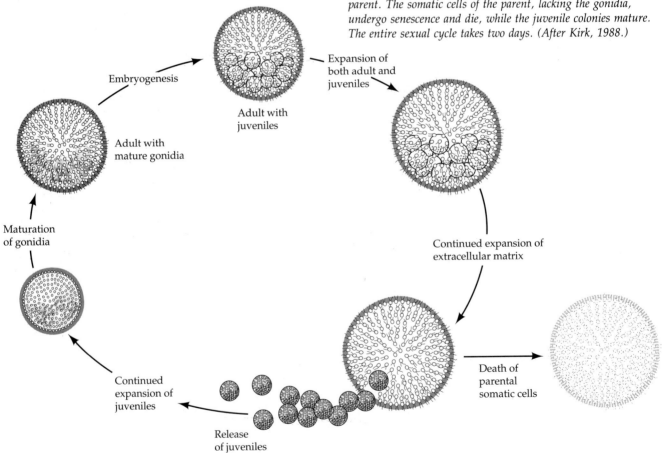

Embryogenesis

Adult with juveniles

Expansion of both adult and juveniles

Adult with mature gonidia

Continued expansion of extracellular matrix

Maturation of gonidia

Death of parental somatic cells

Continued expansion of juveniles

Release of juveniles

cell types. And, as a consequence of this, it is among the simplest organisms to include death as a regular, genetically programmed, part of its life history.

Death and differentiation

Unicellular organisms that reproduce by simple cell division, such as amoebae, are potentially immortal. The amoeba you see today under the microscope has no dead ancestors! When an amoeba divides, neither of the two resulting cells can be considered either as ancestor or offspring; they are siblings. Death comes to an amoeba only if it is eaten or meets with a fatal accident; and when it does, the dead cell leaves no offspring.

Death becomes an essential part of life, however, for any multicellular organism that establishes a division of labor between somatic (body) cells and germ (reproductive) cells. Consider the life history of *Volvox carteri* when it is reproducing asexually (Figure 1.17). Each asexual adult is a spheroid containing some 2000 small, biflagellated somatic cells along its periphery and about 16 large, asexual reproductive cells, called **gonidia**, toward one end of the interior. When mature, each gonidium divides rapidly 11 or 12 times. Certain of these divisions are asymmetric and produce the 16 large cells that will become a new set of gonidia. At the end of cleavage, all the cells that will be present in an adult have been produced from each gonidium. But the embryo is "inside-out": its gonidia are on the outside and the flagella of its somatic cells are pointing toward the interior of the hollow sphere of cells. This predicament is corrected by a process called inversion, in which the embryo turns itself right-side out by a set of cell movements that somewhat resemble the gastrulation movements of animal embryos (Figure 1.18). Clusters of bottle-shaped cells open a hole at one end of the embryo by producing tension on the interconnected cell sheet (Figure 1.19). The embryo everts through this hole and then closes it up. About a day after this is done, the juvenile colonies are enzymatically released from the parent and swim away.

What happens to the somatic cells of the "parent" *Volvox* now that its young have "left home"? Having produced offspring and being incapable of further reproduction, these somatic cells die. Actually, they commit suicide, synthesizing a set of proteins that cause the death and dissolution of the cells that make these proteins (Pommerville and Kochert, 1982). Moreover, in this death, the cells release for the use of others—including their own offspring—all the nutrients that they had stored during life. "Thus emerges," notes David Kirk, "one of the great themes of life on planet Earth: 'Some die that others may live.'"

In *V. carteri*, a specific gene that plays a central role in regulating cell death has been identified (Kirk, 1988). In laboratory strains possessing mutations of this gene, somatic cells abandon their suicidal ways, gain the ability to reproduce asexually, and become potentially immortal (Figure 1.20). The fact that such mutants are never found in nature indicates that cell death most likely plays an important role in the survival of *V. carteri* under natural conditions.

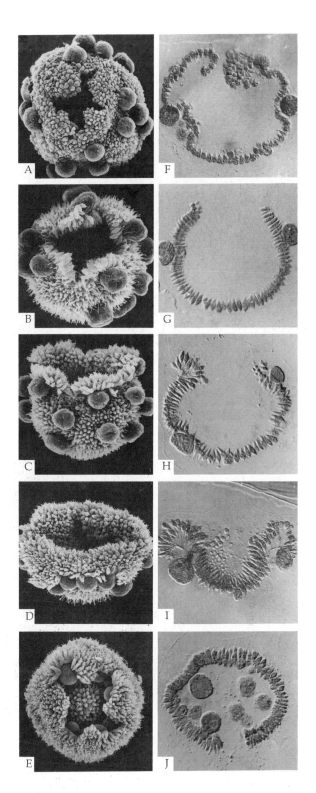

FIGURE 1.18

Inversion of asexually produced embryos of V. carteri. *A–E are scanning electron micrographs of whole embryos. F–J are sagittal sections through the center of the embryo, visualized by differential interference microscopy. Before inversion, the embryo is a hollow sphere of connected cells. When cells change their shape, a hole (the phialopore) opens at the apex of the embryo (A, B, F, G). Cells then curl around and rejoin at the bottom (C–E, H–J). (From Kirk et al., 1982, courtesy of D. Kirk.)*

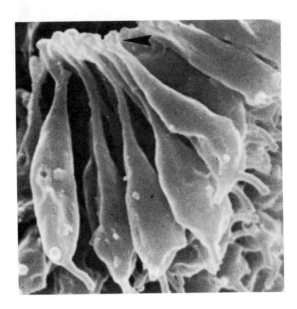

FIGURE 1.19
"Bottle cells" near the opening of the phialopore. These cells remain tightly interconnected through cytoplasmic bridges near their elongated apices, thereby creating the tension that causes the curvature of the interconnected cell sheet. (From Kirk et al., 1982, courtesy of D. Kirk.)

Enter sex

Although *V. carteri* reproduces asexually much of the time, in nature it reproduces sexually once each year. When it does, one generation of individuals passes away and a new and genetically different generation is produced. The naturalist Joseph Wood Krutch (1956) put it more poetically:

> The amoeba and the paramecium are potentially immortal. . . . But for Volvox, death seems to be as inevitable as it is in a mouse or in a man. Volvox must die as Leeuwenhoek saw it die because it had children and is no longer needed. When its time comes it drops quietly to the bottom and joins its ancestors. As Hegner, the Johns Hopkins zoologist, once wrote, 'This is the first advent of inevitable natural death in the animal kingdom and all for the sake of sex.' And he asked: 'Is it worth it?'

For *Volvox carteri*, it most assuredly is worth it. *V. carteri* lives in shallow temporary ponds that fill with spring rains but dry out in the heat of late summer. During most of that time, *V. carteri* swims about, reproducing asexually. These asexual volvoxes would die in minutes once the pond dried out, but *V. carteri* is able to survive by turning sexual shortly before the pond dries up, producing dormant zygotes that survive the heat and drought of late summer and the cold of winter. When rain fills the ponds in spring, the zygotes break their dormancy and hatch out a new generation of individuals to reproduce asexually until the pond is about to dry up once more. How do these simple organisms predict the coming of adverse conditions with sufficient accuracy to produce a sexual generation just in time, year after year?

The stimulus for switching from the asexual to the sexual mode of reproduction in *V. carteri* is known to be a 30,000-Da sexual inducer protein. This protein is so powerful that concentrations as low as 6×10^{-17} M cause gonidia to undergo a modified pattern of embryonic development that results in the production of eggs or sperm, depending on the genetic sex of the individual (Sumper et al., 1993). The sperm are released and swim to a female, where they fertilize eggs to produce the dormant zygotes (Figure 1.21).

What is the source of this sexual inducer protein? Kirk and Kirk (1986) discovered that the sexual cycle could be initiated by heating dishes of *V. carteri* to temperatures that might be expected in a shallow pond in late summer. When this was done, the somatic cells of the asexual volvoxes produced the sexual inducer protein. Since the amount of sexual inducer protein secreted by one individual is sufficient to initiate sexual development in over 500 million asexual volvoxes, a single inducing volvox can convert the entire pond to sexuality. This discovery explained an observation made over 80 years ago that "in the full blaze of Nebraska sunlight, *Volvox* is able to appear, multiply, and riot in sexual reproduction in pools of rainwater of scarcely a fortnight's duration" (Powers, 1908). Thus, in temporary ponds formed by spring rains and dried up by summer's heat, *Volvox* has found a means of survival: it uses the heat to induce the formation of sexual individuals whose mating produces zygotes capable of surviving conditions that kill the adult organism. We see, too, that development is critically linked to the ecosystem wherein the organism has adapted to survive.

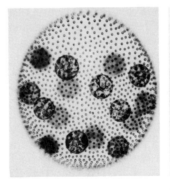 

(A)   (B)

FIGURE 1.20
Mutation of a single gene (called somatic regulator A) abolishes programmed cell death in V. carteri. *The newly hatched volvox carrying this mutation (A) is indistinguishable from the wild-type spheroid. However, shortly before the time when the somatic cells of the wild-type spheroids begin to die, the somatic cells of this mutant redifferentiate as gonidia (B). Eventually, every cell of the mutant will divide to form a new spheroid that will repeat this potentially immortal developmental cycle.*

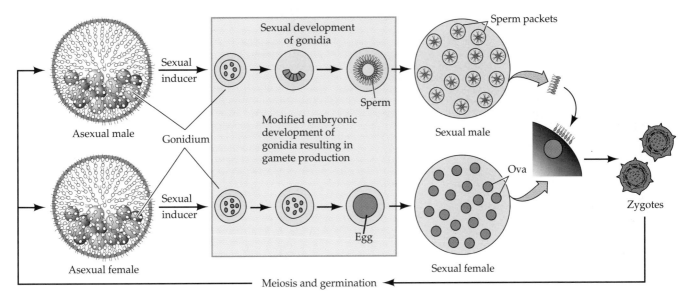

FIGURE 1.21
Sexual reproduction in V. carteri. *Males and females are indistinguishable in their asexual phase. When the sexual inducer protein is present, the gonidia of both mating types undergo a modified embryogenesis that leads to the formation of eggs in the females and sperm in the males. When the gametes are mature, sperm packets (containing 64 or 128 sperm each) are re-* *leased and swim to the females. Upon reaching the female, the sperm packet breaks up into individual sperm, which can fertilize the eggs. The resulting zygote has tough cell walls that can resist drying, heat, and cold. When spring rains cause the zygote to germinate, it undergoes meiosis to produce haploid males and females that reproduce asexually until heat induces the sexual cycle again.*

Differentiation and morphogenesis in *Dictyostelium*

The life cycle of *Dictyostelium*. Another type of multicellular organization derived from unicellular organisms is found in *Dictyostelium discoideum.** The life cycle of this fascinating and unusual organism is illustrated in Figure 1.22. In its vegetative cycle, solitary haploid amoebae (called myxamoebae to distinguish them from the true amoeba species) live on decaying logs, eating bacteria and reproducing by binary fission. When they have exhausted their food supply, tens of thousands of these amoebae join together to form moving streams of cells that converge at a central point. Here they form a conical aggregate, which eventually absorbs all the streaming cells and bends over to produce the migrating slug. The slug (often given the more dignified titles of pseudoplasmodium or **grex**) is usually 2–4 mm long and is encased in a slimy sheath. The grex begins to migrate (if the environment is dark and moist) with its anterior tip slightly raised. When migration ceases, these anterior cells, representing 15–20 percent of the entire cellular population, form a tubed stalk. The stalk begins as some of the central anterior cells, the **prestalk cells,** begin secreting an extracellular coat and extending a tube through the grex. As the prestalk cells differentiate, they form vacuoles and enlarge, lifting the

*This organism is colloquially called a "cellular slime mold." But "*Dictyostelium* is neither slimy nor a mold and is as unrelated to *Physarum* (the true slime mold) as it is to *Acanthamoeba* or *Neurospora*. It is certainly not a filamentous fungi and is just an amoebal organism." (W. Loomis, personal communication.)

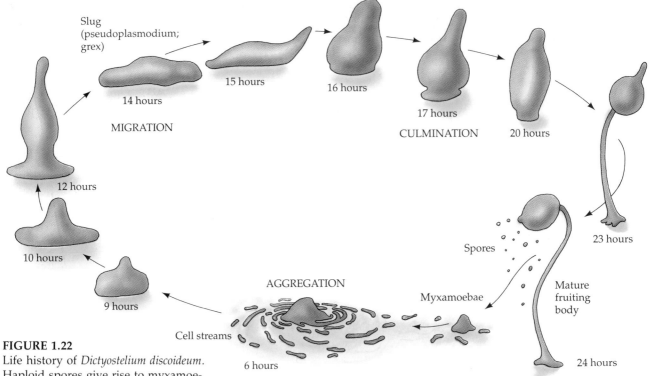

Slug (pseudoplasmodium; grex)

14 hours

15 hours

16 hours

17 hours

20 hours

MIGRATION

CULMINATION

12 hours

10 hours

9 hours

23 hours

Spores

Mature fruiting body

Myxamoebae

AGGREGATION

Cell streams

6 hours

24 hours

FIGURE 1.22

Life history of *Dictyostelium discoideum.* Haploid spores give rise to myxamoebae, which can reproduce asexually to form more haploid myxamoebae. As the food supply diminishes, aggregation occurs at central points and a migrating pseudoplasmodium is formed. Eventually it stops moving and forms a fruiting body that releases more spores. The numbers refer to hours since nutrient dilution begins the developmental sequence.

mass of **prespore cells** that had been in the posterior two-thirds of the grex (Jermyn and Williams, 1991). The stalk cells die, but the posterior cells, elevated above the stalk, become **spore cells.** These spore cells disperse, each one becoming a new myxamoeba.

In addition to this asexual cycle, there is a possibility for sex in *Dictyostelium*. Two amoebae can fuse to create a giant cell, which digests all the other cells of the aggregate. When it has eaten all its neighbors, it encysts itself in a thick wall and undergoes meiotic and mitotic divisions; eventually new myxamoebae are liberated.

Dictyostelium has been a wonderful experimental organism for developmental biologists, for it is an organism wherein initially identical cells are differentiated into one of two alternative cell types, spore or stalk. It is also an organism wherein individual cells come together to form a cohesive structure composed of differentiated cell types, akin to tissue formation in more complex organisms. The aggregation of thousands of amoebae into a single organism is an incredible feat of organization and invites experimentation to answer questions about the mechanisms involved.

Aggregation of *Dictyostelium* cells. The first question is, What causes the amoebae to aggregate? Time-lapse microcinematography has shown that no directed movement occurs during the first 4–5 hours following nutrient starvation. During the next 5 hours, however, the cells are seen moving at about 20 μm/min for 100 seconds. This movement ceases for about 4 minutes, then resumes. Although the movement is directed toward a central point, it is not a simple radial movement. Rather, cells join with each other to form streams; the streams converge into larger streams, and eventually all streams merge at the center. Bonner (1947) and Shaffer (1953) showed that this movement is due to **chemotaxis:** the cells are guided to aggregation centers by a soluble substance. This substance was

(A)

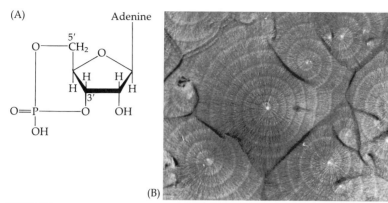

(B)

(C)

FIGURE 1.23

Chemotaxis of *Dictyostelium* amoebae due to spiral waves of cAMP. (A) Chemical structure of cAMP. (B) Visualization of several cAMP "waves" in the medium. Central cells secrete cAMP at regular intervals, and each secretion diffuses outward as a concentric wave. Waves are charted by saturating filter paper with radioactive cAMP and placing it on an aggregating colony. The cAMP from the secreting cells dilutes the radioactive cAMP. When the radioactivity on the paper is recorded (by placing it over X-ray film), the regions of high cAMP concentration in the culture appear lighter than those of low cAMP concentration. (C) Spiral waves of amoebae moving towards the initial source of cAMP. This digitally processed dark-field photomicrograph shows about 10^7 cells. Moving and nonmoving cells scatter light differently, and thereby the photograph reflects cell movement. The bright bands are composed of elongated migrating cells; the dark bands are of cells that have stopped moving and have rounded up. (D) As cells form streams, the spiral of movement can still be seen moving toward the center. (B from Tomchick and Devreotes, 1981, courtesy of P. Devreotes; C and D from Siegert and Weijer, 1989, courtesy of F. Siegert.)

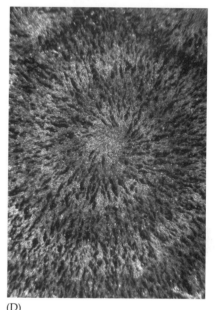

(D)

later identified as **cyclic adenosine 3',5'-monophosphate (cAMP)** (Konijn et al., 1967; Bonner et al., 1969), the chemical structure of which is shown in Figure 1.23A.

Aggregation is initiated as each of the cells begins to synthesize cAMP. There are no "dominant" cells that begin the secretion or control the others. Rather, the sites of aggregation are determined by the distribution of amoebae (Keller and Segal, 1970; Tyson and Murray, 1989). Neighboring cells respond to cAMP in two ways: they initiate a movement toward the cAMP pulse, and they release cAMP of their own (Robertson et al., 1972; Shaffer, 1975). After this, the cell is unresponsive to further cAMP pulses for several minutes. The result is a rotating spiral wave of cAMP that is propagated throughout the population of cells (Figure 1.23B). As each wave arrives, the cells take another step toward the center.*

The differentiation of individual amoebae into either stalk (somatic) or

*The biochemistry of this reaction involves a receptor that binds cAMP. When this binding occurs, specific gene transcription takes place, motility toward the source of the cAMP is initiated, and adenyl cyclase enzymes (which synthesize cAMP from ATP) are activated. The newly formed cAMP activates its own receptors as well as those of its neighbors. The cells in the area remain insensitive to new waves of cAMP until the bound cAMP is removed from the receptors by another cell-surface enzyme, phosphodiesterase (Johnson et al., 1989). The mathematics of such oscillation reactions predict that the diffusion of cAMP would initially be circular. However, as cAMP interacts with the cells that receive and propagate the signal, the cells that receive the front part of the wave begin to migrate at a different rate than the cells behind them. The result is a rotating spiral of cAMP and migration as seen in Figure 1.23. Interestingly, the same mathematical formulas predict the behavior of certain chemical reactions and the formation of new stars in rotating spiral galaxies (Tyson and Murray, 1989).

spore (reproductive) cells is a complex matter. Raper (1940) and Bonner (1957) have demonstrated that in all cases the anterior cells become stalk, while the remaining cells are destined to form spores. Surgically removing the anterior part of the slug does not change this situation. The new anterior cells (which had been destined to produce spores) now become the stalk (Raper, 1940). Somehow a decision is made so that whichever cells are anterior become stalk cells and whichever are posterior become spores. This ability of cells to change their developmental fates according to their location within the whole organism is called **regulation.** We will see this phenomenon in many embryos, including those of mammals.

Cell adhesion molecules in *Dictyostelium.* How do these individual cells stick together to form a cohesive organism? This is the same problem that embryonic cells face, and the solution that evolved in the protists is the same one used by embryos: developmentally regulated **cell adhesion molecules.**

While growing mitotically on bacteria, *Dictyostelium* cells do not adhere to one another. However, once cell division stops, the cells become increasingly adhesive, reaching a plateau of maximum cohesiveness around 8 hours after starvation. The initial cell-to-cell adhesion is mediated by a 24,000-Da glycoprotein that is absent in growing cells but is seen shortly thereafter (Figure 1.24; Knecht et al, 1987; Loomis, 1988). This protein is synthesized from newly transcribed mRNA and becomes localized in the cell membranes of the myxamoebae. If these cells are treated with antibodies that bind to and mask this protein, the cells will not stick to each other and all subsequent development ceases.

Once this initial aggregation has occurred, it is stabilized by a second cell adhesion molecule. This 80,000-Da glycoprotein is also synthesized during the aggregation phase. If it is defective or absent in the cells, small slugs will form, and their fruiting bodies will be only about one-third the normal size. Thus, the second cell adhesion system seems to be needed for retaining a large enough number of cells to form large fruiting bodies (Müller and Gerisch, 1978; Loomis, 1988). In addition, a third cell adhesion system is activated late in development, while the slug is migrating. The protein or group of proteins that mediates the third system may exist only on prespore cells and may be responsible for separating the prespore cells from the prestalk cells (Loomis, personal communication). Thus, *Dictyostelium* has evolved three developmentally regulated systems of cell–cell adhesion that are necessary for the morphogenesis of individual cells into a coherent organism. As we shall see in subsequent chapters, metazoan cells also use a variety of cell adhesion molecules to form the tissues and organs of the embryo.

Dictyostelium is a "part-time multicellular organism" that does not form many cell types (Kay et al., 1989), and the more complex multicellular organisms do not form from the aggregation of formerly independent cells. However, many of the principles found in the development of this

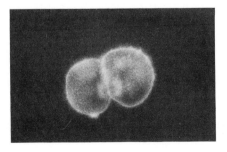

FIGURE 1.24
Dictyostelium cells synthesize an adhesive, 24-kDa glycoprotein shortly after nutrient starvation. *Dictyostelium* cells were stained with a fluorescent antibody that binds to the 24-kDa glycoprotein and were then observed under ultraviolet light. This protein is not seen on amoebae that have just stopped dividing. However, as shown here—10 hours after cell division has ceased—individual amoebae are seen to have this protein in their cell membranes and are capable of adhering together. (Courtesy of W. Loomis.)

"simple" organism are the same as those found in embryos of the more complex phyla. The ability of individual cells to sense a chemical gradient (as in the amoeba's response to cAMP) is very important for cell migration and morphogenesis during animal development. The role of cell-surface proteins for cell cohesiveness is found throughout the animal kingdom, and differentiation-inducing molecules are just beginning to be isolated in metazoan organisms.

SIDELIGHTS & SPECULATIONS

Evidence and antibodies

Biology, like any other science, does not deal with Facts. Rather, it deals with evidence. Several types of evidence will be presented in this book, and they are not all equivalent in strength. As an example, we will use the analysis of cell adhesion in *Dictyostelium*. The first, and weakest, type of evidence is *correlative evidence*. Correlations are made between two or more events, and there is an inference that one event causes the other. As we have seen, fluorescently labeled antibodies to a certain 24-kDa glycoprotein do not label dividing vegetative cells, but they *do* find this protein in myxamoeba cell membranes soon after the cells stop dividing and become competent to aggregate (see Figure 1.24). Thus, there is a correlation between the presence of this cell membrane glycoprotein and the ability to aggregate.

Correlative evidence gives a starting point to investigations, but one cannot say with certainty that one event causes the other based solely upon correlations. It could be equally probable that cell adhesion causes cells to synthesize this new glycoprotein, or that cell adhesion and the synthesis of the 24-kDa glycoprotein are separate events initiated by the same underlying cause. The simultaneous occurrence of the two events could even be coincidental, the events having no relationship to each other.*

How, then, does one get beyond mere correlation? In the study of cell adhesion in *Dictyostelium*, the next step was to use those same antibodies to block the adhesion of myxamoebae. Based on a technique pioneered by Gerisch's laboratory (Beug et al., 1970), Knecht and co-workers (1987) took the antibodies that bound this 24-kDa glycoprotein and isolated their antigen-binding sites (the portions of the antibody molecule that actually recognize the antigen). This was necessary because the whole antibody molecule contains two antigen-binding sites and would therefore artificially cross-link and agglutinate the myxamoebae. When

these antigen-binding fragments (called Fab's) were added to the aggregation-competent cells, the cells could not aggregate. The antibody fragments inhibited the cells' adhering together, presumably by binding to the 24-kDa glycoprotein and blocking its function. This type of evidence is called *loss-of-function evidence*. While stronger than correlative evidence, it still does not make other inferences impossible. For instance, perhaps the antibodies killed the cell (as might have been the case if the 24-kDa glycoprotein were a critical transport channel). This would also stop the cells from adhering. Or perhaps the 24-kDa glycoprotein has nothing to do with adhesion itself but is necessary for the *real* adhesive molecule to function (such as by stabilizing membrane proteins in general). In this case blocking the glycoprotein would similarly cause the inhibition of cell aggregation. Thus, loss-of-function evidence must be bolstered by many controls demonstrating that the agents causing the loss of function specifically knock out the particular function and nothing else.

The strongest type of evidence is *gain-of-function evidence*. Here, the initiation of the first event causes the second event to happen even in instances where neither event usually occurs. Recently, da Silva and Klein (1990) and Faix and co-workers (1990) have obtained such evidence to show that the 80-kDa glycoprotein is an adhesive molecule. They isolated the gene for the 80-kDa protein and modified the gene in a way that would cause it to be expressed all the time. They then placed it back into well-fed, vegetatively growing myxamoebae that do not usually express this protein and that are not usually able to adhere to each other. The presence of this protein on the cell membrane of these dividing cells was confirmed by antibody labeling. Moreover, such cells now adhered to one another even in the vegetative stages when they normally would not. Thus, they had gained an adhesive function solely upon expressing this particular glycoprotein on their cell surfaces. This gain-of-function evidence is more convincing than other types of analysis. Similar experiments have recently been performed on mammalian cells (see Chapter 3) to demonstrate the presence of particular cell adhesion molecules in the developing embryo.

*In a tongue-in-cheek letter spoofing such correlative inferences, Sies (1988) demonstrated a remarkably good correlation between the number of storks seen in West Germany from 1965 to 1980 and the number of babies born during those same years.

Differentiation in *Dictyostelium*. Differentiation into stalk cell or spore cell reflects one of the major phenomena of embryogenesis: the cell's selection of a developmental pathway. Cells are often seen to select a particular developmental fate when alternatives are available. A given bone marrow cell, for instance, can become a lymphocyte or a blood cell; a particular embryonic cell can become either an epidermal skin cell or a neuron. In *Dictyostelium*, we see a simple dichotomous decision, because only two cell types are possible. How is it that a given cell becomes a stalk cell or a spore cell? Although the details are not fully known, a cell's fate appears to be regulated by certain diffusible molecules. The two major candidates are **differentiation-inducing factor (DIF)** and cyclic AMP. DIF appears to be necessary for stalk cell differentiation. This factor, like the sex-inducing factor of *Volvox*, is effective at very low concentrations (10^{-10} M); and, like the *Volvox* protein, it appears to induce the differentiation of a particular type of cell. When added to isolated amoebae or even to prespore (posterior) cells, it causes them to form stalk cells. This low-molecular-weight lipid is genetically regulated, for there are mutant strains of *Dictyostelium* that form only spore precursors and no stalk cells. When DIF is added to these mutant cultures, stalk cells are able to differentiate (Kay and Jermyn, 1983; Morris et al., 1987) and new prestalk-specific mRNAs are seen in the cell cytoplasm (Williams et al., 1987).

While DIF stimulates amoebae to become prestalk cells, the differentiation of prespore cells is most likely controlled by the continuing pulses of cAMP. High concentrations of cAMP initiate the expression of prespore-specific mRNAs in aggregated amoebae. Moreover, when slugs are placed in a medium containing an enzyme that destroys extracellular cAMP, the prespore cells lose their differentiated characteristics (Figure 1.25; Schaap and van Driel, 1985; Wang et al., 1988a,b).

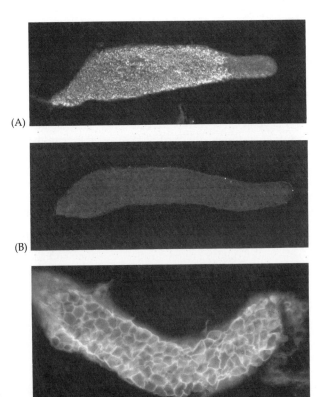

FIGURE 1.25
Chemicals controlling differentiation in *Dictyostelium*. A and B show the effects of placing *Dictyostelium* slugs into a medium containing enzymes that destroy extracellular cAMP. (A) Control grex stained for the presence of a prespore-specific protein (white regions). (B) Similar grex stained after the treatment with cAMP-degrading enzymes. No prespore-specific product is seen. (C) Higher magnification of a slug treated with DIF (in the absence of ammonia). The stain used here binds to the cellulose wall of the stalk cells. All cells of the grex have become stalk cells. (A and B from Wang et al., 1988a; C from Wang and Schaap, 1989; courtesy of the authors.)

How the grex knows which end is up

If all the amoebae of the grex start out equal, how can those cells in the posterior two-thirds of the slug differentiate into spore cells, while those equivalent cells in the anterior third become stalk cells? The answer may lie in the observation that all the original cells are *not* equal. Amoebae that become starved during the early portion of their cell cycle tend to form the anterior portion of the slug, while amoebae starved toward the end of their cell cycle tend to form the posterior (McDonald and Durston, 1984; Weijer et al., 1984). This work has been confirmed and extended by Ohmori and Maeda (1987), who showed that those cells starved in the latter part of the cell cycle respond differently to cAMP and show much higher adhesivity than cells starved just after mitosis. Williams and co-workers (1989) have found that prespore and prestalk cells can be distinguished in early aggregates, and that they are randomly distributed throughout these hemispherical mounds. Thus, the fates of the cells are established even before the grex starts migrating. Within each aggregate, most of the prestalk cells actively migrate to the anterior, while prespore cells remain in what becomes the posterior region of the grex. This migration is probably due to repeated pulses of cAMP that are still emanating from the apical tip of the aggregate. These pulses are chemotactic for prestalk cells but not for prespore cells, so they draw the prestalk cells to the tip of the aggregate (Matsukuma and Durston, 1979; Mee et al., 1986; Takeuchi, 1991; Siegert and Weijer, 1991). Cyclic AMP, then, is seen to play several different roles in *Dictyostelium* development. It aggregates the cells together, it induces prespore cell differentiation, and it directs the migration of prestalk cells to the anterior of the aggregate.

Once the aggregate is complete, it topples over onto its side and forms the migrating grex. Most of the prestalk cells are in the anterior 20 percent of the grex, but there are also some scattered prestalk cells throughout the posterior. Prestalk cells can be distinguished by their secretion of extracellular matrix protein A into the spaces between their cells. Within the center of the anterior portion of the grex, another group of prestalk cells begin secreting a second new protein (extracellular matrix protein B) into their extracellular matrix. These central cells are called prestalk B cells, while the majority of the prestalk cells are denoted as prestalk A cells (Figure 1.26). When the grex finds itself in sunlight, it ceases migrating and undergoes the final differentiation into spores and a stalk. During this process (called culmination), the grex sits upon one end so that the rearguard cells become its base. The prestalk A cells migrate onto the central tube of prestalk B cells, and as they get to the central tube they differentiate into prestalk B cells by synthesizing new extracellular matrix components. The new cells are added to the anterior region of the tube and force the tube farther down into the culminating structure. This tube differentiates to become the stalk. At the same time, the prestalk A cells that had been in the posterior region of the grex migrate to the boundaries of the prespore cells and differentiate into the spore case and the basal disc (Williams and Jermyn, 1991; Harwood et al., 1992). Eventually, the spores are lifted 2 mm off the ground, from which point they can be dispersed.

The trigger for this culmination process appears to be sunlight or low humidity. Recent experiments suggest that these two factors act by causing the diffusion of ammonia from the slug. Ammonia is copiously produced by migrating slugs and represses culmination. Whenever ammonia is depleted (either naturally or experimentally), culmination begins (Schindler and Sussman, 1977; Newell and Ross, 1982; Bonner et al., 1985). Ammonia inhibits the conversion of the prestalk A cells into prestalk B cells and prohibits further stalk formation (Gross et al., 1983; Wang

 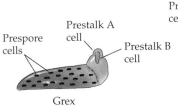 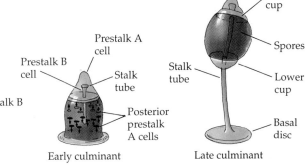

FIGURE 1.26
Regulation of stalk cell differentiation during the culmination phase of Dictyostelium *growth. Schematic representation of cell migration shows that prespore and prestalk cells are usually mixed in the early aggregate stage, but sort out so that most of the prestalk cells are at the anterior of the grex. Prespore cells are shown in color, while stalk cells are shown in shades of gray and black. The prestalk A cells constitute most of the anterior of the grex, with some similar cells in the posterior. Prestalk B cells are seen in the center of the anterior portion of the grex. In the early culmination stages, the prestalk cells in the posterior migrate to form the basal disc and cups of the spore sac; and the anterior prestalk A cells migrate toward the center and become prestalk B cells. This extends the stalk until it lifts the spore case off the surface. (After Harwood et al., 1992.)*

et al., 1990). Bonner and co-workers (1985) have suggested that as light causes more rapid diffusion of ammonia, it would remove the inhibitor and thereby allow culmination to proceed.

Ammonia appears to inhibit stalk production in at least two ways. First, it inhibits the action of DIF (Wang and Schaap, 1989). Second, it inhibits the production of cAMP in the prestalk cells (Schindler and Sussman, 1977; Harwood et al., 1992). This cAMP is needed to activate the enzyme cAMP-dependent protein kinase (PKA). Prestalk cells carrying nonfunctional PKA cannot phosphorylate certain proteins. Neither can these cells migrate into the central anterior region, nor do they differentiate into stalk cells (Firtel and Chapman, 1990; Harwood et al., 1992). The data suggest that when PKA is activated, it phosphorylates a repressor that inhibits the stalk differentiation genes from being expressed. In its phosphorylated state, the repressor is inactive. Once cAMP levels are elevated (by the removal of ammonia), PKA can suppress the inhibitor in the stalk-forming genes (Figure 1.27).

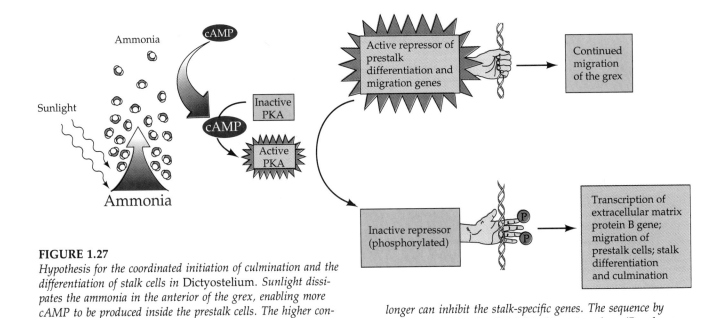

FIGURE 1.27
Hypothesis for the coordinated initiation of culmination and the differentiation of stalk cells in Dictyostelium. *Sunlight dissipates the ammonia in the anterior of the grex, enabling more cAMP to be produced inside the prestalk cells. The higher concentrations of cAMP activate PKA, which phosphorylates an inhibitor of stalk gene expression. The phosphorylated inhibitor no longer can inhibit the stalk-specific genes. The sequence by which spore formation is inhibited is not as clear. (Based on models of Bonner et al., 1985, and Harwood et al., 1992.)*

Developmental patterns among the metazoans

Since the remainder of this book concerns the development of **metazoans**—multicellular organisms that pass through embryonic stages of development—we will present an overview of their developmental patterns. The final chapter of the text will discuss these patterns in more detail. Figure 1.28 illustrates the major evolutionary trends of metazoan development. The most striking observation is that life has not evolved in a straight line, but rather there are several evolutionary paths. We can see that most of the species of metazoans belong to one of two major branches of animals: protostomes and deuterostomes.

The Porifera

The colonial protists are thought to have given rise to two groups of metazoans, both of which pass through embryonic stages of development. One of these groups is the Porifera (sponges). These animals develop in

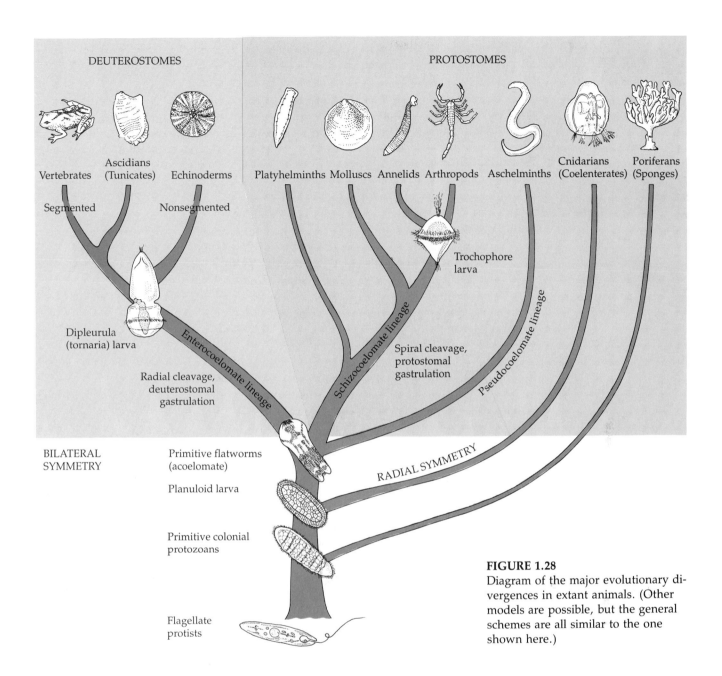

DEUTEROSTOMES

PROTOSTOMES

Vertebrates   Ascidians   Echinoderms   Platyhelminths Molluscs Annelids Arthropods Aschelminths   Cnidarians   Poriferans
               (Tunicates)   (Coelenterates) (Sponges)

Segmented   Nonsegmented

Trochophore
larva

Dipleurula
(tornaria) larva

Enterocoelomate lineage

Schizocoelomate lineage

Spiral cleavage,
protostomal
gastrulation

Pseudocoelomate lineage

Radial cleavage,
deuterostomal
gastrulation

BILATERAL
SYMMETRY

Primitive flatworms
(acoelomate)

Planuloid larva

RADIAL SYMMETRY

Primitive colonial
protozoans

FIGURE 1.28
Diagram of the major evolutionary divergences in extant animals. (Other models are possible, but the general schemes are all similar to the one shown here.)

Flagellate
protists

a manner so different from that of any other animal group that some taxonomists do not consider them metazoans at all. A sponge has three types of somatic cells, but one of these, the **archeocyte,** can differentiate into all the other cell types in the body. The cells of a sponge passed through a sieve can regenerate new sponges from individual cells. Moreover, such reaggregation is species-specific, and if individual sponge cells from different species are mixed together, each of the sponges that reforms contains cells from only one species (Wilson, 1907). It is generally thought that the motile archeocytes collect cells from their own species and not from others (Turner, 1978). Sponges contain no mesoderm, so there are no true organ systems in the Porifera; nor do they have a digestive tube or circulatory system, nerves, or muscles. Thus, even though they pass through an embryonic and a larval stage, sponges are very unlike most metazoans.

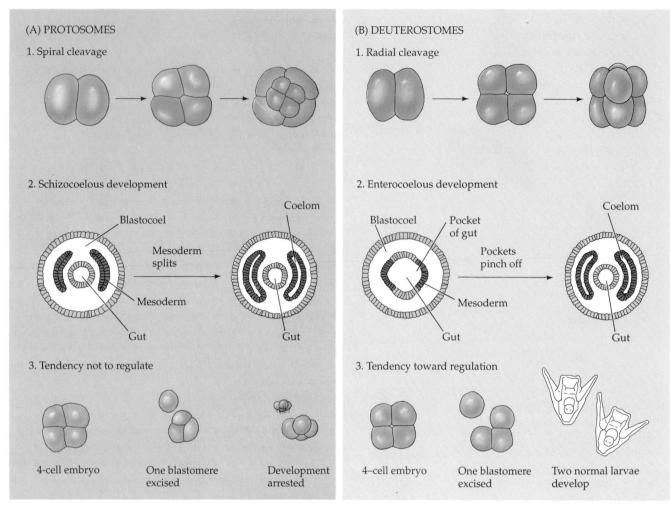

(A) PROTOSOMES

1. Spiral cleavage

2. Schizocoelous development

Blastocoel

Mesoderm splits

Coelom

Mesoderm

Gut

Gut

3. Tendency not to regulate

4-cell embryo

One blastomere excised

Development arrested

(B) DEUTEROSTOMES

1. Radial cleavage

2. Enterocoelous development

Blastocoel

Pocket of gut

Pockets pinch off

Coelom

Mesoderm

Gut

Gut

3. Tendency toward regulation

4–cell embryo

One blastomere excised

Two normal larvae develop

FIGURE 1.29
Principal tendencies of the deuterostomes and protostomes. Exceptions to all these general tendencies have evolved secondarily in certain members of each group. (Mammals and birds, for instance, do not have a strictly enterocoelous formation of the body cavity.)

Protostomes and deuterostomes

The other group of metazoans arising from the colonial protists is characterized by the presence of three germ layers during development. Some members of this group constitute the **Radiata,** so called because they have radial symmetry, like that of a tube or a wheel. The Radiata include the cnidarians (jellyfish, corals, and hydra) and ctenophores (comb jellies). In these animals, the mesoderm is rudimentary, consisting of sparsely scattered cells in a gelatinous matrix. Most metazoans, however, have bilateral symmetry and thus constitute the **Bilateria**. These bilateral phyla are classified as either protostomes or deuterostomes. All Bilateria are thought to have descended from a primitive type of flatworm. These flatworms were the first to have a true mesoderm (although it was not hollowed out to form a body cavity), and they are thought to have resembled the larvae of certain contemporary coelenterates. The differences in the two major divisions of the Bilateria are illustrated in Figure 1.29. **Protostomes** (Gk., "mouth first"), which include the mollusc, arthropod, and worm phyla, are so called because during gastrulation the mouth regions are formed first. The body cavity of these animals forms from the hollowing out of a previously solid cord of mesodermal cells.

The other great division of the Bilateria is the **deuterostome** lineage. Phyla in this division include chordates and echinoderms. Although it may seem strange to classify humans and wolverines in the same group as starfish and sea urchins, certain embryological features stress this kinship. First, in deuterostomes (Gk., "mouth second"), the mouth region is formed after the anal region. Also, whereas protostomes form their body cavity by hollowing out a solid mesodermal block (schizocoelous formation of the body cavity), most deuterostomes form their body cavities from mesodermal pouches extending from the gut (enterocoelous formation of the body cavity).

In addition, protostomes and deuterostomes differ in the way they undergo cleavage. Most deuterostomes undergo cleavage such that the blastomeres are perpendicular or parallel to each other. This is called **radial cleavage.** Protostomes, on the other hand, form blastulae composed of cells that are at acute angles to the polar axis of the embryo. Thus, they are said to undergo **spiral cleavage.** Furthermore, the cleavage stage blastomeres of most deuterostomes have a greater ability to regulate development than do those of protostomes. If a single blastomere is removed from a 4-cell sea urchin or human embryo, that blastomere will develop into an entire organism, and the remaining three-quarters of the embryo can also develop normally. However, if the same operation is performed on a snail or worm embryo, both the single blastomere and the remaining ones develop into partial embryos—each lacking what was formed from the other.

The evolution of organisms depends upon inherited changes in their development. One of the greatest evolutionary advances—the **amniote egg**—occurred among the deuterostomes. This type of egg, exemplified by that of a chicken (Figure 1.30), first appeared in reptiles about 255 million years ago. The amniote egg allowed vertebrates to roam on land, far from existing ponds. Whereas amphibians must return to water to breed and to enable their eggs to develop, the amniote egg carries its own water and food supplies. The egg is fertilized internally and contains **yolk** to nourish the developing embryo. Moreover, it contains two sacs: the **amnion,** which contains the fluid bathing the embryo, and the **allantois,** in which waste materials from embryonic metabolism collect. The entire structure is encased in a shell that allows the diffusion of oxygen but is hard enough to protect the embryo from environmental assaults. A similar development of egg casings enabled arthropods to be the first terrestrial invertebrates. Thus, the final crossing of the boundary between water and land occurred with the modification of the earliest stage in development, the egg.

Developmental biology provides an endless assortment of fascinating animals and problems. It is the author's task to select a small sample of

FIGURE 1.30
Diagram of the amniote egg of the chick, showing the development of membranes enfolding the embryo. (A) Three-day incubation. (B) Seven-day incubation. The origin of the membranes will be detailed in Chapter 9. The yolk is eventually surrounded by membranes derived from the embryo (to absorb nutrients); other membranes extend from the embryo to the shell (where they will exchange oxygen and carbon dioxide and obtain calcium from the shell). The ectoderm eventually covers the outside of the embryo and extends to the shell. The endoderm becomes the gut and encircles the yolk, while the mesoderm extends to cover these layers and provide the blood vessels to bring the extraembryonic material into the embryo.

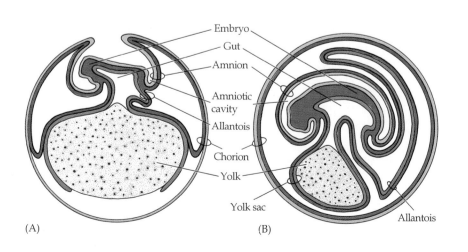

Embryo
Gut
Amnion
Amniotic cavity
Allantois
Chorion
Yolk
Yolk sac
Allantois

(A)   (B)

them to illustrate the major principles of animal development. Yet this sample is an incredibly small collection. We are merely observing the small tidepool within our reach while the whole ocean of developmental principles lies before us. After a brief outline of the genetic and cellular principles relevant to developmental biology, we will investigate the early stages of animal embryogenesis: fertilization, cleavage, gastrulation, and the establishment of the vertebrate body plan. Later chapters will concentrate on the genetic and cellular mechanisms by which animal bodies are constructed. Although an attempt has been made to survey the important variations throughout the animal kingdom, a certain deuterostome chauvinism may be apparent.

LITERATURE CITED

Bergman, K., Goodenough, U. W., Goodenough, D. A., Jawitz, J. and Martin, H. 1975. Gametic differentiation in *Chlamydomonas reinhardtii*. II. Flagellar membranes and the agglutination reaction. *J. Cell Biol.* 67: 606–622.

Beug, H., Gerisch, G., Kempff, S., Riedel, V. and Cremer, G. 1970. Specific inhibition of cell contact formation in *Dictyostelium* by univalent antibodies. *Exp. Cell Res.* 63: 147–158.

Bonner, J. T. 1947. Evidence for the formation of cell aggregates by chemotaxis in the development of the slime mold *Dictyostelium discoideum*. *J. Exp. Zool.* 106: 1–26.

Bonner, J. T. 1957. A theory of the control of differentiation in the cellular slime molds. *Q. Rev. Biol.* 32: 232–246.

Bonner, J. T., Berkley, D. S., Hall, E. M., Konijn, T. M., Mason, J. W., O'Keefe, G. and Wolfe, P. B. 1969. Acrasin, acrasinase and the sensitivity to acrasin in *Dictyostelium discoideum*. *Dev. Biol.* 20: 72–87.

Bonner, J., Hay, A. and John, D. 1985. pH affects fruiting and slug orientation in *Dictyostelium discoideum*. *J. Embryol. Exp. Morphol.* 87: 207–213.

da Silva, A. M. and Klein, C. 1990. Cell adhesion transformed *D. discoideum* cells: Expression of gp80 and its biochemical characterization. *Dev. Biol.* 140: 139–148.

Faix, J., Gerisch, G. and Noegel, A. A. 1990. Constitutive overexpression of the contact A glycoprotein enables growth-phase cells of *Dictyostelium discoideum* to aggregate. *EMBO J.* 9: 2709–2716.

Firtel, R. A. and Chapman, A. L. 1990. A role for cAMP-dependent protein kinase A in early *Dictyostelium* development. *Genes Dev.* 4: 18–28.

Fulton, C. 1977. Cell differentiation in *Naegleria gruberi*. *Annu. Rev. Microbiol.* 31: 597–629.

Fulton, C. and Walsh, C. 1980. Cell differentiation and flagellar elongation in *Naegleria gruberi*: Dependence on transcription and translation. *J. Cell Biol.* 85: 346–360.

Garcia, E. and Dazy, A.-C. 1986. Spatial distribution of poly(A)$^+$ RNA and protein synthesis in *Acetabularia mediterranea*. *Biol. Cell* 58: 23–29.

Goodenough, U. W. and Weiss, R. L. 1975. Gametic differentiation in *Chlamydomonas reinhardtii*. III. Cell wall lysis and microfilament associated mating structure activation in wild-type and mutant strains. *J. Cell Biol.* 67: 623–637.

Gross, J., Bradbury, J., Kay, R. and Peacey, M. 1983. Intracellular pH and the control of cell differentiation in *Dictyostelium discoideum*. *Nature* 303: 244–245.

Hämmerling, J. 1934. Über formbildended Substanzen bei *Acetabularia mediterranea*, ihre räumliche und zeitliche Verteilung und ihre Herkunft. *Wilhelm Roux Arch. Entwicklungsmech. Org.* 131: 1–81.

Hartmann, M. 1921. Die dauernd agame Zucht von *Eudorina elegans*, experimentelle Beiträge zum Befruchtungs-und Todproblem. *Arch. Protistk.* 43: 223–286.

Harwood, A. J., Hopper, N. A., Simon, M.-N., Driscoll, D. M., Veron, M. and Williams, J. G. 1992. Culmination in *Dictyostelium* is regulated by the cAMP-dependent protein kinase. *Cell* 69: 615–624.

Jermyn, K. A. and Williams, J. 1991. An analysis of culmination in *Dictyostelium* using prestalk and stalk-specific cell autonomous markers. *Development* 111: 779–787.

Johnson, R. L., Gundersen, R., Lilly, P., Pitt, G. S., Pupillo, M., Sun, T. J., Vaughan, R. A. and Devreotes, P. N. 1989. G-protein-linked signal transduction systems control development in *Dictyostelium*. *Development* [Suppl.]: 75–80.

Kay, R. R. and Jermyn, K. A. 1983. A possible morphogen controlling differentiation in *Dictyostelium*. *Nature* 303: 242–244.

Kay, R. R., Berks, M. and Traynor, D. 1989. Morphogen hunting in *Dictyostelium*. *Development* [Suppl.]: 81–90.

Keller, E. F. and Segal, L. A. 1970. Initiation of slime mold aggregation viewed as an instability. *J. Theoret. Biol.* 26: 399–415.

Kirk, D. L. 1988. The ontogeny and phylogeny of cellular differentiation in *Volvox*. *Trends Genet.* 4: 32–36.

Kirk, D. L. and Kirk, M. M. 1986. Heat shock elicits production of sexual inducer in *Volvox*. *Science* 231: 51–54.

Kirk, D. L., Viamontes, G. I., Green, K. J. and Bryant, J. L. Jr. 1982. Integrated morphogenetic behavior of cell sheets: Volvox as a model. *In* S. Subtelny and P. B. Green (eds.), *Developmental Order: Its Origin and Regulation.* Alan R. Liss, New York, pp. 247–274.

Kloppstech, K. and Schweiger, H. G. 1975. Polyadenylated RNA from *Acetabularia*. *Differentiation* 4: 115–123.

Knecht, D. A., Fuller, D. and Loomis, W. F. 1987. Surface glycoprotein gp24 involved in early adhesion of *Dictyostelium discoideum*. *Dev. Biol.* 121: 277–283.

Konijn, T. M., van der Meene, J. G. C., Bonner, J. T. and Barkley, D. S. 1967. The acrasin activity of adenosine -3',5'-cyclic phosphate. *Proc. Natl. Acad. Sci. USA* 58: 1152–1154.

Krutch, J. W. 1956. *The Great Chain of Life.* Houghton Mifflin, Boston. [pp. 28–29]

Loomis, W. F. 1988. Cell–cell adhesion in *Dictyostelium discoideum*. *Dev. Genet.* 9: 549–559.

Matsukuma, S. and Durston, A. J. 1979. Chemotactic cell sorting in *Dictyostelium discoideum*. *J. Embryol. Exp. Morphol.* 50: 243–251.

McDonald, S. A. and Durston, A. J. 1984. The cell cycle and sorting out behaviour in *Dictyostelium discoideum*. *J. Cell Sci.* 66: 196–204.

Mee, J. D., Tortolo, D. M. and Coukell, M. B. 1986. Chemotaxis-associated properties of separated prestalk and prespore cells of *Dictyostelium discoideum*. *Biochem. Cell Biol.* 64: 722–732.

Morris, H. R., Taylor, G. W., Masento, M. S., Jermyn, K. A. and Kay, R. R. 1987. Chemical structure of the morphogen differentiation-inducing factor from *Dictyostelium discoideum*. *Nature* 328: 811–814.

Müller, K. and Gerisch, G. 1978. A specific glycoprotein as the target of adhesion blocking Fab in aggregating *Dictyostelium* cells. *Nature* 274: 445–447.

Newell, P. and Ross, F. 1982. Genetic analysis of the slug stage of *Dictyostelium discoideum*. *J. Genet. Microbiol.* 128: 1639–1652.

Ohmori, T. and Maeda, Y. 1987. The developmental fate of *Dictyostelium discoideum* cells depends greatly on the cell–cycle position at the onset of starvation. *Cell Differ.* 22: 11–18.

Pommerville, J. and Kochert, G. 1982. Effects of senescence on somatic cell physiology in the green alga *Volvox carteri*. *Exp. Cell Res.* 14: 39–45.

Powers, J. H. 1908. Further studies on *Volvox*, with description of three new species. *Trans. Am. Micros. Soc.* 28: 141–175.

Raper, K. B. 1940. Pseudoplasmodium formation and organization in *Dictyostelium discoideum*. *J. Elisha Mitchell Sci. Soc.* 56: 241–282.

Robertson, A., Drage, D. J. and Cohen, M. H. 1972. Control of aggregation in *Dictyostelium discoideum* by an external periodic pulse of cyclic adenosine monophosphate. *Science* 175: 333–335.

Schaap, P. and van Driel, R. 1985. The induction of post–aggregative differentiation in *Dictyostelium discoideum* by cAMP. Evidence for involvement of the cell surface cAMP receptor. *Exp. Cell Res.* 159: 388–398.

Schindler, J. and Sussman, M. 1977. Ammonia determines the choice of morphogenetic pathways in *Dictyostelium discoideum*. *J. Mol. Biol.* 116: 161–169.

Shaffer, B. M. 1953. Aggregation in cellular slime molds: In vitro isolation of acrasin. *Nature* 171: 975.

Shaffer, B. M. 1975. Secretion of cyclic AMP induced by cyclic AMP in the cellular slime mold *Dictyostelium discoideum*. *Nature* 255: 549–552.

Siegert, F. and Weijer, C. J. 1989. Digital image processing of optical density wave propagation in *Dictyostelium discoideum* and analysis of the effects of caffeine and ammonia. *J. Cell Science* 93: 325–335.

Siegert, F. and Weijer, C. J. 1991. Analysis of optical density wave propagation and cell movement in the cellular slime mould *Disctyostelium discoideum*. *Physica D* 49: 224–232.

Sies, H. 1988. A new parameter for sex education. *Nature* 332: 495.

Sumper, M., Berg, E., Wenzl, S. and Godl, K. 1993. How a sex pheremone might act at a concentration below 10^{-16} *M. EMBO J.* 12: 831–836.

Strickberger, M. W. 1985. *Genetics*, 3rd Ed. Macmillan, New York.

Takeuchi, I. 1991. Cell sorting and pattern formation in *Dictyostelium discoideum*. *In* J. Gerhart (ed.), *Cell–Cell Interactions in Early Development*. Wiley-Liss, New York, pp. 249–259.

Tomchick, K. J. and Devreotes, P. N. 1981. Adenosine 3′,5′ monophosphate waves in *Dictyostelium discoideum*: A demonstration by isotope dilution-fluorography. *Science* 212: 443–446.

Turner, R. S. Jr. 1978. Sponge cell adhesions. *In* D. R. Garrod (ed.), *Specificity of Embryological Interactions*. Chapman and Hall, London, pp. 199–232.

Tyson, J. J. and Murray, J. D. 1989. Cyclic AMP waves during aggregation of *Dictyostelium amoebae*. *Development* 106: 421–426.

Walsh, C. 1984. Synthesis and assembly of the cytoskeleton of *Naegleria gruberi* flagellates. *J. Cell Biol.* 98: 449–456.

Wang, M. and Schaap, P. 1989. Ammonia depletion and DIF trigger stalk cell differentiation in intact *Dictyostelium discoideum* slugs. *Development* 105: 569–574.

Wang, M., van Driel, R. and Schaap, P. 1988a. Cyclic AMP-phosphodiesterase induces dedifferentiation of prespore cells in *Dictyostelium discoideum* slugs: Evidence that cyclic AMP is the morphogenetic signal for prespore differentiation. *Development* 103: 611–618.

Wang, M. , Aerts, R. J., Spek, W. and Schaap, P. 1988b. Cell cycle phase in *Dictyostelium discoideum* is correlated with the expression of cyclic AMP production, detection and degradation. Involvement of cAMP signalling in cell sorting. *Dev. Biol.* 125: 410–416.

Wang, M., Roelfsema, J. H., Williams, J. G. and Schaap, P. 1990. Cytoplasmic acidification facilitates but does not mediate DIF-induced prestalk gene expression in *Dictyostelium discoideum*. *Dev. Biol.* 140: 182–188.

Weijer, C. J., Duschl, G. and David, C. N. 1984. Dependence of cell-type proportioning and sorting on cell cycle phase in *Dictyostelium discoideum*. *Exp. Cell Res.* 70: 133–145.

Williams, J. G. and Jermyn, K. A. 1991. Cell sorting and positional differentiation during *Dictyostelium* morphogenesis. *In* J. Gerhart (ed.), *Cell–Cell Interactions in Early Development*. Wiley-Liss, New York, pp. 261–272.

Williams, J. G., Ceccarelli, A., McRobbie, S., Mahbubani, H., Kay, R. R., Early, A., Berks, M. and Jermyn, K. A. 1987. Direct induction of *Dictyostelium* pre-stalk gene expression by DIF provides evidence that DIF is a morphogen. *Cell* 49: 185–192.

Williams, J. G., Duffy, K. T., Lane, D. P., McRobbie, S. J., Harwood, A. J., Traynor, D., Kay, R. R. and Jermyn, K. A. 1989. Origins of the prestalk–prespore pattern in *Dictyostelium* development. *Cell* 59: 1157–1163.

Wilson, E. B. 1896. *The Cell in Development and Inheritance*. Macmillan, New York.

Wilson, H. V. 1907. On some phenomena of coalescence and regeneration in sponges. *J. Exp. Zool.* 5: 245–258.

2

Genes and development
Introduction and techniques

"Between the characters that furnish the data for the theory, and the postulated genes, to which the characters are referred, lies the whole field of embryonic development." Here Thomas Hunt Morgan (1926) noted that the only way to get from genotype to phenotype is through developmental processes. In the early twentieth century, embryology and genetics were not considered to be separate sciences. They diverged in the 1920s, when Morgan redefined genetics as the science studying the *transmission* of traits, as opposed to embryology, which was to study the *expression* of these traits. Within the past five years, however, the techniques of molecular biology have effected a *rapprochement* of embryology and genetics. In fact, the two fields have again become linked to a point that necessitates an early discussion of molecular genetics in this text. Problems in animal development that could not be addressed a decade ago are now being solved by a set of techniques involving nucleic acid synthesis and hybridization. This chapter seeks to place these new techniques within the context of the ongoing dialogues between genetics and embryology.

The embryological origins of the gene theory

Mendel called them *Formbildungelementen*, form-building elements; we call them genes. It is in Mendel's term, however, that we see how closely intertwined were the concepts of inheritance and development in the nineteenth century. Up through the first decades of the twentieth century, theories of heredity (such as those proposed by Charles Darwin or August Weismann) had to explain both the transmission and the expression of these inherited form-building elements.

Nucleus or cytoplasm: Which controls heredity?

The gene theory that was to become the cornerstone of modern genetics originated from a controversy within the field of embryology. In the late 1800s, a group of scientists began to study, for its own intrinsic value, how fertilized eggs give rise to adult organisms. Two young American embryologists, Edmund Beecher Wilson and Thomas Hunt Morgan (Figure 2.1), became part of this group of "physiological embryologists," and each became a partisan in the controversy over which of the two com-

(A)

(B)

FIGURE 2.1
(A) E. B. Wilson (1856–1939; shown here around 1899), an embryologist whose work on early embryology and sex determination greatly advanced the chromosomal hypotheses of development. (Wilson was also acknowledged to be among the best amateur cellists in the country.) (B) Thomas Hunt Morgan (1866–1945), who brought the gene theory out of embryology. This photo—taken in 1915, as the basic elements of the gene theory were coming together—shows Morgan sorting flies. (He stood up to do so and used a hand lens.) (A courtesy of W. N. Timmins; B courtesy of G. Allen.)

partments of the fertilized egg—the nucleus or the cytoplasm—controls inheritance.

When Morgan and Wilson entered into this debate, the dispute was already a lively one. One school, associated with Oskar Hertwig, Wilhelm Roux, and Theodor Boveri, proposed that the chromosomes of the *nucleus* contained the form-building elements. This group was challenged by Eduard Pflüger, T. L. W. Bischoff, Wilhelm His, and their colleagues, who believed that no preformed structures could cause such enormous changes during development, but that the inherited patterns of development were caused by the creation of new molecules from the interacting gamete *cytoplasms*. Morgan allied himself with this latter group and obtained data that he interpreted as being consistent with the cytoplasmic model of inheritance. His most crucial experiment was to remove cytoplasm from the newly fertilized ctenophore (comb jelly) egg. "Here, although the entire segmentation nucleus is present, yet by loss of cytoplasm, defects are produced in the embryos." Morgan (1897) concluded that "there seems to be no escape from the conclusion that in the cytoplasm and not in the nucleus lies the differentiating power of the early stages of development."

Wilson, however, became a major proponent of those theories placing the form-building elements on the nuclear chromosomes. He stated this view forcefully in his book *The Cell in Development and Inheritance* (1896): "This fact [of the necessity for the nucleus for protozoan regeneration; see Chapter 1] establishes the presumption that the nucleus is, if not the seat of the formative energy, at least the controlling factor in that energy and hence the controlling factor in inheritance. This presumption becomes a certainty when we turn to the facts of maturation [meiosis], fertilization, and cell-division. All those converge to the conclusion that the chromatin is the most essential element in development." Wilson (1895) did not shrink from the consequences of this conclusion:*

> Now, chromatin is known to be closely similar, if not identical with, a substance known as nuclein . . . which analysis shows to be a tolerably definite chemical composed of a nucleic acid (a complex organic acid rich in phosphorus) and albumin. And thus we reach the remarkable conclusion that inheritance may, perhaps, be effected by the physical transmission of a particular chemical compound from parent to offspring.

Wilson thought that the cytoplasmic material Morgan had removed from the cytoplasm of ctenophore eggs had already been secreted there by the nuclear chromosomes (Wilson, 1894, 1904). To Wilson (1905), "The protoplasmic stuffs appear to be only the immediate means or the efficient cause of differentiation, and we still seek its primary determination in the causes that lie more deeply."

Some of the major support for the chromosomal hypothesis of inheritance was coming from the embryological studies of Theodor Boveri (Figure 2.2A), a researcher at the Naples Zoological Station. Boveri fertilized sea urchin eggs with large concentrations of their sperm and obtained eggs that had been fertilized by two sperm. At first cleavage, these eggs formed four mitotic poles and divided the egg into four cells instead of two (see Chapter 4). Boveri then separated the blastomeres and demonstrated that each cell developed abnormally and in different ways. He showed that these abnormalities were due to each of the cells' having different *types* of chromosomes. Thus, Boveri claimed that each chromosome had an individual nature and controlled different vital processes.

The X chromosome as bridge between genes and development

In addition to Boveri's evidence, E. B. Wilson (1905) and Nettie Stevens (1905a,b) demonstrated a critical correlation between nuclear chromosomes and organismal development. Stevens (Figure 2.2B), a former student of Morgan's, showed that in 92 species of insects (and one primitive chordate), females have two sex-specific chromosomes in each nucleus (XX), while males have only one X chromosome (XY or XO). It appeared that a nuclear structure, the X chromosome, was controlling sexual development.† Morgan disagreed with their interpretation that the chromosomes actually determined sex. Rather, he viewed the assortment of chro-

FIGURE 2.2
Chromosomal uniqueness was shown by Boveri and Stevens. (A) Theodor Boveri (1862–1915) whose work, Wilson (1918) said, "accomplished the actual amalgamation between cytology, embryology, and genetics—a biological achievement which . . . is not second to any of our time." This photograph was taken in 1908, when Boveri's chromosomal and embryological studies were at their zenith. (B) Nettie M. Stevens (1861–1912), who trained with both Boveri and Morgan, seen here in 1904 when she was a postdoctoral student pursuing the research that correlated the number of X chromosomes with sexual development. (A from Baltzer, 1967; B courtesy of the Carnegie Institute of Washington.)

---

*Notice that Wilson is writing about form-building units in chromatin in *1896*—before the rediscovery of Mendel's paper or the founding of the gene theory. For further analysis of the interactions between Morgan and Wilson that led to the gene theory, see Gilbert (1978; 1987) and Allen (1986).

†Wilson was one of Morgan's closest friends, and Morgan considered Stevens his best graduate student at that time. Both were against Morgan on this issue. Even though they disagreed, Morgan wholeheartedly supported Stevens's request for research funds, saying that her qualifications were the best possible. Wilson wrote an equally laudatory letter of support even though she would be a rival researcher (see Brush, 1978).

mosomes as a secondary sexual characteristic that was controlled by some cytoplasmic sex-determining substance.

Morgan's "conversion" to the chromosomal hypothesis came after he obtained data contrary to his theories (see Allen, 1978; Gilbert, 1978; Lederman, 1989). While breeding *Drosophila* for a series of experiments on evolution, Morgan began to obtain several mutations that all correlated with sex. (As Morgan was soon to show, one would expect to find X-linked mutations sooner than those on other chromosomes, since defects on the X chromosome would not be masked by the homologous chromosome in the male.) In 1910, he showed that the traits for both sex and for white eye color are correlated in some way to the presence of a particular X chromosome; but he avoided calling them physically linked. However, by 1911, Morgan had shown that factors regulating eye color, body color, wing shape, and sex all segregate together with the X chromosome. He began to interpret these results in terms of the genes being physically linked to one another on the chromosomes. The embryologist Morgan had shown that nuclear chromosomes are responsible for the development of inherited characters.

The split between embryology and genetics

Morgan's evidence provided a material basis for the concept of the gene. Genetics had mostly been a science of animal and plant breeding, and this gave it a more scientific credibility. Driven both by the urge for progress in animal and plant (and human) breeding and by the ability of geneticists to obtain mathematically verifiable and concrete results within short periods of time, genetics soon became the predominant biological science in the United States (see Allen, 1986; Sapp, 1987; Paul and Kimmelman, 1988). In the 1930s, genetics became its own discipline, developing its own vocabulary, journals, societies, favored organisms, professorships, and rules of evidence. Hostility between embryology and genetics also emerged. Geneticists believed that the embryologists were old-fashioned and that development would be completely explained as the result of gene expression. As Richard Goldschmidt (1938) proclaimed, "Development is, of course, the orderly production of pattern, and therefore after all, genes control pattern." If embryologists were not going to look at embryogenesis in terms of gene activity, the geneticists would.

Conversely, the embryologists thought the geneticists to be irrelevant and uninformed. Embryologists such as Frank Lillie (1927), Ross Granville Harrison (1937), Hans Spemann (1938), and Ernest E. Just (1939) (Figure 2.3) claimed that there could be no genetic theory of development until at least three major challenges had been met by the geneticists.

1. Geneticists had to explain how chromosomes—which were thought to be *identical* in every cell of the organism—direct *different* and *changing* types of cell cytoplasms.
2. Almost all the genes known at the time affected terminal, final modeling steps (eye color, bristle shape, wing veination). Geneticists had to provide evidence that genes control the early stages of embryogenesis. As Just said (quoted in Harrison, 1937), embryologists were interested in how a fly forms its back, not in the number of bristles on its back.
3. Geneticists had to explain phenomena such as sex determination in certain invertebrates (and vertebrates such as reptiles), where the environment determines sexual phenotype.

(A)

(B)

(C)

FIGURE 2.3
Embryologists attempted to keep genetics from "taking over" their field in the 1930s. (A) Frank Lillie headed the Marine Biology Laboratory at Woods Hole and was a leader in fertilization research and reproductive endocrinology. (B) Hans Spemann (left) and Ross Harrison (right) perfected transplantation operations to discover when the body and limb axes are determined. They argued that geneticists had no mechanism for explaining how the same nuclear genes could create different cell types during development. (C) Ernest E. Just distinguished the fast and slow blocks to polyspermic fertilization. He spurned genetics and emphasized the role of the cell membrane in determining the fates of cells. (A courtesy of V. Hamburger; B courtesy of T. Horder; C courtesy of the Marine Biology Laboratory, Woods Hole.)

This debate became quite vehement. In rhetoric reflecting the political anxieties of the late 1930s, Harrison (1937) warned:

> Now that the necessity of relating the data of genetics to embryology is generally recognized and the *Wanderlust* of geneticists is beginning to urge them in our direction, it may not be inappropriate to point out a danger of this threatened invasion. The prestige of success enjoyed by the gene theory might easily become a hindrance to the understanding of development by directing our attention solely to the genom, whereas cell movements, differentiation, and in fact all of developmental processes are actually effected by cytoplasm. Already we have theories that refer the processes of development to gene action and regard the whole performance as no more than the realization of the potencies of genes. Such theories are altogether too one-sided.

Until geneticists could demonstrate the existence of inherited variants during early development and until geneticists had a well-documented theory for how the same chromosomes could produce different cell types, embryologists generally felt no need to ground their embryology in gene action.

Early attempts at developmental genetics

There were some scientists, though, who felt that neither embryology nor genetics was complete without the other. There were some attempts to synthesize these two disciplines, but the first successful reintegration of genetics and embryology came in the late 1930s from two embryologists, Salome Gluecksohn-Schoenheimer (now Gluecksohn Waelsch) and Conrad Hal Waddington. Both were trained in European embryology and had learned genetics in the United States from Morgan's students. Gluecksohn-Schoenheimer and Waddington attempted to find mutations that affected early development and to find the processes these genes affected. Gluecksohn-Schoenheimer (1938, 1940) showed that mutations in the *T*-locus genes of the mouse caused the aberrant development of the posterior portion of the embryo, and she traced the effects of these mutant genes to defects in the axial mesoderm that would normally have helped induce the dorsal axis. Moreover, Gluecksohn-Schoenheimer (1938) reflected that in working with mice, it was not possible to do what experimental embryologists should be doing—changing a structure during its development and seeing what the consequences of that operation were. Rather, a new type of scientist was called for: the **developmental geneticist.**

> While the experimental embryologist carries out a certain experiment and then studies its results, the developmental geneticist first has to study the course of development (that is, the results of the developmental disturbance) and then sometimes draw conclusions on the nature of the "experiment" carried out by the gene.

At the same time, Waddington (1939) isolated several genes that caused wing malformations in *Drosophila*. He, too, analyzed them in terms of how these genes might affect the developmental primordia that give rise to these structures. The *Drosophila* wing, he correctly claimed, "appears favorable for investigations on the developmental action of genes." Thus, one of the main objections of embryologists to the genetic model of development—that genes appear to be working only on the final modeling of the embryo and not on its major outlines—was countered.*

Evidence for genomic equivalence

There still remained the other big objection to a genetically based embryology: How could nuclear genes direct development when the genes were believed to be the same in every cell type? This genomic equivalence was not so much proved as assumed (as each cell is the mitotic descendant of the fertilized egg), and one of the first problems of developmental genetics was to determine whether each cell of an organism had the same genome as every other cell.

Metaplasia

The first evidence for genomic equivalence came after World War II, from embryologists studying the regeneration of excised tissues. The study of salamander eye regeneration has demonstrated that even adult *differentiated* cells can retain their potential to produce other cell types. Therefore,

*Gluecksohn-Schoenheimer's observations took 60 years to be confirmed by DNA hybridization. However, when the T-locus gene was cloned and its expression detected by the in situ hybridization technique (discussed later in this chapter), Wilkinson and co-workers (1990) found that "the expression of the T gene has a direct role in the early events of mesoderm formation and in the morphogenesis of the notochord." While a comprehensive history of early developmental genetics needs to be written, more information on its turbulent origins can be found in Oppenheimer, 1981; Sander, 1986; Gilbert, 1988, 1991; Burian et al., 1991; Harwood, 1993; and Keller, in press).

FIGURE 2.4
Wolffian regeneration of the newt lens from the dorsal margin of the iris. (A) Normal unoperated eye of the larval stage newt *Notophthalmus viridiscens.* (B–G) Regeneration of lens, seen respectively on days 5, 7, 9, 16, 18, and 30. The new lens is complete at day 30. (From Reyer, 1954, courtesy of R. W. Reyer.)

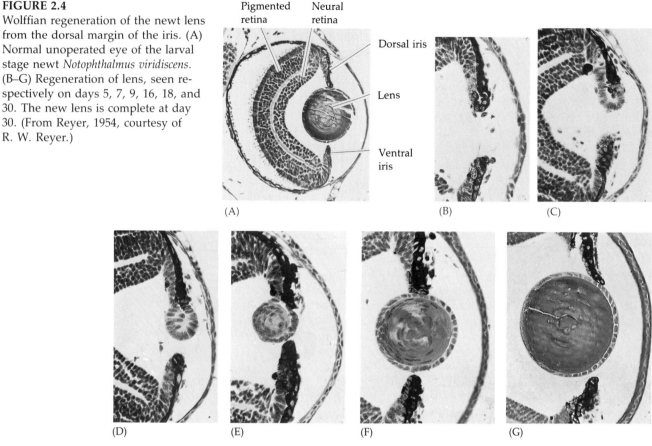

(A)   (B)   (C)

(D)   (E)   (F)   (G)

the genes for these other cell types' products must still be present, even though they are not normally expressed. In the salamander, removal of the neural retina promotes its regeneration from the pigmented retina, and a new lens can be formed from the cells of the dorsal iris. The regeneration of lens tissue from the iris (called Wolffian regeneration after the person who first observed it, in 1894) has been intensively studied. Yamada and his colleagues (Yamada, 1966; Dumont and Yamada, 1972) found that after the removal of a lens, a series of events leads to the production of a new lens from the iris (Figure 2.4). The nuclei of the dorsal side of the iris begin to synthesize enormous amounts of ribosomes, their DNA replicates, and a series of mitotic divisions ensues. The pigmented iris cells then begin to dedifferentiate by throwing out their melanosomes (the pigmented granules that give the eye its color; these melanosomes get ingested by macrophages that enter the wound site). The dorsal iris cells continue to divide, forming a globe of dedifferentiated tissue in the region of the removed lens. These cells then start synthesizing the differentiated products of *lens* cells, the crystallin proteins. These proteins are made in the same order as in normal lens development. Once a new lens has formed, the cells on the dorsal side of the iris cease mitosis.

These events are not the normal route by which the vertebrate lens is formed. As we will detail later, the lens normally develops from a layer of head epithelial cells induced by the underlying retinal precursor cells. The formation of the lens by the differentiated cells of the iris represents **metaplasia** (or **transdifferentiation**), the transformation of one differentiated cell type into another. The salamander iris, then, has not lost any of the genes that are used to differentiate the cells of the lens.

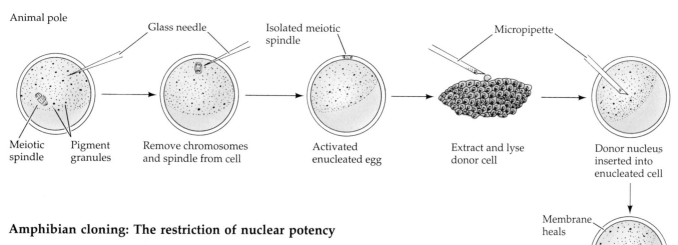

Animal pole

Glass needle

Isolated meiotic spindle

Micropipette

Meiotic spindle | Pigment granules

Remove chromosomes and spindle from cell

Activated enucleated egg

Extract and lyse donor cell

Donor nucleus inserted into enucleated cell

Membrane heals

Amphibian cloning: The restriction of nuclear potency

The ultimate test of whether or not the nucleus from a differentiated cell has undergone any irreversible functional restriction would be to have that nucleus generate every other type of differentiated cell in the body. If each nucleus were identical to the zygote nucleus, each cell's nucleus should be capable of directing the entire development of the organism when transplanted into an activated enucleated egg. Before such an experiment could be done, however, three techniques had to be perfected: (1) a method for enucleating host eggs without destroying them; (2) a method for isolating intact donor nuclei; and (3) a method for transferring such nuclei into the egg without damaging either the nucleus or the oocyte.

These techniques were developed in the 1950s by Robert Briggs and Thomas King. First, they combined the enucleation of the egg with its activation. When an oocyte from the leopard frog (*Rana pipiens*) is pricked with a clean glass needle, the egg undergoes all the cytological and biochemical changes associated with fertilization. The internal cytoplasmic rearrangements of fertilization occur, and the completion of meiosis takes place near the animal pole of the cell. This meiotic spindle can easily be located as it pushes away the pigment granules at the animal pole, and puncturing the oocyte at this site causes the spindle and its chromosomes to flow outside the egg (Figure 2.5). The host egg is now considered to be both activated (in that fertilization reactions necessary to initiate development have been completed) and enucleated. The transfer of nuclei into the eggs is accomplished by disrupting donor cells and transferring a released nucleus into the oocyte through a micropipette. Some cytoplasm accompanies the nucleus to its new home, but the ratio of donor to recipient cytoplasm is only $1:10^5$, and the donor cytoplasm does not seem to affect the outcome of the experiments. In 1952, Briggs and King demonstrated that blastula cell nuclei could direct the development of complete tadpoles when transferred into the oocyte cytoplasm.

What happens when nuclei from more advanced stages are transferred into activated enucleated oocytes? The results of King and Briggs (1956) are outlined in Figure 2.6. Whereas most blastula nuclei can produce entire tadpoles, there is a dramatic decrease in the ability of nuclei from later stages to direct development to the tadpole stage. When nuclei from the *somatic cells* of tailbud-stage tadpoles were used as donors, no nucleus was able to direct normal development. However, *germ cell* nuclei from tailbud-stage tadpoles (which eventually will give rise to a complete organism after fertilization) were capable of directing normal development in 40 percent of the blastulae that developed (Smith, 1956). Thus, somatic cells appear to lose their ability to direct complete development as they become

FIGURE 2.5

Procedure for transplanting blastula nuclei into activated enucleated *Rana pipiens* eggs. The relative dimensions of the meiotic spindle have been exaggerated to show the technique. The handsome *R. pipiens* in the photograph was derived in this way. (After King, 1966; photograph courtesy of M. Di-Berardino and N. Hoffner.)

FIGURE 2.6

Graph of successful nuclear transplants as a function of the developmental age of the nucleus. The abscissa represents the stage at which the donor nucleus (from *R. pipiens*) is isolated and inserted into the activated enucleated oocyte. The ordinate shows the percentage of those transplants capable of producing blastulae that could then direct development to the swimming tadpole stage. (After McKinnell, 1978.)

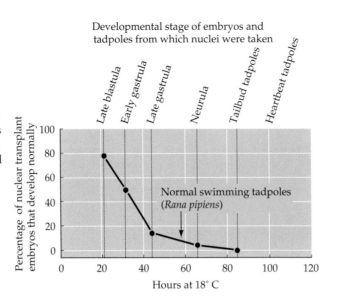

Developmental stage of embryos and tadpoles from which nuclei were taken

determined and differentiated. The progressive restriction of nuclear potency during development appears to be the general rule. It is possible that some differentiated cell nuclei differ from others.

Amphibian cloning: The pluripotency of somatic cells

John Gurdon and his colleagues, using slightly different methods of nuclear transplantation on the frog *Xenopus*, have obtained results suggesting that the nuclei of some differentiated cells may remain totipotent. Gurdon, too, found a progressive loss of potency with increased development, although *Xenopus* cells retained their potencies for a longer period of development (Plate 1). The exceptions to this rule, however, proved very interesting. Gurdon had transferred the intestinal endoderm of feeding *Xenopus* tadpoles into activated enucleated eggs. These donor nuclei contained a genetic marker (one nucleolus per cell instead of the usual two) that distinguished them from host nuclei. Out of 726 nuclei transferred, only 10 (1.4 percent) promoted development to the feeding tadpole stage. Serial transplantation (which here involved placing an intestinal nucleus into an egg and, when the egg had become a blastula, transferring the nuclei of the blastula cells into several more eggs) increased the yield to 7 percent (Gurdon, 1962). In some instances nuclei from tadpole intestinal cells were capable of generating all the cell lineages—neurons, blood cells, nerves, and so forth—of a living tadpole. Moreover, seven of these tadpoles (from two original nuclei) metamorphosed into fertile adult frogs (Gurdon and Uehlinger, 1966); these nuclei were totipotent (Figure 2.7).

King and his colleagues, however, criticized these experiments, pointing out that (1) not enough precautions were taken to make certain that primordial germ cells—which migrate through and often stay in the gut—were not used as sources of nuclei, and (2) the intestinal epithelial cell of such a young tadpole may not qualify as a truly differentiated cell type. Such cells of feeding tadpoles still contain yolk platelets (DiBerardino and King, 1967; McKinnell, 1978; Briggs, 1979). To answer these criticisms, Gurdon and his colleagues cultured epithelial cells from adult frog foot webbing. These cells were shown to be differentiated, because each of them contained keratin, the characteristic protein of adult skin cells. When nuclei from these cells were transferred into activated, enucleated *Xenopus* oocytes, none of the first-generation transfers progressed further than the

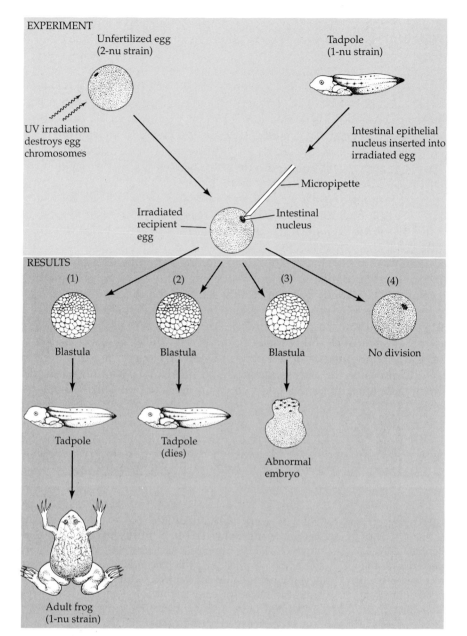

EXPERIMENT

Unfertilized egg
(2-nu strain)

Tadpole
(1-nu strain)

UV irradiation
destroys egg
chromosomes

Intestinal epithelial
nucleus inserted into
irradiated egg

Micropipette

Irradiated
recipient
egg

Intestinal
nucleus

RESULTS

(1)   (2)   (3)   (4)

Blastula   Blastula   Blastula   No division

Tadpole   Tadpole
(dies)

Abnormal
embryo

Adult frog
(1-nu strain)

FIGURE 2.7
Procedure used to obtain mature frogs from the intestinal nuclei of *Xenopus* tadpoles. The wild-type egg (2-nu) is irradiated to destroy the maternal chromosomes, and an intestinal nucleus from a 1-nu tadpole is inserted. In some cases there is no division; in some cases, the embryo is arrested in development; but in other cases, an entire, new frog is formed and has a 1-nu genotype. (After Gurdon, 1968, 1977.)

formation of the neural tube shortly after gastrulation. By serial transplantation, however, numerous tadpoles were generated (Gurdon et al., 1975). Although these tadpoles all died prior to feeding, a single differentiated cell nucleus still retained incredible potencies. A single nucleus derived from an adult frog red blood cell (which neither replicates nor synthesizes mRNA) can undergo over 100 divisions after being transplanted into an activated oocyte and still retain the ability to generate swimming tadpoles (Orr et al., 1986; DiBerardino, 1989). Although DiBerardino (1987) has observed that "to date, no nucleus of a documented specialized cell nor of an adult cell has yet been shown to be totipotent," such a nucleus can still instruct the formation of all the organs of a swimming tadpole.

Some of the differences between the results of Briggs's laboratory and Gurdon's may involve differences in developmental physiology between *Rana* and *Xenopus* frogs. When transferring a nucleus of a differentiated cell into oocyte cytoplasm, one is asking the nucleus to revert back to physiological conditions it is not used to. The cleavage nuclei of frogs divide at a rapid rate, whereas some differentiated cell nuclei divide rarely, if at all. Failure to replicate DNA rapidly can lead to chromosomal breakage, and such chromosomal abnormalities have been seen in many cells of the cloned tadpoles. Sally Hennen (1970) has shown that the developmental success of donor nuclei can be increased by treating such nuclei with spermine and by cooling the egg to give the nucleus time to adapt to the egg cytoplasm. Spermine is thought to remove histones from chromatin and may "reset" the activity of the nuclei. When endoderm nuclei from *Rana pipiens* tailbud-stage tadpoles were treated in this fashion, 62 percent of those nuclei that initiated normal development went on to generate normal tadpoles. In control animals, 0 percent of the nuclei succeeded in generating such tadpoles. Thus, the genes for the development of the entire tadpole do not appear to be lost in the tadpole endoderm cells.

One can look at these amphibian cloning experiments in two ways. First, one can recognize a general restriction of potency concomitant with development. Second, one can readily see that the differentiated cell genome is remarkably potent in its ability to produce all the cell types of the amphibian tadpole. In other words, even if there is debate over the *totipotency* of such nuclei, there is little doubt that they are extremely *pluripotent*. Certainly many unused genes in skin or blood cells can be reactivated to produce the nerves, stomach, or heart of a swimming tadpole. Thus, each nucleus in the body contains most (if not all) of the same genes.

SIDELIGHTS & SPECULATIONS

Cloning mammals for fun and profit

Cloning human beings from previously differentiated cells seems to be the goal of several newspaper editors and novelists. It should be obvious from the preceding discussion, though, that cloning a fully developed individual from differentiated cells is a formidable undertaking. Even in amphibians, the nuclei of differentiated adult cells have not been able to generate adult animals when they are placed into activated enucleated eggs.

Moreover, even if adult frogs could be generated from differentiated cell nuclei, this ability could not be extrapolated to human cells. Besides the ethical and technical diffulties working with the human organism, the human oocyte cytoplasm may not be able to respond to signals from a nucleus of a more advanced-stage cell. Nuclear transplantation has been accomplished in mice by removing the sperm and egg pronuclei (haploid nuclei) from one zygote and then replacing them with pronuclei from other zygotes (Figure 2.8; McGrath and Solter, 1983). These reconstructed zygotes begin dividing and are then implanted into the uterus. The resulting mice display the phenotype of the donor nucleus.

Whereas over 90 percent of enucleated mouse zygotes receiving pronuclei from other zygotes develop successfully to the blastocyst (blastula) stage, not a single embryo (out of 81) developed even this far when nuclei from 4-cell embryos were transferred into the enucleated zygotes (McGrath and Solter, 1984). Similarly, nuclei from 8-cell embryos and the inner cell mass (the blastomeres that form the embryo rather than the placenta*) would not support development. Unlike nuclei from sea urchins and amphibians, the nuclei of early mouse blastomeres (whose cells are known to be totipotent) do not support full development. These experiments probably fail because blastomere nuclei cannot function properly in zygote cytoplasm. Thus, the cloning of Elvis Presley from fully differentiated cells is not something we should count on.

Not all mammalian blastomeres are the same, however, and mammalian species differ greatly in the time of gene activation and implantation into the uterus. Using modifications of the above technique, Willadsen (1986) has produced full-term lambs from the transplanted nuclei of 8-cell-stage blastomeres, and nuclei from the preimplantation embryos of cattle, pigs, and rabbits have been able to direct full development when transplanted into enucleated and activated oocytes (Prather et al., 1987; Stice and Robl, 1988; Prather et al., 1989; Willadsen, 1989). However, the nuclei in all these cases of mammalian cloning came from

*Each blastomere of the inner cell mass is totipotent in that it retains the ability to form any cell type in the body. As we will see in later chapters, the ability of a single inner-cell-mass blastomere to form any cell in the body allows the possibility of twins.

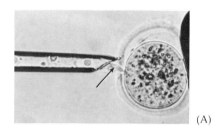

(A)

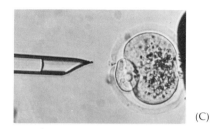

(C)

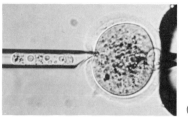

(B)

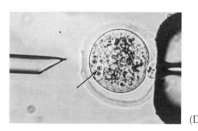

(D)

FIGURE 2.8

Procedure for transferring nuclei into an activated enucleated mammalian egg. A single-cell embryo, incubated in colcemid and cytochalasin to relax the cytoskeleton, is held on a suction pipette. The sperm-derived and egg-derived haploid nuclei have not yet come together. The enucleation pipette pierces the zona pellucida (the protein coat around the egg) and sucks up adjacent cell membrane and the area of the cell containing the pronuclei. The enucleation pipette is then withdrawn (A) and the pronuclei-containing cytoplasm is removed from the egg. The cell membrane is not broken; the continuity of the membrane-bounded cytoplasm is indicated by the arrow. (B) The cell membrane forms a vesicle around the pronuclei within the enucleation pipette. (C) This vesicle is mixed with Sendai virus (which induces the fusion of cell membranes) and is inserted into the space between the zona pellucida and another enucleated egg. (D) The Sendai virus mediates the fusion of the enucleated egg and the membrane-bounded pronuclei, allowing the pronuclei (arrow) to enter the cell. (From McGrath and Solter, 1983, courtesy of the authors.)

preimplantation embryos. In no case has any cloning been achieved from the nuclei of adult differentiated cells. This ability to clone, even from preimplantation embryos, may have important agricultural consequences (see Prather, 1991).

Plant cloning

The only cases where the nuclei of differentiated cells of adult organisms are readily seen to be capable of directing the development of another adult organism occur in plants. This ability has been demonstrated dramatically in cells from carrots and tobacco. In 1958, F. C. Steward and his colleagues established a procedure by which the differentiated tissue of carrot roots could give rise to an entire new plant (Figure 2.9). Small pieces of the phloem tissue were isolated from the carrot and rotated in large flasks containing coconut milk. This fluid (which is actually the liquid endosperm of the coconut seed) contains the factors and nutrients necessary for plant growth and the hormones required for plant differentiation. Under these conditions, the tissues proliferate and form a disorganized mass of

FIGURE 2.9

Steward's experiment demonstrating the totipotency of carrot phloem cells.

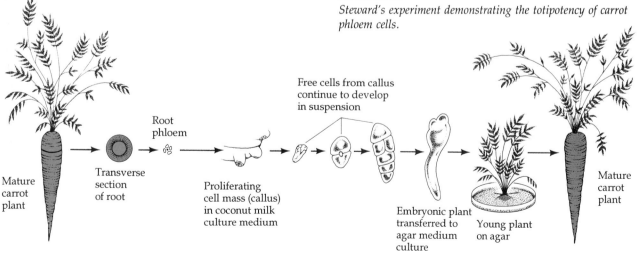

Mature carrot plant

Transverse section of root

Root phloem

Proliferating cell mass (callus) in coconut milk culture medium

Free cells from callus continue to develop in suspension

Embryonic plant transferred to agar medium culture

Young plant on agar

Mature carrot plant

tissue called a **callus.** Continued rotation causes the shearing away of individual cells from the callus and into suspension. These individual cells give rise to rootlike nodules of cells that continue to grow as long as they remain in suspension. When these nodules are placed into a medium solidified with agar, the rest of the plant is able to develop, ultimately forming a complete, fertile carrot plant (Steward et al., 1964; Steward, 1970).

But plants and animals develop differently; the vegetative propagation of plants by cuttings (i.e., portions of the plant that, when nourished, regenerate the missing parts) is a common agricultural practice. Moreover, in contrast to amphibians and mammals (in which germ cells are set off as a distinctive cell lineage early in development), plants normally derive their gametes from somatic cells. It is not overly surprising, then, that a single plant cell can differentiate into other cell types and form a genetically identical clone (Gk. *klon,* "twig").

Of *E. coli* and elephants: The operon model

In most cases studied, the genome is the same from cell to cell within the organism. The genes for globin proteins can be found in skin cells, and the genes for skin keratins can be found in brain neurons. But that still leaves unanswered the other big question posed by the embryologists:* If the nucleus of each cell in an organism has the same genes, how can these genes cause cells to become different? Shortly after World War II, many biologists agreed that

> The widest gap, still to be filled, between two fields of research in biology, is probably the one between genetics and embryology. It is the repeatedly stated—and thus far unsolved problem—of understanding how cells with identical genomes may become differentiated, that of acquiring the property of manufacturing molecules with new or, at least, different specific patterns or configurations.

Interestingly, the above quotation comes from Jacques Monod (1947), a microbial geneticist who had been working on adaptive enzyme synthesis. Adaptive enzymes are those proteins that, although not usually synthesized by the strain of bacteria or yeast, become synthesized when the microorganisms encounter a novel substrate. One example is the ability of the bacterium *Escherichia coli* to synthesize β-galactosidase and other lactose-digesting enzymes when it encounters lactose. If lactose is absent in the cytoplasm, these enzymes are not synthesized. Upon the introduction of lactose into the cytoplasm, this set of new enzymes appears. In microbes, at least, the same genome can produce two functionally different cytoplasmic states after seeing a particular small compound (in this case, lactose). Monod hypothesized that the phenomenon of enzymatic adaptation could be a solution to the problem of how identical genomes can synthesize different "specific" molecules.

Monod was not the only person who felt that unicellular microbes might explain multicellular differentiation. Microbiologist Sol Spiegelman claimed that embryology was being impeded by its own terminology. First, the problem of differentiation had to change from being seen as a structural property of tissues to being seen as a biochemical property of individual

*The big exception to this rule of gene constancy—the immunoglobulin genes—are discussed in Chapter 10. Each cell has all the immunoglobulin gene subuits, but in lymphocytes, some of these subunits get rearranged, and others sometimes get deleted from the genome. The third challenge—the explanation of how the environment can direct development—followed readily once the general explanation of differential gene expression was established. As we will see, the operon model demonstrated how an environmental substance could effect differential gene expression.

cells. Differentiation was to be seen not in terms of tissue anatomy but "as the controlled production of unique enzyme patterns." This redefinition, he stated (1947), would focus attention on "the relationship between the genes in the nucleus and the properties of the cytoplasm." Second, the synthesis of adaptive enzymes in the presence of their substrate was to be discussed as an "induction." This is the technical term in embryology for the ability of one cell to produce a substance that influences the differentiation of another cell. Similarly, the molecular agent responsible for this induction was to be called "the inducer." Thus, Spiegelman saw an essential similarity between the induction of new cell types in the embryo and the induction of new enzymes in the microorganism.

By the late 1950s there were several researchers who felt that microbes were an excellent (and easily studied) model for embryonic differentiation. There was much discussion over the relevance of the microbial model for differentiation, and many microbial geneticists explicitly linked inducible enzymes to embryological concepts. The geneticists felt the extrapolation valid, and they appealed to the unity of nature and the ultimately simple rules that they expected to find. As Monod claimed (see Judson, 1979), if you understand the bacterium, you understand the elephant. Many embryologists, however, remained skeptical of the extrapolation from bacteria to embryos, and they emphasized the complexity of development and the diversity of embryological performances.

By 1961 Jacob and Monod synthesized the data on β-galactosidase induction into the operon model. This model postulated that the small inducer molecule causes the transcription of different genes in *E. coli* (Figure 2.10). In inductive systems, a gene-encoded repressor protein binds at the operator site adjacent to the structural genes. This prevents RNA polymerase from binding to the promoter site and initiating transcription. If the inducer is present, it binds to and alters the conformation of the repressor protein so that it cannot bind to the operator. Thus, the gene becomes able to transcribe mRNA, which can then be translated into proteins. In this way, the same genome could synthesize different enzymes depending on whether or not the inducer were present. In their closing statement of a major 1961 review article, Jacob and Monod emphasized that operon-like control mechanisms may be a universal part of gene regulation. They linked their results to "the fundamental problem of chemical embryology [which] is to understand why tissue cells do not express, all the time, all the potencies inherent in their genome."

The operon model of development was brought into embryology texts immediately by those people who had been looking for a synthesis of genetics and embryology. Waddington's 1962 book, *New Patterns in Genetics and Development*, begins with a chapter relating the Jacob and Monod operon model to the control of gene expression in amphibian development. Waddington especially liked this model because it means that genes are not only active, but reactive. They respond to changes in the cytoplasm. Waddington viewed the genes and the cytoplasm as mutually interacting with each other. This view was also celebrated in *Heredity and Development* (1963), John Moore's synthesis of embryology and genetics, which concluded:

A generation ago, few embryologists or geneticists would have predicted that a synthesis of their fields would be made possible by studies on the bacterium *Escherichia coli*. But this microscopic creature, with no embryology of its own, has shown a way. A decade from now it may be difficult to distinguish between a geneticist and an embryologist, as they advance their science beyond what each might independently achieve.

FIGURE 2.10
Differential gene regulation in *E. coli.* (A–C) In its wild-type, inducible state, no β-galactosidase RNA is transcribed unless lactose is present. (B) When no lactose is available, a repressor protein, made by gene *i*, binds to the operator site (*o*), inhibiting transcription by RNA polymerase from the promoter (*p*). (C) When the inducer, lactose, is present, it combines with the repressor protein, changing its shape. The altered repressor protein can no longer bind to the operator DNA, and transcription ensues. (D) The soluble nature of this repressor protein is shown in studies on mutant *E. coli.* When haploid bacteria cells with a nonfunctional inducer gene (i^-) are made partially diploid with the wild-type *i* gene (i^+), wild-type repressor is manufactured and is able to make the original β-galactosidase gene inducible.

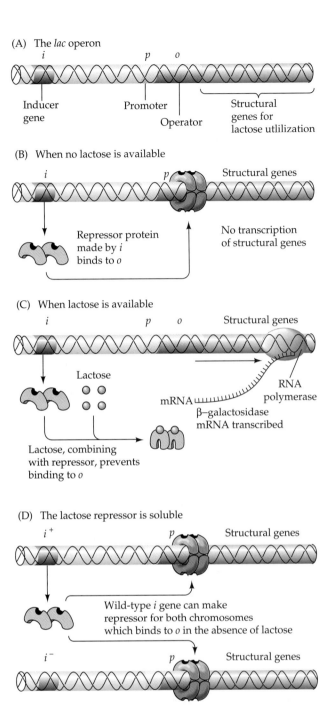

Differential RNA synthesis

The desired unification did not come as quickly as Moore had hoped. However, based on the embryological evidence for genomic equivalence and the operon model of *E. coli,* a consensus emerged in the 1960s that cells regulate development through **differential gene expression.** Since bacteria were the models for such activity, expression usually meant the transcription of mRNA. The three postulates of differential gene expression were:

1. Every cell nucleus contains the complete genome established in the fertilized egg. In molecular terms, the DNAs of all differentiated cells are identical.

2. The unused genes in differentiated cells are not destroyed or mutated, and they retain the potential for being expressed.
3. Only a small percentage of the genome is being expressed in each cell, and a portion of the RNA synthesized is specific for that cell type.

The first two postulates have already been discussed. The third postulate—that only a small portion of the genome is active in making tissue-specific products—was first tested in insect larvae. Upon hatching, an insect larva has two distinct cell populations. About 10,000 cells form the larval tissue. Most of these larval cells have **polytene chromosomes.** Such chromosomes undergo DNA replication in the absence of mitosis and therefore contain 512 (2^9), 1024 (2^{10}), or even more parallel DNA double helices instead of just one (Figure 2.11). These cells do not undergo mitosis, and they grow by expanding to about 150 times their original volume. During metamorphosis, these cells die and are replaced by the nonpolytene diploid cells clustered in certain regions of the larva (Chapter 20). Beermann (1952) showed that the banding patterns of polytene chromosomes were identical throughout the larva, and that no loss or addition of any chromosomal region was seen when different cell types were compared (Figure 2.12). However, Beerman, studying the larval gall midge *Chironomus*, and Becker (1959), studying *Drosophila*, found that there were regions of the chromosomes that were "puffed out." These puffs appeared in different places on the chromosomes in different tissues, and their appearances changed with the development of these cells (Figure 2.13). Furthermore, certain

(A)

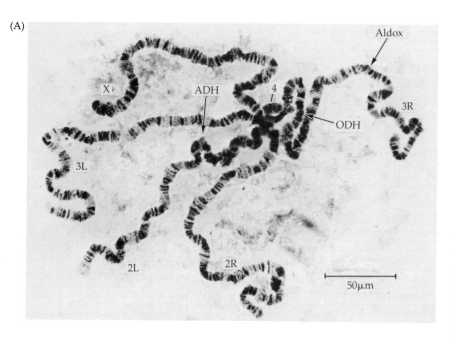

(B)

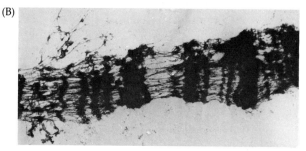

FIGURE 2.11
Polytene chromosomes. (A) Polytene chromosomes from the salivary gland cells of *Drosophila melanogaster*. The four chromosomes are connected at their centromere regions, forming a dense chromocenter. The structural genes for alcohol dehydrogenase (ADH), aldehyde oxidase (Aldox), and octanol dehydrogenase (ODH) have been mapped to the assigned positions on these chromosomes. (B) Electron microscope view of a small region of a *Drosophila* polytene chromosome. The bands (dark) are highly condensed compared with the interband (lighter) regions. (A from Ursprung et al., 1968, courtesy of H. Ursprung; B from Burkholder, 1967, courtesy of G. D. Burkholder.)

FIGURE 2.12
Genomic identity in polytene chromosomes. (A) A region of the polytene chromosome set of the midge *Chironomus tentans*. Note the constancy of band number in different tissues. (B) Hybridization of a yolk protein mRNA to the polytene chromosome of a larval *Drosophila* salivary gland. The dark grains (arrow) show where the radioactive yolk protein message bound to the chromosomes. Note that the gene for the yolk protein is present in the salivary gland chromosomes, even though yolk protein is not synthesized there. (A After Beermann, 1952; B from Barnett et al., 1980; photograph courtesy of P. C. Wensink.)

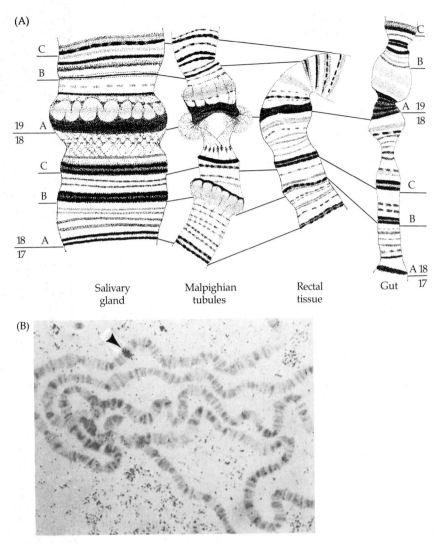

(A)

Salivary gland   Malpighian tubules   Rectal tissue   Gut

(B)

puffs could be induced or inhibited by certain physiological changes caused by heat or hormones (Clever, 1966; Ashburner, 1972; Ashburner and Berondes, 1978).

Beermann (1961) provided evidence that these puffs represent a local loosening of the polytene chromosomes (Figure 2.14) and the first evidence that these puffs are sites of active messenger RNA synthesis. He found two interbreeding species of *Chironomus* that differed: one produced a major salivary protein, whereas the other did not (Figure 2.15). The producers had a large puff (Balbiani ring) at a certain band in the polytene chromosomes in cells of the larval salivary gland; that puff was absent in the nonproducers. When producer was mated with nonproducer, the hybrid larva produced an intermediate amount of salivary protein. When two hybrid flies were mated, the ability to produce salivary protein segregated in proper Mendelian fashion (1 high producer:2 intermediate producers:1 nonproducer). Moreover, whereas high producers were found to have two puffs (one on each homologous chromosome), the intermediate producers had only one puff. The nonproducers lacked any puffs at these sites. Beermann concluded that the genetic information needed for the synthesis of this salivary protein is present in this distal chromosome band and that the synthesis of that product depends on that band's becoming a puff region.

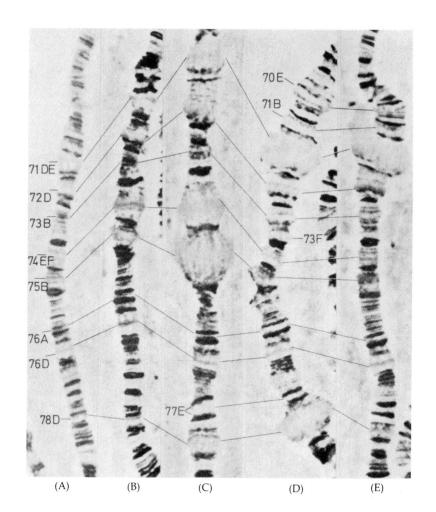

71DE
72D
73B
74EF
75B
76A
76D
78D

70E
71B

73F

77E

(A)   (B)   (C)   (D)   (E)

FIGURE 2.13
Puffing sequence of a portion of chromosome 3 in the larval *Drosophila melanogaster* salivary gland. (A,B) 110-hour larva; (C) 115-hour larva; (D,E) prepupal stage (4 hours apart). Note the puffing and regression of bands 74EF and 75B. Other bands (71DE, 78D) puff later, but most do not puff at all during this time. (Courtesy of M. Ashburner.)

(A)

(B)

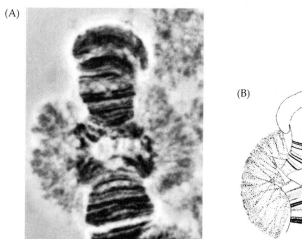
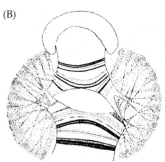

FIGURE 2.14
Proximal end of chromosome 4 from the salivary gland of *Chironomus pallidivitatus,* showing the enormous puff, BR2. (A) Phase-contrast photomicrograph of stained preparation showing extended puff on the polytene chromosome. (B) Diagram of the BR2 region undergoing puffing. (A from Grossbach, 1973, courtesy of U. Grossbach; B after Beermann, 1963.)

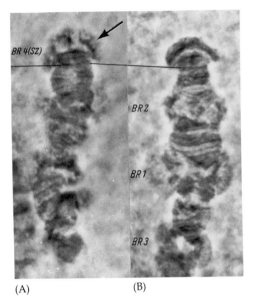

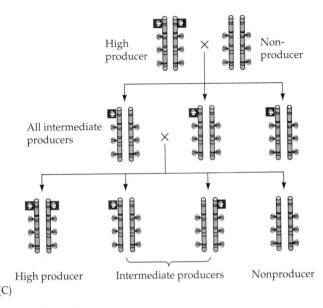

(A)   (B)   (C)

FIGURE 2.15
Correlation of puffing patterns with specialized functions in salivary gland cells of *Chironomus pallidivitatus*. (A) Chromosome 4 from a salivary gland cell producing a granular secretion and showing an additional Balbiani ring [4(SZ)]. (B) Chromosome 4 from a salivary cell, showing only Balbiani rings 1, 2, and 3 (BR1, BR2, BR3). (C) Genetic evidence that the synthesis of a major salivary protein is dependent on the formation of BR4(SZ) puffs. Larvae with high amounts of granular secretions have salivary gland cells with BR4(SZ) puffs on both chromosomes 4 (color), whereas larvae without these secretions have no such puffs. Intermediate producers have only one chromosome 4 with a puffed BR4(SZ) region in each salivary cell making the secretion. (A and B from Beermann, 1961, courtesy of W. Beermann.)

Further proof that chromosomal puffs make mRNA comes from studies of Balbiani ring 2 (BR2) puffs in *Chironomus tentans*. Because of its exceptionally large size, BR2 can be isolated by microdissection and its products can be analyzed by autoradiography (Lambert and Daneholt, 1975). Figure 2.16A,B shows the isolation of BR2 from chromosome 4 of *C. tentans*. BR2 transcription was demonstrated by incubating isolated salivary glands with radioactive RNA precursors. Radioactive RNA could then be extracted from the BR2 portion of the dissected chromosome (Lambert, 1972). This RNA was found to be exceptionally large—about

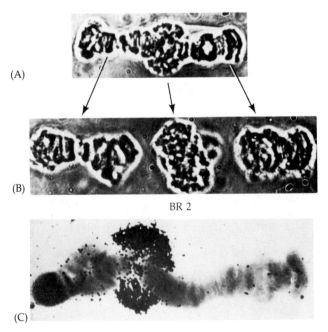

FIGURE 2.16
(A,B) Isolation of BR2 region of *Chironomus tentans* by micromanipulation. The intact chromosome 4 can be divided into three regions, one containing BR2. (C) Transcription from the BR2 region of chromosome 4 of *Chironomus tentans* salivary gland cells. In situ autoradiograph after BR2 RNA was hybridized to the chromosome preparation. (A,B from Lambert and Daneholt, 1975; C from Lambert, 1972; photographs courtesy of B. Lambert.)

50,000 bases. This large radioactive RNA segment hybridized specifically to the BR2 region of the chromosome, thereby showing that the puffed DNA—and no other locus—had been actively transcribing it (Figure 2.16C). This same RNA could also be isolated from protein-synthesizing polysomes, indicating that it is active in protein synthesis (Wieslander and Daneholt, 1977). Thus, an RNA transcribed from a specific band of DNA that puffs in the larval salivary gland is later seen to be making proteins on cytoplasmic ribosomes. Therefore, the puffs on salivary chromosomes are actively making mRNA. In the cells that synthesize this protein, the gene would be activated; in cells that do not use this protein, the gene would remain repressed.

Nucleic acid hybridization

Few genes could be analyzed like those on *Chironomus* polytene chromosome puffs. Moreover, these puff genes were active in cells (such as those of the salivary gland) that had already differentiated; they were not the genes that actually caused cells to differentiate in a certain direction. In order to find the genes that *are* responsible for embryonic development and to analyze how they accomplish their activities, new techniques had to be perfected.

Most techniques of eukaryotic gene analysis are based on **nucleic acid hybridization.** This technique involves annealing single-stranded pieces of RNA and DNA to allow complementary strands to form double-stranded hybrids. For example, if DNA is cut into small pieces and each piece dissociated into two single strands—**denatured**—each strand in the solution should find and reunite with its complementary partner, given sufficient time. The conditions of this renaturation must be such that specific binding between complementary strands is maintained while non-specific matchings are dissociated. This is usually accomplished by varying the temperature or the ionic conditions in the solution in which such renaturation is taking place (Wetmur and Davidson, 1968). Similarly, *RNA synthesized from a particular region of DNA would be expected to bind to the strand from which it was transcribed* (Figure 2.17). Thus, RNA is

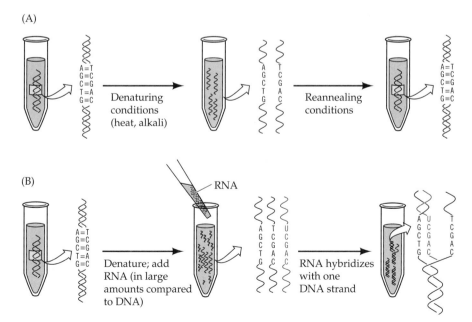

FIGURE 2.17
Nucleic acid hybridization. (A) If the DNA helix is separated into two strands, the strands should reanneal given the appropriate ionic conditions and time. (B) Similarly, if DNA is separated into its two strands, RNA should be able to bind to the genes that encode it. If present in sufficiently large amounts compared with the DNA, the RNA will replace one of the DNA strands in this region.

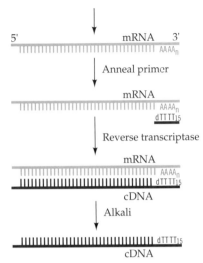

FIGURE 2.18
Method for preparing complementary DNA (cDNA). Most mRNA contains a long stretch of adenosine residues (AAA^n) at the 3' end of the message (to be discussed in Chapter 13); therefore, investigators anneal a primer consisting of 15 deoxythymidine residues (dT_{15}) to the 3' end of the message. Reverse transcriptase then transcribes a complementary DNA strand, starting at the dT_{15} primer. The cDNA can be isolated by raising the pH of the solution, thereby denaturing the double-stranded hybrid and cleaving the RNA.

expected to hybridize specifically with a gene that encodes it. In order to measure this hybridization, one of the nucleic acid strands (the *probe*) is usually radiolabeled. One technical problem that originally plagued nucleic acid hybridization studies was the difficulty of getting enough radioactivity into the RNA molecule. This problem is circumvented by isolating the RNA and making a **complementary DNA (cDNA)** copy in the presence of radioactive precursors. This can be done in a test tube containing the RNA, a short stretch of DNA (called a **primer**), radioactive DNA precursors, and the viral enzyme **reverse transcriptase**. This enzyme is capable of making DNA from an RNA template (Figure 2.18). Because the DNA is synthesized in vitro, one need not worry about the dilution of the radioactive precursor. Furthermore, such a cDNA can hybridize with both the gene that produced the RNA (albeit the other strand) and the RNA itself. It is therefore extremely useful in detecting small amounts of specific RNAs.

Cloning from genomic DNA

As early as 1904, Theodor Boveri despaired that the techniques of his time would ever allow him to study how genes act to create embryos. A particular type of gene amplification technique was needed: "For it is not cell nuclei, not even individual chromosomes, but certain parts of certain chromosomes from certain cells that must be isolated and collected in enormous quantities for analysis; that would be the precondition for placing the chemist in such a position as would allow him to analyse [the hereditary material] more minutely than the morphologists."

However, beginning in the 1970s, nucleic acid hybridization techniques have enabled developmental biologists to do just what Boveri

TABLE 2.1
Commonly used restriction enzymes

Enzyme	Derivation	Recognition and cleavage site[a]
EcoRI	*Escherichia coli*	G▼A A T T C C T T A A▲G
BamHI	*Bacillus amyloliquifaciens*	G▼G A T C C C C T A G▲G
HindIII	*Haemophilus influenzae*	A▼A G C T T T T C G A▲A
SalI	*Streptomyces albus*	G▼T C G A C C A G C T▲G
SmaI	*Serratia marcescens*	C C C▼G G G G G G▲C C C
HhaI	*Haemophilus haemolyticus*	G C G▼C C▲G C G
HaeIII	*Haemophilus aegyptius*	G G▼C C C C▲G G
AluI	*Arthrobacter luteus*	A G▼C T T C▲G A

[a] All restriction enzyme recognition sites have a center of symmetry. The double-stranded sequence read in one direction is identical to the sequence read backwards in the other direction.

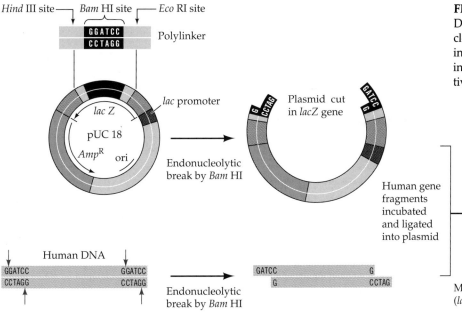

FIGURE 2.19
Diagram of the general protocol for cloning DNA, using as an example the insertion of a human DNA sequence into a plasmid with one *Bam*HI-sensitive site.

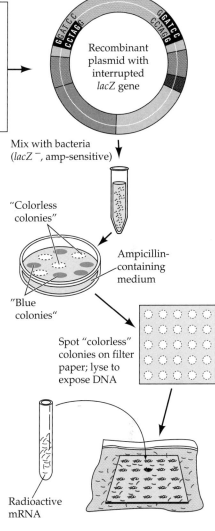

Filter paper incubated with radioactive mRNA from gene to be cloned

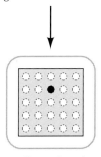

Prepare autoradiograph to show those bacterial clones with a DNA fragment that hybridized to the radioactive DNA

wanted: to isolate and amplify specific regions of the chromosome. The main technique for isolating and amplifying individual genes is called **gene cloning.** The first step in this process involves cutting nuclear DNA into discrete pieces. This is done by incubating the DNA with a **restriction endonuclease** (more commonly called a **restriction enzyme**). These endonucleases are usually bacterial enzymes that recognize specific sequences of DNA and cleave the DNA at these sites (Table 2.1; Nathans and Smith, 1975). For example, when human DNA is incubated with the enzyme *Bam*HI (from *Bacillus amyloliquifaciens* strain H), the DNA is cleaved at every site where the sequence GGATCC occurs. The products are variously sized pieces of DNA, all ending with G on one end and GATCC on the other (Figure 2.19). These pieces are often called **restriction fragments**.

The next step in gene cloning is to incorporate these restriction fragments into **cloning vectors.** These vectors are usually circular DNA molecules that replicate in bacterial cells independently of the bacterial chromosome. Either drug-resistant plasmids or specially modified viruses (which are especially useful for cloning large DNA fragments) are used. Such a vector can be constructed to have only one *Bam*HI-sensitive site. This vector can be opened by incubating it with that restriction enzyme. After being opened, it can be mixed with the *Bam*HI-fragmented human DNA. In numerous cases, the cut DNA pieces will become incorporated into these vectors (because their ends are complementary to the vector's open ends), and the pieces can be joined covalently by placing them in a solution containing the enzyme **DNA ligase.** The whole process yields bacterial plasmids that each contain a single piece of human DNA. These are called recombinant plasmids or, usually, **recombinant DNA** (Cohen et al., 1973; Blattner et al., 1978).

The plasmid illustrated in Figure 2.19 is pUC18, a cloning vector often used by molecular biologists (Vierra and Messing, 1982). It contains (1) a drug-resistance gene, *Ap*^{R}, which makes the bacterium immune to ampicillin and allows researchers to select for those bacteria that have incorporated a plasmid; (2) an origin of DNA replication that enables the plasmid to replicate hundreds of times in each bacterium; and (3) a polylinker, a short, artificial stretch of DNA that contains the restriction enzyme sites for several of these endonucleases. The polylinker resides within a *lacZ* gene that encodes β-galactosidase. The polylinker is short enough (and has the correct number of base pairs) so it does not interfere with the enzymatic activity of the β-galactosidase. The restriction fragments of the nuclear DNA are mixed with the opened pUC18 plasmids and are then ligated shut. The putative recombinant plasmids made in the above manner are then incubated with ampicillin-sensitive *E. coli* that lack a β-galactosidase gene. Even though the bacteria and the plasmids are mixed together under conditions that encourage the bacteria to take in plasmids, not every bacterium incorporates a plasmid. To screen for those bacteria that have incorporated plasmids, the treated *E. coli* are grown on agar containing ampicillin. Only those bacteria that have incorporated a plasmid (with its dominant ampicillin-resistance gene) survive.

But not every plasmid has incorporated a foreign gene (since it is possible for the "sticky ends" of the restriction enzyme site to renature with themselves). To distinguish the bacterial colonies that have incorporated foreign DNA from those that have not, the agar also contains a dye called X-gal. This compound is colorless, but when acted upon by β-galactosidase it forms a blue precipitate.* Thus, if the plasmids have not incorporated a restriction fragment into their restriction enzyme site in the polylinker, the β-galactosidase (*lacZ*) gene is functional, and the resulting β-galactosidase turns the dye blue. The result is the appearance of "blue colonies." However, if the plasmid has taken up a DNA fragment, the β-galactosidase gene is destroyed by the insertion. These bacteria will not turn the dye blue; they produce colorless colonies on the agar.

Colorless colonies are then screened for the presence of the particular gene. Cells from each of these colonies are placed on a nitrocellulose or nylon filter. When these cells are lysed, their DNA gets stuck on the filter. Next, the DNA strands are separated by heating, and the filter is incubated in a solution containing the radioactive RNA (or its cDNA copy) of the gene one wishes to clone. (In some cases, the sequence of the mRNA or the gene is not yet known, and one has to guess the sequence from the amino acid sequence of the protein.) If a plasmid contains that gene, its DNA should be on the paper and only that DNA should be able to bind the radioactive RNA or cDNA probe. Therefore, only those areas will be radioactive. The radioactivity of these regions is detected by **autoradiography.** Sensitive X-ray film is placed over the treated paper. The high-energy electrons emitted by the radioactive RNA sensitize the silver grains in the film, causing them to turn dark when the film is developed. Eventually, a black spot is produced over each colony containing the recombinant plasmid carrying that particular gene (Figure 2.19). This colony is then isolated and grown, producing billions of bacteria, each containing hundreds of identical recombinant plasmids.

The recombinant plasmids can be separated from the *E. coli* chromosome by centrifugation, and incubating the plasmid DNA with *Bam*HI releases the human DNA fragment that contains the gene. This fragment can then be separated from the plasmid DNA, so the investigator has micrograms of purified DNA sequences containing a specific gene. Although this procedure sounds very logical and easy, the numbers of colonies that have to be screened is often astronomical. The number of random fragments that must be cloned in order to obtain the gene you want gets larger with the increasing complexity of the organism's ge-

---

*The dye is 5-bromo-4-chloroindole, and it is blue unless complexed with a molecule such as galactose. The β-galactosidase encoded by the plasmid gene cleaves the galactose off the dye, allowing the dye to achieve its blue conformation.

nome.* To detect a particular gene from a mammalian genome, millions of individual clones must be screened.

DNA hybridization: Within and across species

Clones can be screened by any radioactive stretches of nucleotides. Therefore, the genes cloned from one organism can be probed with radioactive cDNAs derived from the genome of another species. One of the most exciting findings of modern developmental biology has been that genes that are used for specific developmental processes in one organism may be used for similar processes in other organisms. *Drosophila* has been critical in the discovery of these genes. Starting with Morgan, these genes have been mapped, and in the 1960s E. B. Lewis confirmed that some of these genes are responsible for the formation of basic body parts (as we will detail in Chapter 15). One of these, *Antennapedia,* is a gene whose protein product is essential for inhibiting head structures from forming in the thorax. If the gene is missing, antennae grow where the legs should be. If the gene is expressed in the head (as it is in a particular mutant), the fly develops an extra set of legs coming out of its eye sockets (see Figure 15.28). Could such a gene exist in vertebrates?

Evidence for such genes in vertebrates came first from **DNA blots** [sometimes called **Southern blots** after their inventor, E. M. Southern (1975)]. DNA from numerous vertebrate and invertebrate organisms was treated with a restriction enzyme, and the resulting DNA fragments were separated on an electrophoresis gel. The mixtures of fragments are placed into slots on one side of a gel and an electric current is then passed through the gel. The negatively charged DNA fragments migrate toward the positive pole, the smaller fragments moving faster than the larger ones.† However, hybridization cannot be done inside a gel. One has to get the DNA onto a flat surface; this is done by blotting. After denaturing the DNA strands in alkali, investigators returned the gel to a neutral pH, then placed it on wet filter paper atop a plastic support (Figure 2.20; McGinnis et al., 1984; Holland and Hogan, 1986). Nitrocellulose paper (capable of binding single-stranded DNA) was placed directly over the gel and covered with multiple layers of dry paper towels. The filter paper beneath the gel extended into a trough of high-ionic-strength buffer. The buffer traveled through the gel up through the nitrocellulose filter and into the towels. The DNA was brought up through the gel by this wave of buffer, but it was stopped by the nitrocellulose filter; thus the DNA was transferred from the gel to the nitrocellulose paper. After baking the DNA fragments onto the nitrocellulose paper (otherwise it would have come off), the DNA fragments were incubated with radioactive cDNA from a

---

Complexity is the term referring to the number of different types of genes within a nucleus. Although millions of clones must be screened, about 100,000 colonies can now be screened on a single plate. Another common way of screening the clones is to use a plasmid that has its restriction enzyme site near a strong bacterial promoter (such as the one for β-galactosidase). The bacteria will transcribe the gene and translate it into protein. After the bacterial colonies are lysed on filter paper, the proteins stick to the paper and can be found by antibodies directed against that protein. This is called **expression cloning,** and the plasmids are referred to as **expression vectors.**

†Given the same charge-to-mass ratio, smaller fragments obtain a faster velocity than larger ones when propelled by the same energy. This is a function of the kinetic energy equation, $E = \frac{1}{2} mv^2$. Solving for velocity, one finds that it is inversely proportional to the square root of the mass.

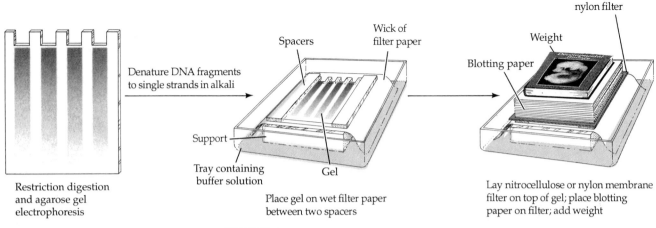

Restriction digestion and agarose gel electrophoresis

Denature DNA fragments to single strands in alkali

Spacers

Wick of filter paper

Support

Tray containing buffer solution

Gel

Place gel on wet filter paper between two spacers

Nitrocellulose or nylon filter

Weight

Blotting paper

Lay nitrocellulose or nylon membrane filter on top of gel; place blotting paper on filter; add weight

FIGURE 2.20
Southern blotting. DNA is treated with restriction enzymes, and the resulting restriction fragments of DNA are placed in a gel. After the fragments of DNA are separated on the gel by electrophoresis, the DNA is denatured into single strands. The gel is then placed on a support on top of a filter paper saturated with high-ionic-strength buffer. Nitrocellulose paper or a nylon filter is placed on the gel, and towels are placed atop the filter. The transfer buffer makes its way through the gel, nitrocellulose paper, and towels by capillary action, taking the DNA with it. The single-stranded DNA is stopped by the nitrocellulose paper. The positions of the DNA in the paper directly reflect the positions of the DNA fragments in the gel.

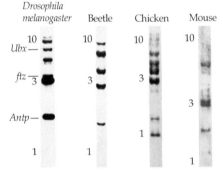

FIGURE 2.21
Southern blots of various organisms' DNA using a radioactive probe from the *Antennapedia* gene of *Drosophila melanogaster*. Since one does not expect the sequences between such diverged species to be perfectly identical, the stringency of the hybridization is lowered by changing the salt conditions. (Such low-stringency blots across phyla are colloquially refered to as "zoo blots," for obvious reasons). Autoradiography shows that *Drosophila* genes contain several portions that are like *Antennapedia* genes in structure, and that many organisms contain several genes that will hybridize this radioactive gene fragment, suggesting that *Antennapedia*-like genes exist in these organisms. The numbers beside the blots indicate size of bands, in kilobases. (From McGinnis et al., 1984, courtesy of W. McGinnis.)

portion of the *Drosophila Antennapedia* gene. An autoradiogram of the nitrocellulose paper showed where the radioactive DNA had found its match. The results from these experiments (Figure 2.21) showed that even vertebrates (mice, humans, and chicks) have genes that hybridize to these sequences. This radioactive section of the *Antennapedia* gene was then used to screen a genomic library of DNA clones derived from the genome of these different species. As we will see in Chapter 17, investigators found clones containing genes that resemble *Antennapedia*, and these genes were revealed to be extremely important in the formation of the vertebrate body axis.

DNA sequencing

Sequence data can tell us the structure of the encoded protein and can identify regulatory DNA sequences that certain genes have in common. The simplicity of the Sanger "dideoxy" sequencing technique (Sanger et al., 1977) has made it a standard procedure in many molecular biology laboratories. One starts with the vector carrying the cloned gene and isolates a single strand of the circular DNA (Figure 2.22). One then anneals a radioactive primer of DNA (about 20 base pairs) complementary to the vector DNA immediately 3' to the cloned gene. (Because these vector sequences are known, oligonucleotide primers can be readily synthesized or purchased commercially.) The primer has a free 3' end to which more nucleotides can be added. One places the primed DNA and all four deoxyribonucleoside triphosphates into four test tubes. Each of the test tubes contains the polymerizing subunit of DNA polymerase and a different *di*deoxynucleoside triphosphate: one tube contains dideoxy-G, one tube

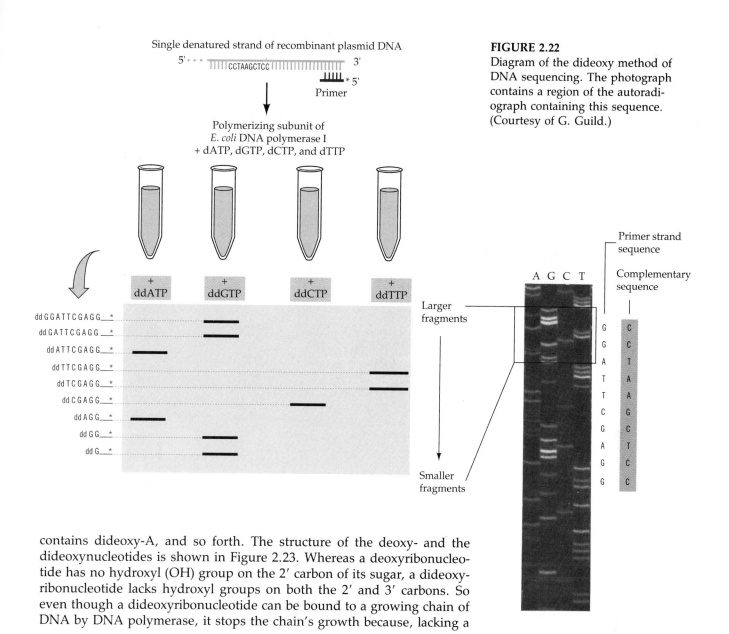

Single denatured strand of recombinant plasmid DNA

5′ ··· CCTAAGCTCC 3′

Primer

↓

Polymerizing subunit of
E. coli DNA polymerase I
+ dATP, dGTP, dCTP, and dTTP

+ ddATP   + ddGTP   + ddCTP   + ddTTP

ddGGATTCGAGG___*
ddGATTCGAGG___*
ddATTCGAGG___*
ddTTCGAGG___*
ddTCGAGG___*
ddCGAGG___*
ddAGG___*
ddGG___*
ddG___*

A G C T

Larger
fragments

Smaller
fragments

Primer strand
sequence

Complementary
sequence

G   C
G   C
A   T
T   A
T   A
C   G
G   C
A   T
G   C
G   C

FIGURE 2.22
Diagram of the dideoxy method of
DNA sequencing. The photograph
contains a region of the autoradi-
ograph containing this sequence.
(Courtesy of G. Guild.)

contains dideoxy-A, and so forth. The structure of the deoxy- and the
dideoxynucleotides is shown in Figure 2.23. Whereas a deoxyribonucleo-
tide has no hydroxyl (OH) group on the 2′ carbon of its sugar, a dideoxyribonucleotide lacks hydroxyl groups on both the 2′ and 3′ carbons. So
even though a dideoxyribonucleotide can be bound to a growing chain of
DNA by DNA polymerase, it stops the chain's growth because, lacking a

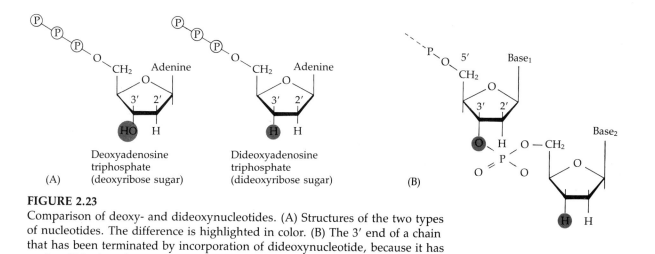

(A)   Deoxyadenosine
triphosphate
(deoxyribose sugar)

Dideoxyadenosine
triphosphate
(dideoxyribose sugar)

(B)

FIGURE 2.23
Comparison of deoxy- and dideoxynucleotides. (A) Structures of the two types
of nucleotides. The difference is highlighted in color. (B) The 3′ end of a chain
that has been terminated by incorporation of dideoxynucleotide, because it has
no free 3′ hydroxyl group for further DNA polymerization.

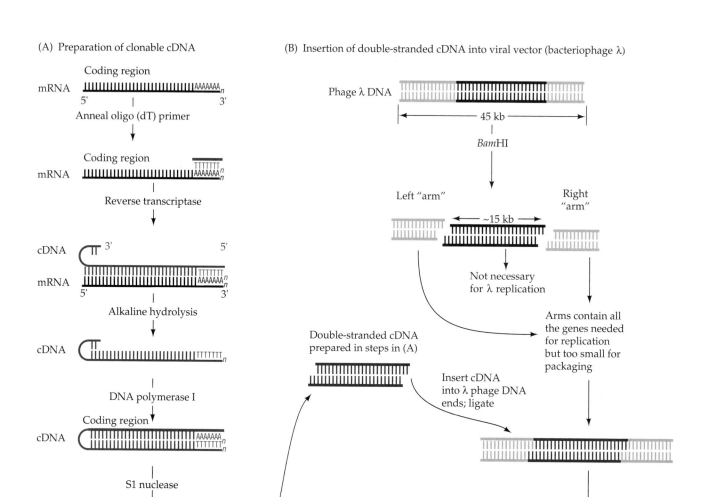

(A) Preparation of clonable cDNA

Coding region

mRNA

Anneal oligo (dT) primer

Coding region

mRNA

Reverse transcriptase

cDNA
mRNA

Alkaline hydrolysis

cDNA

DNA polymerase I

Coding region

cDNA

S1 nuclease

Coding region

Double-stranded cDNA

Add *Bam*HI ends

(B) Insertion of double-stranded cDNA into viral vector (bacteriophage λ)

Phage λ DNA

45 kb

*Bam*HI

Left "arm"   Right "arm"

~15 kb

Not necessary for λ replication

Arms contain all the genes needed for replication but too small for packaging

Double-stranded cDNA prepared in steps in (A)

Insert cDNA into λ phage DNA ends; ligate

cDNA from mRNA now cloned in viral vectors

3′ hydroxyl group, no new nucleotide can bind to it. Thus, when the DNA polymerase is synthesizing DNA from the primer, it will be complementary to the cloned gene. In the tube with dideoxy-A, however, every time the polymerase puts an A into the growing chain, there is a chance that the dideoxy-A will be placed there instead of the deoxy-A. If this happens, the chain stops. Similarly, in the tube with dideoxy-G, the chain has the potential to stop every time a G is inserted. (The process has been likened to a Greek folk dance in which some small percentage of the potential dancers have one arm in a sling.) Because there are millions of chains being made in each tube, each tube will contain a population of chains, some stopped at the first possible site, some at the last, and some at sites in between. The tube with dideoxy-A, for instance, will contain chains of different discrete lengths, each ending at an A residue. The resulting radioactive DNA fragments are separated by electrophoresis. The result is a "ladder" of fragments wherein each "rung" is a nucleotide sequence of a different length. By reading up the ladders, one obtains the DNA sequence complementary to that of the cloned gene.

(C) Preparing library of phage clones

(D) Screening of phage clone library

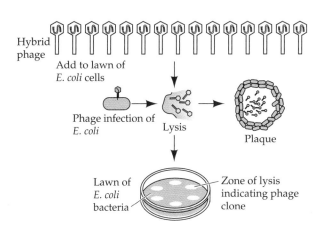

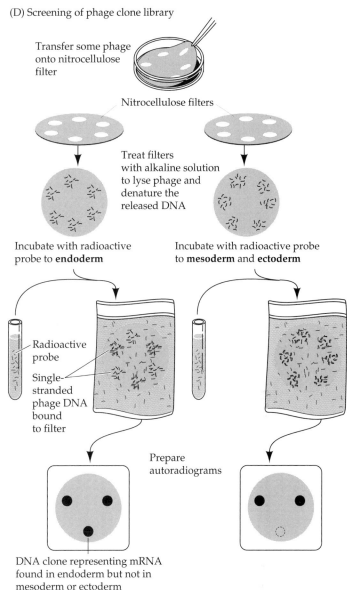

Transfer some phage onto nitrocellulose filter

Nitrocellulose filters

Treat filters with alkaline solution to lyse phage and denature the released DNA

Incubate with radioactive probe to **endoderm**

Incubate with radioactive probe to **mesoderm** and **ectoderm**

Radioactive probe

Single-stranded phage DNA bound to filter

Prepare autoradiograms

DNA clone representing mRNA found in endoderm but not in mesoderm or ectoderm

FIGURE 2.24

Diagram of protocol used to make cDNA libraries. (A) Messenger RNA is isolated and made into cDNA. This cDNA is made double-stranded, and restriction fragment ends are added. (B) The cDNA "genes" can then be inserted into specially modified vectors, in this case bacteriophages. (C) Phages containing the recombinant DNA will lyse *E. coli,* forming plaques. Biochemical techniques can distinguish plaques of recombinant phages from those that lack the inserted gene. (D) The plaques are transferred to nitrocellulose paper and treated with alkali to lyse the phages and denature the DNA in place. These filters are then incubated in radioactive probes (usually cDNA) from a tissue. In the case of the differential cDNA library screening discussed in the text, the same phage library was screened with radioactive probes from two different tissues. This allowed the researchers to look for an mRNA that would be found in one type of tissue but not in the other.

Analyzing mRNA through cDNA libraries

Now we can return to the specificity of mRNA trancription: Can we isolate populations of mRNA that characterize certain cell types and are absent in all others? To find these RNAs, one can "clone" the mRNA from different types of cells and compare them. As shown in Figure 2.24A, this is done by taking the population of messenger RNAs from a cell or tissue and converting them into complementary DNA strands. By taking the procedure a step further (with the aid of DNA polymerase and S1 nuclease), one can make this population of single-stranded cDNA into a population of double-stranded cDNA pieces. These strands of DNA can be inserted into plasmids by adding the appropriate "ends" onto them with DNA ligase. Appending a GATCC/G fragment onto the blunt ends of such a DNA piece creates an artificial *Bam*HI restriction cut and enables the piece to be inserted into a virus or plasmid cut with that enzyme (Figure 2.24B).

Such collections of clones derived from mRNAs are often called **libraries.** Thus, one can have a 16-day embryonic mouse liver library, representing all the genes active in making embryonic liver proteins. One can also have a *Xenopus* vegetal oocyte library, representing messages present only in a particular part of that cell. Genes cloned in this manner are very important because they lack introns. When they are added to bacterial cells, these genes can be transcribed and then translated into the proteins they encode.

Libraries have been extremely useful in studying development. As an example, we will follow the steps used by Wessel and co-workers (1989) to detect differences in the RNAs in different parts of the gastrulating sea urchin embryo. To find endoderm-specific mRNAs in sea urchins, Wessel and co-workers prepared a cDNA library from gastrulating embryos. The mRNA of these samples (most of the RNA of eukaryotic cells is ribosomal) was isolated by running the samples through oligo-dT beads that capture the poly(A) tails of the messages (see legend to Figure 2.19). Then the mRNA population was converted into a cDNA population by using reverse transcriptase (Figure 2.24A). By using *E. coli* polymerase I, the single-stranded cDNA was then made double-stranded. Next, commercially available *Eco*RI "ends" were ligated onto the double-stranded cDNAs. This made them clonable into vectors that were cut with *Eco*RI restriction enzyme. This DNA was then mixed with the arms of a genetically modified λ phage (Figure 2.24B). This phage is so constructed that when grown in a petri dish, the phages that have incorporated the DNA (and thus destroyed the β-galactosidase gene) produce colorless plaques (Figure 2.24C). In this way, approximately 4 million recombinant phages were generated, each containing a cDNA representing an mRNA molecule.

The next steps involved screening these recombinant phages. Which ones might represent mRNAs found in endoderm and not in the other cell layers? Wessel and his colleagues isolated the mRNA populations from mesoderm, ectoderm, and endoderm. They then made labeled cDNAs from each of these mRNA populations using radioactive precursors. They now had three collections of radioactive cDNA molecules, each representing the mRNA population from one of the three germ layers.

The recombinant phages representing the mRNAs of the gastrulating sea urchin embryo were grown, and samples of numerous colonies—each containing thousands of phages—were placed on two nitrocellulose filters (Figure 2.24D). These samples were then placed in alkaline solutions to lyse the phages and make the DNA single-stranded. One of these filter papers was incubated with radioactive cDNA made from the total mRNA of the *endoderm*; the other paper was incubated with radioactive probes to both *mesoderm and ectoderm.* The filters were then washed to remove any unhybridized radioactive cDNA, dried, and exposed on X-ray film. If an mRNA were present in the endoderm but not in either the ectoderm or mesoderm, the recombinant DNA made from that message should bind radioactive cDNA from the endoderm but should not find an mRNA anywhere else. As a result, that spot of recombinant DNA from the endoderm should be radioactive (since it bound radioactive cDNA from the endoderm)—but the same clone should not be radioactive when exposed to ectodermal or mesodermal mRNA. This was found to be the case. One recombinant phage in particular only bound radioactive cDNA made from endodermal mRNA; hence it represented an mRNA found in the endoderm and not in the mesoderm or ectoderm. The phage containing this gene can now be grown in large quantities and characterized.

RNA localization techniques

In situ hybridization

Once one has a cloned a gene, one can determine where it is active by the process of **in situ hybridization.** This technique, developed by Mary Lou Pardue and Joseph Gall (1970), allows one to visualize the positions of specific nucleic acids within cells and tissues. After one has identified a particular clone as being interesting (e.g., the endoderm-specific clone just mentioned), it is grown in large quantities, and the cloned gene is isolated by treating the recombinant vector with restriction enzymes. This gene is then made single-stranded and radioactive. When this radioactive cDNA is added to appropriately fixed cells on a microscope slide, the radioactive cDNA only binds where the complementary mRNA is present. After the unbound cDNA is washed off, the slides are covered with a transparent photographic emulsion for autoradiography. The resulting spots, which are directly above the places where the radioactive cDNA had bound, appear black when viewed directly, or white when viewed under dark-field illumination. Thus, one can visualize those cells (or even regions within cells) that have accumulated a specific type of mRNA. Figure 2.25 shows in situ hybridization using the cDNA found to be specific for endodermal cells. The cDNA finds mRNAs only in the endoderm of the early sea urchin gastrula. As gastrulation continues, the cDNA (and thus the mRNA) becomes localized even more precisely—to the hindgut–midgut region of the endodermal tube.

Northern blots

One can also determine the temporal and spatial expression of RNAs by running an **RNA blot** (often called a **Northern blot**). Whereas Southern blots transfer *DNA* fragments from a gel to paper, Northern blots (not named after their inventor) transfer *RNA* from gel to paper in the same way. An investigator can extract messenger RNA from embryos at various stages of their development and electrophorese these RNAs side by side on a gel. After transferring the separated RNA (by blotting) to nitrocellulose paper or nylon membrane, the RNA-containing paper is incubated in a solution containing a radioactive single-stranded DNA fragment from a particular gene. This DNA sticks only to regions where the complementary RNA is located. Thus, if the mRNA for that gene is present at a certain embryonic stage, the radioactive DNA binds to it and can be detected by autoradiography. Autoradiographs of this type, in which several stages are compared simultaneously, are called **developmental Northern blots.** Figure 2.26A shows a developmental Northern blot for the expression of the above-mentioned endoderm-specific gene during sea urchin development. One can see that mRNA for this protein is first synthesized during the mesenchyme blastula stage and is continually made during the rest of development. The Northern blot shown in Figure 2.26B shows that the accumulation of this mRNA in the prism stage is restricted to the endoderm (Wessel et al., 1989). In situ hybridization and Northern blots provide the best evidence for differential RNA transcription in both space and time. Transcription of certain genes can be tissue-specific and time-specific.

The timing of transcription for several genes can be visualized by **dot blots.** Sargent and Dawid (1983) have isolated from *Xenopus* gastrulae the mRNA that was not present in the egg. To do this, they extracted the gastrula mRNA and made cDNA copies of these messages. They mixed

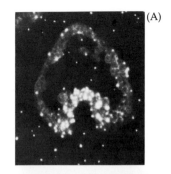
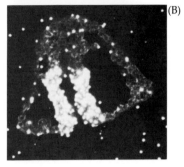

FIGURE 2.25
Dark field photomicrograph of in situ hybridization showing the localization of an endoderm-specific mRNA. The radioactive cDNA used as a probe was prepared from the cloned gene made from endoderm-specific mRNA (Figure 2.24). This radioactive cDNA binds to mRNA in the endoderm of early gastrula-stage sea urchins (A) and in the mid- and hindgut endoderm of late gastrulating sea urchins (B). (From Wessel et al., 1989, courtesy of G. Wessel.)

FIGURE 2.26

Northern blots for an endoderm-specific gene of the sea urchin *Lytechinus variegatus*. (A) Developmental Northern blot showing stage-specific accumulation of the mRNA from this gene. Total mRNA (10 μg per stage) was electrophoresed into an agarose gel. The gel was blotted onto treated paper, and the mRNAs were bound to the paper. The paper was then incubated with radioactive cDNA from the endoderm-specific clone. This mRNA was found to be synthesized during the mesenchyme blastula stage and increased throughout development. (B) Northern blot at the prism stage showing that the mRNA is present in the endoderm (with some mesoderm attached) but not in the ectoderm. Total RNA from the endoderm was electrophoresed (lane 2) next to total mRNA from the rest of the sea urchin (lane 1). Binding of the radioactive cDNA detected mRNA only in the endoderm. (From Wessel et al., 1989, courtesy of G. Wessel.)

these gastrula cDNAs with large quantities of oocyte mRNA. If the oocyte mRNA hybridized with gastrula cDNA, that meant that the cDNA was derived from an mRNA present in both oocyte and gastrula stages. These double-stranded hybrid molecules were removed by filtration, thereby leaving a population of gastrula-specific cDNAs. These cDNAs were then made double-stranded (by DNA polymerase) and inserted into cloning vehicles. This technique is called **subtraction cloning.** Like the dual screening of cDNA libraries, subtraction cloning generates a stage-specific set of clones whose mRNA is found in some stages but not others, or in some tissues but not others (Figure 2.27).

Sargent and Dawid then took embryos from the zygote through the tailbud tadpole stage and separately isolated their RNAs. The RNAs were spotted onto nitrocellulose filters directly (with no prior gel electrophoresis) so that each filter had RNAs from all stages. After the RNAs were baked onto the filter, single-stranded DNA derived from a particular "gastrula" clone was made radioactive and incubated with the filters. If a gene was being transcribed at a given stage, the radioactive cDNA from that gene would find its complement in the mRNAs from that stage on the filter. The nonbound cDNA was then washed off and the binding of radioactive cDNA observed by autoradiography. This technique is called **dot blotting.** Figure 2.28 shows the time courses of gene expression for 17 genes that are active at various stages of gastrulation. None of them is expressed earlier than the midblastula transition at 7 hours. Some genes (DG64, DG39) are expressed immediately afterward, whereas other genes (such as DG72 or DG81) begin to be transcribed at midgastrula, about 7 hours later. Some genes (DG76, DG81) appear to "stay on" once activated, while the activity of other genes (DG56 or DG21) is far more transient.

Finding rare messages by the polymerase chain reaction

The **polymerase chain reaction** (PCR) is a method of in vitro cloning that can generate enormous quantities of a specific DNA fragment from a small amount of starting material (Saiki et al., 1985). It can be used for cloning a specific gene or can be used for determining if a specific gene is actively

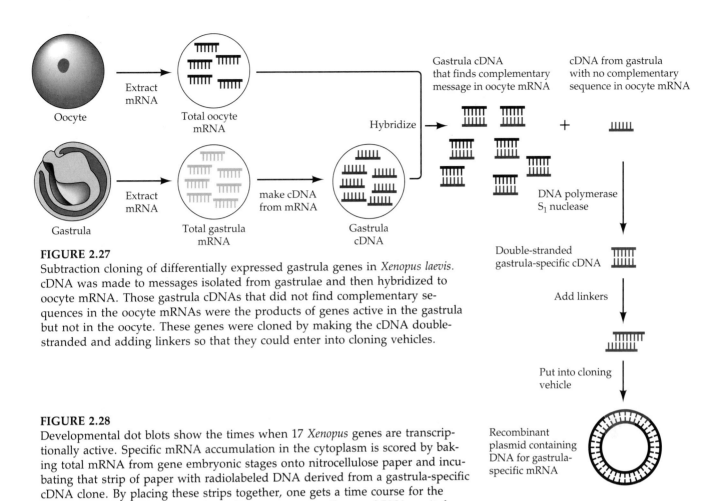

FIGURE 2.27

Subtraction cloning of differentially expressed gastrula genes in *Xenopus laevis*. cDNA was made to messages isolated from gastrulae and then hybridized to oocyte mRNA. Those gastrula cDNAs that did not find complementary sequences in the oocyte mRNAs were the products of genes active in the gastrula but not in the oocyte. These genes were cloned by making the cDNA double-stranded and adding linkers so that they could enter into cloning vehicles.

FIGURE 2.28

Developmental dot blots show the times when 17 *Xenopus* genes are transcriptionally active. Specific mRNA accumulation in the cytoplasm is scored by baking total mRNA from gene embryonic stages onto nitrocellulose paper and incubating that strip of paper with radiolabeled DNA derived from a gastrula-specific cDNA clone. By placing these strips together, one gets a time course for the activity of the specific genes. The r5 line represents a ribosomal RNA control that should always be present. (From Jamrich et al., 1985, courtesy of I. Dawid and M. Sargent.)

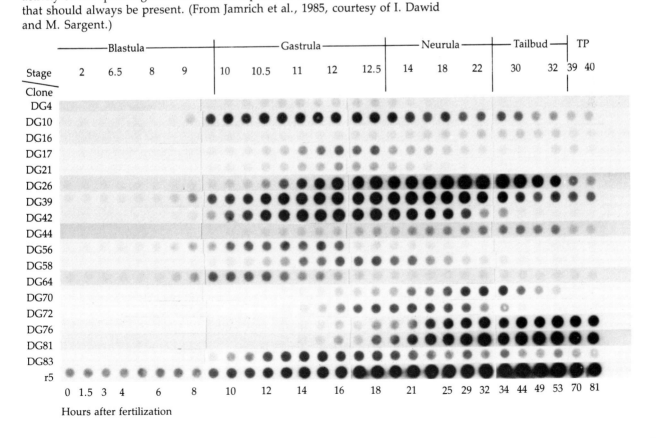

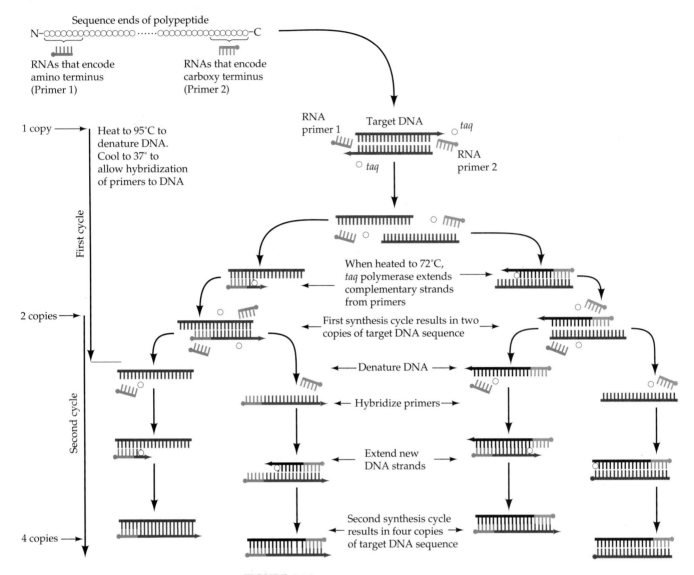

FIGURE 2.29

Protocol for the polymerase chain reaction (PCR). For determining whether a particular type of mRNA is present, all the mRNA is converted to double-stranded DNA by reverse transcriptase and DNA polymerase. This DNA is denatured, and two sets of primers are added. If the specific sequence is present, the primers will hybridize to its opposite ends. (Specific primers are made based on the sequence one is looking for. If one knows only the sequence of the protein encoded by the message, a set of different primers is made, each possibly being complementary to the DNA). Using thermostable DNA polymerase from *T. aquaticus,* each DNA strand synthesizes its complement. These strands are then denatured and the primers are hybridized to them, starting the cycle again. In this way, the number of new strands having the sequence between the two primers increases exponentially.

transcribing RNA in a particular organ or cell type. The standard methods of cloning use living microorganisms to amplify recombinant DNA. PCR, however, can amplify a single DNA molecule several millionfold in a few hours, and it does it in a test tube. This technique has been extremely useful in cases where there is very little nucleic acid to study. Preimplantation mouse embryos, for instance, have very little mRNA, and one cannot obtain millions of such embryos to study. If one wanted to know whether the preimplantation mouse embryo contained the mRNA for a particular protein, it would be very difficult to find out using standard cloning methods. One would have to lyse thousands of mouse embryos to get enough mRNA. However, the PCR technique allows one to find this message in a few embryos by specifically amplifying only that message millions of times (Rappolee et al., 1988).

The use of PCR for finding rare mRNAs is illustrated in Figure 2.29. The mRNAs from a group of cells are purified and converted into cDNA by reverse transcriptase. By using DNA polymerase and S1 nuclease, the population of single-stranded DNAs is then made into double-stranded DNAs. Next one targets a specific DNA for amplification. To do this, one separates the DNA double helices and adds to them two small oligonucleotide primers that are complementary to a portion of the message being looked for. If the oligonucleotides recognize sequences in the DNA, that means that the mRNA was originally present. The oligonucleotides have been made so that they hybridize to opposite strands and opposite sides of the targeted sequence. (If one is trying to isolate the gene or mRNA for a specific protein of known sequence, these flanking regions can be prepared by synthesizing oligonucleotides that would encode the amino end of the protein and oligonucleotides complementary to those that would encode the carboxyl end of the protein.) The 3′ ends of these primers face each other, so that replication is through the target DNA. Once the first primer has hybridized, the DNA polymerase can synthesize a new strand.

This is not normal *E. coli* DNA polymerase, however; it is DNA polymerase from bacteria such as *Thermus aquaticus* or *Thermococcus litoralis*. These bacteria live in hot springs (such as those in Yellowstone National Park) or in submarine thermal vents, where the temperature reaches nearly 90°C. These DNA polymerases can withstand temperatures near boiling; PCR takes advantage of this evolutionary adaptation. Once the second strand is made, it is heat-denatured from its complement. The second primer is added, and now both strands can synthesize new DNA. Repeated cycles of denaturation and synthesis will amplify this region of DNA in geometric fashion. After 20 such rounds, that specific region has been amplified 2^{20} (a little more than a million) times. This material can be used in two ways. First, one can determine if the cells whose mRNA was purified were transcribing that specific gene. To do this, one would take the material and perform a Southern blot on it. If the specific message were present, it would have been made into DNA and amplified a million or so times by the PCR. It would therefore show up on the blot when probed with a radioactive oligonucleotide (such as either primer) only if the mRNA were originally present (Figure 2.30). Second, one can take these amplified copies and place them into cloning vectors and clone them.

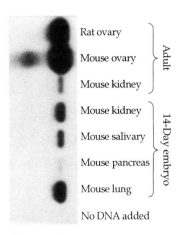

FIGURE 2.30
Evidence from PCR for the synthesis of a growth factor, activin, from embryonic mouse organs. The mRNA from these organs was converted into DNA and amplified through 20 replication cycles. The DNA was electrophoresed and Southern blotted using a radioactive probe to a portion of the activin gene. Activin mRNA was found in the adult mouse ovary (as expected) and also in several embryonic organs. The possible role of activin in these organs will be discussed in Chapter 18. (Courtesy of O. Ritvos.)

Determining the function of a gene: Transgenic cells and organisms

Techniques of inserting new DNA into a cell

While it is nice to know the sequence of a gene and its temporospatial pattern of expression, what's really important is the function(s) of that gene during development. Recent techniques have enabled us to study gene function by taking certain genes into and out of embryonic cells. Cloned pieces of DNA can be isolated, modified (if so desired), and placed into cells by several means. One very direct technique is **microinjection,**

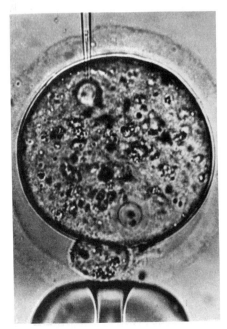

FIGURE 2.31
Injection of DNA (from cloned genes) into a nucleus (in this case, a pronucleus of a mouse egg). (From Wagner et al., 1981, courtesy of T. E. Wagner.)

in which a solution containing the cloned gene is injected very carefully into the nucleus of a cell (Capecchi, 1980). This is an especially useful technique for injecting genes into newly fertilized eggs, since the haploid nuclei of the sperm and egg are relatively large (Figure 2.31). In **transfection,** DNA is incorporated directly into the cell by incubating it in a particular solution that makes the cells drink it in. The chances of such a DNA fragment being incorporated into the chromosomes are relatively small, so the DNA is usually mixed with another gene that enables those rare cells that incorporate them to survive under culture conditions that will kill all the other cells (Perucho et al., 1980; Robins et al., 1981).

Another technique is **electroporation.** Here the DNA is mixed with cells and a high-voltage pulse "pushes" the DNA into the cells. A more "natural" technique to get genes into cells is to put the cloned gene into a **transposable element** or **retroviral vector.** These are naturally occurring mobile regions of DNA that can integrate into the genome. Retroviruses are RNA-containing viruses. Within a host cell, they make a DNA copy of themselves (using their own reverse transcriptase); the copy becomes double-stranded and integrates into a host chromosome. This integration is accomplished as a result of two identical sequences (long terminal repeats) at the ends of the retroviral DNA. Retroviral vectors are made by removing the viral packaging genes (needed for the exit of viruses from the cell) from the center of a mouse retrovirus. This extraction creates a vacant site where other genes can be placed. By using the appropriate restriction enzymes, researchers can remove genes from a cloned phage or plasmid and reinsert the gene into the retroviral vectors. These retroviral vectors infect mouse cells with an efficiency approaching 100 percent. In *Drosophila,* new genes can be carried into a fly via **P elements.** These DNA sequences are naturally occurring transposable elements that can integrate like viruses into any region of the *Drosophila* genome. Moreover, they can be isolated, and cloned genes can be inserted into the center of the P element. When the recombined P element is injected into a *Drosophila* oocyte, it can integrate into the DNA and provide the embryo with this new gene (Spradling and Rubin, 1982).

Chimeric mice

The techniques described have recently been used to transfer genes into every cell of the mouse embryo (Figure 2.32). During mouse development (as will be discussed later), there is a stage when only two cell types are present: the outer cells that will form the fetal portion of the placenta, and the inner cells that will give rise to the embryo itself. These inner cells are called **embryonic stem cells,** in that each of them can (if isolated) generate all the cells of the embryo (Gardner, 1968; Moustafa and Brinster, 1972). These cells can be isolated from a mouse embryo and grown in culture. Once in culture, they can be treated as described so that they will incorporate new DNA. This new embryonic stem cell can then be injected (the entire cell, not just the DNA) into another early-stage mouse embryo, and the treated stem cell will integrate into the host embryo. The result is a **chimeric** mouse.* Some of its cells are derived from the host embryonic stem cells, but some portion of its cells are also derived from the treated stem cell. If the treated cells have become part of the germ line of the

---

*It is critical to note the difference between a chimera and a hybrid. A **hybrid** results from the union of two different *genomes* within the same cell; the offspring of an *AA* genotype parent and an *aa* genotype parent is an *Aa* hybrid. A **chimera** results when *cells* of different genetic constitution appear in the same organism. The term is apt: it refers to a mythical beast with lion's head, goat's body, and serpent's tail.

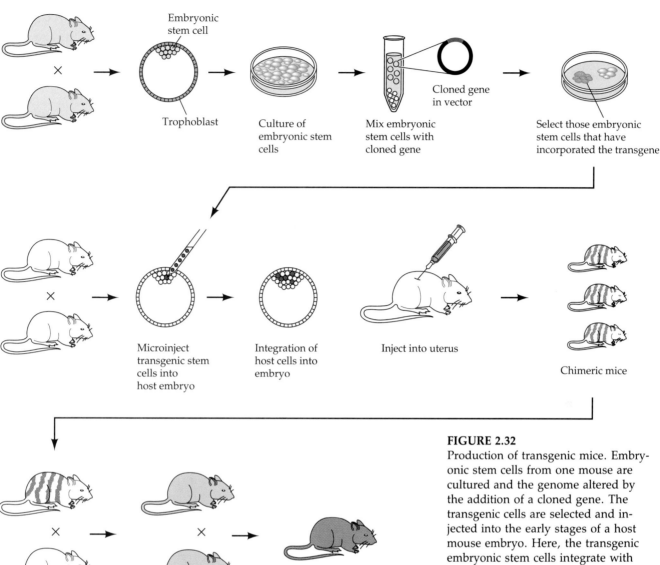

FIGURE 2.32
Production of transgenic mice. Embryonic stem cells from one mouse are cultured and the genome altered by the addition of a cloned gene. The transgenic cells are selected and injected into the early stages of a host mouse embryo. Here, the transgenic embryonic stem cells integrate with the host's embryonic stem cells. This embryo is placed into the uterus of a pregnant mouse and develops into a chimeric mouse. If donor stem cells contributed to the germ line, some of the progeny of that mouse crossed with a wild-type mouse will be heterozygous for the added allele. By mating heterozygotes, a strain of mice can be generated that is homozygous for the added allele. This would be a transgenic strain. The added gene (the *transgene*) can be from any eukaryotic source.

mouse, some of its gametes will be derived from the donor cell. When mated with a wild-type mouse, some of the progeny will therefore carry one copy of the inserted gene. When these heterozygote progeny are mated, about 25 percent of the resulting embryos will carry two copies of the inserted gene in every cell of its body (Gossler et al., 1986). Thus, a gene that had been cloned from some other individual is now part of the genome of the entire mouse embryo. Mice with stable genes derived from other individuals are called **transgenic mice.**

Gene targeting ("knockout") experiments

The analysis of early mammalian embryos has long been hindered by our inability to breed and select mutations that affect early embryonic development. This block has been circumvented by the techniques of **gene targeting** (or, as it is sometimes called, **gene "knockout"**). These techniques are similar to those that generate transgenic mice, but instead of adding genes, gene targeting replaces wild-type alleles with mutant ones. Chisaka and Capecchi (1991) used gene targeting to find the function of the *Hoxa-3* gene in the development of the mouse. *Hoxa-3* is similar to several genes of *Drosophila* that have been found to control segment-specific gene expression in the early embryo, and the protein encoded by the *Hoxa-3* gene binds to DNA, just like its *Drosophila* counterparts. Could *Hoxa-3* similarly regulate spatially specific gene expression in mammals? Chisaka and Capecchi isolated the *Hoxa-3* gene, cut it with a restriction enzyme, and inserted a gene for neomycin resistance into that site (Figure 2.33). In other words, they mutated the *Hoxa-3* gene by inserting into it a large piece of DNA that contained a neomycin-resistance gene. This insertion destroyed the ability of the Hoxa-3 protein to bind to DNA. These mutant *Hoxa-3* genes were electroporated into embryonic stem cells that were sensitive to neomycin. Once inside the nucleus of these cells, the mutated *Hoxa-3* genes replaced a normal allele of *Hoxa-3* by a process called **homologous recombination.** Here the enzymes involved in DNA repair and replication incorporate the mutant gene at the expense of the normal copy. It's a rare event, but such cells can be selected by growing the stem cells in neomycin. Most of the cells die in the drug, but the ones that acquired resistance from the incorporated gene survive. The resulting cells have one normal *Hoxa-3* gene and one mutated *Hoxa-3* gene. The heterozygous stem cells are then microinjected into a mouse blastocyst and integrated into the cells of the embryo. The resulting mouse is a chimera composed of wild-type cells from the host embryo and heterozygous *Hoxa-3* cells from the donor stem cells. The chimeras are mated to wild-type mice, and if some of the donor cells became integrated into the germ cell lineage, some of the progeny will be heterozygous for the *Hoxa-3* gene. These heterozygous mice can be bred with each other, and about 25 percent of their progeny are expected to carry two copies of the mutated *Hoxa-3* gene. These homozygous mutant mice lack thyroid, parathyroid, and thymus glands! In this way, gene targeting can be used to analyze the roles of particular genes during mammalian development.

Determining the function of a message: Antisense RNA

"Antisense" messages can be generated by taking cloned cDNA and recloning it *backward* next to a strong bacterial promoter in another vector; this is the basis of another important technique for analyzing gene function. The bacterial promoter will initiate transcription of the message "in the wrong direction" when it is incubated with RNA polymerase and nucleotide triphosphates. In so doing, it synthesizes a transcript that is complementary to the natural one (Figure 2.34A). This complementary transcript is called **antisense RNA** since it is the reverse of the original message. When large amounts of this antisense RNA are injected or transfected into cells containing the normal mRNA from this gene, the antisense RNA binds to the normal message; the resulting double-stranded polypeptide is degraded. (Cells have enzymes to digest double-stranded nucleic acids in the cytoplasm.) This causes a functional depletion of the message, just as if there were a deletion mutation for that gene.

These results were seen when antisense RNA was made to the *Krüppel* gene of *Drosophila*. *Krüppel* is critical for forming the thorax and abdomen of the fly. If this gene is absent, fly larvae die because they lack thoracic and anterior abdominal segments (Figure 2.34B); a similar situation is

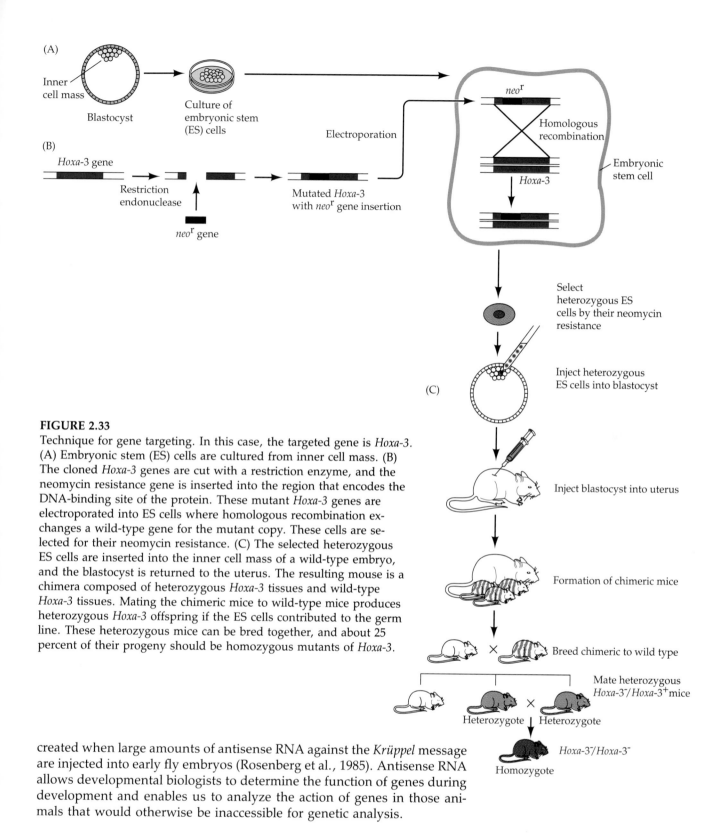

FIGURE 2.33
Technique for gene targeting. In this case, the targeted gene is *Hoxa-3*. (A) Embryonic stem (ES) cells are cultured from inner cell mass. (B) The cloned *Hoxa-3* genes are cut with a restriction enzyme, and the neomycin resistance gene is inserted into the region that encodes the DNA-binding site of the protein. These mutant *Hoxa-3* genes are electroporated into ES cells where homologous recombination exchanges a wild-type gene for the mutant copy. These cells are selected for their neomycin resistance. (C) The selected heterozygous ES cells are inserted into the inner cell mass of a wild-type embryo, and the blastocyst is returned to the uterus. The resulting mouse is a chimera composed of heterozygous *Hoxa-3* tissues and wild-type *Hoxa-3* tissues. Mating the chimeric mice to wild-type mice produces heterozygous *Hoxa-3* offspring if the ES cells contributed to the germ line. These heterozygous mice can be bred together, and about 25 percent of their progeny should be homozygous mutants of *Hoxa-3*.

created when large amounts of antisense RNA against the *Krüppel* message are injected into early fly embryos (Rosenberg et al., 1985). Antisense RNA allows developmental biologists to determine the function of genes during development and enables us to analyze the action of genes in those animals that would otherwise be inaccessible for genetic analysis.

Reinvestigating old problems with new methods

The combination of embryology and molecular biology is enabling developmental biologists to have a new appreciation of how genes work to

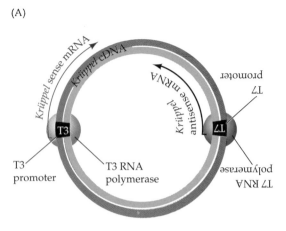

(A)

Krüppel sense mRNA

Krüppel cDNA

Krüppel antisense mRNA

T7 promoter

T3

T7

T3 promoter

T3 RNA polymerase

T7 RNA polymerase

FIGURE 2.34
Production of antisense RNA to examine the roles of genes in development. (A) Production of antisense message (in this case, to the *Krüppel* gene of *Drosophila*) by placing the cloned cDNA fragment for the *Krüppel* message between two strong promoters. The promoters are in opposite orientation with respect to the *Krüppel* cDNA. In this case, the T3 promoter is in normal orientation and the T7 promoter is reversed. These promoters recognize different RNA polymerases (from the T3 and T7 bacteriophages, respectively). T3 polymerase enables the transcription of "sense" mRNA, while T7 polymerase transcribes antisense transcripts. (B) Result of injecting *Krüppel* antisense message into the early embryo (syncytial blastoderm stage) of *Drosophila* before normal *Krüppel* message is produced. The central figure is the wild-type embryo just prior to hatching. To its left is the mutant caused by the lack of *Krüppel* genes. To its right is a wild-type embryo that had been injected with *Krüppel* antisense message in the early embryo stage. Both the mutation and the antisense-treated embryos lack their thoracic and anterior abdominal segments. (B after Rosenberg et al., 1985.)

(B) *Krüppel* mutant embryo

Normal embryo

Normal embryo infected with *Krüppel* "antisense" RNA

construct the organism. As we will see, we are in the middle of a revolution in our understanding of development. One of the first major successes of all this cloning and sequencing was a new "anatomy" of the eukaryotic gene. Although we will be describing gene structure in more detail in Chapter 10, it is important to realize that eukaryotic genes that encode proteins have several regulatory sites (Figure 2.35). One site, the **promoter,** is located directly upstream (before the start) of the gene and is the site where RNA polymerase binds. **RNA polymerase** is the enzyme that will transcribe the DNA into an RNA copy. Located somewhere near the gene (either downstream, upstream, or in an intron within the gene) is a second region called the **enhancer.** Protein factors that bind to the enhancer enable it to interact with the promoter to enable the RNA polymerase to transcribe the gene. Whereas some promoters (such as those used by products involved in the cell's general metabolism) do not need to be activated by enhancers, most genes involved in development are activated at specific times and in specific cells. These genes need to be activated by those factors that bind to the enhancer and the promoter. As we will discuss

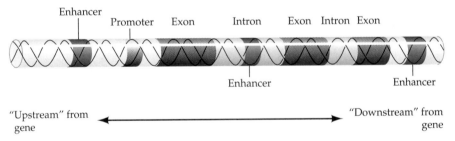

Enhancer   Promoter   Exon   Intron   Exon   Intron   Exon

Enhancer   Enhancer

"Upstream" from gene   "Downstream" from gene

FIGURE 2.35
Basic structure of a developmentally regulated gene. The promoter of most genes encoding proteins is found at the 5′ (upstream) end of the gene. The enhancer is often even farther upstream, but it can also be within an intron of the gene or at the 3′ end. Proteins that bind to the promoter and enhancers interact to regulate the transcription of the gene. (In the ZP3 example below, the OSP-1 site, GATAA, is located in the promoter, about 95 base pairs upstream from the transcription initiation site. An estrogen-sensitive enhancer site is found in the first intron of the *ZP3* gene.)

more fully in Chapter 10, the binding of different transcription factors to the promoters and enhancers of specific genes is one of the mechanisms that control the production of different proteins from identical genomes. One example is the activation of the gene for ZP3.

As we will detail in Chapter 4, ZP3 is the principal sperm-binding protein of the mouse egg surface. It is a glycoprotein synthesized by the oocyte as it matures in the ovum (Roller et al., 1989). A Northern blot shows that the mRNA for this glycoprotein is made only in growing oocytes and cannot be detected in any other cell type tested (Figure 2.36). What enables this gene to become active only in the oocytes? Lira and co-workers (1990) have isolated the gene for ZP3, determined its sequence, and found a promoter site 28 base pairs upstream from the site where transcription of this gene is initiated. They hypothesized that the sequences conferring oocyte-specific activation might exist even farther upstream from the gene. So they used restriction enzymes to isolate the DNA from the 5' upstream region (150 base pairs long) and fused it to the gene for firefly luciferinase. (Needless to say, this light-producing enzyme is not found in mice. It is being used here as a "reporter gene" to monitor where the upstream DNA may cause its expression.) The newly constructed gene containing the upstream region of the *ZP3* gene attached to the *luciferinase* structural gene was injected into mouse zygotes to create transgenic mice that carried the *ZP3* regulatory region–*luciferinase* gene in each nucleus. In female transgenic mice, in situ hybridization located *luciferinase* mRNA in only one cell type: the oocyte (Figure 2.37). Thus, the 150-base-pair sequence of DNA was necessary and sufficient for activating the gene (any gene!) in the oocyte. Within this 150-base pair region (at 99 to 86 base pairs upstream from the *ZP3* structural gene) there is a sequence 5'−GATAA−3', which was found to bind a protein called OSP-1. OSP-1 is found solely in maturing oocytes and it activates the *ZP3* gene by binding to this DNA sequence in the promoter. It appears, then, that ZP3 is made in oocytes because oocytes have OSP-1 protein, which binds to certain sequences of DNA that are part of its promoter (Schickler et al., 1992). Research is now underway to see how the OSP-1-encoding gene is regulated.

FIGURE 2.36
Northern blot of ZP3 mRNA accumulation in the mouse. RNA from various tissues (10 μg per lane) and oocytes (125 ng) was electrophoresed and blotted onto nitrocellulose paper. A radioactively labeled fragment of the *ZP3* gene was used to probe the mRNA. The ZP3 message was found only in the ovary, and especially within the oocytes. (From Roller et al., 1989, courtesy of P. Wassarman.)

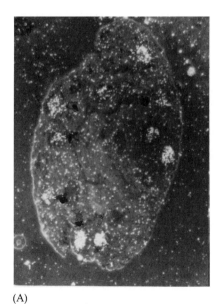
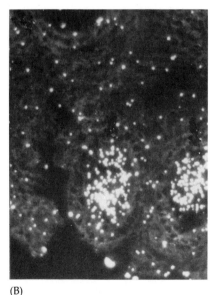

(A) (B)

FIGURE 2.37
In situ hybridization of reporter gene *luciferinase* expression when *luciferinase* was attached to the promoter of the *ZP3* gene. Radioactive probe was against the luciferinase message, which appeared where it was expressed under the direction of the ZP3 promoter. (A) View of entire ovary (60×). (B) Higher magification (160×) of two of the ovarian follicles containing maturing oocytes. (From Lira et al., 1990, courtesy of P. Wassarman.)

A conclusion and a caveat

After nearly a century, we are beginning to understand how cells regulate the differential expression of their genes so that different genes can become active in different cells. This knowledge is helping to explain how that inherited information is utilized to construct the basic body plans and specific cell types of developing organisms.

A word of warning, however. Lest the celebratory tone of this chapter gives you the impression that development is solely a function of gene activity, please recall from Chapter 1 that the distinction between stalk and spore (*Dictyostelium*), amoeboid and flagellate state (*Naegleria*), and sexual and asexual gonidia (*Volvox*) is determined by the *environment*. In later chapters we will see other examples of the environmental control of development: temperature-dependent sex determination in reptiles, diet-dependent development in insects, and the experience-dependent differentiation of neurons and lymphocytes in mammals. In these cases, the organism inherits the *ability* to respond to environmental cues, but there is no predicting the phenotype from the genotype.

LITERATURE CITED

Allen, G. E. 1978. *Thomas Hunt Morgan: The Man and His Science*. Princeton University Press, Princeton, NJ.

Allen, G. E. 1986. T. H. Morgan and the split between embryology and genetics, 1910–1935. *In* T. J. Horder, J. A. Witkowski and C. C. Wylie (eds.), *A History of Embryology*. Cambridge University Press, New York, pp. 113–146.

Ashburner, M. 1972. Patterns of puffing activity in the salivary glands of *Drosophila*. VI. Induction by ecdysone in salivary glands of *D. melanogaster* cultured in vitro. *Chromosoma* 38: 255–281.

Ashburner, M. and Berondes, H. D. 1978. Puffing of polytene chromosomes. In *The Genetics and Biology of Drosophila*, Vol. 2B. Academic Press, New York, pp. 316–395.

Baltzer, F. 1967. *Theodor Boveri: Life and Work of a Great Biologist*. (Trans. D. Rudnick.) University of California Press, Berkeley.

Barnett, T., Pachl, C., Gergen, J. P. and Wensink, P. C. 1980. The isolation and characterization of *Drosophila* yolk protein genes. *Cell* 21: 729–738.

Becker, H. J. 1959. Die Puffs der Speicheldrüsenchromosomen von *Drosophila melanogaster*. I. Beobachtungen zum Verhalten des Puffmusters im Normalstamm und bei zwei Mutanten, giant- und lethal-giant Larvae. *Chromosoma* 10: 654–678.

Beermann, W. 1952. Chromomerenkonstanz und spezifische Modifikationen der Chromosomenstruktur in der Entwicklung und Organdifferenzierung von *Chironomus tentans*. *Chromosoma* 5: 139–198.

Beermann, W. 1961. Ein Balbiani-ring als Locus einer Speicheldrüsen-Mutation. *Chromosoma* 12: 1–25.

Beermann, W. 1963. Cytological aspects of information transfer in cellular differentiation. *Am. Zool.* 3: 23–28.

Blattner, F. R. and eight others. 1978. Cloning human fetal γ-globin and mouse α-type globin DNA: Preparation and screening of shotgun collections. *Science* 202: 1279–1283.

Boveri, T. 1904. *Ergebnisse über die Konstitution der chromatischen Substanz des Zelkerns*. Gustav Fisher, Jena. [p. 123]

Briggs, R. 1979. Genetics of cell type determination. *Int. Rev. Cytol.* [Suppl.] 9: 107–127.

Briggs, R. and King, T. J. 1952. Transplantation of living nuclei from blastula cells into enucleated frogs' eggs. *Proc. Natl. Acad. Sci. USA* 38: 455–463.

Brush, S. 1978. Nettie Stevens and the discovery of sex determination. *Isis* 69: 132–172.

Burian, R., Gayon, J. and Zallen, D. T. 1991. Boris Ephrussi and the synthesis of genetics and embryology. *In* S. Gilbert (ed.), *A Conceptual History of Modern Embryology*. Plenum, New York, pp 207–227.

Burkholder, G. D. 1976. Whole mount electron microscopy of polytene chromosome from *Drosophila melanogaster*. *Can. J. Genet. Cytol.* 18: 67–77.

Capecchi, M. R. 1980. High efficiency transformation by direct microinjection of DNA into cultured mammalian cells. *Cell* 22: 479–488.

Chisaka, O. and Capecchi, M. R. 1991. Regionally restricted developmental defects resulting from targeted disruption of the homeobox gene *hox 1.5*. *Nature* 350: 473–479.

Clever, U. 1966. Induction and repression of a puff in *Chironomus tentans*. *Dev. Biol.* 14: 421–438.

Cohen, S. N., Chang, A. C. Y., Boyer, H. W. and Helling, R. B. 1973. Construction of biologically functional bacterial plasmids in vitro. *Proc. Natl. Acad. Sci. USA* 70: 3240–3244.

DiBerardino, M. A. 1987. Genomic potential of differentiated cells analyzed by nuclear transplantation. *Am. Zool.* 27: 623–644.

DiBerardino, M. A. 1989. Genomic activation in differentiated somatic cells. *In* M. A. DiBerardino and L. D. Etkin (eds.), *Developmental Biology: A Comprehensive Synthesis*. Plenum, New York, pp. 175–198.

DiBerardino, M. A. and King, T. J. 1967. Development and cellular differentiation of neural nuclear transplants of known karyotypes. *Dev. Biol.* 15: 102–128.

Dumont, J. N. and Yamada, T. 1972. Dedifferentiation of iris epithelial cells. *Dev. Biol.* 29: 385–401.

Gardner, R. L. 1968. Mouse chimeras obtained by the injection of cells into the blastocyst. *Nature* 220: 596–597.

Gilbert, S. F. 1978. The embryological origins of the gene theory. *J. Hist. Biol.* 11: 307–351.

Gilbert, S. F. 1987. In friendly disagreement: Wilson, Morgan, and the embryological origins of the gene theory. *Am. Zool.* 27: 797–806.

Gilbert, S. F. 1988. Cellular politics: Ernest Everett Just, Richard B. Goldschmidt, and the attempts to reconcile embryology and genetics. *In* R. Rainger, K. R. Benson and J. Maienschein (eds.), *The American Development of Biology*. University of Pennsylvania Press, Philadelphia, pp. 311–346.

Gilbert, S. F. 1991. Induction and the origins of developmental genetics. *In* S. Gilbert (ed.), *A Conceptual History of Modern Embryology*. Plenum, New York, pp. 181–206.

Gluecksohn-Schoenheimer, S. 1938. The development of two tailless mutants in the house mouse. *Genetics* 23: 573–584.

Gluecksohn-Schoenheimer, S. 1940. The effect of an early lethal (t^o) in the house mouse. *Genetics* 25: 391–400.

Goldschmidt, R. B. 1938. *Physiological Genetics*. McGraw-Hill, New York. [p. 1]

Gossler, A., Doetschman, T., Korn, R., Serfling, E. and Kemler, R. 1986. Transgenesis by means of blastocyst-derived stem cell lines. *Proc. Natl. Acad. Sci. USA* 83: 9065–9069.

Grossbach, U. 1973. Chromosome puffs and gene expressions in polytene cells. *Cold Spring Harbor Symp. Quant. Biol.* 38: 619–627.

Gurdon, J. B. 1962. The developmental capacity of nuclei taken from intestinal epithelial cells of feeding tadpoles. *J. Embryol. Exp. Morphol.* 10: 622–640.

Gurdon, J. B. 1968. Transplanted nuclei and cell differentiation. *Sci. Am.* 219(6): 24–35.

Gurdon, J. B. 1977. Egg cytoplasm and gene control in development. *Proc. R. Soc. Lond.* [B] 198: 211–247.

Gurdon, J. B. and Uehlinger, V. 1966. "Fertile" intestinal nuclei. *Nature* 210: 1240–1241.

Gurdon, J. B., Laskey, R. A. and Reeves, O. R. 1975. The developmental capacity of nuclei transplanted from keratinized cells of adult frogs. *J. Embryol. Exp. Morphol.* 34: 93–112.

Harrison, R. G. 1937. Embryology and its relations. *Science* 85: 369–374.

Harwood, J. 1993. *Styles of Scientific Thought: The German Genetics Community 1900–1933*. The University of Chicago Press, Chicago.

Hennen, S. 1970. Influence of spermine and reduced temperature on the ability of transplanted nuclei to promote normal development in eggs of *Rana pipiens*. *Proc. Natl. Acad. Sci. USA* 66: 630–637.

Holland, P. W. H. and Hogan, B. L. M. 1986. Phylogenetic distribution of *Antennapedia*-like homeoboxes. *Nature* 321: 251–253.

Jacob, F. and Monod. J. 1961. Genetic regulatory mechanisms in the synthesis of proteins. *J. Mol. Biol.* 3: 318–356.

Jamrich, J., Sargent, T. D. and Dawid, I. 1985. Altered morphogenesis and its effects on gene activity in *Xenopus laevis* embryos. *Cold Spring Harbor Symp. Quant. Biol.* 50: 31–35.

Judson, H. F. 1979. *The Eighth Day of Creation*. Simon & Schuster, New York.

Just, E. E. 1939. *The Biology of the Cell Surface*. Blakiston, Philadelphia.

Keller, E. F. In press. Rethinking the meaning of genetic determinism. In *Tanner Lectures 1993*. University of Utah Press, Salt Lake City.

King, T. J. 1966. Nuclear transplantation in amphibia. *Methods Cell Physiol.* 2: 1–36.

King, T. J. and Briggs, R. 1956. Serial transplantation of embryonic nuclei. *Cold Spring Harbor Symp. Quant. Biol.* 21: 271–289.

Lambert, B. 1972. Repeated DNA sequences in a Balbiani ring. *J. Mol. Biol.* 72: 65–75.

Lambert, B. and Daneholt, B. 1975. Microanalysis of RNA from defined cellular components. *Methods Cell Biol.* 10: 17–47.

Lederman, M. 1989. Research note: Genes on chromosomes: The conversion of Thomas Hunt Morgan. *J. Hist. Biol.* 22: 163–176.

Lillie, F. R. 1927. The gene and the ontogenetic process. *Science* 64: 361–368.

Lira, A. A., Kinloch, R. A., Mortillo, S. and Wassarman, P. A. 1990. An upstream region of the mouse *ZP3* gene directs expression of firefly luciferinase specifically to growing oocytes in transgenic mice. *Proc. Natl. Acad. Sci. USA* 87: 7215–7219.

McGinnis, W., Garber, R. L., Wirz, J. Kurioiwa, A. and Gehring, W. J. 1984. A homologous protein-coding sequence in *Drosophila* homeotic genes and its conservation in other metazoans. *Cell* 37: 403–408.

McGrath, J. and Solter, D. 1983. Nuclear transplantation in the mouse embryo by microsurgery and cell fusion. *Science* 220: 1300–1302.

McGrath, J. and Solter, D. 1984. Inability of mouse blastomere nuclei transferred to enucleated zygotes to support development in vitro. *Science* 226: 1317–1319.

McKinnell, R. G. 1978. *Cloning: Nuclear Transplantation in Amphibia*. University of Minnesota Press, Minneapolis.

Monod, J. (1947). The phenomenon of enzymatic adaptation and its bearing on problems of genetics and cellular differentiation. *Growth Symp.* 11: 223–289.

Moore, J. A. (1963). *Heredity and Development*. Oxford University Press, Oxford. [p. 236]

Morgan, T. H. 1897. *The Frog's Egg*. Macmillan, New York. [p. 135]

Morgan, T. H. 1926. *The Theory of the Gene*. Yale University Press, New Haven.

Moustàfa, L. A. and Brinster, R. L. 1972. Induced chimaerism by transplanting embryonic cells into mouse blastocysts. *J. Exp. Zool.* 181: 193–202.

Nathans, D. and Smith, H. O. 1975. Restriction endonucleases in the analysis and restructuring of DNA molecules. *Annu. Rev. Biochem.* 44: 273–293.

Oppenheimer, J. M. 1981. Walter Landauer and developmental genetics. *In* S. Subtelny and U. K. Abbott (eds.), *Levels of Genetic Control in Development*. Alan R. Liss, New York, pp. 1–13.

Orr, N. H., DiBerardino, M. A. and McKinnell, R. G. 1986. The genome of frog erythrocytes displays centuplicate replications. *Proc. Natl. Acad. Sci. USA* 83: 1369–1373.

Pardue, M. L. and Gall, J. G. 1970. Chromosomal localization of mouse satellite DNA. *Science* 168: 1356–1358.

Paul, D. B. and Kimmelman, B. A. 1988. Mendel in America: Theory and practice, 1900–1919. *In* R. Rainger, K. R. Benson and J. Maienschein (eds.), *The American Development of Biology*. University of Pennsylvania Press, Philadelphia, pp. 281–310.

Perucho, M., Hanahan, D. and Wigler, M. 1980. Genetic and physical linkage of exogenous sequences in transformed cells. *Cell* 22: 309–317.

Prather, R. S. 1991. Nuclear transplantation and embryo cloning in mammals. *Int. Lab. Animal Res. News* 33: 62–68.

Prather, R. S., Barnes, F. L., Sims, M. M., Robl, J. M., Eyestone, W. H. and First, N. L. 1987. Nuclear transplantation in the bovine embryo: Assessment of donor nuclei and recipient oocyte. *Biol. Reprod.* 37: 859–866.

Prather, R. S., Sims, M. M. and First, N. L. 1989. Nuclear transplantation in early porcine embryos. *Biol. Reprod.* 41: 414–418.

Rappolee, D. A., Brenner, C. A., Schultz, R., Mark, D. and Werb, Z. 1988. Developmental expression of *PDGF*, *TGF-α* and *TGF-β* genes in preimplantation mouse embryos. *Science* 241: 1823–1825.

Reyer, R. W. 1954. Regeneration in the lens in the amphibian eye. *Q. Rev. Biol.* 29: 1–46.

Robins, D. M., Ripley, S., Henderson, A. S. and Axel, R. 1981. Transforming DNA integrates into the host chromosome. *Cell* 23: 29–39.

Roller, R. J., Kinloch, R. A., Hiraoka, B. Y., Li, S. S.-L. and Wassarman, P. M. 1989. Gene expression during mammalian oogenesis and early embryogenesis: Quantification of three messsenger RNAs abundant in fully grown mouse oocytes. *Development* 106: 251–261.

Rosenberg, U. B., Preiss, A., Seifert, E., Jäckle, H. and Knipple, D. C. 1985. Production of phenocopies by *Krüppel* antisense RNA injection into *Drosophila* embryos. *Nature* 313: 703–706.

Saiki, R. K., Scharf, S., Faloona, F., Mullis, K. B., Horn, G. T., Erlich, H. A. and Arnheim, N. 1985. Enzymatic amplification of β-globin genomic sequences and restriction site analysis for diagnosis of sickle cell anemia. *Science* 230: 1350–1354.

Sander, K. 1986. The role of genes in ontogenesis—evolving concepts from 1883 to 1983 as perceived by an insect embryologist. *In* T. J. Horder, J. A. Witkowski and C. C. Wylie (eds.), *A History of Embryology*. Cambridge University Press, New York, pp. 363–395.

Sanger, F., Nicklen, S. and Coulson, A. R. 1977. DNA sequencing with chain-terminating inhibitors. *Proc. Natl. Acad. Sci. USA* 74: 5463–5467.

Sargent, T. D. and Dawid, I. 1983. Differential gene expression in the gastrula of *Xenopus laevis*. *Science* 222: 135–139.

Schickler, M., Lira, S., Kinloch, R. A. and Wassarman, P. A. 1992. A mouse oocyte-specific protein that binds to a region of mZP3 promoter responsible for oocyte-specific mZP3 gene expression. *Mol. Cell Biol.* 122: 120–127.

Smith, L. D. 1956. Transplantation of the nuclei of primordial germ cells into enucleated eggs of *Rana pipiens. Proc. Natl. Acad. Sci. USA* 54: 101–107.

Southern, E. M. 1975. Detection of specific sequences among DNA fragments separated by gel electrophoresis. *J. Mol. Biol.* 98: 503–517.

Spemann, H. 1938. *Embryonic Development and Induction.* Yale University Press, New Haven.

Spiegelman, S. (1947). Differentiation as the controlled production of unique enzymatic patterns. *In* J. F. Danielli and R. Brown (eds.), *Growth in Relation to Differentiation and Morphogenesis.* Cambridge University Press, Cambridge, p. 287.

Spradling, A. C. and Rubin, G. M. 1982. Transposition of cloned P elements into *Drosophila* germ line chromosomes. *Science* 218: 341–347.

Stevens, N. M. 1905a. A study of the germ cells of *Aphis rosae* and *Aphis oenotherae. J. Exp. Zool.* 2: 371–405; 507–545.

Stevens, 1905b. *Studies in Spermatogenesis with Especial Reference to the "Accessory Chromosome".* Carnegie Institute of Washington, Washington, D.C.

Steward, F. C. 1970. From cultured cells to whole plants: The induction and control of their growth and morphogenesis. *Proc. R. Soc. Lond.* [B] 175: 1–30.

Steward, F. C., Mapes, M. O. and Smith, J. 1958. Growth and organized development of cultured cells. I. Growth and division of freely suspended cells. *Am. J. Bot.* 45: 693–703.

Steward, F. C., Mapes, M. O., Kent, A. E. and Holsten, R. D. 1964. Growth and development of cultured plant cells. *Science* 143: 20–27.

Stice, S. J. and Robl. J. M. 1988. Nuclear reprogramming in nuclear transplant rabbit embryos. *Biol. Reprod.* 39: 657–664.

Ursprung, H., Smith, K. D., Sofer, W. H. and Sullivan, D. T. 1968. Assay systems for the study of gene function. *Science* 160: 1075–1081.

Vierra, J. and Messing, J. 1982. The pUC plasmids, an M13mp7-derived system for insertion mutagenesis and sequencing with synthetic universal primers. *Gene* 19: 259–268.

Waddington, C. H. 1939. Preliminary notes on the development of wings in normal and mutant strains of *Drosophila. Proc. Natl. Acad. Sci. USA* 25: 299–307.

Waddington, C. H. 1962. *New Patterns in Genetics and Development.* Columbia University Press, New York, pp. 14–36.

Wagner, T. E., Hoppe, P., Jollick, J. D., Scholl, D. R., Hodinka, R. L. and Gault, J. B. 1981. Microinjection of rabbit β-globin gene into zygotes and its subsequent expression in adult mice and offspring. *Proc. Natl. Acad. Sci. USA* 78: 6376–6380.

Wieslander, L. and Daneholt, B. 1977. Demonstration of Balbiani ring RNA sequence in polysomes. *J. Cell Biol.* 73: 260–264.

Wessel, G. M., Goldberg, L., Lennarz, W. J. and Klein, W. H. 1989. Gastrulation in the sea urchin is accompanied by the accumulation of an endoderm-specific mRNA. *Dev. Biol.* 136: 526–538.

Wetmur, J. G. and Davidson, N. 1968. Kinetics of renaturation of DNA. *J. Mol. Biol.* 31: 349–370.

Wilkinson, D. G., Bhatt, S. and Herrmann, B. G. 1990. Expression pattern of the mouse *T* gene and its role in mesoderm formation. *Nature* 343: 657–659.

Willadsen, S. M. 1986. Nuclear transplantation in sheep embryos. *Nature* 320: 63–65.

Willadsen, S. M. 1989. Cloning of sheep and cows embryos. *Genome* 31: 956–962.

Wilson, E. B. 1894. The mosaic theory of development. *Biol. Lect. Marine Biol. Lab Woods Hole* 2: 1–14.

Wilson, E. B. 1895. *An Atlas of the Fertilization and Karyogenesis of the Ovum.* Macmillan, New York. [p. 4]

Wilson, E. B. 1896. *The Cell in Development and Inheritance.* Macmillan, New York. [p. 262]

Wilson, E. B. 1904. Experimental studies on germinal localization. I. The germ regions in the egg of *Dentalium. J. Exp. Zool.* 1: 1–72.

Wilson, E. B. 1905. The chromosomes in relation to the determination of sex in insects. *Science* 22: 500–502.

Wilson, E. B. 1918. Theodor Boveri. *In* W. C. Roentgen (ed.), *Erinnerungen an Theodor Boveri.* J. C. B. Mohr, Tübingen, pp. 67–89.

Yamada, T. 1966. Control of tissue specificity: The pattern of cellular synthetic activities in tissue transformation. *Am. Zool.* 6: 21–31.

3

The cellular basis of morphogenesis
Differential cell affinity

A body is not merely a collection of randomly distributed cell types. Development involves not only the *differentiation* of cells, but also their *morphogenesis* into multicellular arrangements such as tissues and organs. When one observes the detailed anatomy of a tissue such as the neural retina, one sees an intricate and precise arrangement of many different types of cells. In this chapter we will introduce the ways by which the cells of the developing embryo change to create the functional organs of the body. There are four major questions that form the framework for discussions of morphogenesis.

- *How are tissues formed from cells?* How do neural retina cells stick to other neural retina cells and not become integrated into the pigmented retina or iris cells next to them? How are the various cell types within the neural retina (the three distinct layers of photoreceptors, bipolar neurons, and ganglion cells) arranged so that the retina is functional?
- *How are organs constructed from tissues?* The retinal cells of the eye form behind the cornea and lens at a precise distance. A retina would not serve any function were it to form behind a bone or at any place where the lens could not focus light on it. Moreover, the neurons from the retina must enter the brain to innervate certain cells of the optic cortex. All these connections must be precisely ordered.
- *How do migrating cells reach their destinations and how do organs form in particular locations?* Eyes develop in the head and nowhere else. What stops eye formation in other parts of the body if all the cells have the same genetic potential? In some cases, such as the precursors of our pigment cells, germ cells, and adrenal gland, the cells have to migrate long distances to reach their final destinations. How are cells instructed to travel along certain routes and to stop when they reach a particular region of the body?
- *How do organs and their cells grow, and how is their growth coordinated throughout development?* The cells of the eye must grow together, and the cells of the retina are hardly ever seen to divide after birth. Our intestine, however, is constantly shedding cells and regenerating new cells throughout its existence. Yet its mitotic rate is carefully

PROCESS	ACTION	MORPHOLOGY	EXAMPLE

MESENCHYMAL CELLS

Condensation	Mesenchyme becomes epithelium	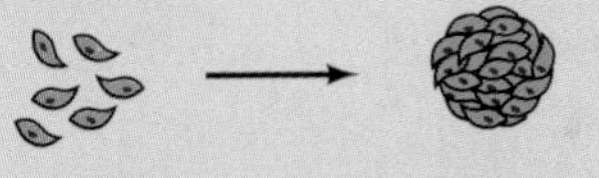	Cartilage mesenchyme
Cell division	Mitosis to produce more cells (hyperplasia)	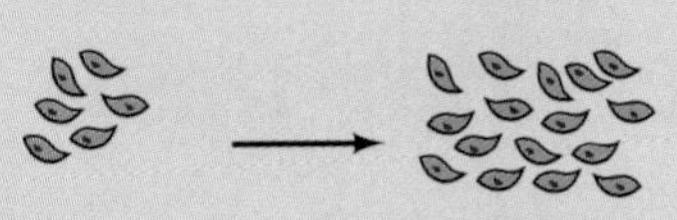	Limb mesenchyme
Cell death	Cells die	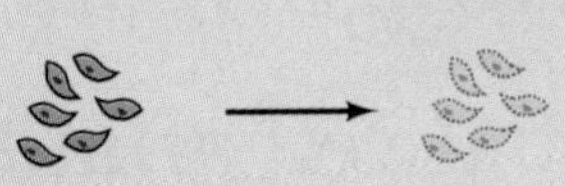	Interdigital mesenchyme
Migration	Cells move at particular times and places	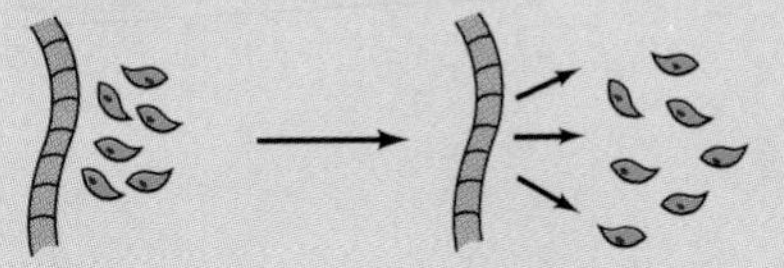	Heart mesenchyme
Matrix secretion and degradation	Synthesis or removal of extracellular layer	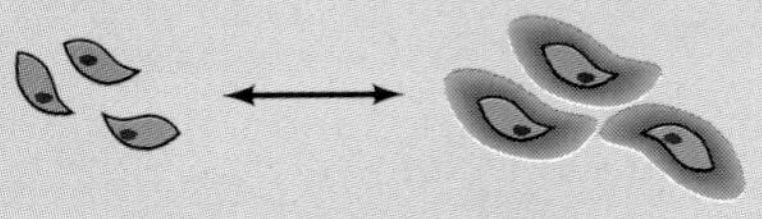	Cartilage mesenchyme
Growth	Cells get larger (hypertrophy)	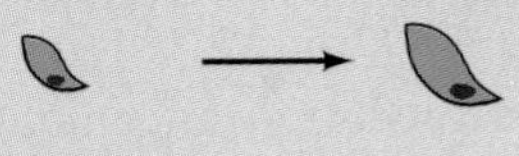	Fat cells

EPITHELIAL CELLS

Dispersal	Epithelium ⟶ Mesenchyme (entire structure)	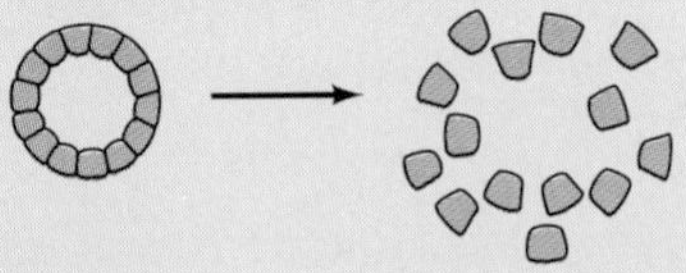	Müllerian duct degeneration
Delamination	Epithelium ⟶ Mesenchyme (part of structure)	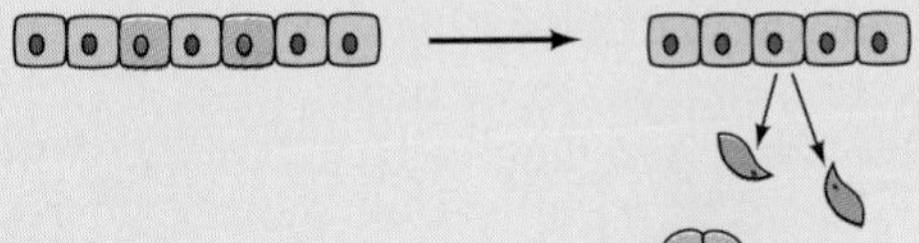	Chick hypoblast
Shape change or growth	Cells remain attached as morphology is altered	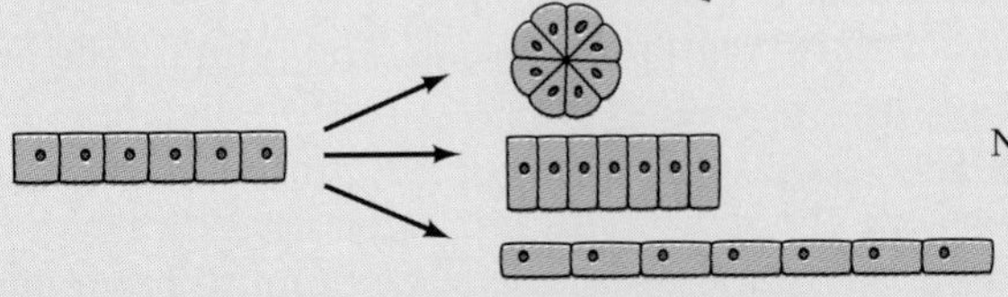	Neurulation
Cell migration (intercalation)	Rows of epithelia merge to form fewer rows		Vertebrate gastrulation
Cell division	Mitosis within row or other direction		Vertebrate gastrulation
Matrix secretion and degradation	Synthesis or removal of extracellular layer	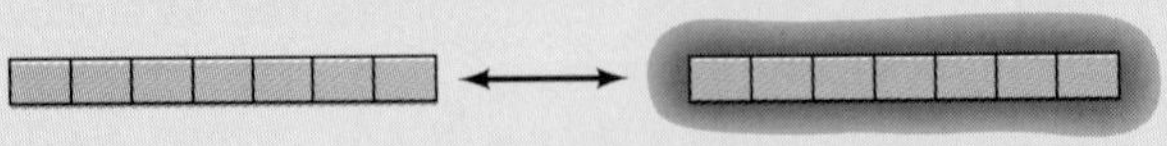	Vertebrate organ formation
Migration	Formation of free edges	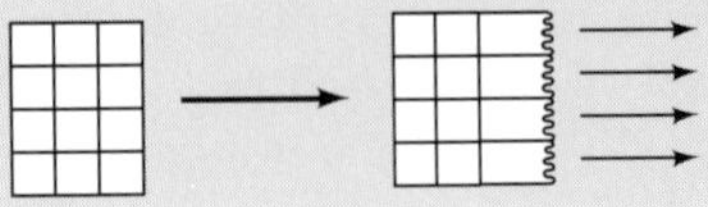	Chick ectoderm

regulated. If it were to generate more cells than it sloughs off, it could produce cancerous outgrowths. If it produced fewer cells than it sloughed off, the intestine could not function to digest food. What controls these differences in growth rate?

All these questions concern aspects of cell behavior. There are two main groups of cells in the embryo: **epithelial cells,** which are tightly connected to one another in sheets or tubes, and **mesenchymal cells,** which are unconnected to each other and operate as individual units. Morphogenesis is brought about through a limited repertoire of cellular processes in these two classes of cells: (1) the direction and number of cell divisions; (2) cell shape changes; (3) cell movement; (4) cell growth; (5) cell death; and (6) changes in the composition of the cell membrane and extracellular matrix. How these processes are accomplished can differ between mesenchymal and epithelial cells (Figure 3.1). Many examples of these diverse morphological changes will be encountered throughout this book.

There appear to be two major ways by which cells communicate with one another to effect morphogenesis. The first is through **diffusible substances** that are made by one type of cell and that change the behavior of other types of cells. These substances include hormones, growth factors, and morphogens; each will be detailed in subsequent chapters. The second method involves **contact** between the surfaces of adjacent cells. During the period of organogenesis, individual cells and groups of cells change their relative positions and become associated with other cell types. Organ and tissue formation involve the selective interactions between neighboring cells to create the proper arrangement of the different cell types.

Cells have the ability to selectively recognize other cells, adhering to some and migrating over others. The molecular events mediating the selective recognition of cells and their formation into tissues and organs occur at the cell surface. Whereas the dominant paradigm of developmental genetics is differential gene expression, the dominant paradigm of morphogenesis involves **differential cell affinities.** These affinities can be for the surfaces of other cells or for the molecules of the extracellular matrix secreted by the cells. In this chapter, we will analyze the ways by which adjacent cell surfaces can interact during development such that cells become localized in their respective sites within tissues and organs.

Differential cell affinity

Just as there was a struggle to show the importance of genes in development, there was also a debate over whether the cell surface plays a role in forming the embryo. The cell surface looks pretty much the same in all cell types, and many investigators thought that the cell surface was not even a living part of the cell. Observations of fertilization and early embryonic development made by E. E. Just (1939) suggested that the cell surface differed between cell types, but the modern analysis of morphogenesis begins with the experiments of Townes and Holtfreter in 1955. Taking advantage of the discovery that amphibian tissues become dissociated into single cells when placed in alkaline solutions, they prepared single-cell suspensions from each of the three amphibian germ layers soon

FIGURE 3.1
Summary of major morphogenetic processes in mesenchymal cells and in epithelial cells.

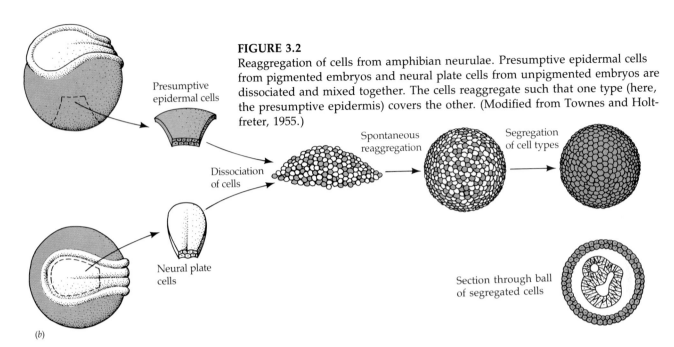

FIGURE 3.2
Reaggregation of cells from amphibian neurulae. Presumptive epidermal cells from pigmented embryos and neural plate cells from unpigmented embryos are dissociated and mixed together. The cells reaggregate such that one type (here, the presumptive epidermis) covers the other. (Modified from Townes and Holtfreter, 1955.)

Presumptive epidermal cells

Dissociation of cells

Neural plate cells

Spontaneous reaggregation

Segregation of cell types

Section through ball of segregated cells

(b)

after the neural tube had formed. Two or more of these single-cell suspensions could be combined in various ways, and when the pH was normalized, the cells adhered to each other, forming aggregates on agar-coated petri dishes. By using the embryos from species having cells of different sizes and colors, Townes and Holtfreter were able to follow the behavior of the recombined cells (Figure 3.2).

The results of their experiments were striking. First, they found that the reaggregated cells became spatially segregated. That is, instead of remaining mixed, each cell type sorted out into its own region. Thus, when epidermal (ectoderm) and mesodermal cells were brought together to form a mixed aggregate, the epidermal cells were found at the periphery of the aggregate and the mesodermal cells were found inside. In no case did the recombined cells remain randomly mixed. In most cases, one tissue type completely enveloped the other.

Second, they found that the final positions of the reaggregated cells reflected their embryonic positions. The mesoderm migrates centrally to the epidermis, adhering to the inner epidermal surface (Figure 3.3A). The mesoderm also migrates centrally with respect to (gut) endoderm (Figure 3.3B). However, when the three germ layer cells are mixed together, the endoderm separates from the ectoderm and mesoderm and is then enveloped by them (Figure 3.3C). In its final configuration, the ectoderm is on the periphery, the endoderm is internal, and the mesoderm lies in the region between them. Holtfreter interpreted this in terms of selective affinity. The inner surface of the ectoderm has a positive affinity for the mesodermal cells, while it has a negative affinity for the endoderm. The mesoderm has positive affinities for both the ectodermal and the endodermal cells. The mimicry of normal embryonic structure by cell aggregates is also seen in the recombination of epidermis and neural plate cells (Figure 3.3D). The presumptive epidermal cells migrate to the periphery as before; the neural plate cells migrate inward, forming a structure reminiscent of the neural tube. When axial mesoderm (notochord) cells are added to the suspension of presumptive epidermal and presumptive neural cells, the cell segregation results in an external epidermal layer, a centrally located neural tissue, and a layer of mesodermal tissue between them (Figure 3.3E). Somehow, the cells are able to sort out into their proper embryological positions.

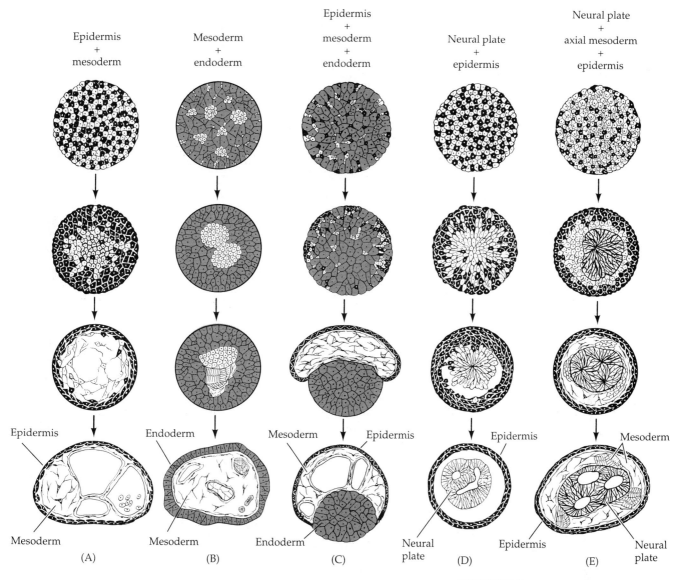

Epidermis
+
mesoderm

Mesoderm
+
endoderm

Epidermis
+
mesoderm
+
endoderm

Neural plate
+
epidermis

Neural plate
+
axial mesoderm
+
epidermis

Epidermis

Mesoderm

(A)

Endoderm

Mesoderm

(B)

Mesoderm   Epidermis

Endoderm

(C)

Epidermis

Neural
plate

(D)

Mesoderm

Epidermis   Neural
plate

(E)

FIGURE 3.3
Sorting out and reorganization of embryonic spatial relationships in aggregates of embryonic amphibian cells. (Modified from Townes and Holtfreter, 1955.)

Such preferential affinities have also been noted by Boucaut (1974), who injected individual cells from specific germ layers back into the cavity of amphibian gastrulas. He found that these cells migrate to their appropriate germ layer. Endodermal cells find positions in the host endoderm, whereas ectodermal cells are found only in host ectoderm. Thus, selective affinity appears to be important for imparting positional information to embryonic cells.

The third conclusion of Holtfreter and his colleagues was that selective affinities change during development. This should be expected, because embryonic cells do not retain a single stable relationship with other cells. For development to occur, cells must interact differently with other cell populations at specific times. Such changes in cellular affinity were dramatically confirmed by Trinkaus (1963), who showed a clear correlation between adhesive changes in vitro and changes in embryonic cell behavior. More recently, the experiments of Fink and McClay (1985) demonstrated this behavior during sea urchin development. In the blastula, all cells seem to have the same affinity for one another. Each cell also has high affinity for the extracellular matrix (hyaline layer) covering of the embryo and a

FIGURE 3.4
Summary of the changes in cell adhesion of skeletal precursor cells (boxed). (A) In the sea urchin blastula, each cell has high affinity for its neighbors and for its substrate, the hyaline layer. (B) As development progresses, changes in the cell surface of these cells cause them to weaken their affinities for surrounding cells and the extracellular matrix and to increase their affinities for the protein inside the blastocoel cavity. The result is that these cells migrate into the blastocoel (arrows) and will form the skeleton.

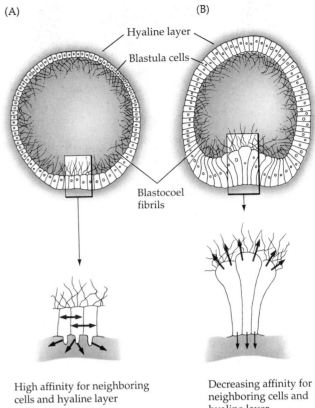

(A)   (B)

Hyaline layer

Blastula cells

Blastocoel fibrils

High affinity for neighboring cells and hyaline layer

Decreasing affinity for neighboring cells and hyaline layer
Increased affinity for blastocoel fibrils

low affinity for the proteins inside the embryonic cavity (blastocoel). However, at the onset of gastrulation, a specific group of cells at the vegetal end of the blastula lose their affinity for neighboring cells and for the external extracellular matrix while simultaneously acquiring affinity for the protein fibrils lining the blastocoel (Figure 3.4). These changes in affinities cause these cells to release their contacts with their neighboring cells and to migrate inside the blastocoel, where they will form the larval skeleton. When they begin to form this skeleton, their adhesion properties have to change again. These cells, which had been "antisocial" toward each other since their ingression into the blastocoel, must now adhere together to form the rudiments of the skeletal ring. These changes in adhesion are temporally specific and are also specific to the skeletal precursor cells (McClay and Ettensohn, 1987). Such changes in cell affinity are extremely important in the processes of morphogenesis.

The reconstruction of aggregates from older embryos of birds and mammals was accomplished by the use of the protease trypsin to dissociate the cells from one another (Moscona, 1952). When the resulting single cells were mixed together in a flask and swirled so that the shear forces would break any nonspecific adhesions, the cells moved around and sorted out according to their cell type. In so doing, they reconstructed the organization of the original tissue (Moscona, 1961; Giudice, 1962). Figure 3.5 shows the "reconstruction" of skin tissue from a 15-day embryonic mouse. The skin cells are separated by proteolytic enzymes and are then aggregated in a rotary culture. The epidermal cells migrate to the periphery and the dermal cells migrate toward the center. By 72 hours, the epidermis has been reconstituted, a keratin layer has formed, and hair follicles are

FIGURE 3.5
Reconstruction of skin from a suspension of skin cells from a 15-day embryonic mouse. (A) Section through the embryonic skin showing epidermis, dermis, and primary hair follicle. (B) Suspension of single skin cells from both the dermis and the epidermis. (C) Aggregates after 24 hours. (D) Section through an aggregate, showing migration of epidermal cells to the periphery. (E) Further differentiation of aggregates (72 hours), showing reconstituted epidermis and dermis, complete with hair follicles and keratinized layer. (From Monroy and Moscona, 1979, courtesy of A. Moscona.)

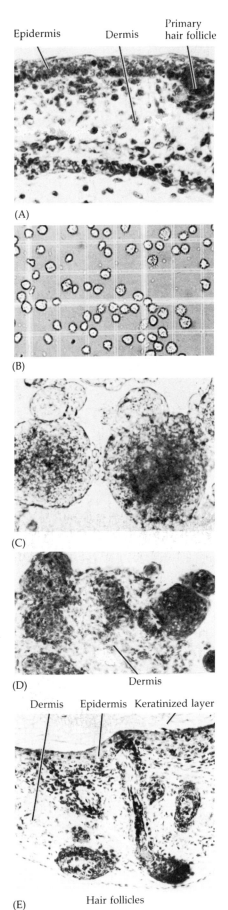

seen in the dermal region. Such reconstruction of complex tissues from single cells is called **histotypic aggregation.**

Thermodynamic model of cell interactions

Cells, then, do not sort randomly but can actively move to create tissue organization. What forces direct cell movement during morphogenesis? In 1964 Malcolm Steinberg proposed a model that explained the directions of cell sorting on thermodynamic principles. Using cells derived from trypsinized embryonic tissues, Steinberg showed that certain cell types always migrate centrally when combined with some cell types, but migrate peripherally when combined with others. Figure 3.6 illustrates the interactions between cultures of pigmented retina cells and neural retina cells. When single-cell suspensions of these two cell types are mixed together, they form aggregates of randomly arranged cells. However, after several hours, the pigmented retina cells are no longer seen on the periphery of these aggregates, and by two days, two distinct layers are seen; the pigmented retina lies internal to the neural retina cells. The same type of interactions can also be seen when spherical aggregates of tissues are

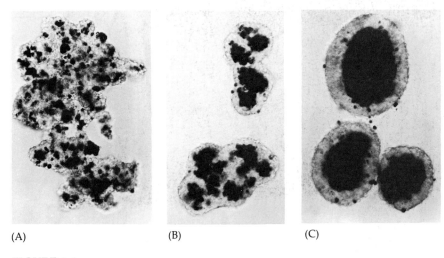

FIGURE 3.6
Aggregates formed by mixing 7-day-old chick embryo neural retina (unpigmented) cells with pigmented retina (dark) cells. (A) 5 hours after the single-cell suspensions were mixed, aggregates of randomly distributed cells are seen. (B) At 19 hours, the pigmented retina cells are no longer seen on the periphery. (C) By two days, a great majority of the pigmented retina cells are located in a central internal mass, surrounded by the neural retina cells. (The scattered pigmented cells are probably dead cells.) (From Armstrong, 1989, courtesy of P. B. Armstrong.)

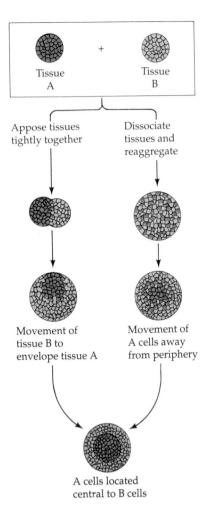

FIGURE 3.7

Spreading of one cell type over another. The final position of aggregates composed of two tissue types is independent of their initial position. An identical final condition is formed whether the tissues are made into single-cell suspensions and then reaggregated or the two tissues are kept intact and brought into contact. (After Armstrong, 1989.)

placed in contact with each other. One of the tissues will envelop the other. The final topography is independent of the starting positions (Figure 3.7).

Moreover, such interactions form a hierarchy (Steinberg, 1970). If the final position of one cell type A is internal to a second cell type B, and the final position of B is internal to a third cell type C, then the final position of A will always be internal to C. For example, pigmented retina cells migrate internally to neural retina cells, and heart cells migrate centrally to pigmented retina. Therefore, heart cells migrate internally to neural retina cells. This observation led Steinberg to propose that the mixed cells interact to form an aggregate with the smallest interfacial free energy (Figure 3.8). In other words, the cells rearrange themselves into the most thermodynamically stable pattern. If cell types A and B have different strengths of adhesion and if the strength of A–A connections is greater than the strength of A–B or B–B connections, sorting will occur, with the A cells becoming central. If the strength of A–A connections is less than or equal to the strength of A–B connections, then the aggregate will remain as a random mix of cells. Last, if the strength of A–A connections is far greater than the strength of A–B connections—in other words, A and B cells show essentially no adhesivity toward one another—then A cells and B cells will form separate aggregates.

All that is needed for sorting out to occur is that cells differ in the strengths of their adhesion. In the simplest form of this model, all cells could have the same type of "glue" distributed on the cell surface. The amount of this cell surface product or the cellular architecture that allows the substance to be differentially concentrated could cause a different amount of stable contacts to be made between cell types. Alternatively, the thermodynamic differences could be caused by several types of adhe-

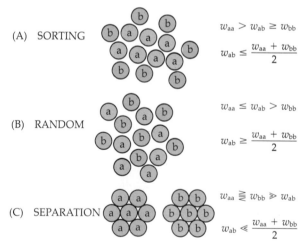

FIGURE 3.8

Sorting out as a process tending toward the maximum thermodynamic stability. (A) Sorting out occurs when the average strength of adhesions between different types of cells (w_{ab}) is less than the average of homotypic (a–a or b–b) adhesive strengths (w_{aa}, w_{bb}). The more adhesive type of cells becomes centrally located. (B) If the strength of the a–b adhesions is greater than or equal to the average of the homotypic adhesions, no sorting will occur, because the system has already achieved a thermodynamic equilibrium, and the mixture of cell types will be random. (C) If the a–b bonds are much weaker than the average of the homotypic adhesions, complete separation will ensue. (This is characteristic of oil and water, for example.)

Labeled blastema

| | Wrist | Elbow | Upper arm |

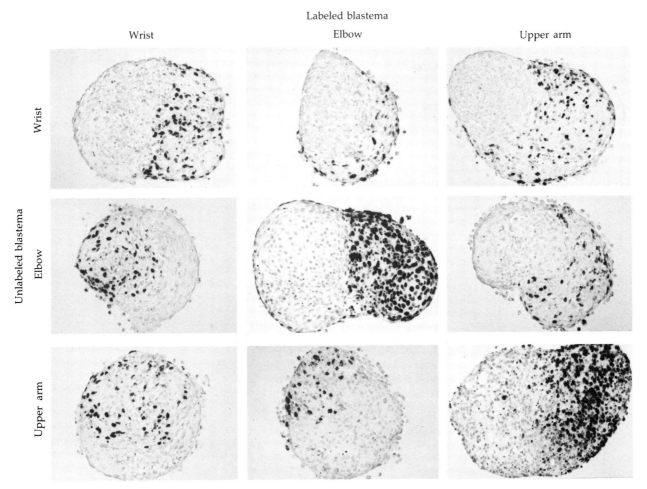

FIGURE 3.9

Sorting when blastemas from the same or different forelimb levels are brought together in culture. (One member of each pair was marked with tritium to distinguish it from the other.) After three days in culture, the aggregates were fixed and sectioned. Blastemas from the same level fused in a straight line. When blastemas were from different levels, the proximal blastema could be seen attempting to surround the more distal cells. (From Nardi and Stocum, 1983, courtesy of D. Stocum.)

sion molecules. This thermodynamic model is called the **differential adhesion hypothesis.** In this hypothesis, the early embryo can be viewed as existing in an equilibrium state until some change in gene activity changes the cell surface molecules. The movements that occur are those that seek to restore the cells to a new equilibrium configuration.

In vivo evidence for the thermodynamic model

Until recently it was extremely difficult to design an experiment to test this model of cell migration in vivo; but evidence for this hypothesis is emerging from studies of limb regeneration in salamanders. As will be detailed in Chapter 19, salamander limbs have certain remarkable attributes. When one amputates a forelimb at the *upper arm*, the remaining stump forms a dedifferentiated mass of cells at its tip (the **regeneration blastema**), which divides and differentiates to form a new limb. The new limb tissue starts at the amputation site, in this case forming the remainder of the limb from the upper arm downward. When the forelimb is amputated at the *wrist*, a similar regeneration blastema forms. However, it does not start making upper arm tissue, elbow, and ulna bones. Rather, it "knows" its location and regenerates only the wrist and digits.

How is this "positional memory" stored? Nardi and Stocum (1983) demonstrated that when two salamander limb blastemas from the same level of origin are placed together they fuse, but neither tissue surrounds the other (Figure 3.9). However, when the blastemas are from different

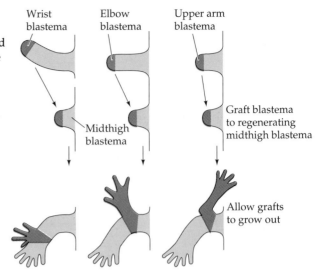

FIGURE 3.10
Sorting out in vivo, whereby regenerating forelimb blastemas (color) grafted into the midthigh blastemas (gray) are displaced to the corresponding region of the regenerating hindlimb (wrist to tarsus; elbow to knee; upper arm to midthigh) and initiate forelimb formation from that point distally. (After Crawford and Stocum, 1988.)

levels, the more proximal (closer to the body) blastema surrounds the more distal one. It appears, then, that the adhesive properties of these cells form a gradient along the proximodistal axis; these properties are greatest at the wrist and lowest at the upper arm.

Crawford and Stocum (1988) were able to relate this in vitro sorting of cells to processes in the living, regenerating limb. Blastemas from the wrist, elbow, or upper arm were grafted into the blastema–stump junction of a hindlimb regenerating from the midthigh. The forelimb blastemas migrated distally to the corresponding level of the host hindlimb and then regenerated a new structure (Figure 3.10). The *upper arm* blastema immediately regenerated a complete arm from the midthigh level. The *elbow* blastema moved to the level of the knee and then formed the remainder of the arm from this point onward; and the *wrist* blastema was displaced to the end of the regenerating hindlimb, where it formed a wrist adjacent to the tarsus of the foot. These data suggest that the hierarchies of cellular sorting out seen in vitro reflect differences that are used by the body in constructing new organs in vivo.

The molecular basis of cell–cell adhesion

The classes of cell adhesion molecules

The formation of tissues and organs is mediated by events occurring at the cell surfaces of adjacent cells. The cell surface includes the cell plasma membrane, the molecules directly beneath the membrane and associated with it, and the molecules found in the extracellular spaces. Eukaryotic cells are surrounded by a complex molecular border called the **plasma** (or cell) **membrane**. Our present understanding of the plasma membrane structure is summarized by the **fluid mosaic model** illustrated in Figure 3.11 (Singer and Nicolson, 1972). First, there are two layers of phospholipids, the polar ends of which are oriented toward the aqueous solutions on either side of the phospholipid bilayer. Some of the proteins traverse the membrane from one side to the other, whereas other proteins extend only part way through the phospholipid bilayer. The distribution of proteins gives rise to the "mosaic" nature of the membrane, and the ability

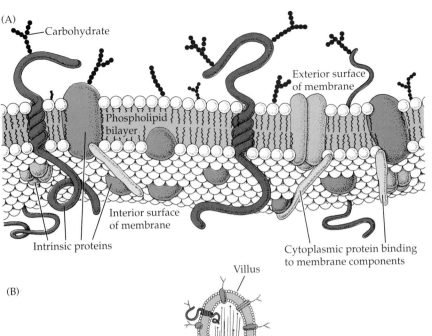

(A)

Carbohydrate

Exterior surface
of membrane

Phospholipid
bilayer

Interior surface
of membrane

Intrinsic proteins

Cytoplasmic protein binding
to membrane components

FIGURE 3.11
The structure of the cell surface. (A)
Fluid mosaic model of the cell membrane. (B) Underlying cytoskeleton.
Contractile fibers consisting of actin
microfilaments and tubulin microtubules are linked to each other and to
the cell membrane. Membrane glycoproteins may be linked by extracellular
proteins as well as by internal fibers.
(C) Quick-frozen, deep-etched scanning electron micrograph showing the
surface of an intestinal cell. Actin extends into the villi and is linked within
the cell by thick and thin filaments directly beneath the cell membrane.
(After Hirokawa et al., 1983, photograph courtesy of N. Hirokawa.)

(B)

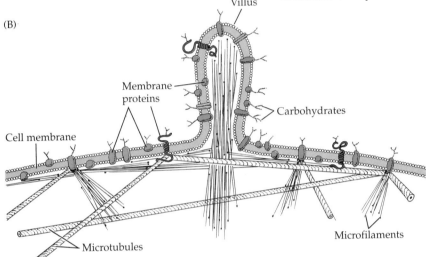

Villus

Membrane
proteins

Carbohydrates

Cell membrane

Microtubules

Microfilaments

(C)

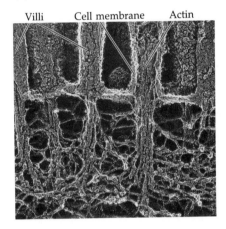

Villi   Cell membrane   Actin

of these proteins to move laterally within the phospholipid sea is evidence of its "fluid" nature.

The cell membrane contains proteins that are capable of interacting with the outside environment. Certain proteins have their active sites pointing outward, toward other cells. The most familiar of these types of proteins are the **transport proteins,** which facilitate the movement of ions and nutrients into the cell, and the **receptor proteins,** which bind hormones and growth factors. These proteins—either directly, or indirectly through other proteins—also serve to transmit information into the cytoplasm and affect a cell's behavior. In addition, there are three classes of cell membrane molecules (usually proteins) that are particularly involved in controlling specific interactions with other cells (Edelman and Thiery, 1985):

- *Cell adhesion molecules.* These proteins are involved in cell–cell adhesion. They can unite cells into epithelial sheets and condense mesenchymal cells into cohesive aggregates. They are critical in separating different tissues from one another.
- *Substrate adhesion molecules.* These molecules are involved in binding cells to their extracellular matrices. They include components of the extracellular matrix and the cell surface receptors for these mole-

cules. Substrate adhesion molecules permit the movement of mesenchymal cells and neurons and allow the spatial separation of epithelial sheets.
- *Cell junctional molecules.* These molecules provide communication
  pathways between the cytoplasm of adjacent cells and provide
  permeability barriers and mechanical strength to epithelial sheets.

The local patterns of expression of these cell surface molecules are thought
to provide a major link between the one-dimensional genetic code and the
three-dimensional organism. By modulating the appearance of these molecules, the genetic potential of the genome can become manifest in the
mechanical processes of morphogenesis.

SIDELIGHTS & SPECULATIONS

Monoclonal antibodies and reverse genetics

Monitoring cell membrane changes through monoclonal antibodies

The expression of membrane components changes in time
and space. Different cell types display different cell surface
components, and these components change as the cell develops. Such tissue-specific membrane components are
often recognized by antisera and are therefore called **differentiation antigens** (Boyse and Old, 1969). Specific differentiation antigens can now be identified by **monoclonal
antibodies** (Figure 3.12). These antibodies are usually made
by injecting foreign cells into mice (or mouse cells of one
strain into mice of another strain). The mouse B lymphocytes will begin producing antibodies against each foreign
component on these cells, with each B lymphocyte producing a single type of antibody. These lymphocytes are
then "immortalized" by fusing them with cultured B-lym-

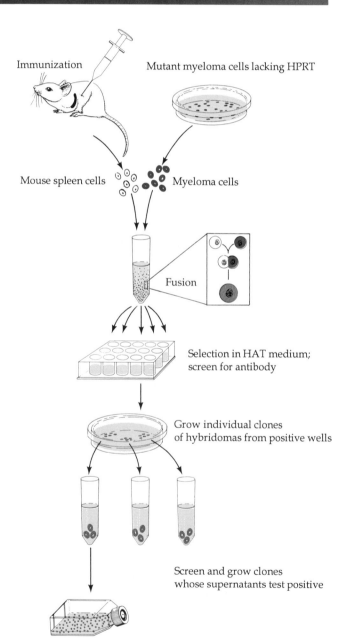

FIGURE 3.12
*Protocol for making monoclonal antibodies. Spleen cells from an
immunized mouse are fused with mutated myeloma cells lacking
the enzyme HPRT. Cells are grown in a medium containing
hypoxanthine, aminopterin, and thymidine (HAT). Unfused
myeloma cells cannot grow in this medium because aminopterin
blocks the only way they have of making purine nucleotides. B
cells die in this medium even though they contain an enzyme
(HPRT) that would allow them to utilize the hypoxanthine
placed in the medium. The fused cells (hybridomas) grow and
divide. The wells in which the hybridomas grow are screened for
the presence of the effective antibody, and the cells from positive
wells are plated at densities low enough to allow individual cells
to give rise to discrete clones. These clones are isolated and
screened for the effective antibody. Such an antibody is monoclonal. The hybridomas producing this antibody can be grown
and frozen. (After Yelton and Scharff, 1980.)*

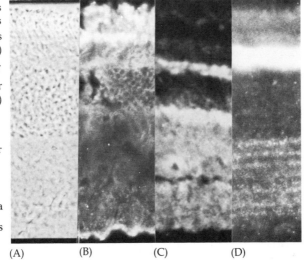

Outer segments
of photoreceptors

Somas of photoreceptors
(outer nuclear layer)

Outer synaptic layer

Inner nuclear layer
(interneuron soma)

Inner synaptic layer

Ganglion cell soma

Ganglion cell axons

(A) (B) (C) (D)

FIGURE 3.13

Cell surface specificity in the chick neural retina. (A) Phase-contrast photograph of a section through newly hatched chick neural retina. (B) Retinal section stained with a fluorescent monoclonal antibody that recognizes retinal (but not other neuronal) cells. (C) Retinal section stained with fluorescent monoclonal antibody that recognizes neuronal processes but not cell bodies in the retina. (D) Retinal section stained with fluorescent monoclonal antibody that recognizes antigens on a subset of nerve cell processes in the outer and inner synaptic layer. (Courtesy of G. Grunwald.)

phocyte tumor cells (myelomas) that have been mutated so that (1) they can no longer synthesize their own antibodies and (2) they lack the purine salvage enzyme hypoxanthine-phosphoribosyltransferase (HPRT). This latter alteration means that the myeloma cells can only make purine nucleotides de novo and cannot utilize purines from the culture medium. After fusion, the cells are then grown in a medium containing aminopterin. This drug inhibits the de novo purine synthetic pathway. Thus, unfused myeloma cells die of purine starvation. They cannot make purine nucleotides using the HPRT-mediated salvage pathway, and the aminopterin blocks the de novo pathway as well. Normal B lymphocytes do not divide in culture, so they, too, die. The fused product of the B lymphocyte and the myeloma cell—a **hybridoma**—proliferates, having the purine salvage enzyme from the B lymphocyte and the growth properties from the tumor. Moreover, each of these hybridomas secretes the specific antibody of the B lymphocyte. The medium in which hybridomas are growing is then tested for antibody that binds to the original population of foreign cells. Such antibody, having a single B lymphocyte as its original source, is called a **monoclonal antibody.** Monoclonal antibodies can be produced in enormous amounts and can recognize antigens (proteins, lipids, or carbohydrates) that are only weakly expressed (Köhler and Milstein, 1975).

Monoclonal antibodies directed against specific types of cells have uncovered numerous differentiation antigens appearing at different times and places during development. Figure 3.13 shows different cell surface molecules at different *spatial* layers of the newly hatched chick neural retina. Each of the monoclonal antibodies recognizes a different molecule in the cell membrane. As is evident from this composite photograph, the membranes of all the cells of the neural retina are not the same. In fact, regions of the same cell membrane can differ; for example, the membranes of the axons and the membranes of the nerve soma contain some different molecules. Figure 3.14 shows *temporal* changes in the cell membrane of a single *Drosophila* epithelial cell as it develops into a retinal photoreceptor. Monoclonal antibodies were obtained after mice had been

injected with homogenates of *Drosophila* head tissue, and a panel of antibodies was tested on the cells of the larval eye imaginal disc that were differentiating into eye structures. As soon as the undifferentiated epithelial cells of the disc show neuronal properties, they express the 22C10 antigen. This antigen is also found in other neuronal cell types. Shortly thereafter, though, the cell begins to express another cell membrane molecule, the 24B10 antigen. This molecule is seen only in those neurons destined to become photoreceptors. At subseqent stages (about 80 hours later) the 21A6 antigen becomes expressed on certain regions of the maturing photoreceptors, and another antigen, 28H9, is characteristic of the terminally differentiated retinal photoreceptor (Zipursky et al., 1984). Thus, cell membranes of different cell types contain different molecules, and these molecules can change during the maturation of the particular cell.

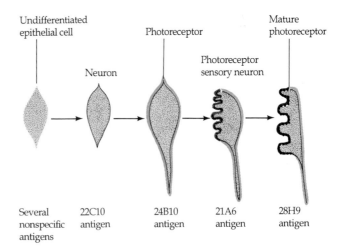

Undifferentiated epithelial cell
Neuron
Photoreceptor
Photoreceptor sensory neuron
Mature photoreceptor

Several nonspecific antigens
22C10 antigen
24B10 antigen
21A6 antigen
28H9 antigen

FIGURE 3.14

Temporal changes in the cell membrane correlated with the morphogenesis of a Drosophila *retinal photoreceptor cell. As differentiation proceeds, different antigens become expressed on the cell membrane. (After Venkatesh et al., 1985.)*

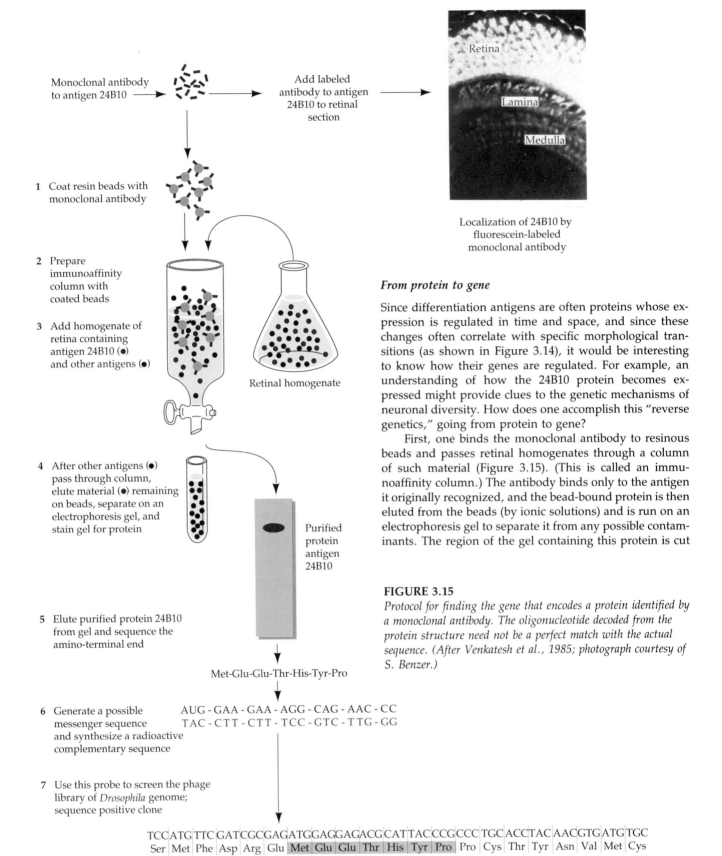

Monoclonal antibody
to antigen 24B10 →

Add labeled
antibody to antigen
24B10 to retinal
section →

Retina

Lamina

Medulla

Localization of 24B10 by
fluorescein-labeled
monoclonal antibody

1 Coat resin beads with
monoclonal antibody

2 Prepare
immunoaffinity
column with
coated beads

3 Add homogenate of
retina containing
antigen 24B10 (●)
and other antigens (●)

Retinal homogenate

4 After other antigens (●)
pass through column,
elute material (●) remaining
on beads, separate on an
electrophoresis gel, and
stain gel for protein

Purified
protein
antigen
24B10

5 Elute purified protein 24B10
from gel and sequence the
amino-terminal end

Met-Glu-Glu-Thr-His-Tyr-Pro

6 Generate a possible
messenger sequence
and synthesize a radioactive
complementary sequence

AUG - GAA - GAA - AGG - CAG - AAC - CC
TAC - CTT - CTT - TCC - GTC - TTG - GG

7 Use this probe to screen the phage
library of *Drosophila* genome;
sequence positive clone

TCC ATG TTC GAT CGC GAG ATG GAG GAG ACG CAT TAC CCG CCC TGC ACC TAC AAC GTG ATG TGC
Ser | Met | Phe | Asp | Arg | Glu | Met | Glu | Glu | Thr | His | Tyr | Pro | Pro | Cys | Thr | Tyr | Asn | Val | Met | Cys

Expected sequence

8 Isolate and characterize gene

From protein to gene

Since differentiation antigens are often proteins whose expression is regulated in time and space, and since these
changes often correlate with specific morphological transitions (as shown in Figure 3.14), it would be interesting
to know how their genes are regulated. For example, an
understanding of how the 24B10 protein becomes expressed might provide clues to the genetic mechanisms of
neuronal diversity. How does one accomplish this "reverse
genetics," going from protein to gene?

First, one binds the monoclonal antibody to resinous
beads and passes retinal homogenates through a column
of such material (Figure 3.15). (This is called an immunoaffinity column.) The antibody binds only to the antigen
it originally recognized, and the bead-bound protein is then
eluted from the beads (by ionic solutions) and is run on an
electrophoresis gel to separate it from any possible contam-
inants. The region of the gel containing this protein is cut

FIGURE 3.15
*Protocol for finding the gene that encodes a protein identified by
a monoclonal antibody. The oligonucleotide decoded from the
protein structure need not be a perfect match with the actual
sequence. (After Venkatesh et al., 1985; photograph courtesy of
S. Benzer.)*

from the gel, and the protein is eluted from the gel matrix and partially sequenced. Radioactive oligonucleotides that would bind to a DNA sequence capable of encoding such a protein are then synthesized. In the case of 24B10, these radioactive probes were used to screen a library of recombinant DNA clones containing regions of the *Drosophila* genome. The *Drosophila* DNA of each positive clone was sequenced to see if it matched the sequence of the original protein isolated by the monoclonal antibody. In this way, one can go from a rare protein identified by a monoclonal antibody to a specific piece of genomic DNA (Zipursky et al., 1984; Venkatesh et al., 1985).

Cell adhesion molecules

Identifying cell adhesion molecules and their roles in development

The sorting-out studies of Holtfreter's and Steinberg's laboratories did not identify the molecules involved in these differential cell adhesions. Roth (1968; Roth et al., 1971) demonstrated that different cell types display selective cell adhesion independent of cell sorting. He modified the rotary aggregation assay by incubating ^{3}H-labeled cartilage cells and ^{14}C-labeled hepatocytes in a rotating solution containing small aggregates of unlabeled cartilage cells. By measuring the ^{14}C- and ^{3}H-labeled cells in these aggregates, he demonstrated that the cartilage aggregates specifically picked up cartilage cells. Similar experiments extended this finding to liver and muscle cells as well (Figure 3.16). These studies demonstrated that different cell types could use different adhesion molecules.

The next task is to identify those molecules mediating cell adhesion and to discover how they accomplish this feat. Several of these **cell adhesion molecules (CAMs)** have been identified and are grouped into two general categories: the **cadherins,** whose cell adhesive properties are dependent on calcium ions, and the **immunoglobulin superfamily CAMs,** whose cell-binding domains resemble those of antibody molecules. Table 3.1 lists some of the recently discovered general CAMs.

Cadherins

One of the approaches to finding cell adhesion molecules used the observation that calcium is often necessary for cell adhesion. Moreover, the presence of calcium ions protects certain adhesive proteins from digestion by trypsin, thereby facilitating their isolation. These calcium-dependent glycoproteins are called **cadherins,** and they appear to be crucial to the spatial segregation of cells and to the organization of animal form (Takeichi, 1987). Cadherins are critical for establishing and maintaining intercel-

FIGURE 3.16
Specificity of cell–cell attachment. Collecting aggregates, each consisting of one type of cell, are placed in a rotating culture containing single cells of both the same (isotypic) and different (heterotypic) types. The single isotypic and heterotypic cells were previously labeled with different radioactive isotopes. After 6 hours, the aggregates are collected, washed, and counted for both isotypic and heterotypic cells that adhered to the aggregate as shown in the table. (Data from Roth, 1968.)

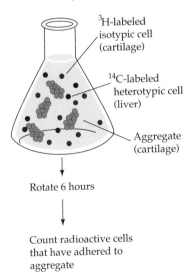

Count of radioactive cells that have adhered to aggregate

Aggregate type	Labeled single cells in suspension*		
	Cartilage	Liver	Pectoral muscle
Cartilage	100	6	48
Liver	10	100	0
Pectoral muscle	38	49	100

*Percentage of mean number of cells collected by isotopic aggregates.

TABLE 3.1
General classification of major cell adhesion molecules (CAMs)

Class	CAM	Cell types
Cadherins (calcium-dependent)	N-cadherin (a.k.a. A-CAM) P-cadherin E-cadherin (a.k.a. L-CAM, uvomorulin)	Nerve, kidney, lens, heart Placenta, epithelia Epithelia, mouse blastula
Immunoglobulin superfamily CAMs (calcium-independent)	N-CAM Ng-CAM (a.k.a. L1, NILE) Neurofascin Cell-CAM LFA-1 CD4 glycoprotein (HIV receptor)	Muscle, nerve, kidney Glia, neurons *Drosophila* neurons Hepatocytes Lymphocytes T-inducer cells

lular connections, and cells with fewer cadherin molecules are generally less adhesive than those cells with many of these molecules in their cell membranes. In mammalian embryos, three major cadherin classes have been identified. (1) **E-cadherin** (epithelial cadherin; also called uvomorulin and L-CAM) is expressed on all early mammalian embryonic cells, even at the one-cell stage. Later, this molecule is restricted to epithelial tissues of embryos and adults. (2) **P-cadherin** (placental cadherin) appears to be expressed primarily on those placental cells of the mammalian embryo that contact the uterine wall (the trophoblast cells) and the uterine wall epithelium itself (Nose and Takeichi, 1986). It is possible that P-cadherin facilitates the connection of the trophoblast with the uterus, since P-cadherin on the uterine cells is seen to contact P-cadherin on the trophoblastic cells of mouse embryos (Kadokawa et al., 1989). (3) **N-cadherin** (neural cadherin) is first seen on the mesodermal cells in the gastrulating embryo as they lose their E-cadherin expression. It also is highly expressed on the cells of the central nervous system (Figure 3.17; Hatta and Takeichi, 1986).

Cadherins adhere cells together by binding to the same type of cadherin on another cell. Thus, cells with E-cadherin stick to other cells having E-cadherin, and they will sort out from cells containing N-cadherin in their membranes. This is called **homophilic binding.** Cells expressing N-cadherin readily sort out from N-cadherin-negative cells in vitro, and Fab antibodies against cadherins will convert a three-dimensional histotypic aggregate of cells into a monolayer (Takeichi et al., 1979). Moreover, when activated E-cadherin genes are transfected into and expressed in cultured mouse fibroblasts (which usually do not express this protein), E-cadherin is seen on their cell surfaces and the treated fibroblasts become tightly connected to each other (Nagafuchi et al., 1987). These cells begin acting like epithelial cells. Cadherin expression is often correlated with aggregation and dispersion. Neural crest cells (which are at the dorsalmost portion of the neural tube) initially express N-cadherin. However, as they begin migrating away from the neural tube as individual cells (to form pigment cells, sensory neurons, and other cell types), these cells lose their N-cadherin expression (Figure 3.17; see also Chapter 7). However, when

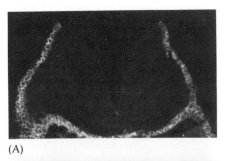

(A)

(B)

FIGURE 3.17
Localization of two different cadherins during the formation of the mouse neural tube. Double immunofluorescent staining was used to localize E- cadherin (A) and N-cadherin (B) in the same transverse section of an 8.5-day embryonic mouse hindbrain. Antibodies to E-cadherin were labeled with one type of fluorescent dye (which fluoresces under one set of wavelengths) while antibodies to N-cadherin were marked with a second type of fluorescent dye (which emits its color at other wavelengths). Photographs taken at the different wavelengths reveal that the outer ectoderm expresses predominantly E-cadherin, while the invaginating neural plate ceases E-cadherin expression and instead expresses N-cadherin. (C) Composite diagram of cadherin expression soon after the formation of the chick neural tube. (Photographs by K. Shimamura and H. Matsunami, courtesy of M. Takeichi; C after Rutishauser, 1990.)

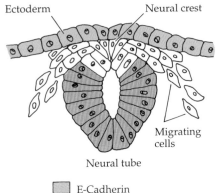

Ectoderm

Neural crest

Migrating cells

Neural tube

▨ E-Cadherin
▨ N-Cadherin

(C)

the migrating cells reach their destination and begin to aggregate together to form nerve ganglia, they re-express N-cadherin (Hatta et al., 1987).

Differential cadherin expression can also explain the homotypic sorting-out data presented earlier. As we discussed, Roth and co-workers had shown that liver cells tend to collect liver cells and that retinal cells collect other retinal cells. Takeichi (1987) demonstrated that the retinal cells express N-cadherin while the liver cells express E-cadherin, and that the sorting out would be expected because of this difference in cadherin expression. He also suggested that the observations of Townes and Holtfreter may likewise be explained by differential cadherin expression. Support for this came from studies wherein different cadherin genes were transfected into mouse fibroblasts that do not usually express either cadherin type. Fibroblasts expressing E-cadherin adhered to other E-cadherin-bearing fibroblasts, while P-cadherin fibroblasts stuck to other fibroblasts expressing P-cadherin. Moreover, when embryonic lung tissue was dissociated and allowed to recombine in the presence of untreated or E-cadherin-bearing fibroblasts, the E-cadherin-expressing fibroblasts became integrated into the epithelial lung tubules (which express E-cadherin), while the untreated fibroblasts became associated with the mesenchymal cells (which do not express any cadherin) (Nose et al., 1988).

All the above-mentioned experiments have been performed on cells in culture. Recently, however, in vivo studies have shown that cadherins may be critical in sorting-out phenomena occurring within the embryo. When the mRNA for chicken N-cadherin is injected into one of the two first-cleavage blastomeres of the *Xenopus* frog embryo, N-cadherin is often expressed on cells that would normally lack it. The embryos that express extra N-cadherin are often characterized by clumps of cells and thickened tissue layers. Normally, the neural tube (which expresses N-cadherin) separates from the cells that would become the epidermis (which express E-cadherin). However, in those embryos wherein both the epidermis and the neural tube expressed this extra N-cadherin, the neural tube would not separate from the epidermis (Detrick et al., 1990; Fujimori et al., 1990). Thus, the cadherins are probably playing major roles in the assortment of cells into tissues.

Molecular regulators of development: The cadherins

The binding specificity of the extracellular region

The cadherin molecules have three major regions. There is an extracellular region that mediates specific adhesion, a transmembrane domain that spans the cell membrane, and a cytoplasmic domain that extends into the cell. The extracellular region is critical for cell–cell binding. When mutant N-cadherin genes that lack the ability to encode the extracellular domains are injected into an early embryonic cell, the descendants of that cell make a defective N-cadherin and cannot properly bind to each other in those tissues where N-cadherin is usually expressed (Figure 3.18; Kintner, 1992). The nervous tissue made by these cells is not organized properly and often migrates into other regions. By exchanging regions between E- and P-cadherins, Nose and colleagues (1990) showed that the amino acids of the cadherin extracellular region involved in cell–cell binding are located in the 113 amino acids of its amino-terminal. This homophilic cadherin recognition site is probably the tripeptide sequence histidine-alanine-valine (HAV): this site is conserved on all major cadherins, mutations in this region severely inhibit cell–cell binding, and synthetic peptides containing this sequence can inhibit cell–cell adhesion by competing with the cell's cadherins for binding (Blaschuk et al., 1990). The amino acids surrounding the HAV sequence appear critical for the *specificity* of this binding (i.e., whether it binds a P-, E-, or N-cadherin). The amino-terminal extracellular region also appears to be responsible for binding the calcium ions.

The intracellular region: Catenins and aggregation

Although it is this extracellular region of the cadherin molecule most distant from the cell that appears to retain the specific binding capacity of the molecule, altering or deleting the amino acids in the E-cadherin *cytoplasmic* domain also prevents cadherin function. The 70 carboxyl-terminal amino acids of cadherins link the cadherin to the cytoskeleton and aggregate the cadherin proteins at the sites of cell–cell attachment (Nagafuchi and Takeichi, 1989). Without this aggregation, cadherin does not seem able to hold the cells in place. The proteins that bind to this carboxyl site on the cadherins and connect them to the cytoskeleton are called **catenins**. The cytoplasmic domain of the cadherins interacts (directly or indirectly) with the three types of catenins (Figure 3.19). The catenin-associated form of E-cadherin can bind to the network of actin microfilaments in the cytoskeleton, but the E-cadherin is unable to be linked to the cytoskeleton if this catenin-binding region is deleted or mutated (Ozawa et al., 1990). Some cells may express a particular cadherin on their cell membrane, but if they do not make the specific catenins as well, the cadherin cannot function (Hirano et al., 1992).

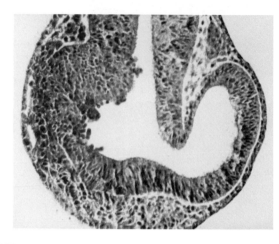

FIGURE 3.18
At the four-cell stage, the blastomeres that form the left side of the frog had been injected with an mRNA for N-cadherin that lacks the extracellular region of the cadherin. The cells with the mutant protein do not form a coherent layer. (From Kintner et al., 1992, courtesy of C. Kintner.)

More cadherins

The "classic" N-, E-, and P-cadherins are not the totality of this family of adhesive glycoproteins. There are several others, including a retinal (R) cadherin and a brain (B) cadherin. And it has recently been shown that some of the molecules involved in forming cell junction adhesions are cadherins (thus obscuring the line between CAMs and cell junctional molecules). The **adherens junction** (*zonulae adherens*) is a beltlike region of cell–cell adhesion and represents the clustering of cadherins, catenins, and the actin cytoskeleton. **Desmosomes** (*macula adherens*; Figure 3.20) are spotlike sites of intercellular contact. Here the intercellular attachment is mediated by *desmoglein I* and *desmocollins I* and *II*. These three calcium-dependent adhesion molecules are also in the cadherin family (Wheeler et al., 1991; Luna and Hitt, 1992). The intracellular regions of these molecules diverge from that of the classic cadherins and enable the desmosome to be linked to the intermediate filament portion of the cytoskeleton, rather than to the actin microfilaments. Epithelial tissues use adherens junctions and desmosomes to separate the extracellular space on one side of an epithelial sheet from the extracellular space on the other side (Figure 3.20B).

Cadherins are extremely important in establishing and maintaining cell–cell interactions between epithelial cells. In adults, they are critical for retaining epithelial integrity. If adult epithelial cells lose E-cadherin, they can become malignant (Vleminckz et al., 1991), and if people make antibodies against a particular epidermal cadherin, they produce a life-threatening disease called pemphigus vulgaris, wherein the epidermal cells lose their cell–cell adhesion, blister, and fall off the body (Amagai et al., 1991).

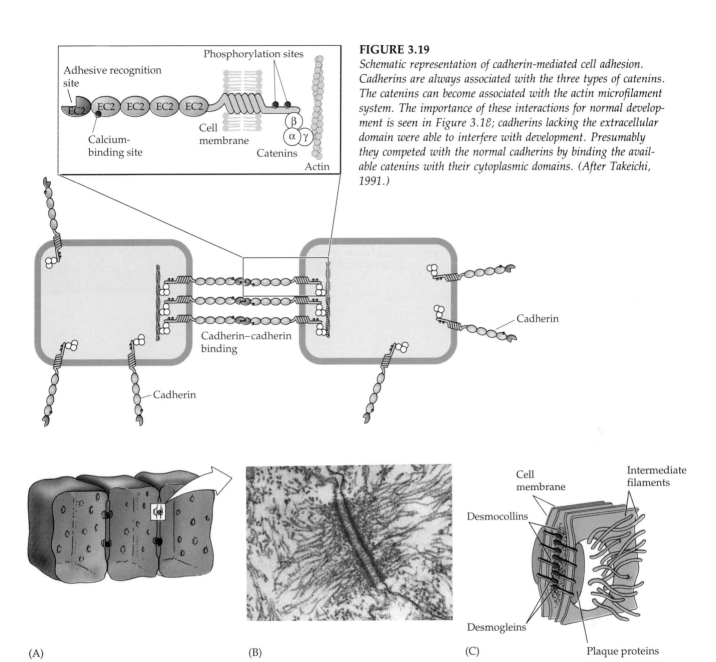

FIGURE 3.19
Schematic representation of cadherin-mediated cell adhesion. Cadherins are always associated with the three types of catenins. The catenins can become associated with the actin microfilament system. The importance of these interactions for normal development is seen in Figure 3.18; cadherins lacking the extracellular domain were able to interfere with development. Presumably they competed with the normal cadherins by binding the available catenins with their cytoplasmic domains. (After Takeichi, 1991.)

(A)   (B)   (C)

FIGURE 3.20
Desmosomes. (A) Cells arranged in an epithelial sheet. (B) Electron micrograph of desmosomes on two adjacent cells from newt epidermis. (C) Integrative model depicting adhesion in the desmosome occurring between cadherin-family molecules (desmogleins and desmocollins) on apposing cell membranes. The plaque-forming molecules may be forms of catenins. (B from Kelly, 1966, courtesy of D. E. Kelly; C after Wolfe, 1993.)

Immunoglobulin superfamily CAMs

As we discussed in Chapter 1, antibodies were first used to identify cell adhesion molecules in *Dictyostelium*. Gerisch and colleagues (Beug et al., 1970) prepared antibodies against *Dictyostelium* and then chemically split the antibodies so that only their monovalent antigen-binding regions—the Fab fragments—remained (Figure 3.21A). (The divalent antibodies had to be split, for if they remained divalent, they might artificially join the cells

(A)

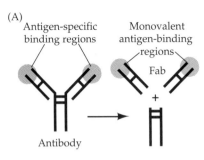

Antigen-specific
binding regions

Antibody

Monovalent
antigen-binding
regions

Fab

+

(B)

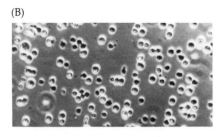

(C)

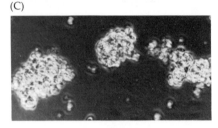

(D)

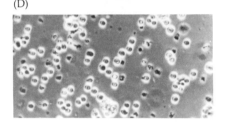

FIGURE 3.21
Aggregation of neural retina cells from 10-day chick embryos. (A) Creation of antibody fragments (Fab). (B) Single cells prior to aggregation. (C) Aggregates produced after 30 minutes of rotation. (D) Inhibition of aggregation when retinal cells were rotated 30 minutes in the presence of Fab antibody fragments to N-CAM. (From Brackenbury et al., 1977, courtesy of G. M. Edelman.)

together, and the effect could not be measured.) This led to the discovery of the 80-kDa glycoprotein that mediates cell–cell adhesion during slime mold aggregation. The same approach was used to study embryonic cell adhesion by Edelman and his colleagues (Brackenbury et al., 1977) and led to the isolation of **neural cell adhesion molecule (N-CAM).** In this study, chick neural retina cells were injected into rabbits, and rabbit antibodies were made to the neural retina cell surface (and to some other cells as well). These antibodies were split into their Fab fragments. When these Fab fragments were added to the culture media, they were able to inhibit the aggregation of the chick neural retina cells (Figure 3.21B). Moreover, when an excess of purified neural retina membranes were added to this mixture, they bound the Fab fragments, thus taking them out of solution and enabling the retina cells to aggregate. By extracting the components of these membranes, Edelman and his colleagues (Thiery et al., 1977) found that the Fab fragments bound to a 140-kDa protein. When this protein was placed into the solution of aggregating retina cells, it alone inhibited the effect of the Fab fragments. It was concluded that the 140-kDa protein, N-CAM, is a cell surface component with a very important role in neuronal aggregation.

N-CAM is a member of a class of CAMs that do not need calcium ions and that share a similar structure (Figure 3.22). This structure, with its extracellular globular domains held in place by disulfide bonds, resembles that of the immunoglobulin molecule, and it is considered likely that the immunoglobulins are derived from this group of CAMs (Williams and Barclay, 1988; Lander, 1989). Thus, these glycoproteins are called the **immunoglobulin superfamily CAMs.***

Although the N-CAM protein is encoded by one gene per haploid genome, it can occur in three sizes, depending on alternative RNA splicing

---

*The designation *superfamily* is often given because the different classes of immunoglobulin molecules themselves constitute a "family." These other members of the superfamily have structures resembling immunoglobulins, but are not exactly "close" family.

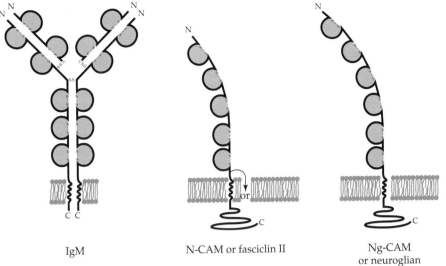

FIGURE 3.22
Three members of the immunoglobulin superfamily. The IgM molecule has two heavy chains, each with five domains, and two light chains, each with two domains. N-CAM is a single polypeptide chain with five domains. Its anchor into the membrane can be either a transmembrane amino acid sequence in the protein or a lipid. Ng-CAM is a transmembrane protein with six globular domains. The insect cell adhesion molecules fasciclin II and neuroglian resemble N-CAM and Ng-CAM, respectively.

IgM

N-CAM or fasciclin II

Ng-CAM
or neuroglian

(as will be shown in Chapter 12). The predominant forms are 180 kDa and 140 kDa in mass and contain three regions: an extracellular region, a membrane-spanning domain, and a cytoplasmic domain. (A 120-kDa form does not span the membrane and appears to consist of an extracellular domain covalently bound to a membrane lipid.)

N-CAM molecules on adjacent cells interact homophilically with each other to construct the bonds between the cells. N-CAM-containing neural retina cells can even bind to lipid vesicles that contain N-CAM in their bilayer (Rutishauser et al., 1982). Electron microscopy confirms that N-CAM molecules can interact with each other to form complex multimeric structures (Hall and Rutishauser, 1987). N-CAM can mediate the adhesion of cells only if the cells all contain N-CAM on their surfaces, and each region of cell–cell contact involves many interactions among N-CAM molecules. N-CAM and N-cadherin are often found on the same cells, but in different regions. On the neuroepithelial cells of the neural tube, N-CAM is concentrated near the outer margin, whereas N-cadherin staining is more intense at the luminal surface (Hatta et al., 1987; Rutishauser and Jessell, 1988).

N-CAM molecules can be made either adhesive or repulsive by modifying them with sialic acid residues. Sialic acid is a complex (10-carbon) saccharide derived from the conjugation of D-mannose with pyruvic acid; it carries a negative charge. Highly glycosylated N-CAM is about 30 percent sialic acid by weight. Less glycosylated N-CAM consists of about 10 percent sialic acid. The difference between the highly and less sialylated molecules appears to be the length of the sialic acid polymers (Hoffman and Edelman, 1983). The sialic acid molecules are covalently linked to the N-CAM protein at three sites on the middle of the molecule, within the extracellular domain (Cunningham et al., 1983). Sialic acid can severely reduce the adhesiveness of one cell toward another. Cells with a relatively low amount of sialic acid on their N-CAM aggregate four times more readily than those with the high levels of sialic acid (Hoffman and Edelman, 1983). As the embryo gets older, most of the N-CAM proteins progress from the high-sialic acid to the low-sialic acid forms (Rieger et al., 1985; Chuong and Edelman, 1985a,b), which may help to stabilize mature tissues. In this manner, N-CAM may play a role both in stimulating two adjacent cells to form regions of contact and in inhibiting such contact. If the neighboring cells both express the low-sialic acid form of N-CAM, strong cell adhesion is promoted. However, if the cells contain the highly

(A)

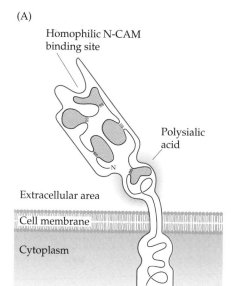

Homophilic N-CAM binding site

Polysialic acid

Extracellular area

Cell membrane

Cytoplasm

FIGURE 3.23
Mechanisms by which N-CAM could promote or inhibit cell–cell adhesions. (A) Structure of the N-CAM molecule. (B) Expression of low sialic acid N-CAM molecules on adjacent cell surfaces leads to the formation of numerous bonds between N-CAM molecules that promote junction formation from subunits (represented as rectangles) on the adjacent cell membranes. (C) Highly sialated forms of N-CAM can inhibit cell–cell interactions that would otherwise have resulted from other adhesive molecules (represented as balls and sockets) on the adjacent cell membranes. When the polysialic acid residues are removed, the two cells can adhere. (After Rutishauser, 1990.)

(B)

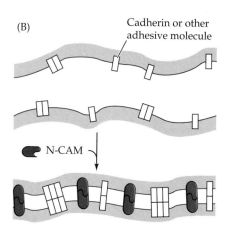

Cadherin or other adhesive molecule

N-CAM

(C)

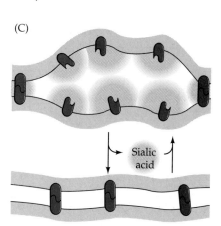

Sialic acid

sialated forms of N-CAM, cell adhesion will be weakened or inhibited (Figure 3.23; Rutishauser et al., 1988).

Differential CAM expression is critical at boundaries between two groups of cells. At such places, the body segregates different cells into different regions. Notochord cells do not enter into the neural tube, nor do dermal cells trespass into the epidermis. Such segregation may be accomplished by the adjacent populations' having different CAMs. For instance, feathers are induced when mesodermally derived mesenchymal cells condense together to form a ball of cells directly beneath the epidermis of the chicken skin. The ectodermal cells are linked together by E-cadherin, while the previously CAM-negative mesenchymal cells begin to express N-CAM and collect to form an aggregate (Figure 3.24). Throughout the developing feather, different groups of cells become separated from one another as a result of their ability to express N-CAM, E-cadherin, or both proteins (Chuong and Edelman, 1985a,b).

In addition to helping generate and maintain boundaries between different cell types, CAMs also appear to play roles in the binding of one cell type to another. This role is especially important in the connections made from an axon to its target organ, and recent studies have shown that N-CAM is needed for the proper attachment of axons to specific target tissue. Tosney and her co-workers (1986) have shown that muscle cells begin to express N-CAM in their cell membranes around the time when

FIGURE 3.24
Distribution of different CAMs at tissue boundaries. As the mesodermal cells collect to induce the feather bud in the ectoderm, the newly aggregated mesenchymal cells express N-CAM (A), while the ectodermal cells express E-cadherin (B) in their cell membranes. (From Chuong and Edelman, 1985a, courtesy of G. M. Edelman.)

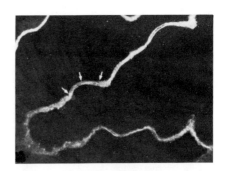

(A)

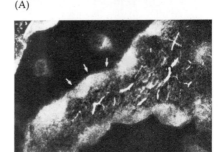

(B)

axons that bear N-CAM are first seen near them. However, once synapses between the nerve and muscle are established, the only regions of the muscle that retain N-CAM are those at the neuromuscular junctions (Covault and Sanes, 1986). When antibodies against N-CAM are placed into specific regions of the developing embryo, the connections between the neuron and its target tissue can be dramatically distorted. Fraser and his colleagues (1988) have shown that antibodies to N-CAM distort the recognition of axons from the frog retina to their specific targets in the optic tectum of the brain. Landmesser and her colleagues (1988) have similarly demonstrated that the innervation of embryonic chick skeletal muscle is affected by anti-N-CAM antibodies. In both cases, the effects can be explained by the ability of the antibodies to interfere with N-CAM's roles in tethering axons together and in maintaining connections between axons and their target tissues.

While some cell adhesion molecules are used in many places in the embryo, others have a much more restricted role. Restricted CAMs are thought to be expressed at very specific times and places during development, and this has made them difficult to identify and characterize. Some of these CAMs are responsible for enabling axons traveling in the same direction to bind together. This binding is called **fasciculation,** and the resulting bundle of axons produces the nerve. (The optic nerve, for instance, is a fasciculated bundle of axons originating in the ganglion cells of the neural retina.) As we will see in Chapter 8, an insect axon can change its direction when it reaches another axon and travel along that axon's surface. This phenomenon is thought to happen only when the two axons contain the same adhesion molecule in their cell membranes. It appears that an insect axon contains many of these CAMs on its cell surface and that the surface molecules differ at different sites along the axon (Bastiani et al., 1985, 1987). These surface compounds have been called **fasciclins** (Figure 3.25) and some of them are also in the immunoglobulin superfamily (Harrelson and Goodman, 1988).

FIGURE 3.25
Fasciclin expression in the developing grasshopper nervous system. (A) Scaffold of fasciculated axons in a grasshopper embryo as seen by Nomarski microscopy. A com and P com are the anterior and posterior commissures whose axons traverse the segment; ISN is the intersegmental neuron, and con is a connective neuron. (B,C) Embryonic nervous system as in (A), but stained with monoclonal antibodies made to the cell surface fasciclin glycoproteins. The antibody on (B) recognizes a subset of axons in the anterior and posterior commissures, while the antibody in (C) binds to a membrane glycoprotein of most longitudinal axon fascicles. The arrows show the same locations on (B) and (C). Note that the antibody stains only a portion of each axon. (From Bastiani et al., 1987, courtesy of C. Goodman.)

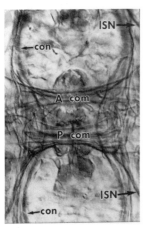

(A)

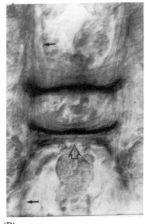

(B)

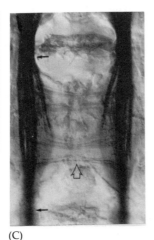

(C)

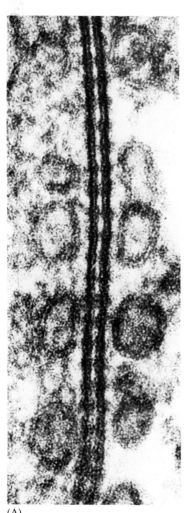

(A)

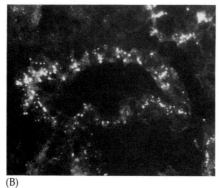

(B)

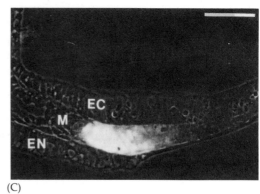

(C)

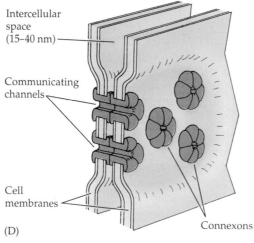

Intercellular
space
(15–40 nm)

Communicating
channels

Cell
membranes

(D)

Connexons

FIGURE 3.26

Gap junction proteins. (A) Electron micrograph of a row of gap junctions connecting two apposed cells. (B) Fluorescence micrograph of gap junctions in a 17-day embryonic mouse kidney tubule. (C) Compartments formed by gap junction proteins between cells that communicate with one another. This compartment in the mouse gastrula can be seen by injecting Lucifer Yellow dye into one cell and seeing it transferred to a small group of cells. (D) Subunit structure of the gap junction. (A from Peracchia and Dulhunty, 1976, courtesy of C. Peracchia; B from Sainio et al., 1992, courtesy of K. Sainio; C from Kalimi and Lo, 1988, courtesy of C. Lo; D after Darnell et al., 1986.)

Gap junctions and their proteins

Gap junctions are specialized intercellular regions where adjacent cells are 15 to 40 nm apart. Through this space, fine connections are seen (Figure 3.26A,B). These connections serve as communication channels between the adjacent cells. Cells so linked are said to be "coupled," and small molecules (MW <1500) and ions can freely pass from one cell to another. In most embryos, at least some of the early blastomeres are connected by gap junctions, thereby enabling ions and small soluble molecules to pass readily between them. The ability of cells to form gap junctions with some cells and not with others enables the formation of physiological "compartments" within a developing embryo (Figure 3.26C).

The importance of gap junctions in development has been demonstrated in amphibian and mammalian embryos (Warner et al., 1984). When antibodies to gap junction proteins were microinjected into one specific cell of an 8-cell *Xenopus* blastula, the progeny of that cell, which are usually coupled through gap junctions, could no longer pass ions or small molecules from cell to cell. Moreover, the tadpoles that resulted from such treated blastulae showed defects specifically relating to the developmental fate of the injected cell (Figure 3.27). The progeny of such a cell did not die, but they were unable to undergo their normal development (Warner et al., 1984). In the mouse embryo, the first eight blastomeres are connected to one another by gap junctions. These eight cells are loosely associated with one another, but they move together to form a compacted

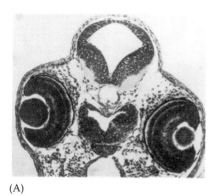
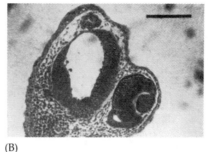

(A) (B)

FIGURE 3.27
Developmental effects of gap junctions. Section through *Xenopus* tadpole in which one of the blastomeres at the eight-cell stage had been injected with (A) a control antibody or (B) antibody against gap junction protein. The side formed by the injected blastomere lacks its eye and has abnormal brain morphology. (From Warner et al., 1984, courtesy of A. E. Warner.)

embryo. If compaction is inhibited, further development ceases. Compaction is inhibited if the gap junctions of uncompacted embryos are blocked by these antibodies. The treated blastomeres continue to divide, but compaction fails to occur (Lo and Gilula, 1979; Lee et al., 1987). If antisense RNA to gap junction messages is injected into one of the blastomeres of a normal mouse embryo, that cell will not form gap junctions and will not be included in the embryo (Bevilacqua et al., 1989).

The gap junction channels are made of **connexin** proteins. In each cell, six identical connexins in the membrane group together to form a transmembrane channel containing a central pore. The gap junction complex of one cell connects to the gap junction complex of another cell, thus enabling the cytoplasms of both cells to be joined (Figure 3.26D). There are several types of connexins, and some of them may be regulated by cadherins. Jongen and co-workers (1991) found that when cells are coupled by E-cadherin, gap junction-mediated communication between those cells is dependent upon cadherin function. Evidence suggests that cadherin works both by keeping the cells in contact and by modifying the connexin protein.

The cell membrane, then, has several mechanisms by which it can form attachments with other cell membranes. It can use immunoglobulin superfamily CAMs, calcium-dependent CAMs, restricted CAMs, and junctional proteins. This does not exhaust its repertoire. As was briefly mentioned above, the cell can also bind specifically to particular components of the extracellular matrix. It is to these components that we turn our attention as we discuss cell migration.

The molecular bases of migrational specificity

There are several mechanisms that enable cells to migrate from one region of the body to another.

Chemotaxis

Chemotaxis is defined as cellular locomotion directed in response to a concentration gadient of a chemical factor in solution (Harris, 1954; Armstrong, 1985). Cells sense the chemical and migrate toward higher concentrations of this substance until they reach the source secreting it. It has been proposed (Champion et al., 1986; Dargemont et al., 1989) that the migration of lymphocyte precursors from the bone marrow into the embryonic thymus (wherein they generate the T cells of the immune system) is an example of chemotaxis. Quail bone marrow cells placed into a 1-mm

FIGURE 3.28
Chemotaxis of lymphocyte precursors to a diffusible compound from the thymus. (A) Cells were placed on a coverslip between two wells of culture medium. The media mixed between the wells, forming a gradient, and the migration of cells was recorded by a video camera. (B) Video camera tracing of cell movement when both wells contained control medium. (C) Tracing when one of the wells contained control (unconditioned) medium and the other contained medium in which embryonic quail thymic epithelium had been grown. (After Champion et al., 1986.)

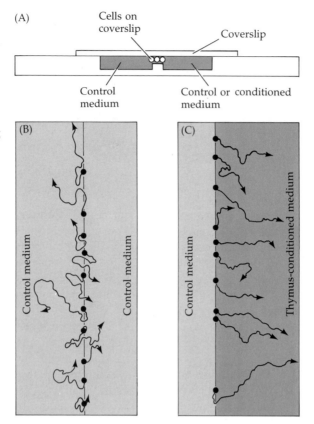

region connecting two wells (Figure 3.28) will migrate into the well containing either thymic epithelial cells or β_2-microglobulin, an 11 kDa peptide produced by these cells.

Haptotaxis

Gradients do not have to be in solution. An adhesive molecule could be present in increasing amounts along an extracellular matrix. A cell that was constantly making and breaking adhesions with such a molecule would move from a region of low concentration to an area where that adhesive molecule was more highly concentrated. Such a phenomenon is called **haptotaxis** (Carter, 1967; Curtis, 1969).

Poole and Steinberg (1982) have evidence that the migration of the pronephric duct cells in salamanders is regulated by haptotaxis. The pronephric duct rudiment is important for kidney and gonad development (Chapter 18). The duct rudiment originates as a solid clump of mesodermal cells near the head of the embryo (Figure 3.29). As the embryo develops, these cells migrate (without cell division) along the ventrolateral border of the dorsal mesoderm, eventually reaching the cloaca (where the urine will be excreted). Thus, the cells of the pronephric rudiment appear to migrate upon a specific path from one place to another along the surface of the embryo. The migration cannot be due to chemotaxis, because the removal of the cloacal region does not stop migration, nor does placing the cloacal region in a different position in the embryo redirect the path of the pronephric rudiment. The migration also cannot be due to the contours of the embryo, and no electrical gradient is observed (see below). When a pronephric rudiment is transplanted from one embryo into the ventral region of another, the secondary duct always travels dorsally across the

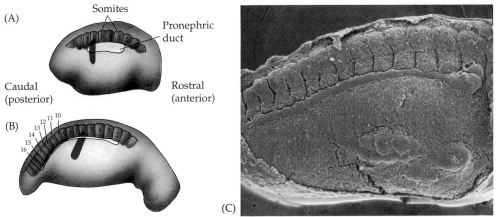

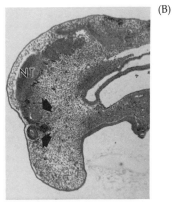

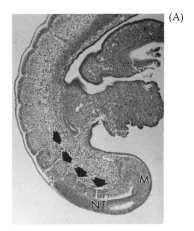

FIGURE 3.29

Elongation of the axolotl (salamander) pronephric duct by cell migration. (A,B) A portion of the embryo is stained with a vital dye (color) across the region of the newly formed pronephric duct rudiment. Several hours later, the dye mark in the pronephric duct has extended into the caudal region of the embryo, away from its original source. (C) Grafted pronephric duct rudiment (bottom) extends a duct dorsocaudally across the host's flank mesoderm to fuse with the host's pronephric duct. (A and B after Poole and Steinberg, 1982; C, courtesy of M. S. Steinberg.)

flank mesoderm to meet the original one and then migrates caudally beneath the dorsal mesoderm toward the cloaca (Figure 3.29C). The transplanted ducts never migrate ventrally or toward the head, regardless of the orientation in which the graft is inserted (Zackson and Steinberg, 1987). It appears that the duct cells are recognizing a gradient of some molecule across the mesodermal surface of the salamander embryo and that this gradient exists ventral-to-dorsal and anterior-to-posterior. The identity of this molecule is still unknown (Thibaudeau et al., 1993).

Galvanotaxis

Charged ions are another possible source of polar gradients in the embryo. Potential differences between cells and their environment are critical in development (as in fertilization), and such voltage differences between embryonic regions may also be important in morphogenesis (Jaffe, 1981; Nuccitelli, 1984). Electrical currents have been detected in early chick embryos (Jaffe and Stern, 1979), and direct current of that magnitude is able to direct cultured mesenchymal cells toward the negative pole (Erickson and Nuccitelli, 1984). Hotary and Robinson (1992) have shown that when the endogenous electrical fields of the chick embryo are disturbed by conductive implants, tail development ceases in more than 80 percent of the cases (Figure 3.30). The internal abnormalities are similar to those of the *rumpless* mutant of chickens, and these mutants were found to have lower electric currents.

Contact guidance and contact inhibition of movement

Physical factors may also play important roles in directing cell migration. Weiss (1934) showed that the physical terrain can influence cell movement and that growing nerve fibers follow the stress contours of a plasma clot rather than move out of shallow grooves (Figure 3.31A). He called this **contact guidance.** Weiss (1955) noted that if the substratum is stressed in

FIGURE 3.30

Sections through the tail region of (A) a control chick embryo or (B) an embryo whose electrical currents were dissipated in the region close to where the hindlimbs form. Mesodermal somites (arrows) can be seen in the control embryo, but do not form normally in the treated ones. NT, neural tube. (From Hotary and Robinson, 1992, courtesy of the authors.)

FIGURE 3.31

Contact guidance of axons by the stress fibers of a plasma clot. (A) Along the random meshes, the axons travel randomly along the clot fibers and branch frequently. Where the fibrous matrix consists of parallel grooves, the axons are made to run parallel to one another. (B) Stress fibers made by a single embryonic chick fibroblast crawling upon a sheet of silicone rubber. The cell is approximately 80 μm wide. (A from Weiss, 1955; B from Harris, 1984, courtesy of A. K. Harris.)

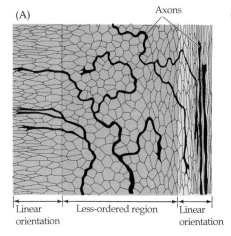

(A)

Axons

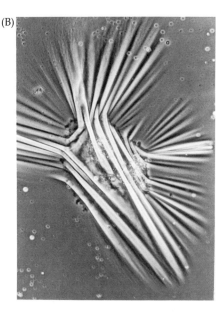

(B)

Linear orientation | Less-ordered region | Linear orientation

certain ways, the channels that are created act as guides for cell migration. At that time, he could not envision what might cause these stresses, but in 1980 Harris showed that fibroblasts can dramatically alter the substrate on which they are cultured. When they are cultured on silicone rubber or collagen, they deform their substrate, generating stress folds practically identical to the ones seen by Weiss (Figure 3.31B). If muscle cells are added to such a substrate, the randomly scattered muscle cells become aligned into well-formed functional units (Stopak and Harris, 1982). Contact guidance is also hypothesized to account for the migration of mesenchymal cells inside the fish fin (Wood and Thorogood, 1987). Cells move by extending a thin process called a **lamellipodium**, and when the lamellipodium of one migrating cell contacts the surface of another cell, the lamellipodium is paralyzed. This process is called **contact inhibition** (Figure 3.32). A new lamellipodium is then formed elsewhere on the cell, thereby taking the cell away from its neighbor (Abercrombie and Ambrose, 1958). The net result is the migration of motile cells away from central masses. (This phenomenon is seen primarily in mesenchymal cells and does not occur when the cells are joined to each other in epithelial sheets, except when a free margin is exposed.) Contact inhibition may provide the stimulus for the migration of the neural crest cells away from the original crest (Rosavio et al., 1983).

FIGURE 3.32

Contact inhibition of movement. The lamellipodium of one fibroblast (A) approaches and (B) contacts the membrane of another fibroblast. While retaining these contacts, the cell on the right (C) passes beneath (underlaps) the other cell and then (D) moves away from it. (From Erickson et al., 1980, courtesy of C. A. Erickson.)

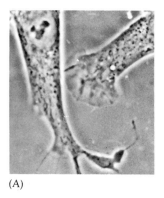

(A)

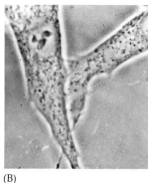

(B)

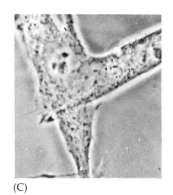

(C)

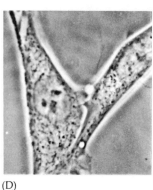

(D)

Differential substrate affinity

This hypothesis postulates that different cells recognize different molecules in various extracellular matrices. Each migrating cell type prefers certain combinations of matrix molecules to others. Weiss (1945) and Tyler (1946) suggested that in some instances, cells can interact with their substrata through lock-and-key interactions between the cell membrane and the extracellular matrix. The relationship between the cell membrane protein and the matrix molecule would resemble that between enzyme and substrate or antibody and antigen. During the past decade, this type of interaction has been shown to be extremely important for cell migration.

The molecular bases for differential substrate affinity

The extracellular matrix

Whether cells are migrating or remaining stationary depends upon their relationship to their immediate **extracellular matrix.** Extracellular matrices consist of macromolecules secreted by the cells into their immediate environment. These molecules interact to form insoluble networks that can play several roles during development. In some instances, they separate two adjacent groups of cells and prevent any interactions. In other cases, the extracellular matrix may serve as the substrate upon which cells migrate, or it may even induce the differentiation of certain cell types. One type of matrix is shown in Figure 3.33. Here, a sheet of epithelial cells is adjacent to a layer of loose mesenchymal tissue. The epithelial cells have formed a tight extracellular layer called the **basal lamina;** the mesenchymal cells secrete a loose **reticular lamina.** Together, these layers constitute the **basement membrane** of the epithelial cell sheet. There are three major components of most extracellular matrices: collagen, proteoglycans, and the large glycoproteins that are called substrate adhesion molecules (Table 3.2).

Collagen. **Collagen** is a family of glycoproteins containing a large percentage of glycine and proline residues. Because collagen is the major structural support of almost every animal organ, it constitutes nearly half the total body protein. Most of this collagen is type I—found in the extracellular matrices of skin, tendons, and bones. This type makes up 90 percent of the collagen in the body. In addition, there are numerous other

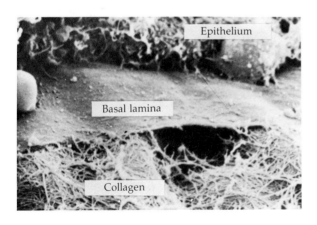

FIGURE 3.33
Location and formation of extracellular matrices in the chick embryo. Scanning electron micrograph showing extracellular matrix at the junction of the epithelial cells (above) and mesenchymal cells (below). The epithelial cells synthesize a tight glycoprotein-based lamina, while the mesenchymal cells secrete reticular lamina made primarily of collagen. (Courtesy of R. L. Trelsted.)

TABLE 3.2
Main constituents of the extracellular matrix

Mesenchymal Extracellular Matrix	Basal Laminae of Epithelial Cells
COLLAGENS Long, thin molecules (most common is type I; types II, III, and V–XIII are also found) that assemble to form fibrils, usually 60–70 nm in diameter. Collagens provide strength and stability to tissues. **MATRIX PROTEOGLYCANS** Made up of proteins and repeating disaccharides (glycosaminoglycans). Glycosaminoglycans include hyaluronic acid, a very large (10^8 Da) molecule that binds large amounts of water. Sulfated proteoglycans each comprise a linear core protein to which are attached chains of one or more of the sulfated glycosaminoglycans (chondroitin, heperan, keratan, and dermatan sulfate). Proteoglycans stimulate and modulate cell movement; their range suggests they may have other properties that are not yet known. **SUBSTRATE ADHESION MOLECULES** Molecules to which cells make the adhesions that allow them to move. They include fibronectin, chondronectin, and tenascin.	**COLLAGEN IV** The major structural component of the basal lamina. Unlike the other collagens, its fibrils are like fine "chicken wire" and assemble into a feltlike substrate. **MATRIX PROTEOGLYCANS** Hyaluronic acid and sulfated proteoglycans are often present in basal laminae; their presence may facilitate the passage of secretory products through the lamina. **SUBSTRATE ADHESION MOLECULES** Laminin is the major functional component of the basal lamina. A glycoprotein trimer with adhesion sites for the cell membrane, collagen IV, and glycosaminoglycans. Basal laminae may contain fibronectin, entactin, nidogen, and other adhesive glycoproteins.

Source: Adapted from Bard, 1990.

TABLE 3.3
Structural and functional heterogeneity of major types of collagen

Collagen type	Characteristics and function	Localization
I	Low carbohydrate content Thick, tightly packed, 67-nm banded fibrils, giving high tensile strength; most abundant collagen	Skin, bone, tendons
II	High carbohydrate content Small 67-nm banded fibrils to form compressible elastic matrix	Cartilage, vitreous body
III	High carbohydrate content Triple-helical, interchain disulfide bonds Small 67-nm banded fibrils to provide support and elasticity	Skin, blood vessels, internal organs (not bone or tendon)
IV	"Chicken wire"-like fibrils act as a cell attachment site and as a selective barrier. Binds to laminin	Basement membranes

Source: Adapted from Vuorio, 1986, and Hostikka, 1990.

types of collagen that serve special functions (Table 3.3). Type II collagen is most evident as the secretion of cartilage cells, but it is also found in the notochord and in the vitreous body of the eye. Type III collagen is most evident in blood vessels, and type IV is found in various basal laminae produced by epithelial cells (Vuorio, 1986). Other types of collagen exist throughout the body and especially in the regions of cartilage. Collagen is important for the formation of basal lamina, and it is also implicated in the branching of epithelial tubules in the salivary glands, lungs, and other organs.

Proteoglycans. These are specific types of glycoproteins in which (1) the weight of the carbohydrate residues far exceeds that of the protein and (2) the carbohydrates are linear chains composed of repeating disaccharides. Usually one of the sugars of the disaccharide has an amino group, so the repeating unit is called a **glycosaminoglycan**. Table 3.4 lists the common glycosaminoglycans; the basic proteoglycan structure is illustrated in Figure 3.34. The interconnection of protein and carbohydrate forms a weblike matrix, and in many motile cell types, the proteoglycan surrounds the cells and is thought to mediate against their coming together

TABLE 3.4
Repeating disaccharide units of the most common glycosaminoglycans of matrix proteoglycans

Glycosaminoglycan	Repeating disaccharide unit[a]	Distribution
Hyaluronic acid	Glucuronic acid-*N*-acetylglucosamine	Connective tissues, bone, vitreous body
Chondroitin sulfate	Glucuronic acid-*N*-acetylgalactosamine sulfate	Cartilage, cornea, arteries
Dermatan sulfate	[Glucuronic or iduronic acid]-*N*-acetylgalactosamine sulfate	Skin, heart, blood vessels
Keratan sulfate	Galactose-*N*-acetylglucosamine sulfate	Cartilage, cornea
Heparan sulfate	[Glucuronic or iduronic acid]-*N*-acetylgalactosamine sulfate	Lung, arteries, cell surfaces

[a]These are the typical repeating units of these glycosaminoglycans. However, regions of each GAG can have slightly modified saccharides.

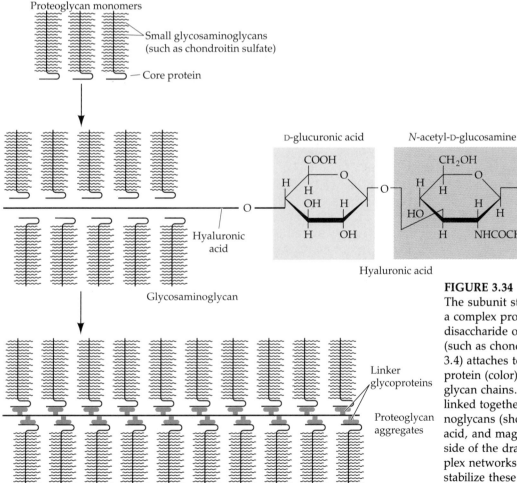

FIGURE 3.34
The subunit structure and assembly of a complex proteoglycan. The repeating disaccharide of the glycosaminoglycan (such as chondroitin sulfate; see Table 3.4) attaches to a relatively small core protein (color) to make up the proteoglycan chains. These chains can be linked together by longer glycosaminoglycans (shown here as hyaluronic acid, and magnified in the right-hand side of the drawing) to produce complex networks. Linker glycoproteins stabilize these latter associations. (Modified from Cheney and Lash, 1981.)

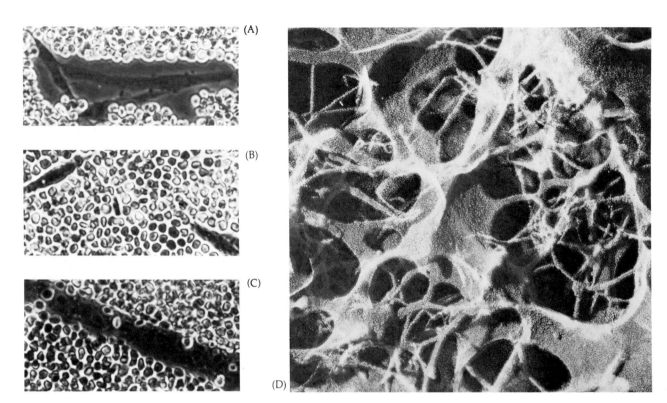

FIGURE 3.35
The proteoglycan coat surrounding
mobile cells. (A) Hyaluronidate coat
surrounds chick myoblasts. Myoblasts
in culture exclude small particles (in
this case fixed red blood cells) for a
significant distance from the cell border. (B) When the myoblasts are
treated with hyaluronidase (which dissolves hyaluronic acid), this extracellular coat vanishes. (C) The coat also
vanishes as the myoblasts cease dividing and join together as they differentiate. (D) Electron micrograph of hyaluronidate in aqueous solution shows
a branching fibrillar network. (A–C
from Orkin et al., 1985, courtesy of B.
Toole; D from Hadler et al., 1982,
courtesy of N. M. Hadler.)

(Figure 3.35). The consistency of the extracellular matrix depends on the
ratio of collagen to proteoglycan. Cartilage, which has a high percentage
of proteoglycans, is soft, whereas tendons, which are predominantly collagen fibers, are tough. In basal laminae, proteoglycans predominate,
forming a molecular sieve in addition to providing structural support.

Proteoglycans have also been shown to be important in mediating the
connections between adjacent tissues of an organ. Here, they can act to
draw cells together* (San Antonio et al., 1987; Thesleff et al., 1989; Vainio
et al., 1989; Bernfield and Sanderson, 1990). In some instances, proteoglycans secreted by one cell type are essential for the growth of neighboring
cells. Axons of the dorsal root ganglia have heparan sulfate proteoglycan
on some of their cell surface proteins, and the removal of this proteoglycan
prevents the associated Schwann cells from proliferating around them
(Ratner et al., 1985). One of the ways that the glycosaminoglycan chains
of the proteoglycans can function is to retain and present **growth factors**
to cellular receptors. Growth factors are hormone-like proteins that cause
mitosis or differentiation when they bind to particular cells. However, the
cell receptor for the growth factor often does not bind the factor with great
affinity. Rather, the growth factor is first bound by the proteoglycan's
carbohydrates, and this concentrates the growth factor so that it can bind
to its receptor (Massagué, 1991; Yayon et al., 1991).

Extracellular glycoproteins. Extracellular matrices contain a variety of
other specialized molecules, such as fibronectin, laminin, and entactin.
These large glycoproteins probably are responsible for organizing the collagen, proteoglycan, and cells into an ordered structure. **Fibronectin** is a

*Heparan sulfate proteoglycans are thought to bring together chondrocytes, the cartilage-
producing cells. Excessive levels of glucose, however, inhibit the synthesis of the core protein
of the proteoglycan, thus inhibiting cartilage formation. Leonard and co-workers (1989)
propose this as a possible mechanism for the skeletal problems frequently seen in children
born to severely diabetic mothers.

(A)

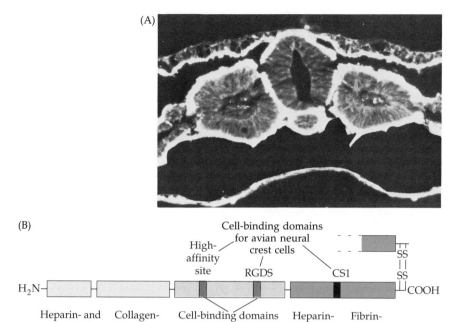

(B)

Cell-binding domains
for avian neural
crest cells

High-
affinity
site   RGDS   CS1

Heparin- and
fibrin-binding
domain

Collagen-
binding
domain

Cell-binding domains
for fibroblasts

Heparin-
binding
site II

Fibrin-
binding
site II

H₂N—

—COOH

SS
||
SS

FIGURE 3.36
Fibronectin in the developing chick embryo. (A) Fluorescent antibodies to fibronectin show that the fibronectin deposition in the 24-hour chick embryo lies along the basal lamina of many organs. (B) Structure and binding domains of fibronectin. The rectangles represent protease-resistant domains. The fibroblast cell binding domain consists of two units, the RGDS site and the high-affinity site, both of which are essential for cell binding. Avian neural crest cells have another site that is necessary for their motility on a fibronectin substrate. Other regions of fibronectin enable it to bind to collagen, heparin,* and other molecules of the extracellular matrix. (A courtesy of J. Lash; B after Dufour et al., 1988.)

very large (460-kDa) glycoprotein dimer synthesized by fibroblasts, chondrocytes, endothelial cells, macrophages, and certain epithelial cells (such as hepatocytes and amniocytes). One of the functions of fibronectin is to serve as a general adhesive molecule linking cells to various substrates such as collagen and proteoglycans. Fibronectin also organizes the extracellular matrix by having several distinct binding sites, whose interaction with the appropriate molecules results in the proper alignment of cells with their extracellular matrix (Figure 3.36).

As we will see in later chapters, fibronectin also has an important role in cell migration. The "roads" over which certain migrating cell types travel are paved with this protein. Mesodermal cell migration during gastrulation is seen on the fibronectin surfaces of many species, and the movement of these cells ceases when fibronectin is locally removed. In chick embryos, the precursors of the heart, the precardiac cells, migrate upon fibronectin to travel from the lateral sides of the embryo to the midline. If chick embryos are injected with antibodies to fibronectin, the precardiac cells fail to migrate to the midline and two separate hearts develop. Fluorescent antibodies to fibronectin have demonstrated a gradient of fibronectin in the path of migration between the endoderm and the mesoderm. If this region is cut and rotated, the heart cells follow the gradient to new positions away from the midline (Linask and Lash, 1988a,b). Thus, fibronectin appears to play a major role in the migration of the precardiac cells to the embryonic midline. Other cell types, such as the germ cell precursors of frog embryos, also travel over cells that secrete fibronectin onto their surfaces (Heasman et al., 1981).

Laminin is a major component of basal laminae. It is made of three peptide chains, and, like fibronectin, it can bind to collagen, glycosaminoglycans, and cells. The collagen bound by laminin is type IV (specific for basal lamina), and the cell-binding region of laminin recognizes chiefly epithelial cells and neurons. The adhesion of epithelial cells to laminin

*Heparin is a portion of the heparin proteoglycan secreted by mast cells and basophils. Heparan and heparan sulfate are names given to similar glycosaminoglycans found in the extracellular matrix or on the cell surface. It is presumed that the binding sites for heparin are also those of heparan sulfate (Bernfield and Sanderson, 1990).

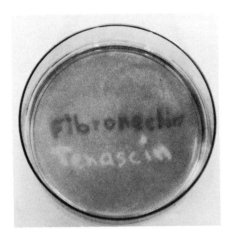

FIGURE 3.37
Inhibition of cell adhesion by tenascin. Fibronectin and tenascin were both plated onto a tissue culture dish in the form of letters. Fibroblasts were added to the dish and allowed to adhere and migrate. The result shows that fibronectin was a much preferred substrate over tissue culture plastic, whereas cells did not adhere to or migrate well upon tenascin. (Courtesy of R. Chiquet.)

(which they sit upon and utilize) is much greater than the affinity of mesenchymal cells for fibronectin (which they have to bind to and release if they are to migrate). Like fibronectin, laminin is seen to play roles in assembling extracellular matrices, promoting cell adhesion and growth, changing cell shape, and permitting cell migrations (Hakamori, 1984).

Not all the large extracellular glycoproteins promote cell adhesion. **Tenascin** (also called **cytotactin**) resembles fibronectin for about half the length of the molecule, and it is found transiently in several extracellular matrices during embryonic development. However, different cells react in different ways to tenascin. Some cells stick to it, while other cells round up and detach from tenascin (Figure 3.37; Spring et al., 1989). Different relative amounts of fibronectin and tenascin may be able to generate substrates of various degrees of adhesiveness. In addition, tenascin appears to increase the synthesis and secretion of proteases from the cells that are on it (Werb et al., 1990). Both these characteristics may be important in generating paths for cell migration and in remodeling the extracellular matrix during development (Tan et al., 1987; Bronner-Fraser, 1988; Wehrle and Chiquet, 1990).

Cell receptors for the extracellular matrix molecules: Integrins

The ability of a cell to bind these adhesive glycoproteins depends on its expressing a cell membrane receptor for the "cell-binding site" of these large molecules. The main fibronectin receptors have been purified by using monoclonal antibodies that block the attachment of cells to fibronectin (Chen et al., 1985; Knudsen et al., 1985). The fibronectin receptor complex was found not only to bind fibronectin on the *outside* of the cell, but also to bind cytoskeletal proteins on the *inside* of the cell. Thus the fibronectin receptor complex appears to span the cell membrane and unite two types of matrices. On the inside of the cell, it serves as the anchorage site for the actin microfilaments that move the cell; meanwhile, on the outside of the cell, it binds to the fibronectin of the extracellular matrix (Figure 3.38). Horwitz and co-workers (1986; Tamkun et al., 1986) have called this family of receptor proteins **integrins** because they integrate the extracellular and intracellular scaffolds, allowing them to work together. Integrin proteins have been found to span the cell membrane of numerous cell types. On the extracellular side, integrin binds to the sequence arginine-glycine-aspartate (RGD) of several adhesive proteins in extracellular matrices, including vitronectin (found in the basal lamina of the eye), fibronectin, and laminin (Ruoslahti and Pierschbacher, 1987). On the cytoplasmic side, integrin binds to **talin** and **α-actinin,** two proteins that connect to actin microfilaments. This dual binding enables the cell to move by contracting the actin microfilaments against the fixed extracellular matrix (see Wang et al., 1993). Different cell types can have different types of integrin molecules with different affinities for extracellular matrix molecules (Hemler et al., 1987; Hemler, 1990). Each integrin protein has two distinct subunits, α and β, and different binary combinations of α and β subunits enable the integrin to bind to particular extracellular molecules. For instance, α2β1 binds to collagen and laminin, while α4β1 binds solely to fibronectin.

While both α and β subunits of integrin are needed for binding to fibronectin or laminin, only the β subunit connects with the internal cytoskeleton. During migration, the bonds linking the β subunit of integrin to the cytoskeleton can be continually made and broken by a protease that cleaves talin and is specifically localized to the cell membrane sites where integrin binds to the substratum. It is possible that this protease could

(A)

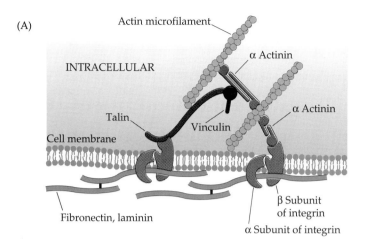

Actin microfilament

INTRACELLULAR

α Actinin

α Actinin

Talin

Vinculin

Cell membrane

β Subunit
of integrin

Fibronectin, laminin

α Subunit of integrin

(B)

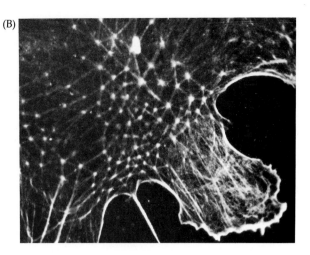

cleave the bridge between the fibronectin receptor and the cytoskeleton (Beckerle et al., 1987).

The importance of integrins for morphogenesis is dramatically illustrated during *Drosophila* embryogenesis. Like vertebrate integrins, *Drosophila* integrins are composed of α and β subunits that span the cell membrane. In the two known *Drosophila* integrins, the β subunits are identical, but the α subunits differ. These two integrins often work together to effect tissue and cell adhesion during development. In the development of the *Drosophila* wing, two epithelial sheets are brought together. The PS1 integrin is found on the basal surface of the presumptive *dorsal* wing epithelium, while the PS2 integrin is on the upper surface of the presumptive *ventral* wing epithelium. During metamorphosis, these two epithelia meet and adhere to form the two-layered wing blade. Mutations in the integrins cause the wing to have regions where the two wing epithelia come apart, as evidenced by bubbles between the two blades (Brower and Jaffe, 1989; Wilcox et al., 1989). Some of the mutants of *Drosophila* integrins are lethal, since integrin is needed to attach muscles to the epidermis and the gut wall. In the lethal mutation *lethal (1) myospheroid*, there is a deficiency in the genes encoding the β subunit of the *Drosophila* integrins. In the absence of this subunit, neither integrin is formed. The somatic muscles are contracted into spheres that lack attachments to the body wall or gut (Leptin et al., 1989).

Integrins are not the only molecules capable of binding to laminin and fibronectin. While the integrin receptor binds to an RGD sequence in the A-chain of laminin, another laminin receptor protein in the cell membrane binds to a different sequence (YIGSR) in the B1 chain (Graf et al., 1987; Yow et al., 1988). The receptors have different affinities for laminin, and these might be important in their function (Horwitz et al., 1985). The α3β1 integrin of fibroblasts, for example, has a relatively low affinity for laminin ($K^{\mathbf{d}} = 10^{-6}\ M$), whereas the affinity for laminin of a laminin receptor from epithelial cells is much higher ($K^{\mathbf{d}} = 2 \times 10^{-9}M$). The receptor that is used may be important in allowing the cells to use laminin either as a basement membrane (in which case the affinity of the receptor would be high) or as a substrate for migration (in which lower-affinity receptors would be used).

Cell receptors for the extracellular matrix molecules: Glycosyltransferases

Another set of proteins that can adhere cells to extracellular matrix proteins are the cell surface **glycosyltransferases.** These membrane-bound enzymes

FIGURE 3.38
Dual function of integrin in binding to extracellular matrices and internal cytoskeleton. (A) Speculative diagram relating the binding of cytoskeleton to the extracellular matrix through the integrin molecule. (B) Indirect immunofluorescence staining the actin microfilaments of a cell extending a lamellipodium. The actin fibers radiate from the ordered cytoskeletal lattice into the lamellipodium. (A after Luna and Hitt, 1992; B from Lazarides, 1976, courtesy of E. Lazarides.)

FIGURE 3.39
Cell surface interactions through glycosyltransferases. (A) The standard glycosyltransferase reaction, in which a sugar is transferred from a nucleoside diphosphate carrier to an acceptor. (B) Interaction between glycosyltransferases and carbohydrate group (acceptor) on extracellular matrix glycoprotein. If the activated sugar is absent, adhesion results (this is thought to occur during fertilization). If the activated sugar is present in low amounts, migration is permitted. (Modified from Pierce et al., 1980.)

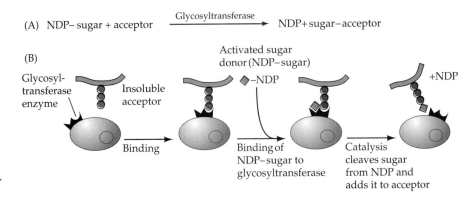

(A) NDP–sugar + acceptor $\xrightarrow{\text{Glycosyltransferase}}$ NDP+sugar–acceptor

(B)
Glycosyltransferase enzyme — Insoluble acceptor
Activated sugar donor (NDP–sugar)
◆–NDP
+NDP

Binding

Binding of NDP–sugar to glycosyltransferase

Catalysis cleaves sugar from NDP and adds it to acceptor

are routinely found in the endoplasmic reticulum and Golgi vesicles, where they are responsible for adding sugar residues onto peptides to make glycoproteins. There are numerous glycosyltransferases, each one specific for a given sugar and some showing substrate specificity as well. Thus, a *galactosyl*transferase is an enzyme capable of transferring *galactose* from an activated donor molecule (UDP-galactose) to an acceptor. There may be several galactosyltransferases with affinities for different acceptor molecules.

Galactosyltransferases are functional cell membrane enzymes, and their adhesion to the extracellular matrix represents a "frustrated" catalysis (Figure 3.39). The enzyme needs two substrates for catalysis to occur, the acceptor carbohydrate and the activated sugar. The membrane glycosyltransferases recognize the carbohydrate acceptors on the extracellular matrix proteins such as laminin. This causes adhesion to occur. When the second substrate appears, these adhesions can be broken by the catalysis. In some instances (such as fertilization in mice, where galactosyltransferase on the sperm cell membrane interacts with carbohydrate components of the extracellular matrix secreted by the egg), adhesion is critical and no catalysis occurs. In migrating cells, both adhesion and catalysis are seen (Figure 3.40; Toole, 1976; Shur, 1977a,b; Turley and Roth, 1979; Eckstein and Shur, 1989).

Two predictions from this model are that (1) the addition of activated sugars should perturb normal development and (2) migrating cells should leave new sugar groups on the substrates over which they migrate. The addition of large concentrations of sugar nucleotides to developing embryos does interfere with normal development (Shur, 1977a,b; Shur et al., 1979). Although the direct cause of these abnormalities has not been ascertained, the simplest explanation is that the sugar nucleotides interfere with the cell surface glycosyltransferases. Turley and Roth (1979) and Runyan and co-workers (1986) have shown that certain migrating cells (such as neural crest cells) can use their glycosyltransferases to bind to glycoprotein components in natural basal laminae.

FIGURE 3.40
Glycosyltransferases on migrating cells. Autoradiograph of a 10-somite chick embryo incubated with UDP-[^{3}H]galactose, or sugar donor. Insoluble radioactivity (black grains) indicates that this radioactive sugar was transferred by the surfaces of migrating cells to other cells or the matrices around them. (From Shur, 1977a, courtesy of B. D. Shur.)

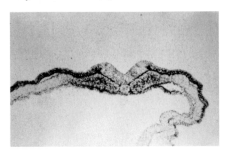

Differential adhesion resulting from multiple adhesion systems

Although we have been discussing these adhesion systems as separate units, it is probable that the morphogenetic processes of cell–cell interaction are brought about by combinations of cell adhesion molecules. For example, the initial attachment of the mouse embryo to the uterine wall appears to be mediated through several adhesion systems. First, the outer cells of the embryo (the trophoblastic cells) have receptors for the collagen and heparan sulfate proteoglycans of the uterine endometrium, and interference with this binding can impede implantation (Farach et al., 1987;

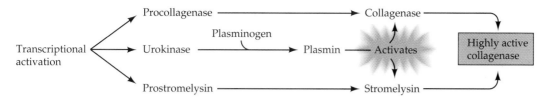

FIGURE 3.41
Cascade of membrane metalloproteinase activation. Urokinase is a plasminogen activator, cleaving plasminogen into plasmin. Plasmin activates the precursor forms of the stromelysins and collagenases to provide a highly active mix of enzymes capable of digesting extracellular matrices. (After Matrisian, 1992.)

Carson et al., 1988, 1993). Second, Dutt and co-workers (1987) have shown that the trophoblast cells can also adhere to uterine cells through cell surface glycosyltransferases. Third, Kadokawa and co-workers (1989) have shown P- and E-cadherin on both the trophoblast and the uterine tissue at the site of implantation. Thus, cells may have many adhesive systems that enable them to bind and/or migrate over specific substrates.

Interestingly, some adhesion can be effected by digesting matrices. For example, when mammalian embryos embed in the uterus, they digest their way through the uterine epithelium and through its basement membrane of laminin, fibronectin, and type IV collagen (Behrendtsen et al., 1992). Bone growth, tadpole tail regression, and branched organ formation (such as in the salivary gland, kidneys, and lung) also require the breakdown of basement membranes. This controlled degradation of extracellular matrix molecules is accomplished by a set of enzymes collectively called **matrix-degrading metalloproteinases** (Matrisian, 1992). These enzymes are secreted directly from the cells onto the matrix to be dissolved and they include (1) collagenases that digest types I, II, and III collagens; (2) gelatinases that digest elastin and collagens IV and V; and (3) stromelysins that digest proteoglycans, fibronectin, and laminin. The activation of these genes is accomplished coordinately, and several members of the metalloproteinases interact to amplify the intensity of the digestive enzymes (Figure 3.41). Soon after the metalloproteinases are activated, the cells activate the genes for inhibitors of these proteins. The controlled production and degradation of extracellular matrices is an essential part of normal development.

In 1782 the French essayist Denis Diderot posed the question of morphogenesis in the fevered dream of a noted physicist. This character could imagine that the body was formed from myriads of "tiny sensitive bodies" that collected together to form an aggregate, but he could not envision how this aggregate could become an animal. Recent studies have shown that this ordering is due to the molecules on the surfaces of these cells. In subsequent chapters we will look more closely at some of these morphogenetic interactions. We are now at the stage where we can begin our study of early embryogenesis and see the integration of the organismal, genetic, and cellular processes of animal development.

LITERATURE CITED

Abercrombie, M. and Ambrose, E. J. 1958. Interference microscope studies of cell contacts in tissue culture. *Exp. Cell Res.* 15: 332–345.

Amagai, M., Klaus-Kovtun, V. and Stanley, J. R. 1991. Autoantibodies against a novel epithelial cadherin in pemphigus vulgaris, a disease of cell adhesion. *Cell* 67: 869–877.

Armstrong, P. B. 1985. The control of cell motility during embryogenesis. *Cancer Metas. Rev.* 4: 59–80.

Armstrong, P. B. 1989. Cell sorting out: The self-assembly of tissues in vitro. *CRC Crit. Rev. Biochem. Mol. Biol.* 24: 119–149.

Bard, J. 1990. *Morphogenesis.* Cambridge University Press, New York.

Bastiani, M. J., Doe, C. Q., Helfand, S. L. and Goodman, C. S. 1985. Neuronal specificity and growth cone guidance in the grasshopper and *Drosophila* embryos. *Trends Neurosci.* 8: 257–266.

Bastiani, M. J., Harrelson, A. L., Snow, P. M. and Goodman, C. S. 1987. Expression of fasciclin I and II glycoproteins on subsets of axon pathways during neuronal development in the grasshopper. *Cell* 48: 745–755.

Beckerle, M. C., Burridge, K., DeMartino, G. N. and Croall, D. E. 1987. Colocalization of calcium-dependent protease II and one of its substrates and sites of adhesion. *Cell* 51: 569–577.

Behrendsten, O., Alexander, C. M. and Werb, Z. 1992. Metalloproteinases mediate extracellular matrix degradation by cells from mouse blastocyst outgrowths. *Development* 114: 447–456.

Bernfield, M. and Sanderson, D. 1990. Syndecan, a morphogenetically regulated cell surface proteoglycan that binds extracellular matrix and growth factors. *Philos. Trans. R. Soc. Lond.* [A] 327: 171–186.

Beug, H., Gerisch, G., Kempff, S., Riedel, V. and Cremer, G. 1970. Specific inhibition of cell contact formation in *Dictyostelium* by univalent antibodies. *Exp. Cell Res.* 63: 147–158.

Bevilacqua, A., Loch-Caruso, R. and Erickson, R. P. 1989. Abnormal development and dye coupling produced by antisense RNA to gap junction protein in mouse preimplantation embryos. *Proc. Natl. Acad. Sci. USA* 86: 5444–5448.

Blaschuk, O. W., Sullivan, R., David, S. and Pouliot, Y. 1990. Identification of a cadherin cell adhesion recognition sequence. *Dev. Biol.* 130: 227–229.

Boucaut, J. C. 1974. Étude autoradiographique de la distribution de cellules embryonnaires isolées, transplantées dans le blastocèle chez *Pleurodeles waltii* Michah (Amphibien, Urodele). *Ann. Embryol. Morphol.* 7: 7–50.

Boyse, E. A. and Old, L. J. 1969. Some aspects of normal and abnormal cell surface genetics. *Annu. Rev. Genet.* 3: 269–289.

Brackenbury, R., Thiery, J.-P., Rutishauser, U. and Edelman, G. M. 1977. Adhesion among neural cells of the chick embryo. I. Immunological assay for molecules involved in cell–cell binding. *J. Biol. Chem.* 252: 6835–6840.

Bronner-Fraser, M. 1988. Distribution of tenascin during cranial neural crest development in the chick. *J. Neurosci. Res.* 21: 135–147.

Brower, D. L. and Jaffe, S. M. 1989. Requirement for integrins during *Drosophila* wing development. *Nature* 342: 285–287.

Carson, D. D., Tang, J.-P. and Gay, S. 1988. Collagens support embryo attachment and outgrowth in vitro: Effects of the Arg-Gly-Asp sequence. *Dev. Biol.* 127: 368–375.

Carson, D. D., Tang, J.-P. and Julian, J. 1993. Heparan sulfate proteoglycan (perlecan) expression by mouse embryos during acquisition of attachment competence. *Dev. Biol.* 155: 97–106.

Carter, S. B. 1967. Haptotaxis and the mechanism of cell motility. *Nature* 213: 256–260.

Champion, S., Imhof, B. A., Savagnier, P. and Thiery, J.-P. 1986. The embryonic thymus produces chemotactic peptides involved in the homing of hemopoietic precursors. *Cell* 44: 781–790.

Chen, W. T., Hasegawa, E., Hasegawa, T., Weinstock, C. and Yamada, K. M. 1985. Development of cell-surface linkage complexes in cultured fibroblasts. *J. Cell Biol.* 100: 1103–1114.

Cheney, C. M. and Lash, J. W. 1981. Diversification within embryonic chick somites: Differential response to notochord. *Dev. Biol.* 81: 288–298.

Chuong, C.-M. and Edelman, G. M. 1985a. Expression of cell adhesion molecules in embryonic induction. I. Morphogenesis of nestling feathers. *J. Cell Biol.* 101: 1009–1026.

Chuong, C.-M. and Edelman, G. M. 1985b. Expression of cell adhesion molecule in embryonic induction. II. Morphogenesis of adult feathers. *J. Cell Biol.* 101: 1027–1043.

Covault, J. and Sanes, J. R. 1986. Distribution of N-CAM in synaptic and extrasynaptic portions of developing and adult skeletal muscle. *J. Cell Biol.* 102: 716–730.

Crawford, K. and Stocum, D. L. 1988. Retinoic acid coordinately proximalizes regenerate pattern and blastema differential affinity in axolotl limbs. *Development* 102: 687–698.

Cunningham, B. A., Hoffman, S., Rutishauser, U., Hemperly, J. J. and Edelman, G. M. 1983. Molecular topography of the neural cell adhesion molecule N-CAM: Surface orientation and location of sialic acid-rich and binding regions. *Proc. Natl. Acad. Sci. USA* 80: 3116–3120.

Curtis, A. S. G. 1969. The measurement of cell adhesiveness by an absolute method. *J. Embryol. Exp. Morphol.* 22: 305–325.

Dargemont, C. and nine others. 1989. Thymotaxin, a chemotactic protein, is identical to β_2-microglobulin. *Science* 246: 803–806.

Darnell, J., Lodish, H. and Baltimore, D. 1986. *Molecular Cell Biology.* Scientific American Books, New York.

Detrick, R. J., Dickey, D. and Kintner, C. R. 1990. The effects of N-cadherin misexpression on morphogenesis in *Xenopus* embryos. *Neuron* 4: 493–506.

Diderot, D. 1782. *D'Alembert's Dream.* Reprinted in J. Barzun and R. H. Bowen (eds.), *Rameau's Nephew and Other Works* (1956). Doubleday, Garden City, NY. [p. 114]

Dufour, S., Duband, J.-L., Humphries, M. J., Obara, M., Yamada, K. M. and Thiery, J. P. 1988. Attachment, spreading and locomotion of avian neural crest cells are mediated by multiple adhesion sites on fibronectin molecules. *EMBO J.* 7: 2661–2671.

Dutt, A., Tang, T.-P. and Carson, D. D. 1987. Lactosaminoglycans are involved in uterine epithelial cell adhesion in vitro. *Dev. Biol.* 119: 27–37.

Eckstein, D. J. and Shur, B. D. 1989. Laminin induces the stable expression of surface glycosyltransferases on lamellipodia of migrating cells. *J. Cell Biol.* 108: 2507–2517.

Edelman, G. M. and Thiery, J.-P. 1985. *The Cell in Contact: Adhesions and Junctions as Morphogenic Determinants.* Wiley, New York.

Erickson, C. A. and Nuccitelli, R. 1984. Embryonic fibroblast motility and orientation can be influenced by physiological electric fields. *J. Cell Biol.* 98: 296–307.

Erickson, C. A., Tosney, K. W. and Weston, J. A. 1980. Analysis of migratory behavior of neural crest and fibroblastic cells in embryonic tissues. *Dev. Biol.* 77: 142–156.

Farach, M. C., Tang, J. P., Decker, G. L. and Carson, D. D. 1987. Heparin/heparan sulfate is involved in attachment and spreading of mouse embryos in vitro. *Dev. Biol.* 123: 401–410.

Fink, R. and McClay, D. R. 1985. Three cell recognition changes accompany the ingression of sea urchin primary mesenchyme cells. *Dev. Biol.* 107: 66–74.

Fraser, S., E., Carhart, M. S., Murray, B. A., Chuong, C.-M. and Edelman, G. E. 1988. Alterations in the *Xenopus* retinotectal projection by antibodies to *Xenopus* N-CAM. *Dev. Biol.* 129: 217–230.

Fujimori, T., Miyatani, S. and Takeichi, M. 1990. Ectopic expression of N-cadherin perturbs histogenesis in *Xenopus* embryos. *Development* 110: 97–104.

Giudice, G. 1962. Restitution of whole larvae from disaggregated cells of sea urchin embryos. *Dev. Biol.* 5: 402–411.

Graf, J., Ogle, R. C., Robey, F. A., Sasaki, M., Martin, G. R., Yamada, Y. and Kleinman, H. K. 1987. A pentapeptide from the laminin B1 chain mediates cell adhesion and binds to the 67000 laminin receptor. *Biochemistry* 26: 6896–6900.

Hadler, N. M., Dourmash, R. R., Nermut, M. V. and Williams, L. D. 1982. Ultrastructure of a hyaluronic acid matrix. *Proc. Natl. Acad. Sci. USA* 79: 307–309.

Hakamori, S., Fukuda, M., Sekiguchi, K. and Carter, W. G. 1984. Fibronectin, laminin, and other extracellular glycoproteins. *In* K. A. Picz and A. H. Reddi (eds.), *Extracellular Matrix Biochemistry.* Elsevier, New York, pp. 229–275.

Hall, A. K. and Rutishauser, U. 1987. Visualization of neural cell adhesion molecule by electron microscopy. *J. Cell Biol.* 104: 1579–1586.

Harrelson, A. L. and Goodman, C. S. 1988. Growth cone guidance in insects: Fasciclin-II is a member of the immunoglobulin superfamily. *Science* 242: 700–708.

Harris, A. K. 1980. Silicone rubber substrate: A new wrinkle in the study of cell locomotion. *Science* 208: 176–179.

Harris, A. K. 1984. Cell traction and the generation of anatomical structure. *In* W. Jäger and J. D. Murray (eds.), *Modeling of Patterns in Time and Space.* Springer-Verlag, Berlin, pp. 103–122.

Harris, H. 1954. Role of chemotaxis in inflammation. *Physiol. Rev.* 34: 529–562.

Hatta, K. and Takeichi, M. 1986. Expression of N-cadherin adhesion molecules associated with early morphogenetic events in chick development. *Nature* 320: 447–449.

Hatta, K. Takagi, S., Fujisawa, H. and Takeichi, M. 1987. Spatial and temporal expression pattern of N-cadherin cell adhesion molecules correlated with morphogenetic processes of chicken embryos. *Dev. Biol.* 120: 215–227.

Heasman, J., Hines, R. D., Swan, A. P., Thomas, V. and Wylie, C. C. 1981. Primordial germ cells of *Xenopus* embryos: The role of fibronectin in their adhesion during migration. *Cell* 27: 437–447.

Hemler, M. E. 1990. VLA proteins in the integrin family: Structures, functions, and their role on leukocytes. *Annu. Rev. Immunol.* 8: 365–400.

Hemler, M. E., Huang, C. and Schwartz, L. 1987. The VLA protein family. Characterization of five distinct cell surface heterodimers each with a common 130,000 molecular weight β subunit. *J. Biol. Chem.* 262: 3300–3309.

Hirano, S., Kimoto, N., Shimoyama, Y., Hirohashi, S. and Takeichi, M. 1992. Identification of a neural α-catenin as a key regulator of cadherin function and multicellular organization. *Cell* 70: 293–301.

Hirokawa, N., Cheney, R. E. and Willard, M. 1983. Localization of a protein of the fodrin-spectrin-TW260/240 family on the mouse intestinal brush border. *Cell* 32: 953–965.

Hoffman, S. B. and Edelman, G. M. 1983. Kinetics of homophilic binding by E and A forms of the neural cell adhesion molecule. *Proc. Natl. Acad. Sci. USA* 80: 5762–5766.

Horwitz, A., Duggan, K., Greggs, R., Decker, C. and Buck, C. 1985. The cell substrate attachment (CSAT) antigen has properties of a receptor for laminin and fibronectin. *J. Cell Biol.* 101: 2134–2144.

Horwitz, A., Duggan, K., Buck, C., Beckerle, M. C. and Burridge, K. 1986. Interaction of plasma membrane fibronectin receptor with talin–α actinin transmembrane linkage. *Nature* 320: 531–533.

Hostikka, S. L. 1990. Human type IV collagen. *Acta Univ. Oulu.*

Hotary, K. B. and Robinson, K. R. 1992. Evidence of a role for endogenous electric fields in chick embryo development. *Development* 114: 985–996.

Jaffe, L. F. 1981. The role of ionic current in establishing developmental pattern. *Philos. Trans. R. Soc. Lond.* [B] 295: 553–566.

Jaffe, L. F. and Stern, C. D. 1979. Strong electrical currents leave the primitive streak of chick embryos. *Science* 206: 569–571.

Jongen, W, M. F. and seven others. 1991. Regulation of connexin 43-mediated gap junction intercellular communication by Ca^{2+} in mouse epidermal cells is controlled by E-cadherin. *J. Cell Biol.* 114: 545–555.

Just, E. E. 1939. *The Biology of the Cell Surface.* Blackiston, Philadelphia.

Kadokawa, Y., Fuketa, I., Nose, A., Takeichi, M. and Nakatsuji, N. 1989. Expression of E- and P-cadherin in mouse embryos and uteri during the periimplantation period. *Dev. Growth Diff.* 31: 23–30.

Kalimi, G. H. and Lo, C. 1988. Communication compartments in the gastrulating mouse embryo. *J. Cell Biol.* 107: 241–255.

Kelly, D. E. 1966. Fin structure of desmosomes, hemidesmosomes and an adepidermal globular layer in developing newt epidermis. *J. Cell Biol.* 28: 51–72.

Kintner, C. 1992. Regulation of embryonic cell adhesion by the cadherin cytoplasmic domain. *Cell* 69: 225–236.

Knudsen, K., Horwitz, A. F. and Buck, C. 1985. A monoclonal antibody identifies a glycoprotein complex involved in cell–substratum adhesion. *Exp. Cell Res.* 157: 218–226.

Köhler, G. and Milstein, C. 1975. Continuous cultures of fused cells secreting antibody of predefined specificity. *Nature* 256: 495–497.

Lander, A. D. 1989. Understanding the molecules of neural cell contacts: Emerging patterns of structure and function. *Trends Neurosci.* 12: 189–195.

Landmesser, L., Dahm, L., Schultz, K. and Rutishauser, U. 1988. Distinct roles for adhesion molecules during innervation of embryonic chick muscle. *Dev. Biol.* 130: 645–670.

Lazarides, E. 1976. Actin, α actinin, and tropomyosin interaction in the structural organization of actin filaments in nonmuscle cells. *J. Cell Biol.* 68: 202–219.

Lee, S., Gilula, N. B. and Warner, A. E. 1987. Gap junctional communication and compaction during preimplantation stages of mouse development. *Cell* 51: 851–860.

Leonard, C. M., Bergman, M., Frenz, D. A., Macreery, L. A. and Newman, S. A. 1989. Abnormal ambient glucose levels inhibit proteoglycan core protein gene expression and reduce proteoglycan accumulation during chondrogenesis: Possible mechanism for teratogenic effects of maternal diabetes. *Proc. Natl. Acad. Sci. USA* 86: 10113–10117.

Leptin, M., Bogaert, T., Lehmann, R. and Wilcox, M. 1989. The function of PS integrins during *Drosophila* embryogenesis. *Cell* 56: 401–408.

Linask, K. L. and Lash, J. W. 1988a. A role for fibronectin in the migration of avian precardiac cells. I. Dose-dependent effects of fibronectin antibody. *Dev. Biol.* 129: 315–323.

Linask, K. L. and Lash, J. W. 1988b. A role for fibronectin in the migration of avian precardiac cells. II. Rotation of the heart-forming region during different stages and its effects. *Dev. Biol.* 129: 324–329.

Lo, C. and Gilula, N. B. 1979. Gap junctional communication in the preimplantation mouse embryo. *Cell* 18: 399–409.

Luna, E. J. and Hitt, A. L. 1992. Cytoskeleton–plasma membrane interactions. *Science* 258: 955–964.

Massagué, J. 1991. A helping hand from proteoglycans. *Curr. Biol.* 1: 117–119.

Matrisian, L. M. 1992. The matrix-degrading metalloproteinases. *BioEssays* 14: 455–463.

McClay, D. R. and Ettensohn, C. A. 1987. Cell recognition during sea urchin gastrulation. *In* W. F. Loomis (ed.), *Genetic Regulation of Development.* Alan R. Liss, New York, pp. 111–128.

Monroy, A. and Moscona, A. A. 1979. *Introductory Concepts in Developmental Biology.* University of Chicago Press, Chicago.

Moscona, A. A. 1952. Cell suspension from organ rudiments of chick embryos. *Exp. Cell Res.* 3: 535–539.

Moscona, A. A. 1961. Rotation-mediated histogenetic aggregation of dissociated cells: A quantifiable approach to cell interaction in vitro. *Exp. Cell Res.* 22: 455–475.

Nagafuchi, A. and Takeichi, M. 1989. Transmembrane control of cadherin-mediated cell adhesion: A 94-kDa protein functionally associated with a specific region of the cytoplasmic domain of E-cadherin. *Cell. Reg.* 1: 37–44.

Nagafuchi, A., Shirayoshi, Y., Okazaki, K., Yasuda, K. and Takeichi, M. 1987. Transformation of cell adhesion properties of exogenously introduced E-cadherin cDNA. *Nature* 329: 341–343.

Nardi, J. B. and Stocum, D. L. 1983. Surface properties of regenerating limb cells: Evidence for gradation along the proximodistal axis. *Differentiation* 25: 27–31.

Nose, A. and Takeichi, M. 1986. A novel cadherin adhesion molecule: Its expression patterns associated with implantation and organogenesis of mouse embryos. *J. Cell Biol.* 103: 2649–2658.

Nose, A., Nagafuchi, A. and Takeichi, M. 1988. Expressed recombinant cadherins mediate cell sorting in model systems. *Cell* 54: 993–1001.

Nose, A., Tsuji, K. and Takeichi, M. 1990. Localization of specificity determining sites in cadherin cell adhesion molecules. *Cell* 61: 147–155.

Nuccitelli, R. 1984. The involvement of transcellular ion currents and electric fields in pattern formation. *In* G. M. Malacinski (ed.), *Pattern Formation: A Primer in Developmental Biology.* Macmillan, New York, pp. 23–46.

Orkin, R. W., Knudson, W. and Toole, P. T. 1985. Loss of hyaluronidate-dependent coat during myoblast fusion. *Dev. Biol.* 107: 527–530.

Ozawa, M., Ringwald, M. and Kemler, R. 1990. Uvomorulin–catenin complex formation is regulated by a specific domain in the cytoplasmic region of the cell adhesion molecule. *Proc. Natl. Acad. Sci. USA* 87: 4246–4250.

Peracchia, C. and Dulhunty, A. F. 1976. Low resistance junctions in crayfish: Structural changes with functional uncoupling. *J. Cell Biol.* 70: 419–439.

Pierce, M., Turley, E. A. and Roth, S. 1980. Cell surface glycosyltransferase activities. *Int. Rev. Cytol.* 65: 1–47.

Pierschbacher, M. D. and Ruoslahti, E. 1984. The cell attachment activity of fibronectin can be duplicated by small synthetic fragments of the molecule. *Nature* 309: 30–33.

Poole, T. J. and Steinberg, M. S. 1982. Evidence for the guidance of pronephric duct migration by a craniocaudally traveling adhesive gradient. *Dev. Biol.* 92: 144–158.

Ratner, N., Bunge, R. P. and Glaser, L. 1985. A neuronal cell surface heparan sulfate proteoglycan is required for dorsal root ganglion neuron stimulation of Schwann cell proliferation. *J. Cell Biol.* 101: 744–754.

Rieger, F., Grument, M. and Edelman, G. M. 1985. N-CAM at the vertebrate neuromuscular junction. *J. Cell Biol.* 101: 285–293.

Rosavio, R. A., Delouvée, A., Yamada, K. M., Timpl, R. and Thiery, J.-P. 1983. Neural crest cell migration: Requirements for exogenous fibronectin and high cell density. *J. Cell Biol.* 96: 462–473.

Roth, S. 1968. Studies on intracellular adhesive selectivity. *Dev. Biol.* 18: 602–631.

Roth, S., McGuire, E. J. and Roseman, S. 1971. An assay for intercellular adhesive specificity. *J. Cell Biol.* 51: 525–535.

Runyan, R. B., Maxwell, G. D. and Shur, B. D. 1986. Evidence for a novel enzymatic mechanism of neural crest cell migration on extracellular glycoconjugate matrices. *J. Cell Biol.* 102: 432–441.

Ruoslahti, E. and Pierschbacher, M. D. 1987. New perspectives in cell adhesion: RGD and integrins. *Science* 238: 491–497.

Rutishauser, U. 1990. Cell adhesion molecules. *In* S. Roth (ed.), *Molecular Approaches to Supracellular Phenomena.* University of Pennsylvania Press, Philadelphia, pp. 175–199.

Rutishauser, U. and Jessell, T. 1988. Cell adhesion molecules in vertebrate neural development. *Physiol. Rev.* 68: 819–857.

Rutishauser, U., Hoffman, S. and Edelman, G. M. 1982. Binding properties of a cell adhesion molecule from neural tissue. *Proc. Natl. Acad. Sci. USA* 79: 685–689.

Rutishauser, U., Acheson, A., Hall, A., Mann, D. M. and Sunshine, J. 1988. The neural cell adhesion molecule (N-CAM) as a regulator of cell–cell interactions. *Science* 240: 53–57.

Sainio, K., Gilbert, S. F., Lehtonen, E., Nishi, M., Kumar, N. M., Gilula, N. B. and Saxén, L. 1992. Differential expression of gap junction mRNAs and proteins in the developing murine kidney and in experimentally induced nephric mesenchymes. *Development* 115: 827–837.

San Antonio, J. D., Winston, B. M. and Tuan, R. S. 1987. Regulation of chondrogenesis by heparan sulfate and structurally related glycosaminoglycans. *Dev. Biol.* 123: 17–24.

Shur, B. D. 1977a. Cell surface glycosyltransferases in gastrulating chick embryos. I. Temporally and spatially specific patterns of four endogenous glycosyltransferase activities. *Dev. Biol.* 58: 23–29.

Shur, B. D. 1977b. Cell surface glycosyltransferases in gastrulating chick embryos. II. Biochemical evidence for a surface localization of endogenous glycosyltransferase activities. *Dev. Biol.* 58: 40–55.

Shur, B. D. 1982. Cell surface glycosyltransferase activities during normal and mutant (*T/T*) mesenchyme migration. *Dev. Biol.* 91: 149–162.

Shur, B. D., Oettgen, P. and Bennett, D. 1979. UDP galactose inhibits blastocyst formation in the mouse: Implications for the mode of action of *T/t*-complex mutations. *Dev. Biol.* 73: 178–181.

Singer, S. J. and Nicolson, G. L. 1972. The fluid mosaic model of the structure of cell membranes. *Science* 175: 720–731.

Spiegel, M. and Spiegel, E. S. 1975. Reaggregation of dissociated embryonic sea urchin cells. *Am. Zool.* 15: 583–606.

Spring, J., Beck, K. and Chiquet-Ehrismann, R. 1989. Two contrary functions of tenascin: Dissection of the active sites by recombinant tenascin fragments. *Cell* 59: 325–334.

Steinberg, M. S. 1964. The problem of adhesive selectivity in cellular interactions. *In* M. Locke (ed.), *Cellular Membranes in Development.* Academic Press, New York, pp. 321–434.

Steinberg, M. S. 1970. Does differential adhesion govern self–assembly processes in histogenesis? Equilibrium configurations and the emergence of a hierarchy among populations of embryonic cells. *J. Exp. Zool.* 173: 395–434.

Stopak, D. and Harris, A. K. 1982. Connective tissue morphogenesis by fibroblast traction. I. Tissue culture observations. *Dev. Biol.* 90: 383–398.

Takeichi, M. 1987. Cadherins: A molecular family essential for selective cell–cell adhesion and animal morphogenesis. *Trends Genet.* 3: 213–217.

Takeichi, M. 1991. Cadherin cell adhesion receptors as a morphogenetic regulator. *Science* 251: 1451–1455.

Takeichi, M., Ozaki, H. S., Tokunaga, K. and Okada, T. S. 1979. Experimental manipulation of cell surface to affect cellular recognition mechanisms. *Dev. Biol.* 70: 195–205.

Tamkun, J. W., DeSimone, D. W., Fonda, D., Patel, R. S., Buck, C, Horwitz, A. F. and Hynes, R. O. 1986. Structure of integrin, a glycoprotein involved in transmembrane linkage between fibronectin and actin. *Cell* 46: 271–282.

Tan, S.-S., Crossin, K. L., Hoffman, S. and Edelman, G. M. 1987. Asymmetric expression of somites of cytotactin and its proteoglycan ligand is correlated with neural crest cell migration. *Proc. Natl. Acad. Sci. USA* 84: 7977–7981.

Thesleff, I., Vainio, S. and Jalkanen, M. 1989. Cell–matrix interactions in tooth development. *Int. J. Dev. Biol.* 33: 91–95.

Thibaudeau, G., Drawbridge, J., Dollarhide, A., Haque, T. and Steinberg, M. 1993. Three populations of migrating amphibian embryonic cells utilize different guidance cues. *Dev. Biol.* 159: 657–668.

Thiery, J.-P., Brackenbury, R., Rutishauser, U. and Edelman, G. M. 1977. Adhesion among neural cells of the chick embryo. II. Purification and characterization of a cell adhesion molecule from neural retina. *J. Biol. Chem.* 252: 6841–6845.

Thiery, J.-P., Delouvée, A., Gallin, W. J., Cunningham, B. A. and Edelman, G. M. 1985. Initial appearance and regional distribution of the neuron–glia cell adhesion molecule of the chick embryo. *J. Cell Biol.* 100: 442–456.

Toole, B. P. 1976. Morphogenetic role of glycosaminoglycans (acid mucopolysaccharides) in brain and other tissues. *In* S. H. Barondes (ed.), *Neuronal Recognition.* Plenum, New York, pp. 276–329.

Tosney, K. W., Watanabe, M., Landmesser, L. and Rutishauser, U. 1986. The distribution of N-CAM in the chick hindlimb during axon outgrowth and synaptogenesis. *Dev. Biol.* 114: 468–481.

Townes, P. L. and Holtfreter, J. 1955. Directed movements and selective adhesion of embryonic amphibian cells. *J. Exp. Zool.* 128: 53–120.

Trinkaus, J. P. 1963. The cellular basis of *Fundulus* epiboly. Adhesivity of blastula and gastrula cells in culture. *Dev. Biol.* 7: 513–532.

Turley, E. A. and Roth, S. 1979. Spontaneous glycosylation of glycosaminoglycan substrates by adherent fibroblasts. *Cell* 17: 109–115.

Tyler, A. 1946. An auto-antibody concept of cell structure, growth, and differentiation. *Growth* 10 (Symposium 6): 7–19.

Vainio, S., Jalkanen, M., Lehtonen, E. and Bernfield, M. 1989. Epithelial–mesenchymal interactions regulate stage-specific expression of a cell surface proteoglycan, syndecan, in the development kidney. *Dev. Biol.* 134: 382–391.

Venkatesh, T. R., Zipursky, S. L. and Benzer, S. 1985. Molecular analysis of the development of the compound eye in *Drosophila. Trends Neurosci.* 8: 251–257.

Vleminckz, K., Vakaet, L., Jr., Mareel, M., Fiers, W. and Van Roy, F. 1991. Genetic manipulation of E-cadherin expression by epithelial tumor cells reveals an invasion suppressor role. *Cell* 66: 107–119.

Vuorio, E. 1986. Connective tissue diseases: Mutations of collagen genes. *Ann. Clin. Res.* 18: 234–241.

Wang, N., Butler, J. P. and Ingber, D. E. 1993. Mechanotransduction across the cell surface and through the cytoskeleton. *Science* 260: 1124–1127.

Warner, A. E., Guthrie, S. C. and Gilula, N. B. 1984. Antibodies to gap junctional protein selectively disrupt junctional communication in the early amphibian embryo. *Nature* 311: 127–131.

Wehrle, B. and Chiquet, M. 1990. Tenascin is accumulated along developing peripheral nerves and allows neurite outgrowth *in vitro. Development* 110: 401–415.

Weiss, P. 1934. In vitro experiments on the factors determining the course of the outgrowing nerve fiber. *J. Exp. Zool.* 68: 393–448.

Weiss, P. 1945. Experiments on cell and axon orientation in vitro: The role of colloidal exudates in tissue organization. *J. Exp. Zool.* 100: 353–386.

Weiss, P. 1947. The problem of specificity in growth and development. *Yale J. Biol. Med.* 19: 235–278.

Weiss, P. 1955. Nervous system. *In* B. H. Willier, P. Weiss and V. Hamburger (eds.), *Analysis of Development.* Saunders, Philadelphia, pp. 346–401.

Werb, Z., Tremble, P. and Damsky, C. H. 1990. Regulation of extracellular matrix degradation by cell–extracellular matrix interaction. *Cell Differ. Dev.* 32: 299–306.

Wheeler, G. N. and eleven others. 1991. Desmosomal glycoprotein DGI, a component of intercellular desmosome junctions, is related to the cadherin family of cell adhesion molecules. *Proc. Natl. Acad. Sci. USA* 88: 4796–4800.

Wilcox, M., DiAntonio, A. and Leptin, M. 1989. The functions of the PS integrins in *Drosophila* wing morphogenesis. *Development* 107: 891–897.

Williams, A. F. and Barclay, A. N. 1988. The immunoglobulin superfamily: Domains for cell surface recognition. *Annu. Rev. Immunol.* 6: 381–405.

Wolfe, S. L. 1993. *Molecular and Cellular Biology.* Wadsworth, Belmont, CA.

Wood, A. and Thorogood, P. 1987. An ultrastructural and morphometric analysis of an in vivo contact guidance system. *Development* 101: 363–381.

Yayon, A., Klagsbrun, M., Esko, J. D., Leder, P. and Ornitz, D. M. 1991. Cell surface heparin-like molecules are required for binding of basic fibroblast growth factor to its high affinity receptor. *Cell* 64: 841–849.

Yelton, D. E. and Scharff, M. D. 1980. Monoclonal antibodies. *Am. Sci.* 68: 510–516.

Yow, H., Wong, J. M., Chen, H. S., Lee, C., Steel, G. D. Jr. and Chen, L. B. 1988. Increased mRNA expression of a laminin-binding protein in human colon carcinoma: Complete sequence of a full-length cDNA encoding the protein. *Proc. Natl. Acad. Sci. USA* 85: 6394–6398.

Zackson, S. L. and Steinberg, M. S. 1987. Chemotaxis or adhesion gradient? Pronephric duct elongation does not depend on distant sources of guidance information. *Dev. Biol.* 124: 418–422.

Zipursky, S. L., Venkatesh, T. R., Teplow, D. B. and Benzer, S. 1984. Neuronal development in the *Drosophila* retina: Monoclonal antibodies as molecular probes. *Cell* 36: 15–26.

II

PATTERNS OF DEVELOPMENT

4

Fertilization

Beginning a new organism

Urge and urge and urge,
Always the procreant urge
of the world.
Out of the dimness opposite equals
advance,
Always substance and increase,
always sex,
Always a knit of identity,
always distinction,
Always a breed of life.
WALT WHITMAN (1855)

The final aim of all love intrigues,
be they comic or tragic, is really of
more importance than all other
ends in human life. What it turns
upon is nothing less than the
composition of the next generation.
A. SCHOPENHAUER
(QUOTED BY C. DARWIN, 1871)

Fertilization is the process whereby two sex cells (gametes) fuse together to create a new individual with genetic potentials derived from both parents. Fertilization, then, accomplishes two separate activities: sex (the combining of genes derived from the two parents) and reproduction (the creation of new organisms). Thus, the first function of fertilization involves the transmission of genes from parent to offspring, and the second function of fertilization is to initiate in the egg cytoplasm those reactions that permit development to proceed.

Although the actual details of fertilization vary enormously from species to species, the events of conception generally consist of four major activities:

- *Contact and recognition between sperm and egg.* In most instances, this insures that the sperm and egg are of the same species.
- *Regulation of sperm entry into the egg.* Only one sperm can ultimately fertilize the egg. This is usually accomplished by allowing only one sperm to enter the egg and inhibiting any others from entering.
- *Fusion of the genetic material of sperm and egg.*
- *Activation of egg metabolism to start development.*

Structure of the gametes

A complex dialogue exists between egg and sperm. The egg is able to activate the sperm metabolism that is essential for fertilization, and the sperm reciprocates by activating the egg metabolism needed for the onset of development. But before proceeding to investigate each of these features of fertilization, we first discuss the structures of the sperm and egg—the two cell types specialized for fertilization.

Sperm

It is only within the past century that the sperm's role in fertilization has been known. Anton van Leeuwenhoek, the Dutch microscopist who co-discovered sperm in 1678, first believed them to be parasitic animals living within the semen (hence the term *spermatozoa*, meaning "sperm animals"). He originally assumed they had nothing at all to do with reproducing the

FIGURE 4.1
The human infant preformed in the sperm as depicted by Nicolas Hartsoeker (1694).

organism in which they were found, but he later came to believe that each sperm contained a preformed embryo. Leeuwenhoek (1685) wrote that sperm were seeds (both *sperma* and *semen* mean "seed") and that the female merely provided the nutrient soil into which the seeds were planted. In this, he was returning to a notion of procreation promulgated by Aristotle 2000 years earlier. Try as he could, Leeuwenhoek was continually disappointed in his inability to find the preformed embryo within the spermatozoa. Nicolas Hartsoeker, the other co-discoverer of sperm, drew what he also hoped to find: a preformed human ("homunculus") within the human sperm (Figure 4.1). This belief that the sperm contained the entire embryonic organism never gained much acceptance as it implied an enormous waste of potential life. Most investigators considered the spermatozoa to be unimportant.

The first evidence suggesting the importance of sperm in reproduction came from a series of experiments performed by Lazzaro Spallanzani in the late 1700s. Spallanzani demonstrated that filtered toad semen devoid of sperm would not fertilize eggs. He concluded, however, that the viscous fluid retained by the filter paper, and not the sperm, was the agent of fertilization. He, too, felt that the spermatic animals were clearly parasitic.

The combination of better microscopic lenses and the cell theory led to a new appreciation of spermatic function. In 1824, J. L. Prevost and J. B. Dumas claimed that sperm were not parasites but rather were active agents of fertilization. They noted the universal existence of sperm in sexually mature males and their absence in immature and aged individuals. These observations, coupled with the known absence of spermatozoa in the sterile mule, convinced them that "there exists an intimate relation between their presence in the organs and the fecundating capacity of the animal." They proposed that the sperm actually entered the egg and contributed materially to the next generation.

These claims were largely disregarded until the 1840s, when A. von Kolliker described the formation of sperm from cells within the adult testes. He ridiculed the idea that the semen could be normal and yet support an enormous number of parasites. Even so, von Kolliker denied that there was any physical contact between sperm and egg. He believed that the sperm excited the egg to develop, much as a magnet communicates its presence to iron. It was only in 1876 that Oscar Hertwig and Herman Fol independently demonstrated sperm entry into the egg and the union of those cells' nuclei. Hertwig had sought an organism suitable for detailed microscopic observations, and he found that the Mediterranean sea urchin, *Toxopneustes lividus*, was perfect. Not only was it common throughout the region and sexually mature throughout most of the year, but its eggs were available in large numbers and were transparent even at high magnifications. After mixing sperm and egg suspensions together, Hertwig repeatedly observed a sperm entering an egg and saw the two nuclei unite. He also noted that only one sperm was seen to enter each egg and that all the nuclei of the embryo were derived from the fused nucleus created at fertilization. Fol made similar observations and detailed the mechanism of sperm entry. Fertilization was at last recognized as the union of sperm and egg, and the union of sea urchin gametes remains the best-studied example of fertilization.

Each sperm is known to consist of a haploid nucleus, a propulsion system to move the nucleus, and a sac of enzymes that enable the nucleus to enter the egg. Most of the cytoplasm of the sperm has been eliminated during the sperm's maturation, leaving only certain organelles that have been modified for their spermatic function (Figure 4.2). During the course of sperm maturation, the haploid nucleus becomes very streamlined and

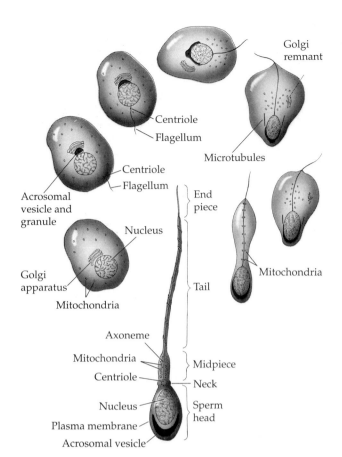

FIGURE 4.2
The modification of a germ cell to form a mammalian sperm. The centriole produces a long flagellum at what will be the posterior end of the sperm, and the Golgi apparatus forms the acrosomal vesicle at the future anterior end. The mitochondria (hollow dots) collect about the flagellum near the base of the haploid nucleus and become incorporated into the midpiece of the sperm. The remaining cytoplasm is jettisoned, and the nucleus condenses. The size of the mature sperm has been enlarged relative to the other figures. (After Clermont and Leblond, 1955.)

its DNA becomes tightly compressed. In front of this compressed haploid nucleus lies the **acrosomal vesicle,** which is derived from the Golgi apparatus and contains enzymes that digest proteins and complex sugars. Thus, it can be considered to be a modified lysosome. These stored enzymes are used to lyse the outer coverings of the egg. In many species, such as sea urchins, a region of globular actin molecules lies between the nucleus and the acrosomal vesicle. These proteins are used to extend a fingerlike process during the early stages of fertilization. In sea urchins and several other species, recognition between sperm and egg involves molecules on this **acrosomal process.** Together, the acrosome and nucleus constitute the head of the sperm.

The means by which sperm are propelled varies according to how the species has adapted to environmental conditions. In some species (such as the parasitic roundworm *Ascaris*), the sperm travel by the amoeboid motion of lamellipodia—local extensions of the cell membrane. In most species, however, each sperm is able to travel long distances by whipping its **flagellum.**

Flagella are complex structures. The major motor portion of the flagellum is called the **axoneme.** An axoneme is formed by the microtubules emanating from the centriole at the base of the sperm nucleus (Figures 4.2 and 4.3). The core of the axoneme consists of two central microtubules surrounded by a row of nine doublet microtubules. Actually, only one microtubule of each doublet is complete, having 13 protofilaments; the other is C-shaped and has only 11 protofilaments (Figure 4.3C). A three-dimensional model of a complete microtubule is shown in Figure 4.3B. Here one can see the 13 interconnected protofilaments, which are made exclusively of the dimeric protein tubulin.

FIGURE 4.3

The motile apparatus of the sperm. (A) Cross section of the flagellum of a mammalian spermatozoon showing the central axoneme and the external fibers. (B) Interpretive diagram of the axoneme, showing the "9 + 2" arrangement of the microtubules and other flagellar components. The schematic diagram shows the association of tubulin protofilaments into a microtubule doublet. The first ("A") portion of the doublet is a normal microtubule comprising 13 protofilaments. The second ("B") portion of the doublet contains only 11 (occasionally 10) protofilaments. (C) A three-dimensional model of the "A" microtubule. The α and β-tubulin subunits are similar but not identical, and the microtubule can change size by polymerizing or depolymerizing tubulin subunits at either end. (A courtesy of D. M. Phillips; B after De Robertis et al., 1975, and Tilney et al., 1973; C from Amos and Klug, 1974, courtesy of the authors.)

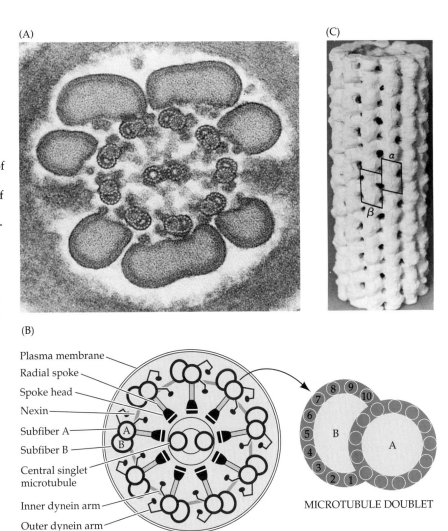

(A)

(C)

(B)

Plasma membrane
Radial spoke
Spoke head
Nexin
Subfiber A
Subfiber B
Central singlet microtubule
Inner dynein arm
Outer dynein arm

AXONEME

MICROTUBULE DOUBLET

Although tubulin is the basis for the structure of the flagellum, other proteins are also critical for flagellar function. The force for sperm propulsion is provided by **dynein,** a protein that is attached to the microtubules (Figure 4.3B). Dynein hydrolyzes molecules of ATP and can convert the released chemical energy into the mechanical energy that propels the sperm (Ogawa et al., 1977). The importance of dynein can be seen in those individuals with the genetic syndrome called Kartagener's triad. These individuals lack dynein on all their ciliated and flagellated cells; therefore, these structures are rendered immotile. Males with this disease are sterile (immotile sperm), are susceptible to bronchial infections (immotile respiratory cilia), and have a 50 percent chance of having their heart on the right-hand side of their body (Afzelius, 1976). Another important flagellar protein appears to be histone H1. This protein is usually seen inside the nucleus, where it folds the chromatin into tight clusters. However, Multigner and his colleagues (1992) find that this same protein stabilizes the flagellar microtubules so that they do not disassemble.

The "9 + 2" microtubule arrangement with the dynein arms has been conserved in axonemes throughout the eukaryotic kingdoms, suggesting that this arrangement is extremely well suited for transmitting energy for movement. The energy to whip the flagellum and thereby propel the sperm comes from rings of mitochondria located in the neck region of the sperm (Figure 4.2). In many species (notably mammals), a layer of dense fibers has interposed itself between the mitochondrial sheath and axoneme. This fiber layer stiffens the sperm tail. Because the thickness of this layer decreases toward the tip, the fibers probably prevent the sperm head from being whipped around too suddenly. Thus, the sperm has undergone extensive modification for the transmission of its nucleus to the egg.

The egg

All the material necessary for the beginning of growth and development must be stored in the mature egg (the **ovum**). Therefore, whereas the sperm has eliminated most of its cytoplasm, the developing egg (called the **oocyte** before it is haploid) not only conserves its material but is actively involved in accumulating more. It either synthesizes or absorbs proteins, such as yolk, that act as food reservoirs for the developing embryo. Thus, birds' eggs are enormous single cells that have become swollen with their accumulated yolk. Even eggs with relatively sparse yolk are comparatively large. The volume of a sea urchin egg is about 2×10^5 μm^3, more than 10,000 times the volume of the sperm. The diagram of the sea urchin egg and sperm in Figure 4.4 shows their relative sizes. The figure also shows the various components of the mature egg. So, while sperm and egg have equal haploid nuclear components, the egg also has a remarkable cytoplasmic storehouse that it has accumulated during its maturation. This cytoplasmic trove includes proteins, ribosomes, transfer RNA (tRNA), mRNA, and morphogenetic factors.

- *Proteins.* It will be a long while before the embryo is able to feed itself or obtain food from its mother. The early embryonic cells need some storable supply of energy and amino acids. In many species, this is accomplished by accumulating yolk proteins in the egg. Many of the yolk proteins are made in other organs (liver, fat body) and travel through the maternal blood to the egg.
- *Ribosomes and tRNA.* As we will soon see, there is a burst of protein synthesis soon after fertilization. This protein synthesis is accom-

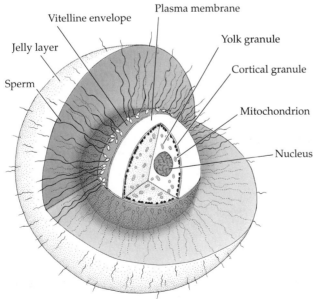

FIGURE 4.4
Structure of the sea urchin egg during fertilization. (After Epel, 1977.)

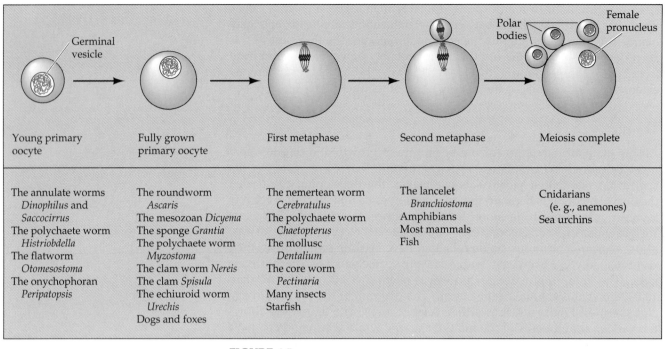

Young primary oocyte	Fully grown primary oocyte	First metaphase	Second metaphase	Meiosis complete
The annulate worms *Dinophilus* and *Saccocirrus* The polychaete worm *Histriobdella* The flatworm *Otomesostoma* The onychophoran *Peripatopsis*	The roundworm *Ascaris* The mesozoan *Dicyema* The sponge *Grantia* The polychaete worm *Myzostoma* The clam worm *Nereis* The clam *Spisula* The echiuroid worm *Urechis* Dogs and foxes	The nemertean worm *Cerebratulus* The polychaete worm *Chaetopterus* The mollusc *Dentalium* The core worm *Pectinaria* Many insects Starfish	The lancelet *Branchiostoma* Amphibians Most mammals Fish	Cnidarians (e. g., anemones) Sea urchins

FIGURE 4.5
Stages of egg maturation at the time of sperm entry in different animals. (After Austin, 1965.)

FIGURE 4.6
The sea urchin egg cell surface. (A) Scanning electron micrograph of an egg before fertilization. The plasma membrane is exposed where the vitelline envelope has been torn. (B) Transmission electron micrograph of an unfertilized egg showing microvilli and plasma membrane, which are closely covered by the vitelline envelope. A cortical granule lies directly beneath the plasma membrane of the egg. (From Schroeder, 1979, courtesy of T. E. Schroeder.)

plished by ribosomes and tRNA, which pre-exist in the egg. The developing egg has special mechanisms to synthesize ribosomes, and certain amphibian oocytes produce as many as 10^{12} ribosomes during their meiotic prophase.

- *Messenger RNA.* In most organisms, the messages for proteins made during early development are already packaged in the oocyte. It is estimated that the eggs of sea urchins contain 25,000 to 50,000 different types of mRNAs. This mRNA, however, remains dormant until after fertilization.

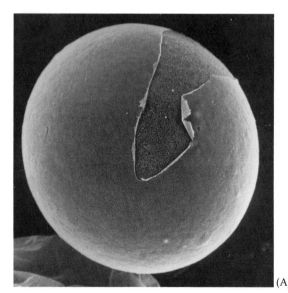

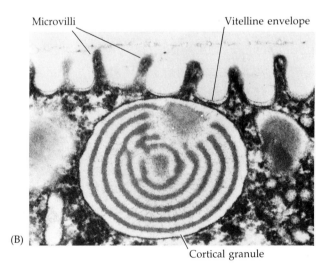

- *Morphogenetic factors.* These are molecules that direct the differentiation of cells into certain cell types. They appear to be localized in different regions of the egg and become segregated into different cells during cleavage (Chapter 14).

Within this enormous volume of cytoplasm resides a large nucleus. In some species (sea urchins, for example) the nucleus is already haploid at the time of fertilization. In other species (including many worms and most mammals) the egg nucleus is still diploid and the sperm enters before the meiotic divisions are completed. The stage of the egg nucleus at the time of sperm entry is illustrated in Figure 4.5.

Enclosing the cytoplasm is the egg **plasma membrane.** This membrane must regulate the flow of certain ions during fertilization and must be capable of fusing with the sperm plasma membrane. Above the plasma membrane is the **vitelline envelope** (Figure 4.6). The major components of this envelope form a fibrous mat above the egg. This mat is supplemented by the extensions of membrane glycoproteins from the plasma membrane and by proteinaceous vitelline posts that adhere the mat to the membrane (Mozingo and Chandler, 1991). The vitelline envelope is essential for the species-specific binding of sperm. In mammals, the vitelline envelope is a separate and thick extracellular matrix called the **zona pellucida.** The mammalian egg is also surrounded by a layer of cells, the **cumulus cells** (Figure 4.7). The cumulus layer represents ovarian follicular cells that were nurturing the egg at the time of its release from the ovary. Mammalian sperm have to get past these cells also to fertilize the egg.*

Lying immediately beneath the plasma membrane of the sea urchin egg is the **cortex.** The cytoplasm in this region is more gel-like than the internal cytoplasm and contains high concentrations of globular actin molecules. During fertilization, these actin molecules polymerize to form long cables of actin known as **microfilaments.** Microfilaments are necessary for cell division, and they also are used to extend the egg surface into the microvilli, which aid sperm entry into the cell (Figure 4.6; also see Figure

*In some mammalian species, the extracellular coverings of the egg are divided into three regions: the zona pellucida, the corona radiata, and the cumulus. The term *corona radiata* then refers to those follicle cells immediately adjacent to the zona pellucida, and the term *cumulus* (or *cumulus oophorus*) is reserved for the more diffuse cells that are farther from the egg and zona. In many mammals (for example, the hamster; Figure 4.7), these two cellular layers are not readily distinguished, and in other species (such as sheep and cattle), these layers are probably shed before fertilization (see Talbot, 1985).

FIGURE 4.7

Hamster eggs immediately before fertilization. (A) The hamster egg, or ovum, is encased in the zona pellucida. This, in turn, is surrounded by the cells of the cumulus. A polar body cell, produced during meiosis, is also within the zona pellucida. (B) At lower magnification, a mouse oocyte is shown in relation to the cumulus. Colloidal carbon particles (India ink) are excluded by the hyaluronidate matrix. (Courtesy of R. Yanagimachi.)

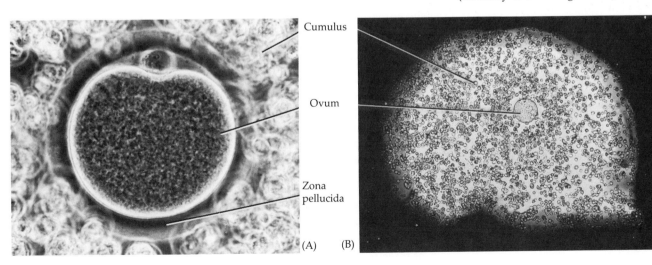

Cumulus

Ovum

Zona pellucida

(A)   (B)

4.18). Also within this cortex are the **cortical granules** (Figures 4.4 and 4.6). These membrane-bound structures are homologous to the acrosomal vesicle of the sperm, being Golgi-derived organelles containing proteolytic enzymes. However, whereas each sperm contains one acrosomal vesicle, each sea urchin egg contains approximately 15,000 cortical granules. Moreover, in addition to containing the digestive enzymes, the cortical granules also contain mucopolysaccharides, adhesive glycoproteins, and **hyaline protein.** The enzymes and mucopolysaccharides are active in preventing other sperm from entering the egg after the first sperm has entered, and the hyaline and adhesive proteins surround the early embryo and provide support for the cleavage-stage blastomeres.

Many types of eggs also secrete an **egg jelly** outside their vitelline envelope (Figure 4.4) This glycoprotein meshwork can have numerous functions. Most often, though, it is used to either attract or activate sperm. The egg, then, is a cell specialized for receiving sperm and initiating development.

Recognition of egg and sperm: Action at a distance

Many marine organisms release their gametes into the local environment. This environment may be as small as a tidepool or as large as the ocean. Moreover, this environment is shared with other species that may shed their sex cells at the same time. These organisms are faced with two problems: (1) How can sperm and eggs meet in such a dilute concentration? (2) What mechanism prevents starfish sperm from trying to fertilize sea urchin eggs? Two major mechanisms have evolved to solve these difficulties: species-specific attraction of sperm and species-specific sperm activation.

Sperm attraction

Species-specific sperm attraction (a type of chemotaxis) has been documented in numerous species, including cnidarians, molluscs, echinoderms, and urochordates (Miller, 1985; Yoshida et al., 1993). In 1978 Miller demonstrated that the eggs of the cnidarian *Orthopyxis caliculata* not only secrete a chemotactic factor but also regulate the timing of its release. Developing oocytes at various stages in their maturation were fixed onto microscope slides, and sperm were released at a certain distance from the eggs. Miller found that when sperm were added to oocytes that had not yet completed their second meiotic division, there was no attraction of sperm to eggs. However, after the second meiotic division was finished and the eggs were ready to be fertilized, the sperm migrated toward them. Thus, these oocytes control not only the type of sperm they attract but also the time at which they attract them.

The mechanisms for chemotaxis are different in other species (see Metz, 1978; Ward and Kopf, 1993).One such chemotactic molecule, a 14-amino acid peptide called **resact,** has been isolated from the egg jelly of the sea urchin *Arbacia punctulata* (Ward et al., 1985). Resact diffuses readily in seawater and has profound effects at very low concentrations when it is added to a suspension of *Arbacia* sperm (Figure 4.8). When a drop of seawater containing *Arbacia* sperm is placed on a microscope slide, the sperm generally swim in circles about 50 μm in diameter. Within seconds of the introduction of a minute amount of resact into the drop, sperm migrate into the region of the injection and congregate there. As resact continues to diffuse from the area of injection, more sperm are recruited

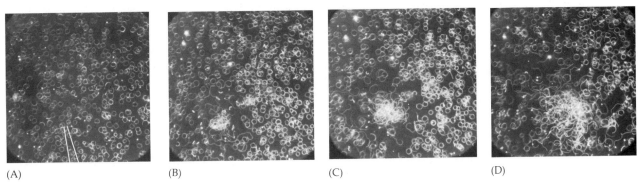

(A) (B) (C) (D)

into the growing cluster. Resact is specific for *A. punctulata* and does not attract sperm of other species. *A. punctulata* sperm bind resact to receptors in their cell membranes (Ramarao and Garbers, 1985; Bentley et al., 1986) and can swim up a concentration gradient of this compound until they reach the egg.

Sperm activation: The acrosome reaction

A second interaction between sperm and egg involves the activation of sperm by egg jelly. In most marine invertebrates, this **acrosome reaction** has two components: the fusion of the acrosomal vesicle with the sperm plasma membrane (an exocytosis that results in the release of the contents of the acrosomal vesicle) and the extension of the acrosomal process (Figure 4.9; Colwin and Colwin, 1963). The acrosome reaction can be initiated

FIGURE 4.8
Sperm chemotaxis in *Arbacia*. One nanoliter of a 10-n*M* solution of resact is injected into a 20-µl drop of sperm suspension. The position of the micropipette is indicated in A. (A) A 1-second photographic exposure showing sperm swimming in tight circles before the addition of resact. (B–D) Similar 1-second exposures showing migration of sperm to the center of the resact gradient 20, 40, and 90 seconds after injection. (From Ward et al., 1985, courtesy of V. D. Vacquier.)

FIGURE 4.9
Acrosome reaction in echinoderm sperm. (A–C) The portion of the acrosomal membrane lying directly beneath the sperm cell membrane fuses with the cell membrane to release the contents of the acrosomal vesicle. (D,E) As the actin molecules assemble to produce microfilaments, the acrosomal process is extended outward. Actual photographs of the acrosome reaction in sea urchin sperm are shown below. (After Summers and Hylander, 1974; photographs courtesy of G. L. Decker and W. J. Lennarz.)

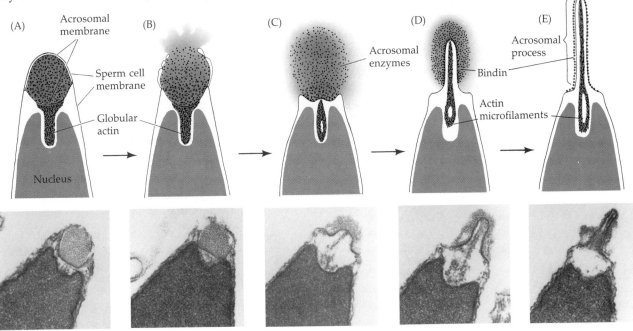

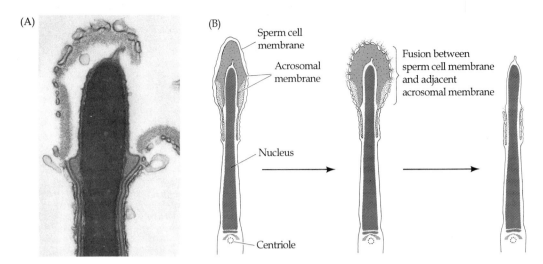

FIGURE 4.10

Acrosome reaction in hamster sperm. (A) Transmission electron micrograph of hamster sperm undergoing acrosomal reaction. The acrosomal membrane can be seen to form vesicles. (B) Interpretive diagram of electron micrographs showing the fusion of the acrosomal and cell membranes in the sperm head. (A from Meizel, 1984, courtesy of S. Meizel; B after Yanagimachi and Noda, 1970.)

by soluble egg jelly, by the egg jelly surrounding the egg, or even by contact with the egg itself in certain species. It can also be activated artificially by increasing the calcium concentration of seawater.

In sea urchins, contact with egg jelly causes the exocytosis of the acrosomal vesicle and release of protein-digesting enzymes that can digest a path through the jelly coat to the egg surface (Dan, 1967; Franklin, 1970; Levine et al., 1978). The sequence of these events is outlined in Figure 4.9. The acrosome reaction is thought to be initiated by a sulfated polysaccharide in the egg jelly that allows calcium and then sodium ions to enter the sperm head and causes potassium and hydrogen ions to leave (SeGall and Lennarz, 1979; Schackmann and Shapiro, 1981). The exocytosis of the acrosomal vesicle is caused by the calcium-mediated fusion of the acrosomal membrane with the adjacent sperm plasma membrane (Figures 4.9 and 4.10). This exocytosis enables the acrosomal vesicle to release its contents at the head of the sperm.*

The second part of the acrosome reaction involves the extension of the acrosomal process (Figure 4.9). This protrusion arises from the polymerization of globular actin molecules into actin filaments, a reaction that appears to be dependent upon the release of hydrogen ions from the sperm head (Schackman et al., 1978; Tilney et al., 1978). Until activation, a regulatory protein may be responsible for blocking actin polymerization, and the rise in intracellular pH accompanying activation might subsequently interfere with this function. In addition to causing the extension of the acrosomal process, this significant increase in pH is also responsible for activating the dynein ATPase in the neck of the sperm. This activation causes a rapid utilization of ATP and a 50 percent increase in mitochondrial respiration. The energy generated is used primarily for flagellar motility (Tombes and Shapiro, 1985). A possible scheme for the events of the acrosome reaction is presented in Figure 4.11.

The egg jelly factors that initiate the acrosomal reactions of sea urchins

*Such exocytotic reactions are seen in the release of insulin from pancreatic cells and in the release of neurotransmitters from synaptic terminals. In all cases, there is a calcium-mediated fusion between the secretory vesicle and the cell membrane. Indeed, the similarity of acrosomal vesicle exocytosis and synaptic vesicle exocytosis may actually be quite deep. Recent studies of acrosome reactions in sea urchins and mammals (González-Martínez et al., 1992; Florman et al., 1992) suggest that the when the receptors for the sperm-activating ligands bind these molecules, they cause a depolarization of the membrane which would open voltage-dependent calcium ion channels in a manner reminiscent of synaptic transmission.

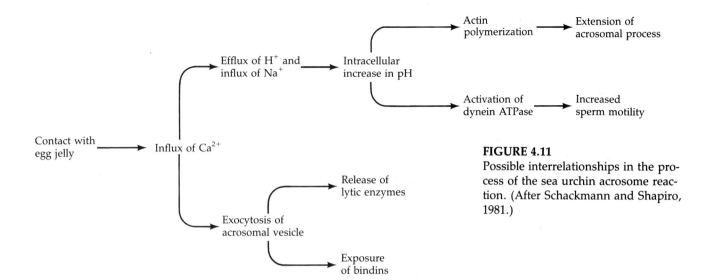

FIGURE 4.11
Possible interrelationships in the process of the sea urchin acrosome reaction. (After Schackmann and Shapiro, 1981.)

are often highly specific. The sperm of sea urchins *Arbacia punctulata* and *Strongylocentrotus drobachiensis* will react only with jelly of their own eggs. However, *S. purpuratus* sperm can also be activated by *Lytechinus variegatus* (but not *A. punctulata*) egg jelly (Summers and Hylander, 1975). Therefore, egg jelly may provide species-specific recognition in some species but not in others.

Recognition of egg and sperm: Contact of gametes

Species-specific recognition in sea urchins

Once the sea urchin sperm has penetrated the egg jelly, the acrosomal process of the sperm contacts the vitelline envelope of the egg (Figure 4.12). A major species-specific recognition step occurs at this point. The

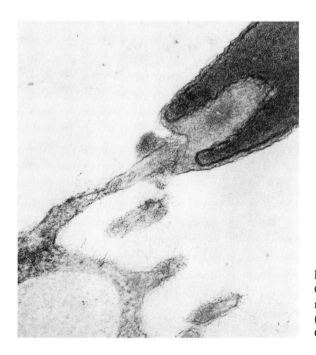

FIGURE 4.12
Contact of a sea urchin sperm acrosomal process with an egg microvillus. (From Epel, 1977, courtesy of F. D. Collins and D. Epel.)

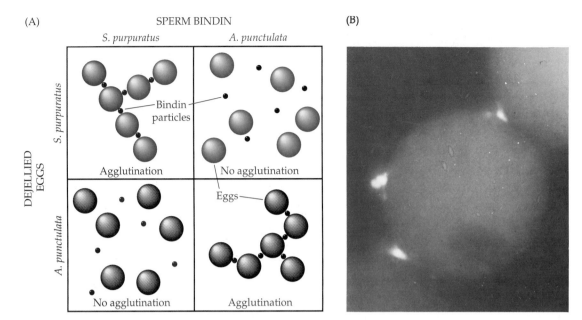

FIGURE 4.13

Species-specific agglutination of dejellied eggs by bindin. (A) Agglutination was promoted by adding 212 μg of bindin to a plastic well containing 0.25 ml of a 2 percent (volume to volume) suspension of eggs. After 2–5 minutes of gentle shaking, the wells were photographed. (B) Fluorescence photomicrograph of *S. purpuratus* eggs bound together by fluorescein-labeled *S. purpuratus* bindin particles. Bindin particles were invariably present at the places where two eggs came together. (A based on photographs of Glabe and Vacquier, 1977; B from Glabe and Lennarz, 1979, courtesy of the authors.)

acrosomal protein mediating this recognition is called **bindin.** In 1977 Vacquier and his co-workers isolated this nonsoluble 30,500-Da protein from the acrosome of *Strongylocentrotus purpuratus*. This protein is capable of binding to dejellied eggs from *S. purpuratus* (Figure 4.13; Vacquier and Moy, 1977). Further, its interaction with eggs is species-specific (Glabe and Vacquier, 1977; Glabe and Lennarz, 1979); bindin isolated from the acrosomes of *S. purpuratus* agglutinates its own dejellied eggs, but not those

FIGURE 4.14

Localization of bindin on the acrosomal process. (A) Immunochemical localization technique places a rabbit antibody wherever bindin is exposed. Rabbit antibodies were made to the bindin protein, and these antibodies were incubated with sperm that had undergone the acrosome reaction. If bindin were present, the rabbit antibodies would remain bound to the sperm. After any unbound antibody was washed off, the sperm were treated with *swine* antibodies that could bind to *rabbit* antibodies. These swine antibodies had been covalently linked to peroxidase enzymes. In such a fashion, peroxidase molecules were placed wherever bindin was present. Peroxidase catalyzes the formation of a dark precipitate from diaminobenzidine (DAB) and hydrogen peroxide. Thus, this precipitate will form only where bindin is present. (B) Localization of bindin to the acrosomal process after the acrosome reaction (×133,200). (C) Localization of bindin to the acrosomal process at the junction of the sperm and the egg. (B and C from Moy and Vacquier, 1979, courtesy of V. D. Vacquier.)

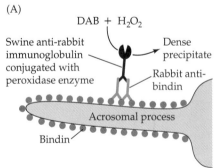

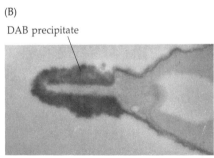

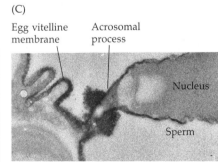

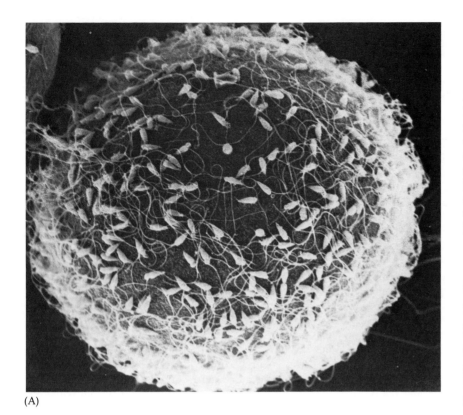

(A)

(B)

FIGURE 4.15
Bindin receptors on the egg. (A) Scanning electron micrograph of sea urchin sperm bound to the vitelline envelope of an egg. (B) Binding of *S. purpuratus* sperm to polystyrene beads that have been coated with purified bindin receptor protein. (A courtesy of C. Glabe, L. Perez, and W. J. Lennarz; B from Foltz et al., 1993.)

of *Arbacia punctulata*. Using immunological techniques, Moy and Vacquier (1979) demonstrated that bindin is located specifically on the acrosomal process—exactly where it should be for sperm–egg recognition (Figure 4.14).

Biochemical studies have shown that the bindins of closely related sea urchin species are indeed different. This finding implies the existence of species-specific bindin receptors on the vitelline envelope. Such receptors were also suggested by the experiments of Vacquier and Payne (1973), who saturated sea urchin eggs with sperm. As seen in Figure 4.15A, sperm binding does not occur over the entire egg surface. Even at saturating numbers of sperm (approximately 1500) there appears to be room on the ovum for more sperm heads, which implies there is a limiting number of sperm-binding sites. A large glycoprotein complex from the vitelline envelopes of sea urchin eggs has been isolated and shown to bind radioactive bindin in a species-specific manner (Glabe and Vacquier, 1978; Rossignol et al., 1984). This glycoprotein is also able to compete with eggs for the sperm of the same species. That is, if *S. purpuratus* sperm are mixed with the bindin receptor from *S. purpuratus* vitelline envelopes, the sperm bind to it and will not fertilize the eggs. The isolated bindin receptor from *S. purpuratus*, however, does not interfere with the fertilization of other related sea urchins. This bindin receptor is a transmembrane glycoprotein of nearly 1300 amino acids (Foltz et al., 1993). The bindin-binding region of this protein extends into the extracellular space and probably becomes a component of the vitelline envelope. These bindin receptors are aggregated into complexes, and hundreds of these complexes are probably needed to tether the sperm to the egg (Figure 4.15B). Thus, species-specific recognition of sea urchin gametes occurs at the levels of acrosome activation and sperm adhesion to the receptors on the egg surface.

Action-at-a-distance between mammalian gametes

It is very difficult to study the interactions that might be occurring between mammalian gametes prior to sperm–egg contact. One obvious reason for this is that mammalian fertilization occurs inside the oviducts of the female. While it is relatively easy to mimic the conditions surrounding sea urchin fertilization (using either natural or artificial seawater), we do not yet know the components of the various natural environments that the mammalian sperm encounter as they travel to the egg. A second reason for this difficulty is that the sperm population ejaculated into the female is probably very heterogeneous, containing spermatozoa at different stages of maturation. Of the 280×10^6 human sperm normally ejaculated into the vagina, only about 200 reach the ampullary region of the oviduct where fertilization takes place (Ralt et al., 1991). Since fewer than 1 in 10,000 sperm get close to the egg, it is difficult to assay those molecules that might enable the sperm to swim toward the egg and become activated. However, the ability to fertilize mammalian eggs in vitro has provided speculations as to what might be occurring naturally.

Capacitation

The reproductive tract of female mammals plays a very active role in the mammalian fertilization process. Newly ejaculated mammalian sperm are unable to undergo the acrosome reaction without residing for some amount of time in the female reproductive tract. This requirement for **capacitation** varies from species to species (Gwatkin, 1976) and can be mimicked in vitro by incubating sperm in tissue culture media or in fluid from the oviducts. The molecular changes that account for capacitation are still unknown (Saling, 1989; Storey and Kopf, 1991), but there are three sets of molecular changes that may be important. First, the fluidity of the sperm cell membrane may be altered by changing its lipid composition. The concentration of cholesterol in the sperm plasma membrane is lowered during sperm capacitation in several species (Davis, 1981), and two proteins found both in serum and in the female reproductive tract (albumin and lipid transport protein 1) have recently been found to remove cholesterol from the human sperm plasma membrane (Langlais et al., 1988; Ravnik et al., 1992). Second, particular proteins or carbohydrates on the sperm surface are lost during capacitation (Poirier and Jackson, 1981; Lopez et al., 1985; Wilson and Oliphant, 1987). It is possible that the moieties lost during capacitation had been blocking zona-binding proteins. Third, phos-

phorylation of certain proteins involved in binding the sperm to the zona pellucida and mediating the exocytosis of the acrosomal vesicle has been observed concomitant with capacitation (Leyton and Saling, 1989a). This phosphorylation may convert the inactive forms of these molecules into functional proteins. However, it is still uncertain to what extent (if any) each of these mechanisms causes capacitation of the sperm. Sperm that are not capacitated are "held up" in the cumulus matrix and so do not reach the egg (Austin, 1960; Corselli and Talbot, 1987).

Hyperactivation and chemotaxis

The different regions of the female reproductive tract may secrete different, regionally specific molecules. These factors may influence sperm mobility as well as capacitation. For instance, when sperm pass from the uterus into the oviducts, they become "hyperactivated," swimming at higher velocities and generating greater force than before. Suarez and co-workers (1991) have shown that while this behavior is not conducive for traveling in low-viscosity fluids, it appears to be extremely well suited for linear sperm movement in the viscous fluid that sperm might encounter in the oviduct.

In addition to increasing the activity of sperm, soluble factors in the oviduct may also provide the directional component of sperm movement. There has been speculation that the ovum (or the cumulus cells or the ruptured ovarian follicle in which the egg had developed) could be secreting chemotactic substances that might attract the sperm toward the egg during the last stages of its migration (see Hunter, 1989). Ralt and colleagues (1991) tested this hypothesis using follicular fluid from human follicles whose eggs were being used for in vitro fertilization. Performing an experiment similar to the one described earlier with sea urchins, they microinjected a drop of follicular fluid into a larger drop of sperm suspension. When they did this, some of the sperm changed their direction of movement to migrate toward the source of follicular fluid. Microinjection of other solutions did not have this effect. These studies did not rule out the possibility that the effect was due to a general stimulation of sperm movement or metabolism. However, these investigations uncovered a fascinating correlation: the fluid from only about half the follicles tested showed a chemotactic effect, and in nearly every case, the egg was fertilizable if and only if the fluid showed chemotactic ability ($P < 0.0001$). It is possible, then, that like certain invertebrate eggs, the human egg secretes a chemotactic factor only when it is capable of fertilization.

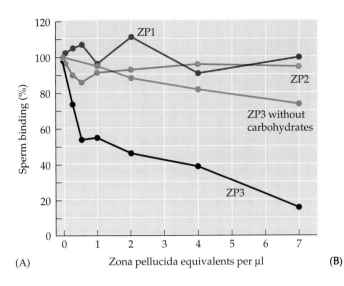

(A)

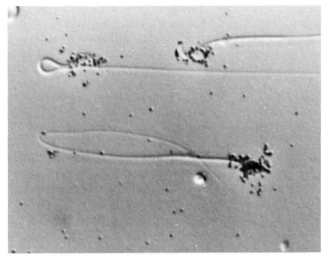

(B)

Gamete binding and recognition in mammals

ZP3: The sperm-binding protein of the mouse zona pellucida. The zona pellucida in mammals plays a role analogous to that of the vitelline envelope in invertebrates. This glycoprotein matrix is synthesized and secreted by the growing oocyte, and it plays two major roles during fertilization: it binds the sperm, and it initiates the acrosome reaction *after* the sperm has bound (Saling et al., 1979; Florman and Storey, 1982; Cherr et al., 1986). The binding of sperm to the zona is relatively, but not absolutely, species-specific (species specificity should not be a major problem when fertilization occurs internally), and the binding of mouse sperm to the mouse zona can be inhibited by first incubating the sperm with zona glycoproteins. Bleil and Wassarman (1980, 1986, 1988) have isolated from the zona pellucida an 83-kDa glycoprotein, **ZP3,** that is the active competitor in this inhibition assay. The other two zona glycoproteins, ZP1 and ZP2, failed to compete for sperm binding (Figure 4.16). Moreover, radiolabeled ZP3 bound to the heads of mouse sperm with intact acrosomes. Thus, ZP3 is the specific glycoprotein in the zona pellucida to which the mouse sperm bind. ZP3 also initiates the acrosome reaction after sperm have bound to it. The mouse sperm can thereby concentrate its proteolytic enzymes directly at the point of attachment at the zona pellucida.

The molecular mechanism by which the zona pellucida and the mammalian sperm recognize each other is presently being studied. The current hypothesis of mammalian gamete binding postulates a set of proteins on the sperm capable of recognizing a specific carbohydrate and protein region on the egg zona ZP3 (Wassarman, 1987; Saling, 1989). Florman and co-workers (1984; Florman and Wassarman, 1985) have shown that the critical moieties for sperm–zona recognition in the mouse are carbohydrate groups on the ZP3 glycoprotein. Removal of these threonine- or serine-linked carbohydrate groups abolishes the ability of ZP3 to compete for sperm binding.

Sperm–zona adhesion proteins. Mouse sperm do not "bore into" the zona. Rather, the sperm approach parallel to the plane of the zona surface and are actively tethered there (Baltz et al., 1988). How is the zona able to bind and retain these wriggling sperm? It appears that ZP3 binds to at least three adhesive proteins on the sperm cell membrane, and thousands of

FIGURE 4.16
Binding of sperm to the zona pellucida. (A) Inhibition assay showing the specific decrease of mouse sperm binding to zonae pellucidae when sperm and zonae are incubated with increasingly large amounts of glycoprotein ZP3. The importance of the carbohydrate portion of ZP3 is also indicated by this figure. (B) Binding of radioactively labeled ZP3 to capacitated mouse sperm. (A after Bleil and Wassarman, 1980, and Florman and Wassarman, 1985; B from Bleil and Wassarman, 1986, courtesy of the authors.)

these sites might be needed to keep these two cells from coming apart. If any of these three adhesive proteins is experimentally inactivated, the sperm cannot adhere to the zona.

The first zona-binding protein of the sperm appears to be a protein that specifically binds to the galactose residues of ZP3. Bleil and Wassarman (1980) have shown that one of the critical carbohydrates of the ZP3 glycoprotein is a terminal galactose group. If this terminal galactose is removed or chemically modified, sperm-binding activity is lost. These researchers later isolated this protein by binding ZP3 covalently to beads and passing the proteins isolated from mouse sperm membranes over them in a column (Bleil and Wassarman, 1990). Most of the proteins passed through the column, but one, a 56-kDa peptide, bound to the ZP3-coated beads. It did not bind to ZP2-coated beads in a similar experiment. This protein was found to be exposed in the sperm membrane, and it bound to galactose residues, strongly suggesting that it is a sperm receptor binding to the carbohydrate moiety of the ZP3 glycoprotein.

The second sperm protein that appears to be important for sperm–zona binding is a sperm cell membrane glycosyltransferase enzyme. Shur's laboratory has shown that this receptor for the zona is an enzyme that recognizes the sugar N-acetylglucosamine on ZP3 (Shur and Hall, 1982a,b; Lopez et al., 1985; Miller et al., 1992). This enzyme, N-acetylglucosamine : galactosyltransferase, is embedded in the sperm plasma membrane, directly above the acrosome, with its active site pointed outward. The enzymatic function of this 60-kDa enzyme would be to catalyze the addition of a galactose sugar (from UDP-galactose) onto a carbohydrate chain terminating with an N-acetylglucosamine sugar (see Chapter 3). However, there are no UDP-galactose residues in the female reproductive tract. Although the enzyme can bind to the N-acetylglucosamine residues of the zona proteins exactly as any enzyme would bind to a substrate, it cannot catalyze the reaction because the second reactant is missing. Therefore, the enzymes (on the sperm) remain bound to their substrate (on the zona).

If this hypothesis were correct, one would expect that sperm–egg binding could be inhibited either by inhibiting the enzyme or by adding the second reactant, UDP-galactose. This is exactly what Shur and co-workers found to be the case. Sperm–zona binding was blocked by (1) addition of UDP-galactose, (2) removal of the N-acetylglucosamine residues from ZP3, (3) addition of antibodies that blocked the activity of the galactosyltransferase, and (4) placement of excess galactosyltransferases in the medium (the excess enzymes would bind to the zona and inhibit the sperm from binding) (Lopez, et al., 1985; Shur and Neely, 1988). Miller and co-workers (1992) have recently shown that the sperm galactosyltransferase will transfer a sugar from UDP-galactose specifically to ZP3. Thus, the sperm surface galactosyltransferase appears to recognize a carbohydrate group on the ZP3 protein of the mouse zona pellucida.

A third sperm protein that binds to the mouse zona pellucida appears to be a 95-kDa transmembrane protein with two functional sites. The extracellular site specifically binds ZP3, while the intracellular site has tyrosine kinase enzymatic activity (Leyton et al., 1992). This enzymatic activity is stimulated when the protein binds ZP3. The ability to phosphorylate the tyrosine residues of proteins is known to be an important way of changing cell metabolism, since several enzymes become active when specific tyrosines in them are phosphorylated (see Chapter 18).

Induction of the mammalian acrosome reaction by ZP3. Once the capacitated sperm has bound to the zona pellucida, how does the mammalian acrosome reaction take place? The reaction is induced by the ZP3 glyco-protein, since modifying the peptide chain of ZP3 can eliminate the ac-rosome reaction even though the binding of sperm to ZP3 is unaffected (Endo et al., 1987a, Leyton and Saling, 1989a). It is thought that ZP3 induces the reaction by clustering together the ZP3 receptors on the sperm cell membrane. If these receptors are artificially crosslinked (either by soluble ZP3 protein or by antibodies to the galactosyltransferase), the acrosome reaction will occur (Leyton and Saling,1989b; Macek et al., 1991). The tyrosine kinase of the 95-kDa ZP3 receptor also appears to be critical, for if its activity is inhibited the acrosome reaction fails to occur after sperm binding (Leyton et al., 1992).

Secondary binding of sperm to zona pellucida. During the acrosome reaction, the anterior portion of the sperm plasma membrane is shed from the sperm (Figure 4.10). This is where the ZP3-binding proteins are lo-cated, yet the sperm must still remain bound to the zona in order to lyse a path through it. In mice, it appears that the secondary binding to the zona is accomplished by proteins in the inner acrosomal membrane that bind specifically to ZP2 (Bleil et al., 1988). Whereas acrosome-intact sperm will not bind to the ZP2 glycoprotein, acrosome-reacted sperm will. More-over, antibodies against the ZP2 protein will not prevent the binding of acrosome-intact sperm to the zona, but will inhibit the attachment of acrosome-reacted sperm. The structure of the zona consists of repeating units of ZP3 and ZP2, occasionally crosslinked by ZP1 (Figure 4.17). It appears that the acrosome-reacted sperm transfer their binding from ZP3 to the adjacent ZP2 molecules. After a mouse sperm has entered the egg, the egg cortical granules release their contents. One of the proteins re-leased by these granules is a protease that specifically alters ZP2 (Moller and Wassarman, 1989). This would inhibit other acrosome-reacted sperm from moving further toward the egg.

It is not known which of the mouse sperm proteins bind to ZP2, and it is possible that one or more of the proteins thought to direct primary binding are actually involved with secondary binding. In porcine sperm, secondary zona binding appears to be mediated by proacrosin. Proacrosin becomes the protease acrosin that has long been known to be involved in digesting the zona pellucida. However, proacrosin is also a fucose-binding protein that maintains the connection between acrosome-reacted sperm and the zona pellucida (Jones et al., 1988). It is possible that proacrosin binds to the zona and is then converted into the active enzyme that locally digests the zona pellucida.

In guinea pigs, secondary binding to the zona is thought to be me-diated by the protein PH-20. Moreover, when this inner acrosomal mem-brane protein was injected into male or female guinea pigs, 100 percent of them became sterile for several months (Primakoff et al., 1988). The blood sera of these sterile guinea pigs had extremely high concentrations of antibodies to PH-20. The antiserum from guinea pigs sterilized by injections of PH-20 not only bound specifically to this protein, but it also blocked sperm–zona adhesion in vitro. The contraceptive effect lasted several months, after which fertility was restored. These animals had been temporarily sterilized by these antibodies. The human analogue of the PH-20 protein is not yet known, but certain sperm antigens display a similar pattern of localization in the sperm. Similarly, the human zona proteins and their functions are not as clearly established as those in the mouse. Nevertheless, these experiments show that the principle of im-munological contraception is well founded.

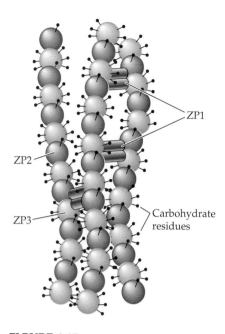

FIGURE 4.17
Diagram of the fibrillar structure of the mouse zona pellucida. The major strands of the zona are composed of repeating dimers of proteins ZP2 and ZP3. These strands are occasionally crosslinked together by ZP1, forming a meshlike network. (After Wassarman, 1989.)

FIGURE 4.18
Scanning electron micrographs of the entry of sperm into sea urchin eggs. (A) Contact of sperm head with egg microvillus through the acrosomal process. (B) Formation of fertilization cone. (C) Internalization of sperm within the egg. (D) Transmission electron micrograph of sperm internalization through the fertilization cone. (A–C from Schatten and Mazia, 1976, courtesy of G. Schatten; D courtesy of F. J. Longo.)

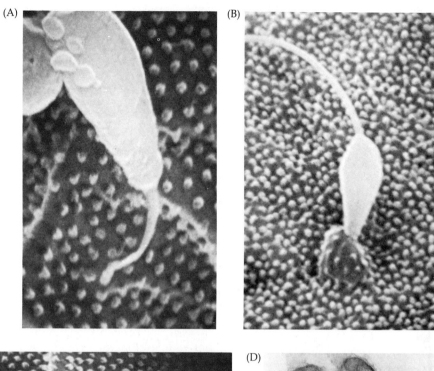

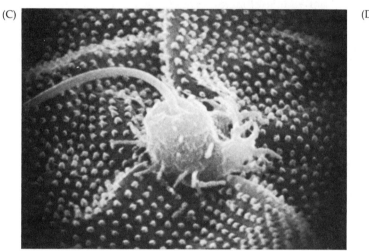
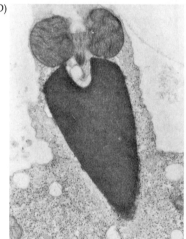

Gamete fusion and the prevention of polyspermy

Fusion between egg and sperm cell membranes

Recognition of sperm by the vitelline envelope or zona is followed by the lysis of that portion of the envelope or zona in the region of the sperm head (Colwin and Colwin, 1960; Epel, 1980). This lysis is followed by the fusion of the sperm cell membrane with the cell membrane of the egg.

The entry of a sperm into a sea urchin egg is illustrated in Figure 4.18. The egg surface is covered with small extensions called microvilli; sperm–egg fusion appears to cause the polymerization of actin and the extension of several microvilli to form the **fertilization cone** (Summers et al., 1975; Schatten and Schatten, 1980, 1983). Homology between the egg and the sperm is again demonstrated, because the transitory fertilization cone, like the acrosomal process, appears to be extended by the polymerization of actin. The sperm and egg membranes join together, and material from the sperm cell membrane can later be found on the egg membrane (Gundersen et al., 1986). The sperm nucleus and tail pass through the cytoplasmic bridge, which is widened by the actin polymerization. Yanagimachi and Noda (1970) have shown a similar process to occur during the fusion of mammalian gametes (Figure 4.19).

In the sea urchin, all regions of the egg are capable of fusing with

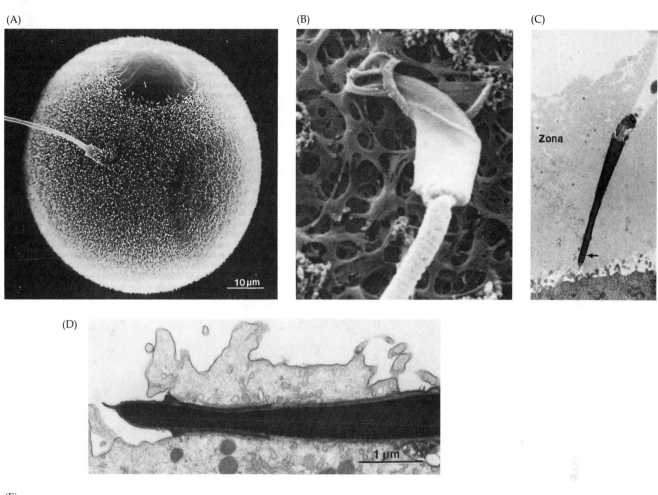

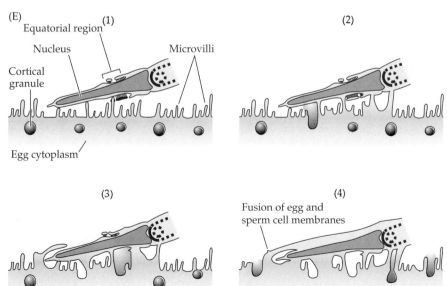

FIGURE 4.19
Entry of sperm into the golden hamster egg. (A) Scanning electron micrograph of sperm fusing with the egg. The "bald" spot (without microvilli) is where the polar body has left. (B) Close-up of sperm-zona binding. (C) Transmission electron micrograph showing the sperm head passing through the zona. (D) Transmission electron micrograph of the hamster sperm fusing parallel to the egg plasma membrane. (E) Diagram of the fusion of the sperm acrosome and plasma membranes with the egg microvilli. (After Yanagimachi and Noda, 1970; Yanagimachi 1988; photographs courtesy of R. Yanagimachi.)

sperm; in several other species, there are specialized regions of the membrane for sperm recognition and fusion (Vacquier, 1979). Fusion is an active process, often mediated by specific "fusogenic" proteins. Proteins such as the HA protein of influenza virus and the F protein of Sendai virus are known to promote cell fusion, and it is possible that bindin is also such a protein. Glabe (1985) has shown that sea urchin bindin will cause phospholipid vesicles to fuse together and that, like the viral fusogenic proteins, bindin contains a long stretch of hydrophobic amino acids near its amino terminus. In abalones, the lysin that dissolves the vitelline envelope has also been found to have fusogenic activity (Hong and Vacquier, 1986), and proteins in the acrosomal membrane of guinea pig and rat sperm are essential for sperm–egg membrane fusion in those species (Primakoff et al., 1987; Blobel et al., 1992; Rochwerger et al., 1992). It appears, then, that one of the proteins on the sperm head is capable of stimulating the fusion of the sperm and egg cell membranes.

Prevention of polyspermy

As soon as one sperm has entered the egg, the fusibility of the egg membrane, which was so necessary to get the sperm inside the egg, becomes a dangerous liability. In sea urchins, as in most animals studied, any sperm that enters the egg can provide a haploid nucleus and a centriole to the egg. In normal **monospermy,** wherein only one sperm enters the egg, a haploid sperm nucleus and a haploid egg nucleus combine to form the diploid nucleus of the fertilized egg (zygote), thus restoring the chromosome number appropriate for the species. The centriole, coming from the sperm, will divide to form the two poles of the mitotic spindle during cleavage.

The entrance of multiple sperm—**polyspermy**—leads to disastrous consequences in most organisms. In the sea urchin, fertilization by two sperm results in a triploid nucleus, wherein each chromosome is represented three times rather than twice. Worse, it means that instead of a bipolar mitotic spindle that separates the chromosomes into two cells, the triploid chromosomes would be divided into as many as four cells. Because there is no mechanism to ensure that each of the four cells receives the proper number and type of chromosomes, the chromosomes would be apportioned unequally. Some cells would receive extra copies of certain chromosomes and other cells would lack them. Theodor Boveri demonstrated in 1902 that such cells either die or develop abnormally (Figure 4.20).

FIGURE 4.20
Aberrant development in a dispermic sea urchin egg. (A) Fusion of three haploid nuclei, each containing 18 chromosomes, and the division of the two sperm centrioles to form four mitotic poles. (B) The 54 chromosomes randomly assort on the four spindles. (C) At anaphase of the first division, the duplicated chromosomes are pulled to the four poles. (D) Four cells containing different numbers and types of chromosomes are formed, thereby causing (E) the early death of the embryo. (After Boveri, 1907.)

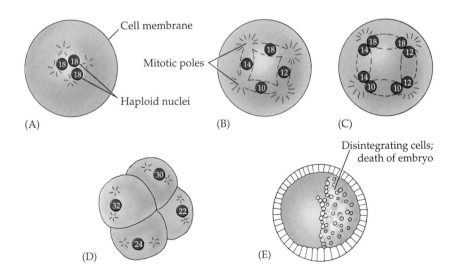

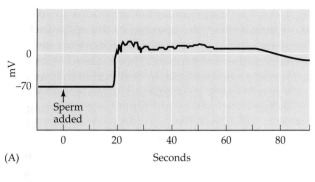

(A)

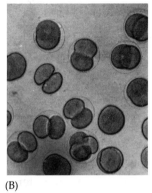

(B)

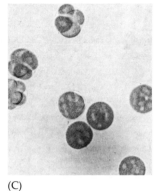

(C)

FIGURE 4.21
Membrane potential of sea urchin eggs before and after fertilization. (A) Before the addition of sperm, the potential difference across the egg cell membrane is about −70 mV. Within 1 second after the fertilizing sperm contacts the egg, the potential shifts in a positive direction. (B) Control eggs developing in 490 mM Na$^+$. (C) Polyspermy in eggs fertilized in 120 mM Na$^+$ (choline was substituted for sodium). The *Lytechinus* eggs were photographed during the first cleavage. (D) Table showing the rise of polyspermy with decreasing sodium ion concentration. (From Jaffe, 1980, photographs courtesy of L. A. Jaffe.)

(D)

[Na](mM)	Percentage of polyspermic eggs
490	22
360	26
120	97
50	100

Species have evolved ways to prevent the union of more than two haploid nuclei. The most common way is to prevent the entry of more than one sperm into the egg. The sea urchin egg has two mechanisms to avoid polyspermy: a fast reaction, accomplished by an electrical change in the egg plasma membrane, and a slower reaction, caused by the exocytosis of the cortical granules.

The fast block to polyspermy. The egg cell membrane is remarkable not only in its ability to fuse with the sperm cell membrane, but also in its ability to become refractile to fusion almost immediately upon the entry of a sperm (Just, 1919). Thus, only one sperm enters the egg cytoplasm. The fast block to polyspermy achieves that goal by changing the electrical potential of the egg membrane.

The cell membrane provides a selective barrier between the egg cytoplasm and the outside environment, and the ionic concentration of the egg is greatly different from that of its surroundings. This concentration difference is especially true for sodium and potassium ions. Seawater has a particularly high sodium ion concentration, whereas the egg cytoplasm has relatively little free sodium. The reverse is the case with potassium ions. This condition is maintained by the cell membrane, which steadfastly inhibits the entry of sodium into the oocyte and prevents potassium ions from leaking into the environment. If an electrode is inserted into the egg and a second electrode is placed outside the oocyte, one can measure the constant potential difference across the egg plasma membrane. This **resting membrane potential** is generally about 70 millivolts, usually expressed as −70 mV because the inside of the cell is negatively charged with respect to the exterior.

Within 1–3 seconds after the binding of the first sperm, the membrane potential shifts to a positive level (Longo et al., 1986). A small influx of sodium ions into the egg is permitted, thereby bringing the potential difference to about +20 mV (Figure 4.21A). Although sperm can fuse with membranes having a resting potential of −70 mV, they cannot readily fuse with membranes having a positive resting potential. It is not known how the binding or entry of a sperm signals the opening of the sodium channels, but Gould and Stephano (1987, 1991) have provided what may be an important clue to understanding this process. They isolated from *Urechis* (a marine worm) sperm an acrosomal protein that is able to open the sodium channels of unfertilized *Urechis* eggs. Moreover, when these eggs are exposed to this protein, the rate of sodium influx and the resulting membrane potential shift are very similar to those produced by live sperm. The opening of the sodium channels in the egg appears to be caused by the binding of the sperm to the egg.

Jaffe and her co-workers have found that polyspermy can be induced in eggs if the eggs are artificially supplied with electrical current that keeps their membrane potential negative. Conversely, fertilization can be prevented entirely by artificially keeping the membrane potential of eggs positive (Jaffe, 1976). The fast block to polyspermy can also be circumvented by lowering the concentration of sodium ions in the water (Figure 4.21B–D). If the sodium ions are not sufficient to cause the positive shift in membrane potential, polyspermy occurs (Gould-Somero et al., 1979; Jaffe, 1980). It is not known how the differences in membrane potential act on the sperm to block secondary fertilization. It is probable that the sperm carry a voltage-sensitive component (possibly a positively charged fusogenic protein) and the insertion of this protein into the egg cell membrane would be regulated by the electrical charge across the membrane (Iwao and Jaffe,1989). An electrical block to polyspermy also occurs in frogs (Cross and Elinson, 1980) but probably not in most mammals (Jaffe and Cross, 1983).

The slow block to polyspermy. Sea urchin eggs (and many others) have a second mechanism to ensure that multiple sperm do not enter the egg cytoplasm (Just, 1919). The fast block to polyspermy is transient, and the

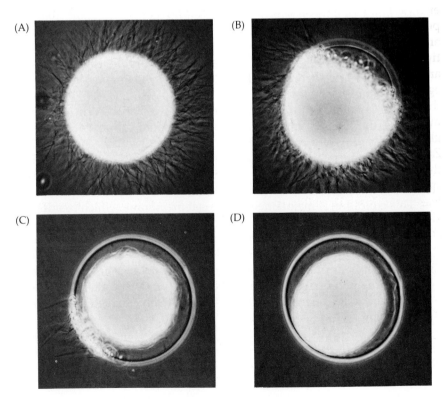

FIGURE 4.22
Formation of the fertilization envelope and removal of excess sperm. Sperm were added to sea urchin eggs, and the suspension was fixed in formaldehyde to prevent further reactions. (A) Ten seconds after sperm addition, sperm are seen surrounding the egg. (B,C) Twenty-five and 35 seconds after insemination, a fertilization envelope forms around the egg, starting at the point of sperm entry. (D) The fertilization envelope is complete and excess sperm are removed. (From Vacquier and Payne, 1973, courtesy of V. D. Vacquier.)

membrane potential of the sea urchin egg remains positive for only about a minute. This brief potential shift may not be sufficient to permanently prevent polyspermy, for Carroll and Epel (1975) have demonstrated that polyspermy can still occur if the sperm bound to the vitelline envelope are not somehow removed. This removal is accomplished by the **cortical granule reaction,** a slower, mechanical, block to polyspermy that becomes active about 1 minute after the first successful sperm–egg attachment.

Directly beneath the sea urchin egg membrane are 15,000 cortical granules, each about 1 μm in diameter (Figure 4.6B). Upon sperm entry, these cortical granules fuse with the egg plasma membrane and release their contents into the space between the cell membrane and the fibrous mat of vitelline envelope proteins. There are several proteins associated with this cortical granule exocytosis. The first are the proteases. These enzymes dissolve the vitelline posts that connect the vitelline envelope proteins to the cell membrane, and they clip off the bindin receptor and any sperm attached to it (Vacquier et al., 1973; Glabe and Vacquier, 1978). Other proteins, mucopolysaccharides released by the cortical granules, produce an osmotic gradient that causes water to rush into the space between the cell membrane and the envelope, causing the vitelline envelope to expand and become the **fertilization envelope** (Figures 4.22 and 4.23). A third protein product of the cortical granules, a peroxidase enzyme, hardens the fertilization envelope by crosslinking tyrosine residues on adjacent proteins (Foerder and Shapiro, 1977; Mozingo and Chandler, 1991). As shown in Figure 4.22, the fertilization envelope starts to form at the site of sperm entry and continues its expansion around the egg. As this envelope forms, sperm are released. This process starts about 20 seconds after sperm attachment and is complete by the end of the first minute of fertilization. Finally, a fourth cortical granule protein, **hyalin,** forms a coating around the egg (Hylander and Summers, 1982). The cell extends elongated microvilli whose tips attach to the hyaline layer. This hyaline layer provides support for the blastomeres during cleavage.

In mammals, the cortical granule reaction does not create a fertilization envelope, but the effect is the same. Released enzymes modify the zona pellucida sperm receptors such that they can no longer bind sperm (Bleil and Wassarman, 1980). This modification process is called the **zona reaction.** During this zona reaction, both ZP3 and ZP2 are modified. Florman and Wassarman (1985) have proposed that the cortical granules of mouse eggs contain an enzyme that clips off the terminal sugar residues of ZP3, thereby releasing bound sperm from the zona and preventing the attachment of other sperm. Cortical granules of mouse eggs contain N-acetylglucosaminidase enzymes capable of cleaving N-acetylglucosamine from ZP3 carbohydrate chains, and Miller and co-workers (1992) demonstrated that after fertilization, ZP3 will not serve as a substrate for the binding of sperm galactosyltransferase. ZP2 is clipped by the cortical granule proteases and loses its ability to bind sperm as well (Moller and Wassarman, 1989). Thus, sperm can no longer initiate or maintain their binding to the zona pellucida and are rapidly shed.

The mechanism for the cortical granule reaction is similar to that of the acrosome reaction. Upon fertilization, the intracellular calcium ion concentration of the egg increases greatly. In the elevated concentration of intracellular free calcium ions, the cortical granule membranes fuse with the egg plasma membrane, thereby causing the exocytosis of their contents (Figure 4.23). Following the fusion of the cortical granules about the point of sperm entry, a wave of cortical granule exocytosis propagates around the cortex to the opposite side of the egg.

The release of calcium from intracellular storage can be monitored

FIGURE 4.23

Cortical granule exocytosis. (A) Schematic diagram showing the events leading to the formation of the fertilization envelope and the hyaline layer. As cortical granules undergo exocytosis, they release proteases that cleave the proteins linking the vitelline envelope to the cell membrane. Mucopolysaccharides released by the cortical granules form an osmotic gradient, thereby causing water to enter and swell the space between the vitelline envelope and the cell membrane. Other enzymes released from the cortical granules harden the vitelline envelope (now the fertilization envelope) and release sperm bound to it. (B,C) Transmission and scanning electron micrographs of the cortex of an unfertilized sea urchin egg. (D,E) Transmission and scanning electron micrographs of the same region of a recently fertilized egg showing the raising of the fertilization envelope and the points at which the cortical granules fused with the plasma membrane of the egg (arrows in D). (A after Austin, 1965. B–E from Chandler and Heuser, 1979, courtesy of D. E. Chandler.)

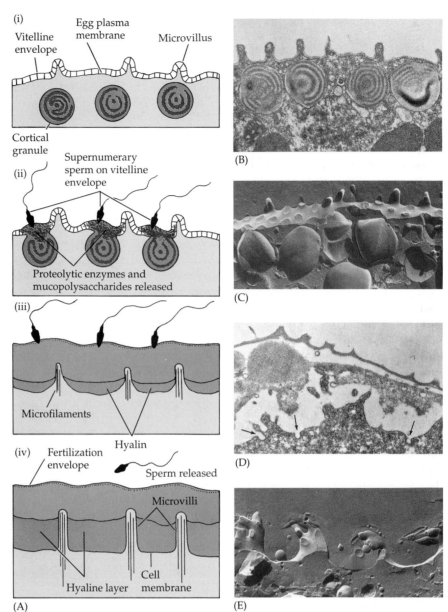

visually by calcium-activated dyes such as aequorin (isolated from luminescent jellyfish) or spectroscopically with dyes such as fura-2. These dyes emit light when they bind free calcium ions. Eggs are injected with the dye and then fertilized. Plate 12 shows the striking wave of calcium release that propagates across the sea urchin egg. Starting at the point of sperm entry, a band of light traverses the cell (Steinhardt et al., 1977; Gilkey et al., 1978; Hafner et al., 1988). As these photographs document, the calcium ions do not merely diffuse across the egg from the point of sperm entry. Rather, the release of calcium ions starts at one end of the cell and proceeds to the other end. The mechanism for this wave of calcium ion release is discussed later in this chapter (Sidelights & Speculations, p. 153). The entire release of calcium ions is complete in roughly 30 seconds in sea urchin eggs, and the free calcium ions are resequestered shortly after they are released. If two sperm enter the egg cytoplasm, calcium ion release can be seen starting at the two separate points on the cell surface (Hafner et al., 1988).

Several experiments have demonstrated that calcium ions are directly responsible for propagating the cortical reaction and that the calcium ions are stored within the egg itself. The first evidence came in 1974 when unfertilized eggs were treated with the ionophore A23187. This drug transports calcium ions across membranes, allowing the cations to traverse the otherwise impermeable barriers. Placing sea urchin eggs into seawater containing A23187 causes the cortical granule reaction and the elevation of the fertilization envelope. Moreover, this reaction occurs in the absence of any calcium ions in the seawater. Therefore, A23187 causes the release of calcium ions already sequestered in organelles within the egg (Chambers et al., 1974; Steinhardt and Epel, 1974). Further studies (Hollinger and Schuetz, 1976; Fulton and Whittingham, 1978; Hamaguchi and Hiramoto, 1981) have shown that calcium ions will initiate cortical granule reactions when injected into sea urchin, mouse, and frog eggs.

The internal calcium ions are probably stored in the endoplasmic reticulum of the egg. In sea urchins and frogs, whose eggs undergo a cortical granule reaction, this reticulum is pronounced in the cortex and surrounds the cortical granules (Figure 4.24; Gardiner and Grey, 1983; Luttmer and Longo, 1985). In the clam *Spisula*, which does not have a cortical granule reaction (and which uses external calcium ions to activate the egg), the cortical endoplasmic reticulum is very sparse. Calcium indicator dyes demonstrate that the endoplasmic reticulum is the source of these calcium ions released during sea urchin fertilization, and when sea urchin eggs were centrifuged to stratify their contents, the source of calcium release was also seen in the region containing the endoplasmic reticulum (Eisen and Reynolds, 1985; Terasaki and Sardet, 1991). In the frog *Xenopus*, the cortical endoplasmic reticulum becomes 10-fold more abundant during the maturation of the egg and disappears locally within a minute after the wave of exocytosis occurs in any particular region of the cortex. Jaffe (1983) likens this calcium-sequestering cortical endoplasmic reticulum to the sarcoplasmic reticulum of skeletal and cardiac muscle. Once initiated, the release of calcium is self-propagating. Free calcium is able to release sequestered calcium from its storage sites, thus causing a wave of calcium ion release and cortical granule exocytosis.

Variations in polyspermy-preventing strategies exist throughout nature. In mammals, polyspermy is minimized by the small number of sperm that reach the site of fertilization (Braden and Austin, 1954). The block to polyspermy in hamsters appears to be controlled solely by the release of

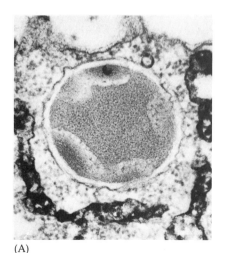

(A)

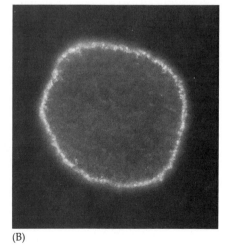

(B)

FIGURE 4.24
Endoplasmic reticulum surrounding cortical granule in sea urchin eggs. (A) Endoplasmic reticulum has been stained with osmium–zinc iodide to allow visualization by transmission electron microscopy. (B) Calcium-dependent calcium release in sea urchin eggs is thought to be effected through calcium ions stored in the endoplasmic reticulum in the egg cortex. Fluorescent antibodies to the calcium-dependent release channel show these channels in the cortical endoplasmic reticulum. (A From Luttmer and Longo, 1985, courtesy of S. Luttmer; B from McPherson et al., 1992, courtesy of F. J. Longo.)

sperm-binding sites on the zona pellucida (Miyazaki and Igusa, 1981; Jaffe and Gould, 1985). Rabbits, however, rely completely on a membrane-level block to polyspermy, and nobody will argue with their success. Lastly, certain animals have defenses to polyspermy about which we know very little. In the yolky eggs of certain birds, reptiles, and salamanders, several sperm actually do enter the egg cytoplasm. In some unknown way, all but one of these sperm are induced to disintegrate in the cytoplasm after the fusion of the egg pronucleus with one of the sperm pronuclei (Ginzburg, 1985; Elinson, 1986). Whatever the mechanism, only one haploid sperm nucleus is allowed to fuse with the haploid nucleus of the egg.

Fusion of the genetic material

In sea urchins, the sperm nucleus enters the egg perpendicular to the egg surface. After fusion of the sperm and egg membranes, the sperm nucleus and its centriole separate from the mitochondria and the flagellum. The mitochondria and the flagellum disintegrate inside the egg, so very few, if any, sperm-derived mitochondria are found in developing or adult organisms (Dawid and Blackler, 1972; Giles et al., 1980). In mice, it is estimated that only 1 out of every 10,000 mitochondria are sperm-derived (Gyllensten et al., 1991). Thus, although each gamete contributes a haploid genome to the zygote, the mitochondrial genome is transmitted primarily from the maternal parent. Conversely, in almost all animals studied (the mouse being the major exception), the centrosome needed to produce the mitotic spindle of the subsequent divisions is derived from the sperm (Sluder et al., 1989, 1993).

The egg nucleus, once it is haploid, is called the **female pronucleus.** Within the egg cytoplasm, the sperm nucleus decondenses to form the **male pronucleus.** Once inside the egg, the male pronucleus undergoes a dramatic transformation. The pronuclear envelope vesiculates into small packets, thereby exposing the compact sperm chromatin to the egg cytoplasm (Longo and Kunkle, 1978). The proteins holding the sperm chromatin in its condensed, inactive, state are exchanged for proteins derived from the egg cytoplasm. This exchange permits the **decondensation** of the sperm chromatin. In sea urchins, decondensation appears to be initiated by the phosphorylation of two sperm-specific histones that bind tightly to the DNA. This process begins when the sperm come into contact with a sulfate-rich compound in the egg jelly that elevates the level of cAMP-dependent protein kinase activity. (Such cAMP-dependent protein kinases were mentioned in Chapter 1). These protein kinases phosphorylate several of the basic residues of the sperm-specific histones and thereby interfere with their binding to DNA (Garbers et al., 1980, 1983; Porter and Vacquier, 1986). This loosening is thought to facilitate the replacement of these sperm-specific histones by other histones that have been stored in the oocyte cytoplasm (Poccia et al., 1981; Green and Poccia, 1985). Once decondensed, the DNA can begin transcription and replication.

In the frog *Xenopus,* the sperm DNA is condensed through its interaction with a protein assembly composed of one molecule each of histone H3, histone H4, and sperm-specific proteins X and Y. The oocyte cytoplasm contains histones H2A and H2B complexed with the protein **nucleoplasmin.** This nucleoplasmin has a greater affinity for the X and Y proteins than it does for histones H2A and H2B, and it trades the histones for the X and Y proteins (Figure 4.25; Philpott and Leno, 1992). This restores the normal chromatin conformation to the male pronuclear DNA.

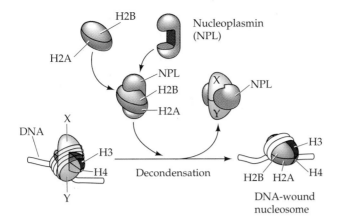

FIGURE 4.25
Schematic model illustrating the re-modeling of sperm chromatin during decondensation. Sperm chromatin is tightly bound through a complex of histones H3, H4, and sperm-specific proteins X and Y. Oocyte nucleoplasmin binds a dimer of histones H2A and H2B and trades them for the X-Y dimer. The resulting chromatin forms the characteristic nucleosome structure containing histones H2A, H2B, H3, and H4, and approximately 200 base pairs of DNA. (After Philpott and Leno, 1992.)

Soon afterward, new membranous vesicles aggregate along the periphery of the chromatin mass and connect with the fragments of the old envelope that have been carried on the ends of the sperm chromosomes.

After the sea urchin sperm enters the egg cytoplasm, the male pronucleus rotates 180° so that the sperm centriole is between the sperm pronucleus and the egg pronucleus. The microtubules of the centriole of the male pronucleus extend and contact the female pronucleus, and the two pronuclei migrate toward each other (Hamaguchi and Hiramoto, 1980; Bestor and Schatten, 1981). Their fusion forms the diploid **zygote nucleus** (Figure 4.26). The initiation of DNA synthesis can occur either in the pronuclear stage (during migration) or after the formation of the zygote nucleus.

In mammals, the process of nuclear fusion takes about 12 hours, compared with less than 1 hour in the sea urchin. The mammalian sperm enters almost tangentially to the surface of the egg rather than approaching it perpendicularly, and it fuses with numerous microvilli (Figure 4.21). The mammalian sperm nucleus also breaks down as its chromatin decondenses and is then reconstructed by coalescing vesicles. Because the decondensation of chromatin can be induced artificially by treating the rabbit sperm nuclei with reagents that disrupt disulfide bonds, it is likely that the oocyte contains a substance that also disrupts these linkages (Calvin and Bedford, 1971; Kvist et al., 1980). The mammalian male pronucleus

FIGURE 4.26
Nuclear events in the fertilization of the sea urchin. (A) Migration of the egg pronucleus and the sperm pronucleus in an egg of *Clypeaster japonicus*. The sperm pronucleus is surrounded by its aster of microtubules. The arrows in the lower portion of the picture indicate the time at which it was taken. (B) Fusion of pronuclei in the sea urchin egg. (A from Hamaguchi and Hiramoto, 1980, courtesy of the authors; B courtesy of F. J. Longo.)

(A)

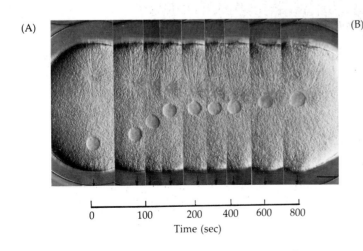

Time (sec)

(B)

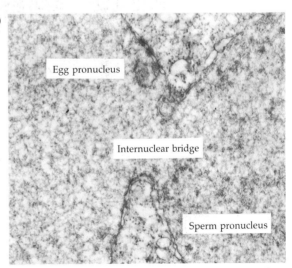

Egg pronucleus

Internuclear bridge

Sperm pronucleus

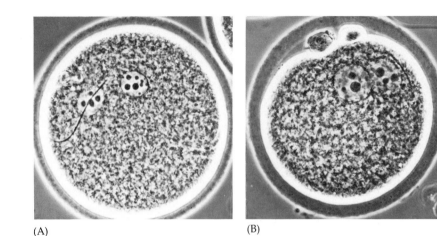

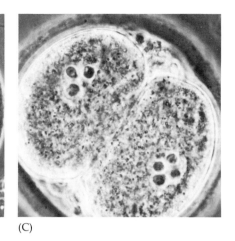

(A)　　　　　　　　　　　　　　(B)　　　　　　　　　　　　　　(C)

FIGURE 4.27

Pronuclear movement in the hamster. (A) Sperm entry into the cell and swelling of the sperm pronucleus. The sperm tail can be seen within egg. (B) Apposition of sperm and egg pronuclei. (C) Two-cell stage, showing two equally sized cells with well-defined nuclei. Debris in perivitelline space is the degenerating polar bodies. (From Bavister, 1980, courtesy of B. D. Bavister.)

enlarges while the oocyte nucleus completes its second meiotic division (Figure 4.27A). Then each pronucleus migrates toward the other, replicating its DNA as it travels. Upon meeting, the two nuclear envelopes break down (Figure 4.27B). However, instead of producing a common zygote nucleus (as happens in sea urchin fertilization), the chromatin condenses into chromosomes that orient themselves upon a common mitotic spindle. Thus, a true diploid nucleus in mammals is first seen not in the zygote, but at the 2-cell stage (Figure 4.27C).

SIDELIGHTS & SPECULATIONS

The nonequivalence of mammalian pronuclei

It is generally assumed that males and females carry equivalent haploid genomes. Indeed, one of the fundamental tenets of Mendelian genetics is that genes derived from the sperm are functionally equivalent to those derived from the egg. However, recent studies show that in mammals the sperm-derived genome and the egg-derived genome may be functionally different and play complementary roles during certain stages of development. The first evidence for this nonequivalence came from studies of a human tumor called a hydatidiform mole. A majority of such moles have been shown to arise from a haploid sperm fertilizing an egg in which the female pronucleus is absent. After entering the egg, the sperm chromosomes duplicate themselves, thereby restoring the diploid chromosome number. Thus, the entire genome is derived from the sperm (Jacobs et al., 1980; Ohama et al., 1981). Here we see a situation in which the cells survive, divide, and have

a normal chromosome number, but development is abnormal. Instead of forming an embryo, the egg becomes a mass of placenta-like cells. Normal development does not occur here when the entire genome comes from the male parent.

Evidence for the nonequivalence of mammalian pronuclei also comes from attempts to get ova to develop in the absence of sperm. This ability to develop an embryo without spermatic contribution is called **parthenogenesis** (Gk., "virgin birth"). The eggs of many invertebrates and some vertebrates are capable of developing normally in the absence of sperm if the egg is artificially activated. In these situations, sperm's contribution to development seems dispensable. Mammals, however, do not exhibit parthenogenesis. Suppression of polar body formation during meiosis produces diploid mouse eggs whose inheritance is derived from the egg alone. These cells divide to form embryos with spinal cords, muscles, skeletons, and organs, including beating hearts. However, development does not continue, and by day 10 or 11 (halfway through the mouse's

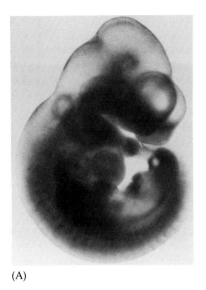
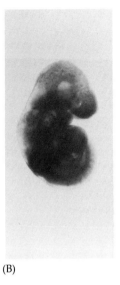

(A)   (B)

FIGURE 4.28
Control (A) and (B) parthenogenetic (two female pronuclei) mouse embryos at 11 days' gestation. The mice were developing in the same female. In addition to being smaller and deteriorating, the parthenogenetic embryos also had much smaller placentas. (From Surani and Barton, 1983, courtesy of the authors.)

TABLE 4.1
Pronuclear transplantation experiments

Class of reconstructed zygotes	Operation	Number of successful transplants	Number of progeny surviving
Bimaternal		339	0
Bipaternal		328	0
Control		348	18

Source: McGrath and Solter, 1984.

gestation), profound differences are observed between the normal and the parthenogenetic embryos (Figure 4.28). The parthenogenetic embryos are deteriorating and are becoming grossly disorganized (Surani et al., 1986). Neither human nor mouse development can be completed solely with egg-derived chromosomes.

The hypothesis that male and female pronuclei are different also gains support from pronuclear transplantation experiments (Surani and Barton, 1983; McGrath and Solter, 1984). Either male or female pronuclei of recently fertilized mouse eggs can be removed and added to other recently fertilized oocytes. (The two pronuclei can be distinguished because the female pronucleus is always the one beneath the polar bodies.) Thus, zygotes with two male or two female pronuclei can be constructed. Although embryonic cleavage occurs, neither of these types of eggs develops to birth, whereas some control eggs (containing one male pronucleus and one female pronucleus from different zygotes) undergoing such transplantation develop normally (Table 4.1). Moreover, the bimaternal or bipaternal embryos cease development at the same time as the parthenogenetic mice. Thus, though the two pronuclei are

equivalent in many animals, in mammals there are important functional differences between them.

The reason for these embryonic deaths is that in some cells, only the maternally derived allele of certain genes is active, while in other cells, only the paternally derived allele of those genes is functional. (To be sure, in most genes, the male-derived and female-derived allele are equivalent and are activated to the same degree in every cell. We are dealing here with the exceptions to that Mendelian rule.) For instance, insulin-like growth factor II (IGF-II) promotes the growth of embryonic and fetal organs. In embryonic mice, the *paternally* derived allele of IGF-II is active throughout the embryo, whereas the *maternally* derived allele is usually inactive (except in a few neural cells). Thus, if a mouse inherits a mutant IGF-II allele from its mother, the mouse will develop to a normal size (since the maternally derived allele is not expressed); but if the same mutant allele is inherited from its father, the mouse will have stunted growth (DeChiara et al., 1991). The opposite pattern of allele expression is found for one of the *receptors* of IGF-II. Here, the *paternal* gene for the receptor is poorly transcribed, while the *maternal* allele is active (Barlow, et al., 1991). The differences between the active and inactive alleles are thought to be caused by modifications of DNA that occur differently in the egg and sperm nuclei, and these will be discussed further in Chapter 11. Since certain developmentally important genes are active only if they come from the sperm and other such genes are active only if they come from the egg, maternal and paternal pronuclei are both necessary for the completion of mammalian development.

Activation of egg metabolism

So far, then, we have discussed the mechanisms by which sperm and egg recognize each other, fuse together, and merge their respective haploid

TABLE 4.2
Events of sea urchin fertilization

Event	Approximate time postinsemination[a]
Sperm–egg bonding	0 seconds
Fertilization potential rise (fast block to polyspermy)	before 3 sec
Sperm–egg membrane fusion	before 30 sec
Calcium increase first detected	30 sec
Cortical vesicle exocytosis (slow block to polyspermy)	30–60 sec
Activation of NAD kinase	starts at 1 min
Increase in NADH and NADPH	starts at 1 min
Increase in O_2 consumption	starts at 1 min
Sperm entry	1–2 min
Acid efflux	1–5 min
Increase in pH (remains high)	1–5 min
Sperm chromatin decondensation	2–12 min
Sperm nucleus migration to egg center	2–12 min
Egg nucleus migration to sperm nucleus	5–10 min
Activation of protein synthesis	starts at 5–10 min
Activation of amino acid transport	starts at 5–10 min
Initiation of DNA synthesis	20–40 min
Mitosis	60–80 min
First cleavage	85–95 min

Source: Whitaker and Steinhardt, 1985.

[a]Approximate times based on data from *S. purpuratus* (15–17°C), *L. pictus* (16–18°C), and *A. punctulata* (18–20°C).

nuclei. But for fertilization to lead to development, changes must occur in the egg cytoplasm.*

The mature sea urchin egg is a metabolically sluggish cell that is reactivated by the entering sperm. This activation is merely a stimulus, however; it sets into action a preprogrammed set of metabolic events. The responses of the egg to the sperm can be divided into those "early" responses that occur within seconds of the cortical reaction and those "late" responses that take place several minutes after fertilization begins (Table 4.2).

Early responses

As we have seen, contact between sea urchin sperm and egg activates the two major blocks to polyspermy: the fast block, initiated by sodium influx into the cell, and the slow block, initiated by intracellular release of calcium ions. The activation of all eggs appears to depend on an increase in the concentration of free calcium ions within the egg. In protostomes such as snails and worms, at least part of the calcium usually enters the egg from outside. In deuterostomes such as fish, frogs, and sea urchins, the activation is accomplished by the release of calcium ions from the endoplasmic

---

*In certain salamanders, this developmental function of fertilization has been totally divorced from the genetic function. The silver salamander (*Ambystoma platineum*) is a hybrid subspecies consisting solely of females. Each female produces an egg with an unreduced chromosome number. This egg, however, cannot develop on its own, so the silver salamander mates with the male Jefferson salamander (*A. jeffersonianum*). The sperm from the male Jefferson salamander only stimulates the egg's development; it does not contribute genetic material (Uzzell, 1964). For details of this complex mechanism of procreation, see Bogart et al., 1989.

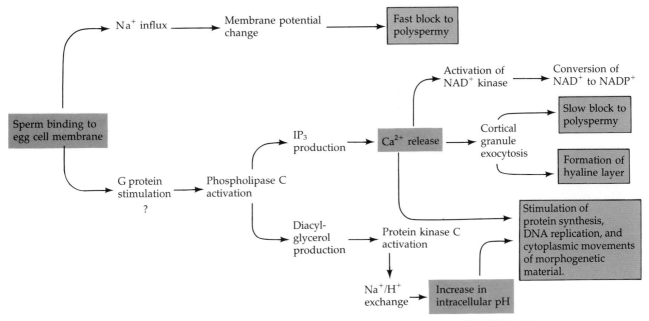

FIGURE 4.29
Model of possible interrelationships in the process of sea urchin fertilization. (After Epel, 1980, and L. A. Jaffe, personal communication.)

reticulum, resulting in a wave of calcium ions sweeping across the egg (Jaffe, 1983; Terasaki and Sardet, 1991).

This release of calcium ions is essential for activating the development of the embryo. If the calcium-chelating chemical EGTA is injected into the eggs, there is no exocytosis of cortical granules, no fertilization potential changes, no sperm decondensation, and no reinitiation of cell division (Kline, 1988). Conversely, eggs can be activated artificially in the absence of sperm by procedures that release free calcium into the oocyte. Steinhardt and Epel (1974) found that micromolar amounts of calcium ionophore A23187 elicit most of the egg responses characteristic of a normally fertilized egg. The elevation of the fertilization envelope, the rise of intracellular pH, the burst of oxygen utilization, and the increases in protein and DNA synthesis are each generated in their proper order. Moreover, this activation takes place in the total absence of calcium ions in the seawater. In most of these cases, development ceases before the first mitosis, because the eggs are still haploid and lack the sperm centriole needed for division.

This calcium release is responsible for activating a series of metabolic reactions (Figure 4.29). One of these is the activation of the enzyme NAD^+ kinase, which converts NAD^+ to $NADP^+$ (Epel et al., 1981). This change may have important consequences for lipid metabolism, since $NADP^+$ (but not NAD^+) can be used as a coenzyme for lipid biosynthesis. Thus, the switch from NAD^+ to $NADP^+$ may be important in the construction of the many new cell membrane components required during cleavage. Another effect of this change involves oxygen consumption. A burst of oxygen reduction is seen during fertilization, and much of this "respiratory burst" is used to crosslink the fertilization membrane. The enzyme responsible for this reduction of oxygen (to hydrogen peroxide) is also NADPH-dependent (Heinecke and Shapiro, 1989).

Late responses

Shortly after the calcium ion levels rise, the intracellular pH also increases. It is thought that these two ionic conditions (higher $[Ca^{2+}]$, lower $[H^+]$) act in concert to give the full spectrum of fertilization events, including

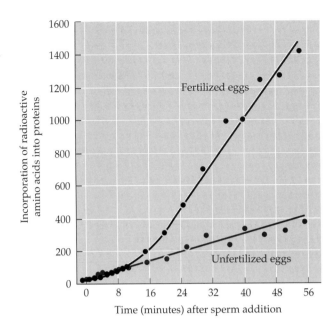

FIGURE 4.30
Protein synthesis in newly fertilized sea urchin eggs. [^{14}C]Leucine incorporation into the proteins of unfertilized eggs or fertilized eggs. (After Epel, 1967.)

FIGURE 4.31
Dual ionic control of protein synthesis during fertilization. (A) Alkalinization of egg by ammonia. Continuous recordings of intracellular pH as an egg is placed into various concentrations of NH$_4$Cl. (B) Protein synthesis as measured by the incorporation of radioactive valine into proteins under particular ionic conditions. The elevation of intracellular pH could be inhibited by fertilizing the eggs in sodium-free seawater or elevated by ammonium ions. The increase in intracellular free calcium concentration could be blocked by EGTA or elevated by A23187. (A after Shen and Steinhardt, 1978; B after Winkler et al., 1980.)

protein synthesis and DNA synthesis (Winkler et al., 1980; Whitaker and Steinhardt, 1982). The rise in intracellular pH begins with a second influx of sodium ions, causing a 1:1 exchange between sodium ions from the seawater and hydrogen ions from the egg. In the sea urchin egg, nearly 75 percent of the hydrogen ions are exchanged. This loss of hydrogen ions causes the pH to rise from 6.8 to 7.2 and brings about enormous changes in egg physiology (Shen and Steinhardt, 1978).

The late responses of fertilization brought about by these ionic changes include the activation of DNA synthesis and protein synthesis. The burst of protein synthesis usually occurs within several minutes after sperm entry and is not dependent on new messenger RNA synthesis (Figure 4.30). Rather, new protein synthesis utilizes mRNAs already present in the oocyte cytoplasm. (Much more will be said about this in Chapter 13.) Such a burst of protein synthesis can be induced artificially by placing unfertilized eggs into a solution containing ammonium ions (Winkler et al., 1980). These ions lose protons outside the cell, diffuse through the plasma membrane as NH$_3$, and once inside the cell, pick up protons to

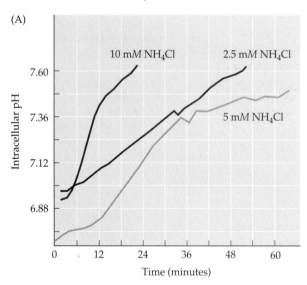

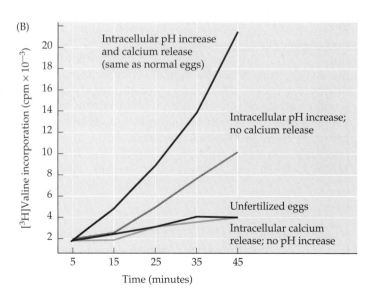

become NH_4^+. The resulting loss of free hydrogen ions from the cytoplasm is reflected in the increase of intracellular pH (Figure 4.31). Thus, bypassing the sodium–hydrogen exchange, ammonium ions are able to raise the pH of the eggs. In so doing, they activate protein synthesis and DNA synthesis. Conversely, agents that block the rise in pH also block these late fertilization events. When newly fertilized eggs are placed into solutions containing low concentrations of sodium ions and amiloride (a drug that inhibits sodium–hydrogen exchange), protein synthesis fails to occur, the movements of the egg and sperm pronuclei are prevented, and cell division fails to take place (Dube et al., 1985).

SIDELIGHTS & SPECULATIONS

Molecular regulators of development: G proteins and phosphoinositides

The activation of both sperm and eggs requires an increase in the concentration of intracellular calcium ions. Moreover, the activation of both gametes includes the calcium-mediated fusion of membrane-enclosed storage organelles (cortical granule, acrosomal vesicle) with the plasma membrane. During the past decade, a biochemical pathway has been discovered that signals calcium release for muscle contractions, cell growth, hormonal secretion, sensory perception, and the release of neurotransmitters (Berridge, 1993). It may also be responsible for the activation of sperm and egg. This sequence of events forms the **G protein–phosphoinositide pathway**.

The first element of the pathway is the transmembrane receptor protein. On one side of the plasma membrane, it specifically binds its ligand (such as acetylcholine, serotonin, or gonadotropin-releasing hormone). On the inside of the plasma membrane, it is connected to a **guanosine nucleotide-binding protein**, or **G protein**. When the ligand binds to its receptor, the receptor activates the G protein (Figure 4.32A). This activation dissociates the G protein into its subunits, which are then able to activate a set of enzymes called **phospholipase C**. These enzymes catalyze the hydrolysis of phosphatidylinositol 4,5-bisphosphate **(PIP$_2$)** into two **second messengers**: inositol 1,4,5-trisphosphate **(IP$_3$)** and diacylglycerol **(DAG)**. IP$_3$ is able to open calcium ion channels, and DAG is able to stimulate those proton pumps that enable the efflux of hydrogen ions from the cells. The result is the elevation of intracellular calcium ions and the increase of intracellular pH.

The question is whether these systems operate in gamete activation. Researchers have found G proteins in the membranes of sperm and eggs, and there is evidence that these proteins may activate the calcium-releasing phosphoinositides.

Activation of the sperm

The binding of the acrosome reaction-activating glycoproteins in sea urchins and mammals causes an influx of calcium ions into the sperm. G proteins have been found in the sperm membranes of all animal species studied (Ward and Kopf, 1993). The G proteins in mammalian acrosomal membranes appear to be activated by the binding of the sperm to ZP3 (Glassner et al., 1991; Ward et al., 1992). If activation of the G protein is inhibited, the sperm can still bind ZP3 (presumably through the carbohydrate chains), but neither the opening of the calcium channels nor the acrosome reaction occurs (Endo et al., 1987b; Wilde and Kopf, 1989; Florman et al., 1992). G proteins have also been found in the membranes of sea urchin sperm, and their activation seems to be required for increased motility and the acrosome reaction (Ramarao and Garbers, 1985; Singh et al., 1988). When sea urchin sperm is stimulated to undergo the acrosome reaction by the egg jelly, their internal pH increases, their internal concentrations of cAMP and IP$_3$ are elevated, and their membrane calcium channels open (Shapiro et al., 1985; Ward and Kopf, 1993). The target enzymes for the activated G proteins of sperm have not yet been identified.

Activation of the egg

"Just how the sperm triggers the explosive release of calcium in the egg is still something of a mystery" (Berridge, 1993). Recent evidence suggests that the ionic activation of the egg may be set off by the same G protein–phosphoinositide cascade. IP$_3$ can release sequestered calcium ions in eggs and in many other cell types (Swann and Whitaker, 1986; Berridge 1993), and increased concentrations of intracellular IP$_3$ are seen within 10 seconds after sea urchin eggs are fertilized (Ciapa and Whitaker, 1986). The release of calcium ions quickly follows the formation of IP$_3$. The importance of IP$_3$ to egg activation was demonstrated by Whitaker and Irvine (1984) and by Busa and his co-workers (1985). When they injected IP$_3$ directly into sea urchin and frog eggs, they activated the release of calcium ions and the cortical granule reaction. These responses were indistinguishable from the events seen during normal fertilization. Moreover, these IP$_3$-mediated effects could be thwarted by pre-injecting the egg with calcium chelating agents (Turner et al., 1986), thereby confirming that IP$_3$ stimulates the release of stored calcium in the egg.

IP$_3$-responsive calcium channels have been found in

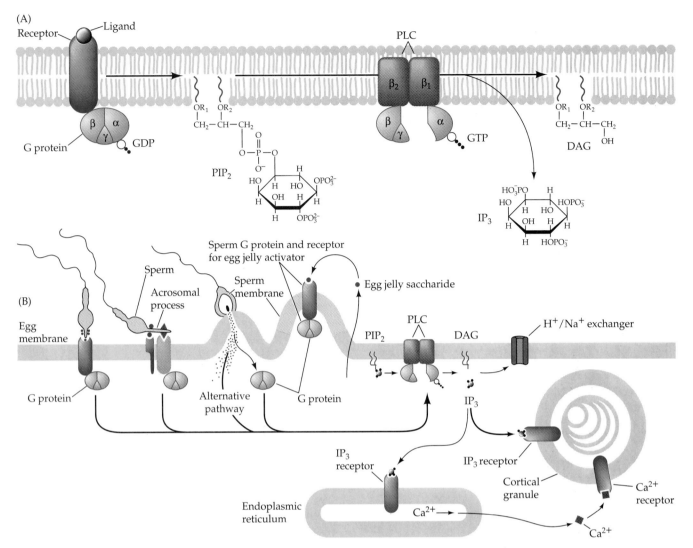

FIGURE 4.32

The G protein–phosphoinositide model of gamete activation. (A) The pathway is initiated when the transmembrane receptor is activated by binding ligand. This activation causes the binding of GTP to the G protein and its dissociation into active subunits. These subunits activate phospholipase C enzymes (PLC), which can catalyze the formation of DAG and IP$_3$. IP$_3$ can bind to a receptor to release calcium ions from the endoplasmic reticulum. The released calcium from the IP$_3$-responsive stores binds to calcium-dependent calcium release channels, which release more calcium ions. These released ions further activate calcium-dependent calcium release channels and propagate the wave of calcium ions across the cortex. Meanwhile, DAG (in the pres- *ence of the released calcium ions) activates protein kinase C. This protein kinase stimulates the sodium/hydrogen transporter to exchange cellular hydrogen ions for extracellular sodium ions, thereby leading to the increase in pH. (B) There are four hypotheses for how the G protein in sea urchin eggs might be activated. (1) The bindin receptor is linked to the G protein, as in (A). (2) The bindin receptor anchors the sperm while another cell membrane component activates a G-protein-linked receptor. (3) Cytoplasmic or membrane components of the sperm activate the egg G protein after cell fusion. (4) The sperm G proteins that had been activated by the egg jelly activate the egg enzymes after cell fusion.*

the egg endoplasmic reticulum. In sea urchin eggs, calcium release occurs in two steps. The first step involves the binding of IP$_3$ to these IP$_3$ receptors in the endoplasmic reticulum near the point of sperm entry (and probably in the cytoplasm just below the cortex). This effects a local release of calcium in the side of the egg near the sperm entry area (Ferris et al., 1989; Furuichi et al., 1989; Terasaki and Sardet, 1991). Once released, these calcium ions initiate the second phase of calcium release. The calcium ions bind

to calcium-sensitive (IP$_3$-*non*responsive) calcium receptors located in the cortical endoplasmic reticulum, adjacent to the egg plasma membrane and surrounding the cortical granule (Figure 4.24; McPherson et al., 1992). The binding of calcium ions to this receptor releases more calcium, and that released calcium can continue the wave by binding to more receptors, and so on. This wave of calcium ions is propagated throughout the cell, starting at the point of sperm entry; and the cortical granules, which will fuse with the cell membrane in the presence of high calcium concentrations, initiate a wave of exocytosis that follows the calcium ions (Vacquier, 1975; Rakow and Shen, 1990; McPherson et al., 1992). IP₃ is similarly found to release calcium ions in vertebrates, and blocking the IP₃ receptor in hamster eggs severely inhibits and retards the release of calcium at fertilization. In vertebrates, unlike sea urchins, waves of IP₃ may mediate calcium release without a separate calcium-responsive channel (Lechleiter and Clapham, 1992; Miyazaki et al., 1992).

The "second" second messenger, DAG, is thought to activate the membrane enzyme **protein kinase C,** which in turn activates the protein that exchanges sodium ions for hydrogen ions (Swann and Whitaker, 1985; Nishizuka, 1986; Shen and Burgart, 1986). Blocking the activity of protein kinase C in sea urchin eggs inhibits the alkalinization of the cytoplasm seen during normal fertilization (Shen and Buck, 1990). The Na^+/H^+ exchange protein also needs calcium ions for activity. Thus, both DAG and IP₃ are involved in the developmental activities of the egg (Figure 4.32A). The key regulatory step is the activation of phospholipase C, the enzyme that produces these two compounds. Jaffe and her co-workers have found G proteins in sea urchin and frog eggs, and when they injected G-protein activators into eggs, the cortical granules underwent exocytosis in the absence of sperm (Turner et al., 1986, 1987; Kline et al., 1991). This activation was inhibited when calcium chelators such as EGTA were added.

It appears, then, that a G protein may be involved in regulating the release of sequestered calcium ions and the exocytosis of the cortical granules. There are several ways that this might be done (Figure 4.32B). First, the binding of sperm to a receptor in the egg plasma membrane might change the conformation of the receptor such that it activates the G protein and initiates the cascade. This is a model seen in the study by Kline and co-workers (1988), who hypothesized that if a G protein mediated the fertilization events by being activated by a sperm-binding receptor, then that same G protein might be activated by a hormone if the egg contained a hormone receptor that could activate the G protein. They injected mRNA for the serotonin receptor or the acetylcholine receptor into frog

eggs. These cell-surface receptors were made and were seen on the egg cell membrane. In these instances, the eggs could be "fertilized" by the neural hormones serotonin and acetylcholine, and the cortical reaction was observed.

Similarly, when mouse oocytes are made to express the acetylcholine receptor on their cell membranes, the addition of acetylcholine causes the complete series of egg activation events, including the early events (calcium release, cortical granule exocytosis, zona modification) and the late events (recruitment and translation of maternal mRNAs, completion of meiosis, pronuclear migration, and DNA synthesis). Blocking the G protein inhibited these events of egg activation (Williams et al., 1992; Moore et al., 1993).

However, while a particular acrosomal peptide is capable of binding to egg membranes and initiating all the fertilization events in the worm *Urechis* (Gould and Stephano, 1991), the only sperm-binding protein so far identified in sea urchins (the bindin receptor mentioned earlier) does not appear to interact with G proteins. This leads to the second model, wherein the bindin receptor merely acts as an anchor to allow the interaction of the proteins that *do* activate the G protein.

In the third model, it is the *fusion* of the sperm and egg membranes that activates the fertilization reaction, not their binding. McCulloh and Chambers (1992) have electrophysiological evidence that sea urchin egg activation does not occur until *after* sperm and egg cytoplasms are joined. They suggest that the egg-activating components are located on the sperm cell membrane or cytoplasm. It is even possible that when the fusion of gamete membranes occurs, the sperm membrane G proteins (activated by the egg jelly to initiate the acrosome reaction) could activate the polyphosphoinositide cascade in the egg (model 4).

This polyphosphoinositide cascade is also seen later in development. It is probably very important in the creation of vertebrate mesodermal cells (Maslanski et al., 1992), controlling the rate of cell growth (Posada and Cooper, 1992), and determining the dorsal–ventral axis (Berridge, 1993). It is still not certain that it is the pathway activating the sperm and egg, but the evidence that it might be is growing.

Rearrangement of egg cytoplasm

Fertilization can initiate radical displacements of the egg's cytoplasmic materials. These rearrangements of oocyte cytoplasm are often crucial for cell differentiation later in development. As we will see in Chapters 14 and 15, the cytoplasm of the egg frequently contains **morphogenetic determinants** that become segregated into specific cells during cleavage. These determinants ultimately lead to the activation or repression of specific genes and thereby confer certain properties to the cells that incorporate them. The correct spatial arrangement of these determinants is crucial for proper development.

In some species, the rearrangement of these determinants into their

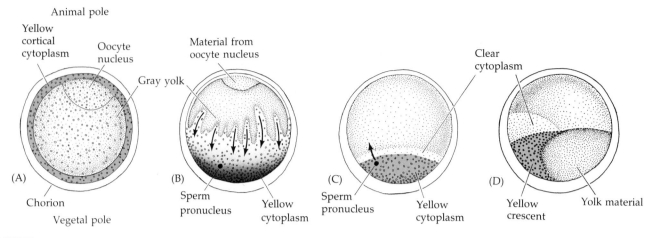

Animal pole

Yellow
cortical
cytoplasm

Oocyte
nucleus

Material from
oocyte nucleus

Clear
cytoplasm

Gray yolk

(A)

Chorion

Vegetal pole

(B)

Sperm
pronucleus

Yellow
cytoplasm

(C)

Sperm
pronucleus

Yellow
cytoplasm

(D)

Yellow
crescent

Yolk material

FIGURE 4.33
Cytoplasmic rearrangement in the egg of the tunicate *Styela partita*. (A) Before fertilization, yellow cortical cytoplasm surrounds gray yolky cytoplasm. (B) After sperm entry, the yellow cortical cytoplasm and the clear cytoplasm derived from the breakdown of the oocyte nucleus stream vegetally toward the sperm. (C) As the sperm pronucleus migrates animally toward the egg pronucleus, the yellow and clear cytoplasms move with it. (D) The final positions of the clear and yellow cytoplasms. These mark the positions where the cells give rise to the mesenchyme and muscles, respectively. (After Conklin, 1905.)

required orientation can be visualized because cytoplasmic pigment granules are present. One such example is the egg of the tunicate *Styela partita* (Conklin, 1905). The unfertilized egg of this animal is seen in Figure 4.33A. A central gray cytoplasm is enveloped by a cortical layer containing yellow lipid inclusions. During meiosis, the breakdown of the nucleus releases a clear substance that accumulates in the animal (upper) hemisphere of the egg. Within 5 minutes of sperm entry, the inner clear and cortical yellow cytoplasms migrate into the vegetal (lower) hemisphere of the egg. As the male pronucleus migrates from the vegetal pole to the equator of the cell along the future posterior side of the embryo, the lipid inclusions migrate with it. This migration forms a **yellow crescent,** extending from the vegetal pole to the equator (Figure 4.33B–D), and brings the yellow cytoplasm into the area where muscle cells will later form in the tunicate larva. The movement of these cytoplasmic regions is dependent upon microtubules that probably are generated by the sperm centriole and by the wave of calcium ions that may contract the animal pole cytoplasm (Sawada and Schatten, 1989; Speksnijder et al., 1990).

Cytoplasmic movement is also seen in amphibian eggs. In frogs, a single sperm can enter anyplace on the animal hemisphere; when it does,

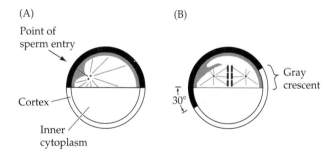

(A)

Point of
sperm entry

Cortex

Inner
cytoplasm

(B)

Gray
crescent

30°

FIGURE 4.34
Reorganization of cytoplasm in the newly fertilized frog egg. (A) Schematic cross section of an egg midway through the first cleavage cycle. The egg has radial symmetry about its animal–vegetal axis. The sperm has entered at one side and the sperm nucleus is migrating inward. The cortex is represented like that of *Rana*, with a heavily pigmented animal hemisphere and a transparent vegetal hemisphere. (B) About 80 percent of the way into first cleavage, the cortical cytoplasm rotates 30° relative to the internal cytoplasm. This rotation will be seen to be important in that gastrulation will begin in that region opposite the point of sperm entry where the greatest displacement of cytoplasm occurs. (After Gerhart et al., 1989.)

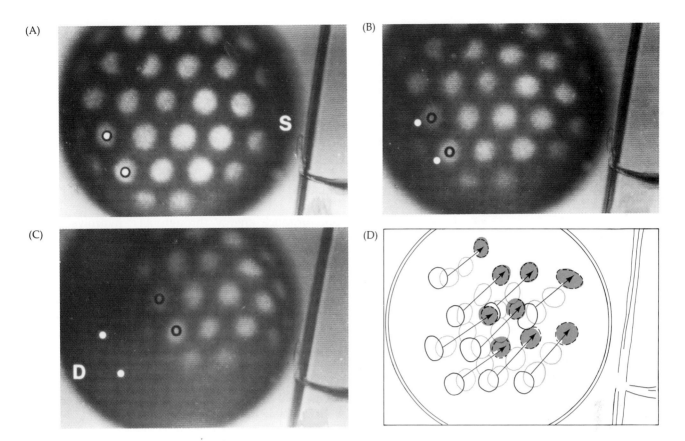

(A)   (B)   (C)   (D)

it changes the cytoplasmic pattern of the egg. Originally, the egg is radially symmetric about the animal–vegetal axis. After sperm entry, however, the cortical (outer) cytoplasm shifts about 30° toward the point of sperm entry, relative to the inner cytoplasm (Manes and Elinson, 1980; Vincent et al., 1986). In some frogs (such as *Rana*), one region of the egg that had formerly been covered by the dark cortical cytoplasm of the animal hemisphere is now exposed (Figure 4.34). This underlying cytoplasm, located near the equator on the side opposite the point of sperm entry, contains diffuse pigment granules and therefore appears gray. Thus, this region has been referred to as the **gray crescent** (Roux, 1887; Ancel and Vintenberger, 1948). As we will see in subsequent chapters, the gray crescent marks the region where gastrulation is initiated in amphibian embryos.

In frogs such as *Xenopus*, where no gray crescent is seen, one can still observe the rotation of the cortical cytoplasm relative to the internal, subcortical, cytoplasm. This movement was demonstrated by Vincent and his colleagues (1986). These investigators imprinted a hexagonal grid of dye (Nile Blue) onto the cytoplasm beneath the cortex while applying another type of dye (a fluorescein-bound lectin) to the egg surface. When the egg was held in one position by embedding it in gelatin, the Nile Blue dots were seen to rotate 30° relative to the fluorescent lectin spots (Figure 4.35). In normal, unembedded eggs, the egg surface is thought to rotate while the subcortical cytoplasm, heavy with yolk platelets, remains stabilized by gravity.

The motor for these cytoplasmic movements in amphibian eggs appears to be an array of parallel microtubules that lie between the cortical and inner cytoplasm of the vegetal hemisphere parallel to the direction of cytoplasmic rotation. These microtubular tracks are first seen immediately before the rotation commences, and they disappear when rotation ceases

FIGURE 4.35
The rotation of subcortical cytoplasm relative to the cell surface cytoplasm. (A) Recently fertilized eggs were imprinted with a hexagonal grid of Nile Blue dye (which stains the lipids in the yolk platelets). The egg was embedded in gelatin, and the original positions of some of the dots were marked on the cell surface with fluorescein (circles in A). The point of sperm entry is marked with an S. (B,C) As the first cell cycle progressed, the dots of the subcortical cytoplasm shifted roughly 30° with respect to the outer immobilized surface of the egg. The site on the egg marking the future dorsal surface of the embryo is marked with a D. (D) Summary of these movements in the vegetal (lower) region of the egg. (From Vincent et al., 1986, photographs courtesy of J. C. Gerhart.)

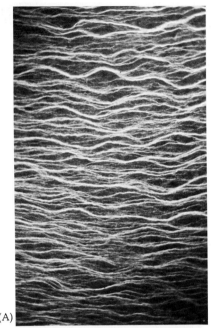

(A)

(B)

(C)

FIGURE 4.36
Parallel arrays of microtubules extend along the vegetal hemisphere along the future dorsal–ventral axis. (A) Parallel arrays of microtubules seen in the second half of the first cell cycle by fluorescent antibodies to tubulin. (B) Prior to cytoplasmic rotation (about midway through the first cell cycle), no array of microtubules can be seen. (C) At the end of cytoplasmic rotation, the microtubules depolymerize. (From Elinson and Rowning, 1988, courtesy of R. Elinson.)

(Figure 4.36; Elinson and Rowning, 1988). Treating the egg with colchicine or ultraviolet radiation at the beginning of rotation stops the formation of these microtubules, thereby inhibiting the cytoplasmic rotations. Using antibodies that bind to the microtubules, Houliston and Elinson (1991a) found that these tracks are formed from sperm- and egg-derived microtubules, and that the sperm centriole directs the polymerization of these microtubules so that they grow into the vegetal region of the egg. Upon reaching the vegetal cortex, these microtubules angle away from the point of sperm entry, toward the vegetal pole. The off-center position of the sperm centriole as it initiates microtubule polymerization provides the directionality to the rotation. The motive force for the rotation is possibly provided by the ATPase **kinesin.** Like dynein and myosin, kinesin is able to attach to fibers and produce energy through ATP hydrolysis. This ATPase is located on the vegetal microtubules and the membranes of the cortical endoplasmic reticulum (Houliston and Elinson, 1991b).

The movement of the cortical cytoplasm with respect to the inner cytoplasm causes profound movement within the inner cytoplasm. Danilchik and Denegre (1991) have labeled yolk platelets with trypan blue and watched their movement by fluorescent microscopy (the bound dye fluoresces red). During the middle part of the first cell cycle, a mass of central egg cytoplasm flows from the presumptive ventral (belly) to the future dorsal (back) side of the embryo (Plate 6). By the end of first division, the cytoplasm of the prospective dorsal side of the embryo is distinctly different from that of the prospective vegetal side. What had been a radially symmetric embryo is now a bilaterally symmetric embryo.

As we will see in Chapters 6 and 16, these cytoplasmic movements initiate a cascade of events that determine the dorsal–ventral axis of the frog. Indeed, the parallel microtubules that allow these rearrangements appear to stretch along the future dorsal–ventral axis. (Klag and Ubbels, 1975; Gerhart et al., 1983).

Preparation for cleavage

The increase in levels of intracellular free calcium ions also sets in motion the apparatus for cell division. The mechanisms by which cleavage is initiated probably differ between species depending upon the stage of meiosis at which fertilization occurs. In all known species, as we will see in the next chapter, the rhythm of cell divisions is regulated by the synthesis and degradation of **cyclin.** Cyclin keeps the cells in metaphase, and the breakdown of cyclin enables the cells to return to interphase. In addition to their other activities, calcium ions also appear to initiate the degradation of cyclin (Watanabe et al., 1991). Once this cyclin is degraded, the cycles of cell division can begin anew.

Cleavage, the processes that create a multicellular organism and prepare the embryo for organogenesis, has a special relationship to these cytoplasmic regions. In tunicate embryos the first cleavage bisects the egg into mirror-image duplicates. From that stage on, every division on one side of that cleavage furrow has a mirror-image division on the opposite side. Similarly, the gray crescent is bisected by the first cleavage furrow in amphibian eggs. Thus, the position of the first cleavage is not random but tends to be specified by the point of sperm entry and the subsequent rotation of egg cytoplasm. The coordination of cleavage plane and cytoplasmic rearrangements is probably mediated through the microtubules of the sperm aster (Manes et al., 1978; Gerhart et al., 1981; Elinson, 1985).

Toward the end of the first cell cycle, then, the cytoplasm is rearranged, the pronuclei have met, DNA is replicating, and new proteins are being translated. The stage is set for the development of a multicellular organism.

Afzelius, B. A. 1976. A human syndrome caused by immotile cilia. *Science* 193: 317–319.

Amos, L. A. and Klug, A. 1974. Arrangement of subunits in flagellar microtubules. *J. Cell Sci.* 14: 523–549.

Ancel, P. and Vintenberger, P. 1948. Recherches sur le determinisme de la symmetrie bilatérale dans l'oeuf des amphibiens. *Bull. Biol. Fr. Belg.* [Suppl.] 31: 1–182.

Austin, C. R. 1960. Capacitation and the release of hyaluronidase. *J. Reprod. Fert.* 1: 310–311.

Austin, C. R. 1965. *Fertilization.* Prentice-Hall, Englewood Cliffs, New Jersey.

Baltz, J. M., Katz, D. F. and Cone, R. A. 1988. The mechanics of sperm–egg interaction at the zona pellucida. *Biophys. J.* 54: 643–654.

Barlow, D. P., Stöger, R., Herrmann, B. G., Saito, K. and Schweifer, N. 1991. The mouse insulin-like growth factor type -2 receptor is imprinted and closely linked to the *Tme* locus. *Nature* 349: 84–87.

Bavister, B. D. 1980. Recent progress in the study of early events in mammalian fertilization. *Dev. Growth Differ.* 22: 385–402.

Bentley, J. K., Shimomura, H. and Garbers, D. L. 1986. Retention of a functional resact receptor in isolated sperm plasma membranes. *Cell* 45: 281–288.

Berridge, M. J. 1993. Inositol triphosphate and calcium signalling. *Nature* 361: 315–325.

Bestor, T. M. and Schatten, G. 1981. Anti-tubulin immunofluorescence microscopy of microtubules present during the pronuclear movements of sea urchin fertilization. *Dev. Biol.* 88: 80–91.

Bleil, J. D. and Wassarman, P. M. 1980. Mammalian sperm and egg interaction: Identification of a glycoprotein in mouse-egg zonae pellucidae possessing receptor activity for sperm. *Cell* 20: 873–882.

Bleil, J. D. and Wassarman, P. M. 1986. Autoradiographic visualization of the mouse egg's sperm receptor bound to sperm. *J. Cell Biol.* 102: 1363–1371.

Bleil, J. D. and Wassarman, P. M. 1988. Galactose at the nonreducing terminus of O-linked oligosaccharides of mouse egg zona pellucida glycoprotein ZP3 is essential for the glycoprotein's sperm receptor activity. *Proc. Natl. Acad. Sci. USA* 85: 6778–6782.

Bleil, J. D. and Wassarman, P. M. 1990. Identification of a ZP3-binding protein on acrosome-intact mouse sperm by photoaffinity crosslinking. *Proc. Natl. Acad. Sci. USA* 87: 5563–5567.

Bleil, J. D., Greve, J. M. and Wassarman, P. M. 1988. Identification of a secondary sperm receptor in the mouse egg zona pellucida: Role in maintenance of binding of acrosome-reacted sperm to eggs. *Dev. Biol.* 28: 376–385.

Blobel, C. P, Wolfsberg, T. G., Turck, C. W., Myles, D. G., Primakoff, P. and White, J. M. 1992. A potential fusion peptide and an integrin domain in a protein active in sperm–egg fusion. *Nature* 356: 248–251.

Bogart, J. P., Elinson, R. P. and Licht, L. E. 1989. Temperature and sperm incorporation in polyploid salamanders. *Science* 246: 1032–1034.

Boveri, T. 1902. On multipolar mitosis as a means of analysis of the cell nucleus. (Translated by S. Glueckshon-Waelsch.) *In* B. H. Willier and J. M. Oppenheimer (eds.), *Foundations of Experimental Embryology.* Hafner, New York, 1974.

Boveri, T. 1907. Zellenstudien VI. Die Entwicklung dispermer Seeigeleier. Ein Beiträge zur Befruchtungslehre und zur Theorie des Kernes. *Jena Zeit. Naturwiss.* 43: 1–292.

Braden, A. W. H. and Austin, C. R. 1954. The number of sperm about the eggs in mammals and its significance for normal fertilization. *Aust. J. Biol. Sci.* 7: 543–551.

Busa, W. B., Ferguson, J. E., Joseph, S. K., Williamson, J. R. and Nuccitelli, R. 1985. Activation of frog (*Xenopus laevis*) eggs by inositol triphosphate. I. Characterization of Ca^{2+} release from intracellular stores. *J. Cell Biol.* 100: 677–682.

Calvin, H. I. and Bedford, J. M. 1971. Formation of disulfide bonds in the nucleus and accessory structures of mammalian spermatozoa during maturation in the epididymis. *J. Reprod. Fertil.* [Suppl.] 13: 65–75.

Carroll, E. J. and Epel, D. 1975. Isolation and biological activity of the proteases released by sea urchin eggs following fertilization. *Dev. Biol.* 44: 22–32.

Chambers, E. L., Pressman, B. C. and Rose, B. 1974. The activity of sea urchin eggs by the divalent ionophores A23187 and X-537A. *Biochem. Biophys. Res. Commun.* 60: 126–132.

Chandler, D. E. and Heuser, J. 1979. Membrane fusion during secretion: Cortical granule exocytosis in sea urchin eggs as studied by quick-freezing and freeze fracture. *J. Cell Biol.* 83: 91–108.

Cherr, G. N., Lambert, H., Meizel, S. and Katz, D. F. 1986. In vitro studies of the golden hamster sperm acrosomal reaction: Completion on zona pellucida and induction by homologous zonae pellucidae. *Dev. Biol.* 114: 119–131.

Ciapa, B. and Whitaker, M. 1986. Two phases of inositol polyphosphate and diacylglycerol production at fertilization. *FEBS Lett.* 195: 347–351.

Clermont, Y. and Leblond, C. P. 1955. Spermiogenesis of man, monkey, and other animals as shown by the "periodic acid-Schiff" technique. *Am. J. Anat.* 96: 229–253.

Colwin, A. L. and Colwin, L. H. 1963. Role of the gamete membranes in fertilization in *Saccoglossus kowalevskii* (Enteropneustra). I. The acrosome reaction and its changes in early stages of fertilization. *J. Cell Biol.* 19: 477–500.

Colwin, L. H. and Colwin, A. L. 1960. Formation of sperm entry holes in the vitelline membrane of *Hydroides hexagonis* (Annelida) and evidence of their lytic origin. *J. Biophys. Biochem. Cytol.* 7: 315–320.

Conklin, E. G. 1905. The orientation and cell-lineage of the ascidian egg. *J. Acad. Nat. Sci. Phila.* 13: 5–119.

Corselli, J. and Talbot, P. 1987. In vivo penetration of hamster oocyte–cumulus complexes using physiological numbers of sperm. *Dev. Biol.* 122: 227–242.

Cross, N. L. and Elinson, R. P. C. 1980. A fast block to polyspermy in frogs mediated by changes in the membrane potential. *Dev. Biol.* 75: 187–198.

Dan, J. C. 1967. Acrosome reaction and lysins. *In* C. B. Metz and A. Monroy (eds.), *Fertilization,* Vol. 1. Academic Press, New York, pp. 237–367.

Danilchik, M. V. and Denegre, J. M. 1991. Deep cytoplasmic rearrangements during early development in *Xenopus laevis. Development* 111: 845–856.

Davis, B. K. 1981. Timing of fertilization in mammals: Sperm cholesterol/phospholipid ratio as determinant of capacitation interval. *Proc. Natl. Acad. Sci. USA* 78: 7560–7564.

Dawid, I. B. and Blackler, A. W. 1972. Maternal and cytoplasmic inheritance of mitochondria in *Xenopus. Dev. Biol.* 29: 152–161.

DeChiara, T. M., Robertson, E. J. and Efstradiatis, A. 1991. Parental imprinting of the mouse insulin-like growth factor II gene. *Cell* 64: 849–859.

De Robertis, E. D. P., Saez, F. A. and De Robertis, E. M. F. 1975. *Cell Biology,* 6th Ed., Saunders, Philadelphia.

Dube, F., Schmidt, T., Johnson, C. H. and Epel, D. 1985. The hierarchy of requirements for an elevated pH during early development of sea urchin embryos. *Cell* 40: 657–666.

Eisen, A. and Reynolds, G. T. 1985. Sources and sinks for the calcium release during fertilization of single sea urchin eggs. *J. Cell Biol.* 100: 1522–1527.

Elinson, R. P. 1985. Changes in levels of polymeric tubulin associated with activation and dorsoventral polarization of the frog egg. *Dev. Biol.* 109: 224–233.

Elinson, R. P. 1986. Fertilization in amphibians: The ancestry of the block to polyspermy. *Int. Rev. Cytol.* 101: 59–100.

Elinson, R. P. and Rowning, B. 1988. A transient array of parallel microtubules in frog eggs: Potential tracks for a cytoplasmic rotation that specifies the dorso-ventral axis. *Dev. Biol.* 128: 185–197.

Endo, Y. G., Kopf, G. S. and Schultz, R. M. 1987a. Effects of phorbol ester on mouse eggs: Dissociation of sperm receptor activity from acrosome reaction-inducing activity of the mouse zona pellucida protein, ZP3. *Dev. Biol.* 123: 574–577.

Endo, Y., Lee, M. A. and Kopf, G. S. 1987b. Evidence for the role of a guanosine nucleotide-binding regulatory protein in the zona pellucida-induced mouse sperm acrosome reaction. *Dev. Biol.* 119: 210–216.

Epel, D. 1967. Protein synthesis in sea urchin eggs: A "late" response to fertilization. *Proc. Natl. Acad. Sci. USA* 57: 899–906.

Epel, D. 1977. The program of fertilization. *Sci. Am.* 237(5): 128–138.

Epel, D. 1980. Fertilization. *Endeavour N.S.* 4: 26–31.

Epel, D., Patton, C., Wallace, R. W. and Cheung, W. Y. 1981. Calmodulin activates NAD kinase of sea urchin eggs: An early response. *Cell* 23: 543–549.

Ferris, C. D., Huganir, R. L., Supattapone, S. and Snyder, S. H. 1989. Purified inositol 1,4,5-trisphosphate receptor mediates calcium flux in reconstituted lipid vesicles. *Nature* 342: 87–89.

Florman, H. M. and Storey, B. T. 1982. Mouse gamete interactions: The zona pellucida is the site of the acrosome reaction leading to fertilization in vitro. *Dev. Biol.* 91: 121–130.

Florman, H. M. and Wassarman, P. M. 1985. O-linked oligosaccharides of mouse egg ZP3 account for its sperm receptor activity. *Cell* 41: 313–324.

Florman, H. M., Bechtol, K. B. and Wassarman, P. M. 1984. Enzymatic dissection of the functions of the mouse egg's receptor for sperm. *Dev. Biol.* 106: 243–255.

Florman, H, M., Corron, M. E., Kim, T. D.-H. and Babcock, D. F. 1992. Activation of voltage-dependent calcium channels of mammalian sperm is required for zona pellucida-induced acrosomal exocytosis. *Dev. Biol.* 152: 304–314.

Foerder, C. A. and Shapiro, B. M. 1977. Release of ovoperoxidase from sea urchin eggs hardens fertilization membrane with tyrosine crosslinks. *Proc. Natl. Acad. Sci. USA* 74: 4214–4218.

Fol, H. 1877. Sur le commencement de l'hémogénie chez divers animaux. *Arch. Zool. Exp. Gén.* 6: 145–169.

Foltz, K. R., Partin, J. S. and Lennarz, W. J. 1993. Sea urchin egg receptor for sperm: Sequence similarity of binding domain to hsp70. *Science* 259: 1421–1425.

Franklin, L. E. 1970. Fertilization and the role of the acrosomal reaction in non-mammals. *Biol. Reprod.* [Suppl.] 2: 159–176.

Fulton, B. P. and Whittingham, D. G. 1978. Activation of mammalian oocytes by intracellular injection of calcium. *Nature* 273: 149–151.

Furuichi, T., Yoshikawa, S., Miyawaki, A., Wada, K., Maeda, N. and Mikoshiba, K. 1989. Primary structure and functional expression of the inositol 1,4,5-trisphosphate-binding protein P400. *Nature* 342: 32–38.

Garbers, D. L., Tubb, D. J. and Kopf, G. S. 1980. Regulation of sea urchin sperm cAMP-dependent protein kinases by an egg associated factor. *Biol. Reprod.* 22: 526–532.

Garbers, D. L., Kopf, G. S., Tubb, D. J. and Olson, G. 1983. Elevation of sperm adenosine 3':5'-monophosphate concentrations by a fucose sulfate-rich complex associated with eggs. I. Structural characterization. *Biol. Reprod.* 29: 1211–1220.

Gardiner, D. M. and Grey, R. D. 1983. Membrane junctions in *Xenopus* eggs: Their distribution suggests a role in calcium regulation. *J. Cell Biol.* 96: 1159–1163.

Gerhart, J., Ubbels, G., Black, S., Hara, K. and Kirschner, M. 1981. A reinvestigation of the role of the grey crescent in axis formation in *Xenopus laevis*. *Nature* 292: 511–516.

Gerhart, J., Black, S., Gimlich, R. and Scharf, S. 1983. Control of polarity in the amphibian egg. *In* W. R. Jeffery and R. A. Raff (eds.), *Time, Space, and Pattern in Embryonic Development*. Alan R. Liss, New York, pp. 261–286.

Gerhart, J., Danilchik, M., Doniach, T., Roberts, S., Rowning, B. and Stewart, R. 1989. Cortical rotation of the *Xenopus* egg: Consequences for the anteroposterior pattern of embryonic dorsal development. *Development* 1989 [Suppl.]: 37–51.

Giles, R. E., Blanc, H., Cann, H. M. and Wallace, D. C. 1980. Maternal inheritance of human mitochondrial DNA. *Proc. Natl. Acad. Sci. USA* 77: 6715–6719.

Gilkey, J. C., Jaffe, L. F., Ridgway, E. G. and Reynolds, G. T. 1978. A free calcium wave traverses the activating egg of the medaka, *Oryzias latipes*. *J. Cell Biol.* 76: 448–466.

Ginzburg, A. S. 1985. Phylogenetic changes in the type of fertilization. *In* J. Mlíkovsky and V. J. A. Novák (eds.), *Evolution and Morphogenesis*. Academia, Prague, pp. 459–466.

Glabe, C. G. 1985. Interaction of the sperm adhesive protein, bindin, with phospholipid vesicles. II. Bindin induces the fusion of mixed–phase vesicles that contain phosphatidylcholine and phosphatidylserine in vitro. *J. Cell Biol.* 100: 800–806.

Glabe, C. G. and Lennarz, W. J. 1979. Species-specific sperm adhesion in sea urchins: A quantitative investigation of bindin-mediated egg agglutination. *J. Cell Biol.* 83: 595–604.

Glabe, C. G. and Vacquier, V. D. 1977. Species-specific agglutination of eggs by bindin isolated from sea urchin sperm. *Nature* 267: 836–838.

Glabe, C. G. and Vacquier, V. D. 1978. Egg surface glycoprotein receptor for sea urchin sperm bindin. *Proc. Natl. Acad. Sci. USA* 75: 881–885.

Glassner, M., Jones, J., Kligman, I., Woolkalis, M. J., Gerton, G. L. and Kopf, G. S. 1991. Immunocytochemical and biochemical characterization of guanine nucleotide-binding regulatory proteins in mammalian spermatozoa. *Dev. Biol.* 146: 438–450.

González-Martínez, M. T., Guerrero, A., Morales, E., de la Torre, L. and Darszon, A. 1992. A depolarization can trigger Ca^{2+} uptake and the acrosome reaction when preceded by a hyperpolarization in *L. pictus* sea urchin sperm. *Dev. Biol.* 150: 193–202.

Gould, M. and Stephano, J. L. 1987. Electrical response of eggs to acrosomal protein similar to those induced by sperm. *Science* 235:1654–1656.

Gould, M. and Stephano, J. L. 1991. Peptides from sperm acrosomal protein that activate development. *Dev. Biol.* 146: 509–518.

Gould-Somero, M., Jaffe, L. A. and Holland, L. Z. 1979. Electrically mediated fast polyspermy block in eggs of the marine worm, *Urechis caupo*. *J. Cell Biol.* 82: 426–440.

Green, G. R. and Poccia, E. L. 1985. Phosphorylation of sea urchin sperm H1 and H2B histones precedes chromatin decondensation and H1 exchange during pronuclear formation. *Dev. Biol.* 108: 235–245.

Gundersen, G. G., Medill, L. and Shapiro, B. M. 1986. Sperm surface proteins are incorporated into the egg membrane and cytoplasm after fertilization. *Dev. Biol.* 113: 207–217.

Gwatkin, R. B. L. 1976. Fertilization. *In* G. Poste and G. L. Nicolson (eds.), *The Cell Surface in Animal Embryogenesis and Development*. Elsevier North-Holland, New York, pp. 1–53.

Gyllensten, U., Wharton, D., Josefson, A. and Wilson, A. 1991. Paternal inheritance of mitochondrial DNA in mice. *Nature* 352: 255–258.

Hafner, M., Petzelt, C., Nobiling, R., Pawley, J. B., Kramp, D. and Schatten, G. 1988. Wave of free calcium at fertilization in the sea urchin egg visualized with Fura-2. *Cell Motil. Cytoskel.* 9: 271–277.

Hamaguchi, M. S. and Hiramoto, Y. 1980. Fertilization process in the heart-urchin, *Clypaester japonicus*, observed with a differential interference microscope. *Dev. Growth Differ.* 22: 517–530.

Hamaguchi, M. S. and Hiramoto, Y. 1981. Activation of sea urchin eggs by microinjection of calcium buffers. *Exp. Cell Res.* 134: 171–179.

Hartsoeker, N. 1694. *Essai de dioptrique.* Paris.

Heinecke, J. W. and Shapiro, B. M. 1989. Respiratory oxygen burst of fertilization. *Proc. Natl. Acad. Sci. USA* 86: 1259–1263.

Hertwig, O. 1877. Beiträge zur Kenntniss der Bildung, Befruchtung, und Theilung des theirischen Eies. *Morphol. Jahr.* 1: 347–452.

Hollinger, T. G. and Schuetz, A. W. 1976. "Cleavage" and cortical granule breakdown in *Rana pipiens* oocytes induced by direct microinjection of calcium. *J. Cell Biol.* 71: 395–401.

Hong, K. and Vacquier, V. D. 1986. Fusion of liposomes induced by a cationic protein from the acrosomal granule of abalone spermatozoa. *Biochemistry* 25: 543–549.

Houliston, E. and Elinson, R. P. 1991a. Patterns of microtubule polymerization relating to cortical rotation in *Xenopus laevis* eggs. *Development* 112: 107–117.

Houliston, E. and Elinson, R. P. 1991b. Evidence for the involvement of microtubules, endoplasmic reticulum, and kinesin in cortical rotation of fertilized frog eggs. *J. Cell Biol.* 114: 1017–1028.

Hunter, R. H. F. 1989. Ovarian programming of gamete progression and maturation in the female genital tract. *Zool. J. Linn. Soc.* 95: 117–124.

Hylander, B. L. and Summers, R. G. 1982. An ultrastructural and immunocytochemical localization of hyaline in the sea urchin egg. *Dev. Biol.* 93: 368–380.

Iwao, Y. and Jaffe, L. A. 1989. Evidence that the voltage-dependent component in the fertilization porcess is contributed by the sperm. *Dev. Biol.* 134: 446–451.

Jacobs, P. A., Wilson, C. M., Sprenkle, J. A., Rosenshein, N. B. and Migeon, B. R. 1980. Mechanism of origin of complete hydatidiform moles. *Nature* 286: 714–716.

Jaffe, L. A. 1976. Fast block to polyspermy in sea urchins is electrically mediated. *Nature* 261: 68–71.

Jaffe, L. A. 1980. Electrical polyspermy block in sea urchins: Nicotine and low sodium experiments. *Dev. Growth Differ.* 22: 503–507.

Jaffe, L. A. and Cross, N. L. 1983. Electrical properties of vertebrate oocyte membranes. *Biol. Reprod.* 30: 50–54.

Jaffe, L. A. and Gould, M. 1985. Polyspermy-preventing mechanisms. *Biol. Fert.* 3: 223–250.

Jaffe, L. F. 1983. Sources of calcium in egg activation: A review and hypothesis. *Dev. Biol.* 99: 265–276.

Jones, R., Brown, C. R. and Lancaster, R. T. 1988. Carbohydrate-binding properties of boar sperm proacrosin and assessment of its role in sperm–egg recognition and adhesion during fertilization. *Development* 102: 781–792.

Just, E. E. 1919. The fertilization reaction in *Echinarachnius parma*. *Biol. Bull.* 36: 1–10.

Klag, J. J. and Ubbels, G. A. 1975. Regional morphological and cytochemical differentiation of the fertilized egg of *Discoglossus pictus* (Anura). *Differentiation* 3: 15–20.

Kline, D. 1988. Calcium-dependent events at fertilization of the frog egg: Injection of a calcium buffer blocks ion channel opening, exocytosis, and formation of pronuclei. *Dev. Biol.* 126: 346–361.

Kline, D., Simoncini, L., Mandel, G., Maue, R., Kado, R. T. and Jaffe. L. A. 1988. Fertilization events induced by neurotransmitters after injection of mRNA into *Xenopus* eggs. *Science* 241: 464–467.

Kline, D, Kopf, G., Muncy, L. F. and Jaffe, L. A. 1991. Evidence for the involvement of a pertussis toxin-insensitive G-protein in egg activation of the frog *Xenopus laevis*. *Dev. Biol.* 143: 218–229.

Kvist, U., Afzelius, B. A. and Nilsson, L. 1980. The intrinsic mechanism of chromatin decondensation and its activation in human spermatozoa. *Dev. Growth Differ.* 22: 543–554.

Langlais, J., Kan, F. W. K., Granger, L., Raymond, L., Bleau, G. and Roberts, K. D. (1988). Identification of sterol acceptors that stimulate cholesterol efflux from human spermatozoa during in vitro capacitation. *Gamete Res.* 20: 185–201.

Leeuwenhoek, A. van. 1685. Letter to the Royal Society of London. Quoted in E. G. Ruestow, 1983, Images and ideas: Leeuwenhoek's perception of the spermatozoa. *J. Hist. Biol.* 16: 185–224.

Lechleiter, J. D. and Clapham, D. E. 1992. Molecular mechanisms of intracellular calcium excitability in *X. laevis* oocytes. *Cell* 69: 283–294.

Levine, A. E., Walsh, K. A. and Fodor, E. J. B. 1978. Evidence of an acrosin-like enzyme in sea urchin sperm. *Dev. Biol.* 63: 299–306.

Leyton, L. and Saling, P. 1989a. 95 kd sperm proteins bind ZP3 and serve as tyrosine kinase substrates in response to zona binding. *Cell* 57: 1123–1130.

Leyton, L. and Saling, P. 1989b. Evidence that aggregation of mouse sperm receptors by ZP3 triggers the acrosome reaction. *J. Cell Biol.* 108: 2163–2168.

Leyton, L., Leguen, P., Bunch, D. and Saling, P. M. 1992. Regulation of mouse gametic interaction by a sperm tyrosine kinase. *Proc. Natl. Acad. Sci. USA* 93: 1164–1169.

Linder, M. E. and Gilman, A. G. 1992. G proteins. *Sci. Am..* (July) 56–65.

Longo, F. J. 1986. Surface changes at fertilization: Integration of sea urchin (*Arbacia punctulata*) sperm and oocyte plasma membranes. *Dev. Biol.* 116: 143–159.

Longo, F. J. and Kunkle, M. 1978. Transformation of sperm nuclei upon insemination. In A. A. Moscona and A. Monroy (eds.), *Current Topics in Developmental Biology*, Vol. 12. Academic Press, New York, pp. 149–184.

Longo, F. J., Lynn, J. W., McCulloh, D. H. and Chambers, E. L. 1986. Correlative ultrastructural and electrophysiological studies of sperm–egg interactions of the sea urchin *Lytechinus variegatus*. *Dev. Biol.* 118: 155–166.

Lopez, L. C., Bayna, E. M., Litoff, D., Shaper, N. L., Shaper, J. H. and Shur, B. D. 1985. Receptor function of mouse sperm surface galactosyltransferase during fertilization. *J. Cell Biol.* 101: 1501–1510.

Luttmer, S. and Longo, F. J. 1985. Ultrastructural and morphometric observations of cortical endoplasmic reticulum in *Arbacia, Spisula*, and mouse eggs. *Dev. Growth Differ.* 27: 349–359.

Macek, M. B., Lopez, L. and Shur, B. D. 1991. Aggregation of β-1,4-galactosyltransferase on mouse sperm induces the acrosome reaction. *Dev. Biol.* 147: 440–444.

Manes, M.E. and Elinson, R.P. 1980. Ultraviolet light inhibits gray crescent formation in the frog egg. *Wilhelm Roux Arch. Dev. Biol.* 189: 73–76.

Manes, M. E., Elinson, R. P. and Barbieri, F. D. 1978. Formation of the amphibian gray crescent: Effects of colchicine and cytochalasin–B. *Wilhelm Roux Arch. Dev. Biol.* 185: 99–104.

Maslanski, J. A., Leshko, L. and Busa, W. B. 1992. Lithium-sensitive production of inositol phosphates during embryonic mesoderm induction. *Science* 256: 243–245.

McCulloh, D. H. and Chambers, E. L. 1992. Fusion of membranes during fertilization. *J. Gen. Physiol.* 99: 137–175.

McGrath, J. and Solter, D. 1984. Completion of mouse embryogenesis requires both the maternal and paternal genome. *Cell* 37: 179–183.

McPherson, S. M., McPherson, P. S., Mathews, L., Campbell, K. P. and Longo, F. J. 1992. Cortical localization of a calcium release channel in sea urchin eggs. *J. Cell Biol.* 116: 1111–1121.

Meizel, S. 1984. The importance of hydrolytic enzymes to an exocytotic event, the mammalian sperm acrosome reaction. *Biol. Rev.* 59: 125–157.

Metz, C. B. 1978. Sperm and egg receptors involved in fertilization. *Curr. Top. Dev. Biol.* 12: 107–148.

Miller, D. J., Macek, M. B. and Shur, B. D. 1992. Complementarity between sperm surface β-1,4-galactosyltransferase and egg-coat ZP3 mediates sperm–egg binding. *Nature* 357: 589–593.

Miller, R. L. 1978. Site–specific agglutination and the timed release of a sperm chemoattractant by the egg of the leptomedusan, *Orthopyxis caliculata*. *J. Exp. Zool.* 205: 385–392.

Miller, R. L. 1985. Sperm chemo-orientation in the metazoa. In C. B. Metz, Jr. and A. Monroy (eds.), *Biology of Fertilization*, Vol. 2. Academic Press, New York, pp. 275–337.

Miyazaki, S. and Igusa, Y. 1981. Fertilization potential in golden hamster eggs consists of recurring hyperpolarizations. *Nature* 290: 702–704.

Miyazaki, S.-I., Yuzaki, M., Nakada, K., Shirakawa, H., Nakanishi, S., Nakade, S. and Mikoshiba, K. 1992. Block of Ca^{2+} wave and Ca^{2+} oscillation by antibody to the inositol 1,4,5-trisphosphate receptor in fertilized hamster eggs. *Science* 257: 251–255.

Moller, C. C. and Wassarman, P. M. 1989. Characterization of a proteinase that cleaves zona pellucida glycoprotein ZP2 following activation of mouse eggs. *Dev. Biol.* 132: 103–112.

Moore, G. D., Kopf, G. S. and Schultz, R. M. 1993. Complete mouse egg activation in the absence of sperm by stimulation of an exogenous G protein-coupled receptor. *Dev. Biol.* 159: 669–678.

Moy, G. W. and Vacquier, V. D. 1979. Immunoperoxidase localization of bindin during the adhesion of sperm to sea urchin eggs. *Curr. Top. Dev. Biol.* 13: 31–44.

Mozingo, N. M. and Chandler, D. E. 1991. Evidence for the existence of two assembly domains within the sea urchin fertilization envelope. *Dev. Biol.* 146: 148–157.

Multigner, L., Gagnon, J., Dorsselaer, A. van and Job, D. 1992. Stabilization of sea urchin flagellar microtubules by histone H1. *Nature* 360: 33–39.

Nishizuka, Y. 1986. Studies and perspectives of protein kinase C. *Science* 233: 305–312.

Ogawa, K., Mohri, T. and Mohri, H. 1977. Identification of dynein as the outer arms of sea urchin sperm axonomes. *Proc. Natl. Acad. Sci. USA* 74: 5006–5010.

Ohama, K., Kajii, T., Okamoto, E., Fukada, Y., Imaizumi, K., Tsukahara, M., Kobayashi, K. and Hagiwara, K. 1981. Dispermic origin of XY hydatidiform moles. *Nature* 292: 551–552.

Philpott, A. and Leno, G. H. 1992. Nucleoplasmin remodels sperm chromatin in *Xenopus* egg extracts. *Cell* 69: 759–767.

Poccia, D., Salik, J. and Krystal, G. 1981. Transitions in histone variants of the male pronucleus following fertilization and evidence for a maternal store of cleavage-stage histones in the sea urchin egg. *Dev. Biol.* 82: 287–296.

Poirier, G. R. and Jackson, J. 1981. Isolation and characterization of two proteinase inhibitors from the male reproductive tract of mice. *Gamete Res.* 4: 555–569.

Porter, D. C. and Vacquier, V. D. 1986. Phosphorylation of sperm histone H1 is induced by the egg jelly layer in the sea urchin *Strongylocentrotus purpuratus*. *Dev. Biol.* 116: 203–212.

Posada, J. and Cooper, J. A. 1992. Molecular signal integration. Interplay between serine, threonine, and tyrosine phosphorylation. *Mol. Biol. Cell* 3: 583–592.

Prevost, J. L. and Dumas, J. B. 1824. Deuxieme mémoire sur la génération. *Ann. Sci. Nat.* 2: 129–149.

Primakoff, P., Hyatt, H. and Tredick-Kline, J. 1987. Identification and purification of a sperm cell surface protein with a potential role in sperm–egg membrane fusion. *J. Cell Biol.* 104: 141–149.

Primakoff, P., Lathrop, W., Woolman, L., Cowan, A. and Myles, D. 1988. Fully effective contraception in male and female guinea pigs immunized with the sperm protein PH-20. *Nature* 335: 543–546.

Rakow, T. L. and Shen, S. S. 1990. Multiple stores of calcium are released in the sea urchin egg during fertilization. *Proc. Natl. Acad. Sci. USA* 87: 9285–9289.

Ralt, D. and eight others. 1991. Sperm attraction to a follicular factor(s) correlates with human egg fertilizability. *Proc. Natl. Acad. Sci. USA* 88: 2840–2844.

Ramarao, C. S. and Garbers, D. L. 1985. Receptor-mediated regulation of guanylate cyclase activity in spermatozoa. *J. Biol. Chem.* 260: 8390–8396.

Ravnik, S. E., Zarutskie, P. W. and Muller, C. H. 1992. Purification and characterization of a human follicular fluid lipid transfer protein that stimulates human sperm capacitation. *Biol. Reprod.* 47: 1126–1133.

Rochwerger, L., Cohen, D. J. and Cuasnicú, P. S. 1992. Mammalian sperm–egg fusion: The rat egg has complementary sites for a sperm protein that mediates gamete fusion. *Dev. Biol.* 153: 83–90.

Rossignol, D. P., Earles, B. J., Decker, G. L. and Lennarz, W. J. 1984. Characterization of the sperm receptor on the surface of eggs of *Strongylocentrotus purpuratus*. *Dev. Biol.* 104: 308–321.

Roux, W. 1887. Beiträge zur Entwicklungsmechanik des Embryo. *Arch. Mikrosk. Anat.* 29: 157–212.

Saling, P. M. 1989. Mammalian sperm interaction with extracellular matrices of the egg. *Oxford Rev. Reprod. Biol.* 11: 339–388.

Saling, P. M., Sowinski, J. and Storey, B. T. 1979. An ultrastructural study of epididymal mouse spermatozoa binding to zonae pellucida in vitro: Sequential relationship to acrosome reaction. *J. Exp. Zool.* 209: 229–238.

Sawada, T. and Schatten, G. 1989. Effects of cytoskeletal inhibitors on ooplasmic segregation and microtubule organization during fertilization and early development in the ascidian *Molgula occidentalis*. *Dev. Biol.* 132: 331–342.

Schackmann, R. W. and Shapiro, B. M. 1981. A partial sequence of ionic changes associated with the acrosome reaction of *Strongylocentrotus purpuratus*. *Dev. Biol.* 81: 145–154.

Schackmann, R. W., Eddy, E. M. and Shapiro, B. M. 1978. The acrosome reaction of *Strongylocentrotus purpuratus* sperm: Ion requirements and movements. *Dev. Biol.* 65: 483–495.

Schatten, G. and Mazia, D. 1976. The penetration of the spermatozoan through the sea urchin egg surface at fertilization: Observations from the outside on whole eggs and from the inside on isolated surfaces. *Exp. Cell Res.* 98: 325–337.

Schatten, G. and Schatten, H. 1983. The energetic egg. *The Sciences* 23(5): 28–35.

Schatten, H. and Schatten, G. 1980. Surface activity at the plasma membrane during sperm incorporation and its cytochalasin-B sensitivity: Scanning electron micrography and time-lapse video microscopy during fertilization of the sea urchin *Lytechinus variegatus*. *Dev. Biol.* 78: 435–449.

Schroeder, T. E. 1979. Surface area change at fertilization: Resorption of the mosaic membrane. *Dev. Biol.* 70: 306–326.

SeGall, G. K. and Lennarz, W. J. 1979. Chemical characterization of the component of the jelly coat from sea urchin eggs responsible for induction of the acrosome reaction. *Dev. Biol.* 71: 33–48.

Shapiro, B. M. Schackmann, R. W., Tombes, R. M. and Kazozoglan, T. 1985. Coupled ionic and enzymatic regulation of sperm behavior. *Curr. Top. Cell. Regul.* 26: 97–113.

Shen, S. S. and Buck, W. R. 1990. A synthetic peptide of the pseudosubstrate domain of protein kinase C blocks cytoplasmic alakalinization during activation of the sea urchin egg. *Dev. Biol.* 140: 272–280.

Shen, S. S. and Burgart, L. J. 1986. 1,2-Diacylglycerols mimic phorbol 12-myristate 13-acetate activation of the sea urchin egg. *J. Cell. Physiol.* 127: 330–340.

Shen, S. S. and Steinhardt, R. A. 1978. Direct measurement of intracellular pH during metabolic depression of the sea urchin egg. *Nature* 272: 253–254.

Shur, B. D. and Hall, N. G. 1982a. Sperm surface galactosyltransferase activities during in vitro capacitation. *J. Cell Biol.* 95: 567–573.

Shur, B. D. and Hall, N. G. 1982b. A role for mouse sperm surface galactosyltransferase in sperm binding for the egg zona pellucida. *J. Cell Biol.* 95: 574–579.

Shur, B. D. and Neely, C. A. 1988. Plasma membrane association, purification, and partial characterization of mouse sperm β1,4-galactosyltransferase. *J. Biol. Chem.* 263: 17706–17714.

Singh, S. and several others 1988. Membrane guanylate cycles is a cell surface receptor with homology to protein kinases. *Nature* 334: 708–712.

Sluder, G., Miller, F. J., Lewis, K., K., Davison, E. D. and Reider, C. L. 1989. Centrosome inheritance in starfish zygotes: Selective loss of the maternal centrosome after fertilization. *Dev. Biol.* 131: 567–579.

Sluder, G., Miller, F. J. and Lewis, K. 1993. Centrosome inheritance in starfish zygotes II. Selective suppression of the maternal centrosome during meiosis. *Dev. Biol.* 155: 58–67.

Speksnijder, J. E., Sardet, C. and Jaffe, L. F. 1990. The activation wave of calcium in the ascidian egg and its role in ooplasmic segregation. *J. Cell Biol.* 110: 1589–1598.

Steinhardt, R. A. and Epel, D. 1974. Activation of sea urchin eggs by a calcium ionophore. *Proc. Natl. Acad. Sci. USA* 71: 1915–1919.

Steinhardt, R., Zucker, R. and Schatten, G. 1977. Intracellular calcium release at fertilization in the sea urchin egg. *Dev. Biol.* 58: 185–196.

Storey, B. T. and Kopf, G. S. 1991. Fertilization in the mouse II. Spermatozoa. *In* B. S. Dunbar and M. G. O'Rand (eds.), *A Comparative Overview of Mammalian Fertilization*. Plenum, New York, pp. 167–216.

Suarez, S. S., Katz, D. F., Owen, D. H., Andrew, J. B. and Powell, R. L. 1991. Evidence for the function of hyperactivated motility in sperm. *Biol. Reprod.* 44: 375–381.

Summers, R. G. and Hylander, B. L. 1974. An ultrastructural analysis of early fertilization in the sand dollar, *Echinarachnius parma*. *Cell Tiss. Res.* 150: 343–368.

Summers, R. G. and Hylander, B. L. 1975. Species-specificity of acrosome reaction and primary gamete binding in echinoids. *Exp. Cell Res.* 96: 63–68.

Summers, R. G., Hylander, B. L., Colwin, L. H. and Colwin, A. L. 1975. The functional anatomy of the echinoderm spermatozoon and its interaction with the egg at fertilization. *Am. Zool.* 15: 523–551.

Surani, M. A. H. and Barton, S. C. 1983. Development of gynocogenetic eggs in the mouse: Implications for parthenogenetic embryos. *Science* 222: 1034–1036.

Surani, M. A. H., Barton, S. C. and Norris, M. L. 1986. Nuclear transplantation in the mouse: Hereditable differences between parental genomes after activation of the embryonic genome. *Cell* 45: 127–136.

Swann, K. and Whitaker, M. 1986. The part played by inositol trisphosphate and calcium in the propagation of the fertilization wave in sea urchin eggs. *J. Cell Biol.* 103: 2333–2342.

Talbot, P. 1985. Sperm penetration through oocyte investments in mammals. *Am. J. Anat.* 174: 331–346.

Talbot, P., DiCarlantonio, G., Zao, P., Penkala, J. and Haimo, L. T. 1985. Motile cells lacking hyaluronidase can penetrate the hamster oocyte cumulus complex. *Dev. Biol.* 108: 387–398.

Terasaki, M. and Sardet, C. 1991. Demonstration of calcium uptake and release by sea urchin egg cortical endoplasmic reticulum. *J. Cell Biol.* 115: 1031–1037.

Tilney, L. G., Bryan, J., Bush, D. J., Fujiwara, K., Mooseker, M. S., Murphy, D. B. and Snyder, D. H. 1973. Microtubules: Evidence for 13 protofilaments. *J. Cell Biol.* 59: 267–275.

Tilney, L. G., Kiehart, D. P., Sardet, C. and Tilney, M. 1978. Polymerization of actin. IV. Role of Ca^{2+} and H^+ in the assembly of actin and in membrane fusion in the acrosomal reaction of echinoderm sperm. *J. Cell Biol.* 77: 536–560.

Tombes, R. M. and Shapiro, B. M. 1985. Metabolite channeling: A phosphocreatine shuttle to mediate high-energy phosphate transport between sperm mitochondrion and tail. *Cell* 41: 325–334.

Turner, P. R., Jaffe, L. A. and Fein, A. 1986. Regulation of cortical vesicle exocytosis in sea urchin eggs by inositol 1,4,5-trisphosphate and GTP-binding protein. *J. Cell Biol.* 102: 70–76.

Turner, P. R., Jaffe, L. A. and Primakoff, P. 1987. A cholera toxin-sensitive G-protein stimulates exocytosis in sea urchin eggs. *Dev. Biol.* 120: 577–583.

Uzzell, T. M. 1964. Relations of the diploid and triploid species of the *Ambystoma jeffersonianum* complex. *Copeia* 1964: 257–300.

Vacquier, V. D. 1975. The isolation of intact cortical granules from sea urchin eggs: Calcium ions trigger granule discharge. *Dev. Biol.* 43: 62–74.

Vacquier, V. D. 1979. The interaction of sea urchin gametes during fertilization. *Am. Zool.* 19: 839–849.

Vacquier, V. D. and Moy, G. W. 1977. Isolation of bindin: The protein responsible for adhesion of sperm to sea urchin eggs. *Proc. Natl. Acad. Sci. USA* 74: 2456–2460.

Vacquier, V. D. and Payne, J. E. 1973. Methods for quantitating sea urchin sperm in egg binding. *Exp. Cell Res.* 82: 227–235.

Vacquier, V. D., Tegner, M. J. and Epel, D. 1973. Protease release from sea urchin eggs at fertilization alters the vitelline layer and aids in preventing polyspermy. *Exp. Cell Res.* 80: 111–119.

Vincent, J. P., Oster, G. F. and Gerhart, J. C. 1986. Kinematics of gray crescent formation in *Xenopus* eggs. The displacement of subcortical cytoplasm relative to the egg surface. *Dev. Biol.* 113: 484–500.

von Kolliker, A. 1841. *Beiträge zur Kenntnis der Geschlectverhältnisse und der Samenflüssigkeit wirbelloser Thiere, nebst einem Versuch Über Wesen und die Bedeutung der sogenannten Samenthiere.* Berlin.

Ward, C. R. and Kopf, G. S. 1993. Molecular events mediating sperm activation. *Dev. Biol.* 158: 9–34.

Ward, G. E., Brokaw, C. J., Garbers, D. L. and Vacquier, V. D. 1985. Chemotaxis of *Arbacia punctulata* spermatozoa to resact, a peptide from the egg jelly layer. *J. Cell Biol.* 101: 2324–2329.

Ward, G. E., Storey, B. T. and Kopf, G. S. 1992. Activation of a G^i protein in cell-free membrane preparations of mouse sperm by the zona pellucida and ZP3, components of the egg's extracellular matrix. *J. Biol. Chem.* 267: 14061–14067.

Wassarman, P. 1987. The biology and chemistry of fertilization. *Science* 235: 553–560.

Wassarman, P. M. 1989. Fertilization in mammals. *Sci. Am.* 256(6): 78–84.

Watanabe, N., Hunt, T., Ikawa, Y. and Sagata, N. 1991. Independent inactivation of MPF and cytostatic factor (Mos) upon fertilization of *Xenopus* eggs. *Nature* 352: 247–249.

Whitaker, M. and Irvine, R. F. 1984. Inositol 1,4,5-triphosphate microinjection activates sea urchin eggs. *Nature* 312: 636–639.

Whitaker, M. and Steinhardt, R. 1982. Ionic regulation of egg activation. *Q. Rev. Biophys.* 15: 593–666.

Whitaker, M. J. and Steinhardt, R. 1985. Ionic signalling in the sea urchin egg at fertilization. *In* C. B. Metz and A. Monroy (eds.), *Biology of Fertilization*, Vol. 3. Academic Press, Orlando, FL, pp. 167–221.

Wilde, M. W. and Kopf, G. S. 1989. Activation of a G-protein in mammalian sperm by an egg-associated extracellular matrix, the zona pellucida. *J. Cell Biol.* 109: 251a.

Williams, C. J., Schultz, R. M. and Kopf, G. S. 1992. Role of G proteins in mouse egg activation: Stimulatory effects of acetylcholine on the ZP2 to ZP2f conversion and pronuclear formation in eggs expressing a functional m1 muscarinic receptor. *Dev. Biol.* 151: 288–296.

Wilson, W. L. and Oliphant, G. 1987. Isolation and biochemical characterization of the subunits of the rabbit sperm acrosome stabilizing factor. *Biol. Reprod.* 37: 159–169.

Winkler, M. M., Steinhardt, R. A., Grainger, J. L. and Minning, L. 1980. Dual ionic controls for the activation of protein synthesis at fertilization. *Nature* 287: 558–560.

Yanagimachi, R. 1988. Mammalian fertilization. *In* E. Knobil and J. Neill (eds.), *The Physiology of Reproduction*. Raven, New York, pp. 135–185.

Yanagimachi, R. and Noda, Y. D. 1970. Electron microscope studies of sperm incorporation into the golden hamster egg. *Am. J. Anat.* 128: 429–462.

Yoshida, M., Inabar, K. and Morisawa, M. 1993. Sperm chemotaxis during the process of fertilization in the ascidians *Ciona savignyi* and *Ciona intestinalis*. *Dev. Biol.* 157: 497–506.

Yudin, A., Cherr, G. and Katz, D. 1988. The structure of the cumulus matrix and zona pellucida in the golden hamster: A new view of sperm interaction with oocyte-associated extracellular matrices. *Cell Tissue Res.* 251: 555–564.

5

Cleavage
Creating multicellularity

To our limited intelligence, it would seem a simple task to divide a nucleus into equal parts. The cell, manifestly, entertains a very different opinion.

E. B. WILSON (1923)

One must show the greatest respect towards any *thing that increases exponentially, no matter how small.*

GARRETT HARDIN (1968)

Remarkable as it is, fertilization is but the initiatory step in development. The zygote, with its new genetic potential and its new arrangement of cytoplasm, now begins the production of a multicellular organism. In all animal species known, this begins by a process called **cleavage,** a series of mitotic divisions whereby the enormous volume of egg cytoplasm is divided into numerous smaller, nucleated cells. These cleavage-stage cells are called **blastomeres.**

In most species (mammals being the chief exception), the rate of cell division and the placement of the blastomeres with respect to one another is completely under the control of the proteins and mRNAs stored in the oocyte by the mother. The zygotic genome, transmitted by mitosis to all the other cells, does not function in early-cleavage embryos. Few, if any, mRNAs are made until relatively late in cleavage, and the embryo can divide properly even when chemicals are used to inhibit transcription. Also, in most species there is no net increase in embryonic volume during cleavage. This differs from most cases of cell proliferation, in which there is a period of cell growth between mitoses: a cell expands to nearly twice its volume, then divides. This growth produces a net increase in the total volume of cells while maintaining a relatively constant ratio of nuclear to cytoplasmic volume. During embryonic cleavage, however, cytoplasmic volume does not increase. Rather, the enormous volume of zygote cytoplasm is divided into increasingly smaller cells. First the egg is divided in half, then quarters, then eighths, and so forth. This division of egg cytoplasm without growth is accomplished by abolishing the growth period between divisions, while the cleavage of nuclei occurs at a rate never seen again (even in tumor cells). A frog egg, for example, can divide into 37,000 cells in just 43 hours. Mitosis in cleavage-stage *Drosophila* occurs every 10 minutes for over 2 hours and in just 12 hours forms some 50,000 cells. This increase in cell number can be appreciated by comparing cleavage with other states of development. Figure 5.1 shows the logarithm of cell number in a frog embryo plotted against the time of development (Sze, 1953). It illustrates a sharp discontinuity between cleavage and gastrulation.

One consequence of this rapid division is that the ratio of cytoplasmic to nuclear volume gets increasingly smaller as cleavage progresses. In many types of embryos, this manifold decrease in the cytoplasmic-to-nuclear volume ratio is crucial in timing the activation of certain genes. For example, in the frog *Xenopus laevis*, transcription of new messages is

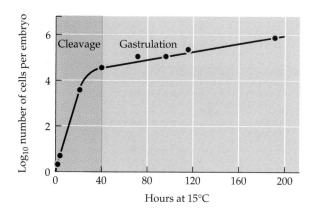

FIGURE 5.1
Formation of new cells during the
early development of the frog *Rana pipiens*. (After Sze, 1953.)

not activated until after 12 divisions. At that time, the rate of cleavage
decreases, the blastomeres become motile, and nuclear genes begin to be
transcribed. It is thought that some factor in the egg is being titrated by
the newly made chromatin, because the time of this transition can be
changed by altering the amount of chromatin in each nucleus: the more
chromatin present in the embryo, the earlier the transition. If the amount
of chromatin present is double the normal amount of chromatin, the
transition occurs one division earlier (Newport and Kirschner, 1982a,b).
Thus, cleavage begins soon after fertilization and ends when the embryo
achieves a new balance between nucleus and cytoplasm.

PATTERNS OF EMBRYONIC CLEAVAGE

Cleavage is a very well coordinated process and is under genetic regulation. The pattern of embryonic cleavage particular to a species is determined by two major parameters: (1) the amount and distribution of yolk
protein within the cytoplasm, and (2) those factors in the egg cytoplasm
influencing the angle of the mitotic spindle and the timing of its formation.

The amount and distribution of yolk determines where cleavage can
occur and the relative size of the blastomeres. When one pole of the egg
is relatively yolk-free, the cellular divisions occur there at a faster rate than
at the opposite pole. The yolk-rich pole is referred to as the **vegetal pole;**
the yolk concentration in the **animal pole** is relatively low. The zygote
nucleus is frequently displaced toward the animal pole. Table 5.1 provides
a classification of cleavage types and shows the influence of yolk on the
cleavage symmetry and pattern. In general, yolk inhibits cleavage. In
zygotes with relatively little yolk (*isolecithal* and *mesolecithal* eggs), cleavage
is **holoblastic,** meaning that the cleavage furrow extends through the
entire egg. Zygotes containing large accumulations of yolk protein
undergo **meroblastic** cleavage, wherein only a portion of the cytoplasm is
cleaved. The cleavage furrow does not penetrate into the yolky portion of
the cytoplasm. Meroblastic cleavage can be **discoidal,** as in birds' eggs, or
superficial, as in insect zygotes, depending on whether the yolk deposit
is located to one side (*telolecithal*) or in the center of the cytoplasm (*centrolecithal*), respectively.

Yolk is an evolutionary adaptation that enables an embryo to develop
in the absence of an external food source. Animals developing without
large yolk concentrations, such as sea urchins, usually form a larval stage
fairly rapidly. This larval stage can feed itself, and development then

TABLE 5.1
Classification of cleavage types

Cleavage pattern	Position of yolk	Cleavage symmetry	Representative animals
Holoblastic (complete cleavage)	Isolecithal (oligolecithal) (sparse, evenly distributed yolk)	Radial	Echinoderms, *Amphioxus*
		Spiral	Most molluscs, annelids, flatworms, and roundworms
		Bilateral	Ascidians
		Rotational	Mammals
	Mesolecithal (moderately telolecithal)	Radial	Amphibians
Meroblastic (incomplete cleavage)	Telolecithal (dense yolk concentrated at one end of egg)	Bilateral	Cephalopod molluscs
		Discoidal (bilateral)	Reptiles, fishes, birds
	Centrolecithal (yolk concentrated in center of egg)	Superficial	Most arthropods

continues from this free-swimming larva. Mammalian eggs, which also lack large quantities of yolk, adopt another strategy, namely, making a placenta. As we will see, the first differentiation of the mammalian embryo sets aside the cells that will form the placenta. This organ supplies food and oxygen for the embryo during its long gestation.

At the other extreme are the eggs of insects, fishes, reptiles, and birds. Most of their cell volumes are yolk. The yolk must be sufficient to nourish these animals, as they develop without a larval stage or placental attachment. The correlation between heavy yolk concentration and lack of larval forms is seen in certain species of frogs. Certain tropical frogs, such as *Eleutherodactylus* and *Arthroleptella*, lack a tadpole stage. Rather, they provision their eggs with an enormous yolk concentration (Lutz, 1947). The eggs do not need to be laid in water because the tadpole stage has been eliminated. This will be discussed further in Chapter 20.

However, the yolk is just one factor influencing a species' pattern of cleavage. There are also inherited patterns of cell division that are superimposed upon the restraints of the yolk. This can readily be seen in isolecithal eggs, in which very little yolk is present. In the absence of a large concentration of yolk, four major cleavage types can be observed: radial holoblastic, spiral holoblastic, bilateral holoblastic, and rotational holoblastic cleavage.

Radial holoblastic cleavage

Radial holoblastic cleavage is the simplest form of cleavage to understand. The furrows in this type of cleavage are oriented parallel to and perpendicular to the animal–vegetal axis of the egg. It is characteristic of echinoderms and the protochordate *Amphioxus*, as well as frogs and salamanders.

The sea cucumber, *Synapta*

The cleavage pattern of the sea cucumber, *Synapta digita*, is illustrated in Figure 5.2. After the union of the pronuclei, the axis of the first mitotic

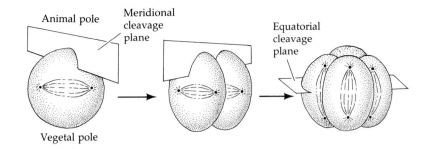

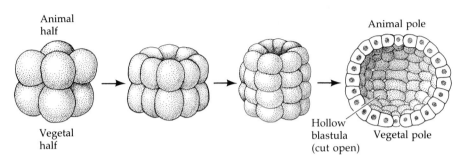

FIGURE 5.2
Holoblastic cleavage in the echinoderm *Synapta digita*, leading to the formation of a hollow blastula, as shown in the cutaway view in the last panel. (After Saunders, 1982.)

spindle is formed perpendicular to the animal–vegetal axis of the egg. Therefore, the first cleavage furrow passes directly through the animal and vegetal poles, creating two equal-sized daughter cells. This cleavage is said to be **meridional** because it passes through the two poles like a meridian on a globe. The mitotic spindles of the second cleavage are at right angles to the first but are still perpendicular to the animal–vegetal axis of the egg. The cleavage furrows appear simultaneously in both blastomeres and also pass through the two poles. Thus, the first two divisions are both meridional and perpendicular to each other. The third division is **equatorial**: the mitotic spindles of each blastomere are now positioned parallel to the animal–vegetal axis, and the resulting cleavage furrow separates the two poles from each other, dividing the embryo into eight equal blastomeres. Each blastomere in the animal half of the embryo is now directly above a blastomere of the vegetal half.

The fourth division is again meridional, producing two tiers of 8 cells each, while the fifth division is equatorial, producing four tiers of 8 cells each. Successive divisions produce 64-, 128-, and 256-cell embryos, with meridional divisions alternating with equatorial divisions. The resulting embryo consists of blastomeres arranged in horizontal rows along a central cavity. At both poles of the embryo, blastomeres move toward each other to create a hollow sphere composed of a single cell layer. This hollow sphere is called the **blastula,** and the central cavity is referred to as the **blastocoel.** At any time during the cleavage of *Synapta*, an embryo bisected through any meridian produces two mirror-image halves. This type of symmetry is characteristic of a sphere or cylinder and is called radial symmetry. Thus, *Synapta* is said to have **radial holoblastic cleavage.**

Sea urchins

Sea urchins also exhibit radial holoblastic cleavage, but with some important modifications. The first and second cleavages are similar to those of *Synapta*; both are meridional and are perpendicular to each other. Similarly, the third cleavage is equatorial, separating the two poles from one another (Figure 5.3). In the fourth cleavage, however, events are very different. The four cells of the animal tier divide meridionally into eight blastomeres, each with the same volume. These cells are called **mesomeres**. The vegetal

FIGURE 5.3
Cleavage in the sea urchin. (A) Planes of cleavage in the first three divisions and the formation of particular tiers of cells in divisions 3–6. (B–D) Photomicrographs of live embryos of the sea urchin *Lytechinus pictus,* looking down upon the animal pole. (B) 2-cell stage; (C) 4-cell stage of sea urchin development. (D) The 32-cell stage, shown without the fertilization membrane in order to allow visualization of the animal pole mesomeres, the central macromeres, and the vegetal micromeres, which angle into the center. (Photographs courtesy of G. Watchmaker.)

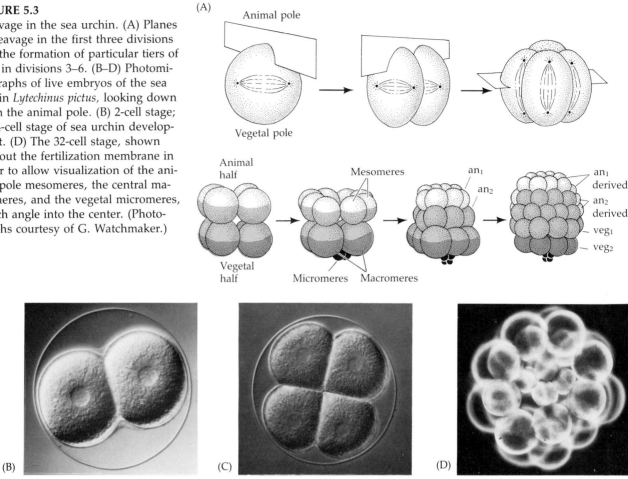

tier, however, undergoes an *unequal* equatorial cleavage to produce four large cells, the **macromeres,** and four smaller **micromeres** at the vegetal pole (Figure 5.4; Summers et al., 1993). As the 16-cell embryo cleaves, the eight mesomeres divide to produce two "animal" tiers, an_1 and an_2, one staggered above the other. The macromeres divide meridionally, forming a tier of eight cells below an_2. The micromeres also divide, producing a small cluster beneath the larger tier. All the cleavage furrows of the sixth division are equatorial; and the seventh cleavage is meridional, producing a 128-cell blastula.

FIGURE 5.4
Formation of the micromeres during the fourth division of sea urchin embryos. The vegetal poles of the embryos are viewed from below. (A) The location and orientation of the mitotic spindle at the bottom portion of the vegetal cells is shown by viewing the living embryo with polarized light. (B) The cleavage through these asymmetrically placed spindles has produced micromeres and macromeres. (From Inoué, 1982, courtesy of S. Inoué.)

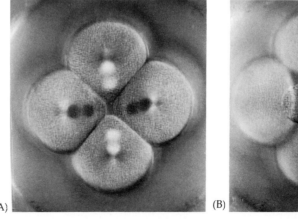
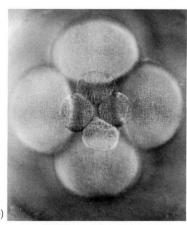

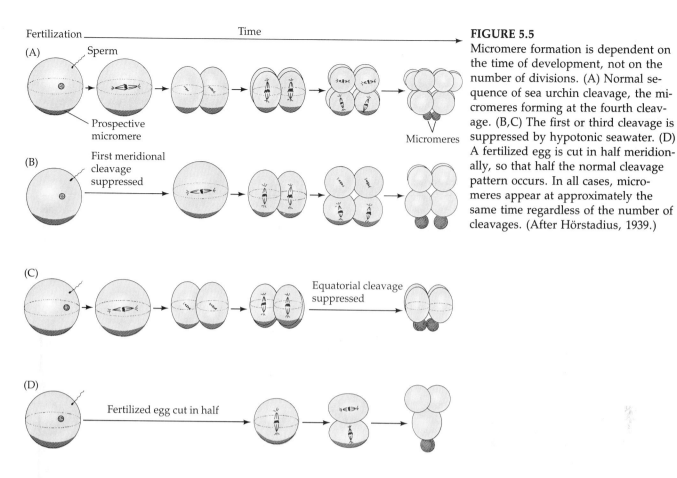

(A)

Sperm

Prospective
micromere

Micromeres

(B)

First meridional
cleavage
suppressed

(C)

Equatorial cleavage
suppressed

(D)

Fertilized egg cut in half

FIGURE 5.5
Micromere formation is dependent on
the time of development, not on the
number of divisions. (A) Normal sequence of sea urchin cleavage, the micromeres forming at the fourth cleavage. (B,C) The first or third cleavage is
suppressed by hypotonic seawater. (D)
A fertilized egg is cut in half meridionally, so that half the normal cleavage
pattern occurs. In all cases, micromeres appear at approximately the
same time regardless of the number of
cleavages. (After Hörstadius, 1939.)

In 1939, Sven Hörstadius performed a simple experiment demonstrating that the timing and placement of each sea urchin cleavage is independent of pre-existing cleavages. He showed that if he inhibited the first
one, two, or three cleavages by shaking the eggs or placing them into
hypotonic seawater, the unequal (fourth) cleavage division that forms the
micromeres would still occur at the appropriate time (Figure 5.5). Thus,
Hörstadius concluded that there are three factors that determine cleavage
in the 8-cell embryo: (1) there are "progressive changes in the cytoplasm
which cause spindles formed after a certain time after fertilization to lie in
a certain direction"; (2) there must be micromere-forming material in the
vegetal cytoplasm; and (3) there must be some mechanism by which the
micromere-forming material is activated at the correct time (Hörstadius,
1973).

The blastula stage of sea urchin development begins at the 128-cell
stage. Here the cells form a hollow sphere surrounding a central blastocoel
(Figure 5.6). By this time, all the cells are the same size, the micromeres

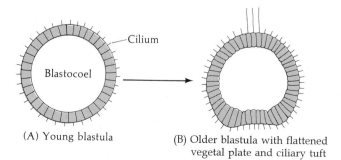

Cilium

Blastocoel

(A) Young blastula

(B) Older blastula with flattened
vegetal plate and ciliary tuft

FIGURE 5.6
Sea urchin blastulae. (A) Early sea urchin blastula shows a single layer of
rounded cells surrounding a large
blastocoel. (B) As division continues,
the cells of the late blastula show differences in shape as the vegetal plate
cells elongate. (After Giudice, 1973.)

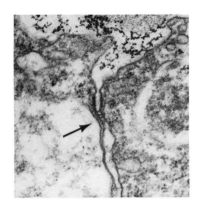

FIGURE 5.7
Area of contact at the junction of two cells of a 1024-cell starfish blastula. Membrane regions cross over at these junctions (arrow) to form tight seals. (From Dan-Sohkawa and Fujisawa, 1980, courtesy of the authors.)

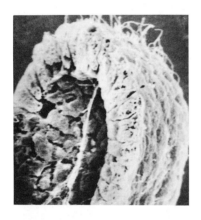

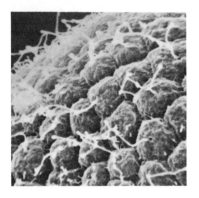

FIGURE 5.8
Ciliated blastula cells. Each cell develops a single cilium. (Courtesy of W. J. Humphreys.)

having slowed down their cell division rate so that they cleave less frequently. Every cell is in contact with the proteinaceous fluid of the blastocoel and with the hyaline layer within the fertilization membrane. During this time, contacts between the cells are tightened. Dan-Sohkawa and Fujisawa (1980) analyzed this process in starfish embryos and showed that the closure of the hollow sphere is contemporaneous with the formation of tight junctions between the blastomeres. These junctions unite the loosely connected cells into a tissue and seal off the blastocoel from the outside environment (Figure 5.7). As the cells continue to divide, the cell layer expands and thins out. During this period, the blastula remains one cell layer thick.

Two theories have been offered to explain the concomitant cell proliferation and blastocoel formation. Dan (1960) hypothesized that the motive force of this expansion is the influx of water into the blastocoel cavity. As the blastomeres secrete proteins into the blastocoel, the blastocoel fluid becomes syrupy. This blastocoel sap absorbs large quantities of water by osmosis, thereby swelling and putting pressure on the blastomeres to expand outward. This pressure would also align the long axis of each cell so that division would never be inward toward the blastocoel. This would create further expansion by having the population oriented in one plane only. Wolpert and Gustafson (1961) and Wolpert and Mercer (1963) proposed that pressure from the blastocoel is not needed to get this effect. They emphasized the role of differential adhesiveness of the cells to one another and to the hyaline layer. They found that as long as the cells remain strongly attached to the hyaline layer, the cells have no alternative but to expand. This expansion creates the blastula rather than the other way around. Certainly the hyaline layer is critical for blastocoel expansion, and if the adhesion of cells to the hyaline layer is inhibited by antibodies to hyalin, then blastocoel expansion ceases (Adelson and Humphreys, 1988). A recent review (Ettensohn and Ingersoll, 1992) concludes that it is likely that both these mechanisms act to expand the blastocoel. During early cleavage, the adhesion to the hyaline layer appears to be the most important factor, while at later stages, the osmotic pressure also seems to play a role.

The blastula cells develop cilia on their outer surfaces (Figure 5.8), thereby causing the blastula to rotate within the fertilization envelope. Soon afterward, the cells of the animal half of the embryo synthesize and secrete a **hatching enzyme** that enables them to digest the fertilization membrane (Lepage et al., 1992). The embryo is now a free-swimming **hatched blastula.**

Amphibians

Cleavage in most frog and salamander embryos is radially symmetrical and holoblastic, just like echinoderm cleavage. The amphibian egg, however, contains much more yolk. This yolk, which is concentrated in the vegetal hemisphere, acts as an impediment to cleavage. Thus, the first division begins at the animal pole and slowly extends down into the vegetal region (Figure 5.9). In the axolotl salamander, the cleavage furrow extends through the animal hemisphere at a rate close to 1 mm/min. The cleavage furrow bisects the gray crescent and then slows down to a mere 0.02–0.03 mm per minute as it approaches the vegetal pole (Hara, 1977).

Figure 5.10A is a scanning electron micrograph showing the first cleavage in a frog egg. One can see the folds in the cleavage furrow and the difference between the furrows in the animal and vegetal hemispheres. Figure 5.10B shows that while the first cleavage furrow is still trying to

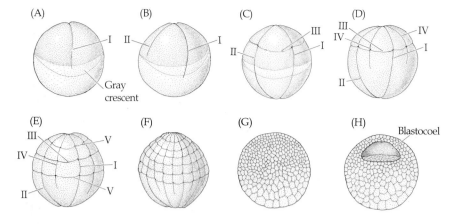

FIGURE 5.9
Cleavage of a frog egg. Cleavage furrows, designated by Roman numerals, are numbered in order of appearance. (A, B) The vegetal yolk impedes the cleavage such that the second division begins in the animal region of the egg before the first division has divided the vegetal cytoplasm. (C) The third division is displaced toward the animal pole (D–H). The vegetal hemisphere ultimately contains longer and fewer blastomeres than the animal half. (After Carlson, 1981.)

cleave the yolky cytoplasm of the vegetal hemisphere, the second cleavage has already started near the animal pole. This cleavage is at right angles to the first one and is also meridional. The third cleavage, as expected, is equatorial. However, because of the vegetally placed yolk, this cleavage furrow in amphibian eggs is much closer to the animal pole. It divides the frog embryo into four small animal blastomeres (micromeres) and four large blastomeres (macromeres) in the vegetal region. This unequal holoblastic cleavage establishes two major embryonic regions: a rapidly dividing region of micromeres near the animal pole, and a more slowly dividing macromere area (Figure 5.10C). As cleavage progresses, the animal region becomes packed with numerous small cells while the vegetal region contains only a relatively small number of large, yolk-laden macromeres (Figure 5.9).

Amphibian embryos containing 16 to 64 cells are commonly called **morulae** (Latin, "mulberry," whose shape they vaguely resemble). At the 128-cell stage, the blastocoel becomes apparent and the embryo is considered to be a blastula. Actually, the formation of the blastocoel has been traced back to the very first cleavage furrow. Kalt (1971) demonstrated that in the frog *Xenopus laevis,* the first cleavage furrow widens in the animal hemisphere to create a small intercellular cavity that is sealed off from the outside by tight intercellular junctions (Figure 5.11). This cavity expands during subsequent cleavages to become the blastocoel.

The blastocoel probably serves two major functions in frog embryos. It is a cavity that permits cell migration during gastrulation; it also prevents the cells beneath it from interacting prematurely with the cells above it. When Nieuwkoop (1973) took embryonic newt cells from the roof of the blastocoel and placed them next to the yolky vegetal cells from the base of the blastocoel, these animal cells became mesoderm tissue instead of ectoderm. Because mesodermal tissue is normally formed from those animal cells that are adjacent to the endoderm precursors, it seems plausible that the vegetal cells influence adjacent cells to differentiate into mesodermal tissues. Thus, the blastocoel appears to preserve the integrity of those cells fated to give rise to the skin and nerves.

FIGURE 5.10
Scanning electron micrographs of the cleavage of a frog egg. (A) First cleavage. (B) Second cleavage (4 cells). (C) Fourth cleavage (16 cells), showing the size discrepancy between the animal and vegetal cells after arising from the third division. (A from Beams and Kessel, 1976, courtesy of the authors; B and C courtesy of L. Biedler.)

FIGURE 5.11
Formation of the blastocoel in a frog egg. (A) First cleavage plane, showing a small cleft, which later develops into the blastocoel. (B) 8-cell embryo showing a small blastocoel (arrow) at the junction of the three cleavage planes. (From Kalt, 1971, courtesy of M. R. Kalt.)

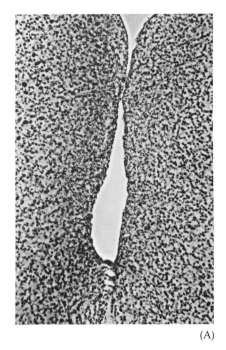

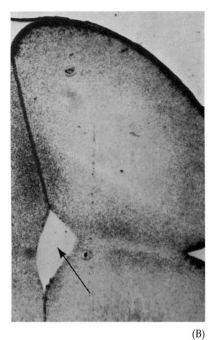

(A)

(B)

Spiral holoblastic cleavage

Spiral cleavage is characteristic of annelid worms, turbellarian flatworms, nemertean worms, and all molluscs except cephalopods. It differs from radial cleavage in numerous ways. First, the eggs do not divide in parallel or perpendicular orientations to the animal–vegetal axis of the egg; rather, cleavage is at oblique angles, forming the "spiral" arrangement of daughter blastomeres. Second, the cells touch each other at more places than do those of radially cleaving embryos. In fact, they take the most thermodynamically stable packing orientation, much like that of adjacent soap bubbles (Figure 5.12). Third, spirally cleaving embryos usually undergo fewer divisions before they begin gastrulation. This makes it possible to know the fate of each individual cell of the blastula. When the fates of the individual cells from annelid, flatworm, and mollusc embryos were compared, the same cells were seen in the same places, and their general fates were identical (Wilson, 1898). The blastulae so produced have no blastocoel and are called **stereoblastulae**.

Figures 5.13 and 5.14 depict the cleavage of mollusc embryos. The first two cleavages are nearly meridional, producing four large macromeres (labeled A, B, C, and D). In many species, the blastomeres are different sizes (D being the largest), a characteristic that allows them to be individually identified. In each successive cleavage, each macromere buds off a small micromere at its animal pole. Each successive quartet of micromeres is displaced to the right or to the left of its sister macromere, creating the spiral relationship characteristic of the cleavage. Looking down on the embryo from the animal pole, the upper ends of the mitotic spindle appear to alternate clockwise and counterclockwise. This causes alternate micromeres to form obliquely to the left and to the right of their macromere. At the third cleavage, the A macromere gives rise to two daughter cells, macromere 1A and micromere 1a. The B, C, and D cells behave similarly, producing the first quartet of micromeres. In most species, the micromeres are to the right of their macromeres (looking down on the animal pole), an arrangement indicating a dextral (as opposed to sinistral) spiral. At the fourth cleavage, the 1A macromere divides to form macromere 2A and

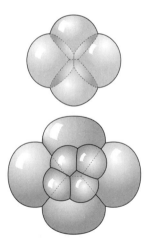

FIGURE 5.12
Diagram showing the arrangement of four and eight soap bubbles in a slightly concave dish. The thermodynamic arrangement maximizes contact and is very reminiscent of spirally cleaving embryos. (After Morgan, 1927.)

(A) View from animal pole

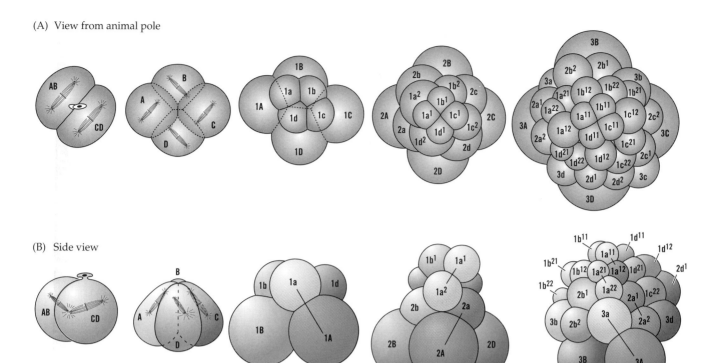

(B) Side view

micromere 2a; and micromere 1a divides to form two more micromeres, 1a[1] and 1a[2]. Further cleavage yields blastomeres 3A and 3a from the 2A macromere; and micromeres such as 1a[2] divide to produce cells such as 1a21 and 1a22.

The orientation of the cleavage plane to the left or to the right is controlled by cytoplasmic factors within the oocyte. This was discovered by analyzing mutations of snail coiling. Some snails have their coils opening to the right of their shells, whereas other snails have their coils opening to the left. Usually the rotation of coiling is the same for all members of a given species. Occasionally, though, mutants are found. For instance, in the species in which the coils open on the right, occasional individuals will be found with coils that open on the left. Crampton (1894) analyzed the embryos of such aberrant snails and found that their early cleavage differed from the norm. The orientation of the cells after second cleavage was different (Figure 5.15). This was the result of a different orientation

FIGURE 5.13
Spiral cleavage of the mollusc *Trochus* viewed (A) from the animal pole and (B) from one side. In (B), the cells derived from the A blastomere are in color. The mitotic spindles, sketched in the early stages, divide the cells unequally and at an angle to the vertical and horizontal axes.

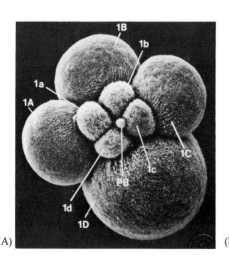

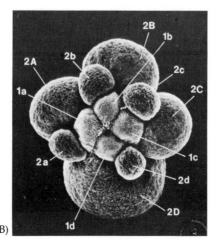

FIGURE 5.14
Spiral cleavage of the snail *Ilyanassa*. The D blastomere is larger than the others, allowing the identification of each cell. Cleavage is dextral. (A) 8-cell stage. (B) Mid-fourth cleavage; the macromeres have already divided into large and small spirally oriented cells. PB, polar body. (From Craig and Morrill, 1986, courtesy of the authors.)

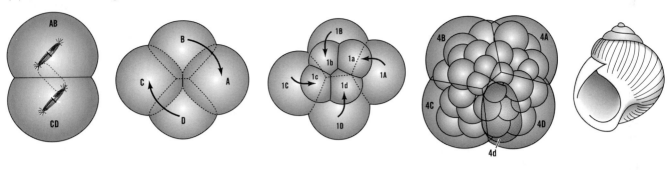

(A) Left-handed coiling

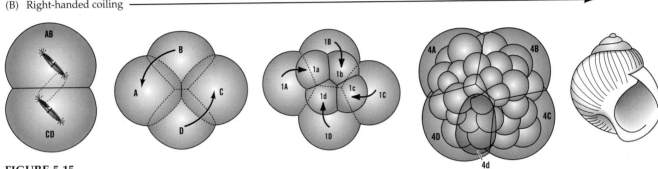

(B) Right-handed coiling

FIGURE 5.15
Looking down upon the animal pole of right-handed and left-handed snails. The origin of right-handed and left-handed snail coiling can be traced to the orientation of the mitotic spindle at second cleavage. The left-handed (A) and right-handed (B) snails develop as mirror images of each other. (After Morgan, 1927.)

of the mitotic apparatus in the sinistrally coiling snails. All subsequent divisions in left-coiling embryos are mirror images of those of dextrally coiling embryos. In Figure 5.15, one can see that the position of the 4d blastomere (which is extremely important, as its progeny will form the mesodermal organs) is different in the two types of spiraling embryos. Eventually, the two snails are formed with their bodies on different sides of the coil opening.

The direction of snail shell coiling is controlled by a single pair of genes (Sturtevant, 1923; Boycott et al., 1930). In the snail *Limnaea peregra*, most individuals are dextrally coiled. Rare mutants exhibiting left-handed coiling were found and mated with wild-type snails. These matings show that there is a "right-handed" allele *D*, which is dominant to the "left-handed" allele *d*. However, the direction of cleavage is determined not by the genotype of the developing snail but by the genotype of the snail's mother. A *dd* female snail can produce only sinistrally coiling offspring, even when the offspring's genotype is *Dd*. A *Dd* individual will coil either left or right, depending on the genome of its mother. The matings produce a chart like this:

		Genotype	Phenotype
$DD♀ \times dd♂$	$\rightarrow$	*Dd*	All right-coiling
$DD♂ \times dd♀$	$\rightarrow$	*Dd*	All left-coiling
Dd	$\times$ *Dd* $\rightarrow$	1*DD*:2*Dd*:1*dd*	All right-coiling

The genetic factors involved in snail coiling are brought to the embryo in the oocyte cytoplasm. It is the genotype of the *ovary* in which the oocyte develops that determines which orientation the cleavage will take. When Freeman and Lundelius (1982) injected a small amount of cytoplasm from dextrally coiling snails into the eggs of *dd* mothers, the resulting embryos coiled to the right. Cytoplasm from sinistrally coiling snails did not affect the right-coiling embryos. This confirmed the view that the wild-type mothers were placing a factor into their eggs that was absent or defective in the *dd* mothers.

Adaptation by modifying embryonic cleavage

Evolution is caused by the hereditary alteration of embryonic development. When we say that the contemporary one-toed horse evolved from a five-toed ancestor, we are stating that during the evolution of the horse, the pattern of embryonic bone formation changed to bring about a one-toed organism. Sometimes we are able to identify a modification of embryogenesis that has enabled the organism to survive in an otherwise inhospitable environment. One such modification, discovered by Frank Lillie in 1898, is brought about by altering the typical pattern of spiral cleavage in the unionid family of clams.

Unlike most clams, *Unio* and its relatives live in swift-flowing streams. Streams usually create a problem for the dispersal of larvae: because the adults are sedentary, the free-swimming larvae would always be carried downstream by the current. These clams, however, have solved this problem by effecting two changes in their development. The first alters embryonic cleavage. In the typical cleavage of molluscs, either all the macromeres are equal in size or the 2D blastomere is the largest cell at that embryonic stage. However, the division of this *Unio* is such that the 2d blastomere gets the largest amount of cytoplasm (Figure 5.16). This cell divides to produce most of the larval structures, including a gland capable of producing a massive shell. These larvae (called *glochidia*) resemble tiny bear traps; they have sensitive hairs that cause the valves of the shell to snap shut when they are touched by the gills or fins of a wandering fish. They "hitchhike" with the fish until they are ready to drop off and metamorphose into adult clams. In this manner, they can spread upstream.

In some species, glochidia are released from the female's brood pouch and merely wait for a fish to come wandering by. Some other species, such as *Lampsilis ven-*

FIGURE 5.17
Phony fish atop the unionid clam Lampsilis ventricosa. *The "fish" is actually the brood pouch and mantle of the clam. (Photograph courtesy of J. H. Welsh.)*

tricola, have increased the chances of its larvae finding a fish by yet another modification of their development (Welsh, 1969). Many clams develop a thin mantle that flaps around the shell and surrounds the brood pouch. In some unionids, the shape of the brood pouch (marsupium) and the undulations of the mantle mimic the shape and swimming behavior of a minnow. To make the deception all the better, they develop a black "eyespot" on one end and a flaring "tail" on the other. The "fish" seen in Figure 5.17 is not a fish at all, but the brood pouch and mantle of the clam beneath it. When a predatory fish is lured within range, the clam discharges the glochidia from the brood pouch.

In such a manner, the modification of existing developmental patterns have permitted unionid clams to survive in environments that would otherwise be inhospitable.

FIGURE 5.16
Formation of glochidium larvae by the modification of spiral cleavage. After the 8-cell embryo is formed (A), placement of the mitotic spindle causes most of the D cytoplasm to enter the 2d blastomere (B). This large 2d blastomere divides (C) to eventually give rise to the large "bear-trap" shell of the larva (D). (After Raff and Kaufman, 1983.)

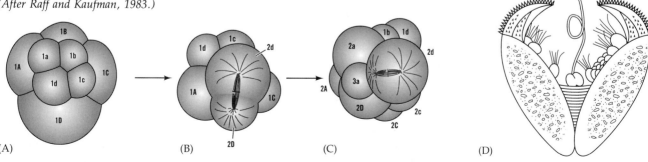

(A)   (B)   (C)   (D)

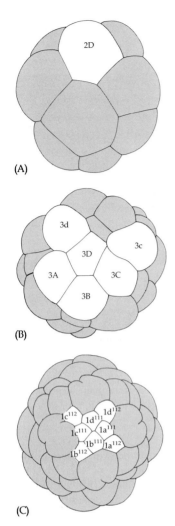

FIGURE 5.18

Intercellular communication during mollusc cleavage. Diagrams show the extent to which the fluorescent dye Lucifer Yellow (white area) has spread after injection into specific macromeres of *Patella vulgata*. (A) Labeling pattern 20 minutes after injection of the 2D macromere of the 16-cell embryo indicates no transfer of dye to adjacent cells. (B) Vegetal view of a 32-cell embryo after the injection of dye into the 3D macromere, showing the spread of the dye to adjacent cells (except 2d22). (C) Animal view of the same embryo about 45 minutes after injection, showing that dye has spread from the 3D macromere to the central group of animal micromeres. (After de Laat et al., 1980.)

Another exciting discovery concerning molluscan cleavage is that certain blastomeres communicate with each other. In those molluscs with equal-sized blastomeres at the 4-cell stage,* the determination of which cell will give rise to the mesodermal precursor cell is accomplished between the fifth and sixth cleavages. At this time, the 3D macromere extends inward and contacts the micromeres of the animal pole. Without this contact, the 4d cell given off by the 3D macromere does not produce mesoderm (Van der Biggelaar and Guerrier, 1979). In 1980, de Laat and co-workers demonstrated that at the time of contact (and not before) small molecules are capable of diffusing between the 3D macromere and the central micromeres. They injected the dye Lucifer Yellow (molecular weight 457) into one of the macromeres. When added before the 32-cell stage, the dye resides only in that single macromere and its progeny. However, when the dye is injected into the 3D blastomere at the beginning of the 32-cell stage, the first tier of micromeres picks up the yellow color (Figure 5.18). Low-molecular-weight material is transferred from one cell to the other at just that time when the micromeres alter the developmental potential of the 3D blastomere. Transmission electron microscopy shows that at this time, gap junctions appear on the surfaces of these cells.

Bilateral holoblastic cleavage

Bilateral holoblastic cleavage is found primarily in ascidians (tunicates). Figure 5.19 shows the cleavage pattern of a tunicate, *Styela partita*. The most striking phenomenon in this type of cleavage is that the first cleavage plane establishes the only plane of symmetry in the embryo, separating the embryo into its future right and left sides. Each successive division orients itself to this plane of symmetry, and the half-embryo formed on one side of the first cleavage is the mirror image of the half-embryo on the other side. The second cleavage is meridional, like the first division, but unlike the first division, it does not pass through the center of the egg. Rather, it creates two large anterior cells (A and D) and two smaller posterior cells (B and C). Each side now has a large and a small blastomere. At the next three divisions, differences in cell size and shape highlight the bilateral symmetry of these embryos. At the 32-cell stage, a small blastocoel is formed and gastrulation begins.

As mentioned in Chapter 4, certain tunicates (including *S. partita*) contain colored cytoplasmic regions. During cleavage, these plasms become partitioned into different cells. Moreover, the type of cytoplasm the cells receive determines their eventual fates. Cells receiving clear cytoplasm become ectoderm; those containing yellow cytoplasm give rise to mesodermal cells; the cells that incorporate the slate-gray inclusions become endoderm; and the light gray cells become the neural tube and notochord. These colored plasms are localized bilaterally around the plane of symmetry, so they are bisected by the first cleavage furrow into the right and left halves of the embryo. The second cleavage causes the prospective mesoderm to lie in the two posterior cells, while the prospective neural and chordamesoderm are to be formed from the two anterior cells. The third division further partitions these cytoplasmic regions such that the mesoderm-forming cells are confined to the two vegetal posterior blastomeres, and the chordamesoderm cells are likewise restricted to the two vegetal anterior cells. The fate of each cell of the early *Styela* embryo has been followed and will be discussed in detail in Chapter 14.

*Don't worry about those mollusc embryos with unequal-sized blastomeres at the 4-cell stage. We will see much of them in Chapter 16.

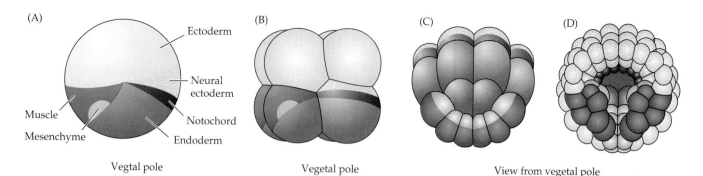

(A) Ectoderm
Neural ectoderm
Muscle
Notochord
Mesenchyme
Endoderm
Vegtal pole

(B) Vegetal pole

(C) (D) View from vegetal pole

FIGURE 5.19
Bilateral symmetry in a tunicate egg. (A) Uncleaved egg. (B) 8-cell embryo, showing the blastomeres and the fates of various cells. It can be viewed as two 4-cell halves; from here on each division on the right side of the embryo has a mirror-image division on the left. (C,D) Views of later embryos from the vegetal pole. (The regions of cytoplasm destined to form particular organs are labeled in panel A and are coded by color throughout the diagram.) (After Balinsky, 1981.)

Rotational holoblastic cleavage

It is not surprising that mammalian cleavage has been the most difficult to study. Mammalian eggs are among the smallest in the animal kingdom, making them hard to manipulate experimentally. The human zygote, for instance, is only 100 μm in diameter—barely visible to the eye, and less than one-thousandth the volume of a *Xenopus* egg. Also, mammalian zygotes are not produced in numbers comparable to sea urchin or frog embryos. Usually, fewer than 10 eggs are ovulated by a female at a given time, so it is difficult to obtain enough material for biochemical studies. As a final hurdle, the development of mammalian embryos is accomplished within another organism rather than in the external environment. Only recently has it been possible to duplicate some of these internal conditions and observe development in vitro.

With all these difficulties, knowledge of mammalian cleavage was worth waiting for, as mammalian cleavage turned out to be strikingly different from most other patterns of embryonic cell division. The mammalian oocyte is released from the ovary and swept into the oviduct (Figure 5.20). Fertilization occurs in the ampulla of the oviduct, a region close to the ovary. Meiosis is completed at this time, and first cleavage begins about a day later. Cleavages in mammalian eggs are among the slowest cleavages occurring in the animal kingdom—about 12 to 24 hours apart. Meanwhile, the cilia in the oviduct push the embryo toward the uterus; the first cleavages occur along this journey.

There are several features of mammalian cleavage that distinguish it

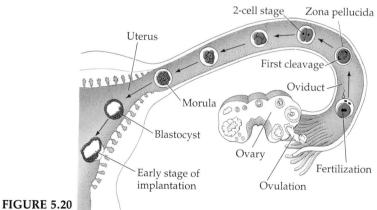

FIGURE 5.20
Development of a human embryo from fertilization to implantation. The egg "hatches" from the zona upon reaching the uterus, and it is probable that the zona prevents the cleaving cells from sticking to the oviduct rather than traveling to the uterus. (After Tuchmann-Duplessis et al., 1972.)

FIGURE 5.21
Comparison of early cleavage in (A) echinoderms (radial cleavage) and (B) mammals (rotational cleavage). Nematodes also have a rotational form of cleavage, but they do not form the blastocyst structure characteristic of mammals. Details of nematode cleavage will be given in Chapter 14. (After Gulyas, 1975.)

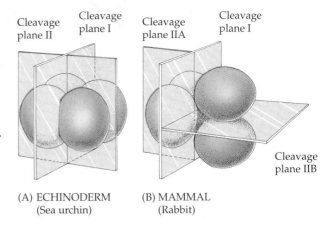

Cleavage plane II   Cleavage plane I   Cleavage plane IIA   Cleavage plane I

Cleavage plane IIB

(A) ECHINODERM
(Sea urchin)

(B) MAMMAL
(Rabbit)

from other cleavage types. The first feature is the relative slowness of the divisions. The second fundamental difference is the unique orientation of mammalian blastomeres with relation to one another. The first cleavage is a normal meridional division; however, in the second cleavage one of the two blastomeres divides meridionally and the other divides equatorially (Figure 5.21). This type of cleavage is called **rotational cleavage** (Gulyas, 1975).

The third major difference between mammalian cleavage and that of most other embryos is the marked asynchrony of early division. Mammalian blastomeres do not all divide at the same time. Thus, mammalian embryos do not increase evenly from 2- to 4- to 8-cell stages, but frequently contain odd numbers of cells. Also, unlike almost all other animal genomes, the mammalian genome is activated during early cleavage and the genome produces the proteins necessary for cleavage to occur. In the mouse and goat, the switch from maternal to zygotic control occurs at the 2-cell stage (Piko and Clegg, 1982; Prather, 1989).

Compaction

Perhaps the most crucial difference between mammalian cleavage and all other types involves the phenomenon of **compaction.** As seen in Figure

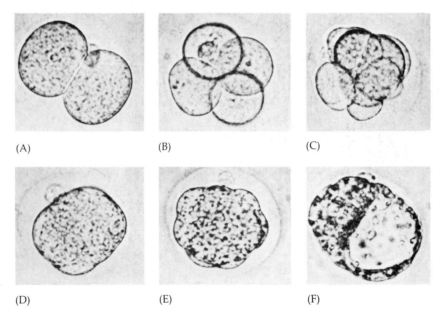

(A)   (B)   (C)

FIGURE 5.22
The cleavage of a single mouse embryo in vitro. (A) 2-cell stage. (B) 4-cell stage. (C) Early 8-cell stage. (D) Compacted 8-cell stage. (E) Morula. (F) Blastocyst. (From Mulnard, 1967, courtesy of J. G. Mulnard.)

(D)   (E)   (F)

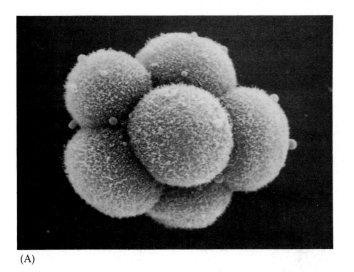

(A)

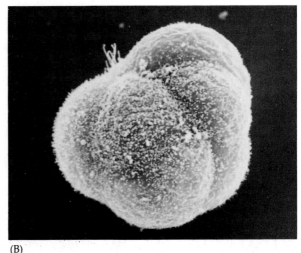

(B)

5.22, mammalian blastomeres through the 8-cell stage form a loose arrangement with plenty of space between them. Following the third cleavage, however, the blastomeres undergo a spectacular change in their behavior. They suddenly huddle together, maximizing their contact with the other blastomeres and forming a compact ball of cells (Figures 5.22C,D and 5.23). This tightly packed arrangement is stabilized by tight junctions that form between the outside cells of the ball, sealing off the inside of the sphere (Figure 5.24). The cells within the sphere form gap junctions, thereby enabling small molecules and ions to pass between the cells.

The cells of the compacted embryo divide to produce a 16-cell morula. This morula consists of a small group of internal cells (1–2 cells) surrounded by a larger group of external cells (Barlow et al., 1972). Most of the descendants of the external cells become the **trophoblast** (or **trophectoderm**) cells. This group of cells produces no embryonic structures. Rather, these cells form the tissue of the chorion, the embryonic portion of the placenta. The chorion enables the fetus to get oxygen and nourishment from the mother. It also secretes hormones so that the mother's uterus will retain the fetus and produces regulators of the immune response so that the mother will not reject the embryo as she would an organ graft. However, trophoblast cells are not able to produce any cell of the embryo itself. They are necessary for implanting the embryo in the uterine wall (Figure 5.25).

The embryo is derived from the descendants of the inner cell(s) of the 16-cell stage, supplemented by an occasional cell dividing from the tro-

FIGURE 5.24
Diagram of compaction and the formation of the blastocyst. (A,B) 8-cell embryo. (C) Morula. (D) Blastocyst.

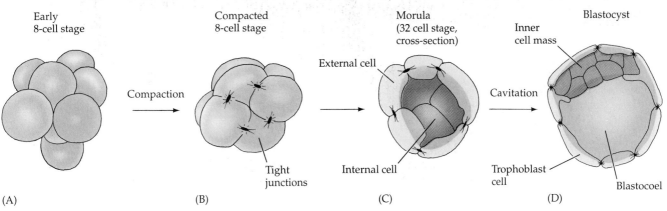

FIGURE 5.25
Implantation of mammalian blastocysts into the uterus. (A) Mouse blastocysts entering uterus. (B) Initial implantation of the blastocyst onto the uterus in a rhesus monkey. (A from Rugh, 1967; B courtesy of the Carnegie Institution of Washington, Chester Reather, photographer.)

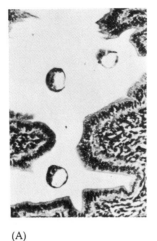

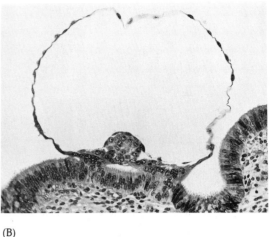

(A)   (B)

phoblast during the transition to the 32-cell stage (Pederson et al., 1986; Fleming, 1987). These cells generate the **inner cell mass** that will give rise to the embryo. These cells not only look different from the trophoblast cells, they also synthesize different proteins at this early developmental stage. By the 64-cell stage, the inner cell mass and the trophoblast cells have become separate cell layers, neither of which contributes cells to the other group (Dyce et al., 1987). Thus, the distinction between trophoblast and inner cell mass blastomeres represents the first differentiation event in mammalian development.

Initially the morula does not have an internal cavity. However, during a process called **cavitation,** the trophoblast cells secrete fluid into the morula to create a blastocoel. The inner cell mass is positioned on one side of the ring of trophoblast cells (Figures 5.22, 5.24, and 5.25). This structure is called the **blastocyst** and is another hallmark of mammalian cleavage.

SIDELIGHTS & SPECULATIONS

The cell surface and the mechanism of compaction

Compaction creates the circumstances that bring about the first differentiation in mammalian development—the separation of trophoblast from inner cell mass. How is this done? There is growing evidence that compaction is mediated by events occurring at the cell surfaces of adjacent blastomeres. First, prior to compaction, each of the eight blastomeres undergoes extensive membrane changes known as **polarization.** Different components of the cell surface migrate to different regions of the cell (Ziomek and Johnson, 1980). This can be seen by tagging certain cell surface molecules with fluorescent dyes. One such tag, which recognizes a class of glycoproteins, shows that at the 4-cell stage these glycoproteins are randomly distributed throughout the membrane (Figure 5.26A). However, at the mid-8-cell stage, these molecules are found predom-

inantly at the poles farthest away from the center of the aggregate (Figure 5.26B). This phenomenon appears to be influenced by cell–cell interactions, because it takes place on pairs of attached blastomeres that are isolated together but not on individual isolated cells.

Second, specific cell surface proteins are seen to play a role in compaction. One such molecule is **E-cadherin,** a 120-kDa adhesive glycoprotein. This protein (also known as uvomorulin) is synthesized at the 2-cell stage and is seen to be uniformly spread throughout the cell membrane. However, as compaction occurs, E-cadherin becomes restricted to those sites on cell membranes that are in contact with adjacent blastomeres. Antibodies to this molecule cause the decompaction of the morula and also inhibit the cell–cell attachments of embryonic stem cells (Figure 5.27; Peyrieras et al., 1983; Johnson et al., 1986). The carbohydrate portion of this glycoprotein may be essential to its function, as tunicamycin (a drug that inhibits the glycosylation of proteins) also prevents compaction.

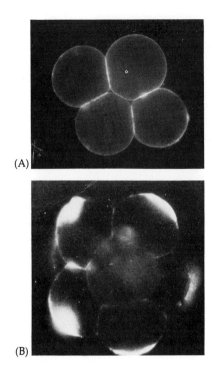

(A)

(B)

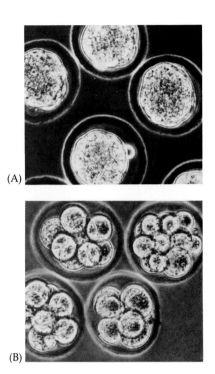

(A)

(B)

FIGURE 5.26
Polarization of membrane components in 8-cell stage mouse blastomeres. (A) Homogeneous, nonpolar distribution of membrane components labeled with fluorescent concanavalin A at the 4-cell stage. (B) Heterogeneous, polar distribution of these components at the 8-cell stage. (A from Fleming et al., 1986; B from Levy et al., 1986. Photographs courtesy of the authors.)

FIGURE 5.27
Prevention of compaction by antiserum directed against the cell surface adhesion glycoprotein E-cadherin. (A) Normal compaction occurring in the absence of antiserum. (B) Proliferation without compaction occurring in the presence of antibodies to E-cadherin. (Photographs courtesy of C. Ziomek.)

Third, recent experiments have shown that the phosphatidylinositol pathway (discussed in Chapter 4 in regard to sperm and egg activation) may also be important for initiating compaction. If 4-cell mouse embryos are placed into media containing drugs that activate protein kinase C, premature compaction occurs. Similarly, diacylglycerides can transiently cause these 4-cell embryos to undergo compaction. When this occurs, the E-cadherin accumulates specifically at the junctions between the blastomeres (Winkel et al., 1990). These results suggest that the activation of protein kinase C may initiate compaction by shifting the localization of E-cadherin.

Fourth, the cell membrane may also be modified during compaction by cytoskeletal reorganization. Microvilli, extended by actin microfilaments, appear on adjacent cell surfaces and attach one cell to the other. These microvilli may be the sites where E-cadherin is functioning to mediate intercellular adhesion. The flattening of the blastomeres against one another may therefore be brought about by the shortening of the microvilli through actin depolymerization (Pratt et al., 1982; Sutherland and Calarco-Gillam, 1983).

Thus, there is growing evidence that compaction is caused by changes in the architecture of the blastomere cell surface. It is not certain, though. how these events relate to one another or how they are coordinated into the integrated network of events that causes compaction to occur.

Formation of the inner cell mass

The creation of an inner cell mass distinct from the trophoblast is *the* crucial process of early mammalian development. How is a cell directed into one or the other of these paths? How is a cell informed that it is either to give rise to a portion of the adult mammal or that it is to give rise to a rather remarkable supporting tissue that will be discarded at birth? Observations of living embryos suggest that this momentous decision is merely a matter of a cell's being in the right place at the right time. Up

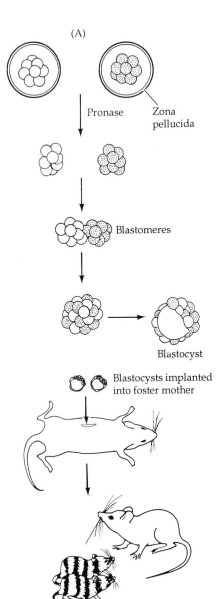

(A)

Pronase

Zona pellucida

Blastomeres

Blastocyst

Blastocysts implanted into foster mother

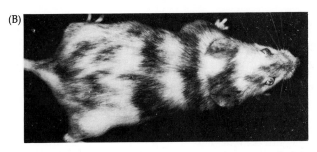

FIGURE 5.28
Production of allophenic mice. (A) The experimental procedures used to produce allophenic mice. Early 8-cell embryos of genetically distinct mice (here, those with coat-color differences) are isolated from mouse oviducts and brought together after their zonae are removed by proteolytic enzymes. The cells form a composite blastocyst, which is implanted into the uterus of a foster mother. (B) An adult allophenic mouse showing contributions from the pigmented (black) and unpigmented (white) embryos. (Photograph courtesy of B. Mintz.)

through the 8-cell stage, there are no obvious differences in the biochemistry, morphology, or potency of any of the blastomeres. However, compaction forms inner and outer cells with vastly different properties. By labeling the various blastomeres, numerous investigators have found that the cells that happen to be on the outside will form the trophoblast whereas the cells that happen to be inside will generate the embryo (Tarkowski and Wróblewska, 1967; Sutherland et al., 1990).* Hillman and co-workers (1972) have shown that when each blastomere of a 4-cell mouse embryo is placed on the outside surface of a mass of aggregated blastomeres, the external, transplanted cells only give rise to trophoblast tissue. Therefore, it seems that whether a cell becomes trophoblast or embryo depends on whether it was an external or an internal cell after compaction.

If most of the cells of the blastocyst give rise to the trophoblast, exactly how many cells actually form the embryo? One way to answer this question is to produce **allophenic mice.** Allophenic mice are the result of two early-cleavage (usually 4- or 8-cell) embryos that have been artificially aggregated to form a composite embryo. As shown in Figure 5.28, the zonae pellucidae of two genetically different embryos are removed and the embryos brought together to form a common blastocyst. These prepared blastocysts are implanted into the uterus of the foster mother. When they are born, the allophenic offspring have some cells from each embryo. This is readily seen when the aggregated blastomeres come from mouse strains that differ in their coat colors. When blastomeres from white and black strains are aggregated, the result is commonly a mouse with black and white bands (Figure 5.28B). If there were only one cell in the blastocyst that gave rise to the embryo, this result would not be possible; the offspring would be either all white or all black. If two cells of the blastocyst were responsible for producing the embryo, we would expect the allophenic pattern to be expressed only half the time (1WW:2WB:1BB). Should there be three embryo-producing cells, the chance that the embryo would have an allophenic pattern would then increase to 75 percent (1WWW:3WWB:3WBB:1BBB), and the 4-cell situation would give an 87.5 percent chance of providing two-color mice. The experimental data of

*The inner cells have been found to come most frequently from the first cell to divide at the 2-cell stage. This cell usually produces the first pair of blastomeres to reach the 8-cell stage, and these cells usually divide so that they are inside the loosely aggregated cluster of blastomeres (Graham and Kelly, 1977).

(A) 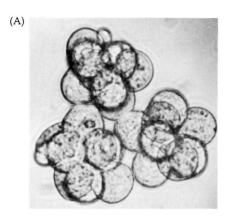 (B) 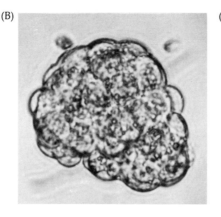 (C)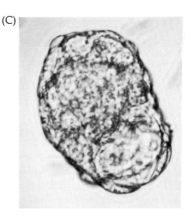

Mintz (1970) are that 73 percent of the double embryos yield allophenic mice, suggesting that three blastomeres of the blastocyst produce the entire embryo. Markert and Petters (1978) have shown that three early 8-cell embryos can unite to form a common compacted morula (Figure 5.29) and that the resulting mouse can have the coat colors of the three different strains (Plate 21). Therefore, although it is not certain that three is the absolute number of blastomeres that form the embryo, we can be fairly certain that the number is not much greater and that most of the cells of the blastocyst never contribute to the adult organism.

FIGURE 5.29
Aggregation and compaction of three 8-cell mouse embryos to form a single compacted morula. Cells from three different embryos (A) are aggregated together to form a morula (B), which undergoes compaction to form a single blastocyst (C). The resulting allophenic mouse is shown in color in Plate 21. (From Markert and Petters, 1978; courtesy of C. Markert.)

Escape from the zona pellucida

While the embryo is moving through the oviduct en route to the uterus, the blastocyst expands within the zona pellucida (the extracellular matrix of the egg that had been essential for sperm binding during fertilization). The cell membranes of the trophectoderm cells contain a sodium pump (a Na^+/K^+–ATPase) facing the blastocoel and transporting sodium ions into the central cavity. This accumulation of sodium ions draws in water osmotically, thus enlarging the blastocoel (Borland, 1977; Wiley, 1984). During this time it is essential that the zona pellucida prevent the blastocyst from adhering to the oviduct walls. When such adherence does take place in humans, it is called an **ectopic** or **tubal pregnancy.** This is a dangerous condition because the implantation of the embryo into the oviduct can cause life-threatening hemorrhage. When it reaches the uterus, however, the embryo must "hatch" from the zona so that it can adhere to the uterine wall.

The mouse blastocyst hatches from the zona by lysing a small hole in it and squeezing through that hole as the blastocyst expands (Figure 5.30). A trypsin-like protease, **strypsin,** is located on the cell membrane and lyses a hole in the fibrillar matrix of the zona (Perona and Wassarman, 1986; Yamazaki and Kato, 1989). Once out, the blastocyst can make direct contact with the uterus. The uterine epithelium "catches" the blastocyst on an extracellular matrix containing collagen, laminin, fibronectin, hyaluronic acid, and heparan sulfate receptors. The trophoblast cells contain the integrins that will bind to the uterine collagen, fibronectin, and laminin, and they synthesize the heparan sulfate proteoglycan dramatically at the time just prior to implantation (see Carson et al., 1993). Once on the uterine epithelial cells, the trophoblast secretes another set of proteases, including **collagenase, stromelysin,** and **plasminogen activator.** These protein-digesting enzymes digest the extracellular matrix of the uterine tissue, enabling the blastocyst to bury itself within the uterine wall (Strickland et al., 1976; Brenner et al., 1989).

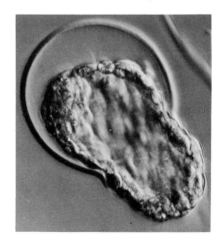

FIGURE 5.30
Mouse blastocyst hatching from the zona pellucida. (Photograph from Mark et al., 1985, courtesy of E. Lacy.)

Twins and embryonic stem cells

Human twins are classified into two major groups: **monozygotic** (one-egg; identical) twins and **dizygotic** (two-egg; fraternal) twins. Fraternal twins are the result of two separate fertilization events, whereas identical twins are formed from a single embryo whose cells somehow dissociated from one another. This fact implies that an isolated mammalian blastomere can give rise to an entire embryo. In 1952 Seidel supported this notion by destroying one cell of a 2-cell rabbit embryo. The resulting blastomere was able to develop into a complete adult. Even a single blastomere of an 8-cell mouse embryo can develop successfully into a complete adult (Kelly, 1977). Gardiner and Rossant (1976) have also shown that if cells of the inner cell mass (but not trophoblast cells) are injected into blastocysts, they contribute to the new embryo. Thus, it is probable that identical twins are produced by the separation of early blastomeres or even by the separation of the inner cell mass into two regions within the same blastocyst.

This seems to be exactly what is happening in roughly 0.25 percent of human births. About 33 percent of identical twins have two complete and separate chorions, indicating that separation occurred before the formation of the trophoblast tissue at day 5 (Figure 5.31A). The remaining identical twins share a common chorion, suggesting that the split occurs within the inner cell mass after the tropho-

blast has formed. By day 9, the human embryo has completed the construction of another extraembryonic layer, the amnion. This tissue forms the amniotic sac (or water sac), which surrounds the embryo with amniotic fluid and protects it from desiccation and abrupt movement (Chapter 6). If the separation of the embryo were to come between the formation of the chorion on day 5 and the amnion on day 9, then the resulting embryos should have one chorion and two amnions (Figure 5.31B). This happens in about two-thirds of human identical twins. A small percentage of identical twins are born within a single chorion and amnion (Figure 5.31C). This means that the division of the embryo came after day 9. Such newborns are at risk of being conjoined ("Siamese") twins.

FIGURE 5.31

Diagram showing the timing of human monozygotic twinning with relation to extraembryonic membranes. (A) Splitting occurs before the formation of the trophectoderm, so each twin has its own chorion and amnion. (B) Splitting occurs after trophectoderm formation but before amnion formation, resulting in twins having individual amniotic sacs but sharing one chorion. (C) Splitting after amnion formation leads to twins in one amniotic sac and a single chorion. (After Langman, 1981).

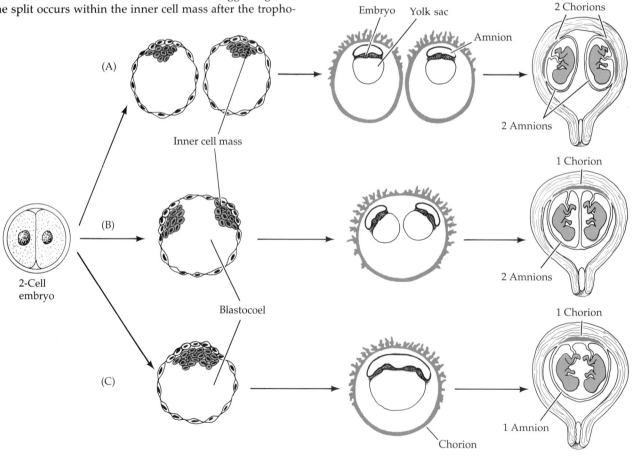

The ability to produce an entire embryo from cells that normally would have contributed to only a portion of the embryo is called *regulation* and is discussed in Chapter 16. Regulation is also seen in the ability of two or more early embryos to form one allophenic mouse rather than twins, triplets, or a multiheaded monster. There is even evidence (de la Chappelle et al., 1974; Mayr et al., 1979) that allophenic regulation can occur in humans. These allophenic individuals have two genetically different cell types (XX and XY) within the same body, each with its own set of genetically defined characteristics. The simplest explanation for the existence of such a phenomenon is that these individuals resulted from the aggregation of two embryos, one male and one female, that were developing at the same time. If this explanation were correct, then two fraternal twins fused to create a single composite individual.

According to our observations of twin formation and allophenic mice, each blastomere of the inner cell mass should be able to produce any cell of the body. This hypothesis has been confirmed, and it will have very important consequences in studying mammalian development. When inner mass cells are isolated and grown under certain conditions, they remain undifferentiated and continue to divide in culture (Evans and Kaufman, 1981; Martin, 1981). These cells are called **embryonic stem cells** (or **ES cells**). As shown in Chapter 2, these cells can be altered when in the Petri dish. Cloned genes can be inserted into their nuclei, or the existing genes can be mutated. When these ES cells are injected into blastocysts of another mouse embryo, the ES cells can integrate into the host inner cell mass. The resulting embryo has cells coming from both the host and the donor tissue. This technique has become extremely important in determining the function of genes during mammalian development.

Meroblastic cleavage

As mentioned earlier in this chapter, yolk concentration plays an important role in cell cleavage. Nowhere is this more apparent than in the meroblastic cleavage types. Here, the large concentrations of yolk prohibit cleavage in all but a small portion of the egg cytoplasm. In **discoidal cleavage,** cell division is limited to a small disc of yolk-free cytoplasm atop a mound of yolk; in **superficial cleavage,** the centrally located yolk permits cleavage only along the peripheral rim of the egg.

Discoidal cleavage

Discoidal cleavage is characteristic of birds, fishes, and reptiles.

Birds. Figure 5.32 shows the cleavage of an avian egg. The bulk of the oocyte is taken over by the yolk, allowing cleavage to occur only in the **blastodisc,** a region of active cytoplasm about 2–3 mm in diameter, at the animal pole of the egg. Because these cleavages do not extend into the yolky cytoplasm, the early-cleavage-stage cells are actually continuous at their bases. The first cleavage furrow appears centrally in the blastodisc, and other cleavages follow to create a single-layered blastoderm. At first, this cellular layer is incomplete, as the cells are still continuous with underlying yolk. Thereafter, equatorial and vertical cleavages divide the blastoderm into a tissue five to six cell layers thick. These cells become linked together with tight junctions (Bellairs et al., 1975; Eyal-Giladi, 1991). Between the blastoderm and the yolk is a space, called the **subgerminal cavity.** This space is created when the blastoderm cells absorb fluid from the albumin ("egg white") and secrete it between themselves and the yolk (New, 1956). At this stage, the deep cells in the center of the blastoderm are shed to create a one-cell thick **area pellucida.** (The shed cells appear to die.) The peripheral ring of blastoderm cells that are not shed constitute the **area opaca.**

By the time the hen has laid the egg, the blastoderm contains some 60,000 cells. Some of these cells are shed into the subgerminal cavity to

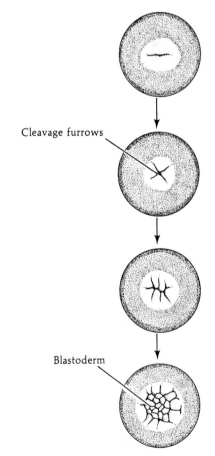

Cleavage furrows

Blastoderm

FIGURE 5.32
Discoidal cleavage in a chick egg, viewed from the animal pole. The cleavage furrows do not penetrate the yolk, and a blastoderm consisting of a single layer of cells is produced.

FIGURE 5.33
Formation of the two layered chick embryo. This sagittal section of the embryo near the posterior margin shows an upper layer consisting of a central epiblast that trails into the cells of Koller's sickle (ks) and the posterior marginal zone (mz). Certain cells from the epiblast fall (delaminate) from the upper layer to form polyinvagination islands (pi) of 5–20 cells each. These cells will be joined by those hypoblast cells (hyp) migrating anteriorly from Koller's sickle to form the lower (hypoblastic) layer. Sc, subgerminal cavity; gwm, germ wall margin. (From Eyal-Giladi et al., 1992, courtesy of H. Eyal-Giladi.)

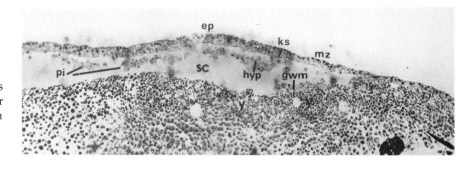

form a second layer (Figure 5.33). Thus, soon after laying, the chick egg contains two layers of cells: the upper **epiblast** and the lower **hypoblast.** Between them lies the blastocoel. We will detail the formation of the hypoblast in the next chapter.

Fishes. The yolky eggs of fishes develop similarly to those of birds, with cell division occurring only in the animal pole blastodisc. Scanning electron micrographs of fish egg cleavage show beautifully the incomplete nature of discoidal cleavage (Figure 5.34). It had been thought that the early blastomeres of fish eggs were like those of mammalian eggs, in that they mixed together such that any blastomere could form any adult structure. Recent evidence, however, shows that the blastomeres are more like those

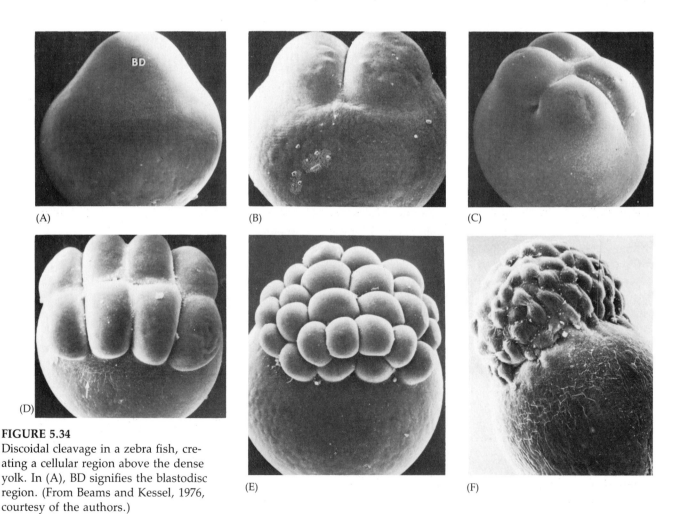

FIGURE 5.34
Discoidal cleavage in a zebra fish, creating a cellular region above the dense yolk. In (A), BD signifies the blastodisc region. (From Beams and Kessel, 1976, courtesy of the authors.)

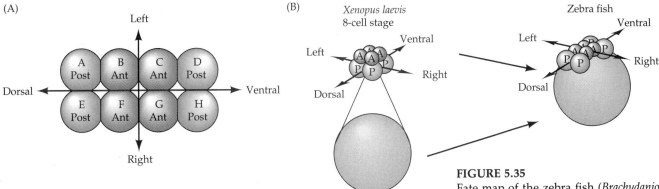

FIGURE 5.35
Fate map of the zebra fish (*Brachydanio rerio*) egg. (A) Schematic depiction of the first three cleavage products and their products. (B) Transformation of the *Xenopus laevis* fate map to the *B. rerio* fate map by imposing a vegetal yolk mass beneath the blastomeres. (After Strehlow and Gilbert, 1993.)

of amphibians, wherein the plane of early cleavage is an indication of cell fate. Strehlow and Gilbert (1993) injected highly fluorescent polymers into individual blastomeres of developing zebra fish and were able to follow their descendants. Instead of mixing, the first eight blastomeres gave rise to specific tissues (Figure 5.35). The first cleavage defined the future dorsal from the future ventral surface; the second division defined the future left-hand side of the embryo from the future right-hand side; and the third cleavage separated the prospective anterior from the posterior of the embryo. The authors noted that the first three divisions of the zebra fish egg are just like those of the *Xenopus* egg if the *Xenopus* egg is viewed as being ventrally compressed against a large yolk mass (Figure 5.35B).

Superficial cleavage

Most insect eggs undergo superficial cleavage, whereby a large mass of centrally located yolk confines cleavage to the cytoplasmic rim of the egg. One of the fascinating details of this cleavage type is that the cells do not form until after the nuclei have divided. Cleavage of an insect egg is shown in Figure 5.36. The zygote nucleus undergoes several mitotic divisions within the central portion of the egg. In *Drosophila*, 256 nuclei are produced by a series of nuclear divisions averaging 8 minutes each. The nuclei then migrate to the periphery of the egg, where the mitoses continue, albeit at a progressively slower rate. The embryo is now called a **syncytial blastoderm,** meaning that all the cleavage nuclei are contained within a common cytoplasm. No cell membranes exist other than that of the egg itself. Those nuclei migrating to the posterior pole of the egg soon become enveloped by new cell membranes to form the **pole cells** of the embryo. These cells give rise to the germ cells of the adult. Thus, one of the first events of insect development is the separation of the future germ cells from the rest of the embryo.

After the pole cells have been formed, the oocyte membrane folds inward between the nuclei, eventually partitioning off each somatic nucleus into a single cell (Figure 5.37). This creates the **cellular blastoderm,** with all the cells arranged in a single-layered jacket around the yolky core of the egg. In *Drosophila* this layer consists of approximately 6000 cells and is formed within 4 hours of fertilization.

Although the nuclei originally divide within a common cytoplasm, this does not mean that the cytoplasm is itself uniform. Karr and Alberts (1986) have shown that each nucleus within the syncytial blastoderm is contained within its own little territory of cytoskeletal proteins. When the nuclei reach the periphery during the tenth cleavage cycle, each nucleus

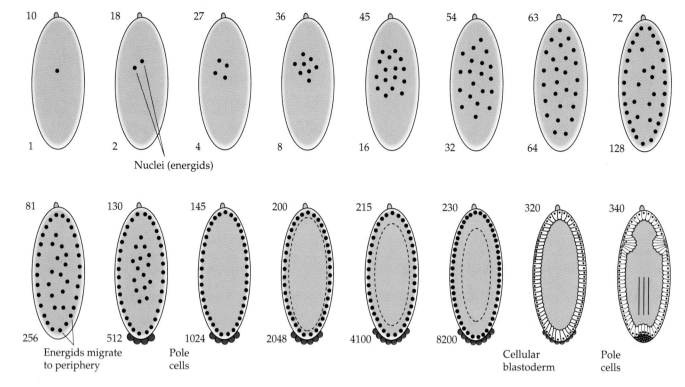

Nuclei (energids)

Energids migrate to periphery

Pole cells

Pole cells

Cellular blastoderm

Pole cells

FIGURE 5.36

Superficial cleavage in a *Drosophila* embryo. The numeral above each embryo corresponds to the number of minutes after deposition of the egg; the numeral at the bottom indicates the number of nuclei present. Pole cells (which will form the germ cells) are seen at the 512-nuclei stage even though the cellular blastoderm does not form until nearly 3 hours later. The times are representative, as the duration of each division cycle depends, in part, on the temperature at which the egg is incubated.

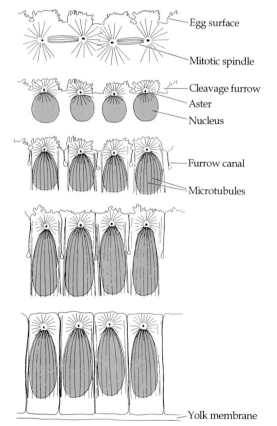

Egg surface

Mitotic spindle

Cleavage furrow

Aster

Nucleus

Furrow canal

Microtubules

Yolk membrane

FIGURE 5.37

Diagram of nuclear elongation and cellularization in *Drosophila* blastoderm. (After Fullilove and Jacobson, 1971.)

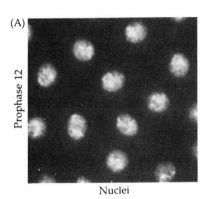

Prophase 12

Nuclei

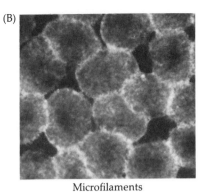

Microfilaments

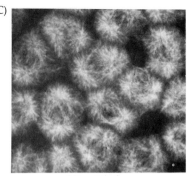

Microtubules

FIGURE 5.38
Localization of the cytoskeleton around nuclei in the syncytial blastoderm of *Drosophila*. A *Drosophila* embryo entering the mitotic prophase of its twelfth division was sectioned and triple-stained. (A) The nuclei were localized by a dye that binds to DNA. (B) Microfilaments were identified using a fluorescent antibody to actin. (C) Microtubules were recognized by a fluorescent antibody to tubulin. Cytoskeletal domains can be seen surrounding each nucleus. (From Karr and Alberts, 1986, courtesy of T. L. Karr.)

becomes surrounded by microtubules and microfilaments. The nuclei and their associated cytoplasmic islands are called **energids.** Figure 5.38 shows the nuclei and their essential microfilament and microtubule domains in prophase of the twelfth mitotic division.

After the nuclei reach the periphery, the time required to complete each of the next four divisions becomes gradually longer. While cycles 1 through 10 are each 8 minutes long, cycle 13, the last cycle in the syncytial blastoderm, takes 25 minutes to complete. The *Drosophila* embryo forms cells in cycle 14 (i.e., after 13 divisions), and cycle 14 is asynchronous. Some groups of cells complete this cycle in 75 minutes, whereas other groups of cells take 175 minutes to complete cycle 14 (Foe, 1989). Transcription from these nuclei (which begins around the eleventh cycle) is greatly enhanced. The slowdown of *Drosophila* nuclear division and the concomitant increase in RNA transcription is often referred to as the **midblastula transition.** Such a transition is also seen in amphibian and echinoderm embryos. The control of this mitotic slowdown (in *Xenopus*, sea urchin, starfish, and *Drosophila* embryos) appears to be effected by the ratio of chromatin to cytoplasm (Newport and Kirschner, 1982a; Edgar et al., 1986). Edgar and his colleagues compared the early development of wild-type *Drosophila* embryos with those of a haploid mutant. The haploid *Drosophila* embryos have half the wild-type quantity of chromatin at each cell division. Hence a haploid embryo at the eighth cell cycle has the same amount of chromatin as a wild-type embryo has at cell cycle 7. These investigators found that whereas wild-type embryos formed their cell layer immediately after the thirteenth division, the haploid embryos underwent an extra, fourteenth, division before cellularization. Moreover, the lengths of cycles 11 through 14 in wild-type embryos corresponded to cycles 12 through 15 in the haploid embryos. Thus, the haploid embryos follow a pattern similar to that of the wild-type embryos, only they lag by one cell division.

If their lagging were due to the haploid mutants' having a chromatin-to-cytoplasm ratio of half the wild-type ratio at any given cycle, then one should be able to accelerate cellularization by tying off (ligating) some of the cytoplasm such that the nuclei divide in a smaller volume. When this ligation was performed, the mitotic pattern of the embryos was accelerated. The final blastoderm division, signaling the end of the cleavage period, is reached when there is one nucleus for each 61 μm^3 of cytoplasm. In *Xenopus,* a similar slowdown in mitotic rate is observed after the twelfth cell division. Here, too, the divisions thereafter become asynchronous. Ligation experiments suggest that the timing of this midblastula transition is also due to the chromatin-to-cytoplasm volume ratio (Newport and Kirschner, 1982a,b).

In both *Drosophila* and *Xenopus*, the initiation of transcription can be induced prematurely by artificially lengthening the cell cycle. When cycloheximide (an inhibitor of protein synthesis) delays cell division, the midblastula transition is induced early in *Xenopus,* and a burst of transcription occurs in *Drosophila* (Edgar et al., 1986; Kimelman et al., 1987).

SIDELIGHTS & SPECULATIONS

Exceptions, generalizations, and parasitic wasp cleavage

What we consider "normal" and what we marginalize as "exceptions" often reflect which animals are most readily accessible to study and most easily domesticated for laboratories. Needless to say, this does not necessarily reflect the condition of the natural world. Rather, our discussions of animal development are often bottlenecked through particular organisms. The varied development of amphibians is generally represented by *Xenopus laevis,* and the mouse and human are the only mammals whose development is usually studied. Similarly, although there are over 800,000 known species of insects, most developmental biologists know only the development of one species: *Drosophila melanogaster. Drosophila* gained preeminence only after it was thought necessary to relate embryological phenomena to particular genes. In 1941, the major compendium of insect development (Johannsen and Butt's *Embryology of Insects and Myriapods*) didn't even mention this species.

Insects are an exceptionally successful and widespread subphylum, however, so it is not surprising to find an enormous amount of variability in their development. The development of the parasitic wasp *Copidosomopsis tanytmemus* differs remarkably from that of the canonical *Drosophila.* Like several other parasitic species, the female *C. tanytmemus* deposits her egg *inside* the egg of another species.

As the host egg (usually that of a moth) is developing, so is the parasite's egg. However, while the host egg begins development in the usual superficial pattern, the wasp egg divides *holoblastically.* Moreover, instead of differentiating a body axis, the cells of the parasitic embryo divide repeatedly to become a mass of undifferentiated cells called a **polygerm.** By two weeks, the growing polygerm is suspended in the host, remaining loosely attached to the larval brain and trachea (Figure 5.39A; Cruz, 1986a).

As the polygerm grows, it splits into dozens (sometimes thousands, depending on the species) of discreet groups of cells. Each of these groups of cells becomes an embryo! This ability of an egg to develop into a mass of cells that routinely forms several embryos is called **polyembryony.** (Polyembryony is characteristic of certain insect groups and of certain mammalian species such as the nine-banded armadillo, whose eggs routinely form identical quadruplets.) Most of these parasitic wasp embryos develop into normal wasp larvae that take about 30 days to develop. A smaller group, about 10 percent of the total number of embryos, become **precocious larvae** (Figure 5.39B), which develop within a week. Not only are they formed earlier, but precocious larvae have very little structure and do not undergo metamorphosis. They are essentially a mobile set of jaws. These larvae do not reproduce, and they die by the time the normal larvae are formed. While they live, however, they go through the host embryo

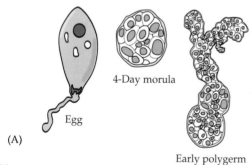

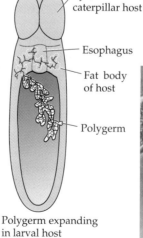

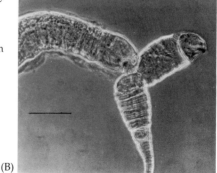

FIGURE 5.39

Development of parasitic wasps (Encyrtidae). (A) Holoblastic cleavage of the Copidosomopsis tanytmemus *egg produces a polygerm of undifferentiated cells. (B) Precocious larvae of a related genus,* Pentalitomastix, *attacks a larva of* Trathala *inside the same host; the photograph taken from a freshly opened host. (A after Cruz, 1986a; B from Cruz, 1981, courtesy of Y. Cruz.)*

Egg   4-Day morula   Early polygerm

Eye/head of caterpillar host
Esophagus
Fat body of host
Polygerm
Polygerm expanding in larval host

(A)   (B)

killing the parasitic larvae of other individuals (of different species and of other clones of the same species). In other words, the precocious larvae are predatory forms that kill possible competitors (Cruz, 1981, 1986b; Grbic et al., 1992).

As the precocious larvae (and their prey) die, the normal larvae emerge from their first molt, and they begin feeding voraciously on the host's larval organs. By 40 days, the parasitic brood has finished eating its host's muscles, fat bodies, gonads, silk glands, gut, nerve cord, and hemolymph, and the host is little more than a sac of skin holding about 70 pupating wasp larvae. After another 5 or 6 days, the new adults gnaw holes in the host's integument and, in a scene repeated in the movie *Alien*, chew their way out of the host's body. These adults then copulate (often on the body of their dead host), find another host in which to deposit an egg, and die shortly thereafter.

Such a life cycle discomforted Charles Darwin and made him question the concept of a benign and all-knowing deity. In 1860 he wrote to the American biologist Asa Gray, "I cannot persuade myself that a benevolent and omnipotent God would have designedly created the Ichneumonidae with the express intention of their feeding within the living bodies of Caterpillars . . ." However, in addition to their usefulness in provoking disquieting notions concerning natural order and the nature of "individuality," parasitic wasps may have important economic consequences. *Macrocentrus grandii* is a polyembryonic wasp that parasitizes the European corn borer. The ability of an insect to form from a holoblastically cleaving embryo should also enable us to appreciate some of the plasticity of nature and make us reluctant to make sweeping generalizations about an entire subphylum of organisms.

MECHANISMS OF CLEAVAGE

Regulating the cleavage cycle

The cell cycle of somatic cells is functionally divided into four stages (Figure 5.40A). After mitosis (M) there is a prereplication gap (G$_1$), after which time the synthesis of DNA takes place (S). After the synthetic period, there is a premitotic gap (G$_2$) which is followed by mitosis. The progression of these phases is regulated by growth factors that will be discussed later. In early-cleavage blastomeres, however, cell division can be much simpler. Early sea urchin blastomeres lack G$_1$, replicating their DNA during the last portion (telophase) of the previous mitosis (Hinegardner et al., 1964). *Xenopus* and *Drosophila* nuclei have eliminated both the G$_1$ and the G$_2$ phases during early cleavage. (*Xenopus* embryos add those phases to the cell cycle sometime after the twelfth cleavage. *Drosophila* adds G$_2$ during cycle 14 and G$_1$ during cycle 17.) For the first 12 divisions, *Xenopus* cells divide synchronously in a biphasic cell cycle: S to M and M to S (Figure 5.40B; Laskey et al., 1977; Newport and Kirschner, 1982a).

The factors regulating this biphasic cycle are located in the cytoplasm. Normal enlarging *Xenopus* oocytes are arrested in first meiotic prophase. They are unable to divide. If nuclei from dividing cells are transplanted into these oocytes, they also cease dividing. When normal oocytes are stimulated by progesterone, they resume their meiotic division and stop at the metaphase of the second meiosis. If nuclei from nondividing cells (such as neurons) are placed into the cytoplasm of progesterone-treated oocytes, they, too, initiate division and stop at metaphase (Gurdon, 1968). The cytoplasm of progesterone-stimulated oocytes still undergoes periodic cortical contractions (characteristic of division), even in the absence of nuclei or centrioles. If cloned DNA fragments are injected into such enucleated embryos, their replication comes under the control of these cycles (Hara et al., 1980; Harland and Laskey, 1980; Karsenti et al., 1984). Thus, the capacity for cell division is regulated by the cytoplasm.

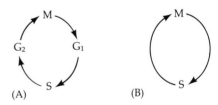

FIGURE 5.40
Cell cycles of somatic cells and early blastomeres. (A) Cell cycle of typical somatic cell. Mitosis (M) is followed by an "interphase" condition. This latter period is subdivided into G$_1$, S (synthesis), and G$_2$ phases. (B) Simpler biphasic cell cycle of the early amphibian blastomeres having only two states, S and M.

Maturation-promoting factor

Some of the factors that govern DNA synthesis and cell division have been identified. The progesterone-induced factor that allows oocyte nuclei to resume their divisions is a two-subunit phosphoprotein called **maturation promoting factor** (MPF; now sometimes called *mitosis-promoting factor*, thus keeping the same initials). MPF was first discovered as the major factor responsible for the resumption of meiotic cell divisions in the ovulated frog egg (Smith and Ecker, 1969; Masui and Markert, 1971). This same factor continues to play a role after fertilization, regulating the biphasic cell cycle of early *Xenopus* blastomeres. Gerhart and co-workers (1984) showed that MPF undergoes cyclic changes in level of activity in mitotic cells. The MPF activity of early frog blastomeres is highest during M and undetectable during S. During S phase, MPF exists in an inactive state. This cyclicity is also seen in enucleated blastomeres. Newport and Kirschner (1984) demonstrated that DNA replication (S) and mitosis (M) are driven solely by the gain and loss of MPF activity, even in the absence of protein synthesis. Cleaving cells can be trapped in S phase by incubating them in an inhibitor of protein synthesis. When MPF is microinjected into these cells, they enter M. Their nuclear envelope breaks down and their chromatin condenses into chromosomes. After an hour, MPF is degraded and the chromosomes return to S phase.

SIDELIGHTS & SPECULATIONS

Molecular regulators of development: MPF and its regulators

The small subunit of MPF: The cdc2 kinase

MPF has a large and a small subunit. The small subunit of MPF is a protein kinase that, when activated, can phosphorylate a variety of proteins. Thus, MPF functions by adding phosphate groups onto specific proteins. One such target is histone H1, which is bound to DNA. The phosphorylation of this protein may bring about chromosomal condensation. Another target is the nuclear envelope. Within 15 minutes after the addition of MPF, the three major proteins (the lamin proteins) of the nuclear envelope become hyperphosphorylated, and within the next 15 minutes, the envelope has depolymerized and is breaking apart (Miake-Lye and Kirschner, 1985; Arion et al., 1988). The purified MPF kinase subunit has been shown to phosphorylate these nuclear envelope proteins and to bring about their depolymerization in vitro (Peter et al., 1990; Ward and Kirschner, 1990). A third target appears to be RNA polymerase (Cisek and Corden, 1989), and the phosphorylation of RNA polymerase may be responsible for the inhibition of transcription seen during mitosis. A fourth target of the kinase appears to be the regulatory subunit of cytoplasmic myosin. When this protein is phosphorylated, it becomes inactive and is unable to function as an ATPase to drive the actin filaments involved in cell division (Satterwhite et al., 1992). The inhibition of this myosin during the early stages of mitosis may prevent the division of the cell until after the chromosomes have separated.

The small subunit of MPF has been remarkably conserved through evolution and is almost identical to a mitosis-inducing phosphoprotein, p34, synthesized by the yeast *cdc2* gene (Dunphy et al, 1988; Gautier et al., 1988). In fact, the human gene that encodes the protein corresponding to the small subunit of *Xenopus* MPF can be inserted into the yeast genome and cause division in *cdc2*-deficient yeast mutants (Lee and Nurse,1987). The p34 protein can exist in phosphorylated and unphosphorylated forms. The active form appears to be phosphorylated on threonine-161 (T-161) and unphosphorylated on tyrosine-15 (Y-15). Both these conditions are important for the kinase activity (Gould and Nurse, 1989; Solomon, 1993).

The large subunit of MPF: Cyclin

How, then, is MPF regulated? Since *Xenopus* cleavage seemed to be regulated by a protein similar to that which regulates yeast cell division, it was thought that whatever regulated the yeast protein might have a counterpart in the animal embryo. One of the most important regulators of the yeast MPF-like protein is the product of the *cdc13* gene, a 56-kDa protein called p56^{cdc13}. This gene was cloned, and its sequence was found to be very similar to that of the cyclin B proteins found in numerous animals (Goebl and Byers, 1988; Solomon et al., 1988). Cyclin B proteins of cleavage-stage cells show a periodic behavior, accumulating during S phase and being degraded during mitosis (Evans et al., 1983; Swenson et al., 1986). Cyclins are often encoded by the mRNAs stored in the oocyte cytoplasm, and if their translation into proteins is selectively inhibited, the

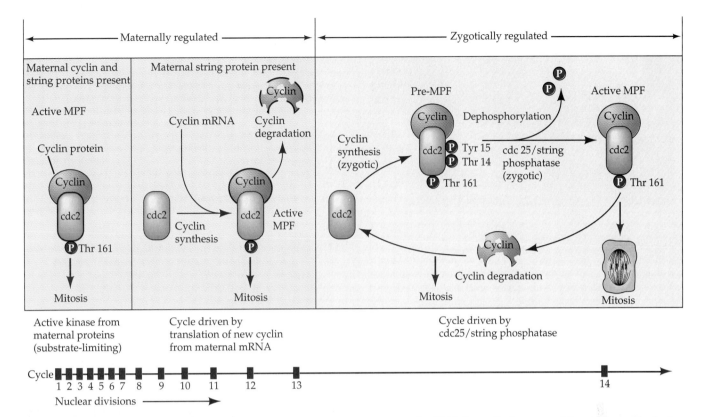

FIGURE 5.41

The development of cell cycle regulation in Drosophila *embryo-genesis. (A) Cyclin and cdc25 (string) protein are both abundant prior to fertilization. Therefore, during the first seven cell cycles, MPF kinase activity is constant and the nuclear divisions proceed as rapidly as the enzymes and substrates function. As cyclin becomes degraded, its synthesis (from mRNAs stored in the cytoplasm) becomes limiting at cycle 8. By cycle 14, the* maternal mRNA *for cyclin is gone and it must be synthesized from nuclear genes. Moreover, the degradation of string protein mandates new synthesis from the nucleus. Pre-MPF accumulates but isn't activated until the string phosphatase cleaves the T-14 and Y-15 phosphates from the cdc2 kinase. The mechanisms that relate MPF activity to the completion of DNA synthesis and the initiation of cytokinesis are currently under active investigation. (After Edgar et al., 1994.)*

cell will not enter mitosis (Minshull et al., 1989). The cyclin B protein combines with the cdc2 kinase of MPF to create the MPF complex. The cyclin enables the cdc2 kinase subunit to become phosphorylated at residues threonine-14 (T-14), tyrosine-15 (Y-15), and threonine-161 (Figure 5.41). The phosphorylation at T-161 is necessary for MPF activity, but phosphorylations at T-14 and Y-15 inhibit it. Thus, when phosphorylated in these positions, the kinase remains inactive, but potentially functional. The supply of potentially functional MPF molecules (pre-MPF) accumulates during the late S period.

The cdc25 phosphatase: Initiator of mitosis

Mitosis begins with the abrupt dephosphorylation of all these MPF kinase subunits at position 15. This is accomplished by the appearance of **cdc25 phosphatase** (Edgar and O'Farrell, 1989; Gautier et al, 1991; Jessus and Beach, 1992; Lee et al., 1992). In this manner, the gradual accumulation of MPF is converted into the sharp burst of kinase activity that initiates mitosis. This phosphatase (which has been found in numerous organisms) is itself developmentally regulated. In *Drosophila*, the cdc25 phosphatase (the product of the *string* gene) is initially synthesized from

stored oocyte mRNA during the first 13 cell cycles. However, during the next cycle, the maternal *string* mRNA is degraded. If the nuclei do not transcribe their own *string* mRNA, then the cells will not divide. Edgar and O'Farrell (1989) have shown that those cells that do divide are synthesizing their own cdc25 phosphatase, while those cells that are not able to join the division cycle are not making it (Figure 5.42). This degradation and need for resynthesis of this protein would explain the switch from cytoplasmic to nuclear control of division seen in cycle 14.

In *Drosophila*, there is a developmental maturation of the regulation of active MPF kinase (Figure 5.41; Edgar et al., 1993). At ovulation, the pre-MPF complexes stored in the egg are dephosphorylated at T-14 and Y-15 by the newly translated string (cdc25) protein. During the first seven nuclear cycles, active MPF remains at high levels and the nucleus divides as fast as the DNA synthetic enzymes permit. During cycles 8–13, cyclin begins to be degraded at metaphase, leading to periodic fluctuations of the MPF kinase activity. Cyclin synthesis from stored oocyte mRNA becomes the rate-limiting step for mitosis. The degradation of the oocyte string protein leads to cell cycle arrest at the interphase of cycle 14. Large concentrations of pre-MPF accumulate. The mitoses for divisions 14, 15, and 16 are

(A)

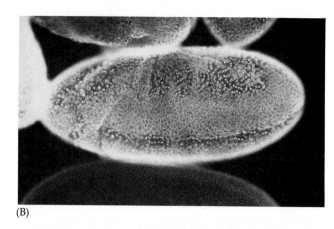

(B)

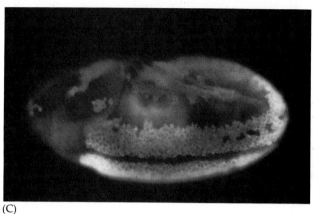

(C)

FIGURE 5.42

Correlation of string *gene expression with cell division in Dro-sophila embryos. (A) In this example, a late stage-14 embryo is stained with a radioactive nucleotide sequence that specifically recognizes and binds* string *mRNA. (B) A slightly older em-bryo is stained with fluorescent antibodies to tubulin to show the microtubules of the mitotic spindles. A comparison of the fluorescence photomicrograph and the autoradiograph obtained from the binding of the radioactive probe shows that only those cells capable of dividing synthesize* string *mRNA. (C) Antibod-ies to the cyclin A protein show that it is degraded after mitosis, and it is not seen in those regions containing string protein. (From Edgar and O'Farrell, 1989, courtesy of B. A. Edgar.)*

triggered only when this pre-MPF is dephosphorylated at position T-14 and Y-15 by string protein. This protein is derived from nuclear transcription at the end of each G_2 period. Mitosis has gone from cytoplasmic to nuclear control.

Cytostatic factor

The synthesis and degradation of MPF leads to the cycling of the cells. However, if cyclin degradation is prevented, MPF remains active and the cell is frozen in metaphase

(Murray et al., 1989). This is apparently what happens during frog oocyte development. The mature frog oocyte stops cell division by producing a protein called **cytostatic factor** (CSF), which keeps the oocytes arrested in meta-phase of the second meiotic division (Figure 5.43). This protein contains the products of the *c-mos* and *cdk-2* genes, and it appears to act by blocking the degradation of cyclin (see Chapter 22). Since cyclin is not degraded, MPF remains active, and the oocyte remains in metaphase. The release of calcium ions during fertilization activates a protease that specifically inactivates CSF (Watanabe et al, 1991). When

FIGURE 5.43

Levels of maturation promoting factor (MPF) during the early development of the frog Xenopus laevis. *The normal matura-tional signal is the hormone progesterone, which stimulates the ovulation of the oocytes and the initiation of meiosis. (After Murray and Kirschner, 1989.)*

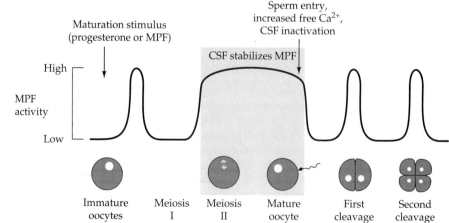

CSF is degraded, the cyclin can then be degraded and the cell can return to S phase. Thus, one of the effects of the calcium ions released at fertilization is to initiate the degradation of cyclin and enable the cell to begin DNA replication. After this, the rhythms of cell division are controlled by MPF activity, which is in turn based on the cyclic rhythms of cyclin synthesis and degradation. While the calcium ions are busy turning off mitosis, the fertilization signals that activate protein kinase C are establishing the conditions of interphase: chromatin decondensation and the re-formation of the nuclear envelope (Bement and Capco, 1991).

The cytoskeletal mechanisms of mitosis

Cleavage is actually the result of two coordinated processes. The first of these cyclic processes is **karyokinesis**—the mitotic division of the nucleus. The mechanical agent of this division is the mitotic spindle, with its microtubules composed of tubulin (the same type of protein that makes the sperm flagellum). The second process is **cytokinesis**—the division of the cell. The mechanical agent of cytokinesis is the **contractile ring** of microfilaments made of actin (the same type of protein that extends the egg microvilli and the sperm acrosomal process). Table 5.2 presents a comparison of these systems of division. The relationship and coordination between these two systems during cleavage is diagrammed in Figure 5.44A, where a sea urchin egg is shown undergoing first cleavage. The mitotic spindle and contractile ring are perpendicular to each other, and the spindle is internal to the contractile ring. The cleavage furrow eventually bisects the plane of mitosis, thereby creating two genetically equivalent blastomeres.

The actin microfilaments are found in the cortex of the egg rather than in the central cytoplasm. Under the electron microscope, the ring of microfilaments can be seen forming a distinct cortical band, 0.1 μm wide (Figure 5.44B). This contractile ring exists only during cleavage and extends 8–10 μm into the center of the egg. It is responsible for exerting the force that splits the zygote into blastomeres, for if it is disrupted, cytokinesis stops. Schroeder (1973) has proposed a model of cleavage wherein the contractile ring splits the egg like an "intercellular purse-string," tightening about the egg as cleavage continues. This tightening of the microfilamentous ring creates the cleavage furrow.

Although karyokinesis and cytokinesis are usually coordinated, they are sometimes separated by natural or experimental conditions. In insect eggs, karyokinesis occurs several times before cytokinesis takes place. Another way to cause this state is to treat embryos with the drug cytochalasin B. This drug inhibits the formation and organization of microfil-

TABLE 5.2
Karyokinesis and cytokinesis

Process	Mechanical agent	Major protein composition	Location	Major disruptive drug
Karyokinesis	Mitotic spindle	Tubulin microtubules	Central cytoplasm	Colchicine, nocodazole[a]
Cytokinesis	Contractile ring	Actin microfilaments	Cortical cytoplasm	Cytochalasin B

[a]Because colchicine has been found to independently inhibit several membrane functions, including osmoregulation and the transport of ions and nucleosides, nocodazole has become the major drug used to inhibit microtubule-mediated processes (see Hardin, 1987).

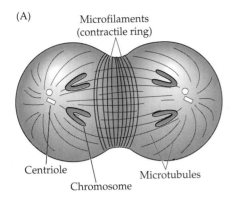

(A)

Microfilaments
(contractile ring)

Centriole

Chromosome

Microtubules

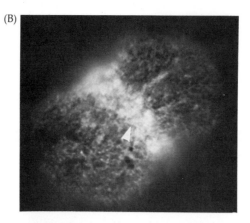

(B)

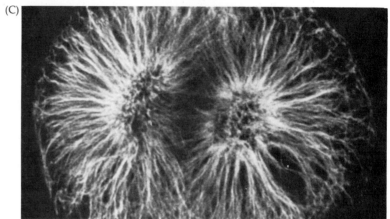

(C)

FIGURE 5.44

Role of microtubules and microfilaments in cell division. (A) Diagram of first-cleavage telophase. The chromosomes are being drawn to the centrioles by microtubules while the cytoplasm is pinched in by the contraction of microfilaments.(B) Localization of actin microfilaments in the cleavage furrow. Fluorescent labeling of the actin microfilaments shows the contractile ring in the first-cleavage furrow (arrow) of a telophase sea urchin egg. (C) Fluorescent labeling of tubulin shows the microtubular asters of a sea urchin egg during telophase of first cleavage. (B from Bonder et al., 1988; C from White et al., 1987.)

aments in the contractile ring, thereby stopping cleavage without stopping karyokinesis (Schroeder, 1972). In some instances, nuclei continue to divide and express developmentally regulated proteins even if cleavage is blocked (Lillie, 1902; Whittaker, 1979).

One of the most intriguing unsolved problems of embryonic cleavage is how cytokinesis and karyokinesis are coordinated with each other. In vitro work suggests that DNA replication may control the phosphorylation of the cdc2 kinase subunit of MPF, and that this phosphorylation may control the ability of the actin to contract (Smythe and Newport, 1992). However, these observations may not be consistent with observations made in early embryonic cells (Ferrell et al., 1991). Meanwhile, the number and location of the cleavage furrows appear to be controlled by the microtubule asters. These asters (Figure 5.44C) are the microtubular "rays" that extend from the poles of the mitotic spindle to the cell periphery. Normal cleavage occurs only if a pair of asters is present (Wilson, 1901), and polyspermic eggs (obtaining centrioles from each sperm) form multiple cleavage furrows within the same egg (see Figure 4.20). In more recent studies, Raff and Glover (1989) showed that if centrioles migrate into the posterior pole of the *Drosophila* embryo, they can form pole cells even in the absence of nuclei. It thus appears that microtubular asters are the *sine qua non* of cleavage.

The second type of evidence linking asters with cleavage furrow formation comes from experiments wherein the direction of cleavage is changed by placing the egg under pressure. Pflüger (1884) discovered that when a frog zygote is gently compressed between two glass plates, the directions of the first three cleavages are all perpendicular to the plane of

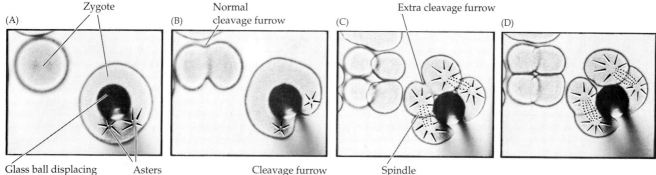

(A) Zygote

(B) Normal cleavage furrow

(C) Extra cleavage furrow

(D)

Glass ball displacing mitotic apparatus

Asters

Cleavage furrow interrupted

Spindle

FIGURE 5.45
Creation of a new cleavage furrow by displacement of the asters. (A,B) By interrupting the cleavage furrow with a glass ball, a horseshoe-shaped cell is created. At the next division (C,D), a new cleavage furrow is created even though there is no mitotic spindle across it. (From Rappaport, 1961, courtesy of R. Rappaport.)

the plates. Both Driesch and Morgan (reviewed in Morgan, 1927) made similar observations with sea urchin embryos. In both cases, the plane of the third cleavage (which is normally parallel to the equator of the oocyte) was displaced by 90°. Thus, by changing the location of the mitotic spindle, one can alter the direction of the cleavage furrow.

Rappaport (1961) has extended this type of experiment by displacing the mitotic spindles to the sides of the cells. In Figure 5.45, a glass ball has been used to displace the asters from the center of the cell toward the periphery. The cleavage furrow that results extends only as far as the ball and does not appear on the other side. Thus, a binucleate, horseshoe-shaped cell is formed. At the next division, two spindle apparatuses form between four asters, but three cleavage furrows are generated! Each arm of the horseshoe has its own mitotic spindle and cleavage furrow as expected, but a third furrow appears between the two asters at the top of the arms (Figure 5.45C). This demonstrates clearly that if two asters are close enough together, their interactions cause the formation of a cleavage furrow even in the absence of a mitotic spindle between them. Here we see again that cell division can occur without nuclear division as long as the asters are present.

The formation of new membranes

One last general consideration of embryonic cleavage involves the formation of new cell membranes. Are these membranes newly synthesized or are they merely extensions of the oocyte cell membrane? The answer is probably that both mechanisms contribute to the internal cell membranes.

Amphibian embryos provide evidence that new membrane components are being synthesized during early cleavage. Figure 5.46A shows the first cleavage furrow of a pigmented frog zygote. Whereas the original membrane has a pigmented cortical region associated with it, the new membrane is white. This new membrane also has electrical conductance properties different from those of the original membrane. Byers and Armstrong (1986) have radiolabeled membrane components of newly fertilized *Xenopus* eggs and followed the redistribution of these molecules through cleavage by autoradiography. During the first cleavage, the membrane of the outer surface of the embryo and the membrane of the leading edge of the cleavage furrow are heavily labeled (showing them to be original membrane regions). Between them is a large region devoid of radioactive label (Figure 5.46B). Thus, the furrow membrane is a mosaic of different parts. The membrane at the leading end of the furrow is derived from the preexisting radiolabeled outer surface of the egg, but most of the mem-

FIGURE 5.46
Formation of new membranes in the first cleavage of a *Xenopus* egg. (A) Old membrane has pigment granules associated with it. The new membrane appears clear, because it does not have these associated granules. (B) Autoradiograph of membrane proteins in the first-cleavage furrow. Surface proteins had been radiolabeled prior to division. White spots indicate regions containing proteins found in the zygote surface before division. (A from de Laat and Bluemink, 1974; B from Byers and Armstrong, 1986, courtesy of the authors.)

(A) New unpigmented membrane

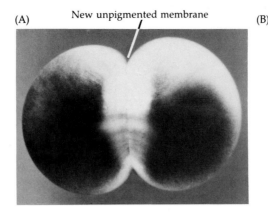

(B)

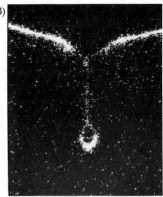

brane of the furrow has been derived from regions that were inaccessible to surface labeling. Byers and Armstrong speculate that the domain of heavily labeled membrane at the leading edge of the furrow contains the membrane anchors for the underlying ring of cortical microfilaments.

Summary

Cleavage is the collection of processes by which the fertilized egg is converted into a multicellular structure. The specific type of cleavage depends upon the evolutionary history of the species and on the ability of the egg to provide the nutritional requirements of the embryo. Some animals need large quantities of yolk to support development to the time when feeding begins, whereas other animals quickly form a feeding larval stage and do not require large amounts of yolk in the egg cytoplasm. In mammals, a peculiar type of cleavage forms the beginnings of the placenta, which will supply nutrients to the embryo. Yet we have seen that underneath the incredible diversity of cleavage types there is an underlying unity of function and mechanism. In all cases, karyokinesis and cytokinesis must be coordinated and the egg divided into cellular regions. During cleavage, the ratio of nucleus to cytoplasm is restored, and developmentally important information is sequestered in various cell regions. In most animals, cleavage is regulated by factors present in the egg cytoplasm and only later does nuclear transcription occur. It appears that the biochemistry regulating these divisions is similar, if not identical, among all the animal phyla, and that the biochemistry of cell division may be the same throughout all eukaryotes. In the next chapter we will see how these blastomeres move about and interact with each other to lay down the framework of the body.

LITERATURE CITED

Adeslon, D. C. and Humphreys, T. 1988. Sea urchin morphogenesisand cell-hyalin adhesion are perturbed by a monoclonal antibody specific for hyalin. *Development* 104: 391–402.

Arion, D., Meijer, L., Brizuela, L. and Beach, D. 1988. cdc2 is a component of the M phase-specific histone H1 kinase: Evidence for identity with MPF. *Cell* 55: 371–378.

Balinsky, B. I. 1981. *Introduction to Embryology,* 5th Ed. Saunders, Philadelphia.

Barlow, P., Owen, D. A. J. and Graham, C. 1972. DNA synthesis in the preimplantation mouse embryo. *J. Embryol. Exp. Morphol.* 27: 432–445.

Beams, H. W. and Kessel, R. G. 1976. Cytokinesis: A comparative study of cytoplasmic division in animal cells. *Am. Sci.* 64: 279–290.

Bellairs, R., Breathnach, A. S. and Gross, M. 1975. Freeze-fracture replication of junctional complexes in unincubated and incubated chick embryos. *Cell Tissue Res.* 162: 235–252.

Bement, W. M. and Capco, D. G. 1991. Parallel pathways of cell cycle control during *Xenopus* egg activation. *Proc. Natl. Acad. Sci. USA* 88: 5172–5176.

Bonder, E. M., Fishkind, D. J., Henson, J. H., Cotran, N. M. and Begg, D. A. 1988. Actin in cytokinesis: Formation of the contractile apparatus. *Zool. Sci.* 5: 699–711.

Borland, R. M. 1977. Transport processes in the mammalian blastocyst. *Dev. Mammals* 1: 31–67.

Boycott, A. E., Diver, C., Garstang, S. L. and Turner, F. M. 1930. The inheritance of sinestrality in *Limnaea peregra* (Mollusca: Pulmonata). *Philos. Trans. R. Soc. Lond.* [B] 219: 51–131.

Brenner, C. A., Adler, R. R., Rappolee, D. A., Pedersen, R. A. and Werb, Z. 1989. Genes for extracellular matrix-degrading metalloproteases and their inhibitor, TIMP, are expressed during early mammalian development. *Genes Dev.* 3: 848–859.

Byers, T. J. and Armstrong, P. B. 1986. Membrane protein redistribution during *Xenopus* first cleavage. *J. Cell Biol.* 102: 2176–2184.

Carlson, B. M. 1981. *Patten's Foundations of Embryology.* McGraw-Hill, New York.

Carson, D. D., Tang, J.-P. and Julian, J. 1993. Heparan sulfate proteoglycan (perlecan) expression by mouse embryos during acquisition of attachment competence. *Dev. Biol.* 155: 97–106.

Cisek, L. J. and Corden, J. L. 1989. Phosphorylation of RNA polymerase by the murine homologue of the cell-cycle control protein *cdc2. Nature* 339: 679–684.

Craig, M. M. and Morrill, J. B. 1986. Cellular arrangements and surface topography during early development in embryos of *Ilyanassa obsoleta. Int. J. Invert. Reprod. Dev.* 9: 209–228.

Crampton, H. E. 1894. Reversal of cleavage in a sinistral gastropod. *Ann. N.Y. Acad. Sci.* 8: 167–170.

Cruz, Y. P. 1981. A sterile defender morph in a polyembryonic hymenopteran parasite. *Nature* 294: 446–447.

Cruz, Y. P. 1986a. Development of the polyembryonic parasite *Copidosomopsis tanytmemus* (Hymenoptera: Encyrtidae). *Ann. Entomol. Soc. Am.* 79: 121–127.

Cruz, Y. P. 1986b. The defender role of the precocious larvae of *Copidosomopsis tanytmemus* Caltagirone (Encyrtidae, Hymenoptera). *J. Exp. Zool.* 237: 309–318.

Dan, K. 1960. Cytoembryology of echinoderms and amphibia. *Int. Rev. Cytol.* 9: 321–367.

Dan-Sohkawa, M. and Fujisawa, H. 1980. Cell dynamics of the blastulation process in the starfish, *Asterina pectinifera. Dev. Biol.* 77: 328–339.

Darwin, C. 1860. Letter to Asa Gray, May 22, 1860. *In* F. Darwin (ed.), *The Life and Letters of Charles Darwin,* Vol. 2. Appleton, New York. [p.105]

de Laat, S. W. and Bluemink, J. G. 1974. New membrane formation during cytokinesis in normal and cytochalasin B-treated eggs of *Xenopus laevis.* II. Electrophysical observations. *J. Cell Biol.* 60: 529–540.

de Laat, S. W., Tertoelen, L. G. J., Dorresteijn, A. W. C and van der Biggelaar, J. A. M. 1980. Intercellular communication patterns are involved in cell determination in early muscular development. *Nature* 287: 546–548.

de la Chappelle, A., Schroder, J., Rantanen, P., Thomasson, B., Niemi, M., Tilikainen, A., Sanger, R. and Robson, E. E. 1974. Early fusion of two human embryos? *Ann. Human Genet.* 38: 63–75.

Dunphy, W. G., Brizuela, L., Beach, D. and Newport, J. 1988. The *Xenopus cdc2* protein is a component of MPF, a cytoplasmic regulator of mitosis. *Cell* 54: 423–431.

Dyce, J., George, M., Goodall, H. and Fleming, T. P. 1987. Do trophectoderm and inner cell mass cells in the mouse blastocyst maintain discrete lineages? *Development* 100: 685–698.

Edgar, B. A. and O'Farrell, P. H. 1989. Genetic control of cell division patterns in the *Drosophila* embryo. *Cell* 57: 177–187.

Edgar, B. A., Kiehle, C. P. and Schubiger, G. 1986. Cell cycle control by the nucleocytoplasmic ratio in early *Drosophila* development. *Cell* 44: 365–372.

Edgar, B., Sprenger, F., Duronio, R. J., Leopold, P. and O'Farrell, P. 1994. MPF regulation during the embryonic cell cycles of *Drosophila. In press.*

Ettensohn, C. A. and Ingersoll, E. P. 1992. Morphogenesis of the sea urchin embryo. *In* E. F. Rossomondo and S. Alexander (eds.), *Morphogenesis.* Marcel Dekker, New York, pp. 189–262.

Evans, M.J. and Kaufman, M. H. 1981. Establishment in culture of pluripotent cells from mouse embryos. *Nature* 292: 154–156.

Evans, T., Rosenthal, E., Youngblom, J., Distel, D. and Hunt, T. 1983. Cyclin: A protein specified by maternal mRNA in sea urchin eggs that is destroyed at each cleavage division. *Cell* 33: 389–396.

Eyal-Giladi, H. 1991. The early embryonic development of the chick, an epigenetic process. *Crit. Rev. Poultry Biol.* 3: 143–166.

Eyal-Giladi, H., Debby, A. and Harel, N. 1992. The posterior section of the chick's area pellucida and its involvement in hypoblast and primitive streak formation. *Development* 116: 819–830.

Ferrell, J. E., Wu, M., Gergart, J. C. and Martin, G. S. 1991. Cell cycle tyrosine phosphorylation of $p34^{cdc2}$ and a microtubule associated protein kinase homologue in *Xenopus* oocytes and eggs. *Mol. Cell. Biol.* 11: 1965–1971.

Fleming, T. P. 1987. Quantitative analysis of cell allocation to trophectoderm and inner cell mass in the mouse embryo. *Dev. Biol.* 119: 520–531.

Fleming, T. P., Pickering, S. J., Qasim, F. and Maro, B. 1986. The generation of cell surface polarity in mouse 8-cell blastomeres: The role of cortical microfilaments analyzed using cytochalasin D. *J. Embryol. Exp. Morphol.* 95: 169–191.

Foe, V. 1989. Mitotic domains reveal early committment of cells in Drosophila embryos. *Development* 107: 1–25.

Freeman, G. and Lundelius, J. W. 1982. The developmental genetics of dextrality and sinistrality in the gastropod *Limnea peregra. Wilhelm Roux Arch. Dev. Biol.* 191: 69–83.

Fullilove, S. L. and Jacobson, A. G. 1971. Nuclear elongation and cytokinesis in *Drosophila montana. Dev. Biol.* 26: 560–577.

Gardiner, R. C. and Rossant, J. 1976. Determination during embryogenesis in mammals. *Ciba Found. Symp.* 40: 5–18.

Gautier, C., Norbury, C., Lohka, M., Nurse, P. and Maller, J. 1988. Purified maturation-promoting factor contains the product of *Xenopus* homolog of the fission yeast cell cycle control gene $cdc2^+$. *Cell* 54: 433–439.

Gautier, J., Solomon, M. J., Booher, R. N., Bazan, J. F. and Kirschner, M. W. 1991. cdc25 is a specific tyrosine phosphatase that directly activates $p34^{cdc2}$. *Cell* 67: 197–211.

Gerhart, J. C., Wu, M. and Kirschner, M. 1984. Cell dynamics of an M–phase-specific cytoplasmic factor in *Xenopus laevis* oocytes and eggs. *J. Cell Biol.* 98: 1247–1255.

Giudice, A. 1973. *Developmental Biology of the Sea Urchin Embryo.* Academic Press, New York.

Goebl, M. and Byers, B. 1988. Cyclin in fission yeast. *Cell* 54: 739–740.

Gould, K. and Nurse, P. 1989. Tyrosine phosphorylation of the fission yeast $cdc2^+$ protein kinase regulates entry into mitosis. *Nature* 342: 39–45.

Graham, C. F. and Kelly, S. J. 1977. Interactions between embryonic cells during early development of the mouse. *In* M. Karkinen-Jaaskelainen, L. Saxén and L. Weiss (eds.), *Cell Interactions in Differentiation.* Academic Press, New York, pp. 45–57.

Grbic, M. and Strand, M. R. 1992. Sibling rivalry and brood sex ratios in polyembryonic wasps. *Nature* 360: 254–256.

Gulyas, B. J. 1975. A reexamination of the cleavage patterns in eutherian mammalian eggs: Rotation of the blastomere pairs during second cleavage in the rabbit. *J. Exp. Zool.* 193: 235–248.

Gurdon, J. B. 1968. Changes in somatic nuclei inserted into growing and maturing amphibian oocytes. *J. Embryol. Exp. Morphol.* 20: 401–414.

Hara, K. 1977. The cleavage pattern of the axolotl egg studied by cinematography and cell counting. *Wilhelm Roux Arch. Entwicklungsmech. Org.* 181: 73–87.

Hara, K., Tydeman, P. and Kirschner, M. W. 1980. A cytoplasmic clock with the same period as the division cycle in *Xenopus* eggs. *Proc. Natl. Acad. Sci. USA* 77: 462–466.

Hardin, J. D. 1987. Archenteron elongation in the sea urchin embryo is a microtubule independent process. *Dev. Biol.* 121: 253–262.

Harland, R. M. and Laskey, R. A. 1980. Regulated replication of DNA microinjected into eggs of *X. laevis. Cell* 21: 761–771.

Hillman, N., Sherman, H. I. and Graham, C. F. 1972. The effects of spatial arrangement of cell determination during mouse development. *J. Embryol. Exp. Morphol.* 28: 263–278.

Hinegardner, R. T., Rao, B. and Feldman, D. E. 1964. The DNA synthetic period during early development of the sea urchin egg. *Exp. Cell Res.* 36: 53–61.

Hörstadius, S. 1939. The mechanics of sea urchin development, studied by operative methods. *Biol. Rev.* 14: 132–179.

Hörstadius, S. 1973. *Experimental Embryology of Echinoderms.* Clarendon Press, Oxford.

Inoué, S. 1982. The role of self–assembly in the generation of biologic form. *In* S. Subtelny and B. P. Green (eds.), *Developmental Order: Its Origin and Regulation,* Alan R. Liss, New York, pp. 35–76.

Jessus, C. and Beach, D. 1992. Oscillation of MPF is accomplished by periodic association between cdc25 and cdc2–cyclin B. *Cell* 68: 323–332.

Johnson, M. H., Chisholm, J. C., Fleming, T. P. and Houliston, E. 1986. A role for cytoplasmic determinants in the development of the early mouse embryo. *J. Embryol. Exp. Morphol.* [Suppl.]: 97–117.

Kalt, M. R. 1971. The relationship between cleavage and blastocoel formation in *Xenopus laevis.* I. Light microscopic observations. *J. Embryol. Exp. Morphol.* 26: 37–49.

Karr, T. L. and Alberts, B. M. 1986. Organization of the cytoskeleton in early *Drosophila* embryos. *J. Cell Biol.* 102: 1494–1509.

Karsenti, E., Newport, J., Hubble, R. and Kirschner, M. 1984. The interconversion of metaphase and interphase microtubule arrays, as studied by the injection of centrosomes and nuclei into *Xenopus* eggs. *J. Cell Biol.* 98: 1730–1745.

Karsenti, E., Bravo, R. and Kirschner, M. W. 1987. Phosphorylation changes associated with early cell cycle in *Xenopus* eggs. *Dev. Biol.* 119: 442–453.

Kelly, S. J. 1977. Studies of the developmental potential of 4- and 8-cell stage mouse blastomeres. *J. Exp. Zool.* 200: 365–376.

Kimelman, D., Kirschner, M. and Scherson, T. 1987. The events of the midblastula transition in *Xenopus* are regulated by changes in the cell cycle. *Cell* 48: 399–407.

Langman, J. 1981. *Medical Embryology,* 4th Ed. Williams & Wilkins, Baltimore.

Laskey, R. A., Mills, A. D. and Morris, N. R. 1977. Assembly of SV40 chromatin in a cell-free system from *Xenopus* eggs. *Cell* 10: 237–243.

Lee, M. and Nurse, P. 1987. Complementation used to clone a human homologue of the fission yeast cell cycle control gene *cdc2*+. *Nature* 335: 251–254.

Lee, M. S., Ogg, S., Xu, M., Parker, M., Donoghue, D. J., Maller, J. L. and Piwnica-Worms, H. 1992. cdc25+ encodes a protein phosphatase the dephosphorylates p34cdc2. *Mol. Biol. Cell* 3: 73–84.

Lepage, T., Sardet, C. and Gache, C. 1992. Spatial expression of the hatching enzyme gene in the sea urchin embryo. *Dev. Biol.* 150: 23–32.

Levy, J. B., Johnson, M. H., Goodall, H. and Maro, B. 1986. The timing of compaction: Control of a major developmental transition in mouse early embryogenesis. *J. Embryol. Exp. Morphol.* 95: 213–237.

Lillie, F. R. 1898. Adaptation in cleavage. *In Biological Lectures of the Marine Biological Laboratory of Woods Hole.* Ginn, Boston, pp. 43–67.

Lillie, F. R. 1902. Differentiation without cleavage in the egg of the annelid *Chaetopterus pergamentaceous. Wilhelm Roux Arch. Entwicklungsmech. Org.* 14: 477–499.

Lutz, B. 1947. Trends towards non-aquatic and direct development in frogs. *Copeia* 4: 242–252.

Mark, W. H., Signorelli, K. and Lacy, E. 1985. An inserted mutation in a transgenic mouse line results in developmental arrest at day 5 of gestation. *Cold Spring Harbor Symp. Quant. Biol.* 50: 453–463.

Markert, C. L. and Petters, R. M. 1978. Manufactured hexaparental mice show that adults are derived from three embryonic cells. *Science* 202: 56–58.

Martin, G. R. 1981. Isolation of a pluripotent cell line from early mouse embryos cultured in medium conditioned by teratocarcinoma stem cells. *Proc. Natl. Acad. Sci. USA* 78: 7634–7638.

Masui, Y. and Markert, C. L. 1971. Cytoplasmic control of nuclear behavior during meiotic maturation of frog oocytes. *J. Exp. Zool.* 177: 129–146.

Mayr, W. R., Pausch, V. and Schnedl, W. 1979. Human chimaera detectable only by investigation of her progeny. *Nature* 277: 210–211.

Miake-Lye, R. and Kirschner, M. W. 1985. Induction of early mitotic events in a cell-free system. *Cell* 41: 165–175.

Minshull, J., Blow, J. J. and Hunt, T. 1989. Translation of cyclin mRNA is necessary for extracts of activated *Xenopus* eggs to enter mitosis. *Cell* 56: 947–956.

Mintz, B. 1970. Clonal expression in allophenic mice. *Symp. Int. Soc. Cell Biol.* 9: 15.

Morgan, T. H. 1927. *Experimental Embryology.* Columbia University Press, New York.

Mulnard, J. G. 1967. Analyse microcinematographique du developpement de l'oeuf de souris du stade II au blastocyste. *Arch. Biol.* (Liege) 78: 107–138.

Murray, A. W. and Kirschner, M. W. 1989. Cyclin synthesis drives the early embryonic cell cycle. *Nature* 339: 275–280.

Murray, A. W., Solomon, M. J. and Kirschner, M. W. 1989. The role of cyclin synthesis and degradation in the control of maturation promoting factor activity. *Nature* 339: 280–286.

New, D. A. T. 1956. The formation of sub-blastodermic fluid in hens' eggs. *J. Embryol. Exp. Morphol.* 43: 221–227.

Newport, J. W. and Kirschner, M. W. 1982a. A major developmental transition in early *Xenopus* embryos: I. Characterization and timing of cellular changes at midblastula stage. *Cell* 30: 675–686.

Newport, J. W. and Kirschner, M. W. 1982b. A major developmental transition in early *Xenopus* embryos: II. Control of the onset of transcription. *Cell* 30: 687–696.

Newport, J. W. and Kirschner, M. W. 1984. Regulation of the cell cycle during *Xenopus laevis* development. *Cell* 37: 731–742.

Nieuwkoop, P. D. 1973. The "organization center" of the amphibian embryo: Its origin, spatial organization, and morphogenetic action. *Adv. Morphogenet.* 10: 1–39.

Pederson, R. A., Wu, K. and Bɛtakier, H. 1986. Origin of the inner cell mass in mouse embryos: Cell lineage analysis by microinjection. *Dev. Biol.* 117: 581–595.

Perona, R. M. and Wassarman, P. M. 1986. Mouse blastocysts hatch *in vitro* by using a trypsin-like proteinase associated with cells of mural trophectoderm. *Dev. Biol.* 114: 42–52.

Peter, M., Nakagawa, J., Dorée, M., Labbé, J. C. and Nigg, E. A. 1990. In vitro disassembly of the nuclear lamina and M phase-specific phosphorylation of lamins by cdc2 kinase. *Cell* 61: 591–602.

Peyrieras, N., Hyafil, F., Louvard, D., Ploegh, H. L. and Jacob, F. 1983. Uvomorulin: A non-integral membrane protein of early mouse embryo. *Proc. Natl. Acad. Sci. USA* 80: 6274–6277.

Piko, L. and Clegg, K. B. 1982. Quantitative changes in total RNA, total poly(A), and ribosomes in early mouse embryos. *Dev. Biol.* 89: 362–378.

Pflüger, E. 1884. Uber die Einwirkung der Schwerkraft und anderer Bedingungen auf die Richtung der Zeiltheilung. *Arch. Ges. Physiol.* 3: 4.

Prather, R. S. 1989. Nuclear transfer in mammals and amphibia. *In* H. Schatten and G. Schatten (eds.), *The Molecular Biology of Fertilization.* Academic Press, New York, pp. 323–340.

Pratt, H. P. M., Ziomek, Z. A., Reeve, W. J. D. and Johnson, M. H. 1982. Compaction of the mouse embryo: An analysis of its components. *J. Embryol. Exp. Morphol.* 70: 113–132.

Raff, J. W. and Glover, D. M. 1989. Centrosomes, not nuclei, initiate pole cell formation in *Drosophila* embryos. *Cell* 57: 611–619.

Raff, R. A. and Kaufman, T. C. 1983. *Embryos, Genes, and Evolution: The Developmental-Genetic Basis of Evolutionary Change.* Macmillan, New York.

Rappaport, R. 1961. Experiments concerning cleavage stimulus in sand dollar eggs. *J. Exp. Zool.* 148: 81–89.

Rugh, R. 1967. *The Mouse.* Burgess, Minneapolis.

Sagata, N., Watanabe, N., Vande Woude, G. F. and Ikawa, Y. 1989. The c-mos proto-oncogene product is a cytostatic factor responsible for meiotic arrest in vertebrate eggs. *Nature* 342: 512–518.

Satterwhite, L. L., Lohka, M. J., Wilson, K. L., Scherson, T. Y., Cisek, L. J. and Pollard, T. D. 1992. Phosphorylation of myosin-II light chain by cyclin-p34^{cdc2}: A mechanism for the timing of cytokinesis. *J. Cell Biol.* 118: 595–605.

Saunders, J. W., Jr. 1982. *Developmental Biology.* Macmillan, New York.

Schroeder, T. E. 1972. The contractile ring. II. Determining its brief existence, volumetric changes, and vital role in cleaving *Arbacia* eggs. *J. Cell Biol.* 53: 419–434.

Schroeder, T. E. 1973. Cell constriction: Contractile role of microfilaments in division and development. *Am. Zool.* 13: 687–696.

Seidel, F. 1952. Die Entwicklungspotenzen einer isolierten Blastomere des Zweizellenstadiums im Säugetierei. *Naturwissenschaften* 39: 355–356.

Smith, L. D. and Ecker, R. E. 1969. Role of the oocyte nucleus in the physiological maturation in *Rana pipiens. Dev. Biol.* 19: 281–309.

Smythe, C. and Newport, J. W. 1992. Coupling of mitosis to the completion of S phase in *Xenopus* occurs via modulation of the tyrosine kinase that phosphorylates p34^{cdc2}. *Cell* 68: 787–797.

Solomon, M. J. 1993. Activation of the various cyclin/cdc2 protein kinases. *Curr. Opin. Cell Biol.* 5: 180–186.

Solomon, M., Booher, R., Kirschner, M. and Beach, D. 1988. Cyclin in fission yeast. *Cell* 54: 738–739.

Strehlow, D. and Gilbert, W. 1993. A fate map for the first cleavages of the zebrafish. *Nature* 361: 451–453.

Strickland, S., Reich, E. and Sherman, M. I. 1976. Plasminogen activator in early embryogenesis: Enzyme production by trophoblast and parietal endoderm. *Cell* 9: 231–240.

Sturtevant, M. H. 1923. Inheritance of direction of coiling in *Limnaea. Science* 58: 269–270.

Summers, R. G., Morrill, J. B., Leith, A., Marko, M., Piston, D. W. and Stonebraker, A. T. 1993. A stereometric analysis of karyogenesis, cytokinesis, and cell arrangements during and following fourth cleavage period in the sea uchin, *Lytechinus variegatus. Dev. Growth Diff.* 35: 41–57.

Sutherland, A. E. and Calarco-Gillam, P. G. 1983. Analysis of compaction in the preimplantation mouse embryo. *Dev. Biol.* 100: 327–338.

Sutherland, A. E., Speed, T. P. and Calarco, P. G. 1990. Inner cell allocation in the mouse morula: The role of oriented division during fourth cleavage. *Dev. Biol.* 137: 13–25.

Swenson, K. L., Farrell, K. M. and Ruderman, J. V. 1986. The clam embryo protein cyclin A induces entry into M phase and the resumption of meiosis in *Xenopus* oocytes. *Cell* 47: 861–870.

Sze, L. C. 1953. Changes in the amount of deoxyribonucleic acid in the development of *Rana pipiens. J. Exp. Zool.* 122: 577–601.

Tarkowski, A. K. and Wróblewska, J. 1967. Development of blastomeres of mouse eggs isolated at the 4- and 8-cell stage. *J. Embryol. Exp. Morphol.* 18: 155–180.

Tuchmann-Duplessis, H., David, G. and Haegel, P. 1972. *Illustrated Human Embryology,* Vol. 1. Springer-Verlag, New York.

van der Biggelaar, J. A. M. and Guerrier, P. 1979. Dorsoventral polarity and mesentoblast determination as concomitant results of cellular interactions in the mollusc *Patella vulgata. Dev. Biol.* 68: 462–471.

Ward, G. E. and Kirschner, M. W. 1990. Identification of cell cycle-regulated phosphorylation cites on nuclear lamin C. *Cell* 61: 561–577.

Watanabe, N., Hunt, T., Ikawa, Y. and Sagata, N. 1991. Independent inactivation of MPF and cytostatic factor (Mos) upon fertilization of *Xenopus* eggs. *Nature* 352: 247–249.

Welsh, J. H. 1969. Mussels on the move. *Nat. Hist.* 78: 56–59.

White, J. C., Amos, W. B. and Fordham, M. 1987. An evaluation of confocal versus conventional imaging of biological structures by fluorescence light microscopy. *J. Cell Biol.* 105: 41–48.

Whittaker, J. R. 1979. Cytoplasmic determinants of tissue differentiation in the ascidian egg. *In* S. Subtelny and I. R. Konigsberg (eds.), *Determinants of Spatial Organization.* Academic Press, New York, pp. 29–51.

Wiley, L. M. 1984. Cavitation in the mouse preimplantation embryo: Na/K ATPase and the origin of nascent blastocoel fluid. *Dev. Biol.* 105: 330–342.

Wilson, E. B. 1898. Cell lineage and ancestral reminiscences. *Biological Lectures of the Marine Biological Laboratory of Woods Hole,* Ginn, Boston, pp. 21–42.

Wilson, E. B. 1901. Experiments in cytology. II. Some phenomena of fertilization and cell division in etherized eggs. III. The effect on cleavage of artificial obliteration of the first cleavage furrow. *Wilhelm Roux Arch. Entwicklungsmech. Org.* 13: 353–395.

Winkel, G. K., Ferguson, J. E., Takeichi, M. and Nuccitelli, R. 1990. Activation of protein kinase C triggers premature compaction in the 4-cell stage mouse embryo. *Dev. Biol.* 138: 1–15.

Wolpert, L. and Gustafson, T. 1961. Studies in the cellular basis of morphogenesis of the sea urchin embryo: The formation of the blastula. *Exp. Cell Res.* 25: 374–382.

Yamazaki, K. and Kato, Y. 1989. Sites of zona pellucida shedding by mouse embryo other than mural trophectoderm. *J. Exp. Zool.* 249: 347–349.

Ziomek, C. A. and Johnson, M. H. 1980. Cell surface interactions induce polarization of mouse 8-cell blastomeres at compaction. *Cell* 21: 935–942.

6

Gastrulation

Reorganizing the embryonic cells

Gastrulation is the process of highly integrated cell and tissue movements whereby the cells of the blastula are dramatically rearranged. The blastula consists of numerous cells, the positions of which were established during cleavage. During gastrulation, these cells are given new positions and new neighbors, and the multilayered body plan of the organism is established. The cells that will form the endodermal and mesodermal organs are brought inside the embryo while the cells that will form the skin and nervous system are spread over the outside surface. Thus, the three germ layers—outer ectoderm, inner endoderm, and interstitial mesoderm—are first produced during gastrulation. In addition, the stage is set for the interactions of these newly positioned tissues.

The movements of gastrulation involve the entire embryo, and cell migrations in one part of the gastrulating organism must be intimately coordinated with other movements occurring simultaneously. Although the patterns of gastrulation vary enormously throughout the animal kingdom, there are relatively few mechanisms involved. Gastrulation usually involves combinations of the following types of movements:

- *Epiboly.* The movement of epithelial sheets (usually of ectodermal cells), which spread as a unit rather than individually, to enclose the deeper layers of the embryo.
- *Invagination.* The infolding of a region of cells, much like the indenting of a soft rubber ball when poked.
- *Involution.* The inturning or inward movement of an expanding outer layer so that it spreads over the internal surface of the remaining external cells.
- *Ingression.* The migration of individual cells from the surface layers into the interior of the embryo.
- *Delamination.* The splitting of one cellular sheet into two more or less parallel sheets.

As we look at the phenomena of gastrulation in different types of embryos, we should keep in mind the following questions (Trinkaus, 1984):

- *What is the unit of migration activity?* Is migration dependent on the movement of individual cells, or are the cells part of a migrating sheet? Remarkable as it first seems, regional migrational properties may be totally controlled by cytoplasmic factors that are independent

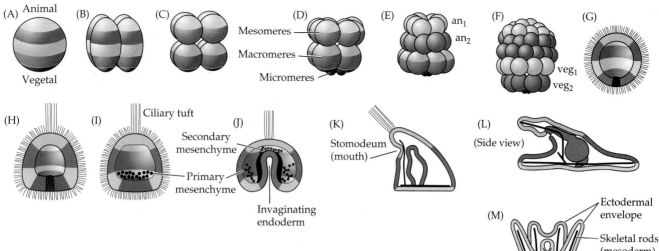

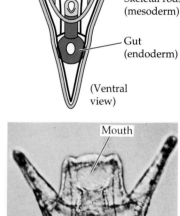

of cellularization. F. R. Lillie (1902) was able to parthenogenetically activate eggs of the annelid *Chaetopterus* and suppress their cleavage. Many events of early development occurred even in the absence of cells. The cytoplasm of the zygote separated into defined regions, and cilia differentiated in the appropriate parts of the egg. Moreover, the outermost clear cytoplasm migrated down over the vegetal regions in a manner specifically reminiscent of the epiboly of animal hemisphere cells during normal development. This occurred at precisely the time that epiboly would have taken place during gastrulation. Thus, epiboly may be (at least in some respects) independent of the cells that form the migrating region.

- *Is the spreading or folding of a cell sheet due to intrinsic factors within the sheet or to extrinsic forces stretching or distorting it?* It is essential to know the answer to this question if we are to understand how the various cell movements of gastrulation are integrated. For instance, do involuting cells pull the epibolizing cells down toward them or are the two movements independent?
- *Is there active spreading of the whole tissue or does the leading edge expand and drag the rest of a cell sheet passively along?*
- *Are changes in cell shape and motility during gastrulation the consequence of changes in cell surface properties, such as adhesiveness to the substrate or to other cells?*

Keeping these questions in mind, we will observe the various patterns of gastrulation found in echinoderms, amphibians, birds, and mammals.*

Sea urchin gastrulation

The sea urchin blastula consists of a single layer of around 1000 cells. These cells, derived from different regions of the zygote, have different sizes and properties. Figures 6.1 and 6.2 show the fates of the various regions of the zygote as it develops through cleavage and gastrulation to the pluteus larva characteristic of sea urchins. The fate of each cell layer can be seen through the layers' movements during gastrulation.

FIGURE 6.1
Normal sea urchin development, following the fate of the cellular layers of the blastula. (A–F) Cleavage through the 60-cell stage (2-cell stage omitted). (G) Early blastula with cilia. (H) Late blastula with ciliary tuft and flattened vegetal plate. (I) Blastula with primary mesenchyme. (J) gastrula with secondary mesenchyme. (K) Prism-stage larva. (L,M) Pluteus larvae. Fates of the zygote cytoplasm can be followed through the variations of shading. (N) Photomicrograph of a live pluteus larva of the sea urchin. (A–M after Hörstadius, 1939; N courtesy of G. Watchmaker.)

---

*Discussion of *Drosophila* gastrulation will be postponed until Chapter 15, where it occurs in the context of axis formation.

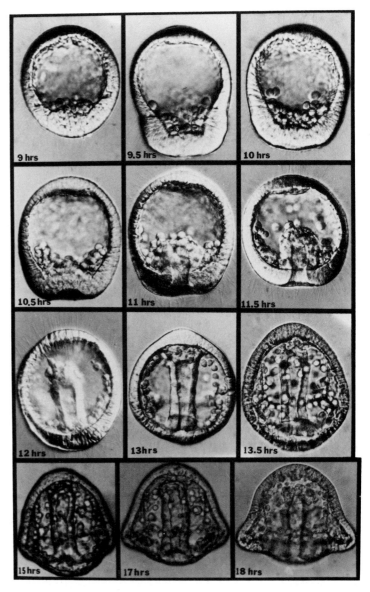

FIGURE 6.2
Entire sequence of gastrulation in *Lytechinus variegatus.* The time shows the
length of development at 25°C. (Courtesy of J. Morrill.)

Ingression of primary mesenchyme

Function of primary mesenchyme cells.   Shortly after the blastula hatches
from its fertilization membrane, the vegetal side of the spherical blastula
begins to thicken and flatten (Figure 6.2, 9 hours). At the center of this
flat **vegetal plate,** a cluster of small cells begins to change. These cells
show pulsating movements on their inner surfaces, extending and contracting long, thin (30 × 5 μm) processes called **filopodia.** The cells then
dissociate from one another and migrate into the blastocoel (Figure 6.2,
9–10 hours). These cells are called the **primary mesenchyme** and are
derived from the micromeres. The 64 or so primary mesenchyme cells of
the sea urchin embryo are the descendants of the four blastomeres that
formed by the asymmetric fourth cleavage.
Gustafson and Wolpert (1961) have used time-lapse films to follow the
microscopic movements of these mesenchyme cells within the blastocoel.

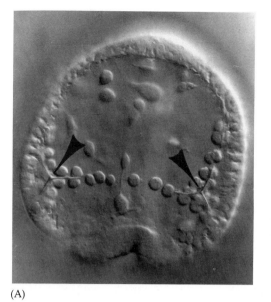

(A)

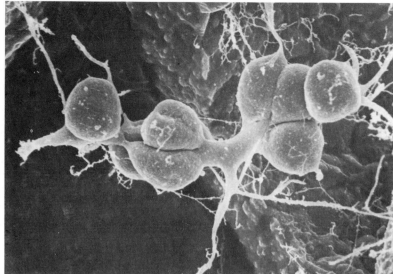

(B)

FIGURE 6.3
Formation of syncytial cables by mesenchymal cells of the sea urchin. (A) Primary mesenchymal cells in the early gastrula align and fuse to lay down the matrix of the calcium carbonate spicule. (B) Scanning electron micrograph of spicules formed by the fusing of primary mesenchyme cells into syncytial cables. (C) Ring of mesenchymal cells around the archenteron (primitive gut). The animal half and the entire archenteron have been removed. (D) Placement of primary mesenchyme cells (color) in the early larva of *Lytechinus variegatus*. (A and D from Ettensohn, 1990; B and C from Morrill and Santos, 1985; all photographs courtesy of the authors.)

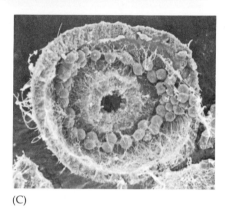

(C)

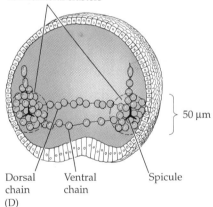

Ventrolateral clusters

50 μm

Dorsal chain   Ventral chain   Spicule

(D)

At first the cells appear to move randomly along the inner blastocoel surface, actively making and breaking filopodial connections to the wall of the blastocoel. Eventually, these cells become localized within the prospective ventrolateral region of the blastocoel, where their adherence is believed to be strongest. Here the primary mesenchymal cells fuse into **syncytial cables,** which will form the axis of the calcium carbonate spicules of the larval skeleton (Figure 6.3). More recent studies (Cherr et al., 1992) suggest that the initial migration of the primary mesenchyme cells is directed by the blastocoel wall and by the parallel fibrils of the extracellular matrix material that pervades the blastocoel. The micromeres appear to migrate along the blastocoel surface and are enmeshed by these fibrils (Figure 6.4).

Importance of extracellular lamina inside the blastocoel. Both cytoplasmic and cell surface events are crucial to the ingression and migration of the primary mesenchymal cells. Gustafson and Wolpert (1967) proposed a model in which the ingression of the micromeres comes about through changes in their adhesion to other cells and to the extracellular matrices that surround them. In 1985, Fink and McClay confirmed Gustafson and Wolpert's speculations by measuring the adhesive strengths of the sea urchin blastomeres for the hyaline layer, for the basal lamina, and for other cells. Originally, all the cells of the blastula are connected on their outer surface to the hyaline layer and on their inner surface by a basal lamina secreted by the cells (see Chapter 3). On their lateral sides, each cell has another cell for a neighbor. Fink and McClay found that the prospective

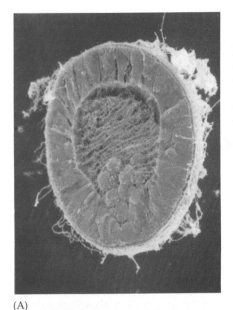

(A)

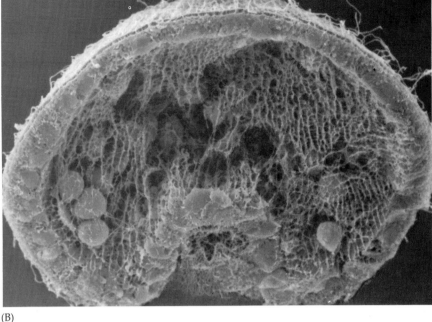

(B)

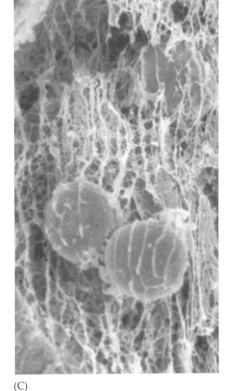

(C)

FIGURE 6.4

Scanning electron micrographs of primary mesenchyme cells within the extracellular matrix fibrils of the blastocoel. (A) Primary mesenchymal cells (PMC) enmeshed in the extracellular matrix of early *Strongylocentrotus* gastrula. (B,C) Gastrula-stage mesenchymal cell migration. The extracellular matrix fibrils of the blastocoel have become parallel to the animal–vegetal axis and are intimately associated with the primary mesenchymal cells. (From Cherr et al, 1992; courtesy of G. Cherr.)

ectoderm and endoderm (descendants of the macromeres and mesomeres, respectively) bind tightly to each other and to the hyaline layer, but they adhere only loosely to the basal lamina (Table 6.1). The micromeres of the blastula originally display a similar pattern of binding. However, the micromere pattern changes at gastrulation. Whereas the other cells retain their tight binding to the hyaline layer and to their neighbors, the primary mesenchyme precursors lose their affinity to these structures (to about 2 percent of its original value) while their affinity to components of the basal lamina and extracellular matrix *increases* 100-fold. This change in affinity causes the micromeres to release their attachments to the external hyaline layer and neighboring cells and, drawn in by the basal lamina, migrate up into the blastocoel (Figure 6.5). The changes in cell affinity have been correlated with changes in cell surface molecules that occur during this time. A cell surface antigen, Meso 1, is detected at the precise time that one first observes mesenchymal movement from the blastula wall (Wessel and McClay, 1985).

As shown in Figure 6.4, there is a heavy concentration of extracellular lamina material around the ingressing primary mesenchymal cells (Galileo and Morrill, 1985; Cherr et al., 1992). Moreover, once inside the blastocoel, the primary mesenchymal cells appear to migrate along the extracellular matrix of the blastocoel wall, extending their filopodia in front of them (Galileo and Morrill, 1985; Karp and Solursh, 1985). The orientation of the fibrils along the animal–vegetal axis may guide the cells in their migration. Two proteins appear to be important in this migration. One is fibronectin, the large (400-kDa) glycoprotein that is a common component of basal

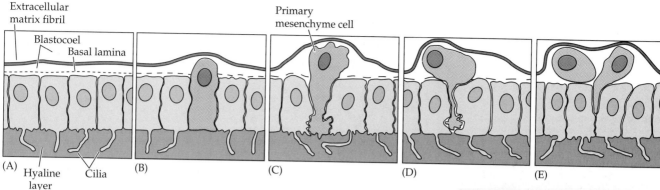

Extracellular matrix fibril

Blastocoel
Basal lamina

Primary mesenchyme cell

(A) Hyaline layer Cilia (B) (C) (D) (E)

FIGURE 6.5

Ingression of primary mesenchymal cells. (A–E) Interpretative diagrams depicting changes in adhesive interactions in the presumptive primary mesenchymal cells (color). These cells lose their affinities for hyaline and for their neighboring blastomeres while they gain an affinity for the basal lamina. Nonmesenchymal blastomeres retain their original high affinities for hyaline and neighboring cells. (F) Scanning electron micrograph showing the ingression of the primary mesenchymal cells of *Lytechinus variegatus*. (A–E after Katow and Solursh, 1980; F courtesy of J. B. Morrill and D. Flaherty.)

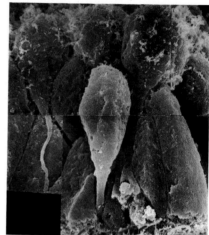

(F)

laminae, including that of the sea urchin blastocoel (Wessel et al., 1984). Fink and McClay (1985) showed that during gastrulation the affinity of the micromeres for this particular molecule increases dramatically. The second set of molecules consists of sulfated proteoglycans found on the cell surface of the ingressing mesenchymal cells (Chapter 3; Sugiyama, 1972; Lane and Solursh, 1991). If the synthesis (or sulfation) of these proteoglycans is inhibited, the mesenchymal cells enter the blastocoel but do not migrate further* (Figure 6.6; Karp and Solursh, 1974; Anstrom et al., 1987).

*In one of the first experiments in chemical embryology, Curt Herbst (1904) found that sea urchin embryos wouldn't gastrulate properly when placed into seawater that lacked sulfate ions. At the time, he couldn't figure out why.

TABLE 6.1

Affinities of mesenchymal and nonmesenchymal cells to cellular and extracellular components[a]

Cell type	Dislodgement force (in dynes)		
	Hyaline	Gastrula cell monolayers	Basal lamina
16-cell stage micromeres	5.8×10^{-5}	6.8×10^{-5}	4.8×10^{-7}
Migratory stage mesenchymal cells	1.2×10^{-7}	1.2×10^{-7}	1.5×10^{-5}
Gastrula ectoderm and endoderm	5.0×10^{-5}	5.0×10^{-5}	5.0×10^{-7}

Source: After Fink and McClay, 1985.
[a]Tested cells were allowed to adhere to plates containing hyaline, extracellular basal lamina, or cell monolayers. The plates were inverted and centrifuged at various strengths to dislodge the cells. The dislodgement force is calculated from the centrifugal force needed to remove the test cells from the substrate.

FIGURE 6.6

Effect of sulfate deprivation on primary mesenchyme movement in the sea urchin *Lytechinus*. (A) Normal gastrula. (B) Abnormal gastrula formed when embryos are grown in sulfate-free seawater. (From Karp and Solursh, 1974, courtesy of M. Solursh.)

(A) 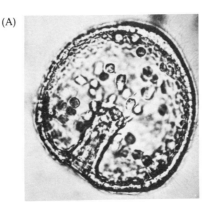 (B)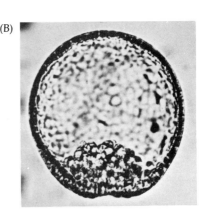

But these guidance cues are not sufficient, since the micromeres "know" when to stop their movement and form spicules near the equator of the blastocoel. Primary mesenchyme cells arrange themselves in a ring at a specific position along the animal–vegetal axis. At two sites near the future ventral side of the larva, many of these primary mesenchyme cells cluster together and initiate spicule formation (Figure 6.3). If a labeled micromere from another embryo is injected into the blastocoel of a gastrulating sea urchin embryo, it migrates to the correct location and contributes to the formation of the embryonic spicules. If primary mesenchyme cells from older embryos are injected into younger gastrulae, they will delay their differentiation, migrate like younger cells, and incorporate normally into the host's mesenchyme. Moreover, if all the host's primary mesenchyme cells are removed before the older mesenchymal cells are injected, the older mesenchyme cells will repeat the earlier stages of their migration and form a normal mesenchymal ring and skeleton (Ettensohn, 1990). It is thought that this positional information is provided by the prospective ectodermal cells and their basal laminae (von Übisch, 1939; Harkey and Whitely, 1980). Only the primary mesenchyme cells (and not other cell types or latex beads) are capable of responding to these patterning cues (Ettensohn and McClay, 1986). Miller and colleagues (1993) have reported the existence of extremely fine (0.3 μm diameter) filopodia on the skeletonogenic mesenchyme; these appear to explore and sense the blastocoel wall.

First stage of archenteron invagination

As the ring of primary mesenchymal cells forms in the vegetal region of the blastocoel, important changes are occurring in the cells that remain at the vegetal plate. These cells remain bound to one another and to the hyaline layer of the egg and move to fill the gaps caused by the ingression of the mesenchyme; therefore, the vegetal plate flattens further. One next sees that the vegetal plate bends inward and extends about one-quarter to one-half of the way into the blastocoel (Figure 6.2, 10.5–11.5 hours; Figure 6.7). Then invagination suddenly ceases. The invaginated region is called the **archenteron** (primitive gut), and the opening of the archenteron at the vegetal region is called the **blastopore**.

What forces work to invaginate these cells? Lane and co-workers (1993) provide evidence that the buckling is similar to that produced by heating a bimetallic metal strip. The hyaline layer is actually made of two laminae, an outer lamina made primarily of hyalin protein and an inner lamina

FIGURE 6.7

Vegetal plate invagination in *Lytechinus variegatus* seen by scanning electron micrography of the external surface of the early gastrula. The blastopore is clearly visible. (From Morrill and Santos, 1985, courtesy of J. B. Morrill.)

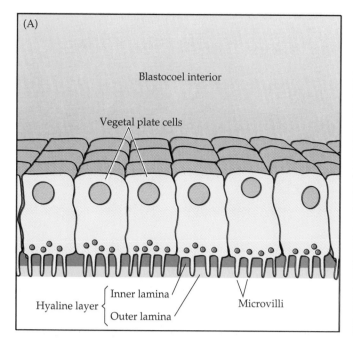

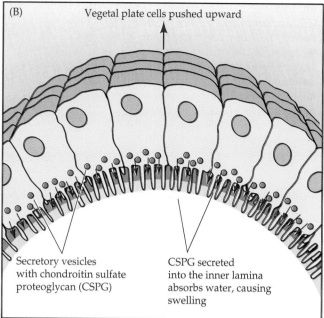

composed of fibropellin proteins[*] (Bisgrove et al., 1991; Hall and Vacquier, 1982). The vegetal plate cells (and only these cells) secrete a chondroitin sulfate proteoglycan into the inner lamina of the hyaline layer directly beneath them. This hygroscopic (water-absorbing) molecule swells the inner lamina, but does not swell the outer lamina. This causes the vegetal region of the hyaline layer to buckle (Figure 6.8). Slightly later, a second force arising from movements of the epithelial cells adjacent to the vegetal plate may facilitate this invagination by drawing the buckled layer inward (Burke et al., 1991).

Second and third stages of archenteron formation

The invagination of the vegetal cells occurs in three discrete stages. After a brief pause, the second phase of archenteron formation begins. During this time, the archenteron extends dramatically, sometimes tripling its length. In the process of extension, the wide, short gut rudiment is transformed into a long, thin tube; yet no new cells are seen to be formed (Figure 6.2, 12 hours; Figure 6.9). To accomplish this extension, the cells of the archenteron rearrange themselves by migrating over one another and by flattening themselves (Ettensohn, 1985; Hardin and Cheng, 1986). This phenomenon, wherein cells intercalate to narrow the tissue and at the same time move it forward, is call **convergent extension**. In at least some species of sea urchins, a third stage of archenteron elongation occurs. This last phase of archenteron elongation is initiated by the tension provided by secondary mesenchymal cells, which form at the tip of the archenteron and remain there (Figure 6.2, 13 hours; Figure 6.10). Filopodia are extended from these cells through the blastocoel fluid to contact the inner surface of the blastocoel wall (Dan and Okazaki, 1956; Schroeder, 1981). The filopodia attach to the wall at the junctions between the blas-

---

[*]Fibropellins are stored in secretory granules within the oocyte. They are secreted from these granules after cortical granule exocytosis releases the hyalin protein. By blastula stage, the fibropellins have formed a meshlike network over the embryo surface.

FIGURE 6.9
Cell rearrangement during the extension of the archenteron in sea urchins embryos. In this species, the early archenteron has 20–30 cells around its circumference. Later in gastrulation, the archenteron has a circumference made by only 6–8 cells. Fluorescently labeled clones can be seen to stretch apically. (After Hardin, 1990.)

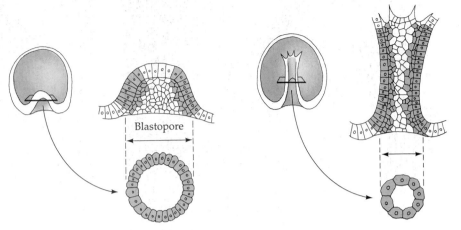

Early gastrulation

Later gastrulation

Blastopore

toderm cells and then shorten, pulling up the archenteron. Hardin (1988) ablated the secondary mesenchyme cells with a laser, with the result that the archenteron could only elongate about two-thirds of the full length. If a few secondary mesenchyme cells were left, elongation continued, although at a slower rate than in controls. The secondary mesenchyme cells, then, play an essential role in pulling the archenteron up to the blastocoel wall during the last phase of invagination.

But can the secondary mesenchyme filopodia attach to any part of the blastocoel wall, or is there a specific target in the animal hemisphere that must be present for this attachment to occur? Is there a region of the blastocoel wall that is already committed to becoming the ventral side of the larva? Studies by Hardin and McClay (1990) show that there is a specific "target" site for the filopodia and that this site differs from other regions of the animal hemisphere. They found that the filopodia of the secondary mesenchyme cells extend, touch the blastocoel wall at random sites, and then retract. However, when the filopodia contact one region of the wall, they remain attached there, flatten out against this region, and pull the archenteron toward it. When Hardin and McClay poked in the other side of the blastocoel wall so that the contacts were made most readily with that region, the filopodia continued to extend and retract after touching it. Only when they found the "target" did they cease these movements. If the gastrula was constricted so that filopodia never reached

FIGURE 6.10
Midgastrula stage of the sea urchin *Lytechinus pictus*, showing filopodial extensions of secondary mesenchyme extending from the archenteron tip to the blastocoel wall. (A) Mesenchymal cells extending filopodia from the tip of the archenteron. (B) Filopodial cables connecting the blastocoel wall to the archenteron tip. The tension of the cables can be seen as they pull on the blastocoel wall at the point of attachment. (Photographs courtesy of C. Ettensohn.)

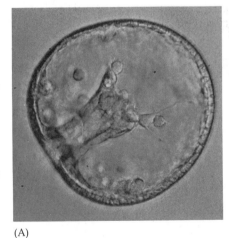

(A)

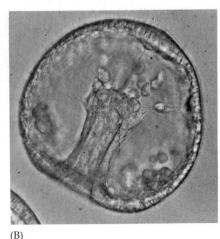

(B)

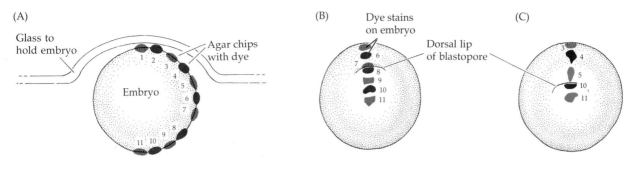

the target area, the secondary mesenchyme cells continued to explore until they eventually moved off the archenteron and found the target tissue as freely migrating cells. There appears, then, to be a target region on what is to become the ventral side of the larva that is recognized by the secondary mesenchyme cells and positions the archenteron in the region where the mouth will form.

As the top of the archenteron meets the blastocoel wall in this region, the secondary mesenchymal cells disperse into the blastocoel, where they proliferate to form the mesodermal organs (Figure 6.2, 13.5 hours). Where the archenteron contacts the wall, a mouth is eventually formed. The mouth fuses with the archenteron to create a continuous digestive tube. Thus, as is characteristic for deuterostomes, the blastopore marks the position of the anus.

Amphibian gastrulation

The study of amphibian gastrulation is both one of the oldest and one of the newest areas of experimental embryology; even though amphibian gastrulation has been extensively studied for the past century, most of our theories concerning the mechanisms of these developmental movements have been revised over the past decade. The study of amphibian gastrulation has been complicated by the fact that there is no one single way that all amphibians gastrulate. Different species employ different means toward the same goal (Smith and Malacinski, 1983; Lundmark, 1986). In recent years the most intensive investigations have focused on *Xenopus*, so we will concentrate on its mode of gastrulation.

Cell movements during amphibian gastrulation

Amphibian blastulae are faced with the same tasks as their echinoderm counterparts, namely, to bring inside those areas destined to form the endodermal organs, to surround the embryo with those cells capable of forming the ectoderm, and to place the mesodermal cells in the proper places between them. The movements whereby this is accomplished can be visualized by the technique of vital dye staining. Vogt (1929) saturated agar chips with dyes, such as Neutral Red or Nile Blue sulfate, that would stain but not damage the embryonic cells. These stained agar chips were pressed to the surface of the blastula, and some of the dye was transferred onto the contacted cells (Figure 6.11). The movements of each group of stained cells were followed throughout gastrulation, and the results were summarized in fate maps. These maps have recently been confirmed and extended by scanning electron microscopy and dye injection techniques (Smith and Malacinski, 1983; Lundmark, 1986).

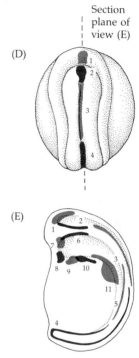

FIGURE 6.11
Vital dye staining of amphibian embryos. (A) Vogt's method for marking specific cells of the embryo surface with vital dyes. (B–D) Surface views of stain on successively later embryos. (E) Newt embryo dissected in medial plane to show stained cells in the interior. (After Vogt, 1929.)

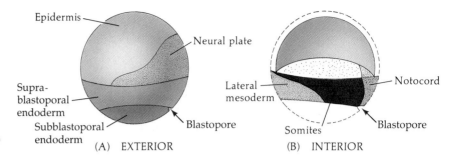

FIGURE 6.12
Fate maps of the blastula of the frog *Xenopus laevis*. Fate maps of (A) the exterior and (B) the interior cells indicate that most of the mesodermal derivatives are formed from the interior cells. In this lateral view, the point at which the dorsal blastopore lip forms is indicated by an arrow. (After Keller, 1976.)

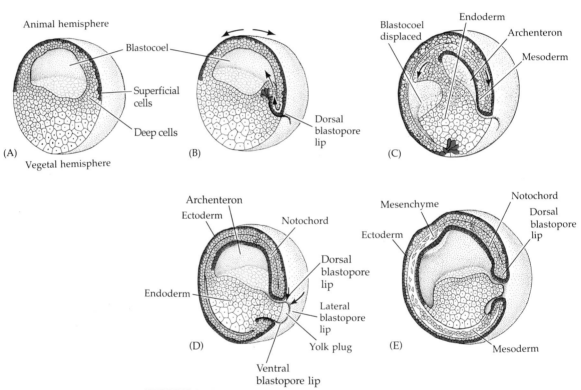

FIGURE 6.13
Cell movements during frog gastrulation. The sections are cut through the middle of the embryo and are positioned so that the vegetal pole is tilted toward the observer and slightly to the left. The major cell movements are indicated by arrows and the superficial animal hemisphere cells are colored so that their movements can be followed. (A,B) Early gastrulation. Bottle cells of the margin move inward to form the dorsal lip of the blastopore, and mesodermal precursors involute under the roof of the blastocoel. (C,D) Midgastrulation. The archenteron forms and displaces the blastocoel, and cells migrate from the lateral and ventral blastopore lips into the embryo. The cells of the animal hemisphere migrate down toward the vegetal region, moving the blastopore to the region near the vegetal pole. (D,E) Toward the end of gastrulation, the blastocoel is obliterated, the embryo becomes surrounded by ectoderm, the endoderm has been internalized, and the mesodermal cells have been positioned between the ectoderm and endoderm. (After Keller, 1986.)

(A)

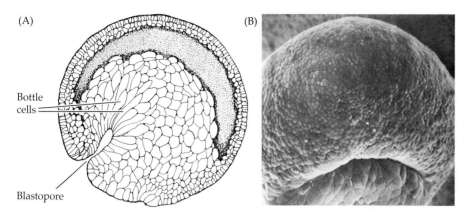

Bottle
cells

Blastopore

(B)

(C)

Vital dye studies by Løvtrup (1975; Landstrom and Løvtrup, 1979) and by Keller (1975, 1976) have shown that cells of the *Xenopus* blastula have different fates depending on whether they are in the deep or the superficial layers of the embryo (Figure 6.12). In *Xenopus*, the mesodermal precursors exist *solely* in the deep layer of cells, while the ectoderm and endoderm arise from the superficial layer on the surface of the embryo. The precursors for the notochord and other mesodermal tissues are located beneath the surface in the equatorial (marginal) region of the embryo. In urodeles (salamanders such as *Triturus* and *Ambystoma*) and in some frogs other than *Xenopus*, the notochord and mesoderm precursors are found in *both* the surface cells and the deep marginal cells (Purcell and Keller, 1993).

Gastrulation in frog embryos is initiated at the future dorsal side of the embryo, just below the equator in the region of the gray crescent (Figure 6.13). Here the prospective local endodermal cells invaginate to form a slitlike blastopore. These cells change their shape dramatically. The main body of each cell is displaced toward the inside of the embryo while it maintains contact with the outside surface by way of a slender neck (Figure 6.14). These **bottle cells** line the initial archenteron. Thus, as in the gastrulating sea urchin, an invagination of cells initiates archenteron formation. However, unlike gastrulation in sea urchins, gastrulation in the frog begins not at the most vegetal region, but in the **marginal zone** near the equator of the blastula, where the animal and vegetal hemispheres meet. Here the endodermal cells are not as large or as yolky as the most vegetal blastomeres.

The next phase of gastrulation involves the involution of the marginal zone cells while the animal cells undergo epiboly and converge at the blastopore. When the migrating marginal cells reach the **dorsal lip of the blastopore,** they turn inward and travel along the inner surface of the outer cell sheets. Thus, the cells constituting the lip of the blastopore are constantly changing. The first cells to compose the dorsal lip are the bottle cells that invaginated to form the leading edge of the archenteron. These cells later become the pharyngeal cells of the foregut. As these first cells pass into the interior of the embryo, the blastopore lip becomes composed of cells that involute into the embryo to become the precursors of the head mesoderm. The next cells involuting into the embryo through the dorsal lip of the blastopore are called the **chordamesoderm** cells. These cells will form the **notochord,** a transient mesodermal "backbone" that is essential for initiating the differentiation of the nervous system.

As the new cells enter the embryo, the blastocoel is displaced to the side opposite the dorsal blastopore lip. Meanwhile, the blastopore lip expands laterally and ventrally as the processes of bottle cell formation and involution continue about the blastopore. The widening blastopore

FIGURE 6.14
Structure of the blastopore lip. (A) Diagram of cells seen in a section of gastrulating salamander embryo, showing the extension of the bottle cells from the blastopore. (B) Surface view of an early dorsal blastopore lip of *Xenopus*. The size difference between the animal and vegetal blastomeres is readily apparent. (C) Close-up of the region where the animal hemisphere cells are involuting through the blastopore lip. (A after Holtfreter, 1943; B and C, scanning electron micrographs courtesy of C. Phillips.)

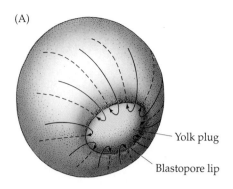

(A)

Yolk plug

Blastopore lip

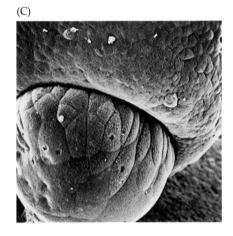

(C)

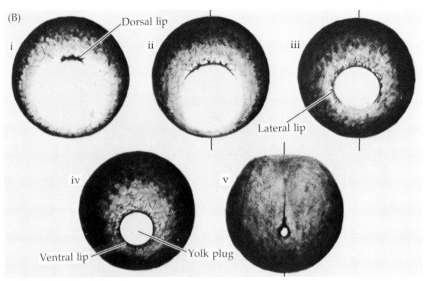

(B)

Dorsal lip

i   ii   iii

Lateral lip

iv   v

Ventral lip   Yolk plug

FIGURE 6.15
Epiboly of the ectoderm. (A) Morphogenetic movements of the cells migrating into the blastopore and then under the surface. (B) Changes in the region around the blastopore as the dorsal, lateral, and ventral lips are formed in succession. When the ventral lip completes the circle, the endoderm becomes progressively internalized. Numbers ii–v correspond to Figures 13B–E, respectively. (C) Scanning electron micrograph of *Xenopus* yolk plug and the epiboly of the ectoderm over it at the blastopore margin. (B from Balinsky, 1975, courtesy of B. I. Balinsky; C courtesy of C. Phillips.)

"crescent" develops lateral lips and finally a ventral lip over which additional mesodermal and endodermal precursor cells pass. With the formation of the ventral lip, the blastopore has formed a ring around the large endodermal cells that remain exposed on the vegetal surface. This remaining patch of endoderm is called the **yolk plug;** it, too, is eventually internalized (Figure 6.15). At that point, all the endodermal precursors have been brought into the interior of the embryo, the ectoderm has encircled the surface, and the mesoderm has been brought between them.

Positioning the blastopore

Having seen the general features of amphibian gastrulation, we can now look at each step in detail. Gastrulation does not exist as an independent process in the life of an animal. In fact, the preparation for gastrulation can be traced back to the literal moment of sperm–egg fusion. The unfertilized egg has a polarity along the animal–vegetal axis. The general fate of these regions can be predicted before fertilization. The surface of the animal hemisphere will become the cells of the ectoderm (skin and nerves), the vegetal hemisphere surface will form the cells of the gut and associated organs (endoderm), and the mesodermal cells will form from the internal cytoplasm around the equator. Thus, the germ layers can be mapped onto the unfertilized ovum; but this tells us nothing about which part of the egg will form the belly and which the back. The dorsal–ventral (back–front), anterior–posterior, and left–right axes have not yet been determined.

The dorsal–ventral and anterior–posterior axes are specified by displacements of the zygote cytoplasm during fertilization. In Chapter 4, we discussed the rotation of the cortical cytoplasm relative to the internal cytoplasm of the frog egg. The internal cytoplasm remains oriented with respect to gravity because of its dense yolk accumulation, while the cortical cytoplasm actively rotates 30° animally ("upward") toward the point of sperm entry (see Figure 4.34). This rotation causes the animal–vegetal axis of the egg surface to be offset 30° relative to the animal–vegetal axis of the internal cytoplasm. In this way, a new state of symmetry is acquired. Whereas the unfertilized egg had been radially symmetrical about the animal–vegetal axis, the fertilized egg now has a dorsoventral axis. It has become bilaterally symmetrical (having right and left sides). The inner cytoplasm moves as well, and fluorescence microscopy of early embryos has shown that the cytoplasmic patterns of presumptive dorsal cells differ from those of the presumptive ventral cells (Plate 6).

These cytoplasmic movements activate the cytoplasm opposite the point of sperm entry to initiate gastrulation (Figure 6.16). The side where the sperm enters marks the future ventral surface of the embryo; the opposite side, where gastrulation is initiated, marks the future dorsum (back) of the embryo (Gerhart et al., 1981, 1986; Vincent et al., 1986). While the sperm is not needed to induce these movements in the egg cytoplasm, it is important in determining the direction of this rotation. If an artificially stimulated egg is enucleated, the cortical rotation still takes place at the correct time. However, the direction of this movement is unpredictable. (In fact, in dispermic eggs, there is still only one direction of rotation.) The sperm appears to provide a spatial cue that orients the autonomous rotation of the cytoplasm, but it is the cytoplasmic rotation that is essential for further development. If the newly fertilized egg is compressed so that the direction of cortical rotation is different from that predicted by the point of sperm entry, the blastopore lip forms in accordance with the direction of cortical rotation (Black and Vincent, 1988). Moreover, if this cortical rotation is blocked, there is no dorsal development, and the embryo dies as a mass of ventral (primarily gut) cells (Vincent and Gerhart, 1987). The direction of cytoplasmic movement determines which side is to be dorsal and which side is to be ventral.

The directional bias provided by the point of sperm entry can be overridden by mechanically redirecting the spatial relationship between the cortical and subcortical cytoplasms. When the egg is prevented from rotating (by immersing it in a polysaccharide that collapses the perivitelline space between the egg and the fertilization envelope), one can turn the egg on its side, 90° so the animal–vegetal axis is horizontal rather than vertical and the point of sperm entry faces upward (Gerhart et al., 1981; Kirschner and Gerhart, 1981; Cooke, 1986). When fertilized eggs are inclined this way for 30 minutes, starting halfway through the first cleavage cycle, the cytoplasm rotates such that almost all the embryos initiate gastrulation on the *same* side as sperm entry (Figure 6.16).

The preceding discussion suggests that one might be able to get two gastrulation initiation sites if one were to combine the sperm-oriented rotation with an artificially induced rotation of the egg. Black and Gerhart (1985) let the initial sperm-directed rotation occur, but they then immobilized these eggs in gelatin and gently centrifuged them so that the internal cytoplasm would go towards the point of sperm entry. When the centrifuged eggs were then allowed to develop in normal water, two sites of gastrulation emerged, leading to conjoined twin larvae (Figure 6.17). Black and Gerhart hypothesize (1986) that such twinning is caused by the formation of two areas of interaction: one axis forms where the normal cortical rotation caused the cytoplasmic interactions in the vegetal region of the cell, the other axis forms where the centrifugation-driven cytoplasm interacts with the vegetal components. Twins can also be produced at normal

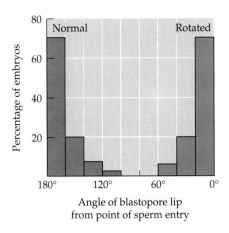

FIGURE 6.16
Relationship between the point of sperm entry and the dorsal blastopore lip in normal and rotated frog eggs. *Xenopus* eggs were fertilized, dejellied, and placed in Ficoll to dehydrate the perivitelline space. The sperm entry point was marked with dye. Rotated eggs were inclined 90°, with the sperm entry point facing upward, 50–80 minutes after fertilization. (After Gerhart et al., 1981.)

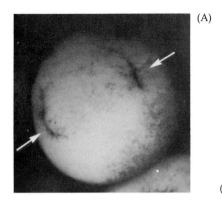

(A)

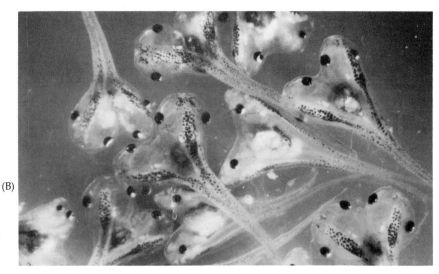

(B)

FIGURE 6.17
Twin blastopores produced by rotating dejellied *Xenopus* eggs ventral side (sperm entry point) up at the time of first cleavage. (A) Two blastopores are instructed to form: the original one (opposite the point of sperm entry) and the new one created by the displacement of cytoplasmic material. (B) These eggs develop two complete axes, which form twin tadpoles, joined ventrally. (Courtesy of J. Gerhart.)

gravity by placing the sperm entry side of the egg uppermost after removing the egg from its fertilization envelope (Gerhart et al., 1981).

The ability to obtain two functional blastopore lips also suggests that there is nothing unique about the gray crescent, where gastrulation is first seen to begin. Rather, the gastrulation-inducing factors appear to be created by the interactions of animal and vegetal cytoplasm, interactions that probably activate some component in the vegetal cytoplasm. Gimlich and Gerhart (1984) performed a series of transplantation experiments that confirmed the hypothesis that the factor(s) that initiate gastrulation originally lie in the deep cytoplasm of the dorsal vegetal cells rather than in the gray crescent. They demonstrated that the three most dorsal vegetal blastomeres of 64-cell *Xenopus* embryos are able to induce the formation of the dorsal lip of the blastopore and of a complete dorsal axis in UV-irradiated recipients (which otherwise would have failed to properly initiate gastrulation; Figure 6.18A). Moreover, these three blastomeres, which underlie the prospective dorsal lip region, can also induce a secondary invagination and axis when transplanted into the ventral side of a normal, unirradiated 64-cell embryo (Figure 6.18B). This small cluster of vegetal blastomeres enables its adjacent marginal cells to invaginate and form the dorsal mesodermal axis of the embryo. Holowacz and Elinson (1993) found that cortical cytoplasm from the dorsal vegetal cells of the 16-cell *Xenopus* embryo was able to induce the formation of secondary axes when injected into vegetal ventral cells. Neither cortical cytoplasm from animal cells nor the deep cytoplasm from ventral cells could induce such axes.

It appears, then, that internal rearrangements of the cytoplasm, probably those normally oriented by the entry of the sperm, are responsible for causing the asymmetric distribution of subcellular factors. This asymmetry creates in the egg a dorsal–ventral distinction that ultimately directs the position of the blastopore above a set of vegetal blastomeres opposite the point of sperm entry. The molecules that may be involved in the formation of the vegetal gastrulation initiation site (the "Nieuwkoop center") will be discussed in Chapter 16.

Cell movements and the construction of the archenteron

The initiation of gastrulation. Amphibian gastrulation is initiated when a group of marginal endoderm cells on the dorsal surface of the blastula

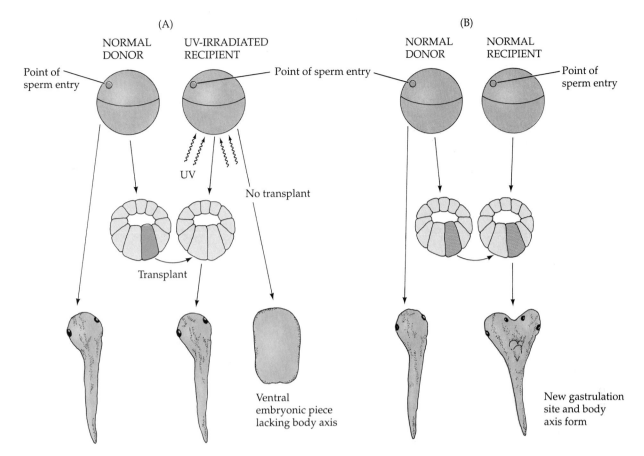

(A)

NORMAL DONOR   UV-IRRADIATED RECIPIENT

Point of sperm entry

Point of sperm entry

UV

No transplant

Transplant

Ventral embryonic piece lacking body axis

(B)

NORMAL DONOR   NORMAL RECIPIENT

Point of sperm entry

New gastrulation site and body axis form

sinks into the embryo. The outer (apical) surfaces of these cells contract dramatically, while their inner (basal) ends expand. The apical–basal length of these cells greatly increases to yield the characteristic "bottle" shape. In salamanders, these cells appear to have an active role in the early movements of gastrulation. Johannes Holtfreter (1943, 1944) found that bottle cells from early salamander gastrulae could attach to glass coverslips and lead the movement of those cells attached to them. Even more convincing were Holtfreter's recombination experiments in which dorsal marginal zone cells (which would give rise to the dorsal blastopore lip) were combined with inner endoderm tissue. When the dorsal marginal zone cells were excised and placed on inner prospective endoderm tissue, the blastopore cell precursors formed bottle cells and sank below the surface of the inner endoderm (Figure 6.19). Moreover, as they sank, they created a depression reminiscent of the early blastopore. Thus, Holtfreter

FIGURE 6.18
Transplantation experiments demonstrating that the vegetal cells underlying the prospective dorsal blastopore lip regions are responsible for causing the initiation of gastrulation. (A) Rescue of irradiated embryos by transplanting the vegetal blastomeres of the most dorsal segment (color) of a 64-cell embryo into a cavity made by the removal of a similar number of vegetal cells. An irradiated zygote without this transplant fails to undergo normal gastrulation. (B) Formation of new gastrulation site and body axis by the transplantation of the most dorsal vegetal cells of a 64-cell embryo into the ventralmost vegetal region of another 64-cell embryo. (After Gimlich and Gerhart, 1984.)

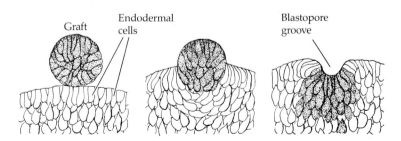

Graft   Endodermal cells   Blastopore groove

FIGURE 6.19
A graft of amphibian cells from the dorsal blastopore lip region sinks into a layer of endodermal cells and forms a blastopore groove. (After Holtfreter, 1944.)

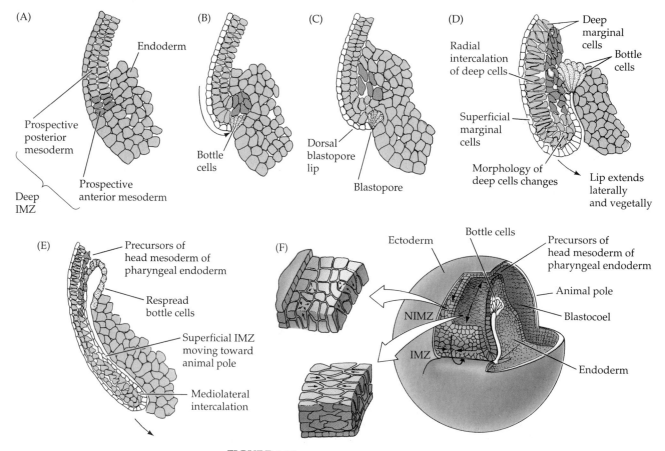

FIGURE 6.20
Integrative model of cell movements during early *Xenopus* gastrulation. (A) Structure of the involuting marginal zone (IMZ) prior to gastrulation. The deep IMZ consists of the prospective anterior mesoderm and the prospective posterior mesoderm. (B) Constriction of the bottle cells pushes up the prospective anterior mesoderm and rotates the IMZ outward. (C) The anterior mesodermal and pharyngeal endoderm precursors lead the movement of the mesoderm into the blastocoel. (D) Radial intercalation (interdigitation) of the deep IMZ cells occurs. The mesoderm moves toward the animal pole, pulling along the superficial cells and the bottle cells by involution. (E) As gastrulation continues, the deep marginal cells flatten and the formerly superficial cells form the wall of the archenteron. (F) Radial intercalation as in (D), looking down at the dorsal blastopore lip from the dorsal surface. In the NIMZ and upper portion of the IMZ, deep (mesodermal) cells are intercalating radially to make a thin band of flattened cells. This thinning of several layers into a few causes extension toward the blastopore lip. Just above the lip, mediolateral intercalation of the cells produces stresses that pull the IMZ over the lip. After involuting over the lip, mediolateral intercalation continues, elongating and narrowing the axial mesoderm. (After Hardin and Keller, 1988; Wilson and Keller, 1991.)

claimed that the ability to invaginate into the deep endoderm is an innate property of the dorsal marginal zone cells.

The situation in the frog embryo is somewhat different. R. E. Keller and his students (Keller, 1981; Hardin and Keller, 1988) have shown that although the bottle cells of *Xenopus* may play a role in *initiating* the involution of the marginal zone as they become bottle-shaped, they are not essential for gastrulation to continue. The peculiar bottle shape of these cells is needed to initiate gastrulation—it is the constriction of the cells that pulls the marginal zone vegetally while pushing the vegetal cells inward (Figure 6.20A,B). The pulling of the marginal zone vegetally enables the ectoderm to expand vegetally and encircle the embryo, while the pushing of the vegetal cells enables these anterior mesodermal precursors to contact the underside of the posterior mesodermal precursors and begin migrating on the roof of the blastocoel (Hardin and Keller, 1988).

However, after starting these movements, the *Xenopus* bottle cells are no longer needed. When bottle cells are removed after their formation, involution and blastopore formation and closure continue. The major factor in the movement of cells into the embryo appears to be the involution of the *subsurface* marginal cells rather than the superficial ones. It appears that these subsurface, **deep involuting marginal zone cells** turn inward and migrate toward the animal pole along the inside surfaces of the remaining deep cells (Figure 6.20C–E) and that the superficial layer forms the lining of the archenteron merely because it is attached to the actively migrating deep cells. The movement of the bottle cells deeper into the embryo depends on their attachment to the underlying deep cells. While removal of the *bottle cells* does not affect the involution of the deep or superficial marginal zone cells into the embryo, the removal of the dorsal marginal zone *deep cells* and their replacement with animal region cells (which do not normally undergo involution) stops archenteron formation.

The formation of mesoderm during **Xenopus** *gastrulation.* Figures 6.20D–F depict the behavior of these involuting marginal zone (IMZ) cells at successive stages of *Xenopus* gastrulation. (Keller and Schoenwolf, 1977; Keller, 1980, 1981; Hardin and Keller, 1988). Shortly before their involution through the blastopore lip, the several layers of deep IMZ cells intercalate radially to form one thin, broad layer. This intercalation further extends the IMZ vegetally. At the same time, the superficial cells spread out by dividing and flattening. When the deep cells reach the blastopore lip, they involute into the embryo and initiate a second type of intercalation. This intercalation causes a convergent extension along the mediolateral axis (Figure 6.20F) that integrates several mesodermal streams to form a long narrow band. The anterior part of this band migrates toward the animal cap. Thus, the mesodermal stream continues to migrate toward the animal pole and the overlying layer of superficial cells (including the bottle cells) is passively pulled toward the animal pole, thereby forming the endodermal roof of the archenteron (Figures 6.13 and 6.20E). Therefore, although the bottle cells may be responsible for creating the initial groove, the motivating force for this involution appears to come from the deep layer of the marginal cells. Furthermore, the radial and mediolateral intercalations of the deep layer of cells appear to be responsible for the continued movement of mesoderm into the embryo.

Migration of the involuting mesoderm

As mesodermal movement progresses, convergent extension continues to narrow and lengthen the involuting marginal zone. The IMZ contains the prospective endodermal roof of the archenteron in its superficial layer (IMZ$_S$) and the prospective mesodermal cells, including those of the notochord, in its deep region (IMZ$_D$). During the middle third of gastrulation, the expanding sheet of mesoderm converges toward the midline of the embryo. This process is driven by the lengthening of each cell along the anterior–posterior axis, thereby further narrowing the band. Toward the end of gastrulation, the central notochord separates from the somitic mesoderm on either side of it and the cells elongate separately (Wilson and Keller, 1991). This convergent extension of the mesoderm appears to be autonomous, because the movements of these cells occur even if this region of the embryo is isolated from the rest of the embryo (Keller, 1986).

During gastrulation, the **animal cap** and **noninvoluting marginal zone cells** expand by epiboly to cover the entire embryo. The dorsal portion of the noninvoluting marginal cells expands more rapidly than the ventral portion, thus causing the blastopore lips to move toward the ventral side. While those mesodermal cells entering through the dorsal blastopore lip give rise to the dorsal axial mesoderm, the remainder of the body mesoderm (which forms the heart, kidneys, blood, bones, and parts of several other organs) enters through the ventral and lateral blastopore lips to create the **mesodermal mantle.** The endoderm is derived from the IMZ_S cells that form the lining of the archenteron roof and from the subblastoporal vegetal cells that become the archenteron floor (Keller, 1986).

SIDELIGHTS & SPECULATIONS

Molecular regulators of development: Fibronectin and the pathways for mesodermal migration

How are the involuting cells informed where to go once they enter the inside of the embryo? In salamanders, it appears that the involuting mesodermal precursors migrate toward the animal pole upon a fibronectin lattice secreted by the cells of the blastocoel roof. Shortly before gastrulation, the presumptive ectoderm of the blastocoel roof secretes an extracellular matrix that contains fibrils of fibronectin (Figure 6.21; Boucaut et al., 1984; Nakatsuji et al., 1985). The involuting mesoderm appears to travel upon these fibronectin fibers. Confirmation of this was obtained by chemically synthesizing a "phony" fibronectin that can compete with the genuine fibronectin of the extracellular matrix. Cells bind to a certain region of the fibronectin protein that contains a three-amino acid sequence (Arg-Gly-Asp; RGD). Boucaut and co-workers injected large amounts of a small peptide containing this sequence into the blastocoels of salamander embryos shortly before gastrulation began. If fibronectin were essential for cell migration, then cells binding this soluble peptide fragment instead of the real cell-bound fibronectin should stop. Unable to find their "road," the mesodermal cells should cease their involution. This is precisely what occurred (Figure 6.22). No migrating cells were seen along the underside of the ectoderm. Instead, the mesodermal precursors remained outside the embryo, forming a convoluted cell mass. Other small synthetic peptides (including other fragments of the fibronectin molecule) did not impede migration. Mesodermal migration can also be arrested by the microinjection of antibodies against either fibronectin or the $\beta 1$ integrin subunit that binds fibronectin to the cell surface (D'Arribère et al., 1988, 1990).

In addition to merely permitting the attachment of mesodermal cells to the fibronectin lattice, this fibronectin-containing extracellular matrix appears to provide the cues for the *direction* of cell migration. Shi and colleagues (1989)

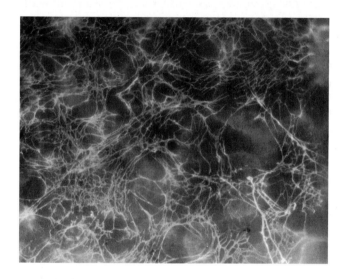

FIGURE 6.21
Fibronectin fibrils in a salamander gastrula, visualized by fluorescent antibodies to fibronectin. Immunofluorescence reveals a fibrillar network of fibronectin on the basal surface of the prospective ectodermal cells lining the blastocoel roof. (From Boucaut et al., 1985, courtesy of J.-C. Boucaut.)

excised the blastocoel roofs from early salamander gastrulae and deposited them on plastic dishes with their extracellular matrices touching the plastic (Figure 6.23A). The axis from the blastopore to the animal pole was marked, and after 2 hours the explant was removed, leaving behind its extracellular matrix. A smaller explant from the dorsal marginal zone (DMZ) was then taken from another early gastrula and placed on the matrix with its own blastopore–animal pole axis perpendicular to that of the matrix. Would the cells of this explant migrate on the matrix and would they migrate in a particular direction? The cells were found to migrate, and the migration could be inhibited by antibodies that block the cells' ability to recognize fibronectin. Moreover, rather than migrating randomly, the DMZ cells

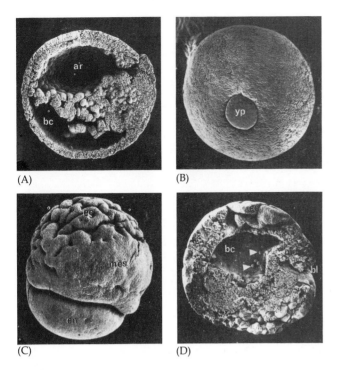

(A)　　　　(B)

(C)　　　　(D)

FIGURE 6.22
Scanning electron micrographs of normal (A,B) and abnormal (C,D) salamander gastrulation. The blastocoel in (C) and (D) was injected with the cell-binding fragment of fibronectin, while the normally gastrulating blastula was injected with a control solution. (A) Section during midgastrulation. (B) The yolk plug towards the end of gastrulation. (C,D) The finishing stages of the arrested gastrulation, wherein the mesodermal precursors, having bound the synthetic fibronectin, cannot recognize the normal fibronectin-lined migration route. The archenteron fails to form and the noninvoluted mesodermal precursors remain on the surface. ar, archenteron; bc, blastocoel; bl, blastopore; ec, ectoderm; en, endoderm; mes, mesoderm; yp, yolk plug. (From Boucaut et al., 1984, courtesy of J.-P. Thiery.)

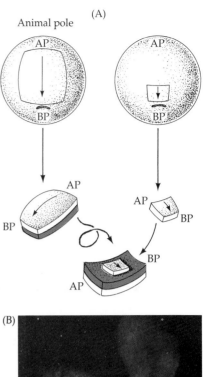

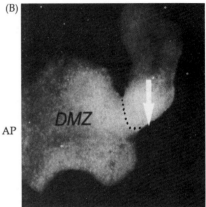

FIGURE 6.23
Direction of salamander dorsal marginal zone (DMZ) cell migration depends on orientation of the extracellular matrix of the blastocoel roof. (A) Explants of blastocoel roof from the blastopore (BP) to the animal pole (AP) were dissected from early gastrula-stage salamander embryos and placed on plastic dishes. The extracellular matrix adhered to the dish, and the tissue was then removed. A smaller explant from an early gastrula, containing DMZ cells, was placed on this matrix with its own axis perpendicular to that of the matrix. (B) DMZ cells from the explant migrate toward the animal pole of the matrix. Dotted line indicates original boundary of the explant, and the white arrow represents its blastopore–animal pole axis. (From Shi et al., 1989, photograph courtesy of the authors.)

migrated to the animal pole of the extracellular matrix that had been absorbed onto the plastic (Figure 6.23B).

In *Xenopus*, convergent extension pushes the migrating cells upward toward the animal pole. However, fibronectin appears to delineate the boundaries within which this movement can occur. The fibronectin of *Xenopus* gastrulae does not form complex lattices. Rather, it is organized into small fibrillar clusters. If fibronectin is synthesized but not organized into these fibrils, the dorsal mesodermal cells will adhere to the basal surface of the presumptive ectoderm but will not migrate (Winkelbauer and Nagel, 1991). The fibronectin fibrils are necessary for the head mesodermal cells to flatten and to extend broad (lamelliform) processes in the direction of migration (Winkelbauer et al., 1991). The importance of these fibronectin fibrils is also seen in interspecific hybrids that arrest at gastrulation. Delarue and colleagues (1985) have shown that certain invia-

ble hybrids between two species of toads arrest during gastrulation because they do not secrete these fibronectin fibrils. It appears, then, that the extracellular matrix of the blastocoel roof, and particularly its fibronectin component, is important in the migration of the mesodermal cells during amphibian gastrulation.

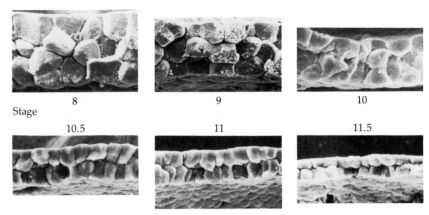

Stage 8   9   10

10.5   11   11.5

FIGURE 6.24

Scanning electron micrographs of the *Xenopus* blastocoel roof showing the changes in cell shape and arrangement. Stages 8 and 9 are blastulae; stages 10 through 11.5 represent progressively later gastrulae. (From Keller, 1980, courtesy of R. E. Keller.)

Epiboly of the ectoderm

While involution is occurring at the blastopore lips, the ectodermal precursors are expanding over the entire embryo. Keller (1980) and Keller and Schoenwolf (1977) have used scanning electron microscopy to observe the changes in both the superficial cells and the deep cells of the animal and marginal regions. The major mechanism of epiboly in *Xenopus* gastrulation appears to be an increase in cell number (through division) coupled with a concurrent integration of several deep layers into one (Figure 6.24). During early gastrulation, three rounds of cell division increase the volume of the deep cells in the animal hemisphere by 45 percent. At the same time, complete integration of the numerous deep cells into one layer occurs. The most superficial layer expands by cell division and flattening. The spreading of cells in the dorsal and ventral marginal zones appears to proceed by the same mechanism, although changes in cell shape appear to play a greater role than in the animal region. The result of these expansions is the epiboly of the superficial and deep cells of the animal and noninvoluting marginal regions over the surface of the embryo (Keller and Danilchik, 1988). Most of the marginal region cells, as previously mentioned, involute to join the mesodermal cell stream within the embryo.

Gastrulation in *Xenopus* is the orchestration of several distinct events. The first indication of gastrulation involves the local *invagination* of endodermal bottle cells in the marginal zone at a precisely defined time and place. Next, the *involution* of marginal cells through the blastopore lip begins the formation of the archenteron. These involuting cells at the leading edge of the mesodermal mantle *migrate* along the inner surface of the blastopore roof, and the prospective chordamesoderm behind them narrows and lengthens posteriorly by convergent extension in the dorsal portion of the embryo. At the same time, the ectodermal precursor cells *epibolize* vegetally by cell division and by the integration of previously independent cell layers. The result of these cell movements is the proper positioning of the three germ layers in preparation for their differentiation into the body organs. We still do not know how the cells are told to initiate and end their movements. Despite our new awareness, Keller (1986) warns: "No morphogenetic process—especially amphibian gastrulation—is satisfactorily understood at the cell or cell population level."

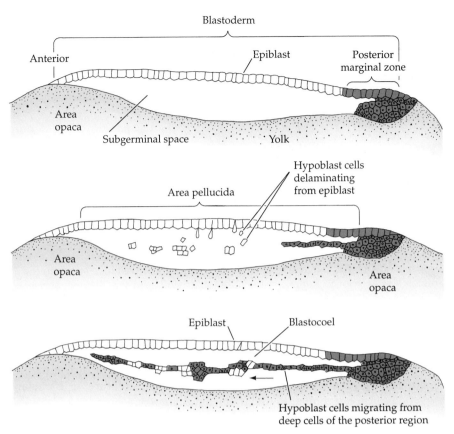

FIGURE 6.25
Formation of the two-layered blasto-derm of the chick embryo. The first hypoblast cells delaminate individually to form islands of cells beneath the epiblast. Cells from the posterior mar-gin (Koller's sickle and the posterior marginal cells behind it) produce a population of cells that migrates be-neath the blastodisc and incorporates the polyinvagination islands. This bot-tom layer becomes the hypoblast. The upper layer is the epiblast. As the hy-poblast moves anteriorly, epiblast cells collect at the region anterior to Koller's sickle to form the primitive streak.

Gastrulation in birds

Overview of avian gastrulation

Cleavage in avian embryos creates a blastodisc above an enormous volume of yolk. This inert, underlying yolk mass imposes severe constraints on cell movements, and avian gastrulation appears at first glance to be very different from that of a sea urchin or a frog. We shall soon see, though, that there are numerous similarities between avian gastrulation and those gastrulations we have already studied. Moreover, we shall see that mam-malian embryos—which do not have yolk—retain gastrulation movements very similar to those of bird and reptile embryos.

Formation of the hypoblast and epiblast. The central cells of the avian blastodisc are separated from the yolk by a subgerminal cavity and appear clear—hence, the center of the blastodisc is called the **area pellucida.** In contrast, the cells at the margin of the area pellucida appear opaque because of their contact with the yolk. They form the **area opaca** (Figure 6.25). Whereas most of the cells remain at the surface, forming the **epi-blast,** certain cells migrate individually into the subgerminal cavity to form the **polyinvagination islands (primary hypoblast),** an archipelago of dis-connected clusters containing 5–20 cells each (Figures 6.25 and 5.33). Shortly thereafter, a sheet of cells from the posterior margin of the blas-toderm (Koller's sickle and the marginal zone behind it) migrates anteriorly to join the polyinvagination islands, thereby forming the **secondary hy-poblast** (Eyal-Giladi et al., 1992). The two-layered blastoderm (epiblast and

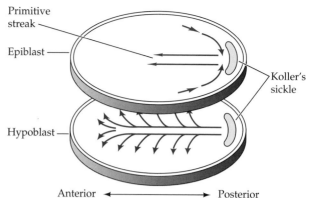

FIGURE 6.26
Cell movement during chick gastrulation. The lower model represents the movements of the central hypoblast cells from their posterior margin into the medial portion of the embryo. These cells first migrate anteriorly and then laterally. The upper model shows movement of the cells that form the primitive streak to the region just anterior to Koller's sickle. The underlying hypoblast induces these cells to become the primitive streak, and the last cells to enter this posterior area become the posteriormost cells of the primitive streak. (After Eyal-Giladi et al., 1992.)

hypoblast) is joined together at the margin of the area opaca, and the space between the layers is a blastocoel. Thus, the structure of the avian blastodisc is not dissimilar to that of the amphibian or echinoderm blastula.

The fate map for the avian embryo is restricted to the epiblast. That is, the hypoblast does not contribute any cells to the developing embryo (Rosenquist, 1966, 1972). Rather, the hypoblast cells form portions of the external membranes, especially the yolk stalk that links the yolk mass to the endodermal digestive tube. All three germ layers of the embryo proper (plus a considerable amount of extraembryonic membrane) are formed from the epiblastic cells. Fate maps of the chick epiblast are shown in Figure 6.26. These maps integrate several types of mapping. Vital dyes and transplants of radioactive cells were useful in mapping out major trends, since they tended to mark groups of cells and diffused as development proceeded. Transplanting genetically marked cells (such as quail cells placed into chick embryos) got around the problem of diffusion but still marked relatively large clusters of cells. Recently, the use of viruses or fluorescent dyes has enabled researchers to follow individual cells throughout development (Schoenwolf, 1991). As can be seen in these figures, there is significant convergent extension as the primitive streak progresses anteriorly. Athough cells in a particular region of the gastrula tend to become specific types of cells, they can still form different cell types if transplanted to another region of the embryo.

Formation of the primitive streak. The major structure characteristic of avian, reptilian, and mammalian gastrulation is the **primitive streak.** This streak is first visible as a thickening of the epiblast cell layer at the posterior region of the embryo, just anterior to Koller's sickle (Figure 6.27). This thickening is caused by the ingression of mesodermal cells from the epiblast into the blastocoel and by the migration of cells from the lateral region of the posterior epiblast toward the center (Figure 6.26B; Vakaet, 1984; Bellairs, 1986; Eyal-Giladi et al., 1992). As the thickening narrows, it moves anteriorly and constricts to form the definitive primitive streak. This streak extends 60–75 percent of the length of the area pellucida and marks the anterior–posterior axis of the embryo (Figure 6.28). As the cells

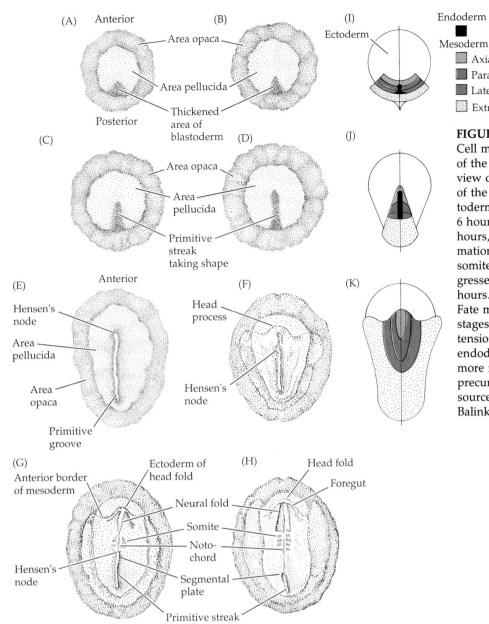

FIGURE 6.27
Cell movements of the primitive streak of the chick embryo. (A–E) Dorsal view of the formation and elongation of the primitive streak. The chick blastoderm is seen at (A) 3–4 hours, (B) 5–6 hours, (C) 7–8 hours, (D) 10–12 hours, and (E) 15–16 hours. (F–H) Formation of notochord and mesodermal somites as the primitive streak regresses. (F) 19–22 hours. (G) 23–24 hours. (H) Four-somite stage. (I–K) Fate maps of the chick epiblast at two stages of gastrulation. Convergent extension is seen in the midline, and the endodermal precursor cells ingress more rapidly than the mesodermal precursor cells. (Adapted from several sources, especially Spratt, 1946, and Balinksy, 1975. I–K after Vakaet, 1985.)

converge to form the primitive streak, a depression forms within the streak. This depression is called the **primitive groove,** and it serves as a blastopore through which the migrating cells pass into the blastocoel. Thus, the primitive groove is analogous to the amphibian blastopore. At the anterior end of the primitive streak is a regional thickening of cells called the **primitive knot,** or **Hensen's node.** The center of this node contains a funnel-shaped depression (sometimes called the primitive pit) through which cells can pass into the blastocoel. Hensen's node is the functional equivalent of the dorsal lip of the amphibian blastopore.

As soon as the primitive streak is formed, epiblast cells begin to migrate over the lips of the primitive streak and into the blastocoel (Figure 6.29). Like the amphibian blastopore, the primitive streak has a continually changing cell population. Those cells migrating through Hensen's node pass down into the blastocoel and migrate anteriorly, forming foregut, head mesoderm, and notochord; those cells passing through the lateral portions of the primitive streak give rise to the majority of endodermal

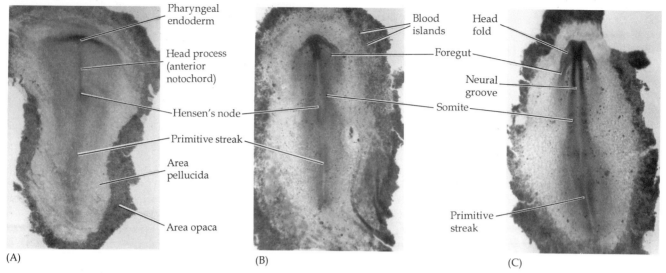

(A)   (B)   (C)

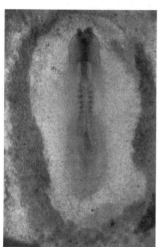

(D)

FIGURE 6.28

Photographs of chick gastrulation from about 24 to about 28 hours. (A) The primitive streak at full extension (24 hours). The head process (anterior notochord) can be seen extending from Hensen's node. (B) Two-somite stage (25 hours). Pharyngeal endoderm is seen anteriorly, while the anterior notochord pushes up the ectoderm above it. The primitive streak is regressing. (C) Four-somite stage (27 hours). (D) At 28 hours, the primitive streak has regressed almost entirely. (Courtesy of K. Linask.)

and mesodermal tissues (Schoenwolf et al., 1992). Unlike the *Xenopus* mesoderm, which migrates as sheets of cells into the blastocoel, cells entering the inside of the avian embryo do so as individuals. Rather than forming a tightly organized sheet of cells, the ingressing population creates a loosely connected mesenchyme. Moreover, there is no true archenteron formed in the avian gastrula.

As the cells enter the primitive streak, the streak elongates toward the future head region. At the same time, the secondary hypoblastic cells are continuing to migrate anteriorly from the posterior margin of the blastoderm. The elongation of the primitive streak appears to be coextensive with the anterior migration of these secondary hypoblast cells.

Migration through the primitive streak: Formation of endoderm and mesoderm. The first cells to migrate through the primitive streak are those destined to become the foregut. This situation is again similar to that seen in amphibians. Once inside the blastocoel, these cells migrate anteriorly and eventually displace the hypoblast cells in the anterior portion of the embryo. The hypoblast cells are confined to a region in the anterior portion of the area pellucida. This region, the **germinal crescent,** does not form any embryonic structures, but it does contain the precursors of germ cells which later migrate through the blood vessels to the gonads. The next cells entering the blastocoel through Hensen's node also move anteriorly, but they do not move as far ventrally as the presumptive endodermal cells. These cells remain between the endoderm and the epiblast to form the head mesoderm and the chordamesoderm (notochordal) cells. These early-ingressing cells have all moved anteriorly, pushing up the anterior midline region of the epiblast to form the **head process.** Meanwhile, cells continue migrating inward through the primitive streak. As they enter the

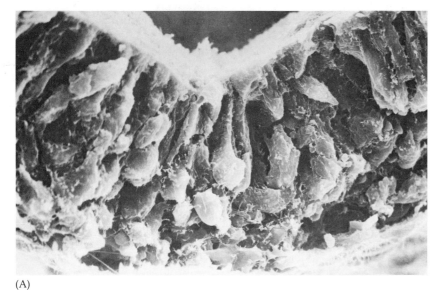

(A)

FIGURE 6.29
Migration of endodermal and mesodermal cells through the primitive
streak. (A) Scanning electron micrograph shows epiblast cells passing into
the blastocoel and extending their apical ends to become bottle cells. (B)
Stereogram of a gastrulating chick embryo, showing the relationship of the
primitive streak, the migrating cells,
and the two original layers of the blastoderm. The lower layer becomes a
mosaic of hypoblast and endodermal
cells; but the hypoblast cells eventually
sort out to form a layer beneath that of
the endoderm and eventually contribute to the yolk sac. (A from Solursh
and Revel, 1978, courtesy of M. Solursh; B after Balinsky, 1975.)

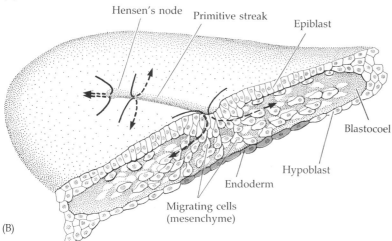

(B)

blastocoel, these cells separate into two streams. One stream moves deeper
and joins the hypoblast along its midline, displacing the hypoblast cells
to the sides. These deep-moving cells give rise to all the endodermal
organs of the embryo as well as to most of the extraembryonic membranes
(the hypoblast forms the rest). The second migrating stream spreads
throughout the blastocoel as a loose sheet, roughly midway between the
hypoblast and the epiblast. This sheet generates the mesodermal portions
of the embryo and extraembryonic membranes. By 22 hours of incubation,
most of the presumptive endodermal cells are in the interior of the embryo,
although presumptive mesodermal cells continue to migrate inward for a
longer time.

Now a new phase of gastrulation begins. While the mesodermal
ingression continues, the primitive streak starts to regress, moving Hensen's node from near the center of the area pellucida to a more posterior
position (Figure 6.30). It leaves in its wake the dorsal axis of the embryo
and the head process. As the node moves further posteriorly, the remaining (posterior) portion of the notochord is laid down. Finally, the node
regresses to its most posterior position, eventually forming the anal region.
By this point, the epiblast is composed entirely of presumptive ectodermal
cells.

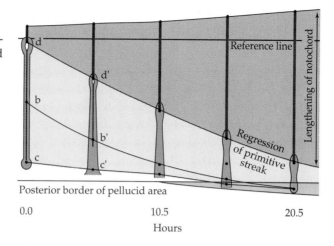

FIGURE 6.30
Regression of the primitive streak, leaving the notochord in its wake. Various points of the streak were followed after it achieved its maximum length. Time represents hours after achieving maximum length. (After Spratt, 1947.)

As a consequence of this two-step gastrulation process, avian (and mammalian) embryos exhibit a distinct anterior-to-posterior gradient of developmental maturity. While cells of the posterior portions of the embryo are undergoing gastrulation, cells at the anterior end are already starting to form organs. For the next several days, the anterior end of the embryo is more advanced in its development (having had a "head start," if you will) than the posterior end.

While the presumptive mesodermal and endodermal cells were moving inward, the ectodermal precursors proliferated to become the only cell population remaining in the upper layer. Moreover, ectodermal cells migrated off the blastodisc to surround the yolk by epiboly. The enclosure of the yolk by the ectoderm (again reminiscent of the epiboly of amphibian ectoderm) is a Herculean task that takes the greater part of 4 days to complete and involves the continuous production of new cellular material and the migration of the presumptive ectodermal cells along the underside of the vitelline envelope. Thus, as avian gastrulation draws to a close, the ectoderm has surrounded the yolk, the endoderm has replaced the hypoblast, and the mesoderm has positioned itself between these two regions.

Mechanisms of avian gastrulation

The role of the hypoblast and the formation of the embryonic axes. The dorsal–ventral (back–belly) axis is critical to the formation of the hypoblast and to the further development of the embryo. This axis is established when the cleaving cells of the blastoderm establish a barrier between the basic (pH 9.5) albumen above the blastodisc and the acidic (pH 6.5) subgerminal space below the disc. Water and sodium ions are transported from the albumen through the cells and into the subgerminal space, creating a membrane potential difference of 25 mV across the cell layer (positive at the ventral side of the cells). This creates two sides to the cells: a side facing the negative and basic albumen and a side facing the acidic and positive subgerminal space fluid. The side facing the albumen becomes dorsal, while the side facing the subgerminal space becomes ventral. This can be reversed either by reversing the pH gradient or by inverting a potential difference across the cell layer (reviewed in Stern and Canning, 1988).

The conversion of the radially symmetrical blastoderm into a bilaterally symmetric structure is determined by gravity. As the ovum rolls down the oviduct, it turns at a rate of 10–15 revolutions per hour. The cytoplasm that is to become the cell layer is always rotating downward but is displaced upward by the denser yolk. Therefore, it is not at the top of the yolk, but is off slightly to the side. That portion of the blastodisc that is highest becomes the caudal (tail) end of the blastoderm, that part where gastrulation begins (Kochav and Eyal-Giladi, 1971). Thus, the anterior–posterior and dorsal-lateral axes are determined prior to gastrulation, while the egg is slowly rolling down the oviduct.

The formation of these two axes enables the formation and migration of the hypoblast. Although the hypoblast does not contribute any cells to the adult chick, it is essential for its proper development. In 1932 C. H. Waddington demonstrated that this lower layer influences the orientation of the chick embryonic axis. He separated the two cellular layers of the chick blastodisc soon after the primitive streak had formed. After bringing the two layers together again so that their original long axes were now nearly perpendicular to each other, he saw that the primitive streak bent according to the long axis of the hypoblast (Figure 6.31). Azar and Eyal-Giladi (1979, 1981) have extended these observations to show that the hypoblast *induces* the formation of the primitive streak. First, removal of the hypoblast stops all further development until the remaining epiblast can regenerate a new lower layer. Second, when the population of hypoblast cells derived from the posterior margin is eliminated, no primitive streak is formed. Rather, the area pellucida develops into a mass of disorganized, mesoderm-like tissue. Thus, it appears that the secondary hypoblast directs the formation and the directionality of the primitive streak.

The blastoderm of the chick embryo acts as a single integrating system to form a single embryo; for if the blastoderm is separated into parts, each having its own marginal zone, each part will form its own embryo (Spratt and Haas, 1960). Control of this field seems to reside in the posterior part of the margin, the cells of which contribute to the hypoblast. These posterior marginal cells not only contribute the inducing cells of the hypoblast, but also prevent other regions of the margin from inducing their own hypoblasts. Khaner and Eyal-Giladi (1989) found that if the posterior margin region is transposed to a lateral margin area (Figure 6.32A), the posterior tear heals and two primitive streaks emerge. Similarly, if a posterior region is reciprocally transposed with a lateral region (Figure 6.32B), only one primitive streak forms, and it arises from the original posterior region. However, if a posterior margin region is placed into an embryo that retains its original posterior margin (Figure 6.32C), only the host's original posterior margin forms the hypoblast that underlies the primitive streak. Khaner and Eyal-Giladi suggest that the marginal zone cells form a gradient of activity whose peak is at the posterior end. These posterior cells will form the hypoblast and at the same time prevent any cells with less activity from forming hypoblasts of their own.

Cell accumulation in the primitive streak. Evidence from the studies of Stern and Canning (1990) suggests that the epiblast is not the homogeneous, undifferentiated tissue that we have long assumed it to be. Rather, there appears to be differentiation in the epiblast cells even before primitive streak formation begins. These studies show that certain cells that are randomly scattered throughout the epiblast can be distinguished by a particular molecule (HNK-1, a sulfated form of glucuronic acid) on their cell surfaces. The cells expressing HNK-1 ingress individually into the blastocoel and migrate to the posterior margin (Figure 6.33). It is probable that the posterior marginal tissue secretes a chemical that attracts the cells expressing HNK-1, while the anterior marginal tissue secretes a repellent

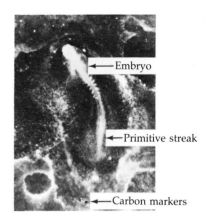

FIGURE 6.31
Rotation of the embryonic axis of a chick embryo, caused by rotating the hypoblast with respect to the developing primitive streak at 20–24 hours incubation. The primitive streak has regressed significantly at this stage. (From Azar and Eyal-Giladi, 1979, courtesy of H. Eyal-Giladi.)

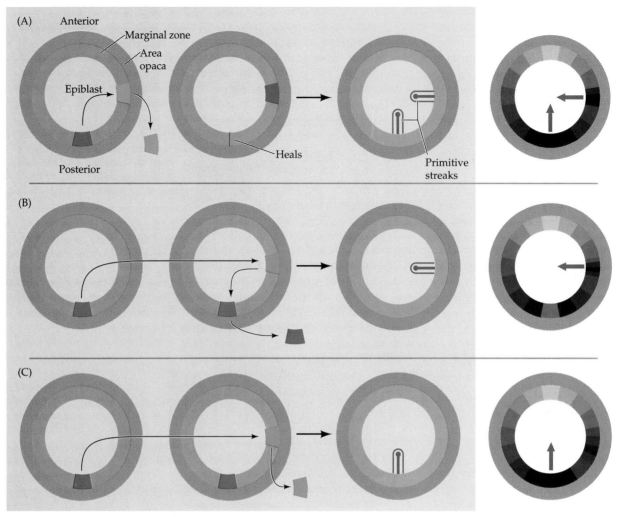

FIGURE 6.32
Experiments of Khaner and Eyal-Giladi demonstrating that the posterior part of the marginal zone contributes to the primitive streak-inducing cells of the hypoblast and prevents other marginal regions from creating their own hypoblasts. (After Khaner and Eyal-Giladi, 1989.)

molecule (Jephcott and Stern, quoted in Stern, 1991). The HNK-1-expressing cells that collect at the posterior margin will give rise to the endoderm and mesoderm, and no cell expressing HNK-1 forms ectodermal derivatives. If the HNK-1 cells are selectively destroyed (by antibodies) while they are still on the epiblast, the embryo will not form any mesoderm or endoderm. These HNK-1-positive cells interact with the epiblast cells above them to form the initial rudiment of the primitive streak. The streak rudiment then undergoes a convergent extension process that narrows and extends the streak. When the streak has reached nearly its full extension, the HNK-1-positive cells dissolve the basal lamina of the central epiblast to form a groove through the primitive streak. This allows epiblast cells (which never had expressed HNK-1) to get recruited into the streak as it extends anteriorly and to contribute (along with HNK-1-positive cells) to the embryonic mesoderm and endoderm.

Movement within the amniote blastocoel is by individual cells, not by an epithelial sheet. But, as in amphibian gastrulation, avian cells passing through the blastopore constrict their apical ends to become bottle cells (Figure 6.29). The breakdown of the basal lamina and the release of these cells from the epiblast may be accomplished by a 190-kDa protein called **scatter factor** (Stern et al., 1990). When resin or plastic containing scatter factor is implanted beneath the epiblast of early gastrulating chick em-

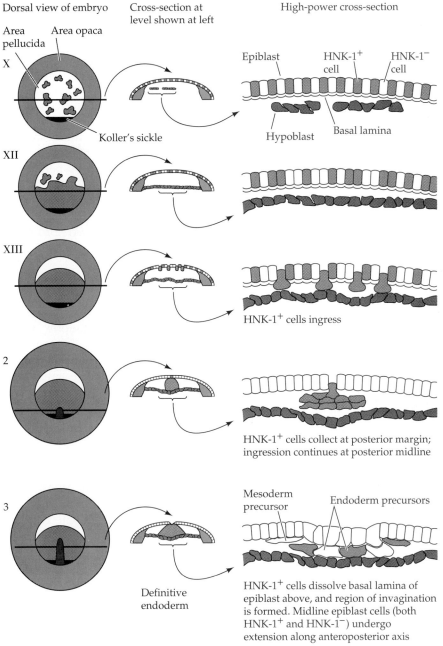

FIGURE 6.33
Model for the role of HNK-1-positive cells during gastrulation. Left-hand figures represent dorsal views looking down upon the blastoderm. Middle and right-hand figures represent low- and high-power cross-sections through the indicated regions of the embryo. The HNK-1$^+$ cells are seen on the high-power cross sections. HNK-1$^+$ cells begin to appear in the epiblast prior to primitive streak formation, when the secondary hypoblast is migrating beneath the epiblast (stage XII). At stage XIII, the secondary hypoblast almost completely underlays the epiblast, and HNK-1$^+$ cells ingress individually into the blastocoel. These cells collect at the posterior marginal zone (stage 2) to form the mesoderm of the initial primitive streak. By stage 3, HNK-1$^+$ cells dissolve the basal lamina of the epiblast above them to allow the invagination of cells from the epiblast into the primitive streak. The cells migrating into the blastocoel do not express HNK-1, and they help form the mesoderm and endoderm of the chick embryo. (After Stern, 1992.)

bryos, new primitive streak regions can be induced. Once cells are released from the primitive streak and enter the blastocoel, they flatten and enter a stream of independent, migrating cells. Extracellular polysaccharides may also play an important role in this migration. One such complex polysaccharide is **hyaluronic acid,** a linear polymer of glucuronic acid and *N*-acetylglucosamine (see Figure 3.35). This compound is made by ectodermal cells and accumulates in the blastocoel, where it coats the surfaces of the incoming cells. Fisher and Solursh (1977) have shown that when this material is digested (by injecting the enzyme hyaluronidase into the blastocoel), the mesenchymal cells clump together and fail to migrate properly. Several studies have shown that hyaluronic acid is important in keeping migrating mesenchymal cells separated from one another. Moreover, hyaluronic acid first accumulates at precisely the time the first cells enter the blastocoel. Hyaluronic acid may be able to keep the cells sepa-

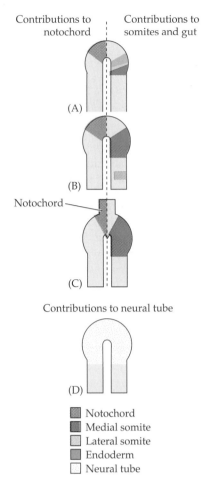

Contributions to notochord | Contributions to somites and gut

(A)

(B)

Notochord —

(C)

Contributions to neural tube

(D)

■ Notochord
■ Medial somite
□ Lateral somite
■ Endoderm
□ Neural tube

FIGURE 6.34
Fate maps of Hensen's node in (A) early, (B) mid, and (C) late gastrulation. Although Hensen's node is bilaterally symmetrical, the left of each diagram shows the region of Hensen's node that contributes cells to the notochord, while the right half of each diagram shows the regions of cells that contribute to the gut and somites. At each time period, Hensen's node contains cells that form part of the neural tube (D). (After Selleck and Stern, 1991.)

rated by its ability to expand in water. Once in an aqueous environment, this polymer can expand to over 1000 times its original volume. Therefore, hyaluronic acid might be an important factor in keeping the mesenchymal cells dispersed during their migration and thereby ensuring that this migration continues.

Hyaluronic acid and other polysaccharides facilitate individual cell migration (Chapter 3), but they do not appear to direct the movement of these cells (Fisher and Solursh, 1979). Rather, cellular movement of these cells is correlated, once again, with the presence of a fibronectin meshwork in the extracellular basal lamina of the epiblast cells. This fibronectin-rich layer appears on the undersurface of the upper layer shortly before the formation of the primitive streak and disappears in the region of the streak. Within the streak, the cells separate and are then seen to migrate laterally along the fibronectin-rich basement membrane of the epiblast (Duband and Thiery, 1982). However, there is no clear evidence that this fibronectin is essential for the directed cell movements of the cells laterally away from the primitive streak.

Commitment of cells in Hensen's node. Hensen's node shares many properties of the dorsal blastopore lip of amphibian embryos. Cells migrating through this site become the gut endoderm and the mesoderm of the dorsal axis, whereas cells migrating through other areas of the primitive streak (analogous to the ventral and lateral lips of the amphibian blastopore) give rise to more lateral mesoderm. Moreover, like the amphibian dorsal blastopore lip, if Hensen's node is explanted and placed under another region of the epiblast, the exised node can organize a secondary dorsal axis. This distinguishes it from any other group of cells in the embryo. (More about these "organizing" properties will be discussed in Chapter 16.) The fate map of Hensen's node has been constructed with more precision as marking techniques become more refined. Originally, large segments of Hensen's node were followed using vital dyes or radioactively labeled grafts. More recently, microinjections of dyes or fluorescent beads enable us to follow single cells to their ultimate destinations. Selleck and Stern (1991) have used these latter techniques to follow cells in various regions of Hensen's node. The anterior region of the primitive streak always contains precursors of the notochord (Figure 6.34). In early stages (Figure 6.34A), endodermal precursor cells are seen at the lateral ends of this prenotochordal wedge. The cells that give rise to the medial (internal) portion of the somites are found lateral to this wedge, and the precursors of the lateral sides of the somites are found further down the primitive streak (Figure 6.34B,C). In all stages, there are neural tube precursor cells throughout the region of Hensen's node.

The cells that are destined to form the notochord are already determined to become notochord cells when they are inside Hensen's node. If they are transplanted to areas of the node that are destined to give rise to somites, these cells will still integrate into the notochord or else form a separate notochord-like structure inside the somite. However, the fate of the presomite cells is still undetermined, for they will form notochordal cells if transplanted into the anterior part of Hensen's node (Selleck and Stern, 1992).

Epiboly of the ectoderm. During gastrulation, ectodermal precursor cells expand outwardly to encircle the yolk. These cells are joined to each other by tight junctions and travel as a unit rather than as individual cells. In birds, the upper surface of the area opaca is seen to adhere tightly to the lower surface of the vitelline envelope and spreads along this inner sur-

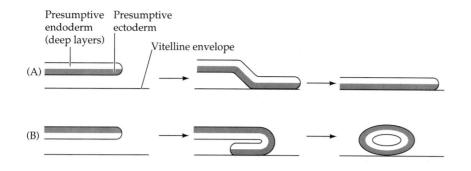

Presumptive endoderm (deep layers) Presumptive ectoderm

Vitelline envelope

(A)

(B)

FIGURE 6.35

Migratory properties of chick ectodermal precursors. (A) When chick blastoderm is placed on vitelline envelope with the ectodermal precursors in contact with the vitelline surface, the marginal cells migrate and cover the vitelline envelope with ectoderm. (B) When the deep layers are placed in contact with the vitelline envelope, the blastoderm layer curls to allow the cells of the superficial layer to adhere to and migrate over the vitelline layer. The result is a closed vesicle. (After New, 1959.)

face. The same behavior is seen in culture. New (1959) demonstrated that isolated blastoderm will spread normally on isolated vitelline envelope, and Spratt (1963) demonstrated that this spreading will not occur on other substrates. These observations suggest that the vitelline envelope is essential for the spreading of the cell sheet. Interestingly, only the marginal cells (i.e., the cells of the area opaca) attach firmly to the vitelline surface. Most of the blastoderm cells adhere loosely, if at all. These marginal cells are inherently different from the other blastoderm cells, as they can extend enormous (500 μm) cytoplasmic processes onto the vitelline envelope. These elongated filopodia are believed to be the locomotor apparatus of the marginal cells.

There are several lines of evidence indicating that the marginal area opaca cells are the agents of ectodermal epiboly. First, the blastoderm spreads only when the margins are expanding. If the marginal cells are removed, the epiboly of the ectoderm stops. Second, when the marginal cells are cut away from the rest of the blastoderm, they continue to expand alone. Thus, it appears that the ectodermal precursor cells are carried along by the actively migrating cells of the area opaca (Schlesinger, 1958). There also exists a specific relationship between the cell membranes of the marginal cells and the lower surface of the vitelline membrane. New (1959) showed that when the blastoderm is placed on a vitelline envelope upside down (deep layers in contact with the vitelline envelope), the edges of the blastoderm curl inward so that the marginal cells of the upper layer are once again contacting the vitelline surface (Figure 6.35). Lash and his co-workers (1990) expanded these results by showing that fibronectin is present on the inner surface of the vitelline envelope. Then, as in the experiment discussed earlier in which Boucaut and co-workers injected the synthetic fibronectin-attachment-site sequence (RGD) into the salamander blastocoel, Lash and colleagues applied the RGD sequence to the vitelline envelope as the cells were migrating on it. This treatment specifically broke the contact between the marginal cells and the vitelline envelope, caused the retraction of the marginal cell filopodia, and stopped the migration of the blastoderm. Thus, there appears to be specific recognition between the marginal cells of the area opaca and the inner surface of the vitelline envelope, a recognition essential for the epibolic migration of the ectoderm that encloses the yolk. It is probable that this attachment is mediated by cell surface receptors for the fibronectin lining the inner vitelline envelope surface.

We have identified many of the processes involved in avian gastrulation, but we are ignorant as to how these processes are carried out. We do not yet know how the subgerminal cavity forms, how certain cells are specified to become hypoblast cells, how certain cells express HNK-1 while their neighbors do not, how HNK-1-expressing cells migrate to the pos-

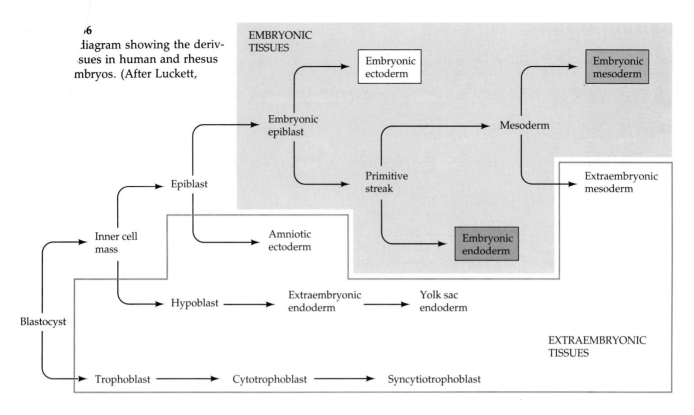

diagram showing the deriv-
,sues in human and rhesus
mbryos. (After Luckett,

EMBRYONIC
TISSUES

Embryonic
ectoderm

Embryonic
epiblast

Embryonic
mesoderm

Mesoderm

Epiblast

Primitive
streak

Extraembryonic
mesoderm

Inner cell
mass

Amniotic
ectoderm

Embryonic
endoderm

Hypoblast

Extraembryonic
endoderm

Yolk sac
endoderm

Blastocyst

EXTRAEMBRYONIC
TISSUES

Trophoblast

Cytotrophoblast

Syncytiotrophoblast

terior margin and how they interact with the epiblast cells there, how the primitive streak is extended and how it retracts, or how the cells are assigned their respective fates. As Gary Schoenwolf (1991) recently remarked, "Despite all that has been written, it is safe to say that what we know about avian gastrulation and neurulation is considerably less than what still remains to be learned."

Gastrulation in mammals

Birds and mammals are both descendants of reptilian species. Therefore, it is not surprising to find that mammalian development parallels that of reptiles and birds. What is surprising is that the gastrulation movements of reptilian and avian embryos, which evolved as an adaptation to yolky eggs, are retained even in the absence of large amounts of yolk in the mammalian embryo. The mammalian inner cell mass can be envisioned as sitting atop an imaginary ball of yolk, following instructions that seem more appropriate to its ancestors.

Modifications for development within another organism

Instead of developing in isolation within an egg, most mammals have evolved the remarkable strategy of developing within the mother herself. The mammalian embryo obtains nutrients directly from its mother and does not rely on stored yolk. This evolution has entailed a dramatic restructuring of the maternal anatomy (such as expansion of the oviduct to form the uterus) as well as the development of a fetal organ capable of absorbing maternal nutrients. This fetal organ—the **placenta**—is derived primarily from embryonic trophoblast cells, supplemented with mesodermal cells derived from the inner cell mass.

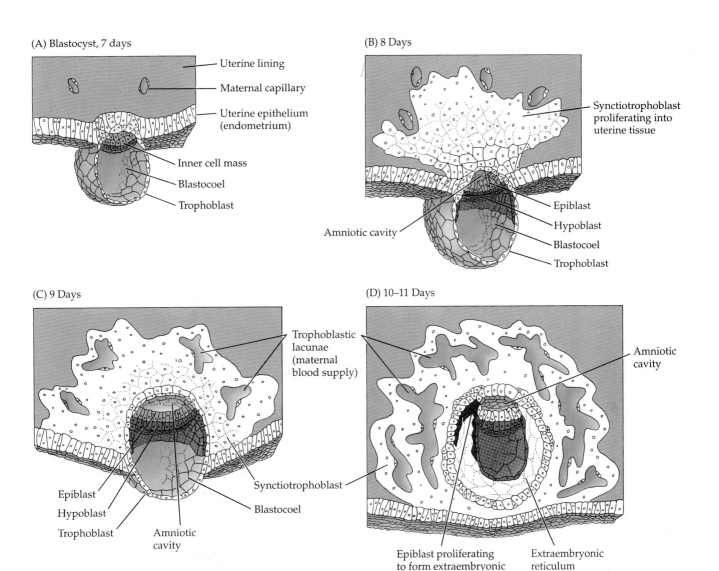

(A) Blastocyst, 7 days

- Uterine lining
- Maternal capillary
- Uterine epithelium (endometrium)
- Inner cell mass
- Blastocoel
- Trophoblast

(B) 8 Days

- Synctiotrophoblast proliferating into uterine tissue
- Amniotic cavity
- Epiblast
- Hypoblast
- Blastocoel
- Trophoblast

(C) 9 Days

- Trophoblastic lacunae (maternal blood supply)
- Epiblast
- Hypoblast
- Trophoblast
- Amniotic cavity
- Synctiotrophoblast
- Blastocoel

(D) 10–11 Days

- Amniotic cavity
- Epiblast proliferating to form extraembryonic mesoderm
- Extraembryonic reticulum

The origins of the various early mammalian tissues are summarized in Figure 6.36. The first segregation of cells within the inner cell mass involves the formation of the **hypoblast** (sometimes called the primitive endoderm) layer (Figure 6.37). These cells separate from the inner cell mass to line the blastocoel cavity, where they give rise to the **yolk sac endoderm.** As in avian embryos, these cells do not produce any part of the newborn organism. The remaining ICM tissue above the hypoblast is now referred to as the **epiblast.** The epiblast cells are split by small clefts that eventually coalesce to separate the embryonic epiblast from the other epiblast cells, which form the lining of the **amnion** (Figures 6.37C and 6.38). Once the lining of the amnion is completed, it fills with a secretion called **amniotic fluid,** which serves as a shock absorber to the developing embryo while preventing its desiccation.

The embryonic epiblast is believed to contain all the cells that will generate the actual embryo, and it is similar in many ways to the avian epiblast. By labeling individual cells of the epiblast with horseradish peroxidase, Kirstie Lawson and her colleagues (1991) were able to construct a detailed fate map of the mouse epiblast (Figure 6.39). Like the chick epiblast cells, the mammalian mesoderm and endoderm migrate through a primitive streak. As they enter the streak, the epiblast cells cease to

FIGURE 6.37

Tissue formation in the human embryo between days 7 and 12. (A,B) Human blastocyst immediately prior to gastrulation. The inner cell mass delaminates hypoblast cells that line the trophoblast, thereby forming the primitive yolk sac and a two-layered (epiblast and hypoblast) blastodisc similar to that seen in avian embryos. The trophoblast in some mammals can be divided into the polar trophoblast, which covers the inner cell mass, and the mural trophoblast, which does not. The trophoblast divides into the cytotrophoblast, which will form the villi, and the syncytiotrophoblast, which will ingress into the uterine tissue. (C) Meanwhile, the epiblast has split into the amniotic ectoderm (which encircles the amniotic cavity) and the embryonic epiblast. The adult mammal forms from the cells of the embryonic epiblast. (D) The extraembryonic endoderm forms the yolk sac. (After Gilbert, 1989 and Larsen, 1993.)

FIGURE 6.38

Amnion structure and cell movements during human gastrulation. (A) Human embryo and uterine connections at 15 days' gestation. In the upper view, the embryo is cut sagitally through the midline; the lower view looks down upon the dorsal surface of the embryo. (B) The movements of the epiblast cells into the primitive streak and Hensen's node and underneath the epiblast are superimposed upon the dorsal surface view. At days 14 and 15, the ingressing epiblast cells are thought to replace the hypoblast cells (which contribute to forming the yolk sac lining), while at day 16, the ingressing cells fan out to form the mesodermal layer. (After Larsen, 1993.)

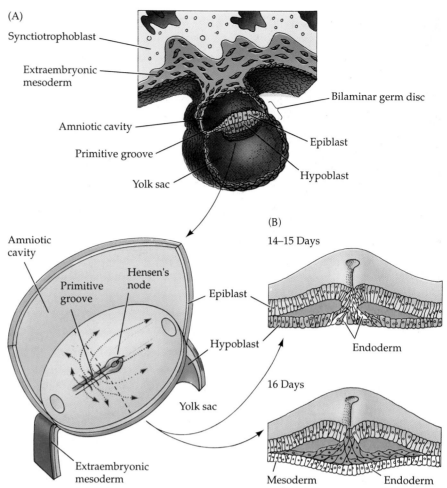

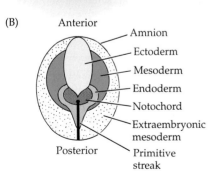

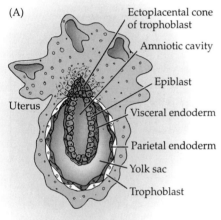

express the E-cadherin that holds cells together, and they migrate as individual cells (Bursdal et al., 1993). Those cells migrating through Hensen's node give rise to the notochord. However, unlike notochord formation in the chick, the cells that form the mouse notochord are thought to become integrated into the endoderm of the primitive gut (Jurand, 1974; K. Sulik, personal communication). These cells can be seen as a band of small, ciliated cells extending rostrally from Hensen's node (Figure 6.40). They form the notochord by converging medially and folding off in a dorsal direction from the roof of the gut.

The ectodermal precursors are located anterior to the fully extended primitive streak, similar to their position in the chick epiblast; but whereas the mesoderm of the chick forms from cells posterior to the furthest extent of the streak, the mouse mesoderm forms from cells anterior to the primitive streak. In some instances, clones of cells give rise to descendants in

FIGURE 6.39

(A) Mouse embryo at the egg cylinder stage, 6 days after fertilization. Note that, unlike chick and human epiblasts, the mouse epiblast is tightly curved. The parietal and visceral endoderm are derived from hypoblast cells, not from the trophectoderm. (B) Fate map of the 7-day mouse epiblast at early gastrula stage. (The mouse fate map has been flattened and should be seen as rolled up with the primitive streak at the edges.) (After Lawson et al., 1991.)

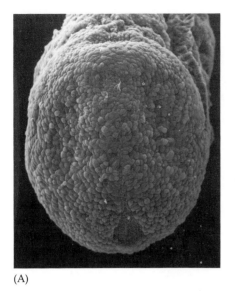

(A)

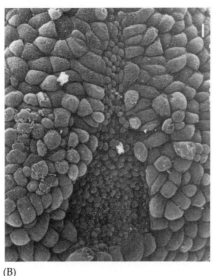

(B)

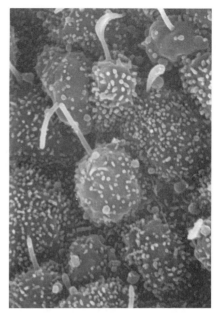

(C)

more than one embryonic layer or to both embryonic and extraembryonic derivatives. Thus, at the epiblast stage, these lineages have not become separate from one another. As in avian embryos, the cells migrating between the hypoblast and epiblast layers appear to be coated with hyaluronic acid, which is first synthesized at the time of primitive streak formation (Solursh and Morriss, 1977). It is thought (Larsen, 1993) that the replacement of the hypoblast cells by endoderm precursors occurs on days 14 to 15 of gestation, while the migration of those cells forming the mesoderm doesn't start until day 16.

Formation of extraembryonic membranes

While the embryonic epiblast is undergoing cell movements reminiscent of those seen in reptilian or avian gastrulation, the extraembryonic cells are making the distinctly mammalian tissues that enable the fetus to survive within the maternal uterus. Although the initial trophoblastic cells of mice and humans appear normal, they give rise to a population of cells wherein nuclear division occurs in the absence of cytokinesis. The initial type of trophoblast cell constitutes a layer called the **cytotrophoblast,** whereas the multinucleated type of cell forms the **syncytiotrophoblast.** The cytotrophoblast adheres to the uterine wall (endometrium) through a series of adhesion molecules that were discussed in Chapter 3. The human cytotrophoblast cells also contain proteolytic enzymes that enable them to enter the uterus and remodel the uterine blood vessels so that the maternal blood bathes fetal blood vessels. The syncytiotrophoblast tissue is thought to further the progression of the embryo into the uterus. This proteolytic activity ceases after the twelfth week of pregnancy (Fisher et al., 1989). The uterus, in turn, sends blood vessels into this area, where they eventually contact the syncytiotrophoblast. Shortly thereafter, mesodermal tissue extends outward from the gastrulating embryo (Figure 6.38). Luckett (1978) has shown that this tissue had migrated through the primitive streak but became extraembryonic rather than embryonic mesoderm. This extraembryonic mesoderm joins the trophoblastic extensions and gives rise to the blood vessels that carry nutrients from the mother to the embryo. The narrow **connecting stalk** of extraembryonic mesoderm that links the embryo to the trophoblast eventually forms the vessels of the umbilical

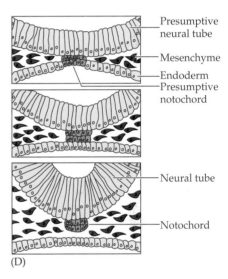

(D)

FIGURE 6.40
Formation of the notochord in the mouse. (A–C) The ventral surface of the 7.5-day mouse embryo, seen by scanning electron microscopy at successively higher magnifications. (Magnification bars: A, 50 μm; B, 10 μm; C, 1 μm.) The presumptive notochord cells are the ciliated cells in the midline that are flanked by the endodermal cells of the primitive gut. (D) The formation of the notochord by the dorsal infolding of the small, ciliated cells. (Courtesy of K. Sulik and G. C. Schoenwolf.)

Labels in figure D:
Presumptive neural tube
Mesenchyme
Endoderm
Presumptive notochord
Neural tube
Notochord

FIGURE 6.41
Human embryo and placenta after 40 days of gestation. The embryo lies within the amnion, and its blood vessels can be seen extending into the chorionic villi. The sphere to the right of the embryo is the yolk sac. (The Carnegie Institution of Washington, courtesy of C. F. Reather.)

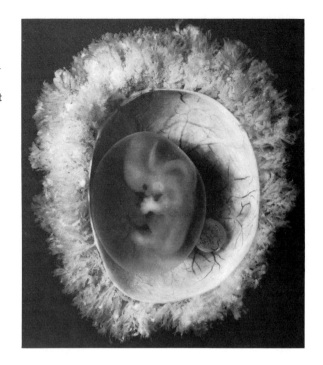

cord. The fully developed organ, consisting of trophoblast tissue and blood vessel-containing mesoderm, is called the **chorion,** and the chorion fuses with the uterine wall to create the placenta. Thus, the placenta has both a maternal portion (the uterine endometrium that is modified during pregnancy) and a fetal component, the chorion. The chorion may be very closely apposed to maternal tissues while still being readily separable (as in the *contact placenta* of the pig), or it may be so intimately integrated that the two tissues cannot be separated without damage to both the mother and the developing fetus (as in the *deciduous placenta* of most mammals, including humans).*

In Figure 6.41, one can observe the relationships between the embryonic and extraembryonic tissues of a 6-week human embryo. The embryo is seen encased in the amnion and is further shielded by the chorion. The blood vessels extending to and from the chorion are readily observable, as are the villi that project from the outer surface of the chorion. These villi contain the blood vessels and allow the chorion to have a large area exposed to the maternal blood. Thus, although fetal and maternal circulatory systems normally never merge, diffusion of soluble substances can occur through the villi (Figure 6.42). In this manner the mother provides the fetus with nutrients and oxygen, and the fetus sends its waste products (mainly carbon dioxide and urea) into the maternal circulation. The blood vessels of the chorionic villi form from extraembryonic mesoderm that enters the mounds of cytotrophoblast tissue called **primary villi** (Figure 6.43). The resulting structures, the **secondary villi,** form during the second week of pregnancy. By the end of the third week, some of this extraembryonic mesoderm has produced blood vessels, and these **tertiary villi** are able to bring nutrients and oxygen from the mother into the embryo.

*There are numerous types of placentas, and the extraembryonic membranes form differently in different orders of mammals (see Cruz and Pedersen, 1991). Although mice and humans gastrulate and implant in a similar fashion, their extraembryonic structures are distinctive. It is very risky to extrapolate developmental phenomena from one group of mammals to another. Even Leonardo da Vinci got caught (Renfree, 1982). His remarkable drawing of the human fetus inside the placenta is stunning art, but poor science: the placenta is that of a cow.

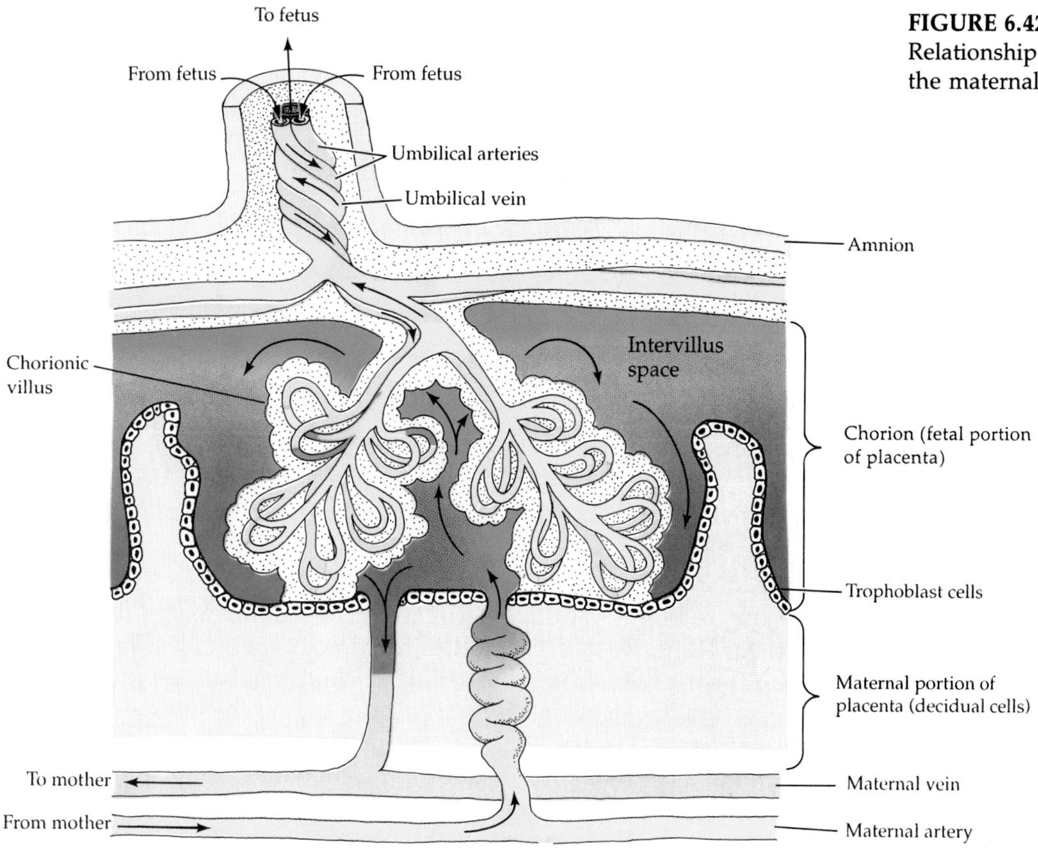

FIGURE 6.42
Relationship of the chorionic villi to the maternal blood in the uterus.

The trophoblast, therefore, is necessary for adhering to and entering the uterine tissues, and the chorion acts to exchange gases and nutrients between mother and fetus. But the chorion has even further importance. It is also an endocrine organ. The syncytiotrophoblast portion of the chorion produces three hormones that are essential for mammalian development. First, it produces **chorionic gonadotropin,** a peptide hormone that is capable of causing other cells in the placenta (and in the maternal ovary) to produce **progesterone.** Progesterone is the steroid hormone that keeps the uterine wall thick and full of blood vessels. In primates, the ovaries can be removed after the first third of pregnancy without harm to the developing fetus because the chorion itself is able to produce the steroid needed to maintain pregnancy (Zander and Münstermann, 1956). Placental progesterone is also used by the fetal adrenal gland as a substrate for the production of biologically important corticosteroid hormones. The third hormone produced by the chorion is **chorionic somatomammotropin** (often called placental lactogen). This hormone is responsible for promoting maternal breast development during pregnancy, thereby enabling milk production later.

Recent studies have indicated that the chorion may have yet another function, namely, the protection of the fetus from the immune response of the mother. A person with a normal immune system recognizes and rejects foreign cells within their body; this fact is demonstrated by the rejection of skin and organ grafts from genetically different individuals. The glycoproteins responsible for this rejection are called the **major histocompatibility antigens,** and these are likely to differ from individual to individual. A human child expresses major histocompatibility antigens from both the mother and the father, and a mother's body may reject her offspring's skin or organs because they contain paternally derived anti-

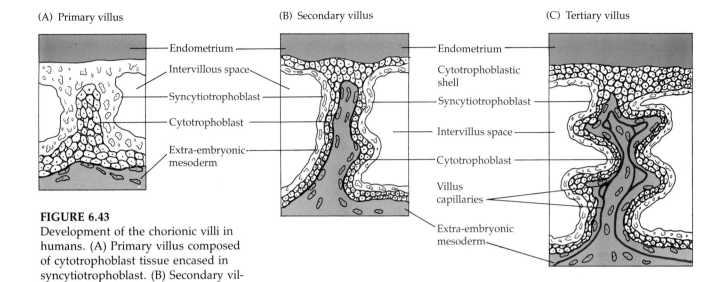

(A) Primary villus   (B) Secondary villus   (C) Tertiary villus

FIGURE 6.43
Development of the chorionic villi in humans. (A) Primary villus composed of cytotrophoblast tissue encased in syncytiotrophoblast. (B) Secondary villus formed when underlying extraembryonic mesoderm penetrates the primary villus. Such secondary villi join adjacent villi to form the cytotrophoblastic shell that will anchor the villi to the endometrium. (C) Within the extraembryonic mesoderm, capillaries form and will connect with branches of the umbilical artery and vein. (After Gilbert, 1989.)

gens. How, then, can a human fetus remain 9 months within the body of its mother? Why doesn't the mother immunologically reject her fetus as she would an organ from that child? It appears that the chorion has evolved several mechanisms by which it can inhibit the immune response against the fetus (Chaouat, 1990). It can secrete soluble proteins that block the production of antibodies, and it can promote the production of certain types of lymphocytes that supress the normal immune response within the uterus. Thus, the functions of the placenta include not only physical support and nutrient exchange, but also regulation of the endocrine and immunological relationships between the mother and the fetus.

SIDELIGHTS & SPECULATIONS

Genes and mammalian gastrulation

Recent experiments have suggested that the maternal and paternal genomes have different roles during mammalian gastrulation. Mouse zygotes can be created that have only sperm-derived chromosomes or only egg-derived chromosomes. The male-derived embryos (androgenones) die with deficiencies in the embryo proper but form well-developed chorions. Conversely, the female-derived embryos (gynogenones) die with deficiencies in their chorions, even though the actual embryo seems normal. It appears that sperm-derived genes are needed for the proper development of the chorion, while egg-derived genes are necessary for the normal development of the embryo itself

(Barton et al., 1984; McGrath and Solter, 1984; Surani et al., 1984). This has been confirmed by making allophenic mice in which blastomeres from a normal 4-cell embryo are aggregated with blastomeres from either androgenetic or gynogenetic embryos. In both cases, cells from both the normal and abnormal embryos were originally seen in all regions of the blastocyst. However, by the end of gastrulation, androgenetic cells are seen almost exclusively in the trophoblast, while gynogenetic cells are hardly ever seen in trophoblast-derived tissues (Thomson and Solter, 1989). This strongly suggests that the maternal and paternal genomes serve distinct functions during early mouse embryogenesis.

In gastrulation, we see an incredibly well coordinated series of cell movements whereby the cleavage-stage blastomeres are rearranged and begin to interact with new neighbors. Moreover, although there are differences between the gastrulation movements of sea urchin, amphibian, avian, and mammalian embryos, certain mechanisms are common to them all. Each group has the problem of bringing the mesodermal and endodermal precursor cells inside the body while surrounding the embryo with ectodermal precursors. Given the different amounts and distributions of yolk, as well as other environmental considerations, each type of organism has evolved a way of accomplishing this goal. The stage is now set for the formation of the first organs.

LITERATURE CITED

Anstrom, J. A., Chin, J. E., Leaf, D. S., Parks, A. L. and Raff, R. A. 1987. Localization and expression of msp130, a primary mesenchyme lineage–specific cell surface protein of the sea urchin embryo. *Development* 101: 255–265.

Azar, Y. and Eyal-Giladi, H. 1979. Marginal zone cells—the primitive streak-inducing component of the primary hypoblast in the chick. *J. Embryol. Exp. Morphol.* 52: 79–88.

Azar, Y. and Eyal-Giladi, H. 1981. Interaction of epiblast and hypoblast in the formation of the primitive streak and the embryonic axis in chick, as revealed by hypoblast rotation experiments. *J. Embryol. Exp. Morphol.* 61: 133–141.

Balinsky, B. I. 1975. *Introduction to Embryology*, 4th Ed. Saunders, Philadelphia.

Barton, S. C., Surani, M. A. and Norris, M. L. 1984. Role of maternal and paternal genomes in mouse development. *Nature* 311: 374–376.

Bellairs, R. 1986. The primitive streak. *Anat. Embryol.* 174: 1–14.

Bisgrove, B. W., Andrews, M. E. and Raff, R. A. 1991. Fibropellins, products of an EGF repeat-containing gene, form a unique extracellular matrix structure that surrounds the sea urchin embryo. *Dev. Biol.* 146: 89–99.

Black, S. D. and Gerhart, J. 1985. Experimental control of the site of embryonic axis formation in *Xenopus laevis* eggs centrifuged before first cleavage. *Dev. Biol.* 108: 310–324.

Black, S. D. and Gerhart, J. 1986. High frequency twinning of *Xenopus laevis* embryos from eggs centrifuged before first cleavage. *Dev. Biol.* 116: 228–240.

Black, S. D. and Vincent, J.-P. 1988. The first cleavage plane and the embryonic axis are determined by separate mechanisms in *Xenopus laevis*. II. Experimental dissociation by lateral compression of the egg. *Dev. Biol.* 128: 65–71.

Boucaut, J.-C., D'Arribère, T., Poole, T. J., Aoyama, H., Yamada, K. M. and Thiery, J. P. 1984. Biologically active synthetic peptides as probes of embryonic development: A competitive peptide inhibition of fibronectin function inhibits gastrulation in amphibian embryos and neural crest cell migration in avian embryos. *J. Cell Biol.* 99: 1822–1830.

Boucaut, J.-C., D'Arribère, T., Li, S. D., Boulekbache, H., Yamada, K. M. and Thiery, J. P. 1985. Evidence for the role of fibronectin in amphibian gastrulation. *J. Embryol. Exper. Morphol.* 89 [Suppl.]: 211–227.

Burke, R. D., Myers, R. L., Sexton, T. L. and Jackson, C. 1991 Cell movements during the initial phase of gastrulation in the sea urchin embryo. *Dev. Biol.* 146: 542–557.

Bursdal, C. A., Damsky, C. H. and Pedersen, R. A. 1993. The role of E-cadherin and integrins in mesoderm differentiation and migration at the mammalian primitive streak. *Development* 118: 829–844.

Chaouat, G. 1990. *The Immunology of the Fetus*. CRC Press, Boca Raton, FL.

Cherr, G. N., Summers, R. G., Baldwin, J. D. and Morrill, J. B. 1992. Preservation and visualization of the sea urchin blastocoelic extracellular matrix. *Microsc. Res. Techn.* 22: 11–22.

Cooke, J. 1986. Permanent distortion of positional system of *Xenopus* embryo by brief early perturbation in gravity. *Nature* 319: 60–63.

Cruz, Y. P. and Pedersen, R. A. 1991. Origin of embryonic and extraembryonic cell lineages in mammalian embryos. In *Animal Applications of Research in Mammalian Development*. Cold Spring Harbor Press, Cold Spring Harbor. Pp. 147–204.

Dan, K. and Okazaki, K. 1956. Cyto-embryological studies of sea urchins. III. Role of secondary mesenchyme cells in the formation of the primitive gut in sea urchin larvae. *Biol. Bull.* 110: 29–42.

D'Arribère, T., Yamada, K. M., Johnson, K. E. and Boucaut, J.-C. 1988. The 140-kD fibronectin receptor complex is required for mesodermal cell adhesion during gastrulation in the amphibian *Pleurodeles waltii*. *Dev. Biol.* 126: 182–194.

D'Arribère, T., Guida, K., Larjava, H., Johnson, K. E., Yamada, K. M., Thiery, J.-P. and Boucaut, J.-C. 1990. In vivo analysis of integrin β1 subunit function in fibronectin matrix assembly. *J. Cell Biol.* 110: 1813–1823.

Delarue, M., D'Arribère, T., Aimar, C. and Boucaut, J.-C. 1985. Bufonid nucleocytoplasmic hybrids arrested at the early gastrula stage lack a fibronectin-containing fibrillar extracellular matrix. *Wilhelm Roux Arch. Dev. Biol.* 194: 275–280.

Duband, J. L. and Thiery, J. P. 1982. Appearance and distribution of fibronectin during chick embryo gastrulation and neurulation. *Dev. Biol.* 94: 337–350.

Ettensohn, C. A. 1985. Gastrulation in the sea urchin embryo is accompanied by the rearrangement of invaginating epithelial cells. *Dev. Biol.* 112: 383–390.

Ettensohn, C. A. 1990. The regulation of primary mesenchyme cell patterning. *Dev. Biol.* 140: 261–271.

Ettensohn, C. A. and McClay, D. R. 1986. The regulation of primary mesenchyme cell migration in the sea urchin embryo: Transplantations of cells and latex beads. *Dev. Biol.* 117: 380–391.

Eyal-Giladi, H., Debby, A. and Harel, N. 1992. The posterior section of the chick's area pellucida and its involvement in hypoblast and primitive streak formation. *Development* 116: 819–830.

Fink, R. D. and McClay, D. R. 1985. Three cell recognition changes accompany the ingression of sea urchin primary mesenchyme cells. *Dev. Biol.* 107: 66–74.

Fisher, M. and Solursh, M. 1977. Glycosaminoglycan localization and role in maintenance of tissue spaces in the early chick embryo. *J. Embryol. Exp. Morphol.* 42: 195–207.

Fisher, M. and Solursh, M. 1979. Influence of local environment on the organization of mesenchymal cells. *J. Embryol. Exp. Morphol.* 49: 295–306.

Fisher, S. J., Cui, T.-Y., Zhang, L., Grahl, K., Guo-Yang, Z., Tarpey, J. and Damsky, C. H. 1989. Adhesive and degradative properties of the human placental cytotrophoblast cells in vitro. *J. Cell Biol.* 109: 891–902.

Galileo, D. S. and Morrill, J. B. 1985. Patterns of cells and extracellular material of the sea urchin *Lytechinus variegatus* (Echinodermata; Echinoidea) embryo, from hatched blastula to late gastrula. *J. Morphol.* 185: 387–402.

Gerhart, J., Ubbels, G., Black, S., Hara, K. and Kirschner, M. 1981. A reinvestigation of the role of the grey crescent in axis formation in *Xenopus laevis*. *Nature* 292: 511–516.

Gerhart, J. and seven others. 1986. Amphibian early development. *BioScience* 36: 541–549.

Gilbert, S. G. 1989. *Pictorial Human Embryology*. University of Washington Press, Seattle.

Gimlich, R. L. and Gerhart, J. C. 1984. Early cellular interactions promote embryonic axis formation in *Xenopus laevis*. *Dev. Biol.* 104: 117–130.

Gustafson, T. and Wolpert, L. 1961. Studies on the cellular basis of morphogenesis in sea urchin embryos: Directed movements of primary mesenchyme cells in normal and vegetalized larvae. *Exp. Cell Res.* 24: 64–79.

Gustafson, T. and Wolpert, L. 1967. Cellular movement and contact in sea urchin morphogenesis. *Biol. Rev.* 42: 442–498.

Hall, H. G. and Vacquier, V. D. 1982. The apical lamina of the sea urchin embryo: Major glycoprotein associated with the hyaline layer. *Dev. Biol.* 89: 168–178.

Hardin, J. 1988. The role of secondary mesenchyme cells during sea urchin gastrulation studied by laser ablation. *Development* 103: 317–324.

Hardin, J. 1990. Context-dependent cell behaviors during gastrulation. *Semin. Dev. Biol.* 1: 335–345.

Hardin, J. D. and Cheng, L. Y. 1986. The mechanisms and mechanics of archenteron elongation during sea urchin gastrulation. *Dev. Biol.* 115: 490–501.

Hardin, J. D. and Keller, R. 1988. The behaviour and function of bottle cells during gastrulation of *Xenopus laevis*. *Development* 103: 211–230.

Hardin, J. and McClay, D. R. 1990. Target recognition by the archenteron during sea urchin gastrulation. *Dev. Biol.* 142: 87–105.

Harkey, M. A. and Whiteley, A. M. 1980. Isolation, culture and differentiation of echinoid primary mesenchyme cells. *Wilhelm Roux Arch. Dev. Biol.* 189: 111–122.

Herbst, C. 1904. Über die zur Entwicklung des Seeigellarven notwendigen anorganischen Stoffe, ihre Rolle und Vertretbarkeit. II. *Wilhelm Roux Arch. Entwicklungsmech. Org.* 17: 306–520.

Holowacz, T. and Elinson, R. P. 1993. Cortical cytoplasm, which induces dorsal axis formation in *Xenopus*, is inactivated by UV irradiation of the oocyte. *Development* 119: 277–285.

Holtfreter, J. 1943. A study of the mechanics of gastrulation: Part I. *J. Exp. Zool.* 94: 261–318.

Holtfreter, J. 1944. A study of the mechanics of gastrulation: Part II. *J. Exp. Zool.* 95: 171–212.

Hörstadius, S. 1939. The mechanics of sea urchin development, studied by operative methods. *Biol. Rev.* 14: 132–179.

Jurand, A. 1974. Some aspects of the development of the notochord in mouse embryos. *J. Embryol. Exp. Morphol.* 32: 1–33.

Karp, G. C. and Solursh, M. 1974. Acid mucopolysaccharide metabolism, the cell surface, and primary mesenchyme cell activity in the sea urchin embryo. *Dev. Biol.* 41: 110–123.

Karp, G. C. and Solursh, M. 1985. Dynamic activity of the filopodia of sea urchin embryonic cells and their role in directed migration of the primary mesenchyme in vitro. *Dev. Biol.* 112: 276–283.

Keller, R. E. 1975. Vital dye mapping of the gastrula and neurola of *Xenopus laevis*. I. Prospective areas and morphogenetic movements of the superficial layer. *Dev. Biol.* 42: 222–241.

Keller, R. E. 1976. Vital dye mapping of the gastrula and neurula of *Xenopus laevis*. II. Prospective areas and morphogenetic movements in the deep layer. *Dev. Biol.* 51: 118–137.

Keller, R. E. 1980. The cellular basis of epiboly: An SEM study of deep cell rearrangement during gastrulation of *Xenopus laevis*. *J. Embryol. Exp. Morphol.* 60: 201–243.

Keller, R. E. 1981. An experimental analysis of the role of bottle cells and the deep marginal zone in the gastrulation of *Xenopus laevis*. *J. Exp. Zool.* 216: 81–101.

Keller, R. E. 1986. The cellular basis of amphibian gastrulation. *In* L. Browder (ed.), *Developmental Biology: A Comprehensive Synthesis*, Vol. 2. Plenum, New York, pp. 241–327.

Keller, R. and Danilchik, M. 1988. Regional expression, pattern and timing of convergence and extension during gastrulation of *Xenopus laevis*. *Development* 103: 193–209.

Keller, R. E. and Schoenwolf, G. C. 1977. An SEM study of cellular morphology, contact, and arrangement as related to gastrulation in *Xenopus laevis*. *Wilhelm Roux Arch. Dev. Biol.* 182: 165–186.

Keller, R. E., Danilchik, M., Gimlich, R. and Shih, J. 1985. Convergent extension by cell intercalation during gastrulation of *Xenopus laevis*. *In* G. M. Edelman (ed.), *Molecular Determinants of Animal Form*. Alan R. Liss, New York, pp. 111–141.

Khaner, O. and Eyal-Giladi, H. 1989. The chick's marginal zone and primitive streak formation. I. Coordinative effect of induction and inhibition. *Dev. Biol.* 134: 206–214.

Kirschner, M. W. and Gerhart, J. C. 1981. Spatial and temporal changes in the amphibian egg. *BioScience* 31: 381–388.

Kochav, S. M. and Eyal-Giladi, H. 1971. Bilateral symmetry in the chick embryo determination by gravity. *Science* 171: 1027–1029.

Lane, M. C. and Solursh, M. 1991. Primary mesenchyme cell migration requires a chondroitin sulfate/dermatan sulfate proteoglycan. *Dev. Biol.* 143: 389–397.

Lane, M. C. Koehl, M. A. R., Wilt, F. and Keller, R. 1993. A role for regulated secretion of apical matrix during epithelial invagination in the sea urchin. *Development* 117: 1049–1060.

Landstrom, U. and Løvtrup, S. 1979. Fate maps and cell differentiation in the amphibian embryo: An experimental study. *J. Embryol. Exp. Morphol.* 54: 113–130.

Larsen, W. J. 1993. *Human Embryology*. Churchill Livingston, New York.

Lash, J. W., Gosfield, E. III, Ostrovsky, D. and Bellairs, R. 1990. Migration of chick blastoderm under the vitelline membrane: The role of fibronectin. *Dev. Biol.* 139: 407–416.

Lawson, K. A., Meneses, J. J. and Pedersen, R. A. 1991. Clonal analysis of epiblast fate during germ layer formation in the mouse embryo. *Development* 113: 891–911.

Lillie, F. R. 1902. Differentiation without cleavage in the egg of the annelid *Chaetopterus pergamentaceous*. *Wilhelm Roux Arch. Entwicklungsmech. Org.* 14: 477–499.

Long, J. A. and Burlingame, P. L. 1937. *A Stereoscopic Atlas of the Chick*. California Laboratory Supply, Berkeley.

Løvtrup, S. 1975. Fate maps and gastrulation in amphibia: A critique of current views. *Can. J. Zool.* 53: 473–479.

Luckett, W. P. 1978. Origin and differentiation of the yolk sac and extraembryonic mesoderm in presomite human and rhesus monkey embryos. *Am. J. Anat.* 152: 59–98.

Lundmark, C. 1986. Roles of bilateral zones of ingressing superficial cells during gastrulation of *Ambystoma mexicanum*. *J. Embryol. Exp. Morph.* 97: 47–62.

McGrath, J. and Solter, D. 1984. Completion of mouse embryogenesis requires both maternal and paternal genomes. *Cell* 37: 179–183.

Miller, J. R., Fraser, S. E. and McClay, D. R. 1993. Extension and dynamics of thin filopodia during gastrulation in the sea urchin embryo. *Mol. Biol. Cell* 4 [Suppl.]: 345a.

Morrill, J. B. and Santos, L. L. 1985. A scanning electron micrographical overview of cellular and extracellular patterns during blastulation and gastrulation in the sea urchin, *Lytechinus variegatus*. *In* R. H. Sawyer and R. M. Showman (eds.), *The Cellular and Molecular Biology of Invertebrate Development*. University of South Carolina Press, pp. 3–33.

Nakatsuji, N., Smolira, M. A. and Wylie, C. C. 1985. Fibronectin visualized by scanning electron microscope immunocytochemistry on the substratum for cell migration in *Xenopus laevis* gastrulae. *Dev. Biol.* 107: 264–268.

New, D. A. T. 1959. Adhesive properties and expansion of the chick blastoderm. *J. Embryol. Exp. Morphol.* 7: 146–164.

Purcell, S. M. and Keller, R. 1993. A different type of amphibian mesoderm morphogenesis in *Ceratophrys ornata*. *Development* 117: 307–317.

Renfree, M. B. 1982. Implantation and placentation. In C. R. Austin and R. V. Short (eds.) *Embryonic and Fetal Development*. Cambridge University Press, Cambridge, pp. 26–69.

Rosenquist, G. C. 1966. A radioautographic study of labeled grafts in the chick blastoderm. Development from primitive-streak stages to stage 12. *Carnegie Inst. Wash. Contr. Embryol.* 38: 31–110.

Rosenquist, G. C. 1972. Endoderm movements in the chick embryo between the short streak and head process stages. *J. Exp. Zool.* 180: 95–104.

Schlesinger, A. B. 1958. The structural significance of the avian yolk in embryogenesis. *J. Exp. Zool.* 138: 223–258.

Schoenwolf, G. G. 1991. Cell movements in the epiblast during gastrulation and neurulation in avian embryos. *In* R. Keller, W. H. Clark, Jr. and F. Griffin (eds.), *Gastrulation: Movements, Patterns, and Molecules.* Plenum, New York, pp. 1–28.

Schoenwolf, G. C., Garcia-Martinez, V. and Diaz, M. S. 1992. Mesoderm movement and fate during amphibian gastrulation and neurulation. *Dev. Dynam.* 193: 235–248.

Schroeder, T. 1981. Development of a "primitive" sea urchin (*Eucidaris tribuloides*): Irregularities in the hyaline layer, micromeres, and primary mesenchyme. *Biol. Bull.* 161: 141–151.

Selleck, M A. J. and Stern, C. D. 1991. Fate mapping and cell lineage analysis of Hensen's node in the chick embryo. *Development* 112: 615–626.

Selleck, M. A. J. and Stern, C. D. 1992. Committment of mesodermal cells in Hensen's node of the chick embryo to notochord and somite. *Development* 114: 403–415.

Shi, D.-L., D'Arribère, T., Johnson, K. E. and Boucaut, J.-C. 1989. Initiation of mesodermal cell migration and spreading relative to gastrulation in the urodele amphibian *Pleurodeles walti. Development* 105: 351–363.

Smith, J. C. and Malacinski, G. M. 1983. The origin of the mesoderm in an anuran, *Xenopus laevis,* and a urodele, *Ambystoma mexicanum. Dev. Biol.* 98: 250–254.

Solursh, M. and Morriss, G. M. 1977. Glycosaminoglycan synthesis in rat embryos during the formation of the primary mesenchyme and neural folds. *Dev. Biol.* 57: 75–86.

Solursh, M. and Revel, J. P. 1978. A scanning electron microscope study of cell shape and cell appendages in the primitive streak region of the rat and chick embryo. *Differentiation* 11: 185–190.

Spratt, N. T., Jr. 1946. Formation of the primitive streak in the explanted chick blastoderm marked with carbon particles. *J. Exp. Zool.* 103: 259–304.

Spratt, N. T., Jr. 1947. Regression and shortening of the primitive streak in the explanted chick blastoderm. *J. Exp. Zool.* 104: 69–100.

Spratt, N. T., Jr. 1963. Role of the substratum, supracellular continuity, and differential growth in morphogenetic cell movements. *Dev. Biol.* 7: 51–63.

Spratt, N. T. Jr. and Haas, H. 1960. Integrative mechanisms in development of early chick blastoderm. I. Regulated potentiality of separate parts. *J. Exp. Zool.* 145: 97–138.

Stern, C. D. 1991. Mesoderm formation in the chick embryo revisited. *In* R. Keller, W. H. Clark, Jr. and F. Griffin (eds.), *Gastrulation: Movements, Patterns, and Molecules.* Plenum, New York, pp. 29–41.

Stern, C.D. 1992. Mesoderm induction and development of the embryonic axis in amniotes. *Trends Genet.* 8: 158–163.

Stern, C. D. and Canning, D. R. 1988. Gastrulation in birds: A model system for the study of animal morphogenesis. *Experientia* 44: 651–657.

Stern, C. D. and Canning, D. R. 1990. Origin of cells giving rise to mesoderm and endoderm in chick embryo. *Nature* 343: 273–275.

Stern, C. D., Ireland, G. W., Herrick, S. E., Gherardi, E., Gray, J., Perryman, M. and Stoker, M. 1990. Epithelial scatter factor and development of the chick embryonic axis. *Development* 110: 1271–1284.

Sugiyama, K. 1972. Occurrence of mucopolysaccharides in the early development of the sea urchin embryo and its role in gastrulation. *Dev. Growth Differ.* 14: 62–73.

Surani, M. A. H., Barton, S. C. and Norris, M. L. 1984. Development of reconstituted mouse eggs suggests imprinting of the genome during gametogenesis. *Nature* 308: 548–550.

Thomson, J. A. and Solter, D. 1989. The developmental fate of androgenetic, parthenogenetic, and gynogenetic cells in chimeric gastrulating mouse embryos. *Genes Dev.* 2: 1344–1351.

Trinkaus, J. P. 1984. *Cells into Organs: The Forces that Shape the Embryo,* 2nd Ed. Prentice-Hall, Englewood Cliffs, NJ.

Vakaet, L. 1984. The initiation of gastrula ingression in the chick blastoderm. *Amer. Zool.* 24: 555–562.

Vakaet, L. 1985. Morphogenetic movements and fate maps in the avian blastoderm. *In* G. M. Edelman (ed.), *Molecular Determinants of Animal Form.* Alan R. Liss, New York, pp. 99–109.

Vincent, J. P. and Gerhart, J. C. 1987. Subcortical rotation in *Xenopus* eggs: An early step in embryonic axis formation. *Dev. Biol.* 123: 526–539.

Vincent, J. P., Oster, G. F. and Gerhart, J. C. 1986. Kinematics of gray crescent formation in *Xenopus* eggs. Displacement of subcortical cytoplasm relative to the egg surface. *Dev. Biol.* 113: 484–500.

Vogt, W. 1929. Gestaltungsanalyse am Amphibienkeim mit ortlicher Vitalfarbung. II. Teil Gastrulation und Mesodermbildung bei Urodelen und Anuren. *Wilhelm Roux Arch. Entwicklungsmech. Org.* 120: 384–706.

von Übisch, L. 1939. Keimblattchimarenforschung an Seeigellarven. *Biol. Rev. Cambr. Phil. Soc.* 14: 88–103.

Waddington, C. H. 1932. Experiments in the development of chick and duck embryos cultivated in vitro. *Philos. Trans. R. Soc. Lond.* [B] 13: 221.

Wessel, G. M. and McClay, D. R. 1985. Sequential expression of germ layer specific molecules in the sea urchin embryo. *Dev. Biol.* 111: 451–463.

Wessel, G. M., Marchase, R. B. and McClay, D. R. 1984. Ontogeny of the basal lamina in the sea urchin embryo. *Dev. Biol.* 103: 235–245.

Wilson, P. and Keller, R. 1991. Cell rearrangement during gastrulation of *Xenopus:* direct observation of cultured explants. *Development* 112: 289–300.

Winkelbauer, R. and Nagel, M. 1991. Directional mesoderm cell migration in the *Xenopus* gastrula. *Dev. Biol.* 148: 573–589.

Winkelbauer, R., Selchow, A., Nagel, M., Stoltz, C. and Angres, B. 1991. Mesodermal cell migration in the *Xenopus* gastrula. *In* R. Keller, W. H. Clark, Jr. and F. Griffin (eds.), *Gastrulation: Movements, Patterns, and Molecules.* Plenum, New York, pp. 147–168.

Zander, J. and von Müstermann, A. M. 1956. Progesteron in menschlichem Blut und Geweben. III. *Klin. Wochenschr.* 34: 944–953.

7

Early vertebrate development
Neurulation and the ectoderm

For the real amazement, if you wish to be amazed, is this process. You start out as a single cell derived from the coupling of a sperm and an egg; this divides in two, then four, then eight, and so on, and at a certain stage there emerges a single cell which has as all its progeny the human brain. The mere existence of such a cell should be one of the great astonishments of the earth. People ought to be walking around all day, all through their waking hours calling to each other in endless wonderment, talking of nothing except that cell.

LEWIS THOMAS (1979)

Even more appealing than virgin forest was the jungle lying before me at that moment: the central nervous system.

RITA LEVI-MONTALCINI (1988)

In 1828, Karl Ernst von Baer, the foremost embryologist of his day,* proclaimed, "I have two small embryos preserved in alcohol, that I forgot to label. At present I am unable to determine the genus to which they belong. They may be lizards, small birds, or even mammals." Figure 7.1 allows us to appreciate his quandary, and it illustrates von Baer's four general laws of embryology. From his detailed study of chick development and his comparison of those embryos with embryos of other vertebrates, von Baer derived four generalizations, illustrated here with some vertebrate examples.

1. *The general features of a large group of animals appear earlier in the embryo than do the specialized features.* All developing vertebrates (fishes, reptiles, amphibians, birds, and mammals) appear very similar shortly after gastrulation. It is only later in development that the special features of class, order, and finally species emerge (Figure 7.1). All vertebrate embryos have gill arches, notochords, spinal cords, and pronephric kidneys.

2. *Less general characters are developed from the more general, until finally the most specialized appear.* All vertebrates initially have the same type of skin. Only later does the skin develop fish scales, reptilian scales, bird feathers, or the hair, claws, and nails of mammals. Similarly, the early development of the limb is essentially the same in all vertebrates. Only later do the differences between legs, wings, and arms become apparent.

3. *Each embryo of a given species, instead of passing through the adult stages of other animals, departs more and more from them.* The visceral clefts of embryonic birds and mammals do not resemble the gill slits of adult fishes in detail. Rather, they resemble those visceral clefts of *embryonic* fishes and other *embryonic* vertebrates. Whereas fishes preserve and elaborate these clefts into true gill slits, mammals convert them into structures such as the eustachian tubes (between the ear and mouth).

4. *Therefore, the early embryo of a higher animal is never like [the adult of] a lower animal, but only like its early embryo.*

---

*K. E. von Baer discovered the notochord, the mammalian egg, and the human egg, as well as making the conceptual advances described here. His work will be discussed further in Chapter 23.

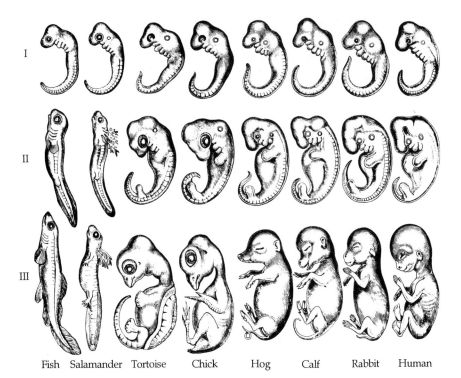

Fish   Salamander   Tortoise   Chick   Hog   Calf   Rabbit   Human

FIGURE 7.1
Illustration of von Baer's law. Early vertebrate embryos exhibit features common to the entire subphylum. As development progresses, embryos become recognizable as members of their class, their order, their family, and finally their species. (After Romanes, 1901.)

Von Baer saw that different groups of animals share certain common features during early embryonic development and that their features become more and more characteristic of the species as development proceeds. Human embryos never pass through a stage equivalent to an adult fish or bird; rather, human embryos initially share characteristics in common with fish and avian embryos. Later, the mammalian and other embryos diverge, none of them passing through the stages of the other.

Von Baer also recognized that there is a common *pattern* to all vertebrate development: the three germ layers give rise to different organs, and this derivation of the organs is constant whether the organism is a fish, a frog, or a chick. **Ectoderm** forms skin and nerves; **endoderm** forms respiratory and digestive tubes; and **mesoderm** forms connective tissue, blood cells, heart, the urogenital system, and parts of most of the internal organs. In this chapter we follow the early development of ectoderm; it and the following chapter focus on the formation of the nervous system in vertebrates. Chapter 9 will follow the early development of endodermal and mesodermal organs.

Neurulation: An overview

"What is perhaps the most intriguing question of all is whether the brain is powerful enough to solve the problem of its own creation." So Gregor Eichele (1992) ended a recent review of research on mammalian brain development. The construction of an organ that perceives, thinks, loves, hates, remembers, changes, and coordinates our conscious and unconscious bodily processes is undoubtedly the most challenging of all developmental enigmas. A combination of genetic, cellular, and organismal approaches is giving us a preliminary understanding of how the basic anatomy of the mammalian brain becomes ordered.

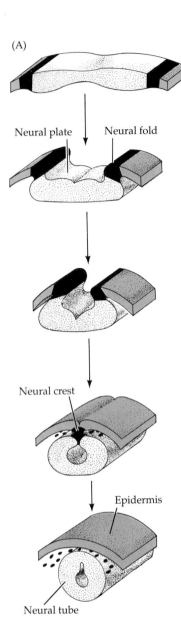

(A)

Neural plate   Neural fold

Neural crest

Epidermis

Neural tube

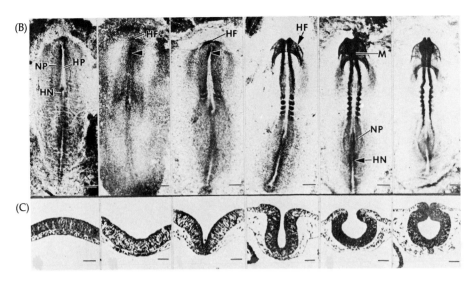

(B)

(C)

FIGURE 7.2

Neurulation in amphibians and amniotes. (A) Diagrammatic representation of neural tube formation. The ectodermal cells are represented either as precursors of the neural crest (black) or as precursors of the epidermis (color). The ectoderm folds in at the most dorsal point, forming an outer epidermis and an inner neural tube connected by neural crest cells. (B) Photomicrographs of neurulation in a 2-day chick embryo. (C) Neural tube formation seen in transverse cross sections of the chick embryo at the region of the future midbrain. Each photograph in C corresponds to the one above it. HF, head fold; HP, head process; HN, Hensen's node; M, midbrain; NP, neural plate. (Photomicrographs courtesy of R. Nagele.)

In vertebrates, gastrulation creates an embryo with an internal endodermal layer, an intermediate mesodermal layer, and an external ectoderm. The interaction between the dorsal mesoderm and its overlying ectoderm is one of the most important interactions of all development, for it initiates **organogenesis,** the creation of specific tissues and organs. In this interaction, the chordamesoderm directs the ectoderm above it to form the hollow **neural tube,** which will differentiate into the brain and spinal cord. The action by which the flat layer of ectodermal cells is transformed into a hollow tube is called **neurulation,** and an embryo undergoing such changes is called a **neurula.** The events of neurulation are diagrammed in Figure 7.2. The original ectoderm is divided into three sets of cells: (1) the internally positioned neural tube, which will form the brain and spinal cord, (2) the externally positioned epidermis of the skin, and (3) the neural crest cells, which migrate from the region that had connected the neural tube and epidermis and which will generate the peripheral neurons and glia, the pigment cells of the skin, and several other cell types. The phenomenon of embryonic induction, which initiates neurulation in the dorsal region of the embryo will be detailed in Chapter 16. In this chapter, we are concerned with the response of the various ectodermal tissues.

FIGURE 7.3

Four views of neurulation in an amphibian embryo, showing early (left), middle (center), and late (right) neurulae in each case. (A) Transverse section through the center of the embryo. (B) The same sequence looking down on the dorsal surface of the whole embryo. (C) Sagittal section through the median plane of the embryo. (D) Computer simulation of the three-dimensional constriction, extension, and rising of the neural plate. (A–C after Balinsky, 1975; D After Jacobson and Gordon, 1976.)

The process of neurulation in frog embryos is depicted in Figure 7.3. The mechanisms of neural tube formation appear to be similar in amphibians, reptiles, birds, and mammals (Gallera, 1971), so we will be considering various groups as we proceed through our survey.* The first indication that a region of ectoderm is destined to become neural tissue is a

*Among the vertebrates, fishes generate their neural tube in a different manner. Fish neural tubes do not form from an infolding of the overlying ectoderm, but rather from a solid cord of cells, which is induced to sink into the embryo. This cord subsequently hollows out (cavitates) to form the neural tube. In birds and mammals, the most posterior portion of the neural tube forms by the aggregation of cells into a medullary cord which then cavitates to form a neural tube that connects to the main neural tube. This is called **secondary neurulation** (see Schoenwolf, 1991b).

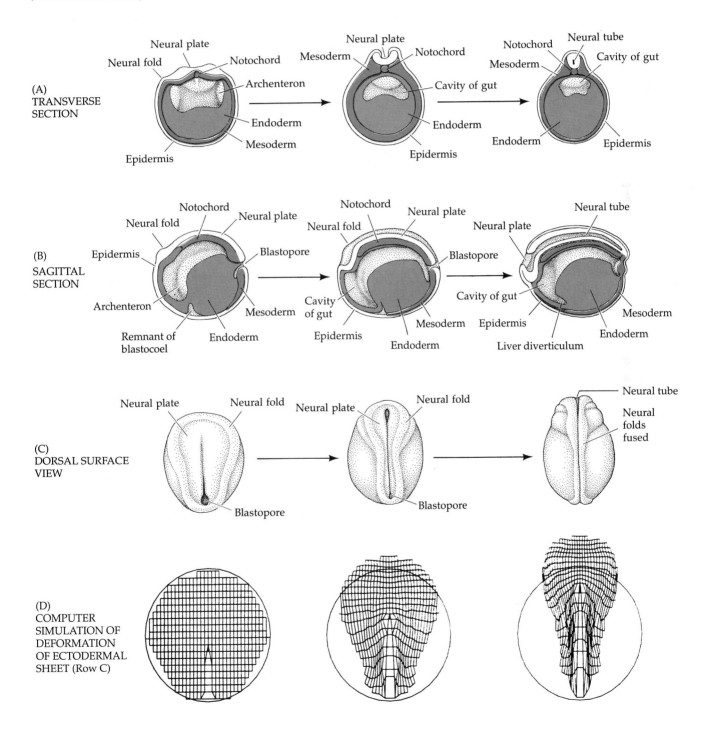

(A) TRANSVERSE SECTION

(B) SAGITTAL SECTION

(C) DORSAL SURFACE VIEW

(D) COMPUTER SIMULATION OF DEFORMATION OF ECTODERMAL SHEET (Row C)

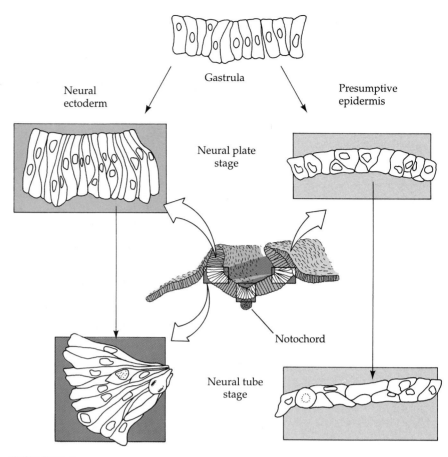

FIGURE 7.4

Shape changes during neurulation in the salamander. At the end of gastrulation, the cells of the ectoderm form a uniform epithelium. During neurulation, neural epithelial cells elongate to form the neural plate. In the three hinge regions (the MHP and the two lateral DHLPs; white areas in center diagram), the neural plate cells change their length and constrict at their apices. This change is fundamental to the folding of the neural tube. Presumptive epidermal cells flatten throughout neurulation. (After Burnside, 1971.)

change in cell shape (Figure 7.4). Midline ectodermal cells become elongated, while the cells destined to form the epidermis become more flattened. The elongation of dorsal ectodermal cells causes these prospective neural regions to rise above the surrounding ectoderm, thereby creating the **neural plate.** As much as 50 percent of the ectoderm is included in this plate. Shortly thereafter, the edges of the neural plate thicken and move upward to form the **neural folds,** while a U-shaped **neural groove** appears in the center of the plate, dividing the future right and left sides of the embryo (Figures 7.3 and 7.5). The neural folds migrate toward the midline of the embryo, eventually fusing to form the neural tube beneath the overlying ectoderm. The cells at the dorsalmost portion of the neural tube become the neural crest cells.

The mechanics of neurulation

Neurulation can be divided into four distinct but spatially and temporally overlapping stages (Schoenwolf, 1991a): (1) the formation of the neural

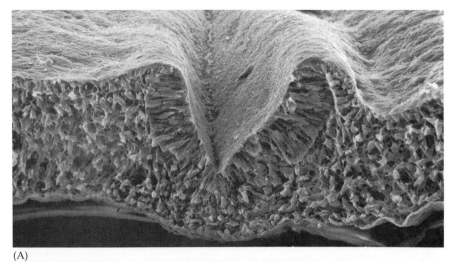

(A)

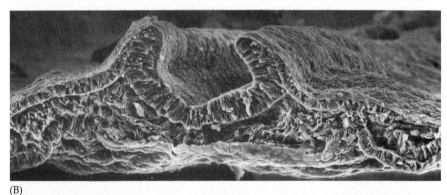

(B)

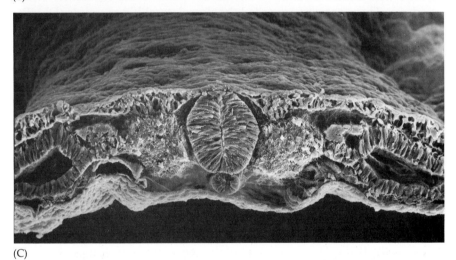

(C)

FIGURE 7.5
Scanning electron micrograph of neural tube formation in the chick embryo. (A) Neural groove surrounded by mesenchymal cells. (B) Elongated neural epithelial cells form a tube as the flattened epidermal cells are brought to the midline of the embryo. The MHP cells form a hinge at the bottom of the tube, while the neural plate cells attached to the basal area of the surface ectoderm form the dorsolateral hinge regions. These three hinges can be seen as furrows. (C) Neural tube formation is completed. The cells that had been the neural plate are now inside the embryo. The presumptive epidermis lies atop the tube, and the neural tube is flanked on its sides by the mesodermal somites and on the bottom by the notochord. (Photographs courtesy of K. W. Tosney.)

plate, (2) the shaping of the neural plate, (3) the bending of the neural plate to form the neural groove, and (4) the closure of the neural groove to form the neural tube.

The neural plate

The *formation* of the neural plate as a region distinct from the other ectodermal cells will be discussed in detail in Chapters 16 and 17. It is generally believed that the underlying dorsal mesoderm (in collaboration with other regions of the embryo) signals the ectodermal cells above it to develop into the columnar neural plate cells (Smith and Schoenwolf, 1989; Keller et al., 1992). As a result of this **neural induction,** the cells of the prospective neural plate are distinguished from the surrounding ectoderm, which is destined to become the epidermis.

The *shaping* of the neural plate involves forces that are intrinsic to the neural plate cells. As the cells become more columnar, they cause a narrowing of the neural plate, but the most important shaping of the neural plate is performed by the cells at the midline of the neural plate that lie directly above the notochord. In amphibians, these cells are called the **notoplate** and are derived from the noninvoluting marginal zone (Jacobson, 1981) In birds and mammals, these midline neural plate cells are called the **median hinge point** (MHP) **cells** and are derived from the anterior midline of Hensen's node and the epiblast region just rostral to it (Schoenwolf, 1991a,b). As Hensen's node regresses, these cells stream caudally to become what will be the ventral floor plate of the neural tube. In both amphibians and amniotes, these midline neural plate cells undergo convergent extension by intercalating several layers of cells into a few layers (Jacobson and Sater, 1988; Schoenwolf and Alvarez, 1989). In so doing, they lengthen and narrow the neural plate (Figure 7.3C).

The *bending* of the neural plate is accomplished by forces both intrinsic to and extrinsic to the neural plate cells. In chicks, the neural plate begins bending even as it is being shaped. The MHP cells become anchored to the notochord beneath them and form a hinge that forms a furrow in the dorsal midline. The notochord induces these cells to decrease their height and to become wedge-shaped (van Straaten et al., 1988; Smith and Schoenwolf, 1989). The cells lateral to the MHP do not undergo such a change (Figures 7.4 and 7.5). Shortly thereafter, two other hinge regions form furrows near the connection of the neural plate to the remainder of the ectoderm. These regions are called the **dorsolateral hinge points** (DLHPs) and they are anchored to the surface ectoderm of the neural fold. These cells *increase* their height and become wedge-shaped. This dorsolateral (but not the medial) wedging is intimately linked to changes in cell shape, and microtubules and microfilaments are both involved in these changes. Colchicine, an inhibitor of microtubule polymerization, inhibits the elongation of these cells, while cytochalasin B, an inhibitor of microfilament formation, prevents the apical constriction of these cells, thereby inhibiting wedge formation (Burnside, 1971, 1973; Karfunkel, 1972; Nagele and Lee, 1980, 1987). After the initial furrowing of the neural plate, the plate bends around these hinge regions. Each hinge acts as a pivot that directs the rotation of the cells around it (Smith and Schoenwolf, 1991).

Meanwhile, extrinsic forces are also at work. The surface ectoderm of the chick embryo pushes towards the center of the embryo, providing another motive force for the bending of the neural tube (Alvarez and Schoenwolf, 1992). This movement of the presumptive epidermis may also be important for ensuring that the neural tube folds *into* the embryo and not outwardly; for if neural plates are isolated from the rest of the embryo, they tend to roll inside out (Schoenwolf, 1991a). Figure 7.5 shows this process of epidermal migration toward the midline of the embryo.

Closure of the neural tube

The neural tube closes as the paired neural folds are brought together at the dorsal midline. The folds adhere to each other, and the cells from the two folds merge. In some species, the cells at this junction form the neural crest cells. In birds, the neural crest cells do not migrate from the dorsal region until after the neural tube has been closed at that site. However, in mammals, the cranial neural crest cells (which form facial and neck structures) migrate while the neural folds are elevating, (i.e., prior to neural tube closure), while in the spinal cord region, the crest cells wait until closure has occurred (Nichols, 1981; Erickson and Weston, 1983).

The formation of the neural tube does not occur simultaneously throughout the ectoderm. This is best seen in those vertebrates (such as birds and mammals) whose body axis is elongated prior to neurulation. Figure 7.6 depicts neurulation in a 24-hour chick embryo. Neurulation in the cephalic (head) region is well advanced while the caudal (tail) region of the embryo is still undergoing gastrulation. Regionalization of the neural tube also occurs as a result of changes in the shape of the tube. In the cephalic end (where the brain will form), the wall of the tube is broad and thick. Here, a series of swellings and constrictions define the various brain

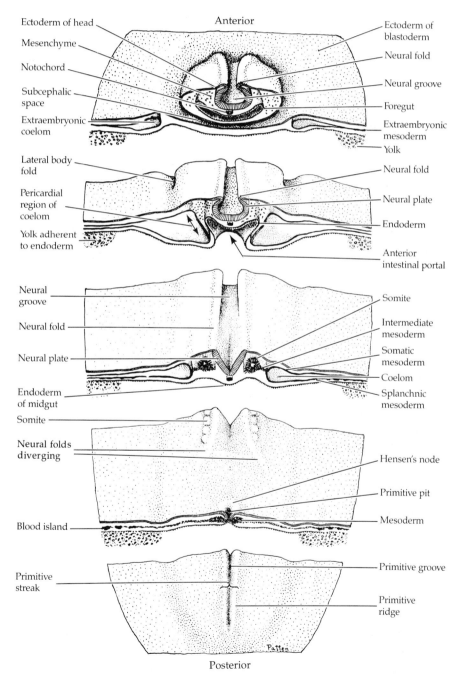

FIGURE 7.6
Stereogram of a 24-hour chick embryo. Cephalic portions are finishing neurulation while the caudal portions are still undergoing gastrulation. (From Patten, 1971; after Huettner, 1949.)

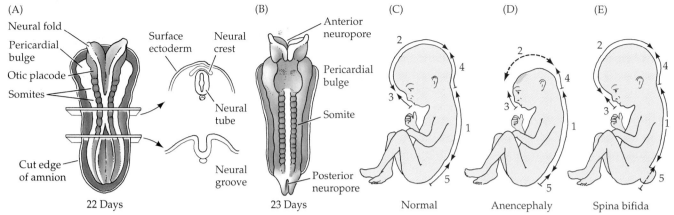

FIGURE 7.7
Neurulation in human embryos. (A) Dorsal and transverse sections of 22-day human embryo initiating neurulation. Both anterior and posterior neuropores are open to the amniotic fluid. (B) Dorsal view of neurulating human embryo a day later. The anterior neuropore region is closing while the posterior neuropore remains open. (C) Regions of neural tube closure postulated by genetic evidence (superimposed on newborn body). (D) Anencephaly due to failure of neural plate fusion in region 2. (E) Spina bifida due to failure of region 5 to fuse (or for the posteriormost neuropore to close). (C–E after Van Allen et al., 1993.)

compartments. Caudal to the head region, however, the neural tube remains a simple tube that tapers off toward the tail. The two open ends of the neural tube are called the **anterior neuropore** and the **posterior neuropore**.

Unlike neurulation in chicks, neural tube closure in mammals is initiated at several places along the anterior–posterior axis (Golden and Chernoff, 1993; Van Allen et al., 1993). Different neural tube defects are caused when various parts of the neural tube fail to close (Figure 7.7). Failure to close the human posterior neural tube regions at day 27 (or the subsequent rupture of the posterior neuropore shortly thereafter) results in a condition called **spina bifida,** the severity of which depends upon how much of the spinal cord remains open. Failure to close the *anterior* neural tube regions results in a lethal condition, **anencephaly.** Here, the forebrain remains in contact with the amniotic fluid and subsequently degenerates. Fetal forebrain development ceases and the vault of the skull fails to form. This abnormality is not that rare in humans, occurring in about 0.1 percent of all pregnancies. Neural tube closure defects can often be detected during pregnancy by various physical and chemical tests.

The neural tube eventually forms a closed cylinder that separates from the surface ectoderm. This separation is thought to be mediated by the expression of different cell adhesion molecules. Although the cells that will become the neural tube originally express E-cadherin, they stop producing this protein as the neural tube forms and instead synthesize N-cadherin and N-CAM (see Figure 3.17). As a result, the two tissues no longer adhere to one another. If the surface ectoderm is made to express N-cadherin (by injecting N-cadherin mRNA into one cell of a 2-cell *Xenopus* embryo), the separation of the neural tube from the presumptive epidermis is dramatically impeded (Detrick et al., 1990; Fujimori et al., 1990).

Differentiation of the neural tube

The differentiation of the neural tube into the various regions of the central nervous system occurs simultaneously in three different ways. On the gross anatomical level, the neural tube and its lumen bulge and constrict to form the chambers of the brain and the spinal cord. At the tissue level, the cell populations within the wall of the neural tube rearrange themselves in various ways to form the different functional regions of the brain and the spinal cord. Finally, on the cellular level, neuroepithelial cells themselves differentiate into the numerous types of neurons and supportive (glial) cells present in the body.

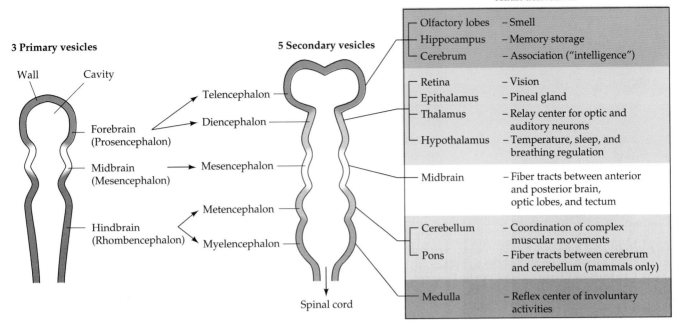

3 Primary vesicles

Wall
Cavity

Forebrain
(Prosencephalon)

Midbrain
(Mesencephalon)

Hindbrain
(Rhombencephalon)

Telencephalon

Diencephalon

Mesencephalon

Metencephalon

Myelencephalon

5 Secondary vesicles

Spinal cord

Adult derivatives

Olfactory lobes	– Smell
Hippocampus	– Memory storage
Cerebrum	– Association ("intelligence")
Retina	– Vision
Epithalamus	– Pineal gland
Thalamus	– Relay center for optic and auditory neurons
Hypothalamus	– Temperature, sleep, and breathing regulation
Midbrain	– Fiber tracts between anterior and posterior brain, optic lobes, and tectum
Cerebellum	– Coordination of complex muscular movements
Pons	– Fiber tracts between cerebrum and cerebellum (mammals only)
Medulla	– Reflex center of involuntary activities

FIGURE 7.8
Early human brain development. The three primary brain vesicles are subdivided as development continues. To the right are the adult derivatives formed by the walls and cavities of the brain. (After Moore and Persaud, 1993.)

Formation of brain regions

The early development of most vertebrate brains is similar, but because the human brain is probably the most organized piece of matter in the solar system and the most interesting organ in the animal kingdom, we will concentrate on the development that is supposed to make *Homo sapient*.

The early mammalian neural tube is a straight structure. However, even before the posterior portion of the tube has formed, the most anterior portion of the tube is undergoing drastic changes. In this anterior region, the neural tube balloons into three primary vesicles (Figure 7.8): forebrain (**prosencephalon**), midbrain (**mesencephalon**), and hindbrain (**rhombencephalon**). By the time the posterior end of the neural tube closes, secondary bulges—the **optic vesicles**—have extended laterally from each side of the developing forebrain.

The forebrain becomes subdivided into the anterior **telencephalon** and the more caudal **diencephalon.** The telencephalon will eventually form the **cerebral hemispheres,** and the diencephalon will form the thalamic and hypothalamic brain regions as well as the region that receives neural input from the eyes. The mesencephalon (midbrain) does not become subdivided, and its lumen eventually becomes the cerebral aqueduct. The rhombencephalon becomes subdivided into a posterior **myelencephalon** and a more anterior **metencephalon.** The myelencephalon eventually becomes the **medulla oblongata,** whose neurons generate the nerves that regulate respiratory, gastrointestinal, and cardiovascular movements. The metencephalon gives rise to the **cerebellum,** the part of the brain responsible for coordinating movements, posture, and balance. The hindbrain develops a segmental pattern that specifies the places where certain nerves originate. Periodic swellings called **rhombomeres** divide the rhombencephalon into smaller compartments. These rhombomeres represent separate developmental "territories" in that cells within each rhombomere can mix freely within the rhombomere, but not with cells from adjacent rhombomeres (Guthrie and Lumsden, 1991). Moreover, each rhombomere has a different developmental fate. This has been most extensively studied in the chick, where the first neurons appear in the even-numbered rhom-

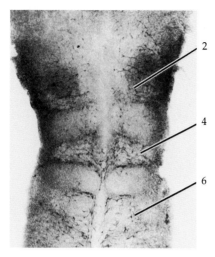

FIGURE 7.9
A 2-day embryonic chick hindbrain splayed to show lateral walls. Neurons were visualized with an antibody staining neurofilament proteins. Rhombomeres 2, 4, and 6 are distinguished by the high density of axons at this early developmental stage. (From Lumsden and Keynes, 1989, courtesy of A. Keynes.)

bomeres, r2, r4, and r6 (Lumsden and Keynes, 1989; Figure 7.9). Neurons from r2 ganglia form the fifth (trigeminal) cranial nerve, those from r4 form the seventh (facial) and eighth (vestibuloacoustic) cranial nerves, while the ninth (glossopharyngeal) cranial nerve exits from r6.

The ballooning of the early embryonic brain is remarkable in its rate, in its extent, and in its being the result primarily of an increase in cavity size, not tissue growth. In chick embryos, the brain volume expands 30-fold between days 3 and 5 of development. This rapid expansion is thought to be caused by positive fluid pressure pressing against the walls of the neural tube by the fluid within it. It might be expected that this fluid pressure might have been dissipated by the spinal cord, but this does not appear to happen. Rather, as the neural folds close in the region between the presumptive brain and spinal cord, the surrounding dorsal tissues push in to constrict the tube at the base of the brain (Figure 7.10; Schoenwolf and Desmond, 1984; Desmond and Schoenwolf, 1986; Desmond and Field, 1992). This occlusion (which also occurs in the human embryo) effectively separates the presumptive brain region from the future spinal cord (Desmond, 1982). If one removes the fluid pressure in the anterior portion of such an occluded neural tube, the chick brain enlarges at a much slower rate and contains many fewer cells than are found in normal, control embryos. The occluded region of the neural tube reopens after the initial rapid enlargement of the brain ventricles.

(A) (B) (D)

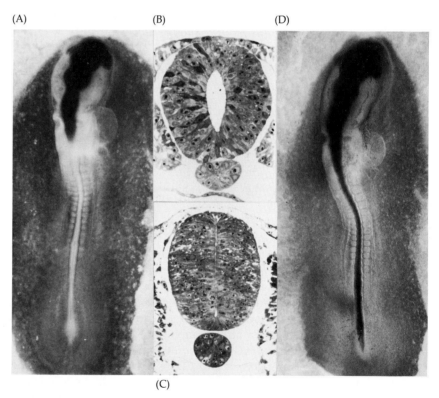

(C)

FIGURE 7.10
Occlusion of the neural tube to allow expansion of the future brain region. (A) Dye injected into the anterior portion of a 3-day chick neural tube fills the brain region but does not pass into the spinal region. (B,C) Section of the chick neural tube at the base of the brain (B) before occlusion and (C) during occlusion. (D) Reopening of occlusion after initial brain enlargement allows dye to pass from the brain region into the spinal cord region. (Courtesy of M. Desmond.)

Molecular regulators of development: The Pax proteins

The molecules involved in the construction of the neural tube are largely unknown, despite the fact that neural tube defects in humans (such as spina bifida and anencephaly) have a combined incidence of about one in every 500 live births. While it is believed that such malformations are due to a combination of environmental and genetic factors, the genetic factors are only now being found. Some of the most important of these genes encode a series of transcription factors known as the **paired-box** (usually abbreviated as **Pax**) **proteins.** These proteins resemble a transcription factor (*paired*) found in the *Drosophila* embryo, and they were found by searching for mammalian genes that were similar to the *Drosophila paired* gene (Bopp et al., 1986; Kessel and Gruss, 1990). Some of these *Pax* genes encode proteins that are expressed along the mouse dorsal axis and whose wild-type form is necessary for proper neural development (Figure 7.11).

The **Pax-3** protein is critical for neural tube closure. The gene for this protein is expressed along the dorsal neural tube and neural crest cells just before neural tube closure. Mice with mutations in both of their *Pax-3* genes

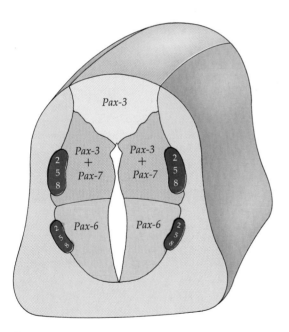

FIGURE 7.11

Pax *genes in the mammalian neural tube. Patterns of* Pax *gene expression in an 11-day mouse embryonic spinal cord. Dorsal and ventral regions are designated by* Pax-3 *and* Pax-6, *respectively. Deficiencies of these genes cause deficiencies in the human and mouse nervous systems. (After Gruss and Walther, 1992.)*

die in utero from a syndrome characterized by the failure of neural tube closure (at both ends) and the failure of neural crest cell migration (Auerbach, 1954; Goulding et al., 1991). The mutations in the various *Pax-3* alleles were in the DNA-binding region of Pax-3 protein, which is thought to be necessary for its function (Epstein et al., 1991). Mice having one mutant allele at this locus have the *Splotch* phenotype characterized by the lack of neural crest cell-derived pigmentation at the extremities of the tail and feet as well as a white belly spot on an otherwise dark mouse. Humans heterozygous for mutations of the *Pax-3* gene have a condition called Waardenburg's syndrome, which is characterized by incomplete pigmentation and congenital deafness (Baldwin et al., 1992; Tassabehji et al., 1992). The **Pax-6** protein is critical for the development of the eyes and nose. When Pax-6 protein is absent in mice and rats, these organs do not form. Humans heterozygous for the *Pax-6* gene lack irises (Hill et al., 1991; Jordan et al., 1992).

In the above cases, developmental anomalies in mammals could be traced back to mutations in genes that are also necessary for *Drosophila* development. (*Drosophila*, it should be noted, does not use the *paired* gene to create a neural tube, and flies have no structure similar to neural crest cells. Nature is using the homologous DNA-binding protein for a different developmental function.) Although finding developmental mutations is difficult in mice, it is now possible (through gene-targeting procedures) to construct mice with specific gene mutations. In *Drosophila*, the *wingless* gene product maintains the boundaries between segments by diffusing to the neighboring cell to activate the *engrailed* gene. Since mice were found to have homologues to both these genes, gene targeting was used to knock out *wnt-1*, a mouse homologue to the *wingless* gene. Mice deficient in *wnt-1* genes lack their midbrain regions as well as the cerebellar portion of their hindbrains (Figure 7.12A,B; McMahon and Bradley, 1990; Thomas and Cappecchi, 1990). However, there is a discrepancy between the places where *wnt-1* is usually expressed and the parts of the brain that fail to form (Figure 7.12C). For instance, the cerebellum does not express *wnt-1*. Why is it missing in the mutants? The answer was found by McMahon and colleagues (1992): the missing cells are those that express the mouse homologue of *engrailed*. Just as in *Drosophila*, it appears that *wnt-1* activates the expression of the *engrailed* gene. In mice (and probably other vertebrates as well), *engrailed* is crucial in forming the cerebellum, and if *engrailed* activity is not turned on in those cells, the cells die. In *wnt-1*-deficient embryos, the entire midbrain–cerebellum domain of *engrailed* expression is deleted (Figure 7.12D,E). The current evidence suggests that the mouse *wnt-1* protein plays a role analogous to the one played in *Drosophila*, where *wingless* protein is needed to activate the *engrailed* gene. It is possible, though, that the wnt-1 protein may act as a specific growth factor for these cells. Either way, it appears that *wnt-1* and probably *engrailed* are needed for

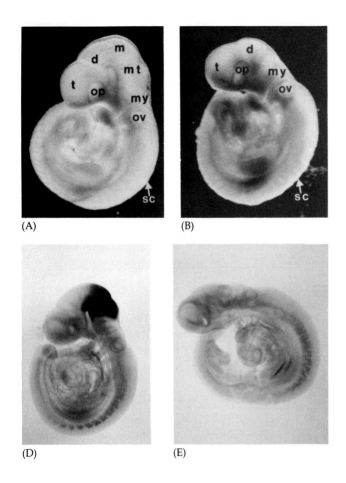

(A)　　　　　　　　(B)

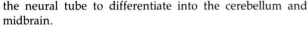

(D)　　　　　　　　(E)

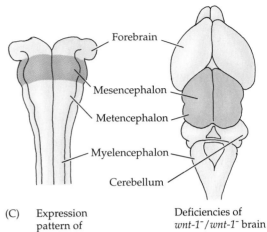

(C)　Expression
　　　pattern of
　　　wnt-1 gene

Deficiencies of
wnt-1⁻/wnt-1⁻ brain

FIGURE 7.12
Wingless *and* engrailed *products interact to create the cerebellum. Comparison of (A) a control mouse embryo and (B) a* wnt-1*-deficient mutant embryo shows that the mutant lacks midbrain and metencephalon structures. (C) Comparing the deleted regions of the* wnt-1 *-deficient brain (gray) with the regions of* wnt-1 *gene expression in the 8.5-day embryonic neural tube (red) indicate that they do not overlap. (D) The expression of the* engrailed (en-2) *gene in wild-type embryos covers those regions missing in the wnt-1-deficient embryos, while (E) engrailed expression fails to occur in this brain region in the* wnt-1*-deficient embryos. (C after Lumsden and Wilkinson, 1990; photographs from McMahon et al., 1992, courtesy of A. McMahon.)*

the neural tube to differentiate into the cerebellum and midbrain.

The transcription of both the *engrailed* and the *wnt-1* genes may be controlled by another *Pax* gene. In zebra fish, *pax[b]* (roughly equivalent to the mouse *Pax-2* gene) is expressed in the posterior midbrain region. If antibodies against pax[b] protein are injected into zebra fish gastrula, the midbrain–hindbrain boundary fails to form, and the cells that would form the posterior hindbrain and cerebel-

lum die. Moreover, in situ hybridization shows that without *pax[b]* expression, the expression of *engrailed* and *wnt-1* mRNAs was drastically reduced in the prospective midbrain region (Krauss et al., 1992). Thus, this Pax protein appears to control the expression of *wnt-1* and *engrailed* genes which are necessary for the proper development of the cerebellum and midbrain structures.

Tissue architecture of the central nervous system

The neurons of the brain cortex are organized into layers, each having different functions and connections. The original neural tube is composed of a germinal neuroepithelium, one cell layer thick. This is a rapidly dividing cell population. Sauer (1935) and others have shown that all these cells are continuous from the luminal edge of the neural tube to the outside edge but that the nuclei of these cells are at different heights, thereby giving the superficial impression that the wall of the neural tube has numerous cell layers. DNA synthesis (S phase) occurs while the nucleus is at the outside edge of the tube, and the nucleus migrates luminally as mitosis proceeds (Figure 7.13). Mitosis occurs on the luminal side of the cell layer. During early mammalian development, 100 percent of the neural tube cells incorporate radioactive thymidine into DNA (Fujita, 1964).

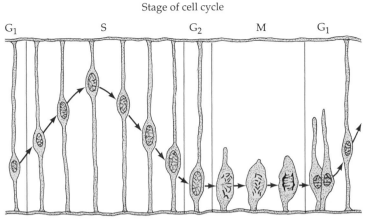

Stage of cell cycle

G_1 S G_2 M G_1

Lumen of neural tube

FIGURE 7.13
Schematic section of a chick embryo neural tube showing the position of the nucleus in a neuroepithelial cell as a function of the cell cycle. Mitotic cells are found near the center of the neural tube, adjacent to the lumen. (After Sauer, 1935.)

Shortly thereafter, certain cells stop incorporating these DNA precursors, thereby indicating that they are no longer participating in DNA synthesis and mitosis. These neuronal and glial cells can then differentiate at the periphery of the neural tube (Figure 7.14; Fujita, 1966; Jacobson, 1968).

If dividing cells are labeled by radioactive thymidine at a single point in their development and their progeny are seen in the outer cortex in the adult brain, then the neurons had to migrate to their cortical positions from the germinal neuroepithelium. The time when its precursor last divides is called the neuron's "birthday," and different types of neuron and glial cells have their birthdays at different times. Labeling at different times during development shows that those cells with the earliest birthdays migrate the shortest distances. The cells with later birthdays migrate through these layers to form the more superficial regions of the cortex. Subsequent differentiation is dependent upon the position these neurons occupy once outside the region of dividing cells (Letourneau, 1977; Jacobson, 1991).

As the cells adjacent to the lumen continue to divide, the migrating cells form a second layer around the original neural tube. This layer becomes progressively thicker as more cells are added to it from the germinal neuroepithelium. This new layer is called the **mantle** (or **intermediate**) **zone,** and the germinal epithelium of the ventricular zone is now called the **ventricular zone** (and, later, the **ependyma**) (Figure 7.15). The mantle zone cells differentiate into both neurons and glia. The neurons make connections among themselves and send forth axons away from the lumen, thereby creating a cell-poor **marginal zone.** Eventually, glial cells cover many of these marginal zone axons in myelin sheaths, thereby giving them a whitish appearance. Hence, the mantle zone, containing the cell bodies, is often referred to as the **gray matter;** and the axonal, marginal layer is often called the **white matter.**

In the spinal cord and medulla, this basic three-zone pattern of ependymal, mantle, and marginal layers is retained throughout development. The gray matter (mantle) gradually becomes a butterfly-shaped structure surrounded by white matter; and both become encased in connective tissue. As the neural tube matures, a longitudinal groove—the **sulcus limitans**—appears to divide it into dorsal and ventral halves. The dorsal portion receives input from sensory neurons, whereas the ventral portion is involved in effecting various motor functions (Figure 7.16). Recent evidence shows that the notochord plays a major role in creating this dorsal-ventral dichotomy. As will be shown in Chapter 16, when the notochord

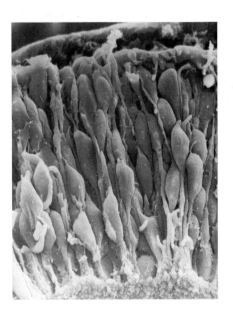

FIGURE 7.14
Scanning electron micrograph of a newly formed chick neural tube showing cells at different stages of their cell cycles. (Courtesy of K. Tosney.)

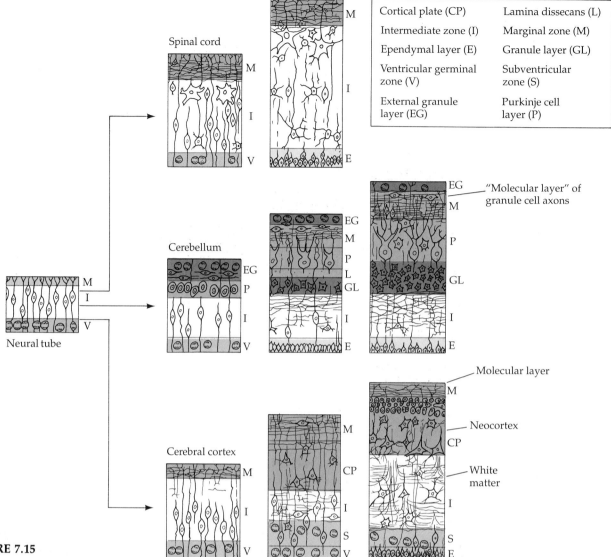

Cortical plate (CP)	Lamina dissecans (L)
Intermediate zone (I)	Marginal zone (M)
Ependymal layer (E)	Granule layer (GL)
Ventricular germinal zone (V)	Subventricular zone (S)
External granule layer (EG)	Purkinje cell layer (P)

FIGURE 7.15

Differentiation of the walls of the neural tube. Section of 5-week human neural tube containing three zones: ependymal, mantle, and marginal. In the spinal cord and medulla (top row), the ependyma remains the sole source of neurons and glial cells. In the cerebellum (middle row), a second mitotic layer, the external granular layer, forms at the region furthest removed from the ependyma. Neuroblasts from this layer migrate back into the intermediate zone to form the granule cells. In the cerebral cortex (bottom row), the migrating neuroblasts and glioblasts form a cortical plate containing six layers. (After Jacobson, 1991.)

contacts the neural cells above it, it induces those medial hinge region cells to become **floor plate cells** of the neural tube. The floor plate then induces cells on both sides of it to become the motor neurons. If the notochord is removed, no floor plate or motor neurons differentiate. Instead, dorsal root ganglion neurons (which are usually formed at the dorsal region of the neural tube) extend out from the ventral portion of the tube (Figure 7.17; Placzek et al., 1991; Yamada et al., 1993).

Cerebellar organization

In the brain, cell migration, differential growth, and selective cell death produce modifications of the three-zone pattern, especially in the cerebellum and cerebrum. Some neurons enter the white matter to differentiate into clusters of neurons called **nuclei.** Each nucleus works as a functional unit, serving as a relay station between the outer layers of the cerebellum and other parts of the brain. In addition, dividing neuronal precursor cells, **neuroblasts,** migrate to the outer surface of the developing cerebellum, forming a new germinal zone, the **external germinal layer,** near the outer

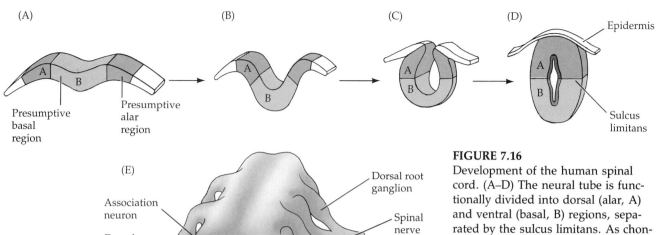

Presumptive basal region

Presumptive alar region

Epidermis

Sulcus limitans

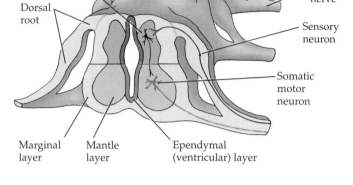

(E)

Dorsal root ganglion

Association neuron

Spinal nerve

Dorsal root

Sensory neuron

Somatic motor neuron

Marginal layer

Mantle layer

Ependymal (ventricular) layer

FIGURE 7.16
Development of the human spinal cord. (A–D) The neural tube is functionally divided into dorsal (alar, A) and ventral (basal, B) regions, separated by the sulcus limitans. As chondroblasts from the sclerotome region of the somite form the spinal vertebrae, the neural tube differentiates into the ependymal, mantle, and marginal zones, and the roof and floor plates become distinguished. (E) A segment of the spinal cord with its sensory (alar) and motor (basal) roots. (After Larsen, 1993.)

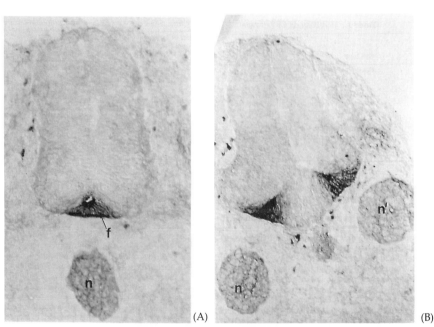

(A)

(B)

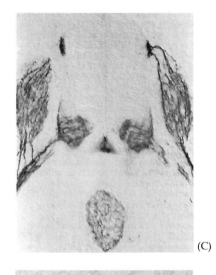

(C)

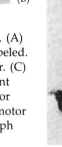

(D)

FIGURE 7.17
Role of the notochord in inducing dorsal–ventral polarity in the spinal cord. (A) Control chick embryo wherein the floor plate neurons (f) are specifically labeled. (B) If another notochord (n) is added, two regions of floor plate cells appear. (C) Control embryo staining for floor plate cells, motor neurons, and the afferent fibers of the dorsal root ganglia (d). (D) If the notochord is removed, no floor plate cells are formed, and the cells that would normally have formed the motor neurons now form dorsal root ganglia. (From Placzek et al., 1991, photograph courtesy of T. Jessell.)

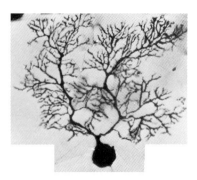

FIGURE 7.18
Normal mouse Purkinje neuron. The large dendritic process of this neuronal type is characteristic. (From Berry et al., 1980, courtesy of M. Berry.)

boundary of the neural tube. At the outer boundary of the external germinal layer (1–2 cells thick), neuroblasts proliferate. The inner compartment of the external germinal layer contains postmitotic neuroblasts that are the precursors of the major neuron of the cerebellar cortex, the **granular cell.** These pregranular cell neurons migrate back into the developing cerebellar white matter to produce a region of granular cell neurons in a region called the **internal granular layer.** Meanwhile, the original ependymal layer of the cerebellum generates a wide variety of neurons and glial cells, including the distinctive and large **Purkinje neurons.** Each Purkinje neuron has an enormous **dendritic apparatus,** which spreads like a fan above a bulblike cell body (Figure 7.18). A typical Purkinje cell may form as many as 100,000 synapses with other neurons, more than any other neuron studied. Each Purkinje neuron also emits a slender axon, which connects to other cells in the deep cerebellar nuclei.

The development of spatial organization is critical for the proper functioning of the cerebellum. All impulses eventually regulate the activity of the Purkinje cells, which are the only output neurons of the cerebellar cortex. For this to happen, the proper cells must differentiate at the appropriate place and time. How is this accomplished?

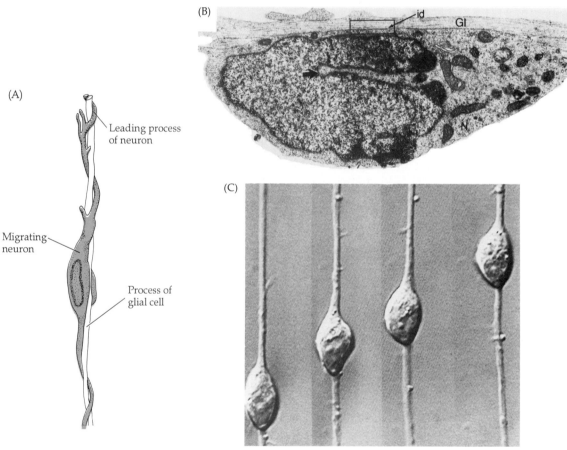

FIGURE 7.19
Neuronal migration upon glial cell processes. (A) Diagram of cortical neuron migrating upon glial cell process. (B) Electron micrograph of the region where the neuron soma adheres to the glial process. (C) Sequential photographs of a neuron migrating upon cerebellar glial process. The leading process has several filopodial extensions. It reaches speeds up to 60 μm/hour as it progresses on the glial process. (A after Rakic, 1975; B from Gregory et al., 1988; C from Hatten, 1990, courtesy of M. Hatten.)

The primary mechanism for positioning young neurons within the developing mammalian brain is glial guidance (Rakic, 1972; Hatten, 1990). Throughout the cortex, neurons are seen to ride "the glial monorail" to their respective destinations. In the cerebellum, the granule cell precursors travel upon the long processes of the **Bergmann glia** (Figure 7.19; Rakic and Sidman, 1973; Rakic, 1975). The neural–glial interaction is a complex and fascinating series of events. First, when the neuronal cell membrane makes contact with the glial membrane, the glial cell stops proliferating and begins to differentiate and extend its glial process. In different regions of the brain, glia generate different morphologies. Some processes are long, others short; some are branched, others are straight. Recombination experiments have shown that the morphologies are glial-specific and are not dependent on the type of neuron interacting with the glial cell (Hatten, 1990). The nerve cell then binds to the glial cell and begins to express calcium ion channels. These channels allow the influx of calcium ions, which are necessary for the motility of the leading process (Komuro and Rakic, 1992). The granule cell neuron wraps this long process around the glial fiber, and as it migrates along the process, it continues to signal the glial cell to retain its differentiated form. The neuron maintains its adhesion to the glial cell through a number of proteins, the most important being an adhesion protein called **astrotactin.** If the astrotactin on the nerve cells is masked by antibodies to that protein, the nerve cell will fail to adhere to the glial processes (Edmondson et al., 1988; Fishell and Hatten, 1991).

Some insight into the mechanisms of spatial ordering has come from the analysis of neurological mutations in mice. Over 30 mutations are known to affect the arrangement of cerebellar neurons. Many of the cerebellar mutants have been found because the phenotype of such mutants (i.e., the inability to keep balance while walking) can be easily recognized. Mice with the *weaver* mutation do not have granule cell neurons in their proper cortical layer and this lack of granular cell neurons prevents the proper development and localization of the Purkinje neurons (Figure 7.20; Goldowitz and Mullen, 1982). This, in turn, leads to the characteristic gait of the mutant mice. The defect in the *weaver* mice is caused by the inability of the granule cell precursors to migrate along the Bergmann glial cells from the outer germinal layer to their proper cortical site. However, the defect does not reside in the neuronal precursor. When *weaver* granule cell precursors are implanted into the cerebellar cortex of wild-type mice, they migrate and differentiate normally (Gao and Hatten, 1993).

It is possible that the defect has to do with neuroblast–neuroblast interactions that prepare the granule cell precursor to "ride" on the glial fiber. Edmondson and colleagues (1988) found that the *weaver* mouse granule cells have little or no functional astrotactin on their surfaces. Astrotactin is synthesized by the granular neuroblasts in response to neuroblast–neuroblast interactions within the external germinal layer, and it is this interaction that is probably deficient in *weaver* mice. If wild-type neuroblasts from this region are added to *weaver* granule cell neuroblasts, the *weaver* neuroblasts begin to synthesize astrotactin and show normal function (Gao et al., 1992). These data strongly suggest that the genetic defect of *weaver* mice involves the interaction among neuroblasts that enables the granule cell precursor to synthesize astrotactin. Astrotactin enables proper granule cell migration; granule cell migration is needed for the proper placement of the Purkinje cells; and proper placement of the Purkinje cells is critical for the functioning of the cerebellar pathways that enable the mouse to balance while walking. Although we've learned a great deal from these mutants, many questions remain. The nature of

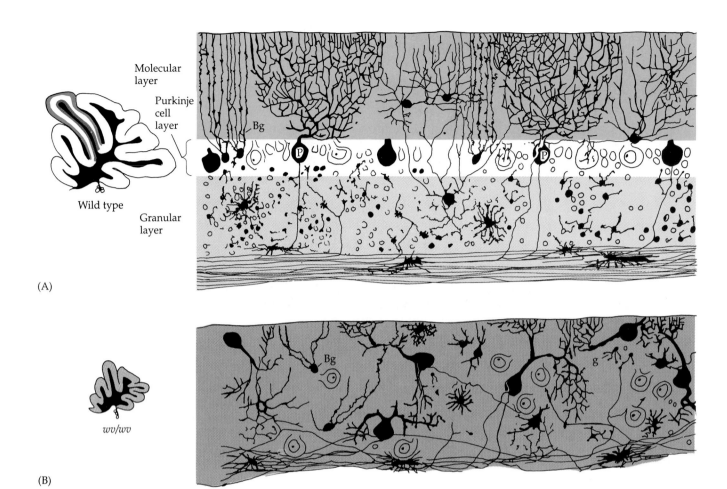

Molecular
layer

Purkinje
cell
layer

Bg

P

P

Wild type

Granular
layer

(A)

wv/wv

(B)

FIGURE 7.20
Abnormal morphogenesis in the brain of the *weaver* mutant mouse. (A) Organization of the cerebellar cortex in wild-type mice. (B) The same region in the cerebellar cortex of mice homozygous for the *weaver* mutation shows a profound deficiency of granule cells, the stunting of dendrites from the Purkinje neurons, and the failure of the neurons to sort themselves into the correct layers. Bg, Bergmann glia; g, granule cells; P, Purkinje neurons. (After Rakic and Sidman, 1973.)

these neuroblast–neuroblast interactions are not yet known, nor are the mechanisms by which the movement of wild-type neurons occurs in only one direction or how the cells are instructed to leave the "glial monorail."

Cerebral organization

In the cerebrum, the three-zone arrangement is also modified. The cerebrum is organized in two distinct ways. First, like the cerebellum, it is organized vertically into layers that interact with each other (Figure 7.15). Certain neuroblasts from the mantle zone migrate upon the glia through the white matter to generate a second zone of neurons. This new mantle zone is called the **neopallial cortex.** This cortex eventually stratifies into six layers of cell bodies, and the adult forms of these neopallial neurons are not completed until the middle of childhood. Each layer of the cerebral cortex differs from the other in its functional properties, the types of neurons found therein, and the sets of connections that emerge. For instance, neurons from layer 4 receive their major input from the thalamus (a region that forms from the diencephalon), while neurons in layer 6 send their major output back to the thalamus. Second, the cerebral cortex is organized horizontally into over 40 regions that regulate geographically distinct processes. For instance, neurons to cortical layer 6 of the visual cortex project axons to the lateral geniculate nucleus of the thalamus, while these layer 6 neurons of the auditory cortex (located more anteriorly than the visual cortex) project axons to the medial geniculate nucleus of the thalamus.

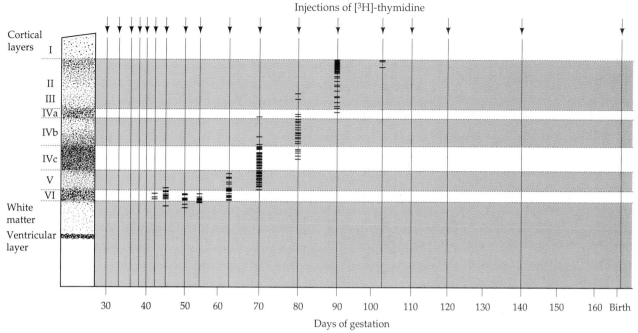

Cortical layers

White matter

Ventricular layer

30 40 50 60 70 80 90 100 110 120 130 140 150 160 Birth

Days of gestation

FIGURE 7.21

"Inside-out" gradient of cerebral cortex formation in the rhesus monkey. The "birthdays" of the cortical neurons were determined by injecting pregnant animals intravenously with [³H]-thymidine at certain times of gestation. The animals were born, and those cells undergoing the S phase of their final cell division cycle were found to be heavily labeled. These cells migrated to various regions and were detected by autoradiography of microscopic slices. The figure represents the position of those neurons in the visual cortex injected once on the day indicated. Full gestation in rhesus monkeys is 165 days. The youngest neurons are at the periphery of the neural tube. (After Rakic, 1974.)

Neither the vertical nor the horizontal organization is clonally specified. Rather, there is a lot of cell movement that mixes the progenies of various precursor cells. After their final mitosis, most of the newly generated neurons migrate radially along glial processes out from the ventricular (ependymal) zone and form the cortical plate below the pia mater of the brain. Like the cerebellar cortex, those neurons with the earliest "birthdays" form the layer closest to the ventricle. Subsequent neurons travel greater distances to form the more superficial layers of the cortex. This forms an "inside-out" gradient of development (Figure 7.21; Rakic, 1974). A single stem cell in the ventricular layer can produce neurons (and glial cells) in any of the cortical layers (Walsh and Cepko, 1988). But how do the cells know which layer to enter? McConnell and Kaznowski (1991) have shown that the determination of laminar identity (i.e., which layer a cell migrates to) is made during the final cell division. Cells transplanted from young brains (where they would form layer 6) into older brains (layer 2) after their last division are committed to their fate and migrate only to layer 6. However, if the cells are transplanted prior to their final division (during mid S phase), they are uncommitted and can migrate to layer 2 (Figure 7.22). We still do not know the nature of the information given to the cell as it becomes committed.

Not all neurons, however, migrate radially. O'Rourke and her colleagues (1992) labeled young neurons with fluorescent dye and followed their migration through the brain. While about 80 percent of the young neurons migrated radially on glial processes from the ventricular zone into the cortical plate, about 12 percent of the new neurons migrated laterally from one functional region of the cortex into another. These observations meshed well with those of Walsh and Cepko (1992), who infected ventricular cells with a retrovirus and were able to stain these cells and their progeny after birth. They found that the neural descendants of a single ventricular cell were dispersed across the functional regions of the cortex. Thus, the specification of these cortical areas into specific functions occurs after neurogenesis.

The development of the human neopallial cortex is particularly strik-

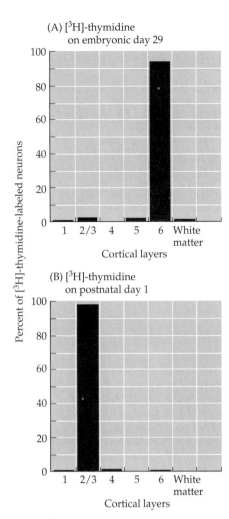

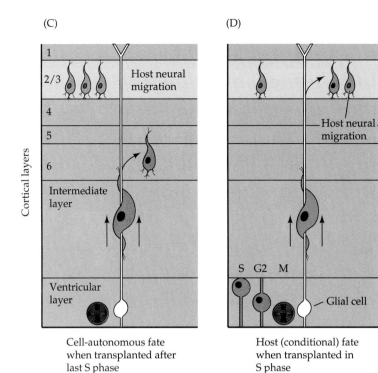

(A) [³H]-thymidine
on embryonic day 29

(B) [³H]-thymidine
on postnatal day 1

(C)

Cell-autonomous fate
when transplanted after
last S phase

(D)

Host (conditional) fate
when transplanted in
S phase

FIGURE 7.22
Determination of laminar identity in the ferret cerebrum. (A) "Early" neuronal precursors (birthdays on embryonic day 29) migrate to layer 6. (B) "Late" neuronal precursors (birthdays on postnatal day 1) migrate farther, into layers 2 and 3. (C) When early precursors (red) are transplanted into older ventricular zones *after* their last mitotic S phase, the neurons they form migrate into layer 6. (D) If these precursors are transplanted *before* or *during* their last S phase, their neurons migrate (with the host neurons) to layer 2. (After McConnell and Kaznowski, 1991.)

ing. The human brain continues to develop at fetal rates even after birth (Holt et al., 1975). Based on morphological and behavioral criteria, Portmann (1941, 1945) suggested that, compared with other primates, human gestation should really last 21 months instead of 9. However, no woman could deliver a 21-month-old infant because the head would not pass through the birth canal; thus, humans give birth at the end of 9 months. Montagu (1962) and Gould (1977) have suggested that during our first year of life we are essentially extrauterine fetuses, and they speculate that much of human intelligence comes from the stimulation of the nervous system as it is forming during that first year.*

Neuronal types

The human brain consists of over 10^{11} nerve cells (neurons) associated with over 10^{12} glial cells. Those cells that remain as integral components of the neural tube lining become **ependymal cells.** These cells can give rise to the precursors of neurons and glial cells. It is thought that the differentiation of these precursor cells is largely determined by the environment that they enter (Rakic and Goldman, 1982), and that, at least in some cases, a given precursor cell can form both neurons and glial cells (Turner and Cepko, 1987). There is a wide variety of neuronal and glial types (as is evident from a comparison of the relatively small granule cell with the enormous Purkinje neuron). The fine extensions of the cell that

*Contrary to the claims of a widely circulated anti-abortion film, the human cerebral cortex has no neuronal connections at 12 weeks' gestation (and therefore cannot move in response to a thought, nor experience consciousness or fear). Measurable electrical activity characteristic of neural cells (the electroencephalogram, or EEG, pattern) is first seen at 7 months' gestation. Morowitz and Trefil (1992) put forth the provocative opinion that since society in the United States has defined death as the *loss* of the EEG pattern, perhaps it should accept the *acquisition* of the EEG pattern as the start of human life.

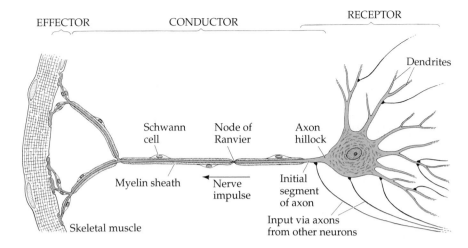

EFFECTOR CONDUCTOR RECEPTOR

Dendrites

Schwann cell

Node of Ranvier

Axon hillock

Myelin sheath

Nerve impulse

Initial segment of axon

Input via axons from other neurons

Skeletal muscle

FIGURE 7.23
Diagram of a motor neuron. Impulses received by the dendrites and the stimulated neuron can transmit electrical impulses through the axon (which may be 2–3 feet long) to its target tissue. The myelin sheath, which provides insulation for the axon, is formed by adjacent Schwann cells. (After Bloom and Fawcett, 1975.)

are used to pick up electrical impulses are called **dendrites** (Figure 7.23). Some neurons develop only a few dendrites, whereas other cells (such as the Purkinje neurons) develop extensive areas for cellular interactions. Very few dendrites can be found on cortical neurons at birth, but one of the amazing things about the first year of human life is the increase in the number of such receptive regions in the cortical neurons. During this year, each cortical neuron develops enough dendritic surface to accommodate as many as 100,000 connections with other neurons. The average cortical neuron connects with 10,000 other neural cells. This pattern of neural connections (synapses) enables the human cortex to function as the center for learning, reasoning, and memory, to develop the capacity for symbolic expression, and to produce voluntary responses to interpreted stimuli.

Another important feature of a developing neuron is its **axon** (sometimes called a neurite). Whereas dendrites are often numerous and do not extend far from the nerve cell body, or **soma,** axons may extend for several feet. The pain receptors on a person's big toe, for example, must transmit their messages all the way to the spinal cord. One of the fundamental concepts of neurobiology is that the axon is a continuous extension of the nerve cell body. At the turn of the twentieth century, there were numerous competing theories of axon formation. Schwann, one of the founders of the cell theory, believed that numerous neural cells linked themselves together in a chain to form an axon. Hensen, the discoverer of the embryonic node, thought that the axon formed around pre-existing cytoplasmic threads between the cells. Wilhelm His (1886) and Santiago Ramón y Cajal (1890) postulated that the axon was indeed an outgrowth (albeit an extremely large one) of the nerve soma. In 1907 Ross Harrison demonstrated the validity of the outgrowth theory in an elegant experiment that founded both the science of developmental neurobiology and the technique of tissue culture. Harrison isolated a portion of neural tube from a 3-mm frog tadpole. At this stage, shortly after the closure of the neural tube, there is no visible differentiation of axons. He placed these neuroblasts in a drop of frog lymph on a coverslip and inverted the coverslip over a depression slide so he could watch what was happening within this "hanging drop." What Harrison saw was the emergence of the axons as outgrowths from the neuroblasts, elongating at about 56 μm/hr.

Such nerve outgrowth is led by the tip of the axon, which is called the **growth cone** (Figure 7.24). This cone does not proceed in a straight line but rather "feels" its way along the substrate. The growth cones move

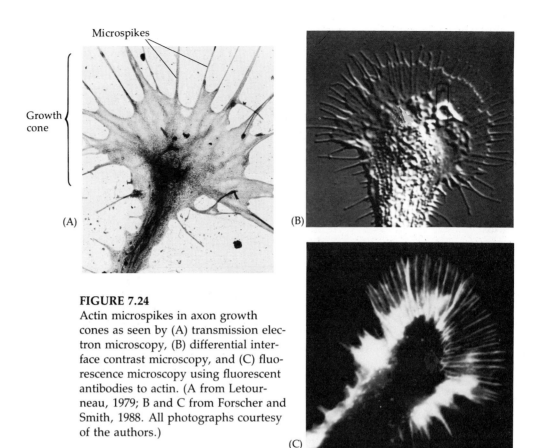

Microspikes

Growth cone

(A)

(B)

(C)

FIGURE 7.24
Actin microspikes in axon growth cones as seen by (A) transmission electron microscopy, (B) differential interface contrast microscopy, and (C) fluorescence microscopy using fluorescent antibodies to actin. (A from Letourneau, 1979; B and C from Forscher and Smith, 1988. All photographs courtesy of the authors.)

by the elongation and contraction of pointed filopodia called **microspikes**. These microspike filopodia contain microfilaments, which are oriented parallel to the long axis of the axon. (This is similar to the situation seen in the filopodial microfilaments of secondary mesenchymal cells in echinoderms.) Treating neurons with cytochalasin B destroys the actin microspikes, inhibiting their further advance (Yamada et al., 1971; Forscher and Smith, 1988). Within the axon itself, structural support is provided by microtubules, and the axon will retract if placed in a solution of colchicine. Thus, the developing neuron retains the same features that we have already noted in the dorsolateral neural tube hinge regions, namely, elongation by microtubules and apical shape changes by microfilaments. As in most migrating cells, moreover, the exploratory filopodia of the axon attach to the substrate and exert a force that pulls the rest of the cell forward. Axons will not grow if the growth cone fails to advance (Lamoureux et al., 1989). In addition to this structural role in axonal migration, filopodia also have a sensory function. Fanning out in front of the growth cone, each filopodium samples the microenvironments and sends signals back to the cell body (Davenport et al., 1993). As we will see in Chapter 8, the filopodia are the fundamental organelles involved in neuronal pathfinding.

Neurons transmit electrical impulses from one region to another. These impulses usually go from the dendrites into the nerve soma, where they are focused into the axon. To prevent dispersion of the electrical signal and to facilitate its conduction, the axon in the central nervous system is insulated at intervals by processes that originate from a type of glial cell called an **oligodendrocyte.** An oligodendrocyte wraps itself around the developing axon. It then produces a specialized cell membrane that is rich

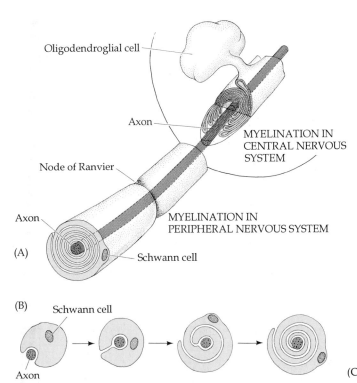

Oligodendroglial cell

Axon

MYELINATION IN
CENTRAL NERVOUS
SYSTEM

Node of Ranvier

Axon

MYELINATION IN
PERIPHERAL NERVOUS SYSTEM

(A)

Schwann cell

(B)

Schwann cell

Axon

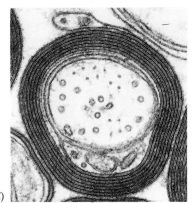

(C)

FIGURE 7.25
Myelination in the central and peripheral nervous systems. (A) In the peripheral nervous system, Schwann cells wrap themselves around the axon; in the central nervous system, myelination is accomplished by the processes of oligodendrocytes. (B) The mechanism of this wrapping entails the production of an enormous membrane complex. (C) Micrograph of an axon enveloped by the myelin membrane of a Schwann cell. (Photograph courtesy of C. S. Raine.)

in myelin basic protein that spirals around the central axon (Figure 7.25). This specialized membrane is called a **myelin sheath.** (In the peripheral nervous system, a glial cell called the **Schwann cell** accomplishes this myelination.) The myelin sheath is essential for proper neural function, and demyelination of nerve fibers is associated with convulsions, paralysis, and several severely debilitating or lethal afflictions. In the mouse mutant *trembler,* the Schwann cells are unable to produce a particular protein component of the myelin sheath, so the myelination is deficient in the peripheral nervous system, but it is normal in the central nervous system. Conversely, in another mouse mutant, *jimpy,* the central nervous system is deficient in myelin, while the peripheral nerves are unaffected (Sidman et al., 1964; Henry and Sidman, 1988).

The axon must also be specialized for secreting a specific neurotransmitter across the small gaps (synaptic clefts) that separate the axon of one cell from the surface of its target cell (the soma, dendrites, or axon of a receiving neuron or a receptor site on a peripheral organ). Some neurons become able to synthesize and secrete acetylcholine, while other neurons develop the enzymatic pathways for making and secreting epinephrine, norepinephrine, octopamine, serotonin, γ-aminobutyric acid, dopamine, or some other neurotransmitter. Each neuron must activate those genes responsible for making the enzymes that can synthesize its neurotransmitter. Thus, neuronal development involves both structural and molecular differentiation.

Development of the vertebrate eye

Dynamics of optic development. An individual gains knowledge of its environment through its sensory organs. The major sensory organs of the head develop from the interactions of the neural tube with a series of

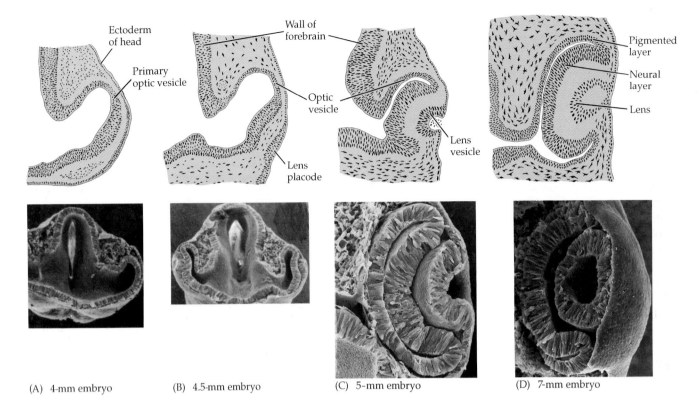

Ectoderm of head

Primary optic vesicle

Wall of forebrain

Optic vesicle

Lens placode

Lens vesicle

Pigmented layer

Neural layer

Lens

(A) 4-mm embryo   (B) 4.5-mm embryo   (C) 5-mm embryo   (D) 7-mm embryo

FIGURE 7.26

Development of the vertebrate eye. (A) Optic vesicle evaginates from the brain and contacts the overlying ectoderm. (B,C) Overlying ectoderm differentiates into lens cells as the optic vesicle folds in on itself and the lens placode becomes the lens vesicle. (C) The optic vesicle becomes the neural and pigmented retina as the lens is internalized. (D) The lens vesicle induces the overlying ectoderm to become the cornea. (Upper illustrations after Mann, 1964; micrographs A–C from Hilfer and Yang, 1980, courtesy of S. R. Hilfer; D courtesy of K. Tosney.)

epidermal thickenings called the **cranial ectodermal placodes.** The most anterior placodes are the two olfactory placodes that form the ganglia for the olfactory nerves, which are responsible for the sense of smell. The auditory placodes similarly invaginate to form the inner ear labyrinth whose neurons form the acoustic ganglion, which enables us to hear. In this section, we will focus on the eye, because this organ, perhaps more than any other organ in the body, must develop with precise coordination of all its components.

The story of optic development begins at gastrulation when the involuting endoderm and mesoderm interact with the adjacent prospective head ectoderm. This interaction gives a lens-forming bias to the head ectoderm* (Saha et al., 1989). But not all parts of the head ectoderm eventually form lenses, and the lens has to be in a precise relationship with the retina. The activation of this latent lens-forming ability and the positioning of the lens in relation to the retina is accomplished by the **optic vesicle.** In humans, the optic vesicles begin as two bulges from the lateral walls of the 22-day embryonic diencephalon (Figure 7.26). These bulges continue to grow laterally from the neural tube, and are connected to the diencephalon by the **optic stalks.** Subsequently, when these vesicles contact the head ectoderm, the ectoderm thickens into the **lens placodes.** The necessity for close contact between the optic vesicles and the surface ectoderm is seen in both experimental cases and in certain mutants. For example, in the mouse mutant *eyeless*, the optic vesicles fail to contact the surface and eye formation ceases (Webster et al., 1984).

Once formed, the lens placode reciprocates and causes changes in the optic vesicle. The vesicle invaginates to form a double-walled **optic cup** (Figure 7.26G–I). As the invagination continues, the connection between the optic cup and the brain is reduced to a narrow slit. At the same time,

*The inductions forming the eye will be detailed in Chapter 18.

the two layers of the optic cup begin to differentiate in different directions. The cells of the outer layer produce pigment and ultimately become the **pigmented retina** (one of the few tissues other than the neural crest cells that can synthesize its own melanin). The cells of the inner layer proliferate rapidly and generate a variety of glia, ganglion neurons, interneurons, and light-sensitive photoreceptor neurons. Collectively, these cells constitute the **neural retina.** The axons from the ganglion cells of the neural retina meet at the base of the eye and travel down the optic stalk. This stalk is then called the **optic nerve.**

Neural retina differentiation. Like the cerebral and cerebellar cortices, the neural retina develops into a layered array of different neuronal types (Figure 7.27). These layers include the light- and color-sensitive photoreceptor (rod and cone) cells, the cell bodies of the ganglion cells, and the bipolar interneurons that transmit the electrical stimulus from the rods and cones to the ganglion cells. In addition, there are numerous glial cells that serve to maintain the integrity of the retina, as well as amacrine neurons, and horizontal neurons that transmit electrical impulses horizontally.

In the early stages of retinal development, cell division from a germinal layer and the migration and differential death of the resulting cells form the striated, laminar pattern of the neural retina. The formation of this highly structured tissue is one of the most intensely studied problems of developmental neurobiology. It has been shown (Turner and Cepko, 1987) that a single retinal neuroblast precursor cell can give rise to at least three types of neurons or to two types of neurons and a glial cell. This analysis was performed using a most ingenious technique to label the cells generated by one particular precursor cell. Newborn rats (whose retinas are still developing) were injected into the back of their eyes with a virus that can integrate into their DNA. This virus contained a β-galactosidase gene (not present in rat retina) that would be expressed only in the infected cells. A month after the rats' eyes were infected, the retinas were removed and stained for the presence of β-galactosidase. Only the progeny of the infected cells should have stained blue. Figure 7.28 shows one of the stripes of cells derived from an infected precursor cell. The stain can be seen in five rods, a bipolar neuron, and a retinal (Müller) glial cell.

Of the three major neuronal types in the retina (ganglion, bipolar, and photoreceptor), the photoreceptive rods and cones are probably the last

FIGURE 7.27
Development of the human retina. Retinal neurons sort out into functional layers during development. (A,B) Initial separation of neuroblasts within the retina. (C) The three layers of neurons in the adult retina and the synaptic layers between them. (D) A functional depiction of the major neuronal pathway in the retina. Light traverses the layers until it is received by the photoreceptors. The axons of the photoreceptors synapse with bipolar neurons that transmit the depolarization to the ganglion neurons. The axons of the ganglion cells join to form the optic nerve that enters the brain. (A and B after Mann, 1964; photograph courtesy of G. Grunwald.)

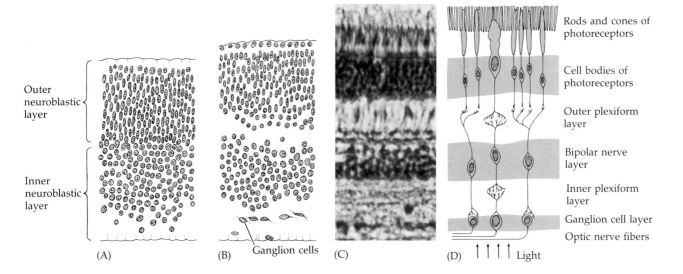

Outer neuroblastic layer

Inner neuroblastic layer

(A)

(B) Ganglion cells

(C)

(D) ↑ ↑ ↑ ↑ Light

Rods and cones of photoreceptors

Cell bodies of photoreceptors

Outer plexiform layer

Bipolar nerve layer

Inner plexiform layer

Ganglion cell layer

Optic nerve fibers

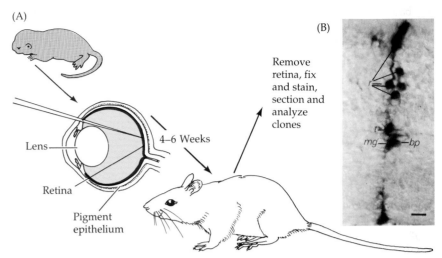

(A)

Lens

Retina

Pigment epithelium

4–6 Weeks

Remove retina, fix and stain, section and analyze clones

(B)

mg — bp

FIGURE 7.28
Determination of the lineage of a precursor cell in the rat retina. (A) Technique whereby a virus containing a functional β-galactosidase gene is injected into the back of the eye to infect some of the retinal precursor cells. After a month to 6 weeks, the eye is removed and the retina is stained for the presence of β-galactosidase. (B) Stained cells forming a strip across the neural retina, including five rods (r), a bipolar neuron (bp), a rod terminal (t), and a Müller glial cell (mg). The identities of these cells were confirmed by Nomarski phase contrast microscopy. Scale bar, 20 μm. (From Turner and Cepko, 1987, photograph courtesy of D. Turner.)

of the neurons to complete their differentiation. As they develop, the cell bodies of these outer neurons produce a bud of cytoplasm that contains several specialized organelles. These organelles elongate the bud and adjust the size and shape of the photoreactive regions of each cell (Detwiler, 1932). The cell membrane of these cells folds back upon itself to form sacs upon which the photoreceptive pigments are placed. Light induces these pigments to undergo chemical changes that ultimately result in a change of membrane potential. This change in membrane potential effects the release of neurotransmitters to a group of bipolar neurons that relay the electrical signal to the ganglion cells. The ganglion cells, whose axons bundle together to form the optic nerve, relay this information to the brain (Fesenko et al., 1985; Stryer, 1986).

SIDELIGHTS & SPECULATIONS

Why babies don't see well

Human newborns see poorly. There may be several reasons for this, but one of the most striking is the immaturity of the retinal photoreceptors. Anatomical studies by Yuodelis and Hendrickson (1986) and physical studies by Banks and Bennett (1988) have shown that the cone photoreceptors of the newborn central retina are over 7.5 μm in diameter, reaching the normal adult width of 2 μm in about 3 years. During this time, the cone density in this region increases from 18 photoreceptors per 100 μm to 42 photoreceptors

per 100 μm, and the photoreceptors develop both their outer segments (that catch the light) and their basal axonal processes. Figure 7.29 highlights the differences between the photoreceptors of neonatal and adult retinas. One can see that the neonatal retina has poorly differentiated photoreceptors, and those photoreceptors it does have are so wide that not many of them can fit into a given area. Banks and Bennett calculate that this enables the central region of the adult retina to absorb light about 350 times more efficiently than the same region of the newborn retina. This low number of photoreceptors per retinal area also causes the babies to be unable to discriminate between two points that are at a distance. This may be why a newborn responds to visual stimuli only when they are brought close to the baby's face. The development of the human retinal photoreceptor provides an excellent example of differentiation that begins early in development but which is not complete until years after birth.

FIGURE 7.29
Development of the cone photoreceptors in the central region of the human retina. Stained light microscope sections were photographed and one cone in each retina outlined for clarity. The pigment epithelium (PE), outer plexiform layer (OPL), Müller glia (M), and outer segments of the photoreceptor (OS) have been labeled. (A) Fetus of 22 weeks' gestation. (B) Newborn 5 days after birth. (C) 72-year-old. The arrow points to the outer limiting membrane, which served as the original border for the retinal neurons. The axon outlined in (C) is actually shorter than normal, thus allowing the synapse with the bipolar neuron to be shown in the picture. The synapse is formed at the cone synaptic pedicle (CP). (From Yuodelis and Hendrickson, 1986, courtesy of A. Hendrickson.)

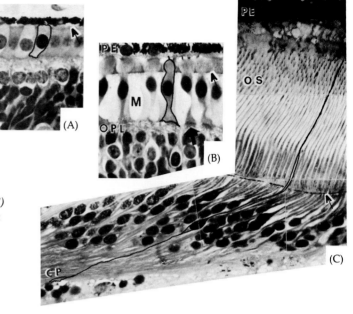

Lens and cornea differentiation. During its continued development into a lens, the lens placode rounds up and contacts the new overlying ectoderm (Figure 7.26D). The lens vesicle then induces the ectoderm to form the transparent cornea. Here, physical parameters play an important role in the development of the eye. Intraocular fluid pressure is necessary for the correct curvature of the cornea so that light can be focused upon the retina. The importance of such ocular pressure can be demonstrated experimentally; the cornea will not develop its characteristic curve when a small glass tube is inserted through the wall of a developing chick eye to drain away intraocular fluids (Coulombre, 1956, 1965). Intraocular pressure is sustained by a ring of scleral bones (probably derived from the neural crest), which act as inelastic restraints.

The differentiation of the lens tissue into a transparent membrane capable of directing light onto the retina involves changes in cell structure and shape as well as synthesis of lens-specific proteins called crystallins (Figure 7.30). Crystallins are synthesized as cell shape changes occur, thereby causing the lens vesicle to become the definitive lens. The cells at the inner portion of the lens vesicle elongate and, under the influence of the neural retina, produce the lens fibers (Piatigorsky, 1981). As these fibers continue to grow, they synthesize crystallins, which eventually fill up the cell and cause the extrusion of the nucleus. The crystallin-synthesizing fibers continue to grow and eventually fill the space between the two layers of the lens vesicle. The anterior cells of the lens vesicle constitute a germinal epithelium, which keeps dividing. These dividing cells move toward the equator of the vesicle, and as they pass through the equatorial region, they, too, begin to elongate (Figure 7.30D). Thus, the lens contains three regions: an anterior zone of dividing epithelial cells, an equatorial zone of cellular elongation, and a posterior and central zone of crystallin-containing fiber cells. This arrangement persists throughout

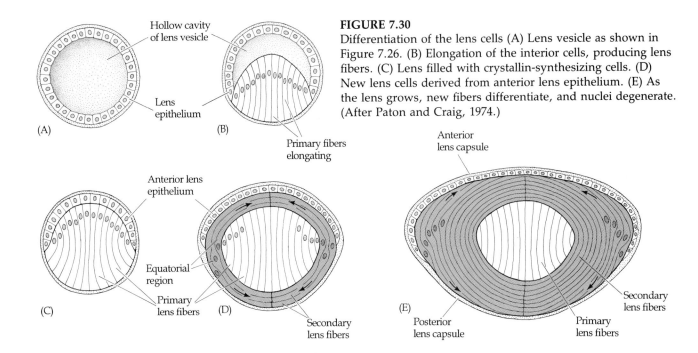

FIGURE 7.30
Differentiation of the lens cells (A) Lens vesicle as shown in Figure 7.26. (B) Elongation of the interior cells, producing lens fibers. (C) Lens filled with crystallin-synthesizing cells. (D) New lens cells derived from anterior lens epithelium. (E) As the lens grows, new fibers differentiate, and nuclei degenerate. (After Paton and Craig, 1974.)

the lifetime of the animal as fibers are continuously being laid down. In the adult chicken, the differentiation from an epithelial cell to a lens fiber takes two years (Papaconstantinou, 1967). The details of lens and cornea formation are discussed in Chapter 18.

Directly in front of the lens is a pigmented and muscular tissue called the iris. These muscles control the size of the pupil (and give an individual his or her characteristic eye color). Unlike the other muscles of the body (which are derived from the mesoderm), the iris is derived from the ectodermal layer. Specifically, the iris develops from a portion of the optic cup that is continuous with the neural retina but does not make photoreceptors.

The neural crest and its derivatives

Although derived from the ectoderm, the neural crest has sometimes been called the fourth germ layer because of its importance. It has even been said, perhaps hyperbolically, that "the only interesting thing about vertebrates is the neural crest" (quoted in Thorogood, 1989). The neural crest cells originate at the dorsalmost region of the neural tube and migrate extensively to generate a bewildering number of differentiated cell types. These include (1) the neurons and glial cells of the sensory, sympathetic, and parasympathetic nervous systems, (2) the epinephrine-producing (medulla) cells of the adrenal gland, (3) the pigment-containing cells of the epidermis, and (4) skeletal and connective tissue components of the head. The fate of the neural crest cells depends to a large degree upon where the cells migrate and settle. The neural crest will be divided into four main functional domains.

- *The cephalic (head) neural crest,* whose cells migrate dorsolaterally to produce the craniofacial mesenchyme that differentiates into the cartilage, bone, cranial neurons, glia, and connective tissues of the face.

These cells also enter the **pharyngeal arches** to give rise to thymic cells, odontoblasts of the tooth primordia, and the cartilage of inner ear and jaw.

- *The trunk neural crest,* whose cells take one of two major pathways. Those neural crest cells that become the pigment-synthesizing **melanocytes** migrate dorsolaterally into the ectoderm and continue their way toward the ventral midline of the belly. However, most of the trunk neural crest cells pass ventrolaterally through the anterior half of each sclerotome. **Sclerotomes** are blocks of mesodermal cells that surround the neural tube and differentiate into the vertebral cartilage of the spine. Those trunk neural crest cells that remain in the sclerotome form the **dorsal root ganglia.** Those cells continuing more ventrally form the **sympathetic ganglia,** the adrenal medulla and the nerve clusters surrounding the aorta.

- *The vagal and sacral neural crest,* whose cells generate the **parasympathetic (enteric) ganglia** of the gut (LeDouarin and Teillet, 1973; Pomeranz et al., 1991). Failure of neural crest cell migration to the colon results in the absence of enteric ganglia and thus to the absence of peristaltic movement in this region. This results in a functional obstruction and the dilation and enlargement of the region above it ("megacolon").

- *A cardiac neural crest* may be located between the cephalic and trunk crests. There is evidence that such neural crest cells exist from the first to the third somites of chick embryos (Kirby 1987; Kirby and Waldo, 1990). These neural crest cells can develop into melanocytes, neurons, cartilage, and connective tissue (of the third, fourth, and sixth pharygeal arches). In addition, this region of the neural crest produces the entire musculoconnective tissue wall of the large arteries as they arise from the heart as well as contributing to the septum that separates the pulmonary circulation from the aorta (LeLièvre and LeDouarin, 1975).

Table 7.1 is a summary of some of the cell types derived from the neural crest.

The trunk neural crest

Migration pathways of trunk neural crest cells

As shown in Figure 7.2, the trunk neural crest is a transient structure, its cells dispersing soon after the neural tube closes. There are two major pathways taken by the migrating neural crest cells (Figure 7.31).

The dorsolateral pathway.　The most obvious pathway for trunk neural crest migration is the dorsolateral pathway by which the melanocyte precursors travel around the periphery of the embryo through the dermal mesoderm underlying the dermis. They enter the ectoderm through minute holes in the basal lamina (which they might make) and colonize the skin and follicles, where they differentiate into melanocytes (Mayer, 1973; Erickson et al., 1992). This pathway was demonstrated in a series of classic experiments by Mary Rawles and others (1948), who transplanted the neural tube and crest from a pigmented strain of chicken into the neural tube of an albino chick embryo. The result was a white chicken with a specific region of colored feathers (Figure 7.32A). The neural crest is re-

TABLE 7.1
Some derivatives of the neural crest

Derivative	Cell type or structure derived
Peripheral nervous system (PNS)	Neurons, including sensory ganglia, sympathetic and parasympathetic ganglia, and plexuses Neuroglial cells Schwann cells
Endocrine and paraendocrine derivatives	Adrenal medulla Calcitonin-secreting cells Carotid body type I cells
Pigment cells	Epidermal pigment cells
Ectomesenchymal derivatives	Facial and anterior ventral skull cartilage and bones
Connective tissue (ectomesenchymally derived)	Corneal endothelium and stroma Tooth papillae Dermis, smooth muscle, and adipose tissue of skin of head and neck Connective tissue of salivary, lachrymal, thymus, thyroid, and pituitary glands Connective tissue and smooth muscle in arteries of aortic arch origin

Source: After Jacobson, 1991, based on multiple sources.

FIGURE 7.31
Neural crest cell migration in the trunk of the chick embryo. Path 1: Cells travel ventrally through the anterior of the sclerotome (that portion of the somite that generates vertebral cartilage). Those cells initially opposite the posterior portions of the sclerotomes migrate along the neural tube until they come to an opposite anterior region. These cells contribute to the sympathetic and parasympathetic ganglia as well as to the adrenal medullary cells and dorsal root ganglia. Path 2: Somewhat later, cells enter a dorsolateral route beneath the ectoderm. These cells become pigment-producing melanocytes.

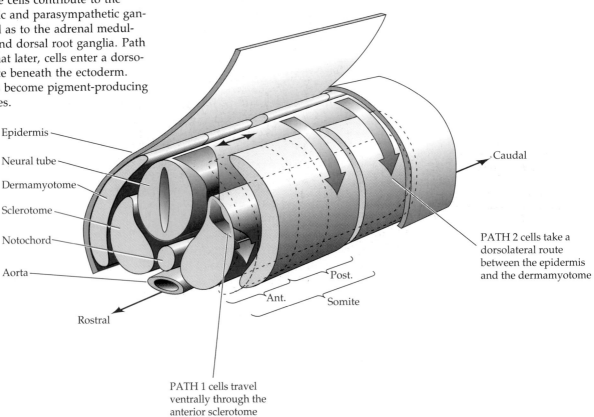

Epidermis
Neural tube
Dermamyotome
Sclerotome
Notochord
Aorta
Rostral
Caudal
Ant. Post. Somite
PATH 2 cells take a dorsolateral route between the epidermis and the dermamyotome
PATH 1 cells travel ventrally through the anterior sclerotome

(A)

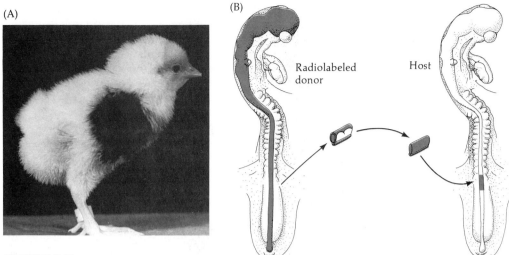

(B)

Radiolabeled donor

Host

FIGURE 7.32

Neural crest cell migration. (A) Chick resulting from the transplantation of a trunk neural crest region from a pigmented strain of chicken into the trunk neural crest region of an unpigmented strain. The crest cells that gave rise to pigment were able to migrate into the wing skin. (B) Grafting technique for mapping neural crest cells. A piece of the dorsal axis is excised from a donor embryo; the neural tube and its associated crest are isolated and implanted into a host embryo whose neural tube and crest have been excised. When the donor crest cells are radiolabeled (with tritiated thymidine) or genetically labeled (from a different species or strain), their descendants can be traced in the host embryo as development proceeds. (C) Autoradiograph showing locations of neural crest cells that have migrated from transplanted radioactive donor neural crest to form melanoblasts (M), sympathetic ganglia (SG), dorsal root ganglia (DRG), and glial cells (G). (A, original photograph from the archives of B. Willier; B after Weston, 1963; C courtesy of J. Weston.)

(C)

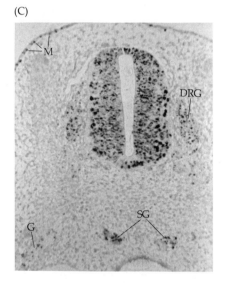

sponsible for the production of all the melanin-containing cells in the organism (with the exception of certain neural derivatives such as the pigmented retina).

The ventral pathway. By grafting a portion of the chick neural tube and its associated crest from radioactively or genetically marked embryos into other embryos (Weston, 1963; LeDouarin and Teillet, 1974), investigators have been able to trace another major route of trunk neural crest cell migration (Figure 7.32B and C). More recent studies have extended these investigations by using fluorescent antibodies, vital dyes, or virally transformed cells to follow *individual* neural crest cells to their destinations. Those cells leaving by the ventral pathway become sensory (dorsal root) and sympathetic neurons, adrenomedullary cells, and Schwann cells. As can be seen in Figure 7.33 and Plate 19, these trunk neural crest cells migrate ventrally through the anterior but not through the posterior section of the sclerotomes (Rickmann et al., 1985; Bronner-Fraser, 1986; Loring and Erickson, 1987; Teillet et al., 1987). Teillet and co-workers combined the antibody approach with transplantation of genetically marked quail neural crest cells into chick embryos. The antibody marker recognizes and labels neural crest cells of both species; the genetic marker enables the investigators to distinguish between quail and chick cells. These studies show that neural crest cells initially opposite the *posterior* regions of the somites migrate anteriorly or posteriorly along the neural tube and then enter the anterior region of their own or adjacent somites. These neural

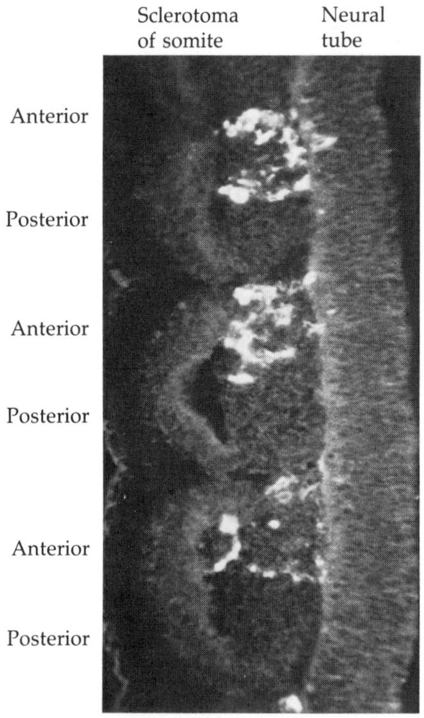

Sclerotoma of somite   Neural tube

Anterior

Posterior

Anterior

Posterior

Anterior

Posterior

FIGURE 7.33
Neural crest cell migration. These fluorescence photomicrographs of longitudinal sections of a 2-day chick embryo are stained with antibody HNK-1, which selectively recognizes neural crest cells. Extensive staining is seen in the anterior, but not in the posterior, half of each sclerotome. (From Bronner-Fraser, 1986, courtesy of M. Bronner-Fraser.)

crest cells join with the neural crest cells that were initially opposite the anterior portion of the somite, and they form the same structures. Thus, each dorsal root ganglion is composed of three neural crest populations: one from the neural crest opposite the anterior portion of the somite, and one from each of the adjacent neural crest regions opposite the posterior portions of the somites. At specific regions of the trunk, crest cells migrating along the same pathway aggregate to form the sympathetic ganglia and the epinephrine-secreting cells of the adrenal medulla. The parasympathetic division of the peripheral nervous system is also formed by neural crest cells migrating by this pathway, but only in the sacral and cervical regions of the embryo.

FIGURE 7.34
Expression of cell adhesion molecule N-cadherin during the initiation and cessation of trunk neural crest cell migration. (A) Graph of relative N-cadherin concentration on the neural crest cell surface. Moderate levels of N-cadherin are seen on neural crest cells before migration, little or no N-cadherin is detected during migration, and N-cadherin is expressed again on the cells derived from the migrating neural crest. (B) As neural crest cells begin to migrate, the dorsalmost cells of the neural tube (as well as the neural crest cells) lose N-cadherin expression, while it is retained by other cells in the neural tube. (C) N-cadherin expression is seen on the newly formed cells of the dorsal root and sympathetic ganglia (as well as on the notochord, neural tube, and other cells. (From Akitaya and Bronner-Fraser, 1992, courtesy of M. Bronner-Fraser.)

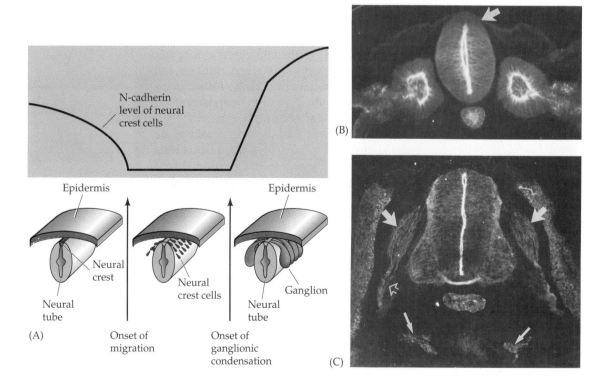

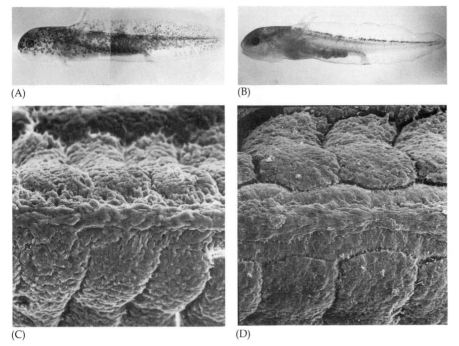

(A)

(B)

(C)

(D)

FIGURE 7.35
Deficiency of neural crest cell migration in the *d/d* mutant of the axolotl. (A) The larvae of wild-type axolotls are characterized by pigment cells throughout the body except for the most ventral portions. (B) In the *d/d* mutant, the neural crest-derived pigment cells form a stripe along the dorsal midline of the larva. (C,D) Scanning electron micrographs of the embryonic neural crest show that (C) the neural crest cells of the wild-type embryos migrate over the neural tube into the somites, while (D) those crest cells of the mutant remain atop the neural tube. (From Löfberg et al, 1989, courtesy of the authors.)

The extracellular matrix and trunk neural crest migration

Any analysis of migration (be it of birds, butterflies, or neural crest cells) has to ask three questions: How is migration initiated? How do the migratory agents know the path on which to travel? and, What signals that the destination has been reached and that migration should end? Although these questions are far from being answered, the migration of neural crest cells appears to be controlled by the substratum over which they travel (Newgreen and Gooday, 1985; Newgreen et al., 1986.) The initiation of migration is probably signaled by a decline in the amount of the adhesion molecule **N-cadherin** on the surface of the neural crest cells (Figure 7.34). Whereas other cells of the neural tube retain this adhesive protein, neural crest cells do not. *Migrating* crest cells have no N-cadherin on their surfaces, but they begin to express it again as they aggregate to form the dorsal root and sympathetic ganglia (Takeichi, 1988; Akitaya and Bronner-Fraser, 1992). At the same time that the neural crest cells lose their N-cadherin and become able to migrate as individual cells, the extracellular surface surrounding them becomes more adhesive (Perris et al., 1990). There appear to be specific paths for the neural crest cells to follow, and when neural crest cells or their derivatives are placed (either by transplantation or by injection) on their normal migration pathway in a host embryo, they migrate along it. (Erickson et al., 1980; Bronner-Fraser and Cohen, 1980).

Evidence from salamander development indicates that the direction of neural crest cell migration may be given by the extracellular matrices over which they travel. In some axolotl salamanders a mutation exists wherein the neural crest forms but its cells fail to migrate along the dorsolateral pathway. This is most readily seen in the lack of pigment cells anywhere except atop the neural tube of these animals (Figure 7.35), and these cells eventually degenerate. When wild-type neural crests are transplanted into mutant embryos, the crest cells are unable to migrate. However, when crests from mutant embryos are transplanted into wild-type embryos, their cells migrate normally (Spieth and Keller, 1984). Thus, the defect in this

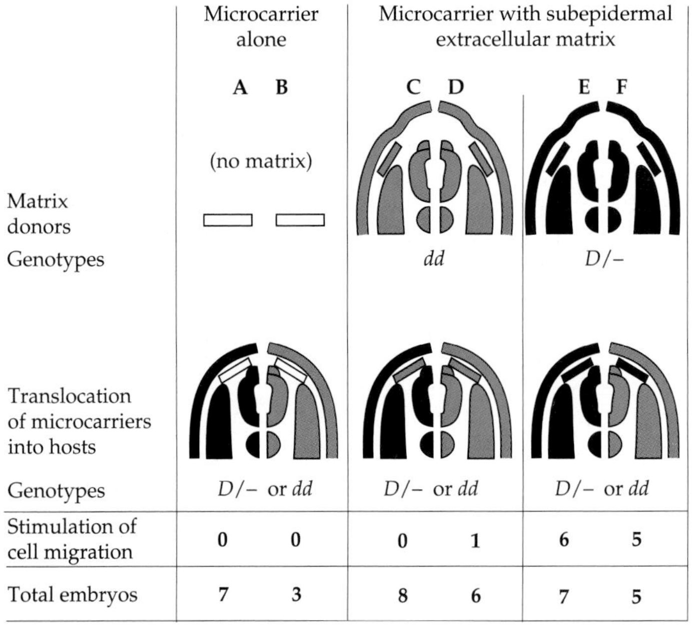

FIGURE 7.36

Importance of the extracellular matrix in neural crest cell migration. These diagrams represent the transplantation experiments wherein extracellular matrix components were carried between wild-type (dark) and mutant (color) donor and host embryos. The upper row represents the donor embryos, while the lower row shows the wild-type and mutant hosts into which the matrix-bearing microcarriers were added. The stimulation denotes the number of times that neural crest cells were seen migrating from the crest in the region of the microcarrier. (After Löfberg et al., 1989.)

	Microcarrier alone		Microcarrier with subepidermal extracellular matrix			
	A	**B**	**C**	**D**	**E**	**F**
Matrix donors	(no matrix)					
Genotypes			*dd*		*D/–*	
Translocation of microcarriers into hosts						
Genotypes	*D/–* or *dd*		*D/–* or *dd*		*D/–* or *dd*	
Stimulation of cell migration	0	0	0	1	6	5
Total embryos	7	3	8	6	7	5

mutant is in the environment that the cells encounter, not in the cells themselves. (The road is deficient, not the vehicle.) Löfberg and co-workers (1989) used this information to show that the extracellular matrix contains critical components regulating neural crest cell migration. They adsorbed, onto membrane microcarriers, extracellular matrix from the subepidermal region of the skin (through which the pigment-forming neural crest cells would migrate). They then placed these microcarriers next to the neural crests of mutant and wild-type embryos just before migration would occur. The microcarriers alone did not stimulate migration from either wild-type or mutant crests (Figure 7.36A,B). Microcarriers containing extracellular matrix material from mutant embryos likewise did not stimulate neural crest cell migration in either mutant or wild-type embryos (Figure 7.36C,D). The microcarriers containing wild-type extracellular matrix, however, were able to stimulate neural crest cell migration from both the mutant and wild-type crests (Figure 7.36E,F), demonstrating the importance of the extracellular matrix in neural crest cell migration.

A similar situation exists in chick embryos, where the mesoderm appears to control where and when the neural crest cells can migrate (Bronner-Fraser and Stern, 1991). When a strip of dorsal mesoderm was inverted so that what was originally anterior was now posterior, the neural crest cells migrated through those cells that would have *originally* produced the anterior half of the sclerotome, not what was the transplanted "anterior" half. Moreover, when host mesoderm was replaced by mesoderm derived solely from cells that would form the anterior portion of the sclerotome, the neural crest cells migrated through the entire graft to produce greatly enlarged ganglia (Goldstein et al., 1990). Thus, the regions allowing neural crest cell migration are actually determined in the mesoderm before migration occurs.

But what are the molecules that enable or forbid trunk neural crest cell migration? The extracellular matrix apposing the migrating neural crest cells is a rich mix of molecules such as fibronectin, laminin, tenascin, various collagen molecules, and proteoglycans. Experiments undertaken to address this issue have to be designed carefully, since neural crest cells may have different migration requirements in different species and even in different parts of the same embryo. One way of finding these answers is to make antibodies to the regions of the extracellular matrix molecules to which cells bind. When such antibodies are injected into the embryo and block these regions of the matrix, do they perturb neural crest cell migration? The migration of *chick cranial* neural crest cells can be severely altered when antibodies to fibronectin, fibronectin receptors, tenascin, or laminin–heparan sulfate proteoglycan are injected into the developing embryo (Poole and Thiery, 1986; Perris and Bronner-Fraser, 1989). However, these antibodies did not severely alter the migration of *chick trunk* neural crest cells.

In 1963, Weston hypothesized that these specific pathway molecules were not needed for trunk neural crest migration; rather, neural crest cells would take advantage of any space open to them, as long as there was no active inhibition of migration. He rotated the neural crest so that it was now on the bottom of the neural tube and noted that when the cells emerged from the ventral region of the neural tube, they migrated in a ventral-to-dorsal fashion (the reverse of normal migration) through the anterior half of the sclerotome. Thus, there seems to be no inherent directionality to the crest cells' migrations. The cells go where there are openings for them. Two molecules may be involved in physically opening or closing channels for trunk neural crest cell migration into the somite. The first of these molecules is hyaluronic acid. Hyaluronic acid may be permissive for neural crest cell migration by causing the formation of cell-free spaces through which neural crest cells can enter (Pratt et al., 1975; Solursh et al., 1979). Meier (1981) has found that hyaluronic acid does indeed accumulate in some regions where neural crest cells migrate. The second molecule is T-cadherin (Ranscht and Bronner-Fraser, 1991). T-cadherin is expressed in the posterior portion of each sclerotome immediately before neural crest cells begin migrating and remains present until migration ceases. This adhesive molecule compresses the caudal portion of the sclerotome, and it is possible that this compression prevents neural crest cells from entering it.

SIDELIGHTS & SPECULATIONS

Spotted mice and growth factors

Two mutations in the mouse have led to the discovery of an important growth factor whose synthesis is necessary for the proliferation of three types of stem cells. Mice having mutations at the *White spotting (W)* and *Steel (Sl)* loci have deficiencies involving three independent cell lineages. Viable homozygotes of these mutations have a severe reduction in their number of melanocytes, hematopoietic stem cells, and primordial germ cells. As a result, these homozygous mutants are white, anemic, and sterile (Sarvella and Russell, 1956; Mintz and Russell, 1957). Heterozygotes of these mutants show fewer than normal cells of these lineages, as is evident from their coat colors. *Steel* heterozygotes are grayish due to their deficiency of melanocytes, while *White spotting* heterozygotes have a white belly spot due to the inability of the melanocytes to migrate from the neural crest into the ventral ectoderm (Figure 7.37).

Transplantation of neural crests and bone marrow cells from wild-type embryos into *White spotting* embryos resulted in the proliferation and migration of the wild-type melanoblasts and hematopoietic stem cells. However, neural crests and bone marrow cells from wild-type embryos did not proliferate or migrate in the *Steel* hosts. Similarly, these stem cells from *White* embryos did not proliferate in wild-type embryos, whereas the *Steel* stem cells did (Bernstein et al., 1968; Mayer and Green, 1968). These studies concluded that the *W* gene is expressed in these three stem cell lineages, while the *Sl* gene affects the microenvironment in which these cells divide and migrate.

In 1988, the wild-type *W* locus was found to encode a transmembrane tyrosine kinase that has the structure of a growth factor receptor (Chabot et al., 1988; Geissler et al., 1988). This tyrosine kinase, c-kit, has been found in the cell membranes of hematopoietic stem cells, primordial germ cells, and presumptive subepidermal melanocytes (Orr-Urteger et al., 1990). Because the product of the wild-

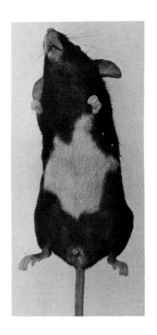

FIGURE 7.37
Ventral surface of a mouse heterozygous for the White *mutation. The mouse has reduced numbers of blood cells, germ cells, and melanocytes. The white spot on the belly is characteristic of heterozygotes since they do not have enough melanocytes to circumvent the mouse. Viable homozygous* White *animals have no pigment in the trunk. (Courtesy of Jackson Laboratories.)*

type *W* allele looked like a growth factor receptor, the search began for its ligand, a presumed growth factor. In 1990, several laboratories (see Witte, 1990) found a new growth factor that specifically binds to the *W*-encoded tyrosine kinase. Moreover, it is encoded by the wild-type *Steel* allele. This protein—stem cell factor—is able to correct the inherited defect of the *Steel* mice.

The cells secreting stem cell factor are associated with the migratory pathways of these three cell lineages (Matsui et al., 1990). Messenger RNA for this protein is seen (1) along the hindgut, mesentery, and genital ridge cells over which the primordial germ cells migrate, (2) in the ectoderm and dorsal region of the somites through which the melanoblasts migrate, and (3) in the yolk sac, embryonic liver, and developing bone marrow that provide the microenvironments for hematopoietic stem cells. In addition, there are regions of the nervous system that express the stem cell growth factor, suggesting that this growth factor may stimulate other cells as well. The identification of the

Steel product as a growth factor for stem cells and the identification of the *White spotting* product as its receptor solves an embryological enigma over 40 years old. This identification also helps us understand how the selective proliferation of these three cell lineages is so well coordinated in the mammalian embryo.

The developmental potency of trunk neural crest cells

Evidence for pluripotency among trunk neural crest cells. One of the most exciting features of neural crest cells is their **pluripotency**. A single neural crest cell can differentiate into several different cell types depending upon its location within the embryo. For example, the parasympathetic neurons formed by the cervical (neck region) neural crest cells produce acetylcholine as their neurotransmitter. They are therefore **cholinergic** neurons. The sympathetic neurons formed by the thoracic (chest) neural crest cells produce norepinephrine; they are **adrenergic** neurons. But when chick cervical and thoracic neural crests are reciprocally transplanted, the former thoracic crest produces the cholinergic neurons of the parasympathetic ganglia, and the former cervical crest forms adrenergic neurons in the sympathetic ganglia (LeDouarin et al., 1975). Kahn and co-workers (1980) found that premigratory trunk neural crest cells from both the thoracic and the cervical regions have the enzymes for synthesizing both acetylcholine and norepinephrine. Thus, the thoracic crest cells are capable of developing into cholinergic neurons when they are placed into the neck, and the cervical crest cells are capable of becoming adrenergic neurons when they are placed in the trunk.

The pluripotency of some neural crest cells is such that regions of the neural crest that never produce nerves in normal embryos can be made to do so under certain conditions. Mesencephalic neural crest cells normally migrate into the eye and interact with the pigmented retina to become scleral cartilage cells (Noden, 1978). However, if this region of the

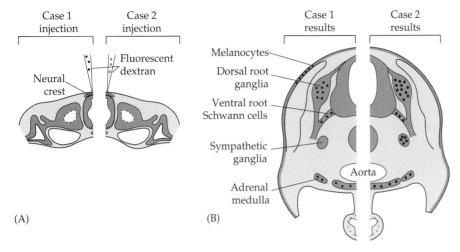

FIGURE 7.38
Pluripotency of trunk neural crest cells. A single neural crest cell is injected with a highly fluorescent dextran molecule. The progeny of this cell will each receive some of these fluorescent molecules. (A) Injection of fluorescent dextran into a neural crest cell shortly before migration of the crest cells is initiated. (B) Two days later, crest-derived tissues contain labeled cells descended from the injected precursor. The figure summarizes data from two different experiments (case 1 and case 2). (After Lumsden, 1988.)

neural crest is transplanted into the trunk region, it can form sensory ganglia neurons, adrenomedullary cells, glia, and Schwann cells (Schweizer et al., 1983).

The above research studied the potential of *populations* of cells. It is still unclear whether most of the cells that leave the neural crest are pluripotent, or whether most have already become restricted to certain fates. Bronner-Fraser and Fraser (1988, 1989) provide evidence that some, if not most, of the *individual* neural crest cells are pluripotential as they leave the crest. They injected fluorescent dextran molecules into individual neural crest cells while the cells were still above the neural tube and then looked to see what types of cells the individual crest cells became after these cells had migrated. The progeny of a single neural crest cell could become sensory neurons, pigment cells, adrenomedullary cells, and glia (Figure 7.38). Moreover, they found that if they labeled individual trunk neural crest cells as they migrated ventrally through the embryo, the label was later found in several cell types, including sensory neurons, sympathetic neurons, and Schwann cells. In mammals, the neural crest cell is similarly seen as a stem cell that can genenerate further multipotent neural crest cells. However, the migrating neural crest cell probably generates derivatives committed to a subset of those potentials (Stemple and Anderson, 1992). Thus, the restriction of the fate of the neural crest cell appears to be late, perhaps as it reaches its final location (Fraser and Bronner-Fraser, 1991).

Evidence for restricted potency among trunk neural crest cells. This does not necessarily mean that *all* individual neural crest cells are so pluripotent. Even at the time of emigration, some neural crest cells may be more determined than others. For one thing, there is a restriction of potency as the neural crest ages. Chick trunk crest cells that migrate *earliest* can form a wide variety of derivatives, including pigmented cells, neurons, and adrenergic cells. Late-migrating cells mostly become melanocytes (Artinger and Bronner-Fraser, 1992). Indeed, if left in the embryo, the late neural crest emigrés usually follow the dorsolateral pathway into the skin (Serbedzija et al., 1989). There is evidence that some restriction in potency exists even in some of the early-emigrating cells. Several investigators (see Sieber-Blum and Sieber, 1984; Stocker et al., 1991; Weston, 1991) have found that a significant number of neural crest cells form clones containing relatively few cell types. In addition, transplantation studies by LeDouarin and Smith (1988) suggest that many of the neural crest-derived cells that form the sensory neurons of the dorsal root ganglia are incapable of forming the autonomic neurons of the sensory ganglia, and vice versa. These studies suggest that some of the neural crest cells have already restricted their potentiality at the time of migration.

Two such committed neural crest cells are hypothesized to be a melanocyte–Schwann cell precursor (Nichols and Weston, 1977; Ciment, 1990) and a sympathoadrenal precursor that can give rise to only sympathetic ganglia and adrenal medullary cells (Landis and Patterson, 1981; Anderson and Axel, 1986). The mechanism for the differential success of these committed cells may involve different requirements for growth factors. The committed dorsal root ganglia neuron precursors appear to need **brain-derived neurotrophic factor** (BDNF), a growth factor produced by the neural tube itself. If a thin, nonpermeable membrane is interposed between the neural tube and the prospective dorsal root ganglia region, the ganglia fail to form. Those neural crest cells that have continued their ventral migration to the sympathetic ganglia regions survive. This inhibition of dorsal root ganglia production can be circumvented by coating the barrier with neural tube extract or with BDNF (Kalcheim et al., 1987; Sieber-Blum, 1991). The neural crest cells committed to form the sympathetic ganglia do not need this growth factor for survival. Rather, their differentiation appears to be spurred by basic fibroblast growth factor (Kalcheim and Neufeld, 1990).

Final differentiation of the neural crest cells. The final differentiation of the autonomic neural crest cells is determined in large part by the environment in which these cells develop. It does not involve the selective death of those cells already committed to secreting another type of neurotransmitter (Coulombe and Bronner-Fraser, 1987). Heart cells, for example, secrete a protein, leukemia inibition factor (LIF) that can convert adrenergic sympathetic neurons into cholinergic neurons without changing their survival or growth (Chun and Patterson, 1977; Fukada, 1980; Yamamori et al., 1989).

Thus, the fate of a committed neural crest cell can be directed by the hormonal milieu of the tissue environment or the substrate upon which it settles. The chick trunk neural crest cells that migrate into the region destined to become the adrenal medulla can differentiate in two directions. Such cells usually differentiate into noradrenergic sympathetic neurons.

FIGURE 7.39
Final differentiation of a neural crest cell committed to become either an adrenal medullary (chromaffin) cell or a sympathetic neuron. Glucocorticoids appear to act at two places. First, they inhibit the actions of those factors that promote neuronal differentiation; and second, they induce those enzymes characteristic of the adrenal cells. Those cells exposed sequentially to basic fibroblast growth factor (bFGF) and nerve growth factor (NGF) differentiate into the sympathetic neurons.

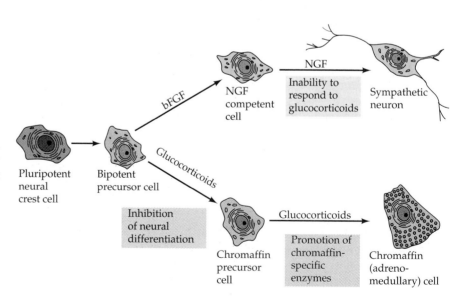

However, if these neural crest cells are given glucocorticoids like those made by the cortical cells of the adrenal gland, they differentiate into adrenal medullary cells (Figure 7.39; Anderson and Axel, 1986; Vogel and Weston, 1990). Adrenergic differentiation is also stimulated by some extracellular matrices but not others (Maxwell and Forbes, 1987). The type of matrix is seen to be important in the differentiation of the salamander neural crest cells, too. If axolotl crest cells are grown on matrices from subepidermal regions (where the pigment cells are found), they become melanocytes. However, if the same crest cells are cultured on matrices from the region of the dorsal root ganglia, they develop a neuronal phenotype (Perris et al., 1988).

The cephalic neural crest

Migratory pathways of the cephalic neural crest cells

The "face" is largely the product of the cephalic (cranial) neural crest, and the evolution of the jaws, teeth, and facial musculature arises from changes in the placement of these cells (see Chapter 23). As mentioned earlier, the hindbrain is segmented along the anterior–posterior axis into rhombomeres. The chick cephalic neural crest cells migrate according to their rhombomeric origin, and there are three major pathways taken by these migrating cells (Figure 7.40; Lumsden and Guthrie, 1991). Cells from rhombomere 2 migrate to the first pharyngeal (mandibular) pouch and also generate the ganglion for the trigeminal nerve; cells from rhombomere 4 populate the second pharyngeal pouch (forming the hyoid cartilage of the neck) and also produce the ganglia for the geniculate and vestibuloacoustic nerves; and cells from rhombomere 6 migrate into the third and fourth pharyngeal pouches to form thymus, parathyroid, and thyroid glands as well as the ganglia of the vagus and glossopharyngeal nerves. If the neural crest is removed from those regions including rhombomere 6, the thymus, parathyroid glands, and thyroid fail to form (Bockman and Kirby, 1984). Neural crest cells from rhombomeres 3 and 5 do not migrate through the mesoderm surrounding them, but either die or enter into the streams of crest cells on either side of them (Graham et al., 1993; Sechrist et al., 1993).

In mammalian embryos, cranial neural crest cells migrate before the neural tube is closed (Tan and Morriss-Kay, 1985) and give rise to the facial mesenchyme (Johnston et al., 1975). The crest cells originating in the forebrain and midbrain contribute to the nasal process, palate, and mesenchyme of the first pharyngeal pouch. This structure becomes part of the gill apparatus in fishes; in humans, it gives rise to the jawbones and to the incus and malleus bones of the middle ear. The neural crest cells originating in the anterior hindbrain region generate the mesenchyme of the second pharyngeal arch, which generates the human stapes bone as well as much of the facial cartilage (Figure 7.40C; Table 7.2). The cervical neural crest cells give rise to the mesenchyme of the third, fourth, and sixth pharyngeal arches (the fifth degenerates in humans), which produce the neck bones and muscles.

As we discussed in Chapter 2, a series of genes appears to specify the fates of these cranial neural crest cells and specifies their migration pathways. Chisaka and Capecchi (1991) "knocked out" the *Hoxa-3* gene from inbred mice and found that these mutant mice had severely deficient or absent thymuses, thyroids, and parathyroid glands, shortened neck ver-

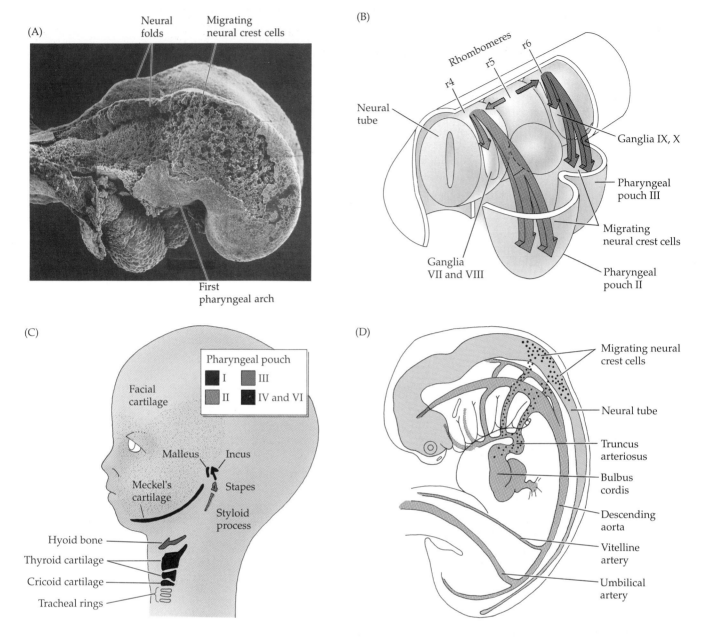

FIGURE 7.40
Neural crest cell migration in the mammalian head. (A) Scanning electron micrograph of a rat embryo with part of the lateral ectoderm removed from the surface. Neural crest migration can be seen over the midbrain, and the column of neural crest cells migrating into the future first pharyngeal arch is evident. (B) Analysis of cranial neural crest cell migration from rhombomeres 4–6 in the mouse suggests that there is a major migration into the pharyngeal arches and a minor migration to form the cranial nerve ganglia. (C) Structures formed in the human face by the ectomesenchymal cells of the neural crest. The cartilaginous elements of the pharyngeal pouches are keyed, and the lightly shaded region indicates the facial skeleton produced by anterior regions of the cephalic crest. (D) Formation of the truncoconal septa (between the aorta and the pulmonary vein) from the neural crest cells of the cardiac crest. Human hindbrain crest cells migrate to pharyngeal arches 4 and 6 during the fifth week of gestation and enter the truncus arteriosus to generate the septa. (A from Tan and Morriss-Kay, 1985, courtesy of S.-S. Tan; B after Sechrist et al., 1993; D after Kirby and Waldo, 1990.)

TABLE 7.2
Some derivatives of the pharyngeal arches

Pharyngeal arch	Skeletal elements (neural crest plus mesoderm)	Arches, arteries (mesoderm)	Muscles (mesoderm)	Cranial nerves (neural tube)
1	Incus and malleus (from neural crest); mandible, maxilla, and temporal bone regions (from crest dermal mesenchyme)	Maxillary branch of the carotid artery (to the ear, nose, and jaw)	Jaw muscles; floor of mouth; muscles of the ear and soft palate	Maxillary and mandibular divisions of trigeminal nerve (V)
2	Stapes bone of the middle ear; styloid process of temporal bone; part of hyoid bone of neck (all from neural crest cartilage)	Arteries to the ear region: cortico-tympanic artery (adult); stapedial artery (embryo)	Muscles of facial expression; jaw and upper neck muscles	Facial nerve (VII)
3	Lower rim and greater horns of hyoid bone (from neural crest)	Common carotid artery; root of internal carotid	Stylopharyngeus (to elevate the pharynx)	Glossopharyngeal nerve (IX)
4	Laryngeal cartilages (from lateral plate mesoderm)	Arch of aorta; right subclavian artery; original spouts of pulmonary arteries	Constrictors of pharynx and vocal cords	Superior laryngeal branch of vagus nerve (X)
6	Laryngeal cartilages (from lateral plate mesoderm)	Ductus arteriosus; roots of definitive pulmonary arteries	Intrinsic muscles of larynx	Recurrent laryngeal branch of vagus nerve (X)

Source: Based on Larsen, 1992.

tebrae, and malformed major heart vessels. It appears the *Hoxa-3* genes are responsible for specifying the cranial neural crest cells that give rise to the neck cartilage and pharyngeal arch derivatives. However, this gene does not control the smaller migration of neural crest cells that form the cranial neural ganglia.

Developmental potency of the cephalic neural crest cells

From the preceding discussion, it would appear that all neural crest cells are originally identical in their potencies. This, however, is not the case. Here, again, cranial crest cells differ from trunk crest cells, for only the cells of the cranial neural crest are able to produce the cartilage of the head. Moreover, when transplanted into the trunk region, the cranial neural crest participates in forming trunk cartilage that normally does not arise from neural crest components. In at least some cases, these cranial neural crest cells are instructed quite early as to what tissues they can form. Noden (1983) removed regions of chick neural crest that would normally seed the second branchial arch and replaced them with cells that would migrate into the first branchial arch. These host embryos developed two sets of lower jaw structures, since the graft-derived cells also produced a mandible. The bases for this instruction will be discussed in Chapter 17.

The cardiac neural crest

As we will detail in Chapter 9, the heart originally forms in the neck region, directly beneath the pharyngeal arches, and it should not be surprising that it acquires cells from the neural crest. However, the contributions of the neural crest to the heart have only recently been appreciated. The caudal region of the cephalic neural crest is sometimes designated the **cardiac neural crest**, since these neural crest cells (and only these particular neural crest cells) can generate the endothelium of the aortic arch arteries and the septum between the aorta and the pulmonary artery (Figure 7.40D). In the chick, the cardiac neural crest lies above the neural tube region from rhombomere 7 through the spinal cord apposing the third somite, and these crest cells migrate into pharyngeal arches 3, 4, and 6. The cardiac neural crest is unique in that if it is removed and replaced by anterior cephalic or trunk neural crests, cardiac abnormalities (notably the failure of aortic–pulmonary separation) occur. Thus, the cardiac crest is already determined to generate cardiac cells, and the other regions of the neural crest cannot substitute for it (Kirby, 1989; Kuratani and Kirby, 1991). Congenital heart defects in humans often occur with defects in the parathyroid, thyroid, or thymus glands. It would not be surprising if these were linked to defects in the migration of cells from the neural crest.

The epidermis and the origin of cutaneous structures

Origin of epidermal cells

The cells covering the embryo after neurulation form the presumptive epidermis. Originally, this tissue is one cell-layer thick, but in most vertebrates this shortly becomes a two-layered structure. The outer layer gives rise to the **periderm,** a temporary covering that is shed once the bottom layer differentiates to form a true epidermis. The inner layer, called the **basal layer** (or **stratum germinativum**), is a germinal epithelium that gives rise to all the cells of the epidermis (Figure 7.41). The basal layer divides to give rise to another, outer population of cells that constitutes the **spinous layer.** These two epidermal layers are referred to as the **Malpighian layer.** The cells of the Malpighian layer divide to produce the granular layer of the epidermis, so called because the cells are characterized by granules of the protein **keratin.** Unlike the cells remaining in the Malpighian layer, the cells of the granular layer do not divide, but begin to differentiate into epidermal skin cells, or **keratinocytes.** The keratin granules become more prominent as the cells of the granular layer age and migrate outward. Here, they form the **cornified layer (stratum corneum)** in which the cells have become flattened sacs of keratin protein. The depth of the cornified layer varies from site to site, but it is usually 10–30 cells thick. The nuclei of these cells have been pushed to one edge of the cell. Shortly after birth, the cells of the cornified layer are shed and are replaced

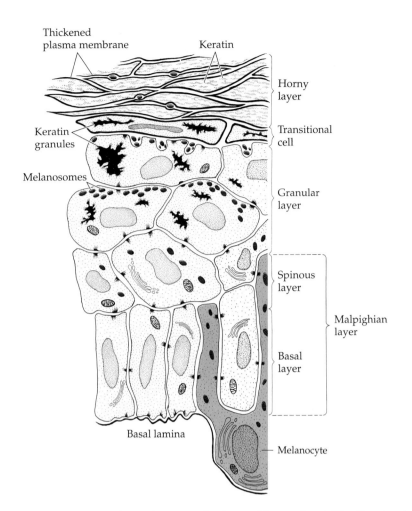

Thickened
plasma membrane

Keratin

Keratin
granules

Melanosomes

Basal lamina

Horny
layer

Transitional
cell

Granular
layer

Spinous
layer

Basal
layer

Malpighian
layer

Melanocyte

FIGURE 7.41
Diagram of the layers of the human epidermis. The basal cells are mitotically active, whereas the fully keratinized cells characteristic of external skin are dead and shed off. The keratinocytes obtain this pigment from the transfer of melanosomes from the processes of melanocytes that reside in the basal layer. (After Montagna and Parakkal, 1974.)

by new cells coming up from the granular layer. Throughout life, the dead keratinized cells of the cornified layer are being shed (we humans lose about 1.5 grams each day*) and are replaced by new cells, the source of which is the mitotic cells of the Malpighian layer. The pigment cells from the neural crest also reside in the Malpighian layer, where they transfer their pigment sacs (melanosomes) to the developing keratinocyte.

The epidermal stem cells of the Malpighian layer are bound to the basement membrane by their integrin proteins. However, as they become committed to differentiate, they downregulate their integrins and eventually lose them as the cells migrate into the spinous layer (Jones and Watt, 1993).

There are two main growth factors stimulating the development of the epidermis. The first of these is **transforming growth factor-α** (TGF-α). TGF-α is made by the basal cells and stimulates their own division. When a growth factor is made by the same cell that receives it, the factor is called an **autocrine growth factor.** Such factors have to be carefully regulated because if their levels are elevated, more cells are rapidly produced. In adult skin, a cell born in the Malpighian layer takes roughly 8 weeks to reach the stratum corneum and remains in the cornified layer about 2 weeks. In individuals with psoriasis, a disease characterized by the exfoliation of enormous amounts of epidermal cells, the time in the cornified layer is only two days (Weinstein and van Scott, 1965; Halprin, 1972). This

---

*Most of this skin becomes "house dust" atop furniture and floors. Should you doubt this, burn some of the dust; it will smell just like singed skin.

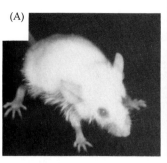
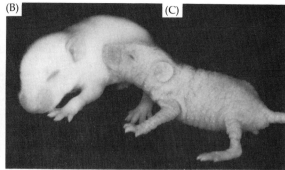

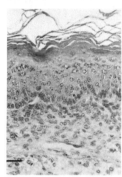
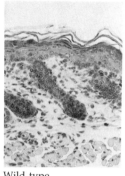
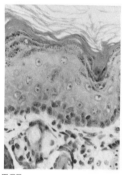

KGF   Wild type   TGF-α

FIGURE 7.42

Growth factors and epidermal proliferation. Upper plate shows (A) a transgenic mouse expressing low levels of KGF in its keratinocytes. Note the sparcity of hair around the legs, eyes, and nose. (B) A wild-type mouse. (C) A mouse (littermate of B) that is expressing high levels of TGF-α in its keratinocytes. It has scaly skin and very little hair. Below each mouse is a cross section through its skin. The mouse expressing excess KGF has no hair follicles and an increased number of basal epidermal cells. The mouse overexpressing TGF-α has very extensive layers of keratinized epithelia, which it sheds. (From Vassar and Fuchs, 1991, and Guo et al., 1993. Photographs courtesy of E. Fuchs.)

condition has been linked to the overexpression of TGF-α (which occurs secondarily to an immune inflammation) (Elder et al., 1989). Similarly, if the *TGF-α* gene is linked to a promoter for keratin 14 (one of the major skin proteins) and inserted into the mouse pronucleus, the transgenic mice activate the *TGF-α* gene in their skin cells and cannot downregulate it. The result is a mouse with scaly skin, stunted hair growth, and an enormous surplus of keratinized epidermis over its single layer of basal cells (Figure 7.42C; Vassar and Fuchs, 1991).

The other growth factor needed for epidermal production is **keratinocyte growth factor** (KGF), which is produced by the fibroblasts of the underlying (mesodermally derived) dermis. When a growth factor is made by a neighboring cell, it is called a **paracrine growth factor.** KGF is received by the basal cells above the dermal fibroblasts and is thought to regulate the proliferation of these basal cells. If the *KGF* gene is fused with the keratin 14 promoter and transgenic mice are made, the KGF becomes autocrine. The resulting transgenic mice (Figure 7.42A) have a thickened epidermis, baggy skin, far too many basal cells, and no hair follicles—not even whisker follicles (Guo et al., 1993). These basal cells are "forced" into the epidermal pathway of differentiation. The alternative for the basal cell is to help generate the hair follicle.

Cutaneous appendages

The epidermis and dermis also interact at specific sites to create the sweat glands and the cutaneous appendages: hair, scales, or feathers (depending on the species). The first indication that a hair follicle will form at a particular place is an aggregation of cells in the basal layer of the epidermis. This aggregation is directed by the underlying dermal cells and occurs at different times and different places in the embryo. The basal cells elongate, divide, and sink into the dermis. The dermal cells respond to this ingression of basal epidermal cells by forming a small node (the **dermal papilla**) beneath the epidermal plug. The dermal papilla then pushes up and

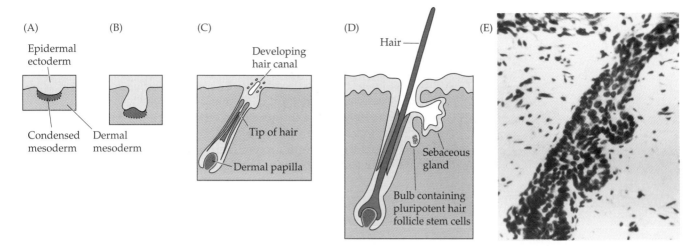

FIGURE 7.43
Development of the hair follicles in fetal human skin. (A) Basal epidermal cells become columnar and bulge slightly into the dermis. (B) Epidermal cells continue to proliferate, and dermal mesenchyme cells collect at the base of the primary hair germ. (C) Hair shaft differentiation begins in the elongated hair germ. (D) The keratinized hair shaft extends from the hair root, the secondary bud forms the sebaceous gland, and beneath it is a region that may contain the hair stem cells for the next cycle of hair production. (E) Photograph of elongated hair germ. (After Hardy, 1992, and Miller et al., 1993. Photograph courtesy of W. Montagna.)

stimulates the basal stem cells to divide more rapidly and to produce postmitotic cells that will differentiate into the keratinized hair shaft (see Hardy, 1992; Miller et al., 1993). Melanoblasts, which were present among the epidermal cells as they ingressed, differentiate into melanocytes and transfer their pigment to the shaft (Figure 7.43). As this is occurring, two epithelial swellings begin to grow on the side of the follicle. The cells of the lower swelling may retain a population of stem cells that will regenerate the hair shaft periodically when the shaft is shed (Pinkus and Mehregan, 1981; Cotsarelis et al., 1990). The cells of the upper bulge will form the **sebaceous glands,** which produce an oily secretion, **sebum.** In many mammals, including humans, the sebum mixes with the desquamated peridermal cells to form the whitish vernix caseosa, which surrounds the fetus at birth.

The first hairs in the human embryo are of a thin, closely spaced type called **lanugo.** This type of hair is usually shed before birth and is replaced (at least in part, by new follicles) by the short and silky **vellus.** Vellus remains on many parts of the human body usually considered hairless, such as the forehead and eyelids. In other areas of the body, vellus gives way to "terminal" hair. During a person's life, some of the follicles that produced vellus can later form terminal hair, and still later revert to vellus production. The armpits of infants, for instance, have follicles that produce vellus until adolescence. At that time, terminal shafts are generated. Conversely, in normal masculine "baldness," the scalp follicles revert back to producing unpigmented and very fine vellus hair (Montagna and Parakkal, 1974). The placement and pattern of hair, feathers, scales, and sweat glands involves the interactions of the dermis and the epidermis, and these will be discussed in more detail in Chapter 18. Just as there is a pluripotent neural stem cell whose offspring become neural and glial cells, so there appears to be a pluripotent epidermal stem cell whose progeny can become epidermis, sebaceous gland, or hair shaft.

Summary

In this chapter we have followed the differentiation of the embryonic ectoderm into a wide variety of tissues. We have seen that the ectoderm produces three sets of cells during neurulation: (1) the neural tube, which gives rise to the neurons, glia, and ependymal cells of the central nervous system; (2) the neural crest cells, which give rise to the peripheral nervous system, pigment cells, adrenal medulla, and certain areas of head cartilage; and (3) the epidermis of the skin, which contributes to the formation of cutaneous structures such as hair, feathers, scales, and sweat and sebaceous glands, as well as forming the outer protective covering of our bodies. We also observed how the interactions of epidermal cells are involved in generating the various tissues of the eye.

Later chapters discuss in more detail the induction of the neural tube and the coordinated development of the eye. Meanwhile, we will discuss the mechanisms by which neurons are directed to travel to specific sites, thus enabling the development of reflexes and behaviors.

LITERATURE CITED

Akitaya, T. and Bronner-Fraser, M. 1992. Expression of cell adhesion molecules during initiation and cessation of neural crest cell migration. *Dev. Dynam.* 194: 12–20.

Alvarez, I. S. and Schoenwolf, G. C. 1992. Expansion of surface epithelium provides the major extrinsic force for bending of the neural plate. *J. Exp. Zool.* 261: 340–348.

Anderson, D. J. and Axel, R. 1986. A bipotential neuroendocrine precursor whose choice of cell fate is determined by NGF and glucocorticoids. *Cell* 47: 1079–1090.

Artinger, K. B. and Bronner-Fraser, M. 1992. Partial restriction in the developmental potential of late emigrating avian neural crest cells. *Dev. Biol.* 149: 149–157.

Auerbach, R. 1954. Analysis of the developmental effects of a lethal mutation in the house mouse. *J. Exp. Zool.* 127: 305–329.

Baldwin, C. T. Hoth, C. F., Amos, J. A., da Silva, E. O. and Milunsky, A. 1992. An exonic mutation in the *HuP2* paired domain gene causes Waardenburg's syndrome *Nature* 355: 637–639.

Balinsky, B. I. 1975. *Introduction to Embryology,* 4th Ed. Saunders, Philadelphia.

Banks, M. S. and Bennett, P. J. 1988. Optical and photoreceptor immaturities limit the spatial and chromatic vision of human neonates. *J. Opt. Soc. Am.* 5: 2059–2079.

Bernstein, S. E., Russell, E. S. and Keighley, G. H. 1968. Two hereditary mouse anemias (*Sl/Sl*d and *W/W*v) deficient in response to erythropoietin. *Ann. N.Y. Acad. Sci.* 149: 475–485.

Berry, M., McConnell, P. and Sievers, J. 1980. Dendritic growth and the control of neuronal form. *Curr. Top. Dev. Biol.* 15: 67–101.

Bloom, W. and Fawcett. D. W. 1975. *Textbook of Histology,* 10th Ed. Saunders, Philadelphia.

Bockman, D. E. and Kirby, M. L. 1984. Dependence of thymus development on derivatives of the neural crest. *Science* 223: 498–500.

Bopp, D., Burri, M., Baumgartner, S., Frigerio, G. and Noll, M. 1986. Conservation of a large protein domain in the segmentation gene *paired* and in functionally related genes of *Drosophila. Cell* 47: 1033–1040.

Bronner-Fraser, M. 1986. Analysis of the early stages of trunk neural crest migration in avian embryos using monoclonal antibody HNK-1. *Dev. Biol.* 115: 44–55.

Bronner-Fraser, M. and Cohen, A. M. 1980. Analysis of the neural crest ventral pathway using injected tracer cells. *Dev. Biol.* 77: 130–141.

Bronner-Fraser, M. and Fraser, S. E. 1988. Cell lineage analysis reveals multipotency of some avian neural crest cells. *Nature* 335: 161–164.

Bronner-Fraser, M. and Fraser, S. 1989. Developmental potential of avian trunk neural crest cells in situ. *Neuron* 3: 755–766.

Bronner-Fraser, M. and Stern, C. 1991. Effects of mesodermal tissues on avian neural crest cell migration. *Dev. Biol.* 143: 213–217.

Burnside, B. 1971. Microtubules and microfilaments in newt neurulation. *Dev. Biol.* 26: 416–441.

Burnside, B. 1973. Microtubules and microfilaments in amphibian neurulation. *Am. Zool.* 13: 989–1006.

Chabot, B., Stephenson, D. A., Chapman, V. M., Besmer, P. and Bernstein, A. 1988. The proto-oncogene c-*kit* encoding a transmembrane tyrosine kinase receptor maps to the mouse *W* locus. *Nature* 335: 88–89.

Chisaka, O. and Capecchi, M. 1991. Regionally restricted developmental defects resulting from targeted disruption of the mouse homeobox gene *Hox1.5. Nature* 350: 473–479.

Chun, L. L. Y. and Patterson, P. H. 1977. Role of nerve growth factor in development of rat sympathetic neurons in vitro. Survival, growth, and differentiation of catecholamine production. *J. Cell Biol.* 75: 694–704.

Ciment, G. 1990. The melanocyte/Schwann cell progenitor: A bipotent intermediate in the neural crest lineage. *Comm. Dev. Neurobiol.* 1: 207–223.

Cotsarelis, G., Sun, T.-T. and Lavker, R. M. 1990. Label-retaining cells reside in the bulge area of pilosebaceous unit: Implications for follicular stem cells, hair cycle and skin carcinogenesis. *Cell* 61: 1329–1337.

Coulombe, J. N. and Bronner-Fraser, M. 1987. Cholinergic neurones acquire adrenergic neurotransmitters when transplanted into an embryo. *Nature* 324: 569–572.

Coulombre, A. J. 1956. The role of intraocular pressure in the development of the chick eye. I. Control of eye size. *J. Exp. Zool.* 133: 211–225.

Coulombre, A. J. 1965. The eye. *In* R. DeHaan and H. Ursprung (eds.), *Organogenesis.* Holt, Rinehart & Winston, New York, pp. 217–251.

Crossin, K. L., Chuong, C.-M. and Edelman, G. M. 1985. Expression sequences of cell adhesion molecules. *Proc. Natl. Acad. Sci. USA* 82: 6942–6946.

Davenport, R. W., Dou, P., Rehder, V. and Kater, S. B. 1993. A sensory role for neuronal growth cone filopodia. *Nature* 361: 721–724.

Desmond, M. E. 1982. A description of the occlusion of the lumen of the spinal cord in early human embryos. *Anat. Rec.* 204: 89–93.

Desmond, M. E. and Field, M. C. 1992. Evaluation of neural fold fusion and coincident initiation of spinal cord occlusion in the chick embryo. *J. Comp. Neurol.* 319: 246–260.

Desmond, M. E. and Schoenwolf, G. C. 1986. Evaluation of the roles of intrinsic and extrinsic factors in occlusion of the spinal neurocoel during rapid brain enlargement in the chick embryo. *J. Embryol. Exp. Morphol.* 97: 25–46.

Detrick, R. J., Dickey, D. and Kintner, C. R. 1990. The effects of N-cadherin misexpression on morphogenesis in *Xenopus* embryos. *Neuron* 4: 493–506.

Detwiler, S. R. 1932. Experimental observations upon the developing retina. *J. Comp. Neural.* 55: 473–492.

Edmondson, J. C. Liem, R. K. H., Kuster, J. C. and Hatten, M. E. 1988. Astrotactin: A novel neuronal cell surface antigen that mediates neuronal–astroglial interactions in cerebellar microcultures. *J. Cell Biol.* 106: 505–517.

Eichele, G. 1992. Budding thoughts. *The Sciences* (Jan., 1992): 30–36.

Elder, J. T. and eight others. 1989. Overexpression of transforming growth factor in psoriatic epidermis. *Science* 243: 811–814.

Epstein, D. J., Vekemans, M. and Gros, P. 1991. *splotch* (Sp[2H]), a mutation affecting development of the mouse neural tube, shows a deletion within the paired homeodomain of *Pax-3. Cell* 67: 767–774.

Erickson, C. A. and Weston, J. A. 1983. An SEM analysis of neural crest migration in the mouse. *J. Embryol. Exp. Morphol.* 74: 97–118.

Erickson, C. A., Tosney, K. W. and Weston, J. A. 1980. Analysis of migrating behaviour of neural crest and fibroblastic cells in embryonic tissues. *Dev. Biol.* 77: 142–156.

Erickson, C. A., Duong, T. D. and Tosney, K. W. 1992. Descriptive and experimental analysis of the dispersion of neural crest cells along the dorsolateral pathway and their entry into ectoderm in the chick embryo. *Dev. Biol.* 151: 251–272.

Fesenko, E. E., Kolenikou, S. S. and Lyubarsky, A. L. 1985. Induction by cyclic GMP of cationic conduction in plasma membranes of retinal rod outer segment. *Nature* 313: 310–313.

Fishell, G. and Hatten, M. E. 1991. Astrotactin provides a receptor system for glia-guided neuronal migration. *Development* 113: 755–765.

Forscher, P. and Smith, S. J. 1988. Actions of cytochalasins on the organization of actin filaments and microtubules in a neural growth cone. *J. Cell Biol.* 107: 1505–1516.

Fraser, S. and Bronner-Fraser, M. 1991. Migrating neural crest cells in the trunk of the avian embryo are multipotent. *Development* 112: 913–920.

Fujimori, T., Miyatani, S. and Takeichi, M. 1990. Ectopic expression of N-cadherin perturbs histogenesis in *Xenopus* embryos. *Development* 110: 97–104.

Fujita, S. 1964. Analysis of neuron differentiation in the central nervous system by tritiated thymidine autoradiography. *J. Comp. Neurol.* 122: 311–328.

Fujita, S. 1966. Application of light and electron microscopy to the study of the cytogenesis of the forebrain. *In* R. Hassler and H. Stephen (eds.), *Evolution of the Forebrain.* Plenum, New York, pp. 180–196.

Fukada, K. 1980. Hormonal control of neurotransmitter choice in sympathetic neuron cultures. *Nature* 287: 553–555.

Gallera, J. 1971. Primary induction in birds. *Adv. Morphogenet.* 9: 149–180.

Gao, W.-Q. and Hatten, M. E. 1993. Neuronal differentiation rescued by implantation of *Weaver* granule cell precursors into wild-type cerebellar cortex. *Science* 260: 367–369.

Gao. W.-Q., Liu, X.-L. and Hatten, M. E. 1992. The *weaver* gene encodes a nonautonomous signal for CNS neuronal differentiation. *Cell* 68: 841–854.

Geissler, E. N., Ryan, M. A. and Housman, D. E. 1988. The dominant white–spotting (*W*) locus of the mouse encodes the *c-kit* proto-oncogene. *Cell* 55: 185–192.

Golden, J. A. and Chernoff, G. F. 1993. Intermittent pattern of neural tube closure in two strains of mice. *Teratology* 47: 73–80.

Goldstein, R. S., Teillet, M.-A. and Kalcheim, C. 1990. The microenvironment created by grafting rostral half-somites is mitogenic for neural crest cells. *Proc. Natl. Acad. Sci. USA* 87: 4476–4480.

Goldowitz, D. and Mullen, R. J. 1982. Granule cell as a site of gene action in the *weaver* mouse cerebellum: Evidence from heterozygous mutant chimera. *J. Neurosci.* 2: 1474–1485.

Gould, S. J. 1977. *Ontogeny and Phylogeny.* Harvard University Press, Cambridge, MA.

Goulding, M. D., Chalepakis, G., Deutsch, U., Erselius, J. and Gruss, P. 1991. Pax-3, a novel murine DNA binding protein expressed during early neurogenesis. *EMBO J.* 10: 1135–1147.

Graham, A., Heyman, I. and Lumsden, A. 1993. Even-numbered rhombomeres control the apoptotic elimination of neural crest cells from odd-numbered rhombomeres of the chick hindbrain. *Development* 119: 233–245.

Guo, L., Yu, Q.-C. and Fuchs, E. 1993. Targetting expression of keratinocyte growth factor to keratinocytes elicits striking changes in epithelial differeentiation in transgenic mice. *EMBO J.* 12: 973–986.

Guthrie, S. and Lumsden, A. 1991. Formation and regeneration of rhombomere boundaries in the developing chick hindbrain. *Development* 112: 221–229.

Gregory, W. A., Edmondson, J. C., Hatten, M. E. and Mason, C. A. 1988. Cytology and neural–glial apposition of migrating cerebellar granule cells in vitro. *J. Neurosci.* 8: 1728–1738.

Gruss, P. and Walther, C. 1992. Pax in development. *Cell* 69: 719–722.

Halprin, K. M. 1972. Epidermal "turnover time"—a reexamination. *J. Invest. Dermatol.* 86: 14–19.

Hardy, M. H. 1992. The secret life of the hair follicle. *Trends Genet.* 8: 55–61.

Harrison, R. G. 1907. Observations on the living developing nerve fiber. *Anat. Rec.* 1: 116–118.

Harrison, R. G. 1910. The outgrowth of a nerve fiber as a mode of protoplasmic movement. *J. Exp. Zool.* 9: 787–846.

Hatten, M. E. 1990. Riding the glial monorail: A common mechanism for glial-guided neuronal migration in different regions of the mammalian brain. *Trends Neurosci.* 13: 179–184.

Henry, E. W. and Sidman, R. L. 1988. Long lives for homozygous *trembler* mutant mice despite virtual absence of peripheral nerve myelin. *Science* 241: 344–346.

Hilfer, S. R. and Yang, J.-J. W. 1980. Accumulation of CPC-precipitable material at apical cell surfaces during formation of the optic cup. *Anat. Rec.* 197: 423–433.

Hill, R. E. and nine others. 1991. Mouse *small eye* results from mutations in a *paired*-like homeobox-containing gene. *Nature* 354: 522–525.

His, W. 1886. Zur Geschichte des menschlichen Rückenmarks und der Nervenwurzeln. *Ges. Wissensch.* BD 13, S. 477.

Holt, A. B., Cheek, D. B., Mellitz, E. D. and Hill, D. E. 1975. Brain size and the relation of the primate to the non-primate. *In* D. B. Cheek (ed.), *Fetal and Postnatal Cellular Growth: Hormones and Nutrition.* Wiley, New York, pp. 23–44.

Huettner, A. F. 1949. *Fundamentals of Comparative Embryology of the Vertebrates,* 2nd Ed. Macmillan, New York.

Hunt, P. and Krumlauf, R., 1991. Deciphering the *Hox* code: Clues to patterning branchial regions of the head. *Cell* 66: 1075–1078.

Jacobson, A. G. 1981. Morphogenesis of the neural plate and tube. *In* I. G. Connely et al. (eds.), *Morphogenesis and Pattern Formation.* Raven Press, New York, pp. 233–263.

Jacobson, A, G. and Gordon, R. 1976. Changes in the shape of the developing vertebrate nervous system analyzed experimentally, mathematically, and by computer simulation. *J. Exp. Zool.* 197: 191–246.

Jacobson, A. G. and Sater, A. K. 1988. Features of embryonic induction. *Development* 104: 341–359.

Jacobson, M. 1968. Cessation of DNA synthesis in retinal ganglion cells correlated with the time of specification of their central connections. *Dev. Biol.* 17: 219–232.

Jacobson, M. 1991. *Developmental Neurobiology,* 2nd Ed. Plenum, New York.

Johnston, M. C., Sulik, K. K., Webster, W. S. and Jarvis, B. L. 1985. Isotretinoin embryopathy in a mouse model: Cranial neural crest involvement. *Teratology* 31: 26A.

Jones, P. H. and Watt, F. M. 1993. Separation of human epidermal stem cells from transit amplifying cells of the basis of differences in integrin function and expression. *Cell* 73: 713–724.

Jordan, T. and seven others. 1992. The human *PAX6* gene is mutated in two patients with aniridia. *Nature Genet.* 1: 328–332.

Kahn, C. R., Coyle, J. T. and Cohen, A. M. 1980. Head and trunk neural crest *in vitro*: Autonomic neuron differentiation. *Dev. Biol.* 77: 340–348.

Kalcheim, C. R. and Neufeld, G. 1990. Expression of basic fibroblast growth factor in the nervous system of early avian embryos. *Development* 109: 203–215.

Kalcheim, C., Barde, Y.-A., Thoenen, H. and LeDourain, N. M. 1987. In vivo effect of brain-derived neurotrophic factor on the survival of neural crest precursor cells of the dorsal root ganglia. *EMBO J.* 6: 2871–2873.

Karfunkel, P. 1972. The activity of microtubules and microfilaments in neurulation in the chick. *J. Exp. Zool.* 181: 289–302.

Keller, R., Shih, J., Sater, A. K. and Moreno, C. 1992. Planar induction of convergence and extension of the neural plate by the organizer of *Xenopus. Dev. Dynam.* 193: 218–234.

Kessel, M. and Gruss, P. 1990. Murine developmental control genes. *Science* 249: 374–379.

Kirby, M. L. 1987. Cardiac morphogenesis: Recent research advances. *Ped. Res.* 21: 219–224.

Kirby, M. L. 1989. Plasticity and predetermination of mesencephalic and trunk neural crest transplanted into the region of the cardiac neural crest. *Dev. Biol.* 134: 401–412.

Kirby, M. L. and Waldo, K. L. 1990. Role of neural crest in congenital heart disease. *Circulation* 82: 332–340.

Komuro, H. and Rakic, P. 1992. Selective role of N-type calcium channels in neuronal migration. *Science* 157: 806–809.

Krauss, S., Maden, Holder, N. and Wilson, S. W. 1992. Zebrafish pax[b] is involved in the formation of the midbrain–hindbrain boundary. *Nature* 360: 87–89.

Kuratani, S. C. and Kirby, M. L. 1991. Initial migration and distribution of the cardiac neural crest in the avian embryo: An introduction to the concept of the circumpharyngeal crest. *Am. J. Anat.* 191: 215–227.

Lamoureux, P., Buxbaum, R. E. and Heidemann, S. R. 1989. Direct evidence that growth cones pull. *Nature* 340: 159–162.

Landis, S. C. and Patterson, P. H. 1981. Neural crest cell lineages. *Trends Neurosci.* 4: 172–175.

Larsen, W. J. 1992. *Human Embryology.* Churchill Livingstone, New York.

LeDouarin, N. and Smith, J. 1988. Development of the peripheral nervous system from the neural crest. *Annu. Rev. Cell Biol.* 4: 375–404.

LeDouarin, N. M. and Teillet, M.-A. 1973. The migration of neural crest cells to the wall of the digestive tract in avian embryo. *J. Embryol. Exp. Morphol.* 30: 31–48.

LeDouarin, N. M. and Teillet, M.-A. 1974. Experimental analysis of the migration and differentiation of neuroblasts of the autonomic nervous system and of neuroectodermal mesenchyme derivatives, using a biological cell marking technique. *Dev. Biol.* 41: 162–184.

LeDouarin, N. M., Renaud, D., Teillet, M.-A. and LeDouarin, G. H. 1975. Cholinergic differentiation of presumptive adrenergic neuroblasts in interspecific chimeras after heterotopic transplantation. *Proc. Natl. Acad. Sci. USA* 72: 728–732.

LeLièvre, C. S. and LeDouarin, N. M. 1975. Mesenchymal derivatives of the neural crest: Analysis of chimaeric quail and chick embryos. *J. Embryol. Exp. Morphol.* 34: 125–154.

Letourneau, P. C. 1977. Regulation of neuronal morphogenesis by cell–substratum adhesion. *Soc. Neurosci. Symp.* 2: 67–81.

Letourneau, P. C. 1979. Cell substratum adhesion of neurite growth cones, and its role in neurite elongation. *Exp. Cell Res.* 124: 127–138.

Löfberg, J., Perris, R. and Epperlin, H. H. 1989. Timing in the regulation of neural crest cell migration: Retarded maturation of regional extracellular matrix inhibits pigment cell migration in embryos of the white axolotl mutant. *Dev. Biol* 131: 168–181.

Loring, J. F. and Erickson, C. A. 1987. Neural crest cell migratory pathways in the trunk of the chick embryo. *Dev. Biol.* 121: 220–236.

Lumsden, A. 1988. Multipotent cells in the avian neural crest. *Trends Neurosci.* 12: 81–83.

Lumsden, A. and Guthrie, S. 1991. Alternating patterns of cell surface properties and neural crest cell migration during segmentation of the chick embryo. *Development* [Supp.] 2: 9–15.

Lumsden, A. and Keynes, R. 1989. Segmental patterns of neuronal development in the chick hindbrain. *Nature* 337: 424–428.

Lumsden, A. and Wilkinson, D. 1990. The promise of gene ablation. *Nature* 347: 335–336.

Mann, I. 1964. *The Development of the Human Eye.* Grune and Stratton, New York.

Matsui, Y., Zsebo, K. M. and Hogan, B. L. M. 1990. Embryonic expression of a maematopoietic growth factor encoded by the *Sl* locus and the ligand for *c-kit. Nature* 347: 667–669.

Maxwell, G. D. and Forbes, M. E. 1987. Exogenous basement membrane-like matrix stimulates adrenergic development in avian neural crest cultures. *Development* 101: 767–776.

Mayer, T. C. 1973. The migratory pathway of neural crest cells into the skin of mouse embryos. *Dev. Biol.* 34: 39–46.

Mayer, T. C. and Green, T. C. 1968. An experimental analysis of the pigment defect caused by mutations at the *W* and *Sl* loci in mice. *Dev. Biol.* 18: 62–75.

McConnell, S. K. and Kaznowski, C. E. 1991. Cell cycle dependence of laminar determination in developing cerebral cortex. *Science* 254: 282–285.

McMahon, A. P. and Bradley, A. 1990. The *Wnt-1* (int-1) proto-oncogene is required for the development of a large region of the mouse brain. *Cell* 62: 1073–1085.

McMahon, A. P., Joyner, A. L., Bradley, A. and McMahon, J. A. 1992. The midbrain–hindbrain phenotype of *wnt-1⁻/wnt-1⁻* mice results from stepwise deletion of *engrailed*-expressing cells by 9.5 days postcoitum. *Cell* 69: 581–595.

Meier, S. 1981. Development of the chick embryo mesoblast: Morphogenesis of the prechordal plate and cranial segments. *Dev. Biol.* 83: 49–61.

Miller, S. J., Lavker, R. M. and Sun, T.-T. 1993. Keratinocyte stem cells of corneal, skin, and hair follicle. *Sem. Dev. Biol.* 4: 217–240.

Mintz, B. and Russell, E. S. 1957. Gene-induced embryological modifications of primordial germ cells in the mouse. *J. Exp. Zool.* 134: 207–237.

Montagna, W. and Parakkal, P. F. 1974. The piliary apparatus. *In* W. Montagna (ed.), *The Structure and Formation of Skin.* Academic Press, New York, pp. 172–258.

Montagu, M. F. A. 1962. Time, morphology, and neoteny in the evolution of man. *In* M. F. A. Montagu (ed.), *Culture and Evolution of Man.* Oxford University Press, New York.

Moore K. L. and Persaud, T. V. N. 1993. *Before We Are Born: Essentials of Embryology and Birth Defects.* W. B. Saunders, Philadelphia.

Morowitz, H. J. and Trefil, J. S. 1992. *The Facts of Life: Science and the Abortion Controversy.* Oxford University Press, New York.

Nagele, R. G. and Lee, H. Y. 1980. Studies on the mechanism of neurulation in the chick: Microfilament-mediated changes in cell shape during uplifting of neural folds. *J. Exp. Zool.* 213: 391–398.

Nagele, R. G. and Lee, H. Y. 1987. Studies in the mechanism of neurulation in the chick. Morphometric analysis of the relationship between regional variations in cell shape and sites of motive force generation. *J. Exp. Biol.* 24: 197–205.

Newgreen, D. F. and Gooday, D. 1985. Control of onset of migration of neural crest cells in avian embryos: Role of Ca⁺⁺ dependent cell adhesions. *Cell Tiss. Res.* 239: 329–336.

Newgreen, D. F., Scheel, M. and Kaster, V. 1986. Morphogenesis of sclerotome and neural crest cells in avian embryos. *In vivo* and *in vitro* studies on the role of notochordal extracellular material. *Cell Tissue Res.* 244: 299–313.

Nichols, D. H. 1981. Neural crest formation in the head of the mouse embryo as observed using a new histôlogical technique. *J. Embryol. Exp. Morphol.* 64: 105–120.

Nichols, D. H. and Weston, J. A, 1977. Melanogenesis in cultures of peripheral nervous tissue. I. The origin and prospective fates of cells giving rise to melanocytes. *Dev. Biol.* 60: 217–225.

Noden, D. M. 1978. The control of avian cephalic neural crest cytodifferentiation. I. Skeletal and connective tissue. *Dev. Biol.* 69: 296–312.

Noden, D. M. 1983. The role of the neural crest in patterning of avian cranial skeletal, connective, and muscle tissues. *Dev. Biol.* 96: 144–165.

Orr-Urteger, A., Avivi, A., Zimmer, Y., Givol, D., Yarden, Y. and Lonai, P. 1990. Developmental expression of *c-kit*, a proto-oncogene encoded by the *W* locus. *Development* 104: 911–924.

O'Rourke, N. A., Dailey, M. E., Smith, S. J. and McConnell, S. K. 1992. Diverse migratory pathways in the developing cerebral cortex. *Science* 258: 299–302.

Papaconstantinou, J. 1967. Molecular aspects of lens cell differentiation. *Science* 156: 338–346.

Paton, D. and Craig, J. A. 1974. Cataracts: Development, diagnosis, and management. *Ciba Clin. Symp.* 26(3): 2–32.

Patten, B. M. 1971. *Early Embryology of the Chick,* 5th Ed. McGraw-Hill, New York.

Perris, R. and Bronner-Fraser, M. 1989. Recent advances in defining the role of the extracellular matrix in neural crest development. *Comm. Dev. Neurobiol.* 1: 61–83.

Perris, R., von Boxburg, Y. and Löfberg, J. 1988. Local embryonic matrices determine region-specific phenotypes in neural crest cells. *Science* 241: 86–89.

Perris, R., Löfberg, J. Fällström, C., von Boxburg, Y., Olsson, L. and Newgreen, D. F. 1990. Structural and compositional divergencies in the extracellular matrix encountered by neural crest cells in the *white* mutant axlotl embryo. *Development* 109: 533–551.

Piatigorsky, J. 1981. Lens differentiation in vertebrates: A review of cellular and molecular features. *Differentiation* 19: 134–153.

Pinkus, H. and Mehregan, A. H. H. 1981. *A Guide to Dermohistopathology*. Appleton Century Crofts, New York.

Placzek, M., Yamada, T., Tessier-Lavigne, M., Jessell. T. and Dodd, J. 1991. Control of dorsoventral pattern in vertebrate neural development: Induction and polarizing properties of the floor plate. *Development* [Suppl. 2]: 105–122.

Pomeranz, H. D., Rothman, T. P. and Gershon, M. D. 1991. Colonization of the postumbilical bowel by cells derived from the sacral neural crest: Direct tracing of cell migration using an intercalating probe and a replication-deficient retrovirus. *Development* 111: 647–655.

Poole, T. J. and Thiery, J. P. 1986. *In* H. C. Slavkin (ed.), *Progress in Clinical and Biological Research*, Vol 217. Alan R. Liss, New York, pp. 235–238.

Portmann, A. 1941. Die Tragzeiten der Primaten und die Dauer der Schwangerschaft beim Menschen: Ein Problem der vergleichen Biologie. *Rev. Suisse Zool.* 48: 511–518.

Portmann, A. 1945. Die Ontogenese des Menschen als Problem der Evolutionsforschung. *Verh. Schweiz. Naturf. Ges.* 125: 44–53.

Pratt, R. M., Larsen, M. A. and Johnston, M. C. 1975. Migration of cranial neural crest cells in a cell-free, hyaluronate-rich matrix. *Dev. Biol.* 44: 298–305.

Rakic, P. 1972. Mode of cell migration to superficial layers of fetal monkey neocortex. *J. Comp. Neurol.* 145: 61–84.

Rakic, P. 1974. Neurons in rhesus visual cortex: Systematic relation between time of origin and eventual disposition. *Science* 183: 425–427.

Rakic, P. 1975. Cell migration and neuronal ectopias in the brain. *In* D. Bergsma (ed.), *Morphogenesis and Malformations of Face and Brain*. Birth Defects Original Article Series II (7): 95–129.

Rakic P. and Goldman, P. S. 1982. Development and modifiability of the cerebral cortex. *Neurosci. Rev.* 20: 429–611.

Rakic, P. and Sidman, R. L. 1973. Organization of cerebellar cortex secondary to deficit of granule cells in *weaver* mutant mice. *J. Comp. Neurol.* 152: 133–162.

Ramón y Cajal, S. 1890. Sur l'origene et les ramifications des fibres neuveuses de la moelle embryonnaire. *Anat. Anz.* 5: 111–119.

Ranscht, B. and Bronner-Fraser, M. 1991. T-cadherin expression alternates with migrating neural crest cells in the trunk of the avian embryo. *Development* 111: 15–22.

Rawles, M. E. 1948. Origin of melanophores and their role in development of color patterns in vertebrates. *Physiol. Rev.* 28: 383–408.

Rickmann, M., Fawcett, J. W. and Keynes, R. J. 1985. The migration of neural crest cells and the growth of motor neurons through the rostral half of the chick somite. *J. Embryol. Exp. Morphol.* 90: 437–455.

Romanes, G. J. 1901. *Darwin and After Darwin*. Open Court Publishing, London.

Saha, M., Spann, C. L. and Grainger, R. M. 1989. Embryonic lens induction: More than meets the optic vesicle. *Cell Differ. Dev.* 28: 153–172.

Sarvella, P. A. and Russell, L. B. 1956. *Steel*, a new dominant gene in the mouse. *J. Hered.* 47: 123–128.

Sauer, F. C. 1935. Mitosis in the neural tube. *J. Comp. Neurol.* 62: 377–405.

Schoenwolf, G. C. 1991a. Cell movements driving neurulation in avian embryos. *Development* 2[Suppl.]: 157–168.

Schoenwolf, G. C. 1991b. Cell movements in the epiblast during gastrulation and neurulation in avian embryos. *In* R. Keller et al. (eds)., *Gastrulation*. Plenum, New York, pp. 1–28.

Schoenwolf, G. C. and Alvarez, I. S. 1989. Roles of neuroepithelial cell rearrangement and division in shaping of the avian neural plate. *Development* 106: 427–439.

Schoenwolf, G. C. and Desmond, N. E. 1984. Descriptive studies of the occlusion and reopening of the spinal canal of the early chick embryo. *Anat. Rec.* 209: 251–263.

Schweizer, G., Ayer-LeLièvre, C. and Le-Douarin, N. M. 1983. Restrictions in developmental capacities in the dorsal root ganglia during the course of development. *Cell Differ.* 13: 191–200.

Sechrist, J., Serbedzija, G. N., Scherson, T., Fraser, S. E. and Bronner-Fraser, M. 1993. Segmental migration of the hindbrain neural crest does not arise from its segmental origin. *Development* 118: 691–703.

Serbedzija, G. N., Bronner-Fraser, M. and Fraser, S. E. 1989. A vital dye analysis of the timing and pathways of avian trunk neural crest cell migration. *Development* 106: 809–816.

Sidman, R. L., Dickie, M. M. and Appel, S. H. 1964. Mutant mice (*quaking* and *jimpy*) with deficient myelination in the central nervous system. *Science* 144: 309–311.

Sieber-Blum, M. 1991. Role of neurotrophic factors BDNF and NGF in the commitment of pluripotent neural crest cells. *Neuron* 6: 949–955.

Sieber-Blum, M. and Sieber, F. 1984. Heterogeneity among early quail neural crest cells. *Dev. Brain Res.* 14: 241–246.

Smith, J. L. and Schoenwolf, G. C. 1989. Notochordal induction of cell wedging in the chick neural plate and its role in neural tube formation. *J. Exp. Zool.* 250: 49–62.

Smith, J. L. and Schoenwolf, G. C. 1991. Further evidence of extrinsic forces in bending of the neural plate. *J. Comp. Neurol.* 307: 225–236.

Solursh, M., Fisher, J. and Singley, C. T. 1979. The synthesis of hyaluronic acid by ectoderm during early organogenesis in the chick embryo. *Differentiation* 14: 77–85.

Spieth, J. and Keller, R. E. 1984. Neural crest cell behavior in white and dark larvae of *Ambystoma mexicanum*: Differences in cell morphology, arrangement, and extracellular matrix as related to migration. *J. Exp. Zool.* 229: 91–107.

Stemple, D. L. and Anderson, D. J. 1992. Isolation of a stem cell for neurons and glia from the mammalian neural crest. *Cell* 71: 973–985.

Stocker, K. M., Sherman, L., Rees, S. and Ciment, G. 1991. Basic FGF and TGF-β1 influence commitment to melanogenesis in neural crest-derived cells of avian embryos. *Development* 111: 635–645.

Stryer, L. 1986. Cyclic GMP cascade of vision. *Annu. Rev. Neurosci.* 9: 87–119.

Takeichi, M. 1988. The cadherins: Cell–cell adhesion molecules controlling animal morphogenesis. *Development* 102: 639–656.

Tan, S.-S. and Morriss-Kay, G. 1985. The development and distribution of the cranial neural crest in the rat embryo. *Cell Tissue Res.* 240: 403–416.

Tassabehji, M., Read, A. P., Newton, V. E., Harris, R., Balling, R., Gruss, P. and Strachan, T. 1992. Waardenburg's syndrome patients have mutations in the human homologue of the *Pax-3* paired bix gene. *Nature* 355: 535–536.

Teillet, M.-A., Kalcheim, C. and LeDouarin, N.M. 1987. Formation of the dorsal root ganglia in the avian embryo: Segmental origin and migratory behavior of neural crest progenitor cells. *Dev. Biol.* 120: 329–347.

Thomas, K. R. and Cappecchi, M. R. 1990. Targeted disruption of the murine *int-1* proto-oncogene resulting in severe abnormalities in midbrain and cerebellar development. *Nature* 346: 847–850.

Thorogood, P. 1989. Review of *Developmental and Evolutionary Aspects of the Neural Crest*. *Trends Neurosci.* 12: 38–39.

Turner, D. L. and Cepko, C. L. 1987. A common progenitor for neurons and glia persists at retina late in development. *Nature* 328: 131–136.

Vogel, K. S. and Weston, J. A. 1990. The sympathoadrenal lineage in avian embryos. II. Effects of glucocorticoids on cultured neural crest cells. *Dev. Biol.* 139: 13–23.

Van Allen, M. I. and fifteen others. 1993. Evidence for multi-site closure of the neural tube in humans. *Am. J. Med. Genet.* 47: 723–743.

Van Straaten, H. W. M., Hekking, J. W. M., Wiertz-Hoessels, E. J. L. M., Thors, F. and Drukker, J. 1988. Effect of the notochord on the differentiation of a floor plate area in the neural tube of the chick embryo. *Anat. Embryol.* 177: 317–324.

Vassar, R. and Fuchs, E. 1991. Transgenic mice provide new insights into the role of TGF-α during epidermal development and differentiation. *Genes Dev.* 5: 714–727.

von Baer, K. E. 1828. *Entwicklungsgeschichte der Thiere: Beobachtung und Reflexion.* Bornträger, Konigsberg.

Walsh, C. and Cepko, C. L. 1988. Clonally related cortical cells show several migration patterns. *Science* 241: 1342–1345.

Walsh, C. and Cepko, C. L. 1992. Widespread dispersion of neuronal clones across functional regions of the cerebral cortex. *Science* 255: 434–440.

Webster, E. H., Silver, A. F. and Gonsalves, N. I. 1984. The extracellular matrix between the optic vesicle and the presumptive lens during lens morphogenesis in an anophthalmic strain of mice. *Dev. Biol.* 103: 142–150.

Weinstein, G. D. and van Scott, E. J. 1965. Turnover times of normal and psoriatic epidermis. *J. Invest. Dermatol.* 45: 257–262.

Weston, J. 1963. A radiographic analysis of the migration and localization of trunk neural crest cells in the chick. *Dev. Biol.* 6: 274–310.

Weston, J. 1970. The migration and differentiation of neural crest cells. *Adv. Morphogenet.* 8: 41–114.

Weston, J. A. 1991. Sequential segregation and fate of developmentally restricted intermediate cell populations in the neural crest lineage. *Curr. Top. Dev. Biol.* 25: 133–153.

Witte, O. N. 1990. *Steel* locus defines new multipotent growth factor. *Cell* 63: 5–6.

Yamada, K. M., Spooner, B. S. and Wessells, N. K. 1971. Ultrastructure and function of growth cones and axons of cultured nerve cells. *J. Cell Biol.* 49: 614–635.

Yamada, T., Pfaff, S. L., Edlund, T. and Jessell, T. M. 1993. Control of cell pattern in the neural tube: Motor neuron induction by diffusible factors from notochord and floor plate. *Cell* 73: 673–686.

Yamamori, T., Fukada, K., Aebersold, R., Korsching, S., Fann, M.-J. and Patterson, P. H. 1989. The cholinergic neuronal differentiation factor from heart cells is identical to leukemia inhibitory factor. *Science* 246: 1412–1416.

Yuodelis, C. and Hendrickson, A. 1986. A qualitative and quantitative analysis of the human fovea during development. *Vision Res.* 26: 847–855.

8

Axonal specificity

Not only do neuronal precursor cells migrate to their place of function, but so do their axons. Unlike most cells whose parts all stay in the same place, the nerve cell is able to elongate axons that may extend meters. The axon has its own locomotory apparatus that resides in the growth cone, and the growth cone can respond to the same types of signals that migrating cells can sense. Thus, the migrating axon can undergo chemotaxis, galvanotaxis, and contact guidance just like migrating cells. The cues for axonal migration, moreover, may be even more specific than those used to get certain cell types to particular areas. The human brain, for instance, is the most ordered piece of matter known. Each of the 10^{11} neurons has the potential to specifically interact with thousands of other cells, and a large neuron (such as a Purkinje cell or motor neuron) can receive input from over 10^5 other cells (Figure 8.1; Gershon et al., 1985). This generation of ordered complexity is one of the greatest challenges to modern science.

Pattern formation in the nervous system

The functioning of the vertebrate brain depends not only on the differentiation and positioning of the neural cells, but also on the specific connections these cells make among themselves and their peripheral targets. In some manner, nerves from a sensory organ such as the eye must connect to specific neurons in the brain that can interpret visual stimuli, and axons from the nervous system must cross large expanses of tissue before innervating the appropriate target tissue. How does the nerve axon "know" to traverse numerous other potential target cells to make its specific connection? Harrison (1910) suggested that the specificity of axonal growth is due to pioneer nerve fibers, which go ahead of other axons and serve as guides for them.* This observation simplifies, but does not solve, the problem of how neurons form the appropriate patterns of interconnections. Harrison also noted, however, that axons must grow upon a solid substrate, and he speculated that differences in the embryonic surfaces might allow axons to travel in certain specified directions. The final

---

*The growth cones of pioneer neurons migrate to their target tissue while embryonic distances are still short and the intervening embryonic tissue is still relatively uncomplicated. Later in development, the other neurons that innervate the target tissue bind (fasciculate) to the pioneer neuron and thereby enter the target tissue. Klose and Bentley (1989) have shown that in some cases, the pioneer neurons die after the other neurons reach their destination. Yet, were that pioneer neuron prevented from differentiating, the other axons would not have reached their target tissue.

FIGURE 8.1
Connections of axons to a cultured hippocampal neuron. The neuron has been outlined by the synaptic protein synaptotagmin, which is present in the terminals of axons that contact the neuron. (Courtesy of M. Matteoli and P. De Camilli.)

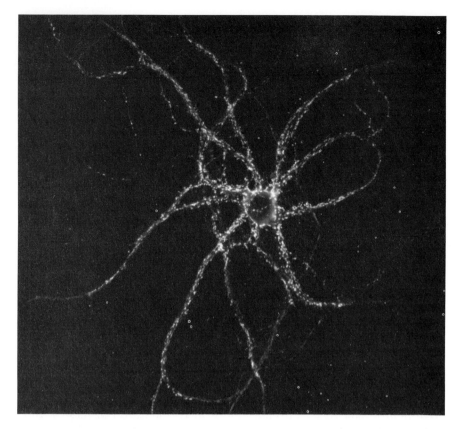

connections would occur by complementary interactions on the cell surface.:

> That it must be a sort of a surface reaction between each kind of nerve fiber and the particular structure to be innervated seems clear from the fact that sensory and motor fibers, though running close together in the same bundle, nevertheless form proper peripheral connections, the one with the epidermis and the other with the muscle. . . . The foregoing facts suggest that there may be a certain analogy here with the union of egg and sperm cell.

Research on the specificity of neuronal connections has focused on two major types of systems: motor neurons, whose axons travel from the nerve to a specific muscle, and the optic system, wherein axons originating in the retina find their way back into the brain. In both cases, the specificity of axonal connections is seen to unfold in three steps (Goodman and Shatz, 1993).

- *Pathway selection*, wherein the axons travel along a route that leads them to a particular region of the embryo.
- *Target selection*, wherein the axons, once they reach the correct area, recognize and bind to a set of cells with which they may form stable connections.
- *Address selection*, wherein the initial patterns are refined such that each axon binds to a small subset (sometimes only one) of its possible targets.

The first two of these processes are independent of neuronal activity. The third process involves the interactions between several active neurons and converts the overlapping projections into a fine-tuned pattern of connections. It has been known since the 1930s that motor axons can find their appropriate muscles even if the neural activity of the axons is blocked. Twitty (who had been Harrison's student) and his colleagues found that the embryos of the newt *Taricha torosa* secreted a toxin, tetrodotoxin, that blocked neural transmission in other species. By grafting pieces of *T. torosa* onto other salamander embryos, they were able to paralyze the host embryos for days while development occurred. About the time the tadpoles were to feed, the toxin wore off, and the salamanders swam and fed normally (Twitty and Johnson, 1934; Twitty, 1937). More recent experiments using zebra fish mutants having nonfunctional neurotransmitter receptors similarly demonstrated that motor neurons establish their normal patterns of innervation in the absence of neuronal activity (Westerfield et al., 1990).

But the question remains, How are the axons instructed where to go? As mentioned in Chapter 3, migratory cells get their cues from diffusible substances, ions, or the extracellular matrix over which they travel. The growth cone is able to respond to the same types of cues and leads the axon from the neural cell soma to its target tissue. It should be remembered from the preceding chapter that the growth cone pulls the axon forward. The axon is not extended by pushes coming from the cell body.

Pathway selection: Guidance by the extracellular matrix

The extracellular matrix can provide information for navigation in many ways. Some ways are more specific than others. For instance, channels and folds in the extracellular matrix can restrict the path of axon growth to a certain region. This is a very crude type of guidance. Second, certain basal lamina proteins may be more adhesive than others and bias axon movement along the basement membrane. Third, molecules in the extracellular matrix may actively repel axons, causing the collapse of the growth cone. As we will see below, each of the mechanisms appears to be working in the embryo.

Guidance by the physical terrain: Contact guidance

One of the first hypotheses to account for the specificity of axonal growth involves **contact guidance,** or **stereotropism.** Here physical cues in the substratum direct neural growth. Harrison developed a technique of growing axons on clotted blood, and using this technique Weiss (1955) noted that the growing axons not only needed a solid substrate on which to migrate but also that the migration tended to follow discontinuities in the clot. When the fibers of the blood clot were randomly oriented, the axons followed this random pattern. But when the clot fibers were made parallel by applying tension to the clot, the nerve axons traveled along these fibers, not veering from the straight and narrow (see Figure 3.31). Singer and his co-workers (1979) found evidence that such physical factors operate in vivo to guide the growth cones. They detected large channels between ependymal cells of the developing newt spinal cord through which growing axons migrated. They hypothesized that these channels provide cues for guiding the axons toward the appropriate regions of the brain. Cellular channels have also been detected in the mouse retina (Silver and Sidman, 1980), and these appear to guide the retinal ganglion cell growth cones into the optic stalk as they develop.

The presence of pre-existing channels is probably not critical for the

FIGURE 8.2
Gradients of extracellular matrix fibrils across the developing wing of the moth *Manduca sexta*. Double-headed arrows represent the proximal–distal axis. (Scanning electron micrographs of the three regions from Nardi, 1983, courtesy of J. Nardi.)

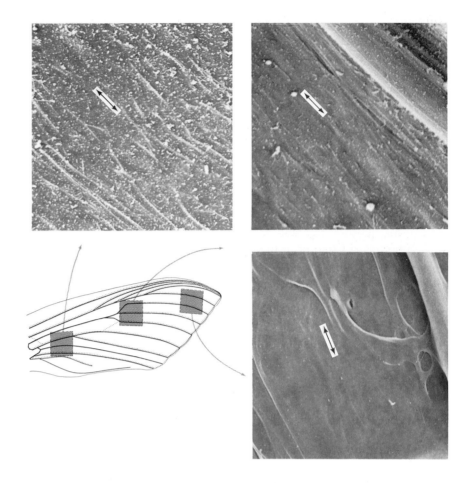

growth of most axons. The growth cone appears to be capable of digesting its own channels through an extracellular matrix by secreting two types of proteolytic enzymes—plasminogen activator and metalloproteases—into its immediate environment (Pittman, 1985).

Guidance by adhesive gradients: Haptotaxis

In addition to providing directional guidance by relatively large channels and grooves, the extracellular matrix can offer a gradient of adhesivity to the growth cone. This would be expected if the extracellular matrix was broad and would thus permit the axon to go in several directions. A gradient would make one direction favored. This type of adhesive gradient has been deduced in certain movements of insect growth axons. Nardi (1983) has seen that when sensory neurons from the tips of pupal moth wings grow toward the central nervous system, they are tightly associated with the extracellular matrix of the upper epithelial layer. When he placed grafts along the route of axon development, Nardi found that the neurons adhered more tightly to proximal tissue (tissue closer to the central nervous system) than to distal tissue. Scanning electron micrographs of the extracellular matrix showed that the matrix changed along the route. It became progressively more "bumpy" and fibrillar as it got closer to the central nervous system (Figure 8.2). Berlot and Goodman (1984) also showed that grasshopper sensory neurons originating in the antennae probably follow a gradient of adhesivity to the central nervous system.

Guidance by differential adhesive specificities

The directional cues need not always be in the form of gradients of one substance. A growing axon's growth cone encounters numerous microenvironments, and some sites may contain molecules that are more adhesive than molecules found in other sites. The growth cone has receptors that recognize proteins that are found in certain basal laminae, and the growth cone leads the axon along the paths coated by these proteins. The **differential adhesive specificity** hypothesis postulates that the growth cone will encounter a patchy environement and that this growth cone can recognize its path by its having particular receptors for certain molecules in the environment. This can be seen in vitro. When placed into culture, a piece of neural retinal tissue will not readily send forth axons onto the plastic dish. However, when the plastic is coated with fibronectin or laminin, long axonal outgrowths are observed (Figure 8.3). Conversely, glycosaminoglycans, another set of molecules associated with extracellular matrices, appear to impede the neural outgrowths (Tosney and Landmesser, 1985).

(A)

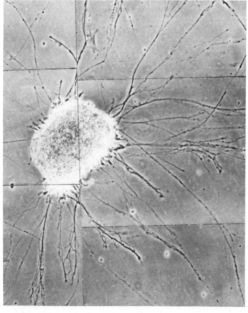

(B)

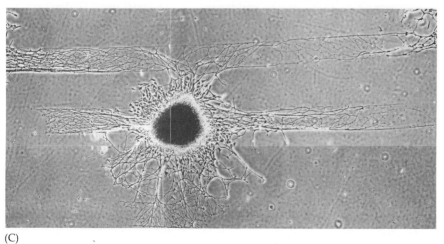

(C)

FIGURE 8.3

Effects of substrate factors on neural outgrowth. (A,B) Effects of fibronectin on neural outgrowth from neural retina aggregates. The aggregate in A was cultured 36 hours on untreated tissue culture plastic. The aggregate in B was cultured on plastic treated with 50 µg fibronectin per milliliter. (C) Outgrowth of sensory neurons placed on patterned substrate consisting of parallel stripes of laminin applied to a background of type IV collagen. (A and B from Akers et al., 1981, courtesy of J. Lilien; C from Gundersen, 1987, courtesy of R. W. Gundersen.)

The presence of such molecules delineates pathways through the embryo (Akers et al., 1981; Gundersen, 1987), and many of the roads upon which axons travel appear to be paved with **laminin.** Letourneau and co-workers (1988) have shown that the axons of certain spinal neurons travel through the neuroepithelium over a transient laminin-coated surface that precisely marks the path of these axons. Similarly, there is a very good correlation between the elongation of retinal axons and the presence of laminin on the neuroepithelial cells and astrocytes in the embryonic mouse brain (Cohen et al., 1986, 1987; Liesi and Silver, 1988). Punctate laminin deposits are seen on the glial cell surfaces along the pathway leading from the retina to the optic tectum, whereas adjacent areas where the optic nerve fails to grow lack these laminin deposits. After the retinal axons have reached the tectum, the glial cells differentiate and lose their laminin. At this point, the retinal ganglial neurons that have formed the optic nerve lose their integrin receptor for laminin. Laminin deposits may also be necessary for the regeneration of neural tissue. Astroglial cells bearing punctate laminin on their surfaces can induce regeneration when placed into embryos where the normal neuronal pathways of the corpus callosum have been severed.

There are at least three regions of the laminin glycoprotein that can support axonal migration and growth (Figure 8.4). First, integrins in the growth cone can bind to the RGD sequence of laminin protein. Second, another receptor in the growth cone can recognize the amino acid sequence YISGR in laminin. The third growth cone receptor for laminin is a glycosyltransferase that recognizes particular carbohydrate side chains on the laminin molecule (Begovac and Shur, 1990; Thomas et al., 1990). These carbohydrates may reside in the "neurite outgrowth" domain of the laminin A chain. In *Caenorhabditis elegans*, the *unc*-6 gene appears to encode a laminin B_2 subunit; the *unc*-5 and *unc*-40 genes are thought to encode receptors for laminin and are expressed in the nerve axons (Hedgecock et al., 1990; Leung-Hagesteijn, 1992). If any of these genes are mutated, the pioneer motor axons cannot connect to their appropriate muscles and the resulting worm shows *un*coordinated movement.

The importance of this pathway in *Caenorhabditis* was shown by Hamelin and his co-workers (1993) who caused the expression of the unc-5 laminin receptor in a set of neurons that does not usually express it. The unc-5 protein is normally expressed in motor neuron growth cones (and mesodermal cells) that move in a *dorsal* direction on the epidermis. This protein is not expressed on those touch receptor neurons whose pioneer axons extend *ventrally* on the epidermis. If the *unc*-5 gene is driven by a promoter that is active in the touch receptor neurons, the unc-5 protein is made in these cells, and the axons extend dorsally on the epidermis. This aberrant dorsal elongation is dependent on the presence of the unc-6 protein, just as it is for the normal elongation of motor neuron axons.

Guidance by specific growth cone repulsion

In addition to specific adhesion, there is also the possibility of *specific repulsion* by the extracellular matrix. Axons from dorsal root ganglion neurons in the trunk only pass through the anterior portion of each somite area (just as the neural crest cells will only migrate through these regions and not the posterior regions) (Figure 8.5A). The cell surface of the posterior portion of the somite may be inhibiting this migration. Davies and colleagues (1990) have shown that membranes isolated from the posterior portion of the somite cause the collapse of growth cones of the dorsal root ganglion neurons (Figure 8.5B and C). Moreover, they have isolated a

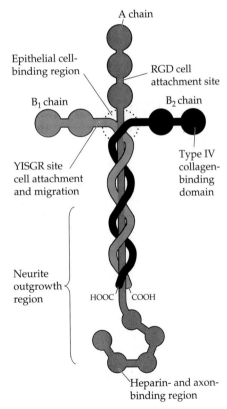

FIGURE 8.4
Structure of laminin and proposed binding regions.

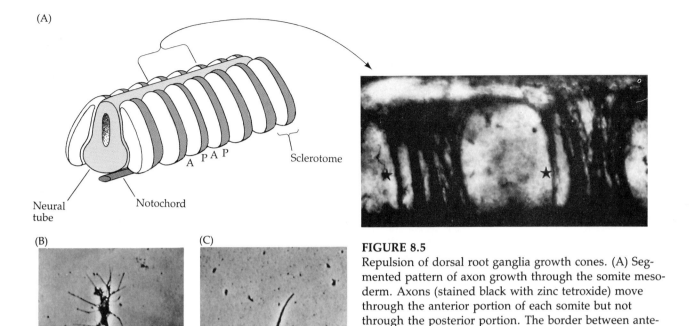

(A)

A P A P

Sclerotome

Neural tube

Notochord

(B)

(C)

FIGURE 8.5
Repulsion of dorsal root ganglia growth cones. (A) Segmented pattern of axon growth through the somite mesoderm. Axons (stained black with zinc tetroxide) move through the anterior portion of each somite but not through the posterior portion. The border between anterior and posterior is marked with a star. (B) Growth cone from an axon of a dorsal root ganglion neuron growing on laminin. Its lamellipodia and filipodia are easily seen. (C) Collapsed growth cone of a dorsal root ganglion neuron when inhibitory protein is added to culture. (A from Keynes and Stern, 1984; B and C from Raper and Kapfhammer, 1990. All photographs courtesy of the authors.)

glycoprotein fraction from chick somites that causes these growth cones to collapse, and components of this fraction are specifically found on the posterior portion of the somites.

SIDELIGHTS & SPECULATIONS

Sex, smell, and specific adhesion

In the late nineteenth century, the Johns Hopkins University professor John Mackenzie (1898), the German psychiatrist Wilhelm Fliess (1897), and the Viennese sexologist Richard von Krafft-Ebing (1886) all shared the mistaken view that there were similarities in the development of the penis and the nose. All three of these investigators used the same case study as evidence: the report of a man who had no sense of smell—no nasal or olfactory nerves—and whose genitals were far smaller than normal.

Such people are now known to have **Kallmann syndrome,** an X-linked disease characterized by anosmia (no sense of smell), small genitalia, and sterile gonads. The anosmia is due to the lack of those neurons in the brain that receive input from the axons coming from nasal neurons. The small gonads and genitalia are the result of a lack of gonadotropin-releasing hormone (GnRH). GnRH is a peptide hormone secreted by the *hypothalamus* that instructs the anterior pituitary to secrete luteinizing hor-

mone, the hormone required for gonadal development and genital maturation. What links these two problems? In 1989, two laboratories (Schwanzel-Fukada and Pfaff, 1989; Wray et al., 1989) made the surprising discovery that the GnRH-secreting neurons do not originate in the hypothalamus. Rather, they originate in the olfactory epithelium (the vomeronasal organ) in the nose rudiment and *migrate* into the hypothalamic region of the brain during fetal development (Figure 8.6). The olfactory receptor neurons of the nose originate from the same place. The axons from the olfactory receptor neurons enter the brain to synapse with the olfactory bulb, while the cell bodies of these neurons remain in the developing nose. Patients with Kallmann syndrome have no olfactory bulb in the brain, as the development of this bulb requires innervation from the olfactory receptor neurons (Stout and Gradziadi, 1980).

The defect in the Kallmann syndrome can be traced to the failure of the GnRH-secreting neurons and the olfactory neuron growth cones to migrate into the brain from the olfactory placode (Schwanzel-Fukada et al., 1989). It is

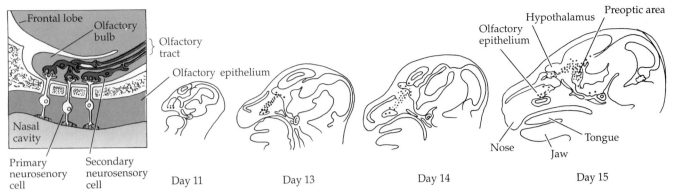

Frontal lobe
Olfactory bulb
Olfactory tract
Olfactory epithelium
Nasal cavity
Primary neurosenory cell
Secondary neurosensory cell

Day 11   Day 13   Day 14   Day 15

Olfactory epithelium
Hypothalamus
Preoptic area
Nose
Jaw
Tongue

FIGURE 8.6

Model for the etiology of Kallmann syndrome. In the left illustration, sensory neurons from the olfactory epithelium extend axons into the olfactory bulb of the brain. In Kallmann syndrome, the olfactory bulb has degenerated, and this loss is thought to be *secondary to the lack of axons from the sensory neurons. The series of sagittal head sections from embryonic mice shows the migration of LHRH-secreting neurons (color) from the nose anlagen into the hypothalamic portion of the brain. This migration does not occur in Kallmann syndrome. (After Calof, 1992.)*

thought that the olfactory axons migrate first and that the GnRH-secreting neurons follow the olfactory nerve fascicles into the brain (Livne et al., 1993). The gene whose absence or abnormality causes the syndrome has been cloned, and its cDNA sequence predicts a cell adhesion protein of the immunoglobulin superfamily class (Franco et al., 1991; Legouis et al., 1991). Members of this class of proteins are known to mediate cell–cell or axon–axon adhesion (Grumet et al., 1991) and they include such proteins as NCAM, Ng-CAM, LFA-1, CD4, fasciclin II, contactin, and neuroglian. The Kallmann syndrome protein also contains regions that resemble the fibronectin molecule, a molecule of the extracellular matrix that is critically important in numerous cell migrations during development. However, the tests to see if this protein is on the tracts followed by the migrating cells and elongating axons from the olfactory epithelium have not yet been done, nor has it been determined that the axons or cells from this region actually bind to this protein.

Guidance by axon-specific migratory cues: The labeled pathway hypothesis

Because extracellular matrix molecules such as laminin and N-CAM are found in several places throughout the embryo, they can usually provide only general cues for growth cone movement. There would be difficulty in using such general molecules to direct the growth cones of several different types of neurons in several different directions. Yet in *Drosophila*, grasshoppers, and *Caenorhabditis* (and probably most invertebrates), the patterning of axonal movement is an astoundingly precise process, and adjacent axons are given different migratory instructions by their environment. For example, within each segment of the grasshopper, 61 neuroblasts emerge (30 on each side and 1 in the center). One of them, the 7-4 neuroblast, is a stem cell and gives rise to a family of six neurons, termed C, G, Q1, Q2, Q5, and Q6. This family of neurons is shown in Figure 8.7 and as the yellow neurons in Plate 20. The axonal growth cones of these neurons reach their targets by following specific pathways formed by other earlier neurons. Q1 and Q2 follow a straight path together, traversing numerous other cells, until they meet the axon from the dorsal midline precursor neuron (dMP2), which they then follow posteriorly. The other four neurons of the 7-4 family migrate across the dMP axon as if it did not exist. Axons from the C and G neurons progress a long way together, but ultimately C follows nerves X1 and X2 toward the posterior of the segment, while G adheres to the P1 and P2 axons (which go posteriorly) and moves anteriorly on their surfaces (Goodman et al., 1984; Taghert et al., 1984).

The G growth cone will have encountered over 100 different surfaces to which it could have adhered, but it is specific for the P neurons. If the P neurons are ablated by laser, the G growth cone acts abnormally, its filopodia searching randomly for its proper migratory surface. If any of the other hundred or so axons are destroyed, the G growth cone behaves normally.

This formulation of axonal pathfinding in insects has been called the **labeled pathways hypothesis** because it implies that a given neuron can specifically recognize the surface of another neuron that has grown out before it. Evidence for this specificity comes from monoclonal antibodies (Bastiani et al., 1987). Neurons aCC and pCC are sibling neurons in the grasshopper (both are derived from neuroblast 1-1) that have very different fates. Moreover, different sets of axons adhere to each of them, creating independent bundles of axons, called **fascicles.** Monoclonal antibodies to the protein **fasciclin I** bound to the two aCC neurons of each segment in the 10-hour embryo. The antibodies did not detect fasciclin I on the pCC neurons. By hour 11, however, other neurons (but not pCC) began to express this cell surface molecule. The nerves were precisely those (RP1, RP2, U1, U2) that fasciculate with aCC. More recent experiments (Zinn et al., 1988; Harreison and Goodman, 1988) have shown that there are at least four fasciclin molecules expressed on different subsets of neurons, and that each of these molecules allows the grown cones of certain neurons to recognize specifically those axons with which they will fasciculate.

In other animals with relatively simple nervous, such as the leech,

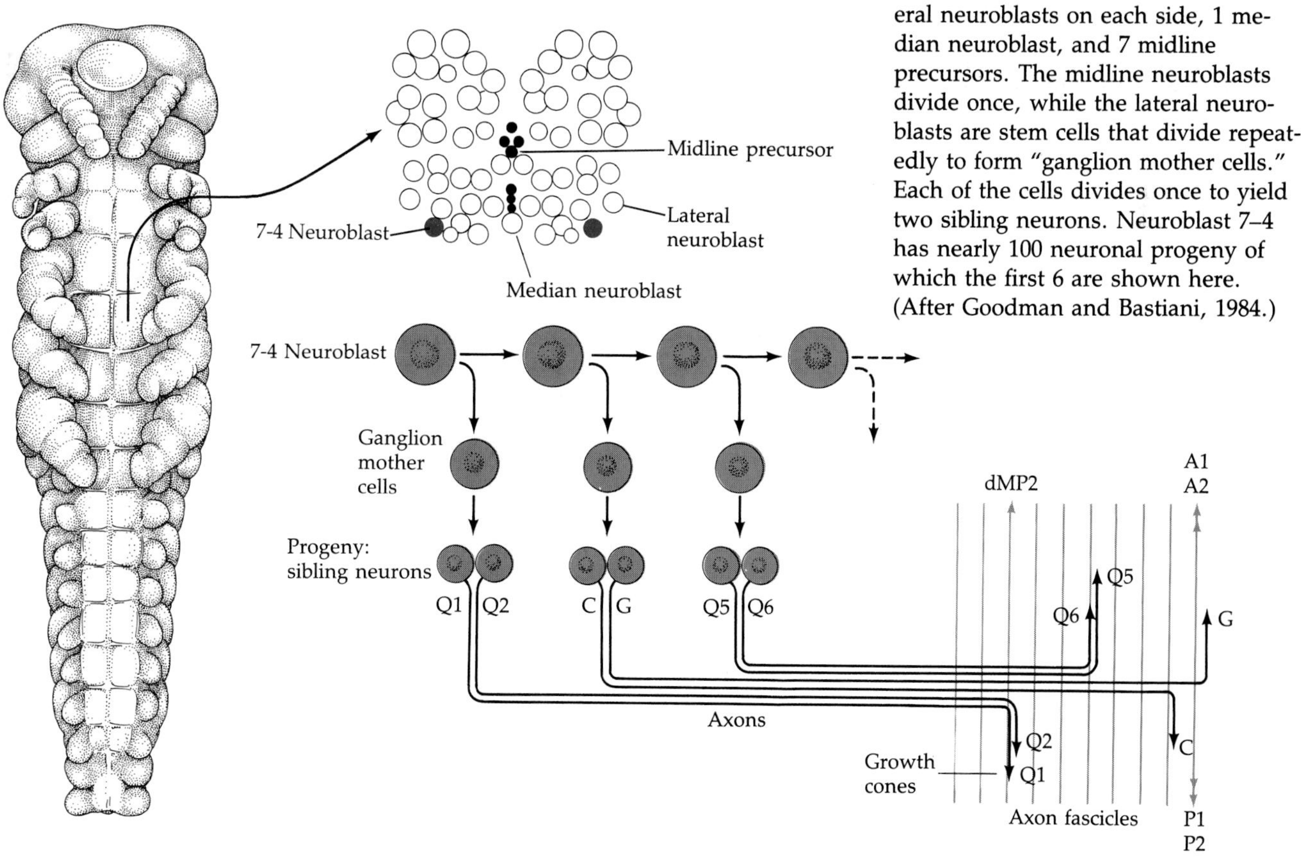

FIGURE 8.7

Each of the 17 segments of the early grasshopper embryo has the same pattern of neuroblasts. There are 30 lateral neuroblasts on each side, 1 median neuroblast, and 7 midline precursors. The midline neuroblasts divide once, while the lateral neuroblasts are stem cells that divide repeatedly to form "ganglion mother cells." Each of the cells divides once to yield two sibling neurons. Neuroblast 7–4 has nearly 100 neuronal progeny of which the first 6 are shown here. (After Goodman and Bastiani, 1984.)

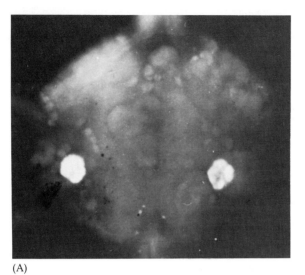

(A)

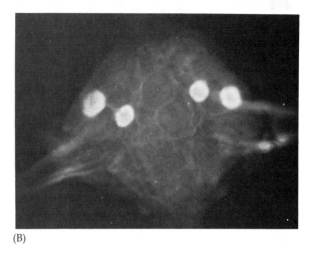

(B)

FIGURE 8.8
Specific functional neurons stained by monoclonal antibodies to cell surface components. (A) Lan 3–1 antibodies recognize a single pair of neurons in a particular ganglion. These neurons function in penile eversion. (B) A set of neurons recognized by Lan 3–2 antibodies; these neurons respond to noxious stimulation of the leech's skin. (From Zipser and McKay, 1981, courtesy of B. Zipser.)

there is evidence that each neuron might have qualitatively different cell surface molecules and that these molecules might be important in synaptic specificity. The nervous system of the leech consists of 34 paired ganglia containing about 400 neurons each. Individual neurons have been identified, and the functions of many of these neurons are known. Zipser and McKay (1981) injected the leech nervous system into mice and obtained hundreds of monoclonal antibodies, which bound to various regions of the nervous system. In some cases, such differences could be correlated with function. Monoclonal antibody Lan 3-1 bound specifically to a single pair of neurons in each of the midbody ganglia (Figure 8.8). These pairs of neurons are known to control the process of penile eversion in mating leeches. Another monoclonal antibody, Lan 3-2, recognized all four neurons in each ganglion that respond to noxious mechanical stimuli. "The situation," according to Zipser and McKay, "seems quite analogous to colour-coded electrical cable containing many wires, where each wire has its own molecule (dye) to facilitate proper recognition and connection at terminals."

Studies on specifically labeled pathways in vertebrates lag far behind those on invertebrates, but recent studies on zebra fish motor neurons suggest that labeled pathways may function here as well. Zebra fish may become the organism of choice in vertebrate developmental neurobiology because their development is very rapid, numerous individuals can be compared, and the embryos are clear. This enables neurobiologists to observe the growth of axons in living embryos. Neurons can be identified by injecting fluorescently labeled substances into the neuronal precursors (Kimmel and Law, 1985) and axon growth can be followed by eye or with a video recorder. Eisen and her colleagues (1986) watched the axonal elongation of three pioneer motor neurons in these embryos. After the axons left the spinal cord, all three followed the same path along a muscle until they reached a particular site in the embryo. At this point they diverged into three neuron-specific pathways that led to the appropriate muscles.

Fetal neurons in adult hosts

In 1976, Lund and Hauschka implanted fetal rat brain tissue into the brain of a newborn rat. The fetal neurons made the appropriate connections within the host brain. This study offered the possibility that transplants of fetal neurons might be able to repair damaged regions in human brains. There are numerous neural degenerative diseases, and Parkinson disease is one of the most prevalent, afflicting about a million Americans. In Parkinson disease, dopamine-producing neurons of the substantia nigra (a cluster of cells in the brainstem) are destroyed, and their axon terminals in the caudate nucleus and putamen (two brain nuclei) degenerate. This leads to muscle tremors, difficulty in initiating voluntary movements, and problems in cognition. The injection of L-dopa (which the body metabolizes into dopamine) relieves the symptoms temporarily, but L-dopa loses its effect with prolonged use, and it sometimes has adverse side effects. In 1990, Lindvall and colleagues implanted human neural cells from the substantia nigra of 8–9 week fetuses into a patient with Parkinson disease. The donor and recipient did not have to be related, since the brain is separated from the immune system by the blood–brain barrier, which shelters transplants of tissue implanted into the brain from rejection by the immune system. Within 5 months the transplant had restored much of the dopamine normally made by the substantia nigra, as well as the patient's capacity for voluntary movement. Two other laboratories have reported similar restoration of function following the transplantation of fetal neurons into such patients (Freed et al., 1992; Spencer et al., 1992). According to Björkland (1987), the optimal donor tissue is that containing presumptive dopamine-secreting neurons that have undergone their last cell division but that have not yet formed extensive synaptic connections. In 1992, Widner and colleagues showed that grafts from fetal mesencephalons were able to restore motor functions to two patients who had destroyed their substantiae nigrae by injecting themselves with a synthetic heroin contaminated with the by-product MPTP. This compound had created a condition that resembled severe Parkinson disease. Since Parkinson disease is progressive, it is not known if the grafted neurons will fall to the same process that destroyed the endogenous neurons. However, it appears likely that human fetal grafts are able to re-establish the synaptic connections that the destroyed neurons once made.

Pathway selection: Guidance by diffusible molecules

The chemotactic hypothesis

The growth of axons from a ganglion to a target tissue may be directed by soluble molecules emanating from the target. The axon's growth cone would be expected to migrate up the concentration gradient of these molecules (chemotaxis). Evidence for such a view comes from explanted ganglia placed close to, but not touching, various potential target tissues. Lumsden and Davies (1983, 1986) have dissected the trigeminal ganglia of sensory neurons out of embryonic mice before the neurons contacted their normal target tissue, the whisker pad. When these explanted ganglia are placed in culture medium near a variety of potential target tissues from the embryonic mouse, the trigeminal neural axons extend only toward the whisker pad tissue. In fact, when cultured between whisker pad epithelial tissue and whisker pad mesenchymal tissue, they only migrate toward the epithelial tissue (Figure 8.9). This observation can best be explained by postulating the secretion of some chemotactic substance by the whisker pad epithelium. The chemotaxis of the trigeminal axons to the whisker pad is not only target-restricted but also neuron-restricted and temporally restricted. The whisker pad tissue does not attract the axons from other sensory ganglia, and it will attract the trigeminal neurons only during a specified time.

   Another example of axon chemotaxis is the growth of spinal commissural neurons toward the neural floor plate (Chapter 3). The axons of these

FIGURE 8.9
Summary of the experiments of Lumsden and Davies showing chemotaxis between neural tissue (trigeminal ganglion) and its target (whisker pad). The chemoattraction is specific for (A) the target and (B) the epithelial cells of the target. Moreover, the chemotactic ability of the whisker pad is specific for the trigeminal neurons (C).

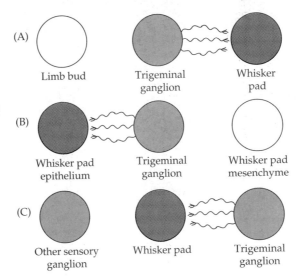

neurons begin growing ventrally down the side of the neural tube. However, about two-thirds of the way down, their direction changes, and they project through the ventrolateral (motor) neuron area of the neural tube toward the floor plate cells (Figure 8.10). The floor plate cells, the most ventral portion of the neural tube, have been shown to synthesize a diffusible protein, **F-spondin,** that is thought to bind to extracellular matrices in the ventral half of the neural tube. When F-spondin binds to these matrices, it may bind to and encourage the growth of these axons (Tessier-Lavigne et al., 1988; Placzek et al., 1990; Klar et al., 1992). Moving the floor plate cells to different positions will change the direction of the commissural axons, and floor plate explants added to cultured spinal cord explants will direct the migration of commissural axons towards them.*

*The binding of a soluble factor to the extracellular matrix makes for an interesting ambiguity between chemotaxis, haptotaxis, and labeled pathways. Nature doesn't necessarily conform to our catagories.

FIGURE 8.10
Trajectory of the commissural axons in the rat spinal cord. Axons from the commissural neurons (color) and association neurons (black) both form in the dorsal region of the spinal cord and begin growing ventrally on embryonic day 11. Association neurons project laterally, while the commissural axons are chemotactically guided ventrally down the lateral margin of the spinal cord toward the floor plate. Upon reaching the floor plate, the commissural axons change their direction due to contact guidance from the floor plate cells. (B) Autoradiographic localization of F-spondin mRNA by in situ hybridization to the hindbrain of a 10-day rat embryo using antisense RNA. Hybridization gives an intense signal from the floor plate neurons. (A after Tessier-Lavigne et al., 1988; B from Klar et al., 1992, courtesy of T. M. Jessell.)

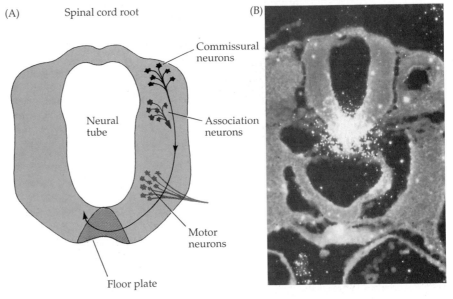

Inhibition of axonal growth by diffusible substances

As mentioned earlier, the pattern of axonal movement can also be influenced by factors that repel growth cones and retard axonal growth. In the snail *Helisoma*, neurotransmitters released by one nerve's growth cone can inhibit or accelerate the elongation of other growth cones. The pattern of neuronal connections is influenced by such interactions. Neuron 5 secretes serotonin, which completely inhibits the axonal growth of neuron 19 by causing the retraction of that growth cone's filopodia (Haydon et al., 1985; Kater, 1985). In this way, neuron 5 prevents neuron 19 from synapsing with it (Figure 8.11). In these snails, nerves can also alter the growth cones of other neurons by electrical activity. Once electrical potentials are established (as at neuromuscular junctions), the action potential can cause the cessation of axonal outgrowth for other neurons (Cohan and Kater, 1986).

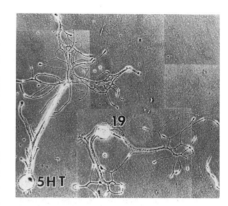

FIGURE 8.11
The growth cones of neuron 19 turn away from serotonin-producing neuron 5 when the two *Helisoma* neurons are cultured together. When another two neurons (such as two neurons 19) are brought together, they attract one another. (From Kater, 1985, courtesy of S. B. Kater.)

Multiple guidance cues

Motor neurons

One of the most important research programs in vertebrate developmental neurobiology concerns the innervation of the limb muscles. Axonal outgrowth of motor neurons occurs very early in development, before the soma of the motor neurons have migrated to their definitive positions in the spinal cord and before the muscles have condensed out of mesenchyme (Landmesser, 1978; Hollyday, 1980). This stage can be seen in Figure 8.12. In order to innervate the limb musculature, the axon extends over hundreds of cells in a complex and changing environment. Recent research has discovered several paths and several barriers that help guide the axons to their appropriate destinations. As mentioned earlier, on either side of the spinal cord are blocks of mesodermal tissue called somites. Just before the axons begin their extension, the somite splits into two types of tissues—the dorsal portion becomes the dermamyotome (which produces the dermis and musculature of the back), while the ventral portion of the

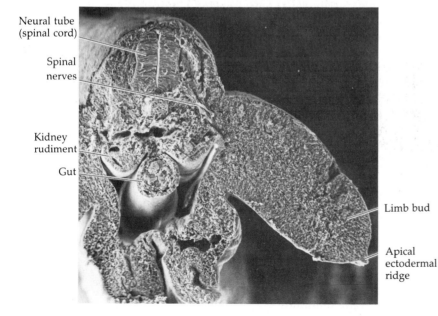

Neural tube
(spinal cord)

Spinal
nerves

Kidney
rudiment

Gut

Limb bud

Apical
ectodermal
ridge

FIGURE 8.12
Scanning electron micrograph of a cross section of a 4-day chick embryo showing the emergence of spinal nerves into the developing limb bud. (From Tosney and Landmesser, 1985, courtesy of K. Tosney.)

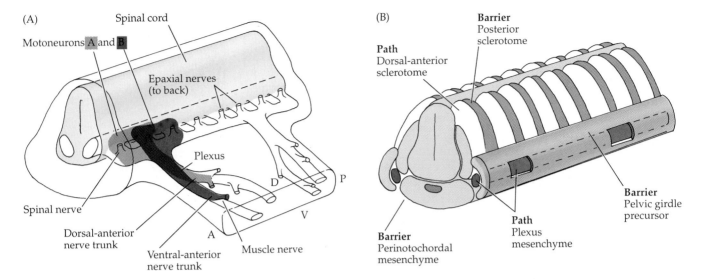

FIGURE 8.13
Motor axon pathways in the hindlimb region of the chick embryo. (A) Diagram of the neural pattern of the hindlimb. Motor neuron axons merge in the plexus and then separate into dorsal and ventral nerve trunks. One plexus is anterior, the other is posterior. (B) Diagram of the environmental components that create the neural patterning. The segmentation of the spinal nerves is created by the sclerotome. Anterior dorsal is permissive to migration, whereas posterior dorsal and all ventral sclerotome (the perinotochordal mesenchyme) are barriers to motor nerve axons. The plexus mesenchyme is permissive, but the pelvic girdle forms a barrier. The two holes in this barrier allow the nerve trunks to extend through them. (After Tosney, 1991.)

somite becomes the sclerotome (which produces the vertebral cartilage). Lateral to the somites, at the base of the limb bud, lies the plexus mesenchyme and the prospective shoulder girdle cells. The cell bodies of the motor neurons lie in the ventrolateral regions of the neural tube as shown in Figure 7.9 diagrammed in Figure 8.13A. Axons from the motor neurons that will innervate the limb muscles are mixed together when they exit from the spinal cord. Axon populations from several segmental levels of spinal cord can form into a common spinal nerve. These spinal nerves meet at a **plexus**. At these plexi, however, the axons from different regions go along different pathways. For example, in this figure, motor neurons for muscles 1 and 2 diverge into the appropriate nerve trunks, and eventually project into single muscles.

By various surgical manipulations of the early chick embryo, some of the environmental cues directing this migration have been discovered. The ventral part of the sclerotome that encircles the notochord forms a barrier against motor axon elongation. Even though cells in this region appear loose and easily circumvented, they repel axons in their vicinity. When the neural tube is rotated such that the motor axons emerge ventrally into this region, they immediately turn to avoid it and travel only through the dorsal anterior sclerotome (Figure 8.13B). Thus, the perinotochordal sclerotome is a barrier to motor axon growth, while the dorsal anterior sclerotome is a pathway (Tosney and Oakley, 1990; Tosney, 1991). Once the axons have progressed through the anterior dorsal sclerotome (along with the neural crest cells that follow the same route), they come to the plexus mesenchyme at the base of the limb bud. This, too, is a permissive environment for axonal growth. However, just past the plexus mesenchyme lie the pelvic girdle precursor cells. These cells inhibit axon growth, and axons turn away from them. There are two holes in the pelvic girdle precursor tissue that fill with plexus mesenchyme. Axons extend through these holes to form the anterior and posterior nerve trunks that enter the limb. If other holes are made experimentally in the pelvic girdle precursor tissue, the axons readily go through them (Tosney and Landmesser, 1984, 1985).

The nerves can even reach their respective destinations if the muscle-forming cells have been removed (Phelan and Hollyday, 1990), and it is probable that the other mesenchymal cells in the limb bud (such as those that form the dermis or cartilage) might be providing the directional cues. These paths to the limb muscle regions appear very well defined. First, if

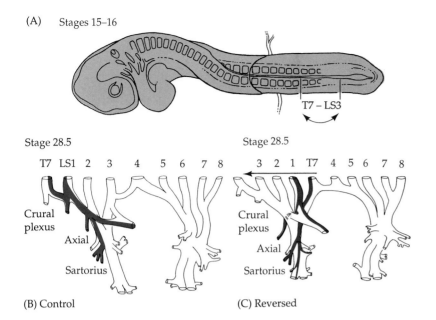

FIGURE 8.14
Compensation for small dislocations of axonal initiation position in the chick embryo. (A) A length of spinal cord comprising several segments T7–LS3 (seventh thoracic to third lumbarsacral segments) is reversed in the 2.5-day embryo. (B) Normal pattern of axon projection into the different muscles at 6 days. (C) Projection of axons in the reversed segment. The ectopically placed neurons eventually found their proper neural pathways and innervated the appropriate muscles. (From Lance-Jones and Landmesser, 1980.)

one redirects axons from a different source (such as a different ganglion) into the limb, they branch like the axons that originally innervated the limb. In other words, the limb is able to dictate the pattern of innervation to a set of axons that normally would not enter the limb (Hamburger, 1939; Hollyday et al., 1977). Moreover, if segments of the chick spinal cord are reversed—so that the motor neurons are in new locations—their axons will find their original targets (Figure 8.14; Lance-Jones and Landmesser, 1980). Yet when limbs develop with duplicated areas (such as two thighs), the neurons innervating the second thigh are not "thigh"-specific neurons, but are neurons that usually innervate the calf (Whitelaw and Hollyday, 1983). These experiments present a paradox that has yet to be resolved: "Particular axons are biased to grow to specific places, yet axons from different motor neuron pools can substitute for one another in the establishment of normal nerve patterns" (Purves and Lichtman, 1985). The most plausible explanation is that several mechanisms act simultaneously to ensure that the axons get to their appropriate places. One of these mechanisms appears to be the guidance from the cell surface of non-muscle-forming mesenchyme cells in the limb bud, while another mechanism probably involves chemotaxis from the limb bud myoblasts (Goodman and Shatz, 1993).

Retinal axons

Multiple guidance cues have also been postulated to explain how individual retinal neurons are able to send axons to the appropriate area of the brain even when transplanted away from the optic nerve (Harris, 1986). This ability to guide the axons of translocated neurons to their appropriate sites implies that the guidance cues are not distributed solely along the normal pathway but exist throughout the embryonic brain. Guiding an axon from the nerve cell body to its destination across the embryo is a complex phenomenon, and several different types of cues might be used simultaneously to ensure that the correct connections get established.

The first steps in getting the retinal axons to their specific regions of the optic tectum take place within the retina (Figure 8.15A). As the retinal ganglion cells differentiate, their position in the inner margin of the retina

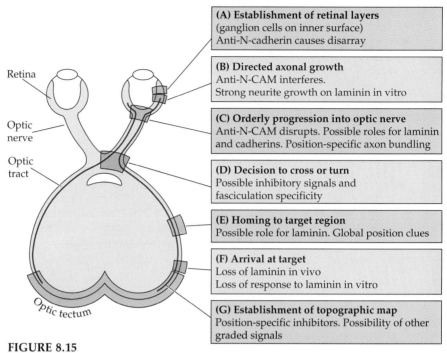

| (A) Establishment of retinal layers
(ganglion cells on inner surface)
Anti-N-cadherin causes disarray |
| (B) Directed axonal growth
Anti-N-CAM interferes.
Strong neurite growth on laminin in vitro |
| (C) Orderly progression into optic nerve
Anti-N-CAM disrupts. Possible roles for laminin
and cadherins. Position-specific axon bundling |
| (D) Decision to cross or turn
Possible inhibitory signals and
fasciculation specificity |
| (E) Homing to target region
Possible role for laminin. Global position clues |
| (F) Arrival at target
Loss of laminin in vivo
Loss of response to laminin in vitro |
| (G) Establishment of topographic map
Position-specific inhibitors. Possibility of other
graded signals |

FIGURE 8.15
Multiple guidance cues direct the movement of retinal ganglia axons to the optic tectum. (After Hynes and Lander, 1992.)

is determined by cadherin molecules (N-cadherin as well as retina-specific R-cadherin) on their cell membranes (Matsunaga et al., 1988; Inuzuka et al., 1991). The axons from these cells grow along the inner surface of the retina toward the optic nerve head (Figure 8.15B). The adhesion and growth of the retinal cell axons along the inner surface of the retina may be governed by the laminin-containing basal lamina. However, the attachment to laminin cannot explain the directionality of the growth. It is possible that a gradient of the inhibitory extracellular matrix molecule chondroitin sulfate proteoglycan plays a role in specifying which way to grow (Hynes and Lander, 1992).

When the axons enter the optic nerve, they grow on glial cells toward the brain. In vitro studies suggest that numerous cell adhesion molecules—N-CAM, cadherins, and integrins—play roles in orienting the axon towards the optic tectum (Neugebauer et al., 1988). Upon their arrival at the optic nerve, the axons fasciculate with those that are already present. N-CAM is critical to this fasciculation, and antibodies against N-CAM cause the axons to enter the optic nerve in a disorderly fashion, which in turn causes them to emerge at the wrong positions in the tectum (Thanos et al., 1984).

Upon entering the brain, mammalian retinal axons reach the **optic chiasm,** where they have to "decide" if they are to continue straight or if they are to turn 90% and enter the other side of the brain (Figure 8.15D). It appears that those axons that are not destined to cross to the other side of the brain are repulsed from doing so when they enter the chiasm (Godement et els, 1990), but the molecular basis of this repulsion is not known. On their way to the optic tectum, the axons travel on a pathway (the **optic tract**) over glial cells whose surfaces are coated with laminin (Figure 8.15E). Very few areas of the brain have laminin, and the laminin in this pathway exists only when the optic nerve fibers are growing on it (Cohen et al., 1987).

The axon migrating from the retina to the tectum encounters numerous other cells and potential targets for innervating. However, a combination of several guidance cues, probably involving both attraction and repulsion, guides the axon along its way. At this point, the retinal axons have reached the optic region of the brain (Figure 8.15F) and target selection begins.

Target selection

When the axons come to the end of this laminin-lined tract, they spread out and find their specific target. Studies on frogs and goldfish (which lack a chiasma) have indicated that each retinal axon sends its impulse to one specific site (a cell or small group of cells) within the tectum (Sperry, 1951). As shown in Figure 8.16, there are two optic tecta in the frog brain. The axons from the right eye enter the left optic tectum, while those from the left eye form synapses with the cells of the right optic tectum.

The map of retinal connections to the frog optic tectum (the retinotectal projection) was detailed by Marcus Jacobson (Jacobson, 1967). Jacobson defined this map by shining a narrow beam of light on a small, limited region of the retina and noting, by means of a recording electrode in the tectum, which tectal cells were being stimulated. The retinotectal projection of *Xenopus laevis* is shown in Figure 8.16. Light illuminating the ventral part of the retina stimulates cells on the lateral surface of the tectum. Similarly, light focused on the posterior part of the retina stimulates cells in the caudal portion of the tectum. These studies demonstrated a point-for-point correspondence between the cells of the retina and the cells of the tectum. When a group of retinal cells is activated, a very small and specific group of tectal cells is stimulated. We also can observe that the points form a continuum; in other words, adjacent points on the retina

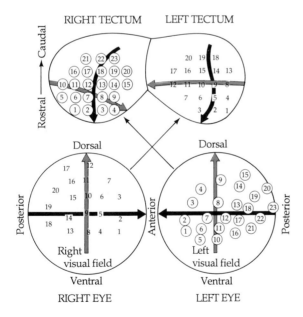

FIGURE 8.16
Map of the normal retinotectal projection in the adult *Xenopus*. The right eye innervates the left tectum, and the left eye innervates the right tectum. The numbers on the visual fields (retina) and the tecta show regions of correspondence; that is, stimulation of spot 15 on the right retina sends electrical impulses to left tectal region 15. The black and colored arrows summarize the pattern of retinotectal connections. (From Jacobson, 1967.)

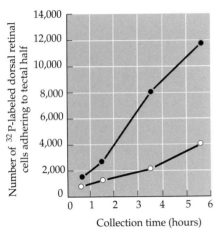

FIGURE 8.17
Differential adhesion of dorsal radioactive chick retinal cells to dorsal and ventral tectal halves. Radioactive cells from the dorsal half of 7-day chick retinas were added to dorsal (color) and ventral (black) halves of 12-day chick optic tecta. The data show the selective adhesion of the dorsal retinal cells to the ventral tectal tissue. (After Roth and Marchase, 1976.)

project onto adjacent points on the tectum. This arrangement enables the frog to see an unbroken image. This intricate specificity caused Sperry (1965) to hypothesize the **chemoaffinity hypothesis:**

> The complicated nerve fiber circuits of the brain grow, assemble, and organize themselves through the use of intricate chemical codes under genetic control. Early in development, the nerve cells, numbering in the millions, acquire and retain thereafter, individual identification tags, chemical in nature, by which they can be distinguished and recognized from one another.

Current theories do not propose that there is a point-to-point specificity between each axon and the nerve that it contacts. Rather, evidence now demonstrates that gradients of adhesivity (especially those involving repulsion) play a role in defining the territories that the axons enter and that activity-driven competition between these neurons determines the final connection of each axon.

Adhesive specificities in different regions of the tectum

There is good evidence that retinal ganglial cells can distinguish between regions of the tectum. Cells prepared from the ventral half of chick neural retina preferentially adhere to dorsal halves of the tectum (Figure 8.17; Roth and Marchase, 1976). Gottlieb and co-workers (1976) found that neurons taken from the dorsalmost part of the chick retina adhere preferentially to the ventralmost portion of the tectum and that the extreme ventral neurons of the retina preferentially adhere to the dorsalmost extremes of the tectum. These results were confirmed in other experimental conditions using axonal tips rather than entire neurons (Halfter et al., 1981).

Gradients of specific compounds have been observed on the cell surface of the neural retina, but their functions are not known. Trisler and his co-workers (1981), using monoclonal antibodies to chick neural retinal cells, found one antibody that localized in a graded fashion across the retinal surface. The antigen identified by this antibody is 35 times more concentrated in the dorsal region of the retina than it is at the ventral margin (Figure 8.18). Moreover, the same cell surface molecule is present in the optic tectum, where the gradient is reversed. Because dorsal retinal cells send axons to the ventral tectum (and ventral retinal cells project to the dorsal tectum), this antigen may be involved in orienting territory-specific axonal connections (Trisler and Collins, 1987). When injected into

FIGURE 8.18
Gradient of antigens on neural retinal cells. (A) Gradient of antibody binding to a cell surface molecule in the 14-day embryonic chick neural retina. Each retina was first divided into eight 45° sections, then into an inner and an outer segment. A radioactive monoclonal antibody was then bound to it. Numbers represent counts per minute per milligram of protein. (B) A gradient across the 13-day embryonic rat neural retina is visualized with a fluorescent monoclonal antibody. (A after Trisler, 1987; B from Constantine-Paton et al., 1986, courtesy of M. Constantine-Paton.)

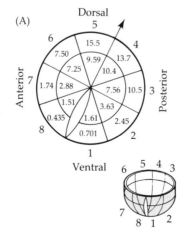

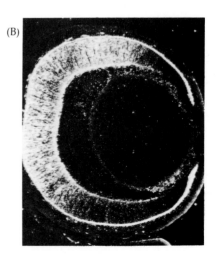

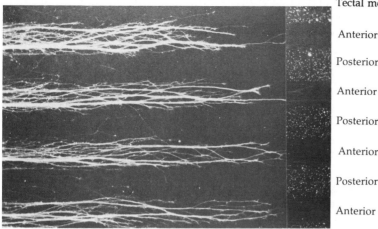

Tectal membranes

Anterior

Posterior

Anterior

Posterior

Anterior

Posterior

Anterior

FIGURE 8.19
Differential repulsion of temporal retinal axons on tectal membranes. Alternating stripes of anterior and posterior tectal membranes were absorbed onto filter paper. When axons from temporal (posterior) retinal ganglial cells grew on such alternating carpets, they preferentially extended axons on the anterior tectal membranes. (From Walter et al., 1987.)

the vitreous humor of 11-day embryonic chick eyes, this antibody inhibits synapse formation (Trisler et al., 1986; Trisler, 1987), further suggesting that this topologically asymmetric antigen may be involved in proper neuronal recognition. In rat embryos, another antigen has been found in a dorsal–ventral gradient across the retina (Figure 8.18B; Constantine-Paton et al., 1986). Its biochemical structure does not seem at all related to the antigen on the chick retinal cells.

One gradient that has been identified functionally is a gradient of "repulsion" that is highest in the posterior tectum and weakest in the anterior tectum. Bonhoeffer and colleagues (Walter et al., 1987) prepared a "carpet" of membranes having alternating stripes derived from the posterior and anterior tecta. They then let cells from the nasal (anterior) or temporal (posterior) regions of the retina extend axons into this carpet. The ganglial cells from the *nasal* portion of the retina extended axons equally well on both the anterior and posterior tectal membranes. The neurons from the *temporal* side of the retina, however, extended axons only on the anterior tectal membranes (Figure 8.19). The basis for this specificity appears to be a repulsive factor on the posterior tectal cell membranes. When the growth cone of a temporal retinal axon contacts a posterior tectal cell membrane, the filopodia of the growth cone withdraw, and the growth cone collapses and retracts (Cox et al., 1990). There appears, then, to be a repulsive factor that steers temporal (posterior) retinal axons away from the posterior tectum. This repulsive factor appears to be a 33-kDa protein that is attached to the plasma membrane of posterior tectal cells (Stahl et al., 1990). If a small amount of this protein is added to the lawn of anterior tectal cell membranes, temporal retinal axons will no longer grow on them. Moreover, a small concentration gradient of this protein is able to guide temporal retinal axon growth in vitro (Baier and Bonhoeffer, 1992). The result of such a gradient in the tectum would be the directing of the temporal retinal axons to the anterior portion of the tectum. Presumably, the nasal retinal axons could be directed (perhaps by exclusion) to the posterior of the tectum.

Address selection: Activity-dependent development

When an axon contacts its "target," usually either a muscle or another neuron, it forms a specialized junction called a **synapse.** Neurotransmitters

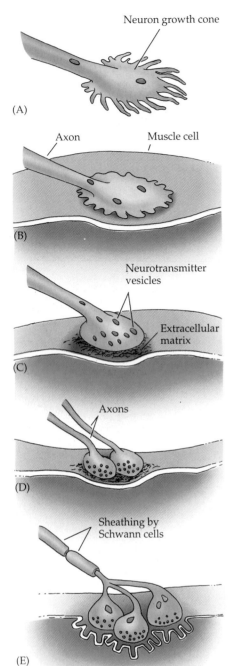

Neuron growth cone

(A)

Axon   Muscle cell

(B)

Neurotransmitter vesicles

Extracellular matrix

(C)

Axons

(D)

Sheathing by Schwann cells

(E)

FIGURE 8.20

Differentiation of motor neuron synapse with muscle. The right column is a magnification of the left. (A) A growth cone approaches a developing muscle cell. (B) The axon stops and forms an unspecialized contact on the myotube surface. (C) Neurotransmitter vesicles enter the axon terminal, and an extracellular matrix connects the axon terminal to the muscle cell as the synapse widens. (D) Other axons converge on the same synaptic site. (E) All axons but one are eliminated. The remaining axon can branch to form a complex junction with the muscle. Each axon terminus is sheathed by a Schwann cell process, and folds form in the muscle cell membrane. (After Hall and Sanes, 1993.)

from the axon terminal are released at these synapses to depolarize or hyperpolarize the membrane of the cell across the synaptic cleft. The construction of the synapse involves several steps (Figure 8.20). When motor neurons in the spinal cord extend axons to muscles, growth cones that contact the newly formed muscle cells migrate over their surfaces. When the growth cone first adheres to the muscle cell membrane, no specializations can be seen in either membrane. However, the axon terminals soon begin to accumulate neurotransmitter-containing synaptic vesicles, the membranes of both cells thicken at the region of contact, and the cleft between the cells fills with extracellular matrix. After this first contact is made, growth cones from other axons converge at this site to form additional synapses. During development, all mammalian muscles studied are seen to be innervated by at least two axons. However, this polyneuronal innervation is transient. During early postnatal life, all but one of these axon branches are retracted. This rearrangement is based on "competition" between the axons (Purves and Lichtman, 1980; Thompson, 1983). When one of the motor neurons is active, it suppresses the synapse of the other neurons (Dan and Poo, 1992). Eventually, the less active synapse is eliminated. The remaining axon terminal expands and is ensheathed by a Schwann cell.

Activity-dependent synapse formation also appears to be involved in the final stages of the retinal projection to the brain. In frog, bird, and rodent embryos treated with tetrodotoxin, axons will grow normally to their respective territories and will make synapses with the tectal neurons. However, the retinotectal map is a coarse one, lacking fine resolution. Just as in the final specification of the motor neuron synapse, neural activity is needed for the point-to-point retinal projection onto the tectal neurons (Harris, 1984; Fawcett and O'Leary, 1985; Kobayashi et al., 1990).

Differential survival after innervation: Neurotrophic factors

Reflecting on his life as an embryo, Lewis Thomas (1992) writes,

> By the time I was born, more of me had died than survived. It is no wonder I cannot remember; during that time I went through brain after brain for nine months, finally contriving the one model that could be human, equipped for language.

Indeed, one of the most puzzling phenomena in the development of the nervous system is neuronal cell death. In many parts of the vertebrate central and peripheral nervous systems, over half the neurons die during the normal course of development. Moreover, there do not seem to be similarities across species. For instance, in the cat retina, around 80 percent of the retinal ganglon cells die, while in the chick retina, this figure is only 40 percent. In the retinas of fishes and amphibians, no retinal ganglion cells appear to die (Patterson, 1992).

The demise of a neuron is not caused by any obvious defect. Indeed, these neurons have differentiated and successfully extended axons to their targets. Rather, it appears that the target tissue is regulating the number of axons innervating it by limiting a supply of some critical survival factor. There seems to be competition for this limited factor. For instance, if more target tissue is grafted into the original target, more of the neurons survive, and if the target tissue is removed before the axons reach it, nearly all the neurons die. These **neurotrophic factors** have been isolated and regulate the survival of different subsets of neurons.

The best-characterized neurotrophic factor is **nerve growth factor** (NGF), a glycoprotein composed of two identical 13-kDa subunits. NGF is necessary for the survival of sympathetic and sensory neurons. Treating mouse embryos with anti-NGF antibodies reduces the number of trigeminal sympathetic and dorsal root ganglia neurons to 20 percent of their control amounts (Levi-Montalcini and Booker, 1960; Pearson et al., 1983). NGF appears to function after innervation has taken place, as NGF is not secreted by the target tissues until after innervation, and growing axons lack the receptors for NGF, so they couldn't respond any earlier (Davies et al., 1987). Removal of these target tissues causes the death of the neurons that would have innervated them, and there is a good correlation between the amount of NGF secreted and the survival of neurons that innervate these tissues (Korsching and Thoenen, 1983; Harper and Davies, 1990).

Other neurotrophic proteins have been characterized. Two of these proteins—**brain-derived neurotrophic factor** (BDNF) and **neurotrophin-3** (NT-3)—share the same basic structure as NGF. However, they favor the survival of somewhat different groups of neurons. While some neurons respond to all three factors, other neurons respond to only one or two (Figure 8.21). NGF supports the growth and differentiation of sympathetic ganglion cells and certain sensory neurons, but it does not appear to influence motor neuron survival. BDNF, however, can rescue fetal motor neurons in vivo from normally occurring cell death and from induced cell death following the removal of its target tissue (Oppenheim et al., 1992). NT-3 made by target tissues can support the survival of visceral neurons that are not responsive to NGF (Hohn et al., 1990; Maisonpierre et al., 1990). BDNF and NT-3 [and two other neurotrophic molecules, neurotrophin-4/5 (NT-4/5) and fibroblast growth factor-5 (FGF-5)] are each synthe-

(A) Sympathetic   (B) Dorsal root   (C) Nodose

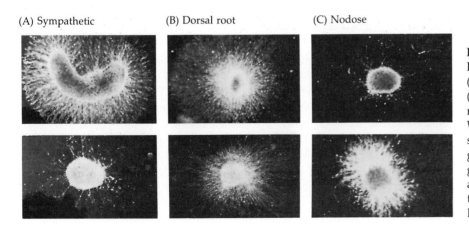

FIGURE 8.21
Effects of NGF (top row) and BDNF (bottom) on neurite outgrowths from (A) sympathetic ganglia, (B) dorsal root ganglia, and (C) nodose ganglia. While both NGF and BDNF had a mild stimulatory effect on dorsal root ganglia axonal outgrowth, the sympathetic ganglia responded to NGF and hardly at all to BDNF, while the converse was true with the nodose ganglia. (From Ibáñez et al., 1991.)

FIGURE 8.22

Interactions between substrate, depolarization, and neurotrophic factor basic FGF (FGF-2) in the survival of ciliary ganglia neurons. Neurons were plated either on laminin (a survival-enhancing substrate) or collagen IV (which does not enhance neuron survival) and observed after 24 hours of culture in the presence or absence of depolarization or FGF-2. When cells were depolarized and grown in the presence of FGF-2, it did not matter on what substrate they grew. However, when FGF-2 was present without polarization, the substrate made a large difference. (From Schmidt and Kater, 1993.)

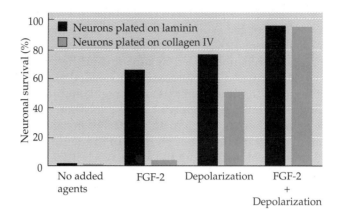

sized in embryonic rat limb muscle cells during the times when motor neuron axons are growing into the muscle and competing for such survival factors. Moreover, BDNF, NT-3, NT-4/5, and FGF prevent the death of the rat embryonic motor neurons when they are placed into culture (Henderson et al., 1993; Hughes et al., 1993). Another newly discovered neurotrophin, glial cell line-derived neurotrophic factor (GDNF) enhances the survival of another group of neurons: the midbrain dopaminergic neurons whose destruction characterizes Parkinson disease (Lin et al., 1993). Yet another, ciliary neurotrophic factor (CNTF), seems to supports the survival of embryonic motor neurons; CNTF is able to prevent the degeneration of motor neurons in a mouse mutant characterized by a progressive loss of motor neurons after birth (Sendtner et al., 1992).

The actual survival of any given neuron in the embryo may be dependent upon a combination of agents. Schmidt and Kater (1993) have shown that neurotrophic factors, depolarization, and interactions with the substrate all combine synergistically to determine neuronal survival. For instance, the survival of chick ciliary ganglion neurons in culture was promoted by FGF, laminin, or depolarization. However, FGF did not promote survival when the laminin was absent, and the combined effects of laminin, FGF, and depolarization were greater than the added effects of each of them (Figure 8.22). The neurotropic factors and the other environmental agents appear to function by suppressing a "suicide program" that would be constitutively expressed unless repressed by these factors (see Chapter 14; Raff et al., 1993). The discovery and purification of these neurotrophic proteins and the analysis of their interactions with substrate and electrical conditions may enable new therapies for neurodegenerative diseases.

The development of behaviors

One of the most fascinating aspects of developmental neurobiology is the correlation of certain neuronal connections with certain behaviors. There are two remarkable aspects of this phenomenon. First there are those cases in which complex behavioral patterns are inherently present in the "circuitry" of the brain at birth. The heartbeat of a 19-day chicken embryo quickens when it hears the distress call, and no other call will evoke this response (Gottlieb, 1965). Furthermore, a newly hatched chick will immediately seek shelter if presented with the shadow of a hawk. The actual hawk is not needed; the shadow cast by a paper silhouette will suffice, and the shadow of no other bird will cause this response (Tinbergen, 1951). There appear, then, to be certain neuronal connections that lead to inherent behaviors in vertebrates.

Equally remarkable are those instances in which the nervous system is seen to be so plastic that new experiences can modify the original set of neuronal connections, causing the creation of new neurons or the formation of new synapses between existing neurons. Since neurons, once formed, do not divide, the "birthday" of a neuron can be identified by treating the organism with radioactive thymidine. Normally, very little radioactive thymidine is taken up into the DNA of a neuron that has already been formed. However, if a new neuron differentiates by cell division during the treatment, it will incorporate radioactive thymidine into its DNA. Such new neurons are seen to be generated when male songbirds learn their songs. Juvenile zebra finches memorize a model song and then learn the pattern of muscle contractions necessary to sing a particular phrase. In this learning and repetition process, new neurons are seen to be generated in the hyperstriatum of the finch's brain. Many of these new neurons send axons to the archistriatum, which is responsible for controlling the vocal musculature (Nordeen and Nordeen, 1988). These changes are not seen in males who are too old to learn the song, nor are they seen in juvenile females (who do not sing these phrases); this will be discussed more fully in Chapter 21.

Even the adult brain is developing in response to new experiences. When adult canaries learn new songs, they generate new neurons whose axons project from one vocal region of the brain to another (Alvarez-Buylla et al., 1990). Similarly, when adult rats learn to keep their balance on dowels, their cerebellar Purkinje cell neurons develop new synapses (Black et al., 1990). Thus, the nervous system continues to develop in adult life, and the pattern of neuronal connections is a product of inherited patterning and patterning produced by experiences. This interplay between innate and experiential development has been detailed most dramatically in studies on mammalian vision.

Experiential changes in inherent mammalian visual pathways

Some of the most interesting research on mammalian neuronal patterning concerns the effects of sensory deprivation on the developing visual system in kittens and monkeys. The paths by which electrical impulses pass from the retina to the brain in mammals are shown in Figure 8.23. Axons from the retinal ganglion cells form the two optic nerves, which meet at the optic chiasm. As in *Xenopus* tadpoles, some fibers go to the opposite (*contralateral*) side of the brain, but unlike most other vertebrates, mammalian retinal cells also send inputs into the same (*ipsilateral*) side of the brain. These nerves end at the two **lateral geniculate nuclei.** Here the input from each eye is kept separate, the uppermost and anterior layers receiving the axons from the contralateral eye, and the middle of the bodies receiving input from the ipsilateral eye. The situation becomes more complicated as neurons from the lateral geniculate nuclei connect with the neurons of the **visual cortex.** Over 80 percent of the neural cells in the cortex receive inputs from both eyes. The result is binocular vision and depth perception. Another remarkable finding is that the retinocortical projection is the same for both eyes. If a cortical neuron is stimulated by light flashing across a region of the left eye 5° above and 1° to the left of the fovea,* it will also be stimulated by a light flashing across a region of the right eye 5 degrees above and 1 degree to the left of the fovea. Moreover, the response evoked in the cortical cell when both eyes are

---

*The fovea is a depression in the center of the retina where only cones are present and the rods and blood vessels are absent. Here it serves as a convenient landmark.

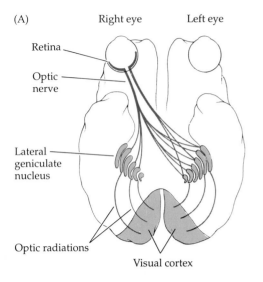

(A) Right eye · Left eye

Retina

Optic nerve

Lateral geniculate nucleus

Optic radiations

Visual cortex

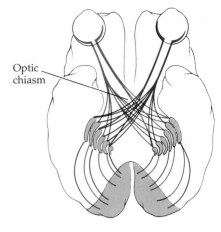

Optic chiasm

Visual pathways from right eye (view is from ventral surface of the brain)

Visual pathways from left eye

Combined left and right visual pathways

FIGURE 8.23
Major pathways of the mammalian visual system. (A) In mammals, the optic nerve from each eye branches, sending nerve fibers to a lateral geniculate nucleus on each side of the brain. On the ipsilateral side, a particular part of the retina goes to a particular part of the lateral geniculate nucleus. On the contralateral side, the lateral geniculate nucleus receives input from all parts of the retina. Neurons from each lateral geniculate nucleus innervate the visual cortex on the same side. (B,C) Isolated (and filleted) retinas showing ipsilateral (B) and contralateral (C) projections from the mouse 16-day embryonic retinal ganglial cells. The fluorescent carbocyanine dye DiI was inserted behind the optic chiasm and the dye was allowed to enter the retinal axons. The dye diffuses along the axons, thereby labeling them to their origin. Ipsilateral projections mostly come from a single part of the retina (in this case, the ventrotemporal region). Contralateral projections to the same site come from all over the retina. (B and C from Colello and Guillery, 1990, courtesy of the authors.)

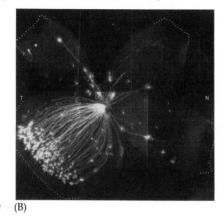

(B)

(C)

stimulated is greater than the response when either retina is stimulated alone.

Hubel, Wiesel, and their co-workers (see Hubel, 1967) demonstrated that the development of the nervous system depends to some degree upon the experience of the individual during a critical period of development. In other words, all of neuronal development is not encoded in the genome: some is learned. Experience appears to strengthen or stabilize some neuronal connections that are already present at birth and to weaken or eliminate other connections. These conclusions come from studies of partial sensory deprivation. Hubel and Wiesel (1962, 1963) sewed shut the right eyelids of newborn kittens and left them closed for 3 months. After this time they unsewed the lids of the right eye. The cortical cells of such kittens could not be stimulated by shining light in the right eye. Almost all the inputs into the visual cortex came from the left eye only. The behavior of the kittens revealed the inadequacy of their right eyes: when only the left eyes of these animals were covered, the kittens became functionally blind. Because the lateral geniculate neurons appeared to be stimulated from both right and left eyes in these kittens, the physiological defect appeared to be between the lateral geniculate nuclei and the visual cortex. In rhesus monkeys, where similar phenomena are observed, the defect has been correlated with a lack of protein synthesis in the lateral geniculate neurons innervated by the covered eye (Kennedy et al., 1981).

Although it would be tempting to conclude that the resulting blindness was due to a failure to form the proper visual connections, this is not the case. Rather, when a kitten is born, axons from lateral geniculate neurons receiving input from each eye overlap extensively in the visual cortex (Hubel and Wiesel, 1963). However, when one eye is covered early in the kitten's life, its connections into the visual cortex are taken over by those of the other eye (Figure 8.24). Competition occurs, and experience plays a role in strengthening and stabilizing the connections from each lateral geniculate nucleus to the visual cortex. Thus, when both eyes of a kitten are sewn shut for 3 months, most cortical cells can still be stimulated by appropriate illumination of one eye or the other. The critical time in kitten development for this validation of neuronal connections begins between the fourth and the sixth week of the kitten's life. Monocular deprivation up to the fourth week produces little or no physiological deficit, but after 6 weeks it produces all the characteristic neuronal changes. If a kitten has had normal visual experience for the first three months, any subsequent monocular deprivation (even for a year or more) is without effect. The synapses have been stabilized.

Two principles, then, can be seen in the patterning of the mammalian visual system. First, neuronal connections involved in vision are present even before the animal sees; and second, experience plays an important role in determining whether or not certain connections remain.* Just as

*Those of you who have studied neurobiology will recall (if properly potentiated) that the concept of the Hebbian synapse is predicated on the notion that experience influences neural pathways. If an axon from neuron A activates neuron B, so that the firing of neuron B is always associated with the firing of neuron A, then the synapse between neuron A and neuron B is strengthened. There are many means by which this strengthening could occur, but most hypotheses focus on changes that would enable calcium ions to more readily enter neuron B. This type of synapse could explain the phenomenon of long-term potentiation, which is thought to be the basis of correlative memory (where one sensation recalls others) (Brown et al., 1990). Such Hebbian mechanisms may mediate the competition between lateral geniculate nuclei axons for cells in the visual cortex (Stent, 1973; Reiter and Stryker, 1988).

FIGURE 8.24
Dark-field autoradiographs of monkey striate cortex 2 weeks after one eye was injected with [³H]proline in the vitreous humor. Each retinal neuron takes up the radioactive label and transfers it to the cells with which it forms synapses. (A) Normal labeling pattern. The white stripes indicate that roughly half the columns took up the label while the other half did not, a pattern reflecting that half the cells were innervated by the labeled eye and half were innervated by the unlabeled eye. (B) Labeling pattern when the unlabeled eye was sutured shut for 18 months. The axonal projections from the normal (labeled) eye take over the regions that would normally have been innervated by the sutured eye. (C,D) Drawings of axons from kitten geniculate nuclei in which one eye was occluded for 33 days. The terminal branching of the axons in the occluded eye (C) were far less extensive than those of the nonoccluded eye (D). (A and B from Wiesel, 1982, courtesy of T. Wiesel; C and D after Antonini and Stryker, 1993.)

(B)

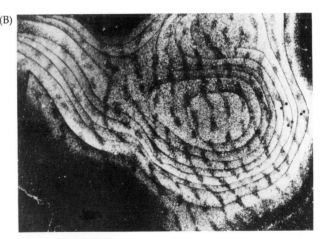

(A)

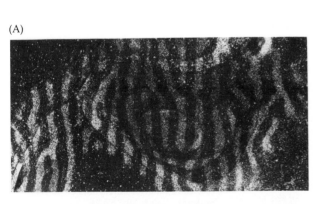

experience refines the original neuromuscular connections, so experience plays a role in refining and improving the visual connections. It is possible, too, that adult functions such as learning and memory arise from the establishment and/or strengthening of different synapses by experience. As Purves and Lichtman (1985) remark, "The interaction of individual animals and their world continues to shape the nervous system throughout life in ways that could never have been programmed. Modification of the nervous system by experience is thus the last and most subtle developmental strategy."

LITERATURE CITED

Akers, R. M., Mosher, D. F. and Lilien, J. E. 1981. Promotion of retinal neurite outgrowth by substratum-bound fibronectin. *Dev. Biol.* 86: 179–188.

Alvarez-Buylla, A., Kirn, J. R. and Nottebohm, F. 1990. Birth of projection neurons in adult avian brain may be related to perceptual or motor learning. *Science* 249: 1444–1446.

Antonini, A. and Stryker, M. P. 1993. Rapid remodeling of axonal arbors in the visual cortex. *Science* 260: 1818–1821.

Baier, H. and Bonhoeffer, F. 1992. Axon guidance by gradients of a target-derived component. *Science* 255: 472–475.

Bastiani, M. J., Harrelson, A. L., Snow, P. M. and Goodman, C. S. 1987. Expression of fasciclin I and II glycoproteins on subsets of axon pathways during neuronal development in the grasshopper. *Cell* 48: 745–755.

Begovac, P. C. and Shur, B. D. 1990. Cell surface galactosyltransferase mediates the initiation of neurite outgrowth from PC12 cells on laminin. *J. Cell Biol.* 110: 461–470.

Berlot, J. and Goodman, C. S. 1984. Guidance of peripheral pioneer neurons in the grasshopper: Adhesive hierarchy of epithelial and neuronal surfaces. *Science* 223: 493–496.

Björkland, A. 1987. Brain implants, transplants. *In* G. Addman (ed.), *Encyclopedia of Neuroscience*, Vol. 1. Birkhauser, Boston, pp. 165–167.

Black, J. E., Issacs, K. R., Anderson, B. J. Alcantara, A. A. and Greenough, W. T. 1990. Learning causes synaptogenesis, whereas motor activity causes angiogenesis, in cerebellar cortex of adult rats. *Proc. Natl. Acad. Sci. USA* 87: 5568–5572.

Brown, T. H., Kairiss, E. W. and Keenan, C. L. 1990. Hebbian synapses: Biophysical mechanisms and algorithm. *Annu. Rev. Neurosci.*. 13: 475–511.

Calof, A. L. 1992. Sex, nose, and genotype. *Curr. Biol.* 2: 103–105.

Cohan, C. S. and Kater, S. B. 1986. Suppression of neurite elongation and growth cone motility by electrical activity. *Science* 232: 1638–1640.

Cohen, J., Burne, J. F., Winter, J. and Bartlett, P. 1986. Retinal ganglial cells lose responsiveness to lamina with maturation. *Nature* 322: 465–467.

Cohen, J., Burne, J. F., McKinlay, C. and Winter, J. 1987. The role of laminin and the laminin/fibronectin receptor complex in the outgrowth of retinal ganglial cell axons. *Dev. Biol.* 122: 407–418.

Colello, R. J. and Guillery, R. W. 1990. The early development of retinal ganglion cells with uncrossed axons in the mouse: retinal position and axon course. *Development* 108: 515–523.

Constantine-Paton, M., Blum A. S., Mendez-Otero, R. and Barnstable, C. J. 1986. A cell surface molecule distributed in a dorsoventral gradient in the perinatal rat retina. *Nature* 324: 459–462.

Cox, E. C., Müller, B. and Bonhoeffer, F. 1990. Axonal guidance in chick visual system: Posterior tectal membranes induce collapse of growth cones from temporal retina. *Neuron* 2: 31–37.

Dan, Y. and Poo, M.-M. 1992. Hebbian depression of isolated neuromuscular synapses in vitro. *Science* 256: 1570–1573.

Davies, A. M., Brandtlow, C., Heumann, R., Korsching, S., Rohrer, H. and Thoenen, H. 1987. Timing and site of nerve growth factor synthesis in developing skin in relation to innervation and expression of the receptor. *Nature* 326: 353–358.

Davies, J. A., Cook, G. W. M., Stern, C. D. and Keynes, R. J. 1990. Isolation from chick somites of a glycoprotein fraction that causes collapse of dorsal root ganglion growth cones. *Neuron* 2: 11–20.

Eisen, J., Meyers, P. Z. and Westerfield, M. 1986. Pathway selection by growth cones of identified motoneurones in live zebra fish embryos. *Nature* 320: 269–271.

Fawcett, J. W. and O'Leary, D. D. M. 1985. The role of electrical activity in the formation of topographic maps in the nervous system. *Trends Neurosci.* 8: 201–206.

Fliess, W. 1897. Quoted in J. Geller, 1992. The cultural construction of the other. *In* H. Eilberg-Schwartz (ed.), *People of the Body.* State University of New York Press, Albany, pp. 243–282.

Franco, B. and fifteen others. 1991. A gene deleted in Kallmann's syndrome shares homology with neural cell adhesion and axonal path-finding molecules. *Nature* 353: 529–536.

Freed, C. R. and eighteen others. 1992. Survival of implanted dopamine cells and neurological iprovement 12 to 46 months after transplantation for Parkinson's disease. *New Eng. J. Med.* 327: 1549–1555.

Gershon, M. D., Schwartz, J. H. and Kandel, E. R. 1985. Morphology of chemical synapses and pattern of interconnections. *In* E. R. Kandel and J. H. Schwartz (eds.), *Principles of Neural Science*, 2nd Ed. Elsevier, New York, pp. 132–147.

Godement, P., Salaun, J. and Mason, C. A. 1990. Retinal axon pathfinding in the optic chiasm: Divergence of crossed and uncrossed fibers. *Neuron* 5: 173–186.

Goodman, C. S. and Bastiani, M. J. 1984. How embryonic nerve cells recognize one another. *Sci. Am.* 251(6): 58–66.

Goodman, C. S. and Shatz, C. J. 1993. Developmental mechanisms that generate precise patterns of neuronal connectivity. *Neuron* 10 [Suppl.]: 77–98.

Goodman, C. S., Bastiani, M. J., Doe, C. Q., du Lac, S., Helfand, S. L., Kuwada, J. Y. and Thomas, J. B. 1984. Cell recognition during neuronal development. *Science* 225: 1271–1287.

Gottlieb, D. I., Rock, K. and Glaser, L. 1976. A gradient of adhesive specificity in developing avian retina. *Proc. Natl. Acad. Sci. USA* 73: 410–414.

Gottlieb, G. 1965. Prenatal auditory sensitivity in chickens and ducks. *Science* 147: 1596–1598.

Grumet, M. 1991. Cell adhesion molecules and their subgroups in the nervous system. *Curr. Opin. Neurobiol.* 1: 370–376.

Gundersen, R. W. 1987. Response of sensory neurites and growth cones to patterned substrata of laminin and fibronectin in vitro. *Dev. Biol.* 121: 423–431.

Hall, Z. W. and Sanes, J. R. 1993. Synaptic structure and development: The neuromuscular junction. *Neuron* 10 [Suppl.]: 99–121.

Halfter, W., Claviez, M. and Schwarz, U. 1981. Preferential adhesion of tectal membranes to anterior embryonic chick retina neurites. *Nature* 292: 67–70.

Hamburger, V. 1939. The development and innervation of transplanted limb primordia of chick embryos. *J. Exp. Zool.* 80: 347–389.

Hamelin, M., Zhou, Y., Su, M.-W., Scott, I. M. and Culotti, J. G. 1993. Expression of the unc-5 guidance receptor in the touch neurons of *C. elegans* steers their axons dorsally. *Nature* 364: 327–330.

Harper, S. and Davies, A. M. 1990. NGF mRNA expression in developing cutaneous epithelium related to innervation density. *Development* 110: 515–519.

Harrelson, A. L. and Goodman, C. S. 1988. Growth cone guidance in insects: Fasciclin II is a member of the immunoglobulin superfamily. *Science* 242: 700–708.

Harris, W. A. 1984. Axonal pathfinding in the absence of normal pathways and impulse activity. *J. Neurosci.* 4: 1153–1162.

Harris, W. A. 1986. Homing behavior of axons in the embryonic vertebrate brain. *Nature* 320: 266–269.

Harrison, R. G. 1910. The outgrowth of the nerve fiber as a mode of protoplasmic movement. *J. Exp. Zool.* 9: 787–848.

Haydon, P. G., Cohan, C. S., McCobb, D. P., Miller, H. R. and Kater, S. B. 1985. Neuron-specific growth cone properties as seen in identified neurons of *Helisoma*. *J. Neurosci. Res.* 13: 135–147.

Hedgecock, E. M., Culotti, J. G. and Hall, D. H. 1990. The *unc-5*, *unc-6*, and *unc-40* genes guide circumferential migrations of pioneer axons and mesodermal cells on the epidermis in *C. elegans*. *Neuron* 2: 61–85.

Henderson, C. E. and twelve others. 1993. Neurotrophins promote motor neuron survival and are present in embryonic limb bud. *Nature* 363: 266–270.

Hohn, A., Leibrock, J., Bailey, K. and Barde, Y.-A. 1990. Identification and characterization of a novel member of the nerve growth factor/brain-derived neurotrophic factor family. *Nature* 344: 339–341.

Hollyday, M. 1980. Motoneuron histogenesis and the development of limb innervation. *Curr. Top. Dev. Biol.* 15: 181–215.

Hollyday, M., Hamburger, V. and Farris, J. M. G. 1977. Localization of motor neuron pools supplying identified muscles in normal and supernumerary legs of chick embryos. *Proc. Natl. Acad. Sci. USA* 74: 3582–3586.

Hubel, D. H. 1967. Effects of distortion of sensory input on the visual system of kittens. *Physiologist* 10: 17–45.

Hubel, D. H. and Wiesel, T. N. 1962. Receptive fields, binocular interaction and functional architecture in the cat's visual cortex. *J. Physiol.* 160: 106–154.

Hubel, D. H. and Wiesel, T. N. 1963. Receptive fields of cells in striate cortex of very young, visually inexperienced kittens. *J. Neurophysiol.* 26: 944–1002.

Hughes, R. A., Sendtner, M., Goldfarb, M., Linholm, D. and Thoenen, H. 1993. Evidence that fibroblast growth factor 5 is a major muscle-derived survival factor for cultured spinal motoneurons. *Neuron* 10: 369–377.

Hynes, R. O. and Lander, A. D. 1992. Contact and adhesive specificities in the associations, migrations, and targeting of cells and axons. *Cell* 68: 303–322.

Ibáñez, C. F., Ebendal, T. and Persson, H. 1991. Chimeric molecules with multiple neurotrophic activities reveal structural elements determining the specificities of NGF and BDNF. *EMBO J.* 10: 2105–2110.

Inuzuka, H., Miyatani, S. and Takeichi, M. 1991. R-cadherin: A novel Ca^{2+}-dependent cell–cell adhesion molecule expressed in the retina. *Neuron* 7: 69–79.

Jacobson, M. 1967. Retinal ganglion cells: Specification of central connections in larval *Xenopus laevis*. *Science* 155: 1106–1108.

Kater, S. B. 1985. Dynamic regulators of neuronal form and conductivity in the adult snail *Helisoma*. In A. I. Selverston (ed.), *Model Neural Networks and Behavior*. Plenum, New York, pp. 191–209.

Kennedy, C., Suda, S., Smith, C. B., Miyaoka, M., Ito, M. and Sokoloff, L. 1981. Changes in protein synthesis underlying functional plasticity in immature monkey visual system. *Proc. Natl. Acad. Sci USA* 78: 3950–3953.

Keynes, R. J. and Stern, C. D. 1984. Segmentation in the vertebrate nervous system. *Nature* 310: 786–787.

Kimmel, C. B. and Law, R. D. 1985. Cell lineage of zebrafish blastomeres I. Cleavage pattern and cytoplasmic bridges between cells. *Dev. Biol.* 108: 78–101.

Klar, A., Baldassare, M. and Jessell, T. M. 1992. F-spondin: A gene expressed at high levels in the floor plate encodes a secreted protein that promotes neural cell adhesion and neurite extension. *Cell* 69: 95–110.

Klose, M. and Bentley, D. 1989. Transient pioneer neurons are essential for formation of an embryonic peripheral nerve. *Science* 245: 982–984.

Kobayashi, T. Nakamura, H. and Yasuda, M. 1990. Disturbance of refinement of retinotectal projection in chick embryos by tetrodotoxin and grayanotoxin. *Dev. Brain Res.* 57: 29–35.

Korsching, S. and Thoenen, H. 1983. Nerve growth factor in sympathetic ganglia and corresponding target organs of the rat: correlation with density of sympathetic innervation. *Proc. Natl. Acad. Sci. USA* 80: 3513–3516.

Krafft-Ebing, R. von. 1886. *Psychopathia Sexualis*. Enke, Stuttgart.

Lance-Jones, C. and Landmesser, L. 1980. Motor neuron projection patterns in chick hindlimb following partial reversals of the spinal cord. *J. Physiol.* 302: 581–602.

Landmesser, L. 1978. The development of motor projection patterns in the chick hindlimb. *J. Physiol.* 284: 391–414.

Legouis, R. *et al.* 1991. The candidate gene for X-linked Kallmann syndrome encodes a protein related to adhesion molecules. *Cell* 67: 423–435.

Letourneau, P., Madsen, A. M., Palm, S. M. and Furcht, L. T. 1988. Immunoreactivity for laminin in the developing ventral longitudinal pathway of the brain. *Dev. Biol.* 125: 135–144.

Leung-Hagesteijn, C., Spence, A. M., Stern, B. D., Zhou, Y., Su, M.-W., Hedgecock, E. M. and Culotti, J. G. 1992. *Unc-5*, a transmembrane protein with immunoglobulin and thrombospondin type 1 domains, guides cell and pioneer axon migrations in *C. elegans*. *Cell* 71: 289–299.

Levi-Montalcini, R. and Booker, B. 1960. Destruction of the sympathetic ganglia in mammals by an antiserum to the nerve growth factor protein. *Proc. Natl. Acad. Sci. USA* 46: 384–390.

Liesi, P. and Silver, J. 1988. Is astrocyte laminin involved in axon guidance in the mammalian CNS? *Dev. Biol.* 130: 774–785.

Lin, L.-F. H., Doherty, D. H., Lile, J. D., Bektesh, S. and Collins, F. 1993. GDNF: A glial cell-line derived neurotrophic factor for midbrain dopaminergic neurons. *Science* 260: 1130–1132.

Lindvall, O., Brundin, P. and Widner, H. 1990. Grafts of fetal dopamine neurons survive and improve motor functions in Parkinson's disease. *Science* 247: 574–577.

Livne, I., Gibson, M. J. and Silverman, A. J. 1993. Biochemical differentiation and intercellular interactions of migratory GnRH cells in the mouse. *Dev. Biol.* 159: 643–666.

Lumsden, A. G. S. and Davies, A. M. 1983. Earliest sensory nerve fibers are guided to peripheral targets by attractants other than nerve growth factor. *Nature* 306: 786–788.

Lumsden, A. G. S. and Davies, A.M. 1986. Chemotropic effect of specific target epithelium in the developing mammalian nervous system. *Nature* 323: 538–539.

Lund, R. D. and Hauschka, S. D. 1976. Transplanted neural tissue develops connection with host rat brain. *Science* 193: 582–584.

Mackenzie, J. L. 1898. The physiological and pathological relations between the nose and the sexual apparatus of man. *Johns Hopkins Hosp. Bull.* 82: 10–17.

Maisonpierre, P. C., Belluscio, L., Squinto, S., Ip, N. Y., Furth, M. E., Lindsay, R. M. and Yancopoulos, G. D. 1990. Neurotrophin-3: A neurotrophic factor related to NGF and BDNF. *Science* 247: 1446–1451.

Matsunaga, M., Hatta, K. and Takeichi, M. 1988. Role of N-cadherin cell adhesion molecules in the histogenesis of neural retina. *Neuron* 1: 289–295.

Nardi, J. B. 1983. Neuronal pathfinding in developing wings of the moth *Manduca sexta*. *Dev. Biol.* 95: 163–174.

Neugebauer, K. M., Tomaselli, K. J., Lilien, J. and Reichardt, L. F. 1988. N-cadherin, N-CAM, and integrins promote retinal neurite outgrowth on astrocytes in vitro. *J. Cell Biol.* 107: 1177–1187.

Nordeen, K. W. and Nordeen, E. J. 1988. Projection neurons within a vocal pathway are born during song learning in zebra finches. *Nature* 334: 149–151.

Oppenheim, R. W., Qin-Wei, Y., Prevette, D. and Yan, Q. 1992. Brain-derived neurotrophic growth factor rescues developing avian motoneurons from cell death. *Nature* 360: 755–757.

Patterson, P. H. 1992. Neuron–target interactions. In Z. Hall, (ed.), *An Introduction to Molecular Neurobiology*. Sinauer Associates, Sunderland, MA., pp. 428–459.

Pearson, J., Johnson, E. M., Jr. and Brandeis, L. 1983. Effects of antibodies to nerve growth factor on intrauterine development of derivatives of cranial neural crest and placode in the guinea pig. *Dev. Biol.* 96: 32–36.

Phelan, K. A. and Hollyday, M. 1990. Axon guidance in muscleless chick wings: The role of muscle cells in motoneural pathway selection and muscle nerve formation. *J. Neurosci.* 10: 2699–2716.

Pittman, R. N. 1985. Release of plasminogen activator and a calcium-dependent metalloprotease from cultured sympathetic and sensory neurons. *Dev. Biol.* 110: 91–101.

Placzek, M., Tessier-Lavigne, M., Jessell, T. and Dodd, J. 1990. Orientation of commissural axons in vitro in response to a floor plate-derived chemoattractant. *Development* 110: 19–30.

Purves, D. and Lichtman, J. W. 1980. Elimination of synapses in the developing nervous system. *Science* 210: 153–157.

Purves, D. and Lichtman, J. W. 1985. *Principles of Neural Development*. Sinauer Associates, Sunderland, MA.

Raff, M. C., Barres, B. A., Burne, J. F., Coles, H. S., Ishizaki, Y. and Jacobson, M. D. 1993. Programmed cell death and the control of cell survival: Lessons from the nervous system. *Science* 262: 695–700.

Raper, J. A. and Kapfhammer, J. P. 1990. The enrichment of a neuronal growth cone collapsing activity from embryonic chick brain. *Neuron* 4: 21–29.

Reiter, H. O. and Stryker, M. P. 1988. Neural plasticity without postsynaptic action potentials: Less-active inputs become dominant when kitten visual cortical cells are phamacologically inhibited. *Proc. Natl. Acad. Sci. USA* 85: 3623–3627.

Roth, S. and Marchase, R. B. 1976. An in vitro assay for retinotectal specificity. *In* S. H. Barondes (ed.), *Neuronal Recognition*. Plenum, New York, pp. 227–248.

Schmidt, M. and Kater, S. B. 1993. Fibroblast growth factors, depolarization, and substrate interact in a combinatorial way to promote neuronal survival. *Dev. Biol.* 158: 228–237.

Schwanzel-Fukada, M. and Pfaff, D. W. 1989. Origin of luteinizing hormone-releasing hormone neurons. *Nature* 338: 161–164.

Schwanzel-Fukada, M., Bick, D. and Pfaff, D. W. 1989. Luteinizing hormone-releasing hormone (LHRH)-expressing cells do not migrate normally in an inherited hypogonadal (Kallman) syndrome. *Mol. Brain Res.* 6: 311–326.

Sendtner, M., Schmalbruch, H., Stöckli, K. A., Carroll, P., Kreutzberg, G. W. and Thoenen, H. 1992. Ciliary neurotrophic factor prevents degeneration of motor neurons in mouse mutant progressive motor neuropathy. *Nature* 358: 502–504.

Silver, J. and Sidman, R.L. 1980. A mechanism for guidance and topologic patterning of retinal ganglion cell axons. *J. Comp. Neurol.* 189: 101–111.

Singer, M., Norlander, R. and Egar, M. 1979. Axonal guidance during embryogenesis and regeneration in the spinal cord of the newt: The blueprint hypothesis of neuronal pathway patterning. *J. Comp. Neurol.* 185: 1–22.

Spencer, D. D. and fifteen others. 1992. Unilateral transplantation of human fetal mesencephalic tissue into the caudate nucleus of patients with Parkinson disease. *New Eng. J. Med.* 327: 1541–1548.

Sperry, R. W. 1951. Mechanisms of neural maturation. *In* S. S. Stevens (ed.), *Handbook of Experimental Psychology*. Wiley, New York, pp. 236–280.

Sperry, R. W. 1965. Embryogenesis of behavioral nerve nets. *In* R. L. DeHaan and H. Ursprung (eds.), *Organogenesis*. Holt, Rinehart & Winston, New York, pp. 161–186.

Stahl, B., Müller, B., von Boxberg, Y., Cox, E. C. and Bonhoeffer, F. 1990. Biochemical characterization of a putative axonal guidance molecule of the chick visual system. *Neuron* 5: 735–743.

Stent, G. S. 1973. A physiological mechanism for Hebb's postulate of learning. *Proc. Natl. Acad. Sci. USA* 70: 997–1001.

Stout, R. P. and Gradziadi, P. P. C. 1980. Influence of the olfactory placode on the development of the brain in *Xenopus laevis* (Daudin). *Neuroscience* 5: 2175–2186.

Taghert, P. H., Doe, C. Q. and Goodman, C. S. 1984. Cell determination and regulation during development of neuroblasts and neurones in grasshopper embryo. *Nature* 307: 163–165.

Tessier-Lavigne, M., Placzek, M., Lumsden, A. G. S., Dodd, J. and Jessell, T. 1988. Chemotropic guidance of developing axons in the mammalian central nervous system. *Nature* 336: 775–778.

Thanos, S., Bonhoeffer, F. and Rutischauser, U. 1984. Fiber–fiber interaction and tectal cues influence the development of the chicken retinotectal projection. *Proc. Natl. Acad. Sci. USA* 81: 1906–1910.

Thomas, L. 1992. *The Fragile Species*. Macmillan, New York.

Thomas, W. A., Schaeffer, A. W. and Treadway, R. M. Jr. 1990. Galactosyl transferase-dependence of neurite outgrowth on substratum-bound laminin. *Development* 110: 1101–1114.

Thompson, W. J. 1983. Synapse elimination in neonatal rat muscle is sensitive to pattern of muscle use. *Nature* 302: 614–616.

Tinbergen, N. 1951. *The Study of Instinct*. Clarendon Press, Oxford.

Tosney, K. W. 1991. Cells and cell interactions that guide motor axons in the developing chick embryo. *BioEssays* 13: 17–23.

Tosney, K. W. and Landmesser, L. T. 1984. Pattern and specificity of axonal outgrowth following varying degrees of limb ablation. *J. Neurosci.* 4: 2158–2527.

Tosney, K. W. and Landmesser, L. T. 1985. Development of the major pathways for neurite outgrowth in the chick hindlimb. *Dev. Biol.* 109: 193–214.

Tosney, K. W. and Oakley, R. A. 1990. Perinotochordal mesenchyme acts as a barrier to axon advance in the chick embryo: Implications for a general mechanism of axon guidance. *Exp. Neurol.* 109: 75–89.

Trisler, D. 1987. Molecular markers of cell position in avian retina are responsible for synapse formation. *Am. Zool.* 27: 189–206.

Trisler, D. and Collins, F. 1987. Corresponding spatial gradients of TOP molecules in the developing retina and optic tectum. *Science* 237: 1208–1210.

Trisler, D., Schneider, M. D. and Nirenberg, M. 1981. A topographic gradient of molecules in retina can be used to identify neuron position. *Proc. Natl. Acad. Sci. USA* 78: 2145–2149.

Trisler, D., Bekenstein, J. and Daniels, M. P. 1986. Antibody to a molecular marker of cell position inhibits synapse formation in the retina. *Proc. Natl. Acad. Sci. USA* 83: 4194–4198.

Twitty, V. C. 1937. Experiments on the phenomenon of paralysis produced by a toxin occuring in *Triturus* embryos. *J. Exp. Zool.* 76: 67–104.

Twitty, V. C. and Johnson, H. H. 1934. Motor inhibition in *Amblystoma* produced by *Triturus* transplants. *Science* 80: 78–79.

Walter, J., Henke-Fahle, S. and Bonhoeffer, F. 1987. Avoidance of posterior tectal membranes by temporal retinal axons. *Development* 101: 909–913.

Weiss, P. 1939. *Principles of Development*. Holt, Rinehart & Winston, New York.

Weiss, P. A. 1955. Nervous system. *In* B. H. Willier, P. A. Weiss and V. Hamburger (eds.), *Analysis of Development*. Saunders, Philadelphia, pp. 346–402.

Westerfield, M., Liu, D. W., Kimmel, C. B. and Walker, C. 1990. Pathfinding and synapse formation in a zebrafish mutant lacking functional acetylcholine receptors. *Neuron* 4: 867–874.

Whitelaw, V. and Hollyday, M. 1983. Position–dependent motor innervation of the chick hindlimb following serial and parallel duplications of limb segments. *J. Neurosci.* 3: 1216–1225.

Widner, H. and eight others. 1992. Bilateral fetal mesencephalic grafts in two patients with parkinsonism induced by 1-methyl-4-phenyl-1,2,3,6 tetrahydropryridine (MPTP). *New Eng. J. Med.* 327: 1556–1563.

Wiesel, T. N. 1982. Postnatal development of the visual cortex and the influence of the environment. *Nature* 299: 583–591.

Wray, S., Grant, P. and Gainer, H. 1989. Evidence that cells expressing luteinizing hormone-releasing hormone mRNA in the mouse are derived from progenitor cells on the olfactory placode. *Proc. Natl. Acad. Sci. USA* 86: 8132–8136.

Yamada, K. M., Spooner, B. S. and Wessells, N. K. 1971. Ultrastructure and function of growth cones and axons of cultured nerve cells. *J. Cell Biol.* 49: 614–635.

Zinn, K., McAllister, L. and Goodman, C. S. 1988. Sequence analysis and neuronal expression of fasciclin I in grasshopper and *Drosophila*. *Cell* 53: 577–587.

Zipser, B. and McKay, R. 1981. Monoclonal antibodies distinguish identifiable neurones in the leech. *Nature* 289: 549–554.

9

Early vertebrate development

Mesoderm and endoderm

*Of physiology from top to toe
 I sing,
Not physiognomy alone or brain
alone is worthy for the Muse,
I say the form complete is worthier
 far,
The Female equally with the Male
 I sing.*
WALT WHITMAN (1867)

In Chapters 7 and 8 we followed the various tissues formed by developing ectoderm. In this chapter, we will follow the early development of the mesodermal and endodermal germ layers. We will see that endoderm forms the lining of the digestive and respiratory tubes, with their associated organs; mesoderm will be seen to generate all the organs between the ectodermal wall and the endodermal tissues.

Theories come and theories go. The frog remains.
JEAN ROSTAND (1960)

MESODERM

The mesoderm of a neurula-stage embryo can be divided into five regions (Figure 9.1). The first region is the **chordamesoderm**. This tissue forms the notochord, a transient organ whose major functions include inducing the formation of the neural tube and establishing the anterior–posterior body axis. As we saw in Chapter 6, the chordamesoderm forms in the center of the embryo on the future dorsal side. The second region is the **somitic dorsal mesoderm**. The term *dorsal* refers to the observation that the tissues developing from this region will be in the *back* of the embryo, along the spine. The cells in this region will form somites, blocks of mesodermal cells on both sides of the neural tube that will produce many of the connective tissues of the back (bone, muscle, cartilage, and dermis). The **intermediate mesoderm** forms the urinary system and genital ducts; we will discuss this region in detail in later chapters. Further away from the notochord, the **lateral plate mesoderm** gives rise to the heart, blood vessels, and blood cells of the circulatory system, as well as to the lining of the body cavities and to all the mesodermal components of the limbs except the muscles. It will also form a series of extraembryonic membranes that are important for transporting nutrients to the embryo. Lastly, the **head mesenchyme** will contribute to the connective tissues and musculature of the face.

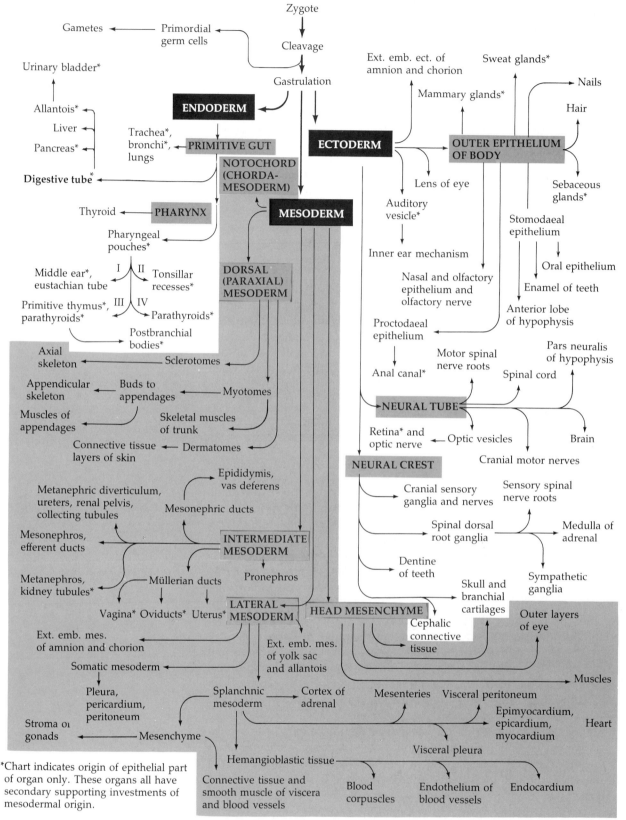

FIGURE 9.1
Chart depicting the lineage of the specialized parts of the amniote body through the three primary germ layers. The germ cells are represented as a line of cells separate from those of the three somatic germ layers because, although the germ cell precursors are located in the presumptive endoderm or mesoderm, they are probably a unique cell type. (After Carlson, 1981.)

Dorsal mesoderm: Differentiation of somites

Paraxial mesoderm

One of the major tasks of gastrulation is to create a mesodermal layer between the endoderm and the ectoderm. As shown in Figure 9.2, the formation of mesodermal and endodermal organs is not subsequent to neural tube formation, but occurs synchronously. Those mesodermal cells not involved in notochord formation have migrated laterally to form thick bands running longitudinally along each side of the notochord and neural tube. These bands of **paraxial mesoderm** are referred to as the *segmental plate* (in birds) and the *unsegmented mesoderm* (in mammals). As the primitive streak regresses and the neural folds begin to gather at the center of the embryo, the paraxial mesoderm separates into blocks of cells called **somites**. Although somites are transient structures, they are extremely important in organizing the segmental pattern of vertebrate embryos. As we saw in the preceding chapter, the somites determine the migration paths of neural crest cells and spinal nerve axons. Somites give rise to the cells that form (1) the vertebrae and ribs, (2) the dermis of the dorsal skin, (3) the skeletal muscles of the back, and (4) the skeletal muscle of the body wall and limbs.

Somitomeres and the initiation of somite formation

The first somites appear in the anterior portion of the embryo, and new somites "bud off" from the rostral end of the paraxial mesoderm at regular intervals (Figures 9.2C,D and 9.3). Because embryos can develop at slightly different rates (as when chicken embryos are incubated at slightly different temperatures), the number of somites present is usually the best indicator how far development has proceeded. The total number of somites formed is characteristic of a species.

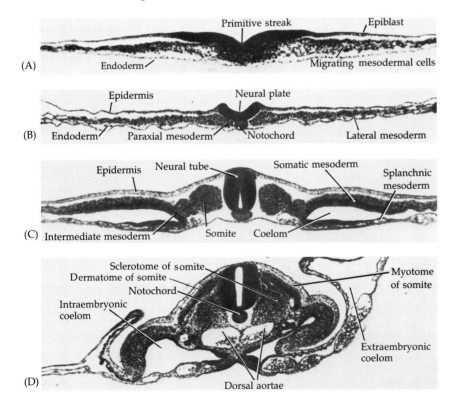

FIGURE 9.2
The progressive development of the chick embryo, focusing on the mesodermal component. (A) Primitive streak region, showing migrating mesodermal and endodermal precursors. (B) Formation of the notochord and paraxial mesoderm. (C,D) Differentiation of the somites, coelom, and the two aortae (which will eventually fuse). A–C, 24-hour embryos; D, 48-hour embryo.

The mechanism for somite formation is not well established, but several studies in chicks have shown that the cells of the segmental plate are organized into whorls of cells called **somitomeres** (Meier, 1979; Packard and Meier, 1983). Conversion from somitomere to somite is seen as the cells of the most anterior somitomere become compacted. Lash and Yamada (1986) have speculated that fibronectin mediates the clumping of the somitomere cells. Their hypothesis is based on two types of data. First, Ostrovsky and co-workers (1984) demonstrated that the amount of fibronectin increases in the anterior region of the presumptive somite at the same time that the cells in the anterior border of this tissue start to compact. The second type of evidence comes from in vitro studies. Lash and co-workers (1984; Cheney and Lash, 1984) found that dissociated cells from the anterior portions of the chick segmental plate (which have just formed somites) readily reassociate into somite-sized clumps. Dissociated cells from the posterior portion of the segmental plate (which has not yet formed somites) do not reassociate. This suggests that the mesodermal cells become more adhesive as they gain the ability to form somites. The posterior cells could be induced to form somite-like masses by adding fibronectin to the culture medium in which they were reaggregating.

Cells of the newly formed somite are organized randomly but soon become organized into a ball of columnar epithelial cells that surround a small cavity filled with loosely connected cells. The epithelial cells become attached to each other by tight junctions. Like the notochord and the neural tube, the somite secretes a **basal lamina** (Figure 9.4) consisting of collagen, fibronectin, laminin, and glycosaminoglycans (GAGs). In fact, GAGs from the neural tube and notochord appear to induce the somites to secrete their own GAGs. If the GAGs are enzymatically removed from the surface of the notochord, GAG synthesis in the somites stops until the notochord once more has elaborated GAGs on its own surface (Kosher and Lash, 1975).

Generation of the somitic cell types

When the somite is first formed, any of its cells can become any of the somite-derived structures. However, as the somite matures, the various regions of the somite become committed to forming only certain cell types. Whether through the action of GAG or some other factor secreted by the notochord or neural tube, the *ventral medial* cells of the somite (those cells located farthest from the back but close to the neural tube) undergo mitosis, lose their round epithelial characteristics, and become mesenchymal cells again. The portion of the somite that gives rise to these cells is called the **sclerotome**, and these mesenchymal cells ultimately become the vertebral **chondrocytes** (Figures 9.2 and 9.5). Chondrocytes are responsible for secreting the special types of collagens and GAGs (such as chondroitin sulfate) characteristic of cartilage. These particular chondrocytes will be responsible for constructing the axial skeleton (vertebrae, ribs, cartilage, and ligaments). The cells of the *lateral* portion of the somite (that region farthest from the neural tube) also disperse. These cells form the precursors of the muscles of the limb and body wall. Ordahl and LeDouarin (1992) have followed these cells by transplanting portions of quail somites into somites of chicken embryos. The chick and quail cells can be distinguished by their nucleolar morphology. They find that whatever somite cells are furthest from the neural tube migrate to form body wall and limb musculature, even if those donor cells were originally from the medial portion of a somite.

Once the cells of the sclerotome and body wall–limb muscle cell pre-

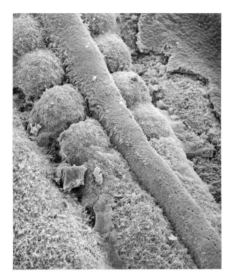

FIGURE 9.3
Neural tube and somites. Scanning electron micrograph showing well-formed somites and paraxial mesoderm (bottom right-hand side) that has not yet separated into distinct somites. A rounding of the paraxial mesoderm into a somitomere can be seen in the lower left side, and neural crest cells can be seen migrating ventrally from the roof of the neural tube. (Courtesy of K. W. Tosney.)

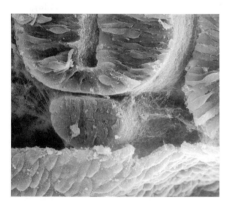

FIGURE 9.4
Scanning electron micrograph of the neural tube–notochord–somite borders. The webbing of collagen and glycosaminoglycan complexes connecting these structures is important in cell migration and the intercellular reactions occurring between them. (Courtesy of K. W. Tosney.)

cursors have migrated away from the somite, the somitic cells closest to the neural tube migrate ventrally down the remaining epithelial portion of the somite to form a solid double-layered epithelium called the **dermamyotome** (Figures 9.2 and 9.5). The dorsal layer of this structure is called the **dermatome**, and it generates the mesenchymal connective tissue of the dorsal skin: the dermis. (The dermis of other areas of the body forms from other mesenchymal cells, not from the somites.) The inner layer of cells is called the **myotome**, and these cells give rise to the vertebral muscles that span the vertebrae and allow the back to bend (Chevallier et al., 1977; Christ et al., 1977). Thus, the somites are critical for forming the back of our bodies: the vertebrae that surround the spinal cord, the muscles and connective tissue that hold the vertebrae together, the dermal underlayer of the skin of our back, and the back musculature. And what happens to the notochord, that central mesodermal structure? After it has provided the axial integrity of the early embryo and has induced the formation of the dorsal neural tube, most of it degenerates. Wherever the sclerotome cells have formed a vertebral body, the notochordal cells die. However, in between the vertebrae, the notochordal cells form the tissue of the intervertebral discs, the **nuclei pulposi**. These are the discs that "slip" in certain back injuries.

Myogenesis: Differentiation of skeletal muscle

A skeletal muscle cell is an extremely large, elongated cell that contains many nuclei. In the mid-1960s developmental biologists debated whether each of these cells (often called myotubes) was derived from the fusion of

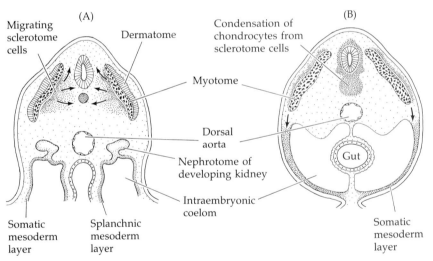

FIGURE 9.5
Diagram of a transverse section through the trunk of (A) an early 4-week and (B) a late 4-week human embryo, showing formation of somite structures. (A) The sclerotome cells are beginning to migrate away from the myotome and dermatome. (B) By the end of the fourth week, the sclerotome cells are condensing to form cartilaginous vertebrae, the dermatome is starting to form the dermis, and the myotome cells are extending ventrally down the walls of the embryo. (C–E) Diagram of the changing structure of the chick somite as cell migrations occur. (A and B after Langman, 1981; C–E after Ordahl, 1993.)

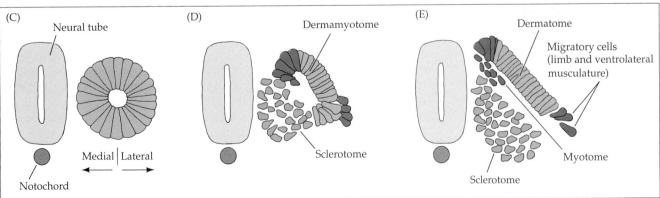

FIGURE 9.6
Quail myoblasts differentiating in culture. All cultures were inoculated, fixed, stained, and photographed on specific days. (A) Day 2 culture consists of unicellular myoblasts (bar represents 0.1 mm). (B) Fusion of myoblasts is first seen in day 3 culture. (C) In day 5 culture, the increase in the number of nuclei within multinucleated myotubes is apparent. (D) Lower magnification of day 5 culture, showing the extensive cell fusion. (E) Higher magnification of multinucleated myotube demonstrating the characteristic cross-striations. Bar, 0.01 mm. (From Buckley and Konigsberg, 1974, courtesy of I. R. Konigsberg.)

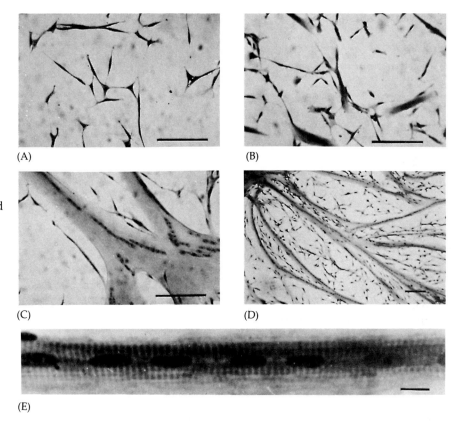

(A)　　　(B)

(C)　　　(D)

(E)

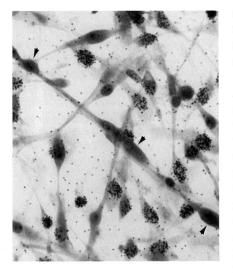

FIGURE 9.7
Autoradiograph showing DNA synthesis in myoblasts. Chick myoblast cells are treated with phospholipase C in culture and then exposed to radioactive thymidine. Unattached myoblasts still divide and incorporate the radioactive thymidine into their DNA. Lined up (but not yet fused) cells (arrows) do not incorporate the label. (From Nameroff and Munar, 1976, courtesy of M. Nameroff.)

several mononucleate muscle precursor cells (myoblasts) or from a single myoblast that undergoes nuclear division without cytokinesis. Evidence from two independent sources demonstrates that the skeletal myotube is derived from the fusion of several mononucleate myoblasts.

In vitro evidence for myoblast fusion. Konigsberg (1963) made a fundamental technical advance in the study of muscle development by defining the culture conditions that would enable myoblasts from certain areas of the body to proliferate in a petri dish and then differentiate into myotubes. (These myoblasts were from the population of cells that migrated away from the somite and were still able to divide prior to their differentiating.) Cells from embryonic regions containing such myoblasts were separated from each other by digesting the extracellular lamina with trypsin. These cells were then placed in collagen-coated petri dishes containing nutrients and serum (Figure 9.6). By 48 hours, the large majority of the cells were rapidly dividing myoblasts. Shortly thereafter, Konigsberg saw that these myoblasts stopped dividing and began to fuse with their neighbors to produce extended myotubes synthesizing muscle-specific proteins. DNA synthesis and nuclear division were not seen in the multinucleated myotubes. As long as there are particular growth factors in the medium (particularly fibroblast growth factor, FGF), the myoblasts will proliferate without differentiating. When these factors are depleted and the cells are prepared to differentiate, the myoblasts secrete fibronectin onto their extracellular matrix and bind to it through their α5β1 integrin, the major fibronectin receptor (Menko and Boettiger, 1987; Menko et al., in press). If this adhesion is blocked, no further muscle development ensues, and it appears that the signal from the integrin–fibronectin attachment is critical for initiating the myoblast to differentiate into muscle cells.

Many of the migrating myoblasts line up in chains and then fuse to form the myotube. Nameroff and Munar (1976) have been able to separate this recognition phenomenon (the making of such chains) from the fusion event by treating the cultured cells with phospholipase C. This enzyme prohibits membrane fusion but permits recognition. Thus, in the presence of the enzyme, chains form but no fusion takes place. Once these cells have aligned themselves, they are no longer dividing. Radioactive thymidine is incorporated into the DNA of the migrating myoblasts but is not incorporated into those myoblasts within the chains (Figure 9.7).

The second stage is the cell fusion event itself. Both the recognition event and the fusion event are dependent upon calcium ions, but during fusion, the calcium ions must get into the myoblasts. Accordingly, fusion can be activated by calcium ionophores, such as A23187, which carry calcium ions across cell membranes (Shainberg et al., 1969; David et al., 1981). While the alignment phenomenon is mediated by cell membrane glycoproteins (Knudsen, 1985: Knudsen et al., 1990), the fusion events appear to be regulated by extensive reorganization of membrane components to form protein-free areas of phospholipids (Kaufman and Foster, 1985; Wakelam, 1985). Kalderon and Gilula (1979) have shown that calcium-mediated membrane fusion occurs in those regions of the myoblast membranes that are enriched for phospholipids. Thus, myoblast cell culture experiments have shown that once the myoblasts stop dividing, they acquire the capacities to fuse and to make muscle-specific proteins. Interestingly, although cell–cell recognition and fusion are essential for myogenesis, species-specificity is not. Myoblasts will fuse only with other myoblasts, but those myoblasts need not be of the same species. Yaffe and Feldman (1965) have shown that embryonic rat and chick myoblasts will readily fuse to form hybrid myotubes.

FIGURE 9.8
The two possible mechanisms of skeletal muscle formation and how to distinguish between them. Allophenic mice are made from the fusion of mouse embryos from two different strains, each strain making a different form of the enzyme isocitrate dehydrogenase. This enzyme is composed of two subunits; one strain of mouse makes AA isocitrate dehydrogenase (indicated in black) and the other makes BB (color). (A) If the enzymes are made in a single cell, or in multinucleate cells arising from the nuclear divisions within a single cell, the enzyme will be purely AA or BB. (B) If there are two different nuclei in the same cell, however, one might code for B subunits while the other might code for A, with the result that some molecules of the enzyme will be hybrid (AB). Electrophoresis can separate these three types of molecules. The presence of the AB molecule in skeletal muscle cells (but not in other cell types) confirms the fusion model. (After Mintz and Baker, 1967.)

(A) Division model

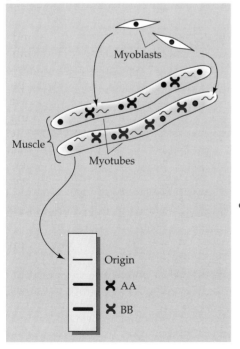

(B) Fusion model

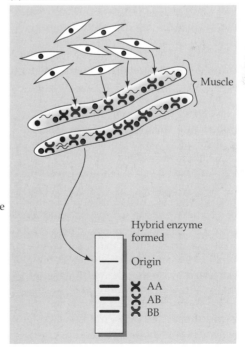

Homogenize and place at origin of electrophoresis plate

Isocitrate dehydrogenase enzymes seen by electrophoresis

- ● Genotype *AA*   ～ Polypeptide A   ✖ Enzyme AA
- ● Genotype *BB*   ～ Polypeptide B   ✖ Enzyme BB
                                        ✖ Enzyme AB

***In vivo* evidence for myoblast fusion.** The second type of evidence for myoblast fusion came from tetraparental (allophenic) mice. These mice can be formed from the fusion of two early embryos, which regulate to produce a single mouse having two distinct cell populations (see Figure 5.28). Mintz and Baker (1967) fused mouse embryos that produce different types of the enzyme isocitrate dehydrogenase. This enzyme, found in all cells, is composed of two identical subunits. Thus, if myotubes are formed from one cell whose nuclei divide without cytokinesis, one would expect to find two distinct forms of the enzyme, that is, the two parental forms, in the allophenic mouse (Figure 9.8). On the other hand, if myotubes are formed by fusion between cells, one would expect to find muscle cells expressing not only the two parental types of enzymes (AA and BB) but also a third class composed of a subunit from each of the parental types (AB). The different forms of isocitrate dehydrogenase can be separated and identified by their electrophoretic mobility. The results clearly demonstrated that although only the two parental types of enzyme were present in all the other tissues of the allophenic mice, the hybrid (AB) enzyme was present in extracts of skeletal muscle tissue. Thus, the myotubes must have been formed from the fusion of numerous myoblasts.

SIDELIGHTS & SPECULATIONS

Molecular regulators of development: Muscle-building and the MyoD *family of transcriptional regulators*

How is an embryonic mesenchymal cell instructed to form a muscle cell instead of a cartilage cell, a fibroblast, or an adipose cell? What molecules commit its fate to one lineage and not another? In 1986, Lassar and co-workers took DNA from myoblast cells and transfected it into a certain type of cell called the C3H10T1/2 cell. This cell has a fibroblast-like appearance, but it resembles mesenchyme since it is able to become either an adipose cell or a muscle cell. When the muscle DNA was added to these cells, the C3H10T1/2 cells were transformed into muscle cells. DNA isolated from fibroblasts or other cell types cannot accomplish this conversion. By subtraction cloning (see Chapter 2), a myoblast-specific mRNA was found that could also effect this change in differentiated phenotype. The myoblast mRNA encoded a protein called myoblast determination 1 protein or, more commonly, **MyoD** (Davis et al., 1987). The *MyoD* gene is expressed only in cells of the muscle lineages. It appears to be a "master switch" gene in that it can convert other cell types into muscles if this gene is active in them. This hypothesis was tested by cloning the *MyoD* gene into a viral vector so that it is under the control of a constitutively active viral promoter (it is always "on"). When this *MyoD* fusion gene was transfected into various cells, pigment cells, nerve cells, fat cells, fibroblasts, and liver cells were converted into muscle-like cells (Figure 9.9; Weintraub et al., 1989). Thus, *MyoD* appears to be sufficient to activate the muscle-specific genes that make up the muscle phenotype.

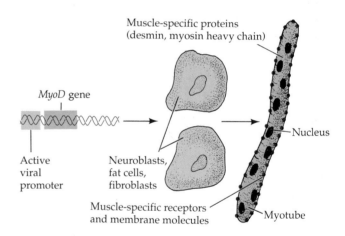

FIGURE 9.9

Summary of several experiments in which the MyoD *gene was activated by a viral promoter and transfected into nonmuscle cells. The* MyoD *protein appears to override the original regulators of the cell phenotype and convert the cells into muscles.*

MyoD encodes a nuclear DNA-binding protein that can bind to regions of the DNA adjacent to muscle-specific genes and activate these genes. For instance, the MyoD protein appears to directly activate the muscle-specific creatine phosphokinase gene by binding to the DNA immediately upstream from it (Lassar et al., 1989). Similarly, there are two MyoD-binding sites on the DNA adjacent to a subunit gene of the chicken muscle acetylcholine receptor (Piette et al., 1990). It also directly activates itself. Once the *MyoD* gene is on, its protein product binds to the DNA

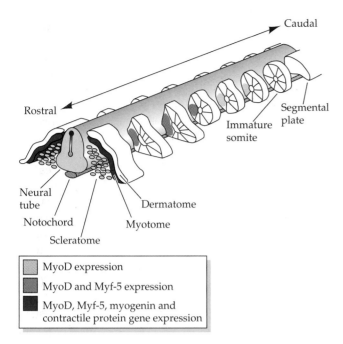

Caudal

Rostral

Segmental plate

Immature somite

Neural tube

Notochord

Dermatome

Scleratome

Myotome

☐ MyoD expression

☐ MyoD and Myf-5 expression

■ MyoD, Myf-5, myogenin and contractile protein gene expression

FIGURE 9.10

Schematic representation of MyoD family myogenic regulatory proteins in the branchial somites of a quail embryo. The figure depicts the expression of MyoD (from quail qmf1 *gene), MyoD plus Myf-5 (from quail* qmf3 *gene), and MyoD, Myf-5, and the myogenin protein (from the quail* qmf2 *gene). The migrating cells that form the limb and body wall musculature do not activate their* MyoD *genes until muscle formation is about to begin in the limb. (After Pownall and Emerson, 1992b.)*

immediately upstream of the *MyoD* gene and keeps this gene from being turned off (Thayer et al., 1989). In other cases, the effects of *MyoD* may be indirect. Not all genes involved in producing the muscle phenotype may be directly activated by MyoD protein. MyoD probably acts indirectly by turning on other regulatory genes, which then activate the structural muscle-specific genes.

MyoD is not the only muscle switch gene. There is a family of MyoD-like proteins that have very similar structures and appear to be able to substitute for each other to a large degree. This family of proteins (sometimes referred to either as the "MyoD family" or the "myfkins") include **myogenin, Myf-5**, and **Myf-6**, and they appear to bind to similar sites on the DNA (to be discussed in Chapter 10). Transfection of any of these myogenic genes into a wide range of cultured cells also converts these cells into muscles. *MyoD* expression leads to *myogenin* expression, and the transfection of *myogenin* genes activates *MyoD* expression. Thus, there is a reciprocal positive feedback loop such that when either *myogenin* or *MyoD* is activated, so is the other gene (Thayer et al., 1989). In the chick and quail, *MyoD* is active in those cells of the early somite that are closest to the neural tube and notochord (Figure 9.10). As the somite matures, Myf-5 also becomes expressed in these cells. After the sclerotome cells have dispersed, the myotome cells express MyoD1, Myf-5, and Myf-6 proteins

(Pownall and Emerson, 1992a,b). It is probable that the neural tube is secreting factors that induce the expression of these proteins in the nearby somitic cells and thereby generate muscle cells from that section of the somite (Packard and Jacobson, 1976; Vivarelli and Cossu, 1986; Rong et al., 1992).

In mice, the order is different, and MyoD is the last of this family to be expressed rather than the first. Myf-5 is the first of these proteins seen in mice, and it is observed in the somitic epithelial cells close to the neural tube (Figure 9.11; Lyons and Buckingham, 1992). Myf-6 expression is seen soon afterward, in the early myotome stage. Myogenin and MyoD proteins are not detected until the myoblasts begin differentiating and the messages for the muscle-specific contractile proteins are being transcribed (Sassoon et al., 1989). Thus, Myf-5 and Myf-6 appear to be involved in the early processes of mouse skeletal muscle development, while MyoD and myogenin appear to be active when myoblasts are fusing.*

These distinctions may not be critical, however. Using a gene-targeting technique (see Chapter 2) that enables researchers to construct mice that lack a particular set of genes, it was shown that some of these myogenic proteins can compensate for the absence of the other members of that family. For instance, when mice lack *MyoD* genes, the expression of the *Myf-5* gene is elevated. Instead of diminishing when MyoD first appears, the *Myf-5* gene stays present and active. The resulting mice have normal muscle development, even though they lack both *MyoD* genes[†] (Rudnicki et al., 1992). When the mice lack their *Myf-5* genes, they also have normal muscle development. Therefore, even though *Myf-5* is the earliest of the myogenic genes expressed, its functions can be taken over by the other myogenic protein genes. However, the absence of Myf-5 protein causes the delay of myotome formation by

---

*As mentioned earlier, dividing myoblasts do not differentiate. This distinction between division and differentiation is characteristic of several cell types that are derived from stem cell populations (Bischoff and Holtzer, 1969; Holtzer, et al., 1975). One of the major growth factors promoting myoblast cell division is basic fibroblast growth factor. The myoblast has receptors for this protein, and the myoblast is stimulated to replicate its DNA and divide when it has bound this factor. FGF is also able to inactivate the MyoD protein. The FGF receptors are lost when the myoblast differentiates into a muscle cell (Olwin and Hauschka, 1988; Moore et al., 1991). FGF simultaneously promotes myoblast cell division and inhibits myoblast differentiation. One of the ways it does the latter is to suppress the transcription of the genes encoding MyoD and myogenin (Vaidya et al., 1989; Brunetti and Goldfine, 1990) while at the same time inactivating the MyoD protein. The mechanism for all this molecular regulation will be discussed in Chapter 10.

†This means that there is some redundancy in the development of skeletal muscles. Such redundancy has long been known to embryologists (Spemann, 1938), but geneticists are rediscovering it (to their dismay, as it muddies up the interpretation of such experiments). Gould (1990) considers developmental redundancy essential for evolution to occur, since one of the redundant partners is free to get a new function while the other partner(s) maintains the original function.

several days, and this causes a failure of the lateral portion of the sclerotome to form properly. Although these mice have normal muscles, their rib cages are distorted and they are unable to breathe (Braun et al., 1992). Recent experiments in Rudolf Jaenisch's laboratory (Rudnicki et al., 1993) show that when the *Myf-5* and *MyoD* genes are both absent from the embryo, no muscles or ribs form! While MyoD and Myf-5 can substitute for one another, there does not

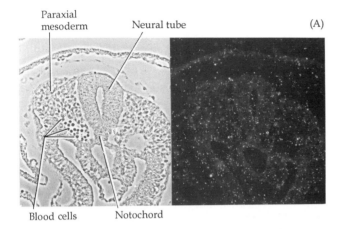

(A)

Paraxial
mesoderm   Neural tube

Blood cells   Notochord

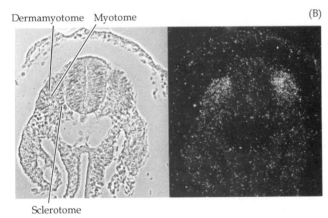

(B)

Dermamyotome   Myotome

Sclerotome

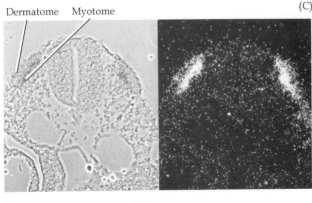

(C)

Dermatome   Myotome

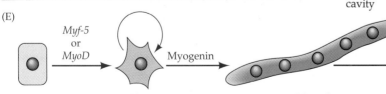

(E)

Myf-5
or
MyoD   Myogenin   *Myf-6*

Cell in somite   Myoblast   Myotube   Myofiber

appear to be redundancy in the functions of myogenin. Mice homozygous for a targeted mutation in the *myogenin* gene die shortly after birth because of defects in their muscle cell formation (Hasty et al., 1993; Nabeshima et al., 1993).

The MyoD family of proteins has two functions. First, they commit an undetermined somite cell to the muscle lineage. Second, they activate the genes that generate the muscle-specific enzymes and contractile fibrils. In the mouse, MyoD and Myf-5 [and perhaps an unrelated protein called myd, which activates *MyoD* and *myogenin* genes (Pinney et al., 1988)] may be responsible for committing the somitic cell to become a myoblast. Myogenin may then be necessary for initiating the differentiation of the myoblast into a myotube, while Myf-6 may be responsible for completing the synthesis of muscle-specific proteins and converting the myotubes into mature myofibrils.

FIGURE 9.11

Expression of Myf-5 *mRNA during somitic muscle formation in the mouse embryo. (A–C) In situ hybridization for* Myf-5 *mRNA in increasingly rostral cross-sections of a 9.5-day embryo. (A) No* Myf-5 *expression is seen in caudal unsegmented paraxial mesoderm. (B) Expression of* Myf-5 *message in the dermamyotome of a newly formed somite. (C) In rostral regions,* Myf-5 *is expressed in the myotome. (D) Expression of* Myf-5 *in a 12-day mouse embryo is seen in the mandibular arch, myotomes, and hindlimb bud (sagittal section). (E) Postulated roles of myogenic proteins during skeletal muscle formation in the mouse. (A–D, photographs courtesy of G. Lyons; E after Rudnicki et al., 1993.)*

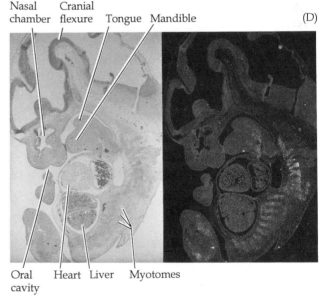

(D)

Nasal   Cranial
chamber flexure   Tongue Mandible

Oral   Heart Liver   Myotomes
cavity

Osteogenesis: The development of bones

Some of the most obvious structures derived from the somitic mesoderm are the bones. In this chapter we can only begin to outline the mechanisms of bone formation, and students wishing further details are invited to consult histology textbooks that devote entire chapters to this topic. There are two major modes of bone formation, or **osteogenesis**, and both involve the transformation of a pre-existing mesenchymal tissue into bone tissue. The direct conversion of mesenchymal tissue into bone is called **intramembranous ossification**. In other cases, the mesenchymal cells differentiate into cartilage, and this cartilage is later replaced by bone. This process by which a cartilage intermediate is formed and replaced by bone cells is called **endochondral ossification**.

Intramembranous ossification. Intramembranous ossification is the characteristic way in which the flat bones of the skull are formed. Mesenchymal cells derived from the neural crest interact with the extracellular matrix of the head epithelial cells to form bone. If the mesenchymal cells do not contact this matrix, no bone will be formed (Tyler and Hall, 1977; Hall, 1988). This was shown in vitro by Hall and colleagues (1983), who isolated head mesenchymal cells and plated them into culture dishes. If no extracellular matrix were present on the surface of these dishes, the cells remained mesenchymal. However, if head epithelial cells had first secreted an extracellular matrix upon the surface, the cells differentiated into bone cells.

The mechanisms responsible for this conversion of mesenchymal cells into bone is still unknown, but recent evidence points to a particular group of molecules at the epithelial–mesenchymal junction. **Bone morphogenetic proteins** can be isolated from adult bone and injected into embryonic muscle or connective tissues. When this is done, cartilage develops from cells within these tissues, and this cartilage is then replaced by bone cells (Syftestad and Caplan, 1984; Urist et al., 1984). Recent research has discovered several of these bone morphogenetic proteins (see Sidelights & Speculations.)

During intramembranous ossification, the mesenchymal cells proliferate and condense into compact nodes. Some of these cells develop into capillaries and others change their shape to become **osteoblasts**, cells capable of secreting the bone matrix. The secreted collagen-glycosaminoglycan matrix is able to bind calcium salts, which are brought to the region through capillaries. In this way, the matrix becomes calcified. In most cases, osteoblasts are separated from the region of calcification by a layer of the prebone (osteoid) matrix they secrete. Occasionally, though, os-

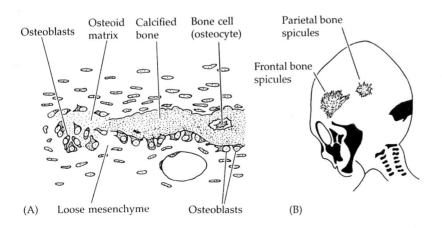

(A) Osteoblasts — Osteoid matrix — Calcified bone — Bone cell (osteocyte) — Loose mesenchyme — Osteoblasts

(B) Parietal bone spicules — Frontal bone spicules

FIGURE 9.12
Schematic diagram of membranous ossification. (A) Mesenchymal cells, probably derived from the neural crest, condense to produce osteoblasts, which deposit osteoid matrix. These osteoblasts become arrayed along the calcified region of the matrix. Osteoblasts that are trapped within the bone matrix become osteocytes. (B) Spread of bone spicules from the primary ossification site in the flat skull bones of a 3-month-old human embryo. The bones shown in black are formed by endochondral ossification. (After Langman, 1981.)

teoblasts become trapped in the calcified matrix and become **osteocytes**, bone cells. As calcification proceeds, the bony spicules radiate out from the center where ossification began (Figure 9.12). Furthermore, the entire region of calcified spicules becomes surrounded by compact mesenchymal cells that form the **periosteum**. The cells on the inner surface of the periosteum also become osteoblasts and deposit bone matrix parallel to that of the existing spicules. In this manner, many layers of bone are formed.

SIDELIGHTS & SPECULATIONS

Molecular regulators of development: The bone morphogenetic proteins

The mass, strength, and shape of a bone are maintained throughout development and adult life through an equilibrium between the forces of bone destruction and those of bone formation. In 1965, Urist demonstrated that the mesodermal cells that form the bone make their own bone-forming factors. Moreover, soluble extracts from demineralized bone tissue could induce the formation of bone in rats when they were implanted into sites that do not usually form bone. This process of ectopic bone formation recapitulated the normal events of endochondral ossification. The implants became invaded by mesenchymal cells that rapidly proliferated into nodules and differentiated in about a week into chondroblasts and chondrocytes. The cartilage then became calcified and was replaced by bone tissue as osteoblasts and osteoclasts appeared along the cartilage model.

The identities of seven of these soluble bone-forming proteins have been elucidated (Wozney et al., 1988; Celeste et al., 1990). These are **bone morphogenetic proteins 1 to 7** (BMP-1–7). Six of these proteins, BMP-2–7, share a very similar structure (Lyons et al., 1991), and they are thought to be 30-kDa glycoprotein dimers containing seven precisely spaced cysteine residues in their carboxyl termini (Figure 9.13). This structure places them into the TGF-β family of growth and differentiation factors, which includes other proteins that induce new tissue formation and regulate mitogenic activity. Unlike the myogenic regulatory proteins mentioned earlier in the chapter, which operate in the nucleus to regulate transcription, the bone morphogenetic proteins travel from cell to cell and appear to function at the membrane of the mesenchymal cells. Purified dimers of human BMP-2, -4, -5, and -7 have the ability to create new bone by the assay mentioned above, and heterodimers composed of BMP-2 and BMP-7 subunits work better than either BMP-2 or BMP-7 alone (Rosen and Thies, 1992; Wozney et al., 1993).

BMPs are thought to function in the production of bone during embryonic development, and they may play different, but overlapping, roles during bone formation. In the early limb bud, BMP-4 is found in the mesenchyme and in the ectodermal ridge at the distal end of the bud.

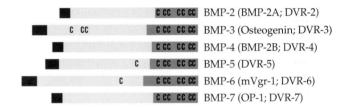

FIGURE 9.13
Structures of BMP-2–7 prior to their secretion from the cell. The black areas represent the hydrophobic leader sequence of each peptide, while the gray regions show the propeptide that is cleaved from the mature protein (color). The seven cysteine residues are labeled (C). The label on the left corresponds to the human protein and lists other names given that molecule. (After Rosen and Thies, 1992.)

BMP-2 is also found in the ridge, but not in the mesenchyme. As the bud grows, BMP-2 and -4 become expressed in the mesenchyme cells between the cartilage cells of the digits, while BMP-6 is expressed in the region of the cartilage that is actively growing. By the end of limb development, BMP-2 is seen in the periostial covering of the bone, while BMP-6 is present in the calcifying cartilage cells (Figure 9.14; Lyons et al., 1989, 1990; Wozney, 1992). The deletion of the BMP-5 genes in mice causes these embryos to have smaller mesenchymal condensations than those of wild-type mice; this leads to shorter ears and sternums, the loss of several small bones, and the deletion of one pair of ribs (Kingsley et al., 1992). In addition, BMP proteins are expressed in numerous nonskeletal regions of the embryo, including the mouse central nervous system, heart, teeth, and whisker follicle precursors (Jones et al., 1991) (see Chapter 18). It is likely, then, that the BMP proteins play several roles during the development of mammalian organs. [For this reason, Lyons and colleagues (1991) have suggested that BMP-2–7 be renamed **d**ecapentaplegic-**V**g-related proteins 2-7, or DVR-2–7, after two of the TGF-β family proteins they most closely resemble.]

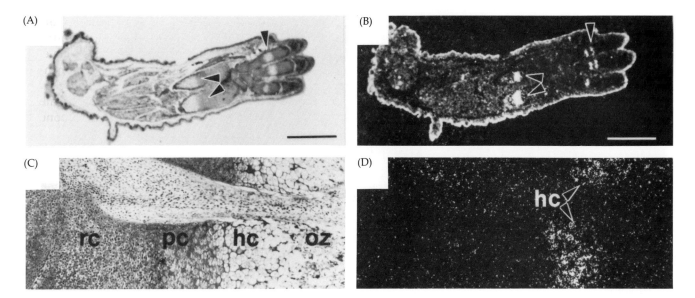

FIGURE 9.14

Localization of BMP mRNA during cartilage and bone formation in the chick embryo. In situ hybridization was performed using radioactive RNA complementary to the mRNA of two BMP proteins. The bright-field photographs on the left correspond to the dark-field autoradiographs on the right. (A,B) A

15.5-day limb hybridized with probe to BMP-2 mRNA shows this message in hypertrophic cartilage of the long bones and digits. (C,D) Higher magnification of a 17.5-day limb shows mRNA for BMP-6 in hypertrophic cartilage. (From Lyons et al., 1990, courtesy of B. Hogan.)

Endochondral ossification.   Endochondral ossification involves the formation of cartilage tissue from aggregated mesenchymal cells and the subsequent replacement of this cartilage tissue by bone (Horton, 1990). The cartilage tissue is a model for the bone that follows. The bones of the vertebral column, the pelvis, and the extremities are first formed of cartilage and are later changed into bone. This remarkable process coordinates **chondrogenesis** (cartilage production) with **osteogenesis** (bone growth), while these skeletal elements are simultaneously bearing a load, growing in width, and responding to local stresses. In humans, the "long bones" of the embryonic limb buds form from mesenchymal cells that form nodules in those regions that will become bone. These cells become **chondrocytes**, and they secrete the cartilage extracellular matrix. The mesenchymal cells around them become the periosteum (Figure 9.15). Soon after the cartilaginous "model" is formed, the cells in the central part of the model become dramatically larger and begin secreting a different type of matrix, one that contains different types of collagen, more fibronectin, and less protease inhibitor. These cells are the **hypertrophic chondrocytes**. Their matrix is more susceptible to invasion by blood vessel cells from the periosteum. A capillary from the periosteum then invades the center of the previously avascular cartilage shaft. As the cartilage matrix is degraded, the hypertrophic cartilage cells die, and **osteoblasts** (bone-forming cells) carried by the blood vessels begin to secrete bone matrix on the partially degraded cartilage. Eventually, all the cartilage is replaced by bone.

As the center of the cartilage model is converted into bone, an ossification front is formed between the newly synthesized bone and the remaining cartilage. The cartilage side of this front contains the hypertrophic cartilage that prepares the shaft for invasion by the blood vessels, and the

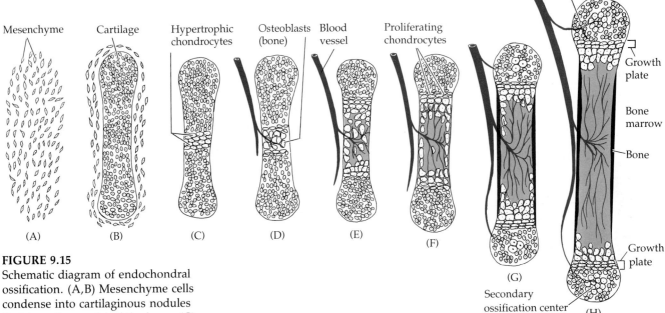

Mesenchyme   Cartilage   Hypertrophic chondrocytes   Osteoblasts (bone)   Blood vessel   Proliferating chondrocytes   Epiphyseal cartilage

Growth plate

Bone marrow

Bone

(A)   (B)   (C)   (D)   (E)   (F)

Growth plate

(G)

Secondary ossification center   (H)

FIGURE 9.15
Schematic diagram of endochondral ossification. (A,B) Mesenchyme cells condense into cartilaginous nodules that form the model of the bone. (C) Chondrocytes in the center of the shaft hypertrophy and change their extracellular matrix, allowing blood vessels to enter. (D,E) Blood vessels bring in osteoblasts, which bind to the degenerating cartilaginous matrix and deposit bone matrix. (F–H) Formation of the epiphyseal growth plates by chondrocytes, which proliferate before undergoing hypertrophy. Secondary ossification centers also form as blood vessels enter near the tips of the bone. (After Horton, 1990.)

bone side contains the osteoblast cells laying down the bone matrix. This front spreads outward in both directions from the center, as more cartilage is turned to bone. If this were all, however, there would be no growth, and our bones would just be as large as the original cartilaginous model. However, as the ossification front nears the ends of the cartilage model, the chondrocytes near the ossification front proliferate prior to undergoing hypertrophy. This pushes out the cartilaginous ends of the bone, providing a source of new cartilage. These cartilaginous regions at the end of the long bones are called **epiphyseal plates**. As this cartilage hypertrophies and the ossification front extends further outward, the remaining cartilage in the epiphyseal plate proliferates. This cartilage forms the growth area of the bone. Thus, the bone keeps growing because of the production of new cartilage cells that undergo hypertrophy, enable the blood vessels to enter, and die as the bony matrix is deposited. As long as the epiphyseal growth plates are able to produce chondrocytes, the bone continues to grow.

The cells of the epiphyseal plate are very responsive to hormones, and their proliferation is stimulated by growth hormone and insulin-like growth factors. Nilsson and colleagues (1986) have recently shown that growth hormone stimulates the production of **insulin-like growth factor I** (IGF-I) in these chondrocytes, and that these chondrocytes respond to it by proliferating. When they added growth hormone to the tibial growth plate of young mice (who could not manufacture their own growth hormone because their pituitaries had been removed), the growth hormone stimulated the formation of IGF-I from the chondrocytes in the proliferative zone (Figure 9.16). The combination of growth hormone and IGF-I appear to provide an extremely strong mitotic signal. The pygmies of the Ituri Forest of Zaire have normal GH and IGF-I levels until puberty. However, at puberty, the pygmies' IGF-I levels fall to about one-third that of other adolescents. It appears that IGF-I is essential for the normal growth spurt at puberty (Merimee et al., 1987). Hormones are also responsible for the cessation of growth. At the end of puberty, high levels of estrogen or testosterone cause the remaining epiphyseal plate cartilage

FIGURE 9.16

Proliferation of cells in epiphyseal plate in response to growth hormone. (A) Cartilaginous region in a young rat that was made growth hormone-deficient by removal of its pituitary. (B) Same region in the rat after injection of growth hormone. (I. Gersh's photographs from Bloom and Fawcett, 1975.)

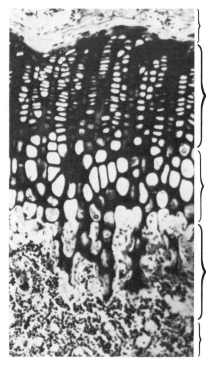

Reserve cartilage

Proliferating cartilage cells

Hypertrophic and calcifying cartilage cells

Zone of cartilage degeneration and ossification

Calcified bone

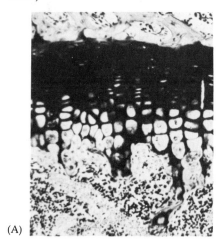

(A)

(B)

to hypertrophy. These cartilage cells grow, die, and are replaced by bone. Without any further cartilage, growth of these bones ceases.

The replacement of chondrocytes by osteoblasts appears to be dependent on the mineralization of the extracellular matrix. In chick embryos, the source of calcium is the calcium carbonate of the eggshell, and during its development the circulatory system of the chick translocates about 120 mg of calcium from the shell to the skeleton (Tuan, 1987). When chick embryos are removed from their shells at day 3 and grown in shell-less culture (in plastic wrap) for the duration of their development, much of the calcium-deficient cartilaginous skeleton fails to mature into bony tissue (Figure 9.17; Tuan and Lynch, 1983). In mammals, calcium is transferred across the placenta and is deposited onto the matrix by the chondrocytes. It has been shown that hypertrophic chondrocytes switch from aerobic to

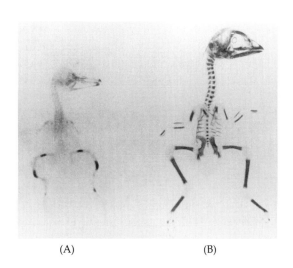

(A)   (B)

FIGURE 9.17

Skeletal mineralization in 17-day chick embryos grown in (A) shell-less and (B) normal culture. The embryos were fixed and stained with Alizarin Red to show the calcified matrix. (From Tuan and Lynch, 1983, courtesy of R. Tuan.)

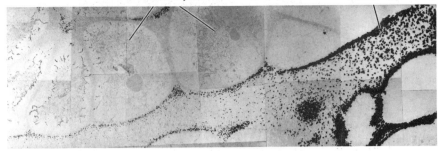

Chondrocytes

Calcium on
extracellular matrix

FIGURE 9.18
Deposition of calcium by chondrocytes in the distal region of the hypertrophic zone. Calcium (stained darkly in this electron micrograph montage) is placed onto the matrix by the enlerging cells. (From Brighton and Hunt, 1974, courtesy of C. T. Brighton.)

FIGURE 9.19
Osteoclast activity on the bone matrix. (A) Electron micrograph of the ruffled membrane of a chick osteoclast cultured on reconstituted bone matrix. (B) Section of ruffled membrane stained for the presence of an ATPase capable of transporting hydrogen ions from the cell. The ATPase is restricted to the membrane of the cell process. (C) Solubilization of inorganic and collagenous matrix components (as measured by the release of [^{45}Ca] and [^{3}H] proline, respectively) by 10,000 osteoclasts incubated on labeled bone fragments. (A and C from Blair et al., 1986; B from Baron et al., 1986, courtesy of the authors.)

anaerobic respiration (Brighton and Hunt, 1974; Brighton, 1984), causing a decrease in cellular ATP and the employment of a phosphocreatine-mediated energy pathway, such as used in oxygen-depleted muscle (Shapiro et al., 1992). By some as yet unknown mechanisms, these metabolic changes are thought to result in the deposition of calcium in the extracellular matrix, within tiny membrane-bound structures known as **matrix vesicles** (Wuthier, 1982). This initiates the process of calcification and allows osteoblasts to bind and initiate bone formation (Figure 9.18).

As new bone material is added peripherally from the internal surface of the periosteum, there is a hollowing out of the internal region to form the bone marrow cavity. This destruction of bone tissue is due to **osteoclasts**, multinucleated cells that enter the bone through the blood vessels (Kahn and Simmons, 1975). Osteoclasts are probably derived from the same precursors as blood cells, and they dissolve both the inorganic and the protein portions of the bone matrix (Ash et al., 1980; Blair et al., 1986). The osteoclast extends numerous cellular processes into the matrix and pumps out hydrogen ions from the osteoclast onto the surrounding material, thereby acidifying and solubilizing it (Figure 9.19; Baron et al., 1985, 1986). The blood vessels also import the blood-forming cells, which will reside in the marrow for the duration of the organism's life.

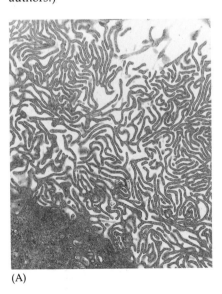

(A)

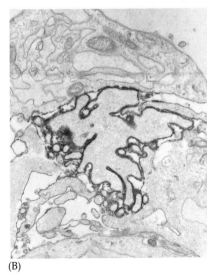

(B)

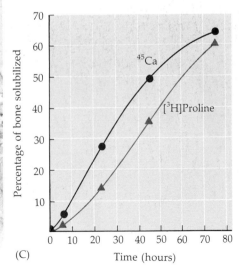

(C)

Lateral plate mesoderm

Not all of the mesodermal mantle is organized into somites. Adjacent to somitic mesoderm is the **intermediate mesodermal** region. This cord of mesodermal cells develops into the pronephric tubule, which is the precursor of kidney and genital ducts. The development of these organ systems will be discussed in detail in Chapters 18 and 21, respectively. Further laterally on each side we come to the **lateral plate mesoderm**. These plates split horizontally into the dorsal **somatic (or parietal) mesoderm**, which underlies the ectoderm, and a **ventral splanchnic (or visceral) mesoderm**, which overlies the endoderm (see Figure 9.2C). Between these layers is the body cavity—the **coelom**—which stretches from the future neck region to the posterior of the body. During later development, the right- and left-side coeloms fuse, and folds extend from the somatic mesoderm, dividing the coelom into separate cavities. In mammals, the coelom is subdivided into the pleural, pericardial, and peritoneal spaces, enveloping the thorax, heart, and abdomen, respectively. The mechanism for creating mesodermal somites and body linings has changed little throughout vertebrate evolution, and the development of the chick mesoderm can be compared with similar stages of frog embryos (Figure 9.20).

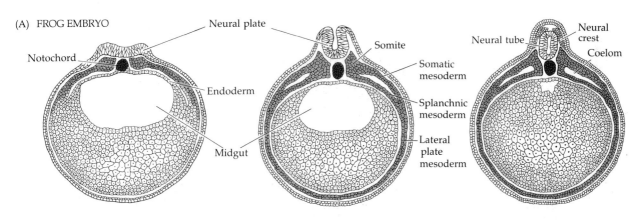

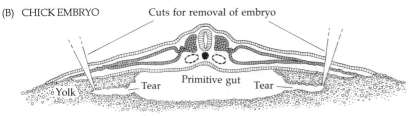

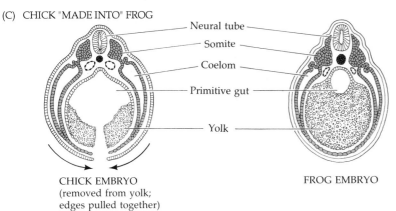

FIGURE 9.20
Comparison of mesodermal development in frog and chick embryos. (A) Neurula-stage frog embryos showing progressive development of the mesoderm and coelom. (B) Transverse section of a chick embryo. (C) When the chick embryo is separated from its enormous yolk mass, it resembles the amphibian neurula at a similar stage. (A after Rugh, 1951; B and C after Patten, 1951.)

Isoforms

Numerous mesodermal cell lineages are characterized by the replacement of one set of structures (cells or molecules) by another during the course of development. The molecules or cells involved in these processes are often related to each other and have been called **isoforms** (Caplan et al., 1983). An example of a cellular isoform set is the chondrocyte and the osteoblast that replaces it. Both types of cells secrete an extensive and calcifiable extracellular matrix. Examples of molecular isoforms include the sequential appearance of embryonic hemoglobin, fetal hemoglobin, and adult hemoglobins.

During mouse development, three isoformic populations of myoblasts are formed. The first myoblasts (myoblast I), formed after the separation of the myotome from the sclerotome, undergo limited fusion to form small myotubes with only 4–6 nuclei per cell. These do not form functional muscles. The myoblast II population soon follows, and these cells fuse into large myotubes (20–100 nuclei) and supplant the myoblast I population (Bonner and Hauschka, 1974). At the same time, the major motor axons begin to enter the limb. These neurons stimulate the production of the third myoblast population (Womble and Bonner, 1980). The myotubes formed by the myoblast III cells form the muscles that characterize the newborn mouse. In the limb, one sees the formation of the characteristic M- and Z-like proteins for the first time in the myoblast III population (Jockusch and Jockusch, 1980).

Within the muscles produced by myoblast III cells, *molecular* isoforms appear and disappear. Consider the heavy (200-kDa) myosin chains. Analyses of these proteins indicate that during fetal development, there is an embryonic myosin heavy chain. This changes at birth to a neonatal myosin heavy chain and later to the adult (fast muscle) myosin heavy chain (Whalen et al., 1979). Thus, there are two types of isoformic replacements—one on the cellular level and one on the molecular level—that are involved in muscle development.

The appearance of these isoforms can be altered by experimentation (as when nerves are prevented from reaching the limb) or by natural mutation. In humans and chicks, there are genetic mutations that cause muscular dystrophy, a disease that eventually leads to the destruction of limb muscle fibers. The muscles of human patients and chicks expressing these mutations have many features characteristic of immature muscles. This suggests a failure of the normal developmental program for muscle development. In both instances the dystrophic muscles continue to make the neonatal form of the myosin heavy chain long after it has disappeared from normal adult muscle (Bandman, 1985; Webster et al., 1988). It should be recognized, then, that the development of a tissue need not be completed as soon as it forms. There is a remodeling that occurs both intercellularly and intracellularly to generate the phenotype of the adult individual.

Formation of extraembryonic membranes

In reptiles, birds, and mammals, embryonic development has taken a new direction. Reptiles evolved a mechanism for laying eggs on dry land, thus freeing them to explore niches that were not close to ponds. To accomplish this, the embryo produced four sets of **extraembryonic membranes** to mediate between it and the environment, and even though most mammals have evolved placentas instead of shells, the basic pattern of extraembryonic membranes remains the same. In developing reptiles, birds, and mammals, there initially is no distinction between embryonic and extraembryonic domains. However, as the body of the embryo takes shape, the border epithelia create a series of body folds, which surround the developing embryo, thereby isolating it from the yolk and delineating which areas are to be embryonic.

The membranous folds are formed by the extension of ectodermal and endodermal epithelium underlaid with mesoderm. The combination of ectoderm and mesoderm is often referred to as the **somatopleure** and forms the amnion and chorion membranes; the combination of endoderm and mesoderm—the **splanchnopleure**—forms the yolk sac and allantois. The endodermal or ectodermal tissue acts as the functioning epithelial cells; and the mesoderm generates the essential blood supply to and from this epithelium. The formation of these folds can be followed in Figure 9.21.

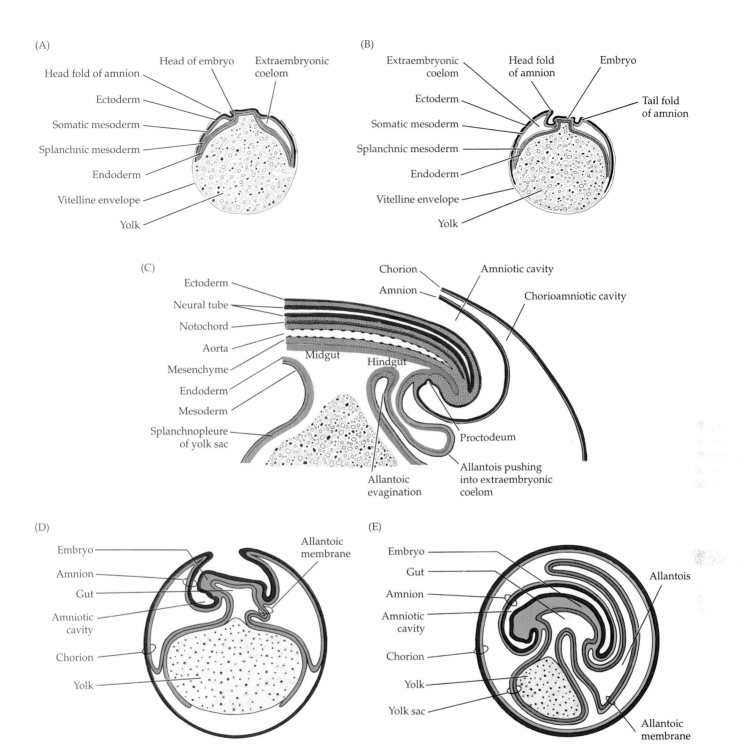

(A)

Head fold of amnion
Head of embryo
Extraembryonic coelom
Ectoderm
Somatic mesoderm
Splanchnic mesoderm
Endoderm
Vitelline envelope
Yolk

(B)

Extraembryonic coelom
Head fold of amnion
Embryo
Ectoderm
Somatic mesoderm
Tail fold of amnion
Splanchnic mesoderm
Endoderm
Vitelline envelope
Yolk

(C)

Ectoderm
Neural tube
Notochord
Aorta
Mesenchyme
Endoderm
Mesoderm
Splanchnopleure of yolk sac
Midgut
Hindgut
Chorion
Amnion
Amniotic cavity
Chorioamniotic cavity
Proctodeum
Allantoic evagination
Allantois pushing into extraembryonic coelom

(D)

Embryo
Amnion
Gut
Amniotic cavity
Chorion
Yolk
Allantoic membrane

(E)

Embryo
Gut
Amnion
Amniotic cavity
Chorion
Yolk
Yolk sac
Allantois
Allantoic membrane

The first problem of a land-dwelling egg is desiccation. Embryonic cells would quickly dry out if they were not in an aqueous environment. This environment is supplied by the **amnion**. The cells of this membrane secrete amniotic fluid; thus embryogenesis still occurs in water. This evolutionary advance is so significant and characteristic that reptiles, birds, and mammals are grouped together as the **amniote vertebrates**.

The second problem of a land-dwelling egg is gas exchange. This exchange is provided for by the **chorion**, the outermost extraembryonic membrane. In birds and reptiles, this membrane adheres to the shell, allowing the exchange of gases between the egg and the environment. In mammals, as we have seen, the chorion has evolved into the placenta, which has many other functions besides respiration.

FIGURE 9.21
Schematic drawings of the extraembryonic membranes of the chick. The embryo is cut longitudinally and the albumen and shell coatings are not shown. (A) 2-day embryo. (B) 3-day embryo. (C) Detailed schematic diagram of the caudal (hind) region of the chick embryo showing the formation of the allantois. (D) 5-day embryo. (E) 9-day embryo. (After Carlson, 1981.)

The **allantois** stores urinary wastes and mediates gas exchange. In reptiles and birds, the allantois becomes a large sac, as there is no other way to keep toxic by-products of metabolism from the developing embryo. The mesodermal layer of allantoic membrane often reaches and fuses with the mesodermal layer of the chorion to create the **chorioallantoic membrane**. This extremely vascular envelope is crucial for chick development and is responsible for transporting calcium from the eggshell into the embryo for bone production (Tuan, 1987). In mammals, the size of the allantois depends upon how well nitrogenous wastes can be removed by the chorionic placenta. In humans the allantois is a vestigial sac, whereas in pigs it is a large and important organ.

The **yolk sac** is the first extraembryonic membrane to be formed, as it mediates nutrition in developing birds and reptiles. It is derived from endodermal cells that grow over the yolk to enclose it. The yolk sac is connected to the midgut by an open tube, the yolk duct, so that the walls of the yolk sac and the walls of the gut are continuous. The blood vessels within the mesoderm of the splanchnopleure transport nutrients from the yolk into the body, for yolk is not taken directly into the body through the yolk duct. Rather, endodermal cells digest the protein into soluble amino acids that can then be passed on to the blood vessels surrounding the yolk sac. Other nutrients, including vitamins, ions, and fatty acids, are stored in the yolk sac and are transported by the yolk sac into the embryonic circulation. In these ways, the four extraembryonic membranes enable the embryo to develop on land.

Heart and circulatory system

The circulatory system is one of the great achievements of the lateral plate mesoderm. Consisting of a heart, blood cells, and an intricate system of blood vessels, the circulatory system provides nourishment to the developing vertebrate embryo. The circulatory system is the first functional unit in the developing embryo, and the heart is the first functional organ.

Fusion of the heart rudiments. In vertebrate embryos, the heart develops in an anterior position, within the splanchnic mesoderm of the neck; only later does it move to the chest. In amphibians, the two presumptive heart-forming regions are initially found at the most anterior position of the mesodermal mantle. While the embryo is undergoing neurulation, these two regions come together in the ventral region of the embryo to form a common **pericardial cavity**. In birds and mammals, the heart also develops by fusion of paired primordia, but the fusion of these two rudiments does not occur until much later in development. In amniote vertebrates, the embryo is a flattened disc, and the lateral plate mesoderm does not completely encircle the yolk sac. The presumptive heart cells originate in the early primitive streak, just posterior to Hensen's node and extending about halfway down its length (Figure 9.22A). These cells migrate through the streak and form two groups of mesodermal cells lateral to (and at the same level as) Hensen's node (Garcia-Martinez and Schoenwolf, 1993). When the embryo is only 18 to 20 hours old, these presumptive heart cells move anteriorly between the ectoderm and endoderm toward the middle of the embryo, remaining in close contact with the endodermal surface (Figure 9.22B; Linask and Lash, 1986). When the cells reach the area where the gut has extended into the anterior region of the embryo, migration ceases. The directionality for this migration appears to be provided by the endoderm. If the cardiac region endoderm is rotated with respect to the rest of the embryo, migration of the precardiac mesoderm cells is reversed. It is

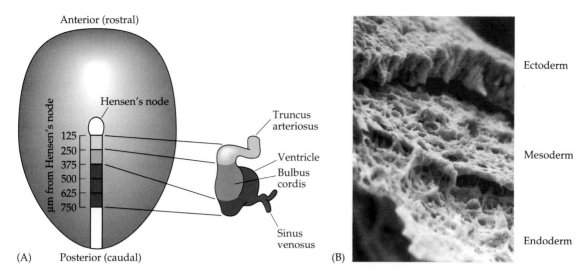

FIGURE 9.22
Heart-forming cells of the chick embryo. (A) Origin of heart cells in the early
(±13 hours) chick embryo. The general anterior–posterior pattern in the primitive streak is seen in the endocardium and epimyocardium of the heart. (B)
Scanning electron micrograph of the heart-forming mesoderm in the 24-hour
chick embryo. The mesoderm is readily separated from the ectoderm but remains in close association with the endoderm. (A after Garcia-Martinez and
Schoenwolf, 1993; B from Linask and Lash, 1986, courtesy of K. Linask.)

thought that the endodermal component responsible for this movement
is an anterior-to-posterior concentration gradient of fibronectin. Antibodies
against fibronectin stop the migration while antibodies against other extracellular matrix components do not (Linask and Lash, 1988a,b). During
this migration, the two heart-forming primordia begin to undergo significant differentiation independently of each other. The presumptive heart
cells of birds and mammals form a double-walled tube consisting of an
inner **endocardium** and an outer **epimyocardium**. The endocardium will
form the inner lining of the heart and the outer layer will form the layer
of heart muscles, which will pump for the lifetime of the organism.

As neurulation proceeds, the foregut is closed by the inward folding
of the splanchnic mesoderm (Figure 9.23). This movement brings the two
tubes together, eventually uniting the epimyocardium into a single tube.
The two endocardia lie within the common chamber for a short while, but
these will also fuse. At this time, the originally paired coelomic chambers
unite to form the body cavity in which the heart resides. The bilateral
origin of the heart can be demonstrated by surgically preventing the
merger of the lateral plate mesoderm (Gräper, 1907; DeHaan, 1959). This
results in a condition called **cardia bifida**, wherein separate hearts form
on each side of the body (Figure 9.24). The next step in heart formation
is the fusion of the endocardial tubes to form a single pumping chamber
(Figure 9.23C,D). This fusion occurs at about 29 hours in chick development or at 3 weeks of human gestation. The unfused posterior portions
of the endocardium become the openings of the **vitelline veins** into the
heart (Figure 9.25). These veins will carry nutrients from the yolk sac into
the **sinus venosus**. The blood then passes through a valvelike flap into
the atrial region of the heart. Contractions of the **truncus arteriosus** speed
the blood into the **aorta**.

Pulsations of the heart begin while the paired primordia are still fusing.
The pacemaker of this contraction is the sinus venosus. The contractions

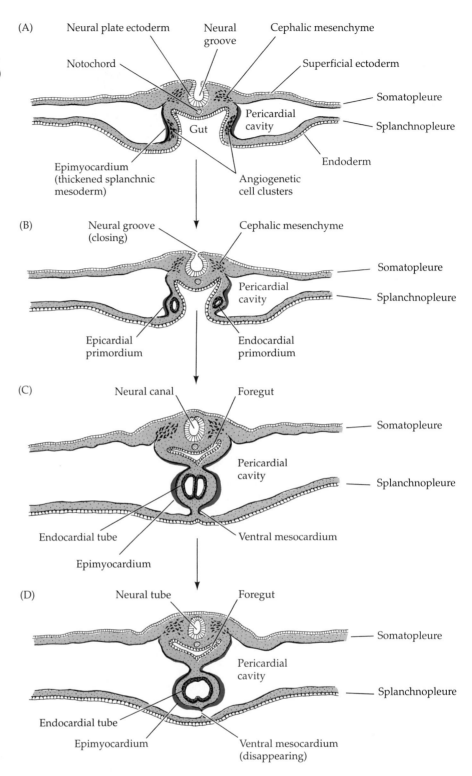

FIGURE 9.23
Formation of the heart. Transverse sections through the heart-forming region of the chick embryo at (A) 25, (B) 26, (C) 28, and (D) 29 hours. (After Carlson, 1981.)

(A)
Neural plate ectoderm
Neural groove
Cephalic mesenchyme
Notochord
Superficial ectoderm
Somatopleure
Pericardial cavity
Splanchnopleure
Gut
Endoderm
Epimyocardium (thickened splanchnic mesoderm)
Angiogenetic cell clusters

(B)
Neural groove (closing)
Cephalic mesenchyme
Somatopleure
Pericardial cavity
Splanchnopleure
Epicardial primordium
Endocardial primordium

(C)
Neural canal
Foregut
Somatopleure
Pericardial cavity
Splanchnopleure
Endocardial tube
Ventral mesocardium
Epimyocardium

(D)
Neural tube
Foregut
Somatopleure
Pericardial cavity
Splanchnopleure
Endocardial tube
Epimyocardium
Ventral mesocardium (disappearing)

begin here, and a wave of muscle contraction is then propagated up the tubular heart. In this way, the heart can pump blood even before its intricate system of valves has been completed. Heart muscle cells have their own inherent ability to contract—and isolated heart cells from 7-day rat or chick embryos will continue to beat in petri dishes (Harary and Farley, 1963; DeHaan, 1967). In the embryo, these contractions become regulated by electrical stimuli from the medulla oblongata via the vagus nerve, and by 4 days, the electrocardiogram of a chick embryo approximates that of an adult.

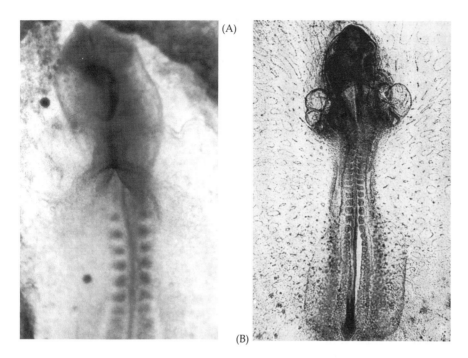

(A)

(B)

FIGURE 9.24
Fusion of the right and left heart rudiments to form a single heart tube. (A) Chick embryo (±30 hours) showing the paired heart primordia meeting at the ventral midlines. (B) Cardia bifida in chick embryo caused by preventing the two heart primordia from fusing. (A courtesy of K. Linask; B courtesy of R. L. DeHaan.)

Formation of the heart chambers. In 3-day chick embryos and 5-week human embryos, the heart is a two-chambered tube, with one atrium and one ventricle. In the chick embryo, the unaided eye can see the remarkable cycle of blood entering the lower chamber and being pumped out through the aorta. The partitioning of this tube into a distinctive atrium and ventricle is accomplished when cells from the myocardium produce a factor (probably transforming growth factor β3) that causes cells from the adjacent endocardium to detach and enter the hyaluronate-rich "cardiac jelly" between the two layers (Markwald et al., 1977; Potts et al., 1991). In

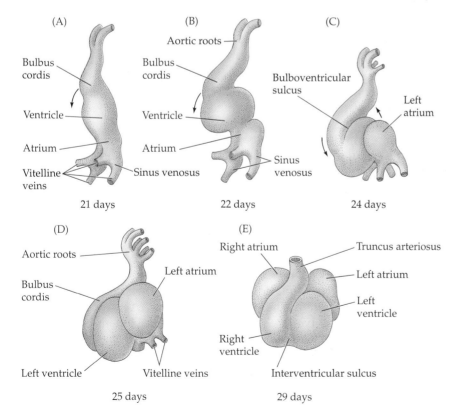

FIGURE 9.25
(A–E) Heart chamber formation during the third week of human development, showing formation of the heart chambers from a simple tube. Views A–D show the developing heart from the left side; E is a frontal view. Although the atria are distinguished externally, they are not separated inside the heart. Note that there are two aortic roots and that these branch out to form the aortic arches (Figure 9.27). (After Langman, 1981, and Larsen, 1993.)

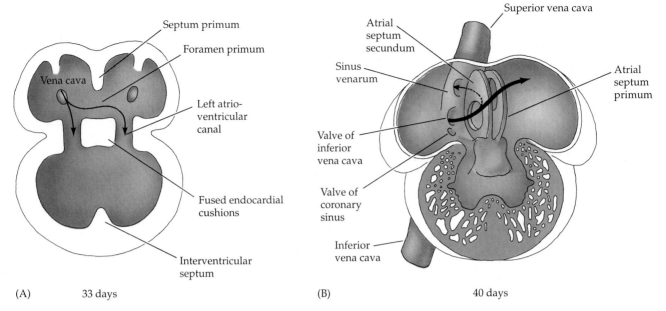

Septum primum

Foramen primum

Vena cava

Left atrio-
ventricular
canal

Fused endocardial
cushions

Interventricular
septum

(A)   33 days

Superior vena cava

Atrial
septum
secundum

Sinus
venarum

Atrial
septum
primum

Valve of
inferior
vena cava

Valve of
coronary
sinus

Inferior
vena cava

(B)   40 days

FIGURE 9.26
Formation of the chambers of the heart. (A) Diagrammatic cross section of the human heart at 4.5 weeks. The atrial and vetricular septa are growing toward the endocardial cushion. (B) Cross section of the human heart just prior to birth. Blood can cross from the right side of the heart to the left side through the openings in the primary and secondary atrial septa. (After Larsen, 1993.)

humans, these cells cause the formation of an **endocardial cushion** that divides the tube into right and left atrioventricular channels (Figure 9.26). Meanwhile, the primitive atrium is partitioned by the growth of two septa that grow ventrally toward the endocardial cushions. The septa, however, have holes in them so blood can still cross from one side into the other. This crossing of blood is needed for survival of the fetus before the circulation to functional lungs has been established. Upon the first breath, however, these holes close and the left and right circulatory loops become established (see Sidelights & Speculations, page 352). The partitioning of the ventricles is accomplished by the growth of the ventricular septum toward the endocardial cushion. With this separation (which usually occurs in the seventh week of human development), the heart is a four-chambered structure with the pulmonary trunk connected to the right ventricle and the aorta connected to the left.

One question that arises in these studies is: How does the left–right polarity emerge in the heart when the sides start off equally? Why should the left side of the heart become different from the right side and vice versa? Studies of fetuses with malformed hearts having either two left or two right sides show a correlation between the presence of the spleen and the left side of the heart. Polysplenia (a spleen in both the left and right sides of the body) is associated with hearts that have two left sides, while asplenia (absence of the spleen) is associated with hearts having two right sides (Anderson et al., 1990; Ho et al., 1991).

Formation of blood vessels

Constraints on how blood vessels may be constructed. There are three major constraints on the construction of blood vessels. The first constraint is *physiological*. Unlike new machines, which do not need to function until they have left the assembly line, new organisms have to function even as they develop. The embryonic cells must obtain nourishment before there is an intestine, use oxygen before they have lungs, and excrete wastes before there are kidneys. Therefore, the developing embryo has needs that are different from those of the adult organism, and its circulatory system reflects those differences. Food is absorbed not through the intes-

346   CHAPTER NINE

FIGURE 9.27
The aortic arches of the human embryo. (A) Originally, the truncus arteriosus pumps blood into the aorta, which branches on either side of the foregut. The six aortic arches take blood from the ventral aorta and allow it to flow into the dorsal aorta. (B) The arches begin to disintegrate or become modified; the dotted lines indicate degenerating structures. (C) Eventually, the remaining arches are modified and the adult arterial system is formed. (After Langman, 1981.)

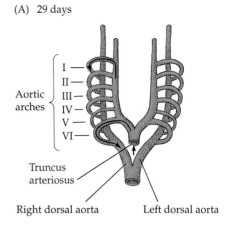

(A) 29 days

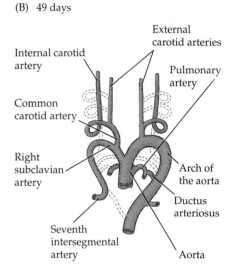

(B) 49 days

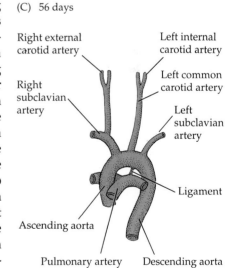

(C) 56 days

tine, but from either the yolk or the placenta, and respiration is not conducted through the gills or lungs but through the chorionic or allantoic membranes. The major embryonic blood vessels must be constructed to serve these extraembryonic structures.

The second constraint is *evolutionary*. The mammalian embryo will extend blood vessels to the yolk sac even though there is no yolk therein. Moreover, the blood leaving the heart loops over the foregut to form the dorsally located aorta. These six pairs of aortic arches loop over the pharynx (Figure 9.27). In primitive fishes, these arches persist and enable the gills to oxygenate the blood. In the adult bird or mammal, where lungs oxygenate the blood, such a system makes little sense, but all six pairs of aortic arches are formed in mammalian and avian embryos before the system eventually becomes simplified into a single aortic arch. Thus, even though our physiology does not require such a structure, our embryonic condition reflects our evolutionary history.

The third set of constraints is *physical*. According to the laws of fluid movement, the most effective transport of fluids is performed by large tubes. As the radius of the blood vessel gets smaller, resistance to flow increases as r^{-4} (Poiseuille's law). A blood vessel that is half as wide as another has a resistance to flow 16 times greater. However, diffusion of nutrients can take place only when blood flows slowly and has access to the membrane. So here is a paradox: The constraints of diffusion mandate that the vessels be small, while the laws of hydraulics mandate that the vessels be large. Living organisms have solved this paradox by evolving circulatory systems with a hierarchy of vessel sizes (LaBarbera, 1990). This hierarchy is formed very early in development, as can be seen in the 3-day chick embryo. In dogs, blood in the large vessels (aorta and vena cava) flows over 100 times faster than it does in the capillaries. By having large vessels specialized for transport and small vessels specialized for diffusion (where the blood spends most of its time), nutrients and oxygen can reach the individual cells of the growing organism. But this is not the entire story, either. If fluid under constant pressure moves directly from a large-diameter pipe into a small-diameter pipe (as in a hose nozzle), the fluid velocity increases. The evolutionary solution to this problem was the emergence of many smaller vessels branching out from a larger one, so the collective cross-sectional area of all the smaller vessels is greater than that of the larger vessel. This relationship (known as Murray's law) is that the cube of the radius of the parent vessel approximates the sum of the cubes of the radii of the smaller vessels. The construction of any circulation system must negotiate among these physical, physiological, and evolutionary constraints.

Vasculogenesis: Formation of blood vessels from blood islands. The major embryonic blood vessels are those that obtain nutrients and bring them to the body and that transport gases to and from the sites of respiratory exchange. In those vertebrates with yolk, the **vitelline (omphalomesenteric)** veins are formed by the aggregation of splanchnic mesodermal cells into **angiogenic clusters**, or **blood islands**, that line the yolk sac (Figure

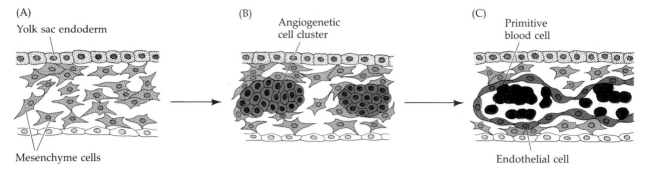

(A) Yolk sac endoderm

Mesenchyme cells

(B) Angiogenetic cell cluster

(C) Primitive blood cell

Endothelial cell

FIGURE 9.28
Vasculogenesis. Blood vessel formation is first seen in the wall of the yolk sac where (A) undifferentiated mesenchyme condenses to form (B) angiogenetic cell clusters. (C) The centers of these clusters form the blood cells, and the outside of the clusters develop into blood vessel endothelial cells. (After Langman, 1981.)

9.28). These cells are first seen in the area opaca at the headfold stage of chick embryogenesis, when the primitive streak is at its fullest extent (Pardanaud et al., 1987). These cords of cells soon hollow out into double-walled tubes analogous to the double tube of the heart. The inner wall becomes the flat **endothelial cell** lining of the vessel, and the outer cells become the smooth muscle. Between these layers is a basal lamina containing a type of collagen specific for blood vessels. It is thought (Murphy and Carlson, 1978; Kubota et al., 1988) that this basal lamina initiates the differentiation of the cell types in the vessel. The central cells of the blood islands differentiate into the embryonic blood cells. As the blood islands grow, they eventually merge to form the capillary network draining into the two vitelline veins, which bring the food and blood cells to the newly formed heart. The creation of blood vessels de novo from the mesoderm is called **vasculogenesis** (Pardanaud et al., 1989).

Two growth factors may be responsible for initiating vasculogenesis. One of these is **basic fibroblast growth factor** (FGF-2). When cells from quail blastodiscs are dissociated in culture, they will not form blood islands or endothelial cells. However, when these cells are cultured in FGF-2, blood islands emerge in culture, and these form endothelial cells (Flamme and Risau, 1992). The second protein is **vascular endothelial growth factor** (VEGF). The latter protein appears to be specific for endothelial cell proliferation and is found in the early mouse embryo. Moreover, the *receptor* for VEGF is found in the blood islands and in other places where VEGF is thought to be active (Millauer et al., 1993).

Vasculogenesis is also seen inside the embryo—notably in the formation of the aortic arches, dorsal aorta, and the blood vessels in the head, lung, and gut. Capillary networks arise independently within the tissues themselves (Auerbach et al., 1985; Pardanaud et al., 1989). In such cases, the capillaries do not arise as smaller and smaller extensions of the major blood vessels growing from the heart. Rather, the mesoderm of each of these organs contains cells called **angioblasts** that arrange themselves into capillary vessels. These organ-specific capillary networks eventually become linked to the extensions of the major blood vessels.

Angiogenesis: Sprouting of blood vessels. Vasculogenesis is not the only way to make blood vessels. In other organs (notably the limb buds, kidney, and brain), existing blood vessels sprout and send endothelial cells into the developing organ (Wilson, 1983; Sariola, 1985). This type of blood vessel formation, in which new vessels emerge from the proliferation of pre-existing blood vessels, is called **angiogenesis**. The organ-forming regions are thought to secrete **angiogenesis factors** that promote the mitosis and migration of endothelial cells into that area. VEGF (mentioned above as a vasculogenesis factor) also promotes the migration of endothelial cells

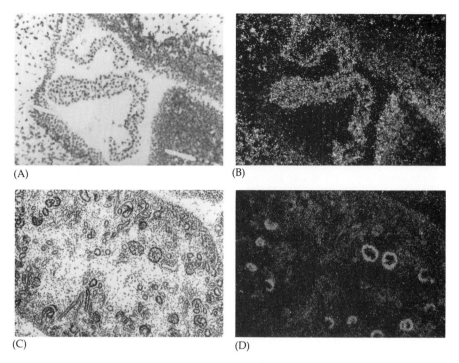

(A)　　　　　　　　　　(B)

(C)　　　　　　　　　　(D)

FIGURE 9.29
Angiogenesis factor production by fetal mouse tissue. In situ hybridization shows that mRNA for secreted VEGF is synthesized by (A,B) the cells of hindbrain in a 6-day embryonic mouse, and by (C,D) the glomeruli of the 15-day fetal mouse kidney. Bright-field micrographs on the left correspond to the dark-field autoradiographs on the right. (From Breier et al, 1992, courtesy of W. Risau.)

into these organs from pre-existing blood vessels on the organs' surface. The temporal and spatial pattern of the expression of this factor correlates well with the times and places where blood vessels enter into the kidney and brain (Figure 9.29; Breier et al., 1992; Millauer et al., 1993).

SIDELIGHTS & SPECULATIONS

Tumor-induced angiogenesis

Judah Folkman (1974) has estimated that as many as 350 billion mitoses occur in each human every day. With each cell division comes the potential that the resulting cells will be malignant. Yet very few tumors actually do develop in any individual. Folkman has suggested that cells capable of forming tumors develop at a certain frequency but that a large majority are never able to form observable tumors. The reason is that solid tumor, like any other rapidly dividing tissue, needs oxygen and nutrients to survive. Without a blood supply, potential tumors either die or remain dormant. Such "microtumors" remain as a stable cell population wherein dying cells are replaced by new cells.

The critical point at which this node of tumorous cells becomes a rapidly growing tumor occurs when the pocket of cells becomes vascularized. The microtumor can expand to 16,000 times its original volume in 2 weeks after vascularization. Without the blood supply, no growth is seen (Folkman, 1974; Ausprunk and Folkman, 1977). To accomplish this vascularization, the original microtumor elaborates substances called tumor angiogenesis factors (TAF). One such TAF, called **angiogenin**, has been isolated from

a human tumor cell population and has been shown to be a single-chain protein of around 14.4 kDa (Fett et al., 1985). Some tumors, such as many gliomas, secrete the VEGF that normally is produced by kidneys and brain cells (Plate et al., 1992; Shweiki et al., 1992), and the inhibition of VEGF-induced angiogenesis suppresses tumor growth in mice (Kim et al., 1993).

TAFs stimulate mitosis in endothelial cells and direct the cells' formation into blood vessels in the direction of the tumor. This phenomenon can be demonstrated by implanting a piece of tumor tissue within the layers of a rabbit or mouse cornea. The cornea itself is not vascularized, but it is surrounded by a vascular border, or *limbus*. Tumor tissue induces blood vessels to form and come toward the tumor (Figure 9.30). Most other adult tissues will not induce these vessels to form. (The exceptions are antigen-stimulated lymphocytes and macrophages, which secrete angiogenesis factors that are probably involved in wound repair.) As little as 3.5 pmol of purified angiogenin can induce extensive capillary formation in the rabbit cornea. Once the blood vessels enter the tumor, the tumor cells undergo explosive growth, eventually bursting the eye.

Folkman and co-workers (1983) have shown that cer-

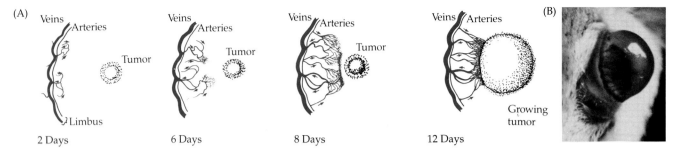

FIGURE 9.30

New blood vessel growth to the site of a tumor in the cornea of an albino mouse. (A) Sequence of events leading to the vascularization of a mammary adenocarcinoma tumor on days 2, 6, *8, and 12. The veins and arteries of the limbus surrounding the cornea both provide vessels. (B) Photograph of living cornea of an albino mouse, with new vessels being constructed to enter the tumor graft. (A from Muthukkaruppan and Auerbach, 1979; B courtesy of R. Auerbach.)*

tain natural substances inhibit tumor-induced angiogenesis. Heparin, a complex polysaccharide found in the matrices of numerous connective tissues, is a potent inhibitor of angiogenesis, especially if administered in the presence of the steroid hormone cortisone. When mice with established tumors received injections of heparin and cortisone, their tumors markedly regressed. The same combination of heparin and cortisone also inhibited normal embryonic angiogenesis in chick embryos. A naturally occurring angiogenesis inhibitor is also produced by cartilage. This 27,600-Da protein keeps cartilage avascular (hence its white, rather than pink, appearance) and is able to inhibit angiogenesis in vivo (Moses et al., 1990).

In addition to supplying the tumor with nutrients and oxygen, the new blood vessels also provide a route that enables tumor cells to migrate to new places, thereby forming secondary tumors. This process is called **metastasis**. Indeed, most victims of cancer do not die from the original tumor but from the numerous secondary tumors spread by metastasis. To enter a blood vessel, a tumor cell has to lyse the collagenous matrix surrounding the new capillaries. This is thought to be done by the secretion of a proteolytic enzyme called **plasminogen activator**. This enzyme is important in the implantation of the blastocyst into the uterus (Chapter 5). Tumor cells use this same enzyme to lyse a hole in the extracellular matrix surrounding the blood vessels. The synthesis of plasminogen activator has been correlated with the metastatic ability of tumors, and antibodies to this enzyme have been found to inhibit human tumor metastasis (Ossowski and Reich, 1983). Once on the capillary endothelial cells, the metastatic tumor cell can squeeze between them and enter the bloodstream (Kramer and Nicolson, 1979). Figure 9.31 shows a metastatic melanoma cell squeezing its way between capillary endothelial cells.

Metastasis requires the ability to travel from one site to another. Like embryonic cells, many tumor cells need a surface upon which to migrate, and some of these cells travel along the extracellular matrix of the blood vessel endothelial cells (Nicolson et al., 1981). This migration, like that of many embryonic cells, appears to be mediated by components of the extracellular basal lamina. When mice are injected with a certain strain of melanoma tumor cells, the cells migrate specifically to the lung, where they form

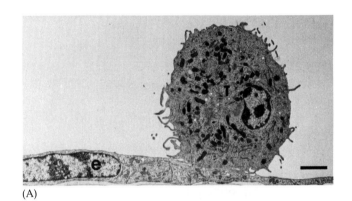

(A)

(B)

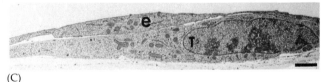

(C)

FIGURE 9.31

(A) Attachment, (B) invasion, and (C) migration of a melanoma tumor cell (T) under a vascular endothelial cell (e). The melanoma cells were placed on the endothelial cells in culture, and photographs were taken at intervals of 0.5, 1, and 3 hours. (From Kramer and Nicolson, 1979, courtesy of G. Nicolson.)

secondary tumors (Figure 9.32). When the cells are inhibited from binding fibronectin or laminin, over 90 percent of the cells fail to reach the lungs (Humphries et al., 1986).

In their rapid division and their secretion of plasminogen activator and angiogenesis factor and in their migration patterns, tumor cells act like normal embryonic cells. There are numerous cases in which tumor cells represent adult cells that have reverted to an embryonic stage of their existence. This finding suggests new approaches to cancer therapy, including the administration of agents that prevent angiogenesis or that actually promote the "differentiation" of the "embryonic" tumor cell back into a normal, "adult" cell (Sachs, 1978; Jimenez and Yunis, 1987).

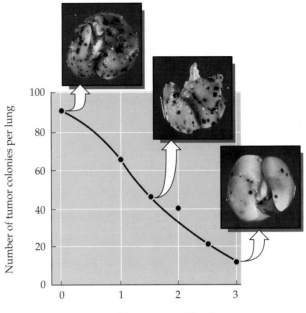

FIGURE 9.32

Metastasis of melanoma cells. The ability of particular melanoma cells to metastasize to the lung is inhibited by the cell-binding regions of fibronectin and laminin. Melanoma cells were injected into the tail vein of mice. Fourteen days later, the animals were killed with ether and the lung metastases were counted. Each point is the average of eight mice. (From Humphries et al., 1986, courtesy of K. M. Yamada.)

Embryonic circulation. The embryonic circulatory system to and from the chick embryo and yolk sac is shown in Figure 9.33. Blood pumped through the dorsal aorta passes over the aortic arches and down into the embryo. Some of this blood leaves the embryo through the vitelline arteries and enters the yolk sac. Nutrients and oxygen are absorbed, and the blood returns through the vitelline veins back into the heart through the sinus

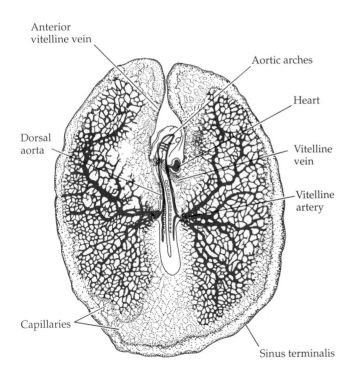

FIGURE 9.33

Circulatory system of a 44-hour chick embryo. This view shows arteries in color; the veins are stippled. The sinus terminalis is the outer limit of the circulatory system and is the site of blood cell generation.

FIGURE 9.34
Circulatory system of a 4-week human embryo. Although at this stage all the major blood vessels are paired left and right, only the right vessels are shown. Arteries are shown in color. (From Carlson, 1981.)

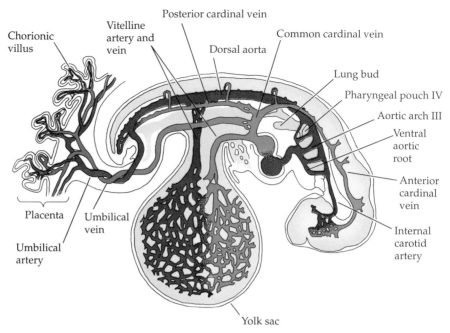

venosus. In mammalian embryos, food and oxygen are obtained from the placenta. Thus, although the mammalian embryo has vessels analogous to the vitelline veins, the main supply of food and oxygen comes from the **umbilical vein**, which unites the embryo with the placenta (Figure 9.34). This vein, which takes the oxygenated and food-laden blood back into the embryo, is derived from what would be the right vitelline vein in birds. The **umbilical artery**, carrying wastes to the placenta, is derived from what would have become the allantoic artery of the chick. It extends from the caudal portion of the aorta and proceeds along the allantois and then out to the placenta.

After entering the embryonic mammalian heart, the blood is pumped into a series of aortic arches, which encircle the pharynx to bring the blood dorsally. In mammals, the left member of the fourth pair of aortic arches is the only one surviving to reach the aorta. The right member of this pair has become the root of the subclavian artery. The third aortic arches have been modified to form the common carotid arteries, which supply blood to the brain and head. The sixth arch is modified to form the pulmonary artery; and the first, second, and fifth arches degenerate. The aorta and pulmonary artery, therefore, have a common opening to the heart for much of their development. Eventually, partitions form within the truncus arteriosus to create two different vessels. Only when the first breath of the newborn animal indicates that the lungs are ready to handle the oxygenation of the blood does the heart become modified to pump blood separately to the pulmonary artery.

SIDELIGHTS & SPECULATIONS

Redirecting blood flow in the newborn mammal

Although the developing fetus shares with the adult the need to get oxygen and nutrients to its tissues, the physi-

ology of the mammalian fetus differs drastically from that of the adult. Chief among these differences is the lack of functional lungs and intestines. All oxygen and nutrients must come from the placenta. This raises two questions. First, how does the fetus obtain oxygen from maternal

FIGURE 9.35

Transfer of oxygen from the mother to the fetus in human embryos. Adult and fetal hemoglobin molecules differ in their protein subunits. The fetal γ chain binds diphosphoglycerol less avidly than does the adult β chain. Consequently, fetal hemoglobin can bind oxygen more efficiently than can adult hemoglobin. In the placenta, there is a net flow (arrow) of oxygen from the mother's blood (which gives oxygen up to the tissue at the lower oxygen pressure) to the fetal blood, which is still picking it up.

FIGURE 9.36

Redirection of blood flow at birth. The expansion of air into the lungs causes pressure changes that redirect the flow of blood in the newborn infant. The ductus arteriosus squeezes shut, breaking off the connection between the aorta and the pulmonary artery, and the foramen ovale, a passageway between the left and right atria, also closes. In this way, pulmonary circulation is separated from systemic circulation.

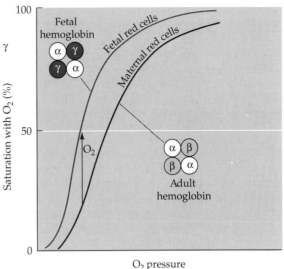

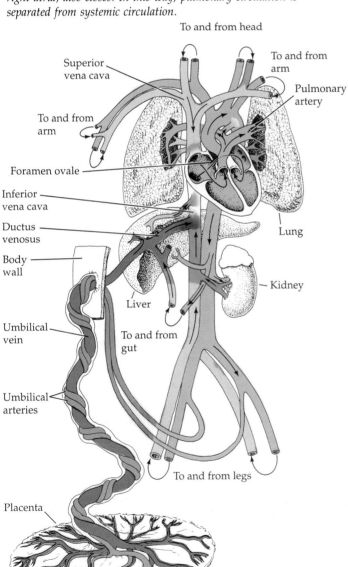

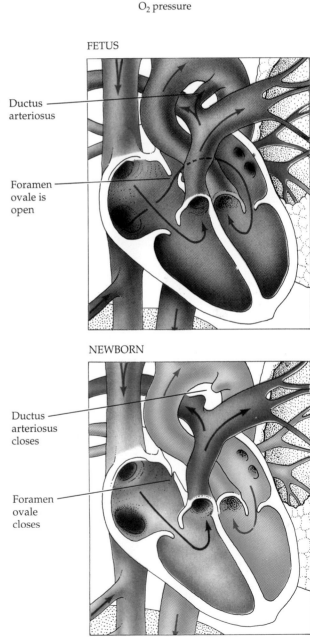

blood? And second, how is blood circulation redirected to the lungs once the umbilical cord is cut and breathing is made necessary?

The solution to the fetus's problems in getting oxygen from its mother's blood involves the development of a fetal hemoglobin. The hemoglobin in fetal red blood cells differs slightly from that in adult corpuscles. Two of the four peptides of the fetal and adult hemoglobin chains are identical—the alpha (α) chains—but adult hemoglobin has two beta (β) chains where the fetus has two gamma (γ) chains (Figure 9.35A). Normal β chains bind the natural regulator diphosphoglycerate, which assists in the unloading of oxygen. The γ-chain isoforms do not bind diphosphoglycerate as well, and therefore they have a higher affinity for oxygen. In the low-oxygen environment of the placenta, oxygen is released from adult hemoglobin. In this same environment, fetal hemoglobin does not give away oxygen, but binds it. This small difference in oxygen affinity mediates the transfer of oxygen from the mother to the fetus (Figure 9.35B). Within the fetus, the myoglobin of the fetal muscles has an even higher affinity for oxygen, so oxygen molecules pass from fetal hemoglobin for storage and use in the fetal muscles. Fetal hemoglobin is not deleterious to the newborn, and, in humans, the replacement of fetal hemoglobin-containing blood cells by adult hemoglobin-containing blood cells is not complete until about six months after birth.*

But once the fetus is not getting its oxygen from the mother, how does it restructure its circulation to get oxygen from its own lungs? During fetal development, an opening—the **ductus arteriosus**—diverts the passage of blood from the pulmonary artery into the aorta (and thus to the placenta). Because blood does not return from the pulmonary vein in the fetus, the developing mammal has to have some other way of getting blood into the left ventricle to be pumped. This is accomplished by the **foramen ovale**, a hole in the septum separating the right and left atria. Blood can enter the right atrium, pass through the foramen to the left atrium, and then enter the left ventricle (Figure 9.36). When the first breath is drawn, the oxygen in the blood causes the muscles surrounding the ductus arteriosus to close the opening. As the blood pressure in the left side of the heart increases, it causes the septa over the foramen ovale to close, thereby separating the pulmonary and systemic circulation.[†] Thus, when breathing begins, the respiratory circulation is shunted from the placenta to the lungs.

*The molecular basis for this switch in globins will be discussed in Chapter 11.

[†]In some infants, the septa fail to close and the foramen ovale is left open. Usually the opening is so small that such children have no physical symptoms, and the foramen eventually closes. However, if the septum secundum fails to form, the atrial septal opening may cause enlargement of the right side of the heart, which can lead to heart failure during early adulthood.

The development of blood cells

The stem cell concept

While many of the cells we have now are the same cells that we acquired as embryos, there are several populations of cells that are constantly regenerating. We lose and replace about 10^{11} red blood cells and small intestinal cells each day, along with about 1.5 grams of epidermal cells. Where are replacement cells coming from? They come from populations of **stem cells**. A stem cell is capable of extensive proliferation, creating more stem cells (self-renewal) as well as more differentiated cellular progeny. Stem cells are, in effect, an embryonic population of cells, continuously producing cells that can undergo further development within an adult organism. Our blood cells, intestinal crypt cells, epidermis, and (in males) spermatocytes are populations in a steady-state equilibrium in which cell production balances cell loss (Hay, 1966). In most cases, stem cells can regulate the production of either more stem cells or more differentiated cells when the equilibrium is stressed by injury or environment. (This is seen by the production of enormous numbers of red blood cells when the body suffers from anoxia.) Stem cells have been identified in all the tissues mentioned earlier, but they are most readily studied in blood cell development.

Potten and Loefffler (1990) present a view in which some stem cells are noncycling *potential* stem cells locked in G_0, while other stem cells are

FIGURE 9.37
Model of the dynamics of stem cell proliferation and differentiation. Proliferation is represented by the horizontal circles, and differentiation is along the vertical axis progressing downward to more differentiated cell types. The initial stem cells can remain quiescent (in G_0 phase) or enter the cell cycle. Stem cells that produce more stem cells remain at one level, but can divide to produce a transition cell type that "falls" to the next level. At each lower level, the probability of falling still lower at the next division is increased. Eventually a mature differentiated cell is generated. (After Potten and Loeffler, 1990.)

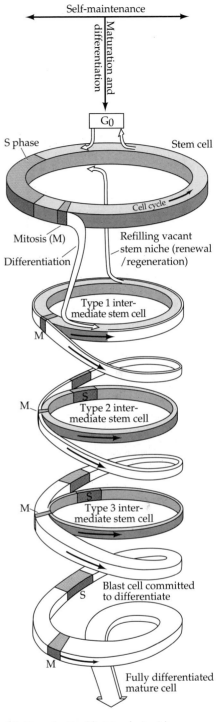

actively in the cell cycle. A cycling stem cell usually divides to create more stem cells, but it can also generate a transitory intermediate stem cell type (T1). A T1 cell can regenerate itself, but usually it proceeds to make a second type of transitory cell, T2. (Under some conditions, a T1 cell can regenerate the original stem cell if the stem cell population is severely depleted.) The T2 cell can also maintain itself, but more usually it divides to create T3 cells. Eventually a transitory cell type is made that *always* matures into a differentiated cell type (Figure 9.37). Thus, the vertebrate body retains populations of stem cells, and these stem cell populations can produce both more stem cells and a population of cells that can undergo further development.

The path of development a stem cell descendant enters depends on the molecular milieu in which it resides. This became apparent when experimental evidence showed that red blood cells (erythrocytes), white blood cells (granulocytes, neutrophils, and platelets), and lymphocytes shared a common precursor—the **pluripotential hematopoietic stem cell**.

Pluripotential stem cells and hematopoietic microenvironments

The CFU-S. The pluripotential hematopoietic stem cell is one of our body's most impressive cells. From it will emerge erythrocytes, neutrophils, basophils, eosinophils, platelets, mast cells, monocytes, tissue macrophages, osteoclasts, and the T and B lymphocytes. The existence of a pluripotential hematopoietic stem cell was shown by Till and McCulloch (1961), who injected bone marrow cells into lethally irradiated mice of the same genetic strain as the marrow donors. (Irradiation kills the hematopoietic cells of the host, enabling one to see the new colonies from the donor mouse.) Some of these donor cells produced discrete nodules on the spleens of the host animals (Figure 9.38). Microscopic studies showed these nodules to be composed of erythrocyte, granulocyte, and platelet precursors. Thus, a single cell from the bone marrow was capable of forming many of the different blood cell types. The cell responsible was called the **CFU-S**, the colony-forming unit of the spleen. Further studies used chromosomal markers to prove that the different types of cells within the colony were formed from the same CFU-S. Here, marrow cells were

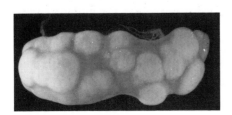

FIGURE 9.38
Isolated blood-forming colonies. When bone marrow containing hematopoietic stem cells is injected into an irradiated mouse, discrete colonies of blood cells are seen on the surface of the spleen of that mouse. (From Till, 1981, courtesy of J. E. Till.)

irradiated so that very few survived. Many of those that did survive had abnormal chromosomes, which could be detected microscopically. When such irradiated CFU-S cells were injected into a mouse whose own blood-forming stem cells had been destroyed, each cell of a spleen colony, be it granulocyte or erythrocyte precursor, had the same chromosomal anomaly (Becker et al., 1963). An important part of the stem cell concept is the requirement that the stem cell be able to form more stem cells in addition to its differentiated cell types. This has indeed been found to be the case. When spleen colonies derived from a single CFU-S are resuspended and injected into other mice, many spleen colonies emerge (Jurśšková and Tkadleček, 1965; Humphries et al., 1979). Thus, we see that a single marrow cell can form numerous different cell types and can also undergo self-renewal; in other words, the CFU-S is a pluripotential hematopoietic stem cell.

The preceding data indicate that although the CFU-S can generate many of the blood cell types, it is not capable of generating lymphocytes. This conclusion is supported by the experiments of Abramson and her colleagues (1977), who have shown that both CFU-S and lymphocytes are derived from yet another stem cell—the pluripotential hematopoietic stem cell. This is sometimes called the **c**olony-forming **u**nit of the **m**yeloid and lymphoid cells, or **CFU-M,L**. When they injected irradiated bone marrow cells into mice having a hereditary deficiency of blood-forming cells, the researchers found the same chromosomal abnormalities in both the spleen colonies and the circulating lymphocytes. This work has been confirmed by studies in which marrow cells are injected with certain viruses that become incorporated into cellular DNA at various random places. The same virally derived genes are seen in the same region of the genome in lymphocytes and in blood cells (Keller et al., 1985; Lemischka et al., 1986). In 1988, Spangrude and colleagues used immunological procedures to isolate a small population of cells (0.05 percent of the total bone marrow cells of mice) that appear to be precursor stem cells of both blood cells and lymphocytes. Any one of these cells is able to form colonies of lymphocytes and blood cells. In fact, the injection of only 30 of these cells into lethally irradiated mice enables half the mice to survive. In these survivors, the millions of lymphocytes and blood cells are all derived from only 30 donor cells.

Blood and lymphocyte lineages. Figure 9.39 summarizes several studies. The first pluripotential hematopoietic stem cell is the CFU-M,L. This cell gives rise to the CFU-S (blood cells) and the CFU-L (lymphocytes). The CFU-S and the CFU-L also are pluripotent stem cells because their progeny can differentiate into numerous cell types. The immediate progeny of the CFU-S, however, are *lineage-restricted* stem cells. Each can produce only one type of cell in addition to renewing itself. The BFU-E (burst-forming unit, erythroid), for instance, is formed from the CFU-S, and it can form only one cell type in addition to itself. This new cell is the CFU-E (colony-forming unit, erythroid), which is capable of responding to the hormone **erythropoietin** to produce the first recognizable differentiated member of the erythrocyte lineage, the **proerythroblast**. Erythropoietin is a glycoprotein that rapidly induces the synthesis of the mRNA for globin (Krantz and Goldwasser, 1965). It is produced predominantly in the kidney, and its synthesis is responsive to environmental conditions. If the level of blood oxygen falls, erythropoietin production is increased, an event leading to the production of more red blood cells. As a red blood cell matures, it becomes an **erythroblast**, synthesizing enormous amounts of hemoglobin. Eventually, the mammalian erythroblast expels its nucleus, becoming

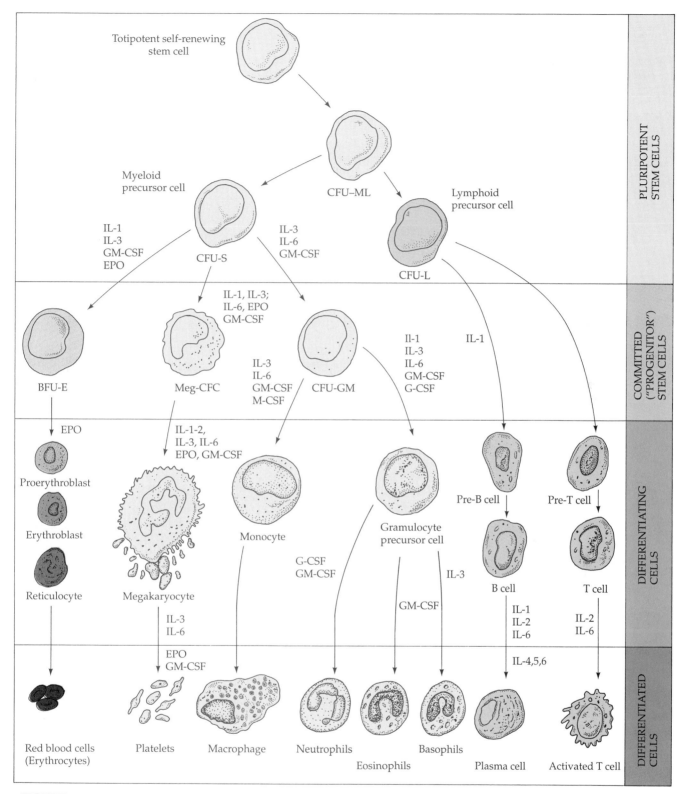

Totipotent self-renewing
stem cell

PLURIPOTENT STEM CELLS

Myeloid
precursor cell

CFU–ML

Lymphoid
precursor cell

IL-1
IL-3
GM-CSF
EPO

CFU-S

IL-3
IL-6
GM-CSF

CFU-L

COMMITTED ("PROGENITOR") STEM CELLS

IL-1, IL-3;
IL-6, EPO
GM-CSF

IL-3
IL-6
GM-CSF
M-CSF

Il-1
IL-3
IL-6
GM-CSF
G-CSF

IL-1

BFU-E

Meg-CFC

CFU-GM

DIFFERENTIATING CELLS

EPO

IL-1-2,
IL-3, IL-6
EPO, GM-CSF

Pre-B cell

Pre-T cell

Proerythroblast

Erythroblast

Monocyte

G-CSF
GM-CSF

IL-3

B cell

T cell

Reticulocyte

Megakaryocyte

Gramulocyte
precursor cell

GM-CSF

IL-1
IL-2
IL-6

IL-2
IL-6

IL-3
IL-6

DIFFERENTIATED CELLS

EPO
GM-CSF

IL-4,5,6

Red blood cells
(Erythrocytes)

Platelets

Macrophage

Neutrophils

Eosinophils

Basophils

Plasma cell

Activated T cell

FIGURE 9.39
A model for the origin of mammalian blood and lymphoid cells. (Other models
are also consistent with the data, and this one summarizes features from several
models.) The growth factors are as listed in Table 9.1. (After Nakauchi and
Gachelin, 1993.)

TABLE 9.1
Some hematopoietic growth factors and their target cells

Regulator	Abbreviation	Responding hematopoietic cells
Erythropoietin	Epo	Erythroid cells, megakaryocytes
Granulocyte colony stimulating factor	G-CSF	Granulocytes, macrophages
Macrophage colony stimulating factor	M-CSF (CSF-1)	Macrophages, granulocytes
Granulocyte-macrophage colony stimulating factor	GM-CSF	Granulocytes, macrophages, eosinophils, megakaryocytes, erythroid cells
Interleukin 1	IL-1	T lymphocytes, stem cells
Interleukin 2	IL-2	T and B lymphocytes
Interleukin 3 (Multipotential colony stimulating factor)	IL-3 (Multi-CSF)	Granulocytes, macrophages, eosinophils, megakaryocytes, stem cells, mast cells
Interleukin 4	IL-4	T and B lymphocytes, granulocytes, macrophages, mast cells
Interleukin 5	IL-5	Eosinophils, B lymphocytes
Interleukin 6	IL-6	B lymphocytes, granulocytes, megakaryocytes, stem cells; osteoclasts
Interleukin 7	IL-7	B and T lymphocytes
Megakaryocyte colony stimulating factor	Meg-CSF	Megakaryocytes
Stem cell factor	SCF	Stem and mast cells, granulocytes, eosinophils, megakaryocytes
Leukemia inhibitory factor	LIF	Megakaryocytes
Macrophage inflammatory protein α	MIP-1α	Stem cells

a **reticulocyte**. Reticulocytes can no longer synthesize globin mRNA but can still translate existing messages into globin. The final stage of differentiation is the **erythrocyte** stage. Here, no division, RNA synthesis, or protein synthesis takes place. The cell leaves the bone marrow to undertake its role of delivering oxygen to the bodily tissues. Similarly, there are lineage-restricted stem cells for platelets and for granulocytes (neutrophils, basophils, and eosinophils) and macrophages.

Some hematopoietic growth factors (such as IL-3—interleukin 3) stimulate the division and maturation of the more primitive stem cells, thus increasing the numbers of all blood cell types. Other factors (such as erythropoietin) are specific for certain cell lineages only. A cell's ability to respond to these factors is dependent upon the presence of receptors for these factors on its surface. The number of these receptors is quite low. There are only about 700 receptors for erythropoietin on a CFU-E, and most other progenitor cells have similar low numbers of growth factor receptors. [The exception is the receptor for macrophage colony-stimulating factor (M-CSF, also known as CSF-1), which can number up to 73,000 per cell on certain progenitor cells.] Table 9.1 shows some of the known hematopoietic growth factors and their cellular targets.

Hematopoietic inductive microenvironments. Some hematopoietic growth factors are made by the stromal cells (fibroblasts and other connective tissue elements) of the bone marrow itself. Other growth factors travel through the blood and are retained by the extracellular matrix of the stromal cells. In the spleen, stem cells are predominantly committed to erythroid development. In the bone marrow, granulocyte development predominates. The developmental path taken by the descendants of a pluripotent stem cell depends on which growth factors it meets, and this is determined by the stromal cells of the bone marrow. Wolf and Trentin (1968) demonstrated that short-range interactions between stromal cells and the stem cells determine the developmental fates of the stem cells' progeny. These investigators placed plugs of bone marrow into the spleen and then injected stem cells. Those colonies in the spleen were predominantly erythroid, whereas those forming in the marrow plugs were predominantly granulocytic. In fact, those colonies that straddled the borders were predominantly erythroid in the spleen and granulocytic in the marrow. The regions of determination are referred to as **hematopoietic inductive microenvironments** (HIM).

The stromal cells of the bone marrow create HIMs by their ability to bind the hematopoietic growth factors (Hunt et al., 1987; Whitlock et al., 1987). GM-CSF and the multilineage growth factor IL-3 both bind to the heparan sulfate glycosaminoglycan of bone marrow stroma (Gordon et al., 1987; Roberts et al., 1988). Moreover, they remain active when bound. In this way, the growth factors may be concentrated and compartmentalized, stimulating stem cells in one area to differentiate into one cell type while allowing the same type of stem cells in another area to differentiate into another cell type. Without these growth factors, the stem cells die.

SIDELIGHTS & SPECULATIONS

Osteoclast development

As we have seen, blood stem cells are influenced by numerous hematopoietic growth factors. Moreover, these factors are themselves influenced by the hormonal milieu of the body. This fact may be extremely important in postmenopausal osteoporosis. Loss of ovarian function in many female mammals causes a loss of bone mass that can often be prevented by giving the individual estrogen, and such bone loss has been associated with the increased production of osteoclasts. It is thought that the osteoclast (the cell responsible for hollowing bones, as described earlier in this chapter) comes from the same stem cell as macrophages and granulocytes, the CFU-GM (Kurihara et al., 1990; Hattersley et al., 1991). The growth factor interleukin-6 stimulates the production of osteoclasts. However, the production of IL-6 is inhibited by estrogen, and when estrogen is added to cultured mouse marrow cells, both IL-6 and os-

teoclast production are inhibited (Girasole et al, 1992a). Jilka and colleagues (1992) showed that the removal of mouse ovaries causes an increase in the number of CFU-GMs, enhanced osteoclast development, and an increase in the number of osteoclasts found in bone. These changes could be prevented by injecting these mice with either estrogen or an antibody to IL-6. This suggests that estrogen usually suppresses IL-6 production and osteoclast formation in female mammals, and that postmenopausal loss of bone mass may be due to the production of new osteoclasts by IL-6.*

*So how come males—who don't have ovaries or much estrogen—do not usually suffer osteoporotic bone loss? It seems that testosterone also suppresses osteoclast development (Girasole et al., 1992b). In human males, testosterone production is usually maintained with age.

Sites of hematopoiesis

In avian and amphibian species, the first blood cells derive from the yolk sac. This cell population, however, is transitory; the hematopoietic stem cells that last the lifetime of the organism are derived from the mesodermal area surrounding the aorta. This was shown in the chick by a series of elegant experiments by Dieterlen-Lièvre, who grafted the blastoderm of chickens onto the yolk of the Japanese quail (Figure 9.40). Chick cells are readily distinguishable from quail cells because the quail cell nucleus stains

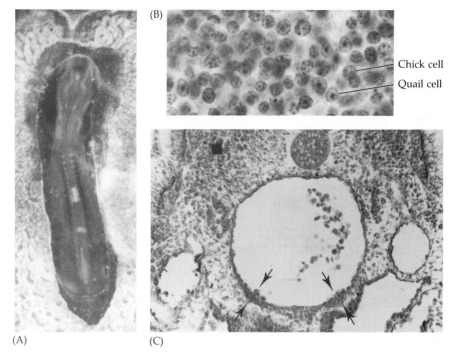

(B)

Chick cell
Quail cell

(A)   (C)

FIGURE 9.40

Blood cell mapping by chick–quail chimeras. (A) Photograph of a "yolk sac chimera," wherein the blastoderm of a quail was transplanted onto the yolk sac of a chick. (B) Photograph of chick and quail cells in the thymus of a chimeric animal showing the difference in the nuclear staining. The lymphoid cells are all chick whereas the structural cells of the thymus are of quail origin. (C) Section through the aorta of a 3-day chick embryo showing the cells (arrows) that give rise to the hematoopoietic stem cells. If cells from this region are taken from quail embryos and placed into chick embryos, the chick embryos have quail blood. (From Martin et al., 1978 and Dieterlen-Lièvre and Martin, 1981, photographs courtesy of F. Dieterlen-Lièvre.)

much more darkly (due to its dense nucleoli), thus providing a permanent marker for distinguishing the two cell types. Using these "yolk sac chimeras," Dieterlen-Lièvre and Martin (1981) showed that the yolk sac stem cells do not contribute cells to the adult animal but that the true stem cells are formed within nodes of mesoderm that line the mesentery and the major blood vessels. In the 4-day chick embryo, the aortic wall appears to be the most important source of new blood cells, and it has been found to contain numerous hematopoietic stem cells (Cormier and Dieterlen-Lièvre, 1988).

In mammals, the situation is more controversial. In the *adult*, the major source of blood cells is the bone marrow (although in some groups, such as mice, the spleen also participates). In the *fetus*, the main hematopoietic organ is the liver, and it is widely thought that stem cells from the fetal liver migrate to the bone marrow. However, two sources of embryonic CFU-S have been found in mice. Moore and Metcalfe (1970) showed that the yolk sac of presomitic mouse embryos forms CFU-S when cultured separately. They speculated that these yolk sac cells were the only hematopoietic lineage in the mouse and that these cells directly seeded the fetal liver at day 11. More recent studies (Kubai and Auerbach, 1983; Godlin et al., 1993; Medvinsky et al., 1993) show that the mesodermal region around the mammalian aorta is also capable of generating CFU-S prior to the presence of CFU-S in the fetal liver. It is not known if one lineage forms prior to the other, or if only one of these two lineages provides the definitive stem cells to the fetal liver. Such questions, though, can now be answered by transplanting cells with lineage-specific markers.

ENDODERM

Pharynx

The function of embryonic endoderm is to construct the linings of two tubes within the body. The first tube, extending throughout the length of the body, is the digestive tube. Buds from this tube form the liver, gallbladder, and pancreas. The second tube, the respiratory tube, forms as an outgrowth of the digestive tube, and it eventually bifurcates into two lungs. The digestive and respiratory tubes share a common chamber in the anterior region of the embryo; this region is called the **pharynx**. Epithelial outpockets of the pharynx give rise to the tonsils, thyroid, thymus, and parathyroid glands.

The respiratory and digestive tubes are both derived from the primitive gut (Figure 9.41). As the endoderm pinches in toward the center of the embryo, the foregut and hindgut regions are formed. At first, the oral end is blocked by a region of ectoderm called the **oral plate**, or **stomodeum**. Eventually (about 22 days in human embryos), the stomodeum breaks, thereby creating the oral opening of the digestive tube. The opening itself is lined by ectodermal cells. This arrangement creates an interesting situation, because the oral plate ectoderm is in contact with the brain ectoderm, which has curved around toward the ventral portion of the embryo. The two ectodermal regions mutually interact with each other. The roof of the oral region forms Rathke's pouch and becomes the glandular part of the pituitary gland. The neural tissue on the floor of the diencephalon gives rise to the infundibular process, which becomes the neural portion of the pituitary. Thus, the pituitary gland has a dual origin; this dual nature of the pituitary is reflected in its adult functions.

The endodermal portion of the digestive and respiratory tubes begins in the pharynx. Here, the mammalian embryo produces four pairs of **pharyngeal pouches** (Figure 9.42). In aquatic vertebrates, these structures produce the gills, but mammalian pharyngeal pouches have been modified for the terrestrial environment. As discussed in Chapter 7, cranial neural crest cells migrate into these pouches to form the cartilagenous or mesenchymal component of these endodermally lined structures. Between these pouches are the **pharyngeal arches**. The first pair of pharyngeal pouches becomes the auditory cavities of the middle ear and the associated eustachian tubes. The second pair of pouches gives rise to the walls of the tonsils. The thymus is derived from the third pair of pharyngeal pouches; it will direct the differentiation of T lymphocytes during later stages of development. One pair of parathyroid glands is also derived from the third pair of pharyngeal pouches, and the other pair is derived from the fourth. In addition to these paired pouches, a small, central diverticulum is formed between the second pharyngeal pouches on the floor of the pharynx. This pocket of endoderm and mesenchyme will bud off from the pharynx and migrate down the neck to become the thyroid gland.

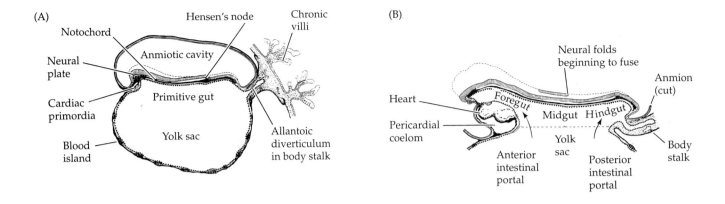

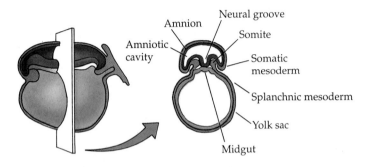

FIGURE 9.41
Formation of the human digestive system, depicted at about (A) 16 days, (B) 18 days, (C) 22 days, and (D) 28 days. (After Crelin, 1961.)

The digestive tube and its derivatives

Posterior to the pharynx, the digestive tube constricts to form the esophagus, which is followed in sequence by the stomach, small intestine, and large intestine. The endodermal cells generate only the lining of the digestive tube and its glands, for mesodermal mesenchyme cells will surround this tube to provide the muscles for peristalsis.

Figure 9.42 shows that the stomach develops as a dilated region close to the pharynx. More caudally, the intestines develop, and the connection between the intestine and yolk sac is eventually severed. At the caudal end of the intestine, a depression forms where the endoderm meets the overlying ectoderm. Here, a thin **cloacal membrane** separates the two tissues. It eventually ruptures, forming the opening that will become the anus. The development of the distinct regions of the digestive tube will be detailed in Chapter 18.

Liver, pancreas, and gallbladder

Endoderm also forms the lining of three accessory organs that develop immediately caudal to the stomach. The **hepatic diverticulum** is the tube of endoderm that extends out from the foregut into the surrounding mesenchyme. The mesenchyme induces the endoderm to proliferate, to branch, and to form the glandular epithelium of the liver. A portion of the hepatic diverticulum (that region closest to the digestive tube) contin-

FIGURE 9.42
Endodermal development of a 6-week human embryo. (A) Sagittal view of 6-week embryo. The stomach region has begun to dilate, and the pancreas is represented by two buds that will eventually fuse. (B–D) Sections through the developing embryo at the plane bisecting (A), showing fates of the pharyngeal clefts and arches. The first cleft forms the external auditory passages, while the second arch expands, eventually covering clefts II, III, and IV. (After Larsen, 1993.)

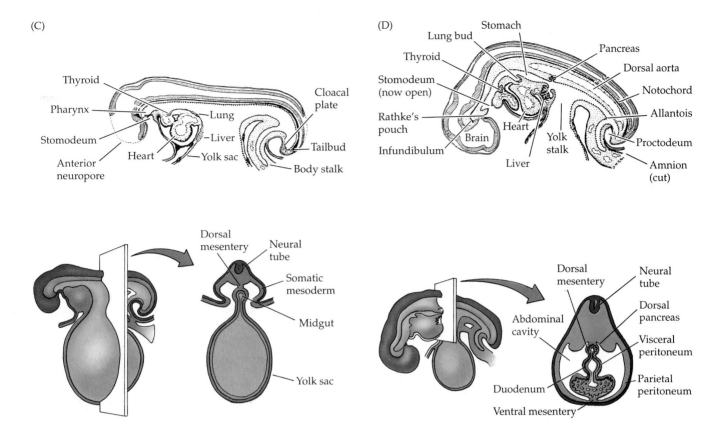

(C)
Thyroid
Pharynx
Stomodeum
Anterior neuropore
Lung
Liver
Heart
Yolk sac
Cloacal plate
Tailbud
Body stalk

(D)
Stomach
Lung bud
Thyroid
Stomodeum (now open)
Rathke's pouch
Infundibulum
Brain
Heart
Liver
Yolk stalk
Pancreas
Dorsal aorta
Notochord
Allantois
Proctodeum
Amnion (cut)

Dorsal mesentery
Neural tube
Somatic mesoderm
Midgut
Yolk sac

Dorsal mesentery
Neural tube
Abdominal cavity
Dorsal pancreas
Visceral peritoneum
Parietal peritoneum
Duodenum
Ventral mesentery

ues to function as the drainage duct of the liver, and a branch from this duct produces the gallbladder (Figure 9.43).

The pancreas develops from the fusion of distinct dorsal and ventral diverticula. Both of these primordia arise from the endoderm immediately caudal to the stomach, and as they grow, they come closer together and eventually fuse. In humans, only the ventral duct survives to carry digestive enzymes into the intestine. In other species (such as the dog), both the dorsal and ventral ducts empty into the intestine.

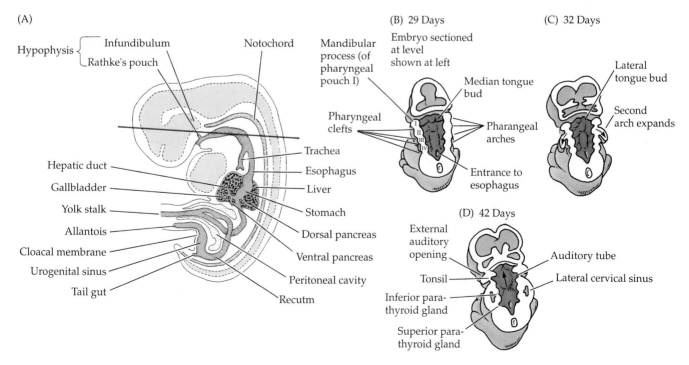

(A)
Hypophysis {
Infundibulum
Rathke's pouch
}
Notochord
Hepatic duct
Gallbladder
Yolk stalk
Allantois
Cloacal membrane
Urogenital sinus
Tail gut
Trachea
Esophagus
Liver
Stomach
Dorsal pancreas
Ventral pancreas
Peritoneal cavity
Recutm

(B) 29 Days
Mandibular process (of pharyngeal pouch I)
Embryo sectioned at level shown at left
Median tongue bud
Pharyngeal clefts
Pharangeal arches
Entrance to esophagus

(C) 32 Days
Lateral tongue bud
Second arch expands

(D) 42 Days
External auditory opening
Tonsil
Inferior parathyroid gland
Superior parathyroid gland
Auditory tube
Lateral cervical sinus

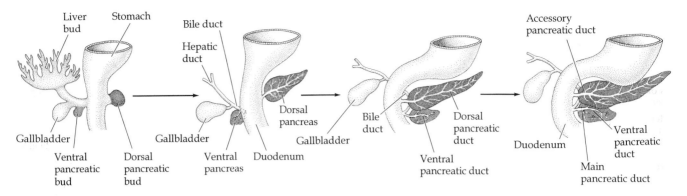

FIGURE 9.43
Pancreatic development in humans. At 30 days (A), the ventral pancreatic bud is close to the liver primordium; by 35 days (B) it begins migrating posteriorly and comes into contact with the dorsal pancreatic bud during the sixth week of development (C). In most individuals, the dorsal pancreatic bud loses its duct into the duodenum. However, in about 10 percent of the population, the dual duct system persists (D). (After Langman, 1981.)

The respiratory tube

The lungs also are a derivative of the digestive tube, even though they serve no role in digestion. In the center of the pharyngeal floor, between the fourth pair of pharyngeal pouches, the **laryngotracheal groove** extends ventrally (Figure 9.44). This groove then bifurcates into the two branches that form the pair of bronchi and lungs. The laryngotracheal endoderm becomes the lining of the trachea, the two bronchi, and the air sacs (alveoli) of the lungs. As we will see in a later chapter, the branching of this endodermal tube depends upon interactions with the different types of mesodermal cells in its path.

The lungs are an evolutionary novelty, and they are among the last of the mammalian organs to fully differentiate. The lungs must be able to draw in oxygen at the baby's first breath. To accomplish this, alveolar cells secrete a *surfactant* into the fluid bathing the lungs. This surfactant, consisting of phospholipids such as sphingomyelin and lecithin, is secreted very late in gestation, and it usually reaches physiologically useful levels around week 34 of human gestation. These compounds enable the alveolar cells to touch one another without sticking together. Thus, infants born prematurely often have difficulty breathing and have to be placed in respirators until their surfactant-producing cells mature.

This concludes our survey of the early features of animal development. We now attend to the *mechanisms* that enable this development to take place. In Part III we focus on the molecular events that direct cell differentiation. In Part IV we see the roles these molecules play in forming the embryonic body axes. Part V will discuss the genetic, cellular, and environmental forces that interact during organ formation.

FIGURE 9.44
Partitioning of the foregut into the esophagus and respiratory diverticulum during the third and fourth weeks of human gestation. (A) Lateral view, end of week 3. (B,C) Ventral views, week 4. (After Langman, 1981.)

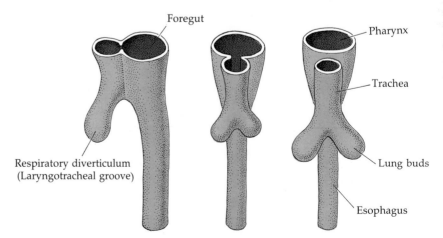

Abramson, S., Miller, R. G. and Phillips, R. A. 1977. The identification in adult bone marrow of pluripotent and restricted stem cells of the myeloid and lymphoid systems. *J. Exp. Med.* 145: 1567–1579.

Anderson, C., Devine, W. A., Anderson, R. H., Debich, D. E. and Zuberbuhler, J. R. 1990. Abnormalities of the spleen in relation to congenital malformation of the heart: A survey of necropsy findings in children. *Br. Heart J.* 63: 122–128.

Ash, P. J., Loutit, J. F. and Townsend, K. M. S. 1980. Osteoclasts derived from haematopoietic stem cells. *Nature* 283: 669–670.

Auerbach, R., Alby, L., Morrissey, L., Tu, M. and Joseph, J. 1985. Expression of organ-specific antigens on capillary endothelial cells. *Microvasc. Res.* 29: 401–411.

Ausprunk, D. H. and Folkman, J. 1977. Migration and proliferation of endothelial cells in preformed and newly formed blood vessels during tumor angiogenesis. *Microvasc. Res.* 14: 53–65.

Bandman, E. 1985. Continued expression of neonatal myosin heavy chain in adult dystrophic skeletal muscle. *Science* 127: 780–782.

Baron, R., Neff, L., Louvard, D. and Courtoy, P. J. 1985. Cell mediated extracellular acidification and bone resorption: Evidence for a low pH in resorbing lacuna and localization of a 100-kD lysosomal membrane protein at the osteoclast ruffled border. *J. Cell Biol.* 101: 2210–2222.

Baron, R., Neff, L., Roy, C., Boisvert, A. and Caplan, M. 1986. Evidence for a high and specific concentration of (Na$^+$,K$^+$) ATPase in the plasma membrane of the osteoclast. *Cell* 46: 311–320.

Becker, A. J., McCulloch, E. A. and Till, J. E. 1963. Cytological demonstration of the clonal nature of spleen cells derived from transplanted mouse marrow cells. *Nature* 197: 452–454.

Bischoff, R. and Holtzer, H. 1969. Mitosis and processes of differentiation of myogenic cells in vitro. *J. Cell Biol.* 41: 188–200.

Blair, H. C., Kahn, A. J., Crouch, E. C., Jeffrey, J. J. and Teitelbaum, S. L. 1986. Isolated osteoclasts resorb the organic and inorganic components of bone. *J. Cell Biol.* 102: 1164–1172.

Bloom, W. and Fawcett, D. W. 1975. *Textbook of Histology*, 10th Ed. Saunders, Philadelphia.

Bonner, P. H. and Hauschka, S. D. 1974. Clonal analysis of vertebrate myogenesis. I. Early developmental events in the chick limb. *Dev. Biol.* 37: 317–328.

Braun, T., Rudnicki, M. A., Arnold, H.-H. and Jaenisch, R. 1992. Targeted inactivation of the muscle regulatory gene *Myf-5* results in abnormal rib development and perinatal death. *Cell* 71: 369–382.

Breier, G., Albrecht, U., Sterrer, S. and Risau, W. 1992. Expression of vascular endothelial growth factor during embryonic angiogenesis and endothelial cell differentiation. *Development* 114: 521–532.

Brighton, C. T. 1984. The growth plate. *Orthoped. Clin. N.A.* 15: 571–594.

Brighton, C. T. and Hunt, R. M. 1974. Mitochondrial calcium and its role in calcification. *Clin. Orthop.* 100: 406–416.

Brunetti, A. and Goldfine, I. D. 1990. Role of myogenin in myoblast differentiation and its regulation by fibroblast growth factor. *J. Biol. Chem.* 265: 5960–5963.

Buckingham, M. 1992. Making muscles in mammals. *Trends Genet.* 8: 144–149.

Buckley, P. A. and Konigsberg, I. R. 1974. Myogenic fusion and the postmitotic gap. *Dev. Biol.* 37: 193–212.

Caplan, A. I., Fiszman, M. Y. and Eppenberger, H. M. 1983. Molecules and cell isoforms during development. *Science* 221: 923–927.

Carlson, B. M. 1981. *Patten's Foundations of Embryology*. McGraw-Hill, New York.

Celeste, A. J., Ianazzi, J. A., Taylor, R. C., Hewick, R. M., Rosen, V., Wang, E. A. and Wozney, J. M. 1990. Identification of transforming growth factor β family members present in bone-inductive protein purified from bovine bone. *Proc. Natl. Acad. Sci. USA* 87: 9843–9847.

Cheney, C. M. and Lash, J. W. 1984. An increase in cell–cell adhesion in the chick segmental plate results in meristic pattern. *J. Embryol. Exp. Morphol.* 79: 1–10.

Chevallier, A., Kieny, M., Mauger, A. and Sengel, P. 1977. Developmental fate of the somitic mesoderm in the chick embryo. In D. A. Ede, J. R. Hinchliffe and M. Balls (eds.), *Vertebrate Limb and Somite Morphogenesis*. Cambridge University Press, Cambridge, pp. 421–432.

Christ, B., Jacob, H. J. and Jacob, M. 1977. Experimental analysis of the origin of the wing musculature in avian embryos. *Anat. Embryol.* 150: 171–186.

Cormier, F. and Dieterlen-Lièvre, F. 1988. The wall of the chick aorta harbours M-CFC, G-CFC, GM-CFC and BFU-E. *Development* 102: 279–285.

Crelin, E. S. 1961. Development of the gastrointstinal tract. *Clinic. Symp.* (CIBA) 13: 68–82.

David, J. D., See, W. M. and Higginbotham, C. A. 1981. Fusion of chick embryo skeletal myoblasts: Role of calcium influx preceding membrane union. *Dev. Biol.* 82: 297–307.

Davis, R. L., Weintraub, H. and Lassar, A. B. 1987. Expression of a single transfected cDNA converts fibroblasts into myoblasts. *Cell* 51: 987–1000.

DeHaan, R. 1959. *Cardia bifida* and the development of pacemaker function in the early chicken heart. *Dev. Biol.* 1: 586–602.

DeHaan, R. L. 1967. Regulation of spontaneous activity and growth of embryonic chick heart cells in tissue culture. *Dev. Biol.* 16: 216–249.

Dieterlen-Lièvre, F. and Martin, C. 1981. Diffuse intraembryonic hemopoiesis in normal and chimeric avian development. *Dev. Biol.* 88: 180–191.

Fett, J. W., Strydom, D. J., Lubb, R. R., Alderman, E. M., Bethune, J. L., Riordan, J. F. and Vallee, B. L. 1985. Isolation and characterization of angiogenin, an angiogenic protein from human carcinoma cells. *Biochemistry* 24: 5480–5486.

Flamme, I. and Risau, W. 1992. Induction of vasculogenesis and hematogenesis in vitro. *Development* 116: 435–439.

Folkman, J. 1974. Tumor angiogenesis. *Adv. Cancer Res.* 19: 331–358.

Folkman, J., Langer, R., Linhardt, R. J., Haudenschild, C. and Taylor, S. 1983. Angiogenesis inhibition and tumor regression caused by heparin or heparin fragment in the presence of cortisone. *Science* 221: 719–725.

Foltz, C. 1989. *Cytokine Pathways in the Immune System*. Genzyme, Boston.

Garcia-Martinez, V. and Schoenwolf, G. C. 1993. Primitive-streak origin of the cardiovascular system in avian ambryos. *Dev. Biol.* 159: 706–719.

Girasole, G., Jilka, R. L., Passeri, G., Boswell, S., Boder, G., Williams, D. C. and Manolanas, S. C. 1992a. 17-β-Estradiol inhibits interleukin-6 production by bone marrow derived stromal cells and osteoblasts in vitro: A potential mechanism for the antiosteoporotic effect of estrogens. *J. Clin. Invest.* 89: 883–891.

Girasole, G., Passeri, G., Knutson, S., Manolanas, S. C. and Jilka, R. L. 1992b. Up-regulation of osteoclastic potential of the marrow is induced by orchiectomy and is reversed by testosterone replacement. *J. Bone Min. Res.* 7: S96.

Godlin, I. E., Garcia-Porrero, J. A., Coutinho, A., Dieterlen-Lièvre, F. and Marcos, M. A. R. 1993. Para-aortic splanchnopleura from early mouse embryos contain B1a cell progenitors. *Nature* 364: 67–70.

Gordon, M. Y., Riley, G. P., Watt, S. M. and Greaves, M. F. 1987. Compartmentalization of a haematopoietic growth factor (GM-CSF) by glycosaminoglycans in the bone marrow microenvironment. *Nature* 326: 403–405.

Gould, S. J. 1990. An earful of jaw. *Nat. Hist.* 1990(3): 12–23.

Gräper, L. 1907. Untersuchungen über die Herzbildung der Vögel. *Wilhelm Roux Arch. Entwicklungsmech. Org.* 24: 375–410.

Hall, B. K. 1988. The embryonic development of bone. *Am. Sci.* 76: 174–181.

Hall, B. K., van Exan, R. J. and Brunt, S. L. 1983. Retention of epithelial basal lamina allows isolated mandibular mesenchyme to form bone. *J. Craniofac. Gen. Dev. Biol.* 3: 253–267.

Harary, I. and Farley, B. 1963. *In vitro* studies on single beating rat heart cells. II. Intercellular communication. *Exp. Cell Res.* 29: 466–474.

Hasty, P., Bradley, A., Morris, J. H., Edmondson, D. G., Venuti, J. M., Olson, E. and Klein, W. H. 1993. Muscle deficiency and neonatal death in mice with a targeted mutation in the *myogenin* gene. *Nature* 364: 501–506.

Hattersley, G., Kirby, J. A., Chambers, T. J. 1991. Identification of osteoclast precursors in multilineage hematopoietic colonies. *Endocrinology* 128: 259–262.

Hay, E. 1966. *Regeneration*. Holt, Rinehart & Winston, New York.

Ho, S. Y., Cook, A., Anderson, R. H., Allan, L. D. and Fagg, N. 1991. Isomerism of the atrial appendages in the fetus. *Pediatr. Pathol.* 11: 589–608.

Hofman, F. and Globerson, A. 1973. Graft-versus-host response induced in vitro by mouse yolk sac cells. *Eur. J. Immunol.* 3: 179–181.

Holtzer, H., Rubinstein, N., Fellini, S., Yeoh, G., Chi, J., Birbaum, J. and Okayama, M. 1975. Lineages, quantal cell cycles, and the generation of cell diversities. *Q. Rev. Biophys.* 8: 523–557.

Horton, W. A. 1990. The biology of bone growth. *Growth Genet. Horm.* 6(2): 1–3.

Humphries, M. J., Oldern, K. and Yamada, K. M. 1986. A synthetic peptide from fibronectin inhibits experimental metastasis in murine melanoma cells. *Science* 233: 467–470.

Humphries, R. K., Jacky, P. B., Dill, F. J., Eaves, A. C. and Eaves, C. J. 1979. CFUs in individual erythroid colonies derived in vitro from adult mature mouse marrow. *Nature* 279: 718–720.

Hunt, P., Robertson, D., Weiss, D., Rennick, D., Lee, F. and Witte, O. N. 1987. A single bone marrow stromal cell type supports the in vitro growth of early lymphoid and myeloid cells. *Cell* 48: 997–1007.

Jilka, R. L. and eight others. 1992. Increased osteoclast development after estrogen loss: Mediation by interleukin-6. *Science* 257: 88–91.

Jimenez, J. J. and Yunis, A. A. 1987. Tumor cell rejection through terminal cell differentiation. *Science* 238: 1278–1280.

Jockusch, M. and Jockusch, B. M. 1980. Structural organization of the Z-line protein, α-actinin, in developing skeletal muscle cells. *Dev. Biol.* 75: 231–238.

Jones, C. M., Lyons, K. M. and Hogan, B. L. M. 1991. Involvement of bone morphogenetic protein-4 (BMP-4) and Vgr-1 in morphogenesis and neurogenesis in the mouse. *Development* 111: 531–542.

Jurśšková V. and Tkadleček, L. 1965. Character of primary and secondary colonies of haematopoiesis in the spleen of irradiated mice. *Nature* 206: 951–952.

Kahn, A. J. and Simmons, D. J. 1975. Investigation of cell lineage in bone using a chimaera of chick and quail embryonic tissue. *Nature* 258: 325–327.

Kalderon, N. and Gilula, N. B. 1979. Membrane events involved in myoblast fusion. *J. Cell Biol.* 81: 411–425.

Kaufman, S. J. and Foster, R. F. 1985. Remodeling of the myoblast membrane accompanies development. *Dev. Biol.* 110: 1–14.

Keller, G., Paige, C., Gilboa, E. and Wagner, E. F. 1985. Expression of a foreign gene in myeloid and lymphoid cells derived from multipotent hematopoietic precursors. *Nature* 318: 149–154.

Kim, K. J. and seven others. 1993. Inhibition of vascular endothelial growth factor-induced angiogenesis suppresses tumour growth in vivo. *Nature* 362: 841–844.

Kingsley, D. M., Bland, A. E., Grubber, J. M., Marker, P. C., Russell, L. B., Copeland, N. G. and Jenkins, N. A. 1992. The mouse *short ear* skeletal morphogenesis locus is associated with defects in a bone morphogenetic member of the TGF-β superfamily. *Cell* 71: 399–410.

Knudsen, K. A. 1985. The calcium-dependent myoblast adhesion that precedes cell fusion is mediated by glycoproteins. *J. Cell Biol.* 101: 891–897.

Knudsen, K. A., McElwee, S. A. and Myers, L. 1990. A role for the neural cell adhesion molecule, N-CAM, in myoblast interaction during myogenesis. *Dev. Biol.* 138: 159–168.

Konigsberg, I. R. 1963. Clonal analysis of myogenesis. *Science* 140: 1273–1284.

Kosher, R. A. and Lash, J. W. 1975. Notochord stimulation of in vitro somite chondrogenesis before and after enzymatic removal of perinotochordal materials. *Dev. Biol.* 42: 362–378.

Kramer, R. H. and Nicolson, G. L. 1979. Interaction of tumor cells with vascular endothelial cell monolayers: A model for metastatic invasion. *Proc. Natl. Acad. Sci. USA* 76: 5704–5708.

Kramer, T. C. 1942. The partitioning of the truncus and conus and the formation of the membranous portion of the intraventricular septum in the human heart. *Am. J. Anat.* 71: 343–370.

Krantz, S. B. and Goldwasser, E. 1965. On the mechanism of erythropoietin induced differentiation. II. The effect on RNA synthesis. *Biochim. Biophys. Acta* 103: 325–332.

Kubai, L. and Auerbach, R. 1983. A new source of embryonic lymphocytes in the mouse. *Nature* 301: 154–156.

Kubota, Y., Kleinman, H. K., Martin, G. R. and Lawley, T. J. 1988. Role of laminin and basement membrane in the morphological differentiation of human endothelial cells into capillary-like structures. *J. Cell Biol.* 107: 1589–1598.

Kurihara, N., Chenu, C., Miller, M., Civin, C. and Roodman, G. D. 1990. Identification of committed mononuclear precursors for osteoclast-like cells in long term human marrow cultures. *Endocrinology* 126: 2733–2741.

LaBarbera, M. 1990. Principles of design of fluid transport systems in zoology. *Science* 249: 992–1000.

Langman, J. 1981. *Medical Embryology*, 4th Ed. Williams & Wilkins, Baltimore.

Larsen, W. J. 1993. *Human Embryology*. Churchill-Livingstone, New York.

Lash, J. W. and Yamada, K. M. 1986. The adhesion recognition signal of fibronectin: A possible trigger mechanism for compaction during somitogenesis. In R. Bellairs, D. H. Ede and J. W. Lash (eds.), *Somites in Developing Embryos*. Plenum, New York, pp. 201–208.

Lash, J. W., Seitz, A. W., Cheney, C. M. and Ostrovsky, D. 1984. On the role of fibronectin during the compaction stage of somitogenesis in the chick embryo. *J. Exp. Zool.* 232: 197–206.

Lassar, A. B., Paterson, B. M. and Weintraub, H. 1986. Transfection of a DNA locus that mediates the conversion of 10T1/2 fibroblasts into myoblasts. *Cell* 47: 649–656.

Lassar, A. B., Buskin, J. N., Lockshon, D., Davis, R. L., Apone, S., Hauschka, S. D. and Weintraub. H. 1989. MyoD is a sequence-specific DNA binding protein requiring a region of *myc* homology to bind to the muscle creatine kinase enhancer. *Cell* 58: 823–831.

Lemischka, I. R., Raulet, D. H. and Mulligan, R. C. 1986. Developmental potential and dynamic behavior of hematopoietic stem cells. *Cell* 45: 917–927.

Linask, K. K. and Lash, J. W. 1986. Precardiac cell migration: Fibronectin localization at mesoderm–endoderm interface during directional movement. *Dev. Biol.* 114: 87–101.

Linask, K. K. and Lash, J. W. 1988a. A role for fibronectin in the migration of avian precardiac cells I. Dose-dependent effects of fibronectin antibody. *Dev. Biol.* 129: 315–323.

Linask, K. K. and Lash, J. W. 1988b. A role for fibronectin in the migration of avian precardiac cells II. Rotation of the heart-forming region during different stages and their effects. *Dev. Biol.* 129: 324–329.

Liu, C.-P. and Auerbach, R. 1991. In vitro development of murine T cells from prethymic and preliver embryonic yolk sac hematopoietic stem cells. *Development* 113: 1315–1323.

Lyons, G. E. and Buckingham, M. E. 1992. Developmental regulation of myogenesis in the mouse. *Semin. Dev. Biol.* 3: 243–253.

Lyons, K. M., Pelton, R. W. and Hogan, B. L. M. 1989. Patterns of expression of murine *Vgr-1* and *BMP-2α* RNA suggest that transforming growth factor β-like genes coordinately regulate aspects of embryonic development. *Genes Dev.* 3: 1657–1668.

Lyons, K. M., Pelton, R. W. and Hogan, B. L. M. 1990. Organogenesis and pattern formation in the mouse: RNA distribution patterns suggest a role for bone morphogenetic protein-2A. *Development* 109: 833–844.

Lyons, K. M., Jones, C. M. and Hogan, B. L. M. 1991. The DVR gene family in embryonic development. *Trends Genet.* 7: 408–412.

Markwald, R. R., Fitzharris, T. P. and Manasek, J. J. 1977. Structural development of endocardial cushions. *Am. J. Anat.* 148: 85–120.

Martin, C., Beaupain, D. and Dieterlen-Lièvre, F. 1978. Developmental relationships between vitelline and intraembryonic haemopoiesis studied in avian yolk sac chimeras. *Cell Differ.* 7: 115–130.

Medvinsky, A. L., Samoylina, N. L., Müller, A. M. and Dzierzak, E. A. 1993. An early pre-liver intraembryonic source of CFU-S in the developing mouse. *Nature* 364: 64–67.

Meier, S. 1979. Development of the chick mesoblast: Formation of the embryonic axis and the establishment of the metameric pattern. *Dev. Biol.* 73: 25–45.

Menko, A. S. and Boettiger, D. 1987. Occupation of the extracellular matrix integrin is a control point for myogenic differentiation. *Cell* 51: 51–57.

Menko, A. S., George-Weinstein, M. and Boettiger, D. 1993. Fibronectin, the extracellular regulator of myogenic differentiation. In press.

Merimee, T. J., Zapf, J., Hewlett, B. and Cavalli-Sforza, L. L. 1987. Insulin-like growth factors in pygmies. The role of puberty in determining final stature. *N. Engl. J. Med.* 316: 906–911.

Millauer, B., Wizigmann-Voos, Schnürch, H., Martinez, R., Müller, N. P. H., Risau, W. and Ullrich, A. 1993. High-affinity VEGF binding and developmental expression suggest *flk-1* as a major regulator of vasculogenesis and angiogenesis. *Cell* 72: 835–846.

Mintz, B. and Baker, W. W. 1967. Normal mammalian muscle differentiation and gene control of isocitrate dehydrogenase synthesis. *Proc. Natl. Acad. Sci. USA* 58: 592–598.

Moore, J. W., Dionne, C., Jaye, M. and Swain, J. 1991. The mRNAs encoding acidic FGF, basic FGF, and FGF receptor are coordinately downregulated during myogenic differentiation. *Development* 111: 741–748.

Moore, M. A. S. and Metcalfe, D. 1970. Ontogeny of the haemopoietic system: Yolk sac origin of in vivo and in vitro colony forming cells in the developing mouse embryo. *Br. J. Haematol.* 18: 279–296.

Moses, M. A., Sudhalter, J. and Langer, R. 1990. Identification of an inhibitor of neovascularization from cartilage. *Science* 248: 1408–1410.

Murphy, M. E. and Carlson, E. C. 1978. Ultrastructural study of developing extracellular matrix in vitelline blood vessels of the early chick embryo. *Am. J. Anat.* 151: 345–375.

Muthukkaruppan, V. R. and Auerbach, R. 1979. Angiogenesis in the mouse cornea. *Science* 205: 1416–1418.

Nabeshima, Y., Hanaoka, K., Hayasaka, M., Esumi, E., Li, S., Nonaka, I. and Nabeshima, Y. 1993. *Myogenin* gene disruption results in perinatal lethality because of severe muscle defect. *Nature* 364: 532–535.

Nakauchi, H. and Gachelin, G. 1993. Les cellules souches. *La Recherche* 254: 537–541.

Nameroff, M. and Munar, E. 1976. Inhibition of cellular differentiation by phospholipase C. II. Separation of fusion and recognition among myogenic cells. *Dev. Biol.* 49: 288–293.

Nicolson, G. L., Irimura, T., Gonzalez, R. and Ruoslahti, E. 1981. The role of fibronectin in adhesion of metastatic melanoma cells to endothelial cells and their basal lamina. *Exp. Cell Res.* 135: 461–465.

Nilsson, A., Isgaard, J., Lindahl, A., Dahlström, A., Skottner, A. and Isaksson, O. G. P. 1986. Regulation by growth hormone of number of chondrocytes containing IGF-I in rat growth plate. *Science* 233: 571–574.

Ordahl, C. P. 1993. Myogenic lineages within the developing somite. In M. Bernfield (ed.), *Molecular Basis of Morphogenesis.* Wiley-Liss, New York, pp. 165–170.

Ordahl, C. P. and LeDouarin, N. 1992. Two myogenic lineages within the developing somite. *Development* 114: 339–353.

Olwin, B. B. and Hauschka, S. D. 1988. Cell surface fibroblast growth factor and epidermal growth factor receptors are permanently lost during skeletal muscle terminal differentiation in culture. *J. Cell Biol.* 107:761–769.

Ossowski, L. and Reich, E. 1983. Antibodies to plasminogen activator inhibit tumor metastasis. *Cell* 35: 611–619.

Ostrovsky, D., Cheney, C. M., Seitz, A. W. and Lash, J. W. 1984. Fibronectin distribution during somitogenesis in the chick embryo. *Cell Differ.* 13: 217–223.

Packard, D. S., Jr. and Jacobson, A. G. 1976. The influence of axial structures on chick somite formation. *Dev. Biol.* 53: 36–48.

Packard, D. S., Jr. and Meier, S. 1983. An experimental study of somitomeric organization of the avian vegetal plate. *Dev. Biol.* 97: 191–202.

Pardanaud, L., Altmann, C., Kitos, P., Dieterlen-Lièvre, F. and Buck, C. 1987. Vasculogenesis in the early quail blastodisc as studied with a monoclonal antibody recognizing endothelial cells. *Development* 100: 339–349.

Pardanaud, L., Yassine, F. and Dieterlen-Lièvre, F. 1989. Relationship between vasculogenesis, angiogenesis, and hemopoiesis during avian ontogeny. *Development* 105: 473–485.

Patten, B. M. 1951. *Early Embryology of the Chick,* 4th Ed. McGraw-Hill, New York.

Piette, J., Bessereau, J.-L., Huchet, M. and Changeaux, J.-P. 1990. Two adjacent MyoD1-binding sites regulate expression of the acetylcholine receptor α-subunit gene. *Nature* 345: 353–355.

Pinney, D. F., Pearson-White, S. H., Konieczny, S. J., Latham, K. E. and Emerson, C. P., Jr. 1988. Myogenic lineage determination and differentiation: Evidence for a regulatory gene pathway. *Cell* 53: 781–793.

Plate, K. H., Breier, G., Weich, H. A. and Risau, W. 1992. Vascular endothelial growth factor is a potential tumour angiogenesis factor in human gliomas in vivo. *Nature* 359: 845–848.

Potten, C. S. and Loeffler, M. 1990. Stem cells: attributes, spirals, pitfalls, and uncertainties. Lessons for and from the Crypt. *Development* 110: 1001–1020.

Potts, J. D., Dagle, J. M., Walder, J. A., Weeks, D. L. and Runyon, R. B. 1991. Epithelial–mesenchymal transformation of embryonic cardiac endothelial cells is inhibited by a modified antisense oligodeoxynucleotide to transforming growth factor b3. *Proc. Natl. Acad. Sci. USA* 88: 1516–1520.

Pownall, M. E. and Emerson, C. E., Jr. 1992a. Sequential activation of three myogenic regulatory genes during somite morphogenesis in quails. *Dev. Biol.* 151: 67–79.

Pownall, M. E. and Emerson, C. E., Jr. 1992b. Molecular and embryological studies of avian myogenesis. *Semin. Dev. Biol.* 3: 229–241.

Risau, W. 1986. Developing brain produces an angiogenesis factor. *Proc. Natl. Acad. Sci. USA* 83: 3855–3859.

Roberts, R., Gallagher, J., Spooncer, E., Allen, T. D., Bloomfield, F. and Dexter, T. M. 1988. Heparan sulphate-bound growth factors: A mechanism for stromal cell mediated haemopoiesis. *Nature* 332: 376–378.

Rong, P. M., Teillet, M.-A., Ziller, C. and LeDouarin, N. M. 1992. The neural tube/notochord complex is necessary for vertebral but not limb and body wall striated muscle differentiation. *Development* 115: 657–672.

Rosen, V. and Thies, S. 1992. The BMP proteins in bone formation and repair. *Trends Genet.* 8: 97–102.

Rudnicki, M. A., Braun, T., Hinuma, S. and Jaenisch, R. 1992. Inactivation of *MyoD* in mice leads to up-regulation of the myogenic HLH gene *Myf-5* and results in apparently normal muscle development. *Cell* 71: 383–390.

Rudnicki, M. A., Schnegelsberg, P. N. J., Stead, R. H., Braun, T., Arnold, H.-H. and Jaenisch, R. 1993. MyoD or Myf-5 is required in a functionally redundant manner for the formation of skeletal muscle. *Cell* 75: 1351–1359.

Rugh, R. 1951. *The Frog: Its Reproduction and Development.* Blakiston, Philadelphia.

Sachs, L. 1978. Control of normal cell differentiation and the phenotypic reversion of malignancy in myeloid leukemias. *Nature* 274: 535–539.

Sariola, H. 1985. Interspecies chimeras: An experimental approach for studies on embryonic angiogenesis. *Med. Biol.* 6: 43–65.

Sassoon, D., Lyons, G., Wright, W. E., Lin, V., Lassar, A., Weintraub, H. and Buckingham, M. E. 1989. Expression of two myogenic regulatory factors myogenin and myoD1 during mouse embryogenesis. *Nature* 341: 303–307.

Shainberg, A., Yagil, G. and Yaffe, D. 1969. Control of myogenesis in vitro by Ca^{2+} concentration in nutritional medium. *Exp. Cell Res.* 58: 163–167.

Shapiro, I., DeBolt, K., Funanage, V., Smith, S. and Tuan, R. 1992. Developmental regulation of creatine kinase activity in cells of the epiphyseal growth plate. *J. Bone Min. Res.* 7: 493–500.

Shweiki, D., Itin, A., Soffer, D. and Keshet, E. 1992. Vascular endothelial growth factor induced by hypoxia may mediate hypoxia-initiated angiogenesis. *Nature* 359: 843–845.

Spangrude, G. J., Heimfeld, S. and Weissman, I. 1988. Purification and characterization of mouse hematopoietic stem cells. *Science* 241: 58–62.

Spemann, H. 1938. *Embryonic Development and Induction*. Yale University Press, New Haven.

Sutherland, W. M. and Konigsberg, I. R. 1983. CPK accumulation in fusion-blocked quail myocytes. *Dev. Biol.* 99: 287–297.

Syftestad, G. T. and Caplan, A. I. 1984. A fraction from extracts of demineralized adult bone stimulates conversion of mesenchymal cells into chondrocytes. *Dev. Biol.* 104: 348–356.

Thayer, M. J., Tapscott, S. J., Davis, R. L., Wright, W. E., Lassar, A. B. and Weintraub, H. 1989. Positive autoregulation of the myogenic determination gene *MyoD1*. *Cell* 58: 241–248.

Till, J. E. 1981. Cellular diversity in the blood-forming system. *Am. Sci.* 69: 522–527.

Till, J. E. and McCulloch, E. A. 1961. A direct measurement of the radiation sensitivity of normal mouse bone marrow cells. *Radiat. Res.* 14: 213–222.

Tuan, R. 1987. Mechanisms and regulation of calcium transport by the chick embryonic chorioallantoic membrane. *J. Exp. Zool.* [Suppl.] 1: 1–13.

Tuan, R. S. and Lynch, M. H. 1983. Effect of experimentally induced calcium deficiency on the developmental expression of collagen types in chick embryonic skeleton. *Dev. Biol.* 100: 374–386.

Tyler, M. S. and Hall, B. K. 1977. Epithelial influence on skeletogenesis in the mandible of the embryonic chick. *Anat. Rec.* 206: 61–70.

Urist, M. R. 1965. Bone: Formation by autoinduction. *Science* 150: 893–899.

Urist, M. R. and eight others. 1984. Purfication of bovine bone morphogenetic protein by hydroxyapatite chromatography. *Proc. Natl. Acad. Sci. USA* 81: 371–375.

Vaidya, T. B., Rhodes, S. J., Taparowsky, E. J. and Konieczny, S. F. 1989. Fibroblast growth factor and transforming growth factor-β repress transcription of the myogenic regulatory gene MyoD1. *Mol. Cell Biol.* 9: 3576–3579.

Vivarelli, E. and Cossu, G. 1986. Neural control of early somite differentiation in cultures of mouse somites. *Dev. Biol.* 117: 319–325.

Wakelam, M. J. O. 1985. The fusion of myoblasts. *Biochem. J.* 228: 1–12.

Webster, C., Silberstein, L., Hays, A. P. and Blau, H. M. 1988. Fast muscle fibers are preferentially affected in Duchenne muscular dystrophy. *Cell* 52: 503–513.

Weintraub, H., Tapscott, S. J., Davis, R. L., Thayer, M. J., Adam, M. A., Lassar, A. B. and Miller, D. 1989. Activation of muscle-specific genes in pigment, nerve, fat, liver, and fibroblast cell lines by forced expression of MyoD. *Proc. Nat. Acad. Sci. USA* 86: 5434–5438.

Whalen, R. G., Schwartz, K., Bouveret, P., Sell, S. M. and Gros, F. 1979. Contractile protein isozymes in muscle development: Identification of an embryonic form of myosin heavy chain. *Proc. Natl. Acad. Sci. USA* 76: 5197–5201.

Whitlock, C. A., Tidmarsh, G. F., Muller-Sieburg, C. and Weissman, I. L. 1987. Bone marrow stromal cell lines with lymphopoietic activity express high levels of a pre-B neoplasia-associated molecule. *Cell* 48: 1009–1021.

Wilson, D. 1983. The origin of the endothelium in the developing marginal vein of the chick wing bud. *Cell Differ.* 13: 63–67.

Wolf, N. S. and Trentin, J. J. 1968. Hemopoietic colony studies. V. Effect of hemopoietic organ stroma on differentiation of pluripotent stem cells. *J. Exp. Med.* 127: 205–214.

Womble, M. D. and Bonner, P. H. 1980. Developmental fate of a distinct class of chick myoblasts after transplantation of cloned cells into quail embryos. *J. Embryol. Exp. Morphol.* 58: 119–130.

Wozney, J. M. and seven others. 1988. Novel regulators of bone formation: Molecular clones and activities. *Science* 242: 1528–1534.

Wozney, J. M., Capparella, J. and Rosen, V. 1993. Bone morphogenesis proteins in cartilage and bone development. *In* M. Bernfield (ed.), *Molecular Basis of Morphogenesis*. Wiley-Liss, New York, pp. 221–230.

Wuthier, R. 1982. A review of the primary mechanism of endochondral ossification with special emphasis on the role of cells, mitochondria, and matrix vesicles. *Clin. Orthoped. Rel. Res.* 169: 219–242.

Yaffe, D. and Feldman, M. 1965. The formation of hybrid multinucleated muscle fibres from myoblasts of different genetic origin. *Dev. Biol.* 11: 300–317.

MECHANISMS OF CELLULAR
DIFFERENTIATION

10

Transcriptional regulation of gene expression

Transcription factors and the activation of specific promoters

But whatever the immediate operations of the genes turn out to be, they most certainly belong to the category of developmental processes and thus belong to the province of embryology. This central problem of fundamental biology at the present time is being attacked from many sides, both by physiologists and biochemists and by geneticists; but it is essentially an embryological problem.
C. H. WADDINGTON (1956)

We have entered the cell, the mansion of our birth, and have started the inventory of our acquired wealth.
ALBERT CLAUDE (1974)

The major paradigm of developmental genetics is *differential gene expression from the same nuclear repertoire.* Different cell types make different sets of proteins even though their genomes are identical. This regulation of gene expression can be accomplished at several levels:

- *Differential gene transcription,* regulating which of the nuclear genes are transcribed into RNA
- *Selective nuclear RNA processing,* regulating which of the transcribed RNAs get into the cytoplasm to become messenger RNAs
- *Selective messenger RNA translation,* regulating which of the mRNAs in the cytoplasm get translated into protein
- *Differential protein modification,* regulating which proteins are allowed to remain or function in the cell

Some genes (such as those coding for the globin proteins of hemoglobin) are regulated at each of these levels. This and the following chapter will discuss the mechanisms of differential gene transcription: how different genes are activated in different types of cells at particular times. We discussed the basic phenomena of differential gene transcription in Chapter 2. The puffs of the polytene chromosomes represent the activation of groups of genes in response to a hormone produced in the larval insect. Similarly, the expression of endoderm-specific genes in sea urchin larvae was controlled at the level of gene transcription. In these chapters, we will be discussing the mechanisms whereby nuclei in different cells are able to activate and repress different genes as the cells differentiate.

Exons and introns

When one looks at genes, the first thing that becomes apparent is that most eukaryotic genes are not like most prokaryotic genes. Eukaryotic genes are not colinear with their peptide products. Rather, the 5′ and 3′ ends of eukaryotic mRNA come from noncontiguous regions on the chro-

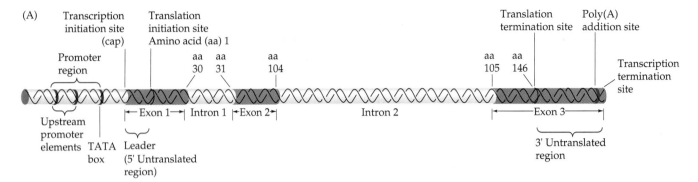

(A)

Transcription initiation site (cap)

Translation initiation site Amino acid (aa) 1

Promoter region

aa 30   aa 31

aa 104

Translation termination site

Poly(A) addition site

aa 105   aa 146

Transcription termination site

Upstream promoter elements   TATA box   Leader (5' Untranslated region)

← Exon 1 → Intron 1 ⊢Exon 2⊣

Intron 2

⊢———— Exon 3 ————⊣

3' Untranslated region

FIGURE 10.1

Nucleotide sequence of the human β-globin gene. (A) Schematic representation of the locations of the promoter region, transcription initiation (cap) site, leader sequence, exons and introns of the β-globin gene. Exons are in color; the numbers flanking them indicate the amino acid positions they encode in β-globin. (B) The nucleotide sequence of the β-globin gene, shown from the 5' to the 3' end of the RNA. The promoter sequences are boxed, as are the translation initiation and termination codes ATG and TAA. The large capital letters boxed in color correspond to exons, and the amino acids for which they code are abbreviated above them. The small capital letters are the bases of the intervening sequences. The codons represented by capital letters after the translation terminator are in the globin mRNA but are not translated into proteins. Within this group is the sequence thought to be needed for polyadenylation. A G in the first intron (arrow) is mutated to an A in one form of β^+-thalassemia. (Sequence from Lawn et al., 1980.)

mosome. Between the regions of DNA coding for the proteins—**exons**—are intervening sequences—**introns**—that have nothing whatever to do with the amino acid sequence of the protein.* The structure of the human β-globin gene is shown in Figure 10.1. This gene consists of the following elements:

1. A **promoter region** responsible for the binding of RNA polymerase and for the subsequent initiation of transcription. This promoter region of the human β-globin gene has (as will be shown later) three distinct units and extends from 95 to 26 base pairs before ("upstream from") the transcription initiation site (i.e., from -95 to -26).

2. The sequence ACATTTG, where *transcription* is initiated. This is often called the **cap sequence** because it represents the 5' end of the RNA, which will receive a "cap" of modified nucleotides soon after it is transcribed (as we will see later). The specific cap sequence varies among genes.

3. The ATG codon for the initiation of *translation*. This codon is located 50 base pairs after the initiation point of transcription (although this distance differs greatly in different genes). The intervening sequence of 50 nucleotide pairs between the initiation points of transcription and translation is called the **leader sequence**. The leader sequence can determine the rate at which translation is initiated.

4. The first exon containing 90 base pairs coding for amino acids 1–30 of human β-globin.

5. An intron containing 130 base pairs with no coding sequences for hemoglobin. The structure of this intron is important in enabling the RNA to be processed into messenger RNA and to exit from the nucleus.

6. An exon containing 222 base pairs coding for amino acids 31–104.

7. A large intron—850 base pairs—having nothing to do with the globin protein structure.

*The term *exon* has taken on two overlapping meanings. In the original sense, it is anatomically defined as a nucleotide sequence whose RNA "exits" the nucleus. It has taken on the functional definition of referring to a protein-encoding nucleotide sequence. For discussion here, we will use the former definition, and we will define the leader sequences and 3' untranslated sequences as exons that are not translated. Certain eukaryotic genes (such as the genes for histones) lack intervening sequences, and any hypothesis concerning intron function must take these exceptions into account. By convention, upstream, downstream, 5', and 3' directions are specified in relation to the RNA. Thus the promoter is *upstream* of the gene, near its 5' end.

(B)

ccctgtgga**gccacaccc**tagggttgg**ccaat**ctactcccaggagcagggagggcaggagccagggctggg**cataaaa**

gtcaggcagagccatctattgctt`ACATTTGCTTCTGACACAACTGTGTTCACTAGCAACCTCAAACAGACACC`ATG

`ValHisLeuThrProGluGluLysSerAlaValThrAlaLeuTrpGlyLysValAsnValAspGluValGlyGlu`
`GTGCACCTGACTCCTGAGGAGAAGTCTGCCGTTACTGCCCTGTGGGGCAAGGTGAACGTGGATGAAGTTGGTGGTGAG`

`AlaLeuGlyArg`
`GCCCTGGGCAGG`TTGGTATCAAGGTTACAAGACAGGTTTAAGGAGACCAATAGAAACTGGGCATGTGGAGACAGAGAAG

⬇

ACTCTTGGGTTTCTGATAGGCACTGACTCTCTCTGCCTATTGGTCTATTTTCCCACCCTTAGG`CTGCTGGTGGTCTAC`
`LeuLeuValValTyr`

`ProTrpThrGlnArgPhePheGluSerPheGlyAspLeuSerThrProAspAlaValMetGlyAsnProLysValLys`
`CCTTGGACCCAGAGGTTCTTTGAGTCCTTTGGGGATCTGTCCACTCCTGATGCTGTTATGGGCAACCCTAAGGTGAAG`

`AlaHisGlyLysLysValLeuGlyAlaPheSerAspGlyLeuAlaHisLeuAspAsnLeuLysGlyThrPheAlaThr`
`GCTCATGGCAAGAAAGTGCTCGGTGCCTTTAGTGATGGCCTGGCTCACCTGGACAACCTCAAGGGCACCTTTGCCACA`

`LeuSerGluLeuHisCysAspLysLeuHisValAspProGluAsnPheArg`
`CTGAGTGAGCTGCACTGTGACAAGCTGCACGTGGATCCTGAGAACTTCAGG`GTGAGTCTATGGGACCCTTGATGTTTT

CTTTCCCCTTCTTTTCTATGGTTAAGTTCATGTCATAGGAAGGGGAGAAGTAACAGGGTACAGTTTAGAATGGGAAAC

AGACGAATGATTGCATCAGTGTGGAAGTCTCAGGATCGTTTTAGTTTCTTTTATTTGCTGTTCATAACAATTGTTTTC

TTTTGTTTAATTCTTGCTTTCTTTTTTTTTCTTCTCCGCAATTTTTACTATTATACTTAATGCCTTAACATTGTGTAT

AACAAAAGGAAATATCTCTGAGATACATTAAGTAACTTAAAAAAAAAAACTTTACACAGTCTGCCTAGTACATTACTATT

TGGAATATATGTGTGCTTATTTGCATATTCATAATCTCCCTACTTTATTTTCTTTTATTTTTAATTGATACATAATCA

TTATACATATTTATGGGTTAAAGTGTAATGTTTTAATATGTGTACACATATTGACCAAATCAGGGTAATTTTGCATT

TGTAATTTTTAAAAAATGCTTTCTTCTTTTAATATACTTTTTTGTTTATCTTATTTCTAATACTTTCCCTAATCTCTTT

CTTTCAGGGCAATAATGATACAATGTATCATGCCTCTTTGCACCATTCTAAAGAATAACAGTGATAATTTCTGGGTTA

AGGCAATAGCAATATTTCTGCATATAAATATTTCTGCATATAAATTGTAACTGATGTAAGAGGTTTCATATTGCTAA

TAGCAGCTACAATCCAGCTACCATTCTGCTTTTATTTTATGGTTGGGATAAGGCTGGATTATTCTGAGTCCAAGCTAG

GCCCTTTTGCTAATCATGTTCATACCTCTTATCTTCCTCCCACAG`CTCCTGGGCAACGTGCTGGTCTGTGTGCTGGCC`
`LeuLeuGlyAsnValLeuValCysValLeuAla`

`HisHisPheGlyLysGluPheThrProProValGlnAlaAlaTyrGlnLysValValAlaGlyValAlaAsnAlaLeu`
`CATCACTTTGGCAAAGAATTCACCCCACCAGTGCAGGCTGCCTATCAGAAAGTGGTGGCTGGTGTGGCTAATGCCCTG`

`AlaHisLysTyrHis`
`GCCCACAAGTATCAC`TAA`GCTCGCTTTCTTGCTGTCCAATTTCTATTAAAGGTTCCTTTGTTCCCTAAGTCCAACTAC`

`TAAACTGGGGGATATTATGAAGGGCCTTGAGCATCTGGATTCTGCCT`AATAAA`AACATTTATTTTCATTGC`aatgat

gtatttaaattatttctgaatatttttactaaaaagggaatgtgggaggtcagtgcatttaaaacataaagaaatgatg

agctgttcaaaccttgggaaaatacactatatcttaaactccatgaaagaaggtgaggctgcaaccagctaatgcaca

ttggcaacagcccctgatgcctatgccttattcatccctcagaaaaggattcttgtagaggcttgatttgcaggttaa

agttttgctatgctgtattttacattacttattgttttagctgtcctcatgaatgtctttttcactacccatttgctta

tcctgcatctctctcagccttgact ...

8. An exon containing 126 base pairs coding for amino acids 105–146.
9. A **translation termination codon**, TAA.
10. A **3' untranslated region** that, although transcribed, is not translated into protein. This region includes the sequence AATAAA, which is needed to place a "tail" of some 200 to 300 adenylate residues on the RNA transcript. This poly(A) tail confers stability and translatability on the mRNA, and it is inserted into the RNA about 20 bases downstream of the AAUAAA sequence. However, transcription continues beyond the AATAAA site for about 1000 nucleotides before being terminated. Within the 3' transcribed but untranslated sequence (about 600–900 base pairs from the AATAAA site) is a DNA sequence that serves as an enhancer. This sequence is necessary for the temporal and tissue-specific expression of the β-globin gene in adult red blood cell precursors (Trudel and Constantini, 1987).

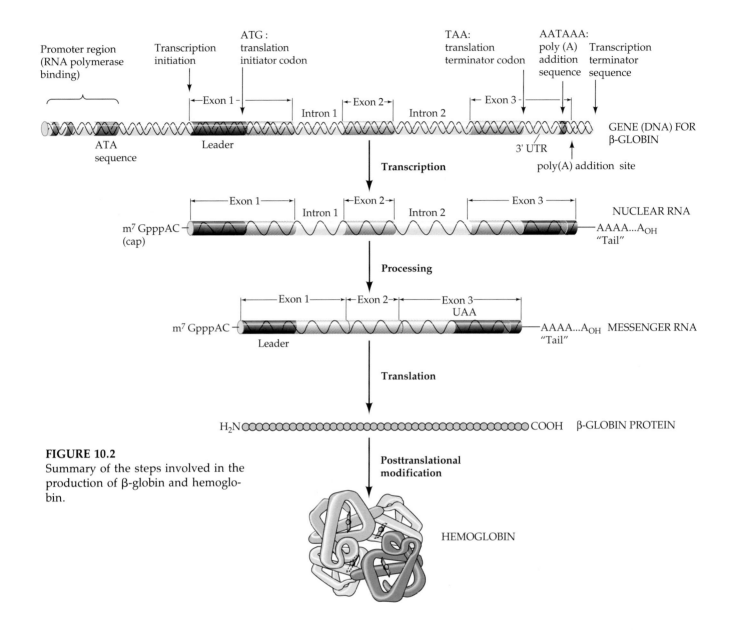

FIGURE 10.2
Summary of the steps involved in the production of β-globin and hemoglobin.

The original nuclear RNA transcript for such a gene contains the capping sequence, the leader sequence, the exons, the introns, and the 3' untranslated region (Figure 10.2). In addition, both its ends become modified. A cap consisting of methylated guanosine is placed on the 5' end of the RNA in opposite polarity to the RNA itself. Thus, whereas all the bases in the message precursor are linked 5' to 3', the cap structure is linked 5' to 5'. This means that there is no free 5' phosphate group on the nuclear RNA (Figure 10.3). Messenger RNA molecules are likewise "capped," although it is not certain whether the mRNA cap is the original one it received in the nucleus. The 5' cap is necessary for the binding of mRNA to the ribosome and for subsequent translation (Shatkin, 1976).

The 3' terminus is usually modified in the nucleus by having roughly 200 adenylate residues added on as a tail. These adenylic acid residues are put together enzymatically and are added to the transcript. They are not part of the gene sequence. Both the 5' and 3' modifications may protect the RNA from exonucleases (Sheiness and Darnell, 1973; Gedamu and Dixon, 1978), thereby stabilizing the message and its precursor.

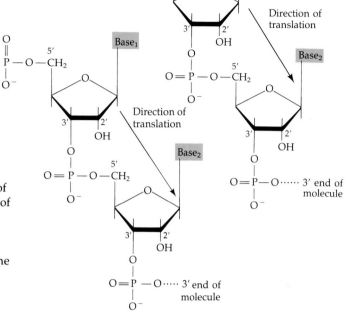

AFTER CAPPING

7-methyl guanosine

FIGURE 10.3
Capping the 5' end of a eukaryotic mRNA. A cap of 7-methylguanylate is linked 5' to 5' with the first base of the newly transcribed mRNA. The original 5' terminus of the mRNA had three phosphate groups. The capping mechanism ligates GTP with this terminus, using one phosphate group from GTP and two phosphate groups from mRNA. Later, an enzyme methylates the guanosine at position 7; the first and second bases of the original mRNA molecule are often methylated as well. (After Rottman et al., 1974.)

Promoters and enhancers

In addition to the gene structure discussed above, there are regulatory sequences that can be on either end of the gene (or even within it). These sequences, the promoters and the enhancers (introduced in Chapter 2), are necessary for controlling where and when a particular gene is transcribed.

Two types of regulatory elements are needed to effect transcription at the proper sites. The first set of regulatory elements are called *cis* **regulators**. These represent specific DNA sequences on a given chromosome. *Cis* regulators act only on adjacent genes. The second group of regulatory elements are called *trans* **regulators**. These are soluble molecules (including proteins and RNAs) that are made by one gene and interact with genes on the same or different chromosomes. If one recalls gene induction in the *lac* operon in *E. coli*, one will remember that a repressor gene makes a repressor protein that interacts with the operator sequence of the *lac* operon genes. In this case, the operator DNA is a *cis*-regulatory element because it controls only the adjacent *lac* operon on its own chromosome. The repressor protein, however, is a *trans* regulator, because it can be made by one chromosome and bind to the *cis*-regulatory operator on another chromosome.

In eukaryotic genes that encode messenger RNA, two types of *cis*-regulatory DNA sequences have been discovered that influence which genes become transcribed in which cells. These are the promoters and the enhancers. **Promoters** are typically located immediately upstream from the site where transcription begins and are generally hundreds of base pairs long. The promoter site is required for the binding of RNA polymerase II and the accurate initiation of transcription. Eukaryotic RNA polymerases require additional protein factors to bind efficiently to the promoter. The **enhancer** activates the utilization of the promoter, controlling the efficiency and rate of transcription from that particular promoter. Enhancers can activate only *cis*-linked promoters (i.e., promoters on the same chromosome), but they can do so at great distances (some as great as 50 kilobases away from the promoter). Moreover, enhancers do not need to be at the 5' ("upstream") side of the gene. They can be at the 3' end, in the introns, or even on the complementary DNA strand (Maniatis et al., 1987). Like the promoter, enhancers function by binding specific *trans*-regulatory proteins called **transcription factors**.

One type of enhancer is a negative enhancer, also called a **silencer**. When transcription factors bind to silencers, they inhibit the transcription from *cis*-linked promoters. Some sequences can act as positive enhancers in some cells and negative enhancers in others, depending on the other transcription factors present in the cell.

Promoter structure

Promoters of genes that transcribe relatively large amounts of mRNA have similar structures. They have a TATA sequence (sometimes called the **TATA box** or the **Goldberg-Hogness box**) about 30 base pairs upstream from the site where transcription begins, and one or more promoter elements further upstream (Figure 10.4; Grosschedl and Birnstiel, 1980; McKnight and Tjian, 1986). The "functional anatomy" of a promoter region can be analyzed by determining which of its bases are necessary for efficient transcription. Cloned genes can be accurately transcribed when they are placed into the nuclei of frog oocytes or fibroblasts or when they are incubated with RNA polymerase in the presence of nucleotides and nuclear extracts (Wasylyk et al., 1980). After the transcription of a gene is confirmed, one uses restriction enzymes to make specific deletions in the gene or in the regions surrounding it. One can then see whether such a modified gene will still be accurately transcribed. Such studies on β-globin genes (Grosveld et al., 1982; Dierks et al., 1983) showed that the first 109 base pairs preceding the cap site were sufficient for the correct initiation of β-globin gene transcription by RNA polymerase.

Myers and co-workers (1986) refined this analysis by cloning the region of a mouse globin gene from 106 base pairs upstream from the start of transcription (−106) through the first 475 base pairs (+475) of the first exon. These clones were subjected to in vitro mutagenesis (wherein specific mutations can be placed into a cloned gene). In this way 130 different single-base substitutions were introduced into the promoter region of the globin gene. These cloned genes were placed into plasmids containing an enhancer from a gene normally expressed in all tissues. The recombinant plasmids were then introduced by transfection into cultured cells that do not usually produce globin. Would the cells transcribe a truncated globin message (475 bases) from the clones? Figure 10.5 shows the results. In most cases, mutating a base in the 5' flanking region did not affect the efficiency of globin gene transcription. But there were three clusters of nucleotides wherein mutations drastically reduced transcription. One cluster was in the TATA box, another was in the CAAT upstream promoter element, and a third was in the CACCC region, about 95 to 87 base pairs upstream from the cap site.

The CAAT and TATA boxes have been found to be critical elements in numerous eukaryotic promoters (Efstratiadis et al., 1980), but the

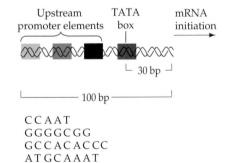

FIGURE 10.4
Typical promoter region for a protein-coding eukaryotic gene. The gene diagrammed here contains a TATA box and three upstream promoter elements. Examples of such upstream elements are shown beneath the diagram. (After Maniatis et al., 1987.)

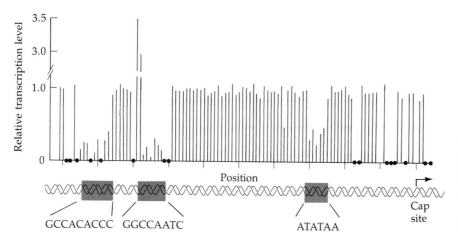

FIGURE 10.5
The effect of specific point mutations in the mouse β-globin promoter on the ability of that promoter to initiate transcription. Each line represents the transcription level of a mutant promoter relative to the transcription level of a wild-type globin promoter assayed concurrently. The black dots represent nucleotides for which no mutation had been generated. Below the histogram is a diagram showing the position of the TATA box and the two upstream promoter elements of the mouse β-globin gene. (After Myers et al., 1986.)

CACCC sequence is seldom seen except in the β-globin gene promoters in several species. In humans, this sequence appears to be critical. A naturally occurring mutation in this sequence causes a total loss of β-globin gene transcription (Orkin and Kazazian, 1984), and this sequence is recognized by an erythroid-specific transcription factor (Mantovani et al., 1988). Two mutations, at positions −78 and −79, have actually increased transcription to three times the wild-type level. It is thought that these changes facilitate the interaction of the promoter with the *trans*-regulatory proteins.

Promoter function

Promoters can function not only to bind RNA polymerase, but also specify the places and times that transcription can occur from that gene. This function of promoters can be vividly demonstrated in certain transgenic animals. Here, a new gene is constructed wherein the normal promoter of a particular gene is replaced with the promoter of some other gene, and the fused gene is placed into a pronucleus of a mammalian zygote. Palmiter and co-workers (1982) isolated the rat growth hormone gene and deleted its 5′ promoter region. Into this space, they substituted the promoter sequence of another gene—*MT-1*, for mouse metallothionein 1, a small protein involved in regulating serum zinc levels. This hybrid gene is shown in Figure 10.6A. The *MT-1* gene can be induced by the presence of heavy metals such as zinc or cadmium, and the sequences responsible for this induction are in the promoter of this gene. By fusing this metallothionein promoter region to the rat growth hormone gene (*rGH*), the rat growth hormone gene is placed under the control of the metallothionein promoter. In this case, rat growth hormone message should be made when the *MT-1* promoter is activated by the presence of zinc or cadmium.

A plasmid containing this fused gene was grown in bacteria, the *MT-1/rGH* piece was isolated, and about 600 copies of this fragment were injected into pronuclei of recently fertilized mouse eggs. DNA hybridization showed that many of these newborn mice had incorporated numerous copies of the rat growth hormone gene into their chromosomes. These transgenic mice were then fed a diet supplemented with zinc. The zinc induced large amounts of rat growth hormone to be made from the livers of these mice. (The liver is where metallothionein is usually made. Growth hormone is usually secreted from the pituitary gland.) The amount of growth hormone secreted correlated with the size of these mice. The transgenic mice became enormous, up to 80 percent larger than their

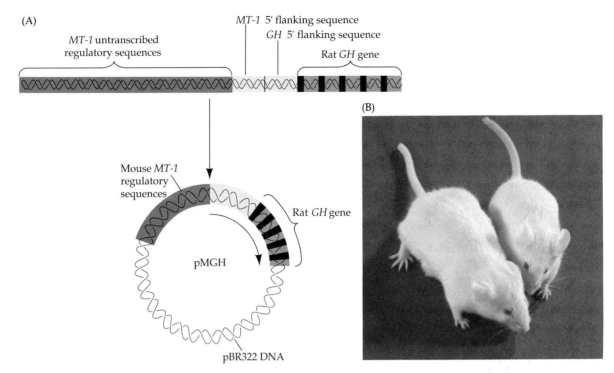

FIGURE 10.6
Promoter function seen in transgenic mice. (A) Recombinant plasmid containing rat growth hormone structural gene, mouse metallothionein regulatory region, and bacterial plasmid pBR322. The plasmid, p*MGH*, was injected into the mouse oocytes. The dark boxes on the injected plasmid correspond to the exons of the *GH* gene. The direction of transcription is indicated by an arrow. (B) A mouse derived from the eggs injected with p*MGH* (left) and a normal littermate (right). (From Palmiter et al., 1982, photograph courtesy of R. L. Brinster.)

normal littermates* (Figure 10.6B). The metallothionein promoter regulated the synthesis of growth hormone in these transgenic mice.

This strategy is presently being used by pharmaceutical companies to make large quantities of protein products such as peptide hormones, α_1-antitrypsin (used for patients with emphysema), or blood clotting factors. Pronuclei from cows, sheep, and goats have been injected with recombinant DNA containing the gene sequence for the desired protein fused with the promoters of genes for casein, lactalbumin, or β-lactoglobulin (three major milk proteins). Lactating cows synthesize enormous quantities of milk proteins, and much of this production is regulated by the transcription of new messages. The hope is that when the animal transcribes the genes for casein or lactalbumin (in response to the hormone prolactin), the genes for these therapeutic proteins will be transcribed and synthesized as well. For instance, in one case, a human gene for α_1-antitrypsin protein was fused to a β-lactoglobulin promoter and injected into sheep zygote pronuclei. One of those sheep embryos developed into a female whose milk contained 35 grams of human α_1-antitrypsin protein per liter[†] (Figure 10.7; Wright et al., 1991). Promoters, then, play a role in specifiying which gene is transcribed in which cells during development.

*Two other things grew with this experiment. The first is our potential ability to cure genetic disease by fertilizing eggs in vitro and injecting a normal gene into a pronucleus. These eggs can begin their development and then be returned to the woman's uterus. The second thing that grew was our responsibility (which usually is proportional to our power, whether we like it or not).

[†]Most secretion in transgenic animal milk is not nearly this high, probably because the genes are not linked to their appropriate enhancers, as we will see later.

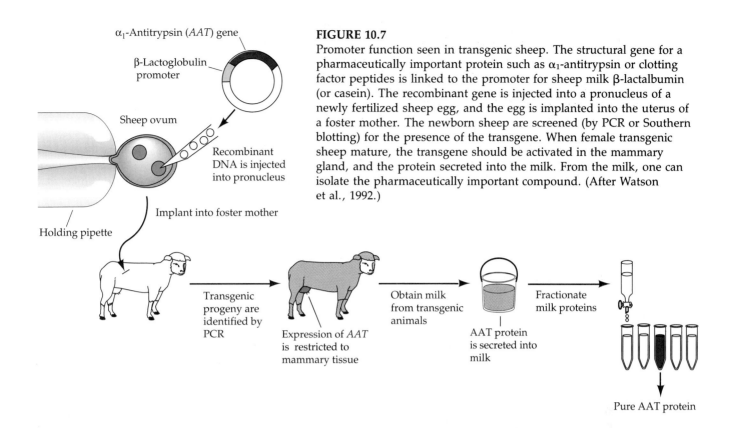

α₁-Antitrypsin (*AAT*) gene

β-Lactoglobulin promoter

Sheep ovum

Recombinant DNA is injected into pronucleus

Holding pipette

Implant into foster mother

Transgenic progeny are identified by PCR

Expression of *AAT* is restricted to mammary tissue

Obtain milk from transgenic animals

AAT protein is secreted into milk

Fractionate milk proteins

Pure AAT protein

FIGURE 10.7
Promoter function seen in transgenic sheep. The structural gene for a pharmaceutically important protein such as α₁-antitrypsin or clotting factor peptides is linked to the promoter for sheep milk β-lactalbumin (or casein). The recombinant gene is injected into a pronucleus of a newly fertilized sheep egg, and the egg is implanted into the uterus of a foster mother. The newborn sheep are screened (by PCR or Southern blotting) for the presence of the transgene. When female transgenic sheep mature, the transgene should be activated in the mammary gland, and the protein secreted into the milk. From the milk, one can isolate the pharmaceutically important compound. (After Watson et al., 1992.)

SIDELIGHTS & SPECULATIONS

Molecular regulators of development: RNA polymerase and the trans-*regulatory factors at the promoter*

Transcription requires the interaction of RNA polymerase with promoter DNA. In eukaryotic cells, there are three different types of RNA polymerases, each having particular functions and properties (Rutter et al., 1976). **RNA polymerase I** is found in the nucleolar region of the nucleus and is responsible for transcribing the large ribosomal RNAs; **RNA polymerase II** transcribes messenger RNA precursors; and **RNA polymerase III** transcribes small RNAs such as transfer RNA, 5 S ribosomal RNA, and other small DNA sequences.* None of the eukaryotic RNA polymerases can bind efficiently to DNA. Rather, there are families of DNA-binding proteins that first bind to DNA and, once bound, interact with the RNA polymerase to initiate RNA synthesis.

How the TATA element binds RNA polymerase II

The classic diagram of transcription shows that DNA, in the presence of RNA polymerase and ribonucleoside triphosphates, transcribes RNA molecules. But this simple

*In most cells, ribosomal and transfer RNAs are constitutively synthesized. However, animals have evolved remarkable mechanisms for up-regulating rRNA synthesis in their oocytes. We will therefore postpone our discussion of RNA polymerases I and III until we detail the events of oogenesis in Chapter 22.

scheme does not account for the difficulties in (1) getting the RNA to start in the correct place and (2) having transcription of a particular gene occur only at particular times and in particular types of cells. There have to be factors that enable RNA polymerase to bind solely to the promoters of particular genes. Here, we will discuss those proteins and DNA sequences that localize the RNA polymerase to the promoter sites. The enzyme responsible for the transcription of *messenger* RNAs is *RNA polymerase II*. However, accurate transcription of cloned genes in vitro will not occur if the genes are incubated with purified RNA polymerase II and nucleoside triphosphates. Nuclear extracts must be added if accurate transcription is to commence. What are these factors that allow transcription to be initiated? Recent studies (Buratowski et al., 1989; Sopta et al., 1989) have shown that at least six nuclear proteins are necessary for the proper initiation of transcription by RNA polymerase II (Figure 10.8).

TFIID and TFIIA. * In the first step of mRNA transcription, the **TFIID complex** binds to the TATA box. This was shown by DNase protection experiments in which TFIID was added to cloned genes and the DNA was then digested with DNase. The only way to "rescue" the DNA from digestion was for the TFIID to bind to the DNA, thereby preventing the DNase from reaching it. In this way Sawadogo and Roeder (1985) showed that TFIID specifically

*TF stands for *t*ranscription *f*actor; *II* indicates that the factor was first found to be needed for RNA polymerase *II*; and the letter designations refer to fractions from phosphocellulose columns that had the activity.

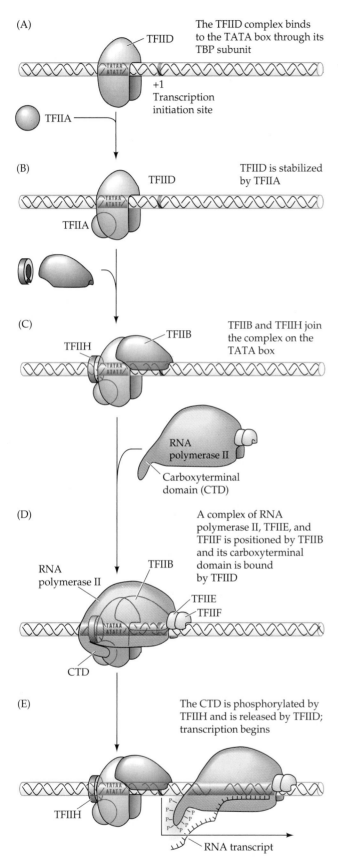

(A)

TFIID

The TFIID complex binds to the TATA box through its TBP subunit

+1
Transcription initiation site

TFIIA

(B)

TFIID

TFIID is stabilized by TFIIA

TFIIA

(C)

TFIIH

TFIIB

TFIIB and TFIIH join the complex on the TATA box

RNA polymerase II

Carboxyterminal domain (CTD)

(D)

RNA polymerase II

TFIIB

TFIIE
TFIIF

A complex of RNA polymerase II, TFIIE, and TFIIF is positioned by TFIIB and its carboxyterminal domain is bound by TFIID

CTD

(E)

The CTD is phosphorylated by TFIIH and is released by TFIID; transcription begins

TFIIH

RNA transcript

FIGURE 10.8

The formation of the active eukaryotic initiation complex. The diagrams represent the complexes formed on the TATA box by the transcription factors and RNA polymerase II. (A) The TFIID complex binds to the TATA box through its TBP subunit. (B) TFIID is stabilized by TFIIA. (C) TFIIB and TFIIH join the complex on the TATA box while TFIIE and TFIIF associate with RNA polymerase II. (D) RNA polymerase is positioned by TFIIB, and its carboxy-terminal domain (CTD) is bound by TFIID. (E) The CTD is phosphorylated by TFIIH and is released by TFIID. The RNA polymerase II is now competent to transcribe mRNA from the gene.

1991). The TFIID complex has several activities; the first is to bind the TATA sequence and to serve as the foundation for the transcriptional complex. Another role of TFIID is to prevent the stabilization of nucleosomes in the promoter region. When promoter-containing DNA is incorporated into nucleosomes, these genes are unable to be transcribed when TFIID, RNA polymerase II, and other factors are added later. However, when TFIID is added to the genes before or during nucleosome formation, the resulting chromatin is transcriptionally active (Workman and Roeder, 1987). In blocking nucleosome production, TFIID appears to act antagonistically to histone H1. Histone H1 (as we will see in the subsequent chapter) stabilizes nucleosomes and prevents transcription in the region where it binds. The addition of histone H1 prevents TFIID from finding the TATA sites and prevents transcription; however, this inhibition is overcome if TFIID is added first (Laybourn and Kadonga, 1991). The binding of TFIID is facilitated and stabilized by the transcription factor TFIIA (Buratowski et al., 1989; Maldonado et al., 1990). Thus, the decision to transcribe or not to transcribe a particular gene often depends on the balance between inhibitory factors (such as histones) and TFIID and TFIIA.

TFIIB and RNA polymerase II. The TFIID/TFIIA complex cannot form a stable complex directly with RNA polymerase II. Rather, TFIID binds factor **TFIIB**. The binding of TFIIB to TFIID appears to be the key rate-limiting step in the transcription of numerous genes. This rate can be dramatically increased by the proximity of certain promoter- and enhancer-binding transcription factors. These transcription factors are sequence-specific and can regulate which genes will be transcribed. The *activator domain* of these transcription factors binds directly to TFIIB and facilitates its assembly with TFIID (Lin and Greene, 1991; Lin et al., 1991). Once TFIIB is in place, it can bind RNA polymerase II. Most of the RNA polymerase is positioned by its interaction with TFIIB, but the carboxy-terminal tail of the large subunit of RNA polymerase II interacts directly with TFIID (Figure 10.8D). In this way, RNA polymerase II is placed at the promoter.

TFIIE/F and TFIIH. Either directly before or during its binding to TFIIB, RNA polymerase II becomes associated with TFIIF and TFIIE (Buratowski et al., 1991; Conaway et al., 1991). **TFIIF** has an enzymatic activity needed to unwind the DNA helix. **TFIIE** is a DNA-dependent ATPase and is probably needed for generating the energy for transcription (Bunick et al., 1982; Sawadogo and Roeder, 1984). But what good is all this if the RNA polymerase remains

bound to the TATA region of the genes. TFIID is a multimeric protein, one of whose components—the **TATA-binding protein (TBP)**—binds directly into the minor groove of the TATA sequence (Lee et al., 1991; Starr and Hawley,

bound to this complex on the TATA box? For transcription to occur, the RNA polymerase must be released from the promoter region. This releasing activity appears to be the function of **TFIIH**. The RNA polymerase is tightly bound by its carboxy-terminal domain (CTD) to TFIID. However, TFIID will only bind the *unphosphorylated* form of the CTD. In mammals, the CTD contains 52 repeats of the seven-amino acid sequence YSPTSPS. When the initiation complex is formed, the completed complex activates the protein kinase activity of TFIIH. TFIIH then phosphorylates each of the 52 repeats (Figure 10.8E; Koleske et al., 1992; Lu et al., 1992; Usheva et al., 1992). TFIID cannot bind this heavily phosphorylated region and releases the RNA polymerase. Transcription of messenger RNA by RNA polymerase II can begin.

TAFs and the activation of basal transcription

TFIID is a multimeric protein, only one of its subunits actually binding to the TATA sequence. Some of the other subunits are called **TATA-binding protein-associated factors** (TAFs). TAFs are needed to mediate enhancer-regulated transcription (Dynlacht et al., 1991). The TAFs may function as bridges linking the transcriptional assembly complex to the enhancer regions that stimulate transcription.

For instance, one of the most common transcription factors is Sp1. This non-TAF protein binds to the promoter or enhancer GGGCGG sequences through its carboxyl end but regulates transcriptional activity via its amino terminus

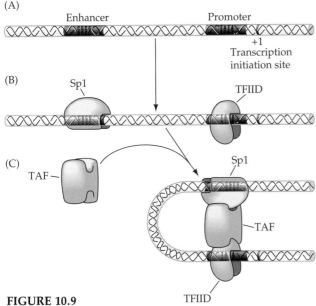

FIGURE 10.9
Possible mechanism whereby Sp1 (in either the enhancer region or an upstream promoter region) comes into contact with TFIID through a TAF that binds to both of them. (A) The enhancer and promoter regions of a gene. (B) The enhancer region binds Sp1, the promoter region binds TFIID. (C) The TAF binds to TFIID and enables Sp1 (and other proteins at different sites in the chromatin) to interact with the transcriptional complex.

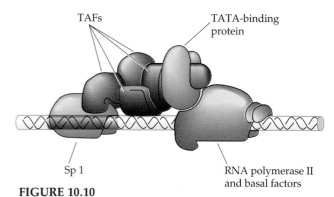

FIGURE 10.10
Possible configuration for transcription factors mediating RNA polymerase II binding to a TATA-less promoter containing an Sp1 binding site. (After Pugh and Tjian, 1991; Comai et al., 1992.)

(Dynan and Tjian, 1985; Kadonaga et al., 1988). It is probable that this factor is found in all cells, and so cannot regulate differential gene expression. However, it appears to be involved with the interactions between the promoter region and the enhancer region in ways that do result in the differential transcription of particular genes in particular cells. Hoey and co-workers (1993) found that a *Drosophila* TAF binds specifically to Sp1 and to TFIID and can form a bridge between them. Moreover, they demonstrated that the strength of the interactions between these proteins correlates with the activation of transcription. The TAF thus appears to bring the Sp1 protein (with its transcriptional activating regions) into the proximity of TFIID. Here, the Sp1 protein may enable TFIID to strengthen its association with the TATA box or to more readily associate TFIIB. In this model, TAFs would enable the looping of DNA so that Sp1 elements in the enhancer would meet the TFIID protein at the promoter (Figure 10.9).

Another TAF has been identified as topoisomerase I. This protein binds to the TATA-binding protein and prevents basal transcription from promoters bound by this protein. However, when TBP is activated by other transcription factors, topoisomerase I stimulates the activated transcription (Merino et al., 1993).

Promoters lacking TATA elements

There are many genes (mostly those encoding general metabolic proteins and not cell-specific proteins) that use RNA polymerase II, but whose promoters lack the TATA sequence. In these cases, some other protein binds to the promoter region. These are usually general promoter-binding proteins such as Sp1. The Sp1 protein on the GC-rich promoter element then binds TFIID either directly or through a TAF. The TFIID is now able to begin the cascade of factors that will form the transcription initiation complex and bind an RNA polymerase II protein to the promoter region (Figure 10.10; Pugh and Tjian, 1991; Rigby, 1993). So even though these promoters lack a TATA box sequence, TFIID is still the deciding factor in regulating whether transcription occurs.

Enhancer structure and function

In addition to the promoters, enhancers are also important in regulating the transcription of nearby genes. One of the first cellular enhancers found was seen to control the cell specificity of immunoglobulin gene transcription. B cells are the only cells in the body that produce immunoglobulin (antibody) protein. Gillies and his co-workers (1983) transfected a cloned immunoglobulin heavy chain gene into cultured B lymphocyte tumor cells that had lost the ability to make their own heavy chain. These transfected cells were then able to synthesize the heavy chain encoded by the incorporated gene. However, if these researchers added the same gene—but lacking a small region of a particular intron—into these defective B cells, they observed very little transcription of the inserted gene. There was an enhancer region within the intron that was necessary for transcription (Figure 10.11).

Enhancers are often the primary elements responsible for tissue-specific transcription: the cloned immunoglobulin genes are *not* transcribed when inserted into the nuclei of cells other than B cells (Banerji et al., 1983; Gillies et al., 1983). Moreover, when the enhancer region of the immunoglobulin heavy chain is inserted into a cloned gene for β-globin, it stimulates the transcription of that hemoglobin gene only when the gene is inserted into a B cell.

FIGURE 10.11
Tissue specificity of enhancer element effect. The immunoglobulin heavy chain gene was isolated and cloned from an IgG-producing myeloma cell line. Some of these clones were kept intact, while in others various regions of the intron between the VDJ and the C^γ exons (which will be discussed later) were excised by restriction enzymes. DNA from the resulting clones was transfected into myeloma cells that had lost their own immunoglobulin genes. The accumulated mRNA from the transfected cells was isolated and separated by electrophoresis on a polyacrylamide gel, along with mRNA from a normal mouse fibroblast and from the original myeloma cell. The RNA was blotted onto nitrocellulose paper and hybridized with a radioactive restriction enzyme fragment of the C_γ region. If the C_γ region was being transcribed from these cloned genes, the radioactive probe should bind to it. The probe detected C_γ message only from those clones that contained a certain region of the intron (indicated by the colored bar within the intron). When transfected into mouse fibroblast cells, however, even the entire cloned gene would not transcribe C_γ mRNA. (Protocol and data of Gillies et al., 1983.)

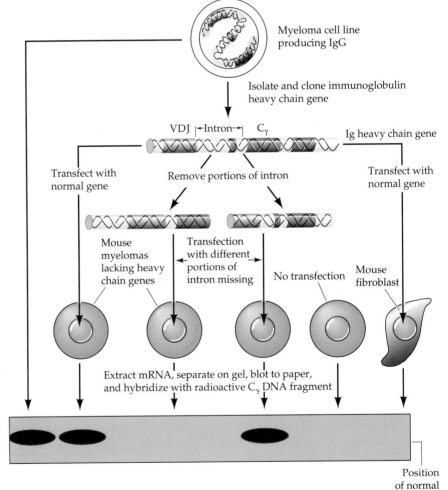

Enhancers regulate temporal and spatial patterns of transcription

Enhancers can regulate the temporal and tissue-specific expression of all differentially regulated genes, and genes active in adjacent cell types are seen to have different enhancers. In the pancreas, for instance, the exocrine protein genes (chymotrypsin, amylase, and trypsin) have enhancers different from that of the endocrine protein insulin. Both these enhancers lie in the 5′ flanking sequences of their respective genes. Walker and colleagues (1983) placed these flanking regions onto the gene for bacterial chloramphenicol acetyltransferase (CAT), a gene whose enzyme product is not found in mammalian cells. CAT activity is easy to assay in mammalian cells and is used as a "reporter gene" to tell investigators whether a particular enhancer is functioning. They then transfected these hybrid genes into (1) ovary cells (which do not secrete insulin or chymotrypsin), (2) an insulin-secreting cell line, and (3) an exocrine cell line, and measured the activity of the marker enzyme in each of these cells. As shown in Figure 10.12, neither enhancer sequence caused the enzyme to be made in the ovarian cells. In the insulin-secreting cell, however, the 5′ flanking region of the insulin gene enabled the chloramphenicol acetyltransferase gene to be expressed, but the 5′ flanking region of the chymotrypsin gene did not. Conversely, when the clones were placed into the exocrine pancreatic cell line, the chymotrypsin 5′ flanking sequence allowed CAT ex-

FIGURE 10.12
Tissue specificity of pancreatic gene enhancers. The 5′ flanking regions of the insulin gene (I) and the chymotrypsin gene (C) were separately inserted next to the gene for bacterial CAT. As a positive control, the enhancer from Rous sarcoma virus (V), which appears to operate in all cell types, was also placed by the CAT gene. The three clones were transfected into three types of cells, and the CAT activity assayed on the cell lysates. The inserts show typical autoradiograms of the CAT assay wherein radioactive chloramphenicol (the substrate of the CAT reaction) can be separated from chloramphenicol monoacetate (the product of the CAT reaction) by chromatography. (After Walker et al., 1983.)

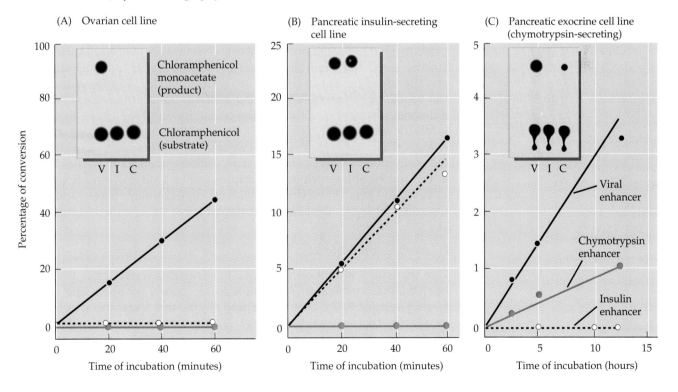

pression, while the insulin enhancer did not. The enhancers for ten exocrine proteins share a 20-base pair consensus sequence, suggesting that these similar sequences play a role in activating these genes in the exocrine cells of the pancreas (Boulet et al., 1986). Thus, the expression of genes in exocrine and endocrine cells of the pancreas appears to be controlled by different enhancers.

Enhancers are critical in the regulation of normal development, and over the past decade, four generalizations that emphasize their importance for differential gene expression have been made:

1. Most genes require enhancers for their transcription.
2. Enhancers are the major determinant of differential transcription in space (cell type) and time.
3. Having the enhancer at a relatively large distance from the promoter means that there can be multiple signals to determine whether a given gene is transcribed. A given gene can have several enhancer sites linked to it, and each enhancer can be bound by more than one factor (which may regulate whether it stimulates or inhibits transcription).
4. The interaction between the proteins bound to the enhancer sites with the transcription apparatus assembled at the promoter is thought to regulate transcription. The mechanism of this association is not fully known, nor do we comprehend how the promoter integrates all these signals.

Transcription factors: The *trans*-regulators of promoters and enhancers

Transcription factors are proteins that bind to the enhancer or promoter regions and interact such that transcription occurs from only a small group of promoters in any cell. Most transcription factors can bind to specific DNA sequences, and these *trans*-regulatory proteins can be grouped together in families based on similarities in structure. Within such a family, proteins share a common framework structure in their respective DNA binding sites, and slight differences in the amino acids at the binding site can alter the sequence of the DNA to which it binds. In addition to having this sequence-specific **DNA-binding domain**, transcription factors contain a domain involved in activating the transcription of the gene whose promoter or enhancer it has bound. Usually, this *trans*-**activating domain** enables that transcription factor to interact with proteins involved in binding RNA polymerase. This interaction often enhances the efficiency by which the basal transcriptional complex can be built and bind RNA polymerase II. There are several families of transcription factors, and those discussed here are just some of the main types.

Homeodomain proteins

One extremely important family of *trans*-regulatory factors is the set of **homeodomain proteins**. These proteins are critical for specifying the anterior–posterior body axes throughout the animal kingdom; they will be detailed further in Chapters 15 and 17. The homeodomain consists of 60 amino acids arranged in a helix–turn–helix, such that the third helix extends into the major groove of the DNA that it recognizes. Amino acids in the amino-terminal portion of the homeodomain also contact the bases

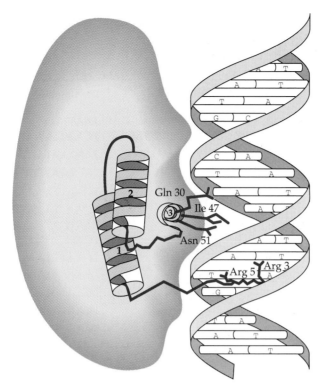

FIGURE 10.13
The homeodomain of the engrailed protein binds to a particular site in the DNA. Helix 3 contacts the base pairs in the major groove, while the amino-terminal portion of the homeodomain enters the minor groove (After Pabo and Sauer, 1992.)

in the minor groove (Figure 10.13). This homeodomain was first seen in proteins that specify segment identity in *Drosophila*. Mutations of these proteins caused one body segment to be transformed into another (a transformation known as homeosis, which will be discussed in detail in Chapter 15). Several of the homeodomain proteins of *Drosophila melanogaster* have been cloned, sequenced, and tested for their ability to regulate transcription. Table 10.1 shows nine homeodomain-containing *Drosophila* proteins and the DNA sequences they recognize. The recognition of specific promoters by homeodomain-containing proteins has been shown to be essential for *Drosophila* development. The bicoid protein, for instance, is a homeodomain transcription factor that binds to the promoters on the *hunchback* gene. This binding activates transcription of the *hunchback* gene; the resulting hunchback protein is also a transcription factor, and it binds to the enhancers of those genes necessary for *Drosophila* head and thorax formation (Driever and Nüsslein-Volhard, 1989; Struhl et al., 1989). Small changes in the amino acid composition of the DNA-binding site can change the DNA sequence recognized by the protein. Treisman and her colleagues (1989) found that by altering a single amino acid in the homeodomain, one could change the promoters that this protein would activate.

The POU transcription factors

Some transcription factors have both a homeodomain and a *second* DNA-binding region (Figure 10.14). In some instances, this region that comprises the homeodomain and the second DNA-binding region is called the **POU domain** (Herr et al., 1988). The initials are taken from the four proteins first seen to have such domains: **Pit-1** (also called GHF-1). a pituitary-

TABLE 10.1
Major homeodomain proteins of *Drosophila melanogaster* and their DNA-binding sites

Protein	DNA-binding site(s)
Abdominal B	TAATTTGCAT
	TCAATTAAAT
Antennapedia	TAATAATAATAATAA
bicoid	TCCTAATCCC
engrailed	TCAATTAAAT
even-skipped	TCAATTAAAT
	TAATAATAATAATAA
	TCAGCACCG
fushi tarazu	TCAATTAAAT
	TAATAATAATAATAA
paired	TCAATTAAAT
Ultrabithorax	TAATAATAATAATAA
zerknült	TCAATTAAAT

specific factor that activates the genes encoding growth hormone, prolactin, and other pituitary proteins; Oct-1, a ubiquitous protein that recognizes a certain eight-base-pair sequence called the octa box, and Oct-2, the B-cell-specific protein that recognizes the octa box and activates immunoglobulin genes; and unc-86, a nematode gene product involved in determining neuronal cell fates. The homeodomain of Pit-1 recognizes the sequence ATATTCAT, while the homeodomain of Oct-2 recognizes the similar sequence ATTTGCAT. If the DNA element recognized by Pit-1 is altered in two places, it becomes an Oct-2 binding site, and the prolactin gene becomes expressed in B lymphocytes (Elsholtz et al., 1990). Thus, a two-base change in the enhancer recognized by a POU protein can convert pituitary-specific transcription into lymphocyte-specific transcription.

The **Pit-1** protein is found only in pituitary cells. When growth hormone genes (which are active in the pituitary) are cloned and placed in nuclear extracts of nonpituitary cells, these genes are not transcribed, whereas transcription will take place if the growth hormone genes are placed into nuclear extracts of anterior pituitary cells. Moreover, the addition of Pit-1 to the nonpituitary nuclear extract will enable it to transcribe

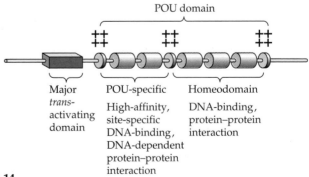

FIGURE 10.14
The domains of POU family transcription factors.

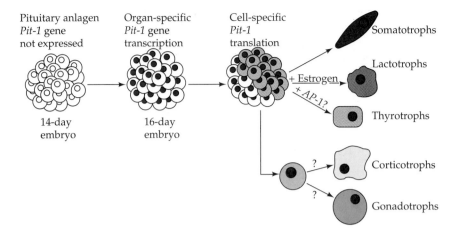

Pituitary anlagen
Pit-1 gene
not expressed

Organ-specific
Pit-1 gene
transcription

Cell-specific
Pit-1
translation

Somatotrophs

Lactotrophs

+ Estrogen
+ AP-1?

Thyrotrophs

14-day
embryo

16-day
embryo

?

Corticotrophs

?

Gonadotrophs

FIGURE 10.15
Determination of cell type by combinations of *trans*-regulatory proteins. The mRNA for the Pit-1 transcription factor is transcribed in all cell types destined to reside in the anterior pituitary. However, Pit-1 protein is only translated in thyrotroph, somatotroph, and lactotroph cells. Somatotrophs synthesize growth hormone upon the translation of *Pit-1*. Lactotrophs synthesize some prolactin concomitant with *Pit-1* translation but produce significant prolactin only when costimulated with estrogen (and its *trans*-regulatory receptor protein). The mechanism distinguishing thryroid-stimulating hormone transcription is thought to involve the silencer region (Figure 10.16). (After Simmons et al., 1990.)

the growth hormone gene (Bodner and Karin, 1987). It appears, then, that the specific expression of the growth hormone gene in the anterior pituitary is mediated by a tissue-specific *trans*-regulating protein. Conversely, when other genes (such as globin) are cloned next to the Pit-1-binding sequences and are used to construct transgenic mice, these genes show pituitary-specific transcription (Behringer et al., 1988).

The Pit-1 enhancer-binding protein is itself developmentally regulated (Simmons et al., 1990; Dollé et al., 1990). Transcription of the mouse *Pit-1* gene is detectable within a day following the histological appearance of Rathke's pouch, the anlage of the anterior pituitary, but translation of this mRNA into protein does not occur for another 2–3 days. Growth hormone mRNA is first detected when the *Pit-1* gene is translated. Interestingly, in two types of anterior pituitary cells, the corticotrophs that synthesize corticotropin (also known as adrenocorticotropic hormone, or ACTH), and the gonadotrophs that synthesize gonadotropins, *Pit-1* mRNA is made but remains untranslated (Figure 10.15). Only in the pituitary cells of somatotrophs (growth hormone-producing), lactotrophs (prolactin-producing), and thyrotrophs (which synthesize thyroid-stimulating hormone) is the *Pit-1* message translated into the DNA-binding nuclear protein. The Pit-1 enhancer-binding protein not only mediates the transcription of the differentiated products of these cells, but it is necessary for the formation of the anterior pituitary cells in Rathke's pouch. Two *dwarf* mutations in mice are caused by a mutation in Pit-1 protein. These mice lack the thyrotroph, lactotroph, and somatotroph cells of the anterior pituitary (Li et al., 1990).

Combinatorial interactions and looping in regulating *prolactin* gene transcription. The activity of the Pit-1 protein on the prolactin gene illustrates many of the characteristics of transcription factors. First, the enhancer of the *prolactin* gene binds several different factors whose interactions regulate transcription. The *prolactin* gene is activated during pregnancy to make the pituitary hormone (prolactin) that stimulates milk production; this gene is maximally stimulated by the *combination* of Pit-1 and estrogen. This combination regulates the place (the pituitary gland) and the time (pregnancy and shortly thereafter) of prolactin synthesis. Simmons and colleagues (1990) showed that this synergism occurs in the enhancer region of the gene. The Pit-1 protein binds to one region of the enhancer, while the estrogen binds through its receptor protein to another region of the enhancer. When these factors are present together, transcription is much greater than when either is added separately (Figure 10.16A). Moreover, there appear to be silencer regions flanking the enhancer that are needed for turning off the prolactin gene in thyrotrophs (which could otherwise

FIGURE 10.16
Combinatorial synergism in the prolactin enhancer. (A) Synergism between sites of the rat prolactin enhancer. The prolactin enhancer was fused to a reporter (*luciferinase*) gene, and the level of reporter gene activity assayed when the fused gene was added to cultured cells. When either Pit-1 or estrogen was present in these cells, there was only minor increase in the level of transcription. However, the addition of *both* substances caused a 1400-fold increase in the level of transcription. (B) Synergism between enhancer and promoter sites of the mouse prolactin gene. Fused genes were made carrying the reporter gene and (1) the entire 5′ prolactin enhancer region, flanking sequences, and prolactin promoter, (2) just the tissue-specific enhancer and the prolactin promoter, (3) just the prolactin promoter, (4) the prolactin enhancer plus the promoter of a *thymidine kinase* (*TK*) gene, and (5) just the *TK* gene promoter. These constructs were placed into mouse zygotes, and the reporter gene expression was monitored in lactotrophic and thyrotrophic cells of the pituitary. (C) Model for the regulation of prolactin gene expression. The promoter and enhancer both have four Pit-1 binding sites. The estrogen receptor binds to the estrogen-reponsive element (ERE) of the enhancer region. The regions that inhibit prolactin transcription in thryotrophs and somatotrophs are in black. (A after Simmons et al., 1990; B after Crenshaw et al., 1989.)

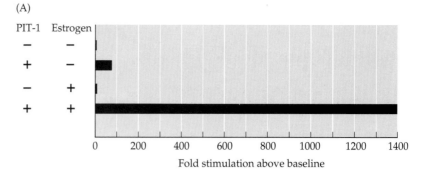

(A)

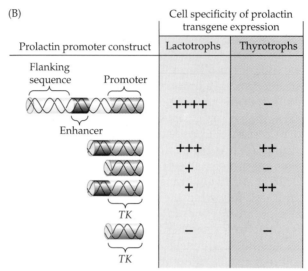

(B)

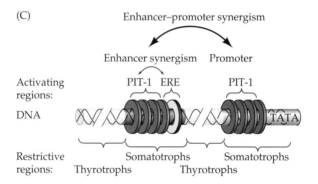

(C)

activate the prolactin gene) (Figure 10.16B,C; Crenshaw et al, 1989). Thus, Pit-1 acts in a combinatorial way with other transcription factors to regulate its target genes.

Second, there is synergism between the enhancer and the promoter of the prolactin gene. The *prolactin* gene enhancer will not stimulate the promoter of another gene nearly as efficiently as it will stimulate its own promoter (Figure 10.16B,C; Crenshaw et al., 1989). This synergism between promoter and enhancer appears to be caused by DNA looping between the enhancer and the promoter sites. In the rat prolactin gene, the enhancer is located over 1300 base pairs upstream from its promoter. Using an assay that would fuse DNA brought together by protein–protein interactions, Cullen and colleagues (1993) showed that the promoter and enhancer regions are brought together only when Pit-1 and estrogen are both present. It appears that the hormone-bound estrogen receptor on the enhancer is able to stabilize the interaction between the enhancer and the promoter regions, thus allowing the interaction between the enhancer-bound proteins (Pit-1 and the estrogen receptor) with the transcription apparatus of the promoter.

Third, the Pit-1 protein positively regulates its own synthesis. One of the targets for the Pit-1 protein is the enhancer of the *Pit-1* gene itself (Rhodes et al., 1993). Once the *Pit-1* gene has been activated (by other transcription factors), Pit-1 protein binds to its own enhancer and maintains *Pit-1* gene transcription. This type of positive autoregulation is important as a mechanism for committing a cell to a given pathway of development. Thus, the *Pit-1* gene, once active, maintains the pituitary phenotype. Such autoregulation is also seen for the MyoD protein (which commits cells to the path of muscle cell development) and for various *Drosophila* proteins that maintain the individual's sex and segment-specific boundaries.

SIDELIGHTS & SPECULATIONS

Regulation of transcription from immunoglobulin light chain genes

The ability to obtain clonal lines of cells "frozen" at a particular stage in their development enables us to sample the transcription factors present during that particular time. Such clones are obtainable from tumors of certain tissues, and leukemias of the B lymphocytes (B cells) of the immune system have enabled researchers to identify many of the the *cis-* and *trans*-regulatory elements necessary for the development of the B-cell lineage.

The structure of the immunoglobulin genes

So far, our model has been that every cell in the body contains the same exact genes. This is probably true for most cell types, but the lymphocytes are different. In each B cell, the genome has been rearranged so that each B cell is able to make only one type of antibody (out of a possible repertoire of over 10 million antibodies).

Antibodies are produced when a foreign substance— the antigen—comes into contact with B cells, which reside in the lymph nodes and spleen. Even before contact with an antigen, each of the resting B cells makes immunoglobulin proteins but does not secrete them. Rather, the immunoglobulin molecules are inserted into the B-cell membranes and are used as antigen receptors. These receptors can signal the cell to divide and further differentiate when they bind antigens. Each B cell makes an antibody that recognizes one and only one antigenic shape. Therefore, an antibody recognizing the protein shell of poliovirus would not be expected to recognize cholera toxin, *E. coli* membranes, or zebra dander. All antibody proteins on the B-cell membrane have a very similar structure. Each consists of two pairs of polypeptide subunits. There are two identical heavy chains and two identical light chains; the chains are linked together by disulfide bonds (Figure

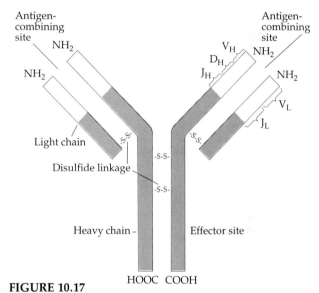

FIGURE 10.17
Structure of a typical immunoglobulin (antibody) protein. Two identical heavy chains are connected by disulfide linkages. The antigen-combining site is composed of the variable regions (white) of the heavy and light chains, whereas the effector site of the antibody (which controls whether it agglutinates antigens, binds to macrophages, or enters into mucous secretions) is determined by the amino acid sequence of the heavy chain constant (color) region.

10.17). The *specificity* of the antibody molecule (i.e., whether it will bind to a poliovirus, an *E. coli* cell, or some other molecule) is determined by the amino acid sequence of the *variable region*. This region is made up of the amino-terminal ends of the heavy and light chains. The variable regions of the antibody molecules are attached to *constant regions* that give the antibody its effector properties. For

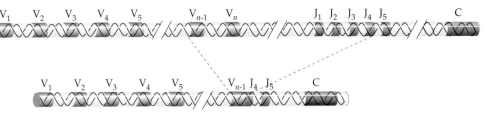

Original gene organization

V_1 V_2 V_3 V_4 V_5 V_{n-1} V_n J_1 J_2 J_3 J_4 J_5 C

V_1 V_2 V_3 V_4 V_5 V_{n-1} J_4 J_5 C

FIGURE 10.18

Rearrangement of the light chain genes during B-lymphocyte development. While the developing B cell is still maturing in the bone marrow, one of the 300 or more V gene segments combines with one of the five J gene segments and moves closer to the constant (C) gene segment.

example, the heavy chain constant region of cell-surface immunoglobulin molecules anchors the protein into the cell membrane. For decades, immunologists puzzled over how the immune system could possibly generate so many different types of antibodies. Could all the 10^7 different types of antibody proteins be encoded in the genome? This would take up an enormous amount of chromosomal space. Moreover, how could the immune system "know" how to make an antibody to some molecule that isn't even found outside the laboratory? Surprisingly, it was discovered that the production of specific antibody molecules involves the creation of *new genes* during B cell differentiation.

Creation of antibody light chain genes.

Antibody heavy and light chain genes are organized in segments. Mammalian light chain genes contain three segments (Figure 10.18). The first gene segment, V, encodes the first 97 amino acids of the light chain variable region. There are about 300 different V sequences linked tandemly on the mouse genome. The second segment, J, consists of four or five possible DNA sequences for the last 15–17 residues of the variable region of the antibody light chain. The third segment is the constant (C) region of the light chain. During B-cell differentiation, one of the 300 V segments and one of the five J segments combine to form the

FIGURE 10.19

Protocol and results of the Hozumi and Tonegawa experiment. DNAs from mouse embryo cells and B-cell tumors (myelomas) were separately digested in BamHI, separated by electrophoresis, and eluted from the gels. After being denatured, each eluted DNA sample was hybridized to either radioactive mRNA coding for the V and C regions of the immunoglobulin light chain (whole) or a fragmented radioactive mRNA encoding only the C region of that light chain protein (3' half). For the embryonic DNA, the V and C regions of the light chain protein were found on two different pieces of DNA (the V region on a piece having a molecular weight of 3.9×10^6, the C region on a DNA fragment of MW 6×10^6). In the lymphocyte tumor, the V and C regions were found together on a single DNA fragment of MW 2.4×10^6.

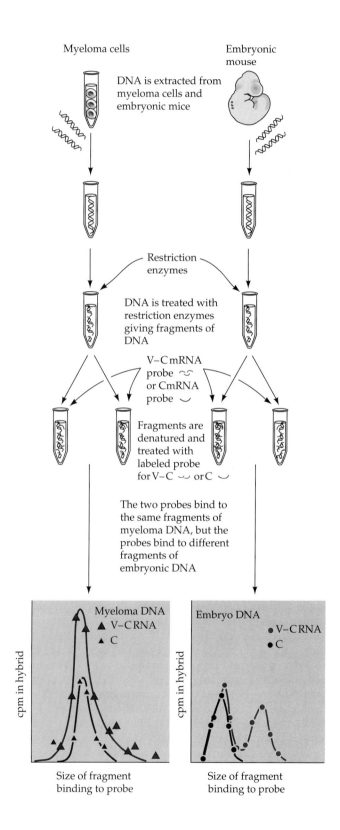

Myeloma cells   Embryonic mouse

DNA is extracted from myeloma cells and embryonic mice

Restriction enzymes

DNA is treated with restriction enzymes giving fragments of DNA

V–C mRNA probe ∿ or C mRNA probe ⌣

Fragments are denatured and treated with labeled probe for V–C ∿ or C ⌣

The two probes bind to the same fragments of myeloma DNA, but the probes bind to different fragments of embryonic DNA

Myeloma DNA
▲ V–C RNA
▲ C

Embryo DNA
● V–C RNA
● C

cpm in hybrid

Size of fragment binding to probe

Size of fragment binding to probe

variable region of the antibody gene. This is done by moving a V-segment sequence to a J-segment sequence, a rearrangement that eliminates the intervening DNA.

This gene rearrangement was first shown by Hozumi and Tonegawa (1976) who isolated DNA from a mouse embryo and from a light chain-secreting B-cell tumor.* They digested these two DNAs separately with the restriction enzyme *Bam*HI (see Chapter 2), which cleaves DNA wherever it locates the sequence GGATCC. The result was a series of DNA fragments, the size of each fragment determined by the length of the DNA molecule between two cleavage sites. These DNA fragments were placed into a well at one end of a gelatinous slab and electrophoresed through the gel. As the DNA migrated toward the positive electrode, the smaller fragments moved faster than the larger ones, effectively separating the fragments by size. The gel, with the DNA spread throughout it, was cut into several pieces, each slice containing pieces of DNA of a certain size. The DNA from each slice of the gel was eluted and denatured into single strands. Part of this DNA was hybridized with radioactive RNA that coded for the entire light chain and had been isolated from the original B-cell tumor. The other part was hybridized with radioactive RNA coding only for the C region of the light chain (the 3′ half of the RNA message). The DNA from the embryo bound the light chain mRNA in two slices. The DNA of the first slice had a molecular weight (MW) of about 6 million; the DNA of the second band was of MW 3.9 million. When the mouse embryo DNA was hybridized with the light chain C-region mRNA, only the 6 million-MW DNA bound the RNA. Thus, in the mouse *embryo*, the C region was coded within DNA fragments having a molecular weight of 6 million (between *Bam*HI sites), while the V region was coded within a region of MW 3.9 million (Figure 10.19).

The lymphocyte tumor DNA, however, gave a very different result. The only lymphocyte DNA that bound the light chain mRNA had a molecular weight of 2.4 million. Moreover, it bound the light chain C-region mRNA segment as well. Both the C and the V regions were found to be coded on the same fragment of DNA! The simplest explanation (and one confirmed by numerous other laboratories and methods; see Bernard et al., 1978; Brack et al., 1978) was that two gene fragments, one encoding the light chain C region and one encoding a specific light chain V region, had fused together *to form a new gene* during the development of the lymphocyte. Their model for such gene synthesis is presented in Figure 10.20.

Creation of antibody heavy chain genes

The heavy chain genes of antibodies contain even more segments than the light chain genes. Heavy chain gene segments include a V segment (200 different sequences for the first 97 amino acids), a D segment (10–15 different sequences encoding 3–14 amino acids), and a J segment (4 sequences for the last 15–17 amino acids of the V region). The next segment codes for the C region. The heavy chain variable region is formed by adjoining one V segment and one D segment to one J segment (Figure 10.21A,B). This VDJ variable region sequence is now adjacent to the first constant region of the heavy chain genes—the C_μ region specifically for antibodies that can be inserted into the plasma membrane. Thus, an immunoglobulin molecule is formed from two genes created during the antigen-independent stage of B lymphocyte development. About 10^3 light chain genes and about 10^4 heavy chain genes can be formed. Since each is formed independently of the other, about 10^7 types of antibody can be created from the union of the light chain and the heavy chain within a cell. Each cell makes only one of these 10^7 antibody types and places the antibodies in the cell membrane to be used as antigen receptors. The genome of each lymphocyte clone can be markedly different from that of any other cell.

Later in adult development, some of the B cells undergo another rearrangement called **class switching**. Here, the entire VDJ sequence of the heavy chain gene is transferred from the C_μ region to another C region. The new segment will have slightly different properties. In Figure 10.21C and D, the VDJ sequence is transferred to a C_α region. The C_α sequence encodes the constant region that enables the antibodies to be secreted into the mucosa and digestive system, where they protect the body against airborne and food-borne antigens.

*B cell tumors (myelomas) were used because they produce an enormous amount of one specific immunoglobulin (and of the mRNA for that immunoglobulin).

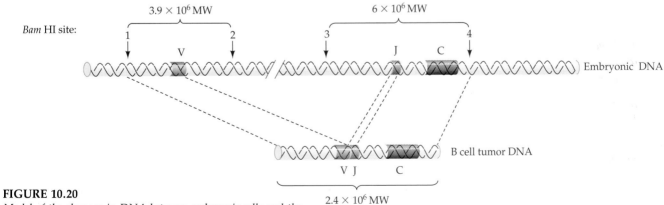

FIGURE 10.20
Model of the changes in DNA between embryonic cells and the B lymphocyte, according to the data of Hozumi and Tonegawa (1976).

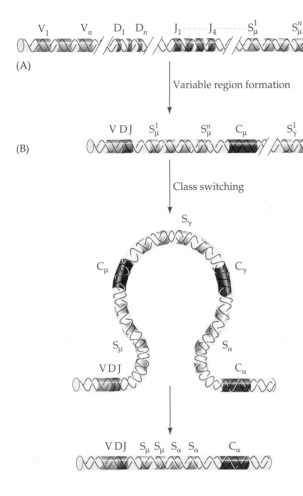

(A)

V_1 V_n D_1 D_n J_1 —— J_4 - - - S_μ^1 S_μ^n C_μ S_γ^1 - - - S_γ^n C_γ S_α^1 - - - S_α^n C_α

Variable region formation

(B)

V D J S_μ^1 S_μ^n C_μ S_γ^1 S_γ^n C_γ S_α^1 S_α^n C_α

Class switching

S_γ

C_μ C_γ

S_μ S_α

V D J C_α

V D J S_μ S_μ S_α S_α C_α

FIGURE 10.21

Variable region gene formation and class switching in the production of immunoglobulin heavy chains. (A) A heavy chain gene contains three segments (V, D, and J) that come together to form the variable (V) region, and the constant (C) region. The four major classes of antibodies are classified on the basis of the constant region (IgA contains C_α; IgM, C_μ; IgG, C_γ). (B) Before antigen presentation, the variable region forms from the union of a V, a D, and a J segment. This VDJ gene segment is adjacent to the C_μ region, and the resulting antibody is placed in the cell membrane. After antigen presentation, a region of DNA can be looped out in such a way that the VDJ segment becomes adjacent to another C region (in this case, the C_α region, which enables the antibodies to enter mucous secretions). This class switching is mediated by a series of switch (S) sequences adjacent to each of the constant regions. (After Davis et al., 1980a,b.)

B cells are not the only cell type that alters its genome during differentiation. The other major cell of the immune system, the T lymphocyte, also deletes a portion of its genome in the construction of its antigen receptor (Fujimoto and Yamagishi, 1987). The enzymes responsible for mediating the DNA recombination events appear to be the same in the B-and T-cell lineages. Called **recombinases** (Schatz et al., 1989; Oettinger et al., 1990), these two proteins recognize the signal regions of DNA immediately upstream from the recombinable DNA. Moreover, the genes for these enzymes are active only in the pre-B cells and pre-T cells wherein the genes are being recombined. These recombinase genes are not active in mature B and T cells, nor are they seen in most other cell types.*

Cis-regulatory regions of the immunoglobulin genes

The discovery of V(D)J joining and class switching solved most of the problems concerning the origin of antibody

---

*Until recently, recombinase proteins were thought to be found solely in lymphocytes, but recent evidence (Chun et al., 1991; Matsuoka et al., 1991) has shown that recombination events and the recombinases exist in brain tissue as well. It is not known what their function might be in the neural cells, but it is fascinating to speculate that some of the receptors that bind a nerve cell axon to its specific target might be made by recombination of several gene regions.

diversity. However, it created some other problems. Each V segment has its own promoter attached to it. Why isn't every V segment gene expressed? Deletion mapping of cloned genes has shown that the promoter of the immunoglobulin light chain contains several critical regions for transcription. One of these upstream promoter elements is called an **octa box** (Bergman et al., 1984; Parslow et al., 1984) because of its eight base pairs: ATTTGCAT. This octa sequence has been found in every light chain immunoglobulin promoter studied. The heavy chain immunoglobulin gene promoters have an *inverted* octa box: ATGCAAAT. When the octa box is placed upstream from a globin gene in a cultured B lymphocytes, the globin gene transcription increases 11 to 18 times. This increase is seen only in lymphoid cells and has not been observed in fibroblasts (Wirth et al., 1987).

The *enhancer* sequence of the light chain immunoglobulin genes is located within the first intron between the VJ sequence and the C region (Queen and Baltimore, 1983; Bergman et al., 1984). When this sequence is translocated onto cloned globin genes, globin gene transcription can also occur specifically in lymphocytes (Picard and Schaffner, 1984). For transcription to occur in the lymphocyte, the promoter must be brought into proximity of the enhancer. All the V gene segments carry a promoter, but only the V segment brought near the constant region (with its enhancer) will be activated (Mather and Perry, 1982). This placement occurs during the construction of the immunoglobulin gene. The gene rearrangements bring a particular promoter into the proximity of the enhancer. A similar

(A) Germ-line gene: No transcription

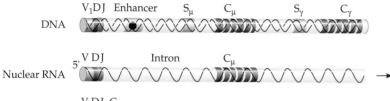

V_1 V_n D region   J region   Enhancer S_μ C_μ S_γ C_γ

(B) Rearranged gene: Transcription of immunoglobulin

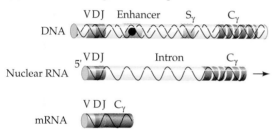

DNA V_1DJ Enhancer S_μ C_μ S_γ C_γ

Nuclear RNA   5′ V D J   Intron C_μ →

mRNA   V D J C_μ

(C) Class switched gene: Transcription of new class of immunoglobulin

DNA   V D J Enhancer S_γ C_γ

Nuclear RNA   5′ VDJ   Intron C_γ →

mRNA   V D J C_γ

FIGURE 10.22

Model for immunoglobulin enhancer activity. (A) Immunoglobulin heavy chain gene enhancer region appears to involve sequences between the J gene segments and the switch sequences (S_μ) preceding C_μ. If the enhancer is deleted, transcription is greatly diminished. The 5′ promoter precedes each of the V region gene segments and is originally very distant from the enhancer. (B) The VDJ gene rearrangement brings one promoter near the enhancer and allows transcription to take place. (C) During class switching, the enhancer stays with the VDJ segments as they are placed near a new constant region (C_γ).

event occurs for the heavy chain gene, and during class switching (when a region of DNA is looped out and deleted), the enhancer region remains by the VDJ piece (Figure 10.22).

Trans-regulation of immunoglobulin synthesis

Gene rearrangement itself, however, is not sufficient for gene activation, since a rearranged immunoglobulin gene will not actively transcribe when placed in a fibroblast or liver cell. There have to be cell-specific *trans*-regulatory

factors. Two promoter-binding factors were identified by Staudt and co-workers in 1986. They incubated a small piece of DNA containing the octa box in nuclear extracts from various cells. They then ran the resulting products on a gel. If the nuclear extract did not have a protein that bound to this DNA, the small DNA fragment should migrate rapidly through the gel. If a protein *did* bind to this DNA, however, the migration should be impeded. This **mobility shift assay** (Figure 10.23) demonstrated that every nucleus had at least one factor capable of binding to the DNA fragment. However, the B-cell lineage (pre-B cells, B cells, and plasma cells) contained, in addition, another factor capable of binding specifically to the octa box DNA.

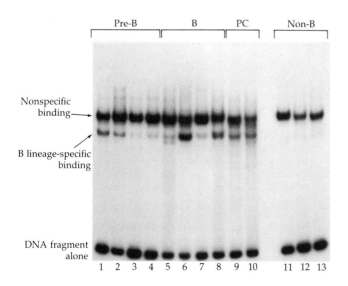

Nonspecific binding →

B lineage-specific binding →

DNA fragment alone

Pre-B   B   PC   Non-B

1   2   3   4   5   6   7   8   9   10   11   12   13

FIGURE 10.23

Gel mobility shift assay. Nuclear extracts from B-cell lineage cells [pre-B, B, and plasma cell (PC)] and non-B cells (cervical, fibroblast, and red blood cell precursor cell lines) were mixed with a small segment of DNA containing the octamer. After incubation, the mixtures were electrophoretically separated on a gel, blotted to nitrocellulose paper, and hybridized with radioactive DNA complementary to the octamer sequence. Without any binding, the octamer-containing fragment rapidly moves to the bottom of the gel. All nuclei contain a protein that nonspecifically binds to the octamer and greatly impedes migration. The nuclei from B-cell-lineage cells, however, also contain another protein that inhibits migration by binding to the octamer sequence. (From Staudt et al., 1986, courtesy of D. Baltimore.)

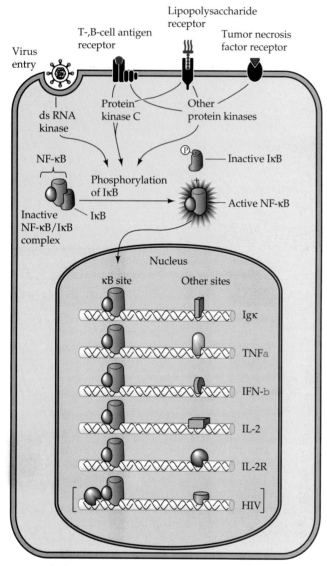

Virus
entry

T-,B-cell antigen
receptor

Lipopolysaccharide
receptor

Tumor necrosis
factor receptor

ds RNA
kinase

Protein
kinase C

Other
protein kinases

NF-κB

Phosphorylation
of IκB

Inactive IκB

Active NF-κB

Inactive
NF-κB/IκB
complex

IκB

Nucleus

κB site   Other sites

Igκ

TNFa

IFN-b

IL-2

IL-2R

HIV

FIGURE 10.24
Regulation of NF-κB by IκB. IκB binds to the larger subunit of NF-κB and prevents the complex from entering the nucleus. IκB can be phosphorylated by several kinases that are activated by replicating viruses, antigens, lipopolysaccharides, or tumor necrosis factor α. The phosphorylation of IκB releases NF-κB which can then enter the nucleus and bind to those promoter and enhancer sites it recognizes. Such genes include those encoding the kappa immunoglobulin light chain, tumor necrosis factor, interleukin 2, and the receptor for interleukin 2. The human immunodeficiency virus also has sites for NF-κB binding.

These binding proteins were isolated, and the lymphocyte-specific protein was termed Oct-2 (NF-A2).

Similar assays were used to find a nuclear protein restricted to B-cell lineage cells that binds specifically to the enhancer sequences of light chain immunoglobulin genes (Sen and Baltimore, 1986a; Atchinson and Perry, 1987).

One such protein, **NF-κB**, was found only in mature B cells and plasma cells and recognized the sequence 5′−GGGACTTTCC−3′ in the light chain enhancer. Moreover, when pre-B cells are induced to become B cells, active NF-κB protein appears.*

This brings the problem of differential gene activity to another level: What controls the synthesis of the B cell-specific *trans*-regulatory factors such as Oct-2 or NF-κB? (That is, to explain how immunoglobulins are produced in a cell-specific manner, we now have to explain how these nuclear proteins are generated in a cell-specific manner.) It appears that the NF-κB protein is actually present in numerous cell types, but is bound by a 65-kDa protein called **IκB** (inhibitor of kappa). In its bound state, NF-κB cannot enter the nucleus and bind to DNA (Figure 10.24; Henkel et al., 1992). Thus, NF-κB can recognize its enhancer region only in those cells that either do not synthesize or do not activate their IκB—that is, in mature lymphocytes (Sen and Baltimore, 1986b; Baeuerle and Baltimore, 1988). The synthesis of immunoglobulin light chains is probably initiated when a signal on the cell surface activates protein kinases that can phosphorylate IκB. This phosphorylation is thought to release NF-κB and allow it to enter the nucleus (Ghosh and Baltimore, 1990; Kerr et al., 1991).

In addition to the positive regulation of immunoglobulin gene transcription seen in B cells, there is also negative regulation of immunoglobulin production in non-B cells. There appear to be several sites flanking the immunoglobulin genes that are bound by proteins that inhibit the genes' transcription (Calame, 1989). Proteins capable of binding to these silencer regions of immunoglobulin genes have been seen in non-B cells. Thus, transcription may be stimulated or inhibited by *trans*-regulatory proteins depending upon a cell's developmental history.

The analysis of cell-specific transcription of immunoglobulin light chain genes has moved, then, to a level beyond looking at the terminal differentiated cell product. We know that the transcription of this immunoglobulin gene depends upon the prior activity of two nuclear proteins, Oct-2 and NF-κB, whose activity is seen only in the B-cell lineage.

*B cells are the only cells that synthesize immunoglobulins. The *pre-B* cell can make the heavy chain but not the light chain of the immunoglobulin protein. Active NF-κB has also been found in activated (but not unactivated) T cells. T cell-specific genes such as those encoding interleukin 2 (T-cell growth factor) and its receptor have enhancers that bind NF-κB. This NF-κB responsiveness may be important in the propagation of the human immune deficiency virus (HIV). When HIV infects a T cell, it induces the formation of active NF-κB (Sen and Baltimore, 1986b; Lenardo and Baltimore, 1989). The production of active NF-κB stimulates the T-cell genes that have enhancer sites for this protein. Thus, the T cell is stimulated to proliferate. At the same time, HIV also has NF-κB enhancer elements that enable it to transcribe its products rapidly, too. It is obvious that NF-κB plays a very important role in normal and altered lymphocyte development.

One might wonder what the nonspecific octa box-binding protein is doing in non-B cells. Although Oct-1 and Oct-2 proteins can bind to the same octa box, they exert their effects by interacting with other proteins at certain promoters (Tanaka et al., 1992).

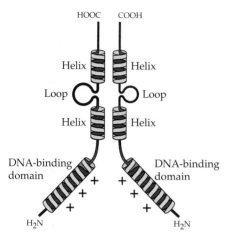

FIGURE 10.25
Domains of the basic helix-loop-helix transcription factors.

Basic helix-loop-helix transcription factors

Another prominent arrangement seen in enhancer and promoter DNA-binding proteins is the **basic helix-loop-helix** (bHLH) motif. The muscle-specific transcription factors MyoD and myogenin (discussed in Chapter 9) contain this motif, as do several *Drosophila* proteins that determine the cells of *Drosophila's* peripheral nervous system: the products of the *daughterless, achaete-scute,* and *extramacrochaetae* genes. As we will see in Chapter 21, the genes that determine the sex of *Drosophila* also contain the bHLH pattern. The bHLH proteins bind to DNA through a region of basic amino acids (typically 10–13 residues) that precedes the first α-helix (Figure 10.25). The helices contain hydrophobic amino acids at every third or fourth position so that the helix presents a surface of hydrophobic residues to the environment. This enables the protein to pair by hydrophobic interaction with the same protein or with a related protein that displays such a surface (Jones, 1990).

Recent studies have shown that homodimers (between two identical bHLH proteins) do not bind well to DNA. Rather, the bHLH proteins recognize their promoter sequences according to the following paradigm (Table 10.2). There is a ubiquitous bHLH protein synthesized in most cells that can form a dimer with either of two potential partners. One of these possible partners is the *positive* regulator (which stimulates transcription); the other partner is the negative regulator. When the positive regulator dimerizes with the ubiquitous protein, it forms an activator complex that stimulates transcription from the genes it recognizes. When the negative regulator dimerizes with the ubiquitous bHLH protein, it forms an inhibitory complex that represses transcription from those same genes. For instance, the MyoD family of proteins are active in promoting myogenesis when complexed to either the E12 or E47 proteins—two ubiquitous bHLH

TABLE 10.2
Developmental bHLH dimers

Dimer	*Drosophila* neurogenesis	Mammalian myogenesis	Mammalian cell division
Ubiquitous protein	daughterless	E12, E47	Max
Positive regulator	achaete-scute	MyoD family	myc
Negative regulator	extramacrochaetae	Id	Mad or Max

proteins (French et al., 1991; Lassar et al., 1991). Muscle development is inhibited, though, when the MyoD, E12, or E47 proteins are bound to the **Id** (inhibitor of **d**ifferentiation) protein. The Id protein contains the HLH motif, but *lacks* the basic region that binds to the DNA. Dimerization of Id with MyoD, E12, or E47 interferes with the ability of these proteins to bind DNA, and the expression of Id in cells prevents the activity of the MyoD proteins (Benezra et al., 1990).

SIDELIGHTS & SPECULATIONS

Regulating the switch between the growth and proliferation of muscle cells

There are ways to regulate myogenic bHLH proteins other than dimerizing them with Id. It has long been known (Stockdale and Holtzer, 1961; Bischoff and Holtzer, 1969) that muscle cells do not generally become differentiated until after they have finished proliferating. Proliferating muscle cells do not express the muscle-specific phenotype, while differentiated muscles no longer divide. Growth and differentiation are considered mutually exclusive states in muscle development, and once the muscle cell has left the cell cycle, it cannot return even if growth factors are provided (Konigsberg et al., 1960; Nadal-Ginard, 1978). Such mutual exclusivity between differentiation and proliferation is also seen in the development of neurons, adipocytes, blood cells, and skin keratinocytes. The mechanisms for this switch during myogenesis appear to involve the regulation of myogenic bHLH proteins. The first of these mechanisms is responsible for preventing premature muscle differentiation when the myogenic bHLH proteins first appear. Myogenic bHLH proteins are extremely sensitive to growth factors. As long as growth factors are present to stimulate mitosis, myogenesis does not occur, even if MyoD or myf5 proteins are present in the cells (Olson, 1992). The inactivation of these bHLH proteins is associated with their inability to bind the CANNTG (where N is any base) sequence of the DNA (Brennan et al., 1991). Why can't these myogenic bHLH proteins function? Growth factors such as fibroblast growth factor (FGF) not only stimulate Id transcription but also activate protein kinase C. This kinase induces the phosphorylation of the myogenic bHLH proteins right in their DNA-binding sites (Li et al., 1992). When this site is phosphorylated, the myogenic bHLH will not bind DNA. Thus, as long as the growth factors are present and able to be received, myogenesis will not occur.

The second mechanism initiates and maintains the differentiated state, preventing the differentiated muscle cell from rejoining the cell cycle. This involves the interaction between MyoD and two transcription factors that regulate cell division. The retinoblastoma protein, RB, is a 105-kDa DNA-binding phosphoprotein that regulates cell division in numerous cell types (Cavanee et al., 1983; Friend et al., 1986; Horowitz et al., 1990). The cell cycle is regulated, in part, by the phosphorylation state of RB. The RB protein is located in the nucleus and is unphosphorylated in resting (G_0 or G_1) cells. However, during the $G_1 \rightarrow S$ transition, almost all the RB protein becomes phosphorylated at several serine and threonine sites (Lee et al., 1987; Buchkovich et al., 1989; Chen et al., 1989; DeCaprio et al., 1989). The unphosphorylated form of the RB protein blocks the entry of the cell into S phase, and the phosphorylation of the protein takes away this inhibition. (Similarly, the absence of the RB protein would take away this inhibition totally and would promote unregulated cell division.) The myogenic bHLH proteins appear to take the cell out of the cell cycle by binding to RB and keeping it in the unphosphorylated state. Myogenic bHLH proteins may bind to the site in RB where the growth factor-dependent phosophorylation would normally occur, thereby making the RB unresponsive to growth factor stimulation (Gu et al., 1993). MyoD has also been found to bind to the c-jun subunit of mitosis-promoting transcription factor AP-1. When this binding occurs, the two proteins mutually inactivate each other. Thus, when AP-1 is abundant, MyoD is inhibited from functioning, and when MyoD is present in abundance, AP-1 is inactivated (Bengal et al., 1992). Thus, the myogenic bHLH proteins mediate the mutual exclusion of muscle cell division and differentiation. As long as growth factors are present, myogenic bHLH proteins cannot function because of (1) the presence of Id and (2) their being phosphorylated at their DNA-binding site. However, when the growth factors no longer signal cell division, more myogenic bHLH proteins are made, they bind to E12 or E47 to promote differentiation, and they bind to pRB and AP1 to prevent their activating cell division.

Basic leucine zipper transcription factors

The structure of **basic leucine zipper** (bZip) transcription factors is very similar to that of the bHLH proteins. The bZip proteins are dimers, each of whose subunits contain a basic DNA-binding domain at the carboxyl end, followed closely by an α-helix containing several leucine residues. These leucines are placed in the helix such that they interact with similarly spaced leucine residues on other bZip proteins to form a "leucine zipper" between them, causing dimers to form. This domain is followed by a regulatory domain that can interact with the promoter to stimulate or repress transcription (Landschulz et al., 1988; Pathak and Sigler, 1992). The C/EBP, AP-1, and yeast GCN4 transcription factors are members of the bZip family. Genetic and X-ray crystallographic methods have converged on a model of DNA binding shown in Figure 10.26 (Vinson et al., 1989; Pu and Struhl, 1991). In the figure the two α-helices containing the DNA-binding region are inserted into the major groove of the DNA, each helix finding an identical DNA sequence. The resulting binding looks like that of a scissors or hemostat.

There appear to be several bZIP proteins that can bind to the sequence CCAAT; one of the most important is called the **CCAAT enhancer-binding**

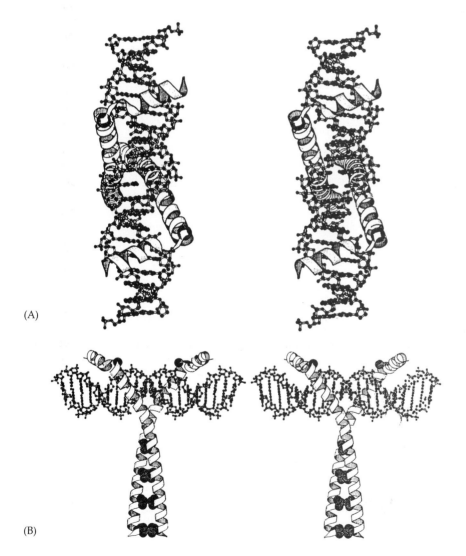

(A)

(B)

FIGURE 10.26
Stereoscopic representation of bZip protein C/EBP DNA-binding region interacting with 20 base pairs of DNA containing the CCAAT sequence. (A) "Dorsal view," looking down at a DNA double helix and parallel to the leucine zipper. (B) "Side view," at right angles to the upper diagram and perpendicular to the DNA axis. Leucine residues connecting the two subunits can be seen at the bottom, as can the "scissors grip" in the DNA. If you aren't used to crossing your eyes to see the composite stereoimage, please borrow a stereopticon. (From Pathak and Sigler, 1992.)

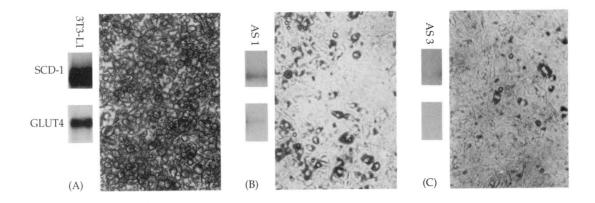

FIGURE 10.27

Adipogenesis (fat cell formation) mediated through the C/EBP transcription factor. (A) Normal adipogenesis of the 3T3-L1 pre-adipocyte cell line in culture. The *SCD-1* and *GLUT4* genes are activated and the cells synthesize and accumulate large amounts of triglycerides. (B,C) Two lines of 3T3-L1 cells that have been transfected with antisense RNA against the C/EBP message. Neither gene is well expressed, and triglyceride levels are 15 and 5 percent of normal. (From Lin and Lane, 1992.)

protein (C/EBP). C/EBP plays a role in adipogenesis similar to that of the myogenic bHLH proteins in myogenesis. Precocious expression of C/EBP in dividing, pre-adipose cells causes the cessation of cell division and the initiation of the adipose phenotype (Umek et al., 1991). (Unlike the myogenic bHLH proteins, which can convert nerve cells and fibroblasts into muscles, C/EBP does not appear to convert other types of cells into the adipocyte lineage.) The C/EBP bZIP protein binds to the enhancers of numerous adipose-specific genes when adipogenesis is initiated in culture (Figure 10.27; Christy et al., 1989; Kaestner et al., 1990). Antisense mRNA against C/EBP suppresses the coordinate expression of adipocyte-specific messages and the differentiation of preadipocytes into adipocytes (Samuelsson et al., 1991; Lin and Lane, 1992).

C/EBP is also enriched in liver cells, and it is one of the major regulators of liver-specific gene expression. In mouse hepatocytes, other transcription factors bind to the promoter and enhancer regions of the liver-specific genes during development. However, these genes do not transcribe large amounts of proteins (such as albumin) until C/EBP is expressed in these cells immediately prior to birth (Milos and Zaret, 1992). Another liver-specific gene activated by the C/EBP protein is blood-clotting factor IX. Mutations in this clotting factor gene cause hemophilia B. In some patients, the cause of this disease has been traced to mutations in the C/EBP binding site in the promoter region of the factor IX gene. These mutations prevent the binding of C/EBP to this gene (Crossley and Brownlee, 1990).

SIDELIGHTS & SPECULATIONS

Enhancers and cancers

One critical transcription factor that is very important in regulating cell division is the **c-myc** protein. It is a member of a class of DNA-binding proteins that have incorporated a leucine zipper and a basic helix–loop–helix motif. The c-myc proteins behave similarly to the bHLH and bZIP proteins in that they form heterodimers that bind DNA (see Table 10.2). The c-myc forms an activator complex when joined to the ubiquitous protein, Max. The complex between Max and an inhibitory protein, Mad, creates the Mad-Max repressor protein (which binds to the same site as the c-myc/Max complex, CACGTG (Ayer et al., 1993).

The c-*myc* gene is the cellular homologue to the cancer-producing gene, or **oncogene**, v-*myc* of the avian myelocytomatosis virus (Donner et al., 1982). The c-*myc* gene synthesizes very short-lived mRNA and protein products when stimulated by a variety of growth factors (Kelly et al., 1983). These c-*myc* gene products appear suddenly as cells are induced from the G_0 state into G_1, and they are degraded shortly thereffer. The c-myc proteins signal cell division, and if they are not degraded rapidly, the cell will continue proliferating, thereby increasing the risk of tumor formation.

Given that enhancers are able to control the expression of reporter genes that are unrelated to it, it would appear

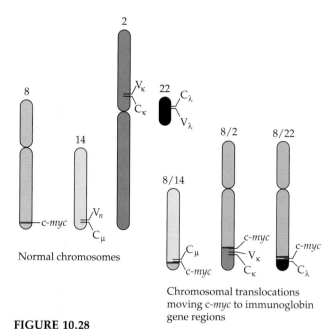

Normal chromosomes

Chromosomal translocations
moving c-*myc* to immunoglobin
gene regions

FIGURE 10.28

*Three chromosomal translocations that independently can give
rise to Burkitt lymphoma of B cells. In each case, the c-myc
gene (on chromosome 8) is translocated to the enhancer region
of an immunoglobulin gene on chromosome 2, 14, or 22. C_μ is
the heavy chain gene; and C_λ and C_κ are the genes for the two
types of immunoglobulin light chains.*

that these sequences are very powerful regulators of the
specificity of gene transcription. What would happen if a
spontaneous chromosomal translocation brought an enhancer for one protein adjacent to a structural gene for
another protein? In most cases, it probably wouldn't matter. Even if the enhancer did stimulate the expression of
the gene in the "wrong" cell, the product of only one cell
would be abnormally regulated. Perhaps that cell would
die and be replaced by another. Perhaps the descendants

of that cell would form a clone of cells expressing a protein
that other cells in the tissue did not make. However, if the
translocation brought the c-*myc* gene next to an enhancer
for a very actively transcribed gene, the c-*myc* gene would
be activated to transcribe large amounts of message as the
cell differentiated. In such cases, the single cell could give
rise to a tumor.

Chromosomal translocations involving the c-*myc* gene
appear to be responsible for tumors of the immunoglobu-
lin-synthesizing B cells of our immune system (Croce,
1987). Here, the c-*myc* gene on the end of the small arm of
human chromosome 8 has been translocated to chromosome 14, 22, or 2 (Figure 10.28). These three chromosomes
contain the genes for the immunoglobulin proteins, and
the c-*myc* gene has been translocated to the region of the
immunoglobulin gene enhancers (Leder et al., 1983; Croce
et al., 1985). The amount of c-*myc* mRNA transcribed from
these translocated chromosomes correlates with the activation of the immunoglobulin genes. Thus, when the enhancers of the immunoglobulin genes are activated (during
B cell development), they turn on the adjacent gene—
which is now c-*myc*. The c-*myc* mRNA is made in enormous
amounts and is translated into the c-*myc* transcription factor. This factor instructs the cells to divide. In the continuous presence of c-*myc* protein, the cells keep on dividing,
thus forming a tumor called Burkitt lymphoma (Nishikura
et al., 1983; Croce et al., 1984). In these situations, the
translocation of the c-*myc* gene to an immunoglobulin enhancer in a single cell can cause the entire leukemia. Indeed, most all leukemias are seen to result from a single
cell.

Several types of leukemias are caused when other transcription factor genes have been translocated to the regions
of the immunoglobulin gene enhancers. These types of
rearrangements between transcription factor encoding regions and lymphocyte-specific regulatory regions may be
errors caused by the VDJ recombinases that are specific to
lymphocytes and would explain why these translocations
are seen in so many leukemias (Rabbitts, 1991).

Zinc finger transcription factors

Another type of DNA binding domain is the **zinc finger** motif. Zinc finger
proteins include WT-1 (a very important transcription factor in the kidney
and gonads); the ubiquitous transcription factor Sp1; *Xenopus* 5S rRNA
transcription factor TFIIIA; Krox 20 (a protein that regulates gene expression in the developing hindbrain); Egr-1 (which commits white blood cell
development to the macrophage lineage); Krüppel (a protein that specifes
abdominal cells in *Drosophila*); and numerous steroid-binding transcription
factors. Each of these proteins has two or more "DNA-binding fingers,"
α-helical domains whose central amino acids tend to be basic. These
domains are linked together in tandem and are each stabilized by a centrally located zinc ion coordinated by two cysteines (at the base of the
helix) and two internal histidines (Figure 10.29). The crystal structure
shows that the zinc fingers bind in the major groove of the DNA.

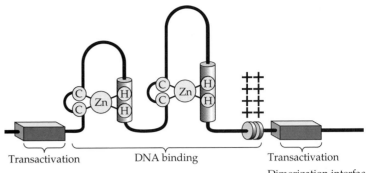

Transactivation

DNA binding

Transactivation

Dimerization interface, HSP90 binding, inhibitory function, transactivation

FIGURE 10.29
Functional domains of zinc finger transcription factors.

The WT-1 protein contains four zinc finger regions, and it is usually expressed in the fetal kidney and gonads. In humans, people with one mutant *WT-1* allele (usually a deletion of the gene or of a zinc finger region) have urogenital malformations and develop Wilms tumor of the kidney (Haber et al., 1990; Bruening et al., 1992; see Chapter 18). In mice, *both WT-1* genes can be deleted by gene targeting, and the resulting mice die in utero, having neither kidney nor gonads (Kriedberg et al., 1993). The WT-1 factor binds to the regulatory regions of several genes that are active during kidney development and also is thought to inhibit the expression of certain growth factors (especially insulin-like growth factor II) in the developing kidney (Drummond et al., 1992).

Nuclear hormone receptors and their hormone-responsive elements

Specific steroid hormones are known to increase the transcription of specific sets of genes. Once the hormone has entered the cell, it binds to its specific receptor protein, converting that receptor into a conformation that is able to enter the nucleus and bind particular DNA sequences (Miesfeld et al., 1986; Green and Chambon, 1988). The family of steroid hormone receptors includes those proteins that recognize estrogen, progesterone, testosterone, and cortisone as well as nonsteroid lipids such as retinoic acid, thyroxin, and vitamin D. The DNA sequences capable of binding nuclear hormone receptors are called **hormone-responsive elements**, and they can be in either enhancers or promoters. One set of steroids includes the glucocorticoid hormones (cortisone, hydrocortisone, and the synthetic hormone dexamethasone). These bind to the glucocorticoid hormone receptors and enable the bound receptors to bind to the glucocorticoid-responsive elements in the chromosomes (Figure 10.30).

The steroid hormone-responsive elements are extremely similar to one another and are recognized by closely related proteins. These steroid receptor proteins each contain three functional domains: (1) a **hormone-binding domain**, (2) a **DNA-binding domain** that recognizes the hormone-responsive element, and (3) a *trans*-activation domain, which is involved in mediating the signal to initiate transcription. These functions can overlap, and all domains appear to have some role in activating transcription (Beato, 1989). For transcriptional activation to occur, the receptor has to enter the nucleus and dimerize with a similar hormone-binding protein. The binding of hormone to the hormone-binding domain appears to be necessary for dimerization, translocation into the nucleus, and the ability of the DNA-binding region to recognize the hormone-responsive element (Kumar et al., 1987).

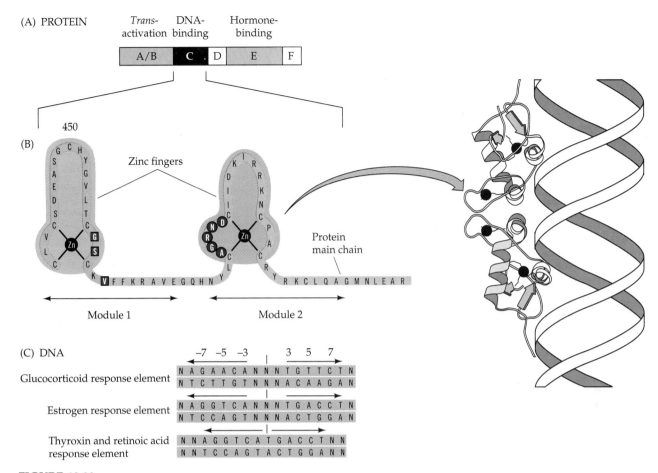

(A) PROTEIN

Trans-activation | DNA-binding | Hormone-binding

A/B | C . D | E | F

(B)

450

Zinc fingers

Module 1

Module 2

Protein main chain

(C) DNA

−7 −5 −3 | 3 5 7

Glucocorticoid response element

N A G A A C A N N N T G T T C T N
N T C T T G T N N N A C A A G A N

Estrogen response element

N A G G T C A N N N T G A C C T N
N T C C A G T N N N A C T G G A N

Thyroxin and retinoic acid response element

N N A G G T C A T G A C C T N N
N N T C C A G T A C T G G A N N

FIGURE 10.30

Structural organization of hormone receptor DNA-binding proteins. (A) Generalized structure of a steroid hormone binding protein. The functions ascribed to each fraction have been determined by analyzing the effects of mutations in each of these regions and by making chimeric protein molecules having regions derived from different receptor proteins. A similar structure is seen in the retinoic acid receptor and the thyroid hormone receptor. (B) Zinc finger DNA-binding region of the glucocorticoid receptor. Boxed residues in module 1 discriminate between estrogen- and glucocorticoid-responsive elements. Circled residues are involved in dimerization. At right, the zinc finger region of the glucocorticoid receptor is shown bound to its response element. (C) The DNA sequences for the response element. Note that they are inverted palindromes so that each dimer is exposed to the same site. N, any base. The distinction between binding glucocorticoids or progesterone receptors or binding thyroxin or retinoic acid receptors is determined by the spacing of the response elements, by limiting amounts of receptors, and by other interacting *cis* elements. (After Kaptein, 1992.)

The hormone-responsive elements within the DNA were originally discovered by binding competition assays (Figure 10.31). For instance, to identify the glucocorticoid responsive element, Karin and co-workers (1984) fixed double-stranded calf thymus DNA (which is presumed to have many glucocorticoid-responsive elements) to cellulose filters and incubated these filters in solutions of glucocorticoid receptors carrying radioactive hormone. The radioactive hormone–protein complexes were allowed to bind to the DNA, and the bound radioactivity was counted (Figure 10.31A). Next, the same assay was run with an excess of some other DNA in solution (Figure 10.31B). If this competitor DNA had the

FIGURE 10.31

The protocol for determining the site of the glucocorticoid enhancer.

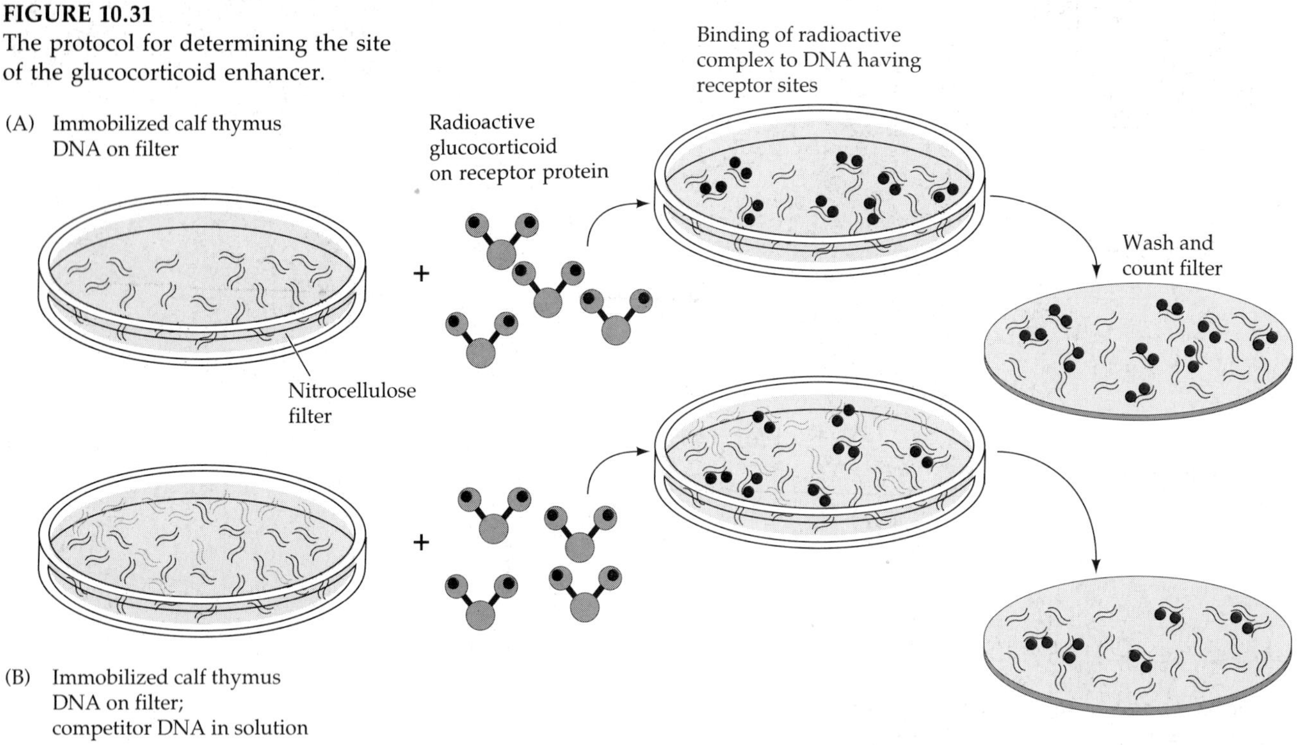

(A) Immobilized calf thymus DNA on filter

Radioactive glucocorticoid on receptor protein

Binding of radioactive complex to DNA having receptor sites

Wash and count filter

Nitrocellulose filter

(B) Immobilized calf thymus DNA on filter; competitor DNA in solution

glucocorticoid-responsive element, it would compete for the radioactive complex and less radioactivity would be bound to the calf thymus DNA (Pfahl,1982). By using various restriction enzyme-derived fragments of DNA and comparing the sequences of numerous glucocorticoid-responsive elements, it was found that the consensus sequence of the glucocorticoid responsive element is AGAACA*NNN*TGTTCT (where *N* can be any base). These glucocorticoid-binding sequences were shown to act as enhancers: when the glucocorticoid response element was ligated onto genes that are not usually hormone-dependent, those genes gained the responsiveness to the glucocorticoid (Figure 10.32; Chandler et al., 1983).

The binding of the receptor protein to the hormone-responsive enhancer element is accomplished by a zinc finger region in the DNA-binding domain (Green et al., 1988). When chimeric proteins are made wherein the zinc finger domain of the estrogen receptor substitutes for the same region of a glucocorticoid receptor, the protein recognizes the DNA that contains estrogen-responsive elements and causes the gene to be responsive to glucocorticoids. The critical amino acids appear to reside at the "knuckle" of the zinc finger (Danielsen et al., 1989; Umesono and Evans, 1989). Changing as few as two amino acids at the knuckle of the zinc finger region will change the specificity of the binding protein. Thus, while these DNA-binding domains of the hormone receptor proteins are very similar, they can distinguish subtle differences in the enhancer sequences. For example, the (palindromic) sequence 5'-GGTCACTGTGACC-3' is a strong estrogen-responsive enhancer element that will bind the estrogen-containing receptor protein. Two symmetrical mutations in this sequence, making it 5'-GG**A**CACTGTG**T**CC-3', convert this DNA into a glucocorticoid-responsive enhancer (Martinez et al., 1987; Klock et al., 1987). Given the similarities among hormone receptor proteins and the similarities among hormone-responsive elements, it is likely that each steroid hormone mediates its transcriptional activation by the same general mechanism.

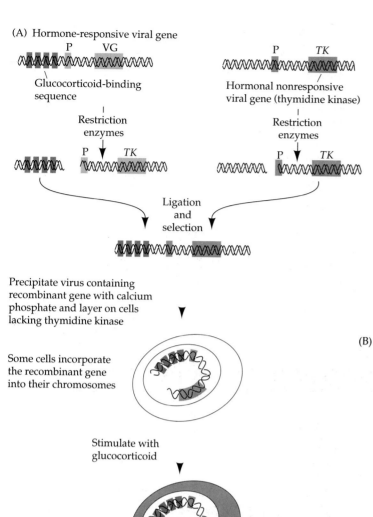

(A) Hormone-responsive viral gene

P   VG

Glucocorticoid-binding
sequence

P   TK

Hormonal nonresponsive
viral gene (thymidine kinase)

Restriction
enzymes

Restriction
enzymes

P   TK

P   TK

Ligation
and
selection

Precipitate virus containing
recombinant gene with calcium
phosphate and layer on cells
lacking thymidine kinase

Some cells incorporate
the recombinant gene
into their chromosomes

Stimulate with
glucocorticoid

Thymidine kinase induced

FIGURE 10.32

Test for the glucocorticoid enhancer sequence. (A) A recombinant virus containing the glucocorticoid-responsive enhancer of the mouse mammary tumor virus and the *thymidine kinase* gene of *Herpes simplex* virus can integrate into the genome of a cell lacking *thymidine kinase* gene. Upon treatment with glucocorticoids, the recombinant gene transcribes viral thymidine kinase. P, promoter region; *TK*, *thymidine kinase* gene; VG, viral gene of mouse mammary tumor virus. (B) Electron micrograph of glucocorticoid enhancer elements, showing hormone receptor bound to that region of the gene. (A modified from Chandler, 1983; B from Payvar et al., 1983, courtesy of K. R. Yamamoto.)

(B)

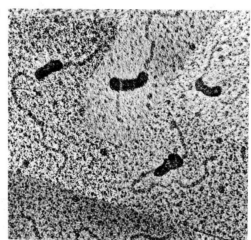

Positive and negative control by interactions between transcription factors

The hormone receptor proteins interact with other *trans*-regulatory proteins on the DNA. This interaction is seen to occur when the glucocorticoid receptor binds to DNA near a CACCC box (which binds other nuclear factors). First, removal of a CACCC element from the promoter of the tryptophan oxygenase gene abolishes its ability to be induced by glucocorticoids (Danesch et al., 1987). Second, placing the marker gene choline acetyltransferase (CAT) on a plasmid containing the glucocorticoid-responsive element and wild-type and mutant CACCC elements shows that both wild-type elements are needed for maximal glucocorticoid enhancement (Figure 10.33; Schüle et al., 1988; Strähle et al., 1988). Thus (as discussed earlier with the interaction between Pit-1 and the estrogen receptor protein), the hormone receptor protein interacts synergistically with other *trans*-regulatory proteins to stimulate the transcription of hormone-responsive genes. The hormone receptors can also interact between themselves to synergistically activate a promoter. In several genes, there are multiple hormone-responsive elements linked together. When more than one receptor binds, the transcription complex is stabilized, sometimes over

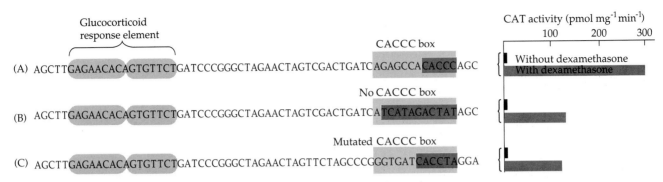

Glucocorticoid
response element

CACCC box

CAT activity (pmol mg^{-1} min^{-1})
100   200   300

(A) AGCTTGAGAACACAGTGTTCTGATCCCGGGCTAGAACTAGTCGACTGATCAGAGCCACACCCAGC

☐ Without dexamethasone
■ With dexamethasone

No CACCC box

(B) AGCTTGAGAACACAGTGTTCTGATCCCGGGCTAGAACTAGTCGACTGATCATCATAGACTATAGC

Mutated CACCC box

(C) AGCTTGAGAACACAGTGTTCTGATCCCGGGCTAGAACTAGTTCTAGCCCGGGTGATCACCTAGGA

FIGURE 10.33

Cooperativity between the glucocorti-coid-responsive element and the CACCC box. Plasmids were constructed containing a glucocorticoid-responsive element (colored box) and (A) a wild-type CACCC box, (B) an unrelated DNA sequence, or (C) a mutated CACCC box. These plasmids were separately transfected into cultured human cells, and their inducibility was tested by adding the glucocorticoid dexamethasone to the cells' medium. The result shows that, while the presence of the glucocorticoid-responsive element was sufficient for some inducibility, the combination of glucocorticoid-responsive element and a functional CACCC box gave three times as much stimulation. (After Schüle et al., 1988.)

100-fold. This tremendously increases the efficiency of transcription from that promoter (Martinez et al., 1987; Tsai et al., 1989). In other cases, hormone-binding receptors may interact in order to remove nucleosomes from an adjacent site that can then become bound by some other transcriptional activator. This other transcriptional factor may then activate transcription from the promoter (Lucas and Granner, 1992).

The interactions between steroid receptors and other transcriptional regulatory proteins can determine whether the steroid's effect is positive or negative. Diamond and co-workers (1990) demonstrated that the effect of glucocorticoid hormones on the transcription of the mouse *proliferin* gene can be positive or negative depending on the prior physiological state of the cell. A 25-base-pair sequence upstream from the *proliferin* gene can bind the glucocorticoid receptor and the bZip dimer of either c-jun and c-jun or c-jun and c-fos. The binding of c-jun to this site is necessary for the function of the glucocorticoid receptor. If the c-jun–c-jun dimer were present at this site (without the glucocorticoid receptor), very little transcription would occur. This transcription is dramatically *enhanced* by the addition of glucocorticoids (Figure 10.34). If the c-jun–c-fos dimer were present at this site, however, it could direct extremely efficient transcription of the *proliferin* gene. This transcription is *inhibited* by the presence of glucocorticoids. Thus, whether the glucocorticoid has a stimulatory or inhibitory effect on *proliferin* transcription depends on the prior physiological

FIGURE 10.34

Alternative enhancing and silencing effects of the glucocorticoid response element upstream from the mouse *proliferin* gene. The effect of the glucocorticoid depends on the prior condition of the cell, i.e., whether high concentrations of c-fos were being synthesized. The arrows represent transcription of the *proliferin* gene. The large circle represents the hormone-bound glucocorticoid receptor. The c-jun protein is represented as a small circle, while the c-fos protein is a small square. (After Diamond et al., 1990.)

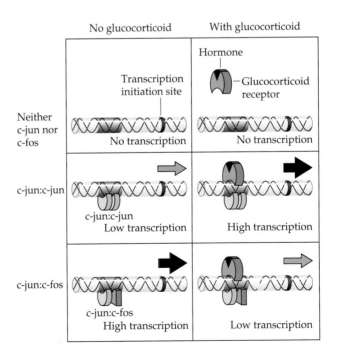

state of the cell. A single DNA sequence binding a particular hormone receptor may be both an enhancer and a silencer for the same protein.

Regulation of transcription factor activity

Transcriptional regulation

If transcription factors are proteins that regulate the expression of particular genes, then how are the transcription factors themselves regulated? One obvious way is to regulate the synthesis of transcription factors by other transcription factors. This method is used in *Drosophila* development, in which there is a cascade of transcription factor synthesis. For instance, the bicoid protein activates the *hunchback* gene. The newly synthesized hunchback protein activates the *Krüppel* gene. The Krüppel protein activates the *odd-skipped* gene, and so on. In such cascades, one is eventually led back to a maternally produced factor that is placed into the oocyte cytoplasm. We will detail this series of events in Chapter 15, because it is critical for the formation of the *Drosophila* anterior–posterior body axis.

In mammals, several transcription factors (such as MyoD and Pit-1) are regulated by the synthesis of other transcription factors. Moreover, such transcription factors often have very complex enhancers and promoters that enable them to be expressed only in certain cells. The mouse *myogenin* gene, for instance, is expressed in the myotome, pharyngeal arches, and limb buds. There appear to be at least three separable sites in the upstream regulatory region for this gene. The closest site is necessary for the transcription of this gene in the limb buds. If this site is mutated, the *myogenin* gene is not transcribed there. A second site, further upstream, is needed for expression of myogenin in the limb buds, pharyngeal arches, and central cells of the posterior somites (Figure 10.35; Cheng et al., 1993; Yeo and Rigby, 1993). A third site, still further upstream, is necessary for the increased efficiency of transcription of the gene. These three sites bind different transcription factors. A similar situation appears to exist for *myf-5,* wherein different regions of DNA regulate the different expression pattern elements (cover photo; Patapoutian et al., 1993).

Protein modification and RNA splicing

Another mechanism of regulating transcription factor activity is by phosphorylation. In one set of cases, the transcription factor protein is present, but inactive, and phosphorylation activates the dormant protein. As we have discussed earlier, the phosphorylation of a sequestered transcription

FIGURE 10.35
Myogenin expression in the 10.5-day mouse embryo. A β-galactosidase reporter gene was linked to the upstream regulatory sequences of the *myogenin* gene, and this was used to make transgenic mice. The transgenic embryos were stained for the presence of bacterial β-galactosidase at 10.5 days. (A) Wild-type *myogenin* promoter region, showing all the places where the *myogenin* gene is usually expressed. (B) Expression from a *myogenin* promoter with a mutation in a site close to the *myogenin* gene. There is no transcription from this gene in the limb buds. (C) Expression from a *myogenin* promoter with a mutation in a site further upstream from the gene. No transcription from this promoter is seen in the pharyngeal arches, limbs, or central posterior myotome cells. (From Cheng et al., 1993.)

(A)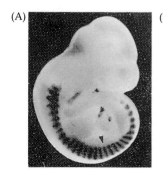

(B)

(C)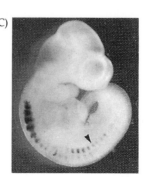

FIGURE 10.36

Regulation of transcription factor activity by phosphorylation. The three subunits of the ISG3 transcription factor are found in the cytoplasm. When interferon binds to its cell surface receptor, the receptor's protein tyrosine kinase is activated. This will directly or indirectly cause the phosphorylation of these proteins. Once phosphorylated, these proteins enter the nucleus and join with a fourth protein to construct the functional transcription factor that activates interferon-responsive genes. (After Marx, 1992.)

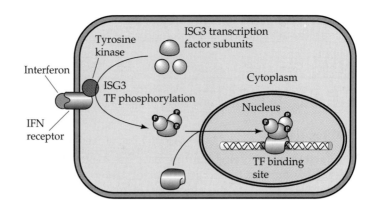

factor in its inhibitor (such as IκB) can release the inhibition and enable the transcription factor (in this case, NF-κB) to enter the nucleus and bind its DNA sequence. Phosphorylation can also work more directly. When interferon-α binds to its receptor at the cell surface, it activates a tyrosine kinase which can phosphorylate three proteins (the ISGF3-α proteins) that reside in the cytoplasm. Once phosphorylated, they join together to create a transcription factor capable of entering the nucleus, joining with a fourth protein, and transcribing the interferon-responsive genes (Figure 10.36; Schindler et al., 1992). Phosphorylation can also be used to repress transcription factors, as when the binding of DNA by Pit-1, Oct-1, or myogenin is inhibited by their being phosphorylated (Hunter and Karin, 1992).

As we will detail in Chapter 12, numerous proteins can be generated from the same gene by splicing alternative exons together. What is an exon in one cell type may be an intron in some other type of cell. TFE3 is a bHLH–bZip transcription factor that binds to the promoter and enhancer of heavy chain immunoglobulin genes. In most cells, however, a truncated form of this protein is also made. This protein comes from an alternatively spliced RNA wherein an exon in the activation domain has been eliminated (i.e., made into an intron and not put into the mRNA). The truncated protein can dimerize with the active form and inhibit it (Roman et al., 1991).

And so we see that the activity of transcription factors can be regulated at several levels. In some cases, a single transcription factor can be regulated in numerous ways. Given that each gene is often regulated by several transcription factors, the cell is given many options on how to express certain genes in only certain cell types. Within the past five years, our knowledge of transcription factors has progressed enormously and has given us a new, dynamic, view of gene expression. The gene itself is no longer seen as an independent entity controlling the synthesis of proteins. Rather, the gene both *directs* and *is directed by* protein synthesis. Angier (1992) writes that

> A series of new discoveries suggests that DNA is more like a certain type of politician, surrounded by a flock of protein handlers and advisers that must vigorously massage it, twist it and, on occasion, reinvent it before the grand blueprint of the body can make any sense at all.

Certainly, the interactions between DNA and its transcription factors are taking the interactive relationship of nucleus and cytoplasm to new and splendidly complex levels. So far, we have been focusing our attention on the relationship of the transcription factors to DNA. But transcription factors do not see mere DNA. Rather, they see a highly structured DNA–protein complex called chromatin. To initiate transcription, one also must deal with the higher-order structures of the cell, and we will continue our discussion of the transcriptional regulation of development into the next chapter.

LITERATURE CITED

Angier, N. 1992. A first step in putting genes into action: Bend the DNA. *New York Times*, August 4, 1992, pp. C1, C7.

Atchinson, M. L. and Perry, R. P. 1987. The role of κ-enhancer and its binding factor NF-κB in the developmental regulation of κ gene transcription. *Cell* 48: 121–128.

Ayer, D. E., Kretzner, L. and Eisenman, R. N. 1993. Mad: A heterodimeric partner for Max that antagonizes Myc transcriptional activity. *Cell* 72: 211–222.

Baeuerle, P. A. and Baltimore, D. 1988. IκB: A specific inhibitor of the NF-κB transcription factor. *Science* 242: 540–545.

Banerji, J., Olson, L. and Schaffner, W. 1983. A lymphocyte-specific cellular enhancer is located downstream of the joining region in immunoglobulin heavy chain genes. *Cell* 33: 729–740.

Beato, M. 1989. Gene regulation by steroid hormones. *Cell* 56: 335–344.

Behringer, R. R., Mathews, L. S., Palmiter, R. D. and Brinster, R. L. 1988. Dwarf mice produced by genetic ablation of growth hormone-expressing cells. *Genes Dev.* 2: 453–461.

Benezra, R., Davis, R. L., Lockshon, D., Turner, D. L. and Weintraub, H. 1990. The protein Id: A negative regulator of helix–loop–helix DNA binding proteins. *Cell* 61: 49–59.

Bengal, E. Ransone, L., Scharfmann, R., Dwarki, V. J., Tapscott, S. J., Weintraub, H. and Verma, I. M. 1992. Functional antagonism between c-jun and MyoD proteins: A direct physical association. *Cell* 68: 507–519.

Bergman, Y., Rice, D., Grosschedl, R. and Baltimore, D. 1984. Two regulatory elements for immunoglobulin κ light chain gene expression. *Proc. Natl. Acad. Sci. USA* 81: 7041–7045.

Bernard, O., Hozumi, N. and Tonegawa, S. 1978. Sequences of mouse immunoglobulin light chain genes before and after somatic change. *Cell* 15: 1133–1144.

Bischoff, R. and Holtzer, H. 1969. Mitosis and the processes of differentiation of myogenic cells in vitro. *J. Cell Biol.* 41: 188–200.

Bodner, M. and Karin, M. 1987. A pituitary-specific *trans*-acting factor can stimulate transcription from the growth hormone promoter in extracts of non-expressing cells. *Cell* 50: 267–275.

Borst, P. and Greaves, D. R. 1987. Programmed gene rearrangements altering gene expression. *Science* 235: 658–667.

Boulet, A. M., Erwin, C. R. and Rutter, W. J. 1986. Cell-specific enhancers in the rat exocrine pancreas. *Proc. Natl. Acad. Sci. USA* 83: 3599–3603.

Brack, C., Hirama, M., Lenhard-Schuller, R. and Tonegawa, S. 1978. A complete immunoglobulin gene is created by somatic recombination. *Cell* 15: 1–14.

Brennan, T., Edmondson, D. G., Li, L. and Olson, E. N. 1991. Transforming growth factor β represses the actions of myogenin through a mechanism independent of DNA binding. *Proc. Natl. Acad. Sci. USA* 88: 3822–3826.

Bruening, W. and seven others. 1992. Germline intronic and exonic mutations in the Wilms' tumor gene (*WT1*) affecting urogenital development. *Nature Genet.* 1: 144–148.

Buchkovich, K., Duffy, L. A. and Harlow, E. 1989. The retinoblastoma gene is phosphorylated during specific phases of the cell cycle. *Cell* 58: 1097–1105.

Bunick, D., Zandomeni, R., Ackerman, S. and Weinmann, R. 1982. Mechanism of RNA polymerase II-specific initiation of transcription in vitro: ATP requirement and uncapped runoff transcripts. *Cell* 29: 877–886.

Buratowski, S., Hahn, S., Guarente, L. and Sharp, P. A. 1989. Five initiation complexes in transcription initiation by RNA polymerase II. *Cell* 56: 549–561.

Buratowski, S., Sopta, M., Greenblatt, J. and Sharp, P. 1991. RNA polymerase II-associated proteins are required for a DNA conformation change in the transcription initiation complex. *Proc. Natl. Acad. Sci. USA* 88: 7509–7513.

Calame, K. L. 1989. Immunoglobulin gene transcription: Molecular mechanisms. *Trends Genet.* 5: 395–399.

Cavanee, W. K. and eight others. 1983. Expression of recessive alleles by chromosomal mechanisms in retinoblastoma. *Nature* 305: 779–784.

Chandler, V. L., Maier, B. A. and Yamamoto, K.R. 1983. DNA sequences bound specifically by glucocorticoid receptor in vitro render a heterologous promoter hormone responsive in vivo. *Cell* 33: 489–499.

Chen, P.-L., Scully, P., Shew, J.-Y., Wang, J. Y. J. and Lee, W.-H. 1989. Phosphorylation of the retinoblastoma gene product is modulated during the cell cycle and cellular differentiation. *Cell* 58: 1193–1198.

Cheng, T.-C., Wallace, M., Merlie, J. P. and Olson, E. N. 1993. Separable regulatory elements govern myogenin transcription in mouse embryogenesis. *Science* 261: 215–218.

Christy, R. J. and seven others. 1989. Differentiation-induced gene expression in 3T3-L1 preadipocytes: CCAAT/enhancer binding protein interacts with and activates the promoter of two adipocyte-specific genes. *Genes Dev.* 3: 1323–1335.

Chun, J. J. M., Schatz, D. G., Oettinger, M. A., Jaenisch, R. and Baltimore, D. 1991. The recombination activating gene 1 (*RAG-1*) is present in the murine central nervous system. *Cell* 64: 189–200.

Comai, L., Tanese, N. and Tjian, R. 1992. The TATA-binding protein and associated factors are integral components of the RNA polymerase I transcription factor, SL1. *Cell* 68: 965–976.

Conaway, R. C., Pfeil-Garrett, K., Hanley, J. P. and Conaway, J. W. 1991. Mechanism of promoter selection by RNA polymerase II: Mammalian transcription factors a and bg promote entry of polymerase into the preinitiation complex. *Proc. Natl. Acad. Sci. USA* 88: 6205–6209.

Crenshaw, E. B., Kalla, K., Simmons, D. M., Swanson, L. W. and Rosenfeld, M. G. 1989. Cell-specific expression of the prolactin gene in transgenic mice is controlled by synergistic interactions between promoter and enhancer elements. *Genes Dev.* 3: 959–972.

Croce, C. M. 1985. Chromosomal translocations, oncogenes, and B-cell tumors. *Hosp. Pract.* 20(1): 41–48.

Croce, C. M. 1987. Role of chromosome translocations in human neoplasia. *Cell* 49: 155–156.

Croce, C. M., Thierfelder, W., Erikson, J., Nishikura, K., Finan, J., Lenoir, G. M. and Nowell, P. C. 1984. Transcriptional activation of an unarranged and untranslocated c-*myc* oncogene by translocation of a C_λ locus in Burkitt lymphoma cells. *Proc. Natl. Acad. Sci. USA* 80: 6922–2926.

Crossley, M. and Brownlee, G. G. 1990. Disruption of a C/EBP binding site in the factor IX promoter is associated with haemophilia B. *Nature* 345: 444–446.

Cullen, K. E., Kladde, M. P. and Seyfred, M. A. 1993. Interaction between transcriptional regulatory regions of prolactin chromatin. *Science* 261: 203–206.

Danesch, U., Gloss, B., Schmid, W., Schötz, G. and Renkawitz, R. 1987. Glucocorticoid induction of the rat tryptophan oxygenase gene is mediated by two widely separated glucocorticoid-responsive elements. *EMBO J.* 6: 625–630.

Danielsen, M. Hinck, L. and Ringold,G. M. 1989. Two amino acids within the knuckle of the first zinc finger specify DNA response element activation by the glucocorticoid receptor. *Cell* 57: 1131–1138.

Davis, M. M., Calame, K., Early, P. W., Livant, D. L., Joho, R., Weissman, I. L. and Hood, L. 1980a. An immunoglobulin heavy chain gene is formed by at least two recombinational events. *Nature* 283: 733–739.

Davis, M. M., Kim, S. K. and Hood, L. 1980b. Immunoglobulin class switching: Developmentally regulated DNA rearrangements during differentiation. *Cell* 22: 1–2.

DeCaprio, J. A. and seven others. 1989. The product of the retinoblastoma susceptibility gene has properties of a cell cycle regulatory element. *Cell* 58: 1085–1095.

Diamond, M. I., Miner, J. N., Yoshinaga, S. K. and Yamamoto, K. R. 1990. Transcription factor interactions: Selectors of positive and negative regulation from a single DNA element. *Science* 249: 1266–1272.

Dierks, P., van Ooyen, A., Chochran, M. D., Dobkin, C., Reiser, J. and Weissman, C. 1983. Three regions upstream from the cap site are required for efficient and accurate transcription of the rabbit β-globin gene in mouse 3T6 cells. *Cell* 32: 695–706.

Dollé, P., Castrillo, J.-L., Theill, L. E., Deerinck, T., Ellisman, M. and Karin, M. 1990. Expression of GHF-1 protein in mouse pituitaries correlates both temporally and spatially with the onset of growth hormone gene activity. *Cell* 60: 809–820.

Driever, W. and Nüsslein-Volhard, C. 1989. The bicoid protein is a positive regulator of *hunchback* transcription in the early *Drosophila* embryo. *Nature* 337: 138–143.

Drummond, I. A., Madden, S. L., Rohwer, N. P., Bell, G. I., Sukhatme, V. P. and Rauscher, F. III. 1992. Repression of the insulin-like growth factor II gene by the Wilms' tumor suppressor gene *WT1*. *Science* 257: 674–678.

Dynan, W. S. and Tjian, R. 1985. Control of eukaryotic messenger RNA synthesis by sequence-specific DNA-binding proteins. *Nature* 316: 774–778.

Dynlacht, B. D., Hoey, T. and Tjian, R. 1991. Isolation of cofactors associated with the TATA-binding protein that mediate transcriptional activation. *Cell* 66: 563–576.

Efstratiadis, A. and fourteen others. 1980. The structure and evolution of the human β-globin gene family. *Cell* 21: 653–668.

Elsholtz, H. P., Albert, V. R., Treacy, M. N. and Rosenfeld, M. G. 1990. A two-base change in a POU factor-binding site switches pituitary specific to lymphoid-specific gene expression. *Genes Dev.* 4: 43–51.

Fletcher, C., Heintz, N. and Roeder, R. 1987. Purification and characterization of OTF-1, a transcription factor regulating cell cycle expression of a human histone H2β gene. *Cell* 51: 773–781.

French, B. A., Chow, K.-L., Olson, E.N. and Schwartz, R. J. 1991. Heterodimers of myogenic regulatory factors and E12 bind a complex element governing myogenic induction of the avian α-actin promoter. *Mol. Cell. Biol.* 11: 2439–2450.

Friedman, A. D., Landschulz, W. H. and McKnight, S. L. 1989. CCAAT/enhancer binding protein activates the promoter of the serum albumin gene in cultured hepatoma cells. *Genes Dev.* 3: 1314–1322.

Friend, S. H., Bernards, R., Rogelj, S., Weinberg, R. A., Rapaport, J. M., Albert, D. M. and Dryja, T. P. 1986. A human DNA segment with properties of the gene that predisposes to retinoblastoma and osteosarcoma. *Nature* 323: 643–646.

Fujimoto, S. and Yamagishi, H. 1987. Isolation of an excision product of T cell receptor α-chain gene rearrangements. *Nature* 327: 242–243.

Gedamu, L. and Dixon, G. H. 1978. Effect of enzymatic decapping on protamine messenger RNA translation in wheat-germ S-30. *Biochem. Biophys. Res. Commun.* 85: 114–124.

Ghosh, S. and Baltimore, D. 1990. Activation in vitro of NF-κB by phosphorylation of its inhibitor, IκB. *Nature* 344: 678–682.

Gillies, S. D., Morrison, S. L., Oi, V. T. and Tonegawa, S. 1983. A tissue-specific transcription enhancer element is located in the major intron of a rearranged immunoglobulin heavy chain gene. *Cell* 33: 717–728.

Green, S. and Chambon, P. 1988. Nuclear receptors enhance our understanding of transcriptional regulation. *Trends Genet.* 4: 309–314.

Green, S., Kumar, V., Thenlaz, I., Wahli, W. and Chambon, P. 1988. The *N*-terminal DNA binding zinc finger of the oestrogen and glucocorticoid receptors determines target gene specificity. *EMBO J.* 7: 3037–3044.

Grosschedl, R. and Birnstiel, M. L. 1980. Spacer DNA upstream from the TATAATA sequence are essential for promotion of H2A histone gene transcription in vivo. *Proc. Natl. Acad. Sci. USA* 77: 7102–7106.

Grosveld, F. , de Boer, E., Shewmaker, C. K. and Flavell, R. A. 1982. DNA sequences necessary for the transcription of the rabbit β-globin gene in vivo. *Nature* 295: 120–126.

Gu, W. Schneider, J. W., Condorelli, G., Kaushal, S., Mahdavi, V. and Nadal-Ginard, B. 1993. Interaction of myogenic factors and the retinoblastoma protein mediates muscle cell commitment and differentiation. *Cell* 72: 309–324.

Haber, D. A and seven others. 1990. An internal deletion within an 11p13 zinc finger gene contributes to the development of Wilms' tumor. *Cell* 61: 1257–1269.

Henkel, T., Zabel, U., van Zee, K., Müller, J. M., Fanning, E. and Baeuerle, P. A. 1992. Intramolecular masking of the nuclear location signal and dimerization domain in the precursor for the p50 NFνκB subunit. *Cell* 68: 1121–1133.

Herr, W. and eleven others. 1988. The POU domain: A large conserved region in the mammalian *pit-1*, *oct-1*, *oct-2*, and *Caenorhabditis elegans unc-86* gene products. *Genes Dev.* 2: 1513–1516.

Hoey, T., Weinzierl, R. O. J., Gill, G., Chen, J.-L., Dynlacht, B. D., Tjian, R. 1993. Molecular cloning and functional analysis of *Drosophila* TAF110 reveal properties expected of coactivators. *Cell* 72: 247–260.

Horowitz, J. M. and eight others. 1990. Frequent inactivation of the retinoblastoma anti-oncogene is restricted to a subset of human tumor cells. *Proc. Natl. Acad. Sci. USA* 87: 2775–2779.

Hozumi, N. and Tonegawa, S. 1976. Evidence for somatic rearrangement of immunoglobulin genes coding for variable and constant regions. *Proc. Natl. Acad. Sci. USA* 73: 3628–3632.

Hunter, T. and Karin, M. 1992. The regulation of transcription by phosphorylation. *Cell* 70: 375–387.

Jones, N. 1990. Transcriptional regulation by dimerization: Two sides to an incestuous relationship. *Cell* 61: 9–11.

Kadonaga, J. T., Courey, A. J., Ladika, J. and Tjian, R. 1988. Distinct regions of Sp1 modulate DNA binding and transcriptional activation. Science 242: 1566–1570.

Kaestner, K. H., Christy, R. J. and Lane, M. D. 1990. Mouse insulin-responsive glucose transporter gene: Characterization of the gene and *trans*-activation by the CCAAT/enhancer binding protein. *Proc. Natl. Acad. Sci. USA* 87: 251–255.

Kaptein, R. 1992. Zinc-finger structures. *Curr. Opin. Struct. Biol.* 2: 109–115.

Karin, M., Haslinger, A., Holtgreve, H., Richards, R. I., Krautner, P., Westphal, H. M. and Beato, M. 1984. Characterization of DNA sequences through which cadmium and glucocorticoid hormones induce human metallothionein-II gene. *Nature* 308: 513–519.

Kelly, K., Cochrane, B. H., Stiles, C. D. and Leder, P. 1983. Cell-specific regulation of the c-*myc* gene by lymphocyte mitogens and platelet-derived growth factor. *Cell* 35: 603–610.

Kerr, L. D., Inoue, J.-I., Davis, N., Link, E., Baeurle, P. A., Bose, H. R. Jr., and Verma, I. M. 1991. The rel-associated pp40 protein prevents DNA binding at rel and NF-κB: Relationship with IκB and regulation by phosphorylation. *Genes Devel.* 5: 1464–1476.

Klock, G., Strähle, U. and Schüutz, G. 1987. Oestrogen and glucocorticoid responsive elements are closely related but distinct. *Nature* 329: 734–736.

Koleske, A. J., Buratowski, S., Nonet, M. and Young, R. A. 1992. A novel transcription factor reveals a functional link between the RNA polymerase II CTD and TFIID. *Cell* 69: 883–894.

Konigsberg, I. R., McElvain, N., Tootle, M. and Herrmann, H. 1960. The dissociability of deoxyribonucleic acid synthesis from the development of multinuclearity of muscle cells in culture. *J. Biophys. Biochem. Cytol.* 8: 333–343.

Kreidberg, J. A., Saviola, H., Loring, J. M., Maeda, M., Pelletier, J., Housman, D. and Jaenisch, R. 1993. WT-1 is required for early kidney development. *Cell* 74: 679–691.

Kumar, V., Green, S., Stack, G., Berry, M., Jin, J.-R. and Chambon, P. 1987. Functional domains of the human estrogen receptor. *Cell* 51: 941–951.

Landschulz, W. H., Johnson, P. F. and McKnight, S. L. 1988. The leucine zipper: A hypothetical structure common to a new class of DNA-binding proteins. *Science* 240: 1759–1764.

Lassar, A. B. and seven others. 1991. Functional activity of myogenic HLH proteins requires hetero-oligomerization with E12/E47-like proteins in vivo. *Cell* 66: 305–315.

Lawn, R. M., Efstratiadis, A., O'Connell, C. and Maniatis, T. 1980. The nucleotide sequence of the human β-globin gene. *Cell* 21: 647–651.

Laybourn, P. J. and Kadonaga, J. T. 1991. Role of nucleosome cores and histone H1 in regulation of transcription by RNA polymerase II. *Science* 254: 238–245.

Leder, P. and seven others. 1983. Translocations among antibody genes in human cancer. *Science* 222: 765–771.

Lee, D. K., Horikoshi, M. and Roeder, R. G. 1991. Interaction of TFIID in the minor groove of the TATA element. *Cell* 67: 1241–1250.

Lee, E. and seven others. 1987. The retinoblastoma susceptibility gene encodes a nuclear phosphoprotein associated with DNA binding activity. *Nature* 329: 642–645.

Lenardo, M. J. and Baltimore, D. 1989. NF-κB: A pleiotropic mediator of inducible and tissue-specific gene control. *Cell* 58: 227–229.

Li, L., Zhou, J., Guy, J., Heller-Harrison, R., Czech, M. P. and Olson, E. N. 1992. FGF inactivates myogenic helix–loop–helix proteins through phosphorylation of a conserved protein kinase C site in their DNA-binding domains. *Cell* 71: 1181–1194.

Li, S., Crenshaw, E. B. III, Rawson, E. J., Simmons, D. M., Swanson, L. W. and Rosenfeld, M. G. 1990. *Dwarf* locus mutants lacking three pituitary cell types result from mutations in the POU-domain gene *pit-1*. *Nature* 347: 528–533.

Lin, F.-T. and Lane, M. D. 1992. Antisense CCAAT/enhancer-binding protein RNA suppresses coordinate gene expression and triglyceride accumulation during differentiation of 3T3-L1 pre-adipocytes. *Genes Dev.* 6: 533–544.

Lin, Y.-S. and Green, M. R. 1991. Mechanism of action of an acidic transcriptional activator *in vitro*. *Cell* 64: 971–981.

Lin, Y.-S., Ha, I., Maldonado, E., Reinberg, D. and Green, M. R. 1991. Binding of general transcription factor TFIIB to an acidic activating region. *Nature* 353: 569–571.

Lu, H., Zawel, L., Fisher, L., Egly, M. and Reinberg, D. 1992. Human general transcription factor IIH phosphorylates the C-terminal domain of RNA polymerase II. *Nature* 358: 641–645.

Lucas, P. C. and Granner, D. K. 1992. Hormone responsive domains in gene transcription. *Annu. Rev. Biochem.* 61: 1131–1173.

Maldonado, E., Ha, I., Cortes, P., Weis, L. and Reinberg, D. 1990. Factors involved in specific transcription by mammalian RNA polymerase II: Role of transcription factors IIA, IID, and IIB during formation of a transcription competent complex. *Mol. Cell Biol.* 10: 6335–6347.

Maniatis, T., Goodbourn, S. and Fischer, J. A. 1987. Regulation of inducible and tissue-specific gene expression. *Science* 236: 1237–1245.

Mantovani, R. and eight others. 1988. An erythroid-specific nuclear factor binding to the proximal CACCC box of the β-globin gene promoter. *Nucl. Acid Res.* 16: 4299–4313.

Martinez, E., Givel, F. and Wahl, W. 1987. An estrogen-responsive element as an inducible enhancer: DNA sequence requirements and conversion to a glucocorticoid-responsive element. *EMBO J.* 6: 3719–3727.

Marx, J. 1992. Taking a direct path to the genes. *Science* 257: 744–745.

Mather, E. L. and Perry, R. P. 1982. Transcriptional regulation of the immunoglobulin V genes. *Nucleic Acids Res.* 9: 6855–6867.

Matsuoka, M., Nagawa, F., Okazaji, K., Kingsbury, L., Yoshida, K., Muller, U., Larue, D.T., Winer, J. A. and Sakano, H. 1991. Detection of somatic DNA recombination in the transgenic mouse brain. *Science* 254: 81–86.

McKnight, S. and Tjian, R. 1986. Transcriptional selectivity of viral genes in mammalian cells. *Cell* 46: 795–805.

Merino, A., Madden, K. R., Lane, W. S., Champoux, J. J. and Reinberg, D. 1993. DNA topoisomerase I is involved in both repression and activation of transcription. *Nature* 365: 227–232.

Miesfeld, R. and seven others. 1986. Genetic complementation of a glucocorticoid receptor deficiency by a cloned receptor cDNA. *Cell* 46: 389–399.

Milos, P. M. and Zaret, K. S. 1992. A ubiquitous factor is required for C/EBP-related proteins to form stable transcription complexes on an ovalbumin promoter segment in vitro. *Genes Dev.* 6: 183–196.

Myers, R. M., Tilly, K. and Maniatis, T. 1986. Fine structure genetic analysis of a β-globin promoter. *Science* 232: 613–618.

Nadal-Ginard, B. 1978. Commitment, fusion, and biochemical differentiation of a myogenic cell line in the absence of DNA synthesis. *Cell* 15: 855–866.

Nishikura, K., Rushdim, A., Erikson, J., Watt, R., Rovera, G. and Croce, C. M. 1983. Differential expression of the normal and of the translocated human c-*myc* oncogenes in B cells. *Proc. Natl. Acad. Sci. USA* 80: 4822–4286.

Oettinger, M. A., Schatz, D. G., Gorka, C. and Baltimore, D. 1990. *RAG-1* and *RAG-2*, adjacent genes that synergistically activate V(D)J recombination. *Science* 248: 1517–1522.

Olson, E. N. 1992. Interplay between proliferation and differentiation within myogenic lineage. *Dev. Biol.* 154: 261–272.

Orkin, S. and Kazazian, H. H. 1984. The mutation and polymorphism of the human β-globin gene and its surrounding DNA. *Annu. Rev. Genet.* 18: 131–171.

Pabo, C. O. and Sauer, R. T. 1992. Transcription factors: Structural families and principles of DNA recognition. *Annu. Rev. Biochem.* 61: 1053–1095.

Palmiter, R. D., Brinster, R. L., Hamm, R. E., Trumbauer, M. E., Rosenfeld, M. G., Birnberg, N. C. and Evan, R. M. 1982. Dramatic growth of mice that develop from eggs microinjected with metallothionein growth hormone fusion genes. *Nature* 300: 611–615.

Parslow, T. G., Blair, D. L., Murphy, W. J. and Granner, D. K. 1984. Structure of the 5'ends of immunoglobulin genes: A novel conserved sequence. *Proc. Natl. Acad. Sci. USA* 81: 2650–2654.

Patapoutian, A., Miner, J. H., Lyons, G. E. and Wold, B. 1993. Isolated sequences from the linked *Myf-5* and *MRF-4* genes drive distinct patterns of muscle-specific expression in transgenic mice. *Development* 118: 61–69.

Pathak, D. and Sigler, P. B. 1992. Updating structure–function relationships in the bZip family of transcription factors. *Curr. Opin. Struct. Biol.* 2: 116–123.

Payvar, F. and seven others. 1983. Sequence-specific binding of glucocorticoid receptor to MTV DNA at sites within and upstream of the transcribed region. *Cell* 35: 381–392.

Pfahl, M. 1982. Specific binding of the glucocorticoid–receptor complex to the mouse mammary tumor proviral promoter region. *Cell* 31: 475–482.

Picard, D. and Schaffner, W. 1984. A lymphocyte-specific enhancer in the mouse immunoglobulin κ gene. *Nature* 307: 80–82.

Pu, W.T. and Struhl, K. 1991. The leucine zipper symmetrically positions the adjacent regions for specific DNA binding. *Proc. Natl. Acad. Sci. USA* 88: 6901–6905.

Pugh, B. F. and Tjian, R. 1991. Transcription from a TATA-less promoter requires a multisubunit TFIID complex. *Genes Dev.* 5: 1935–1944.

Queen, C. and Baltimore, D. 1983. Immunoglobulin gene transcription is activated by downstream sequence elements. *Cell* 33: 741–748.

Rabbitts, T, H, 1991. Translocations, master genes, and the differences between the origins of acute and chronic leukemias. *Cell* 67: 641–644.

Rhodes, S. J. and seven others. 1993. A tissue-specific enhancer confers Pit-1-dependent morphogen inducibility on the *pit-1* gene. *Genes Dev.* 7: 913–932.

Rigby, P. W. J. 1993. Three in one and one in three: It all depends on TBP. *Cell* 72: 7–10.

Roman, C., Dohn, L. and Calame, K. 1991. A dominant negative form of transcription activator mTFE3 created by differential splicing. *Science* 254: 94–97.

Rottman, F. A., Shatkin, A. J. and Perry, R. P. 1974. Sequences containing methylated nucleotides at the 5' termini of messenger RNAs: Possible applications for processing. *Cell* 3: 197–199.

Rutter, W., Jr., Valenzuela, P., Ball, G. E., Holland, M., Hager, G. L., Degennero, L. J. and Bishop, R. J. 1976. The role of DNA-dependent RNA polymerase in transcriptive specificity. *In* E. M. Bradbury and K. Jaucherian (eds.), *The Organization and Expression of the Eukaryotic Genome*. Academic Press, New York, pp. 279–293.

Samuelsson, L., Strömberg, K., Vikman, K., Bjursell, G. and Enerbäck, S. 1991. The CCAAT/enhancer binding protein and its role in adipocyte differentiation: Evidence for direct involvement in terminal adipocyte development. *EMBO J.* 10: 3787–3793.

Sawadogo, M. and Roeder, R. G. 1984. Energy requirement for specific transcription by the human RNA polymerase II system. *J. Biol. Chem.* 259: 5321–5326.

Schatz, D. G., Oettinger, M. A. and Baltimore, D. 1989. The V(D)J recombination activating gene, *RAG-1. Cell* 59: 1035–1048.

Schindler, C., Shuai, K., Prezioso, V. R. and Darnell, J. E., Jr. 1992. Interferon-dependent tyrosine phosphorylation of a latent cytoplasmic transcription factor. *Science* 257: 809–815.

Schüle, R., Muller, M., Kaltschmidt, C. and Renkawitz, R. 1988. Many transcription factors interact synergistically with steroid receptors. *Science* 242: 1418–1420.

Sen, R. and Baltimore, D. 1986a. Multiple nuclear factors interact with the immunoglobulin enhancer sequences. *Cell* 46: 705–716.

Sen, R. and Baltimore, D. 1986b. Inducibility of κ immunoglobulin enhancer-binding protein NF-κB by a posttranslational mechanism. *Cell* 47: 921–928.

Shatkin, A. J. 1976. Capping of eucaryotic mRNAs. *Cell* 9: 645–653.

Sheiness, D. and Darnell, J. E. 1973. Polyadenylic segment in mRNA becomes shorter with age. *Nature New Biol.* 241: 265–268.

Simmons, D. M., Voss, J. W., Ingraham, H. A., Holloway, J. M., Broide, R. S., Rosenfeld, M. G. and Swanson, L. W. 1990. Pituitary cell phenotypes involve cell-specific *Pit-1* mRNA translation and synergistic interactions with other classes of transcription factors. *Genes. Dev.* 4: 695–711.

Sopta, M., Burton, Z. F. and Greenblatt, J. 1989. Structure and associated DNA helicase activity of a general transcription factor that binds to RNA polymerase II. *Nature* 341: 410–414.

Starr, D. B. and Hawley, D. K. 1991. TFIID binds in the minor groove of the TATA box. *Cell* 67: 1231–1240.

Staudt, L. M., Singh, H., Sen, R., Wirth, T., Sharp, P. A. and Baltimore, D. 1986. A lymphoid-specific protein binding to the octamer motif of immunoglobulin genes. *Nature* 323: 640–643.

Stockdale, F. E. and Holter, H. 1961. DNA synthesis and myogenesis. *Exp. Cell Res.* 24: 508–520.

Strähle, U., Schmid, W. and Schütz, G. 1988. Synergistic action of the glucocorticoid receptor with transcription factors. *EMBO J.* 7: 3389–3395.

Struhl, G., Struhl, K. and Macdonald, P. M. 1989. The gradient morphogen *bicoid* is a concentration–dependent transcriptional activator. *Cell* 57: 1259–1273.

Tanaka, M., Lai, J.-S. and Herr, W. 1992. Promoter-specific activation domains in *Oct-1* and *Oct-2* direct differential activation of an snRNA and mRNA promoter. *Cell* 68: 755–767.

Treisman, J., Gönczy, P., Vashishta, M., Harris, E. and Desplan, C. 1989. A single amino acid can determine the DNA binding specificity of homeodomain proteins. *Cell* 59: 553–562.

Trudel, M. and Constantini, F. 1987. A 3' enhancer contributes to the stage-specific expression of the human β-globin gene. *Genes Dev.* 1: 954–961.

Tsai, S. Y., Tsai, M.-J., and O'Malley, B. W. 1989. Cooperative binding of steroid hormone receptors contributes to transcriptional synergism at target enhancer elements. *Cell* 57: 443–448.

Umek, R. M., Friedman, A. D. and McKnight, S. L. 1991. CCAAT–enhancer binding protein: A component of a differentiation switch. *Science* 251: 288–292.

Umesono, K. and Evans, R. M. 1989. Determinants of target gene specificity for steroid/thyroid hormone receptors. *Cell* 57: 1139–1146.

Usheva, A., Maldonado, E., Goldring, A., LuH., Houbavi, C., Reinberg, D. and Aloni, Y. 1992. Specific interaction between the nonphosphorylated form of RNA polymerase II and the TATA-binding protein. *Cell* 69: 871–881.

Vinson, C. R., Sigler, P. B. and McKnight, S. L. 1989. Scissors–grip model for DNA recognition by a family of leucine zipper proteins. *Science* 246: 911–916.

Walker, M. D., Edlund, T., Boulet, A. M. and Rutter, W. J. 1983. Cell-specific expression controlled by the 5' flanking region of the insulin and chymotrypsin genes. *Nature* 306: 557–561.

Wasylyk, B., Kedinger, C., Corden, J., Brison, D. and Chambon, P. 1980. Specific in vitro initiation of transcription on conalbumin and ovalbumin genes and comparison with adenovirus 2 early and late genes. *Nature* 285: 367–373.

Watson, J. D., Gilman, M., Witkowski, J., and Zoller, M. 1992. *Recombinant DNA*, 2nd Ed. Scientific American Books, New York.

Wirth, T., Staudt, L. and Baltimore, D. 1987. An octamer oligonucleotide upstream of a TATA motif is sufficient for lymphoid-specific promoter activity. *Nature* 329: 174–178.

Workman, J. L. and Roeder, R. G. 1987. Binding of transcription factor TFIID to the major late promoter during in vitro nucleosome assembly potentiates subsequent initiation by RNA polymerase II. *Cell* 51: 613–622.

Wright, G. and eight others. 1991. High level expression of active human α-1-antitrypsin in the milk of transgenic sheep. *BioTech.* 9: 830–834.

Yeo, S. P. and Rigby, P. W. J. 1993. The regulation of *myogenin* gene expression during embryonic development of the mouse. *Genes Dev.* 7: 1277–1287.

11

Transcriptional regulation of gene expression

The activation of chromatin

Up to this point, we have limited our discussion of messenger RNA transcription to the structure of the gene itself. But genes do not exist in an uncovered state within the nucleus, readily accessible to RNA polymerase or to any enhancer- or promoter-binding protein. Rather, eukaryotic chromosomes contain as much protein (by weight) as nucleic acid, and this DNA-protein complex is called **chromatin**. The most abundant of the chromatin proteins are basic polypeptides called **histones**, which are organized into **nucleosomes**.

The nucleosome is the basic unit of chromatin structure. It is composed of a histone octamer (two molecules each of histones H2A-H2B and histones H3-H4) wrapped about with two loops of approximately 140 base pairs of DNA (Figure 11.1; Kornberg and Thomas, 1974). Chromatin can thus be visualized as a string of nucleosome beads linked by from 10 to 100 base pairs of DNA. While classical geneticists have likened genes to "beads on a string," molecular geneticists liken genes to "string on the beads."

Transcription factors must be able to find the DNA sequences despite the fact that most of the DNA is packaged in nucleosomes. It is currently thought that rendering a gene competent to transcribe RNA involves (1) the binding of transcription factors to the DNA and (2) the exclusion of nucleosomes from the promoter region of the gene. The interactions between specific transcription factors and the DNA they bind cause the phenomenon of tissue-specific and temporally specific gene transcription.

Nucleosomes aren't the only structural impediment to the binding of transcription factors to their DNA sequences, because the nucleosomes are themselves wound into tight "solenoids" that are stabilized by histone H1. Histone H1 is found in the 60 or so base pairs of "linker" DNA between the nucleosomes (Figures 11.1 and 11.2; Weintraub, 1984). This H1-dependent conformation of nucleosomes inhibits the transcription of genes in somatic cells by packing adjacent nucleosomes together into tight arrays that prohibit the access of transcription factors and RNA polymerases (Thoma et al., 1979; Schlissel and Brown, 1984).

FIGURE 11.1
Nucleosome and chromatin structure.
(A) Model of nucleosome structure as
seen by X-ray crystallography at a resolution of 0.33 nm. The central bicon-
cave protein unit (white) is the H3-H4
tetramer. Each of the two dark ovoids
flanking the tetramer is an H2A-H2B
dimer. (B) Relationship of histone H1
to the core nucleosome (containing
two copies each of histones H2A, H2B,
H3, and H4). (C) H1 can link the DNA
into compact forms and can draw nucleosomes together. About 140 base
pairs of DNA encircle the histone octamer, and about 60 base pairs of DNA
link the nucleosomes together. (D)
Model for the arrangement of nucleosomes into the highly compacted, sole-
noidal chromatin structure. (A from
Burlingame et al., 1985; B–D after
Wolfe, 1993.)

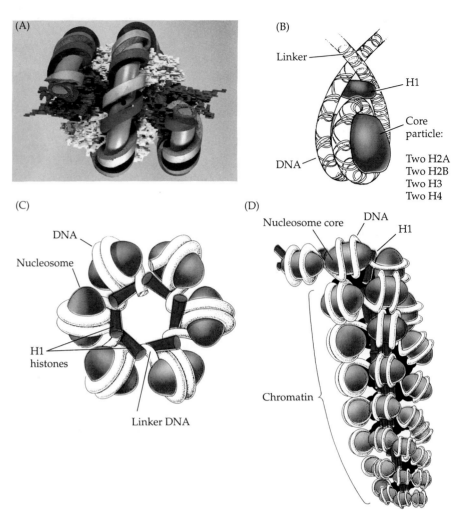

Activating repressed chromatin

Accessibility to *trans*-regulatory factors

It is incredible that DNA can become accessible to *trans*-regulatory factors
at all. There is enough DNA in a single human body to extend the diameter
of the solar system (Crick, 1966), and this enormous length must be packed
up tightly into the nuclei of our cells. Yet our genetic library can be
accessed specifically in each cell type. Solution hybridization experiments
suggest that there is a minimum of 10,000 tissue-specific genes in the
genome of most vertebrates; so it is not surprising that in any given cell
type most of these genes are repressed. It is generally thought, then, that
the "default" condition of chromatin is a repressed state and that tissue-
specific genes become activated by locally interrupting the repressive factors (Weintraub, 1985). As mentioned above, the major mechanism of
general gene repression is probably the compaction of DNA into clusters
of nucleosomes, and the initiation of transcription depends on removing
nucleosomes from the promoter region of the gene. There are three ways
that this might be accomplished. First, during DNA synthesis (S phase in
the cell cycle), nucleosomes are removed from a strand of the DNA and
are replaced shortly thereafter. At this time of replacement, there could
be competition for the promoter sites between histone H1 and transcription factors such as TATA-binding TFIID. Second, in cells that are not

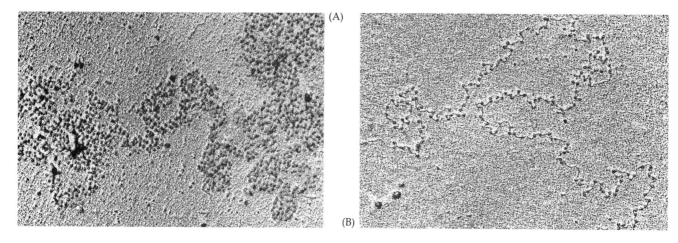

(A)

(B)

FIGURE 11.2
The role of H1 in compaction of chromatin. (A) Chicken liver chromatin observed in the electron microscope. The beads represent the nucleosomes. (B) The same chromatin after the removal of histone H1 by salt elution. The chromatin has become far less compact. (From Oudet et al., 1975; photograph courtesy of P. Chambon.)

dividing there may be an equilibrium between histone H1 and various transcription factors. This equilibrium could be shifted by varying the concentrations of transcription factors or by synthesizing factors that stabilize these transcriptional regulators (Laybourne and Kadonaga, 1991; Felsenfeld, 1992). Third, there appear to be certain transcriptional activators (such as the glucocorticoid receptor) that can bind to existing nucleosomes and disrupt them (Rigaud et al., 1991; Adams and Workman, 1993). Once the nucleosomes are dissociated in the promoter region, other transcription factors can bind (Figure 11.3).

The ability of transcription factors to remove nucleosomes from active

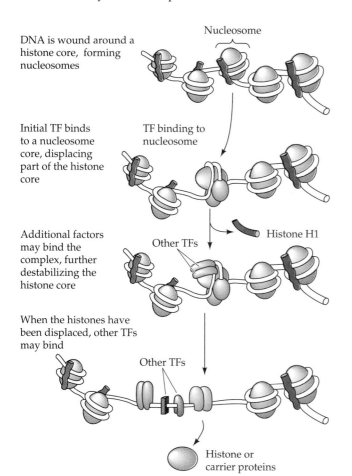

DNA is wound around a histone core, forming nucleosomes

Nucleosome

Initial TF binds to a nucleosome core, displacing part of the histone core

TF binding to nucleosome

Additional factors may bind the complex, further destabilizing the histone core

Other TFs

Histone H1

When the histones have been displaced, other TFs may bind

Other TFs

Histone or carrier proteins

FIGURE 11.3
Binding of a transcription factor (TF) to the nucleosome can destabilize the nucleosome, enable the histones to be removed, and open the region for other transcription factors. (After Adams and Workman, 1993).

FIGURE 11.4
Protocol for determining the specificity of DNase I digestion of chromatin. See Table 1 for results of digestion experiment.

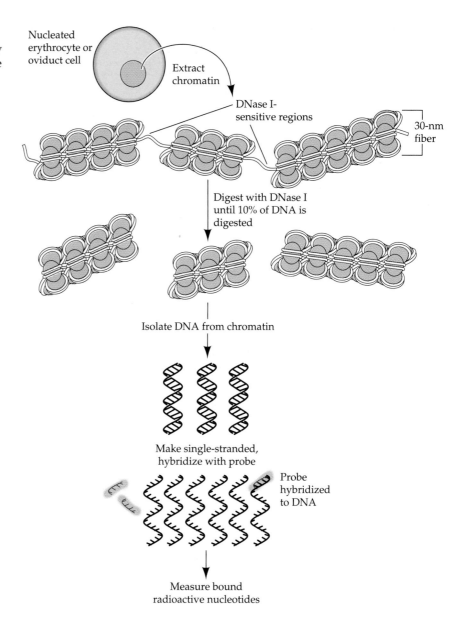

genes and their promoters can be seen by nuclease experiments. The accessibility of a gene to nuclear proteins can be detected by treating the chromatin of a tissue with small amounts of **DNase I**. This pancreatic DNase digests accessible regions of DNA, but the DNA covered by nucleosomes is protected. After digestion, the DNA of the treated chromatin is extracted and mixed with radioactive cDNA for a particular gene (Figure 11.4). If the cDNA finds sequences to bind to, then the gene has been protected from digestion by chromatin proteins—that is, it was not accessible to the DNase, and it probably would not be accessible to transcription factors or RNA polymerase, either. However, if the cDNA probe does not find sequences to bind, then the gene has been exposed to the DNase and probably would be accessible to RNA polymerase and *trans*-regulatory factors.

The DNase I sensitivity of a given gene was found to be dependent upon the cell type in which it resides (Table 11.1; Weintraub and Groudine, 1976). When chromatin from developing chick red blood cells was treated with DNase I and its DNA extracted and mixed with radioactive globin cDNA, the globin cDNA found little with which to bind. The globin genes

TABLE 11.1
Binding studies with DNase I-treated chromatin

Source (chick)	DNase treatment	Radioactive cDNA probe	Percent maximum binding of radioactive cDNA to DNA extracted from treated chromatin
Red blood cell DNA	–	Globin cDNA	94
Brain cell chromatin	+	Globin cDNA	90–100
Fibroblast chromatin	+	Globin cDNA	90–100
Red blood cell chromatin	+	Globin cDNA	25
Red blood cell chromatin	+	Ovalbumin cDNA	90–100

Source: After Weintraub and Groudine, 1976.

in the chromatin had been digested by the small amounts of DNase I. However, treating brain cell chromatin with the same amounts of DNase I did not destroy the globin genes. Therefore, the globin gene was accessible to outside enzymes in developing red blood cell chromatin but not in brain cell chromatin. Similarly, the ovalbumin (egg white) gene is susceptible to DNase I digestion in oviduct chromatin but not in red blood cell chromatin. When the chromatin is treated with DNase I and the DNA is extracted, ovalbumin cDNA is able to find sequences in the erythrocyte preparation but not in the DNA from the treated oviduct chromatin. Here, then, we have clear correlation of differential gene regulation and chromatin structure.

DNase-hypersensitive sites

The DNase I-sensitive sites include large stretches of DNA that contain the genes active in the particular cell. These regions usually contain nucleosomes, but the nucleosomes are not packed together. It is thought that as RNA polymerase passes through a region of the gene containing a nucleosome, the nucleosome comes apart so that the polymerase can continue transcription. The nucleosome will reform on the DNA behind the polymerase (Clark and Felsenfeld, 1992). So the complete absence of nucleosomes in the structural gene is not required for transcription to occur. However, the complete absence of nucleosomes in the *promoter* region of the gene may be essential.

The correlation between nucleosome-free promoter regions and active transcription was shown by looking more closely at the regions digested by DNase I. In addition to these DNase I-*sensitive* sites, there exist **DNase I-*hypersensitive* sites**. These sites, identified on Southern blots by small radioactive DNA fragments, are destroyed by extremely minute amounts of DNase I, indicating that they are extremely accessible to outside molecules. This accessibility appears to arise from the nearly total absence of nucleosomes in this region of DNA (Elgin, 1988). The DNase I-hypersensitive sites mark regions of the chromatin, such as active promoters and enhancers, where DNA-binding proteins are bound. DNase I-hypersensitive regions are therefore associated with tissue-specific developmentally regulated genes (Elgin, 1981; Conklin and Groudine, 1984). For example, globin genes in red blood cells and their immediate precursors contain DNase I-hypersensitive sites, but globin genes in other cells do not (Stalder et al., 1980; Groudine et al., 1983). The 5′ flanking region of the chick vitellogenin gene contains several hypersensitive sites in the chromatin from the liver of laying hens; but these sites are not present in male liver chromatin, embryonic liver chromatin, or brain or lymphocyte chromatin (Burch and Weintraub, 1983). When a DNase I-hypersensitive site was deleted from the 5' region of a developmentally regulated salivary protein gene in *Drosophila*, no transcripts were detectable, and no RNA polymerase molecules bound to that gene's promoter (Shermoen and Beckendorf, 1982; Steiner et al., 1984). Similarly, the gene for serum albumin is transcribed in mouse liver at rates over 1000 times greater than in any other mouse tissue. In liver only, there are seven DNase I-hypersensitive sites that map to the promoter and enhancer regions. Using a gel retardation assay, Liu and co-workers (1988) found that liver nuclei contain *trans*-regulatory factors that bind specifically to the DNA sequences within these regions.

DNase I hypersensitive sites often reside within or adjacent to sites that have enhancer functions, and certain *trans*-regulatory factors are able to induce the formation of these hypersensitive sites. Zaret and Yamamoto (1984), studying the glucocorticoid-responsive enhancer of the mouse mammary tumor virus, demonstrated that before the addition of the hormone to cells containing the virus, this enhancer sequence showed no special DNase I sensitivity. After the hormone was administered, a discrete DNase I-hypersensitive site developed in this region. The formation of the hypersensitive site coincided with the initiation of viral gene transcription; when the hormone was withdrawn, both the hypersensitive site and viral gene transcription disappeared. Zaret and Yamamoto speculated that the interaction between the glucocorticoid receptor complex and the enhancer DNA alters the chromatin configuration to facilitate transcription from the nearby promoter. This would be the case if, as mentioned above, the glucocorticoid receptor could remove nucleosomes from the region of its DNA-binding sequence.

In some instances, the binding of a factor to one hypersensitive site can cause the formation of new hypersensitive sites and facilitate transcription. This is seen to occur in the chick oviduct, where the progesterone receptor appears to bind to sites opened by the estrogen receptor. The administration of estrogen to immature chicks causes the formation of four new DNase I-hypersensitive sites in the promoter region of the ovalbumin gene. These sites occur only in the oviduct cells (which will synthesize and secrete ovalbumin) and not in the ovalbumin genes of any other tissue. If these sites are not formed, transcription does not occur. One of these DNase I-hypersensitive sites contains the TATA sequence of the promoter where the transcription apparatus is initiated (Kaye et al., 1986). When the estrogen is withdrawn, three of these sites disappear. The promoters and enhancers of numerous tissue-specific genes are seen to lie within DNase I-hypersensitive sites that allow for the binding of constitutive and tissue-specific transcription factors.

Thus, some *trans*-regulatory factors appear to compete with nucleosome formation and create regions of DNA near the promoter that are devoid of nucleosomes. These enable other *trans*-regulatory factors to bind. Other studies also indicate that most (if not all) of the DNase I-hypersensitive regions are devoid of nucleosomes. Rose and Garrard (1984) have shown that the immunoglobulin light chain genes become DNase I-sensitive as they proceed from pre-B cells to B cells. Moreover, at the same time, the repeating nucleosome pattern changes, causing large regions of the immunoglobulin gene to be relatively free of nucleosomes. The nucleosomes that do remain appear to be devoid of histone H1 and to contain instead nonhistone chromatin proteins HMG 14 and 17 proteins.

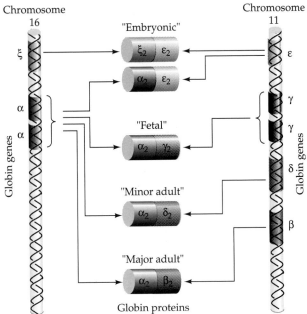

FIGURE 11.5
Sequential gene activation in hemoglobin synthesis during development.

Locus control regions and globin gene transcription

The activation of the globin genes in developing red blood cells involves some of the most interesting and complex forms of transcriptional regulation. In many species, including chicks and humans, the embryonic or fetal hemoglobin differs from that found in adult red blood cells. A schematic diagram of human hemoglobin types and the genes that code for them is shown in Figure 11.5. Human **embryonic hemoglobin** consists largely of two zeta (ζ) globin chains, two epsilon (ϵ) globin chains, and four molecules of heme. During the second month of human gestation, ζ- and ϵ-globin synthesis abruptly ceases, while alpha (α) and gamma (γ) globin synthesis increase (Figure 11.6). The association of two γ-globin chains with two α-globin chains produces **fetal hemoglobin**. At 3 months' gestation, the beta (β) globin and delta (δ) globin genes begin to be active, and their products slowly increase while γ-globin levels gradually decline.

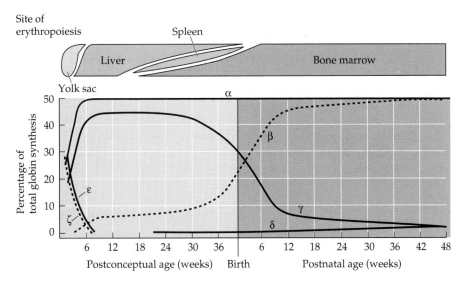

FIGURE 11.6
Percentages of hemoglobin polypeptide chains as a function of human development. The physiological importance of the γ-globin chain in fetal hemoglobin was examined in Chapter 9. (After Karlsson and Nienhaus, 1985.)

FIGURE 11.7

Diagram of the human β-globin family of genes on chromosome 11. (A) The erythroid-specific LCR region is located 6–22 kilobases upstream of the ε-globin gene. The four DNase I-hypersensitive sites within this region are marked by arrows. A fifth DNase I-hypersensitive site downstream from the β-globin gene is also marked, and a deletion of this region causes the persistence of γ-globin gene transcription. Downstream from the ε-globin (embryonic) gene are two nearly identical γ-globin (fetal) genes. These are followed by the adult δ- and β-globin genes. (B) One possible model for LCR activity. Transcription factors binding to the globin promoters are stabilized at the replication fork by binding to the LCR. The complex would not be dissociated and the regions associated with the LCR would remain devoid of nucleosomes. (A after Ryan et al., 1989; B after Felsenfeld, 1992.)

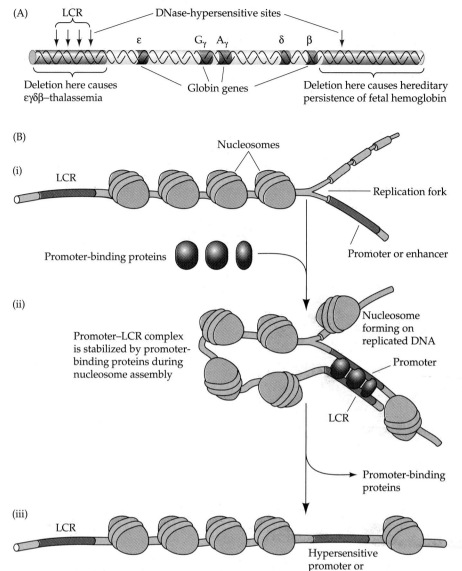

This switchover is greatly accelerated after birth, and fetal hemoglobin is replaced by **adult hemoglobin**: $\alpha_2\beta_2$. The normal adult hemoglobin profile is 97 percent $\alpha_2\beta_2$, 2-3 percent $\alpha_2\delta_2$, and 1 percent $\alpha_2\gamma_2$.

In humans, the ζ- and α-globin genes are on chromosome 16, but the ε-, γ-, δ-, and β-globin genes are linked together, in order of appearance, on chromosome 11. It appears, then, that there is a mechanism that directs the sequential switching of the chromosome 11 genes from embryonic, to fetal, to adult globins.

In the chromosome containing the human β-globin gene family, there are DNase-hypersensitive regions in several areas containing promoters and enhancers. Moreover, some of these regions become more sensitive to DNase during the switch from fetal to adult hemoglobin (Groudine et al., 1983). There is also a region of DNA that is exceptionally sensitive to DNase I that is found 6–22 kilobases upstream of the ε-globin gene (Figure 11.7A). This **locus control region** (LCR; formerly called the dominant control region, master enhancer, or locus-activating region) contains four sites that are DNase-hypersensitive only in erythroid precursor cells and that appear to be necessary for activating high levels of erythroid cell-specific transcription from the entire β-globin gene family (ε-, γ-, β-, and δ-globins) on human chromosome 11 (Grosveld et al., 1987). Deletion of the LCR causes the silencing of all these genes. Conversely, if the LCR is placed adjacent to genes that are not usually expressed in red blood cells (such as the T-cell-specific *thy-1* gene) and then transfected into erythroid precursor cells, these new genes are expressed in red blood cells. This effect is specific for red blood cell precursors, since only they would have the appropriate *trans*-regulatory factors to bind to this region (Blom van Assendelft et al., 1989).

Whereas typical enhancers and promoters activate (or repress) a single gene, the LCR is responsible for permitting gene expression in an entire region. Moreover, the LCR enables the transcription of high levels of these globin genes no matter where in the chromosome the globin gene complex (including the LCR) may happen to lie. Ryan and co-workers (1989) constructed transgenic mice containing the human β-globin gene and its immediate promoters and enhancers. These transgenic mice made only small amounts of human β-globin (less that 0.3 percent of the total cell β-globin). However, when the researchers added the LCR, human β-globin accounted for over half the total globin in these mice. This result explains medical observations that patients who lacked this region had deficiencies of ε-, γ-, δ-, and β-globins, even though their genes for these proteins were intact and the globin genes on the other chromosome functioned normally (Tuan et al., 1987).

The locus control region is crammed with DNA binding sites for known *trans*-regulatory factors. As Gary Felsenfeld (1992) observed, "The domains look as though they were put together by an overenthusiastic student determined to construct a powerful *cis*-acting element." He suggests that one of the functions of the LCR is to loop around to the promoter region during DNA replication and bind to it in a manner that prevents nucleosomes from forming on the globin promoters (Figure 11.7B). Indeed, the globin promoters are not DNase I-hypersensitive except in the presence of the LCR.

SIDELIGHTS & SPECULATIONS

Globin gene switching

Although it is obvious that globin gene transcription switches from embryonic to fetal to adult isoforms during development, we do not know the mechanism by which this takes place. Recent models of globin switching have focused on competition and cooperation between enhancers, promoters, and the locus control region. Before encountering the complicated regulatory network for human globin transcription, we should first observe a slightly simpler globin gene switching system—that of the chick erythrocyte.

Chick globin gene switching

In chicks, the transition from embryonic to adult globin synthesis (there being no "fetus" stage in chickens) appears to involve competetion for the enhancer region between promoters of the embryonic and adult globin genes. In

chick erythroid cells, the transcription factor NF-E4 is found only in definitive "adult" red blood cells. This protein binds to a DNA sequence in the β-globin gene *promoter*. NF-E4 also binds to two *trans*-regulatory *proteins* that are bound to the β-globin gene *enhancer*. Without this factor, the β-globin gene does not form a stable structure; but once NF-E4 binds the promoter and enhancer regions together, the gene is ready for transcription (Figure 11.8; Gallarda et al., 1989). The contact between enhancer and promoter may be critical for the transcriptional activation of developmentally regulated genes.

Choi and Engel (1988) found that in the absence of the β-globin gene, ε-globin genes were continually active, even in the adult chick. Furthermore, the replacement of the ε-globin promoter by the β-globin promoter resulted in the expression of ε-globin only in the adult stages. These results suggested that the β-globin promoter contained a region that was critical for stage-specific regulation. Their data also suggested that there is competition between the

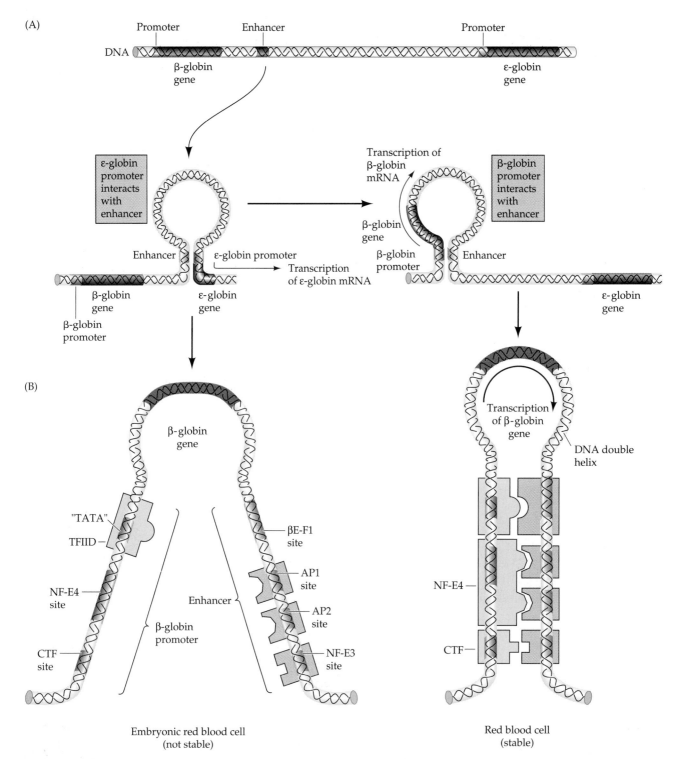

(A)

Promoter   Enhancer   Promoter

DNA

β-globin   ε-globin
gene   gene

ε-globin
promoter
interacts
with
enhancer

Transcription of
β-globin
mRNA

β-globin
promoter
interacts
with
enhancer

Enhancer   ε-globin promoter

β-globin
gene   Transcription
of ε-globin mRNA

β-globin
gene

β-globin
gene   Enhancer

ε-globin
gene

β-globin
promoter   ε-globin
gene

β-globin
promoter

(B)

Transcription
of β-globin
gene

DNA double
helix

β-globin
gene

"TATA"

TFIID

βE-F1
site

AP1
site

NF-E4
site   Enhancer   AP2
site

NF-E4

β-globin
promoter

CTF
site   NF-E3
site

CTF

Embryonic red blood cell
(not stable)

Red blood cell
(stable)

FIGURE 11.8

Model for promoter-enhancer interaction in regulating the transcription of the chick β-globin gene. (A) Competition model, whereby the ε-globin and β-globin promoters compete for interaction with the same enhancer. (B) Structure of the β globin promoter region and the enhancer. The globin gene in the embryonic red blood cell precursor has three major 5' promoter elements that can interact with four DNA elements in the 3' enhancer. Some of these sites are bound by trans-*regulatory fac-* *tors while others are not. Transcription is not seen from the β-globin gene. As erythroid development continues, new* trans *factors are synthesized that bind to the promoter and enhancer and that lead to protein–protein interactions between the enhancer and promoter regions. In the definitive red blood cell stage (where hemoglobin synthesis is greatly increased), new* trans-*regulatory factors, NF-E4 and βE-F1, stabilize the promoter–enhancer interaction and permit efficient globin synthesis from the β-globin gene. (After Gallarda et al., 1989.)*

β and ε promoters for the same enhancer. Analyzing the proteins that bind to the enhancer and promoters, Gallarda and colleagues (1989) have proposed that in *embryonic* red blood cells, the interaction between the ε-globin promoter and the enhancer is stabilized by ε-globin-promoter-binding factors that can interact with the enhancer-binding proteins. The β-globin promoter lacks such binding proteins and thus cannot interact with the enhancer. Therefore, the ε-globin gene is transcribed. In *immature* red blood cells, the developing erythrocyte transcribes two proteins that stabilize the interactions between the β-globin promoter and the enhancer. As a result, the enhancer is thermodynamically more stable when joined with the β-globin promoter than with the ε-globin promoter. Transcription of ε-globin ceases. However, the transcription of β-globin does not begin at high levels until NF-E4 is produced in the *mature* red blood cell. This protein completes the stabilization of the enhancer with the β-globin gene promoter. Foley and Engel (1992) have shown that mutations in the NF-E4 binding site that prohibit NF-E4 binding prevent the synthesis of β-globin and enable the continued synthesis of ε-globin.

If the promoters were "competing" for the same enhancer, then the addition of a second enhancer might enable both promoters to function. Indeed, when a second enhancer was added, both the ε *and* the β genes were transcribed in both the embryonic and the fetal cells. This result suggests that normally the two globin gene promoters "compete" for the same enhancer and that their ability to retain the enhancer depends upon the transcription factors present in the nucleus.*

*If you think things are getting complicated, you're right. Harold Weintraub, one of the leading figures of chromatin research, notes: "A complex enhancer can contain 10 binding sites, an LCR perhaps as many, a transcription complex may contain 15 proteins, and RNA polymerase maybe 12. What a mess!" (H. Weintraub, personal communication). We don't know why there are so many different factors regulating the transcription of these genes.

Human globin gene switching

The gene expression switching system for human globins seems even more complicated. There are several *cis*-regulatory elements for β-globin. We already discussed the β-globin promoter region and the locus control region that keeps the entire region of the chromosome relatively free of nucleosomes. In addition, there is a 3′ enhancer that regulates the temporal expression of the β-globin gene, and there is another enhancer that regulates the tissue specificity of β-globin gene expression. This latter enhancer is actually located *within* the third exon of the β-globin gene itself (Behringer et al., 1987; Trudel and Constantini, 1987) and allows the transcription of β-globin genes solely in erythroid cells. As shown in Figure 11.9, these *cis*-regulatory regions have numerous sites for ubiquitous and erythroid-specific transcription factors. One of the most important of the erythroid-specific factors is a zinc-finger protein, **GATA-1** (Orkin, 1992). This factor binds to GATA sequences, which are found throughout the LCR as well as in the promoters and enhancers of numerous genes that are expressed in red blood cells (including some of the genes for globin, heme synthesis, and the erythropoietin receptor). Gene targeting experiments show that mice without the gene for GATA-1 produce no erythroid cell lineage (Pevny et al., 1991). The second critical transcription factor appears to be **NF-E2**. This erythroid-specific bZIP transcription factor binds to areas of the LCR and may mediate the communication between the LCR and the promoter regions (perhaps by binding to GATA-1) (Talbot and Grosveld, 1991; Gong and Dean, 1993).

Hereditary persistence of fetal hemoglobin

Whereas most all persons switch from fetal to adult globins around the time of birth, there are some people who do not. These individuals retain transcription from their γ-globin gene and are said to have **hereditary persistence of**

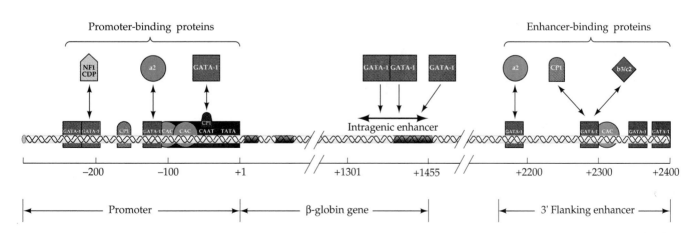

FIGURE 11.9
Schematic representation of the human β-globin gene and its regulatory regions. The shaded shapes represent different tran- *scription factors, and the double arrows indicate that more than one factor can bind at that site. (After Ottolenghi, 1992.)*

fetal hemoglobin (HPFH). It does them no harm.* The mutations that give rise to HPFH cluster in the *cis*-regulatory regions. Deletion mutations that remove the β-globin enhancer or promoter regions suffice to elevate γ-globin levels in adult cells. Point mutations in the promoters for either of the γ-globin genes can also cause HPFH (Martin

*Individuals with HPFH are phenotypically normal and are identified through the screening of populations for other globin abnormalities (such as thalassemia and sickle-cell anemia). Researchers would love to know how to safely reactivate the γ-globin gene in people suffering from β-thalassemia, sickle-cell anemia, and other diseases of β-globin. If the γ-globin gene were reactivated to even a small extent, it could alleviate many of the symptoms of these diseases. Recent studies suggest that administration of butyrate or the combination of intravenous hydroxyurea and erythropoietin may be efficacious in elevating fetal hemoglobin in newly generated red blood cells (Perrine et al., 1993; Rodgers et al., 1993). Studies are underway to evaluate these procedures.

et al., 1989). One of these mutations creates a new binding site for GATA-1, whereas another creates a strong binding site for the ubiquitous factor Sp1 (Ottolenghi, 1992). Thus, there appears to be competition between the promoters of the γ-globin genes and the promoters of the β-globin genes (Enver et al., 1990), and this competition is based on presence of enhancing and repressing *trans*-regulatory factors. There are also point mutations that cause HPFH by preventing the binding of a negative regulator to the γ-globin promoter in adult cells (Berry et al., 1992).

The LCR and globin switching in humans

It appears that the competition is not for the 3' enhancer (which may function locally), but for the LCR. Interestingly, the *distance* between the LCR and the globin genes affects their activation (Hanscombe et al., 1991). When linked closely to the LCR, the human β-globin gene becomes expressed in transgenic mouse embryonic cells. Its

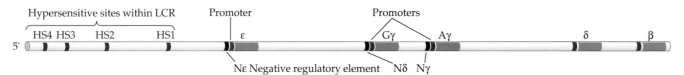

(A) Germy line environment (inactive)

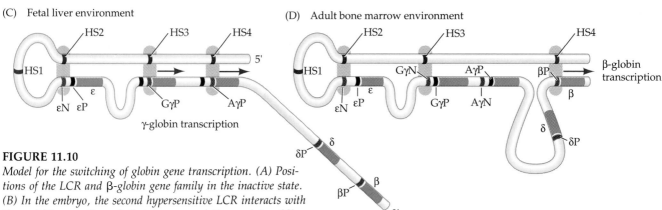

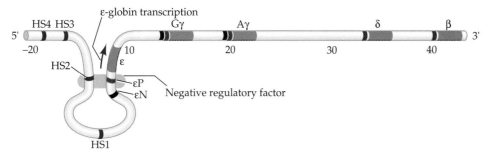

FIGURE 11.10

Model for the switching of globin gene transcription. (A) Positions of the LCR and β-globin gene family in the inactive state. (B) In the embryo, the second hypersensitive LCR interacts with the ε-globin gene through their **trans**-*regulating proteins. (C) In the fetal stage, the synthesis of new transcription factors enable the third and fourth hypersensitive regions of the LCR to interact with the γ-globin gene promoters. Meanwhile, negative regulatory factors bind to silencer on the ε-globin gene, inhibiting its interaction with hypersensitive region 2. (D) In the*

adult stage, the fourth hypersensitive region of the LCR interacts with the β-globin gene promoters, while the 3' enhancer is also activated. Negative regulatory factors suppress the LCR interactions with the γ and ε gene promoters. (After Engel, 1993.)

correct activation (in adult cells only) is restored only when it is placed farther away from the LCR. Similarly, the human γ-globin gene is repressed earlier (like the normal β-globin gene) when it is further separated from the LCR. This suggests that the interaction between the LCR and the globin genes is polarized (Figure 11.10; Hanscombe et al., 1991): those globin genes closest to the LCR are turned on earliest, while those genes more distal are turned on later. Presumably, there is physical contact between the LCR and the gene-specific promoters and enhancers. The mechanisms by which distance from the LCR might regulate the activation of different promoters at different times remain to be explained, but Fraser and colleagues (1993) have shown that each of the four DNase I-hypersensitive sites in the LCR plays a different role in regulating globin gene expression. By making transgenic mice containing human γ- and β-globin genes linked to specific hypersensitive regions of the LCR, they showed that the third site is responsible for activating gene expression during the embryonic and fetal periods, while the fourth hypersensitive site activates gene expression at the adult stage. The binding of β-globin promoter to the fourth hypersensitive site may be in competition with the binding of the γ-globin gene promoter to the third hypersensitive site. Whichever binds most effectively would be given the cooperation of the LCR for transcription. The transcription factors involved in the HPFH syndromes might be those that mediate the interactions between the promoters and the hypersensitive sites of the LCR. In any event, the interactions of transcription factors between the enhancers, promoters, and LCR will provide a fascinating story of differential gene expression in human cells.

DNA methylation and gene activity

It is often assumed that a gene contains exactly the same nucleotides whether it is active or inactive. A β-globin gene in a red blood cell precursor should have the same nucleotides as the β-globin gene in a fibroblast or retinal cell of the same animal. There may be, however, a subtle difference in the DNA in the promoter region. In 1948, R. D. Hotchkiss discovered a "fifth base" in DNA, **5-methylcytosine**. In eukaryotes, this base is made enzymatically after DNA is replicated, and about 5 percent of the cytosines in mammalian DNA are converted to 5-methylcytosine. This conversion can only occur when the cytosine residue is followed by a guanosine (CpG). Recent studies have shown that the degree to which the cytosines of a gene are methylated may also control the gene's transcription. In other words, DNA methylation might change the structure of the gene and, in so doing, regulate its activity.

There are three areas in which DNA methylation is thought to contribute to differential gene activity. First, the methylation of promoter sequences contributes to the temporal and spatial regulation of genes encoding tissue-specific proteins. This is seen primarily in vertebrates. Second, DNA methylation is thought to be responsible for distinguishing between certain egg-derived and sperm-derived genes in mammals, thereby allowing only one of them to be expressed during early development. Third, DNA methylation is thought to be responsible for the continued repression of the genes on one of the two X chromosomes in each female mammalian cell.

Correlations between promoter methylation and gene inactivity

The first evidence that DNA methylation helps regulate gene activity comes from studies showing a correlation between gene activity and low amounts of cytosine methylation (hypomethylation), especially in the promoter region of the gene. In developing human and chick red blood cells, the DNA involved in globin synthesis is completely (or nearly completely) unmethylated, whereas the same genes are highly methylated in cells that do not produce globin (Figure 11.11). Fetal liver cells that produce hemoglobin in early development have unmethylated genes for fetal hemo-

FIGURE 11.11

Methylation of globin genes in human
embryonic blood cells. The activity of
the globin genes correlates inversely
with the methylation of their promoters. (After Mavilio et al., 1983.)

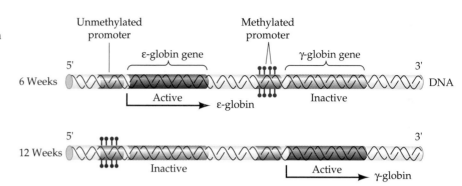

globin. These genes become methylated in the adult tissue (van der Ploeg
and Flavell, 1980; Groudine and Weintraub, 1981; Mavilio et al., 1993).

Organ-specific methylation patterns are also seen in the chick ovalbumin gene; the gene is unmethylated in the oviduct cells but is methylated in other chick tissues (Mandel and Chambon, 1979). Demethylation
accompanies class switching in immunoglobulin synthesis (Rogers and
Wall, 1981) and correlates with the ability of murine lymphocytes to produce the metal-binding protein metallothionein I (Compere and Palmiter,
1981). Thus, the absence of DNA methylation correlates well with the
tissue-specific expression of certain genes.

The second type of evidence for DNA methylation as a regulatory
process comes from experiments in which the expression of cloned genes
is altered by introducing or removing methyl groups from their cytosine
residues. When Busslinger and co-workers (1983) added cloned globin
genes to cells (by coprecipitation with calcium phosphate), the cells ingested the DNA and, in many cases, incorporated the DNA into the
nucleus. In such cases, the cloned globin gene was transcribed. By protecting certain regions of the cloned globin genes from methylation before
adding them to the cells, it is possible to create clones in which the globin
genes have identical sequences but different methylation patterns. A completely unmethylated gene is transcribed, whereas a completely methylated gene (methyl groups on every appropriate C residue) is not transcribed. Using partially methylated clones, Busslinger and co-workers
showed that methylation in the 5′ region of the globin gene (nucleotides
−760 to +100) prevents transcription. Thus, methylation in the 5′ end of
a gene appears to play a direct role in the regulation of gene expression.
In general, methylation of the promoter region inhibits the transcription
of genes.

Methylation and the maintenance of transcription patterns

Methylation differences may be responsible for *maintaining* (as opposed to
initiating) a pattern of transcriptional activity through several cell generations (Holliday, 1987). During replication, each DNA strand serves as a
template for its complementary strand. In regions of methylation, the
methyl groups are usually on both strands of the double helix, since a CG
on one side is mirrored by an antiparallel CG on the other. If the C on
one strand is methylated, so is the C on the other strand (Figure 11.12).
During replication, one strand of the DNA (the template strand) would
have the methylation pattern, while the newly synthesized strand would
not. However, the enzyme DNA (cytosine-5)-methyltransferase has a
strong preference for DNA that has one methylated strand, and when it
sees a methyl-CpG on one side of the DNA, it methylates the new C on
the other side (Gruenbaum et al., 1982; Bestor and Ingram, 1983).

It is doubtful that methylation differences actually *initiate* gene inactivation. Since the methylation pattern should be inherited after each cell division, something else must be able to recognize the genes for differentiated cells in their methylated state and subsequently demethylate them. This has been shown by transfecting a methylated α-actin gene into cultured myoblast cells (which normally transcribe that gene). When transfected into myoblasts, this gene was demethylated and transcribed. However, if transferred into other cell types, this muscle-specific gene remained methylated. New DNA synthesis was not required, but certain *cis*-DNA sequences were necessary for the muscle-specific demethylation to occur (Yisraeli et al., 1986; Paroush et al., 1990). A similar situation has been seen in the demethylation of the immunoglobulin and vitellogenin (yolk protein) genes (Frank et al., 1990; Jost, 1993). Therefore, methylation may be needed to stabilize the pattern of gene transcription, but the gene is probably activated initially by other factors.

How does methylation prohibit transcription? One possibility is that transcription factors cannot find their enhancer or promoter sequences if the DNA is methylated (Iguchi-Ariga and Schaffner, 1989). Another possibility is that certain proteins specifically recognize the methylated DNA and compete against the transcription factors for these sites. Boyse and Bird (1991, 1992) have provided evidence for the second model by showing that methylated promoter sequences are bound by a **methyl-CpG-binding protein**. This protein appears to compete with the binding of transcription factors and thereby reduces transcription from these sites.

The ability to propagate the methylation pattern may cause the propagation of a nucleosome spacing pattern, as well. Keshet and co-workers (1986) have found that methylation affects chromatin structure and suggest that demethylation creates DNase I-hypersensitive sites. When they transfected *un*methylated globin genes into mouse fibroblast nuclei, the genes became packaged into DNase-sensitive chromatin (regardless of the transcriptional ability of the gene). When the same genes were methylated at every CpG site, the DNase-sensitive regions failed to form, presumably because their methylation caused them to be packaged into an inaccessible form. It is possible that when *trans*-regulating factors remove the nucleosomes from the DNA, these regions become demethylated. The demethylation may be necessary to stabilize these regions of activity. The methyl groups would interact with the histones to enable nucleosomes to form only on methylated DNA and not on unmethylated DNA, leaving the active regions free of nucleosomes (Keshet et al., 1986). Once these regions were established, it would become easier for other *trans*-regulatory elements to find these nucleosome-free regions.

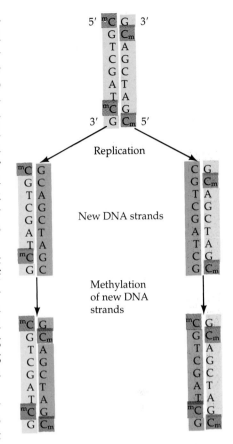

FIGURE 11.12
Model for the propagation of methylation patterns. When DNA replicates, only one of the two strands (the "old" strand) retains the original methylation pattern. The other strand (the "new" strand) is unmethylated. A CG-specific methylating enzyme would be able to bind to CG pairs where a C residue was methylated and then methylate the C residue on the complementary strand. (After Browder, 1984.)

SIDELIGHTS & SPECULATIONS

Methylation and gene imprinting

In Chapter 4, we saw that the genomes of the mammalian sperm and egg are not equivalent. Zygotes do not develop properly if their nuclei are derived from two egg pronuclei or from two sperm pronuclei. This inability to develop is probably due to certain genes that are active only if they are derived from the sperm or from the egg. Recent studies have documented rare conditions in mice and humans in which a mutant gene causes severe or lethal defects if inherited from one parent, but does not produce these deficiencies if inherited from the other. For instance in mice, the gene for insulin-like growth factor-2 (*Igf-2*) on chromosome 7 is active in early embryos only on the chromosome transmitted from the father. Conversely, the gene for the receptor of this growth factor (*Igf-2r*) on chromosome 17 is active only when transmitted from the mother (Barlow et al., 1991; DeChiara et al., 1991; Bartolomei and

TABLE 11.2
Evidence that genomic imprinting affects phenotype in human gene disorders at chromosome 15 (locus 11q13)

Parental origin		
Mother	Father	Phenotype
Normal allele	Mutant allele	Prader-Willi syndrome
Mutant allele	Normal allele	Angelman syndrome
Two copies of allele	Allele absent	Prader-Willi syndrome
Allele absent	Two copies of allele	Angelman syndrome

Source: After Nicholls et al., 1993

Tilghman, 1992). A mouse pup that inherits a deletion of the *Igf-2r* gene from its father is normal, but if the same deletion is inherited from the mother, the pup's growth is stunted. In humans, the loss of a particular gene in the long arm of chromosome 15 results in different phenotypes depending on whether the loss is in the male-derived or the female-derived chromosome (Table 11.2). If the defective or missing gene comes from the father, the child is born with Prader-Willi syndrome, a disease associated with mild mental retardation, obesity, small gonads, and short stature. If the defective or missing gene comes from the mother, the child has Angelman syndrome, characterized by severe mental retardation, lack of speech, and inappropriate laughter (Knoll et al., 1989; Nicholls et al., 1989).

It is now generally thought that most, if not all, of the differences between the mammalian male and female pronuclei involve differences in their DNA methylation patterns. The distribution of the methylated and unmethylated CG doublets can be analyzed by cutting the DNA with two restriction enzymes, *Hpa*II and *Msp*I (McGhee and Ginder, 1979). Both these enzymes cut at the same site—CCGG—but *Hpa*II will not cut DNA if the central C is methylated, whereas *Msp*I cuts whether the sequence is methylated or not. Therefore, DNA from a particular cell type can be digested separately with *Hpa*II and *Msp*I and the cleaved DNA fragments Southern blotted and hybridized with a radioactive gene-specific probe (see Chapter 2). Differences in band patterns in the autoradiograms of the *Msp*I-cleaved and *Hpa*II-cleaved fragments can then be ascribed to methylation differences.

Figure 11.13 shows the results of an experiment in which DNA from sperm was isolated and treated with *Hpa*II or *Msp*I. The probe was a radioactive DNA from the second exon of the β-globin gene. The autoradiogram of fragments from the *Msp*I digestion shows that this probe binds to DNA fragments that have 1400 base pairs between CCGG sites. The *Hpa*II digestion autoradiogram shows that in the sperm these sites (and probably numerous others) are methylated, and that this DNA sequence now resides in a 25,000-base-pair stretch of DNA wherein all CCGG sites are methylated (Groudine and Conklin, 1985).

This technique showed that the primordial germ cell nuclei of both male and female mammals are strikingly hypomethylated (Monk et al., 1987; Driscoll and Migeon, 1990), but both the sperm and egg genes undergo extensive methylation as the gametes mature. It appears that during germ cell formation, previous information is erased, and then, during meiosis, new information is introduced onto the genome. The pattern of methylation on a given gene can differ between egg and sperm, and these gene-specific methylation differences are seen in the chromosomes of embryonic cells (Sanford et al., 1987; Reik et al., 1987; Sapienza et al., 1987; Chaillet et al., 1991; Kafri et al., 1992). Thus, methylation differences between sperm and egg genes may specify whether a gene came from the father or mother. This maternal and paternal imprinting adds additional information to the inherited genomes, information that may regulate spatial and temporal gene activity and chromosome behavior.

Swain and co-workers (1987) were able to follow these events by looking at a specific gene that undergoes differential methylation in the sperm and egg. They constructed a strain of transgenic mice in which a particular gene, c-*myc*, was inserted into the mouse genome. When this gene was inherited from the male parent, it was transcribed specifically in the heart and in no other tissue. When it was inherited from the female parent, it was not expressed at all. The pattern of expression correlated with the degree of methylation; this gene is methylated during egg maturation but remains hypomethylated during sperm formation. In animals that inherit the transgene from males, the gene is unmethylated and is expressed in the heart. In animals that acquire this transgene from their mothers, the

FIGURE 11.13
Detection of methylation sites on DNA. DNA was isolated from chicken sperm and digested with either MspI *(lane 1) or* HpaII *(lane 2). The fragments were separated by electrophoresis, blotted onto paper, and hybridized to a radioactive DNA probe from the second exon of the β-globin gene. This probe bound to a fragment 1400 bases long in the MspI digest, but to a fragment about 25,000 bases long in the HpaII digest. (From Groudine and Conklin, 1985; photograph courtesy of M. Groudine.)*

gene is methylated and silent. In both the males and females, the pattern of methylation is erased in the germ cells (Chaillet et al., 1991; Kafri et al., 1992). In mice, methylation differences from the gametes are also seen in the imprinted genes for Igf-2r and H19 (Ferguson-Smith et al., 1993; Stöger et al., 1993). Moreover, if these genes are placed in a mutant mouse strain that lacks the enzyme capable of methylating the CpG sites, transcription occurs from the previously silent alleles (Li et al., 1993). Such gamete-specific methylation differences provide a plausible explanation for the failure of parthenogenetic mammals to develop and for the need for both male and female pronuclei in the zygote. It also provides a reminder that the organism cannot be explained solely by its genes. One needs knowledge of developmental parameters as well as genetic ones.

Mammalian X-chromosome dosage compensation

In animals as diverse as *Drosophila* and humans, females are characterized as having two X chromosomes per cell, while males are characterized as having a single X chromosome per cell. Unlike the Y chromosome, the X chromosome has on it thousands of genes that are essential for cell activity. Yet, despite the female cell's having double the number of X chromosomes as the male's, male and female cells have approximately equal amounts of X chromosome-encoded enzymes. This equalization is called **dosage compensation**. The transcription rates of the X chromosomes have been altered so that male and female cells transcribe the same amount of RNAs from their X chromosomes. In *Drosophila*, both X chromosomes in the female are active, but there is increased transcription from the male X chromosome so that the single X chromosome of the male cells produces as much product as the two X chromosomes in the female cells (Lucchesi and Manning, 1987). This is accomplished by the binding of particular trancription factors to hundreds of sites along the male X chromosome (Kuroda et al., 1991).

In mammals, X chromosome dosage compensation occurs by inactivating one X chromosome in each female cell. Thus, each mammalian somatic cell, whether male or female, has only one functioning X chromosome. This phenomena is called **X chromosome inactivation**. The chromatin of the inactive X chromosome is converted into **heterochromatin**—chromatin that remains condensed throughout most of the cell cycle and replicates after most of the chromatin (the **euchromatin**) of the nucleus (Figure 11.14). This heterochromatin, in a formation called the **Barr body**, is often seen on the nuclear envelope of female cells (Barr and Bertram, 1949). X chromosome inactivation has to occur early in development. Using a mutated X chromosome that would not inactivate, Tagaki and Abe (1990) showed that the expression of two X chromosomes per cell in mouse embryos leads to ectodermal cell death and the absence of mesoderm formation, eventually causing embryonic death at day 10 of gestation.

FIGURE 11.14
Nuclei of human oral epithelial cells stained with Cresyl Violet. (A) Cell from a normal XY male, showing no Barr body. (B) Cell from a normal XX female, showing a single Barr body (arrow). (C) Cell from a female with three X chromosomes. Two Barr bodies can be seen, and only one X chromosome per cell is active. (From Moore, 1977.)

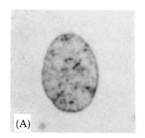
(A)

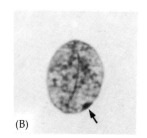
(B)

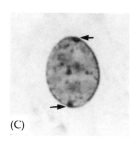
(C)

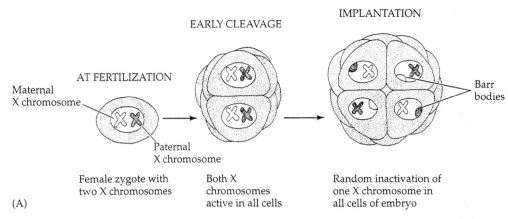

EARLY CLEAVAGE

IMPLANTATION

AT FERTILIZATION

Maternal
X chromosome

Paternal
X chromosome

Barr
bodies

Female zygote with
two X chromosomes

Both X
chromosomes
active in all cells

Random inactivation of
one X chromosome in
all cells of embryo

(A)

FIGURE 11.15
X chromosome inactivation in mammals. (A) Schematic diagram illustrating random X chromosome inactivation. The inactivation is believed to occur at about the time of implantation. (B) A female mouse heterozygous for the X-linked coat color gene *dappled*. Distinctly different pigmented regions are seen. (Photograph courtesy of M. F. Lyon.)

(B)

The early inactivation of one X chromosome per cell has important phenotypic consequences. One of the earliest analyses of X chromosome inactivation was performed by Mary Lyon (1961), who observed coat color patterns in mice. If a mouse is heterozygous for an *autosomal* gene controlling hair pigmentation, then the mouse resembles one of the two parents or has a color intermediate between the two. In either case, the mouse is a single color. But if a female mouse is heterozygous for a pigmentation gene on the *X chromosome*, a different result is seen: patches of one parental color alternate with patches of the other parental color (Figure 11.15). Lyon proposed the following hypothesis to account for these results:

1. Very early in the development of female mammals, both X chromosomes are active.
2. As development proceeds, one X chromosome is turned off in each cell.
3. This inactivation is random. In some cells, the paternally derived X chromosome is inactivated; in other cells, the maternally derived X chromosome is shut off.
4. This process is irreversible. Once an X chromosome has been inactivated, the same X chromosome is inactivated in all that cell's progeny. (The areas of pigment in these mice are large patches, not a "salt and pepper" pattern.) Thus, all tissues in female mammals are mosaics of two cell types.

Some of the most impressive evidence for this model comes from biochemical studies on clones of human cells. In humans, there is a genetic disease—Lesch-Nyhan syndrome—that is characterized by the lack of the X-linked enzyme hypoxanthine phosphoribosyltransferase (HPRT). Lesch-

Nyhan syndrome is transmitted through the X chromosome—that is, males who have this mutation in their one X chromosome suffer (and die) from the disease. In females, however, the presence of the mutant HPRT gene can be masked by the other X chromosome, which carries the wild-type allele. A woman who has sons with this disease is said to be a carrier, as she has a mutant HPRT gene on one chromosome and a wild-type HPRT gene on the other X chromosome. If the Lyon hypothesis were correct, each cell from such a woman should be making either the active or the inactive HPRT, depending on which X chromosome is active. Barbara Migeon (1971) tested this prediction by taking individual skin cells from a woman heterozygous for the HPRT gene and placing them in culture. Each of these cells divided to form a clone of cells. When Migeon stained the clones for the presence of wild-type HPRT, approximately half of the clones had the enzyme and the other half did not (Figure 11.16).

The Lyon hypothesis of X chromosome inactivation provides an excellent account of differential gene inactivation at the level of transcription. Some interesting exceptions to the general rules further show its importance. First, X chromosome inactivation holds true only for *somatic* cells, not *germ* cells. In female germ cells, the inactive X chromosome is reactivated shortly before the cells enter meiosis (Gartler et al., 1973, 1980; Migeon and Jelalian, 1977; Kratzer and Chapman, 1981). Thus, in mature oocytes, both X chromosomes are active. In each generation, X chromosome inactivation has to be established anew.

Second, there are some exceptions to the rule of randomness in the inactivation pattern. The first X chromosome inactivation in the mouse is seen in the trophectoderm, where the *paternally* derived X chromosome is specifically inactivated (Tagaki, 1974; West et al., 1977). Third, X chromosome inactivation does not extend throughout the entire length of the human X chromosome. Almost all known X chromosome genes map to the long arm of the X chromosome. However, on the short arm of the X chromosome, there are several genes (such as that for steroid sulfatase) that do not undergo this dosage-related inactivation (Mohandas et al., 1980; Brown and Willard, 1990). Thus, in humans, heterochromatization does not extend throughout the entire X chromosome.

The fourth exception really ends up proving the rule. There are a few male mammals with coat color patterns we would not expect to find unless the animals exhibited X chromosome inactivation. Male calico and tortoise-shell cats are among these examples. These spotted coat patterns are normally seen in females and are thought to result from random X chromosome inactivation. But rare males exhibit these coat patterns as well. How can this be? It turns out that these cats are XXY. The Y chromosome makes them male (see Chapter 21), but one X chromosome undergoes inactivation, just as in females, so there is only one active X per cell (Centerwall and Benirschke, 1973). Thus, these cats have cells with a Barr body and random X chromosome inactivation. It is clear, then, that one mechanism for transcription-level control of gene regulation is to make a large number of genes heterochromatic and, thus, transcriptionally inert.

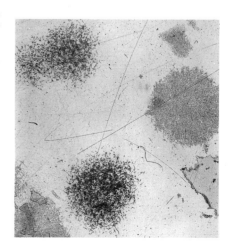

FIGURE 11.16

Retention of X chromosome inactivation. Approximately 30 cells from a woman heterozygous for HPRT deficiency were placed into a petri dish and allowed to grow. The cells were visualized by autoradiography after incubation in a medium containing radioactive hypoxanthine. Cells with HPRT incorporate the radiolabeled compound into their RNA and darken the photographic emulsion placed over them. The clones of cells without HPRT appear lighter because their cells cannot incorporate the radioactive compound. (From Migeon, 1971, courtesy of B. Migeon.)

The mechanism of X chromosome inactivation

The mechanism of X chromosome inactivation is still poorly understood, but new research is giving us some indication of the factors that may be involved in initiating and maintaining a heterochromatic X chromosome.

Initiation of X chromosome inactivation: The **Xist** gene

The first clues for an initiator of X chromosome inactivation came from genetic studies wherein certain mutant X chromosomes in mice were unable to become inactive (Russell, 1963; Cattanach et al., 1969; Mattei et al., 1981). These chromosomes lacked a particular region thereafter called the X chromosome inactivation center (XIC). In 1991, Brown and her colleagues found an RNA transcript that was made solely from the *inactive* X chromosome of humans. (In those loci that escape X chromosome inactivation in humans, both X chromosomes synthesize the transcripts. Here, the transcript was coming only from the "inactive" X chromosome.) This transcript, *Xist*, was being made by a gene within the XIC region. Moreover, this transcript does not appear to encode a protein. It stays within the nucleus and interacts with the inactive X-chromatin Barr body (Brown et al., 1992). A similar situation exists in the mouse, wherein the *Xist* gene of the inactive X chromosome synthesizes a nuclear RNA whose sequence cannot encode a protein (Borsani et al., 1991; Brockdorrf et al., 1992). The transcripts from the *Xist* gene are seen in mouse embryos prior to X chromosome inactivation, which would be expected if this gene plays a role in initiating X chromosome inactivation (Kay et al., 1993). Therefore, in addition to tRNA, mRNA, and rRNA, there is now evidence for X-inactivation RNA. It is still not known what this *Xist* RNA is doing to inactivate the chromosome.

Forbidding transcription: The unacetylated nucleosome

What is preventing the DNA of the inactive X chromosome from being transcribed like the active one? A recent study by Jeppeson and Turner (1993) suggests that the inactive and active X chromosomes differ from one another by the acetylation of their respective histones H4. One of the best ways to release protein from DNA is to add negative charges to the protein. This can be accomplished by adding phosphate or acetate groups to the DNA binding regions of the protein. Acetylation of histone H4 has been correlated with actively transcribing genes. Nucleosomes from nonmethylated CpG promoter regions have highly acetylated H4 proteins, and the acetylation of histone H4 correlates with the activation of certain genes (Chahal et al., 1980; Tazi and Bird, 1990). Moreover, whereas transcription factor TFIIIA cannot bind to the 5S RNA gene if the gene is wrapped up in a nucleosome, that factor can bind to the gene if the nucleosome's histones have been acetylated

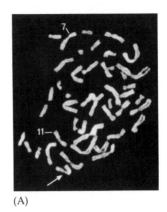
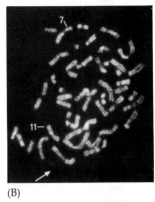

(A)   (B)

FIGURE 11.17
The inactive X chromosome of human female cells contains underacetylated histone H4. (A) Metaphase spread of a human female fibroblast cell stained with Hoechst 33258, which stains chromatin. (Chromosomes 7 and 11 are numbered, and an arrow points to the inactive X). (B) Same preparation stained with fluorescent antibody to acetylated histone H4. While all the other chromosomes are clearly visible, the inactive X is not. (From Jeppesen and Turner, 1993.)

(Lee et al., 1993). Histone acetylation does not appear to be obligatory for gene transcription, but it may facilitate transcription in several systems (Turner, 1991). Using an antibody that recognizes acetylated histone H4 (but not unacetylated histone H4), Jeppesen and Turner found that the active X chromosomes of human and mice have about as much acetylated histone H4 as most other chromosomes. However, the inactive X chromosome has hardly any acetylated histone H4 (Figure 11.17). It is not known how *Xist* expression would cause the occurrence of unacetylated H4 in the nucleosomes of the inactive X chromosomes.

Locking in transcription patterns: DNA methylation

The locking in of the transcriptionally inactive state is done by methylation. The first evidence for such *"cis"* differences between active and inactive X-chromosomal DNA came when Liskay and Evans (1980) transfected the X-linked gene for HPRT into HPRT-deficient mouse cells in culture. When the DNA came from a clone of cells in which the gene for HPRT was on the *inactive* X chromosome, the DNA was unable to make HPRT in the HPRT-deficient host cell. However, when the DNA was derived from a clone of cells expressing the HPRT gene on its *active* X chromosome, the transfected cells made HPRT from this gene. Shortly thereafter, Mohandas and colleagues (1981) found that 5-azacytidine (a drug that inhibits cytosine methyltransferase) could locally reactivate these genes on the inactive X chromosome. Further investigations, using restriction enzymes and cDNA probes, showed that the CpG "islands" at the promoter sites of several genes are methylated in the inactive X chromosome and unmethylated in the active X chromosome (Wolf et al., 1982; 1984; Keith et al., 1986). These methylation patterns are removed during germ cell formation, thereby enabling a new pattern of X chromosome inactivation to occur in the next generation. During the first trimester of human development, the adult pattern of methylation and X chromosome inactivation is established anew* (Migeon et al., 1991).

We are still ignorant of the mechanisms by which the *Xist* transcript regulates the state of the chromatin and how the spreading of inactivation occurs. We are still ignorant of the ways in which Xist transcription, nucleosome modification, and DNA methylation are linked to the heterochromatization of an X chromosome. We don't yet know how the choice between the two X chromosomes is originally made or how the *Xist* RNA is transcribed from a region surrounded by inactivated genes. There is still much to learn about this critical mammalian phenomenon.

*As mentioned in earlier chapters, it is difficult to extrapolate from one mammalian group to another. This is certainly the case with X chromosome inactivation. Just because X chromosome inactivation happens this way in the mouse placenta does not mean it happens this way in the placenta of all mammals. In human chorionic villi, some cells contain two active X chromosomes, and those X chromosomes that are inactivated can be reactivated (Migeon et al., 1985, 1986). Moreover, X chromosome inactivation in the *human* placenta appears to be random; either the paternally or maternally derived X chromosome can be turned off. In marsupials, the paternally derived X chromosome is preferentially inactivated throughout the embryo (Sharman, 1971; Cooper et al., 1971; Samollow et al., 1987). Moreover, in humans, there are obvious regions of the X chromosome that escape inactivation. The somatic differences between humans with XX and XO karyotypes also predict that there would be some X-linked genes that would be needed in two doses for normal female development. In the mouse, X-chromosome inactivation appears to extend throughout the entire chromosome (Ashworth et al., 1991).

Association of active DNA with the nuclear matrix

Attachment of active chromatin to a nuclear matrix

Within the nucleus, the replicative enzymes must somehow find their sites for initiating DNA synthesis; the transcriptional factors and polymerases must find their promoters and enhancers; the RNA processing factors must find their RNA splicing sites; and the messenger RNA must efficiently find the pores through which to exit the nucleus. This is a great deal to ask of molecules in solution. We would have to expect the various factors involved in transcription to be floating around in the nuclear sap, bumping randomly into DNA. The RNA so formed would then be spliced and bump around the nucleus until it found a nuclear pore through which to leave.

An alternative model suggests that RNA is transcribed on a solid substrate wherein all the enzymes needed for transcription, processing, and transport are collected together. There are precedents for thinking in such terms. The electron transport chain of the mitochondria is such an ordered aggregate, and the DNA-synthetic enzymes of bacteria have long been known to reside on the inner surface of the cell membrane. What one has to ask, then, are the following questions: Is there a nuclear reticulum wherein such enzymes might be found? If such a network exists, are the transcriptionally active genes located there? If such a network exists, are the RNA-synthesizing enzymes found there?

A nuclear matrix can be isolated by dissolving nuclei in lipid detergents and solubilizing most of the DNA with DNases (Berezney and Coffey, 1977; Capco et al., 1982). Transmission electron microscopy of such complexes shows a meshwork of proteins that extends throughout the nucleus and connects to the cytoskeleton at the nuclear envelope (Figure 11.18). This matrix is seen in all eukaryotic nuclei thus far examined (Wilson, 1895; Nelson et al., 1986).

When one isolates such a matrix, DNase has removed about 98 percent of the DNA. Is the DNA that is still bound to this matrix (and presumably

FIGURE 11.18

Transmission electron micrograph ($\times$47,000) of a portion of nuclear matrix (left) and surrounding cytoplasm. Cytoskeletal filaments are clearly visible. The mouse fibroblast cells were extracted with detergent to remove lipids and further treated with DNase I. In 1895, E. B. Wilson, looking through the light microscope, reported that the nucleus was traversed by fibers that were continuous with those of the cytoplasmic reticulum and that surrounded the chromatin. (From Capco et al., 1982; courtesy of S. Penman.)

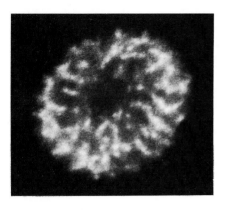

FIGURE 11.19
Presence of active chromatin along the nuclear periphery and channels. Erythrocyte nuclei were treated with DNase I, which is thought to nick actively transcribing chromatin regions. This nick was "healed" by nick translation within the nucleus in the presence of nucleotides whose presence could be detected by fluorescence. The labeled nucleotides were found at the periphery of the nucleus and along structures leading inward from the nuclear envelope. (From Hutchinson and Weintraub, 1985, courtesy of N. Hutchinson.)

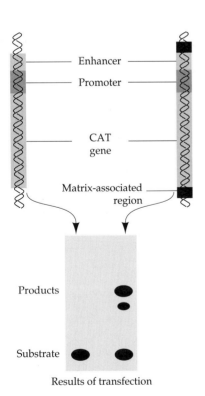

Enhancer

Promoter

CAT gene

Matrix-associated region

Products

Substrate

Results of transfection

protected from the DNase by being so tightly associated with it) enriched for actively transcribing genes? There is evidence that this is so for at least some genes. The ovalbumin gene is preferentially associated with the nuclear matrix in adult chick oviduct cells, but not in chick liver or erythrocyte cells. The globin genes, moreover, are not associated with the nuclear matrix of the oviduct cells (Robinson et al., 1982; Thorburn and Knowland, 1993). Ciejek and co-workers (1983) confirmed and extended these observations, showing that the entire hormone-inducible transcription unit of the ovalbumin gene is bound to the nuclear matrix. No other genes within 100,000 base pairs of these are matrix-associated. Moreover, when estrogen was withdrawn from the animals, the specific attachment of these genes to the nuclear matrix was abolished. It appears that these genes are bound to the nuclear matrix only when activated.

In 1985, Hutchinson and Weintraub showed that DNase I-sensitive sites are not found uniformly throughout the nucleus. They treated nuclei with DNase I, then repaired the nicks with radioactive nucleotides. The labeled DNA should represent just the actively transcribing (i.e., DNase I-sensitive) genes. The results of such treatment showed the DNase I-sensitive DNA existed at the periphery of the nuclei and along channels or fibers that connect to the nuclear envelope (Figure 11.19). It is possible, then, that active genes are specifically associated with the nuclear envelope or matrix.

The third type of evidence for nuclear matrix participation in transcription involves the finding that most (some report 95 percent) of the newly synthesized RNA appears to be attached to the nuclear matrix (Herman et al., 1978; Miller et al., 1978; van Eekelen and van Venrooij, 1981; Mariman et al., 1982). This binding appears to be mediated by a particular nuclear matrix protein. Given that active genes, RNA polymerase, and nascent transcripts appear to be bound to a nuclear matrix, Jackson and Cook (1985) have proposed that transcription does not occur by a mobile polymerase traveling down the length of a gene. Rather, they envision the RNA polymerase as tethered to the nuclear matrix, with the DNA traveling through it.

There is also some evidence that active DNA may be attached to the nuclear matrix by AT-rich DNA sequences called **matrix-associated regions** (MARs), or scaffold-associated regions (Gasser and Laemmli, 1986). Most of these *cis*-regulatory elements map near or within enhancers or promoters. The importance of these regions was shown by Stief and co-workers (1989), who identified two MARs of the chick lysozyme gene. In this case, the MARs were not in the enhancer and so could be separated from it. When they fused the chick lysozyme enhancer and promoter to the reporter gene CAT and transfected the clone into lysozyme-producing cells, it did not produce much CAT protein. They then made a similar gene containing the promoter, enhancer, and CAT sequences, and flanked by the two MARs. When this clone was transfected into the lysozyme-producing cells, CAT synthesis was enormously enhanced (Figure 11.20).

Three proteins have recently been implicated in binding DNA to the nuclear envelope. Ludérus and co-workers (1992) found that MARs are

FIGURE 11.20
Importance of matrix-associated regions in transcription. When clones consisting of lysozyme promoter, enhancer, and the CAT gene are transfected into a lysozyme-secreting cell line, very little CAT protein is made, as assayed by CAT enzyme activity. However, if the two MARs are also included in the cloned gene, much more CAT protein can be found in these cells. (After Stief et al., 1989.)

bound by lamin B$_1$, a major component of the nuclear envelope. Dickinson and colleagues (1992) isolated a thymus-specific MAR-binding protein that unwinds the DNA adjacent to its binding site (Dickinson et al., 1992). Sukegawa and Blobel (1993) isolated a nuclear pore protein that contains a DNA-binding zinc-finger region that faces the nucleoplasm. The sequence of the DNA recognized by this protein is not yet known.

Topoisomerases and gene transcription

In several studies, matrix-associated regions have been seen to contain or be adjacent to a DNA sequence that is recognized by an enzyme—**topoisomerase II**—that might be essential for transcription (Cockerill and Garrard, 1986; Adachi et al., 1989; Scheuermann and Chen, 1989). Recent studies have suggested that the unwinding of the DNA helix is very important in facilitating transcription. Transcriptionally active chromatin has to be twisted to enable the strands to unwind (Ryoji and Worcel, 1984), and this twisting is accomplished by supercoiling the DNA helix (Figure 11.21). Villeponteau and co-workers (1984) have shown that DNase I-sensitive sites in active genes are formed only when the genes are under torsional strain. Topoisomerase II is the enzyme responsible for twisting the DNA and allowing the strands to separate. Using antibodies to this protein, Berrios and colleagues (1985) demonstrated that topoisomerase II is located on the nuclear matrix–nuclear envelope complex.

The proximity of the MAR and the topoisomerase II binding regions suggests that the anchoring of the DNA to the matrix may be necessary to prevent free rotation of the DNA so as to permit the matrix-bound topoisomerases to twist the chromatin. Bode and co-workers (1992) showed that the MARs associated with the interferon 2 and immunoglobulin heavy chain genes have a strong affinity for the nuclear matrix and enable their respective promoters to become unwound. If these sequences were mutated, they no longer had a strong attachment to the nuclear matrix nor did they enhance promoter activity by unwinding the helix.

If binding to the nuclear matrix is essential for RNA transcription, it is possible that negative regulatory proteins such as silencers may inhibit this association. This possibility was suggested by Scheuermann and Chen (1989), who isolated a protein that inhibits the transcription of the heavy chain immunoglobulin gene. This protein, NF-μNR, is expressed in non-B cells and in early stages of B cell development, but is absent from the mature B cells that transcribe large amounts of immunoglobulin genes. This protein binds to four places flanking the heavy chain enhancer: two MAR consensus sequences and two topoisomerase II consensus sequences (Figure 11.22). It is possible, then, that when NF-μNR is present in nuclei, it would bind to the enhancer flanking regions and prevent the association of the heavy chain immunoglobulin gene with the nuclear matrix and topoisomerase. When this protein was not present, these associations would occur and transcription of this gene would result.

FIGURE 11.21
Supercoiling of DNA during transcription. Topoisomerase II joins together two regions of DNA and introduces supercoiling by transiently breaking and recombining the DNA strands. As a result of the strain, a portion of the double helix separates into two strands, enabling RNA polymerase (and presumably other *trans*-regulatory factors) to initiate transcription. The topoisomerase-binding sites have been found on matrix-bound DNA (Cockerill and Garrard, 1986). (After Darnell et al., 1986.)

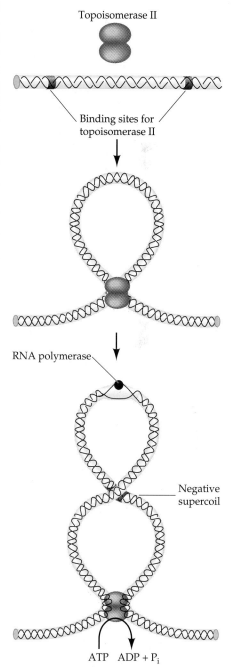

Topoisomerase II

Binding sites for topoisomerase II

RNA polymerase

Negative supercoil

ATP   ADP + P$_i$

FIGURE 11.22

One of four regions of the heavy chain immunoglobulin gene enhancer protected by the NF-μNR protein. NF-μNR was added to the enhancer region DNA and the DNA was digested with DNase. Only those sequences covered by NF-μNR would be preserved. The protected region (gray) includes a matrix-associated region sequence and a topoisomerase II binding site (color). (After Scheuermann and Chen, 1989.)

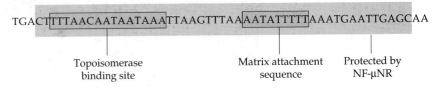

TGACTTTTAACAATAATAAATTAAGTTTAAAATATTTTTAAATGAATTGAGCAA

Topoisomerase
binding site

Matrix attachment
sequence

Protected by
NF-μNR

Gilbert 3/E, Figure#12.42, G#572 Black, Focus: 100%

Differential gene transcription is a major way to regulate development. The *cis-* regulatory regions in the DNA and the *trans*-regulatory proteins that activate and repress transcription are being identified, and their mechanisms of action are being delineated. It appears that certain transcription factors are able to disrupt or prevent nucleosome formation in enhancers and promoter regions, thereby enabling RNA polymerase II to bind to the promoter and transcribe the gene. Certain transcription factors are able to stimulate transcription by interacting with the transcriptional complex and to accelerate its formation. Demethylation and the unwinding of genetic regions on the nuclear matrix are also probably involved in regulating gene expression, as well. As Albert Claude said, we are just beginning to appreciate our acquired wealth.

LITERATURE CITED

Adachi, Y., Käs, E. and Laemmli, U. 1989. Preferential, cooperative binding of DNA topoisomerase II to scaffold-associated regions. *EMBO J.* 8: 3997–4006.

Adams, C. C. and Workman, J. L. 1993. Nucleosome displacement in transcription. *Cell* 72: 305–308.

Asworth, A., Rastan, S., Lovell-Badge, R. and Kay, G. F. 1991. X inactivation may explain the difference in viability of XO humans and mice. *Nature* 351: 406–408.

Barlow, D. P., Stoger, R., Herrmann, B. G., Saito, K. and Schweifer, N. 1991. The mouse insulin-like growth factor type-2 receptor is imprinted and closely linked to the *Tme* locus. *Nature* 349: 84–87.

Barr, M. L. and Bertram, E. G. 1949. A morphological distinction between neurones of the male and female, and the behavior of the nucleolar satellite during accelerated nucleoprotein synthesis. *Nature* 163: 676.

Bartolomei, M. S. and Tilghman, S. M. 1992. Parental imprinting of mouse chromosome 7. *Semin. Dev. Biol.* 3: 107–117.

Behringer, R., Hammer, R., Brinster, R., Palmiter, R. and Townes, T. 1987. Two 3′ sequences direct adult erythroid-specific expression of human β-globin genes in transgenic mice. *Proc. Natl. Acad. Sci. USA* 84: 7056–7060.

Berezney, R. and Coffey, D. S. 1977. Nuclear matrix: Isolation and characterization of a framework structure from rat liver nuclei. *J. Cell Biol.* 73: 616–637.

Berrios, M., Osheroff, N. and Fisher, P. A. 1985. *In situ* localization of DNA topoisomerase II, a major polypeptide component of the *Drosophila* nuclear matrix fraction. *Proc. Natl. Acad. Sci. USA* 82: 4142–4146.

Berry, M., Grosveld, F. and Dillon, N. 1992. A single point mutation is the cause of the Greek form of hereditary persistence of fetal hemoglobin. *Nature* 358: 499–502.

Bestor, T. H. and Ingram, V. M. 1983. Two DNA methyltransferases from murine erythroleukemia cells: Purification, sequence specificity, and mode of interaction with DNA. *Proc. Natl. Acad. Sci. USA* 82: 2674–2678.

Blom van Assendelft, G., Hanscombe, O., Grosveld, F. and Greaves, D. R. 1989. The β-globin dominant control region activates homologous and heterologous promoters in a tissue-specific manner. *Cell* 56: 969–977.

Bode, J., Kohwi, Y., Dickinson, L., Joh, T., Klehr, D., Mielke, C. and Kohwi-Shigematsu, T. 1992. Biological significance of unwinding capability of nuclear matrix-associated DNA. *Science* 255: 195–197.

Borsani, G. and thirteen others. 1991. Characterization of a murine gene expressed from the inactive X chromosome. *Nature* 351: 325–329.

Boyse, J. and Bird, A. 1991. DNA methylation inhibits transcription indirectly via a methyl-CpG binding protein. *Cell* 64: 1123–1134.

Boyse, J. and Bird, A. 1992. Repression of genes by DNA methylation depends upon CpG density and promoter strength: Evidence for involvement of a methyl-CpG binding protein. *EMBO J.* 11: 327–333.

Brockendorrf, N. and seven others. 1992. The product of the mouse *Xist* gene is a 15-kb inactive X-specific transcript containing no conserved ORF and located in the nucleus. *Cell* 71: 515–526.

Browder, L. W. 1984. *Developmental Biology*, 2nd Ed. Saunders, Philadelphia.

Brown, C. J. and Willard, H. F. 1990. Localization of a gene that escapes inactivation to the X chromosome proximal short arm: Implications for X inactivation. *Am. J. Hum. Genet.* 46: 273–279.

Brown, C. J. and six others. 1991a. A gene from the region of the human X inactivation center is expressed exclusively from the inactive X chromosome. *Nature* 349: 38–44.

Brown, C. J. and nine others. 1991b. Localization of the X chromosome inactivation center on the human X chromosome. *Nature* 349: 82–84.

Brown, C. J., Hendrich, B. D., Rupert, J. L., Lafreniere, R. G., Xing, Y., Lawrence, J. and Willard, H. F. 1992. The human *XIST* gene: Analysis of a 17-kb inactive X-specific RNA that contains conserved repeats and is highly localized within the nucleus. *Cell* 71: 527–542.

Burch, J. B. and Weintraub, H. 1983. Temporal order of chromatin structural changes associated with activation of the major chick vitellogenin gene. *Cell* 33: 64–76.

Burlingame, R. W., Love, W. E., Wang, B.-C., Hamlin, R., Xuong, N.-H. and Moudrianakis, E. N. 1985. Crystallographic structure of the octameric histone core of the nucleosome at a resolution of 3.3 Å. *Science* 228: 246–253.

Busslinger, M., Hurst, J. and Flavell, R. A. 1983. DNA methylation and the regulation of globin gene expression. *Cell* 34: 197–206.

Capco, D. G., Wan, K. M. and Penman, S. 1982. The nuclear matrix: Three-dimensional architecture and protein composition. *Cell* 29: 847–858.

Cattanach, B. M., Pollard, C. E. and Perez, J. N. 1969. Controlling elements in the mouse X chromosome. 1. Interaction with the X-linked genes. *Genet. Res.* 14: 223–235.

Centerwall, W. R. and Benirschke, K. 1973. Male tortoiseshell and calico (T–C) cats. *J. Hered.* 64: 272–278.

Chaillet, J. R., Vogt, T. F., Beier, D. R. and Leder, P. 1991. Parental-specific methylation of an imprinted transgene is established during gametogenesis and progressively changes during embryogenesis. *Cell* 66: 77–83.

Chalal, S. S., Matthews, H. R. and Bradbury, E. M. 1980. Acetylation of histone H4 and its role in chromatin structure and function. *Nature* 287: 76–79.

Choi, O.-R. and Engel, J. D. 1988. Developmental regulation of β-globin gene switching. *Cell* 55: 17–26.

Ciejek, E. M., Tsai, M.-J. and O'Malley,B. W. 1983. Actively transcribed genes are associated with the nuclear matrix. *Nature* 306: 607–609.

Clark, D. J. and Felsenfeld, G. 1992. A nucleosome core is transferred out of the path of a transcribing polymerase. *Cell* 71: 11–22.

Cockerill, P. N. and Garrard, W. T. 1986. Chromosome loop anchorage of the κ immunoglobulin gene occurs next to the enhancer in a region containing topoisomerase II sites. *Cell* 44: 273–282.

Compere, S. J. and Palmiter, R. D. 1981. DNA methylation controls the inducibility of the mouse metallothionein-I gene in lymphoid cells. *Cell* 25: 233–240.

Conklin, K.F. and Groudine, M. 1984. Chromatin structure and gene expression. *In* A. Razin, H. Cedar and A. D. Riggs (eds.), *DNA Methylation.* Springer-Verlag, New York, pp. 293–351.

Cooper, D. W., Vandeberg, J. L., Sharmen, G. B. and Poole, W. E. 1971. Phosphoglycerate kinase polymorphism in kangaroos provides further evidence for paternal X inactivation. *Nature New Biol.* 230: 155–157.

Crick, F. H. C. 1966. *Of Molecules and Men.* University of Washington Press, Seattle.

Darnell, J., Lodish, H. and Baltimore, D. 1986. *Molecular Cell Biology.* Scientific American Books, New York.

DeChiara, T. M., Robertson, E. J. and Efstratiadis, A. 1991. Parental imprinting of the mouse insulin-like growth factor II gene. *Cell* 64: 849–859.

Dickinson, L. A., Joh, T., Kohwi, Y. and Kohwi-Shigematsu, T. 1992. A tissue-specific MAR/SAR DNA-binding protein with unusual binding site recognition. *Cell* 70: 631–645.

Driscoll, D. J. and Migeon, B. R. 1990. Sex difference in methylation of single-copy genes in human meiotic germ cells: Implications for X chromosome inactivation, parental imprinting, and the origin of PGC mutations. *Somatic Cell Molec. Genet.* 16: 267–268.

Elgin, S. 1981. DNase I-hypersensitive sites of chromatin. *Cell* 27: 413–415.

Engel, J. D. 1993. Developmental regulation of human β-globin gene transcription: A switch of loyalties? *Trends Genet.* 9: 304–309.

Enver, T., Raich, N., Ebens, A. J., Papayannopoulou, T., Costantini, F. and Stamatoyannopoulos, G. 1990. Developmental regulation of human fetal-to-adult globin gene switching in transgenic mice. *Nature* 344: 309–312.

Felsenfeld, G. 1992. Chromatin as an essential part of the transcriptional mechanism. *Nature* 355: 219–224.

Ferguson-Smith, A. C., Sasaki, H., Cattanach, B. M. and Surani, M. A. 1993. Parental origin-specific epigenetic modification of the mouse H19 gene. *Nature* 362: 751–755.

Foley, K. P. and Engel, J. D. 1992. Individual stage selector element mutations lead to reciprocal changes in β- vs ε-globin gene transcription: Genetic confirmation of promoter competition during globin gene switching. *Genes Dev.* 6: 730–744.

Frank, D., Lichtenstein, M., Paroush, Z., Bergmann, Y., Shani, M., Razin, A. and Ceder. H. 1990. Demethylation of genes in animal cells. *Philos. Trans. R. Soc. Lond.* [B] 326: 241–251.

Fraser, P., Pruzina, S., Antoniou, M. and Grosveld, F. 1993. Each hypersensitive site of the human β-globin locus control region confers a different developmental pattern of expression on the globin genes. *Genes Dev.* 7: 106–113.

Garcia, J. V., Bich-Thuy, L. T., Stafford, J. and Queen, C. 1986. Synergism between immunoglobulin enhancers and promoters. *Nature* 322: 383–385.

Gallarda, J. L., Foley, K. P., Yang, Z. and Engel, J. D. 1989. The β-globin stage-selector element is erythroid-specific promoter/enhancer-binding protein NF-E4. *Genes Dev.* 3: 1845–1859.

Gartler, S. M., Liskay, R. M. and Grant, N. 1973. Two functional X chromosomes in human fetal oocytes. *Exp. Cell Res.* 82: 464–466.

Gartler, S. M., Rivest, M. and Cole, R. E. 1980. Cytological evidence for an inactive X chromosome in murine oogonia. *Cytogenet. Cell Genet.* 28: 203–207.

Gasser, S. M. and Laemmli, U. K. 1986. Cohabitation of scaffold binding regions with upstream/enhancer elements of three developmentally regulated genes in *D. melanogaster. Cell* 46: 521–530.

Gong, Q. and Dean, A. 1993. Enhancer-dependent transcription of the ε-globin promoter requires promoter-bound GATA-1 and enhancer bound AP-1/NF-E2. *Mol. Cell Biol.* 13: 911–917.

Grosveld, F., Blom van Assendelft, G., Greaves, D. R. and Kollins, G. 1987. Position-dependent high-level expression of the human β-globin gene in transgenic mice. *Cell* 51: 975–985.

Groudine, M. and Conklin, K. F. 1985. Chromatin structure and *de novo* methylation of sperm DNA: Implications for activation of the paternal genome. *Science* 228: 1061–1068.

Groudine, M. and Weintraub, H. 1981. Activation of globin genes during chick development. *Cell* 24: 393–401.

Groudine, M., Kohwi-Shigematsu, T., Gelinas, R.,Stamatoyannopoulos, G. and Papayannopoulo, T. 1983. Human fetal to adult hemoglobin switching: Changes in chromatin structure of the β-globin gene locus. *Proc. Natl. Acad. Sci. USA* 80: 7551–7555.

Gruenbaum, Y., Ceder, H. and Razin, A. 1982. Substrate and sequence specificity of a eukaryotic DNA methylase. *Nature* 295: 620–622.

Hanscombe, O., Whyall, D., Fraser, P., Yannoutsos, N., Greaves, D., Dillon, N. and Grosveld, F. 1991. Importance of globin gene order for correct developmental expression. *Genes Dev.* 5: 1387–1394.

Herman, R., Weymouth, L. and Penman, S. 1978. Heterogeneous nuclear RNA–protein fibers in chromatin depleted nuclei. *J. Cell Biol.* 78: 663–674.

Holliday, R. 1987. The inheritance of epigenetic defects. *Science* 238: 163–170.

Hotchkiss, R. D. 1948. The quantitative separation of purines, pyrimidines, and nucleosides by paper chromatography. *J. Biol. Chem.* 175: 315–332.

Hutchinson, N. and Weintraub, H. 1985. Localization of DNase I-sensitive sequences to specific regions of interphase nuclei. *Cell* 43: 471–482.

Iguchi-Ariga, S. M. M. and Schaffner, W. 1989. CpG methylation of the cAMP-responsive enhancer/promoter sequence TGACGTCA abolishes specific factor binding as well as transcriptional activation. *Genes Dev.* 3: 612–619.

Jackson, D. A. and Cook, P. R. 1985. Transcription occurs at a nucleoskeleton. *EMBO J.* 4: 919–925.

Jeppesen, P. and Turner, B. M. 1993. The inactive X chromosome in female mammals is distinguished by a lack of histone H4 acetylation, a cytogenetic marker for gene expression. *Cell* 74: 281–289.

Jost, J. P. 1993. Nuclear extracts of chicken embryos promote an active demethylation of DNA by excision repair of 5-methyldeoxycytidine. *Proc. Natl. Acad. Sci. USA* 90: 4684–4688.

Kafri, T. and seven others. 1992. Developmental pattern of gene-specific DNA methylation in the mouse embryo and germ line. *Genes. Dev.* 6: 705–714.

Karlsson, S. and Nieuhaus, A. W. 1985. Developmental regulation of human globin genes. *Annu. Rev. Biochem.* 54: 1071–1108.

Kay, G. F., Penny, G. D., Patel, D., Ashworth, A., Brockdorrf, N. and Rastan, S. 1993. Expression of *Xist* during mouse development suggests a role in the initiation of X chromosome inactivation. *Cell* 72: 171–182.

Kaye, J. S., Pratt-Kaye, S., Bellard, M., Dretzen, G., Bellard, F. and Chambon, P. 1986. Steroid hormone dependence of four DNase I-hypersensitive regions located within the 7000-bp 5' flanking segment of the ovalbumin gene. *EMBO J.* 5: 277–285.

Keith, D. H., Singersam, J. and Riggs,A. D. 1986. Active X-chromosome DNA is unmethylated at eight CCGG sites clustered in a guanine-plus-cytosine-rich island at the 5' end of the gene for phosphoglycerate kinase. *Mol. Cell Biol.* 6: 4122–4125.

Keshet, I., Lieman-Hurwitz, J. and Cedar, H. 1986. DNA methylation affects the formation of active chromatin. *Cell* 44: 535–543.

Knoll, J. H. M., Nicholls, R. D., Magenis, R. E., Graham, J. M., Jr., Lalande, M. and Latt, S. A. 1989. Angelman and Prader-Willi syndromes share a common chromosome 15 deletion but differ in the parental origin of the deletion. *Am. J. Med. Genet.* 32: 285–290.

Kornberg, R. D. and Thomas, J.D. 1974. Chromatin structure: Oligomers of histones. *Science* 184: 865–868.

Kratzer, P. G. and Chapman, V. M. 1981. X–chromosome reactivation in oocytes of *Mus caroli*. *Proc. Natl. Acad. Sci. USA* 78: 3093–3097.

Kuroda, M. I., Kernan, M. J., Kreber, R., Ganetzky, B. and Baker, B. S. 1991. The maleless protein associates with the X chromosome to regulate dosage compensation in *Drosophila*. *Cell* 66: 935–947.

Laybourne, P. J. and Kadonaga, J. T. 1991. Role of nucleosome cores and histone H1 in regulation of transcription by RNA polymerase II. *Science* 254: 238–245.

Lee, D. Y., Hayes, J. J., Pruss, D. and Wolffe, A. P. 1993. A positive role for histone acetylation on transcription factor access to nucleosomal DNA. *Cell* 72: 73–84.

Li, E., Beard, C. and Jaenisch, R. 1993. The role of DNA methylation in genomic imprinting. *Nature* 36: 362–365.

Liskay, R. M. and Evans, R. 1980. Inactive X chromosome DNA does not function in DNA-mediated cell transformation for the hypoxanthine phosphoribosyltransferase gene. *Proc. Natl. Acad. Sci. USA* 77: 4895–4898.

Liu, J.-K., Bergman, Y. and Zaret, K. S. 1988. The mouse albumin promoter and a distal upstream site are simultaneously DNase I hypersensitive in liver chromatin and bind similar liver-abundant factors in vitro. *Genes Dev.* 2: 528–541.

Lucchesi, J. C. and Manning, J. E. 1987. Gene dosage and compensation in *Drosophila melanogaster*. *Adv. Genet.* 24: 371–429.

Ludérus, L. A., M. E. E., de Graaf, A., Mattia, E., den Blaauwen, J. L., Grande, M. A., de Jong, L. and van Driel, R. 1992. Binding of matrix attachment regions to lamin B$_1$. *Cell* 70: 949–959.

Lyon, M. F. 1961. Gene action in the X chromosome of the mouse (*Mus musculus* L.) *Nature*: 190: 372–373.

Mandel, J. L. and Chambon, P. 1979. DNA methylation differences: Organ-specific variations in methylation pattern within and around ovalbumin and other chick genes. Nucleic Acid Res. 7: 2081–2103.

Mariman, E. C. M., van Eekelen, C. A. G., Reinders, R. J., Berns, A. J. M. and van Venrooji, W. J. 1982. Adenoviral heterogenous nuclear RNA is associated with the host nuclear matrix during splicing. *J. Mol. Biol.* 154: 103–119.

Martin, D. I. K., Tsai, S.-F. and Orkin, S. H. 1989. Increased γ-globin expression in a nondeletion HPFH mediated by an erythroid-specific DNA-binding factor. *Nature* 338: 435–437.

Mattei, M. G., Mattei, J. F., Vidal, I. and Giraud, F. 1981. Structural anomalies of the X chromosome and activation center. *Hum. Genet.* 56: 401–408.

Mavilio, F. and nine others. 1983. Molecular mechanisms for human hemoglobin switching: Selective undermethylation and expression of globin genes in embryonic, fetal, and adult erythroblasts. *Proc. Natl. Acad. Sci. USA* 80: 6907–6911.

McGhee, J. D. and Ginder, G. D. 1979. Specific DNA methylation sites in the vicinity of the chick β-globin genes. *Nature* 280: 419–420.

Migeon, B. R. 1971. Studies of skin fibroblasts from ten families with HGPRT deficiency, with reference to X–chromosomal inactivation. *Am. J. Hum. Genet.* 23: 199–209.

Migeon, B. R. and Jelalian, K. 1977. Evidence for two active X chromosomes in germ cells of female before meiotic entry. *Nature* 269: 242–243.

Migeon, B. R., Wolf, S. F., Axelman, J., Kaslow, D.C. and Schmidt, M. 1985. Incomplete X chromosome dosage compensation in chorionic villi of human placenta. *Proc. Natl. Acad. Sci. USA* 82: 3390–3394.

Migeon, B. R., Schmidt, M., Axelman, J. and Cullen, C. R. 1986. Complete reactivation of X chromosomes from human chorionic villi with a switch to early DNA replication. *Proc. Natl. Acad. Sci. USA* 83: 2182–2186.

Migeon, B. R., Holland, M. M., Drscoll, D. J. and Robinson, J. C. 1991. Programmed demethylation in CpG islands during human fetal development. *Somatic Cell Molec. Genet.* 17: 159–168.

Miller, T. E., Huang, C.-Y. and Pogo, A. O. 1978. Rat liver nuclear skeleton and ribonucleoprotein complexes containing hnRNA. *J. Cell Biol.* 76: 675–691.

Mohandas, T., Sparkes, R. S., Hellkuhl, B., Brzeschik, K. H. and Shapiro, L. J. 1980. Expression of an X-linked gene from an inactive human X chromosome in mouse–human hybrid cells: Further evidence for the non-inactivation of the steroid sulfatase locus in man. *Proc. Natl. Acad. Sci. USA* 77: 6759–6763.

Mohandas, T., Sparkes, R. S. and Shapiro, L. J. 1981. Reactivation of an inactive human X chromosome: Evidence for X inactivation by DNA methylation. *Science* 211: 393–396.

Monk, M., Boubelik, M. and Lehnert, S. 1987. Temporal and regional changes in DNA methylation in the embryonic, extraembryonic, and germ cell lineages during mouse embryo development. *Development* 99: 371–382.

Moore, K. L. 1977. *The Developing Human.* Saunders, Philadelphia.

Nelson, W. G., Pienta, K. J., Barrack, E. R. and Coffey, D. S. 1986. The role of the nuclear matrix in the organization and function of DNA. *Annu. Rev. Biophys. Biophysical Chem.* 15: 457–475.

Nicholls, R. D. 1993. Genomic imprinting and uniparental disomy in Angelman and Prader-Willi syndromes. *Am. J. Med. Genet.* 46: 16–25.

Nicholls, R. D., Kroll, J. H. M., Butler, M. G., Karma, S. and Lalande, M. 1989. Genetic imprinting suggested by maternal heterodisomy in non-deletion Prader-Willi syndrome. *Nature* 342: 281–285.

Orkin S. H. 1992. GATA-binding transcription factor in hematopoietic cells. *Blood* 80: 575–581.

Ottolenghi, S. 1992. Developmental regulation of human globin genes: A model for cell differentiation in the hematopoietic system. *In* V. E. Russo et al., (eds.), *Development: The Molecular Genetic Approach.* Springer-Verlag, New York, pp. 519–536.

Oudet, P., Gross-Bellard, M. and Chambon, P. 1975. Electron microscope and biochemical evidence that chromatin structure is a repeating unit. *Cell* 4: 281–300.

Paroush, Z., Keshet, I., Yisraeli, J. and Cedar, H. 1990. Dynamics of demethylation and activation of the α-actin gene in myoblasts. *Cell* 63: 1229–1337.

Perrine, S. P. and eight others. 1993. A short-term trial of butyrate to stimulate fetal globin gene expression in the β-globin disorders. *New Engl. J. Med.* 328: 81–86.

Pevny, L. and seven others. 1991. Erythroid differentiation in chimeric mice blocked by a targeted mutation in the gene for transcription factor GATA-1. *Nature* 349: 257–261.

Reik, W., Collick, A., Norris, M. L., Barton, S. C. and Surani, M. A. 1987. Genomic imprinting determines methylation of paternal alleles in transgenic mice. *Nature* 328: 248–250.

Rigaud, G., Roux, J., Pictet, R. and Grange, T. 1991. In vivo footprinting of rat TAT gene: Dynamic interplay between glucocorticoid receptor and a liver-specific factor. *Cell* 67: 977–986.

Robinson, S. I., Nelkin, B. D. and Vogelstein, B. 1982. The ovalbumin gene is associated with the nuclear matrix of chicken oviduct cells. *Cell* 28: 99–106.

Rodgers, G. P., Dover, G. J., Vyesaka, N., Noguchi, C. T., Schecter, A. N., and Nieuhuis, A. W. 1993. Augmentation by erythropoietin of the fetal hemoglobin response to hydroxyurea in sickle cell disease. *New Engl. J. Med.* 328: 73–80.

Rogers, J. and Wall, R. 1981. Immunoglobulin heavy–chain genes: Demethylation accompanies class switching. *Proc. Natl. Acad. Sci. USA* 78: 7497–7501.

Rose, S. and Garrard, W. T. 1984. Differentiation-dependent chromatin alterations precede and accompany transcription of immunoglobulin light chain genes. *J. Biol. Chem.* 259: 8534–8544.

Russell, L. B. 1963. Mammalian X chromosome action: Inactivation limited in spread and region of origin. *Science* 140: 976–978.

Ryan, T. M., Behringer, R. B., Martin, N. C., Townes, T. M., Palmiter, R. D. and Brinster, R. L. 1989. A single erythroid-specific DNase I super-hypersensitivity site activates high levels of human β-globin gene expression in transgenic mice. *Genes Dev.* 3: 314–323.

Ryoji, M. and Worcel, A. 1984. Chromatin assembly in *Xenopus* oocytes: In vitro studies. *Cell* 37: 21–32.

Samollow, P. B., Ford, A. L. and Vande-Berg, J. L. 1987. X–linked gene expression in the Virginia opossum: Differences between the paternally derived *Gpd* and *Pgk-A* loci. *Genetics* 115: 185–195.

Sanford, J. P., Clark, H. J., Chapman, V. M. and Rossant, J. 1987. Differences in DNA methylation during oogenesis and spermatogenesis and their persistence during early embryogenesis in the mouse. *Genes Dev.* 1: 1039–1046.

Sapienza, C., Peterson, A. C., Rossant, J. and Balling, R. 1987. Degree of methylation of transgenes is dependent on gamete of origin. *Nature* 328: 251–254.

Scheuermann, R. H. and Chen, U. 1989. A developmental-specific factor binds to suppressor sites flanking the immunoglobulin heavy-chain enhancer. *Genes Dev.* 3: 1255–1266.

Schlissel, M. S. and Brown, D. D. 1984. The transcriptional regulation of *Xenopus* 5S RNA genes in chromatin: The roles of active stable transcription complex and histone H1. *Cell* 37: 903–913.

Sharman, G. B. 1971. Late DNA replication in the paternally derived X chromosome of female kangaroos. *Nature* 230: 231–232.

Shen, C. K. J. and Maniatis, T. 1980. Tissue-specific DNA methylation in a cluster of rabbit β-like globin genes. *Proc. Natl. Acad. Sci. USA* 77: 6634–6638.

Shermoen, A. W. and Beckendorf, S. K. 1982. A complex of DNase I-hypersensitive sites near the *Drosophila* glue protein gene, *Sgs-4*. *Cell* 29: 601–607.

Stalder, J., Larsen, A., Engel, J. D., Dolan, M., Groudine, M. and Weintraub, H. 1980. Tissue-specific DNA cleavages in the globin chromatin domain introduced by DNase I. *Cell* 20: 451–460.

Steiner, E. K., Eissenbert, J. C. and Elgin, S. C. R. 1984. A cytological approach to the orderly events in gene activation using the *Sgs-4* locus of *Drosophila melanogaster*. *J. Cell Biol.* 99: 233–238.

Stief, A., Winter, D. M., Strätling, W. H. and Sippel. A. E. 1989. A nuclear DNA attachment element mediates elevated and position-dependent gene activity. *Nature* 341: 343–345.

Stöger, R., Kublicka, P, Liu, C.-G., Kafri, T., Razin, A., Cedar, H. and Barlow, D. P. 1993. Maternal-specific methylation of the imprinted mouse *Igf2r* locus identifies the expressed locus as carrying the imprinting signal. *Cell* 73: 61–71.

Sukegawa, J. and Blobel, G. 1993. A nuclear pore complex protein that contains zinc finger motifs, binds DNA, and faces the nucleoplasm. *Cell* 72: 29–38.

Swain, J. L., Stewart, T. A. and Leder, P. 1987. Parental legacy determines methylation and expression of an autosomal transgene: A molecular mechanism for parental imprinting. *Cell* 50: 719–727.

Tagaki, N. 1974. Differentiation of X chromosomes in early female mouse embryos. *Exp. Cell Res.* 86: 127–135.

Tagaki, N. and Abe, K. 1990. Detrimental effects of two active X chromosomes on early mouse development. *Development* 109: 189–201.

Talbot, D. and Grosveld, F. 1991. The 5′ HS2 of the globin locus control region enhances transcription through the interaction of a multimeric complex binding at two functionally distinct NF-E2 binding sites. *EMBO J.* 10: 1391–1398.

Tazi, J. and Bird, A. P. 1990. Alternative chromatin structure at CpG islands. *Cell* 60: 909–920.

Thoma, F., Koller, T. and Klug, A. 1979. Involvement of histone H1 in the organization of the nucleosome and of the salt-dependent superstructures of chromatin. *J. Cell Biol.* 83: 403–427.

Thorburn, A. and Knowland, J. 1993. Attachment of vitellogenin genes to the nucleoskeleton accompanies their activation. *Biochem. Biophys. Res. Communic.* 191: 308–313.

Trudel, M. and Constantini, F. 1987. A 3′ enhancer contributes to the stage-specific expression of the human β-globin gene. *Genes Dev.* 1: 954–961.

Tuan, D., Abeliovich, A., Lee-Oldham, M. and Lee, D. 1987. Identification of regulatory elements in human β-like globin genes. *In* G. Stamatoyannopoulos and A. W. Nienhuis (eds.), *Developmental Control of Globin Gene Expression.* Alan R. Liss, New York, pp. 211–220.

Turner, B. M. 1991. Histone acetylation and control of gene expression. *J. Cell Sci.* 99: 13–20.

van der Ploeg, L. H. T. and Flavell, R. D. 1980. DNA methylation in the human γ-δ-β globin locus in erythroid and non-erythroid cells. *Cell* 19: 947–958.

van Eekelen, C. A. G. and van Venrooij, W. J. 1981. HnRNA and its attachment to a nuclear protein matrix. J. Cell Biol. 88: 554–563.

Villeponteau, B., Lundell, M. and Martinson, H. 1984. Torsional stress promotes the DNase hypersensitivity of active genes. Cell 39: 469–478.

Weintraub, H. 1984. Histone H1-dependent chromatin superstructures and the suppression of gene activity. *Cell* 38: 17–27.

Weintraub, H. 1985. Assembly and propagation of repressed and derepressed chromosomal states. *Cell* 42: 705–711.

Weintraub, H. and Groudine, M. 1976. Chromosomal subunits in active genes have an altered configuration. *Science* 193: 848–856.

West, J. D., Frels, W. I., Chapman, V. M. and Papaioannou, V. E. 1977. Preferential expression of the maternally derived X chromosome in the mouse yolk sac. *Cell* 12: 873–882.

Wilson, E. B. 1895. *An Atlas of the Fertilization and Karyogenesis of the Ovum.* Macmillan, New York.

Wolf, S. F., Jolly, D. J., Lunnen, K., Friedmann, T. and Migeon, B. R. 1982. Methylation of the hypoxanthine phosphoribosyltransferase locus on the human X chromosome: Implications for X chromosome inactivation. *Proc. Natl. Acad. Sci. USA* 81: 2806–2810.

Wolf, S. F., Dintgis, S., Toniolo, D., Persico, G., Lunnen, K. D., Axelman, J. and Migeon, B. R. 1984. Complete concordance between glucose-6-phosphate dehydrogenase activity and hypomethylation of 3′ CpG clusters: Implication for X chromosome dosage compensation. *Nucleic Acid Res.* 12: 9333–9348.

Wolfe, S. L. 1993. *Molecular and Cellular Biology.* Wadsworth, Belmont, CA.

Yisraeli, J., Adelstein, R. S., Melloui, D., Nudel, U., Yaffe, D. and Ceder, H. 1986. Muscle-specific activation of a methylated chimeric actin gene. *Cell* 46: 409–416.

Zaret, K. S. and Yamamoto, K. R. 1984. Reversible and persistent changes in chromatin structure accompany activation of a glucocorticoid-dependent enhancer element. *Cell* 38: 29–38.

12

Control of development by RNA processing

It may be in the interpretation and analysis of differentiation that the new concepts derived from the study of microorganisms will prove of the greatest value. . . . Eventually, however, differentiation will have to be studied in differentiated cells.

J. MONOD AND F. JACOB (1961)

*Between the conception
And the creation . . .
Between the potency
And the existence
Between the essence
And the descent
Falls the Shadow.*

T. S. ELIOT (1936)

The essence of differentiation is the production of different sets of proteins in different types of cells. In bacteria, differential gene expression can be effected at the levels of transcription, translation, and protein modification. In eukaryotes, however, another possible level of regulation exists, namely, control at the level of RNA processing and transport. This chapter will present some of the recent evidence suggesting (1) that this type of regulation is crucial for development, (2) that different cells can process the same transcribed RNA in different ways, thus creating different populations of cytoplasmic messenger RNAs from the same set of nuclear transcripts, and (3) that some cells allow a nuclear RNA transcript to be processed into mRNA while other cells cause the same nuclear RNA to be degraded before being processed into mRNA.

During the 1960s, developmental biologists had a strong bias toward explaining differentiation in terms of transcriptional regulation of gene expression. First, research on polytene chromosomes and their products indicated that transcriptional regulation occurs during insect development. Second, DNA–RNA hybridization studies showed that whereas the DNA of different cell types is identical, their mRNAs are different (McCarthy and Hoyer, 1964). Third, gene expression could be studied more readily in bacteria than in eukaryotic cells, and this research demonstrated the importance of transcriptional regulation in producing different physiological states in *Escherichia coli*. The *lac* operon of *E. coli*, for instance, transcribes mRNA only in the presence of a specific inducer molecule (in this case, lactose). Transcriptional gene expression in unicellular prokaryotes was a powerful paradigm for differential gene expression in developing eukaryotic cells.

Heterogeneous nuclear RNA

It soon became apparent, however, that messenger RNA is not the primary transcription product of eukaryotic cells. Rather, there is *nuclear* RNA that was much larger than mRNA and has a much shorter half-life than the cytoplasmic messages. Because they vary widely in size, these nuclear RNA molecules are called **heterogeneous nuclear RNA** (hnRNA). The hnRNA molecules are approximately 5 to 20×10^3 nucleotides long, compared with the 2×10^3 bases usually found in mRNA, and they decay with a half-life of minutes instead of hours (see Lewin, 1980). In sea urchin embryos, the half-life of mRNA is about 5 hours whereas the half-life of nuclear RNA is about 20 minutes. The relationship between hnRNA and cytoplasmic mRNA was disputed until 1977, at which time several laboratories found mRNA sequences within the large nuclear RNA molecules. Bastos and Aviv (1977) isolated mouse globin message and made complementary DNA from it by using reverse transcriptase (Figure 12.1). They then conjugated this cDNA to cellulose beads. After nucleated hemoglobin-synthesizing cells were grown with radioactive uridine, the radiolabeled RNA was extracted and passed over this column. Any RNA sequence coding for globin should bind to the complementary DNA on the column. No other RNA sequence should be "fished out" in such a manner. The RNA bound to the column was eluted from the column by washing it with low-ionic-strength solutions, which destabilized the hydrogen bonds holding the complementary molecules together. When cells were labeled for 5 minutes, the radioactive RNA binding to the column had an average size of 5000 nucleotides, nearly eight times that of the cytoplasmic globin message.

However, when the labeling was stopped and the RNA extracted 5 minutes later, the globin sequences were found in a smaller RNA. Finally, the globin-coding RNA was found in a 600-nucleotide molecule characteristic of the cytoplasmic message. Thus, hnRNA appears to contain the precursors of cytoplasmic globin mRNA. Furthermore, studies of β^+-thalassemia, a disease characterized by low levels of β-globin, showed that the basis of this disease is the faulty processing and transport of the globin message precursor inside the nucleus (Kantor et al., 1980; Marquat et al., 1980). Thus, at least some of the hnRNA contains precursor mRNA.

Sequence complexity of nuclear and messenger RNAs

The next question involved the number of different types of RNA sequences found in the nucleus and cytoplasm. Does the nucleus contain a greater diversity of sequences than the cytoplasm, or are they about the same? The answer to this question will tell us whether much more of the genome is transcribed into hnRNA than the amount we measure in mRNA. The diversity of nucleotide sequences is called **complexity**. If one were to take one copy of each different mRNA and lay them end to end, and express the total length in nucleotides, that is the complexity of the mRNA.

The analysis of complexity is often graphed on C_0t curves. C_0 refers to the original concentration of labeled nucleic acid (in moles of nucleotide per liter) and t refers to the time (in seconds). The extent of reassociation is a function of both concentration and time. Thus, the percentage of renatured DNA is plotted against the C_0t value (Figure 12.2). This type of analysis has revealed that the genome contains some DNA sequences represented millions of times (the highly repetitive fraction), some sequences represented thousands of times (the moderately repetitive fraction), and some DNA (such as those genes coding for enzymes) that are essentially represented once per haploid genome (Figure 12.2B). This last fraction is referred to as **single-copy DNA** (Britten and Kohne, 1968).

To measure complexity of RNA, RNA is isolated from the tissue of interest and a small amount of radioactive single-copy DNA is added to a great abundance of the RNA. The fraction of DNA that finds its "partner" is the fraction of the genome that is transcribed into RNA (after being

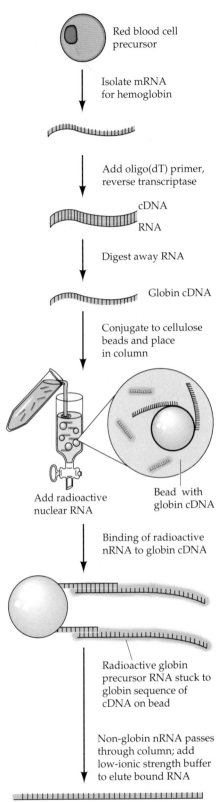

Red blood cell precursor

Isolate mRNA for hemoglobin

Add oligo(dT) primer, reverse transcriptase

cDNA

RNA

Digest away RNA

Globin cDNA

Conjugate to cellulose beads and place in column

Add radioactive nuclear RNA

Bead with globin cDNA

Binding of radioactive nRNA to globin cDNA

Radioactive globin precursor RNA stuck to globin sequence of cDNA on bead

Non-globin nRNA passes through column; add low-ionic strength buffer to elute bound RNA

Radioactive nRNA eluted containing globin mRNA sequences

FIGURE 12.1
Protocol for isolating heterogeneous nuclear RNA containing globin-coding sequences.

FIGURE 12.2

Reassociation of nucleic acids from various sources. The curve represents the normal kinetics expected when each gene is present in the same frequency. Because the hybridization reaction demands that two complementary sequences come together, the rate of hybridization can be described as $dC/dt = -kC^2$, where C is the concentration of single-stranded sequences present at time t, and k is the association constant. Note that the ordinate is plotted with 0 percent at the top and that the C_0t axis is represented logarithmically. (A) Reassociation of DNAs or artificial RNA with little or no repetitive sequences. According to these data, the DNA of *E. coli* has about 100 times as many sequences as the DNA of the T4 bacteriophage has. (B) Reassociation of DNA from a typical eukaryotic organism. DNA is sheared to about 500 nucleotide pairs and denatured in alkali. When the DNA is allowed to renature, a fraction of highly repetitive DNA reanneals almost immediately. Then a fraction of moderately repetitive DNA renatures. Eventually the "single-copy" region of the genome—the portion that encodes the proteins—renatures. According to these data, some sequences are present a million times per haploid genome, whereas most sequences are present only once. (A after Britten and Kohne, 1968; B after Hood et al., 1975.)

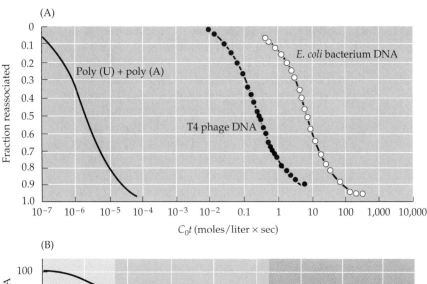

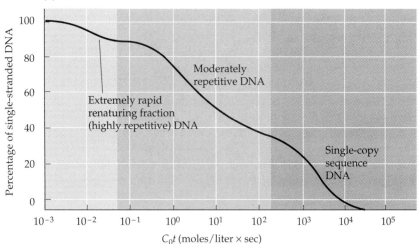

multiplied by 2 to account for the presence of both strands in the single-copy DNA). By this method the hnRNA complexity was determined to be more than tenfold greater than that of mRNA from the same cells. About 90 percent of the hnRNA contains sequences that do not enter the cytoplasm.

Control of early development by hnRNA processing

Many of these nonmessage RNA sequences that are present in the nucleus but not in the cytoplasm are the introns of the mRNA prescursors. However, studies in sea urchins suggest that entire transcripts are being degraded in some cells and processed into mRNA in others. In other words, these studies suggest that different cell types may be transcribing the same type of nuclear RNA, but that different subsets of this population are being processed to mRNA in different types of cells.

In 1977, Kleene and Humphreys showed that nuclear RNA from whole pluteus larvae and from blastulae were (within experimental error) identical to each other. Both RNAs bound to the same 30 percent of the genome. When the RNA complexity was analyzed, hnRNA from blastula cells was found to bind to 15 percent of this DNA (i.e., to 30 percent of the genomic single copy DNA). Similarly, pluteus-stage hnRNA, even when present in great excess, also bound to 30% of the single-copy DNA.

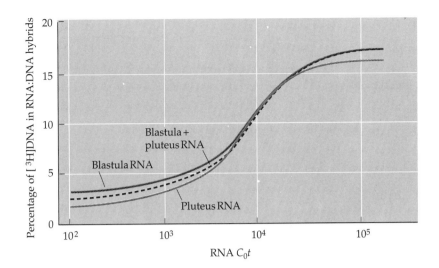

FIGURE 12.3

Hybridization of nuclear RNA from sea urchin embryos with single-copy [³H]DNA. Radioactive single-copy DNA was mixed with either blastula RNA, pluteus RNA, or a mixture of blastula and pluteus RNAs. The mixtures were incubated to allow all complementary sequences to pair. In all three cases, about 15 percent of the DNA hybridized to the RNA. (After Kleene and Humphreys, 1977.)

Are these two sets of DNA sequences the same or are they different? This question was approached by mixing blastula and pluteus nuclear RNAs and adding them to the denatured single-copy DNA. If the sequences were totally different, one would expect 30 percent of the DNA to be bound (i.e., 60 percent of the genome would be coding for the combined set of blastula and pluteus messages). If they were identical, one would expect 15 percent of the DNA to be bound. The result is shown in Figure 12.3. The mixture bound to only 15 percent of the DNA. The hnRNA sequences of the blastula and pluteus bound to the same DNA. Within experimental error, the hnRNA was identical in blastula and pluteus cells.

To confirm this unexpected observation, Kleene and Humphreys (1977, 1985) isolated radioactive single-copy DNA and reassociated it with pluteus or blastula RNA to a high C_0t value (so that all DNA sequences could bind the RNA if their complementary RNA sequences were present). They then isolated those DNA fragments that would not bind to blastula hnRNA (blastula-null DNA) and those DNA fragments that would not bind to the pluteus hnRNA (pluteus-null DNA). When blastula-null DNA

FIGURE 12.4

Hybridization of "pluteus-plus" DNA and "pluteus-null" DNA by pluteus and blastula RNA. (A) Both pluteus and blastula RNA bound "pluteus-plus" DNA to the same extent. (B) Neither RNA bound significantly to the "pluteus-null" DNA. (After Kleene and Humphreys, 1977.)

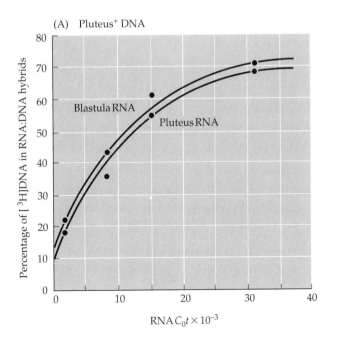

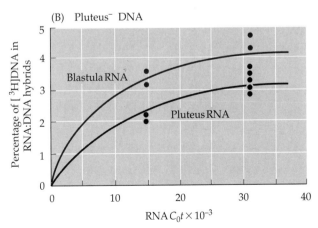

was denatured and mixed to new samples of *pluteus* hnRNA and when pluteus-null DNA was mixed with *blastula* hnRNA, no hybridization above background levels was observed (Figure 12.4). Therefore, the DNA sequences transcribed during the blastula stage are the same as DNA sequences transcribed during the pluteus stage. Wold and her colleagues (1978) extended these observations by showing that sequences present in blastula *messenger* RNA (isolated from translating polysomes) but absent in gastrula and adult tissue mRNA were nonetheless present in the *nuclear*

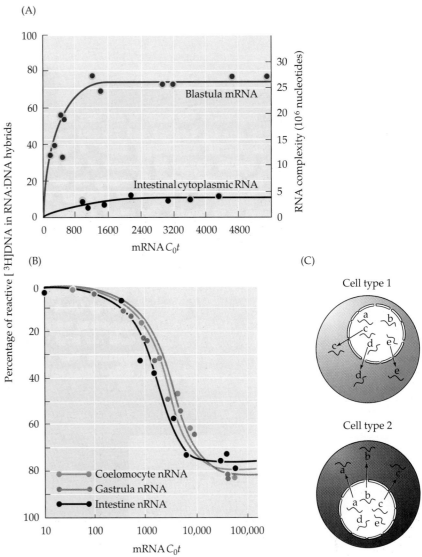

FIGURE 12.5
Sequences found in the nuclear RNA of several cell types not found in the mRNA. (A) Specificity of sea urchin blastula messenger cDNA. Hybridization of blastula messenger cDNA (cDNA to blastula mRNA) with blastula mRNA and intestinal cytoplasmic RNA shows that the mRNAs are very different. (B) Hybridization of the blastula messenger cDNA to the nuclear RNAs (nRNAs) of gastrulae and adult coelomocytes and intestinal cells suggests the identity of all nuclear RNAs. (C) Speculative model based on differential RNA processing. In both cell types, the same RNAs (a, b, c, d, e) are transcribed, but in one cell type, sequences c, d, and e are processed to cytoplasmic mRNA, whereas in another cell type sequences a, b, and c are processed into the cytoplasm. (A and B after Wold et al., 1978.)

TABLE 12.1
Intertissue comparisons of structural gene sequences in messenger RNA and nuclear RNA

Reference tracer complementary to	Normalized reaction with parent mRNA		Normalized reaction with other mRNA		Normalized reaction with nRNA	
	mRNA	%	mRNA	%	nRNA	%
SEA URCHIN						
Blastula mRNA (single copy DNA)	Blastula	100	Intestine	12	Intestine	97
			Coelomocyte	13	Coelomocyte	101
MOUSE						
Brain mRNA (total cDNA)	Brain	100	Kidney	78	Kidney	102
Brain mRNA (cDNA representing rare messages)	Brain	100	Kidney	56	Kidney	100

Source: Davidson and Britten, 1979.

RNA of the gastrula and adult tissues. These results were interpreted to indicate that more genes are transcribed in the nucleus than are allowed to become mRNAs in the cytoplasm (Figure 12.5; Table 12.1).

This post-transcriptional control of RNA processing has been confirmed for specific messages by ribonuclease protection assays and nuclear run-on transcription assays. The **ribonuclease protection assay** is a sensitive test for determining the presence (or absence) of a particular sequence in a population of RNAs (Figure 12.6). One makes a relatively small radioactively labeled RNA that is complementary to a specific sequence in the RNA one wishes to detect. This RNA "probe" is mixed with cellular RNA that one is testing for the presence of the particular sequence. If the sequence is present, the probe binds to it. Then ribonuclease (RNase, an enzyme that digests single-stranded RNA) is added to the mixture to cleave away all the RNA that is not hybridized. However, the double-stranded RNA formed by the hybridization of radioactive and cellular RNAs is not cleaved. This double-stranded RNA can be run on a gel and autoradiographed. If the RNA sequence is present, a band of a particular size should show up on the autoradiograph. Gagnon and his colleagues (1992) performed such an analysis on the transcripts from the *Spec1* and *CyIIIa* genes of the sea urchin *Strogylocentrotus purpuratus*. These genes encode calcium-binding and actin proteins, respectively, that are expressed only in the aboral ectoderm of the pluteus larva. Using probes that bound to an exon and to an intron, they found that these genes were being transcribed not only in the ectoderm cells, but also in the mesoderm and endoderm. The analysis of the *CyIIIa* gene showed that the level of *introns* was the same in both the gastrula ectoderm and in the mesoderm/endoderm samples, suggesting that this gene was being transcribed at the same rate in all cell types (Figure 12.7A). The *exon*, however, accumulated in the ectoderm (which expresses the protein), but not in the mesoderm or endoderm (which do not). Thus, while the genes appear to be transcribed at a similar rate in the ectoderm and in other tissues, the messenger RNA for these proteins (represented by the exons) accumulates only in the ectoderm.

This conclusion was confirmed when nuclei were isolated from the gastrula tissues. Radioactive UTP enabled the researchers to follow whatever RNAs were being transcribed at the time the nuclei were isolated.

FIGURE 12.6
Assays for detecting the accumulation of a message in the cytoplasm. (A) Ribonuclease protection assay. RNA is isolated and purified from embryonic tissue. A radioactive RNA probe is synthesized that is complementary to a small stretch of the RNA one is assaying. If the specific RNA is present, the radioactive probe will bind to it. RNase is then added, destroying all the RNA except for the double-stranded region that contains the radioactive oligonucleotide. This can be electrophoresed on a gel and autoradiographed. (B) Nuclear run-on assay. Nuclei are isolated from embryonic tissue. Radioactive UTP is added to the nuclei. The mRNA that is being synthesized incorporates radioactive label as it continues being transcribed. The mRNA can be isolated and hybridized to complementary DNA sequences immobilized on paper. If the radioactive transcript binds, it is detected by autoradiography.

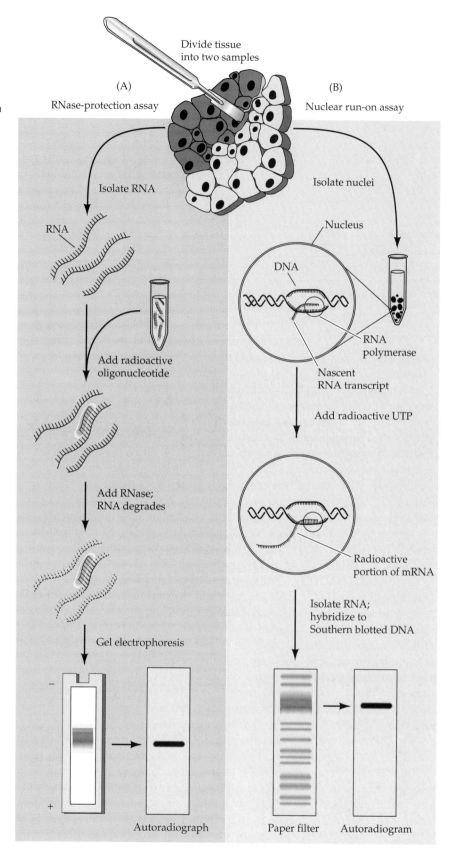

FIGURE 12.7

Regulation of ectoderm-specific gene expression by RNA processing. (A) Autora-diograms of ribonuclease protection assay. The left-hand column represents RNA isolated from the gastrula ectodermal tissue; the right-hand column represents RNA isolated from the endoderm and mesodermal tissues. The upper band is the RNA protected by a probe that binds to an intron sequence (which should be found only in the nucleus) of *CyIIIa*. The lower band represents the RNA protected by a probe complementary to an exon sequence. (B) Results of a nuclear transcription run-on assay. Radioactive RNA synthesized in vitro by ecto-dermal nuclei (left) and endodermal/mesodermal nuclei (right) was hybridized to an intron of the *Spec1* gene affixed to a filter. Both ectodermal and endoderm/mesoderm nuclei were transcribing this gene in the sea urchin gastrula, even though the *Spec1* message is seen only in the ectodermal cells. (From Gagnon et al., 1992, courtesy of R. and L. Angerer.)

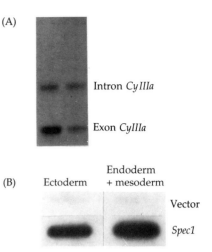

(A)

Intron *CyIIIa*

Exon *CyIIIa*

(B)   Ectoderm   Endoderm + mesoderm

Vector

Spec1

Both ectoderm and endoderm/mesoderm nuclei were transcribing the *Spec1* gene at the gastrula stage (Figure 12.7B). Thus, the expression of the *Spec1* and *CyIIIa* genes is at the level of RNA processing in the gastrula. Later in development (by the time of the pluteus stage), these genes come under transcriptional control, where transcription of the genes stops in the cells not expressing these proteins. It appears, then, that RNA processing plays a major role in the control of sea urchin gene expression in early embryos.

SIDELIGHTS & SPECULATIONS

Molecular regulators of development: Spliceosomes

How are the message precursors spliced? How might different mRNAs be generated from the same nuclear transcript? What distinguishes an exon from an intron? Studies on the biochemistry of RNA processing are just beginning to bring light to this relatively unknown area.

Splice sites

Comparison of the DNA sequences of many different genes revealed certain similarities at the intron–exon junctions (see Jacob and Gallinaro, 1989). The base sequence of the intron usually begins with GT, and the base sequence of the exon usually ends with AG . In fact, the consensus sequence for the **5′ splice site** is (in the language of RNA): (A or C)AG/GUAAGU (i.e., most exons end in AG, and the intron begins with GU). The **3′ splice site** sequence of most introns is similar to the sequence $(C/U)_{>10}N(C/T)AG/G$ such that most introns end in AG after a long stretch of pyrimidines. These sites appear to be very important for intron processing. The mutations that cause β-thalassemia are not in the exons coding for the globin protein, but in the *introns* or other regulatory regions. In one of these thalassemia mutations (depicted by the arrow in Figure 10.1), the sequence TTGGTCT in the first intron is mutated to TTAGTCT. However, the usual 3′ splicing site for this

intron is TTAG/GCT. Thus, the mutation makes an artificial splicing site, creating the possibility for erroneous splicing (Spritz et al., 1981). In fact, this thalassemic β-globin precursor splices at the wrong junction over 90 percent of the time (Busslinger et al., 1981).

Branch sites

In addition to the splicing junctions, there is one other sequence necessary for proper processing of the introns. This is the **branch site**. This site is located within the intron (generally 20–50 nucleotides upstream from the 3′ splice site) and has the consensus sequence UAUAAC. In most cases, U can be replaced by C and A can be replaced by G. The penultimate (boldface) A residue, however, is invariant. This A residue is critical, since it will create a phosphodiester "branch" with the phosphate of the 5′ splice site (Figure 12.8). It does this by using its free 2′ hydroxyl group to combine with the phosphate at the 5′ splice site, breaking the phosphodiester bond at that splice site, and making a 2′−5′ phosphodiester bond with the end of the intron. In effect, the GMP phosphate, instead of holding on to the first exon, releases it and hooks on to the 2′ hydroxyl group of the branch site AMP. This maneuver releases the first exon and makes a "lariat" structure as an intermediate.

The free 3′ OH terminus of the first exon then binds to the phosphate at the 3′ splice junction downstream from

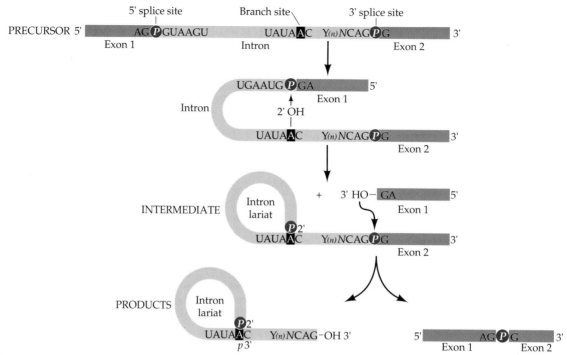

FIGURE 12.8

Hypothetical mechanism for splicing an mRNA precursor. The model precursor RNA is drawn with its intron spanning from the 5' splice site to the 3' splice site, and the consensus sequences at these sites are indicated. The branch site and its surrounding nucleotides are also shown. Splicing occurs in two steps. The first involves the creation of a lariat structure as the phosphate of the 5' splice site forms a phosphodiester linkage with the 2' OH group of the branch site AMP. This lariat formation leads to detachment of the first exon. In the second step, the terminal OH group of the first exon replaces the intron at the 5' end of the second exon. The phosphate groups involved in the reactions are identified with circles. (Modified from Sharp, 1987.)

this lariat, thereby substituting itself for the intron. The result is the linkage of the two exons with the AG group coming from the first exon and the G coming from the second (Sharp, 1987).

The lariat-shaped intermediate with its 2'–5' phosphodiester bond was discovered by using a cloned fragment of the human β-globin gene containing the first two exons and the surrounded intron (Ruskin et al., 1984). The RNA transcribed from such a clone was placed into an extract of human cell nuclei capable of splicing this "precursor" at the appropriate sites (Krainer et al., 1984). (When the products of this reaction were run on gels, the lariat intermediates were noticable because they did not migrate according to their expected molecular weights. The lariat and circle structures impeded their migration during electrophoresis.) Chemical analysis of these intermediates confirmed that the branch site A has two phosphodiester linkages, one from its 2' OH group and one from its 3' OH group. Shortly thereafter, Zeitlin and Efstratiadis (1984) identified similar intermediates for β-globin transcripts being processed in vivo in fetal rabbits.

The roles of small nuclear ribonucleoprotein particles

In metazoan nuclei, these splicing reactions do not happen spontaneously.* Rather, they are directed by catalytic particles called **spliceosomes** that are composed of small nuclear RNAs (snRNAs) and proteins. These small ribonucleoprotein particles are called **snRNPs** ("snurps"). There are five major snRNAs—U1, U2, U4, U5, and U6—each present between 2×10^5 and 10^6 copies per nucleus, and each containing from 56 to 217 nucleotides. In addition, numerous other snRNAs are present in smaller amounts. Most snRNPs contain one snRNA molecule complexed with approximately 10 proteins. The U4, U5, and U6 snRNPs interact to form a single particle called the U4-U5-U6 tri-snRNP (Maniatis and Reed, 1987; Wassarman and Steitz, 1992).

The first suggestion that snRNPs were involved in splicing came from the observation that the 5' terminal region of U1 snRNA was complementary to the 5' splice site of the introns (Figure 12.9; Lerner et al., 1980; Rogers and Wall, 1980). Further experiments (Mount et al., 1983; Tatei, 1984) have shown that purified U1 snRNP binds specifically to the 5' splicing site sequences in the RNA precursors (Figure 12.9B). The U2 snRNP binds specifically to the branch site by base pairing (Wu and Manley, 1989; Zhuang and Weiner, 1989). If U2 snRNP is mixed with mRNA precursors and then RNase is added, the only RNA protected from digestion (since it is covered by the U2 snRNP) is the branch site and the nucleotides surrounding it (Black et al., 1985). The binding of the U2 snRNP to the branch site is mediated through the U2 snRNP auxiliary

*Remarkably enough, such spontaneous splicing does occur in removing the introns of yeast mitochondrial genes and the introns of the ribosomal RNA in the protist *Tetrahymena* (see Sharp, 1987; Cech, 1992).

(A)

U1 snRNA

CAU*U*CAUU A^{m} ^{m} pppG^{m} 5'

5' AG P GUG A GUA — AAC Y(n)NCAG P G 3'

Exon 1 | | Exon 2
5' splice site | Branch site | 3' splice site

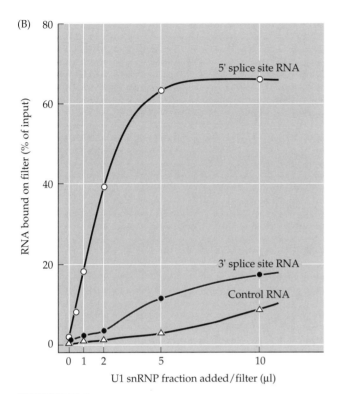

(B)

FIGURE 12.9

Binding of U1 snRNA to 5′ splice region of an intron. (A) Consensus sequence of the 5′ exon–intron boundary is complementary to the 5′ end of the U1 snRNA. The m indicates a methyl group; the asterisk marks a posttranscriptional modification of uridine. (B) Specific binding of the U1 snRNP to purified RNA fragments. Little specific binding is seen when the U1 snRNP is combined with bulk RNA or with the 3′ splicing junction. When combined with fragments containing the 5′ splicing junction, this snRNP binds to the RNA. (A after Padgett et al., 1985; B after Tatei et al., 1984.)

factor (U2AF). U2AF binds to the 3′ splice site and positions the U2 snRNP on the branch point. (Ruskin et al., 1988). The U5 snRNP binds to the 3′ splice site, since this snRNP is able to protect the 3′ splicing site from RNase digestion (Chabot et al., 1985). This interaction is thought to be mediated by the recognition of the 3′ splice sequence by the protein rather than by the RNA portion of the snRNP (Tazi et al., 1986).

Splicing occurs, however, only after snRNPs interact with one another to form a spliceosome on the message precursor. Such particles have been isolated from live cells and from in vitro assemblages of snRNPs and pre-mRNAs (Figure 12.10A). Spliceosome assembly is probably initiated

by the binding of U1 snRNP to the 5′ end splice junction of the intron (Figure 12.10B). Although the details may differ among phyla, it is generally thought that the U2 snRNP binds next, selecting the branch site (Nelson and Green, 1989). The U4-U5-U6 tri-snRNP causes the U5 snRNA to form a loop whose end can base-pair with both the 5′ and 3′ splice sites (Figure 12.10C; Pikielny et al., 1989; Newman and Norman, 1992). The completed spliceosome, like the ribosome, is involved in bringing different RNAs into proximity with one another and synthesizing new covalent bonds. By cross-linking the RNAs together with psoralen (a reagent that will covalently bind apposing uridine bases together when exposed to ultraviolet light), Wassarman and Steitz (1992) delineated the temporal changes in spliceosome assembly and proposed a model to explain how the spliceosome can cleave the message precursor and religate their exons (Figure 12.10C–E). The base pairing between the U1 snRNA and the 5′ splice site is unwound, allowing U5 snRNA to shift and to pair with the first bases of the intron. This changes the configuration within the tri-snRNP such that the U6 snRNA is released from its base pairing with U4 snRNA, enabling the U6 snRNA to pair to a conserved intron sequence at the 5′ splice site. Cleavage occurs first at the 5′ splice site, followed soon afterwards by cleavage at the 3′ splice site and ligation of the two exons (Figure 12.10E).

Spliceosomes do not meet "naked" mRNA precursors. These pre-mRNAs are bound by highly conserved basic proteins, which fold the RNA into specific structures (Skoglund et al., 1983, 1986). The globin hnRNA-protein complex, for instance, is folded in such a way that certain regions of the message precursor are more exposed than others (Patton and Chae, 1985; Eperon et al., 1988). Thus, the folding of RNA by these proteins may allow certain splice sites to be recognized more efficiently than other possible sites (a property that becomes important when introns extend for 10,000 or more nucleotides). Two of these serine- and arginine-rich basic proteins (SR proteins) appear to play a role in alternative RNA processing (discussed in the next section). In many mRNA precursors, there is a choice of potential exons to include in the mRNA. One or the other exon can be ligated to a 3′ splice site, and this creates two different, but related messenger RNAs. In at least some cases, the choice between the alternative 5′ splice sites is influenced by the ratio of proteins SF2 and hnRNP-A1. In general, an excess of SF2 results in the utilization of the proximal (closer) 5′ splice site, while an excess of hnRNP-A1 results in the spliceosome's utilizing the distal (further) 5′ splice site (Mayeda and Krainer, 1992). The regulation of the SF2/hnRNP-A1 system may be crucial in regulating the tissue-specific expression of some genes by alternative RNA processing.

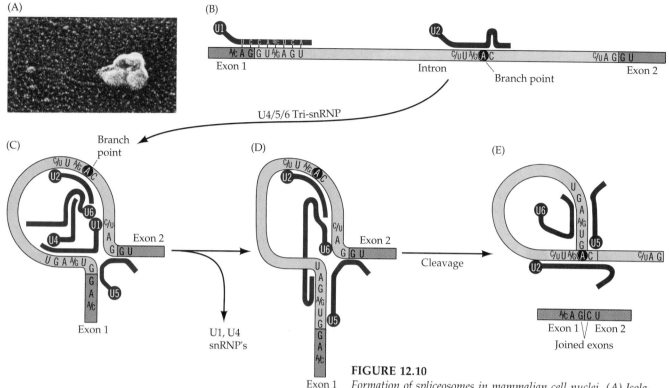

FIGURE 12.10

Formation of spliceosomes in mammalian cell nuclei. (A) Isolation of spliceosomes bound to a fragment of human β-globin pre-mRNA. (B) U1 snRNA binds to the 5' splice site, while U2 snRNA binds to the branch site. Intron sequences are depicted by lower case letters; exons are in boxes. (C–E) A proposed model for spliceosome activity. (C) The U4–U5–U6 tri-snRNP binds so that the U5 snRNA binds to both the 5' and 3' splice sites. (D) The U5 snRNA shifts from the 5' splice site to the intron while the U6 snRNA binds to the 5' splice site. (E) Cleavage brings together the two exons. It is possible for three RNA helices (such as U5, U6, and the 5' end splice site) to interact through their base pairings (Parker and Siliciano, 1993). (A from Reed et al., 1988, courtesy of the authors; B after Wassarman and Steitz, 1992.)

The SR proteins may also be used to determine which pre-mRNAs become spliced. One of these basic proteins, SC35, binds to the globin pre-mRNA before any of the spliceosome components; once it binds, it commits the RNA to be spliced. SC35 is necessary for the splicing of the globin pre-mRNA, but it does not need to be present for the splicing of other nuclear RNAs. Other SR proteins, however, commit other hnRNAs to be spliced. Thus, it is possible that different basic hnRNA-binding proteins commit different nuclear transcripts into the splicing pathway (Fu, 1993).

Differential RNA processing: Creating alternative proteins from the same gene

One gene, many related proteins

Differential RNA processing has been found to control the alternative forms of expression of over 100 proteins. The deletion of certain potential exons in some cells but not in others enables one gene to create a family of closely related proteins. Instead of one gene: one polypeptide, one can have one gene: one family of proteins. For instance, the mRNA precursor for the cell adhesion molecule N-CAM can be alternatively processed into over 100 different forms depending on which exons are included in the mRNA. Although only four major forms of this protein are usually seen in any embryo, some of the minor forms are seen in the brain and heart

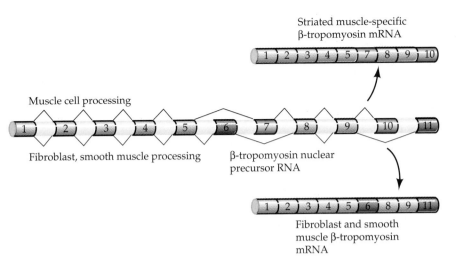

Striated muscle-specific
β-tropomyosin mRNA

Muscle cell processing

Fibroblast, smooth muscle processing

β-tropomyosin nuclear
precursor RNA

Fibroblast and smooth
muscle β-tropomyosin
mRNA

FIGURE 12.11
Schematic diagram of alternative RNA splicing in the β-tropomyosin mRNA precursor. In the skeletal muscle, possible exons 6 and 11 are skipped (and made into introns); while in fibroblasts and smooth muscle cells, possible exons 7 and 10 are made into introns, and 6 and 11 are used as exons.

(Zorn and Krieg, 1992). Similarly, alternative RNA splicing enables the β-tropomyosin gene to encode both the skeletal muscle and fibroblast forms of this protein. The nuclear RNA for β-tropomyosin contains 11 exons. Exons 1–5, 8, and 9 are common to all mRNAs expressed from this gene. Exons 6 and 11 are also used in fibroblasts and smooth muscle cells, while exons 7 and 10 are used in the synthesis of skeletal muscle β-tropomyosin (Figure 12.11). In smooth muscles and fibroblasts, a protein is made that blocks spliceosomes from forming at the the skeletal muscle-specific splice sites (Guo et al., 1991; d'Orval et al., 1991).

In some cases, the properties of the alternatively spliced proteins may have important consequences during development. The five different human fibronectin proteins are generated from one pair of identical fibronectin genes. The diverse (and in some instances organ-specific) forms of fibronectin come from different mRNAs generated by splicing together different exons of the fibronectin mRNA precursor (Tamkun et al., 1984; Hynes, 1987). Some forms of fibronectin are found on the pathways over which embryonic cells migrate whereas other forms are not, suggesting that the alternatively spliced forms of fibronectin have different embryonic functions (ffrench-Constant and Hynes, 1989). Another adhesion molecule, the CD44 glycoprotein, is expressed on lymphocytes. However, when these cells begin to differentiate in response to antigen stimulation, they produce an alternatively spliced variant of this protein, CD44v. This variant molecule enables the lymphocytes to adhere to regions of the veins and leave the blood circulation (presumably to fight the infection). If this variant is generated in lymphocyte *tumor* cells, the cells become metastatic and are able to leave the circulation to seed new colonies of tumors (Arch et al., 1992).

In some instances, RNA splicing may coordinate the production of different sets of proteins in different cell types. The processing of a particular mRNA precursor in certain thyroid cells yields the message for the hormone calcitonin. The same mRNA precursor in nerve cells, however, is processed into the message for a neuropeptide, CGRP (Amara et al., 1982; Crenshaw et al., 1987). This phenomenon is interesting in its own right and also because another alternatively spliced transcript generates substance P in neurons and substance K in the same thyroid cells that make calcitonin. Crenshaw and co-workers speculate that there may be neuron-specific splice-regulating factors that could function on a variety of alternately spliceable precursors. This would enable the generation of a tissue-specific array of proteins.

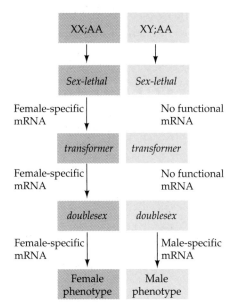

FIGURE 12.12

Sex determination in *Drosophila*. This simplified scheme shows that the X-to-autosome ratio is monitored by the *Sex-lethal* gene. If this gene is active, it processes the *transformer* pre-mRNA into a functional female-specific message. In the presence of the female-specific transformer protein, the *doublesex* gene transcript is processed in a female-specific fashion, leading to the production of the female phenotype. If the *transformer* gene does not make a female-specific product (that is, if the *Sex-lethal* gene is not activated), the *doublesex* transcript is spliced in the male-specific manner, leading to the realization of the male phenotype. (The details of this pathway will be discussed in Chapter 21.)

Alternative RNA processing and *Drosophila* sex determination

In addition to generating the products of differentiated cells, alternative RNA splicing is also used for earlier events that determine cell fate. Recent studies, for instance, demonstrate that differential RNA processing plays a major role in determining the sexual phenotype of *Drosophila* (Baker et al., 1987; Boggs et al., 1987).

As we will see in Chapter 21, the development of the sexual phenotype in *Drosophila* is mediated by a series of genes that convert the X chromosome to autosome ratio into either a male or a female cell. When the X: autosome ratio is 1 (i.e., when there are two X chromosomes per diploid cell), the embryo develops into a female fly. When the ratio is 0.5 (i.e., when the fly is XY with only one X chromosome per diploid cell), the embryo develops into a male (Figure 12.12).

One of the key genes of this pathway is *transformer* (*tra*). This gene is necessary for the production of females, and its loss results in male flies, irrespective of the chromosomal ratio. Throughout the larval period, the *tra* gene actively synthesizes a transcript that is processed into a general mRNA (found in both females and males) or into a female-specific mRNA (Figure 12.13A). Only females contain the alternatively spliced message. The general mRNA found in both males and females contains a stop codon (UGA) early in the second exon, and the small protein produced by this mRNA is not functional. Therefore, the general nonspecific transcript has no bearing on sex determination (Belote et al., 1989). However, in the female-specific message, this UGA codon is in an *intron* that is spliced out during mRNA formation and does not interfere with the translation of the message. In other words, the female transcript is the only functional transcript of this gene. In fact, when the cDNA of this female-specific transcript is incorporated into the genomes of XY flies, these flies become female. The protein encoded by the female-specific mRNA appears to be an arginine-rich peptide 196 amino acids long (Boggs et al., 1987).

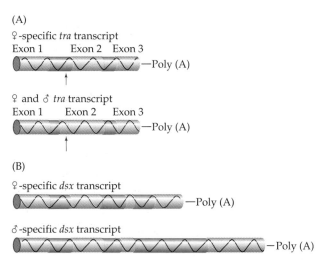

FIGURE 12.13

Sex-specific transcripts during *Drosophila* development. (A) Female-specific and general mRNAs generated from the *transformer* (*tra*) gene by alternative RNA splicing. The arrows point to the UGA stop codon, which is present in the general transcript exons but not included (because it is in an intron) in the female-specific transcript from the same gene. The colored sequences represent exons; the light gray sequences are introns. (B) Alternative RNA splicing in the *doublesex* (*dsx*) gene. During pupation, the *doublesex* precursor RNA is spliced to make male-specific or female-specific gene products depending on whether the *tra* and *tra-2* genes are active. (A after Boggs et al., 1987; B after Baker et al., 1987.)

What makes the *transformer* gene process a female-specific transcript in XX cells and not in XY cells? It appears that sex-specific alternative splicing of *tra* involves competition between two possible 3' (acceptor) splice sites in the intron. Sosnowski and her colleagues (1989) have provided evidence that this competition is altered by the presence or absence of a functional *Sex-lethal* gene product. *Sex-lethal* (*Sxl*) is one of the first genes on the sex phenotype pathway, and it acts prior to *transformer*. If the X:autosome ratio is 1, the first gene to be activated appears to be *Sxl*.* This gene does not appear to be active in XY embryos or larvae. When the *Sex-lethal* gene is active, the *transformer* gene makes both the general and the female-specific transcript and the fly becomes female. If the *Sex-lethal* gene is deleted or mutated, the *transformer* gene makes only the nonfunctional general transcript and the fly becomes male. It appears that the product of the *Sex-lethal* gene is controlling which of the 3' splice sites is being used.

There are two major ways that the *Sex-lethal* gene could control which 3' splice site is used (Figure 12.14). One way is to block the use of the general acceptor site so that only the alternative, female-specific, acceptor site can be used. The other model is to activate the female-specific acceptor site in a positive manner. Valcárcel and colleagues (1993) have shown that the Sex-lethal protein inhibits splicing at the general (non-female-specific) acceptor site by specifically binding to its polypyrimidine tract. This blocks the binding of a splicing factor, U2AF, to the general site and causes it to activate the lower-affinity female-specific acceptor site. It appears, then, that if the general 3' acceptor splice site of *transformer* is blocked (either by mutation or by the product of the *Sex-lethal* gene), the alternative, female-specific, acceptor site is used. The result is a female fly.

*As we will see in Chapter 21, the maintenance of *Sex-lethal* gene activity is also a matter of differential RNA splicing, but we will save that discussion until then.

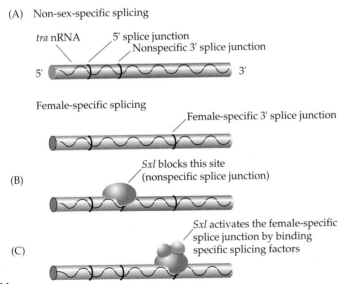

FIGURE 12.14
Two ways in which the product of the *Sex-lethal* (*Sxl*) gene might cause the female-specific 3' intron splice site of *tra-1* nRNA to be utilized. (A) The two alternative splicing modes of the *tra-1* gene: the processing of an inactive transcript in both males and females, and the active transcript processed only in females. (B) Possible mechanism 1: Block of the general splice site. (C) Possible mechanism 2: Activation of the female-specific splice site. (After Belote et al., 1989.)

In the pathway of *Drosophila* sex determination, the *transformer* gene is thought to control the expression of a pivotal gene called *doublesex* (*dsx*). This gene is needed for the production of either sexual phenotype, and mutations of *dsx* can reverse the expected sexual phenotype, causing XX embryos to become males or XY embryos to become females. During pupation, *doublesex* makes a transcript that can be processed in two alternative ways. It can generate a female-specific mRNA or a male-specific mRNA (Figure 12.13B; Nagoshi et al., 1988). In females and males, the first three exons are the same. However, the fourth exons are different. The male-specific RNA deletes a large section of the precursor RNA that includes the female-specific exon.

Tian and Maniatis (1992) have shown that sex-specific processing of *dsx* pre-mRNA involves the activation of the female-specific 3′ splice site by the products of the *transformer* and *transformer-2* genes. The male 3′ splice site of *doublesex* is much more efficient than the female site, since the polypyrimidine (U/C) region female 3′ splice site is interrupted by a tract of purine residues. However, the tra-2 protein binds specifically to a 13-nucleotide region that is repeated six times in the female-specific exon. This region has been seen to be critical for female-specific splicing. It is thought that tra-2 and tra proteins act by directing the splicing complex to the adjacent 3′ splice junction If the *tra* genes are absent (as in mutations) or not activated (as in XY flies; see Chapter 21), the *dsx* transcript is processed in the male-specific manner and male flies are generated. Research into *Drosophila* sex determination shows that differential RNA processing plays enormously important roles throughout development.

Determining the 3′ end of the RNA

The 3′ end of most eukaryotic mRNAs is created by two sequential chemical reactions. First, pre-mRNA is cleaved at a specific site to leave a 3′ hydroxyl group. After this 3′ end is generated, about 200 adenylate molecules—poly(A)—are added to that end. This 3′ end cleavage and polyadenylation are done prior to the completion of splicing, so the splicing reactions are performed on polyadenylated mRNA precursors. The 3′ processing reactions are regulated by both *cis*- and *trans*-acting factors. The first *cis*-acting element is the AAUAAA sequence. The primary RNA transcript contains a 3′ untranslated sequence that often extends far beyond the point at which the translation terminates. Within this trailer is the sequence AAUAAA, which is essential for the cleavage of the message 10–30 bases downstream from this site (Proudfoot and Brownlee, 1976). Mutations of this sequence prevent the 3′ end formation of mRNA (Wickens and Stephenson, 1984; Orkin et al., 1985). If the AATAAA site of human β-globin is mutated to AACAAA, the 3′ terminus of the new RNA is located 900 nucleotides downstream from the normal poly(A) addition site, within 15 nucleotides of the first AAUAAA in the 3′ flanking region of the transcript. The poly(A) tail appears to be added in a biphasic manner (Sheets and Wickens, 1989). The addition of the first 10 adenylate residues is relatively slow and dependent upon the AAUAAA site. The addition of the remaining residues, however, is rapid and does not depend on AAUAAA, but on the presence of those 10 adenylate residues that start the poly(A) tail.

Another *cis*-acting element is a GU- or U-rich sequence usually located further downstream (3′) from the cleavage. This sequence appears to be critical for the efficient cleavage of the hnRNA at the 3′ processing site (McDevitt et al., 1984; Christofori and Keller, 1988).

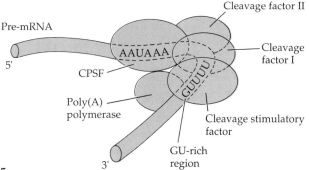

FIGURE 12.15
A model for the formation of the 3′ end of mRNA and its polyadenylation. The *cis*-element AAUAAA on the pre-mRNA is recognized by the nuclear protein cleavage and polyadenylation specificity factor (CPSF). Five other nuclear proteins form around this complex to splice the pre-mRNA (cleavage factors I and II and cleavage stimulatory factor) and to synthesize the polyadenylate tail [poly(A) polymerase]. (After Takagaki et al., 1989.)

Trans-acting factors

The *cis*-acting elements are binding sites for *trans*-acting proteins that will cleave the message precursor and polyadenylate its 3′ end. The fractionation of proteins from human cancer cell nuclei has resulted in the isolation of six nuclear proteins that are necessary, and perhaps sufficient, to cleave and polyadenylate the mRNA precursor (Christofori and Keller, 1988; Takagaki et al.,1989). The first factor is a 290-kDa protein, **cleavage and polyadenylation specificity factor** (CPSF), which recognizes the *cis*-AAUAAA sequence (Figure 12.15; Keller et al., 1991; Wahle and Keller, 1992). The second nuclear protein is the **poly(A) polymerase** enzyme that synthesizes the poly(A) tail. Two other proteins, **cleavage factors I** and **II**, are the enzymes thought to be responsible for actually cleaving the pre-mRNA, but they cannot synthesize the poly(A) tail, nor can they bind specifically to the AAUAAA signal sequence. (They probably do stabilize the interaction between specificity factor and the AAUAAA sequence, however.) When these four factors are mixed together, they can cut and polyadenylate pre-mRNAs in vitro. A fifth nuclear protein, **cleavage stimulatory factor**, can significantly increase the efficiency of this reaction. Once the 3′ end is cleaved and given a short (10 nucleotides) poly(A) tail, **poly(A)-binding protein II** can bind to this tail and stimulate poly(A) polymerase to rapidly elongate the tail to about 200 adenylate residues (Wahle, 1991).

There are some instances in which transcription is terminated prematurely, and the resulting transcript does not receive its poly(A) tail. This deficiency causes the truncated message to be quickly degraded. At least two proteins involved in regulating cell proliferation appear to be regulated in this manner. In nongrowing cells, transcription of these genes is initiated, but the transcript is not completed. Only in growing cells is the complete RNA made (Bentley and Groudine, 1987; Bender et al., 1987). In at least one of these cases, mutations near the 3′ end of the first exon (where the premature termination can occur) cause the completion of transcription in nuclei that normally would prematurely terminate the transcript (Cesarman et al., 1987). In such cases, the mitosis-promoting protein is made and the cells become malignant.

In some instances, the 3′ processing of message precursors occurs by some other mechanism that does not use polyadenylation. This is routinely seen in the processing of the histone transcripts that contain neither introns nor poly(A) tails. Here, the participation of U7 snRNP has been demonstrated in the relatively simple cleavage reaction (Strub and Birnstiel, 1986; Mowry and Steitz, 1987).

Alternative 3′ end formation: Regulation of the immunoglobulin message precursor

It should be remembered that the mechanisms by which the RNA precursors are cleaved and the poly(A) tail attached have yet to be explained. Moreover, just as there are alternative splicing sites in many message precursors, there are sometimes alternative 3′ cleavage sites to end the message. Differential 3′ cleavage and polyadenylation have been found to be extremely important in the generation of different immunoglobulin (antibody) proteins during the differentiation of the B lymphocyte. The expression of antibody-coding genes during the development of lymphocytes and their responses to antigen involve numerous events of selection and rearrangement of gene segments. Much of this selection and rearrangement occurs at the DNA level, as we saw in Chapters 10 and 11. But several aspects of the regulation of antibody gene expression take place at the level of RNA processing.

During the development of a B lymphocyte, before it becomes an antibody-secreting cell, it produces samples of the type of antibody it will be capable of secreting and inserts them into the plasma membrane. Here they function as antigen receptors. Depending on its stage of development and whether or not the cell has been exposed to the type of antigen with which its antibodies will react, the surface immunoglobulins may be of either the IgM or the IgD class. Initially they are of the IgM class; after the first antigen stimulation, both IgM and IgD may be made simultaneously; and later only IgD may be on the cell surface. But the variable (antigen-binding) portion of the immunoglobulin remains the same throughout the life history of the specific lymphocyte and its progeny cells. It has been shown (Maki et al., 1981) that switching between IgM and IgD occurs at the RNA processing level. A single RNA transcript is synthesized that contains the exons coding for the variable (antigen-binding) region plus the μ (IgM) and δ (IgD) constant regions, separated by various introns. Whether the mRNA that reaches the cytoplasm codes for IgM or IgD depends on which of two alternative cleavage and polyadenylation pathways is used. If the message precursor is cleaved and polyadenylated after the μ exons, and the δ exons are downstream from this cleavage site, IgM will be made. However, If the μ exons are removed by splicing, cleavage and polyadenylation can occur after the δ exons, and IgD will be made. During transient periods when both splicing mechanisms are being utilized, both classes of immunoglobulins are produced (Figure 12.16).

The decision to make IgM or IgD is not the only decision mediated by alternative cleavage and polyadenylation. The placement of these proteins—whether they are to be inserted into the plasma membrane or secreted—is also determined in this manner. IgD is confined to the cell membrane, but IgM can exist in two functionally and structurally distinct forms, membrane-bound (mIgM) and secreted (sIgM). Membrane-bound IgM acts as an antigen receptor, whereas secreted IgM is the first type of antibody found in the blood when an organism is exposed to an antigen for the first time. Several investigators have found that sIgM and mIgM differ from each other at their carboxyl ends. The molecules are identical except that the membrane-bound form contains a hydrophobic "tail" that

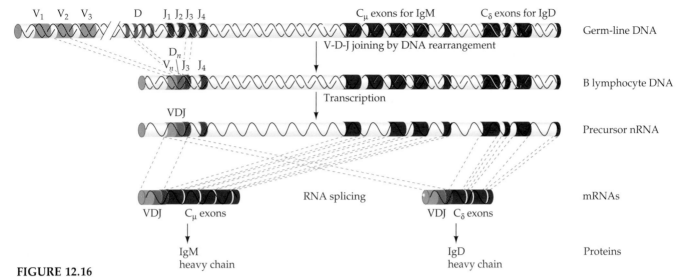

FIGURE 12.16
Proposed model for the expression of IgM and IgD in B lymphocytes. The heavy chain gene is rearranged during lymphocyte development by the translocation of V, D, and J segment sequences to a position adjacent to the constant region sequences. The transcript of this gene includes both the μ (IgM) and δ (IgD) constant regions along with the newly constructed variable region. Alternative pathways of processing can delete μ constant region (C_μ) exons or δ constant region (C_δ) exons. (After Liu et al., 1980.)

keeps it inserted in the membrane (Figure 12.17). The mRNAs for membrane and secreted forms of IgM also differ. Although both of them are found to be transcribed from the same C_μ gene, they are processed differently (Alt et al., 1980; Early et al., 1980; Rogers et al., 1980). The secreted IgM contains regions encoded by the *VDJ* genes and exons $C_\mu 1$, $C_\mu 2$, $C_\mu 3$, and $C_\mu 4$. It also contains a terminal portion that allows it to be secreted. The membrane-bound IgM contains the same arrangement except that instead of the "secretion" exon, it has added a portion encoded by two more exons, $C_\mu 5$ and $C_\mu 6$, which gives it a hydrophobic tail that can integrate into the lymphocyte membrane. Thus, cleavage and polyadenylation of the $C_\mu 4$ exon yield the secretory form of IgM, while cleavage and polyadenylation of the $C_\mu 6$ exon (and the differential splicing of $C_\mu 4$) yield membrane-bound IgM (Figure 12.18). Peterson and Perry (1989) have suggested that there is competition between the two sites for the cleavage and polyadenylation factors and that the biases in this competition change as the B cell matures.

Transport out of the nucleus

One of the largest mysteries of RNA processing is how it is connected with the transport of the message out of the nucleus. The proper 3′ end appears to be essential for RNA transport from the cytoplasm, and adding a "termination casette" (containing AAUAAA and G/U sequences with the proper distance between them) to bacterial mRNAs enables them to be efficiently transported from nucleus to cytoplasm (Eckner et al., 1991). It has long been assumed that mRNA exits through the nuclear pores, and recent evidence supports this view. By microinjecting gold-labeled

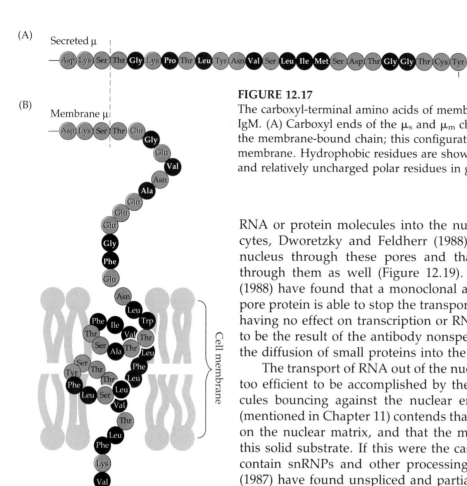

(A)

Secreted μ

—Asp–Lys–Ser–Thr–Gly–Lys–Pro–Thr–Leu–Tyr–Asn–Val–Ser–Leu–Ile–Met–Ser–Asp–Thr–Gly–Gly–Thr–Cys–Tyr—COOH

(B)

Membrane μ

Cell membrane

FIGURE 12.17
The carboxyl-terminal amino acids of membrane (m) and secreted (s) forms of IgM. (A) Carboxyl ends of the μ$_s$ and μ$_m$ chains. (B) Possible configuration of the membrane-bound chain; this configuration anchors the molecule in the cell membrane. Hydrophobic residues are shown in black, charged residues in color, and relatively uncharged polar residues in gray. (After Rogers et al., 1980.)

RNA or protein molecules into the nucleus or cytoplasm of *Xenopus* oocytes, Dworetzky and Feldherr (1988) have shown that RNA exits the nucleus through these pores and that protein can enter the nucleus through them as well (Figure 12.19). Featherstone and her co-workers (1988) have found that a monoclonal antibody directed against a nuclear pore protein is able to stop the transport of RNAs out of the nucleus while having no effect on transcription or RNA processing. This does not seem to be the result of the antibody nonspecifically occluding the pores, since the diffusion of small proteins into the nucleus remains unaffected.

The transport of RNA out of the nucleus and into the cytoplasm seems too efficient to be accomplished by the simple diffusion of mRNA molecules bouncing against the nuclear envelope. One possible alternative (mentioned in Chapter 11) contends that RNA is transcribed and processed on the nuclear matrix, and that the mRNA is delivered to the pores on this solid substrate. If this were the case, then the nuclear matrix should contain snRNPs and other processing proteins. Zeitlin and co-workers (1987) have found unspliced and partially spliced message precursors for rabbit β-globin on the nuclear matrix, and Ciejek and co-workers (1982) have found that partially spliced ovalbumin message precursors are bound to the matrix of chick oviduct nuclei. Studies in Lawrence's laboratory (Xing et al., 1993; Carter et al., 1993) show that the nuclear RNA does not

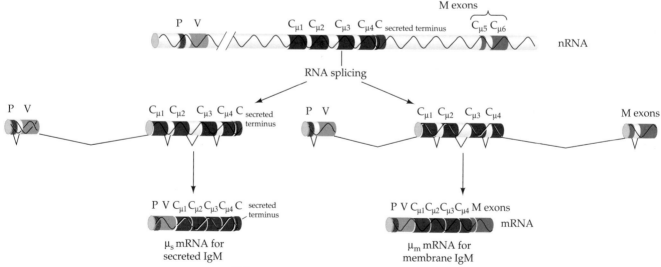

FIGURE 12.18
Proposed alternative splicing patterns for secreted μ and membrane μ mRNA. P includes the leader sequence; V encodes the variable (VDJ) region; the possible constant region exons of the μ (IgM) constant region follow downstream. The spliced RNAs (introns) are shown as thin lines. (After Early et al., 1980.)

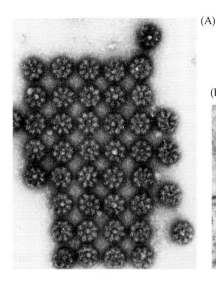

(A)

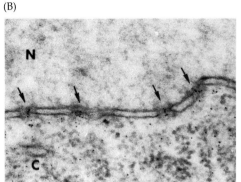

(B)

N

C

FIGURE 12.19

Nuclear pores and RNA transport. (A) Nuclear pores from *Xenopus laevis* oocytes. The ring of the pore is formed by eight units. (B) When RNA coated with gold particles is injected into the *Xenopus* oocyte, its export from the nucleus (N) to the cytoplasm (C) can be monitored by electron microscopy. (Other polymers complexed with gold are not exported.) This figure shows gold-coated tRNA molecules being transported through the channels of the nuclear pores. (A from Unwin and Mulligan, 1982; B from Dworetzky and Feldherr, 1988; photographs courtesy of the authors.)

diffuse far from the gene and that the splicing of fibronectin and neurotensin pre-mRNAs occurs on the matrix tracks leading to the nuclear envelope.

Research on viral mRNAs also provides evidence for the active translocation of pre-mRNAs to the nuclear envelope. Adenovirus message is transcribed on the nuclear matrix and undergoes translocation to the nuclear envelope before entering the cytoplasm. A particular viral protein is necessary for this translocation to occur, and if the viral gene encoding this protein is mutated, viral mRNAs accumulate in the nucleus without reaching the nuclear envelope (Leppard and Shenk, 1989). Similarly, the Rev protein of HIV-1 encodes a factor that binds to the RNA for envelope glycoprotein and transports it to the nuclear envelope (Heaphy et al., 1990; Lawrence et al., 1992).

The mechanism by which mRNAs get into the cytoplasm is not well characterized. The cytoplasmic side of the nuclear envelope is often seen to be studded with ribosomes. Could the ribosomes grab the message as it comes out of the pores and pull it from the nucleus as they translate it? While we do not know the answer to this, there is at least one piece of evidence in its favor. In a certain type of human β-thalassemia, there is a mutation in codon 39 such that it now codes for the termination of translation. Therefore, it is understandable that no β-globin protein is made from this gene. But this does not explain why there should be a deficiency of β-globin *mRNA* in the cytoplasm. Studies by Humphries and co-workers (1984) show that when this message is made in vitro from a cloned gene, it is not less stable than wild-type β-globin mRNA, and Orkin and Kazazian (1984) speculate that translation may be needed on the cytoplasmic side of the nuclear envelope in order to get the globin message out. When this translation ceases (because of the termination codon), so would the transport of message out of the nucleus.

Summary

RNA processing must be viewed as a major means of regulating differential gene expression. Evidence indicates that it determines the cell-specific mRNA population in developing sea urchins and that it may do likewise in mice and rats. DNA and RNA sequencing have shown that differential RNA processing can create different proteins in different cell types (calcitonin or CGRP; IgM or IgD) or at different times within the same cell lineage (membrane-bound and secreted IgM). The regulation of differential RNA processing is also required for determining the sexual phenotype of *Drosophila*. We still know relatively little about the mechanisms of alternative mRNA processing or about the ways in which some cells process transcripts that other cells do not. The mechanism underlying such differential RNA processing may give us insights into the very core of cell differentiation and embryonic determination.

LITERATURE CITED

Alt, F. W., Bothwell, A. L. M., Knapp, M. N., Siden, E., Mather, E., Koshland, M. and Baltimore, D. 1980. Synthesis of secreted and membrane-bound immunoglobulin μ heavy chains is directed by mRNAs that differ at their 3′ ends. *Cell* 20: 293–301.

Amara, S. G., Jonas, V., Rosenfeld, M. G., Ong, E. S. and Evans, R. M. 1982. Alternative RNA processing in calcitonin gene expression generates mRNAs encoding different polypeptide products. *Nature* 298: 240–244.

Arch, R., Wirth, K., Hofmann, M., Ponta, H., Matzku, S., Herrlich, P. and Zöller, M. 1992. Participation in normal immune responses of a metastasis-inducing splice variant of CD44. *Science* 257: 682–685.

Axel, R., Feigleson, P. and Schutz, G. 1976. Analysis of the complexity and diversity of mRNA from chicken liver and oviduct. *Cell* 7: 247–254.

Baker, B., Nagoshi, R. N. and Burtin, K. C. 1987. Molecular genetic aspects of sex determination in *Drosophila*. *BioEssays* 6: 66–70.

Bastos, R. N. and Aviv, H. 1977. Globin RNA precursor molecules: Biosynthesis and processing erythroid cells. *Cell* 11: 641–650.

Belote, J. M., McKeown, M., Boggs, R. T., Ohkawa, R. and Sosnowski, B. A. 1989. Molecular genetics of *transformer*, a genetic switch controlling sexual differentiation in *Drosophila*. *Dev. Genet.* 10: 143–154.

Bender, T.P., Thompson, C.B. and Kuehl, W.M. 1987. Differential expression of c-*myb* mRNA in murine B lymphomas by a block to transcription elongation. *Science 237*: 1473–1476.

Bentley, D. L. and Groudine, M. 1987. A block to elongation is largely responsible for decreased transcription of c-*myc* in differentiated HL60 cells. *Nature* 321: 702–706.

Black, D. L., Charbot, B. and Steitz, J. 1985. U2 as well as U1 small nucleus ribonucleoproteins are involved in pre-mRNA splicing. *Cell* 42: 737–750.

Boggs, R. T., Gregor, P., Idriss, S., Belote, J. M. and McKeown, M. 1987. Regulation of sexual differentiation in *D. melanogaster* via alternative splicing of RNA from the *transformer* gene. *Cell* 50: 739–747.

Britten, R. J. and Kohne, D. E. 1968. Repeated sequences in DNA. *Science* 161: 529–540.

Busslinger, M., Moschonas, N. and Flavell, R. A. 1981. β⁺ Thalassemia: Aberrant splicing results from a single point mutation in an intron. *Cell* 27: 289–298.

Carter, K. C., Bowman, D., Carrington, W., Fogarty, K., McNeil, J. A., Fay, F. S. and Lawrence, J. B. 1993. A three-dimensional view of precursor messenger RNA metabolism within the mammalian nucleus. *Science* 259: 1330–1335.

Cech, T. R. 1992. RNA catalysis by group I ribozyme: Developing a model for transition-state stabilization. *J. Biol. Chem.* 267: 17479–17482.

Cesarman, E., Dalla-Favera, R., Bentley, D. and Groudine, M. 1987. Mutations in the first exon are associated with altered transcription in c-*myc* in Burkitt lymphoma. *Science* 238: 1272–1275.

Chabot, B., Black, D. L., LeMaster, D. M. and Steitz, J. A. 1985. The 3′ splice site of pre-messenger RNA is recognized by a small ribonucleoprotein. *Science* 230: 1344–1349.

Christofori, G. and Keller, W. 1988. 3′ cleavage and polyadenylation of mRNA precursors in vitro requires a poly(A) polymerase, a cleavage factor, and a snRNP. *Cell* 54: 875–889.

Ciejek, E. M., Nordstrom, J. L., Tsai, M. J. and O'Malley, B. W. 1982. Ribonucleic acid precursors are associated with the chick oviduct nuclear matrix. *Biochemistry* 21: 4945–4953.

Crenshaw, E. B., III, Russo, A. F., Swanson, L. W. and Rosenfeld, M. G. 1987. Neuron-specific alternative RNA processing in transgenic mice expressing a metallothionein–calcitonin fusion gene. *Cell* 49: 389–398.

Davidson, E. H. and Britten, R. J. 1979. Regulation of gene expression: Possible role of repetitive sequences. *Science* 204: 1052–1059.

d'Orval, B. C., Carafa, Y. d'A., Sirand-Pugnet, P., Gallego, M., Brody, E. and Marie, J. 1991. RNA secondary structure repression of a muscle-specific exon in HeLa cell nuclear extracts. *Science* 252: 1823–1828.

Dworetzky, S. and Feldherr, C. 1988. Translocation of RNA-coated gold particles through the nuclear pores of oocytes. *J. Cell Biol.* 106: 575–584.

Early, P., Rogers, J., Davis, M., Dalame, K., Bond, M., Wall, R. and Hood, L. 1980. Two mRNAs can be produced from a single immunoglobulin gene by alternative RNA processing pathways. *Cell* 20: 313–319.

Eckner, R., Ellmeier, W. and Birnstiel, M. L. 1991. Mature mRNA 3′ end formation stimulates RNA export from the nucleus. *EMBO J.* 10: 3513–3522.

Eperon, L. P., Graham, I. R., Griffiths, A. D. and Eperon, I. C. 1988. Effects of RNA secondary structure on alternative splicing of pre-mRNA: Is folding limited to a region behind the transcribing RNA polymerase? *Cell* 54: 393–401.

Featherstone, C., Darby, M. K. and Gerace, L. 1988. A monoclonal antibody against the nuclear pore complex inhibits nucleocytoplasmic transport of protein and RNA in vivo. *J. Cell Biol.* 107: 1289–1297.

ffrench-Constant, C. and Hynes, R. O. 1989. Alternative splicing of fibronectin is temporally and spatially regulated in the chicken embryo. *Development* 106: 375–388.

Fu, X.-D. 1993. Specific commitment of different pre-mRNAs to splicing by single SR proteins. *Nature* 365: 82–85.

Gagnon, M. L., Angerer, L. M. and Angerer, R. C. 1992. Posttranscriptional regulation of ectoderm-specific gene expression in early sea urchin embryos. *Development* 114: 457–467.

Guo, W., Mulligan, G. J., Wormsley, S., Helfman, D. M. 1991. Alternative splicing of β-tropomyosin pre-mRNA: Cis-acting elements and cellular factors that block the use of a skeletal muscle exon in nonmuscle cells. *Genes Dev.* 5: 2096–2107.

Heaphy, S. and eight others. 1990. HIV-1 regulator of virion expression (Rev) protein binds to an RNA stem–loop structure located within the Rev response element. *Cell* 60: 685–693.

Hood, L. E., Wilson, J. H. and Wood, W. B. 1975. *The Molecular Biology of Eukaryotic Cells.* Benjamin, New York.

Humphries, R. K., Ley, T. J., Anagnou, N. P., Baur, A. W. and Nienhuis, A. W. 1984. β⁰-39 thalassemia gene: A premature termination codon causes β-mRNA deficiency without affecting cytoplasmic β-mRNA stability. *Blood* 64: 23–32.

Hynes, R. O. 1987. Fibronectins: A family of complex and versatile adhesive glycoproteins derived from a single gene. *Harvey Lect.* 81: 133–152.

Jacob, M. and Gallinaro, H. 1989. The 5' splice site: Phylogenetic evolution and variable geometry of association with U1RNA. *Nucl. Acid Res.* 17: 2159–2180.

Kantor, J. A., Turner, P. H. and Nienhuis, A. W. 1980. β Thalassemia: Mutations which affect processing of the β-globin mRNA precursor. *Cell* 21: 149–157.

Keller, W., Bienroth, S., Lang, K. M. and Christofori, G. 1991. Cleavage and polyadenylation factor CPF specifically interacts with the pre-mRNA 3' processing signal AAUAAA. *EMBO J.* 10: 4241–4249.

Kleene, K. C. and Humphreys, T. 1977. Similarity of hnRNA sequences in blastula and pluteus stage sea urchin embryos. *Cell* 12: 143–155.

Kleene, K. C. and Humphreys, T. 1985. Transcription of similar sets of rare maternal RNAs and rare nuclear RNAs in sea urchin blastulae and adult coelomocytes. *J. Embryol. Exp. Morphol.* 85: 131–149.

Krainer, A. R., Maniatis, T., Ruskin, B. and Green, M. R. 1984. Normal and mutant human β-globin pre-messenger RNAs are faithfully and efficiently spliced in vitro. *Cell* 36: 993–1005.

Lawrence, J. B. and Cochrane, A. W., Johnson, C. V., Perkin, A. and Rosen, C. A. 1992. The HIV-1 Rev protein: A model system for coupled RNA transport and translation. *New Biol.* 3: 1220–1232.

Leppard, K. N. and Shenk, T. 1989. The adenovirus E1B 55-kd protein influences mRNA transport via an intranuclear effect on RNA metabolism. *EMBO J.* 8: 2329–2336.

Lerner, M. R., Boyle, J. A., Mount, S. M., Wolin, S. L. and Steitz, J. A. 1980. Are snRNPs involved in splicing? *Nature* 283: 220–224.

Lewin, B. 1980. *Gene Expression 2*, 2nd Ed. Wiley–Interscience, New York, pp. 694–760.

Liu, C.-P., Tucker, P. W., Mushinski, F. and Blattner, F. R. 1980. Mapping of heavy chain genes for mouse immunoglobulins M and G. *Science* 209: 1348–1353.

Maki, R., Roeder, W., Traunecker, A., Sidman, C., Wabl, M., Rasjhke, W. and Tonegawa, S. 1981. The role of DNA rearrangement and alternate mRNA processing in the expression of immunoglobulin delta genes. *Cell* 24: 353–365.

Maniatis, T. and Reed, R. 1987. The role of small nuclear ribonucleoprotein particles in pre-mRNA splicing. *Nature* 325: 673–678.

Maquat, L. E., Kinniburgh, A. J., Beach, L. R., Honig, G. R., Lazerson, J., Ershler, W. B. and Ross, J. 1980. Processing of human β-globin mRNA precursor to mRNA is defective in three patients with β$^+$ thalassemias. *Proc. Natl. Acad. Sci. USA* 77: 4287–4291.

Mayeda, A. and Krainer, A. R. 1992. Regulation of alternative pre-mRNA splicing by hnRNP A1 and splicing factor SF2. *Cell* 68: 367–375.

McCarthy, B. J. and Hoyer, B. H. 1964. Identity of DNA and diversity of RNA in normal mouse tissues. *Proc. Natl. Acad. Sci. USA* 52: 915–922.

McDevitt, M. A., Imperiale, M. J., Ali, H. and Nevins, J. R. 1984. Requirement of a downstream sequence for the generation of poly(A) addition site. *Cell* 37: 329–338.

Mount, S. M., Pettersson, I., Hinterberger, M., Karmas, A. and Steitz, J. 1983. The U1 small nuclear RNA–protein complex selectively binds a 5' splice site in vitro. *Cell* 33: 509–518.

Mowry, K. L. and Steitz, J. A. 1987. Identification of the human H7 snRNP as one of several factors involved in the 3' end maturation of histone premessenger RNAs. *Science* 238: 1682–1687.

Nagoshi, R. N., McKeown, M., Burtis, K. C., Belote, J. M. and Baker, B. S. 1988. The control of alternative splicing at genes regulating sexual differentiation in *D. melanogaster*. *Cell* 53: 229–236.

Nelson, K. K. and Green, M .R. 1989. Mammalian U2 snRNP has a sequence-specific RNA–binding activity. *Genes Devel.* 3: 1562–1571.

Newman, A. J. and Norman, C. 1992. U5 snRNA interacts with exon sequences at 5' and 3' splice sites. *Cell* 68: 743–754.

Orkin, S. H. and Kazazian, H. H. 1984. The mutation and polymorphism of the human β-globin gene and its surrounding DNA. *Annu. Rev. Genet.* 18: 131–171.

Orkin, S. H., Cheng, T.-C., Antonarakis, S. E. and Kazazian, H. H., Jr. 1985. Thalassemia due to a mutation in the cleavage–polyadenylation signal of the human β-globin gene. *EMBO J.* 4: 453–456.

Padgett, R. A., Grabowski, P. J., Konarska, M. M. and Sharp, P. A. 1985. Splicing messenger RNA precursors: Branch sites and lariat RNAs. *Trends Biochem. Sci.* 10: 154–157.

Parker, R. and Siliciano, P. G. 1993. Evidence for an essential non-Watson–Crick interaction between the first and last nucleotides of a nuclear pre-mRNA intron. *Nature* 361: 660–662.

Patton, J. R. and Chae, C.-B. 1985. Specific regions of the intervening sequences of β-globin RNA are resistant to nuclease in 50S heterogeneous nuclear RNA–protein complexes. *Proc. Natl. Acad. Sci. USA* 82: 8414–8418.

Peterson, M. L. and Perry, R. 1989. The regulated production of μ_m and μ_s mRNA is dependent upon the relative efficiencies of the poly(A) site usage and the $C_\mu4$-to-M1 splice. *Mol. Cell Biol.* 9: 726–738.

Pikielny, C. W., Bindereif, A. and Green, M. R. 1989. In vitro reconstitution of snRNPs: a reconstituted U4/U6 snRNP participates in splicing complex formation. *Genes Dev.* 3: 479–487.

Proudfoot, N. J. and Brownlee, G. G. 1976. 3' Non-coding region sequences in eukaryotic messenger RNA. *Nature* 263: 211–214.

Reed, R., Griffith, J. and Maniatis, T. 1988. Purification and visualization of native spliceosomes. *Cell* 53: 949–961.

Rogers, J. and Wall, R. 1980. A mechanism for RNA splicing. *Proc. Natl. Acad. Sci. USA* 77: 1877–1879.

Rogers, J., Early, P., Carter, C., Calame, K., Bond, M., Hood, L. and Wall, R. 1980. Two mRNAs with different 3' ends encode membrane-bound and secreted forms of immunoglobulin μ chain. *Cell* 20: 303–312.

Ruskin, B., Krainer, A. R., Maniatis, T. and Green, M. R. 1984. Excision of an intact intron as a novel lariat structure during pre-mRNA splicing in vitro. *Cell* 38: 317–331.

Ruskin, B., Zamore, P. D. and Green, M. R. 1988. A factor, U2AF, is required for U2 snRNP binding and splicing complex assembly. *Cell* 52: 207–219.

Sharp, P. A. 1987. Splicing of messenger RNA precursors. *Science* 235: 766–771.

Sheets, M. D. and Wickens, M. 1989. Two phases in the addition of a poly(A) tail. *Genes Dev.* 3: 1401–1412.

Skoglund, U., Andersson, K., Bjorkroth, B., Lamb, M. M. and Daneholt, B. 1983. Visualization of the formation and transport of a specific hnRNP particle. *Cell* 34: 847–855.

Skoglund, U., Andersson, K., Strandberg, B. and Dancholt, B. 1986. Three–dimensional structure of a specific pre-messenger RNP particle established by electron microscope tomography. *Nature* 319: 560–564.

Sosnowski, B. A., Belote, J. M. and McKeown, M. 1989. Sex-specific alternative splicing of RNA from the *transformer* gene results from sequence-specific splice site blockage. *Cell* 58: 449–459.

Spritz, R.A. and nine others. 1981. Base substitution in an intervening sequence of a β$^+$-thalassemic human globin gene. *Proc. Natl. Acad. Sci. USA* 78: 2455–2459.

Strub, K. and Birnstiel, M. L. 1986. Genetic complementation in the *Xenopus* oocyte: Co-expression of sea urchin histone and U7 RNAs restores 3' processing of H3 pre-mRNA in the oocyte. *EMBO J.* 5: 1675–1682.

Takagaki, Y., Ryner, L. C. and Manley, J. L. 1989. Four factors required for 3'-end cleavage of pre-mRNAs. *Genes Dev.* 3: 1711–1724.

Tamkun, J. W., Schwartzbauer, J. E. and Hynes, R. O. 1984. A single rat fibronectin gene generates three different mRNAs by alternative splicing of a complex exon. *Proc. Natl. Acad. Sci. USA* 81: 5140–5144.

Tatei, K., Takemura, K., Mayeda, A., Fujiwara, Y., Tanaka, H., Ishimura, A. and Oshima, Y. 1984. U1 RNA–protein complex preferentially binds to both 5' and 3' splice junction sequences in RNA or single-stranded DNA. *Proc. Natl. Acad. Sci. USA* 81: 6281–6285.

Tazi, J., Alibert, C., Temsamani, J., Reveillaud, I., Cathala, G., Brunel, C. and Jeanteur, P. 1986. A protein that specifically recognizes the 3' splice site of mammalian pre-mRNA introns is associated with a small ribonucleoprotein. *Cell* 47: 755–766.

Tian, M. and Maniatis, T. 1992. Positive control of pre-mRNA splicing *in vitro*. *Science* 256: 237–240.

Tsai, S. Y., Tsai, M.-J., Lin, C.-T. and O'Malley, B. W. 1979. Effect of estrogen on ovalbumin gene expression in differentiated nontarget tissues. *Biochemistry* 18: 5726–5731.

Unwin, P. N. T. and Mulligan, R. A. 1982. A large particle is associated with the perimeter of the nuclear pore complex. *J. Cell Biol.* 93: 63–75.

Valcárcel, J., Singh, R., Zamore, P. D. and Greene, M. R. 1993. The protein Sex-lethal antagonizes the splicing factor U2AF to regulate alternative splicing of transformer pre-mRNA. *Nature* 362: 171–175.

Wahle, E. 1991. A novel poly(A)-binding protein acts as a specificity factor in the second phase of messenger RNA polyadenylation. *Cell* 66: 759–768.

Wahle, E. and Keller, W. 1992. The biochemistry of 3'-end cleavage and polyadenylation of messenger RNA precursors. *Annu. Rev. Bioch.* 61: 419–440.

Wassarman, D. A. and Steitz, J. A. 1992. Interactions of small nuclear RNAs with precursor messenger RNA during in vitro splicing. *Science* 257: 1918–1925.

Wickens, M. and Stephenson, P. 1984. Role of the conserved AAUAAA sequence: Four AAUAAA point mutants prevent 3' end formation. *Science* 226: 1045–1051.

Wold, B. J., Klein, W. H., Hough-Evans, B. R., Britten, R. J. and Davidson, E. H. 1978. Sea urchin embryo mRNA sequences expressed in nuclear RNA of adult tissues. *Cell* 14: 941–950.

Wu, J. and Manley, J. L. 1989. Mammalian pre-mRNA branch site selection by U2 snRNP involves base pairing. *Genes Dev.* 3: 1553–1561.

Xing, Y., Johnson, C. V., Dobner, P. R. and Lawrence, J. B. 1993. Higher level organization of individual gene transcription and RNA splicing. *Science* 259: 1326–1330.

Zeitlin, S. and Efstratiadis, A. 1984. In vivo splicing of the rabbit β-globin pre-mRNA. *Cell* 39: 589–602.

Zeitlin, S., Parent, A., Silverstein, S. and Efstratiadis, A. 1987. Pre-mRNA splicing and the nuclear matrix. *Mol. Cell. Biol.* 7: 111–120.

Zhuang, Y. and Weiner, A. M. 1989. A compensatory base change in human U2 snRNA can suppress a branch site mutation. *Genes Dev.* 3: 1545–1552.

Zorn, A. M. and Krieg, P. A. 1992. Developmental regulation of alternative splicing in the mRNA encoding *Xenopus laevis* neural cell adhesion molecule (N-CAM). *Dev. Biol.* 149: 197–205.

13

Translational regulation of developmental processes

We must not conceal from ourselves the fact that the causal investigation of organisms is one of the most difficult, if not the most difficult problem which the human intellect has attempted to solve, and that this investigation, like every causal science, can never reach completeness, since every new cause ascertained only gives rise to fresh questions concerning the cause of this cause.
WILHELM ROUX (1894)

There is no rest for the messenger til the message is delivered.
JOSEPH CONRAD (1920)

After a messenger RNA has been transcribed, processed, and exported from the nucleus, it still needs to be translated in order to form the protein encoded in the genome. In this chapter we will see that regulation at the level of translation is an extremely important mechanism in the control of gene expression. In such cases, the message is already present but may or may not be translated, depending on certain cellular conditions. Thus, translational control of gene expression can be used when a burst of protein synthesis is needed immediately (as we will find to be the case in the newly fertilized egg), or it can be used as a fine-tuning mechanism to ensure that a very precise amount of protein is made from the available supply of messages (as is the case in hemoglobin synthesis). We will also see that there are several ways to effect translational control and that different cells have evolved different means to do so.

Mechanisms of eukaryotic translation

Translation is the process by which the information contained in the nucleotide sequence of mRNA instructs the synthesis of a particular polypeptide. This process, outlined in Figure 13.1, has been divided into three phases—initiation, elongation, and termination—and it is regulated by soluble proteins called (appropriately) initiation factors, elongation factors, and termination factors (Hershey, 1989; Safer, 1989).

Initiation consists of the reactions wherein the first aminoacyl-transfer RNA and the mRNA are bound to the ribosome. The only transfer RNA (tRNA) capable of initiating translation is a special initiator tRNA (tRNA$_i$), which carries the amino acid methionine. As shown in Figure 13.2, the first reactions involve the formation of an **initiation complex** consisting of methionyl-initiator tRNA bound to a 40S ("small") ribosomal subunit. This reaction is catalyzed by the active form of **eukaryotic initiation factor 2** (eIF2-GTP), which binds the initiator Met-tRNA to the 40S ribosomal subunit. Note that this binding occurs in the absence of mRNA. The mRNA is added next. First, a **cap-binding protein—eIF4E**—binds to the 7-methylguanosine cap at the 5′ end of the message. Without this cap, the binding of mRNA to the ribosomal subunit is not often completed (Shatkin, 1976, 1985), and eIF4E is critical for the translation to proceed. However, there is less eIF4E than the number of messages in the cell, so it is thought that

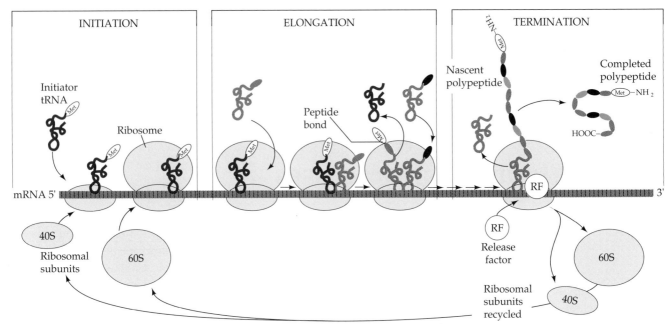

FIGURE 13.1

Schematic representation of the events of eukaryotic translation. The initiation steps bring together the 40S and 60S ribosomal subunits, mRNA, and the initiator tRNA, which is complexed to the amino acid methionine (Met). During elongation, amino acids are brought to the polysome and peptide bonds are formed between the amino acids. The sequence of amino acids in the growing protein is directed by the sequence of nucleic acid codons in the mRNA. After the last peptide bond of the protein has been made, one of the codons UAG, UGA, or UAA signals the termination of translation. The ribosomal subunits and message can be reutilized.

each mRNA has to compete for this cap-binding protein (Thach, 1992). Initiation factor 4A then complexes with eIF4E and positions itself on a helical hairpin loop in the leader sequence of the mRNA. The eIF4A (stimulated by eIF4B and ATP) unwinds the helix. This step can be rate-limiting if the hairpin loop helix is hidden by some other stable secondary structure. The 40S ribosomal subunit then travels down the message until it reaches an AUG codon in the proper context. Kozak (1986) has shown that just any AUG will not do. For the 40S ribosomal subunit to stop and initiate translation, the nucleotides around AUG are also important. By mutating cloned genes and assaying the translation of their RNAs, Kozak found the "optimum" sequence to be ACC**AUG**G. Mutations in the flanking nucleotides could reduce translation 20-fold. The importance of the flanking sequence also has been seen in vivo. Morle and co-workers (1985) have reported a patient whose α-thalassemia (deficiency of the α-globin subunit of hemoglobin) was due to a change in this sequence from ACCAUGG to CCCAUGG. The binding of the 40S subunit to the AUG of the message positions the initiator tRNA over the AUG codon. Only after the mRNA has been properly positioned on the small ribosomal subunit can the 60S ("large") ribosomal subunit bind. This completes the initiation reaction. During this process, the GTP on eIF2 is hydrolyzed to GDP. In order for the eIF2 to pick up a new initiator tRNA, it must be regenerated to eIF-GTP by eIF2B.

Elongation involves the sequential binding of aminoacyl-tRNAs to the ribosome and the formation of peptide bonds between the amino acids as they sequentially relinquish their tRNA carriers (Figure 13.1). As amino acids are joined together, the ribosome travels down the message, thereby exposing new codons for tRNA binding. This allows another ribosome to initiate on the 5' end of the message and begin its traveling. Thus, any mRNA usually will have several ribosomes attached to it. This structure is then called a polyribosome—or, more commonly, a polysome (Figure 13.3). The **termination** of protein synthesis takes place when one of the mRNA codons UAG, UAA, or UGA is exposed on the ribosome. These nucleotide triplets (called *termination codons*) are not recognized by tRNAs

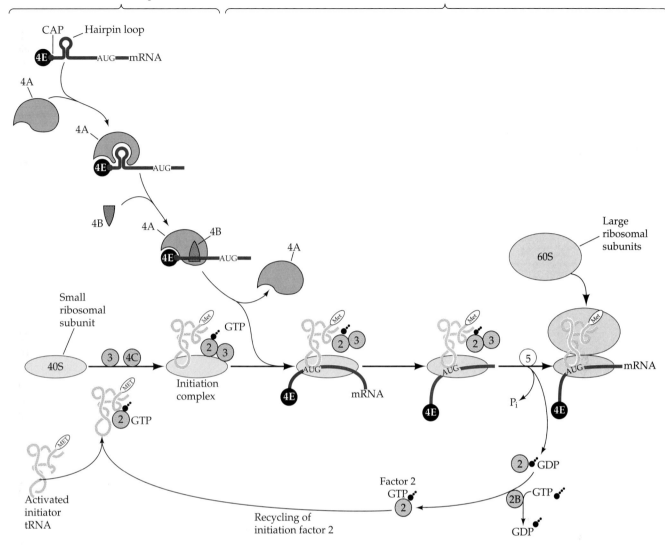

and hence do not code for any amino acids. Rather, they are recognized by release factors, which hydrolyze the peptide from the last tRNA, freeing it from the ribosome. The ribosome separates into its two subunits, and the cycle of translation begins anew.

Control of protein synthesis by differential longevity of mRNA

One of the chief ways of regulating gene expression at the translational level involves the selective degradation or stabilization of the mRNA. If the mRNA were degraded rapidly after it entered the cytoplasm, it could generate very few proteins. However, if a message with a relatively short half-life were selectively stabilized in certain cells at certain times, then it would make large amounts of that particular protein only at certain times and places.

FIGURE 13.2
Initiation phase of eukaryotic translation. All the initiation factors are represented as circles. The first complex is made by the union of the 40S ribosomal subunit with initiator-tRNA. The initiator-tRNA has been complexed with the active (GTP) form of initiation factor 2. After this complex is formed, the mRNA is positioned with the aid of cap-binding protein (eIF4E) and other eIF4 subunits. Once the mRNA is in place, eIF5 mediates the joining of the 60S ribosomal subunit and the release of the previous initiation factors. The eIF2, now in its inactive (GDP) form, is reactivated by eIF2B. (After Hershey, 1989; Thach, 1992.)

FIGURE 13.3
Individual polysome transcribing the giant mRNA from the BR2 puff of *Chironomus tentans*. (A) Electron micrograph of a polysome containing 74 ribosomes. The nascent proteins can be seen extending from the ribosomes and growing as the ribosomes move from the 5′ end of the message to the 3′ end. Near the 3′ end are ribosomes from which the protein has detached. (B) Higher magnification of such a polysome; the polysome has been stretched during the preparation of the specimen. The relationship of the mRNA to the ribosomal subunits and the nascent polypeptide can be seen. (From Francke et al., 1982; photograph courtesy of J. E. Edstrom.)

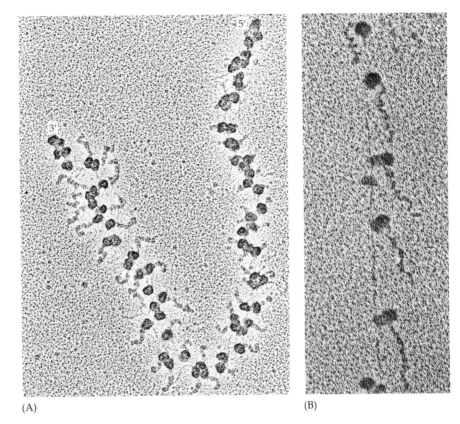

(A)   (B)

Selective degradation of mRNAs

Longevity of an mRNA is encoded in its 3′ untranslated region. Not all mRNAs have the same stability within a cell. A stable message such as β-globin has a half-life of around 17 hours, while the mRNAs for various growth factors have half-lives of less than 30 minutes. Thus, the amount of protein made from a single globin message should be much greater than that from a single growth factor message. The major element regulating message half-life appears to reside in the **3′ untranslated region** (3′ UTR) of the mRNA, and the ephemeral RNA species contain one or more AU-rich sequences in this region. Shaw and Kamen (1986) inserted a 51-base-pair AT-rich region from the 3′ UTR of the gene for growth factor GM-CSF into the 3′ UTR of the rabbit β-globin gene (Figure 13.4). The resulting globin message had a half-life of less than 30 minutes. A similar sequence, but containing 14 G and C residues, was inserted into another β-globin gene as a control. Its globin message had the normally long half-life.

The ability to differentially degrade mRNAs is critical for cell function. For example, the c-*fos* gene encodes a nuclear protein necessary for normal fibroblast cell division (Holt et al., 1986). Like the GM-CSF growth factor message, the mRNA for c-*fos* contains a large 3′ untranslated region rich in AU sequences. If this region is deleted (either experimentally or by natural mutation), the message gains a longer half-life. Consequently there is more c-*fos* protein being made, and the cell is continuously signaled to divide. The result is a tumor of those cells having c-*fos* genes that lack the AU-rich 3′ UTR (Meijlink et al., 1985). Wilson and Treisman (1988) found that this region stimulates the removal of the poly(A) tail when that message is translated. If the AU-rich region was deleted or replaced by some other sequence, the poly(A) tail remained and the message had a

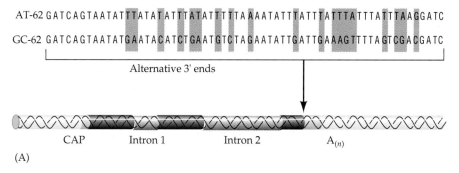

AT-62 GATCAGTAATATTTATATATTTATATTTTTAAAATATTTATTTATTTATTTATTTATTTAAGGATC

GC-62 GATCAGTAATATGAATACATCTGAATGTCTAGAATATTGATTGAAAGTTTTAGTCGACGATC

Alternative 3′ ends

CAP   Intron 1   Intron 2   A$_{(n)}$

(A)

WT   AT   GC

← β-globin RNA

← β$_2$ microglobulin RNA control

1   2   3

(B)

FIGURE 13.4
Regulation of mRNA longevity by a sequence in the 3′ untranslated region. (A) The 3′ end of the rabbit β-globin gene was altered by the insertion of a 62-base pair fragment derived from the 3′ end of the human *GM-CSF* gene or a related sequence in which several of the AT pairs had been replaced by GC pairs (indicated in color). (B) The clones were injected into cultured mouse cells, and the presence of the message after 30 hours was measured by incubating cell extracts with a ^{32}P-labeled DNA complementary to the 5′ end of the message. If the rabbit β-globin message still existed, the radioactive cDNA would bind to it and therefore be resistant to S1 nuclease (which destroys single-stranded nucleic acids only). If the message were not present, the added S1 nuclease would digest the probe to mononucleotides and no radioactivity would be bound. The resultant solutions were run on a gel and autoradiographed. Lane 1: Extract from cells incorporating the wild-type (WT) cloned rabbit β-globin gene. Lane 2: Extract from cells incorporating the rabbit β-globin gene with the AT-rich 3′ end (showing no mRNA after 30 hours). Lane 3: Extract from cells incorporating the rabbit β-globin gene with the GC-substituted 3′ end (showing stable mRNA after 30 hours). The gene and probe for β$_2$-microglobulin (producing a long-lived mRNA) was used as a control. (After Shaw and Kamen, 1986.)

longer half-life. It is possible that the AU-rich region undergoes extensive base-pairing with the poly(A) tail to cause a large double-stranded stem to form. This stem, they predict, is recognized by an RNase on the ribosome that would snip it off, thus removing the poly(A) tail that would otherwise stabilize the message. The stability of certain mRNAs in oocytes and embryos is similarly controlled by sites within the 3′ UTR (Brown and Harland, 1990).

Differential poly(A) tail shortening plays a decisive role in the life cycle of slime mold *Dictyostelium*. In the slime mold, a new set of messages are transcribed during the switch from vegetative growth (amoeba) to development (grex). At the same time, the poly(A) tails of the existing vegetative-stage mRNAs are dramatically shortened. As a result, the newly transcribed messages are translated whereas the pre-existing messages are not (Palatnik et al., 1984). This mechanism has been seen in larval *Drosophila* salivary glands (Restifo and Guild, 1986).

Hormonal stabilization of specific messenger RNAs. Differentiated gene products are often synthesized in response to hormonal induction. In some of these cases, hormones do not increase the transcription of certain messages but act at the level of translation. One such case involves the synthesis of casein by lactating mammals. Casein is the major phosphoprotein of milk and is therefore a differentiated product of the mammary gland. As will be discussed more fully in Chapter 20, the mammary gland is prepared by the sequential actions of several hormones. Prolactin, however, is the hormone responsible for lactation—that is, actual milk pro-

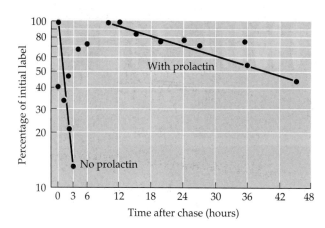

FIGURE 13.5
Degradation of casein mRNA in the presence and absence of prolactin. Cultured mammary cells were given radioactive RNA precursors (pulse) and, after a given time, were washed and fed nonradioactive precursors (chase). The casein RNA synthesized during the pulse time was then isolated and counted. In the absence of prolactin, the newly synthesized casein mRNA decayed rapidly, with a half-life of 1.1 hours. When the same experiment was done in medium containing prolactin, the half-life was extended to 28.5 hours. (After Guyette et al., 1979.)

duction. Prolactin augments the transcription of casein messages only about twofold; its major effect appears to be the stabilization of casein mRNA (Guyette et al., 1979). Prolactin somehow increases the longevity of the casein message so it exists 25 times longer than most other messages in the cell (Figure 13.5). Consequently, each casein mRNA can be used for more rounds of translation. In this way, a greater than normal number of casein molecules can be synthesized from each casein message. Table 13.1 summarizes these data and shows that other hormones also increase the stability of specific messenger RNAs.

Translational control of oocyte messages

Evidence for maternal regulation of early development

In most animal species, the diploid nucleus is not immediately expressed. Evidence that early development is controlled by factors stored in or made by the oocyte came from several experiments at the turn of the century (reviewed in Davidson, 1976). These experiments clearly demonstrated the dominance of maternal traits during the initial stages of embryogenesis and a switch to paternal or hybrid characteristics only later in develop-

TABLE 13.1
Stabilization of specific messenger RNAs by hormones

mRNA	Cell or tissue	Regulatory effector	Half-life (hours)	
			+ Effector	− Effector
Vitellogenin	*Xenopus* liver	Estrogen	500	16
Albumin	*Xenopus* liver	Estrogen	10	3
Vitellogenin	Avian liver	Estrogen	22	~2.5
Apo VLDL II	Avian liver	Estrogen	26	3
Casein	Rat mammary gland	Prolactin	92	5
Growth hormone	Rat pituitary cell cultures	Dexamethasone and thyroxine	20	2
Insulin	Rat pancreatic islet cells	Glucose	77	29
Ovalbumin	Hen oviduct	Estrogen, progesterone	~24	2–5

Source: After Shapiro et al., 1987.

TABLE 13.2
Driesch's data on the maternal control of primary mesenchyme cell number in hybrid sea urchin embryos

Egg		Sperm	Average number of primary mesenchyme cells (range in parentheses)
Echinus	×	*Echinus*	55 (±4)
Spherechinus	×	*Spherechinus*	33 (±4)
Spherechinus	×	*Echinus*	35 (±5)
Strongylocentrotus	×	*Strongylocentrotus*	49 (±3)
Spherechinus	×	*Strongylocentrotus*	33 (±3)

Source: After Davidson, 1976.

ment. Such far-reaching maternal effects have already been alluded to in our discussion of the cleavage orientation in snail embryos, in which the oocyte cytoplasm contains a factor that directs the rotations of the cleavage planes in a dextral or sinistral direction.

In 1898, Hans Driesch crossed two species of sea urchins and found that the average number of primary mesenchyme cells in the hybrid depended solely on which species provided the egg (Table 13.2). In a related experiment, Tennant (1914) fertilized eggs from the sea urchin *Cidaris* with sperm from the sea urchin *Lytechinus*. He showed that all the chromosomes were retained in the resulting blastomeres and that the maternal pattern of development was followed completely through the invagination of the archenteron and the timing of mesenchyme formation (Table 13.3). The paternal genes were first seen to be involved in the placement of the skeletal cells.

A second type of evidence for maternal regulation of early development comes from enucleation studies. If one could enucleate an egg after fertilization, one should be able to observe how far development could proceed in the absence of new mRNA synthesis. This enucleation has been accomplished by physical, chemical, and genetic means. Physical enucleation of the oocyte was accomplished by E. B. Harvey (1940), who removed the nuclei from sea urchin eggs and chemically activated the enucleated oocytes. These activated enucleated oocytes cleaved and

TABLE 13.3
Tennant's data on the control of early gastrulation events in hybrid sea urchin embryos

Species of sea urchin	Archenteron invagination (hours)	Mesenchyme formation (hours)	Site of origin of primary mesenchyme cells
Cidaris (♀)	20–33	23–26 (follows invagination)	Archenteron tip
Lytechinus (♂)	9	8 (precedes invagination)	Archenteron base and sides
Hybrid (*Cidaris* ♀ × *Lytechinus* ♂)	20	24 (follows invagination)	Archenteron base and sides

Source: After Davidson, 1976.

formed abnormal blastulae (that lacked a blastocoel). Frog eggs, too, are capable of development through the midblastula stage, even in the absence of a nucleus. In both instances, the embryos have a burst of protein synthesis as they begin their development. In sea urchin eggs, this burst of protein synthesis occurs during fertilization (or artificial activation); while in the frog, the stimulus for the increased rate of protein synthesis is the meiotic maturation of the oocyte preparatory to ovulation. In activated enucleated frog eggs, the proteins synthesized shortly after fertilization are not only synthesized in the proper amounts, they are also the same type of protein that is normally made (Smith and Ecker, 1965; Ecker and Smith, 1971).

Stored messenger RNAs

Evidence that the oocyte controls early development by storing mRNAs was first obtained by Brachet and co-workers (1963) and by Denny and Tyler (1964). The results of these investigations demonstrated that enucleated and activated sea urchin half-eggs contain RNA and can synthesize proteins at a rate comparable to that of normally fertilized eggs. Because there can be no transcription from these cells, stored mRNA is implicated. Craig and Piatigorsky (1971) demonstrated that this increase in protein synthesis could not have come from the transcription of mitochondrial DNA in the oocyte. It seemed that the oocyte had stored messenger RNAs that were not translated until after fertilization.

A gentler means of enucleation was made possible when it was discovered that the drug *actinomycin D* inhibits RNA synthesis; transcription could be eliminated by placing the newly fertilized egg in a solution of this drug. Gross and Cousineau (1964) found that when sea urchin eggs were treated with enough actinomycin D to shut down 94 percent of their RNA synthesis, the embryos still became blastulae and synthesized protein at the same rate as control eggs (Figure 13.6). That this amount of development is dependent on stored messages and not on pre-formed proteins was shown by treating the fertilized eggs with emetine or cycloheximide. These drugs inhibit translation, and embryos fertilized in the presence of these inhibitors did not develop at all (Figure 13.6C).

Although the initial burst of protein synthesis was identical in the actinomycin D-treated and control embryos, the treated embryos did not

FIGURE 13.6
Effect of protein synthesis inhibitors on the development of the sea urchin. (A) Control pluteus larva. (B) Development arrested at blastula stage when actinomycin D prevents the transcription of new RNAs. (C) Development arrested at a single-cell stage when emetine or cycloheximide prevents translation of new proteins. (D) Protein synthesis in embryos of the sea urchin *Arbacia punctulata* fertilized in the presence or absence of actinomycin D. For the first few hours, there is no significant difference in the rate of new protein synthesis. A burst of new protein synthesis beginning in the midblastula stage (colored curve) represents the translation of newly transcribed messages and therefore is not seen in the embryos growing in actinomycin D. (D after Gross et al., 1964.)

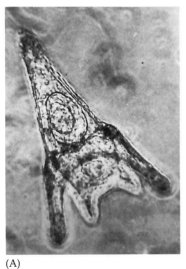

(A)

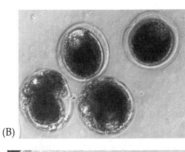

(B)

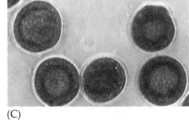

(C)

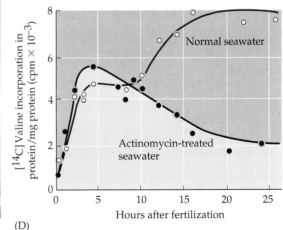

(D)

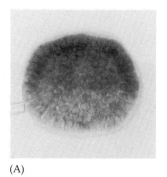 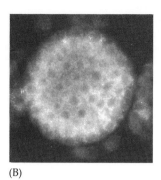 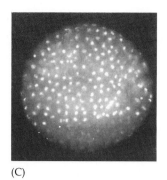 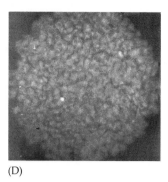

(A)   (B)   (C)   (D)

produce a second burst of protein synthesis at the blastula stage (Figure 13.6D). The message for this second burst of protein synthesis comes from nuclear transcription. One of these newly transcribed genes is that encoding the hatching enzyme that digests the fertilization envelope surrounding the blastula. In the presence of actinomycin D, this enzyme is not made from the ectodermal cells of the 220-cell blastula (Figure 13.7; Lepage et al., 1992). Thus, the actinomycin D-treated embryos can develop up to the blastula stage, but they proceed no further. The stored mRNAs of the oocyte are sufficient to take development only through the hatched blastula stage.

Physical, chemical, and genetic enucleations of the egg make it clear that stored messages do indeed exist in the cytoplasm of the oocyte and that the products of these messages support the early development of the embryo. Not all the mRNAs present in the oocyte cytoplasm are immediately translated. Instead, some reside dormant in the cytoplasm until fertilization or later, when they become translated. Just as most nuclei transcribe only a small subset of their many genes, so oocytes and early embryos translate only some of the many messenger RNAs present in their cytoplasm.

Characterization of stored oocyte messenger RNAs

Types of stored oocyte mRNAs. Davidson and his colleagues have estimated the complexity of the oocyte mRNA in a manner similar to their analysis of nuclear RNA complexity (Chapter 12). RNA (in large excess) was hybridized to denatured DNA, and the half C_0t value of the hybridization was found to be proportional to the amount of different RNA sequences present. By this analysis, they estimated that each oocyte (in numerous phyla) had enough different nucleotide sequences to account for roughly 1600 copies each of 20,000 to 50,000 RNA types (Galau et al., 1976; Hough-Evans et al., 1977). This is the greatest message complexity of any known cell type, and it reflects the enormous developmental potential of the oocyte. Relatively few of these messages have been characterized. Several of these mRNAs are not used in the oocyte but are stored there and translated after fertilization. This was first shown by using protein synthesis inhibitors, but more recently, PCR and RNase protection assays have also shown the existence of mRNAs that are stored in the oocyte and first translated during oocyte maturation (immediately prior to and during ovulation), fertilization, or in the early embryo. Table 13.4 gives a partial list of these stored mRNAs. Some of these mRNAs are for proteins that will be needed during cleavage, when the embryo makes enormous amounts of chromatin, cell membranes, and cytoskeletal components. Other stored messages encode proteins that determine the fate

FIGURE 13.7
Transcription and translation of the hatching enzyme gene during the midblastula stages of the sea urchin *Paracentrotus lividus*. (A) The hatching enzyme is made from newly transcribed mRNA that is transcribed in the animal hemisphere cells of the midblastula. (B) The enzyme is not seen before the 120-cell stage and is expressed most abundantly from (C) the ectodermal precursors of the 350-cell (9-hour) blastula prior to hatching. (D) As hatching takes place in the 500-cell embryos (around 13 hours), the hatching enzyme is degraded. (A) is an in situ hybridization using digoxygenin-labeled cDNA probe coupled with an alkaline-phosphatase-conjugated antibody to digoxygenin. In (B–D), hatching enzyme was detected by fluorescent antibodies against it. (From Lepage et al., 1992, courtesy of T. Lepage and C. Sardet.)

TABLE 13.4
Some mRNAs stored in oocyte cytoplasm and translated at or near fertilization

mRNAs encoding	Function(s)	Organism(s)
Cyclins	Cell division regulation	Sea urchin, clam, starfish, frog
Actin	Cell movement and contraction	Mouse, starfish
Tubulin	Form mitotic spindles, cilia, flagella	Clam, mouse
Small subunit of ribonucleotide reductase	DNA synthesis	Sea urchin, clam, starfish
Hypoxanthine phosphoribosyl-transferase	Purine synthesis	Mouse
Vg1	Mesodermal determination(?)	Frog
Histones	Chromatin formation	Sea urchin, frog, clam
Cadherins	Blastomere adhesion	Frog
Metalloproteinases	Implantation in uterus	Mouse
Growth factors	Cell growth; uterine cell growth(?)	Mouse
Sex determination factor fem-3	Sperm formation	*C. elegans*
par gene products	Segregate morphogenetic determinants	*C. elegans*
skn-1 protein	Blastomere fate determination	*C. elegans*
bicoid morphogen	Anterior fate determination	*Drosophila*
nanos morphogen	Posterior fate determination	*Drosophila*
germ cell-less protein	Germ cell determination	*Drosophila*
oskar protein	Germ cell localization	*Drosophila*
Ornithine transcarbamylase	Urea cycle	Frog
Elongation factor 1a	Protein synthesis	Frog
Ribosomal proteins	Protein synthesis	Frog, *Drosophila*

Sources: Compiled from numerous sources, including Raff, 1980; Shiokawa et al., 1983; Rappollee et al., 1988; Brenner et al., 1989; Standart, 1992.

of cells. One of the most remarkable instances is the storage of the information needed to make ribonucleotide reductase. The large subunit is stored as a protein in the oocyte cytoplasm. The small subunit is stored as an untranslated maternal message. Only after fertilization, when the mRNA for the small subunit is translated, can the newly synthesized small subunit combine with the preformed larger subunit to generate the functional enzyme (Standart et al., 1986).

Another remarkable finding was that the rate and pattern of cell divisions during early sea urchin development do not require a nucleus. Rather, they require continued protein synthesis from stored maternal mRNAs (Wagenaar and Mazia, 1978). The reason for this dependency on stored message was shown in 1983, when Evans and colleagues found a class of proteins they called called *cyclins*. These proteins regulate cell division (as discussed in Chapter 5) and are encoded by maternal mRNAs. What is striking about cyclins is that they are destroyed upon cell division and have to be resynthesized anew from the stored messages after the completion of each cleavage. Cyclin synthesis from stored messages is seen to decline as the embryo nears the end of the blastula stage.

Localization of stored mRNAs. Some stored mRNAs are not uniformly distributed throughout the oocyte. We have already seen that certain RNA sequences show cytoplasmic localization in snail and tunicate eggs, but such localizations have also been observed in sea urchin eggs and frog eggs (Rodgers and Gross, 1978). Rebagliati and co-workers (1985) have shown that whereas most of the maternal messages are found uniformly throughout the unfertilized *Xenopus* egg, some stored mRNAs are localized in the animal or vegetal poles of the cytoplasm. They extracted poly(A)-containing RNA from oocytes and used reverse transcriptase and DNA polymerase to convert the RNAs into a population of double-stranded DNAs. These double-stranded DNAs were then inserted into cloning vectors and grown separately in *E. coli*. About 2 million clones were derived in this manner. The DNA from these clones (the *Xenopus* oocyte library) was then transferred to two pieces of filter paper and denatured under conditions yielding single-stranded DNA. The investigators then cut the animal or vegetal poles from the egg and extracted the poly(A)-containing RNA from these regions. Radioactive cDNAs were made from the RNAs, and one group of the DNA-containing filters was incubated in the cDNAs from the animal messages while the other group was incubated in the cDNAs from the vegetal messages. When the binding of radioactive cDNAs was measured, most clones bound equal amounts of cDNA from the animal and vegetal poles, indicating that those messages were equally distributed. However, about 1.2 percent of the clones only bound cDNA made from animal pole messages, and about 0.2 percent of the clones only bound to the cDNA derived from vegetal pole mRNA.

The DNA of the clones specific for animal or vegetal messages could then be used to identify the localized mRNAs. RNA was extracted from whole eggs or their animal and vegetal poles and run on gels. The electrophoretically separated RNAs were blotted onto nitrocellulose paper (Northern blots) and probed with radioactive DNA from each of the region-specific clones. Two of these results are shown in Figure 13.8. The cytoplasmic localization of preformed oocyte messages can be seen in both mosaic and regulative eggs.

FIGURE 13.8
Demonstration of localized messages in the animal and vegetal poles of the *Xenopus* oocyte. RNA was obtained from the entire egg (T), the animal cap (A), or the vegetal cap (V) and electrophoretically separated on gels. The RNA was transferred to paper by the Northern blot procedure, and the paper was incubated with radioactive DNA from clones derived from cDNA complementary to oocyte messages. The radioactive DNA from clone *An*2 hybridizes to a message present in the animal pole but not in the vegetal pole. The reverse distribution is seen for the messages hybridizing to the DNA from clone *Vg*1. (B) In situ hybridization showing the *Vg*1 message at different stages of localization in the *Xenopus* oocyte. In the mature oocyte, it resides solely in the vegetal cortex. (The *Vg*1 RNA has recently been shown to encode a critical factor for determining the dorsal–ventral axis in vertebrates and will be discussed more fully in Chapter 16.) (A from Rebagliati et al., 1985, courtesy of D. Melton; B from Melton, 1987.)

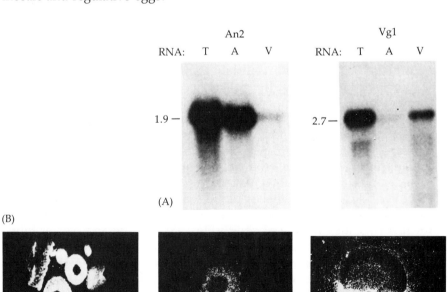

An2

RNA: T A V

Vg1

RNA: T A V

1.9 —

2.7 —

(A)

(B)

Mechanisms for translational control of oocyte messages

There are at present at least five hypotheses for the regulation of oocyte mRNA translation. Three of them involve the availability of mRNAs, whereas the other two involve the efficiency of mRNA translation. Although these hypotheses may be seen as competing with one another, it is probable that most species use more than one of these mechanisms to regulate oocyte mRNA translation.

The fundamental question involves *how the messenger RNAs are recruited into the polysomes.* It is known that whereas the unfertilized sea urchin egg has less than 1 percent of its ribosomes incorporated into polysomes, the cleavage-stage blastomeres have nearly 20 percent of their ribosomes in such structures (Figure 13.9). Since new mRNA is not being made at that time, it appears that pre-existing messages are being recruited into polysomes. This recruitment has been confirmed by numerous experiments demonstrating that the mRNA from the *oocyte ribonucleoprotein* (RNP) becomes incorporated into *embryonic polysomes.* Young and Raff (1979) showed that radioactively labeled RNA that is originally in the sea urchin oocyte RNPs later becomes associated with blastomere polysomes. Rosenthal and his co-workers (1980) found that when protein-free mRNAs from surf clam oocytes and embryos are placed in cell-free translation systems, they both code for the same proteins. In other words, the oocytes and embryos contain identical sets of mRNA. However, they found different subsets of messages in the *polysomes* in oocytes and embryos. Messages for three prominent embryo-specific proteins (A, B, and C) are seen as untranslated RNPs in the oocyte cytoplasm, whereas they are seen as polysomal mRNA in embryonic blastomeres. The recruitment of mRNA can be seen from the untranslated RNP to the translationally active polysomes. Conversely, messages for three characteristic oocyte proteins (X, Y, and Z) are found on the polysomes of oocytes but not on those of embryos. The recruitment of stored mRNAs into the polysomes of early embryonic cells has also been seen in *Drosophila* (Mermod et al., 1980) and *Xenopus* (Taylor and Smith, 1985). The question then becomes, How do mRNAs that are lying dormant in the cytoplasm of the oocyte suddenly acquire their competence to be translated?

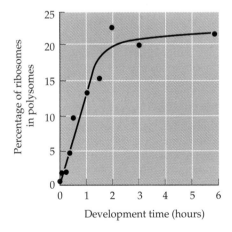

FIGURE 13.9
Increase in the percentage of ribosomes incorporated into polysomes during early sea urchin development. (After Humphreys, 1971.)

The masked maternal message hypothesis

This hypothesis contends that the oocyte messages are physically masked by proteins so that the mRNA cannot attach to ribosomes. Upon oocyte maturation or fertilization, the masking protein would fall away, enabling the mRNA to be translated. Messenger RNA is never found devoid of proteins. However, the type of protein associated with the RNA can vary. In 1966, Spirin proposed that the mRNA of the oocyte is stored in *informosomes,* ribonucleoprotein complexes wherein the mRNA is masked. The masked messages would be unable to bind to the ribosomes and thus would not be translated. At fertilization, the proteins masking the message would be released (possibly because of the ionic changes occurring during fertilization) and the message would be free to initiate translation. Support for this hypothesis followed shortly. In 1968, Infante and Nemer found that the unfertilized sea urchin egg contains RNP particles that sediment more slowly than ribosomes, and Gross and co-workers (1973) found that these particles contain various mRNAs.

Support for the masked maternal message hypothesis came from experiments showing that while the mRNA of unfertilized eggs stored in

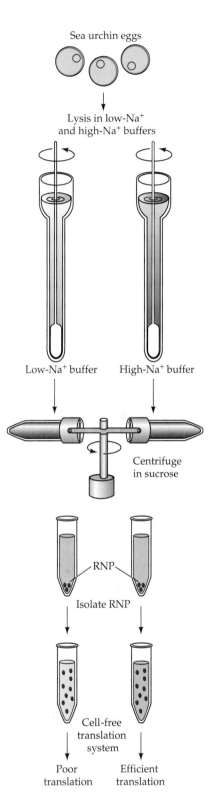

FIGURE 13.10
Demonstration of the presence of mRNA in RNP. Protein-bound RNA cannot be translated in low-sodium buffer, but it can be translated into proteins in buffer containing high concentrations of sodium ions, which cause the mRNA to dissociate from the RNP.

RNPs cannot be translated, the same RNAs can be translated if their RNPs are placed in solutions mimicking the changed ionic state of the egg after fertilization (Figure 13.10; Jenkins et al., 1978). It was proposed that the influx of sodium during fertilization might destabilize the RNP particle, thereby allowing its mRNA to be translated (Raff, 1980). A similar masking might be occurring in the surf clam *Spisula*. Here, mRNAs encoding the small subunit of ribonucleotide reductase and cyclin A are severely repressed in oocytes and are not translated until fertilization. However, two procedures can "unmask" the messages. First, high salt concentrations (0.5 *M* KCl) allow these mRNAs to make proteins; so does the removal of a particular sequence of bases in the 3′ untranslated regions of these messages (Figure 13.11; Standart et al., 1990). There are regions in the 3′ UTRs of both these messages that are very similar and may constitute binding sites for an 82-kDa protein that binds to the 3′ UTR of these mRNAs (Standart, 1992). Upon fertilization, this protein becomes phosphorylated. It is possible that the phosphorylated form of this protein can no longer block translation and that one of the kinases activated at fertilization is responsible for releasing the block to translation.

Another support for the masked maternal message hypothesis comes from studies of ribosomal RNA. In *Xenopus*, 5S rRNA is stored in an RNP complex with TFIIIA, its transcription factor (Wolffe and Brown, 1988; Allison et al., 1991). This association stores the 5S RNA in an inactive form until it is later incorporated into new ribosomes. Amphibian oocytes contain specific proteins that bind to some mRNAs and not others (Richter and Smith, 1984; Audet et al., 1987; Swiderski and Richter, 1988), but it is not known if these proteins functionally mask endogenous RNAs.

The poly(A) tail hypothesis

The developing oocyte provides embryologists with a remarkable display of translational regulation of gene expression. Indeed, frog and mouse oocytes each store numerous different types of messages in their respective cytoplasms, but they translate these mRNAs at different times during their maturation. Recent studies have demonstrated that altered polyadenylation is critical to the timing of oocyte mRNA translation and that this altered polyadenylation is regulated by the 3′ untranslated region.

The 3′ UTR can regulate the translational efficiency of oocyte messages by controlling the size of the poly(A) tail. In oocytes, the shortening of the poly(A) tail does not doom the message to extinction. Rather, it merely represses its ability to be translated (Hyman and Wormington, 1988). This repression is often temporary. In the mouse oocytes, those mRNAs that are being used for oocyte growth and metabolism retain their long poly(A) tails and are immediately translated. However, those mRNAs that are to be stored in the oocyte for translation at meiotic maturation (just prior to ovulation) or at fertilization tend to lose most of their poly(A) tails upon entering the cytoplasm. These mRNAs retain only from 15 to 90 adenylate residues (Figure 13.12). At meiotic maturation, an inversion occurs. Those mRNAs that had been actively translated lose their poly(A) tails and no longer function, while those poorly adenylated mRNAs that had been

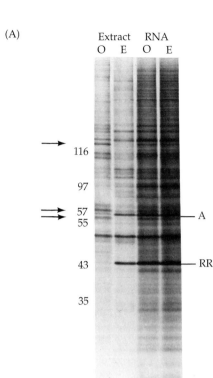

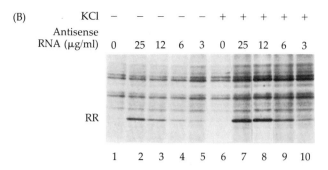

FIGURE 13.11

Unmasking the ribonucleotide reductase (RR) small subunit mRNA in clam oocytes. (A) Clam RR message is present but not translated in oocytes. Extracts from oocytes (lane 1) or activated eggs (lane 2) were mixed with ribosomes, translation factors, and radioactive amino acids and translated in vitro. The proteins were run on a gel and autoradiographs were made. The RR protein is made in the egg extract but not in the oocyte extract. When the mRNA of the oocytes (lane 3) and the eggs (lane 4) were isolated and separated from all proteins (by phenol extraction), RR message was translated in both cases. (B) The degree of unmasking is dependent on the presence of a high salt concentration or the addition of antisense mRNA that blocks the 3′ UTR protein-binding site. (From Standart et al., 1990.)

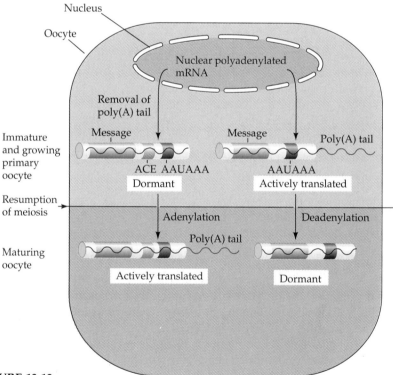

FIGURE 13.12

Model for the translational regulation of oocyte mRNAs in the mouse. Those mRNAs to be used in oocyte metabolism have polyadenylation sequences in their 3′ UTRs and retain their poly(A) tails. These mRNAs are translated until meiotic maturation (just prior to ovulation), when they lose their poly(A) tails. Those mRNAs that remain translationally dormant until meiotic maturation have adenylation control elements (ACEs) as well as the polyadenylation sequences, and they lose their poly(A) tails in the cytoplasm of the immature oocyte. When meiotic maturation begins, the tails are restored and translation of these messages is initiated.

stored rapidly gain long (150–600 adenylates) poly(A) tails and are translated into proteins (Vassalli et al., 1989; Huarte et al., 1992).

In mammals, the messages that are translated in the immature oocyte have a standard AAUAAA polyadenylation sequence. These messages retain their poly(A) tails until the resumption of meiotic maturation. At that time, their tails are deadenylated, and they become translationally inactive. Those mRNAs that will be stored in the immature oocyte cytoplasm for translation after oocyte maturation get their poly(A) tails clipped off immediately after they leave the nucleus. These messages have *two* signals in their 3′ UTR: the polyadenylation sequence AAUAAA, and a sequence known as the **adenylation control element** (ACE; sometimes called the cytoplasmic polyadenylation element, CPE), whose consensus sequence in mice and frogs is UUUUUAU (Fox et al., 1989; Bachvarova, 1992; Huarte et al., 1992). Upon the resumption of oocyte maturation, these stored transcripts are polyadenylated again (probably by the same polyadenylation enzyme as was found in the nucleus) and become translationally active. The acquisition of a long tail is seen to be critical for the onset of translation from the stored oocyte mRNA, and the control of this lengthening depends upon the presence or absence of an ACE.

In *Xenopus* oocytes, the story is similar, but with some variations. Like the mammalian oocyte mRNAs, a long poly(A) tail is necessary for a message to be translated (Fox and Wickens, 1990; Varnum and Wormington, 1990). A switch in RNA translation occurs during maturation. When the germinal vesicle (the haploid nucleus) breaks down to begin meiotic division, it releases deadenylation factors. Those mRNAs without ACEs are deadenylated, while those messages containing ACEs are able to become polyadenylated (Varnum et al., 1992). The polyadenylation sequence and the ACE are both needed for translational activation of these messages, but the presence per se of a poly(A) tail is not sufficient. Rather, the *process* of polyadenylation is critical for the message to be translated. That is, an mRNA injected with a pre-existing poly(A) tail will not be translated. It is possible that the process of polyadenylation also removes an inhibitor protein that otherwise masks the message (Fox et al., 1989; McGrew et al., 1989). There are some differences among the ACEs and these differences may cause different patterns of polyadenylation in the messages containing them (Paris and Richter, 1990). For instance, a particular ACE with a stretch of 12 U bases inhibits the polyadenylation of those mRNAs containing it during the time of oocyte maturation. However, after fertilization, the mRNAs containing this ACE are polyadenylated and translated into proteins (Simon et al., 1992). This suggests that there are specific factors that bind to these ACEs at various times. Paris and co-workers (1991) demonstrated that a 58-kD oocyte protein binds to that ACE and can be phosphorylated by cdc2 kinase. This phosphorylation is required for the polyadenylation of a set of mRNAs during oocyte maturation. The cdc2 kinase is activated in the cascade initiated by progesterone to stimulate the oocyte to resume meiosis prior to fertilization. This demonstrates that developmental cues can regulate which set of mRNAs become functional.*

*Despite these similarities, the functions of the poly(A) sequences and ACEs differ between mouse and frog oocytes. In frog oocytes, the deadenylation that occurs at maturation is the "default state", and deadenylation and translational inactivation occur unless an ACE is present. Polyadenylation will activate a masked message and maintain the translation of those mRNAs associated with polysomes. In mouse oocytes, the ACE controls both polyadenylation and deadenylation. In *immature* oocytes, those messages lacking the ACE are translated immediately, while ACE-containing mRNAs are deadenylated and translationally inactivated. At maturation, the mouse system becomes similar to that of *Xenopus*, and the ACE-containing RNAs are now polyadenylated and active in translation (Huarte et al., 1992).

Molecular regulators of development: The 3' untranslated region

The polyadenylation hypothesis is probably the most robust of the ideas concerning the regulation of stored mRNA in the oocyte. Moreover, interest in the 3' untranslated region is blossoming. This region of the message, "long viewed as a wasteland of genetic information—is in fact populated by critical regulatory elements that help orchestrate early development" (Wickens, 1992). In addition to regulating mRNA recruitment in newly fertilized eggs, the 3' UTR is being seen in several other developing systems where mRNA is regulated in space and time.

Gamete determination in C. elegans

A particularly dramatic role for the 3' UTR in masked mRNA is seen in *Caenorhabditis elegans*. This nematode worm has a female body but is hermaphroditic, producing both sperm and eggs at different times. The first germ cells to differentiate in the nematode become sperm, which are stored in the uterus for later use. After the fourth molt (from larva to adult), the germ cells cease making sperm and begin to make eggs. These eggs will eventually become fertilized by the stored sperm. The process determining which path the germ cell follows—to sperm or to egg—depends on the translational repression of different messages. The initiation of sperm formation is achieved by the repression of the *tra-2* message. The tra-2 protein is essential for the development of eggs and female body cells, and repression of *tra-2* mRNA translation in germ cells causes them to become sperm. The 3' UTR of this message contains two regions of 28 nucleotides, each of which appears to bind a putative repressor protein that is synthesized during the larval stages associated with spermatogenesis. If these regions are mutated, the translation of the *tra-2* mRNA is not repressed, no sperm is made, and the nematode is functionally female instead of hermaphroditic (Evans et al., 1992).

The story does not end here. The switchover from spermatogenesis to oogenesis also requires suppressing the translation of the *fem-3* gene through its 3' UTR. The fem-3 protein is critical for specifying male body cells and sperm production. Transcription of the *fem-3* gene is inhibited by tra-2, but the repression of existing *fem-3* messages is also needed. This translational repression appears to be effected by the binding of a translational inhibitor by the 3' UTR of the *fem-3* mRNA (Ahringer and Kimble, 1991; Evans et al., 1992).* Thus, the initiation of spermatogenesis in hermaphroditic nematodes and the transition from spermatogenesis to oogenesis appears to be regulated by translational repression through the 3' UTR.

*The translational repressors of the *fem-3* and *tra-2* messages have not yet been isolated, so it is premature to discuss how they work. It sounds like they might "mask" the message, but it is also possible they could act by interfering with polyadenylation.

Regulation of larval gene expression

Caenorhabditis elegans lives up to its name, having evolved a particularly elegant solution to the problem of controlling larval gene expression (Lee et al., 1993). High levels of lin-14 transcription factor specify the protein synthesis of early larval organs. Thereafter, the lin-14 protein is not seen anymore, although *lin-14* messages are detected throughout development. *C. elegans* is able to inhibit the synthesis of lin-14 from its mRNA by activating the *lin-4* gene; this gene encodes RNA that binds to the 3' UTR of the *lin-14* message and blocks its translation. (This RNA does not appear to encode any protein.) In loss-of-function *lin-14* mutations, the lin-14 protein is synthesized continuously, and the nematode is arrested in early development.

Localization of mRNA in oocytes

Not only is the translational activity of the stored maternal messages regulated by their 3' UTRs; so is their location within the oocyte. The *Vg1* mRNA of the *Xenopus* egg (see Figure 13.8) is localized into the vegetal pole via a 340-nucleotide sequence in its 3' UTR that binds to the oocyte cytoskeleton (Mowry and Melton, 1992). If this sequence is missing, the *Vg1* message will not go to the vegetal pole, and if this sequence is placed on other mRNAs, those messages will be translocated to the bottom of the egg. In *Drosophila*, the proper placement of several mRNAs is critical to normal anterior–posterior polarity (Chapters 15 and 22). *Bicoid* mRNA must be placed in the anterior pole of the oocyte, while *oskar*, *nanos*, and *cyclin B* must each be placed in the posterior pole. In all these cases, the messages are directed to their appropriate position by their respective 3' UTRs (Wharton and Struhl, 1991; Macdonald, 1992). In the case of *bicoid*, we know from genetic studies that the *exuperantia* (*exu*) gene is needed in order to localize *bicoid* message to the anterior of the oocyte. The *exu* gene appears to make a protein that binds the 3' UTR of the *bicoid* message to the anterior microtubules (Pokrywka and Stephenson, 1991).

Maintaining the differentiated state

The 3' UTR may also be necessary for the maintenance of differentiation in some tissues. In muscle cells, the mRNAs for troponin I, tropomyosin, and α-actin not only encode their respective muscle-specific proteins, but also block the ability of these developing muscle cells to return to a dedifferentiated, proliferative state. Rastinejad and Blau (1993) have shown that the 3' UTR of these messages is critical in maintaining muscle differentiation in cultured cells, and they propose that these 3' UTRs may be binding (and thereby sequestering) proteins that promote growth or suppress differentiation. Thus, while transcription factors such as the MyoD family may be responsible for *initiating* muscle differentiation, the 3' UTRs on muscle-specific mRNAs may be necessary for *maintaining* it.

The translational efficiency hypothesis

The preceding models of translational regulation assumed that the translational apparatus is capable of efficiently translating any message, but that mRNA and the ribosomes are kept apart by physical or chemical means. This need not be the case. The initial low pH of the oocyte is itself able to impede protein synthesis. As discussed in Chapter 4, there is an dramatic release of hydrogen ions during sea urchin fertilization, resulting in an elevation of cytoplasmic pH. Winkler and Steinhardt (1981) prepared an in vitro cell-free translation system from unfertilized sea urchin eggs. The pH of the resulting suspension was then retained or altered by dialysis, and protein synthesis was measured by the incorporation of radioactive valine into proteins. No exogenous mRNA was added. Figure 13.13 shows the results of one such experiment. Increasing the pH from oocyte levels (pH 6.9) to zygote levels (pH 7.4) occasioned a burst of protein synthesis mimicking that seen during fertilization.

Hille and her colleagues (1985; Danilchik et al., 1986) have suggested that the change in pH activates the translational apparatus of the egg. Ribosomes and initiation factors derived from unfertilized eggs were less active in translation than ribosomes and factors derived from fertilized eggs. Moreover, the injection of exogenous globin message into unfertilized eggs did not increase the amount of protein being synthesized. The globin mRNA was translated at the expense of other messages, suggesting that there is a limiting amount of some portion of the translational apparatus. The limiting factor is probably a translational initiation factor. The addition of eIF2B (the GTP-binding recycling factor) or eIF4F (which contains cap-binding protein) to a lysate prepared from unfertilized eggs increased the translational efficiency of that lysate (Colin et al., 1987; Lopo et al., 1988). It is probable that the alkalinization of egg cytoplasm serves both to unmask the mRNA (either physically or through polyadenylation) and to activate initiation factors. Support for this notion also comes from Winkler and his colleagues (1985; Kelso-Winemiller and Winkler, 1991), who have seen a threefold increase in the binding of mRNA to ribosomes once the pH has been raised.

Other activation systems: Uncapped messages and sequestered messages

Uncapped mRNA. The 5' and 3' ends of messenger RNA are necessary for efficient translation. We have already seen how differences in the length of the 3' poly(A) tail can effect differential RNA translation in *Xenopus* and *Spisula* oocytes. Certain moths use a mechanism for translational control involving changes at the 5' cap (Kastern et al., 1982). In order to be efficiently translated, almost all eukaryotic messages need a 7-methylguanosine "cap" on their 5' ends (Shatkin, 1976). The stored messages of the tobacco hornworm moth oocyte have a nonmethylated cap. The guanosine is present, but the methyl group has not been added to it. Such messages are not translated into proteins in a cell-free system. However, at fertilization, there is a burst of methylation in these oocytes and the caps are completed. The mRNAs with the completed caps are then able to bind to the ribosomes and initiate translation. Data on artificial messages suggest that secondary structures (such as hairpin loops) in the 5' untranslated region can also regulate the timing of RNA translation in the oocyte (Fu et al., 1991).

Sequestered mRNA. Some studies have disputed the evidence that all oocyte RNPs are untranslatable. Rather, it is possible that the protein

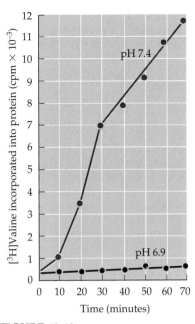

FIGURE 13.13
Evidence for the inefficiency of protein synthesis at prefertilization pH levels. The in vitro translation system made from unfertilized eggs is kept at pH 6.9 or dialyzed to pH 7.4. Endogenous messages are translated much more efficiently at the postfertilization pH. (After Winkler and Steinhardt, 1981.)

FIGURE 13.14
Sequestering of the sea urchin oocyte histone messages. A cDNA probe recognizing histone message is hybridized to sea urchin eggs fixed at various times after fertilization. Autoradiography shows the message to be sequestered in the maternal pronucleus until its breakdown 80–90 minutes after sperm entry. (From DeLeon et al., 1983, courtesy of L. and R. Angerer.)

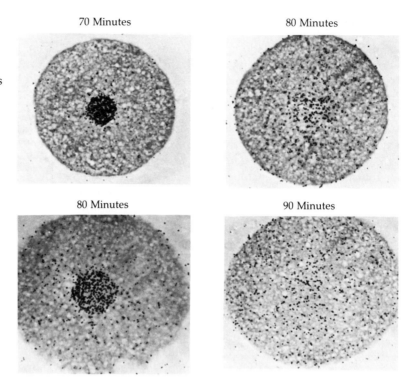

70 Minutes
80 Minutes
80 Minutes
90 Minutes

synthetic apparatus is compartmentalized, thereby preventing the mRNA (within the RNP) from getting close to the ribosomes (Moon et al., 1982). The histone mRNAs of sea urchin oocytes seem to be regulated by this type of restriction. The histone messages of the oocyte are not found in the cytoplasm. Rather, they are localized in the large pronucleus of the unfertilized egg. It is only when the pronucleus breaks down, at the end of fertilization, that the histone mRNA gets into the cytoplasm (Figure 13.14; DeLeon et al., 1983). This may not be the case for other messages. Less than 0.1 percent of the total mRNA of the unfertilized egg is found in pronuclei (Angerer and Angerer, 1981), and those RNPs containing the messages for actin and tubulin are predominantly cytoplasmic (Showman et al., 1982). The observation that some maternal messages as well as individual ribosomes are bound to the cytoskeleton (Moon et al., 1983) suggests that the cytoskeleton may also act to separate mRNAs from ribosomes. It is possible that all these possible mechanisms of translational control are utilized, even within the same oocyte. The egg has evolved numerous ways of regulating the translation of its stored mRNA, and species are able to use several of these mechanisms at once.

SIDELIGHTS & SPECULATIONS

The activation of the embryonic genome

The timing of developmental events differs enormously among animal species. At 24 hours after fertilization, *Drosophila* larvae have hatched and are busily eating; amphibian embryos are either at late gastrula or early neurula stages; and the sea urchin embryo is a late blastula or early gastrula with hundreds of cells. Mammalian embryos take their time. At 24 hours of development, a mouse zygote has divided only once, and a human egg still has about six hours until its first cleavage. Organisms also differ in the timing and the abruptness of the transition from the cytoplasmic control of development to the regulation of development by nuclear transcription. In most species studied,

TABLE 13.5
Activation of embryonic genomes and duration of functional maternal mRNA

Organism	Time of first observable transcription from nucleus[a]	Time of major new nuclear transcription[a]	Longevity of functional maternal message
Mammal (*Mus musculus*)	Late 1-cell stage (11–17 hr)	8- to 16-cell early morula (day 3)	4-cell cleavage stage (day 2–4)
Amphibian (*Xenopus laevis*)	Early cleavage (≤32-cell) midblastula (3 hr)	12th cleavage (4000-cell) midblastula (7 hr)	Neurula stage (15–30 hr)
Echinoderm (*S. purpuratus* and other sea urchins)	Zygote (pronuclear stage) (≤0.5 hr)	Midblastula (~128 cells) (11 hr)	Late blastula (15 hr)
Insect (*Drosophila melanogaster*)	Syncytial blastoderm after 10th nuclear division (2.5 hr)	Cellular blastoderm after 14th nuclear division (3.5 hr)	Mid-organogenesis (~15 hr)

Source: Adapted from Wilt, 1964; Woodland and Ballantine, 1980; Clegg and Piko, 1983; Gilbert and Solter, 1985; Poccia et al., 1985; Weir and Kornberg, 1985; Davidson, 1986; Edgar and Schubiger, 1986; Shiokawa et al., 1989.
[a]Times indicate incubation at an appropriate temperature.

the early embryo is an "RNA world" where the genome counts for naught (Wickens, 1992). In other embryos, nuclear transcription begins immediately upon fertilization, and new gene products are seen during the first cell cycle (Table 13.5).

Xenopus embryos appear to develop through the cleavage stage without the need for nuclear transcription. As mentioned in Chapter 5, the nucleus is essentially inactive until the "midblastula transition" at the end of the twelfth cell division (Figure 13.15; Newport and Kirschner, 1982). Until this midblastula transition, development proceeds on the materials stored in the oocyte cytoplasm. Eventually, transcription is initiated in the nuclei of the embryo. After the midblastula transition, different genes get turned on at different times; but the genes that become activated earliest may be activated by the maternal factors in the oocyte. The OZ-1 protein is a transcription factor that is made in the developing oocyte. It binds to a 14-base pair DNA sequence found in the promoters of several genes that are turned on at or shortly after the midblastula transition (Ovsenek et al., 1992). If other genes (such as the *Xenopus* β-globin gene) are linked to this sequence, they become expressed at the midblastula transition; but if the sequence is mutated, they are not expressed well (Figure 13.16). It is possible that this OZ-1 protein is, itself, inactive until some other factor (perhaps associated with the lengthening of the cell cycle) activates it.

The concept that maternal proteins may activate the genome during the midblastula transition is supported by the investigations of the *o* mutant of the axolotl salamander. This is a maternal effect mutation wherein homozygous females produce eggs that are successfully fertilized and are totally normal until the late cleavage and early blastula stages (Briggs and Cassens, 1966). At midblastula, eggs shed by an *o/o* female have slower mitoses and go on to form a dorsal blastopore lip, but always arrest in gastrulation. Malacinski (1971) and Carroll (1974) have shown that in embryos from wild-type females, new RNA and

protein synthesis starts at this midblastula stage. However, the midblastulae from eggs of *o/o* mothers do not undergo this burst of protein synthesis and have a pattern of proteins identical to that produced by enucleated zygotes (Figure 13.17). Briggs and Cassens (1966) demonstrated that

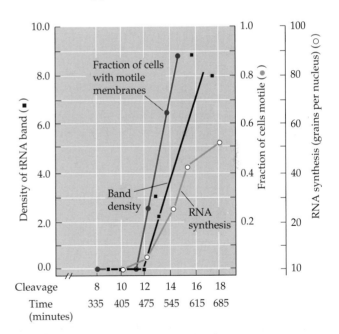

FIGURE 13.15

Activation of transcriptions and membrane motility in Xenopus after the twelfth cell division. Transcription was scored autoradiographically by the number of reduced silver grains over the nuclei of embryos immersed in radioactive uridine and by the activation of a cloned tRNA gene whose radioactive product could be assayed by measuring the density of the band of the autoradiographed gel. Motility was scored by determining the fraction of cells showing pseudopodia or blebbing in video recordings. (After Davidson, 1986.)

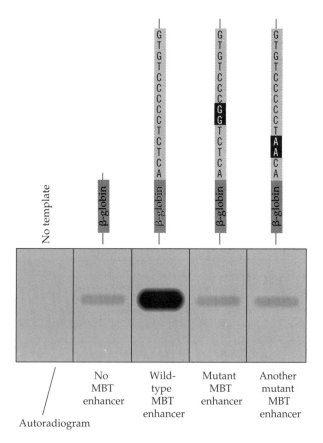

FIGURE 13.16
Effect of the OZ-1-binding midblastula transition (MBT) enhancer. The DNA sequence that activates transcription at the midblastula transition for the GS17 gene of Xenopus laevis was placed on a β-globin gene and injected into Xenopus oocytes. This globin construct became expressed at the midblastula stage. Globin genes without this enhancer or with a mutated MBT enhancer did not show significant expression at this stage. (After Ovsenek et al., 1992.)

the embryos from *o/o* mothers lacked a factor that activates the midblastula nuclear genome. In the absence of such a factor, the only development that occurs is that which can be supported by the stored oocyte mRNA. In amphibians, there is enough stored oocyte material to enable the embryo to enter gastrulation. Without new RNA synthesis, however, no further development can occur.

In *Drosophila* there also appears to be a midblastula transition from the RNAs and proteins of the oocyte cytoplasm to nuclear transcription. This transition is first seen after the tenth nuclear division. This is the first cycle with a G₂ phase, and the G₂ phase increases in length from 10 minutes after the tenth cycle to 60 minutes following the fourteenth. In the G₂ phase of the fourteenth cycle, the genome is transcribing at the highest level of activity seen during embryogenesis (Anderson and Lengyel, 1979; Weir and Kornberg, 1985). Edgar and Schubiger (1986) have shown that *Drosophila* nuclei become competent to transcribe at cycle 10 but that most genes need a longer G₂ phase to become activated. The high transcriptional activity

of cycle 14 embryos can be induced prematurely by artificially extending the G₂ period of younger embryos with cycloheximide. This activation can be accomplished with embryos as young as the tenth cycle but not earlier. It appears, then, that most genes become capable of activation during cycle 10, but that they do not initiate their transcription until cycle 14.

Sea urchins lack a stark midblastula transition. Although their enucleated eggs can develop through blastula stages, and although there is certainly a burst of nuclear transcription from midblastula nuclei, there seems to be no time in sea urchin development when the embryonic nucleus is not functioning. Low levels of transcription (including new histone messages) can be seen from pronuclei even before they fuse (Poccia et al., 1985). These newly transcribed messages join the larger pool of maternal mRNA. The chromatin of the first four cleavages is made primarily with histones stored in the oocyte cytoplasm and with histones synthesized from maternal messages. From the 16-cell stage onward, however, most histones are synthesized from messages transcribed from embryonic cell nuclei (Goustin and Wilt, 1981). This pattern is in marked contrast to that of *Xenopus* embryos, wherein a large pool of maternally stored histone protein and a large supply of stored oocyte histone message are utilized by thousands of cells.

Mammalian, ascidian, nematode, and molluscan embryos also appear to initiate transcription within the first cell cycle (Schauer and Wood, 1990). However, as in so many developmental events, mammals cannot be said to have evolved a uniform strategy. In the most-studied mammal group, mice, the embryonic genome is extremely active during the 2-cell stage. Between the 1-cell stage and the 2-cell embryo, over two-thirds of the proteins have a greater

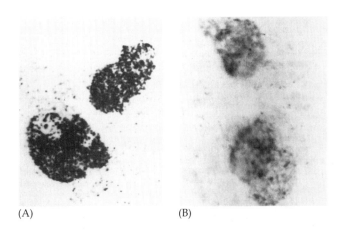

(A)   (B)

FIGURE 13.17
Incorporation of [³H]uridine into RNA of wild-type and o/o-mutant axolotl embryos. Blastula-stage embryos were incubated in the radioactive RNA precursor for 3 hours, washed, fixed, stained, and observed by autoradiography. (A) Normal embryo cells showing intense radioactivity indicating RNA synthesis. (B) Embryo from an o/o female. Stain is present, but no significant labeling is seen, indicating that little or no transcription has occurred. (From Carroll, 1974.)

than fivefold alteration of their synthesis (Latham et al., 1991, 1992). If cultured with the transcriptional inhibitor α-amanitin (which blocks RNA polymerase II), mouse eggs are blocked at the two-cell stage (Flach et al., 1982). In mice, the maternal mRNAs last about two days—roughly the same amount of time as those of the other phyla—and then, during the second day, the maternal messages are rapidly degraded (Clegg and Piko, 1983; Paynton et al., 1988). As the gene products encoded by the maternal messages decay, they are replaced by new proteins made from mRNA that is newly transcribed from the nucleus. In most instances, the sperm-derived chromosomes are probably activated simultaneously with the egg-derived chromosomes (Gilbert and Solter, 1985). Latham and colleagues have transplanted nuclei into different cytoplasms and demonstrated that the cytoplasm changes during the latter part of the 1-cell stage. The cytoplasm in the early 1-cell embryo will not support the transcription of genes from the nuclei of later embryos. However, the cytoplasm of late

1-cell embryos will. Since inhibitors of the cAMP-dependent protein kinase (PKA) inhibit the competence of the cytoplasm to support transcription, it is possible that the activation of PKA is essential for the cytoplasm's acquiring its transcriptionally permissive state. Other mammals do not necessarily follow the same schedule. Human mRNA synthesis is first seen at the 4-cell stage, and transcriptional inhibitors block development at the 4- to 8-cell stages. In cows and sheep, transcriptional activity is seen at the 8- to 16-cell stages (Braude et al., 1988; Telford et al., 1990).

In all animal species observed, there is a period of time when the phenomena of early development are controlled by messages and proteins stored in the oocyte cytoplasm. In most species (mammals being the exception), the nuclear genome is activated long before the maternal messages are degraded, so that both sets of mRNAs are being translated simultaneously. Eventually, as the maternal messages are degraded on day 1 or 2, the transcripts from the embryonic genome become more important.

Translational control of coordinated protein synthesis: Hemoglobin production

One of the major problems in genetic regulation is the coordinated production of several products from different regions of the genome. When a developing red blood cell synthesizes hemoglobin, it must ensure that the α-globin chains, β-globin chains, and heme molecules are in a 2:2:4 ratio (Figure 13.18). Any major deviation from this ratio results in severely debilitating diseases.

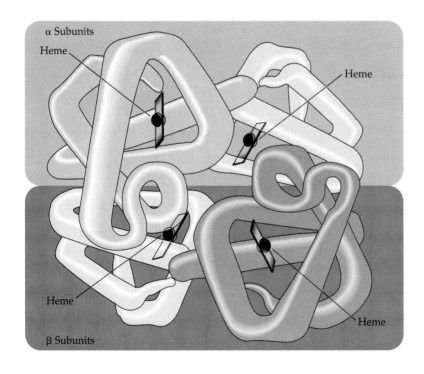

FIGURE 13.18
The structure of adult human hemoglobin, which has four polypeptide chains (two α, two β) and four heme molecules.

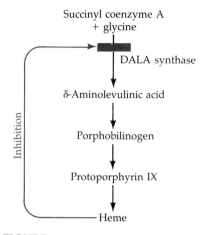

Succinyl coenzyme A
+ glycine

DALA synthase

δ-Aminolevulinic acid

Porphobilinogen

Protoporphyrin IX

Heme

Inhibition

FIGURE 13.19
Feedback regulation of heme synthesis. (After Harris, 1975.)

Recent evidence has shown that the heme molecule regulates the proportional synthesis of the hemoglobin components. It accomplishes this feat in two ways. First, excess heme (i.e., heme that has not been bound to protein such as globin) shuts off its own synthesis (Karibian and London, 1965). It does this by inactivating δ-aminolevulinate synthase (DALA synthetase), the first enzyme in the pathway for the production of heme (Figure 13.19). Thus, when there is more heme present than there are molecules to bind it, no more heme will be produced. Second, excess heme stimulates the production of globin proteins (Gribble and Schwartz, 1965; Zucker and Schulman, 1968). When heme (as its oxidized form, hemin) is added to a cell-free translation system that includes all the factors needed to translate mRNAs (Table 13.6), the synthesis of globin is greatly enhanced (Figure 13.20A). Therefore, if there is no globin to bind the heme, the excess heme shuts off its own synthesis and stimulates the production of more globin.

Several laboratories have investigated how a molecule as small as heme could regulate protein synthesis. In 1972, Adamson and colleagues demonstrated that the stimulatory effect of heme on globin synthesis could be mimicked by adding to the translation system those proteins that are loosely associated with the ribosomes. Because such solutions are rich in translation initiation factors, each factor was tested separately. It was found that eukaryotic initiation factor 2 (eIF2) restored protein synthesis to heme-deficient lysates in the translation system (Figure 13.20B). This initiation factor is responsible for combining with the initiator tRNA and complexing it to the 40S ribosomal subunit.

What, then, is the relationship between heme and eIF2? To answer this, London and his co-workers (Levin et al., 1976; Ranu et al., 1976; Ramaiah, et al., 1992) added heme-deficient lysates to heme-supplemented translation systems. They found that a portion of the heme-deficient lysate could actually depress the synthesis of globin in the translation system to which it was added. This finding indicated that an inhibitor was present. Furthermore, this inhibiting fraction had enzymatic activity. It contained a kinase capable of phosphorylating eIF2.

Phosphorylated eIF2 eventually halts translation. Normally, once the ribosomal subunits come together, eIF2 is released as a complex with GDP (Raychaudhury et al., 1985). For eIF2 to be used again in initiation, it must complex with eIF2B (recycling factor). This eIF2B acts to exchange GTP for GDP (see Figure 13.2), and the resulting eIF2-GTP complex is able to enter another round of initiation. However, if the alpha subunit of the eIF2 is

TABLE 13.6
Components of the in vitro translation system containing rabbit reticulocyte lysate

Component	Concentration (in 100 μl)	Component	Concentration (in 100 μl)
Reticulocyte lysate (1:1)	50 μl	KCl	76 mM
Tris-HCl (pH 7.6) buffer	10 mM	Proportional amino acid mixture	6–170 μM
ATP	1 mM	[^{14}C]Leucine	0.8 μCi
GTP	0.2 mM	"Cold" leucine	26 μM
Creatine phosphate	5 mM	Hemin	10–30 μM
Creatine phosphokinase	10 μg	H$_2$O to bring the total reaction	
Magnesium acetate	2 mM	to 100 μl volume	

Source: After London et al., 1976.

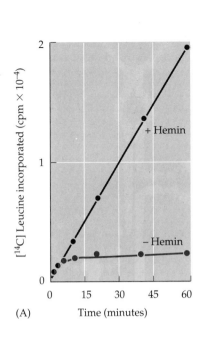

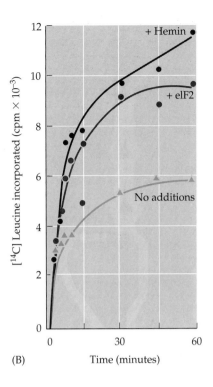

FIGURE 13.20
Translational regulation by hemin and eukaryotic initiation factor 2. (A) Translation of the globin mRNA in the rabbit reticulocyte in vitro protein synthesis system. Inclusion of hemin causes a dramatic elevation in protein synthesis. (B) Effect of the addition of additional eukaryotic initiation factor 2 to the rabbit reticulocyte in vitro translation system. The additional eIF2 elevated the level of protein synthesis close to that of the hemin-stimulated system. (A after London et al., 1976; B after Clemens et al., 1974.)

phosphorylated, the eIF2B recycling factor binds but cannot let go (Thomas et al., 1985; Gross et al., 1985). Eventually, all the eIF2B (whose concentration is 10- to 20-fold lower than that of eIF2) gets bound on these complexes, and translation stops. Addition of eIF2B to heme-deficient lysates restores protein synthesis to the levels of heme-supplemented systems (Grace et al., 1984).

To summarize, heme appears to regulate globin synthesis in the following manner (Figure 13.21):

1. In the absence of heme, a specific protein kinase phosphorylates eIF2.
2. Phosphorylated eIF2 binds to its GTP-recycling factor (eIF2B) and will not release it. Eventually, all the eIF2B is immobilized in these complexes, and translation comes to a halt since eIF2 remains in its inactive (GDP-bound) state.
3. Excess heme is able to bind to the protein kinase, inactivating it (Fagard and London, 1981). Inactivated kinase will not phosphorylate eIF2, so translation proceeds. Thus, as long as heme is present, globin synthesis continues.

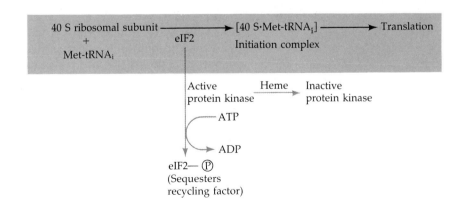

FIGURE 13.21
Scheme for the translational control of globin synthesis. As a result of inactivation by protein kinase, eIF2 is depleted, unless heme inactivates the protein kinase.

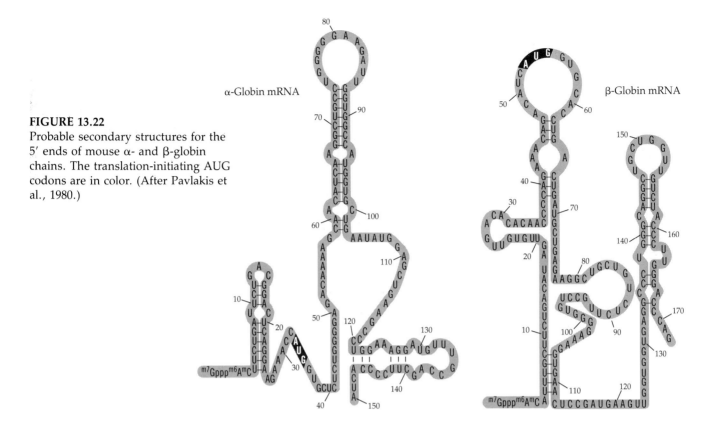

FIGURE 13.22
Probable secondary structures for the 5' ends of mouse α- and β-globin chains. The translation-initiating AUG codons are in color. (After Pavlakis et al., 1980.)

α-Globin mRNA

β-Globin mRNA

The story of the translational control of globin synthesis does not end here. As we discussed in Chapter 11, there are four active α-globin genes per diploid cell and only two active β-globin genes. If each gene were transcribed and translated at the same rate, one would expect twice as many α-globin molecules as β-globin molecules. This, of course, is not the case. One finds a 1.4:1 ratio of α:β mRNA, but 1:1 ratio of the proteins (Lodish, 1971). The equalization of the proteins appears to involve translational regulation.

Kabat and Chappell (1977) have suggested that the equalization is done at the initiation step of translation. They showed that the α-globin mRNA competes with the β-globin message for initiation factors and that the β-globin message appears to be the better competitor. The β-globin message is recognized more efficiently by the initiation factors and is thus translated more frequently. When the two mRNAs were present in equal amounts with a severely limiting supply of initiation factors, only 3 percent of the resulting protein was α-globin. However, when the unfractionated mRNA (α- and β-globin messages from lysed cells) was added to an excess of such initiation factors, all mRNAs were translated with equal efficiency and the resulting α to β ratio was 1.4:1. Recent experiments (Ray et al., 1983; Sarkar et al., 1984) have implicated the cap binding protein as the initiation factor responsible for discriminating between the two types of globin message. While it is not yet known how this discrimination occurs, it is known that the secondary structure of the 5' leader sequence affects the efficiency of translation (Pelletier and Sonnenberg, 1985). As is seen in Figure 13.22, the 5' ends of α- and β-globin messages differ significantly. Thus, at the initiation step of translation, the proper ratios of α-globin, β-globin, and heme are established. Although hemoglobin synthesis involves regulation at the transcriptional and RNA-processing levels, the final molecule is constructed through fine-tuned coordination at the level of translation.

The widespread use of translational regulation

Translational control is certainly a major mechanism of developmental regulation. We have seen its importance in the burst of protein synthesis following fertilization in many species, in balancing the synthesis of the components of hemoglobin, and in the hormone-induced synthesis of casein. There are many more cases of translational regulation throughout the animal kingdom. The reactivation of dehydrated brine shrimp embryos is also regulated at the level of translation. These embryos (often sold as "sea monkeys") begin development and can then lie dormant for long periods of time. During this dormancy, no protein synthesis occurs. It has been found that an inhibitor of protein synthesis exists in these embryos and that upon reactivation, a 20-fold increase in the level of functional eIF2 occurs (Filipowicz et al., 1975). Moreover, Sierra et al. (1977) have shown that this inhibitor may be a protein kinase. Thus, the same mechanism that regulates hemoglobin synthesis may also be used by brine shrimp embryos.

The coordinate and opposite regulation of the two major mammalian iron-binding proteins, ferritin and the transferrin receptor, has recently been elucidated (see Klausner and Harford, 1989; Klausner et al., 1993). Both the ferritin and the transferrin receptor mRNAs contain regions that bind an iron-responsive binding protein (IRE-BP). The ferritin message has this sequence on its leader sequence (5′ to the protein-encoding region), while the transferrin receptor message contains two of these sequences on its 3′ untranslated region. When cellular iron is in low supply, the iron-binding protein cannot bind iron and is in a conformation that binds to these mRNAs. When it binds to the leader sequence of the ferritin message, it blocks its translation, thus preventing the synthesis of this iron-storage protein. Simultaneously, the binding protein binds to the 3′ end of the transferrin receptor message, thereby stabilizing it against degradation and enabling more transferrin receptors to be produced. The transferrin receptors act to bring more iron into the cell (Figure 13.23).

RNA editing

One of the most unexpected mechanisms of translational control has recently been seen in the regulation of the apolipoprotein-B proteins. Apo-B proteins are components of serum lipid-carrier proteins and are thought to play a major role in the genesis of atherosclerosis. Apo-B48 (48 kDa) is synthesized in the intestines and becomes part of the chylomicron complexes necessary for the absorption and transport of dietary cholesterol and triglycerides. Apo-B100 (100 kDa) is made in the liver and is the major protein component of the very low, low, and intermediate density lipid carrier proteins. The apo-B100 and apo-B48 proteins are transcribed from the same gene, and no differential RNA processing is seen to occur to generate different mRNAs for these two proteins. Analysis of apo-B cDNAs indicates that the apo-B message in the *liver* encodes the entire apo-B100 peptide. The *intestinal* message, however, differs from the liver message by only one base. A C-to-U transition has occurred, changing a normal glutamine codon (CAA) into a terminator codon (UAA) at codon 2153. This difference results in the formation of the shorter apo-B48 protein in the intestine (Chen et al., 1987; Powell et al., 1987). This **RNA editing** is an instance where a specific base change is made in an existing RNA, thereby changing the message. The primary transcript of the *apo-B* gene does not appear to be edited, and the C-to-U editing can be accomplished by a factor contained within the nucleus. Therefore, Lau and colleagues

FIGURE 13.23
Coordinate and opposite translational regulation of ferritin and the transferrin receptor. Both messages contain regions that are recognized by an iron-binding regulatory protein (IRE-BP). In the absence of intracellular iron, this protein binds to these messages, inhibiting the translation of ferritin mRNA and stabilizing the mRNA for the transferrin receptor. (After Klausner and Harford, 1989.)

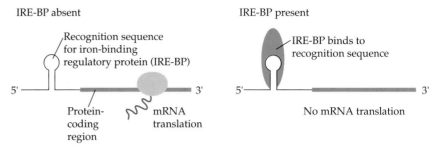

FERRITIN mRNA

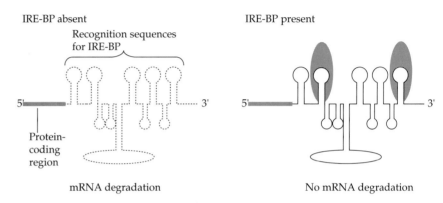

TRANSFERRIN RECEPTOR mRNA

(1991) conclude that this RNA editing is done during the RNA-processing steps. The nuclear factor responsible for this editing has not yet been identified, but by altering the sequence structure of the RNA near the edited cytosine, Chen and co-workers (1990) discovered two regions that are critical for editing. One is a region of nucleotides conserved between various mammalian species, and the other is a species-specific sequence further downstream. They postulate an enzyme that recognizes these two regions and places its catalytic site upon the particular cytosine. The deamination of the cytosine converts it into a uridine residue (Figure 13.24). Up to this time, it had been an axiom of molecular biology that the nucleotide sequence of a message, once transcribed, could not be altered. Such RNA editing is an exceptionally rare event and is usually seen only in certain organelle messages (see Chan, 1993). However, Sommer and colleagues (1991) proposed that regulated RNA editing can alter the cal-

FIGURE 13.24
Model of an enzymatic mechanism that could enable deamination of a specific cytosine of apoB mRNA. Two regions are needed for the RNA editing: a region that is conserved in several mammals and a species-specific element that has a hairpin loop structure that might be recognized by the enzyme. (After Chan, 1993.)

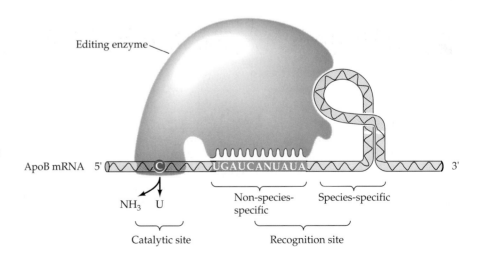

cium ion permeability of certain glutamate-gated ion channels during mammalian brain development.

Translational control, then, is an important and widely used mechanism for regulating gene expression in development. It can be used to activate a certain set of existing mRNAs at a certain time or to regulate the ratio at which different, competing, mRNAs can be translated. Animals have evolved several mechanisms whereby mRNAs can be stored in the oocyte for later use during early embryogenesis. The molecular bases for these translational regulatory mechanisms are now being studied.

Epilogue: Posttranslational regulation

When the peptide is synthesized, the story is still not over. Once a protein is made, it becomes part of a larger level of organization. It may become part of the structural framework of the cell or it may become involved in one of the myriad enzymatic pathways for the synthesis or breakdown of cellular metabolites. In either case, the individual protein is now part of a complex "ecosystem" that integrates it into a relationship with numerous other proteins. Thus, several changes can still take place that determine whether or not the protein is active. First, some newly synthesized proteins are inactive without further modifications. These modifications can involve the cleaving away of certain inhibitory sections of the protein or may involve the binding of a small compound to enhance its activity. Second, some proteins may be selectively inactivated. In some cases, inactivation involves the degradation of the protein itself; in other cases, inactivation may be brought about by the binding of an inhibitory ligand. Third, some proteins must be "addressed" to their specific intracellular destinations. The cell is not merely a sack of enzymes: proteins are often sequestered in certain regions, such as membranes, lysosomes, nuclei, or mitochondria. Fourth, some proteins need to assemble with other proteins to form a functional unit. The hemoglobin protein, the microtubule, and the ribosome are all examples of numerous proteins joining together to form a functional unit. Therefore, the expression of genetic information can still be influenced at the posttranslational level. Some of these cases (such as the phosphorylation of mitosis-promoting factor) have already been discussed, while others will be discussed as they appear. At this point we will leave our discussion of the molecular aspects of gene expression and return to the dynamics of the developing embryo. We can now look at early developmental processes to study the molecular mechanisms for the determination of cell fate and tissue structure.

LITERATURE CITED

Adamson, S. D., Yau, P. M. P., Herbert, E. and Zucker, W. V. 1972. Involvement of hemin, a stimulatory fraction from ribosomes, and a protein synthesis inhibitor in the regulation of hemoglobin synthesis. *J. Mol. Biol.* 63: 247–264.

Ahringer, J. and Kimble, J. 1991. Control of the sperm–oocyte switch in *Caenorhabditis elegans* by the *fem-3* 3′ untranslated region. *Nature* 349: 346–348.

Allison, L., Romaniuk, P. J. and Bakken, A. H. 1991. RNA–protein interactions of stored 5S RNA with TFIIIA and ribosomal protein L5 during *Xenopus oogenesis. Dev. Biol.* 144: 129–144.

Anderson, K. W. and Lengyel, J. A. 1979. Rates of synthesis of major classes of RNA in *Drosophila* embryos. *Dev. Biol.* 70: 217–231.

Angerer, L. M. and Angerer, R. C. 1981. Detection of poly A⁺ RNA in sea urchin eggs and embryos by quantitative in situ hybridization. *Nucleic Acids Res.* 9: 2819–2840.

Audet, R. G., Goodchild, J. and Richter, J. D. 1987. Eukaryotic initiation factor 4A stimulates translation in microinjected *Xenopus* oocytes. *Dev. Biol.* 121: 58–68.

Bachvarova, R. F. 1992. A maternal tail of poly(A): The long and the short of it. *Cell* 69: 895–897.

Bachvarova, R. DeLeon, V., Johnson, A., Kaplan, G. and Paynton, B. V. 1985. Changes in total RNA, poly(A) RNA and actin mRNA during meiotic maturation of mouse oocytes. *Dev. Biol.* 108: 325–331.

Brachet, J., Decroly, M., Ficq, A. and Quertier, J. 1963. Ribonucleic acid metabolism in fertilized and unfertilized sea urchin eggs. *Biochim. Biophys. Acta* 72: 660–662.

Braude, P., Bolton, V. and Moore, S. 1988. Human gene expression first occurs between the four- and eight-cell stages of preimplantation development. *Nature* 332: 459–461.

Brenner, C. A., Adler, R. R., Rappolee,D. A., Pedersen, R. A. and Werb, Z. 1989. Genes for extracellular matrix-degrading metalloproteases and their inhibitor, TIMP, are expressed during early mammalian development. *Genes Dev.* 3: 848–859.

Briggs, R. and Cassens, G. 1966. Accumulation in the oocyte nucleus of a gene product essential for embryonic development beyond gastrulation. *Proc. Natl. Acad. Sci. USA* 55: 1103–1109.

Brown, B. D. and Harland, R. M. 1990. Endonucleolytic cleavage of a maternal homeobox mRNA in *Xenopus* oocytes. *Genes Dev.* 4: 1925–1935.

Carroll, C. R. 1974. Comparative study of the early embryonic cytology and nucleic acid synthesis of *Ambystoma mexicanum* normal and *o* mutant embryos. *J. Exp. Zool.* 187: 409–422.

Chan, L. 1993. RNA editing: Exploring one mode with apolipoprotein B mRNA. *BioEssays* 15: 33–41.

Chen, S.-H. and 12 others. 1987. Apolipoprotein B48 is the product of a messenger RNA with an organ-specific in-frame stop codon. *Science* 238: 363–366.

Chen, S.-H., Li, X., Liao, W. S. L., Wu, J. H. and Chan, L. 1990. RNA editing of apolipoprotein B mRNA: Sequence specificity determined by in vitro coupled transcription–editing. *J. Biol. Chem.* 265: 6811–6816.

Clegg, K. B. and Piko, L. 1983. Poly(A) length, cytoplasmic adenylation, and synthesis of poly(A)+ RNA in early mouse embryos. *Dev. Biol.* 95: 331–341.

Clemens, M. J., Henshaw, E. C., Rahaminoff, H. and London, I. M. 1974. Met–tRNAfmet binding to 40S ribosomal units: A site for the regulation of initiation of protein synthesis by hemin. *Proc. Natl. Acad. Sci. USA* 71: 2946–2950.

Colin, A. M., Brown, B. D., Dholakia, J. N., Woodley, C. L., Wahba, A. J. and Hille, M. B. 1987. Evidence for simultaneous derepression of messenger RNA and the guanine nucleotide exchange factor in fertilized sea urchin eggs. *Dev. Biol.* 123: 354–363.

Craig, S. P. and Piatigorsky, J. 1971. Protein synthesis and development in the absence of cytoplasmic RNA synthesis in non-nucleate egg fragments and embryos of sea urchins: Effect of ethidium bromide. *Dev. Biol.* 24: 213–232.

Danilchik, M. V., Yablonka-Reuveniz, Z., Moon, R. T., Reed, S. K. and Hille, M. B. 1986. Separate ribosomal pools in sea urchin embryos: Ammonia activates a movement between pools. *Biochemistry* 25: 3696–3702.

Davidson, E. H. 1986. *Gene Activity in Early Development*, 3rd Ed. Academic Press, New York.

DeLeon, C. V., Cox, K. H., Angerer, L. M. and Angerer, R. C. 1983. Most early variant histone mRNA is contained in the pronucleus of sea urchin eggs. *Dev. Biol.* 100: 197–206.

Denny, P. C. and Tyler, A. 1964. Activation of protein synthesis in non-nucleate fragments of sea urchin eggs. *Biochem. Biophys. Res. Commun.* 145: 245–249.

Driesch, H. 1898. Über rein-mütterliche Charaktere und Bastardlarven von Echiniden. *Wilhelm Roux Arch. Entwicklungsmech. Org.* 7: 65–102.

Dworkin, M. B. and Dworkin-Rastl, E. 1985. Changes in the RNA titers and polyadenylation during oogenesis and oocyte maturation in *Xenopus laevis. Dev. Biol.* 112: 451–457.

Ecker, R. E. and Smith, L. D. 1971. The nature and fate of *Rana pipiens* proteins synthesized during maturation and early cleavage. *Dev. Biol.* 24: 559–576.

Edgar, B. A. and Schubiger, G. 1986. Parameters controlling transcriptional activation during early *Drosophila* development. *Cell* 44: 871–877.

Evans, T. C., Goodwin, E. B. and Kimble, J. 1992. Translational regulation of development and maternal RNAs in *Caenorhabditis elegans. Semin. Dev. Biol.* 3: 381–389.

Fagard, R. and London, I. M. 1981. Relationship between phosphorylation and activity of heme-regulated eukaryotic initiation factor 2 kinase. *Proc. Natl. Acad. Sci. USA* 78: 866–870.

Filipowicz, W., Sierra, J. J. and Ochoa, S. 1975. Polypeptide chain initiation in eukaryotes: Initiation factor MP in *Artemia salina* embryos. *Proc. Natl. Acad. Sci. USA* 72: 3947–3951.

Flach, G., Johnson, M. H., Braude, P. R., Taylor, R. A. S. and Bolton, V. N. 1982. The transition from maternal to embryonic control in the 2-cell mouse embryo. *EMBO J.* 1: 681–686.

Fox, C. A. and Wickens, M. 1990. Poly(A) removed during oocyte maturation: A default reaction selectively prevented by specific sequences in the 3' UTR of certain maternal mRNAs. *Genes Dev.* 4: 2287–2298.

Fox, C. A., Sheets, M. D. and Wickens, M. 1989. Poly(A) addition during maturation of frog oocytes: distinct nuclear and cytoplasmic activities and regulation by the sequence UUUUUAU. *Genes Dev.* 3: 2151–2162.

Francke, C., Edstrom, J. E., McDowell, A. W. and Miller, O. L. 1982. Microscopic visualization of a discrete class of giant translation units in salivary gland cells of *Chironomus tentans. EMBO J.* 1: 59–62.

Fu, L., Ye, R., Browder, L. and Johnston, R. 1991. Translational potentiation of mRNA with secondary structure in *Xenopus. Science* 251: 807–810.

Galau, G., Kelin, W. H., Davis, M. M., Wold, B., Britten, R. J. and Davidson, E. H. 1976. Structural gene sets active in embryos and adult tissues of the sea urchin. *Cell* 7: 487–505.

Gilbert, S. F. and Solter, D. 1985. Onset of paternal and maternal *Gpi-2* expression in preimplantation mouse embryos. *Dev. Biol.* 109: 515–517.

Goustin, A. S. and Wilt, F. H. 1981. Protein synthesis, polyribosomes, and peptide elongation in early development of *Strongylocentrotus purpuratus. Dev. Biol.* 82: 32–40.

Grace, M. and eight others. 1984. Protein synthesis in rabbit reticulocytes: Characteristics of the protein factor RF that reverses inhibition of protein synthesis in heme-deficient reticulocyte lysates. *Proc. Natl. Acad. Sci. USA* 79: 6517–6521.

Gribble, T. J. and Schwartz, H. C. 1965. Effect of protoporphyrin on hemoglobin synthesis. *Biochim. Biophys. Acta* 103: 333–338.

Gross, K. W., Jacobs-Lorena, M., Baglioni, G. and Gross, P. R. 1973. Cell-free translation of maternal messenger RNA from sea urchin eggs. *Proc. Natl. Acad. Sci. USA* 70: 2614–2618.

Gross, M., Redman, R. and Kaplansky, D. A. 1985. Evidence that the primary effects of phosphorylation of eukaryotic initiation factor 2a in rabbit reticulocyte lysate is inhibition of the release of eukaryotic initiation factor 2-GDP from 60S ribosomal subunits. *J. Biol. Chem.* 260: 9491–9500.

Gross, P. R. and Cousineau, G. H. 1964. Macromolecular synthesis and the influence of actinomycin D on early development. *Exp. Cell Res.* 33: 368–395.

Gross, P. R., Malkin, L. I. and Moyer, W. A. 1964. Templates for the first proteins of embryonic development. *Proc. Natl. Acad. Sci. USA* 51: 407–414.

Guyette, W. A., Matusik, R. J. and Rosen, J. M. 1979. Prolactin-mediated transcriptional and post–transcriptional control of casein gene expression. *Cell* 17: 1013–1023.

Harris, H. 1975. *Principles of Human Biochemical Genetics*. Elsevier North-Holland, New York.

Harvey, E. B. 1940. A comparison of the development of nucleate and non-nucleate eggs of *Arbacia punctulata. Biol. Bull.* 79: 166–187.

Hershey, J. W. B. 1989. Protein phosphorylation controls translation rates. *J. Biol. Chem.* 264: 20823–20826.

Hille, M. B., Danilchik, M. V., Colin, A. M. and Moon, R. T. 1985. Translational control in echinoid eggs and early embryos. *In* R. H. Sawyer and R. M. Showman (eds.), *The Cellular and Molecular Biology of Invertebrate Development*. University of South Carolina Press, pp. 91–124.

Holt, J. T., Gopal, T. V., Moulton, A. D. and Nienhuis, A. W. 1986. Inducible production of c-*fos* antisense RNA inhibits 3T3 cell proliferation. *Proc. Natl. Acad. Sci. USA* 83: 4794–4798.

Hough-Evans, B. R., Wold, B. J., Ernst, S. G., Britten, R. J. and Davidson, E. H. 1977. Appearance and persistence of maternal RNA sequences in sea urchin development. *Dev. Biol.* 260: 258–277.

Huarte, J. and seven others. 1992. Transient translational silencing by reversible mRNA deadenylation. *Cell* 69: 1021–1030.

Humphreys, T. 1971. Measurements of messenger RNA entering polysomes upon fertilization in sea urchins. *Dev. Biol.* 26: 201–208.

Hyman, L. E. and Wormington, W. M. 1988. Translational inactivation of ribosomal protein messenger RNAs during *Xenopus* oocyte maturation. *Genes Dev.* 2: 598–605.

Infante, A. and Nemer, M. 1968. Heterogeneous RNP particles in the cytoplasm of sea urchin embryos. *J. Mol. Biol.* 32: 543–565.

Jenkins, N. A., Kaumeyer, J. R., Young, E. M. and Raff, R. A. 1978. A test for masked message: The template activity of messenger ribonucleoprotein particles isolated from sea urchin eggs. *Dev. Biol.* 63: 279–298.

Kabat, D. and Chappell, M. R. 1977. Competition between globin messenger ribonucleic acids for a discriminating initiation factor. *J. Biol. Chem.* 252: 2684–2690.

Karibian, D. and London, I. M. 1965. Control of heme synthesis by feedback inhibition. *Biochem. Biophys. Res. Commun.* 18: 243–249.

Kastern, W. H., Swindlehurst, M., Aaron, C., Hooper, J. and Berry, S. J. 1982. Control of mRNA translation in oocytes and developing embryos of giant moths. I. Functions of the 5′ terminal "cap" in the tobacco hornworm *Manduca sexta. Dev. Biol.* 89: 437–449.

Kelso-Winemiller, L. and Winkler, M. M. 1991. "Unmasking" of stored maternal mRNAs and the activation of protein synthesis at fertilization in sea urchins. *Development* 111: 623–633.

Klausner, R. D. and Harford, J. B. 1989. *Cis–trans* models for post-transcriptional gene regulation. *Science* 246: 870–872.

Klausner, R. D., Rouault, T. A. and Harford, J. B. 1993. Regulating the fate of mRNA: The control of cellular iron metabolism. *Cell* 72: 19–28.

Kozak, M. 1986. Point mutations define a sequence flanking the AUG initiator codon that modulates translation by eukaryotic ribosomes. *Cell* 44: 283–292.

Latham, K. E., Garrels, J. I., Chang, C. and Solter, D. 1991. Quantitative analysis of protein synthesis in mouse embryos. I. Extensive reprogramming at the one- and two-cell stages. *Development* 112: 821–932.

Latham, K. E., Solter, D. and Schultz, R. M. 1992. Acquisition of a transcriptionally permissive state during the 1-cell stage of mouse embryogenesis. *Dev. Biol.* 149: 457–462.

Lau, P. P., Xiong, W., Zhu, H.-J., Chen, S.-H. and Chan, L. 1991. Apolipoprotein B mRNA editing is an intranuclear event that occurs posttranscriptionally coincident with splicing and polyadenylation. *J. Biol. Chem.* 266: 20550–20554.

Lee, R. C., Feinbaum, R. L. and Ambros, V. 1993. The *C. elegans* heterochromatic gene *lin-4* encodes small RNAs with antisense complementarity to *lin-14. Cell* 75: 843–854.

Lepage, T., Sardet, C. and Gache, C. 1992. Spatial expression of the hatching enzyme gene in the sea urchin embryo. *Dev. Biol.* 150: 23–32.

Levin, D., Ranu, R., Ernst, V. and London, I. M. 1976. Regulation of protein synthesis in reticulocyte lysates: Phosphorylation of methionyl–tRNA$_f$ binding factor by protein kinase activity of translational inhibitor isolated from heme-deficient lysates. *Proc. Natl. Acad. Sci. USA* 73: 3112–3116.

Lodish, H. F. 1971. Alpha and beta globin messenger ribonucleic acid. Different amounts and rates of translation. *J. Biol. Chem.* 246: 7131–7138.

London, I. M., Clemens, M. J., Ranu, R. S., Levin, D. H., Cherbas, L. F. and Ernst, V. 1976. The role of hemin in the regulation of protein synthesis in erythroid cells. *Fed. Proc.* 35: 2218–2222.

Lopo, A. C., MacMillan, S. and Hershey, J. W. B. 1988. Translational control in early sea urchin embryogenesis: Initiation factor eIF4F stimulates protein synthesis in lysates from unfertilized eggs of *S. purpuratus. Biochemistry* 27: 351–357.

Macdonald, P. M. 1992. The means to the ends: localization of maternal messenger RNAs. *Semin. Dev. Biol.* 3: 413–424.

Malacinski, G. M. 1971. Genetic control of qualitative changes in protein synthesis during early amphibian (Mexican axolotl) embryogenesis. *Dev. Biol.* 26: 442–451.

McGrew, L., Dworkin-Rastl, E., Dworkin, M. B. and Richter, J. D. 1989. Poly(A) elongation during *Xenopus* oocyte maturation is required for translational recruitment and is mediated by a short sequence element. *Genes Dev.* 3: 803–815.

Meijlink, F., Curran, T., Miller, A. D. and Verma, I. M. 1985. Removal of a 67-base pair sequence in the non-coding region of protooncogene *fos* converts it to a transforming gene. *Proc. Natl. Acad. Sci. USA* 82: 4987–4991.

Melton, D. 1987. Translocation of a localized maternal mRNA to the vegetal pole of *Xenopus* oocytes. *Nature* 328: 80–82.

Mermod, J. J., Schatz, G. and Croppa, M. 1980. Specific control of messenger translation in *Drosophila* oocytes and embryos. *Dev. Biol.* 75: 177–186.

Moon, R. T., Danilchik, M. V. and Hille, M. 1982. An assessment of the masked messenger hypothesis: Sea urchin egg messenger ribonucleoprotein complexes are efficient templates for in vitro protein synthesis. *Dev. Biol.* 93: 389–403.

Moon, R. T., Nicosia, R. F., Olsen, C., Hille, M. B. and Jeffery, W. R. 1983. The cytoskeletal framework of sea urchin eggs and embryos: Developmental changes in the association of messenger RNA. *Dev. Biol.* 95: 447–458.

Morle, F., Lopez, B., Henni, T. and Godet, J. 1985. α-Thalassaemia associated with the deletion of two nucleotides at positions −2 and −3 preceding the AUG codon. *EMBO J.* 4: 1245–1250.

Mowry, K. and Melton, D. 1992. Vegetal messenger RNA localization directed by a 340 nucleotide RNA sequence element in *Xenopus* oocytes. *Science* 255: 991–994.

Newport, J. and Kirschner, M. 1982. A major developmental transition in early *Xenopus* embryos. II. Control of the onset of transcription. *Cell* 30: 687–696.

Ovsenek, N., Zorn, A. M. and Krieg, P. A. 1992. A maternal factor, OZ–1, activates embryonic transcription of the *Xenopus laevis GS17* gene. *Development* 115: 649–655.

Palatnik, C. M., Wilkins, C. and Jacobson, A. 1984. Translational control during early *Dictyostelium* development: Possible involvement of poly(A) sequences. *Cell* 36: 1017–1025.

Paris, J. and Richter, J. D. 1990. Maturation-specific polyadenylation and translational control: Diversity of polyadenylation elements, influence of poly(A) tail size, and the formation of stable polyadenylation complexes. *Mol. Cell. Biol.* 10: 5634–5645.

Paris, J., Swensen, K., Piwnica-Worms, H. and Richter, J. D. 1991. Maturation-specific polyadenylation: In vitro activation by p34^{cdc2} and phosphorylation of a 58-kD CPE-binding protein. *Genes Dev.* 5: 1697–1708.

Pavlakis, G. N., Lockard, R. E., Vamvakopolous, N., Rieser, L., Rajbhandary, U. L. and Vournakis, J. N. 1980. Secondary structure of mouse and rabbit α- and β-globin mRNAs: Differential accessibility of initiator AUG codons towards nucleases. *Cell* 19: 91–102.

Paynton, B. V., Rempel, R. and Bachvarova, R. 1988. Changes in states of adenylation and time course of degradation of maternal mRNAs during oocyte maturation and early embryonic development in the mouse. *Dev. Biol.* 129: 304–314.

Pelletier, J. and Sonenberg, N. 1985. Insertional mutagenesis to increase secondary structure within the 5′ noncoding region of a eukaryotic mRNA reduces translational efficiency. *Cell* 40: 515–526.

Poccia, D., Wolff, R., Kragh, S. and Williamson, P. 1985. RNA synthesis in male pronuclei of the sea urchin. *Biochim. Biophys. Acta* 824: 349–356.

Pokrywka, N. and Stephenson, E. 1991. Microtubules mediate the localization of *bicoid* mRNA during *Drosophila* oogenesis. *Development* 113: 55–66.

Powell, L. M., Wallis, S. C., Pease, R. J., Edwards, Y. H., Knott, T. J. and Scott, J. 1987. A novel form of tissue-specific RNA processing produces apolipoprotein-B48 in intestine. *Cell* 50: 831–840.

Raff, R. A. 1980. Masked messenger RNA and the regulation of protein synthesis in eggs and embryos. *In* D. M. Prescott and L. Goldstein (eds.), *Cell Biology: A Comprehensive Treatise,* Vol. 4. Academic Press, New York, pp. 107–136.

Ramaiah, K. V. A., Dhindsa, R. S., Chen, J. J., London, I. M. and Levin, D. 1992. Recycling and phosphorylation of eukaryotic initiation factor 2 on 60S subunits of 70S initiation complexes and polysomes. *Proc. Natl. Acad. Sci. USA* 89: 12063–12067.

Ranu, R. S., Levin, D. H., Delaunay, J., Ernst, U. and London, I. M. 1976. Regulation of protein synthesis in rabbit reticulocyte lysates: Characteristics of inhibition of protein synthesis by a translational inhibitor from heme-deficient lysates and its relationship to the initiation factor which binds Met-tRNA$_f$. *Proc. Natl. Acad. Sci. USA* 73: 2720–2726.

Rappollee, D., Brenner, C. A., Schultz, R., Mark, D. and Werb, Z. 1988. Developmental expression of PDGF, TGF-α, and TGF-β genes in preimplantation mouse embryos. *Science* 241: 1823–1825.

Rastinejad, F. and Blau, H. M. 1993. Genetic complementation reveals a novel regulatory role for 3' untranslated regions in growth and differentiation. *Cell* 72: 903–917.

Ray, B. K. and eight others. 1983. Role of mRNA competition in regulating translation: Further characterization of mRNA discriminatory initiation factors. *Proc. Natl. Acad. Sci. USA* 80: 663–667.

Raychaudhury, P., Chaudhuri, A. and Maitra, U. 1985. Formation and release of eukaryotic initiation factor 2 GDP complex during eukaryotic ribosomal polypeptide chain initiation complex formation. *J. Biol. Chem.* 260: 2140–2145.

Rebagliati, M. R., Weeks, D. L., Harvey, R. P. and Melton, D. A. 1985. Identification and cloning of localized maternal RNAs from *Xenopus* eggs. *Cell* 42: 769–777.

Restifo, L. L. and Guild, G. M. 1986. Poly(A) shortening of coregulated transcripts in *Drosophila*. *Dev. Biol.* 115: 507–510.

Richter, J. D. and Smith, L. D. 1984. Reversible inhibition of translation by *Xenopus* oocyte-specific proteins. *Nature* 309: 378–380.

Rodgers, W. H. and Gross, P. R. 1978. Inhomogeneous distribution of egg RNA sequences in the early embryo. *Cell* 14: 279–288.

Rosenthal, E. T. and Ruderman, J. V. 1987. Widespread change in the translation and adenylation of maternal messenger RNA following fertilization of *Spisula* oocytes. *Dev. Biol.* 121: 237–246.

Rosenthal, E., Hunt, T. and Ruderman, J. V. 1980. Selective translation of mRNA controls the pattern of protein synthesis during early development of the surf clam, *Spisula solidissima*. *Cell* 20: 487–494.

Safer, B. 1989. Nomenclature of initiation, elongation and termination factors for translation in eukaryotes. *Eur. J. Biochem.* 186: 1–3.

Sarkar, G., Edery, I., Gallo, R. and Sonenberg, N. 1984. Preferential stimulation of rabbit α-globin mRNA translation by a cap-binding protein complex. *Biochim. Biophys. Acta* 783: 122–129.

Schauer, I. E. and Wood, W. B. 1990. Early *C. elegans* embryos are transcriptionally active. *Development* 110: 1303–1317.

Shapiro, D. J., Blume, J. E. and Nielsen, D. A. 1987. Regulation of messenger RNA stability in eukaryotic cells. *BioEssays* 6: 221–226.

Shatkin, A. J. 1976. Capping of eukaryotic mRNAs. *Cell* 9: 645–653.

Shatkin, A. J. 1985. mRNA cap binding proteins: Essential factors for initiating translation. *Cell* 40: 223–224.

Shaw, G. and Kamen, R. 1986. A conserved AU sequence from the 3' untranslated region of GM–CSF mRNA mediates selective mRNA degradation. *Cell* 46: 659–667.

Shiokawa, K., Tashiro, K., Oka, T. and Yamana, K. 1983. Contribution of maternal mRNA for maintenance of Ca^{2+}-dependent reaggregating activity in dissociated cells of *Xenopus laevis* embryos. *Cell Differ.* 13: 247–255.

Shiokawa, K., Misumi, Y., Nakamura, N., Yamana, K. and Oh-uchida, M. 1989. Changes in the patterns of RNA synthesis in early embryogenesis of *Xenopus laevis*. *Cell Differ. Dev.* 28: 17–26.

Showman, R. M., Wells, D. E., Anstrom, J., Hursh, D. A. and Raff, R. A. 1982. Message-specific sequestration of maternal histone mRNA in the sea urchin egg. *Proc. Natl. Acad. Sci. USA* 79: 5944–5947.

Sierra, J. M., de Haro, C., Datta, A. and Ochoa, S. 1977. Translational control by protein kinases in *Artemia salina* and wheat germ. *Proc. Natl. Acad. Sci. USA* 74: 4356–4359.

Smith, L. D. and Ecker, R. C. 1965. Protein synthesis in enucleated eggs of *Rana pipiens*. *Science* 150: 777–779.

Sommer, B., Köhler, M., Sprengel, R. and Seeburg, P. H. 1991. RNA editing in brain controls a determinant of ion flow in glutamate-gated channels. *Cell* 67: 11–19.

Spirin, A. S. 1966. On "masked" forms of messenger RNA in early embryogenesis and in other differentiating systems. *Curr. Top. Dev. Biol.* 1: 1–38.

Standart, N. 1992. Masking and unmasking of maternal mRNAs. *Semin. Dev. Biol.* 3: 367–379.

Standart, N., Hunt, T. and Ruderman, J. V. 1986. Differential accumulation of ribonucleotide reductase subunits in clam oocytes: The large subunit is stored as a polypeptide, the small subunit as untranslated mRNA. *J. Cell Biol.* 103: 2129–2136.

Standart, N., Dale, M., Stewart, E. and Hunt, T. 1990. Maternal mRNA from clam oocytes can be specifically unmasked in vitro by antisense RNA complementary to the 3' untranslated region. *Genes Dev.* 4: 2157–2168.

Swiderski, R. E. and Richter, J. D. 1988. Photocrosslinking of proteins to maternal mRNA in *Xenopus* oocytes. *Dev. Biol.* 128: 349–358.

Taylor, M. A. and Smith, L. D. 1985. Quantitative changes in protein synthesis during oogenesis in *Xenopus laevis*. *Dev. Biol.* 110: 230–237.

Telford, N. A., Watson, A. J. and Schultz, G. A. 1990. Transition from maternal to embryonic control in early mammalian development: A comparison of several species. *Mol. Reprod. Dev.* 26: 90–100.

Tennant, D. H. 1914. The early influence of spermatozoa upon the characters of echinoid larva. *Carnegie Inst. Wash. Publ.* 182: 127–138.

Thach, R. E. 1992. Cap recap: The involvement of eIF-4F in regulating gene expression. *Cell* 69: 177–180.

Thomas, N. S. B., Matts, R. L., Levin, D. H. and London, I. M. 1985. The 60S ribosomal subunit as a carrier of eukaryotic initiation factor 2 and the site of reversing factor activity during protein synthesis. *J. Biol. Chem.* 260: 9860–9866.

Varnum, S. M. and Wormington, W. M. 1990. Deadenylation of maternal mRNAs during *Xenopus* oocyte maturation does not require *cis* sequences: A default mechanism for translational control. *Genes Dev.* 4: 2278–2286.

Varnum, S., Hurney, C. A. and Wormington, W. M. 1992. Maturation-specific deadenylation in *Xenopus* oocytes requires nuclear and cytoplasmic factors. *Dev. Biol.* 153: 283–290.

Vassalli, J. D. and seven others. 1989. Regulated polyadenylation controls mRNA translation during meiotic maturation of mouse oocytes. *Genes Dev.* 3: 2163–2171.

Wagenaar, E. B. and Mazia, D. 1978. The effect of emetine on the first cleavage division of the sea urchin, *Strongylocentrotus purpuratus*. In E. R. Dirksen, D. M. Prescott and L. F. Fox (eds.), *Cell Reproduction: In Honor of Daniel Mazia*. Academic Press, New York, pp. 539–545.

Weir, M. P. and Kornberg, T. 1985. Patterns of *engrailed* and *fushi tarazu* transcripts reveal novel intermediate stages of *Drosophila* segmentation. *Nature* 318: 433–439.

Wharton, R. P. and Struhl, G. 1991. RNA regulatory elements mediate control of *Drosophila* body pattern by the posterior morphogen, nanos. *Cell* 67: 955–967.

Wickens, M. 1992. Introduction: RNA and the early embryo. *Semin. Dev. Biol.* 3: 363–365.

Wilson, T. and Treisman, R. 1988. Removal of poly(A) and consequent degradation of c-*fos* mRNA facilitated by 3' AU-rich sequences. *Nature* 336: 396–399.

Wilt, F. H. 1964. Ribonucleic acid synthesis during sea urchin embryogenesis. *Dev. Biol.* 9: 299–313.

Winkler, M. M. and Steinhardt, R. A. 1981. Activation of protein synthesis in a sea urchin cell-free system. *Dev. Biol.* 84: 432–439.

Winkler, M. M., Nelson, E. M., Lashbrook, C. and Hershey, J. W. B. 1985. Multiple levels of regulation of protein synthesis at fertilization in sea urchin eggs. *Dev. Biol.* 107: 290–300.

Wolffe, A. P. and Brown, D. D. 1988. Developmental regulation of two 5S ribosomal RNA genes. *Science* 241: 1626–1632.

Woodland, H. R. and Ballantine, J. E. M. 1980. Paternal gene expression in developing hybrid embryos of *Xenopus laevis* and *Xenopus borealis*. *J. Embryol. Exp. Morphol.* 60: 359–372.

Young, E. M. and Raff, R. A. 1979. Messenger ribonucleoprotein particles in developing sea urchin embryos. *Dev. Biol.* 72: 24–40.

Zucker, W. V. and Schulman, H. M. 1968. Stimulation of globin-chain initiation by hemin in the reticulocyte cell-free system. *Proc. Natl. Acad. Sci. USA* 59: 582–589.

IV

SPECIFICATION OF CELL FATE AND THE EMBRYONIC AXES

14

Autonomous cell specification by cytoplasmic determinants

Each metazoan organism is a complex assortment of specialized cell types. For example, red and white blood cells differ not only from each other but also from the heart cells that propel them through the body. They also differ from the outstretched neurons that conduct neural impulses from the brain to the heart, and from the glandular cells, which secrete hormones into the blood. Table 14.1 presents a very incomplete list of specialized cell types, their characteristic products, and their functions.

I hold it probable that in the germ cells there exist fine internal differences which predetermine the subsequent transformation to a determinant substance; not differences which are mere potencies present in the germ cells, but actual material differences so fine that we have not as yet been able to demonstrate them.
R. VIRCHOW (1858)

Studying the period of cleavage we approach the source whence emerge the progressively branched streams of differentiation that end finally in almost quiet pools, the individual cells of the complex adult organism.
E. E. JUST (1939)

Cell commitment and differentiation

The development of specialized cell types from the single fertilized egg is called **differentiation**. This overt change in cellular biochemistry and function is preceded by a process involving the covert commitment of cells to a particular fate or set of fates. Here, the cell does not *appear* phenotypically different from its uncommitted state, but somehow its developmental fate has become restricted. Although embryologists have routinely used the word *determination* to describe this hidden commitment, a particular tissue might be classified as determined or undetermined depending on which assay for determination was used (see Harrison, 1933). Slack (1991) has therefore divided this commitment into two stages, **specification** and **determination**. A cell or tissue is said to be *specified* when it is capable of differentiating autonomously when placed in a neutral environment such as a petri dish. (The environment is "neutral" with respect to the developmental pathway.) A cell or tissue is said to be *determined* when it is capable of differentiating autonomously even when placed in another region of the embryo. If it is able to differentiate according to its original fate even when placed in another region of the embryo, it is assumed that the commitment is irreversible.

We know of three major ways by which this commitment can take place (Table 14.2). The first mechanism of commitment involves the cytoplasmic segregation of determinative molecules during embryonic cleavage, wherein the cleavage planes separate qualitatively different regions of the zygote cytoplasm into different daughter cells. Each cell becomes specified by the type of cytoplasm it acquires during cleavage, and cell fate is thereby determined without any reference to neighboring cells. This

TABLE 14.1
Some differentiated cell types and their major products

Cell type	Differentiated cell product	Specialized function
Keratinocyte (skin cell)	Keratin	Protection against abrasion, desiccation
Erythrocyte (red blood cell)	Hemoglobin	Transport of oxygen
Lens cell	Crystallins	Transmission of light
B lymphocyte	Immunoglobulins	Antibody synthesis
T lymphocyte	Cell surface antigens (lymphokines)	Destruction of foreign cells; regulation of immune response
Melanocyte	Melanin	Pigment production
Pancreatic islet cells	Insulin	Regulation of carbohydrate metabolism
Leydig cell (♂)	Testosterone	Male sexual characteristics
Chondrocyte (cartilage cell)	Chondroitin sulfate; type II collagen	Tendons and ligaments
Osteoblast (bone-forming cell)	Bone matrix	Skeletal support
Myocyte (muscle cell)	Muscle actin and myosin	Contraction
Hepatocyte (liver cell)	Serum albumin; numerous enzymes	Production of serum proteins and numerous enzymatic functions
Neurons	Neurotransmitters (acetylcholine, epinephrine, etc.)	Transmission of electrical impulses
Tubule cell (♀) of hen oviduct	Ovalbumin	Egg white proteins for nutrition and protection of embryo
Follicle cell (♀) of insect oviduct	Chorion proteins	Eggshell proteins for protection of embryo

mechanism of committing cell fate is called **autonomous specification**, because the cells are specified by their own internal cytoplasmic components (Davidson, 1991). Thus, if a particular blastomere were removed early in development, that blastomere would produce the same cells that it would have if it were still part of the larger embryo, and the remaining embryo would lack those cells (and only those cells) that would have been formed by the missing cell (see Figure 1.29). Autonomous specification gives rise to a pattern of embryogenesis referred to as **mosaic development**, since the embryo appears to be constructed like a tile mosaic of self-differentiating parts.

A second way of committing cell fate involves interactions with neighboring cells. Here, the cells originally have the ability to follow more than one path of differentiation, and the interaction of these cells with other cells or tissues restricts the fates of one or both of the participants. This type of cell fate determination is sometimes called **conditional specification**, because the fate of a cell depends upon the conditions in which it finds itself. If a blastomere were removed from an early embryo of an organism having the conditional specification of its cells, the remaining embryonic cells could alter their normal fates so that the roles of the

TABLE 14.2
Modes of cell type specification and their characteristics

I. Autonomous specification

Characteristic of most invertebrates.
Specification by acquisition of certain cytoplasmic molecules present in the egg
Invariant cleavages produce the same lineages in each embryo of the species. Blastomere fates are generally invariant.
Lineage "founder cells" are usually specified autonomously at poles of embryonic axes.
Cell type specification precedes any large-scale embryonic cell migration.
Produces "mosaic" ("determinative") development: cells cannot change fate if a blastomere is lost.

II. Conditional specification

Characteristic of all vertebrates and few invertebrates.
Specification by interactions between cells. Relative positions are important.
Variable cleavages produce no invariant fate assignments to cells.
Massive cell rearrangements and migrations precede or accompany specification.
Capacity for "regulative" development: allows cells to acquire different functions.

III. Syncytial specification

Characteristic of most insect classes.
Specification of body regions by interactions between cytoplasmic regions prior to cellularization of the blastoderm.
Variable cleavage produces no rigid cell fates for particular nuclei.
After cellularization, conditional specification is most often seen.

Source: After Davidson, 1991.

missing cell could be taken over. Thus, conditional specification gives rise to a pattern of embryogenesis called **regulative development**. As we will see, all organisms use both autonomous and conditional means to specify different cell types, and there is a spectrum ranging from mosaic to regulative development. However, most invertebrates have a predominantly autonomous mode of cell type specification, whereas vertebrates are characterized by their extensive use of conditional specification.

Many insects utilize a third means to determine cell fate. Here, interactions between maternal components within the syncytial blastoderm occur before the cell membranes separating nuclei have formed. In **syncytial specification**, major cell fate decisions are made even before cells have formed. This chapter will focus on experiments that demonstrate autonomous specification, while the following chapters will cover conditional and syncytial modes of cell commitment during early embryogenesis.

Preformation and epigenesis

Any explanation of the differentiation of the various bodily cells from the fertilized egg has to explain (1) the constant morphology of each species (i.e., that chickens beget only chickens, not crocodiles), and (2) the diversity among the bodily parts of each organism. Indeed, one of the major characteristics of development is that each species reproduces a characteristic developmental pattern. Development involves the expression of the inherited properties of the species.

In the seventeenth century, a union of development and inheritance was achieved in the hypothesis of **preformationism**. According to this view, all the organs of the adult were prefigured in miniature within the sperm or (more usually) the ovum. Organisms were not seen to be "developed," but were "unrolled." This hypothesis had the backing of both science and philosophy (Gould, 1977; Roe, 1981). First, because all organs were prefigured, embryonic development merely required the growth of existing structures, not the formation of new ones. No extra mysterious force was needed for embryonic development. Second, just as the adult organism was prefigured in the germ cells, another generation already existed in a prefigured state within the germ cells of the first prefigured generation. This corollary, called *embôitment* (encapsulation), assured that the species would always remain constant. Although certain microscopists claimed to see fully formed human miniatures within the sperm or egg, the major proponents of this hypothesis—Albrecht von Haller and Charles Bonnet—knew that organ systems develop at different rates and that embryonic structures need not be in the same place as those in the newborn.

The preformationists had no cell theory to provide a lower limit to the size of their preformed organisms, nor did they view mankind's tenure on Earth as potentially infinite. Rather, said Bonnet (1764), "Nature works as small as it wishes," and the human species existed in that finite time spanning the creation and the resurrection. This was in accord with the best science of its time, conforming to the French mathematician-philosopher René Descartes' principle of the infinite divisibility of a mechanical nature initiated, but not interfered with, by God.

Preformation was a conservative theory, emphasizing the lack of change between generations. Its principal failure was its inability to account for the variations known by the limited genetic evidence of the time. It was known, for instance, that matings between white and black parents produced children of intermediate skin color, an impossibility if inheritance and development were solely through either the sperm or the egg. In more controlled experiments, the German botanist Joseph Kölreuter (1766) had produced hybrid tobacco plants having the characteristics of both species. Moreover, by mating the hybrid to either the male or female parent, Kölreuter was able to "revert" the hybrid back to one or the other parental type after several generations. Thus, inheritance seemed to arise from a mixture of parental components. In addition, preformationism could not explain the generation of "monstrosities" and distinct deviations such as hexadactylism (six fingers per hand) when both parents were normal.

There developed, then, an alternative hypothesis: **epigenesis**. According to this hypothesis, each adult organism develops anew from an undifferentiated condition. This view of development, having philosophical roots as far back as Aristotle, was revived by a German embryologist working in St. Petersburg, Kaspar Friedrich Wolff. By carefully observing the development of chick embryos, Wolff demonstrated that the embryonic parts develop from tissues that have no counterpart in the adult organism. The heart and blood vessels (which, according to preformationism, had to be present from the beginning in order to ensure embryonic growth) could be seen to develop anew in each embryo. Similarly, the intestinal tube was seen to arise by the folding of an originally flat tissue. This latter observation was explicitly detailed by Wolff, who proclaimed (1767), "When the formation of the intestine in this manner has been duly weighed, almost no doubt can remain, I believe, of the truth of epigenesis." However, to create an organism anew each generation, Wolff had to postulate an unknown force, the *vis essentialis* ("essential force"), which, acting like gravity or magnetism, would organize embryonic development.

Preformationism best explained the continuity between generations, whereas epigenesis best explained variation and the direct observations of organ formation. A reconciliation of sorts was attempted by the German philosopher Immanuel Kant (1724–1804) and his colleague, biologist Johann Friedrich Blumenbach (1752–1840). Attempting to construct a scientific theory of racial descent, Blumenbach postulated a mechanical, goal-directed force called the *Bildungstrieb* ("development force"). Such a force, he said, was not theoretical but could be shown to exist by experimentation. A *Hydra*, when cut, regenerates its amputated parts from the rearrangement of existing elements. Some purposive organizing force could be observed in operation, and this force was a property of the organism itself. This *Bildungstrieb* was thought to be inherited through the germ cells. Thus, development could proceed epigenetically through a predetermined force inherent in the matter of the embryo (Cassirer, 1950; Lenoir, 1980). Moreover, such a force was believed to be susceptible to change; the left-handed variant of snail coiling was used as an example of such modifications in the organizing force. In this hypothesis, wherein epigenetic development is directed by preformed instructions, we are not far from the view held by some modern biologists that "the complete description of the organism is already written in the egg" (Brenner, 1979). However, until the rediscovery of Mendel's work at the beginning of the twentieth century, there was no consistent genetic theory in which to place such ideas of inherited variation, and each scientist was free to speculate on the mechanisms by which developmental patterns are inherited.

The French teratologists

The attempts to find a hypothesis that would explain species constancy and epigenetic development led to the creation of modern embryology. The searches for such a hypothesis were undertaken in two different intellectual traditions. One, centered in France, stressed the anatomical aspects of teratogenesis. It sought to discover the mechanisms by which embryological errors cause some infants to be born with developmental abnormalities. The second search for such a hypothesis was centered in Germany and focused on the physiology of developmental processes. Both research traditions began manipulating embryos to see how the developing organism would respond to these perturbations (Churchill, 1973; Fischer and Smith, 1984).

The French teratological experiments began in the 1820s with the studies of Etienne Geoffrey Saint-Hilaire and his son, Isadore. These investigators attempted to show that anomalous births were the products of disrupted fetal development rather than preformed aberrations. They sought to artificially produce developmental anomalies by altering the incubation conditions of chick egg development. Though largely unsuccessful in these attempts (their crude techniques either allowed normal development to continue or killed the embryos), they set the stage for Dareste's more refined analysis in 1877. Dareste performed thousands of experiments and traced developmental anomalies in chicks back to early stages of their development.

But the chick embryo was a poor choice of organism for studying the earliest stages of embryogenesis. If one wanted to examine whether perturbations at the earliest stages of development affected adult structures, one would have to use another organism. In 1886, a French medical student, Laurent Chabry, began studying teratogenesis in the more readily accessible tunicate embryo. This was a fortunate choice, because these embryos develop rapidly into larvae with relatively few cell types. Chabry set out to produce specific malformations by lancing specific blastomeres of the cleaving tunicate embryo. He discovered that each blastomere was responsible for producing a particular set of larval tissues. In the absence of those cells, the larva lacked just those structures normally formed by those cells. Moreover, he observed that when particular cells were isolated from the rest of the embryo, they formed their characteristic structure apart from the context of the other cells. Thus, each of the tunicate cells appeared to be developing autonomously.* As we discussed earlier, this ability of each cell to develop independently from the other embryonic cells is often referred to as *autonomous* or *mosaic development*, because the embryo seems to be a mosaic of self-differentiating parts.

*This was not the answer Chabry expected, or the one he had hoped to find. In nineteenth century France, conservatives favored preformationist views, which were interpreted to support hereditary inequalities between members of a community. What you were was determined by your lineage. Liberals, especially Socialists, favored epigenetic views, which were interpreted to indicate that everyone started off with an equal hereditary endowment, and no one had a "right" to a higher position than any other person. Chabry, a Socialist who hated the inherited rights of the aristocrats, took pains not to extrapolate his data to anything beyond tunicate embryos (see Fischer, 1991).

FIGURE 14.1
Segregation of cytoplasmic determinants at the time of fertilization. (A) Fate map of the cytoplasmic regions of the tunicate *Phallusia mammillata* shortly after the cytoplasmic movements of fertilization have ended. Anterior is to the left, posterior to the right. (B) The organs of the tunicate larva. (A after Nishida, 1987.)

(A) ANIMAL VIEW

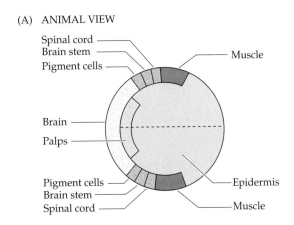

LATERAL VIEW

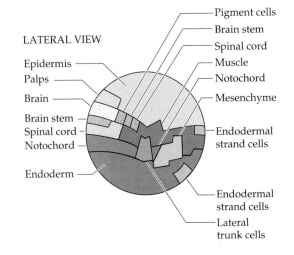

VEGETAL VIEW

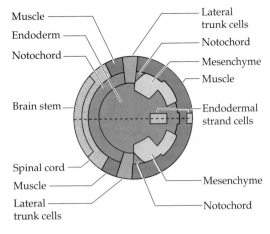

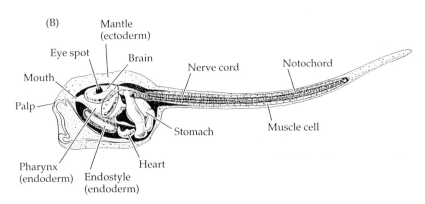

Autonomous specification in tunicate embryos

More recent studies have shown that the tunicate embryo does indeed approximate a "mosaic of self-differentiated parts" constructed from information stored in the oocyte cytoplasm. As the embryo divides, different cells incorporate different regions of cytoplasm. These different cytoplasmic regions are thought to contain **morphogenetic determinants** that control the commitment of the cell to a particular cell type. Studies of tunicate cell determination have been helped considerably by the eggs of certain species that segregate their cytoplasm into a series of colored regions immediately after fertilization (Plate 11).

In 1905, E. G. Conklin described how these colored plasms become apportioned into various blastomeres. The first cleavage separates the egg into right and left mirror images. From then on, each cell division on one side parallels a similar division on the other. By following the fate of *each* blastomere of the tunicate *Styela partita*, Conklin came to the astonishing conclusion that each of the colored regions of cytoplasm delineates a specific embryonic fate (Figure 14.1). The yellow cytoplasmic crescent gives rise to the muscle cells; the gray equatorial crescent produces the notochord and the neural tube; the clear animal cytoplasm becomes the larval epidermis; and the yolky gray vegetal region gives rise to the larval gut.

Reverberi and Minganti (1946) analyzed tunicate determination in a series of isolation experiments, and they, too, observed the self-differentiation of each isolated blastomere and the remaining embryo. The results of one of these experiments is shown in Figure 14.2. When the 8-cell embryo is separated into its four doublets (the right and left sides being

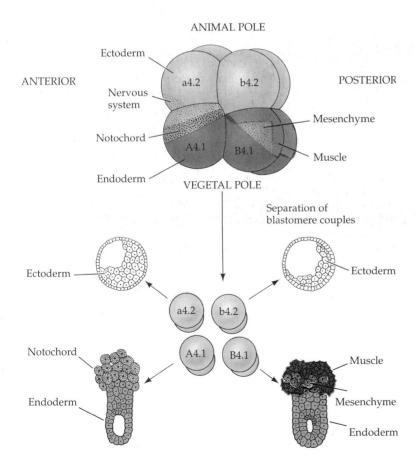

FIGURE 14.2
Mosaic determination in tunicates. When the four blastomere couples of the 8-cell embryo are dissociated, they develop as indicated, each forming separate structures. (After Reverberi and Minganti, 1946.)

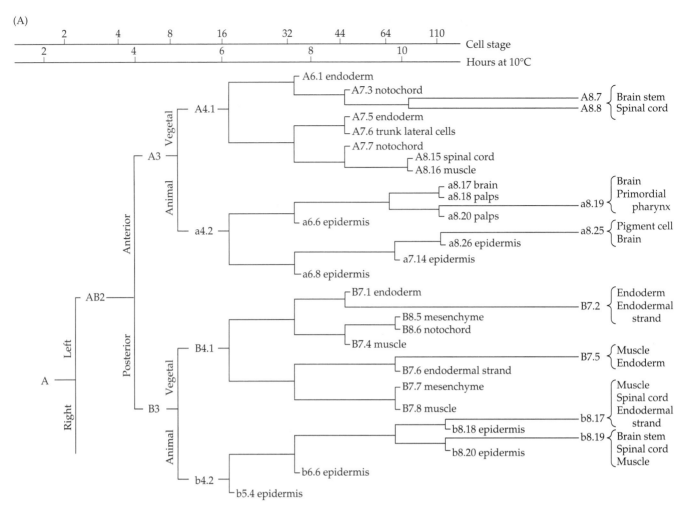

FIGURE 14.3
Determinative lineage of the tunicate blastomeres. (A) Lineage fate chart of embryonic development in the tunicate *H. roretzi*. Because both right and left halves develop identically, only one half of the embryo is represented here. (B) Muscle cell lineages. (A after Nishida, 1987; B after Nishida, 1992a.)

equivalent), mosaic determination is the rule. The animal posterior pair of blastomeres gives rise to the ectoderm; the vegetal posterior pair produces endoderm, mesenchyme, and muscle tissue, just as expected from the fate map. Neural development, however, is an exception. The nerve-producing cells are generated from both the animal and the vegetal anterior quadrants, yet neither produces them alone. When these anterior pairs are reunited, though, the brain and palp tissues arise. Even in an embryo as strictly determined as that of the tunicates, some inductive interactions take place between blastomeres. In fact, Ortolani (1959) has shown that this region of ectoderm is not determined for "neuralness" until the 64-cell stage, right before gastrulation. Thus, although most tissues are determined immediately by segregation of the egg cytoplasm, certain tissues in these embryos have a progressive determination by cell–cell interaction.

In 1973, J. R. Whittaker provided dramatic biochemical confirmation of the cytoplasmic segregation of tissue determinants. Whittaker (1973) stained cells for the presence or absence of the enzyme acetylcholinesterase. This enzyme is found only in the larval muscle tissue and is involved in enabling muscles to respond to repeated nerve impulses. From the cell

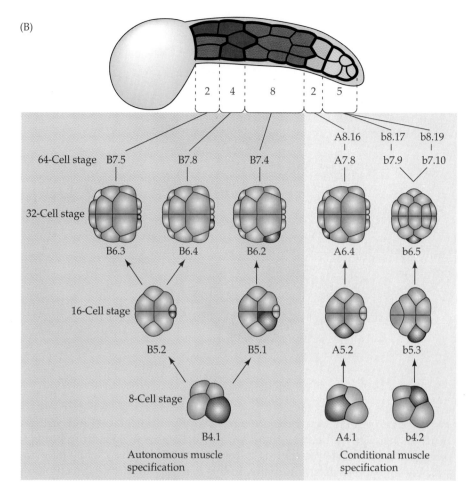

(B)

64-Cell stage B7.5   B7.8   B7.4   A8.16   b8.17   b8.19

                                                          A7.8   b7.9   b7.10

32-Cell stage

              B6.3   B6.4   B6.2   A6.4   b6.5

16-Cell stage

              B5.2   B5.1   A5.2   b5.3

8-Cell stage

              B4.1   A4.1   b4.2

        Autonomous muscle   Conditional muscle
        specification   specification

lineage studies of Conklin and others (Figures 14.2 and 14.3), it was known that only one pair of blastomeres (posterior vegetal; B4.1) in the 8-cell embryo is capable of producing tail muscle tissue. When Whittaker removed these two cells and placed them in isolation, they produced muscle tissue that stained positively for the presence of acetylcholinesterase (Figure 14.4). No other cell was able to form muscles when separated. (The b4.2 and A4.1 blastomeres can generate muscles, but through interaction with other cells, so they are not autonomous; Meedel et al., 1987; Nishida, 1987, 1990.)

FIGURE 14.4
Acetylcholinesterase development in the progeny of the muscle lineage blastomeres (B4.1) isolated at the 8-cell stage. (A) Diagram of the isolation procedure. (B) Localization of acetylcholinesterase in the tail muscles of the intact tunicate larva. Enzyme presence is demonstrated in the progeny of the B4.1 blastomere pair (C), but not in the remaining 6/8 of the embryo (D) when incubated for the length of time it normally takes to form a larva. (From Whittaker et al., 1977, courtesy of J. R. Whittaker.)

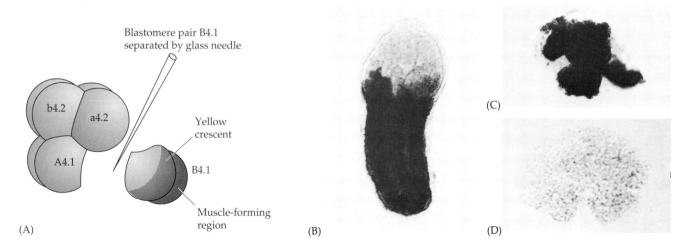

(A)

b4.2

a4.2

A4.1

Blastomere pair B4.1
separated by glass needle

Yellow
crescent

B4.1

Muscle-forming
region

(B)

(C)

(D)

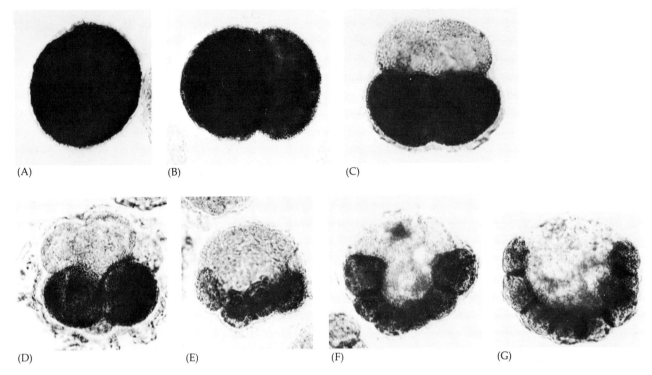

(A)   (B)   (C)

(D)   (E)   (F)   (G)

FIGURE 14.5
Acetylcholinesterase localization in tunicate embryos whose cytokinesis has been arrested at various times by cytochalasin B. (A) 1-cell, (B) 2-cell, (C) 4-cell, (D) 8-cell, (E) 16-cell, (F) 32-cell, and (G) 64-cell stage. (From Whittaker, 1973a, courtesy of J. R. Whittaker.)

Whittaker then took tunicate embryos of various stages and arrested further cleavage by treating them with cytochalasin B. This drug binds to microfilaments, thus preventing cytokinesis (cell division) while allowing nuclear division to occur normally. In this manner, all further development occurs within the population of cells present at the time when cytochalasin was added. After the embryos had their further cleavages blocked, they were allowed to develop and then were stained for the presence of acetylcholinesterase. A comparison of these results (Figure 14.5) with the cell lineage chart (Figure 14.3) shows a striking concordance. The ability to produce muscle cells was originally present in both of the 2-cell blastomeres. However, by the 4-cell stage, the ability to produce acetylcholinesterase-synthesizing cells is limited to the vegetal blastomeres. In the 8-cell embryo, only the two posterior vegetal blastomeres can give rise to such cells. These are the cells known to form most of the tail muscles of the tunicate larva (Meedel et al., 1987; Nishida, 1987).

The cells whose descendants were shown to synthesize acetylcholinesterase are precisely those cells destined to produce muscle autonomously. The production of that enzyme in the isolated cells occurs at exactly the time when it appears in normal embryos (Satoh, 1979). When cytoplasm is transferred from the B4.1 (muscle-forming) blastomere to the b4.2 (ectoderm-forming) blastomere of 8-cell tunicate embryos, the ectoderm-forming blastomere generates muscle cells as well as its normal ectodermal progeny (Whittaker, 1982). Moreover, cytoplasm from the yellow plasm area of the fertilized egg is also able to cause the b4.2 blastomere to express muscle-specific proteins (Figure 14.6; Nishida, 1992a). Conversely, Tung and colleagues (1977) have shown that when larval *nuclei* are transplanted into enucleated tunicate egg fragments, the newly formed cells show the structures typical of those cells providing the cytoplasm, not of those cells providing the nuclei. We can conclude, then, that certain determinants that exist in the cytoplasm cause the formation of certain tissues. These morphogenetic determinants appear to work by selectively activating (or inactivating) specific genes. The determination of the blas-

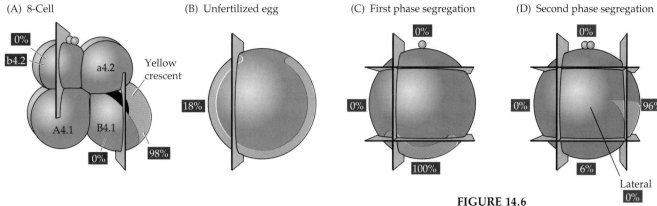

(A) 8-Cell

0%
b4.2
a4.2
Yellow crescent
A4.1
B4.1
0%
98%

(B) Unfertilized egg

18%

(C) First phase segregation

0%
0%
100%

(D) Second phase segregation

0%
0%
96%
6%
Lateral
0%

tomeres and the activation of certain genes are controlled by the spatial localization of morphogenetic determinants within the egg cytoplasm.

The nature of tunicate morphogenetic determinants

The autonomous determination of cellular phenotypes appears to be accomplished by two types of mechanisms. The first mechanism involves the localization and partitioning of morphogenetic factors responsible for activating the *transcription* of certain genes in those cells that have incorporated it. This type of factor would act as the MyoD family of transcription factors does when it generates muscle cells. Once partitioned into those cells, these morphogenetic factors would enter the nuclei and activate the genes that create the particular cellular phenotype.

Transcription factors of this type appear to activate the acetylcholinesterase gene. When tunicate embryos are grown in the presence of the transcription inhibitor actinomycin D, no acetylcholinesterase activity emerges, suggesting that the genes for acetylcholinesterase production fail to be activated. Moreover, when RNA is purified from *Ciona* embryos of various stages and injected into frog oocytes, the oocytes translate the newly inserted messages. By this technique, Meedel and Whittaker (1983, 1984) found that there is no detectable mRNA for acetylcholinesterase until the gastrula stage, when it appears only in the descendants of the B4.1 blastomeres. Therefore, new RNA synthesis has to occur for the production of this enzyme. Crowther and Whittaker (1984) have demonstrated that most of the cell-specific markers of tunicate development are sensitive to transcription inhibitors, so morphogenetic determinants may act generally to activate specific transcription. Despite numerous attempts to find them (see Jeffery, 1985; Nishida, 1987), these tunicate cell transcription factors have not yet been isolated.

The second mechanism of autonomous cell specification involves the localization and partitioning of the messages and proteins that are characteristic of a particular cell type. This situation is seen when one looks at the synthesis of intestinal alkaline phosphatase in tunicate embryos. This enzyme is synthesized by the larval endoderm and becomes segregated into those cells that form the gut (Figure 14.7; Whittaker, 1977; Nishida, 1993). Actinomycin D has no effect on the synthesis of this enzyme, although inhibitors of translation prevent its expression. Therefore, mRNA for intestinal alkaline phosphatase must already exist in the oocyte cytoplasm. Somehow, the mRNA is sequestered into the appropriate blastomere, and recent studies suggest that the cytoskeleton may be responsible (see Sidelights & Speculations). The alkaline phosphatase enzyme is not thought to cause the creation of the endodermal cells. Rather, it is a component of the endoderm cells that is placed in the right area at the right time.

FIGURE 14.6
Location of muscle-forming cytoplasm during early ascidian development. Regions of cytoplasm were transferred into the b4.2 (presumptive epidermis) blastomere and screened for muscle-specific proteins made from the b4.2-derived cells. The colored region represents the "yellow crescent" material thought to contain muscle-forming determinants. Percentages indicate the fraction of the specimen showing muscle gene expression. (A) 8-cell embryo. (B) Unfertilized egg. (C) Fertilized egg in the first phase of cytoplasmic movements. (D) Fertilized egg in the second phase of cytoplasmic movements. (After Nishida, 1992b.)

FIGURE 14.7
Alkaline phosphatase localization in cytochalasin B-arrested *Ciona* embryos. (A) 2-cell, (B) 4-cell, (C) 8-cell, and (D) 16-cell stage. Staining reactions 14–16 hours after fertilization reveal the localization of alkaline phosphatase determinants. (From Whittaker, 1977, courtesy of J. R. Whittaker.)

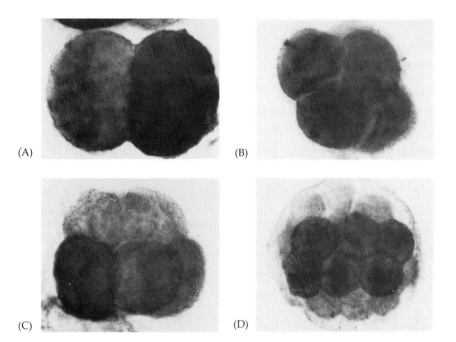

(A)　　　　　(B)

(C)　　　　　(D)

SIDELIGHTS & SPECULATIONS

Intracellular localization and movements of morphogenetic determinants

Whereas the various inclusions of the fertilized egg (including yolk, pigment, and soluble proteins) can easily be displaced by centrifugation, such displacement does not usually affect embryogenesis (reviewed in Morgan, 1927). It appears, then, that either the determinants are too small to be moved by centrifugation or they are somehow anchored within the egg. The lack of diffusion manifest in the cytoplasmic localization of these determinants weighs against the first possibility. Most likely, the determinants are attached to insoluble material, probably the cytoskeletal framework of the cell. This infrastructure of filaments and tubules is particularly prominent in the oocyte cortex, but it extends throughout the cell. Cervera et al. (1981) have reported that most cellular RNA in cultured cells is asso-

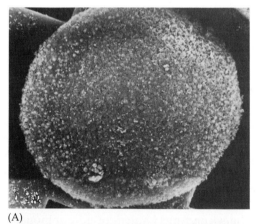

(A)

FIGURE 14.8
Scanning electron micrographs of tunicate eggs undergoing the segregation of morphogenetic determinants. The eggs have been extracted with Triton X-100 detergent to solubilize the membrane and to allow the observation of cytoskeletal components. (A) Unfertilized Styela *egg; yellow pigment granules and the cytoskeleton can be seen around the entire surface. (B) A newly fertilized* Boltenia *zygote segregating its yellow cytoplasm. The region containing the yellow pigment granules is elevated and consists of a plasma membrane lamina and a deeper network of filaments. The presumed direction of the pigment migration is shown by the arrow. (From Jeffery and Meier, 1983, courtesy of S. Meier.)*

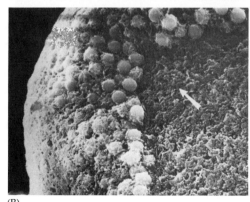

(B)

ciated with the cytoskeletal framework. Thus, the cytoskeleton might be a means of specifically localizing cytoplasmic determinants.

The cytoskeleton can be isolated by extracting cells with nonionic detergents such as Triton X-100. The detergent solubilizes lipids, tRNA, and monoribosomes. The remaining cytoskeleton contains microtubules, microfilaments, intermediate filaments, and roughly 200 proteins, including one that is capable of binding the 5' cap of mRNAs (Zumbe et al., 1982; Moon et al., 1983). In the tunicates *Styela* and *Boltenia*, the muscle-forming yellow crescent is characterized by an actin-containing cytoskeletal domain. This domain is originally coextensive with the unfertilized egg. After fertilization, however, the actin microfilaments contract and become segregated into those blastomeres fated to form muscle cells, taking with them the yellow pigment granules and a set of mRNAs (Jeffery and Meier, 1983; Jeffery, 1984). Figure 14.8 shows that the cytoskeletal framework contains the yellow pigment gran-

ules and that these granules are given their intracellular localization by the movements of the oocyte cytoplasm during fertilization. The proteins of this yellow crescent region differ significantly from the proteins of the rest of the egg, whereas the mRNAs in this region are not seen to be specific to the yellow crescent (Jeffery, 1985). One of these cytoskeletal proteins is a 58-kDa peptide that is localized in the periphery of the unfertilized egg, segregates to the yellow crescent during fertilization, and enters the tail muscle cells during subsequent development (Swalla et al., 1991). Thus, this cytoskeletal protein appears to follow the same route as the myoplasm during development. Moreover, in those species of ascidians that have direct development and lack a tailed tadpole stage, this protein is absent. The 58-kDa cytoskeletal protein is likely to be involved in myogenic determination in tunicates, perhaps as the protein that binds and moves the muscle cell determinant of ascidians.

Cytoplasmic localization in mollusc embryos

The mosaic type of differentiation is widespread throughout the animal kingdom, especially in protostomal organisms such as ctenophores (comb jellies), annelids, nematodes, and molluscs, all of which initiate gastrulation at the future anterior end after only a few cell divisions. Molluscs provide some of the most impressive examples of "mosaic" development and of the phenomenon of **cytoplasmic localization**, wherein the morphogenetic determinants are found in a specific region of the oocyte. Moreover, these cytoplasmic factors become actively moved so that a blastomere having these factors can restrict their transmission to only one of its two daughter cells. The fate of the two daughter cells is thus changed by which one of them gets the morphogenetic determinant.

E. B. Wilson, the outstanding American embryologist at the turn of the century, isolated early blastomeres from embryos of the mollusc *Patella coerulea* and compared their development with that of the same cells left within other *Patella* embryos. Figure 14.9 shows one set of results published by Wilson in 1904. Not only did the isolated blastomeres follow their normal developmental fates (in this case, to produce the ciliated trochoblast cells), but they also completed the normal number of cell divisions at precisely the same time as those cells remaining within the embryo. Their cleavages were in the correct orientation, and the derived cells became ciliated at the appropriate time. Wilson concluded from these experiments that these cells possess within themselves all the factors that determine the form and rhythm of cleavage and the characteristic and complex differentiation that they undergo, wholly independent of their relation to the remainder of the embryo.

The polar lobe

In his next experiment, Wilson was able to demonstrate that such development is predicated on the segregation of specific morphogenetic determinants into specific blastomeres. Certain spirally cleaving embryos

FIGURE 14.9

(A–C) Differentiation of trochoblast cells in the normal embryo of the mollusc *Patella*. (A) 16-cell stage seen from the side; the presumptive trochoblast cells are shaded. (B) 48-cell stage. (C) Ciliated larval stage, seen from the animal pole. Cilia are seen on trochoblast cells. (D–G) Differentiation of *Patella* trochoblast cells isolated and culture in vitro. (D) Isolated trochoblast cell. (E,F) Results of first and second divisions in culture. (G) Ciliated product of F. Even in isolated culture, cells become ciliated at the correct time. (After Wilson, 1904.)

Normal development of *Patella*

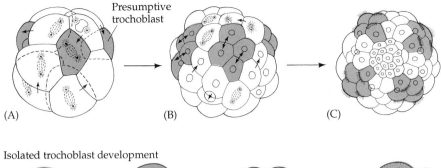

Isolated trochoblast development

(mostly in the mollusc and annelid phyla) extrude a bulb of cytoplasm immediately before first cleavage (Figure 14.10). This protrusion is called the **polar lobe**. In certain species of snails, the region uniting the polar lobe to the rest of the egg becomes a fine tube. The first cleavage splits the zygote asymmetrically, so the polar lobe is connected only to the CD blastomere. In several species, nearly one-third of the total cytoplasmic volume is present in these anucleate lobes, giving them the appearance of another cell. This three-lobed structure is often referred to as the trefoil-stage embryo (Figure 14.11). The CD blastomere then absorbs the polar lobe material, but extrudes it again prior to second cleavage (Figure 14.10). After this division, the polar lobe is attached only to the D blastomere, which absorbs its material. Thereafter, no polar lobe is formed.

Wilson showed that if one removes the polar lobe at the trefoil stage, the remaining cells divide normally. However, instead of producing a normal trochophore (snail) larva, they produce an incomplete larva, wholly lacking its mesodermal organs—muscles, mouth, shell gland, and

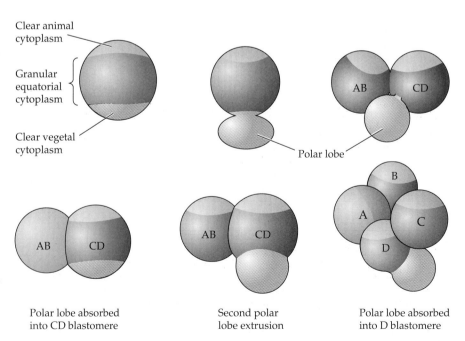

FIGURE 14.10

Cleavage in the mollusc *Dentalium*. Extrusion and reincorporation of the polar lobe occur twice. (After Wilson, 1904.)

foot.* Moreover, Wilson demonstrated that the same type of abnormal embryo can be produced by removing the D blastomere from the 4-cell embryo. Wilson concluded that the polar lobe cytoplasm contains the mesodermal determinants and that these determinants give the D blastomere its mesoderm-forming capacity. Wilson also showed that the localization of the mesodermal determinants is established shortly after fertilization, thereby demonstrating that a specific cytoplasmic region of the egg, destined for inclusion into the D blastomere, contains whatever factors are necessary for the special cleavage rhythms of the D blastomere and for the differentiation of the mesoderm.

The morphogenetic determinants sequestered within the polar lobe are probably located in the cytoskeleton or cortex and not in the diffusible cytoplasm of the embryo. Evidence for this came from the studies of A. C. Clement (1968). When the animal hemisphere is separated from the vegetal hemisphere in the snail *Ilyanassa obsoleta*, the animal hemisphere forms ectodermal organs that resemble those embryos formed from lobeless eggs. Clement took those embryos that had begun resorbing their second polar lobe and placed them into gelatin slabs. He then placed these embedded embryos in centrifuge tubes and centrifuged the fluid, yolky, cytoplasm from the vegetal part of the cell into the animal hemisphere. By centrifuging these embryos in a second, viscous, medium, he caused the separation of the animal and vegetal hemispheres. The animal halves from such centrifuged embryos did not develop any more mesodermal and endodermal structures than those of uncentrifuged eggs. Thus, the determinants of the polar lobe were not transferred to the animal hemisphere in the fluid contents of the vegetal hemisphere. Van den Biggelaar obtained similar results when he removed the cytoplasm from the polar lobe with a micropipette. Cytoplasm from other regions of the cell flowed into the polar lobe, replacing the portion that he had removed. The subsequent development of these embryos was normal. In addition, when he added the soluble polar lobe cytoplasm to the B blastomere, duplications of structures were not seen (Verdonk and Cather, 1983). Therefore, the diffusible part of the cytoplasm does not contain these morphogenetic determinants. They probably reside in the nonfluid cortical cytoplasm or on the cytoskeleton.

Clement also analyzed the further development of the D blastomere in order to observe the further appropriation of these determinants. The development of the D blastomere is illustrated in Figure 14.12. This macromere, having received the contents of the polar lobe, is larger than the other three. When one removes the D blastomere or its first or second macromere derivatives (1D or 2D), one obtains an incomplete larva, lacking heart, intestine, velum (the ciliated border of the larva), shell gland, eyes, and foot. When one removes the 3D blastomere (*after* the division of the 2D cell to form the 3d blastomere), one obtains an almost-normal embryo, having eyes, feet, velum, and some shell gland, but no heart or intestine (Figure 14.13). Therefore, some of the morphogenetic determinants originally present in the D blastomere were apportioned to the 3d cell. After the 4d cell is given off (by the division of the 3D blastomere), removal of the D derivative (the 4D cell) produces no qualitative difference in development. In fact, all the essential determinants for heart and intestine formation are now in the 4d blastomere, and removal of *that* cell results in a heartless and gutless larva (Clement, 1986). The 4d blastomere is

---

*The shell gland is an ectodermal organ formed through induction by mesodermal cells. Without the mesoderm, no cells are present to induce the competent ectoderm. Again, we see some limited induction within a mosaic embryo.

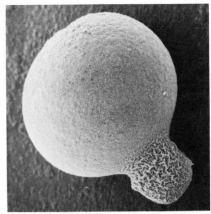

(A)

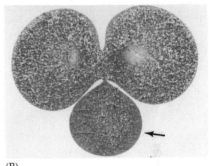

(B)

FIGURE 14.11
Polar lobes of molluscs. (A) Scanning electron micrograph of the extending polar lobe in the uncleaved egg of *Buccinum undatum*. The surface ridges are confined to the polar lobe region. (B) Section through the first cleavage, or trefoil-stage, embryo of *Dentalium*. The arrow points to the large polar lobe. (Courtesy of M. R. Dohmen.)

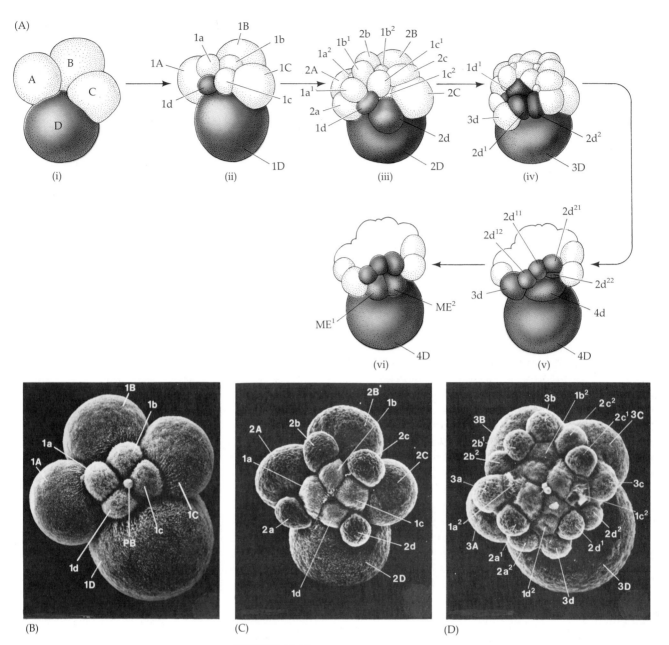

FIGURE 14.12

Development of the D blastomere. (A) Schematic diagrams of the lineage of the D blastomere in *Ilyanassa* embryos. (i) 4-cell embryo. (ii) 1D and 1d blastomeres at 8-cell stage. (iii) 16-cell stage containing 2D and 2d blastomeres (derived from 1D). The D-derived cells (color) often divide later than others. (iv) Division of the 2D macromere to generate 3D and 3d cells while the 2d cell divides into 2d1 and 2d2. (v) 64-cell stage. The 3D blastomere produces the 4D and 4d cells. (vi) The 4d blastomere divides symmetrically to produce the two mesentoblasts ME1 and ME2. (B) 8-cell embryo. (C) 12-cell embryo (1a–1d have not divided yet). (D) 32-cell embryo. The small PB cell is the polar body and is not part of the embryo. (A after Clement, 1962; photographs from Craig and Morrill, 1986, courtesy of the authors.)

responsible for forming (at its next division) the two mesentoblasts, the cells that give rise to both the mesodermal (heart) and endodermal (intestine) organs.

The material in the polar lobe is also responsible for organizing the dorsal–ventral (back–belly) polarity of the embryo. When polar lobe material is allowed to pass into the AB blastomere as well as into the CD cell, twin larvae are formed that are joined at their ventral surfaces (Figure 14.14; Guerrier et al., 1978; Henry and Martindale, 1987).

Thus, experiments have demonstrated that the nondiffusible polar lobe cytoplasm is extremely important for normal mollusc development because:

1. It contains the determinants for the proper rhythm and cleavage orientation of the D blastomere.
2. It contains certain determinants (those entering the 4d blastomere and hence leading to the mesentoblasts) for mesodermal and intestinal differentiation.
3. It is responsible for permitting the inductive interactions (through the material entering the 3d blastomere) leading to the formation of the shell gland and eye.
4. It contains determinants needed for specifying the dorsal-ventral axis of the embryo.

Although the polar lobe is clearly important for normal snail development, we still do not know the mechanisms of its effects. There appear to be no major differences in mRNA or protein synthesis between lobed and lobeless embryos (Brandhorst and Newrock, 1981; Collier, 1983, 1984). One possible clue has been provided by Atkinson (1987), who has observed differentiated cells of the velum, digestive system, and shell gland within the lobeless embryo. Lobeless embryos can produce these cells, but they appear unable to organize them into functional tissues and organs. Tissues of the digestive tract can be found, but they are not connected; myocytes are scattered around the lobeless larva but are not organized into a functional muscle tissue. Thus, the developmental functions of the polar lobe may be very complex.

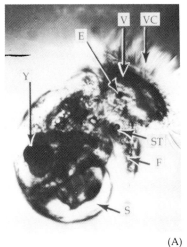

(A)

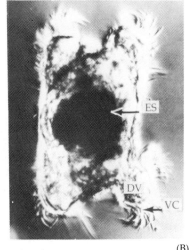

(B)

FIGURE 14.13
Importance of polar lobe in the development of *Ilyanassa*. (A) Normal veliger larva. (B) Aberrant larva, typical of those produced when the polar lobe of the D blastomere is removed. E, eye; F, foot; S, shell; ST, statocyst (balancing organ); V, velum; VC, velar cilia; Y, residual yolk; ES, everted stomodeum; DV, disorganized velum. (From Newrock and Raff, 1975, courtesy of K. Newrock.)

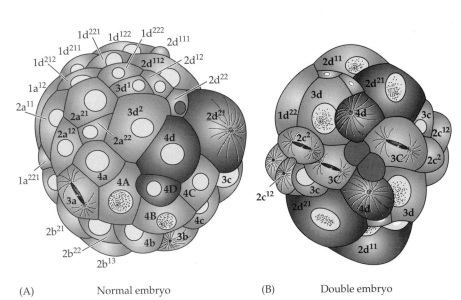

(A)   Normal embryo   (B)   Double embryo

FIGURE 14.14
Formation of twin embryos by the suppression of polar lobe formation in *Dentalium*. (A) Normal embryo at sixth-cleavage stage. (B) Twin embryo formed when low concentrations of cytochalasin inhibit polar lobe formation, and the polar lobe material is distributed to both the AB and CD blastomeres. (After Guerrier et al., 1978.)

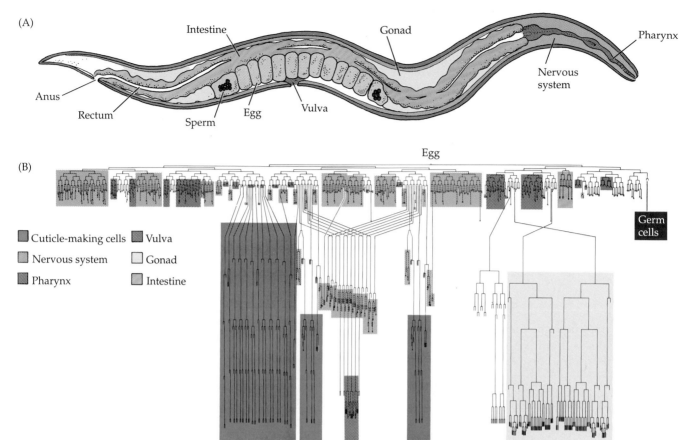

FIGURE 14.15

Caenorhabditis elegans. (A) Side view of adult hermaphrodite. Early in its development, sperm are formed. These sperm are stored during later stages so that a mature egg passes through the sperm on its way to the vulva. In this manner, the hermaphrodite unites its own sperm with its own eggs. (B) Entire cell lineage chart for *C. elegans*. Each vertical line represents a cell; each horizontal line represents a cell division. (After Pines, 1992, based on Sulston and Horvitz, 1977, and Sulston et al., 1983.)

In figure (A), labels: Intestine, Gonad, Pharynx, Anus, Rectum, Sperm, Egg, Vulva, Nervous system.

In figure (B), labels: Egg, Germ cells. Legend: Cuticle-making cells, Vulva, Nervous system, Gonad, Pharynx, Intestine.

Cell specification in the nematode *Caenorhabditis elegans*

The ability to analyze development requires appropriate organisms. Sea urchins have long been a favorite organism of embryologists because their gametes are readily obtainable in large numbers, their eggs and embryos are transparent, and fertilization and development can occur under laboratory conditions. But sea urchins are difficult to rear in the laboratory for more than one generation, making their genetics difficult to study. Geneticists, on the other hand (at least those working with multicellular eukaryotes), favor *Drosophila*. The rapid life cycle, the readiness to breed, and the polytene chromosomes of the fly larva (which allow gene localization) make this animal superbly suited for hereditary analysis. But *Drosophila* development is very complex and difficult to study. A research program spearheaded by Sidney Brenner (1974) was set up to identify an organism wherein it might be possible to identify each gene involved in development as well as to trace the lineage of every single cell. Such an organism is *Caenorhabditis elegans*, a small (1 millimeter long), free-living soil nematode (Figure 14.15A). It has a rapid (about 16 hours) period of embryogenesis, which it can accomplish in petri dishes, and relatively few cell types. Moreover, its predominant form is hermaphroditic, each individual containing both eggs and sperm. These roundworms can reproduce either by self-fertilization or by cross-fertilization with the infrequently occurring males. The body of a hermaphroditic *C. elegans* contains exactly 959 somatic cells, whose entire lineage has been mapped through its transparent cuticle (Figure 14.15; Sulston and Horvitz, 1977; Kimble and

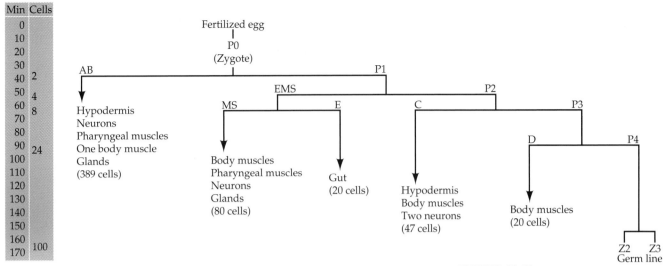

Min	Cells
0	
10	
20	
30	
40	2
50	4
60	8
70	
80	
90	24
100	
110	
120	
130	
140	
150	
160	
170	100

FIGURE 14.16

Abbreviated cell lineage chart for *C. elegans*, emphasizing the germ line precursors (P cells P0–P4) that receive the P granules. The number of cells (in parentheses) refers to those present in the newly hatched larva. Some of these continue to divide to produce the 959 somatic cells of the adult. (From Strome and Wood, 1983, courtesy of W. Wood.)

Hirsh, 1979). Furthermore, unlike vertebrate cell lineages, the cell lineage of *C. elegans* is almost entirely invariant from one individual to the next. There is little room for randomness (Sulston et al., 1983). (This is a consequence of the spatial ordering of cytoplasmic segregation.) *Caenorhabditis* also has a small number of genes for a multicellular organism, about 15,000 (Sulston et al., 1992).

The pattern of *C. elegans* division (Figure 14.16) resembles that of stem cell lines, in that during early cleavage, asymmetric divisions produce one "differentiating" daughter cell (collectively called the **founder cells** and denoted AB, MS, E, C, and D) and another stem cell (the P1–P4 lineage). The localization of cytoplasmic substances into specific blastomeres has been elegantly demonstrated in these asymmetric divisions. Within the egg is the set of **germ line granules**, or **P granules**, that are redistributed in the zygote shortly after fertilization and become restricted to those cells capable of forming gametes. Using fluorescent antibodies to a component of the P granules, Strome and Wood (1983) discovered that during the pronuclear migration in the zygote, the randomly scattered P granules become localized in the posterior end of the zygote (toward the site of sperm entry), so that they only enter the blastomere (P1) formed from the posterior cytoplasm (Figure 14.17 and Plate 10). Following cleavage, the P granules disperse throughout the P1 blastomere until the start of mitosis, when they once again migrate to the posterior end of the cell. Here they become apportioned to the P2 blastomere. Eventually the P granules will reside in the P4 cell, whose progeny become the sperm and eggs of the adult. The localization of the P granules requires microfilaments but can occur in the absence of microtubules. Treating the zygotes with cytochalasin D (a microfilament inhibitor) prevents the segregation of these granules to the posterior of the cell, whereas demecolcine (a colchicine-like microtubule inhibitor) fails to stop this movement (Strome and Wood, 1983). Once within the posterior region of the zygote, P granules remain there even if microfilaments are then disrupted (Hill and Strome, 1987, 1990).

The mechanisms for the movement and anchoring of these cytoplasmic granules remain unknown, but they are probably regulated by the *par* **genes** that control cytoplasmic partitioning during the first cleavages of *C. elegans*. Mutations in five *par* genes are expressed as maternal effect mutants where microfilament distributions are aberrant, and the P granules

FIGURE 14.17

Asymmetric localization of P granules during fertilization and first cleavage. Left-hand figures are stained to show DNA; right-hand figures are the same cells stained with fluorescent antibodies to P-granule protein. (A) A zygote prior to pronuclear migration shows random dispersal of P granules. (B) As pronuclei come together, granules become localized to the posterior periphery of the zygote. (C) Two-cell embryo in which P1 is entering mitotic prophase; the P granules are now positioned at the posterior periphery to be transmitted to the P2 cell. (From Strome and Wood, 1983, courtesy of S. Strome.)

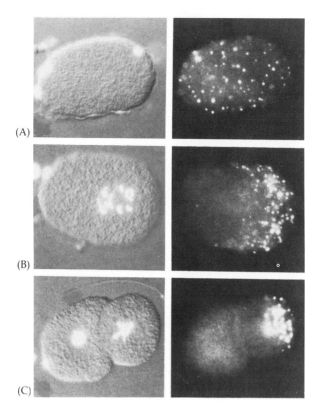

are distributed abnormally (Kemphues et al., 1988; Kirby et al., 1990). Early cleavages in these mutant embryos are symmetric and synchronous, and the P granules are found in several blastomeres (Figure 14.18). The phenotypes of the *par-2* and *par-3* mutants resemble what occurs when wild-type embryos are exposed to a microfilament inhibitor for a 10-minute period during the first cell cycle (Hill and Strome, 1990), and it is likely that these genes encode proteins that regulate the cytoskeletal apparatus that anchors or translocates the P granules in the early embryo.

FIGURE 14.18

Aberrant actin and P-granule distribution in the *par-3* mutant. Distribution of cytoplasmic actin in the wild-type embryo (A) and in the embryo of a *par-3* deficient female (B). Distribution of P granules is asymmetric in the wild-type embryo (C), but is symmetric in the *par-3* deficient embryo (D). (E) In the 4-cell mutant embryo, P granules can be seen in all four cells. (From Kirby, 1992, courtesy of C. M. Kirby.)

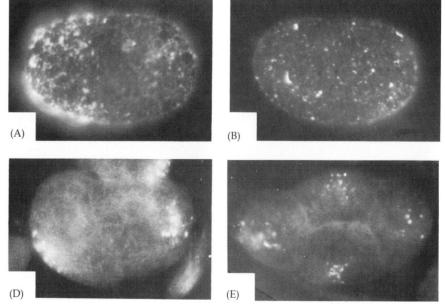

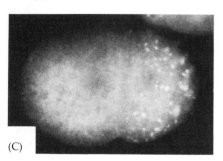

Maternal control of blastomere identity: The genetic control of the pharyngeal progenitor cells of *C. elegans*

The determination for much of the *C. elegans* embryo is autonomous, the cell fates being determined by internal, cytoplasmic factors rather than by interactions with neighboring cells. It is thought that the protein factors might determine cell fate by entering the nuclei of the particular blastomeres and activating or repressing specific fate-determining genes. Have any transcription factors been found in autonomously determined cell lineages? While the P granules of *C. elegans* are localized in a way consistent with a role as morphogenetic determinant, they do not enter the nucleus, and their role in development is still unknown. However, the **skn-1 protein** of *C. elegans* embryos is a very promising candidate for a transcription factor morphogen.

The skn-1 protein is a maternally specified polypeptide that may control the fate of the MS blastomere, the cell that generates the posterior pharynx. After first cleavage, only the posterior blastomere, P1, has the ability to autonomously produce pharyngeal cells when isolated. After P1 divides, only the EMS is able to generate pharyngeal muscle cells even when isolated from the other cells of the body (Priess and Thomson, 1987). Similarly, when the EMS cell divides, only one of its progeny, MS, has the intrinsic ability to generate pharyngeal tissue. This suggests that pharyngeal cell fate may be determined autonomously by maternal factors residing in the cytoplasm that get parceled to these particular cells. Bowerman and his co-workers (1992) searched for maternal effect mutants that lack pharyngeal cells, and they isolated a mutation called *skn-1*. Embryos from homozygous *skn-1* deficient mothers lack both pharyngeal and intestinal derivatives of EMS (Figure 14.19). Instead of making the normal intestine and pharyngeal structures, these embryos seem to make extra hypodermal (skin) tissue where their intestine and pharynx should be. In addition, the progeny of the ABa cell, which is usually induced by the EMS cell to produce the anterior pharynx musculature, stops dividing in the affected embryos. In other words, only those cells that are destined

FIGURE 14.19
Deficiencies of intestine and pharynx in *skn-1* mutants. Embryos derived from wild-type females (A,C) and females homozygous for mutant *skn-1* (B,D) were tested for the presence of pharyngeal muscles (A,B) and gut-specific granules (C,D). The pharyngeal muscle-specific antibody labels the pharynx musculature of those embryos derived from wild-type females but does not bind to any structure in the embryos from *skn-1* mutant females. Similarly the birefringent gut-granules characteristic of embryonic intestines are absent from the embryos derived from the *skn-1* mutant females. (From Bowerman et al., 1992, courtesy of B. Bowerman.)

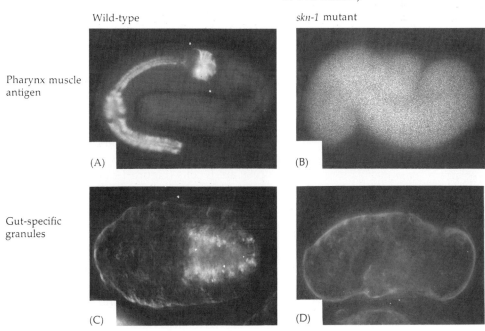

Wild-type   *skn-1* mutant

Pharynx muscle antigen

(A)   (B)

Gut-specific granules

(C)   (D)

Wild-type   mex-1

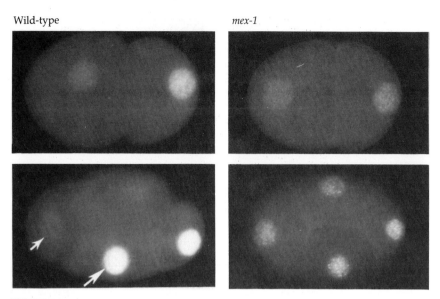

FIGURE 14.20
Cytoplasmic localization of the skn-1 protein. Antibodies to the skn-1 protein find that it is present predominantly in the P1 cell nucleus after first division. After second division, this protein accumulates in both cells derived from P1, but not in those cells derived from AB (compare intensities of the nuclei marked by arrows). In *mex-1* mutants, the skn-1 protein is given equally to all blasto-meres. (From Bowerman et al., 1993, courtesy of B. Bowerman.)

to form pharynx or intestine are affected by this mutation. Moreover, the protein that would be encoded by this message has a DNA-binding-site motif similar to that seen in the bZIP family transcription factors.

Bowerman et al. (1993) showed that the skn-1 protein is made during gametogenesis and is put into the egg cytoplasm. However, after first cleavage, this protein enters the P1 nucleus to a much higher level than it enters the AB nucleus (Figure 14.20). After the second division, both P1 derivatives receive the skn-1 protein in their nuclei. Thus, the skn-1 protein might be a morphogen that activates certain genes in the P1 cell and its descendants.

This hypothesis is supported by studies involving maternal effect mu-tants that fail to form the correct numbers of pharyngeal cells. Mutations of the *pie-1* and *mex-1* genes alter the determination of cells in the eight-cell *C. elegans* embryo such that several additional cells in the embryo become determined as MS cells (Mello et al., 1992). In embryos derived from *pie-1*-deficient females, the P3 and C sister blastomeres are converted into E and MS blastomeres, respectively, while in embryos from *mex-1*-deficient females, all the descendants of the AB cell are redefined as MS cells. Females simultaneously deficient in both of these genes produce embryos in which the anterior six cells are MS cells and the posterior two cells are E cells (Figure 14.21). Thus, in all cases, the wild-type *skn-1* gene was needed for the extra MS cells to form. This links the mex-1 and pie-1 proteins to the placement of skn-1 and is consistent with the hypothesis wherein (1) the *skn-1* gene product would activate all or part of the MS pattern of development, and (2) the wild-type *pie-1* and *mex-1* gene prod-ucts would be required to either localize or regulate the wild-type *skn-1* product. Bowerman and colleagues (1993) have shown that the wild-type *mex-1* gene product is needed to restrict the skn-1 determinant to the P1 cell during first cleavage and that the wild-type pie-1 product is critical for restricting its activity to the EMS cell during second cleavage. Figure 14.21

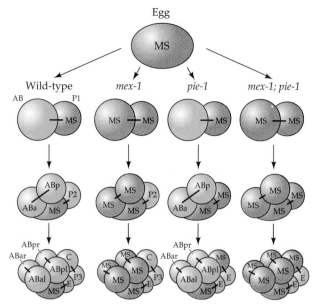

FIGURE 14.21

Schematic model of MS cell determination in wild-type and mutant embryos. The MS determinant (thought to be the product of the *skn-1* gene) is present in an inactive state within the egg. During first division in wild-type embryos, the MS determinant is localized into the posterior (P1) blastomere, and at second division it becomes further localized into the EMS cell. At third division, the EMS cell divides into the MS cell (where the factor is activated) and the E cell. In embryos derived from *mex-1* deficient females, the factor does not segregate at first division, but the segregation from P1 is normal at the next division. In embryos derived from *pie-1* deficient females, the initial segregation of the MS determinant into P1 is normal, but the second asymmetric distribution of the determinant (into the EMS cell) is defective. In the combined mutant, the patterns are superimposed such that all four cells of the 4-cell embryo have the inactive MS determinant. (After Mello et al., 1992.)

shows the pattern of skn-1 protein distribution in the first two cleavages of wild-type and of *mex-1* deficient *C. elegans* embryos. Thus, the skn-1 protein, placed into the egg cytoplasm, may be a transcription factor that activates specific genes in the MS blastomere to determine its fate.

Regulation in *C. elegans*

The development of *C. elegans* is largely autonomous, but regulatory interactions between cells are also important for specifying cell fate. If the EMS blastomere is separated from all the other cells at the four-cell stage shortly after its formation, it will not form gut-specific rhabditin granules. If it is recombined with the P2 blastomere, however, it will form such granules; but it will not do so when combined with ABa, ABp, or both AB derivatives (Figure 14.22; Goldstein, 1992). Cellular interactions are needed for this stage of intestinal determination.

Since the nematode has invariant cell lineages, it also has invariant cell–cell interactions. In the 4-cell embryo, sister blastomeres ABa (anterior) and ABp (posterior) have different developmental fates. ABa makes neurons, hypodermis, and the anterior pharynx cells, while ABp makes only neurons and hypodermal cells. However, if one experimentally reverses their positions, their fates are similarly reversed, and a normal embryo is formed. In other words, ABa and ABp are equivalent cells whose fate is determined by their positions within the embryo (Priess and Thomson,

FIGURE 14.22

Results of isolation and recombination experiments showing that cellular interactions are required for the EMS cell to form intestine lineage determinants. (A) When separated shortly after formation of the EMS cell, the EMS blastomere cannot produce gut-specific granules. If left there for longer periods, it can. (B) If the EMS cells are recombined with either or both derivatives of the AB blastomere, it will not form gut-specific granules. (C) If recombined with the P2 blastomere, the EMS cell will give rise to gut-specific structures. (After Goldstein, 1992.)

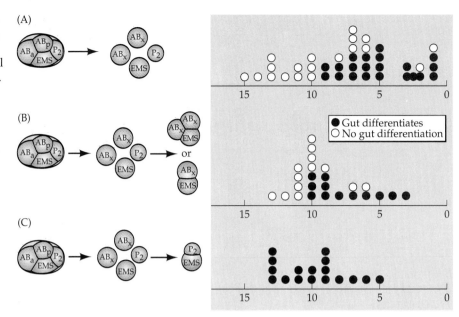

Time of separation
(min before EMS cleavage)

1987). Under normal circumstances, however, the invariant embryonic cleavage pattern dictates that the descendants of ABa, not ABp, make 19 pharyngeal cells. ABa daughters cells differentiate into these pharyngeal muscles because of their interaction with the EMS blastomere or its descendents (which autonomously produce 18 pharyngeal muscle cells). If the P1 cell is removed, the isolated AB cell does not generate muscle cells. By separating the muscle-forming progeny of the AB cell from the adjacent progeny of the P1 cell (the EMS blastomere), Priess and Thomson showed that the cell contact is needed between the ABa lineage and the EMS cell lineages between the 4- and 28-cell stages.

Genetic studies have identified one of the components of this interaction. If the maternal product of the *glp-1* gene is absent from the embryo, the P1 lineages are unaffected, and they generate the normal amount of muscle cells. However, the AB blastomere no longer produces its pharyngeal muscle cells (Preiss et al., 1987). The development of the AB blastomere in the *glp-1*-deficient embryos closely resembles that of the development of the AB blastomere that had been physically separated from P1. First, like the physically isolated AB cell, it produces neurons and hypodermal cells, but not muscles. Second, the critical period of the *glp-1* gene product appears to be between the 4- and 28-cell stages.* This was shown by using temperature-sensitive mutants. Temperature-sensitive mutants are important for developmental studies because they enable investigators to find the time when a gene product is functional. The mutant protein is functional at one temperature but not at another. By shifting the temperature at various stages of development, one can determine the stage when the gene product must be active (Figure 14.23). Thus, as we have seen in tunicates and snails, some regulation and induction still occur in organisms whose early embryonic development is characterized by an invariant determinism.

Cellular interactions have also been seen later in *C. elegans* develop-

*The *glp-1* gene is also active in regulating postembryonic cell–cell interactions. It is used later by the distal tip cell of the gonad to control the number of germ cells entering meiosis (Chapters 18 and 22; Austin and Kimble, 1987).

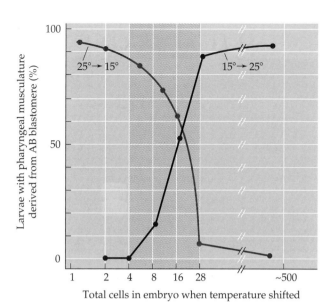

FIGURE 14.23
Temperature shift experiment ⟨to deter⟩mine at what stage the matern⟨al⟩ gene product is active. The glp⟨-1 pro⟩tein in this mutant functions at ⟨15°C,⟩ but does not function at 25°C. By shifting the temperature at different embryonic stages, it was found that the glp-1 protein was needed between the 4- and the 28-cell stage. (After Priess et al., 1987.)

ment. Kimble (1981) has used focused laser beams to selectively kill specific cells whose progeny would have formed the vulva (the passage through which the eggs are laid). One of the cells in this region becomes the anchor cell, which connects the overlying gonad to the vulva. If this cell is destroyed, the hypodermal (skin) cells will not divide to form the vulva. However, as long as the anchor cell is near the hypodermis, the vulva forms, indicating that the anchor cell induces vulva formation (Figure 14.24). Moreover, the six vulval precursor cells influenced by the anchor cell form an **equivalence group**. Here, each cell is originally able to be induced by the anchor cell and can substitute for any of the other cells in the group. However, their fate is determined by their positions relative to the anchor cell. During normal development, the three central cells divide to form the vulva while the three outer cells divide once to form six hypodermal cells. (If the anchor cell is destroyed, all the six cells of the equivalence group divide once to form a dozen hypodermal cells.) If the three central cells are destroyed, the three outer cells, which normally form hypodermal cells, generate vulval cells instead. This type of induction will be discussed more fully in Chapter 18.

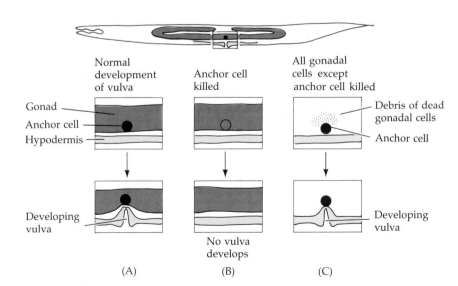

FIGURE 14.24
Induction in *C. elegans*. Schematic representation of experiments demonstrating that the anchor cell is necessary for vulva development. (After Alberts et al., 1983.)

"To be or not to be: That is the phenotype"

Granted we all are poised at life-or-death decisions, but this existential dichotomy is rarely as stark as that seen in the cell lineage of *C. elegans*. During the normal development of *C. elegans*, 131 cells kill themselves. This programmed cell death, or **apoptosis**, is an active event initiated by two genes, *ced-3* and *ced-4*. When these genes are expressed, the cells expressing them die. Loss-of-function mutations in either of these genes allows the survival of cells that normally would have undergone apoptosis. The products of *ced-3* and *ced-4* are thought either to be toxic to the cell or to cause the formation of toxic compounds from other metabolites (Ellis and Horvitz, 1986; Yuan and Horvitz, 1990).

What determines which cells shall live and which shall die? Studies in Robert Horvitz's laboratory (Hengartner et al., 1992) have demonstrated that the **ced-9 gene** inhibits the activities of *ced-3* and *ced-4*. Mutations that inactivate the *ced-9* protein cause numerous cells that normally survive to activate their *ced-3* and *ced-4* genes and die. This leads to the death of the entire embryo. Conversely, gain-of-function mutants of *ced-9* prevent those cells that normally undergo apoptosis from dying. (These are the same cells that survive in the *ced-3* and *ced-4* mutants.) Thus, the normal function of *ced-9* is to prevent those cells that should live from initiating programmed cell death (Figure 14.25). The *ced-9* gene appears to function as a binary switch regulating the choice between life and death. This decision is made independently in every cell of the embryo, and it is possible that every cell of the embryo is poised to die, and those cells that do survive do so because an active *ced-9* gene prevents programmed cell death from occurring.

It is not known how *ced-9* functions to prevent cell death. Nor is it known how it becomes differentially regulated, although certain other genes produce products that turn it on. Similar genes are now being described in mammals, as well. The gene *bcl-2* encodes an intracellular membrane protein that prevents or delays normal apoptosis of human neurons and lymphocytes (Hockenbery et al., 1990; Williams et al., 1990; Allsopp et al., 1993). Most lympho-

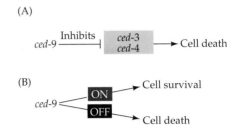

FIGURE 14.25

Model for the function of the ced-9 *gene in* C. elegans. *(A)* ced-9 *acts as a negative regulator of* ced-3 *and* ced-4, *the two genes whose activities cause cell death. (B) The binary switch effected by* ced-9. *When on, the activities of* ced-3 *and* ced-4 *are inhibited and the cell survives. If* ced-9 *is not on, the* ced-3 *and* ced-4 *gene products kill the cell. (From Hengartner et al., 1992.)*

cytes and their precursors die during their maturation, and those that survive have a limited life span. They are prevented from dying by certain growth factors that appear to activate the *bcl-2* gene. Moreover, if *bcl-2* genes on a constitutively active promoter are transferred into cells that are soon going to die, bcl-2 protein is made and the life spans of the cells are significantly increased (Nuñez et al., 1990). This is done naturally by the Epstein-Barr virus (which causes mononucleosis). Lymphocytes infected with Epstein-Barr virus do not undergo their usual cell death because one of the viral proteins induces the activity of the *bcl-2* gene (Henderson et al., 1991). The similarities between *ced-9* and *bcl-2* are so impressive that if an active human *bcl-2* gene is placed into *C. elegans* embryos, it prevents the normally occurring cell deaths (Vaux et al., 1992). This suggests that it functions in humans by acting upon the human analogues of *ced-3* and *ced-4*. It is possible that certain degenerative disorders (such as stroke or other neurodegenerative disorders) may arise from the inactivation or circumvention of genes such as *bcl-2* and *ced-9*. If this is shown to be the case, those genes active in preventing cell death in development may be among the most medically important genes in our bodies.

Cytoplasmic localization of germ cell determinants

Cytoplasmically localized determinants are found throughout the animal kingdom. The most frequently observed determinants are those responsible for the determination of germ cell precursors, that is, those cells that give rise to gametes. Even in many embryos in which other aspects of early development are regulative, those cells containing a certain region of egg cytoplasm are destined to become germ cell precursors.

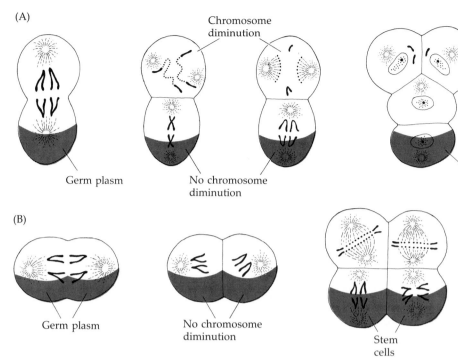

(A)

Chromosome diminution

Germ plasm

No chromosome diminution

Stem cell

(B)

Germ plasm

No chromosome diminution

Stem cells

FIGURE 14.26

Distribution of germ plasm (color) during cleavage of (A) normal and (B) centrifuged zygotes of *Parascaris*. (A) The germ plasm is normally conserved in the most vegetal blastomere, as shown by the lack of chromosomal diminution in that particular cell. Thus, at the 4-cell stage, the embryo has one stem cell for its gametes. (B) When the first cleavage is displaced 90° by centrifugation, both resulting cells have vegetal germ plasm and neither cell undergoes chromosome diminution. After the second cleavage, these two cells give rise to germinal stem cells. (After Waddington, 1966.)

Germ cell determination in nematodes

Theodor Boveri (1862–1915) was the first person to look at an organism's chromosomes throughout its development. In so doing, he discovered a fascinating feature in the development of the roundworm *Parascaris aequorum* (formerly *Ascaris megalocephala*). This nematode has only two chromosomes per haploid cell, thus allowing detailed observations of the individual chromosomes. The cleavage plane of the first embryonic division is unusual in that it is equatorial, separating the animal from the vegetal half of the zygote (Figure 14.26A). More bizarre, however, is the behavior of the chromosomes in the subsequent division of these first two blastomeres. The ends of the chromosomes in the animal-derived blastomere fragment into dozens of pieces just before cleavage of this cell. This phenomenon is called **chromosome diminution**, because only a portion of the original chromosome survives. Numerous genes are lost in these cells when the chromosome fragments, and these genes are not included in the newly formed nuclei (Tobler et al., 1972). Meanwhile, in the vegetal blastomere, the chromosomes remain normal. During the second division, the animal cell splits meridionally while the vegetal cell again divides equatorially. Both vegetally derived cells have normal chromosomes. However, the chromosomes of the more animally located of these two vegetal blastomeres fragment before third division. Thus, at the 4-cell stage, only one cell—the most vegetal—contains a full set of genes. At successive cleavages, somatic nuclei are given off from this vegetalmost line, until at the 16-cell stage there are only two cells with undiminished chromosomes. One of these two blastomeres gives rise to the germ cells; the other eventually undergoes chromosome diminution and forms somatic cells. The chromosomes are kept intact only in those cells destined to form the germ line. If this were not the case, the genetic information would degenerate from one generation to the next. The cells that have undergone chromosome diminution generate the somatic cells.

Boveri has been called the last of the great "observers" of embryology and the first of the great experimenters. Not content with observing the retention of the full chromosome complement solely by the germ cell

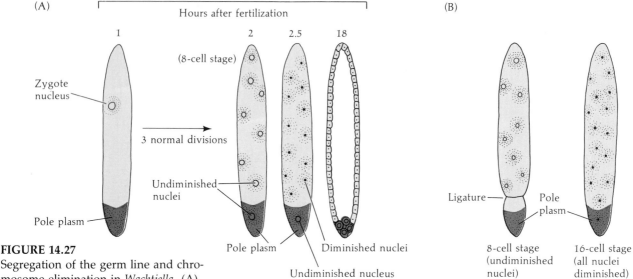

FIGURE 14.27
Segregation of the germ line and chromosome elimination in *Wachtiella*. (A) Chromosome elimination depends on location. The first three divisions are normal. Thereafter all nuclei except a single nucleus at the posterior pole lose 32 of their 40 chromosomes. (B) When a ligature is tied around the 8-cell zygote near its posterior end so that no nucleus enters the pole region, all nuclei undergo elimination at the 16-cell stage, even if the connection to the pole plasm is reestablished. (After Geyer-Duszynska, 1959.)

precursors, he set out to test whether a specific region of cytoplasm protects the nuclei within it from diminution. If so, any nucleus happening to reside in this region should be protected. Boveri (1910) tested this by centrifuging *Parascaris* eggs shortly before their first cleavage. This treatment shifted the orientation of the mitotic spindle. When the spindle forms perpendicular to its normal orientation, both resulting blastomeres should contain some of the vegetal cytoplasm (Figure 14.26). Indeed, Boveri found that after the first division neither nucleus underwent chromosomal diminution. However, the next division was equatorial along the animal–vegetal axis. Here the resulting animal blastomeres both underwent diminution, whereas the two vegetal cells did not. Boveri concluded that the vegetal cytoplasm contains a factor (or factors) that protects nuclei from chromosomal diminution and determines them to be germ cells.

Germ cell determination in insects

Certain insect eggs also contain a germ plasm that appears to act very similarly to the one observed in *Parascaris*. In the midge *Wachtiella persicariae*, most nuclei lose 32 of their original 40 chromosomes! However, two undiminished nuclei are found at the posterior pole of the egg and do not divide for a period of time (Figure 14.27). These two nuclei eventually give rise to the germ cells.* When nuclei are prevented by a ligature from migrating into the posterior pole region, every nucleus undergoes diminution, and the resulting midge is sterile. When the ligature is loosened and a diminished nucleus enters the posterior pole, germ cells are made but never differentiate into functional gametes (Geyer-Duszynska, 1959). Kunz and co-workers (1970) have shown that the eliminated chromatin contains genes that are active during germ cell production.

The germinal cytoplasm of insects is different from that of any other cytoplasm in the egg. Hegner (1911) found that when he removed or destroyed this region of beetle eggs before pole cell formation had occurred the resulting embryos had no germ cells and were sterile. Geigy (1931)

*This and the *Parascaris* example sound like perfect evidence for Weismann's hypothesis of the segregation of nuclear determinants (to be discussed in Chapter 16). While these cases of chromosome diminution and elimination are exceptions to the general rule that the nuclei of differentiated cells retain unused genes, there is no evidence that different somatic cells in *Wachtiella* or *Parascaris* retain different parts of the genome.

(A)

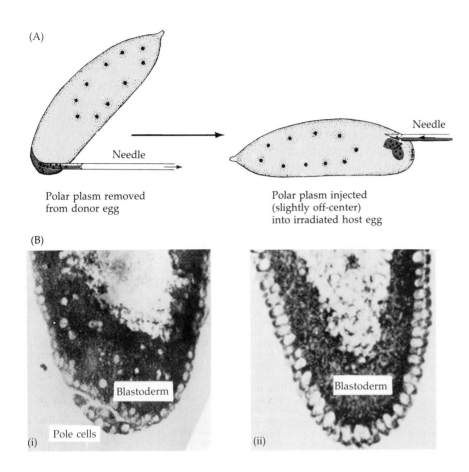

Polar plasm removed
from donor egg

Needle

Polar plasm injected
(slightly off-center)
into irradiated host egg

Needle

(B)

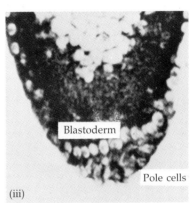

(i) Blastoderm Pole cells

(ii) Blastoderm

(iii) Blastoderm Pole cells

FIGURE 14.28
Ability of pole plasm to correct radiation-induced sterility. (A) Technique of pole plasm transplantation from unirradiated donor to irradiated host. (B) Longitudinal sections of the posterior portion of the *Drosophila* embryo fixed at the completion of cleavage. (i) Normal embryo with complete blastoderm and pole cells. (ii) Embryo irradiated during early cleavage. Blastoderm has formed, but pole cells are absent. (iii) Embryo irradiated during early cleavage but subsequently injected with pole plasm from normal embryos. Pole cells and blastoderm are both seen. (From Okada et al., 1974, courtesy of M. Okada.)

showed that irradiating *Drosophila* egg pole plasm with ultraviolet light produced sterile flies; Okada and co-workers (1974) extended this line of experimentation by showing that the addition of pole plasm from unirradiated donor embryos can cure the sterility of irradiated eggs (Figure 14.28). No other part of the cytoplasm could accomplish this reversal of sterility. This posterior pole plasm is conveniently marked with **polar granules** (Figure 14.29). Although their role in germ cell determination is not known, their constant association with the pole plasm and the **pole cells** derived from it makes them a convenient marker for this region. This region of pole cells is easily identifiable under a scanning electron microscope (Figure 14.30).

Recent work on the pole cell cytoplasm has focused primarily on *Drosophila* embryos. As detailed in Chapter 3, the *Drosophila* embryo develops as a syncytium the first 2 hours after fertilization. The nuclei divide without any corresponding cellular division, forming a blastoderm layer that contains about 3500 nuclei. Each nucleus is eventually enclosed in a cellular membrane, transforming the syncytial blastoderm into a cellular blastoderm. Transplantation experiments (Zalokar, 1971; Illmensee, 1968, 1972) have shown that all syncytial nuclei are equivalent and totipotent. The cells of the cellular blastoderm, however, are precisely determined. A similar conclusion was reached by Schubiger and Wood (1977), who ligated *Drosophila* eggs during various stages of their development. *Drosophila* nuclei do not undergo diminution, and any syncytial-stage nucleus can give rise to germ cells or to somatic cells. Their determination depends

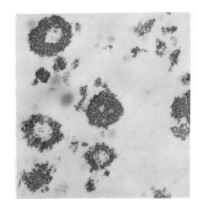

FIGURE 14.29
Electron micrograph of polar granules from particulate fraction from *Drosophila* pole cells. (Courtesy of A. P. Mahowald.)

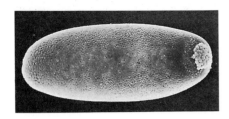

FIGURE 14.30
Scanning electron micrograph of the pole cells of a *Drosophila* embryo just prior to completion of cleavage. The pole cells can be seen at the right of this picture. (Courtesy of A. P. Mahowald.)

upon the region of the egg into which they migrate. One of these regions is the posterior pole plasm. Again, this pole plasm appears to contain the morphogenetic determinants for germ cell production.

The autonomy of this cytoplasmic region and its ability to determine any *Drosophila* nucleus was shown in 1974 by the ingenious experiments of Karl Illmensee and Anthony Mahowald. In these experiments (outlined in Figure 14.31), an incredibly small amount (5–100 picoliters) of anucleate pole plasm was transferred from wild-type *Drosophila* eggs into the *anterior* pole of genetically marked eggs prior to their cellularization. These donor eggs carried the chromosomal mutations *multiple wing hair* and *ebony*. After the cellular blastoderm had formed, the cells in the anterior pole of the recipient embryo resembled normal posterior pole cells, having incorpo-

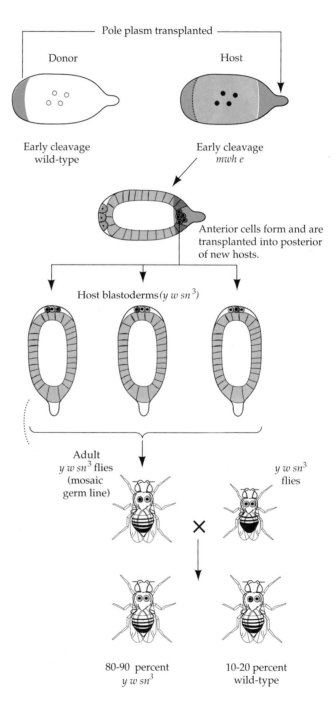

FIGURE 14.31
Ability of germ plasm to determine the fate of cells that contain it. Pole plasm from wild-type *Drosophila* eggs is transplanted into the anterior (but not the posterior) pole of genetically marked (mutant) embryos. Subsequently, the cells in the anterior pole resemble the normal germ cell precursors seen at the posterior pole. These anterior pole cells are then transplanted into host posterior regions (marked with different mutations) so that their descendants can migrate to the gonads. When these flies are mated to other flies with the second series of mutations (which did not have transplanted cells), some of the progeny are wild-type, indicating that the germ cells of these progeny came from a nonparental strain of fly. (After Illmensee and Mahowald, 1974.)

rated the polar granules and developed a typical pole cell morphology. To test whether or not these cells had become functioning germ cell precursors, Mahowald and Illmensee transplanted these modified anterior cells into the posterior region of cleaving embryos containing their own genetically marked pole cells.* These new host embryos were marked by different mutations (recessives *yellow*, *white*, and *singed*). When these embryos developed, they all became flies bearing the mutations *yellow*, *white*, and *singed*. These flies were mated to other flies carrying the same mutations. In most cases (88/92), these matings produced individuals identical to both parents. In four cases, however, wild-type progeny emerged, indicating that some of the germ cells in these flies derived from the transplanted cells; the germ plasm from one embryo was able to cause the anterior nuclei of another embryo to develop into functioning germ cells! This technique has also proved useful in showing the time of the germ cell determinant's localization. Illmensee and his colleagues (1976) found that this determinant is able to function before fertilization and becomes localized in the developing oocyte at about the same time that the yolk reaches the posterior end of the egg.

Nature has also provided confirmation of the importance of both pole plasm and its polar granules. Female *Drosophila* homozygous for the *grandchildless* mutation produce normal but sterile offspring [$GG \, \male \times gg \, \female \to Gg$ (sterile)]. Mahowald and colleagues (1979) have shown that when such females are mated with normal males, the nuclei of the resulting embryos never migrate into the pole plasm of the egg. No pole cells are formed, and the resulting adults have no primordial germ cells with which to produce gametes. Another maternal effect mutation—*agametic*—causes the absence of germ cells in about half the gonads of offspring derived from homozygous female flies. Here, the normal number of pole cells form, but the polar granules degenerate shortly after fertilization (Engstrom et al., 1982). Transplantation experiments demonstrate that the defect is in the polar cytoplasm and not in the ovarian environment. Thus, we now have fairly strong evidence that the pole plasm is directly concerned with germ cell determination.

Components of the *Drosophila* pole plasm

What are the determinants of the *Drosophila* germ plasm and how do they become localized in the posterior of the embryo? *Drosophila* polar granules have been isolated and appear to be composed of both protein and RNA (Mahowald, 1971a,b; Waring et al., 1978), but the identity of these macromolecules (and there are still some whose identity is unknown) was not uncovered until the genetic approach was used. One of the components of the germ plasm is the mRNA of the ***germ cell-less (gcl)* gene**. This gene was discovered by Jongens and his colleagues (1992) when they mutated *Drosophila* and screened for those females who did not have "grandchildren" ("grandoffspring"?). The rationale was that if a female did not place functional germ plasm in her eggs, she could still have offspring, but the offspring would be sterile (since they would lack germ cells). The wild-type *gcl* gene is transcribed in the nurse cells of the fly's ovary, and its mRNA is transported through the ring canals into the egg. Once inside the egg, it becomes transported to the posteriormost portion and resides within what will become the pole plasm (Figure 14.32A,B).

*The reason for doing this was so that these cells might be incorporated into the developing gonads. There is evidence from other organisms that the pole plasm also contains determinants for the proper *migration* of germ cells (Züst and Dixon, 1977; Ikenishi and Kotani, 1979).

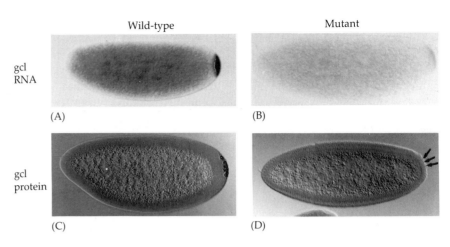

FIGURE 14.32

Localization of *germ cell-less* gene products in the posterior of the egg and embryo. The *gcl* mRNA can be seen in the posterior pole of early-cleavage embryos produced from wild-type females (A), but not in those embryos produced by *gcl*-deficient mutant females (B). Antibodies against the protein encoded by the *gcl* gene can be detected in the cellular blastoderm stage of those embryos produced by wild-type females (C) but not in the embryos from mutant females (D). (From Jongens et al., 1992, courtesy of T. A. Jongens.)

This message gets translated into protein during the early stages of cleavage (Figure 14.32C,D). The *gcl*-encoded protein appears to enter the nucleus, and it is essential for pole cell production. Flies mutant for this gene lack germ cells, and when antisense RNA against the *gcl* message is placed into the embryo, the ability to make germ cells is similarly destroyed (Figure 14.33).

A second candidate for a germ plasm component was a big surprise: **mitochondrial large ribosomal RNA**. Using the assay system of ultraviolet-irradiated eggs, Kobayashi and Okada (1989) showed that the injection of mitochondrial large ribosomal RNA (mtlrRNA) restores the ability of these irradiated embryos to form pole cells. Moreover, in normal fly eggs, mtlrRNA is located outside the mitochondria solely in the pole plasm of cleavage-stage embryos. Here it appears as a component of the polar granules (Kobayashi et al., 1993). While the mtlrRNA is involved in directing the formation of the pole cells, it does not enter into them.

What directs *germ cell-less* mRNA and mtlrRNA (and presumably other pole plasm molecules) to the posterior of the egg? There are at least seven

FIGURE 14.33

Migration of germ cells in the embryos produced by wild-type females and in those embryos produced by mutants that cannot synthesize the *germ cell-less* protein. Staining of the germ cells is accomplished by antibodies directed against vasa, a polar granule component that is not mutated in either type of *Drosophila*. (A) An early blastoderm-stage embryo from wild-type females has pole cells at the posterior pole. (B) Embryos from *gcl*-mutant females do not. In embryos from wild-type females, these pole cells can be moved into the posterior midgut primordium (C) from whence they migrate into the gonads (E). Such cells are not seen in the embryos from mothers that lack *germ cell-less* activity (D,F). (From Jongens et al., 1992, courtesy of T. A. Jongens.)

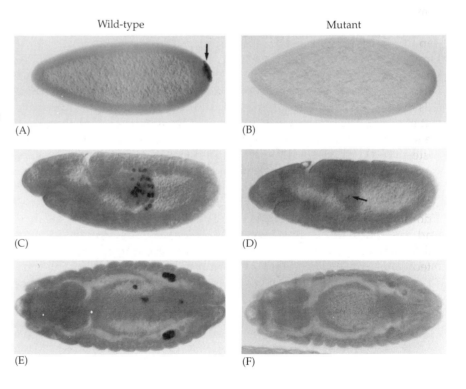

(A)

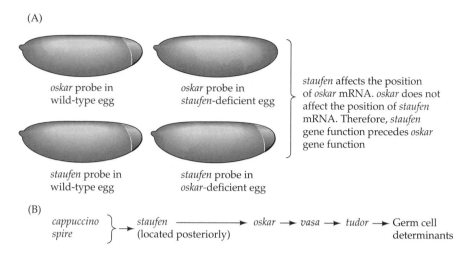

oskar probe in
wild-type egg

oskar probe in
staufen-deficient egg

staufen probe in
wild-type egg

staufen probe in
oskar-deficient egg

staufen affects the position
of oskar mRNA. oskar does not
affect the position of staufen
mRNA. Therefore, staufen
gene function precedes oskar
gene function

(B)

cappuccino
spire
} → staufen ——————————→ oskar → vasa → tudor → Germ cell
(located posteriorly)   determinants

FIGURE 14.34
(A) Diagram of the rationale for determining the steps of the genetic pathway leading to the posterior localization of the germ cell determinants. (B) Summary of those steps.

other mutants that have the inability to form germ cells, and these mutants also have poorly formed abdomens. These mutants are in the genes for *cappuccino*, *spire*, *staufen*, *oskar*, *vasa*, *valois*, and *tudor*. Each of these genes is active in the ovary and places a gene product into the growing oocyte. By probing for the localization of the mRNA or protein for one gene in a mutant that lacks another gene, one can place the actions of these genes in a defined order (Figure 14.34). These studies (reviewed by Strome, 1992; Ephrussi and Lehmann, 1992) show that two proteins, those made by the *cappuccino* and *spire* genes, are needed for the posterior localization of the staufen protein. (That is, the staufen protein will not be placed in the posterior of eggs from mothers whose ovaries cannot make cappuccino or spire). The staufen protein is needed for the posterior localization of *oskar* mRNA. The protein made by the *oskar* message is critical for the posterior localization of the vasa protein, another component of the polar granules. Mutants of *tudor* and *valois* do not affect the positioning of vasa, but appear to be critical for the maintainance of the pole plasm once it has formed (Hay et al., 1990; Lasko and Ashburner, 1990).

The *oskar* gene product appears to be the rate- and position-limiting step in pole cell formation. Embryos derived from females with only one copy of the *oskar* gene produce from 10 to 15 pole cells at the cellular blastoderm stage, whereas those with two copies produce around 35 pole cells. If one raises the number of *oskar* genes to four copies, about 50 pole cells are formed. Moreover, Ephrussi and Lehmann (1992) demonstrated that germ cells will form wherever the *oskar* message is localized, and that the steps preceding it are crucial solely in getting the *oskar* mRNA to the posterior pole of the egg. If the *oskar* message becomes localized to the anterior of the embryo (which can be arranged experimentally), the germ plasm and germ cells will form in the anterior. The oskar protein probably creates the first part of the scaffold of the polar granules. The vasa and tudor proteins bind to oskar to create a more complex scaffold that can bind the germ cell determinants. The localization of *gcl* mRNA and mtlrRNA to the posterior pole of the egg is wiped out by any of the above-listed mutations. In *valois* and *tudor* mutants, small amounts of *gcl* message can be seen localized to the posterior plasm in early cleavage embryos, but this localization is lost by late cleavage (Jongens et al., 1992). Thus, the polar granules include both the germ cell determinants and the scaffolding that keeps them at the posterior of the egg and embryo. The scaffold will bind the *germ cell-less* mRNA (and presumably gene products for other germ cell determinants). These messages become translated into

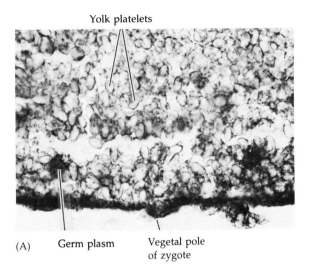

(A)   Germ plasm   Vegetal pole
                          of zygote

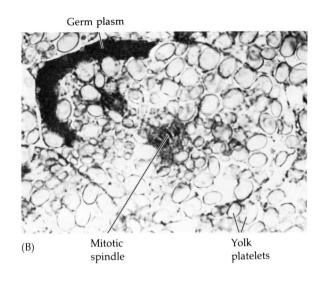

(B)   Mitotic   Yolk
            spindle   platelets

(C)   Germ cell

FIGURE 14.35
Germ plasm of frog embryos. (A) Germ plasm (dark regions) near the ventral pole of a newly fertilized zygote. (B) Germ plasm-containing cell in endodermal region of blastula in mitotic anaphase. Note the germ plasm entering into only one of the two yolk-laden daughter cells. (C) Primordial germ cell and somatic cells near the floor of the blastocoel in early gastrula. (Courtesy of A. Blackler.)

protein during early cleavage, enter the nucleus of the pole cells, and (in some as yet unknown way) determine the fate of these cells to be germ cells.

Germ cell determination in amphibians

Cytoplasmic localization of germ cell determinants has also been observed in vertebrate embryos. Bounoure (1934) showed that the vegetal region of fertilized frog eggs contains a material with staining properties similar to those of *Drosophila* pole plasm (Figure 14.35). He was able to trace this cortical cytoplasm into the few cells in the presumptive endoderm that would normally migrate into the genital ridge. Blackler (1962) showed that these cells are the primordial germ cell precursors. He removed the endodermal region from the neurula of one frog and inserted it into the neurula endoderm of another (Figure 14.36). The donor frogs were genetically distinguishable because their cells contained only one nucleolus instead of the usual two. The host frogs were wild-type (i.e., two nucleoli per nucleus). After the operation, the donor frogs were sterile, indicating that the neurula could not regulate for the production of germ cells and that the correct region had been excised. The host frogs, however, were fertile. Moreover, when these frogs were mated to normal adults, the offspring were a mixture of one-nucleolus and two-nucleoli frogs. Therefore, some of the gametes came from the endodermal region of the *donor* embryo. (If none had come from the donor, one would get 100 percent two-nucleoli frogs. If every germ cell were derived from the donor, one would expect a 1:1 ratio, as is shown in Figure 14.36).

The early movements of the germ plasm have been analyzed in detail by Savage and Danilchik (1993), who labeled the germ plasm with fluorescent dye. They found that the germ plasm of unfertilized eggs consisted of tiny "islands" that appear to be tethered to the yolk mass near the vegetal cortex. These germ plasm islands move with this vegetal yolk mass during the cortical rotation of fertilization. After the rotation, the

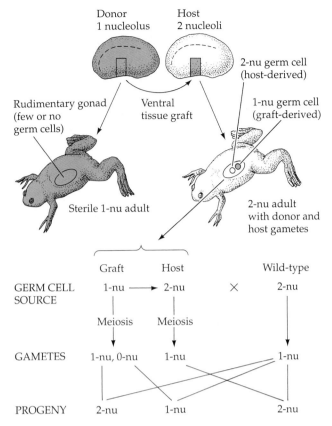

FIGURE 14.36
Demonstration of primordial germ cells in early tadpole endoderm. A block of
ventral tissue from a strain of frog with one nucleolus per cell is transplanted
into the analogous region in a host tadpole with two nucleoli per cell. The donor
tadpole metamorphoses into a sterile adult, while the host tadpole develops into
a frog with both host and donor gametes. This fact can be demonstrated by
crossing that frog with a normal (two-nucleoli) adult; the results of such a cross
are shown in the flow chart. Some of the offspring have only one nucleolus per
cell, indicating that they received one nucleolus from one parent (the two-nucleoli frog) and no nucleolus from the other (the two-nucleoli frog with some
one-nucleolus germ cells). (After Blackler, 1966.)

islands are released from the yolk mass and begin fusing together and
migrating to the vegetal pole. This aggregation depends upon microtubules, and the movement of these clusters to the vegetal pole is dependent
on peristalsis-like contractions of the vegetal surface. These periodic surface contractions of the vegetal cell surface also appear to push this germ
plasm along the furrows of the newly formed blastomeres, enabling it to
enter inside the embryo.

When ultraviolet light is applied to the vegetal surface (but nowehere
else) of the frog embryo, the resulting frogs are normal but lack germ cells
in their gonads (Bounoure, 1939; Smith, 1966). Very few primordial germ
cells reach the gonads, and those that do have about one-tenth the volume
of normal primordial germ cells and have aberrantly shaped nuclei (Züst
and Dixon, 1977). Savage and Danilchik (1993) found that UV light prevents the vegetal surface contractions and inhibits the migration of germ
plasm to the vegetal pole. Without this region of cytoplasm, germ cells
fail to form. So, like the *Drosophila* pole plasm, the cytoplasm from the
vegetal region of frog zygotes contains the determinants for germ cell
formation.

Summary

We have evidence in certain organisms that the determination of a cell's fate is due to the portion of egg cytoplasm that it acquires during cleavage. Such a cell differentiates independently of other cells, and organisms that utilize such a mechanism tend toward a mosaic, or determinate, type of development. Embryos exhibiting this form of development include those of molluscs, tunicates, and nematodes. The placement of morphogenetic determinants within the egg cytoplasm, their redistribution during egg development and fertilization, and the patterns of cell cleavage are important for determining which cell acquires which fate. Each of these phenomena is a function of the egg. Although most of the development of these organisms follows the mosaic pattern, some interactive determination is also seen. In tunicates, the nervous system and some of the muscles are formed by inductive interactions between blastomeres, and snails and nematodes also have certain organs formed in this interactive manner. In the next chapter, we will look at certain organisms in which interactions between molecules in the syncytial blastoderm of insect eggs constitute the primary mechanism of determining cell fate.

LITERATURE CITED

Alberts, B., Bray, D., Lewis, J., Raff, M., Roberts, K. and Watson, J. D. 1983. *Molecular Biology of the Cell.* Garland, New York.

Allsopp, T. E., Wyatt, S., Paterson, H. F. and Davies, A. M. 1993. The proto-oncogene *bcl-2* can selectively rescue neurotrophic factor-dependent neurons from apoptosis. *Cell* 73: 295–307.

Atkinson, J. W. 1987. An atlas of light micrographs of normal and lobeless larvae of the marine gastropod *Ilyanassa obsoleta. Int. J. Invert. Reprod. Dev.* 9: 169–178.

Austin, J. and Kimble, J. 1987. glp-1 is required in the germ line for regulation of the decision between mitosis and meiosis in *C. elegans. Cell* 51: 589–600.

Blackler, A. W. 1962. Transfer of primordial germ cells between two subspecies of *Xenopus laevis. J. Embryol. Exp. Morphol.* 10: 641–651.

Blackler, A. W. 1966. The role of a "germinal plasm" in the formation of primordial germ cells in *Rana pipiens. Dev. Biol.* 14: 330–347.

Bonnet, C. 1764. *Contemplation de la Nature.* Marc-Michel Ray, Amsterdam.

Bounoure, L. 1934. Recherches sur la lignée germinale chez la grenouille rousse aux premiers stades du développement. *Ann. Sci. Natl. 10e Wer.* 17: 67–248.

Bounoure, L. 1939. *L'origine des Cellules Reproductries et le Probleme de la Lignée Germinale.* Gauthier-Villars, Paris.

Boveri, T. 1910. Über die Teilung centrifugierter Eier von *Ascaris megalocephala. Wilhelm Roux Arch. Entwicklungsmech. Org.* 30: 101–125.

Bowerman, B., Eaton, B. A. and Priess, J. R. 1992. *skn-1*, a maternally expressed gene required to specify the fate of ventral blastomeres in the early *C. elegans* embryo. *Cell* 68: 1061–1075.

Bowerman, B., Draper, B. W., Mello, C. C. and Priess, J. R. 1993. The maternal gene *skn-1* encodes a protein that is distributed unequally in early *C. elegans* embryos. *Cell* 74: 443–452.

Brandhorst, B. P. and Newrock, K. M. 1981. Post-transcriptional regulation of protein synthesis in *Ilyanassa* embryos and isolated polar lobes. *Dev. Biol.* 83: 250–254.

Brenner, S. 1974. The genetics of *Caenorhabditis elegans. Genetics* 77: 71–94.

Brenner, S. 1979. Cited in H. F. Judson, *The Eighth Day of Creation.* Simon and Schuster, New York, p. 219.

Cassirer, E. 1950. Developmental mechanics and the problem of cause in biology. In E. Cassirer (ed.), *The Problem of Knowledge.* Yale University Press, New Haven.

Cervera, M., Dreyfuss, G. and Penman, S. 1981. Messenger RNA is translated when associated with the cytoskeletal framework in normal and VSV–infected HeLa cells. *Cell* 23: 113–120.

Chabry, L. M. 1887. Contribution a l'embryologie normale tératologique des ascidies simples. *J. Anat. Physiol. Norm. Pathol.* 23: 167–321.

Churchill, F. B. 1973. Chabry, Roux, and the experimental method in nineteenth century embryology. In R. N. Giere and R. S. Westfall (eds.), *Foundations of Scientific Method: The Nineteenth Century.* Indiana University Press, Bloomington, pp. 161–205.

Clement, A. C. 1962. Development of *Ilyanassa* following removal of the D macromere at successive cleavage stages. *J. Exp. Zool.* 149: 193–215.

Clement, A. C. 1968. Development of the vegetal half of the *Ilyanassa* egg after removal of most of the yolk by centrifugal force, compared with the development of animal halves of similar visible composition. *Dev. Biol.* 17: 165–186.

Clement, A. C. 1986. The embryonic value of the micromeres in *Ilyanassa obsoleta*, as determined by deletion experiments. III. The third quartet cells and the mesentoblast cell, 4d. *Int. J. Invert. Reprod. Dev.* 9: 155–168.

Collier, J. R. 1983. The biochemistry of molluscan development. In N. H. Verdonk, J. A. M. van den Biggelaar and A. S. Tompa (eds.), *The Mollusca*, Vol. 3. Academic, New York, pp. 215–252.

Collier, J. R. 1984. Protein synthesis in normal and lobeless gastrulae of *Ilyanassa obsoleta. Biol. Bull.* 167: 371–377.

Conklin, E. G. 1905a. The organization and cell lineage of the ascidian egg. *J. Acad. Nat. Sci. Phila.* 13: 1–119.

Conklin, E. G. 1905b. Organ–forming substances in the eggs of ascidians. *Biol. Bull.* 8: 205–230.

Conklin, E. G. 1905c. Mosaic development in ascidian eggs. *J. Exp. Zool.* 2: 145–223.

Craig, M. M. and Morrill, J. B. 1986. Cellular arrangements and surface topology during early development in embryos of *Ilyanassa obsoleta. Int. J. Invert. Reprod. Dev.* 9: 209–228.

Crowther, R. J. and Whittaker, J. R. 1984. Differentiation of histotypic ultrastructural features in cells of cleavage-arrested early ascidian embryos. *Wilhelm Roux Arch. Dev. Biol.* 194: 87–98.

Dareste, C. 1877. *Recherches sur la Production Artificielle des Monstruosités ou Essais de Tératogenie Experimentale.* Reinwald, Paris.

Davidson, E. H. 1991. Spatial mechanisms of gene regulation in metazoan embryos. *Development* 113: 1–26.

Deno, T., Nishida, H. and Satoh, N. 1984. Autonomous muscle cell differentiation in partial ascidian embryos according to the newly verified cell lineages. *Dev. Biol.* 104: 322–328.

Ellis, R. E. and Horvitz, H. R. 1986. Genetic control of programmed cell death in the nematode *C. elegans*. *Cell* 44: 817–829.

Engstrom, L., Caulton, J. H., Underwood, E. M. and Mahowald, A. P. 1982. Developmental lesions in the agametic mutant of *Drosophila melanogaster*. *Dev. Biol.* 91: 163–170.

Ephrussi, A. and Lehmann, R. 1992. Induction of germ cell formation by *oskar*. *Nature* 358: 387–392.

Fischer, J.-L. 1991. Laurent Chabry and the beginnings of experimental embryology in France. *In* S. Gilbert (ed.), *A Conceptual History of Modern Embryology*. Plenum, New Yor, pp. 31–41.

Fischer, J.-L. and Smith, J. 1984. French embryology and the "mechanics of development" from 1887 to 1910: L. Chabry, Y. Delage and E. Bataillon. *Hist. Phil. Life Sci.* 6: 25–39.

Geigy, R. 1931. Action de l'ultra-violet sur le pôle germinal dans l'oeuf de *Drosophila melanogaster* (castration et mutabilité). *Rev. Suisse Zool.* 38: 187–288.

Geyer-Duszynska, I. 1959. Experimental research on chromosome diminution in *Cecidomiidae* (Diptera). *J. Exp. Zool.* 141: 381–441.

Goldstein, B. 1992. Induction of gut in *Caenorhabditis elegans* embryos. *Nature* 357: 255–257.

Gould, S. J. 1977. *Ontogeny and Phylogeny*. Belknap Press, Cambridge, MA.

Guerrier, P., van den Biggelaar, J. A. M., Dongen, C. A. M. and Verdonk, N. H. 1978. Significance of the polar lobe for the determination of dorsoventral polarity in *Dentalium vulgare* (da Costa). *Dev. Biol.* 63: 233–242.

Harrison, R. G. 1933. Some difficulties of the determination problem. *Am. Natur.* 67: 306–321.

Hay, B., Jan, L. Y. and Jan, Y. N. 1990. Localization of vasa, a component of *Drosophila* polar granules, in maternal effect mutants that alter embryonic anteroposterior polarity. *Development* 109: 425–433.

Hegner, R. W. 1911. Experiments with chrysomelid beetles. III. The effects of killing parts of the eggs of *Leptinotarsa decemlineata*. *Biol. Bull.* 20: 237–251.

Henderson, S. and seven others. 1991. Induction of *bcl-2* expression by Epstein-Barr virus latent membrane protein 1 protects infected B cells from programmed cell death. *Cell* 65: 1107–1115.

Hengartner, M. O., Ellis, R. E. and Horvitz, H. R. 1992. *Caenorhabditis elegans* gene *ced-9* protects cell from programmed cell death. *Nature* 356: 494–499.

Henry, J. J. and Martindale, M. Q. 1987. The organizing role of the D quadrant as revealed through the phenomenon of twinning in the polychaete *Chaetopterus variopedatus*. *Wilhelm Roux Arch. Dev. Biol.* 196: 449–510.

Hill, D. P. and Strome, S. 1987. An analysis of the role of microfilaments in the establishment and maintainance of asymmetry in *Caenorhabditis elegans* zygotes. *Dev. Biol.* 125: 75–84.

Hill, D. P. and Strome, S. 1990. Brief cytochalasin-induced disruption of microfilaments during a critical interval in 1-cell *C. elegans* embryos alters the positioning of developmental instructions to the 2-cell embryo. *Development* 108: 159–172.

Hockenbery, D. M., Nuñez, G., Milliman, C., Schreiber, R. D. and Korsmeyer, S. J. 1990. Bcl-2 is an inner mitochondrial membrane protein that blocks programmed cell death. *Nature* 348: 334–336.

Ikenishi, K. and Kotani, M. 1979. Ultraviolet effects on presumptive primordial germ cells (pPGCs) in *Xenopus laevis* after the cleavage stage. *Dev. Biol.* 69: 237–246.

Illmensee, K. 1968. Transplantation of embryonic nuclei into unfertilized eggs of *Drosophila melanogaster*. *Nature* 219: 1268–1269.

Illmensee, K. 1972. Developmental potencies of nuclei from cleavage, preblastoderm, and syncytial blastoderm transplanted into unfertilized eggs of *Drosophila melanogaster*. *Wilhelm Roux Arch. Entwicklungsmech. Org.* 170: 267–298.

Illmensee, K. and Mahowald, A. P. 1974. Transplantation of posterior polar plasm in *Drosophila*. Induction of germ cells at the anterior pole of the egg. *Proc. Natl. Acad. Sci. USA* 71: 1016–1020.

Illmensee, K., Mahowald, A. P. and Loomis, M. R. 1976. The ontogeny of germ plasm during oogenesis in *Drosophila*. *Dev. Biol.* 49: 40–65.

Jeffery, W. R. 1984. Spatial distribution of messenger RNA in the cytoskeletal framework of ascidian eggs. *Dev. Biol.* 103: 482–492.

Jeffery, W. R. 1985. Identification of proteins and mRNAs in isolated yellow crescents of ascidian eggs. *J. Embryol. Exp. Morphol.* 89: 275–287.

Jeffery, W. R. and Meier, S. 1983. A yellow crescent cytoskeletal domain in ascidian eggs and its role in early development. *Dev. Biol.* 96: 125–143.

Jongens, T. A., Hay, B., Jan, L. Y. and Jan, Y. N. 1992. The *germ cell–less* gene product: A posteriorly localized component necessary for germ cell development in *Drosophila*. *Cell* 70: 569–584.

Kemphues, K. J., Priess, J. R., Morton, D. G. and Cheng, N. 1988. Identification of genes required for cytoplasmic localization in early *C. elegans* embryos. *Cell* 52: 311–320.

Kimble, J. 1981. Lineage alterations after ablation of cells on the somatic gonad of *Caenorhabditis elegans*. *Dev. Biol.* 87: 286–300.

Kimble, J. and Hirsch, D. 1979. The postembryonic cell lineages of the hermaphrodite and male gonads in *Caenorhabditis elegans*. *Dev. Biol.* 70: 396–417.

Kirby, C. M. 1992. Cytoplasmic reorganization and the generation of asymmetry in *Caenorhabditis elegans*, with an emphasis on *par-3*, a maternal-effect gene essential for both processes. PhD dissertation, Cornell University, Ithaca, NY.

Kirby, C. M., Kusch, M. and Kemphues, K. 1990. Mutations in the *par* genes of *Caenorhabditis elegans* affect cytoplasmic reorganization during the first cell cycle. *Dev. Biol.* 142: 203–215.

Kobayashi, S. and Okada, M. 1989. Restoration of pole-cell forming ability to UV-irradiated *Drosophila* embryos by injection of mitochondrial lrRNA. *Development* 107: 733–742.

Kobayashi, S., Amikura, R. and Okada, M. 1993. Presence of mitochondrial large ribosomal RNA outside mitochondria in germ plasm of *Drosophila melanogaster*. *Science* 260: 1521–1524.

Kölreuter, J. G. 1761–1766. *Vorläufige Narchicht von einigen das Geschlecht der Pflanzen beteffenden Versuchen und Beobachtungen, nebst Fortsetzugen 1, 2 und 3*. Leipzig.

Kunz, W., Trepte, H. H. and Bier, K. 1970. On the function of the germ line chromosomes in the oogenesis of *Wachtiella persiariae* (Cecidomyiidae). *Chromosoma* 30: 180–192.

Lasko, P. F. and Ashburner, M. 1990. Posterior localization of vasa protein correlates with, but is not sufficient for, pole cell development. *Genes Dev.* 4: 905–921.

Lenoir, T. 1980. Kant, Blumenbach, and vital materialism in German biology. *Isis* 71: 77–108.

Mahowald, A. P. 1971a. Polar granules of *Drosophila*. III. The continuity of polar granules during the life cycle of *Drosophila*. *J. Exp. Zool.* 176: 329–343.

Mahowald, A. P. 1971b. Polar granules of *Drosophila*. IV. Cytochemical studies showing loss of RNA from polar granules during early stages of embryogenesis. *J. Exp. Zool.* 176: 329–343.

Mahowald, A. P., Caulton, J. H. and Gehring, W. J. 1979. Ultrastructural studies of oocytes and embryos derived from female flies carrying the *grandchildless* mutation in *Drosophila subobscura*. *Dev. Biol.* 69: 118–132.

Meedel, T. H. and Whittaker, J. R. 1983. Development of transcriptionally active mRNA for larval muscle acetylcholinesterase during ascidian embryogenesis. *Proc. Natl. Acad. Sci. USA* 80: 4761–4765.

Meedel, T. H. and Whittaker, J. R. 1984. Lineage segregation and developmental autonomy in expression of functional muscle acetylcholinesterase in the ascidian embryo. *Dev. Biol.* 105: 479–487.

Meedel, T. H., Crowthier, R. J. and Whittaker, J. R. 1987. Determinative properties of muscle lineages in ascidian embryos. *Development* 100: 245–260.

Mello, G. C., Draper, B. W., Krause, M., Weintraub, H. and Priess, J. R. 1992. The *pie-1* and *mex-1* genes and maternal control of blastomere identity in early *C. elegans* embryos. *Cell* 70: 163–176.

Moon, R. T., Nicosia, R. F., Olsen, C., Hille, M. and Jeffery, W. R. 1983. The cytoskeletal framework of sea urchin eggs and embryos: Developmental changes in the association of messenger RNA. *Dev. Biol.* 95: 447–458.

Morgan, T. H. 1927. *Experimental Embryology*. Columbia University Press, New York.

Newrock, K. M. and Raff, R. A. 1975. Polar lobe specific regulation of translation in embryos of *Ilyanassa obsoleta*. *Dev. Biol.* 42: 242–261.

Nishida, H. 1987. Cell lineage analysis in ascidian embryos by intracellular injection of a tracer enzyme. III. Up to the tissue restricted stage. *Dev. Biol.* 121: 526–541.

Nishida, H. 1990. Determinative mechanisms in secondary muscle lineages of ascidian embryos: Development of muscle-specific features in isolated muscle progenitor cells. *Development* 108: 559–568.

Nishida, H. 1992a. Determination of developmental fates of blastomeres in ascidian embryos. *Dev. Growth Differ.* 34: 253–262.

Nishida, H. 1992b. Regionality of egg cytoplasm that promotes muscle differentiation in embryo of the ascidian *Halocynthia roretzi*. *Development* 116: 521–529.

Nishida, H. 1993. Localized regions of egg cytoplasm that promote expression of endoderm-specific alkaline phosphatase in embryos of the ascidian *Halocynthia roretzi*. *Development* 118: 1–7.

Nishikata, T., Mita-Miyazawa, I., Deno, T. and Satoh, N. 1987. Monoclonal antibodies against components of the myoplasm of the ascidian *Ciona intestinalis* partially block the development of muscle-specific acetylcholinesterase. *Development* 100: 577–586.

Nuñez, G., London, L., Hockenbury, D., Alexander, M., McKearn, J. P. and Korsmeyer, S. J. 1990. Deregulation of *blc-2* gene expression selectively prolongs survival of growth factor-deprived hemopoietic cells. *J. Immunol.* 144: 3602–3610.

Okada, M., Kleinman, I. A. and Schneiderman, H. A. 1974. Restoration of fertility in sterilized *Drosophila* eggs by transplantation of polar cytoplasm. *Dev. Biol.* 37: 43–54.

Ortolani, G. 1959. Richerche sulla induzione del sistema nervoso nelle larve delle Ascidie. *Boll. Zool.* 26: 341–348.

Pines, M. (ed.). 1992. *From Egg to Adult*. Howard Hughes Med. Inst., Bethesda, pp. 30–38.

Priess, R. A. and Thomson, J. N. 1987. Cellular interactions in early *C. elegans* embryos. *Cell* 48: 241–250.

Priess, R. A., Schnabel, H. and Schnabel, R. 1987. The *glp-1* locus and cellular interactions in early *C. elegans* embryos. *Cell* 51: 601–611.

Reverberi, G. and Minganti, A. 1946. Fenomeni di evocazione nello sviluppo dell'uovo di Ascidie. Risultati dell'indagine spermentale sull'uovo di *Ascidiella aspersa* e di *Ascidia malaca* allo stadio di 8 blastomeri. *Pubbl. Staz. Zool. Napoli* 20: 199–252. (Quoted in Reverberi, 1971, p. 537.

Roe, S. 1981. *Matter, Life, and Generation: Eighteenth-Century Embryology and the Haller–Wolff Debate*. Cambridge University Press, Cambridge.

Satoh, N. 1979. On the "clock" mechanism determining the time of tissue-specific development during ascidian embryogenesis. I. Acetylcholinesterase development in cleavage-arrested embryos. *J. Embryol. Exp. Morphol.* 54: 131–139.

Savage, R. M. and Danilchik, M. V. 1993. Dynamics of germ plasm localization and its inhibition by ultraviolet irradiation in early cleavage *Xenopus* eggs. *Dev. Biol.* 157: 371–382.

Schubiger, G. and Wood, W. J. 1977. Determination during early embryogenesis in *Drosophila melanogaster*. *Am. Zool.* 17: 565–576.

Slack, J. M. W. 1991. *From Egg to Embryo: Regional Specification in Early Development*. Cambridge University Press, New York.

Smith, L. D. 1966. The role of a "germinal plasm" in the formation of primordial germ cells in *Rana pipiens*. *Dev. Biol.* 14: 330–347.

Strome, S. 1992. The germ of the issue. *Nature* 358: 368–369.

Strome, S. and Wood, W. B. 1983. Generation of asymmetry and segregation of germ-like granules in early *Caenorhabditis elegans* embryos. *Cell* 35: 15–25.

Sulston, J. and Horvitz, H. R. 1977. Postembryonic cell lineages of the nematode *Caenorhabditis elegans*. *Dev. Biol.* 56: 110–156.

Sulston, J. E., Schierenberg, J., White, J. and Thomson, N. 1983. The embryonic cell lineage of the nematode *Caenorhabditis elegans*. *Dev. Biol.* 100: 64–119.

Sulston, J. and eighteen others. 1992. The *C. elegans* genome sequencing project: A beginning. *Nature* 356: 37–42.

Swalla, B. J., Bladgett, M. R. and Jeffery, W. R. 1991. Identification of a cytoskeletal protein localized in the myoplasm of ascidian eggs: Localization is modified during anuran development. *Development* 111: 425–423.

Tobler, H., Smith, K. D. and Ursprung, H. 1972. Molecular aspects of chromatin elimination in *Ascaris lumbricoides*. *Dev. Biol.* 27: 190–203.

Tung, T. C., Wu, S. C., Yel, Y. F., Li, K. S. and Hsu, M. C. 1977. Cell differentiation in ascidians studied by nuclear transplantation. *Scientia Sinica* 20: 222–233.

Vaux, D. L., Weissman, I. L. and Kim, S. K. 1992. Prevention of programmed cell death in *Caenorhabditis elegans* by human *bcl-2*. *Science* 258: 1955–1957.

Verdonk, N. H. and Cather, J. N. 1983. Morphogenetic determination and differentiation. *In* N. H. Verdonk, J. A. M. van den Biggelaar and A. S. Tompa (eds.), *The Mollusca*. Academic Press, New York, pp. 215–252.

Waddington, C. H. 1966. *Principles of Development and Differentiation*. Macmillan, New York.

Waring, G. L., Allis, C. D. and Mahowald, A. P. 1978. Isolation of polar granules and the identification of polar granule-specific protein. *Dev. Biol.* 66: 197–206.

Whittaker, J. R. 1973. Segregation during ascidian embryogenesis of egg cytoplasmic information for tissue-specific enzyme development. *Proc. Natl. Acad. Sci. USA* 70: 2096–2100.

Whittaker, J. R. 1977. Segregation during cleavage of a factor determining endodermal alkaline phosphatase development in ascidian embryos. *J. Exp. Zool.* 202: 139–153.

Whittaker, J. R. 1982. Muscle cell lineage can change the developmental expression in epidermal lineage cells of ascidian embryos. *Dev. Biol.* 93: 463–470.

Whittaker, J. R., Ortolani, G. and Farinella–Ferruzza, N. 1977. Autonomy of acetylcholinesterase differentiation in muscle lineage cells in ascidian embryos. *Dev. Biol.* 55: 196–200.

Williams, G. T., Smith, C. A., Spooncer, E., Dexter, T. M. and Taylor, D. R. 1990. Haemopoietic colony stimulating factors promote cell survival by suppressing apoptosis. *Nature* 343: 76–79.

Wilson, E. B. 1904. Experimental studies on germinal localization. I. The germ regions in the egg of *Dentalium*. II. Experiments on the cleavage-mosaic in *Patella* and *Dentalium*. *J. Exp. Zool.* 1: 1–72.

Wolff, K. F. 1767. De formatione intestinorum praecipue. *Novi Commentarii Academie Scientarum Imperialis Petropolitanae*.

Yuan, J. Y. and Horvitz, H. R. 1990. The *Caenorhabditis elegans* genes *ced-3* and *ced-4* act cell autonomously to cause programmed cell death. *Dev. Biol.* 138: 33–41.

Zalokar, M. 1971. Transplantation of nuclei in *Drosophila melanogaster*. *Proc. Natl. Acad. Sci. USA* 68: 1539–1541.

Zumbe, A., Stahli, C. and Trachsel, H. 1982. Association of a M_r 50,000 cap-binding protein with the cytoskeleton in baby hamster kidney cells. *Proc. Natl. Acad. Sci. USA* 79: 2927–2931.

Züst, B. and Dixon, K. E. 1977. Events in the germ cell lineage after entry of the primordial germ cells into the genital ridges in normal and UV-irradiated *Xenopus laevis*. *J. Embryol. Exp. Morphol.* 41: 33–46.

15

The genetics of axis specification in *Drosophila*

The Library is limitless and periodic. If an eternal voyager were to traverse it in any direction, he would find, after many centuries, that the same volumes are repeated in the same disorder (which repeated, would constitute an order: Order itself). My solitude rejoices in this elegant hope.

J. L. BORGES (1962)

Those of us who are at work on Drosophila *find a particular point to the question. For the genetic material available is all that could be desired, and even embryological experiments can be done. . . . It is for us to make use of these opportunities. We have a complete story to unravel, because we can work things from both ends at once.*

JACK SCHULTZ (1935)

In the last chapter, we discussed the specification of early embryonic cells by their acquiring different cytoplasmic morphogens that had been stored in the oocyte. The cell membranes establish the region of cytoplasm incorporated into each cell, and it is thought that the morphogenetic determinants then direct differential gene expression in these blastomeres. During *Drosophila* development, cellular membranes do not form until after the thirteenth nuclear division. Prior to this time, all the nuclei share a common cytoplasm, and material can diffuse through the embryo. In these embryos, the specification of the cell types along anterior–posterior and dorsal–ventral axes is accomplished by the interactions of cytoplasmic materials within the single, multinucleated cell.

A summary of *Drosophila* development

As discussed in Chapter 5, *Drosophila* embryos develop very rapidly through a series of nuclear divisions that form a syncytial blastoderm. During the ninth division cycle, about five nuclei reach the surface of the posterior pole of the embryo. These nuclei become enclosed by cell membranes and generate the pole cells that give rise to the gametes of the adult. Most of the other nuclei arrive at the periphery of the embryo at cycle 10 and then undergo four more divisions at progressively slower rates. Following cycle 13, cell membranes grow between the nuclei to form a cellular blastoderm of about 6000 cells (Turner and Mahowald, 1977; Foe and Alberts, 1983). At cycle 14, the level of transcription, which had been very low, increases dramatically. At the same time, the cells become motile, and the 2- to 3-hour embryo begins gastrulation (Edgar and Schubiger, 1986).

The first movements of *Drosophila* gastrulation segregate the presumptive mesoderm, endoderm, and ectoderm (Figure 15.1). The prospective mesoderm—about 1000 cells constituting the ventral midline—folds in to produce the **ventral furrow**. This furrow eventually pinches off from the surface to become the ventral tube within the embryo. It then flattens to form a layer of mesodermal tissue beneath the ventral ectoderm. The prospective endoderm invaginates as two pockets at the anterior and posterior ends of the ventral furrow. The pole cells are internalized along with the endoderm. At this time, the embryo bends to form the **cephalic furrow** and the **anterior** and **posterior transverse folds**.

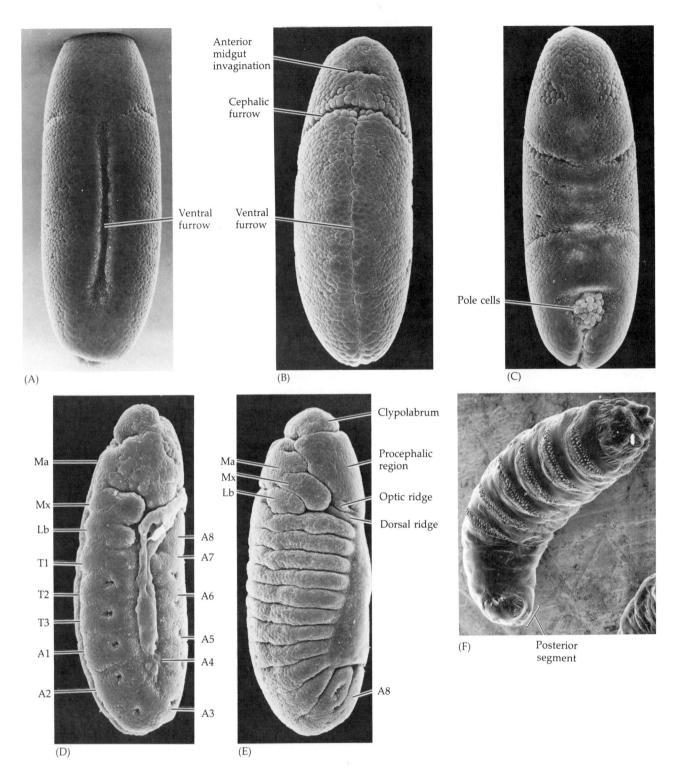

FIGURE 15.1
Gastrulation in *Drosophila*. (A) Ventral furrow beginning to form as cells migrate toward ventral midline. (B) Invagination of the ventral furrow. (C) Dorsal view of a slightly older embryo showing the migration of the pole cells and posterior endoderm into the embryo. (D) Lateral view showing fullest migration of germ band. Incipient segments of the germ band are labeled: Ma, Mx, and Lb correspond to the mandibular, maxillary, and labial head segments; T1–T3, thoracic segments; A1–A8, abdominal segments. (E) Germ band reversing direction. The clypolabrum, procephalic region, optic, and dorsal ridges of the head are distinguished. (F) Newly hatched first instar larva. (Courtesy of F. R. Turner.)

The cells remaining on the surface migrate toward the ventral midline to form the **germ band**. The cells of the germ band migrate posteriorly and then around the dorsal surface so that at the end of germ band formation, the cells destined to form the most posterior larval structures are located immediately behind the future head region. At this time, the body segments begin to appear, dividing the ectoderm and mesoderm. The germ band then contracts, placing the presumptive posterior segments into the posterior tip of the embryo. It is also during this time that the imaginal discs are set aside from the ectoderm and that the nervous system forms from two strips of ectodermal cells along the ventral midline. Unlike the neural tube in vertebrate nervous systems, the insect neural tube is formed ventrally, and it arises from neuroblasts differentiating from the ventral neurogenic ectoderm of each segment and from the procephalic neurogenic ectoderm of the head. Some of the genes involved in this differentiation (such as *Notch*) and patterning will be discussed in subsequent chapters.

THE ORIGINS OF ANTERIOR–POSTERIOR POLARITY

Overview

Development can be analyzed biochemically, anatomically, cellularly, genetically, and even evolutionarily. The analysis of *Drosophila* embryonic axis formation shows the power of the genetic approach to development. The *Drosophila* egg, embryo, larva, and adult are polar structures with respect to the anterior–posterior axis. The oocyte develops in such a way that one end (the future anterior) connects directly with the cytoplasm of the ovarian nurse cells, which transport proteins, ribosomes, and mRNA into the egg. The opposite end of the egg (the future posterior) accumulates the pole plasm that gives rise to the germ cells of the adult. The embryo, larva, and adult fly have a distinct head end and a distinct tail end, between which are repeating segmental units (Figure 15.2). Three of these segments form the thorax, while another eight segments form the abdomen. Each segment of the adult fly has its own identity. The first thoracic segment, for example, has only legs; the second thoracic segment contains legs and wings. The third thoracic segment has legs and halteres

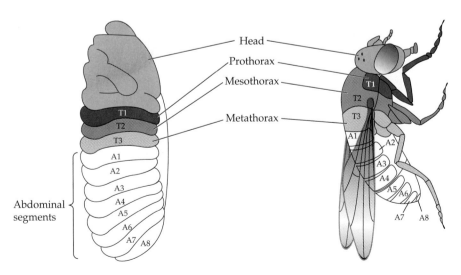

FIGURE 15.2

Comparison of larval and adult segmentation in *Drosophila*. The three thoracic segments can be distinguished by their appendages: T1 (prothoracic) has legs only; T2 (mesothoracic) has wings and legs; T3 (metathoracic) has halteres and legs.

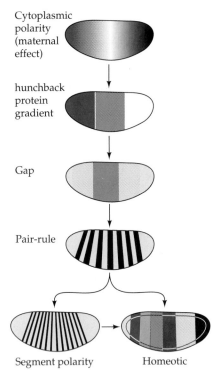

FIGURE 15.3
Generalized model of *Drosophila* pattern formation. The pattern is established by maternal effect genes that form gradients and regions of morphogenetic proteins. These morphogenetic determinants create a gradient of hunchback protein that differentially activates the gap genes that define broad territories of the embryo. The gap genes enable the expression of pair-rule genes, each of which divides the embryo into regions about two segment primordia wide. The segment polarity genes then divide the embryo into segment-sized units along the anterior–posterior axis. The combination of these genes defines the spatial domains of the homeotic genes that define the identities of each of the segments.

(balancers). Thoracic and abdominal segments can also be distinguished by their cuticle.

How does the polarity of the *Drosophila* egg give rise to the polarity of the fly body with its repetitive yet individuated segments? During the past decade, a model has emerged that synthesizes many of the data in this field (Figure 15.3). First, **maternal effect genes** in the fly ovaries produce proteins and messenger RNAs that are placed into different regions of the egg. These encode transcriptional and translational regulatory proteins that diffuse through the syncytial blastoderm and activate or repress the expression of certain zygotic genes. One of these proteins, bicoid, regulates the production of anterior structures, while another maternally specified protein, the product of *nanos* mRNA, regulates the formation of the posterior part of the embryo. Second, the zygotic genes regulated by these maternal factors are expressed in certain broad (about three segments wide), partially overlapping domains. These genes are called **gap genes** (because mutations in them cause gaps in the segmentation pattern), and they are among the first genes transcribed in the embryo. One of these gap genes, the *hunchback* gene, is directly regulated by the maternal factors placed into the egg and forms a gradient of hunchback protein. This directs the expression of the other gap genes. Third, the different concentrations of the gap gene proteins cause the **pair-rule genes** to be transcribed in the primordia of each alternate segment. The transcription pattern of each of these pair-rule genes gives a striped pattern of seven vertical bands along the anterior–posterior axis. The stripes of the pair-rule gene proteins activate the transcription of the **segment polarity genes**. Their mRNA and protein products form 14 bands that divide the embryo into segmentwide units. At the same time, proteins of the gap, pair-rule, and segment polarity genes interact to regulate another class of genes, the **homeotic genes**, whose transcription determines the developmental fate of each segment.

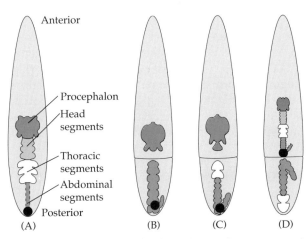

FIGURE 15.4
Sander's ligature experiments on the embryo of the leafhopper insect *Euscelis*. (A) Normal embryo seen in ventral view. The black ball at the bottom represents a cluster of symbiotic bacteria that marks the posterior pole. (B) After ligating the early embryo, partial embryos form, but the head and thoracic segments are missing from both embryos. (C) When ligated later (at the blastoderm stage), more of the missing segments are formed, but the embryos still lack the most central segments. (D) When the posterior pole cytoplasm is transplanted into an embryo ligated at the blastoderm stage, a small but complete embryo forms in the anterior half, while the posterior half forms an inverted partial embryo. These results can be explained in terms of gradients at the poles of the embryo that turn on one set of structures and repress the formation of others. (After Sander, 1960, and French, 1988.)

The maternal effect genes

Embryological evidence of polarity regulation by oocyte cytoplasm

Classic embryological experiments demonstrated that there are at least two "organizing centers" in the insect egg. One is the anterior organizing center, the other the posterior organizing center. Sander (1975) postulated that these two organizing areas form two gradients, one initiated at the anterior end and the other at the posterior end. Each of these gradients forms its own structures at the poles and interacts with the other gradient to form the central portion of the embryo. Sander based this model on experiments that involved ligating the embryo at various times during development and transplanting regions of polar cytoplasm from one region of the egg to another (Figure 15.4). First, if he moved cytoplasm from the posterior pole more anteriorly, he obtained a small embryo anterior to the posterior pole plasm, while extra segments, not organized into an embryo, formed behind it. Second, if he ligated the egg early in development, separating the anterior from the posterior region, one half developed into an anterior embryo and one half developed into a posterior embryo, but neither half contained the middle segments of the embryo. The later in development the ligature was made, the fewer middle segments were missing. Thus, it appeared that there were indeed gradients emanating from the two poles during cleavage and that these gradients interacted to produce the positional information determining the identity of each segment.

The possibility that mRNA is responsible for generating the anterior gradient was suggested in a series of experiments by Kalthoff and Sander (1968). They found that when the anterior portion of the *Smittia* (midge) egg was exposed to ultraviolet light at wavelengths capable of inactivating RNA (265 and 285 nm), the resulting embryo lacked its head and thorax. Instead, the embryos developed two abdomens with mirror-image symmetry (Figure 15.5). Further evidence that RNA is important in specifying the anterior portion of the fly embryo was obtained by Kandler-Singer and Kalthoff (1976), who submerged *Smittia* eggs in solutions containing various enzymes and then punctured the eggs in specific regions. Double abdomens resulted when RNase was permitted to enter the anterior end. Other enzymes did not cause this abnormality, nor did RNase effect this change when it entered other regions of the egg. Thus, it seemed like the egg sequestered an RNA that generated a gradient of anterior-forming material.

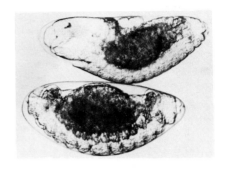

FIGURE 15.5
Normal and irradiated embryos of the midge *Smittia*. The normal embryo (top) shows a head on the left and abdominal segments on the right. The UV-irradiated embryo has no head region but has abdominal segments at both ends. (From Kalthoff, 1969, courtesy of K. Kalthoff.)

SIDELIGHTS & SPECULATIONS

Gradient models of positional information

How can cells be informed of their position in the embryo and then use that information to differentiate into the appropriate cell type? One major type of explanation proposes gradients of morphogenic substances (Boveri, 1901; Child, 1941; Wolpert, 1971). In these models, a soluble substance (morphogen) is posited to diffuse from a *source* (where it is produced) to a *sink* (where it is degraded), establishing a continuous range of concentrations within that region. Theoretical considerations (see Crick, 1970) suggest that such gradients could only function over relatively small distances, less than 100 cell diameters. In gradient models, the concentration of the morphogen changes over distance, the highest concentrations being near the

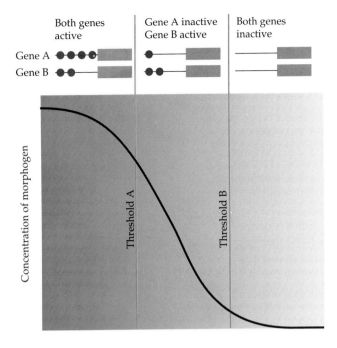

Both genes active | Gene A inactive Gene B active | Both genes inactive

Gene A
Gene B

Concentration of morphogen

Threshold A

Threshold B

FIGURE 15.6

A hypothetical model for gradients establishing positional information. The concentration of the morphogen drops from the source. In this diagram, the receptors for the morphogen are enhancer elements of two genes that control cell fate, but the receptors could also be cytoplasmic receptors or membrane receptors. One of the receptors (in this case, the enhancer on gene A) needs a high concentration of morphogen in order to act. At high concentrations of morphogen, both genes A and B are active. In moderate concentrations, only gene B is active. Where the morphogen concentration falls below another threshold, neither gene is active. (After Wolpert, 1978.)

source of the morphogen. The cells would have to have "sensors" that would respond differently to different concentrations of the gradient. This could be accomplished by having enhancer or promoter elements that could bind the morphogen at different strengths (Figure 15.6). For example, if a morphogen were being made at the anterior of the body, the genes responsible for organizing head development might have an enhancer that bound the morphogen poorly. Only when there was a large concentration of the morphogen present would that gene be active. The gene(s) responsible for thorax formation, on the other hand, might have an enhancer that bound the morphogen rather well, enabling it to respond to relatively low levels of that morphogen. The cells of the head would express *both* these genes, while the cells of the thorax would express only that gene whose enhancer could bind low amounts of the morphogen. The cells in the posterior portion of the body would not see any of this morphogen, and neither of these genes would be activated. In this way, cells could sense the presence of a morphogen and respond differentially, depending on the morphogen concentration. The sensor would not have to be an enhancer; it could just as well be a cell receptor in the cytoplasm or cell membrane.

There is evidence that certain morphogens may be

used for different processes within an embryo. In such cases, a piece of tissue transplanted from one region to another retains its organ-specific identity but can differentiate according to its new position within a field. Therefore, if cells that would normally become the middle of a *Drosophila* leg are removed from a late leg imaginal disc and placed into the center of an antennal disc (where they would become the end of the antenna), they differentiate into claws (the end of the *foot*) at the tip of the antenna. Specification of position (middle or tip) is separate from the determination of cell type (leg or antenna).

Most gradient models assume that all the cells that can respond to a gradient are equivalent. They all interpret the morphogen signal in the same way, and the concentration of morphogen that they receive determines their identity. However, the interpretation of gradients does not have to be linear. Take, for example, a series of exam grades that stretches uniformly from 100 to 60. In one scheme (a "linear" reading), a grade between 100 and 90 is A, 89–80 is B, 79–70 is C, and 69–60 is D. In another class (using a "curved" reading), 100–95 is A, 95–85 is B, 84–70 is C, and 69–60 is D. Nijhout (1981) has used a two-gradient model to explain the development of "eyespot" patterns on butterfly wings. One gradient consists of a linear diffusion of a morphogen. The second gradient involves the interpretation of this morphogen; in other words, the sensitivity threshold of the cells involved differs at different regions of the wings. The existence of the second gradient gives rise to an elliptical spot, not the circular spot that would result if the sensitivity gradient were absent (Figure 15.7).

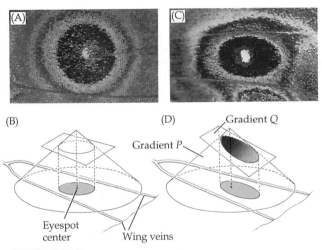

(A) | (C)

(B) | (D) Gradient *Q*

Gradient *P*

Eyespot center | Wing veins

FIGURE 15.7

Gradient model of positional information proposed to explain butterfly wing spots. (A) Photograph of an eyespot on the wing of Morpho peleides. *(B) Diagram of a two-gradient model that may explain the way the spot was generated. The origin of the morphogen is at the center of the spot and corresponds to the apex of a cone, the height of which reflects its concentration. Concentration Q represents the level of morphogen needed to reach the threshold sensitivity for the formation of color in those wing cells. (C) Photograph of the wing of* Smyrna blomfildia, *in which the eyespots are elliptical. (D) Different orientations of the sensitivity gradient Q could result in such elliptical eyespots. (After Nijhout, 1981, courtesy of H. F. Nijhout.)*

The anterior organizing center: The gradient of bicoid protein

In 1988, the gradient hypothesis was united with a genetic approach to the study of *Drosophila* embryogenesis. If there were gradients, what were the morphogens whose concentrations changed over space? What were the genes that shaped these gradients? And did these substances act by activating or inhibiting certain genes in the areas where they were concentrated? Christiane Nüsslein-Volhard led a research program that found that one set of genes encoded gradient morphogens for the anterior part of the embryo, another set of genes encoded the morphogens responsible for organizing the abdominal region of the embryo, and a third set of genes encoded proteins that produced the terminal regions at both ends of the embryo (Figure 15.8; Table 15.1).

In *Drosophila*, the phenotype of the *bicoid* mutant is very interesting if one is thinking in terms of gradients. Instead of having anterior structures (acron, head, thorax) followed by abdominal structures and telson, the structure of the *bicoid* mutant is telson, abdomen, telson (Figure 15.9). It would appear that these embryos lack whatever morphogen is needed for anterior stuctures. Moreover, one could postulate that the substance these mutants lack is the one postulated by Sander and Kalthoff to turn on genes for anterior structures and to turn off genes for the telson structures.

Further studies have strengthened the view that the product of the wild-type *bicoid* (*bcd*) gene is the morphogen that controls anterior development. First, *bicoid* is a maternal effect gene. The messenger RNA from the mother's *bicoid* genes are placed into the embryo by the mother's ovarian cells (Frigerio et al., 1986; Berleth et al., 1988). The *bicoid* mRNA is strictly localized in the anterior portion of the oocyte (Figure 15.10A). This would not be compatible with its proposed role as a morphogen

TABLE 15.1
Maternal effect genes that effect the anterior–posterior polarity of the *Drosophila* embryo

Gene	Phenotype	Proposed function and structure
ANTERIOR GROUP		
bicoid (bcd)	Head and thorax deleted, replaced by inverted telson	Graded anterior morphogen; contains homeodomain
exuperantia (exu)	Anterior head structures deleted	Anchors *bicoid* mRNA
swallow (swa)	Anterior head structures deleted	Anchors *bicoid* mRNA
POSTERIOR GROUP		
nanos (nos)	No abdomen	Posterior morphogen
tudor (tud)	No abdomen, no pole cells	Localization of nanos
oskar (osk)	No abdomen, no pole cells	Localization of nanos
vasa (vas)	No abdomen, no pole cells; oogenesis defective	Localization of nanos
valois (val)	No abdomen, no pole cells; cell arization defective	Stabilization of the nanos localization complex
pumilio (pum)	No abdomen	Transports posterior signal to abdomen
TERMINAL GROUP		
torso (tor)	No terminals	Possible morphogen for terminals
trunk (trk)	No terminals	Transmits *torsolike* signal to *torso*
fs(1)Nasrat[fs(1)N]	No terminals; collapsed eggs	Transmits *torsolike* signal to *torso*
fs(1)pole hole[fs(1)ph]	No terminals; collapsed eggs	Transmits *torsolike* signal to *torso*

Source: After Anderson, 1989.

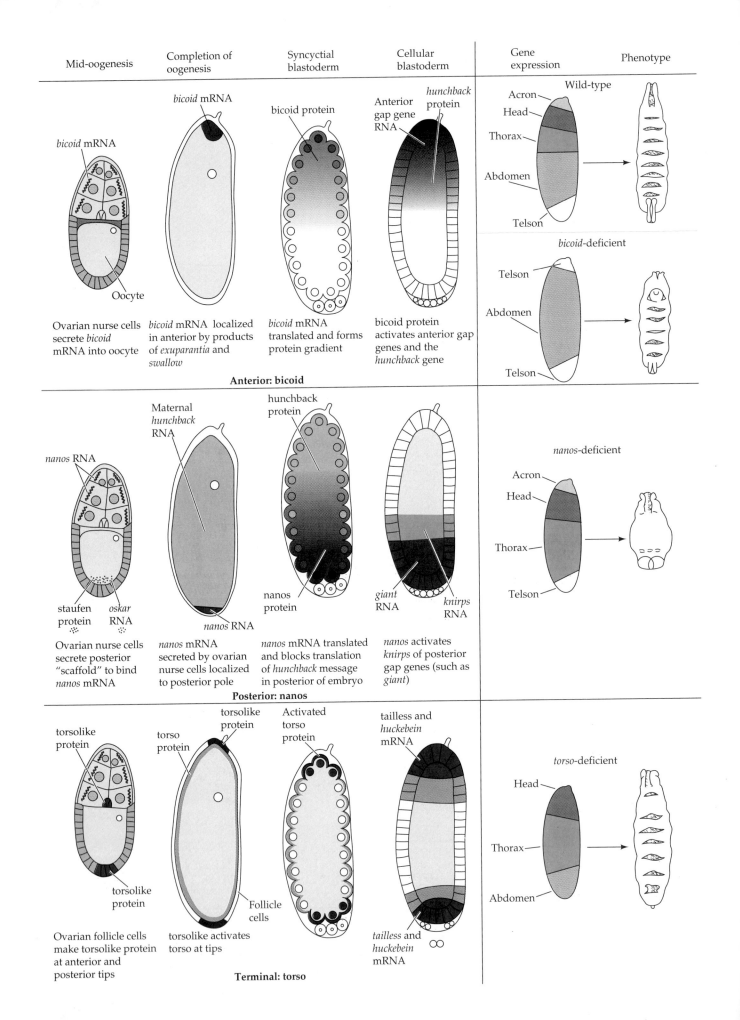

Mid-oogenesis	Completion of oogenesis	Syncytial blastoderm	Cellular blastoderm	Gene expression	Phenotype

Anterior: bicoid

bicoid mRNA

Ovarian nurse cells secrete *bicoid* mRNA into oocyte

Oocyte

bicoid mRNA

bicoid mRNA localized in anterior by products of *exuparantia* and *swallow*

bicoid protein

bicoid mRNA translated and forms protein gradient

Anterior gap gene RNA

hunchback protein

bicoid protein activates anterior gap genes and the *hunchback* gene

Wild-type

Acron
Head
Thorax
Abdomen
Telson

bicoid-deficient

Telson
Abdomen
Telson

Posterior: nanos

nanos RNA

staufen protein

oskar RNA

Ovarian nurse cells secrete posterior "scaffold" to bind *nanos* mRNA

Maternal *hunchback* RNA

nanos RNA

nanos mRNA secreted by ovarian nurse cells localized to posterior pole

hunchback protein

nanos protein

nanos mRNA translated and blocks translation of *hunchback* message in posterior of embryo

giant RNA

knirps RNA

nanos activates *knirps* of posterior gap genes (such as *giant*)

nanos-deficient

Acron
Head
Thorax
Telson

Terminal: torso

torsolike protein

torsolike protein

Ovarian follicle cells make torsolike protein at anterior and posterior tips

torso protein

torsolike protein

Follicle cells

torsolike activates torso at tips

Activated torso protein

tailless and *huckebein* mRNA

tailless and *huckebein* mRNA

nanos activates

torso-deficient

Head
Thorax
Abdomen

FIGURE 15.8
Three independent genetic pathways interact to form the anterior–posterior axis of the *Drosophila* embryo. In each case, the initial asymmetry is established during oogenesis, and the pattern is organized by the maternal products soon after fertilization. The realization of the pattern comes about when the localized maternal products activate or repress specific zygotic genes in different regions of the embryo. (After St. Johnston and Nüslein-Volhard, 1992.)

whose influence extends as a gradient over half the length of the egg. However, Driever and Nüsslein-Volhard (1988a) have shown that when bicoid protein is translated from this mRNA during early cleavage, it forms a gradient with its highest concentration in the anterior of the egg and reaches background levels in the posterior third of the egg. Moreover, this protein soon becomes concentrated in the embryonic *nuclei* in the anterior portion of the embryo (Figure 15.10B,C; Plate 14A).

Further evidence that bicoid protein is the anterior morphogen came from experiments that altered the steepness of the gradient. Two genes, *exuperantia* and *swallow*, are responsible for keeping the *bicoid* message at the anterior pole of the egg. In their absence, the *bicoid* message diffuses further into the posterior of the egg, and the gradient of bicoid protein is shallower (Driever and Nüsslein-Volhard, 1988b). The phenotype produced by these two mutants is similar to that of bicoid-deficient embryos, but is less severe. These embryos lack their most anterior structures and have an extended mouth and thoracic region. Thus, by altering the gradient of bicoid protein, one correspondingly alters the fate of the embryonic regions.

Confirmation that the bicoid protein alone is the morphogen responsible for initiating head and thorax formation came from the experiments wherein purified *bicoid* mRNA was injected into early-cleavage embryos (Figure 15.11; Driever, et al., 1990). When injected into the anterior of *bicoid*-deficient embryos (whose mother lacked *bicoid* genes), the *bicoid* mRNA rescued the embryos and caused them to have normal anterior–posterior polarity. Moreover, any location in the embryos where the *bicoid* messages were injected became the head. If *bicoid* mRNA was injected into the center of the embryo, that middle region became the head and the regions on either side of it became thorax structures. If *bicoid* mRNA was placed in the *posterior* pole of a wild-type embryo (with its own endogenous *bicoid* message in its anterior pole), two heads emerged, one at either end. Therefore, the *bicoid* gene is now thought to encode the anterior morphogen of the *Drosophila* embryo.

(A)

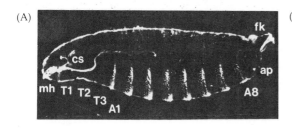

(B)

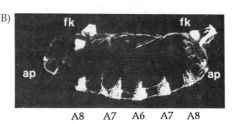

FIGURE 15.9

Phenotype of a strongly affected embryo from a female deficient in the *bicoid* gene. (A) Wild-type cuticle pattern. (B) *bicoid* mutant. The head and thorax have been replaced by a second set of posterior telson structures. Abbreviations: fk, filzkörper; ap, anal plates (both telson structures). T1–T3, thoracic segments; A1, A8, the two terminal abdominal segments; mh, cs, head structures. (From Driever et al., 1990, courtesy of W. Driever.)

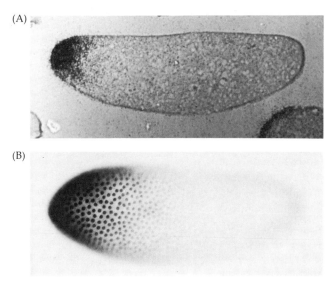

(A)

(B)

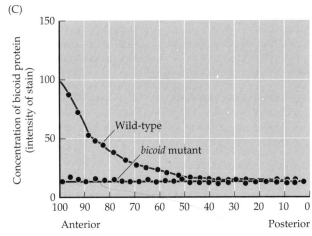

(C)

FIGURE 15.10

Gradient of bicoid protein in the early *Drosophila* embryo. (A) Localization of *bicoid* mRNA to the anterior tip of the embryo. (B) Gradient of bicoid protein shortly after fertilization. Note that the concentration is greatest anteriorly and trails off posteriorly. Notice also that bicoid protein is concentrated in the nuclei of the embryo. (C) Densitometric scan of the bicoid protein gradient. The upper curve represents the gradient of bicoid protein in wild-type embryos. The lower curve represents bicoid protein in embryos of *bicoid*-deficient mothers. (A from Kaufman et al., 1990; B and C from Driever and Nüsslein-Volhard, 1988a. Photographs courtesy of the authors.)

FIGURE 15.11

Schematic representation of the experiments demonstrating that the *bicoid* gene encodes the morphogen responsible for head structures in *Drosophila*. The phenotypes of the *bicoid*-deficient and wild-type embryos are shown at the sides. When *bicoid*-deficient embryos are injected with *bicoid* mRNA, the point of injection forms the head structures. When the posterior pole of an early-cleavage wild-type embryo is injected with *bicoid* mRNA, head structures form at both poles. (After Driever et al., 1989.)

The next question then emerged: How might a gradient in bicoid protein control the determination of the anterior–posterior axis? One clue was that the bicoid protein is found in nuclei and contains a homeodomain DNA-binding region (see Chapter 10). These facts suggested that bicoid protein could bind DNA and regulate gene expression. The earliest zygotic genes expressed in *Drosophila* are the gap genes. One of them, *hunchback* (*hb*) is active in the anterior half of the embryo—the region where bicoid protein is seen. Mutants lacking hunchback protein lack mouthparts and the thorax structures. In the late 1980s, two laboratories independently demonstrated that bicoid protein binds to and activates the *hunchback* gene

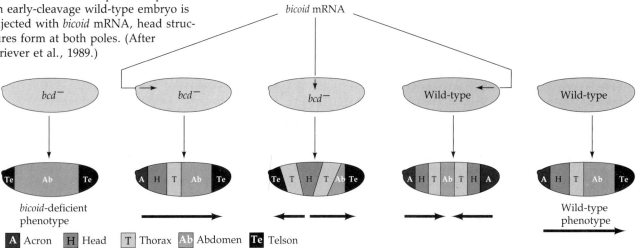

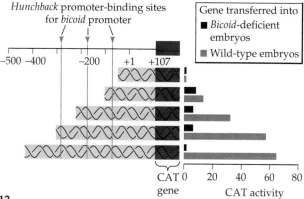

FIGURE 15.12
Influence of bicoid protein in activating the *hunchback* gene. Different regions of the *hunchback* promoter were fused to the CAT reporter gene and injected into either wild-type embryos or embryos from *bicoid*-deficient mothers. The more bicoid-binding sites there were in the promoter region, the more effective its expression in wild-type embryos. In embryos without bicoid protein, no transcription resulted from any of the *hunchback* promoter-driven genes. (After Driever and Nüsslein-Volhard, 1989.)

(Driever and Nüsslein-Volhard, 1989; Struhl et al., 1989). The *hunchback* gene is thought to repress abdominal-specific genes, thereby allowing the region of *hunchback* expression (hence, the region of *bicoid* expression) to form the head and thorax. Using DNase footprinting (proteins are bound to a segment of DNA, DNase is added, and the only DNA that remains is that protected by the DNA-binding protein), the researchers found that bicoid protein binds to five sites in the upstream promoter region of the *hunchback* gene. These sites all have the consensus sequence 5′−TCTAATCCC−3′.

But binding does not necessarily mean activation. The activation of this gene by bicoid protein was shown by fusing these *hunchback* promoter sites to the chloramphenicol acetyltransferase (CAT) reporter gene and injecting these genes into early *Drosophila* embryos. In all cases, bicoid protein was needed to activate the reporter genes. If injected into *bicoid*-deficient embryos, no CAT was produced (Figure 15.12). It was also shown that while some activation is seen when only one of the five bicoid protein-binding sequences was present, the full expression of the reporter gene (and presumably of *hunchback*) came when three of the five sites were present. Thus, the bicoid protein gradient probably functions by activating *hunchback* gene transcription in the anterior portion of the embryo.

Driever and co-workers (1989) predicted that at least one other anterior gene besides *hunchback* must be activated by bicoid. First, deletions of *hunchback* produce only some of the defects seen in the *bicoid* mutant phenotype. Second, as we saw in the *swallow* and *exuparentia* experiments, only moderate levels of bicoid protein are needed to activate thorax formation (i.e., *hunchback* gene expression), but head formation needs higher concentrations. Driever et al. (1989) predicted that the promoters of such a head-specific gap gene would have low-affinity binding sites for bicoid protein. This gene would be activated only at extremely high concentrations of bicoid protein—that is, near the anterior tip of the embryo. Since then, three gap genes of the head have been discovered that are dependent on very high concentrations of bicoid protein for their expression (Cohen and Jürgens, 1990; Finkelstein and Perrimon, 1990). The *buttonhead (btd)*, *empty spiracles (ems)*, and *orthodenticle (otd)* genes are needed to specify the progressively anterior regions of the head.

FIGURE 15.13
Schematic model for the action of bicoid and nanos proteins in specifying the anterior–posterior axis of the embryo. In the anterior region (color), bicoid protein activates *hunchback* transcription, and nanos protein is not there to stop the translation of the *hunchback* transcripts into protein. In the posterior region (gray), nanos represses *hunchback* translation, thereby allowing abdominal genes to be expressed.

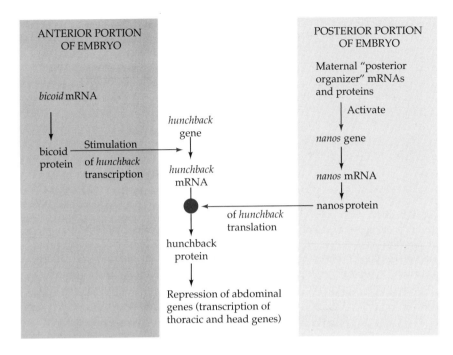

The posterior organizing center: Activating and transporting the *nanos* product

The posterior organizing center is defined by the activities of the *nanos* gene (Lehmann and Nüsslein-Volhard, 1991; Wang and Lehmann, 1991; Wharton and Struhl, 1991). The *nanos* mRNA is produced in the ovary and is transported into the egg, where it becomes bound in the posterior region (farthest away from the ovarian nurse cells). The products of several other genes [*oskar*, *valois*, *vasa*, *staufen*, and *tudor*—the same gene products that place the germ plasm determinant into the posterior pole plasm (Chapter 14)] are needed to place the *nanos* mRNA into the posterior part of the egg.* If *nanos* or any other of these maternal effect genes are absent in the mother, no embryonic abdomen forms (Lehmann and Nüsslein-Volhard, 1986; Schüpbach and Wieschaus, 1986).

The *nanos* message is translated into protein soon after fertilization, just as the *bicoid* message is. Tautz (1988) has shown that during normal abdomen formation, the protein product of the *nanos* gene represses the *translation* of *hunchback* mRNA (Figure 15.13). This *hunchback* mRNA is initially present throughout the embryo, although more can be made from zygotic nuclei if they are activated by bicoid protein. Thus, the combination of bicoid and nanos protein causes a gradient of hunchback protein across the egg (Figure 15.14). The bicoid protein activates *hunchback* gene transcription in the anterior part of the embryo, while the nanos protein inhibits the translation of *hunchback* mRNA in the posterior part of the embryo. If the *nanos* gene product were not present, hunchback protein would be made throughout the embryo, and presumably would inhibit the abdomen-generating gap genes such as *knirps* (Hülskamp et al., 1989; Irish et al., 1989; Struhl, 1989). The *hunchback* gene, therefore, appears to

*Like the placement of the *bicoid* message, the location of the *nanos* message is determined by its 3′ untranslated region. If the bicoid 3′ UTR is placed onto the protein-encoding region of *nanos* mRNA, the *nanos* message gets placed in the anterior of the egg. When the RNA is translated, the nanos protein inhibits the translation of *hunchback* mRNA, and the embryo forms two abdomens—one in the anterior and one in the posterior of the embryo (Gavis and Lehmann, 1992).

FIGURE 15.14

Formation and function of the hunchback protein gradient. (A) Maternal mRNAs for bicoid and nanos are placed in the anterior and posterior of the egg, respectively. The concentration of maternal *hunchback* message is constant throughout the length of the embryo. (B) Upon fertilization, *bicoid* and *nanos* messages are translated into their respective proteins. The bicoid protein stimulates new hunchback synthesis; the nanos protein inhibits the translation of *hunchback* message. This leads to a gradient of hunchback protein. (C) The concentration of hunchback protein specifies the positions where the gap genes *giant*, *Krüppel*, and *knirps* are transcribed. (After Struhl et al., 1992.)

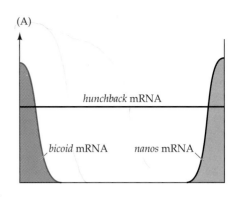

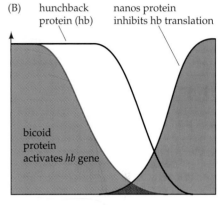

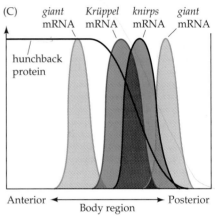

be the focal point, under the regulation of both the anterior and the posterior organization centers long known to exist in insect development. These studies on *nanos* and *bicoid* expression can now explain why destroying *bicoid* mRNA (for example, with UV light or RNase) allows the formation of a second abdomen, and why ligation procedures stop the formation of the central segments of the fly embryo.

Although *nanos* mRNA is present at the extreme posterior pole, its activity is somewhat anterior in the abdomen. This movement appears to be the function of the *pumilio* gene product (Lehmann and Nüsslein-Volhard, 1987). If pole plasm is transferred from one *pumilio* mutant embryo into the *abdominal region* of another *pumilio* embryo, the host mutant embryo is rescued and forms an abdomen. Moreover, if one genetically deletes a number of segments between the posterior pole and the abdomen region (thereby shortening the distance between the posterior organizing center and the abdomen), abdominal segmentation is restored in *pumilio* mutant embryos. The product of the *nanos* gene is considered to be the posterior morphogen of the *Drosophila* anterior–posterior axis.

The terminal gene group

If both the anterior and the posterior organizing centers are nonfunctional, an embryo still can develop some anterior–posterior pattern (Nüsslein-Volhard et al., 1987). When females are made doubly mutant for both the anterior and the posterior morphogens, their embryos produce two telsons, one at each end of the embryo. Thus, there exists a third set of maternal effect genes that help to create the extremes of the anterior–posterior axis. Mutations in these *terminal genes* result in the loss of the unsegmented extremities of the organism: the anterior acron and the posterior telson. Therefore, the terminal gene set defines the boundaries of the segmented parts of the body. In the absence of these gene products, the segmented portion of the embryo expands to the extremities (Degelmann et al., 1986; Klingler et al., 1988).

The critical gene here appears to be *torso* (see Figure 15.8). The *torso* mRNA is synthesized by the ovarian cells before fertilization, but it is not translated in the oocyte until after fertilization. The torso protein becomes localized in the cell membrane throughout the egg (Casanova and Struhl, 1989). In the absence of *torso* product, there is neither acron nor telson. Rather, the segmented portion of the embryo takes up all the egg. Conversely, in a dominant mutation of this gene, the entire anterior half of the embryo is converted into acron and the entire posterior half into telson. Stevens and her colleagues (1990) have shown that torso protein is probably activated by follicle cells at either pole of the embryo. The activator of the torso protein is *torsolike*, and a negative mutation in the *torsolike* gene creates a phenotype almost identical to *torso*. The *torsolike* gene is expessed in the follicle cells that surround the egg. By somatic mutation,

Stevens and colleagues showed that if the follicle cells at the poles of the egg chamber are deficient in the *torsolike* gene (even if the other follicle cells express the wild-type allele of this gene), the resulting embryo will have a phenotype similar to *torso*.

Activation of *torso* initiates a pathway that activates the *tailless* and *huckebein* gap genes. These genes cause the differentiation of the acron- and telson-specific genes. If the terminal genes act alone, they activate the *tailless* gap gene and cause the formation of a telson. However, if the *bicoid* gene product is also present when *tailless* is functioning, the region forms an acron (Pignoni, et al., 1992).

The anterior–posterior axis of the embryo is seen, therefore, to be specified by three sets of genes: those that define the anterior organizing center, those that define the posterior organizing center, and those that define the terminal boundary region. The anterior organizing center is located at the anterior end of the embryo and acts through a gradient of bicoid protein that activates anterior-specific gap genes and suppresses posterior-specific gap genes. The posterior organizing center is located at the posterior pole and acts through the formation of nanos protein, which gets transported into the abdominal region. Here, nanos inhibits the inhibitor of abdominal-specific gene expression and activates those genes that form the abdomen. The boundaries of the acron and telson are defined by the *torso* gene product, which is activated at the tips of the embryo.

The segmentation genes

The commitment of cell fate in *Drosophila* appears to be a two-step process: specification and determination (Slack, 1983). Early in development, the fate of a cell is dependent upon environmental cues such as those provided by the gradients mentioned above. This **specification** of cell fate is flexible, and it can still be altered in response to environmental signals. Eventually, the cells undergo a transition from this loose type of commitment to an irreversible **determination**. Here, the fate of a cell has become cell-intrinsic.* The transition from specification to determination in *Drosophila* is mediated by the **segmentation genes**. Segmentation genes divide the early embryo into a repeating series of segmental primordia along the anterior–posterior axis. Mutations in segmentation genes cause the embryo to lack certain segments or parts of segments, and these mutations show the existence of three classes of segmentation genes (Table 15.2). Often these mutations affect **parasegments**, regions of the embryo that are separated by mesodermal thickenings and ectodermal grooves, which divide the embryo into 14 regions (Martinez-Arias and Lawrence, 1985). The parasegments of the embryo do not become the segments of the larva or adult. Rather, they include the posterior part of an anterior segment and the anterior portion of the segment behind it (Figure 15.15).

The **gap genes** are activated or repressed by the maternal effect genes and divide the embryo into broad regions containing several segmental primordia. The *Krüppel* gene, for example, is expressed primarily in parasegments 4–6 in the center of the *Drosophila* embryo (Plate 14B; Figures

Early embryo (normal)   Later embryo (normal)   Larva (normal)   Larva (lethal mutant)

Area of gene action   Area of gene action   Denticle bands

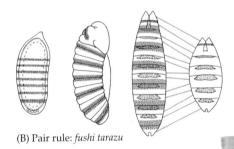

(A) Gap: *Krüppel*

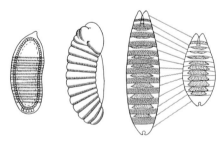

(B) Pair rule: *fushi tarazu*

(C) Segment polarity: *engrailed*

FIGURE 15.15
Three types of segmentation pattern mutants. The left panel shows the cleavage-stage embryo, with the region where the particular gene is normally transcribed in wild-type embryos shown in color. In the three panels to the right, the areas shown in color are deleted when these mutants develop. (After Mange and Mange, 1990.)

*Aficionados of information theory will recognize that the process by which the anterior–posterior information in morphogenetic gradients is transferred to discrete domains of homeotic selector genes represents a transition from analogue to digital specification. Specification is analogue, determination digital. This enables the transient information of the gradients in the syncytial blastoderm to be stabilized so that it can be utilized much later in development (Baumgartner and Noll, 1990).

TABLE 15.2
Major loci affecting segmentation pattern in *Drosophila*

Category	Loci	Category	Loci
Gap genes	*Krüppel (Kr)*	Pair-rule genes (secondary)	*fushi tarazu (ftz)*
	knirps (kni)		*odd-paired (opa)*
	hunchback (hb)		*odd-skipped (sdd)*
	giant (gt)		*sloppy-paired (slp)*
	tailless (tll)		*paired (prd)*
	huckebein (hkb)	Segment polarity genes	*engrailed (en)*
	buttonhead (btd)		*wingless (wg)*
	empty spiracles (ems)		*cubitus interruptus^D (cb^D)*
Pair-rule genes (primary)	*orthodenticle (otd)*		*hedgehog (hh)*
	hairy (h)		*fused (fu)*
	even-skipped (eve)		*armadillo (arm)*
	runt (run)		*patched (ptc)*
			gooseberry (gsb)

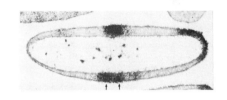

FIGURE 15.16
Expression of the *Krüppel* gene in the center and posterior of the *Drosophila* embryo (arrows). A 2.5-hour-old embryo was hybridized with cDNA that recognized *Krüppel* mRNA accumulations. (From Levine and Harding, 1989, courtesy of M. Levine.)

15.15A and 15.16); the absence of *Krüppel* causes the embryo to lack these regions. Next, the **pair-rule genes** subdivide the broad gap gene domains into segments. Mutations of the pair-rule genes (as in *fushi tarazu*; Plate 14C) usually delete portions of every other segment. Figures 15.15B and 15.18 compare the morphology of the wild-type embryo with that of the *fushi tarazu* mutant. Finally, the **segment polarity genes** are responsible for maintaining certain repeated structures within each segment. Mutations in this group of genes cause a portion of each segment to be deleted and replaced by a mirror-image structure of another portion of the segment. For instance, the *engrailed* gene is needed for the maintainance of the anterior–posterior boundary between segments. In *engrailed* mutants (Figure 15.15C; Plate 14D), adjacent segments fuse together.

The regulation of these different genes is structured as a hierarchy, as shown in Figure 15.3. A cascade is initiated by the maternal effect genes that control the activation of the gap genes. The gap genes interact to control their own transcription, and acting as a group they control the patterns of pair-rule gene expression. The pair-rule genes interact among themselves to form the repetitive segmental divisions of the body, and they also control the pattern of segment polarity gene expression. The

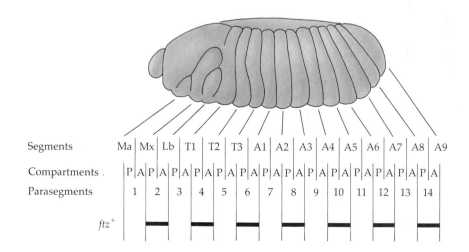

FIGURE 15.17
Segments and parasegments. A and P represent the anterior and posterior compartments of the segments. The parasegments are shifted one compartment forward. Ma, Mx, and Lb represent the three head segments (mandibular, maxillary, and labial), T segments are thoracic, and A segments are abdominal. The parasegments are numbered 1 through 14. Underneath the map are the boundaries of gene expression observed by the in situ hybridization of radioactive cDNA from the pair-rule gene *fushi tarazu* (*ftz*). (After Martinez-Arias and Lawrence, 1985.)

FIGURE 15.18
Defects seen in the *ftz⁻* embryo. (A) Scanning electron micrograph of wild-type embryo seen in lateral view. (B) Same stage of an *ftz⁻* embryo. The white lines connect the homologous portions of the segmented germ band. (C) Diagram of wild-type embryonic segmentation. The shaded regions show the parasegments of the germ band that are missing in the *ftz⁻* embryo. (After Kaufman et al., 1990, photographs courtesy of T. Kaufman.)

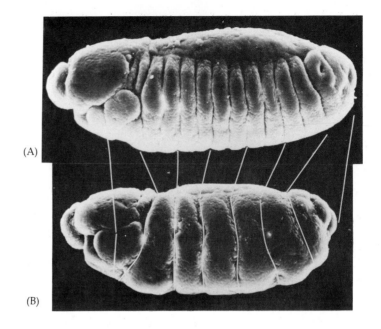

(A)

(B)

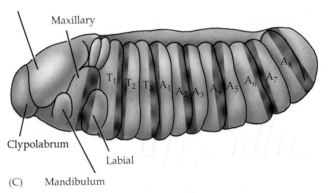

Maxillary

T$_1$ T$_2$ T$_3$ A$_1$ A$_2$ A$_3$ A$_4$ A$_5$ A$_6$ A$_7$ A$_8$

Clypolabrum

Labial

(C)   Mandibulum

pair-rule and gap genes also interact to regulate the **homeotic genes** that determine the structure of each segment. By the end of the cellular blastoderm stage, each segment primordium has been given an individual identity by its unique constellation of gap, pair-rule, and homeotic gene products (Levine and Harding, 1989).

The gap genes

The gap genes were originally defined by a series of mutations whose embryos lacked groups of consecutive segments (Nüsslein-Volhard and Weischaus, 1980). As shown in Figure 15.19, deletions caused by the *hunchback* (*hb*), *Krüppel* (*Kr*), and *knirps* (*kni*) genes span the entire segmented region of the *Drosophila* embryo. The *giant* (*gt*) gap gene overlaps with these three, and the phenotypes of the *tailless* and *huckebein* mutants delete portions of the unsegmented termini of the embryo.

The expression of these genes is a dynamic situation. There is usually a low level of transcriptional activity across the entire embryo that becomes defined into discrete regions of high activity as cleavage continues (Jäckle et al., 1986). The critical element appears to be the expression of the hunchback protein, which is stimulated by bicoid protein and inhibited by nanos protein. By the end of nuclear cycle 12, the expression of hunchback protein is high across the anterior part of the embryo and then forms

FIGURE 15.19
Segment deletions in gap gene mutants. The table below the photographs indicates missing segmental regions with white bars. In *hunchback* mutants, the region is extended (gray area) if the mother as well as the zygote lacks *hunchback* gene activity. (After Gaul and Jäckle, 1990; photographs courtesy of E. Wieschaus.)

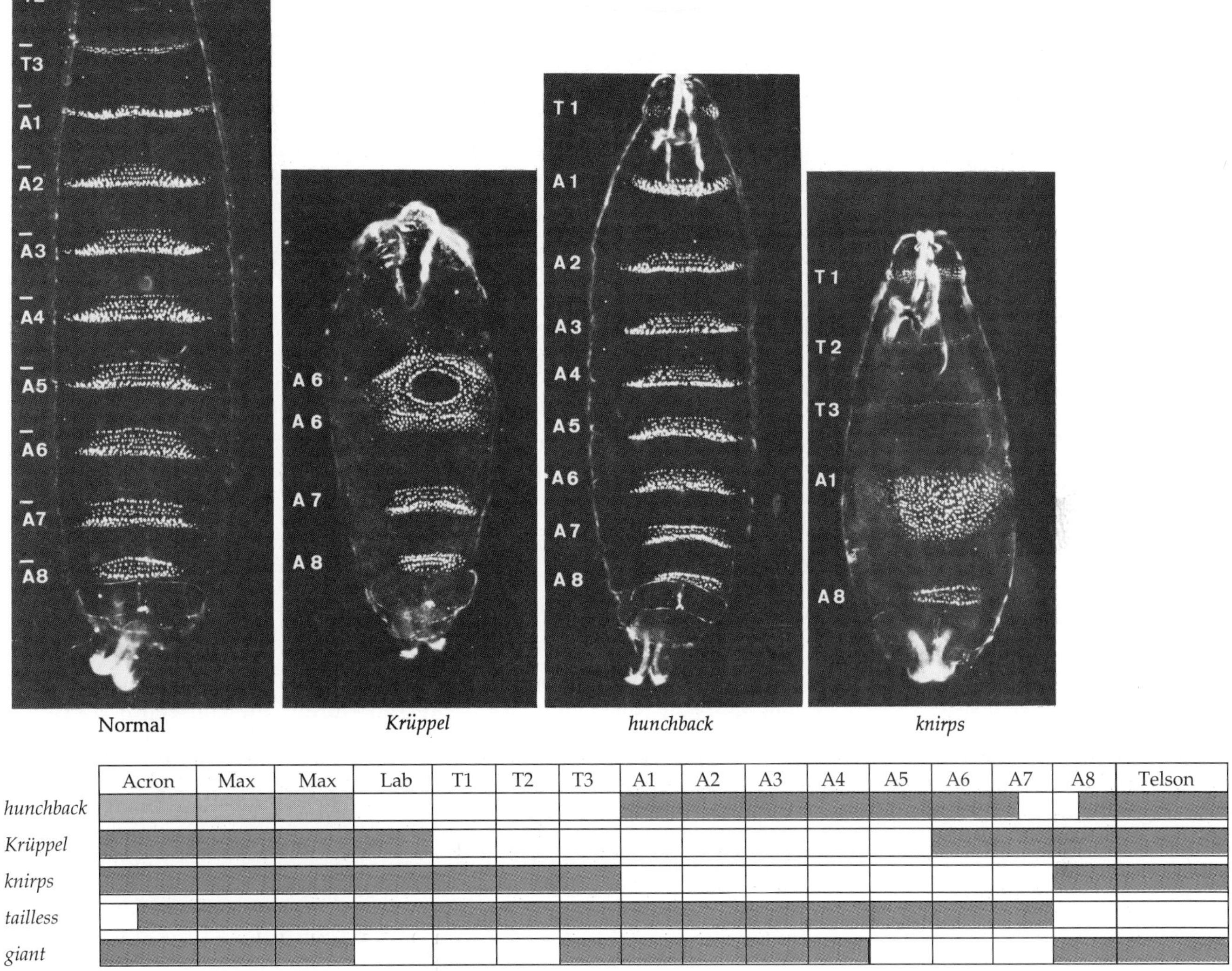

T1
T2
T3
A1
A2
A3
A4
A5
A6
A7
A8
Normal
A 6
A 6
A 7
A 8
Krüppel
T 1
A 1
A 2
A 3
A 4
A 5
A 6
A 7
A 8
hunchback
T 1
T 2
T 3
A 1
A 8
knirps
Acron
Max
Max
Lab
T1
T2
T3
A1
A2
A3
A4
A5
A6
A7
A8
Telson
hunchback
Krüppel
knirps
tailless
giant

a steep gradient through about 15 nuclei, such that only the last third of the embryo has undetectable hunchback expression. The transcription patterns of the gap genes are initiated by the different concentrations of the hunchback protein. High levels of hunchback protein induce the expression of *giant,* while the *Krüppel* transcript appears over the region where hunchback declines (Figure 15.14C). The *knirps* gene gets turned on at even lower concentrations of hunchback protein. Finally, when there is no hunchback protein detectable, the *giant* gene is activated again, this time by proteins from the posterior terminus (Struhl et al., 1992).

The original placement of these proteins is dependent on hunchback protein concentration, but they become stabilized and maintained by interactions between the different gap genes themselves.* For instance, *Krüppel* gene expression is negatively regulated on its anterior boundary

*These interactions between genes and gene products are facilitated by the fact that all these reactions occur within a syncytium. The cell membranes have not yet formed.

by hunchback protein and on its posterior boundary by knirps and tailless proteins (Jäckle et al., 1986; Harding and Levine, 1988; Hoch et al., 1992). If *hunchback* activity is lacking, the domain of *Krüppel* expression extends anteriorly. If *knirps* activity is lacking, *Krüppel* gene expression extends more posteriorly. The boundaries between the regions of gap gene transcription are probably created by mutual repression. Just as the *giant* and *hunchback* proteins can control the anterior boundary of *Krüppel*, so can *Krüppel* determine the posterior boundaries of *giant* and *hunchback* transcription. If an embryo lacks the *Krüppel* gene, *hunchback* transcription continues into the area usually alloted to *Krüppel* (Jäckle et al., 1986; Kraut and Levine, 1991). These boundary-forming inhibitions are thought to be directly mediated by the gap gene products, because all four major gap genes (*hb, gt, Kr,* and *kni*) encode DNA-binding proteins that can activate or repress transcription (Knipple et al., 1985; Gaul and Jäckle, 1990; Capovilla et al., 1992).

Moreover, these interactions are highly specific, and the product of one gap gene can bind to the promoters of other gap genes. DNase I footprinting shows that the protein encoded by the wild-type *Krüppel* gene binds to the promoter region of the *hunchback* gene (which it inhibits) and to the promoter region of the *knirps* gene (which it stimulates). The *knirps* promoter region is also recognized by the protein product of the *tailless* gene that inhibits *knirps* transcription. Hunchback protein (in addition to recognizing the *Krüppel* promoter) also recognizes its own promoter, suggesting that *hunchback* is involved in regulating its own expression (Pankratz et al., 1990; Štanojević et al., 1989; Treisman and Desplan, 1989).

The pair-rule genes

The first indication of repeated segmentation in the fly embryo comes when the pair-rule genes are expressed during the thirteenth division cycle. The transcription patterns of these genes are striking in that they each divide the embryo into the areas that are the precursors of the segmental body plan. As can be seen in Figure 15.20 and Plate 14C, one vertical stripe of nuclei (the cells are just beginning to form) expresses this gene, then another stripe of nuclei does not express it, and then another stripe of nuclei expresses this gene. The result is a "zebra stripe" pattern along the anterior–posterior axis, dividing the axis into 15 subunits (Hafen

FIGURE 15.20
Relationship between three pair-rule genes. (A) The transcription pattern of the *even-skipped* gene. (B) Double-labeling of the *even-skipped* and *fushi tarazu* genes showing that *fushi tarazu* is expressed between the domains of *even-skipped* transcription. (C) Diagram of the expression of three pair-rule genes—*even-skipped, hairy,* and *fushi tarazu*—in individual nuclei at the cellular blastoderm stage. (After Carroll et al., 1988; photographs courtesy of B. Thalley and S. Carroll.)

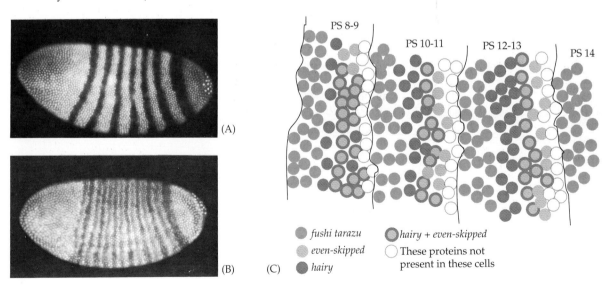

(A)

(B)   (C)

PS 8-9   PS 10-11   PS 12-13   PS 14

○ *fushi tarazu*   ◑ *hairy + even-skipped*
○ *even-skipped*   ○ These proteins not present in these cells
● *hairy*

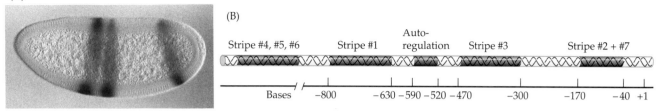

(A)

(B)

| Stripe #4, #5, #6 | Stripe #1 | Auto-regulation | Stripe #3 | Stripe #2 + #7 |

Bases −800 −630 −590 −520 −470 −300 −170 −40 +1

FIGURE 15.21

Specific promoter regions of the *even-skipped* gene control specific transcription bands in the embryo. The wild-type transcription pattern of seven bands (plus a band in the head region) in the blastoderm is shown in Figure 15.20. A reporter β-galactosidase gene was fused to regions of the even-skipped promoter and inserted into the fly genome. The full promoter region produces the seven normal transcription stripes. (A) If only the most proximal 480 base pairs of the promoter are present, only stripes 2, 3, and 7 form. (B) Partial map of the *eve* promoter showing the regions responsible for the various stripes and for autoregulation. (From Harding et al., 1989, courtesy of M. Levine.)

et al., 1984). Eight genes are presently known to be capable of dividing the early embryo in this fashion; they are listed in Table 15.2. It is important to note that not all nuclei express the same pair-rule genes. In fact, in each parasegment, each row of nuclei probably has its own constellation of pair-rule genes that distinguishes it from any other row. Figure 15.20 shows the relationship between three pair-rule genes, *even-skipped*, *hairy*, and *fushi tarazu*.

How are some nuclei of the *Drosophila* embryo told to transcribe a particular gene while their neighbors are told not to transcribe it? The answer appears to come from the distribution of the protein products of the gap genes. Whereas the *mRNA* of each of the gap genes has a very discrete distribution that defines abutting or slightly overlapping regions of expression, the *protein* products of these genes extend more broadly. In fact, they overlap by at least 8–10 nuclei (which at this stage accounts for about 2–3 segment primordia). This was demonstrated in a striking manner by Štanojević and co-workers (1989). They fixed cellularizing blastoderms, stained the hunchback protein with an antibody carrying a red dye, and simultaneously stained the Krüppel protein with an antibody carrying a green dye. Cellularizing regions that contained *both* proteins bound both antibodies and were stained bright yellow (Plate 14B). Similarly, Krüppel protein overlaps with knirps protein in the posterior region (Pankratz et al., 1990).

Three genes are known to be the **primary pair-rule genes**. These genes—*hairy*, *even-skipped*, and *runt*—are essential for the formation of the periodic pattern, and they are the genes directly controlled by the gap proteins. The promoters of the primary pair-rule genes are recognized by gap gene proteins, and it is thought that the different concentrations of gap gene proteins determine whether the gene is active or not. The promoters of these genes are often modular, containing a different region for each transcriptional stripe. For example, one particular deletion in the promoter region of the *even-skipped* gene prevents the formation of the seventh *even-skipped* stripe, while a deletion slightly more downstream causes the loss of the second *even-skipped* stripe (Figure 15.21), DNase I footprinting of this latter region shows that it contains six binding sites for Krüppel protein, three binding sites for hunchback protein, three binding sites for giant protein, and five binding sites for bicoid protein. Genetic evidence shows that if some of these sites are deleted, the position of the second stripe moves. Štanojević and colleagues (1991) have shown that the second *even-skipped* stripe is repressed by both giant and Krüppel proteins and is activated by hunchback protein at low concentrations of bicoid. Their model is shown in Figure 15.22. The region responsible for the third stripe of *even-skipped* transcription contains 20 hunchback protein-binding sites and not one site for the Krüppel protein (Štanojević et al., 1989). This situation would enable the site to respond to very low levels of the *hunchback* gene product. Gap proteins activate transcription of some pair-rule genes while repressing the transcription of others. The result is the pattern of transcription stripes that emerges as the embryos develop.

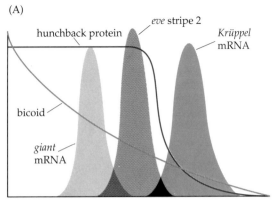

(A)

hunchback protein

eve stripe 2

Krüppel mRNA

bicoid

giant mRNA

931

(B)   1600

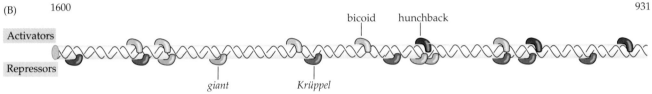

Activators

Repressors

bicoid   hunchback

giant   *Krüppel*

FIGURE 15.22

Hypothesis for the formation of the second stripe of transcription from the *even-skipped* gene. (A) Limits of *even-skipped* transcription are restricted by negative regulation from giant and Krüppel gap gene proteins. The bicoid and hunchback proteins activate transcription where possible. (B) Enhancer element for stripe 2 regulation, containing binding sequences for Krüppel, giant, bicoid, and hunchback proteins. Note that nearly every activator site is closely linked to a repressor site, suggesting competitive interactions at these positions. (After Ŝtanojević et al., 1991.)

Once initiated by the gap proteins, the transcription pattern of the primary pair-rule genes becomes stabilized by their interactions among themselves (Levine and Harding, 1989). The primary pair-rule genes also form the context that allows or inhibits the expression of the later-acting **secondary pair-rule genes**. One such secondary pair-rule gene is *fushi tarazu* (*ftz*; Japanese, "too few segments"). Early in cycle 14, *ftz* mRNA and protein are seen throughout the segmented portion of the embryo. However, as the proteins from the primary pair-rule genes begin to interact with the *ftz* promoter, the *ftz* gene is repressed in certain bands of nuclei to create the interstripe regions. Meanwhile, the ftz protein interacts with its own promoter to stimulate more transcription of the *ftz* gene (Figure 15.23; Edgar, et al., 1986; Karr and Kornberg, 1989; Schier and Gehring, 1992).

The segment polarity genes

So far our discussion has identified interactions between molecules within the syncytial embryo. But once cells begin to form, interactions take place between the *cells* containing certain segmentation genes. The interactions between these cells establish the fate of the cell within the segment. The segment polarity genes (Table 15.2) divide the domains of the pair-rule genes, forming 14 bands of transcription, each 1–2 nuclei wide (Plate 14D). One of the best studied interactions between the segment polarity genes is that between *engrailed* and *wingless* (see Peifer and Bejsovec, 1992; Siegfried et al., 1992). The *engrailed* gene is needed to ensure the anterior–posterior boundary of the segment. In its absence, portions of the posterior part of each segment are replaced by duplications of anterior regions of the next adjoining segment. The *wingless* gene is needed for the proper expression of different regions of each segment.

The initiation of the segment polarity gene transcription pattern is regulated by the pair-rule genes. The transcription of the *wingless* gene is repressed by both fushi tarazu and even-skipped proteins, and it is stimulated by odd-paired. Meanwhile, the *engrailed* gene is active in those cells containing the ftz protein (which stimulates engrailed transcription) and lacking odd-skipped (which inhibits its transcription). This causes *wingless* to be transcribed solely in the cell directly anterior to the cells where *engrailed* is transcribed (Figure 15.24A).

(A)

(B)

(C)

(D)

1 2 3 4 5 6 7

FIGURE 15.23

Transcription of the *ftz* gene. At the beginning of cycle 14, there is low-level transcription in each of the nuclei in the segmented region of the *Drosophila* embryo. Within the next 30 minutes, the expression pattern alters as *ftz* transcription is enhanced in certain regions (which form the stripes) and repressed in the interstripe regions. (After Karr and Kornberg, 1989.)

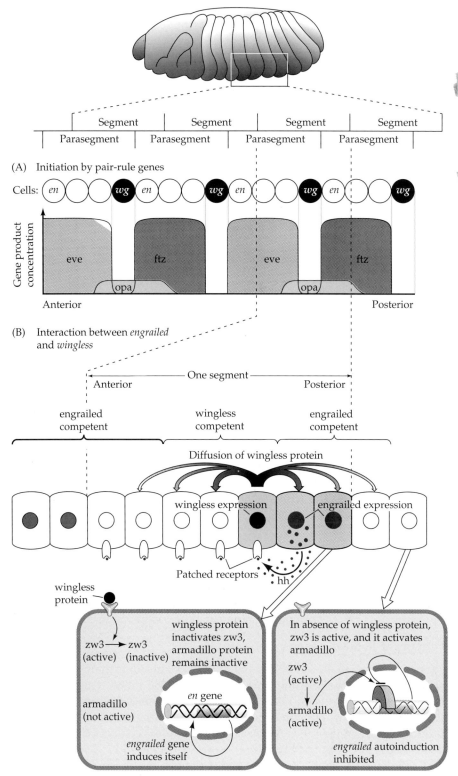

FIGURE 15.24
Model for the transcription of segment polarity genes *engrailed* (*en*) and *wingless* (*wg*). (A) The initiation of *wg* and *en* expression is initiated by pair-rule genes. The *engrailed* gene is expressed when the cells contain high concentrations of either even-skipped or fushi tarazu proteins. The *wingless* gene is transcribed when neither of these two genes is active, but a third gene (probably *odd-skipped*) is present. (B) The continued expression of *wg* and *en* is maintained by interactions between the *engrailed*- and *wingless*-expressing cells. The wingless protein diffuses into the surrounding cells. In those cells competent to express engrailed (having eve or ftz proteins), wingless protein blocks the inhibition of engrailed autoinduction by zeste-white 3 (zw3). The engrailed protein is synthesized. The cells synthesizing engrailed also synthesize hedgehog, a protein that diffuses and binds to the patched protein. This product of the *patched* gene inhibits the *wg*-competent cells from synthesizing the wingless protein unless the *patched* signal is inhibited by the hedgehog protein that is secreted by *engrailed*-transcribing cells. In this manner, engrailed and wingless transcription patterns are mutually stabilizing. (After Levine and Hardin, 1989; Peifer and Bejsovec, 1992; Ingham, 1993; Siegfried et al., 1994.)

The *maintainance* of these patterns is regulated by interactions between the cells expressing *wingless* and those expressing *engrailed*. The wingless protein is secreted from the cell synthesizing it and enters the adjacent cells. In these neighboring cells, the zest-white protein (zw3) is constitutively active and blocks the ability of engrailed protein to stimulate its own transcription. The wingless protein, in a complex pathway involving the

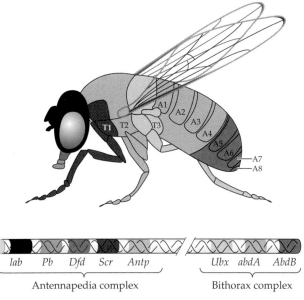

FIGURE 15.25
The functional domains of the Bithorax complex and Antennapedia complex genes in *Drosophila*. The bithorax complex has been divided into the three lethal complementation groups identified by E. B. Lewis. The antennapedia complex genes are *labial* (*lab*), *Deformed* (*Dfd*), *Sex comb reduced* (*Scr*), and *Antennapedia* (*Antp*). (After Dessain et al., 1992.)

products of the *armadillo* and *disheveled* genes, inactivates the zw3 repression and allows the expression of *engrailed* (Siegfried et al., 1992). In addition to the wingless protein's enabling the transcription of *engrailed* in neighboring cells, a reciprocal interaction occurs whereby the engrailed-synthesizing cells secrete the hedgehog protein which (by binding to the patched protein on the wingless-secreting cells) maintains the expression of the *wingless* gene* (Heemskerk et al., 1991; Ingham et al., 1991; Mohler and Vani, 1992). In this way, the transcription pattern of these two cells is stabilized.

The homeotic selector genes

Patterns of homeotic gene expression

After the segmental boundaries have been established, the characteristic structures of each segment are specified. This specification is accomplished by the **homeotic selector genes** (Lewis, 1978). There are two regions of *Drosophila* chromosome 3 that contain most of these homeotic genes (Figure 15.25). One region, the **Antennapedia complex**, contains the homeotic genes *labial* (*lab*), *Antennapedia* (*Antp*), *Sex comb reduced* (*Scr*), *Deformed* (*Dfd*), and *proboscipedia* (*pb*). The *labial* and *Deformed* genes specify the head segments, while *Sex comb reduced* and *Antennapedia* contribute to giving the thoracic segments their identities. The *proboscipedia* gene appears to act

---

*There may even be a further step in specifying which cells express the *engrailed* gene. Eventually, the synthesis of engrailed no longer depends on autoactivation from the engrailed protein. Rather, the chromatin appears to be fixed in a conformation that is inherited by the descendants of these cells. The products of the *Polycomb* gene appear to stabilize repressed chromatin, while the proteins of the *trithorax* gene seem to stabilize active chromatin conformations (Ingham and Whittle, 1980; Paro, 1990.).

only in adults, but in its absence, the labial palps of the mouth are transformed into legs (Wakimoto et al., 1984; Kaufman et al., 1990). The second region of homeotic genes is the **bithorax complex** (Lewis, 1978). There are three protein-coding genes found in this complex: *Ultrabithorax* (*Ubx*), which is required for the identity of the third thoracic segment; and *abdominal A* (*abdA*) and *Abdominal B* (*AbdB*) which are responsible for the segmental identities of the abdominal segments (Sánchez-Herrero et al., 1985). The lethal phenotype of the triple-point mutant Ubx^-, $abdA^-$, $AbdB^-$ is identical to that of a deletion of the entire bithorax complex (Casanova et al., 1987). An additional homeotic gene, *caudal*, is in neither of these regions and is involved in specifying the telson (MacDonald and Struhl, 1986; Mlodzik and Gehring, 1987).

Because these genes are responsible for the specification of body parts, mutations in them lead to bizarre phenotypes. In 1894 William Bateson called these organisms "homeotic mutants," and they have fascinated developmental biologists for decades. The *Antennapedia* gene, for instance, is thought to specify the identity of the second thoracic segment. In the *dominant* mutation of *Antennapedia*, this gene is expressed in the head as well as in the thorax, and the imaginal discs of the head region are specified as thoracic. Thus, legs rather than antennae grow out of the head sockets (Figure 15.26). In the *recessive* mutant of *Antennapedia*, the gene fails to be expressed in the second thoracic segment, and antennae sprout out of the leg positions (Struhl, 1981; Frischer et al., 1986; Schneuwly et al., 1987). Likewise, when the Ultrabithorax complex is deleted, the third thoracic segment (which is characterized by halteres) becomes transformed into another *second* thoracic segment. The result (Figure 15.27) is a fly with four wings—an embarassing situation for a classic dipteran.*

These major homeotic selector genes have been cloned and their expression analyzed by in situ hybridization (Harding et al., 1985; Akam, 1987). A summary of these experiments is shown in Figure 15.28. Transcripts from each locus are detected in specific regions of the embryo and are especially prominent in the central nervous system. In homeotic mutants, this normal expression is altered. For instance, in the dominant

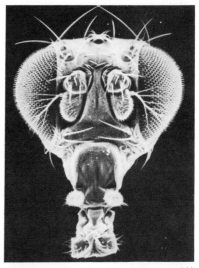

(A)

(B)

FIGURE 15.26
(A) Head of a wild-type fly. (B) Head of a fly containing the *Aristopedia* mutation that converts antennae into legs. (From Kaufman et al., 1990, courtesy of T. C. Kaufman.)

---

*Dipterans (two-winged insects such as flies) are thought to have evolved from normal four-winged insects; it is possible that this change arose via alterations in the bithorax complex. Chapter 23 includes more speculation on the relationship between *bithorax* genes and evolution.

FIGURE 15.27
This four-winged fruit fly was constructed by putting together three mutations in *cis* regulators of the *Ultrabithorax* gene. These mutations effectively transform the third thoracic segment into another second thoracic segment (i.e., halteres into wings). (Courtesy of E. B. Lewis.)

Antennapedia alleles mentioned above, the *Antennapedia* gene has been inverted on the chromosome, so that it has lost its own promoter and is under the control of a different promoter that is active in the head. This causes the ectopic expression of *Antp* in the head. Similarly, if the *Ultrabithorax* gene is placed on a new promoter and expressed in the head region, the antennae are transformed into legs (Mann and Hogness, 1990).

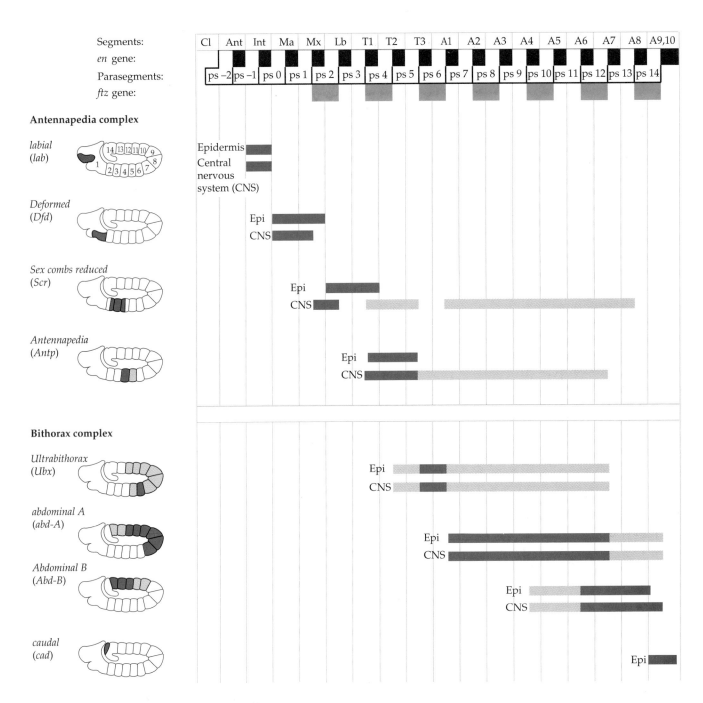

FIGURE 15.28
Regions of homeotic gene expression (both mRNA and protein) in the blastoderm of the *Drosophila* embryo and (a few hours later) in its central nervous system. The darker shaded areas are those segments or parasegments with the most product. The diagram beneath the chart represents gene expression within the parasegmental boundaries. (After Kaufman et al., 1990.)

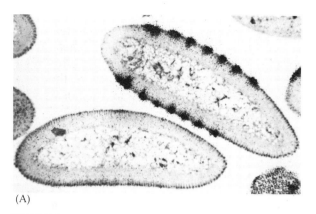

(A)

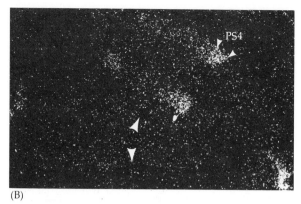

(B)

FIGURE 15.29

Dependence of a homeotic gene on the proper expression of the *fushi tarazu* segmentation gene. In situ hybridization of various radioactive DNA probes was performed on adjacent microscopic sections from wild-type (lower) and *ftz⁻* (upper) *Drosophila* embryos. Both embryos are at the cellularizing blastoderm stage. (A) Autoradiograph of section hybridized with radioactive *ftz* DNA. Seven areas of *ftz* RNA accumulation are seen in the wild-type embryo, but none is seen in the *ftz⁻* mutant. (B) Dark-field image of the autoradiograph using radioactive DNA of the *Antennapedia* gene shows wild-type gene expression in the fourth parasegment (PS4 and small arrows) but no transcription in the same region (large arrows) in the *ftz⁻* mutant. (From Ingham and Martinez-Arias, 1986, courtesy of the authors.)

Initiating the patterns of homeotic gene expression

The initiation of the homeotic gene domains is influenced by the pair-rule genes and the gap genes. The ftz protein has a positive effect on the transcription rates of the *Scr, Antp,* and *Ubx* genes; therefore, these genes are not transcribed efficiently in *ftz⁻* embryos. The early expression of these three genes is usually confined to parasegments 2, 4, and 6, respectively; these are the sites corresponding to *ftz* gene expression bands 1, 2, and 3 (Figure 15.29; Ingham and Martinez-Arias, 1986). Conversely, ftz protein has a negative effect on *Dfd* transcription. In wild-type embryos, the anterior margin of an ftz protein band forms the posterior border of the *Dfd* territory. In *ftz⁻* embryos, the posterior of *Dfd* extends posteriorly. Other pair-rule genes, such as *even-skipped* and *odd-paired*, stimulate *Dfd* expression (Jack et al., 1988).

But the pair-rule gene products alone cannot account for the boundaries of the homeotic genes. (Otherwise the homeotic genes would show periodicity across the entire anterior–posterior axis.) Since each homeotic gene is usually expressed in contiguous segments (or parasegments), the gap genes are obvious candidates for establishing the limits of homeotic gene expression. Much evidence has accumulated for this view. The *Krüppel* gene, for instance, activates *Antp* transcription and inhibits *AbdB* transcription (Figure 15.30). The gap gene *knirps* also is required for delimiting the initial domains of *Antp* and *AbdB*, but it acts to inhibit both genes. Thus, the presence of high concentrations of Krüppel protein in the central

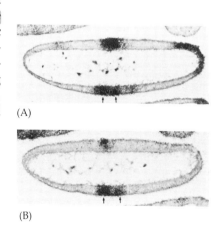

(A)

(B)

FIGURE 15.30

The initial expression of the homeotic gene *Antennapedia* (B) is predicated upon the prior expression of *Krüppel* (A) in the same area. If the placement of *Krüppel* expression is altered, so is the expression of *Antennapedia*. (From Levine and Harding, 1989, courtesy of the authors.)

region of the embryo enhances the transcription of *Antennapedia* while it keeps the *Abdominal B* gene from being transcribed (Harding and Levine, 1988).

Maintaining the patterns of homeotic gene expression

The expression of homeotic genes is also a dynamic process. The *Antp* gene, for instance, although initially expressed in presumptive parasegment 4, soon appears in parasegment 5. As the germ band expands, *Antp* gene expression is seen in the presumptive neural tube as far posterior as parasegment 12. During further development, the pattern contracts again, and *Antp* transcripts are localized strongly to parasegments 4 and 5. *Antp* transcription is negatively regulated by all the homeotic gene products posterior to it. In other words, each of the bithorax complex genes represses the transcription of *Antennapedia*. If *Ultrabithorax* is deleted, *Antp* activity extends through the region that would normally have expressed *Ubx* and stops where the *Abd* region begins. (This allows the third thoracic segment to form wings like the second thoracic segment, as in Figure 15.27.) If the entire bithorax complex is deleted, *Antp* expression extends throughout the abdomen. (The larva doesn't survive, but the cuticle pattern throughout the abdomen is that of the second thoracic segment.) In general, the anteriorly expressed homeotic genes are inhibited by those homeotic proteins more posterior to them (Harding and Levine, 1989). Thus, *Antp* is repressed by *Ubx*, *Ubx* is repressed by *abdA*, and *abdA* is repressed by *AbdB*.

Realisator genes. The search is now on for the "realisator genes," those genes that are the targets of the homeotic genes and function to form the specified tissue or organ primordia. One approach to finding such genes has been sequencing. Gene sequencing has shown that some genes have enhancer elements that bind homeotic genes, thereby causing them to be regulated by homeotic gene expression patterns. The *Distal-less* gene (itself a homeobox-containing gene) is necessary for limb development and is active solely in the thorax. *Distal-less* expression is repressed in the abdomen, probably by a combination of Ubx and abdA proteins that can bind to its enhancer and block transcription (Vachon et al., 1992).

Another technique, pioneered by Gould and co-workers (1990), involves antibody precipitation. Precipitating Ubx-bound chromatin with antibodies directed against the Ubx protein should identify those genes regulated by the Ultrabithorax protein during development. One of the genes precipitated in this fashion turns out to encode *connectin*, a cell adhesion molecule involved in connecting motor neurons to muscle cells. Like *Distal-less*, this gene is transcribed primarily in the thorax, and it is repressed by the combination of abdA and Ubx proteins (Gould and White, 1992).

A third method, pioneered in Walter Gehring's laboratory, has used "enhancer traps" to detect those genes regulated by *Antennapedia*. Here, a transposon containing a β-galactosidase reporter gene is linked to a weak promoter and is introduced randomly into the genome of different *Drosophila*. The expression of β-galactosidase (which can be readily detected by staining) becomes controlled by the enhancers in the vicinity of the promoter. If the enhancer is regulated by Antennapedia protein (which is present in the thoracic region of the embryo but not in the head), then β-galactosidase activity should be different when thoracic and head tissues are compared. Using this approach, Wagner-Bernholz and colleagues (1991) found what may be a critical gene that is regulated by *Antennapedia*.

(A)

(B)

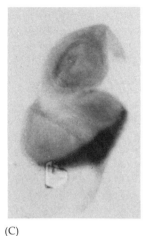
(C)

FIGURE 15.31
The "enhancer trap" transposon carries a β-galactosidase gene that becomes activated when placed near an enhancer. In one strain, the transposon became incorporated near a gene that was differentially regulated in the head and in the thorax. (A) Leg imaginal discs from wild-type larvae (at third instar stage just prior to pupation) do not express a particular gene, *salm*. (B) The antennal discs from the same larva do express *salm*. (C) Antennal discs from an *Antennepedia* mutant show that this gene is repressed in this mutant. (From Wagner-Bernholz et al., 1991, courtesy of W. J. Gehring.)

This gene, *salm*, is not active in leg imaginal discs from the thorax, but is expressed in the antennal imaginal disc (Figure 15.31). Thus, *salm* appears to be a gene that is repressed by the Antennapedia protein. In those mutants of *Antennapedia* that cause its expression in the head and cause legs to evert from the antennal sockets, the antennal disc lacks this protein's expression. The repression of the *salm* gene may be critical for the formation of leg tissue, rather than antennal tissue, from the thoracic imaginal discs.

SIDELIGHTS & SPECULATIONS

Molecular regulation of development: The homeodomain proteins

The homeodomain

Homeodomain proteins are a family of transcription factors characterized by a 60-amino acid domain that binds to certain regions of DNA. The homeodomain was first seen in those proteins whose absence or misregulation causes homeotic transformations of *Drosophila* segments. It is thought that homeodomain proteins activate batteries of genes that specify the particular properties of that segment. Such proteins include the products of the eight homeotic genes of the Antennapedia–bithorax complex, as well as other proteins such as fushi tarazu, caudal, and bicoid. Homeodomain transcription factors are important in determining the anterior–posterior axes of both invertebrates and vertebrates. In *Drosophila*, the presence of certain homeodomain-containing proteins is also necessary for the determination of specific neurons. Without these transcription factors, the fates of these neuronal cells are altered (Doe et al., 1988).

The homeodomain is encoded by the 180-base pair **homeobox** (see Chapter 10). The homeodomains appear to specify the binding sites for these proteins and are critical in specifying cell fate. For instance, if a chimeric protein is constructed mostly of Antennapedia but with the carboxy terminus (including the homeodomain) of Ultrabithorax, the protein can substitute for Ultrabithorax and specify the appropriate cells as parasegment 6 (Mann and Hogness, 1990). The isolated homeodomain of Antennapedia will bind to the same promoters as the entire Antennapedia protein, indicating that the binding of this protein is dependent upon its homeodomain (Müller et al., 1988).

The homeodomain folds into three α-helices, the latter two folding into a helix-turn-helix conformation that is characteristic of a family of transcription factors that bind DNA in the major groove of the double helix (Otting et al., 1990; Percival-Smith et al., 1990). The TAA(A)T motif is conserved in nearly all sites recognized by homeodomains; it probably serves to distinguish those sites to which homeodomain proteins might bind. The terminal T appears to be critical in this recognition, as mutating it destroys all homeodomain binding. Mutation studies have shown that bicoid and Antennapedia-class homeodomain proteins use lysine or glutamine, respectively, at position 9 to distinguish related recognition sites. If that lysine is replaced by glutamine, a bicoid protein will recognize Antennapedia-binding sites (Hanes and Brent, 1989, 1991). The amino acid at position 9 in the homeodomain helix recognizes the seventh base pair of the DNA recognition sequence. The lysine of the bicoid homeodomain recognizes the G of CG

(A)

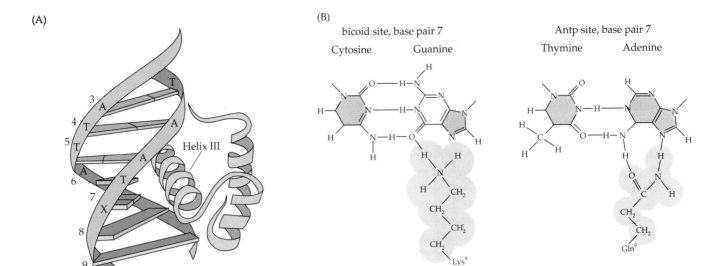

(B)

bicoid site, base pair 7

Cytosine   Guanine

Antp site, base pair 7

Thymine   Adenine

FIGURE 15.32

Homeodomain–DNA interactions. (A) Homeodomain helix-turn-helix sequence within major groove of the DNA. (B) Proposed pairing between the bicoid homeodomain lysine and the CG base pair of the recognition sequence, and between the glutamine of the Antennapedia homeodomain and the TA base pair of its recognition sequence. In both cases, the ninth amino acid of the helix bonds with the base pair immediately following the TAAT sequence. (A after Riddihough, 1992; B after Hanes and Brent, 1991.)

pairs, while the glutamine of the Antennapedia homeodomain recognizes the A of an AT pair (Figure 15.32; Hanes and Brent, 1991). Other homeodomain proteins show a similar pattern where one portion of the homeodomain recognizes the common sequence, while another portion recognizes a specific structure close to the TAAT.

Multiple regulatory modes for homeodomain proteins

The structure of the homeotic genes points toward complex modes of regulation. Some of the genes have multiple promoters and multiple sites for transcription initiation. The *Antennapedia* gene, for instance, has two transcription products, one of them initiating within an intron of the other (Figure 15.33; Frischer et al., 1986). Transcription from the first promoter is activated by Krüppel protein and

repressed by Ultrabithorax protein, while transcription from the second promoter is activated by hunchback and fushi tarazu proteins and inhibited by oskar protein (Boulet and Scott, 1988; Irish et al., 1989; Krasnow et al., 1989). Although the protein produced from both these promoters is the same, the promoters cause it to be expressed in different cells. The transcript of the first promoter is necessary for dorsal thorax development, whereas the transcript of the second promoter is required for ventral thorax (leg) development and for embryonic viability (Bermingham et al., 1990). Hogness and his co-workers (Beachy et al., 1988; Krasnow et al., 1989) used DNA footprinting to show that Ultrabithorax protein binds to a series of sites in the first promoter region of the *Antennapedia* gene. By fusing the *Antennapedia* first promoter region to the CAT reporter gene and placing the fused gene into cultured cells that do not usually transcribe any of the homeotic genes (Figure 15.34), they showed that this binding represses *Antennapedia* gene activity. The *Antennapedia* gene was activated by the presence of ftz protein (supplied by adding the *ftz* gene on a strong constitutive promoter). However, when a third gene was transfected into the cells—*Ubx* on a strong constitutive promoter—the *Antennapedia* gene was repressed.

Another complication of the homeotic genes is that several of them can produce families of related proteins by alternative RNA splicing. The *Ultrabithorax* gene produces several proteins by such a mechanism, and these proteins

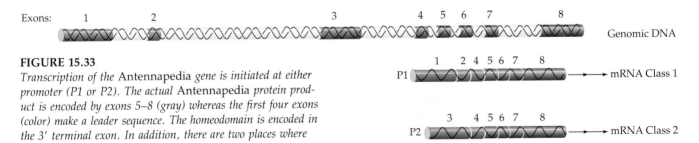

FIGURE 15.33

Transcription of the Antennapedia *gene is initiated at either promoter (P1 or P2). The actual* Antennapedia *protein product is encoded by exons 5–8 (gray) whereas the first four exons (color) make a leader sequence. The homeodomain is encoded in the 3' terminal exon. In addition, there are two places where poly(A) addition can terminate the transcript (double arrowheads). (After Frischer et al., 1986; Kaufman et al., 1990.)*

(A)

Fusion of *CAT* gene and *Antp* promoter and transfection in *Drosophila* cells

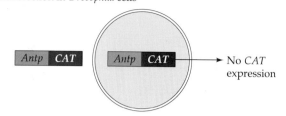

→ No *CAT* expression

(B)

Addition of *ftz* on a strong constitutive promoter

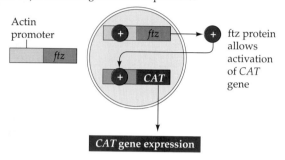

ftz protein allows activation of *CAT* gene

(C)

Addition of *Ubx* suppresses *CAT* expression again

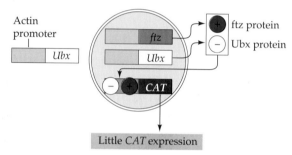

ftz protein

Ubx protein

FIGURE 15.34

Regulation of Antennapedia *expression positively by* fushi tarazu *and negatively by* Ultrabithorax. *(A) Cultured Drosophila cells will not express the* CAT *gene when it is fused with an* Antp *promoter and transfected into the cell. (B) However, when an* ftz *gene is also transfected into these cells and allowed to express its product, the homeotic promoters are activated and the* CAT *gene is expressed. (C) When an active* Ubx *gene is also present, the* CAT *gene is not expressed.*

(A) *Ultrabithorax* gene

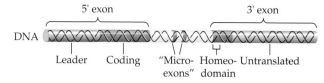

(B) Splicing in early embryogenesis, epidermis, mesoderm

(C) Splicing in CNS including 1 or 0 "microexons"

FIGURE 15.35

Ultrabithorax *mRNA structures generated by differential RNA processing. (A)* Ultrabithorax *gene. (B) The mRNAs that include two "microexons." The 5' splicing site of the first exon can vary within this group. (C) The mRNA family that includes one or no microexons. The homeodomain of both sets of* Ultrabithorax *proteins is encoded in the 3' exon. (After Beachy, 1990.)*

have different, although overlapping, specificities (Figure 15.35). All of them transform antennae into legs if they are expressed at the time of the second-to-third larval molt. However, in the early embryo, one of the Ubx proteins is important in determining the parasegmental identity of the peripheral nervous system, while other Ubx proteins are not (Mann and Hogness, 1990).

Another striking thing about the homeotic genes is that some of them have enormous introns. One of the several *Antennapedia* introns is about 57,000 base pairs long—over 10 times the combined length of all its exons! This extra length may be important in temporal regulation of the gene's expression (Kornfeld, et al. 1989). In *Drosophila* embryogenesis, events happen quickly. The gap, pair-rule, and homeotic selector genes in the embryo are each active for only about three hours. Since it is estimated that *Drosophila* genes are transcribed at a rate of 1000 nucleotides per minute at 25°C (Ashburner, 1990), this huge *Antennapedia* intron adds nearly 1 hour to the time lag before the protein is expressed.

Cis-regulatory elements and the bithorax complex

The three bithorax complex genes are obviously critical to the fly's development, but how can these three genes establish the segment identities of the third thoracic segment and the 10 abdominal segments? One possibility is that differential concentrations of the three proteins are expressed differently in each of the segments. Another model involves *cis*-regulator sites in these proteins that are segment-specific. This model was proposed on the basis of genetic studies by Lewis (1978, 1985). Lewis noted that when the entire bithorax complex was deleted, all remaining segments of

FIGURE 15.36

Representation of Lewis's model for gene regulation of the bithorax complex. The second thoracic segment is considered the ground state. For each segment after the second thoracic segment, another gene gets activated. In the last segment, all the genes are active. (After Spierer and Goldschmidt-Clermont, 1985.)

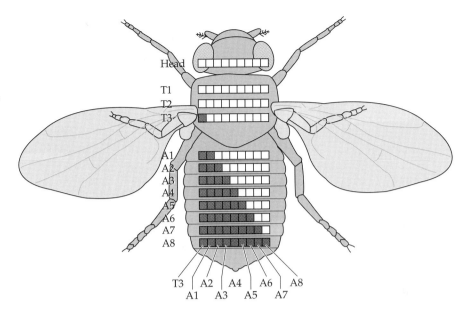

the dying embryo resembled the second thoracic segment. He postulated that this segment was the "baseline" from which all the other segments would be modified.

Lewis then postulated that the bithorax complex should contain at least one gene for each segment below the second thoracic level. In other words, the development of T3 would involve the turning on of all the "ground-level" T2 genes plus the gene(s) responsible for T3 characteristics (Figure 15.36). A mutation in these T3 genes would return the segment to the T2 state. Similarly, the development of the first abdominal segment would require that the T2 and T3 genes be expressed in addition to the A1 gene. Mutations of the A1 gene would cause the segment to develop T3 structures. This activation of different genes in different segments could be explained by postulating a gradient of a repressor molecule that inhibits the expression of all these genes. The repressor would be concentrated in the anterior segments, where fewer genes are turned on, and would be at a low concentration in the posterior segments. Furthermore, the genes controlling the development of the most posterior segments would have a higher affinity for the repressor molecule. In the anterior segments, therefore, these genes would be more readily turned off. They would be active only in the posterior segments, where repressor concentration is lowest.

But how can Lewis's model work if there are only three protein-encoding genes within the bithorax complex? Lewis and his colleagues (Bender et al., 1983; Karch et al., 1985) have identified other regions of the bithorax complex that produce homeotic transformations. The *anterobithorax* (*abx*) and *bithorax* (*bx*) mutants cause the anterior compartment of the third thoracic segment (anterior balancers) to assume the identity of the anterior compartment of the *second* thoracic segment (anterior wings). Similarly, the *posterobithorax* (*pbx*) and *bithoraxoid* (*bxd*) mutants cause the *posterior* compartment of the third thoracic segment to resemble that of the second thoracic segment. The combination of *abx*, *pbx*, and *bxd* mutations in a single embryo causes the entire transformation of the third thoracic segment into another second thoracic segment. (The result is the fly shown in Figure 15.27.) Whereas these mutations were originally considered to be on separate genes, it now appears that they are mutations of enhancer elements, which enable the position-specific expression of the *Ubx* gene (Lewis, 1985; Peifer et al., 1987).

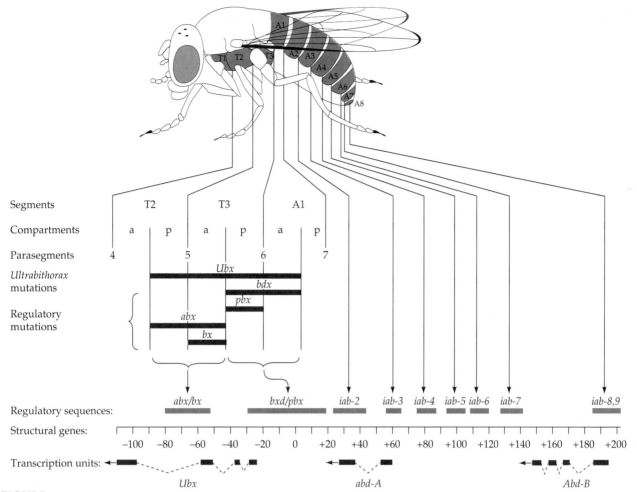

FIGURE 15.37
Regulatory mutations in the bithorax complex. The schematic adult fly is divided into segments and anterior and posterior compartments. The regulatory regions of the *Ultrabithorax* gene are shown below the fly. The shaded areas represent the region specified by the particular regulatory domain. The continuous line below that represents the 300,000-base-pair region of the complex. The three transcription units that encode the three homeotic proteins of the bithorax complex are shown in relation to the regulatory loci. Each of these genes is transcribed right to left. The exons are shown as shaded boxes, the introns as dashed lines. Above the line are the regulatory sequences defined by genetic mutations. (After Peifer et al., 1987, and Beachy, 1990.)

The relationship between the *cis*-regulatory mutations and the three transcription units of the bithorax complex is shown in Figure 15.37. The protein-encoding regions of the bithorax complex take up less than one-tenth of the DNA in this complex. The regulatory mutations generally map to the flanking regions of these three genes or to introns within the genes. Further evidence that *abx*, *bx*, and *bxd* are *cis*-regulatory elements comes from analyzing specific mutations and deletions. The deletion of the *Ubx* gene results in the homeotic transformation of parasegment 5 (posterior T2 and anterior T3) and parasegment 6 (posterior T3 and anterior A1) into copies of parasegment 4 (posterior T1 and anterior T2). Such a transformation is lethal; the embryo dies before hatching. In *abx* and *bx* mutants, however, only parasegment 5 is transformed into parasegment 4, as *Ubx* expression is reduced in parasegment 5 (Casanova et al., 1985; Peifer and Bender, 1986). Hence the anterior wing emerges in what would

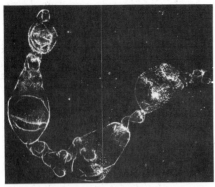

(A)

(B)

FIGURE 15.38
Rescue of larva by injection of wild-type mRNA into eggs destined to have the *snake* phenotype. (A) Deformed larva consisting entirely of dorsal cells. Larvae like these developed from eggs of a female homozygous for the *snake* allele. (B) Wild-type appearance of such larvae developing from *snake* eggs that received injections of mRNA from wild-type eggs. (From Anderson and Nüsslein-Volhard, 1984, courtesy of C. Nüsslein-Volhard.)

otherwise have been anterior haltere. Similarly, the *bxd* mutations reduce *Ubx* expression in parasegment 6 (Peifer et al., 1987). The *bithorax* regulatory element for Ubx contains an enhancer that binds the proteins encoded by the *tailless, fushi tarazu,* and *hunchback* segmentation genes (Qian et al., 1991). In the abdominal region, the *cis*-regulatory sequences *intra-abdominal* (*iab*) 2–9 direct the expression of *abdA* or *AbdB* in the various segments (Sánchez-Herrero, 1991; Boulet et al., 1991).

THE GENERATION OF DORSAL–VENTRAL POLARITY IN *DROSOPHILA*

In 1936, embryologist E. E. Just criticized those geneticists who sought to explain development by looking at specific mutations affecting eye color, bristle number, and wing shape. He said that he wasn't interested in the development of the bristles on a fly's back; rather, he wanted to know how the fly embryo makes the back itself. Fifty years later, embryologists and geneticists are finally answering that question.

Dorsal protein: Morphogen for dorsal–ventral polarity

The dorsal–ventral axis is specified by nearly 20 products placed into the oocyte by the maternal genome during egg development. Anderson and Nüsslein-Volhard (1984) have isolated 11 maternal effect genes, each of whose absence is associated with a lack of ventral structures (Figure 15.38). In addition, the absence of another maternal effect gene, *cactus*, causes the ventralization of all cells. In some cases, the maternal product is a protein, while in other cases the female fly places a specific mRNA into the egg.

The specification of the dorsal–ventral axis can be divided into several steps. The critical step is the translocation of the **dorsal protein** from the cytoplasm into the nuclei of the ventral cells during the fourteenth division cycle. The events preceding this step specify that the signal to translocate the dorsal protein will be given only by the ventral cells. The steps after the translocation of the dorsal protein concern what the dorsal protein does to specify the different regions of the embryo.

Translocation of dorsal protein

The actual protein that distinguishes dorsum from ventrum is the protein product of the *dorsal* gene. The mRNA from the mother's *dorsal* genes is put into the egg by the mother fly's ovarian cells. However, dorsal protein is not synthesized from its maternal message until about 90 minutes after fertilization. When this protein is translated, it is found *throughout* the embryo, not just on the ventral or dorsal side. How, then, can this protein act as a morphogen if it is located everywhere the embryo? In 1989, the surprising answer was found (Roth et al., 1989; Rushlow et al., 1989; Steward, 1989). While dorsal protein can be found throughout the syncytial blastoderm of the early *Drosophila* embryo, it is transported into cell nuclei only in the *ventral* part of the embryo (Figure 15.39A,B). Here, dorsal protein binds to certain nuclear genes to activate or suppress their transcription. If it does not enter the nucleus, the ventralizing genes of

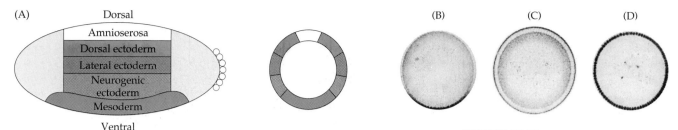

(A) Dorsal / Amnioserosa / Dorsal ectoderm / Lateral ectoderm / Neurogenic ectoderm / Mesoderm / Ventral

Transverse section

FIGURE 15.39

Inclusion of dorsal protein into the nuclei of ventral, but not lateral or dorsal, nuclei. (A) Fate map of the mid-*Drosophila* embryo. The most ventral part becomes the mesoderm; the next higher portion becomes the neurogenic (ventral) ectoderm. The lateral and epidermal ectoderm can be distinguished in the cuticle, and the dorsalmost region becomes the amnioserosa, the extraembryonic layer that surrounds the embryo. (B–D) Cross sections of embryos stained with antibody to show the presence of dorsal protein. In all cases, the dark stain represents the dorsal protein. (B) A wild-type embryo, showing dorsal protein in the ventralmost nuclei. (C) A dorsalized mutant, showing no localization of dorsal protein in any nucleus. (D) A ventralized mutant; dorsal protein has entered the nucleus of every cell. (A from Rushlow et al., 1989; B–D from Roth et al., 1989, courtesy of the authors.)

the nucleus (*snail* and *twisted*) are not activated, the dorsalizing genes of the nucleus (*decapentaplegic* and *zerknüllt*) are not repressed, and the region of the embryo develops into dorsalized cells. This hypothesis that the dorsal–ventral axis of *Drosophila* is specified by the selective transport of the dorsal morphogen protein into the nucleus is strengthened by the analysis of mutations having an entirely dorsalized or an entirely ventralized phenotype (Figure 15.39C,D). In those mutants wherein all the cells are dorsalized (as is evident by their dorsal cuticle), dorsal protein does not enter the nucleus in any cell. Conversely, in those mutants wherein all cells have a ventral phenotype, dorsal protein is found in every cell nucleus.

Providing the asymmetric signal for dorsal protein translocation

Signal from the oocyte nucleus to the follicle cells

If dorsal protein is found throughout the embryo but gets translocated into the nucleus only of ventral cells, something else must be providing the asymmetric cues (Figure 15.40). It appears that the signal is mediated through a complex interaction between the oocyte and its surrounding follicle cells. The original signal appears to be the dorsal location of the oocyte nucleus during mid-oogenesis (Montell et al., 1991; Schüpbach et al., 1991). Soon afterward, the follicle cells above the nucleus assume a more columnar shape than the other follicle cells. These differences in shape and packing become accentuated as the egg matures, eventually distinguishing dorsal and ventral follicle cells. The dorsalizing signal appears to be made by the products of the *gurken* and *cornichon* genes* (Schüpbach, 1987; Forlani et al., 1993). Mutations of these genes in the oocyte cause the ventralization of both the embryo and its surrounding follicle cells. (If the mutation is in the follicle cells and not in the egg, the embryo is normal).

The dorsalizing signal appears to be received by the follicle cells through a receptor encoded by the *torpedo* gene. This gene encodes a protein very similar in structure to the epidermal growth factor receptor of mammals (Price et al., 1989). Maternal deficiency of this gene causes the ventralization of the embryo. Moreover, the *torpedo* gene is active in

---

*A clue to the phenotype of these mutants is that they are the German and French terms, respectively, for small pickles. Recall also that *Drosophila* genes are often named for their *mutant* phenotypes. Hence dorsal protein is needed in order to produce *ventral* cells (i.e., if the protein is deficient, the cells become dorsal).

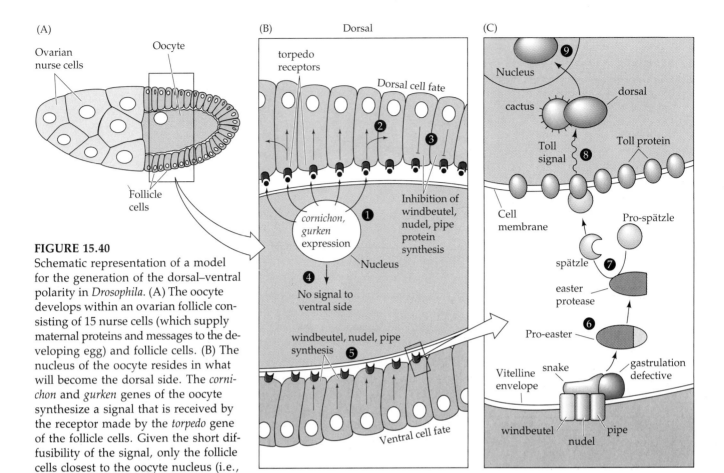

(A)

Ovarian nurse cells

Oocyte

Follicle cells

(B) Dorsal

torpedo receptors

Dorsal cell fate

2

3

Inhibition of windbeutel, nudel, pipe protein synthesis

cornichon, gurken expression **1**

Nucleus

4

No signal to ventral side

windbeutel, nudel, pipe synthesis **5**

Ventral cell fate

Ventral

(C)

Nucleus **9**

cactus

dorsal

Toll signal **8**

Toll protein

Cell membrane

Pro-spätzle

spätzle **7**

easter protease

Pro-easter **6**

gastrulation defective

Vitelline envelope

snake

windbeutel

nudel

pipe

FIGURE 15.40
Schematic representation of a model for the generation of the dorsal–ventral polarity in *Drosophila*. (A) The oocyte develops within an ovarian follicle consisting of 15 nurse cells (which supply maternal proteins and messages to the developing egg) and follicle cells. (B) The nucleus of the oocyte resides in what will become the dorsal side. The *cornichon* and *gurken* genes of the oocyte synthesize a signal that is received by the receptor made by the *torpedo* gene of the follicle cells. Given the short diffusibility of the signal, only the follicle cells closest to the oocyte nucleus (i.e., the dorsal follicle cells) receive this signal. The signal from the torpedo receptor causes the follicle cells to differentiate a characteristic dorsal follicle morphology that inhibits the synthesis of windbeutel, nudel, and pipe proteins; therefore these proteins are made only by the ventral follicle cells. (C) The three ventral follicle proteins are thought to become incorporated into the vitelline membrane, but only on the ventral side. They split the products of the *snake* and *gastrulation defective* genes to create an active enzyme that will split the zymogen form of the easter protein into an active easter protease. The easter protease splits the spätzle protein into a form that can bind to the Toll receptor (which is found throughout the cell membrane). Thus, only the ventral side receives the Toll signal. This signal is responsible for removing the cactus protein from the amino acids of dorsal that enable dorsal protein to be translocated into the nucleus. The dorsal protein enters the nucleus and ventralizes the cells. (After Schüpbach et al., 1991, and Hecht and Anderson, 1992.)

1. *cornichon* and *gurken* genes synthesize signal received by torpedo receptors during mid-oogenesis

2. torpedo signal causes follicle cells to differentiate to a dorsal morphology

3. Synthesis of windbeutel, nudel and pipe inhibited

4. cornichon and gurken proteins do not diffuse to ventral side

5. Ventral follicle cells synthesize ventral proteins windbeutel, nudel, and pipe

6. Ventral follicle proteins absorb snake and gastrulation-defective proteins to effect splitting of easter zymogen, making active easter protease only on ventral side

7. easter splits spätzle, which binds to Toll receptor protein

8. Toll signal splits cactus and dorsal proteins

9. dorsal protein enters the nucleus and ventralizes the cell

the ovarian follicle cells, not in the embryo. This was discovered by making germ line chimeras. Schüpbach (1987) transplanted germ cell precursors from wild-type embryos to embryos whose mothers carried the *torpedo* mutation. Conversely, she transplanted those germ cells from the *torpedo* embryos to wild-type embryos (Figure 15.41). The results were surprising in that the wild-type eggs produced ventralized embryos when these eggs had developed within *torpedo* mutant follicles. The *torpedo* mutant eggs were able to produce normal embryos if the eggs developed within a wild-type ovary. Thus, unlike *gurken* and *cornichon* gene products, the wild-type *torpedo* gene is needed in the follicle cells, not in the egg itself. It is probable that *gurken* and *cornichon* are genes that produce the ligand that binds to the *torpedo*-encoded receptor (Neuman-Silberberg and Schüpbach, 1993.).

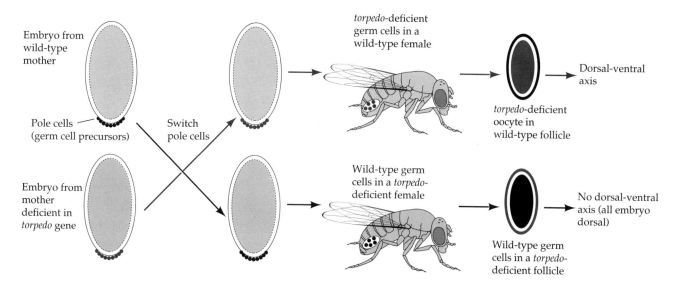

Embryo from wild-type mother

Pole cells (germ cell precursors)

Switch pole cells

Embryo from mother deficient in *torpedo* gene

torpedo-deficient germ cells in a wild-type female

Wild-type germ cells in a *torpedo*-deficient female

torpedo-deficient oocyte in wild-type follicle

Dorsal-ventral axis

Wild-type germ cells in a *torpedo*-deficient follicle

No dorsal-ventral axis (all embryo dorsal)

FIGURE 15.41
Germ line chimeras made by interchanging pole cells (germ cell precursors) between wild-type embryos and embryos from mothers homozygous for the *torpedo* gene. These transplants produce wild-type females whose embryos come from eggs from the mutant mothers, and *torpedo*-deficient embryos with wild-type eggs. The eggs from the *torpedo*-deficient mothers produce normal embryos if the eggs develop in the wild-type ovary, while the wild-type eggs produce ventralized embryos if the eggs develop in the mutant ovary.

Signal from the follicle cells to the oocyte cytoplasm

The genes *nudel (nd), pipe (pip),* and *windbeutel (wind)* are also needed in the follicle cell and not in the oocyte. If the mother lacks any of these three genes, the embryo forms a totally dorsalized phenotype. These genes are turned off by the activation of the *torpedo* receptor (Stein et al., 1991). If they are allowed to be active (as is normally the case in the ventral follicle cells), their proteins are thought to be incorporated into the ventral portion of the vitelline envelope that is secreted around the egg by the follicle cells (Hecht and Anderson, 1992; Stein and Nüsslein-Volhard, 1992). In this way, an asymmetric signal is now present on the envelope adjacent to the egg and separated from the egg by a perivitelline fluid. The complex formed by the nudel, pipe, and windbeutel proteins is thought to activate three serine proteases that are secreted by the embryo into the perivitelline fluid (Figure 15.40). These three serine proteases are the products of the *snake (snk), gastrulation-defective (gd),* and *easter (ea)* genes. As with most extracellular proteases, they are probably secreted in an inactive form and become activated by peptide cleavage. The snake and gastrulation-defective proteases appear to be activated by the asymmetric complex (snake, easter, and gastrulation-defective proteins), and they activate the easter protein from its precursor form (Chasan et al., 1992). The easter protease, now activated, cleaves the protein product of the *spätzle* gene. Cleaved spätzle protein is now able to bind to a receptor in the oocyte cell membrane, the product of the *Toll* gene.

Toll protein is also a maternal product evenly distributed throughout the cell membrane of the egg (Hashimoto et al., 1988, 1991). The recessive mutation *Toll* has a similar dorsalized phenotype, and injections of mRNA from wild-type eggs will restore the dorsal–ventral polarity of eggs laid by *Toll⁻/Toll⁻* mothers. But unlike the case of *snake* or the other 10 maternal genes, the site of the injection is important. Whichever part of the egg is injected becomes the ventral region of the rescued embryo (Anderson et al., 1985). This suggests that the *Toll⁻/Toll⁻* eggs lack a dorsal–ventral axis (whereas in *snake,* the ventral region occurs in its normal place). In normal development, the Toll receptor is located throughout the oocyte cell membrane, but it becomes activated only by binding the spätzle protein, which is produced on the ventral side of the egg. In this way, the Toll receptor on the ventral side of the egg is transducing a signal into the egg, while the Toll receptor on the dorsal side of the egg is not.

FIGURE 15.42

Model of a conserved pathway for regulating nuclear transport of transcription factors in *Drosophila* and mammals. (A) In *Drosophila*, the Toll protein binds the signal from the spätzle protein and activates the kinase region of the pelle protein. The pelle protein phosphorylates cactus, allowing it to separate from the dorsal protein. The dorsal protein can then enter the nucleus and regulate the transcription of ventrally specific genes. (B) In mammalian lymphocytes, the IL-1 receptor can cause the phosphorylation of IκB (through a protein kinase that has not yet been identified). This enables the NF-κB protein to enter the nucleus and effect the transcription of several lymphocyte-specific genes. (After Shelton and Wasserman, 1993.)

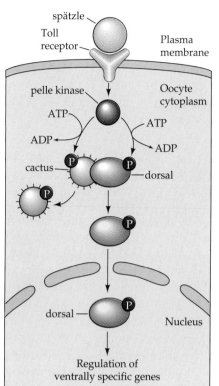

(A) *Drosophila* embryo

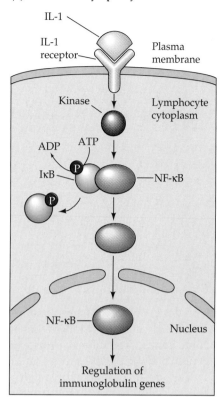

(B) Mammalian lymphocyte

Separation of the dorsal and cactus proteins

The products of the *tube* (*tub*) and *pelle* (*pll*) genes somehow translate the signal from Toll receptor into a gradient of dorsal nuclear localization, and the spread of this signal appears to be restricted by the *cappuccino* and *spire* gene products. What might the signal be doing? It appears that the cactus protein is sitting on the portion of the dorsal protein that enables the dorsal protein to get into nuclei. As long as this cactus protein is bound to it, dorsal protein remains in the cytoplasm. Thus, this entire complex signaling system is organized to split the cactus protein from the dorsal protein in the ventral part of the egg. Toll protein, in conjunction with tube protein, activates pelle protein. The pelle protein is a protein kinase that can phosphorylate dorsal protein. Once phosphorylated, dorsal can no longer bind to the cactus protein and can enter the nucleus (Kidd, 1992; Shelton and Wasserman, 1993; Whalen and Steward, 1993). The result is a gradient of dorsal localization into the ventral cells of the embryo.

The process described for the translocation of dorsal protein into the nucleus is very similar to the process described in Chapter 10 for the translocation of the NF-κB transcription factor into the nucleus. In fact, there is substantial homology between the NF-κB and dorsal and between IκB and cactus. Indeed, a further homology exists between the Toll protein and the interleukin-1 receptor. Thus, the biochemical pathway used to specify the dorsal–ventral polarity in *Drosophila* appears to be formally the same as that used to differentiate lymphocytes in mammals (Figure 15.42).

Targets for binding the dorsal protein. What does the dorsal protein do once it is located in the nuclei of the ventral cells? Looking at the fate map of the cross section through the middle of the *Drosophila* embryo at the

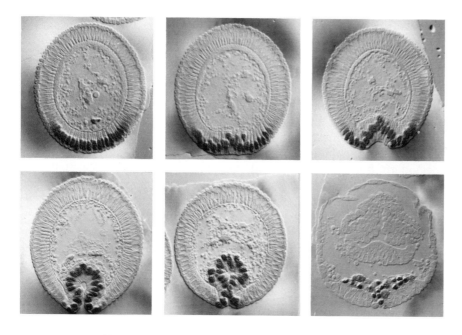

FIGURE 15.43
Gastrulation in *Drosophila*. The meso-dermal cells at the ventral portion of the embryo buckle inward, forming a tube which then generates the meso-dermal organs. The nuclei are stained with antibody to the twist protein. (From Leptin, 1991b, courtesy of M. Leptin.)

fourteenth cell nuclear division cycle (Figure 15.39), it is obvious that the 16 cells with the highest concentration of dorsal protein are those that generate the mesoderm. The next cell up from this region generates the specialized glial and neural cells of the midline. The next two cells are those that give rise to the ventral epidermis and ventral nerve cord, while the nine cells above them produce the dorsal epidermis. The most dorsal group of six cells do not divide; they generate the amnioserosal covering of the embryo (Ferguson and Anderson, 1991).

Researchers have interpreted this fate map to be produced by a gradient of dorsal protein in the nuclei. Large amounts specify the cells to be ventral cells, while lesser amounts specify the cells to be glial or ectodermal (Jiang and Levine, 1993). The first morphogenetic event of *Drosophila* gastrulation is the invagination of the 16 ventralmost cells of the embryo (Figure 15.43). All the mesodermal derivatives of the body—muscles, fat bodies, gonads—derive from these cells (Foe, 1989). Identification of mutants having aberrant gastrulation has resulted in identification of three targets of the dorsal protein: the *twist*, *snail*, and *rhomboid* genes (Figure 15.44). These genes are transcribed only in the ventral cell nuclei that have

FIGURE 15.44
Subdivision of the dorsal–ventral axis by the gradient of dorsal protein in the nuclei. The dorsal protein activates the zygotic genes *rhomboid*, *twist*, and *snail*, depending on its nuclear concentration. The snail protein, formed most ventrally, inhibits the transcription of rhomboid protein. The dorsal protein inhibits the expression of *tolloid*, *dpp*, and *zen* in the ventral region. Differing concentrations of zerknüllt protein determine the fates of the dorsal cells. (After Steward and Govind, 1993.)

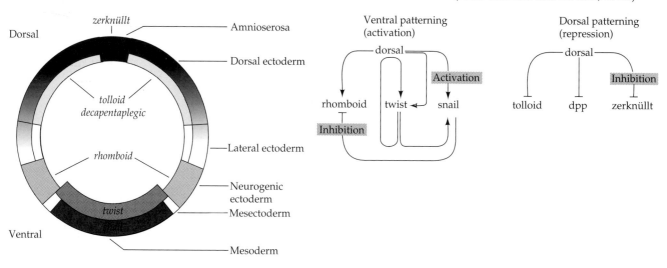

received the dorsal protein, and the dorsal protein binds to four sites in the *twist* promoter. These binding sites do not bind dorsal protein with a very high affinity, so it takes a high dorsal protein concentration to activate the *twist* gene (Jiang, et al., 1991; Pan et al., 1991; Thisse et al., 1988, 1991). The *twist* gene is required for activating and maintaining the expression of further mesodermal genes, including itself. The *snail* gene is also activated by high concentrations of the dorsal protein. Instead of activating further mesoderm-specific genes, the snail protein turns off the *rhomboid* gene, enabling *rhomboid* to be expressed in the presumptive neuroectoderm but not in the mesoderm (Jiang and Levine, 1993). Both *snail* and *twist* are needed to get the entire mesodermal phenotype and proper gastrulation (Leptin et al., 1991a). The sharp border between the mesodermal cells and those cells adjacent to them that generate glial cells is produced by the presence of the *snail* and *twist* gene products in the ventralmost cells, but of only the *twist* gene product in the next dorsal cell (Kosman et al., 1991). In mutants of *snail*, the ventralmost cells still have the *twist* gene activated, and they resemble the more lateral cells (Nambu et al., 1990).

In addition to activating *twist* and snail *gene* expression, the dorsal protein inhibits the expression of the *zerknüllt* (*zen*) and *decapentaplegic* (*dpp*) genes. Mutants of dorsal express *dpp* and *zen* genes throughout the embryo (Rushlow et al., 1987), and embryos deficient in *dpp* and *zen* fail to form dorsal structures (Irish and Gelbart, 1987). The result is that mesodermal precursors express *twist* and *snail* (but not *zen* or *dpp*). Precursors of the dorsal epidermis and amnioserosa express *zen* and *dpp* but not *twist* or *snail*. Glial (mesectoderm) precursors express only *snail*, while the lateral neuroectodermal precursors do not express any of these four genes (Kosman et al., 1991; Ray et al., 1991). Thus, the responses to the dorsal protein gradient pattern the dorsal–ventral polarity of the embryo and specify the various organ primordia of the *Drosophila* body.

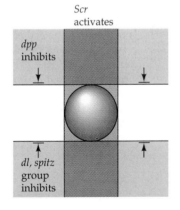

FIGURE 15.45
Cartesian coordinate system for the expression of those genes giving rise to salivary glands. The genes are activated by the protein product of the *Sex combs reduced* homeotic gene along the anterior–posterior axis, and they are inhibited in the regions marked by *decapentaplegic* and *dorsal* gene products along the dorsal–ventral axis. This allows salivary glands to form in the midline of the embryo in the second parasegment. (After Panzer et al., 1992.)

AXES AND ORGAN PRIMORDIA

The Cartesian coordinate model

The anterior–posterior and dorsal–ventral axes of *Drosophila* embryos form a coordinate system that can be used to specify positions within the embryo. Theoretically, cells that are initially equivalent in developmental potential can respond to their coordinates by expressing different sets of genes. This has been seen to be the case in the formation of the salivary gland rudiments (Panzer et al., 1992). First, salivary glands only form in the strip of cells defined by the *Sex combs reduced* (*Scr*) gene along the anterior–posterior axis (parasegment 2). No salivary glands form in *Scr*-deficient mutants. Moreover, if *Scr* is caused to function throughout the embryo, salivary gland genes are expressed in a ventrolateral stripe along most of the length of the embryo. The position of the salivary gland along the dorsal–ventral axis is repressed by both *decapentaplegic* and by *dorsal*. These proteins inhibit salivary gland formation both dorsally and ventrally. Thus, the salivary gland forms at the intersection of the vertical *Scr* expression band (second parasegment) and the horizontal region in the middle of the embryo's circumference that has neither *decapentaplegic* nor *dorsal* gene products (Figure 15.45). The cells that form the salivary gland are directed to do so by the intersecting gene activities of the anterior–posterior and dorsal–ventral axes.

A similar situation is seen with tissues that are found in every segment of the fly. Neuroblasts arise from 10 clusters of 4–6 cells each that form twice in every segment in the strip of neuroectoderm at the midline of the embryo (Skeath and Carroll, 1992). The potential to form neurals cells is conferred on these cells by the expression of proneural genes from the achaete-scute gene complex: *achaete (ac)*, *scute (sc)*, and *lethal of scute (l'sc)*. The cells in each cluster will interact (in ways that will be discussed in a later chapter) to generate a single neural cell from the cluster. Skeath and colleagues (1993) have shown that the pattern of *acheate* and *scute* transcription is imposed by a coordinate sytem. Their expression is repressed by decapentaplegic and snail proteins along the dorsal–ventral axis, while positive enhancement by pair-rule genes along the anterior–posterior axis cause their repetition in each half-segment. The enhancer recognized by these axis-specifying proteins lies between the *achaete* and *scute* genes and appears to regulate both of them. It is very likely, then, that the positions of organ primordia are specified throughout the fly through a two-dimensional coordinate system based on the intersection of the anterior–posterior and dorsal–ventral axes.

Homologous specification

The final step in the realization of the *Drosophila* phenotype involves the activation of specific batteries of genes (presumably by the homeotic selector proteins). This is an exceedingly complicated event that involves two processes. One is differentiation, wherein cells are instructed as to what proteins to make. But before differentiation occurs, the relative *positions* of the cells must first be specified. For instance, is this particular group of cells in the second thoracic segment going to be near the beginning of a leg (such as the coxa or trochanter structures) or at the end of the leg (tarsal or claw structures)? The types of cells in the structures may be the same, but their placement is different.

It appears that the positional specification of the cells—their commitment to form a particular part of the structure within a morphogenetic field—is quite independent of their specification to be different cell types. For example, the determination of each imaginal disc is different. Wing discs give rise to wings, leg discs to legs. The regional specifications of the different discs, however, appear to be identical. This can be seen with certain homeotic mutants such as *Antennapedia*, wherein antennal structures are transformed into legs (Postlethwait and Schneiderman, 1971). Occasionally, the entire antenna becomes an entire leg, but it is more common that only a portion of the antenna becomes leglike. In the latter cases, the replacement is absolutely position-specific. The cells of the antennal disc that normally would have formed the distal tip of the antenna (arista) are transformed into the most distal portion of the leg (claw); cells specified to give rise to the second portion of the antenna are transformed into the second portion (trochanter) of the leg. The corresponding parts of the two structures are shown in Figure 15.46. It is apparent, then, that the two differently determined discs must use a common mechanism for the specification of cell fates within the respective discs.

Summary

We are beginning to learn how the genome influences the construction of the organism. The genes regulating pattern formation in *Drosophila* operate

FIGURE 15.46
Correspondence between portions of the antenna and portions of the leg. In the mutant *Antennapedia*, regions of the antenna are transformed into leg structures. The arrows show the portions of the antenna that form specific corresponding portions of the leg. Such correspondence has also been seen in the transcription patterns of genes such as *salm*. (After Postlethwait and Schneiderman, 1971.)

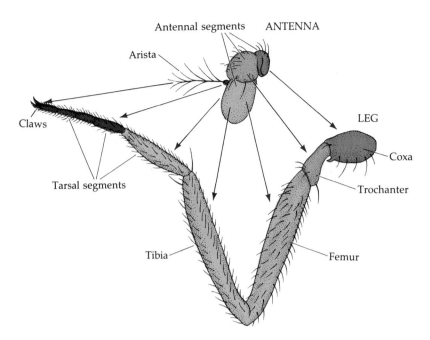

according to certain principles. First, there is a temporal order wherein the different classes of genes are transcribed. Second, the genes are transcribed in a precise spatial arrangement that is brought about by regulatory interactions with previously acting genes (and gene products) and with other genes of the same class. Third, the spatial patterns of gene expression provide a unique identity to every cell along the anterior–posterior axis. When this information is combined with the dorsal–ventral gradient, this constellation of gene products is thought to specify the identity of each cell.

Genetic studies on the *Drosophila* embryo have uncovered numerous genes that are responsible for the specification of the anterior–posterior and the dorsal–ventral axes. We are far from a complete understanding of *Drosophila* pattern formation, but we are much more aware of its complexity than we were five years ago. The mutations of *Drosophila* have given us our first glimpses of the multiple levels of pattern regulation in a complex organism and have enabled the isolation of these genes and their gene products. Moreover, as we will see in the forthcoming chapters, these genes may provide clues to a general mechanism of pattern formation used throughout the animal kingdom.

LITERATURE CITED

Akam, M. E. 1987. The molecular basis for metameric pattern in the *Drosophila* embryo. *Development* 101: 1–22.

Anderson, K. V. 1989. *Drosophila*: The maternal contribution. *In* D. M. Glover and B. D. Hames (eds.), *Genes and Embryos*. IRL, New York, pp. 1–37.

Anderson, K. V. and Nüsslein-Volhard, C. 1984. Information for the dorsal–ventral pattern of the *Drosophila* embryo is stored as maternal mRNA. *Nature* 311: 223–227.

Anderson, K., Bokla, L. and Nüsslein-Volhard, C. 1985. Establishment of dorsal–ventral polarity in the *Drosophila* embryo: The induction of polarity by the *Toll* gene product. *Cell* 42: 791–798.

Ashburner, M. 1990. Puffs, genes, and hormones revisited. *Cell* 61: 1–3.

Bateson, W. 1894. *Materials for the Study of Variation*. Macmillan, London.

Baumgartner, S. and Noll, M. 1990. Networks of interaction among pair–rule genes regulating *paired* expression during primordial segmentation of *Drosophila*. *Mech. Dev.* 1: 1–18.

Beachy, P. A. 1990. A molecular view of the *Ultrabithorax* homeotic gene of *Drosophila*. *Trends Genet.* 6(2): 46–51.

Beachy, P. A., Krasnow, M. A., Gavis, E. R. and Hogness, D. S. 1988. An *Ultrabithorax* protein binds sequences near its own and the *Antennapedia* P1 promoters. *Cell* 55: 1069–1081.

Bender, W. and seven others. 1983. Molecular genetics of the bithorax complex in *Drosophila melanogaster*. *Science* 221: 23–29.

Berleth, T. and seven others. 1988. The role of localization of bicoid RNA in organizing the anterior pattern of the *Drosophila* embryo. *EMBO J.* 7: 1749–1756.

Bermingham, J. R. Jr., Martinez-Arias, A., Pettit, M. G. and Scott, M. 1990. Different patterns of transcription from the two *Antennapedia* promoters during *Drosophila* embryogenesis. *Development* 109: 553–566.

Boulet, A. M. and Scott, M. P. 1988. Control elements of the P2 promoter of the *Antennapedia* gene. *Genes Dev.* 2: 1600–1614.

Boulet, A. M., Lloyd, A., Sakonju, S. 1991. Molecular definition of the morphogenetic and regulatory functions and the *cis*-regulatory elements of the *Drosophila Abd-B* gene. *Development* 111: 393–405.

Boveri, T. 1901. Über die Polarität des Seeigeleiers. *Eisverh. Phys. Med. Ges. Würzburg* 34: 145–175.

Capovilla, M., Eldon, E. D. and Pirrotta, V. 1992. The *giant* gene of *Drosophila* encodes a b-ZIP DNA-binding protein that regulates the expression of other segmentation gap genes. *Development* 114: 99–112.

Carroll, S. B., Laughton, A. and Thalley, B. S. 1988. Expression, function, and regulation of the hairy segmentation protein in the *Drosophila* embryo. *Genes Dev.* 2: 883–890.

Casanova, J. and Struhl, G. 1989. Localized surface activity of *torso*, a receptor tyrosine kinase, specifies body pattern in *Drosophila*. *Genes Dev.* 3: 2025–2038.

Casanova, J., Sanchez-Herrero, E. and Morata, G. 1985. Prothoracic transformation and functional structure of the *Ultrabithorax* gene of *Drosophila*. *Cell* 42: 663–669.

Casanova, J., Sanchez-Herrero, E., Busturia, A. and Morata, G. 1987. Double and triple mutant combination of the bithorax complex of *Drosophila*. *EMBO J.* 6: 3103–3109.

Chasan, R., Jin, Y. and Anderson, K. V. 1992. Activation of the *easter* zymogen is regulated by five other genes to define dorsal–ventral polarity in the *Drosophila* embryo. *Development* 115: 607–615.

Child, C. M. 1941. *Patterns and Problems of Development*. University of Chicago Press, Chicago.

Cohen, S. M. and Jürgens, G. 1990. Mutations of *Drosophila* head development by gap-like segmentation genes. *Nature* 346: 482–485.

Crick, F. H. C. 1970. Diffusion in embryogenesis. *Nature* 225: 420–422.

Degelmann, A., Hardy, P. A., Perrimon, N. and Mahowald, A. P. 1986. Developmental analysis of the *torso-like* phenotype in *Drosophila* produced by a maternal-effect locus. *Dev. Biol.* 115: 479–489.

Dessain, S., Gross, C. T., Kuziora, M. A. and McGinnis, W. 1992. *Antp*-type homeodomains have distinct DNA-binding specificities that correlate with their different regulatory functions in embryos. *EMBO Journal* 11: 991–1002.

Doe, C. Q., Hiromi, Y., Gehring, W. J. and Goodman, C. S. 1988. Expression and function of the segmentation gene *fushi tarazu* during *Drosophila* neurogenesis. *Science* 239: 170–175.

Driever, W. and Nüsslein-Volhard, C. 1988a. A gradient of bicoid protein in *Drosophila* embryos. *Cell* 54: 83–93.

Driever, W. and Nüsslein-Volhard, C. 1988b. The bicoid protein determines position in the *Drosophila* embryo in a concentration-dependent manner. *Cell* 54: 95–104.

Driever, W. and Nüsslein-Volhard, C. 1989. The bicoid protein is a positive regulator of *hunchback* transcription in the early *Drosophila* embryo. *Nature* 337: 138–143.

Driever, W., Thoma, G. and Nüsslein-Volhard, C. 1989. Determination of spatial domains of zygotic gene expression in the *Drosophila* embryo by the affinity of binding sites for the bicoid morphogen. *Nature* 340: 363–367.

Driever, W., Siegel, V. and Nüsslein-Volhard, C. 1990. Autonomous determination of anterior structures in the early *Drosophila* embryo by the bicoid morphogen. *Development*. 109: 811–820.

Edgar, B. A. and Schubiger, G. 1986. Parameters controlling transcriptional activation during early *Drosophila* development. *Cell* 44: 871–877.

Edgar, B. A., Weir, M. P., Schubiger, G. and Kornberg, T. 1986. Repression and turnover pattern of *fushi tarazu* RNA in the early *Drosophila* embryo. *Cell* 47: 747–754.

Ferguson, E. L and Anderson, K. V. 1991. Dorsal–ventral pattern formation in the *Drosophila* embryo The role of zygotically active genes. *Curr. Top. Dev. Biol.* 25: 17–43.

Finkelstein, R. and Perrimon, N. 1990. The *orthodenticle* gene is regulated by bicoid and torso and specifies *Drosophila* head development. *Nature* 346: 485–488.

Foe, V. E. 1989. Mitotic domains reveal early committment of cells in *Drosophila* embryos. *Development* 107: 1–22.

Foe, V. E. and Alberts, B. M. 1983. Studies of nuclear and cytoplasmic behavior during the five mitotic cycles that precede gastrulation in *Drosophila* embryogenesis. *J. Cell Sci.* 61: 31–70.

Forlani, S., Ferrandon, D., Saget, O. and Mohier, E. 1993. A regulatory function for K10 in the establishhment of dorsoventral polarity in the *Drosophila* egg and embryo. *Mech. Dev.* 41: 109–120.

French, V. 1988. Gradients and insect segmentation. *In* V. French, P. Ingham, J. Cooke and J. Smith (eds.), *Mechanisms of Segmentation*. Company of Biologists, Cambridge, pp. 3–16.

Frigerio, G., Burri, M., Bopp, D., Baumgartner, S. and Noll, M. 1986. Structure of the segmentation gene *paired* and the *Drosophila* PRD gene set as part of a gene network. *Cell* 47: 735–746.

Frischer, L. E., Hagen, F. S. and Garber, R. L. 1986. An inversion that disrupts the *Antennapedia* gene causes abnormal structure and localization of RNAs. *Cell* 47: 1017–1023.

Gaul, U. and Jäckle, H. 1990. Role of gap genes in early *Drosophila* development. *Annu. Rev. Genet.* 27: 239–275.

Gavis, E. R. and Lehmann, R. 1992. Localization of *nanos* RNA controls embryonic polarity. *Cell* 71: 301–313.

Gould, A. P. and White, R. A. H. 1992. Connectin, a target of homeotic gene control in *Drosophila*. *Development* 116: 1163–1174.

Gould, A. P., Brookman, J. J., Strutt, D. I. and White, R. A. H. 1990. Targets of homeotic gene control in *Drosophila*. *Nature* 348: 308–312.

Hafen, E., Levine, M. and Gehring, W. J. 1984. Regulation of *Antennapedia* transcript distribution by the bithorax complex in *Drosophila*. *Nature* 307: 287–289.

Hanes, S. D. and Brent, R. 1989. DNA specificity of the bicoid activator protein is determined by homeodomain recognition helix residue 9. *Cell* 57: 1275–1283.

Hanes, S. D. and Brent, R. 1991. A genetic model for interaction of the homeodomain recognition helix with DNA. *Science* 251: 426–430.

Harding, K. and Levine, M. 1988. Gap genes define the limits of *Antennapedia* and *Bithorax* gene expression during early development in *Drosophila*. *EMBO J.* 7: 205–214.

Harding, K. and Levine, M. 1989. *Drosophila*: The zygotic contribution. *In* D. M. Glover and B. D. Hames (eds.), *Genes and Embryos*. IRL, New York, pp. 38–90.

Harding, K., Wedeen, C., McGinnis, W. and Levine, M. 1985. Spatially regulated expression of homeotic genes in *Drosophila*. *Science* 229: 1236–1242.

Harding, K., Hoey, T., Warrior, R. and Levine, M. 1989. Autoregulatory and gap gene response elements of the *even-skipped* promoter of *Drosophila*. *EMBO J.* 8: 1205–1212.

Hashimoto, C., Hudson, K. L. and Anderson, K. V. 1988. The *Toll* gene of *Drosophila*, required for dorsal–ventral embryonic polarity, appears to encode a transmembrane protein. *Cell* 52: 269–279.

Hashimoto, C., Gerttula, S. and Anderson, K. V. 1991. Plasma membrane localization of the *Toll* protein in the syncytial *Drosophila* embryo: importance of transmembrane signalling for dorsal–ventral pattern formation. *Development* 11: 1021–1028.

Hecht, P. M. and Anderson, K. V. 1992. Extracellular proteases and embryonic pattern formation. *Trends Cell Biol.* 2: 197–202.

Heemskerk, J., DiNardo, S., Kostriken, R. and O'Farrell, P. H. 1991. Multiple modes of *engrailed* regulation in the progression towards cell fate determination. *Nature* 352: 404–410.

Hoch, M., Gerwin, N., Taubert, H. and Jäckle, H. 1992. Competition for overlapping sites in the regulatory region of the *Drosophila* gene *Krüppel*. *Science* 256: 94–97.

Howard, K. and Ingham, P. 1986. Regulatory interactions between the segmentation genes *fushi tarazu*, *hairy*, and *engrailed* in the *Drosophila* blastoderm. *Cell* 44: 949–957.

Hülskamp, M., Schröder, C., Pfeifle, C., Jäckle, H. and Tautz, D. 1989. Posterior segmentation of the *Drosophila* embryo in the absence of a maternal posterior organizer gene. *Nature* 338: 629–632.

Ingham, P. W. 1993. Localized *hedgehog* activity controls spatially restricted transcription of *wingless* in the *Drosophila* embryo. *Nature* 366: 560–562.

Ingham, P. W. and Martinez-Arias, A. 1986. The correct activation of *Antennapedia* and bithorax complex genes requires the *fushi tarazu* gene. *Nature* 324: 592–597.

Ingham, P. W. and Whittle, R. 1980. *Trithorax*: A new homeotic mutation of *Drosophila* causing transformations of abdominal and thoracic imaginal segments. I. Putative role during embryogenesis. *Mol. Gen. Genet.* 179: 607–614.

Ingham, P. W., Taylor, A. M. and Nakano, Y. 1991. Role of *Drosophila patched* gene in positional signalling. *Nature* 353: 184–187.

Irish, V. F. and Gelbart, W. M. 1987. The *decapentaplegic* gene is required for dorsal–ventral patterning of the *Drosophila* embryo. *Genes and Dev.* 1: 868–879.

Irish, V., Lehmann, R. and Akam, M. 1989. The *Drosophila* posterior-group gene *nanos* functions by repressing hunchback activity. *Nature* 338: 646–648.

Irish, V. F., Martinez-Arias, A. and Akam, M. 1989. Spatial regulation of the *Antennapedia* and *Ultrabithorax* genes during *Drosophila* early development. *EMBO J.* 8: 1527–1537.

Jack, T., Regulski, M. and McGinnis, W. 1988. Pair-rule segmentation genes regulate the expression of the homeotic selector gene, *Deformed*. *Genes Dev.* 2: 592–597.

Jäckle, H., Tautz, D., Schuh, R., Seifert, E. and Lehmann, R. 1986. Cross-regulatory interactions among the gap genes of *Drosophila*. *Nature* 324: 668–670.

Jiang, J. and Levine, M. 1993. Binding affinities and cooperative interactions with bHLH activators delimit threshold responses to the dorsal gradient morphogen. *Cell* 72: 741–752.

Jiang, J., Kosman, D., Ip, Y. T. and Levine, M. 1991. The *dorsal* morphogen gradient regulates the mesoderm determinant *twist* in early *Drosophila* embryos. *Genes and Dev.* 5: 1881–1891.

Kalthoff, K. 1969. Der Einfluss vershiedener Versuchparameter auf die Häufigkeit der Missbildung "Doppelabdomen" in UV-bestrahlten Eiern von *Smittia* sp. (Diptera, Chironomidae). *Zool. Anz. Suppl.* 33: 59–65.

Kalthoff, K. and Sander, K. 1968. Der Enwicklungsgang der Missbildung "Doppelabdomen" im partiell UV-bestrahlten Ei von *Smittia parthenogenetica* (Diptera, Chironomidae). *Wilhelm Roux Arch. Entwicklungsmech. Org.* 161: 129–146.

Kandler-Singer, I. and Kalthoff, K. 1976. RNase sensitivity of an anterior morphogenetic determinant in an insect egg (*Smittia* sp., Chironomidae, Diptera). *Proc. Natl. Acad. Sci. USA* 73: 3739–3743.

Karch, F. and seven others. 1985. The abdominal region of the bithorax complex. *Cell* 43: 81–96.

Karr, T. L. and Kornberg, T. B. 1989. *fushi tarazu* protein expression in the cellular blastoderm of *Drosophila* detected using a novel imaging technique. *Development* 105: 95–103.

Kaufman, T. C., Seeger, M. A. and Olsen, G. 1990. Molecular and genetic organization of the *Antennapedia* gene complex of *Drosophila melanogaster*. *Adv. Genet.* 27: 309–362.

Kidd, S. 1992. Characterization of the *Drosophila cactus* locus and analysis of interactions between cactus and dorsal proteins. *Cell* 71: 623–635.

Klingler, M., Erdélyi, M., Szabad, J. and Nüsslein-Volhard, C. 1988. Function of *torso* in determining the terminal anlagen of the *Drosophila* embryo. *Nature* 335: 275–277.

Knipple, D. C., Seifert, E., Rosenberg, U. B., Preiss, A. and Jäckle, H. 1985. Spatial and temporal patterns of *Krüppel* gene expression in early *Drosophila* embryos. *Nature* 317: 40–44.

Kornfeld, K., Saint, R. B., Beachy, P. A., Harte, P. J., Peattie, D. A. and Hogness, D. S. 1989. Structure and expression of a family of Ultrabithorax mRNAs regulated by alternative splicing and polyadenylation in *Drosophila*. *Genes Dev.* 3: 243–258.

Kosman, D., Ip, Y. T., Levine, M. and Arora, K. 1991. Establishment of the mesoderm-neuroectoderm boundary in the *Drosophila* embryo. *Science* 254: 118–122.

Krasnow, M. A., Saffman, E. E., Kornfeld, K. and Hogness, D. S. 1989. Transcriptional activation and repression by Ultrabithorax proteins in cultured *Drosophila* cells. *Cell* 57: 1031–1043.

Kraut, R. and Levine, M. 1991. Mutually repressive interactions between the gap genes *giant* and *Krüppel* define middle body regions of the *Drosophila* embryo. *Development* 111: 611–621.

Lehmann, R. and Nüsslein-Volhard, C. 1987. Involvement of the *pumilio* gene in the transport of the abdominal signal in the *Drosophila* embryo. *Nature* 329: 167–170.

Lehmann, R. and Nüsslein-Volhard, C. 1991. The maternal gene *nanos* has a central role in posterior pattern formation of the *Drosophila* embryo. *Development* 112: 679–691.

Leptin, M. 1991a. *twist* and *snail* as positive and negative regulators during *Drosophila* mesoderm development. *Genes Dev.* 5: 1568–1576.

Leptin, M. 1991b. Mechanics and genetics of cell shape changes during *Drosophila* ventral furrow formation. *In* R. Keller et al. (eds.), *Gastrulation: Movements, Patterns, and Molecules*. Plenum, New York, pp. 199–212.

Levine, M. S. and Harding, K. W. 1989. *Drosophila*: The zygotic contribution. *In* D. M. Glover and B. D. Hames (eds.), *Genes and Embryos*. IRL, New York, pp. 39–94.

Lewis, E. B. 1978. A gene complex controlling segmentation in *Drosophila*. *Nature* 276: 565–570.

Lewis, E. B. 1985. Regulation of the genes of the bithorax complex in *Drosophila*. *Cold Spring Harbor Symp. Quant. Biol.* 50: 155–164.

MacDonald, P. M. and Struhl, G. 1986. A molecular gradient in early *Drosophila* embryos and its role in specifying the body pattern. *Nature* 324: 537–545.

Mange, A. P. and Mange, E. J. 1990. *Genetics: Human Aspects*. Sinauer Associates, Sunderland, MA.

Mann, R. S. and Hogness, D. S. 1990. Functional dissection of *Ultrabithorax* proteins in *D. melanogaster*. *Cell* 60: 597–610.

Martinez-Arias, A. and Lawrence, P. A. 1985. Parasegments and compartments in the *Drosophila* embryo. *Nature* 313: 639–642.

Mlodzik, M. and Gehring, W. J. 1987. Expression of the *caudal* gene in the germ line of *Drosophila*: Formation of an RNA and protein gradient during early embryogenesis. *Cell* 48: 465–478.

Mohler, J. and Vani, K. 1992. Molecular organization and embryonic expression of the *hedgehog* gene involved in cell–cell communication in segmental patterning in *Drosophila*. *Development* 115: 957–971.

Montell, D. J., Keshishian, H. and Spradling, A. C. 1991. Laser ablation studies of the role of the *Drosophila* oocyte nucleus in pattern formation. *Science* 254: 290–293.

Müller, M. Affolter, M., Leupin, W., Otting, G., Wüthrich, K. and Gehring, W. J. 1988. Isolation and sequence-specific DNA binding of the *Antennapedia* homeodomain. *EMBO J.* 7: 4299–4304.

Nambu, J. R., Franks, R. G., Hong, S. and Crews, S. 1990. The *single-minded* gene of *Drosophila* is required for the expression of genes important for the development of CNS midline cells. *Cell* 63: 63–75.

Neuman-Silberberg, F. S. and Schüpbach, T. 1993. The *Drosophila* dorsoventral patterning gene *gurken* produces a dorsally localized RNA and encodes a TGF-α-like protein. *Cell* 75: 165–174.

Nijhout, H. F. 1981. The color patterns of butterflies and moths. *Sci. Am.* 245 (5): 140–151.

Nüsslein-Volhard, C. and Wieschaus, E. 1980. Mutations affecting segment number and polarity in *Drosophila*. *Nature* 287: 795–801.

Nüsslein-Volhard, C., Frohnhöfer, H. G. and Lehmann, R. 1987. Determination of anterioposterior polarity in *Drosophila*. *Science* 238: 1675–1681.

O'Farrell, P. H. and seven others. 1985. Embryonic pattern in *Drosophila*: The spatial distribution and sequence-specific DNA binding of engrailed protein. *Cold Spring Harbor Symp. Quant. Biol.* 50: 235–242.

Otting, G., Qian, Y. Q., Billeter, M., Müller, M., Affolter, M., Gehring, W. J. and Wüthrich, K. 1990. Protein–DNA contacts in the structure of a homeodomain–DNA complex determined by nuclear magnetic resonance spectroscopy in solution. *EMBO J.* 9: 3085–3092.

Pan, D., Huang, J.-D. and Courey, A. J. 1991. Functional analysis of the *Drosophila twist* promoter reveals a *dorsal*-binding ventral activator region. *Genes and Dev.* 5: 1892–1901.

Pankratz, M. J., Seifert, E., Gerwin, N., Billi, B., Nauber, U. and Jäckle, H. 1990. Gradients of *Krüppel* and *knirps* gene products direct pair-rule gene stripe patterning in the posterior region of the *Drosophila* embryo. *Cell* 61: 309–317.

Panzer, S., Weigel, D. and Beckendorf, S. K. 1992. Organogenesis in *Drosophila melanogaster*: embryonic salivary gland determination is controlled by homeotic and dorsoventral patterning genes. *Development* 114: 49–57.

Paro, P. 1990. Imprinting a determined state into the chromatin of *Drosophila*. *Trends Genet.* 6: 416–421.

Peifer, M. and Bender, W. 1986. The *anterobithorax* and *bithorax* mutations of the bithorax complex. *EMBO J.* 5: 2293–2303.

Peifer, M. and Bejsovec, A. 1992. Knowing your neighbors: Cell interactions determine intrasegmental patterning in *Drosophila*. *Trends Genet.* 8: 243–249.

Peifer, M., Karch, F. and Bender, W. 1987. The bithorax complex: Control of segmental identity. *Genes Dev.* 1: 891–898.

Percival-Smith, A., Müller, M., Affolter, M. and Gehring, W. J. 1990. The interaction with DNA of wild-type and mutant *fushi tarazu* homeodomains. *EMBO J.* 9: 3967–3974.

Pignoni, F., Steingrímsson, E. and Lengyel, J. A. 1992. *bicoid* and the terminal system activate *tailless* expression in the early *Drosophila* embryo. *Development* 115: 239–251.

Postlethwait, J. H. and Schneiderman, H. A. 1971. Pattern formation and determination in the antenna of the homeotic mutant *Antennapedia* of *Drosophila melanogaster*. *Dev. Biol.* 25: 606–640.

Price, J. V., Clifford, R. J. and Schüpbach, T. 1989. The maternal ventralizing gene *torpedo* is allelic to *faint little ball*, an embryonic lethal, and encodes the *Drosophila* EGF receptor homolog. *Cell* 56: 1085–1092.

Qian, S., Capovilla, M. and Pirrotta, V. 1991. The *bx* region enhancer, a distant *cis*-control element of the *Drosophila Ubx* gene and its regulation by *hunchback* and other segmentation genes. *EMBO J.* 10: 1415–1425.

Ray, R. Arora, K., Nüsslein-Volhard, C. and Gelbart, W. M. 1991. The control of cell fate along the dorsal–ventral axis of the *Drosophila* embryo. *Development* 113: 35–54.

Riddihough, G. 1992. Homing in on the homeobox. *Nature* 357: 643–644.

Roth, S., Stein, D., Nüsslein-Volhard, C. 1989. A gradient of nuclear localization of the *dorsal* protein determines dorsoventral pattern in the *Drosophila* embryo. *Cell* 59: 1189–1202.

Rushlow, C., Frasch, J., Doyle, H. and Levine, M. 1987. Maternal regulation of a homeobox gene controlling differentiation of dorsal tissues in *Drosophila*. *Nature* 330: 583–586.

Rushlow, C. A., Han, K., Manley, J. L. and Levine, M. 1989. The graded distribution of the *dorsal* morphogen is initiated by selective nuclear transport in *Drosophila*. *Cell* 59: 1165–1177.

Sánchez-Herrero, E. 1991. Control of the expression of the bithorax complex genes *abdominal-A* and *Abdominal-B* by *cis*-regulatory regions in *Drosophila* embryos. *Development* 111: 437–449.

Sánchez-Herrero, E., Verños, I., Marco, R. and Morata, G. 1985. Genetic organization of *Drosophila* bithorax complex. *Nature* 313: 108–113.

Sander, K. 1960. Analyse des ooplasmatischen Reaktionssystems von Euscelis plebajus Fall (Circadina) durch Isolieren und Kombinieren von Keimteilen. II. Die Differenzierungsleistungen nach Verlagern von Hinterpolmaterial. *Wilhelm Roux Arch. Entwicklungsmech. Org.* 151: 660–707.

Sander, K. 1975. Pattern specification in the insect embryo. In *Cell Patterning. CIBA Foundation Symp.* 29: 241–263.

Schier, A. F. and Gehring, A. J. 1992. Direct homeodomain-DNA interaction in the autoregulation of the *fushi tarazu* gene. *Nature* 356: 804–807.

Schneuwly, S., Kuroiwa, A. and Gehring, W. J. 1987. Molecular analysis of the dominant homeotic *Antennapedia* phenotype. *EMBO J.* 6: 201–206.

Schüpbach, T. 1987. Germ line and soma cooperate during oogenesis to establish the dorsoventral pattern of egg shell and embryo in *Drosophila melanogaster*. *Cell* 49: 699–707.

Schüpbach, T. and Wieschaus, E. 1986. Maternal effect mutations altering the anterior–posterior pattern of the *Drosophila* embryo. *Roux Arch. Dev. Biol.* 195: 302–317.

Schüpbach, T., Clifford, R. J., Manseau, L. J. and Price, J. V. 1991. Dorsoventral signaling processes in *Drosophila* oogenesis. *In* J. Gerhart, (ed.) *Cell–Cell Interactions in Early Development*. Wiley-Liss, New York pp. 163–174.

Shelton, C. A. and Wasserman, S. A. 1993. *pelle* encodes a protein kinase required to establish dorsoventral polarity in the *Drosophila* embryo. *Cell* 72: 515–525.

Siegfried, E., Wilder, E. L. and Perrimon, N. 1994. Components of *wingless* signalling in *Drosophila*. *Nature* 367: 76–80.

Simon, J., Peifer, M., Bender, W. and O'Connor, M. 1990. Regulatory elements of the bithorax complex that control expression along the anterior–posterior axis. *EMBO J.* 9: 3945–3956.

Skeath, J. B. and Carroll, S. B. 1992. Regulation of proneural gene expression and cell fate during neuroblast segregation in the *Drosophila* embryo. *Development* 114: 939–946.

Skeath, J. B., Panganiban, G., Selegue, J. and Carroll, S. B. 1993. Gene regulation in two dimensions: The proneural achaete and scute genes are controlled by combinations of axis-patterning genes through a common intergenic control region. *Genes Dev.* 6: 2606–2619.

Slack, J. M. W. 1983. *From Egg to Embryo: Determinative Events in Early Development*. Cambridge University Press, Cambridge.

Spierer, P. and Goldschmidt-Clermont, M. 1985. La génétique du développement de la mouche. *La Récherche* 16: 452–461.

St. Johnston, D. and Nüsslein-Volhard, C. 1992. The origin of pattern and polarity in the *Drosophila* embryo. *Cell* 68: 201–219.

Štanojević, D., Hoey, T. and Levine, M. 1989. Sequence-specific DNA-binding activities of the gap proteins encoded by *hunchback* and *Krüppel* in *Drosophila*. *Nature* 341: 331–335.

Štanojević, D., Small, S. and Levine, M. 1991. Regulators of a segmentation stripe by overlapping activators and repressors in the *Drosophila* embryo. *Science* 254: 1385–1387.

Stein, D. and Nüsslein-Volhard, C. 1992. Multiple extracellular activities in *Drosophila* egg perivitelline fluid are required for establishment of embryonic dorsal–ventral polarity. *Cell* 68: 429–440.

Stein, D., Roth, S., Vogelsang, E., Nüsslein-Volhard, C. 1991. The polarity of the dorsoventral axis in the *Drosophila* embryo is defined by an extracellular signal. *Cell* 65: 725–735.

Stevens, L. M., Frohnhöfer, H. G., Klingler, M. and Nüsslein-Volhard, C. 1990. Localized requirement for *torso-like* expression in follicle cells for development of terminal anlagen of the *Drosophila* embryo. *Nature* 346: 660–662.

Steward, R. 1989. Relocalization of the dorsal protein from the cytoplasm to the nucleus correlates with its function. *Cell* 59: 1179–1188.

Steward, R. and Govind, S. 1993. Dorsal–ventral polarity in the *Drosophila* embryo. *Curr. Opin. Genet. Dev.* 3: 556–561.

Struhl, G. 1981. A homeotic mutation transforming leg to antenna in *Drosophila*. *Nature* 292: 635–638.

Struhl, G. 1989. Differing strategies for organizing anterior and posterior body pattern in *Drosophila* embryos. *Nature* 338: 741–744.

Struhl, G., Struhl, K. and Macdonald, P. M. 1989. The gradient morphogen *bicoid* is a concentration-dependent transcriptional activator. *Cell* 57: 1259–1273.

Struhl, G., Johnson, P. and Lawrence, P. 1992. Control of a *Drosophila* body pattern by the hunchback morphogen gradient. *Cell* 69: 237–249.

Tautz, D. 1988. Regulation of the *Drosophila* segmentation gene *hunchback* by two maternal morphogenetic centers. *Nature* 332: 281–284.

Thisse, B., Stoetzel, C., Gorostiza-Thisse, C. and Perrin-Schmidt, F. 1988. Sequence of the *twist* gene and nuclear localization of its protein in endomesodermal cells of early *Drosophila* embryos. *EMBO J.* 7: 2175–2183.

Thisse, C, Perrin-Schmidt, F., Stoetzel, C. and Thisse, B. 1991. Sequence-specific *trans* activation of the *Drosophila twist* gene by the *dorsal* gene product. *Cell* 65: 1191–1201.

Treisman, J. and Desplan, C. 1989. The products of the *Drosophila* gap genes *hunchback* and Krüppel bind to the *hunchback* promoters. *Nature* 341: 335–336.

Turner, F. R. and Mahowald, A. P. 1977. Scanning electron microscopy of *Drosophila melanogaster* embryogenesis. *Dev. Biol.* 57: 403–416.

Vachon, G., Cohen, B., Pfeifle, C., McGuffin, M. E., Botas, J. and Cohen, S. M. 1992. Homeotic genes of the bithorax complex repress limb development in the abdomen of the *Drosophila* embryo through the target gene *Distal-less*. *Cell* 71: 437–450.

Wagner-Bernholz, J. T., Wilson, C., Gibson, G., Schuh, R. and Gehring, W. J. 1991. Identification of target genes of the homeotic gene *Antennapedia* by enhancer detection. *Genes Dev.* 5: 2467–2480.

Wakimoto, B. T., Turner, F. R. and Kaufman, T. C. 1984. Defects in embryogenesis in mutants associated with the *Antennapedia* gene complex of *Drosophila melanogaster*. *Dev. Biol.* 102: 147–172.

Wang, C. and Lehmann, R. 1991. *Nanos* is the localized posterior determinant in *Drosophila*. *Cell* 66: 637–647.

Whalen, A. M. and Steward, R. 1993. Dissociation of the dorsal–cactus complex and phosphorylation of the dorsal protein correlate with the nuclear localization of dorsal. *J. Cell. Biol.* 123: 523–534.

Wharton, R. P. and Struhl, G. 1991. RNA regulatory elements mediate control of *Drosophila* body pattern by the posterior morphogen *nanos*. *Cell* 67: 955–967.

Wolpert, L. 1971. Positional information and pattern formation. *Curr. Top. Dev. Biol.* 6: 183–224.

Wolpert, L. 1978. Pattern formation in biological development. *Sci. Am.* 239(4): 154–164.

16

Specification of cell fate by progressive cell–cell interactions

The study of the role of genes in ontogeny is a field for developmental physiology. Not that the geneticist will be excluded from the working out of this problem—he will become a genetic experimental embryologist. After a long journey which took him sometimes out of sight of his fellow biologists, he returns home again with some new concepts and tools.
CURT STERN (1936)

We are standing and walking with parts of our body which could have been used for thinking had they developed in another part of the embryo.
HANS SPEMANN (1943)

In the last chapter we saw that cell determination and axis specification can be caused by the interactions of specific cytoplasmic substances within a syncytial cell. We also saw that homeotic genes play a major role in the regional specification of the axes. Whereas axis specification and cell determination in *Drosophila* occur by interactions between cellular compartments, we will now see this interactive mode of determination occurring between blastomeres.

Regulative development

In deuterostomes such as sea urchins and vertebrates, the fate of a cell depends on its position in the embryo, rather than on what piece of cytoplasm it has acquired. Sidney Brenner (quoted in Wilkins, 1993) has remarked that animal development proceeds in either of two ways. Some organisms are specified predominantly by the "European style"—that is, each cell is determined by who its ancestors were. The lineage is the important thing. Conversely, the blastomeres of most vertebrates are specified predominantly in the "American style": there is a great deal of mixing between cells, and each cell is determined by who its neighbors are. Each cell starts off with similar potentials and develops according to whom it meets. In these embryos, for at least part of cleavage, each cell is able to develop into the entire embryo if it is separated from the other cells, and the remaining cells are able to alter their fates to produce a whole embryo (as in twin formation). This type of commitment is called conditional (or dependent) specification and it gives rise to **regulative development**.

August Weismann: The germ plasm theory

The discovery of regulative determination has its roots in the failure of the mosaic theories of development formulated in Germany at the end of the nineteenth century. In 1883, August Weismann began proposing a theory that integrated such diverse biological phenomena as heredity, development, regeneration, sexual reproduction, and evolution by natural selection. This mechanical model for cellular differentiation was called the

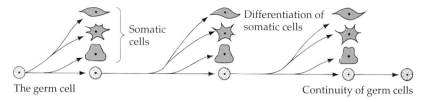

Somatic cells

Differentiation of somatic cells

The germ cell

Continuity of germ cells

FIGURE 16.1

Weismann's theory of inheritance. The germ cell gives rise to the differentiating somatic cells of the body (indicated in color), as well as to new germ cells. (From Wilson, 1896.)

germ plasm theory. Based on the scant knowledge of fertilization available at that time, Weismann boldly proposed that the sperm and egg provided equal chromosomal contributions, both quantitatively and qualitatively, to the new organism. Moreover, chromosomes were postulated to carry the inherited potentials of this new organism and were said to be the basis for the continuity between generations.* However, not all the determinants on the chromosomes were thought to enter every cell of the embryo; instead of dividing equally, the chromosomes were hypothesized to divide in such a way that different nuclear determinants entered different cells. Whereas the fertilized egg would carry the full complement of all determinants, certain cells would retain the "blood-forming" determinants while other cells would retain the "muscle-forming" determinants. Only in the nuclei of those cells destined to become gametes (the germ cells) were all types of determinants thought to be retained. The nuclei of all other cells would have only a fraction of the original determinant types.

Weismann's hypothesis proposed the continuity of the germ plasm and the diversity of the somatic lines. Differentiation was due to the "segregation of the nuclear determinants" into the various cell types. The chromosomes, while appearing equal in all cells, would be unequal in their qualities. Only the germ cell line would keep all the determinants, and this cell lineage would be totally *independent* of the somatic cells. Hence, there could be no inheritance of characteristics acquired by the somatic cells. Weismann obtained support for this model by cutting off the tails of newborn mice for 19 generations. The mice of each succeeding generation had tails of normal length, indicating that the germ line was insulated from insults to the somatic tissue.[†]

Weismann's germ plasm theory is depicted in Figure 16.1. It emphasizes the continuity and immortality of the germ line contrasted with the temporary nature of the adult organism, showing, as physiologist Michael Foster noted, that "the animal body is in reality a vehicle for ova." E. B. Wilson, who claimed that his remarkable textbook *The Cell in Development*

---

*Embryologists were thinking in these terms some 15 years before the rediscovery of Mendel's work. Weismann (1892, 1893) also speculated that these nuclear determinants of inheritance functioned by elaborating substances that became active in the cytoplasm!

[†]At this time, the major alternative view was that of pangenesis. This theory, advocated as a "provisional hypothesis" by Charles Darwin, posited that each somatic cell contained particles (*pangenes*) that migrated back into the sex cells to provide for the transmission of that cell's characteristics. According to this theory, Weismann should have obtained mice with shorter tails. More recently, Thomas Jukes, commenting on these results, quoted Hamlet's intuition that "there's a divinity that shapes our ends, rough-hew them how we will." It should be noted that the independence of the germ line from the somatic line is not absolute in all organisms. In sponges, flatworms, hydroids, and colonial tunicates, germ cells can develop from somatic tissues, and genetic changes made to those somatic tissues can be inherited (Berrill and Liu, 1948; Buss, 1987).

and Inheritance (1896) grew out of Weismann's hypothesis, also saw the implications of Weismann's scheme:

> The death of the individual involves no breach of continuity in the series of cell divisions by which the life of the race flows onwards. The individual dies, it is true, but the germ-cells live on, carrying with them, as it were, the traditions of the race from which they have sprung, and handing them on to their descendants.

Wilhelm Roux: Mosaic development

Weismann had intuited that chromosomes are the bearers of the inherited information for development. More importantly, though, he had proposed a hypothesis of development that could be tested immediately. Weismann had claimed that when the first cleavage division separated the future right half of the embryo from the future left half, there would be a separation of "right" determinants from "left" determinants in the resulting blastomeres. This assertion was tested by Wilhelm Roux, a young German embryologist. In 1888, Roux published the results of a series of experiments in which he took 2- and 4-cell frog embryos and destroyed some of the cells of each embryo with a hot needle. Weismann's hypothesis predicted the formation of right- or left half-embryos; Roux obtained half-morulae, just as Weismann predicted (Figure 16.2). These developed into half-neurulae having a complete right or left side, with one medullary fold, one ear pit, and so on. He therefore concluded that the frog embryo was a mosaic of self-differentiating parts and that it was likely that each cell received a specific set of determinants and differentiated accordingly. With this series of experiments, Roux inaugurated his program of developmental mechanics (*Entwicklungsmechanik*), an experimental and physiological approach to embryology (see Sander, 1991a,b). No longer, insisted Roux, would embryology merely be the servant of evolutionary studies. Rather, embryology would assume its role as an independent experimental science.

Regulation in sea urchin embryos

Hans Driesch: Regulative development

Nobody appreciated the experimental approach to embryology more than Hans Driesch. Driesch's goal was to explain development in terms of the

FIGURE 16.2
Roux's attempt to show mosaic development. Destroying one cell of a 2-cell frog embryo results in the development of only one half of the embryo.

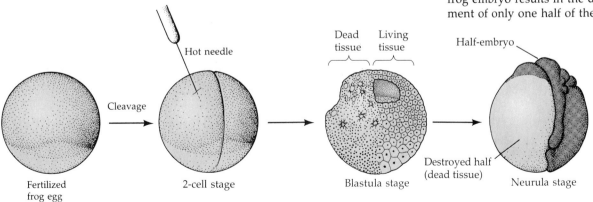

Fertilized frog egg — Cleavage → 2-cell stage → Hot needle — Dead tissue / Living tissue — Blastula stage → Half-embryo / Destroyed half (dead tissue) — Neurula stage

FIGURE 16.3

Driesch's demonstration of regulative development. (A) A normal pluteus larva. (B) Smaller, but normal, plutei that each developed from one blastomere of a dissected 4-cell embryo. (All larvae are drawn to the same scale.) (After Hörstadius and Wolsky, 1936.) Note that the four larvae derived in this way are not identical, despite their ability to generate all the necessary cell types. Such variations are also seen in those adult sea urchins formed in this way (Marcus, 1979).

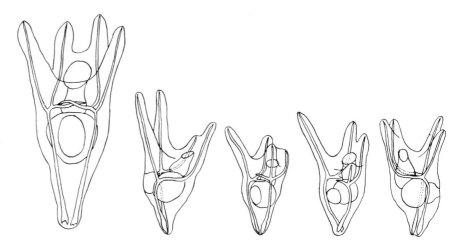

(A) Normal pluteus larva

(B) Plutei developed from single cells of 4-cell embryo

laws of physics and mathematics. His initial investigations were similar to those of Roux. Roux's experiments were, technically, *defect* studies that answered the question of how the remaining blastomeres of an embryo would develop when a subset of them was destroyed. Driesch (1892) sought to extend this research by performing *isolation* experiments. Sea urchin blastomeres were separated from each other by vigorous shaking (or, later, by placing them in calcium-free seawater). To Driesch's surprise, each of the blastomeres from a 2-cell embryo developed into a complete larva. Similarly, when Driesch separated the blastomeres from 4- and 8-cell embryos, some of the cells produced entire pluteus larvae (Figure 16.3). Here was a result drastically different from that predicted by Weismann or Roux. Rather than self-differentiating into its future embryonic part, each blastomere could regulate its development so as to produce a complete organism. Here was the first experimentally observable instance of regulative development.

Regulative development was also demonstrated in another experiment by Driesch. In sea urchin eggs, the first two cleavage planes are meridional, passing through the animal and vegetal poles, whereas the third division is equatorial, dividing the embryo into four upper and four lower cells (see Figure 5.3). Driesch (1893) changed the direction of the third cleavage by gently compressing the early embryos between two glass plates, thus causing the third division to be meridional like the preceding two cleavages. After he released the pressure, the fourth division was equatorial. This procedure reshuffled the nuclei, causing a nucleus that normally would be in the region destined to form endoderm to now be in the presumptive ectoderm region. Some nuclei that would normally have produced dorsal structures were now found in the ventral cells (Figure 16.4). If the segregation of nuclear determinants had occurred (as had been proposed by Weismann and Roux), the resulting embryo should be strangely disordered. However, Driesch obtained normal larvae from these embryos. He concluded, "The relative position of a blastomere within the whole will probably in a general way, determine what shall come from it."

The consequences of these experiments were momentous both for embryology and for Driesch personally. First, Driesch had demonstrated that the "prospective potency" of an isolated blastomere (those cell types

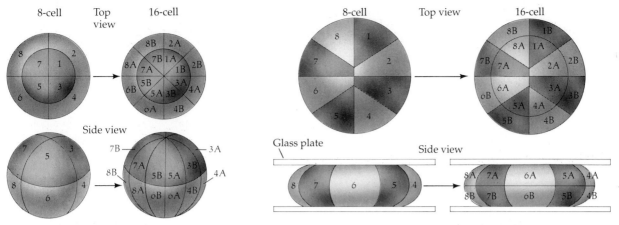

8-cell   Top view   16-cell

Side view

(A) NORMAL CLEAVAGE

8-cell   Top view   16-cell

Glass plate   Side view

(B) CLEAVAGE UNDER PRESSURE

FIGURE 16.4
Driesch's pressure-plate experiment for altering the distribution of nuclei. (A)
Normal cleavage from 8- to 16-cell sea urchin embryos seen from the animal pole
(upper sequence) and from the side (lower sequence). (B) Abnormal cleavage
planes formed under pressure, as seen from the animal pole and from the side.
(After Huxley and deBeer, 1934.)

it was possible for it to form) is greater than its "prospective fate" (the cell
types it would normally give rise to over the unaltered course of its
development). According to Weismann and Roux, the prospective potency
and the prospective fate of a blastomere should be identical. Second,
Driesch concluded that the sea urchin embryo is a "harmonious equipotential system," because all these potentially independent parts functioned
together to form a single organism. Third, he concluded that the fate of a
nucleus depended solely on its location in the embryo. Driesch (1894)
hypothesized a series of events wherein development proceeded by the
interactions of the nucleus and cytoplasm:

> Insofar as it contains a nucleus, every cell, during ontogenesis, carries the
> totality of all primordia; insofar as it contains a specific cytoplasmic cell body,
> it is specifically enabled by this to respond to specific effects only. . . . When
> nuclear material is activated, then, under its guidance, the cytoplasm of its
> cell that had first influenced the nucleus is in turn changed, and thus the
> basis is established for a new elementary process, which itself is not only the
> result but also a cause.

This strikingly modern concept of nuclear–cytoplasmic interaction and
nuclear equivalence eventually caused Driesch to abandon science. He
could no longer envision the embryo as a physical machine, because it
could be subdivided into parts that were each capable of reforming the
entire organism. In other words, Driesch had come to believe that development could not be explained by physical forces. He was driven to invoke
a vital force, *entelechy* ("internal goal-directed force"), to explain how development proceeds. Essentially, he believed the embryo was imbued with
an internal psyche and wisdom to accomplish its goals despite the obsta-
cles embryologists placed in its path. Unable to explain his results in terms
of the physics of his day, Driesch renounced the study of developmental
physiology and became a philosophy professor, proclaiming vitalism until
his death in 1941. Others, especially Oscar Hertwig (1894), were able to

TABLE 16.1
Experimental procedures and results of Roux and Driesch

Investigator	Organism	Type of experiment	Conclusion	Interpretation concerning potency and fate
Roux (1888)	Frogs (*Rana fusca*)	Defect	Mosaic development (autonomous)	Prospective potency equals prospective fate
Driesch (1892)	Sea urchin (*Echinus microtuberculatus*)	Isolation	Regulative development (conditional)	Prospective potency is greater than prospective fate
Driesch (1893)	Sea urchin (*Echinus* and *Paracentrotus*)	Recombination	Regulative development (conditional)	Prospective potency is greater than prospective fate

incorporate Driesch's experiments into a more sophisticated experimental embryology.*

The differences between Roux's experiments and those of Driesch are summarized in Table 16.1. The difference between isolation and defect experiments and the importance of the interactions provided by the destroyed blastomeres were highlighted in 1910, when J. F. McClendon showed that isolated frog blastomeres behave just like separated sea urchin cells. Therefore, the mosaic-like development of the first two frog blastomeres in Roux's study was an artifact of the defect experiment. Something in or on the dead blastomere still informed the live cells that it existed. We have already seen that early mammalian blastomeres have a regulative type of development. As we discussed in Chapter 5, each isolated blastomere of a mouse inner cell mass is capable of generating an entire fertile mouse. The ability of two or more early mouse embryos to fuse into one normal embryo (Figure 5.28) and the phenomenon of identical twins (Figure 5.31) also attest to the regulative ability of mammalian blastomeres. Therefore, even though Weismann and Roux pioneered the study of developmental physiology, their proposition that differentiation is caused by the segregation of nuclear determinants was shortly shown to be incorrect.

Sven Hörstadius: Potency and oocyte gradients

But Driesch was not 100 percent correct either. As we saw in the previous chapter, there are numerous animals that do develop largely as a mosaic of self-differentiating parts. More importantly, though, even the sea urchin

*These experiments strengthened within embryology a type of mechanistic philosophy called *holist organicism*. This philosophy refers to the views that (1) the properties of the whole cannot be predicted solely from the properties of the component parts, and (2) the properties of the parts are informed by their relationship to the whole. As an analogy, the meaning of a sentence obviously depends on the meanings of its component parts, words. However, the meaning of each word depends on the entire sentence. In the sentence, "The party leaders were split on the platform," the possible meanings of each noun and verb are limited by the meaning of the entire sentence and by the relationships to other words within the sentence. Similarly, a cell in the embryo develops its phenotype depending on its interactions within the entire embryo. The opposite materialist view is *reductionism*, which maintains that the properties of the whole can be known if all the properties of the parts are known. Embryology has traditionally espoused holist organicism, while genetics has been characterized as being a reductionistic discipline (Haraway, 1976; Roll-Hansen, 1978; Allen, 1985; Tauber and Sarkar, 1992; Gilbert, in press).

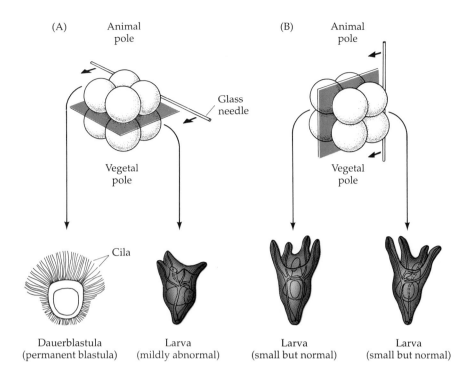

(A) Animal pole

Glass needle

Vegetal pole

Cila

Dauerblastula (permanent blastula)

Larva (mildly abnormal)

(B) Animal pole

Vegetal pole

Larva (small but normal)

Larva (small but normal)

FIGURE 16.5
Early asymmetry in the sea urchin embryo. (A) When the four animal blastomeres are separated from the four vegetal blastomeres and each half is allowed to develop, the animal cells form a ciliated dauerblastula and the vegetal cells form a larva with an expanded gut. (B) When the 8-cell embryo is split so that each half contains animal and vegetal cells, small, normal-appearing larvae develop.

embryo is not a collection of completely equipotential cells. In a series of experiments conducted from 1928 to 1935, Swedish biologist Sven Hörstadius separated various layers of early sea urchin embryos with fine glass needles and observed their subsequent development (Hörstadius, 1928, 1939). When the 8-cell embryo was divided meridionally through the animal and vegetal poles, both halves produced pluteus larvae, just as Driesch had foretold. But when embryos at the same stage were split equatorially (separating animal and vegetal poles), neither part developed into a complete larva (Figure 16.5). Rather, the animal half became a hollow ball of ciliated epidermal cells (called a **dauerblastula**), and the vegetal half developed into a slightly abnormal embryo with an expanded gut. Hörstadius was able to duplicate these results by cutting unfertilized sea urchin eggs in half and fertilizing the halves separately. In sea urchins, egg fragments (**merogones**) can divide and develop even if they have only a haploid nucleus. If a sperm enters the half that lacks the haploid egg nucleus, the merogone will still develop (Figure 16.6). When the egg has been split meridionally, normal embryos formed from both halves of the egg. However, when the oocyte has been equatorially cut, fertilization produced either the ciliated animal ball or an embryo with expanded vegetal gut. Therefore, even in sea urchin embryos, there appears to be some degree of mosaicism, at least along the animal–vegetal axis. This has been confirmed by Maruyama and co-workers (1985), who similarly split unfertilized sea urchin eggs either meridionally or equatorially. They found that when they separated the animal half from the vegetal half, only the fertilized vegetal half was able to form micromeres and to gastrulate. Therefore, the determinants that enable micromere formation and gastrulation appear to be localized in the vegetal portion of the egg.

Regulation of cell fate in late-cleavage embryos. These observations led Hörstadius to perform some of the most exciting experiments in the history of embryology. First, Hörstadius (1935) traced the normal development of each of the six tiers of cells of the 64-cell sea urchin embryo. As shown in

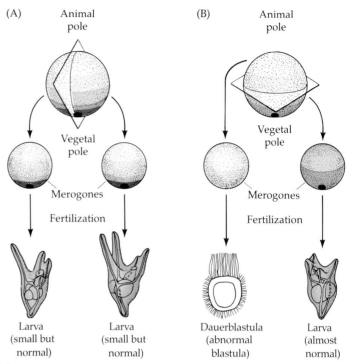

FIGURE 16.6
Asymmetry in the sea urchin egg. (A) When Hörstadius split the sea urchin egg meridionally so that both merogones contained animal and vegetal cytoplasm, small, normal-appearing plutei developed. (B) When the sea urchin egg was divided into animal and vegetal halves (merogones) and the halves were allowed to be fertilized by sperm, the animal half developed into a ciliated dauerblastula and the vegetal half produced a pluteus with an expanded gut.

Figure 16.7A, the animal cells and the first vegetal layer normally produce ectoderm; the second vegetal layer gives rise to endoderm and some larval mesoderm; and the micromeres generate the mesodermal skeleton.

Next, Hörstadius removed the fertilization membrane from the 64-cell embryos, separated the tiers with fine glass needles, and recombined them in various ways. The isolated animal hemisphere, alone, became a ball of ciliated ectoderm cells (Figure 16.7B). Such a ciliated dauerblastula was said to be "animalized." When Hörstadius recombined an isolated animal hemisphere with the veg_1 tier (Figure 16.7C), the resulting larva was less animalized. Ciliary development was suppressed and a portion of the gut was formed. However, when the animal hemisphere was combined with the veg_2 tier (Figure 16.7D), a normal-looking pluteus larva developed. In this combination, the veg_2 cells, which normally form only the archenteron and its derivatives, now formed skeletal structures as well. Similarly, when the animal half was recombined with just the micromeres (Figure 16.7E), a small, normal-looking pluteus was formed, but in this case the endoderm was completely derived from the animal cells. Here, the gut was formed by cells that would normally have created the ciliated ectoderm. These experiments showed that the animal cells have the genetic potential to be gut cells even at the 64-cell stage.

Eric Davidson: The transcription factor and induction model of sea urchin cell determination

It should be noted that the biochemical nature of the vegetalizing signal(s) has not yet been elucidated. Hörstadius interpreted these experiments in

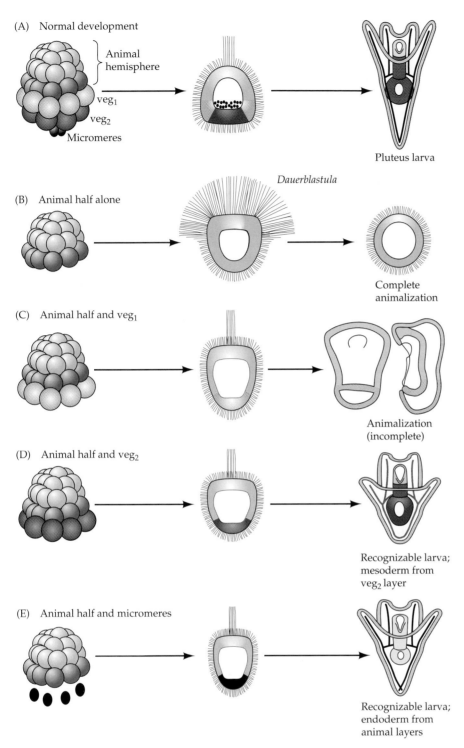

(A) Normal development

Animal
hemisphere

veg₁

veg₂

Micromeres

Pluteus larva

Dauerblastula

(B) Animal half alone

Complete
animalization

(C) Animal half and veg₁

Animalization
(incomplete)

(D) Animal half and veg₂

Recognizable larva;
mesoderm from
veg₂ layer

(E) Animal half and micromeres

Recognizable larva;
endoderm from
animal layers

FIGURE 16.7
Hörstadius's demonstration of regulation in sea urchins. (A) Fate of each cell layer of the 64-cell sea urchin embryos through blastula to pluteus stage. The different cell layers are marked as in Figure 6.1. (B) Fate of the isolated animal half. (C) Recombination of animal half plus the veg₁ tier of cells. (D) Recombination of animal half plus the veg₂ tier of cells. (E) Recombination of animal half plus the micromeres. In each case, the original fates of the cells have been altered by the new neighbors. (After Hörstadius, 1939.)

terms of gradients of animalizing and vegetalizing substances. More recently, Davidson (1989) sought to explain these phenomena in terms of transcriptional regulatory proteins (such as those that bind to promoters or enhancers) that are localized throughout the egg and that become activated by inductive interactions between adjacent cells. One of these factors, the one in the micromere precursors of the primary mesenchyme cells, is thought to become active autonomously, early in cleavage (Figure 16.8). This factor would cause the determination of the micromeres, alter

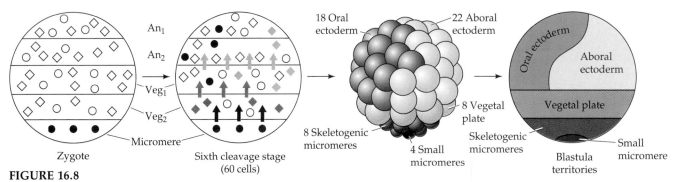

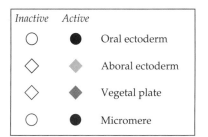

FIGURE 16.8

Davidson's model of cell fate determination in sea urchins. In the zygote, the animal–vegetal axis is specified. Along this axis are inactive DNA-binding regulatory factors, represented by open circles. The filled circles indicate that the factor is active (unless attached to a specific inhibitor). In the zygote, only the skeletogenic mesenchyme factors are active. During cleavage, the cell surfaces of the skeletogenic micromeres induce the vegetal cells above them to activate their vegetal plate factors only. The cell membranes of the vegetal plate cells then induce the cells above them to activate the ectoderm factors. The oral–aboral axis appears to be specified by the end of the second cleavage. This axis might form a gradient of some substance that releases the inhibition of aboral ectoderm regulatory factor. In so doing, the future oral ectoderm would be distinguished from the future aboral ectoderm. (After Davidson, 1989; Ransick and Davidson, 1993.)

their cell membranes so they could react with the cells above them, and initiate the cascade of determination. By the end of blastulation, there would be five territories that would constitute both the precursor cells for certain regions of the pluteus larva and the regions of specific gene activation. For example, the cytoskeletal actin gene, which is expressed only in the aboral ectoderm cells, cannot be activated anywhere else in the embryo. Similarly, if one injects the gene for the skeletal matrix protein anywhere else in the embryo except the region of skeletogenic mesenchyme, it will not be activated (Hough-Evans et al., 1987; Sucov et al., 1988). Therefore, each territory would constitute a region of differential gene expression that would be reflected in different parts of the organism being formed.

However, if the embryo is perturbed, the neighbors would differ. New fates would ensue as the new vegetalmost cells activated their skeletogenic mesenchyme determinants and started activation of their neighbors. To see if this was indeed the case, Ransick and Davidson (1993) transplanted micromeres into the animal cap of an 8-cell sea urchin embryo. These skeletogenic mesenchyme precursor cells induced the formation of the archenteron and the expression of the archenteron-specific genes (Figure 16.9). The fluorescently labeled micromeres did not contribute structurally to the secondary archenteron, but provided a short-range inductive signal that respecified the animal cap cells.

Forming an integrated organism: Restricting the potency of neighboring cells

Driesch referred to the embryo as an "equipotential harmonious system" because each of the composite cells had surrendered most of its potential in order to form part of a single complete organism. Each cell could have become a complete animal on its own, yet didn't. What made the cells cooperate instead of becoming autonomous entities? In the case of snails and tunicates, the answer was simple. The maternal cytoplasm does not

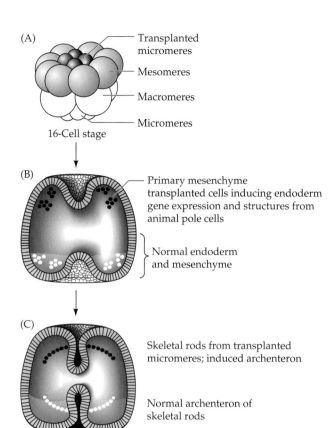

(A)

Transplanted micromeres

Mesomeres

Macromeres

Micromeres

16-Cell stage

(B)

Primary mesenchyme transplanted cells inducing endoderm gene expression and structures from animal pole cells

Normal endoderm and mesenchyme

(C)

Skeletal rods from transplanted micromeres; induced archenteron

Normal archenteron of skeletal rods

FIGURE 16.9
(A) Fluorescently labeled micromeres are transplanted from the vegetal pole of a 16-cell embryo to the animal pole of a host 16-cell embryo. (B) The transplanted micromeres invaginate into the blastocoel to become primary mesenchymal cells, and they induce the animal cells next to them to become vegetal plate endoderm cells. (C) The transplanted micromeres differentiate into skeletal cables while the induced animal cap cells form the archenteron. Meanwhile, gastrulation proceeds normally at the original vegetal plate. Eventually, the two archenterons fuse. (After Ransick and Davidson, 1993.)

permit each cell to become autonomous; each cell can only develop into a portion of the embryo. In sea urchins and other embryos that show regulation, the answer is more complex.

Recent evidence suggests that the "harmonious equipotential system" is caused by negative induction events that mutually restrict the fate of neighboring cells. Jon Henry and colleagues in Rudolf Raff's laboratory (1989) have shown that if one isolates pairs of cells from the animal cap of a 16-cell sea urchin embryo, these cells can give rise to both ectodermal and mesodermal components. However, their capacity to form mesoderm is severely restricted if they are aggregated with other animal cap pairs. Thus, the presence of neighbor cells, even of the same kind, restricts the potencies of both partners. Ettensohn and McClay (1988) have shown that potency is also restricted when a cell is combined with its neighbors along the animal–vegetal axis. First, they demonstrated that the number of primary mesenchyme cells appears to be fixed and can be regulated by changes in the macromeres. If all the 60 primary mesenchymal cells of *Lytechinus variegatus* are removed from the early gastrula, an equal number of secondary mesenchymal cells (from the archenteron that had been vegetal plate macromeres) convert into primary mesenchyme and start forming spicules. If one removes 10, 20, 30, or 40 primary mesenchymal cells, about 50, 40, 30, or 10 secondary mesenchyme cells become spicule-forming primary mesenchymal cells (Figure 16.10). Thus, the primary mesenchyme cells have a restrictive influence preventing the formation of new primary mesenchyme cells from the archenteron, and there is negative induction occurring here. We do not yet know the mechanism by which the primary mesenchyme cells stops the archenteron from forming primary mesenchyme and sets a limit for the number of such cells in the blastocoel.

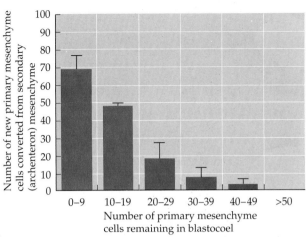

FIGURE 16.10

Results of experiments showing that primary mesenchyme cells (PMCs) restrict the potency of archenteron cells. When all 60 of the PMCs are sucked out of the blastocoel, an equal number of PMCs are induced from the archenteron. When a fraction of the PMCs are removed, the archenteron replaces that number of cells. When extra PMCs are added, regulation occurs so that two sets of spicules are formed, as is normal. They just have more cells forming each set of spicules.

By recombining cells from various layers, Khaner and Wilt (1990, 1991) found that in most all cases, a cell from one layer *restricts* the ability of a cell from another layer to express all its potential fates (Figure 16.11). The major exception—as mentioned above—is the recombining of animal pole mesomere cells with certain vegetal pole micromeres to form gut tissue from the mesomeres. However, in normal sea urchin development, these cells never associate with each other.

Regulation during amphibian development

Hans Spemann: Progressive determination of embryonic cells

In the previous sections, we saw evidence for regulative development. We noted that the two major aspects of regulation—first, that an isolated blastomere has a potency greater than its normal embryonic fate, and second, that a cell's fate is determined by interactions between neighboring cells—hold true during the early stages of sea urchin cleavage. Eventually, however, the blastomeres become committed to certain fates. In 1918, Hans Spemann of the University of Freiburg discovered that a similar situation exists in the salamander egg. The experiments by which he and his colleagues analyzed this phenomenon over the next 20 years form the basis for much of our knowledge of embryonic physiology and won Spemann a Nobel prize in 1935.

Spemann, like Roux and Driesch, sought to test Weismann's hypothesis, and, by an ingenious method, he demonstrated that early newt blastomeres have identical nuclei, each capable of producing an entire larva. Shortly after fertilizing a newt egg, Spemann took a baby's hair and lassoed the zygote in the plane of the first cleavage. He then partially constricted the egg, causing all the nuclear divisions to remain on one side of the constriction. Eventually, often as late as the 16-cell stage, a nucleus would escape across the constriction into the non-nucleated side.

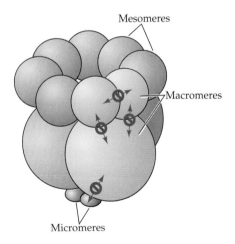

FIGURE 16.11

Summary of inhibitory inductions in the sea urchin blastula. Double-headed arrows illustrate the mutually restrictive interactions between adjacent cells. (After Henry et al., 1989.)

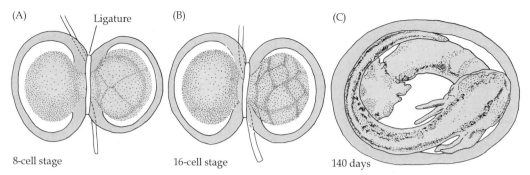

(A) Ligature   8-cell stage

(B) 16-cell stage

(C) 140 days

FIGURE 16.12
Spemann's demonstration of nuclear equivalence in newt cleavage. (A) When the fertilized egg of the newt *Triturus taeniatus* was constricted by a ligature, the nucleus was restricted to one-half of the embryo. The cleavage on that side of the embryo reached the 8-cell stage while the other side remained undivided. (B) At the 16-cell stage, a single nucleus entered the as yet undivided half, and the ligature was constricted to complete the separation of the two halves. (C) After 140 days, each side had developed into a normal embryo. (After Spemann, 1938.)

Cleavage then began on this side, too, whereupon the lasso was tightened until the two halves were completely separated. Twin larvae developed, one slightly older than the other (Figure 16.12). Spemann concluded from this that early amphibian nuclei are genetically identical and each is capable of giving rise to an entire organism. In this respect, amphibian blastomeres were similar to those of the sea urchin.

Moreover, when Spemann performed a similar experiment with the constriction still longitudinal but perpendicular to the plane of the first cleavage (separating the future dorsal and ventral regions rather than right and left sides), he obtained a different result altogether. The nuclei continued to divide on both sides of the constriction, but only one side—the future *dorsal* side of the embryo—would give rise to a normal larva. The other side would produce an unorganized tissue mass of ventral cells, which Spemann called the *Bauchstück*—the belly piece. This tissue mass was a ball of epidermal cells (ectoderm) containing blood and mesenchyme (mesoderm) and gut cells (endoderm) but no dorsal structures such as nervous system, notochord, or somites (Figure 16.13).

Why should these two experiments give different results? Could it be

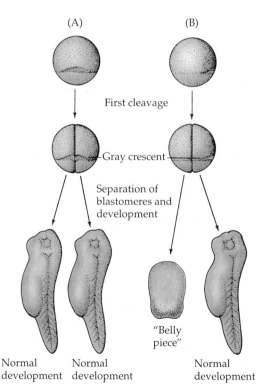

(A)   (B)

First cleavage

Gray crescent

Separation of blastomeres and development

Normal development   Normal development

"Belly piece"   Normal development

FIGURE 16.13
Asymmetry in an amphibian egg. (A) When the plane of first cleavage divides the egg into two blastomeres that each get one-half of the gray crescent, each experimentally separated cell develops into a normal embryo. (B) When only one of the two blastomeres receives the entire gray crescent, it alone forms a normal embryo. The other piece lacks dorsal structures and remains a mass of unorganized tissues. (After Spemann, 1938.)

FIGURE 16.14

Determination of ectoderm during newt gastrulation. Presumptive neural ectoderm from one newt embryo is transplanted into a region in another embryo that normally becomes epidermis. (A) When the transfer is done in the early gastrula, the presumptive neural tissue develops into epidermis and only one neural plate is seen. (B) When the same experiment is performed on late gastrula tissues, the presumptive neural cells form neural tissue, thereby causing two neural regions to form on the host. (After Saxén and Toivonen, 1962.)

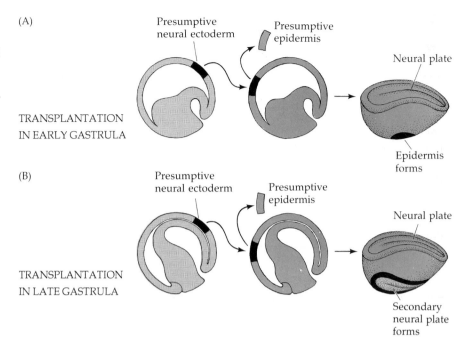

(A)

Presumptive neural ectoderm

Presumptive epidermis

Neural plate

TRANSPLANTATION IN EARLY GASTRULA

Epidermis forms

(B)

Presumptive neural ectoderm

Presumptive epidermis

Neural plate

TRANSPLANTATION IN LATE GASTRULA

Secondary neural plate forms

that when the egg is divided perpendicular to the first cleavage plane, some cytoplasmic substance is not equally distributed into the two halves? Fortunately, the salamander egg was a good place to look for answers. As we have seen in Chapters 4 and 6, there are dramatic movements in the cortical cytoplasm following the fertilization of amphibian eggs, and in some amphibians these movements expose a gray, crescent-shaped area of cytoplasm in the region directly opposite the point of sperm entry. Moreover, the first cleavage plane normally bisects this region equally into the two blastomeres. If these cells are then separated, two complete larvae develop. However, should this cleavage plane be aberrant (either in the rare natural event or in an experiment in which an investigator constricts a hair-loop lasso perpendicular to the normal cleavage plane), the gray crescent material passes into only one of the two blastomeres. Spemann found that when these two blastomeres are separated, only the blastomere containing the gray crescent develops normally.

It appears, then, that something in the gray crescent region is essential for proper embryonic development. But how does it function? What role does it play in normal development? The most important clue came from the fate map of this area of the egg, for it showed that the gray crescent region gives rise to the cells that initiate gastrulation. These cells form the dorsal lip of the blastopore. As was shown in Chapter 6, the cells of the dorsal blastopore lip are somehow committed to invaginate into the blastula, thus initiating gastrulation and the formation of the archenteron. Because all future amphibian development depends on the interaction of cells rearranged during gastrulation, Spemann speculated that the importance of gray crescent material lies in its ability to initiate gastrulation and that crucial developmental changes occur during gastrulation.

In 1918, Spemann demonstrated that enormous changes in cell potency do indeed take place during gastrulation. He found that the cells of the *early* gastrula are uncommitted with respect to their eventual differentiation, but that the fates of the *late* gastrula cells are fixed. Spemann exchanged tissues between the early gastrulae of two differently pigmented species of newts (Figure 16.14). When a region of prospective epidermal cells was transplanted to an area where the neural plate formed,

TABLE 16.2
Results of tissue transplantation during early- and late-gastrula stages in the newt

Donor region	Host region	Differentiation of donor tissue	Conclusion
EARLY GASTRULA			
Prospective neurons	Prospective epidermis	Epidermis	Dependent (conditional) development
Prospective epidermis	Prospective neurons	Neurons	Dependent (conditional) development
LATE GASTRULA			
Prospective neurons	Prospective epidermis	Neurons	Independent (autonomous) development (determined)
Prospective epidermis	Prospective neurons	Epidermis	Independent (autonomous) development (determined)

the transplanted cells gave rise to neural tissue. When prospective neural plate cells were transplanted to the region fated to become belly skin, these cells became epidermal (Table 16.2). Thus, these early newt gastrula cells were not yet committed to a specific type of differentiation. Their prospective potencies were still greater than their prospective fates. Such cells are said to exhibit conditional (or regulative, or dependent) development because their ultimate fates depend on their location in the embryo. However, when the same heteroplastic (interspecies) transplantation experiments were performed on *late* gastrulae, Spemann obtained completely different results. Rather than regulating their differentiation in accordance with their new location, the transplanted cells exhibited autonomous (or independent, or mosaic) development. Their prospective fate was fixed, and the cells developed independently of their new embryonic location. Specifically, prospective neural cells now developed into brain tissue even when placed in the region of prospective epidermis, and prospective epidermis formed skin even in the region of the prospective neural tube. Within the time separating early and late gastrulation, the potencies of these groups of cells had become restricted to their eventual paths of differentiation. Such cells are said to be determined: they can no longer regulate their differentiation into other cell types. It should be noted that the criteria for determination are completely operational. There are no obvious changes occurring in the cells, and no overt differentiation can be seen. The molecular basis of determination remains one of the major unsolved puzzles of development.

Hans Spemann and Hilde Mangold: Primary embryonic induction

The most spectacular transplantation experiments were published by Hans Spemann and Hilde Mangold in 1924. They showed that when tissues are placed in new locations, the dorsal lip of the blastopore is the only self-differentiating region in the early gastrula. When dorsal blastopore lip tissue from an early gastrula was transplanted into the ventral ectoderm of another gastrula, it not only continued to be blastopore lip, but it also initiated gastrulation and embryogenesis in the surrounding tissue. In

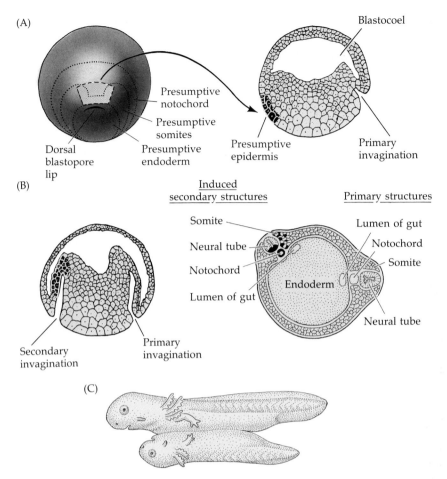

FIGURE 16.15
Self-differentiation of the dorsal blastopore lip tissue. (A) Dorsal blastopore lip from early gastrula is transplanted into another early gastrula in the region that normally becomes ventral epidermis. (B) Tissue invaginates and forms a second archenteron and then a second embryonic axis. Both donor and host tissues are seen in the new neural tube, notochord, and somites. (C) Eventually, a second embryo forms that is joined to host.

these experiments, Spemann and Mangold used differently pigmented embryos from two species of newt, the darkly pigmented *Triturus taeniatus* and the nonpigmented (clear) *Triturus cristatus*. When Spemann and Mangold prepared these heteroplastic transplants, they were able to readily identify host and donor tissues on the basis of color. The dorsal blastopore lips (the dorsal marginal zone tissue) of early *T. cristatus* gastrulae were removed and implanted into the regions of early *T. taeniatus* gastrulae fated to become ventral epidermis (Figure 16.15). Unlike the other early-gastrula tissues, which developed according to their new location, the donor blastopore lip did not become belly skin. Rather, it invaginated just as it would normally have done (showing self-determination) and disappeared beneath the vegetal cells. The light-colored donor tissue then continued to self-differentiate into the chordamesoderm and other mesodermal structures that constituted the original fate of that blastopore tissue. As the axis was formed, host cells began to participate in the production of the new embryo, becoming organs that normally they never would have formed. Thus, a somite could be seen containing both colorless (donor) and pigmented (host) tissue. What was even more spectacular,

(A)

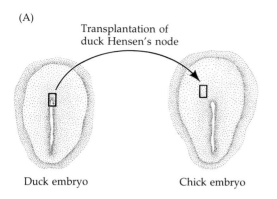

Transplantation of
duck Hensen's node

Duck embryo

Chick embryo

(B)

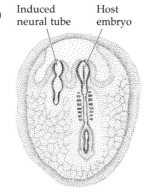

Induced
neural tube

Host
embryo

FIGURE 16.16
Induction of a new embryonic axis by
Hensen's node. (A) Hensen's node tissue is removed from a duck embryo
and implanted into a host chick embryo. (B) An accessory neural tube is
induced at the graft site. (After Waddington, 1933.)

the dorsal blastopore lip cells were able to interact with the host tissues
to form a complete neural plate. Eventually, a secondary embryo formed,
face-to-face with its host (Figure 16.15; Plate 3). These technically difficult experiments have recently been repeated using nuclear markers, and
the results of Spemann and Mangold have been confirmed (Gimlich and
Cook, 1983; Smith and Slack, 1983; Jacobson, 1984; Recanzone and Harris,
1985).*

Spemann (1938) referred to the dorsal blastopore lip cells as the **organizer** because (1) they induced the host's ventral tissues to change their
fates to form a neural tube and dorsal mesodermal tissue, and (2) they
organized these host and donor tissues into a secondary embryo with clear
anterior–posterior and dorsal–ventral axes. He proposed that during normal development, these cells would organize the dorsal ectoderm into a
neural tube and transform the flanking mesoderm into the body axis. It is
now known (thanks largely to Spemann and his students) that the interaction of the chordamesoderm and ectoderm is not sufficient to "organize"
the entire embryo. Rather, it initiates a series of sequential inductive
events. The process by which one embryonic region interacts with a second region to influence that second region's differentiation or behavior is
called **induction**. Because there are numerous inductions during embryonic development, this key induction wherein the blastopore lip cells
induce the dorsal axis and the neural tube is traditionally called **primary
embryonic induction.**[†]

We also know that the dorsal blastopore lip is active in organizing
secondary embryos in *Amphioxus*, cyclostomes, and a variety of amphibians. In birds and mammals, the organizer originates at Koller's sickle (the
posterior margin of the embryo), and Hensen's node acts as the dorsal
blastopore lip. Cells migrating through Hensen's node become the head
endoderm and chordamesoderm, while cells migrating through other portions of the primitive streak become lateral and ventral mesodermal cells.
When Hensen's nodes from young gastrula are transplanted into epiblasts
of other young gastrula, they will induce a complete secondary axis to
form (Figure 16.16; Waddington, 1933; Storey et al., 1992).

*Hilde Proescholdt Mangold died in a tragic accident when her gasoline heater exploded. At
the time she was 26 years old, and her paper was just being published. The experiment,
which formed the basis for her doctoral thesis, is one of the very few doctoral theses in
biology that have been directly concerned with the awarding of a Nobel prize. For more
information about Hilde Mangold and her times, see Hamburger, 1984.

[†]This classic term has been a source of confusion, because the induction of the neural tube
by the notochord is no longer considered to be the first inductive process in the embryo. We
will shortly discuss inductive events that precede this "primary" induction.

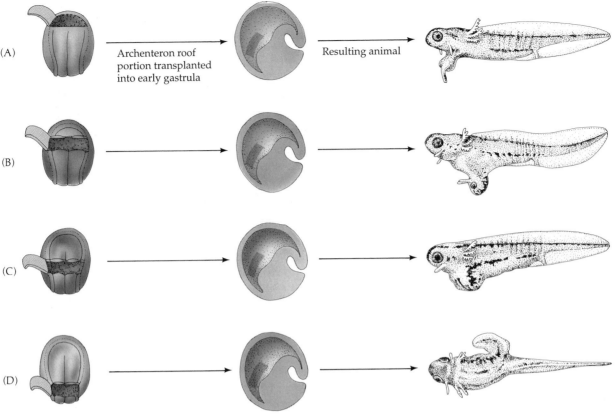

FIGURE 16.17
Regional specificity of induction can be demonstrated by implanting different regions (color) of the archenteron roof into early *Triturus* gastrulae. The resulting animals have secondary parts. (A) Head with balancers; (B) head with balancers, eyes, and forebrain; (C) posterior part of head, deuterence-phalon, and otic vesicles; and (D) trunk-tail segment. (After Mangold, 1933.)

Regional specifity of induction

The determination of regional differences

One of the most fascinating phenomena in neural induction is the regional specificity of the neural structures that are produced. Forebrain (archencephalic), hindbrain (deuterencephalic), and spinocaudal regions of the neural tube must all be properly organized in an anterior-to-posterior direction. Thus, the organizer tissue not only induces the neural tube but also specifies the *regions* of the neural tube. This region-specific induction was shown by Otto Mangold (1933) in a series of experiments wherein various regions of the *Triturus* (newt) archenteron roof were transplanted into early-gastrula embryos (Figure 16.17). After the superadjacent neural plate was removed, four successive sections of the archenteron roof were excised from embryos that had just completed gastrulation and were placed into the blastocoels of early gastrulae. The most anterior portion of the archenteron roof induced balancers and portions of the oral apparatus (Figure 16.17A); the next most anterior section induced the formation of various head structures, including nose, eyes, suckers, and otic (hearing organ) vesicles (Figure 16.17B). The third section induced the otic vesicles without any other head organs (Figure 16.17C); and the most posterior segment induced the formation of dorsal trunk and tail mesoderm (Figure 16.17D). [The induction of dorsal mesoderm—rather than the dorsal ectoderm of the nervous system—by the posterior end of the notochord was confirmed by Bijtel (1931) and Spofford (1945), who showed that the posterior fifth of the neural plate gives rise to tail somites and the posterior portions of the pronephric kidney duct.]

(A) Transplantation of young gastrula dorsal lip

(B) Transplantation of advanced gastrula dorsal lip

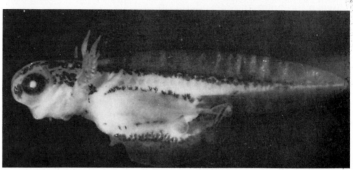

Moreover, when dorsal blastopore lips from *early* salamander embryos (early gastrulae) were placed into other early salamander embryos, they formed secondary heads. When dorsal lips from *later-stage* embryos were transplanted into early salamander embryos, they induced the formation of secondary tails (Figure 16.18; Mangold, 1933). This means that the first cells of the organizer to enter the embryo induce the formation of brains and heads, while those cells that form the dorsal blastopore lip of later-stage embryos induce the cells above them to become spinal cords and tails. A similar phenomenon occurs in chick embryos (Storey at al., 1992).

The double gradient model

To further study the organizer phenomenon, Holtfreter (1936) made a blastopore "sandwich," enclosing the dorsal blastopore lip between slices of undifferentiated ectoderm. The blastopore lips from the *early* gastrula stages induced the most archencephalic (anterior) structures, whereas the blastopore lips of *later* embryos caused the differentiation of the more posterior neural elements. There are two general models to explain this. One model holds that the early and late blastopore lips differ in the *quantity* of some substance. Large quantities of that substance produce either anterior or posterior neural structures, while small quantities produce the other. Midbrain regions would have an intermediate quantity of this substance. The second model holds that the early and late blastopore lips produce *qualitatively* different substances. Regional specification would be caused by the interaction of two substances secreted by the cells of the chordamesoderm. A high concentration of one would give forebrain development, while a high concentration of the other would elicit the formation of spinal cord and trunk structures. Mixtures of the two substances would elicit the midbrain and hindbrain regions.

Evidence for the latter model came from studies involving artificial tissue-specific inducers. Guinea pig bone marrow, for instance, was found to induce mesodermal structures only. Pellets of guinea pig liver, however, could only cause the induction of forebrain structures. Toivonen and Saxén (1955) implanted these inducers together within the blastocoel of the same

FIGURE 16.18
Regionally specific inducing action of the dorsal blastopore lip. (A) Young dorsal blastopore lips (which will form the anterior portion of the dorsal mesoderm) induce anterior structures when placed into young newt gastrulae. (B) Older dorsal blastopore lips placed into similar newt gastrulae produce more posterior structures. (From Saxén and Toivonen, 1962, photographs courtesy of L. Saxén.)

FIGURE 16.19

Evidence for the dual-gradient model of induction. (A) Simultaneous implantation of a neuralizing inducer (guinea pig liver) and a mesodermalizing inducer (guinea pig bone marrow) into the blastocoel of an early newt gastrula. (B) Results of such an implantation. Hindbrain and spinal cord structures, which were intermediate between forebrain and mesoderm on the fate map of the neural plate, were not induced well by either inducer. When the two inducers were implanted together, these structures were produced. (After Toivonen and Saxén, 1955.)

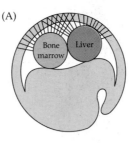
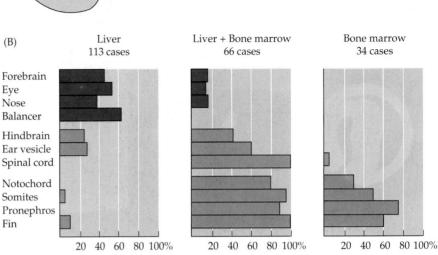

early gastrula. Whereas the liver would only have induced forebrain and the bone marrow would only have induced mesoderm, the two together induced all the normal forebrain, hindbrain, spinal cord, and trunk mesoderm. Thus, the regional specificity of neural induction may be due to opposing gradients of forebrain-inducing and spinal cord-inducing substances (Figure 16.19).

The integration of the double gradient model of regionalization into the two-step model of neural induction

In 1940, C. H. Waddington had hypothesized that neural induction occurred by the temporally separated processes: evocation and individuation. Evocation determined that the ectodermal tissue would become a neural tube, while individuation determined the regionally specific type of neural tube produced. Nieuwkoop (1952) and co-workers provided evidence for this two-step model. They inserted folds of competent ectoderm at various positions along the anterior–posterior axis of the host gastrulae. The proximal parts of these folds produced structures typical of the host's region of insertion, while the more distal part of the fold developed into neural structures of a more anterior nature than that of the insertion (Figure 16.20). These results were interpreted as implying that the induction is a two-step process: an "activation" that leads to the formation of forebrain dervatives, followed by a "transformation" that produces a quantitatively more caudal series of structures.

The merging of these two hypotheses was accomplished by a series of experiments coming from Helsinki. In 1964, Saxén and colleagues wrapped competent ectoderm around a forebrain inducer or around a spinocaudal (mesodermal) inducer (Figure 16.21). After an initial 24-hour period of induction (which is a sufficient duration for induction to occur), the inducers were removed and the ectodermal cells were disaggregated into single-cell suspensions. These ectodermal suspensions were reaggre-

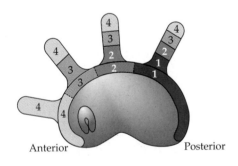

FIGURE 16.20
Evidence for a two-step model of neural induction: activation and transformation. A fold of gastrula ectoderm was implanted to a region of the neural plate. The more anterior structures are on the left-hand side, and 1–4 represent different neural structures. Folds of unspecified gastrula ectoderm tended to differentiate into an anterior neural structures, but were posteriorized by material coming from the posterior of the embryo. (After Doniach, 1993.)

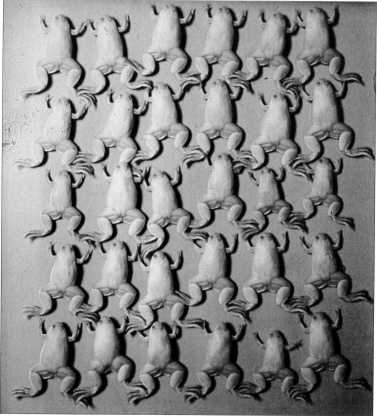

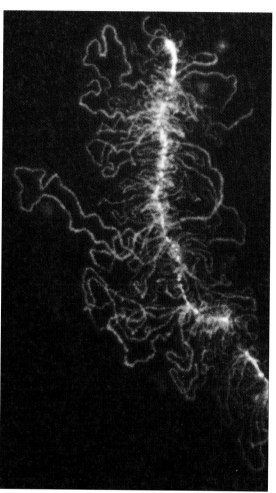

PLATE 1
A clone of *Xenopus* frogs
The nuclei for all the members of this clone came from a single individual—a tailbud-stage tadpole whose parents were both marked with an albino gene. The nuclei were transferred into activated enucleated unfertilized eggs from a wild-type female (upper panel). The resulting frogs were all female and albino (lower panel). Chapter 2. (Photograph courtesy of J. Gurdon.)

PLATE 2
Active transcription units
on a newt chromosome
Oocytes from amphibians such as *Notophthalmus viridescens* have lampbrush chromosomes where the active RNA-synthesizing genes loop out. The DNA axis of these loops is stained with a white dye. The red stain is from an antibody that binds to RNA-binding proteins. Chapter 22. (Photograph courtesy of M. B. Roth and J. Gall.)

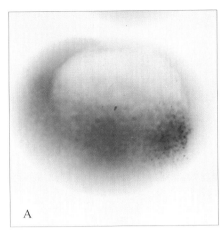

A

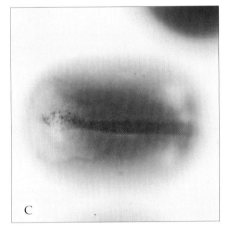

B

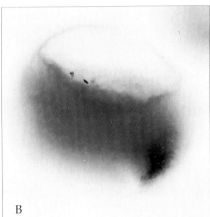

C

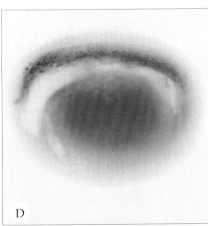

D

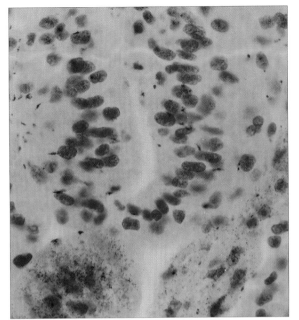

PLATE 3
Ability of dorsal blastopore lip to generate secondary neural axis in amphibians
The dorsal blastopore lip of amphibian embryos can organize a second embryonic axis when transplanted to the ventral side of another gastrula. This photograph is taken from an actual slide prepared by Hilde Mangold and shows that the secondary dorsal structures contain both host (unpigmented) and donor (pigmented) tissues. Chapter 16. (Photograph courtesy of P. Fässler and K. Sander.)

PLATE 4
Rescue of dorsal structures by the noggin protein
The noggin protein may be critical for inducing the dorsal mesoderm and the neural tube. When *Xenopus* eggs are exposed to UV irradiation prior to first cleavage, no dorsal structures form (top). If an early cell of such an embryo is injected with *noggin* mRNA, the embryos form dorsal structures. If too much *noggin* message is injected, the embryos produce too much dorsal anterior tissue (bottom.) Chapter 16. (Photographs courtesy of R. M. Harland.)

PLATE 5 *(left)*
The *noggin* gene is transcribed in the dorsal mesoderm and endoderm tissue
The *noggin* mRNA accumulates in the region of the dorsal marginal zone (A) and is seen in the dorsal blastopore lip (B). When these cells involute, *noggin* expression is seen in the notochord and pharyngeal endoderm (C), which extend anteriorly in the center of the embryo (D). Chapter 16. (Photographs courtesy of R. M. Harland.)

PLATE 6 *(below, left)*
Cytoplasmic rearrangements in *Xenopus laevis*
(A) The unfertilized *Xenopus laevis* egg is radially symmetric. (B) Cytoplasmic movements are seen as the egg starts cleaving, 90 minutes after fertilization. The cytoplasm of the future dorsal side (right) differs from that of the future ventral side (left). These differences can be seen throughout embryonic cleavage (C,D) and result in the positioning of the dorsal morphogenetic determinants in the side of the embryo opposite the point of sperm entry. Chapter 4. (Photographs courtesy of M. V. Danilchik.)

PLATE 7 *(left)*
Fibroblast growth factor is essential for ventral and lateral mesoderm production in *Xenopus*
When *Xenopus* eggs are injected with a dominant negative mutant receptor for fibroblast growth factor (FGF), the embryo is unable to respond to FGF. In the absence of the FGF signal, the ventral and lateral mesoderm fail to form, and the embryo lacks trunk and tail. Chapter 16. (Photograph courtesy of M. Kirschner.)

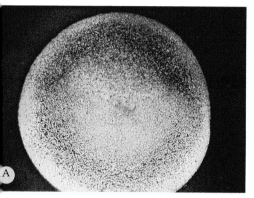

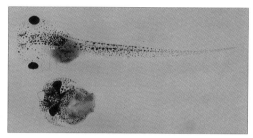

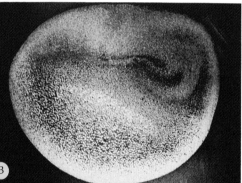

PLATE 8 *(right)*
Localization of a specific mRNA in a region of the egg
The *vg1* mRNA, which encodes a mesodermal growth factor, is found by in situ hybridization to reside solely in the vegetal region of the *Xenopus* egg. The white crescent at the bottom of the egg is due to the radioactivity of the probe recognizing the mRNA; the rest of the egg is green due to staining with Giemsa dye. Chapters 16. (Photograph courtesy of D. A. Melton.)

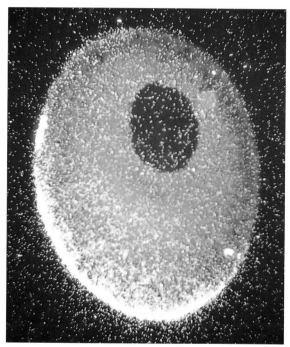

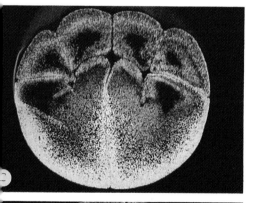

PLATE 9 *(right)*
Inducation of a secondary body axis in *Xenopus*
The normal *Xenopus* tadpole is seen on top. When mRNA for the activin protein is injected into a single blastomere on the ventral side of a 32-stage embryo, it induces a new axis to form. Chapter 16. (Photograph courtesy of G. Thomsen, M. Whitman, and D. A. Melton.)

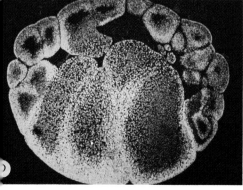

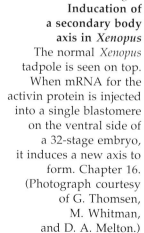

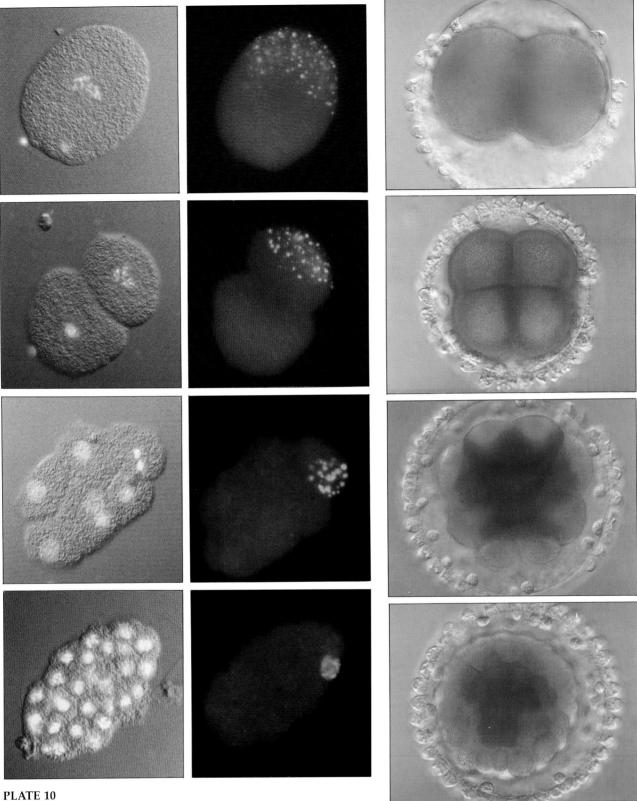

PLATE 10
Progressive cytoplasmic localization
The segregation of certain cytoplasmic granules ("P granules") is seen to progress into the posteriormost cells of the *Caenorhabditis elegans* embryo. These cells generate the sperm and egg of the nematode. When the pronuclei meet during fertilization, the P granules move to the posterior portion of the cell. This movement continues until they are found only in the P cell that gives rise to the gametes. The left-hand column is stained to show the position of the nuclei, while the right-hand column is stained to show the P granules. Chapter 14. (Photographs courtesy of S. Strome.)

PLATE 11
Cytoplasmic localization in tunicate embryos
Cleavage separates regions of cytoplasm into particular cells. The yellow crescent of the *Styela* embryo becomes localized into a small group of cells that will generate the larval musculature. This figure shows the 2-, 4-, 16-, and 64-cell stages. Chapter 14. (Photographs courtesy of J. R. Whittaker.)

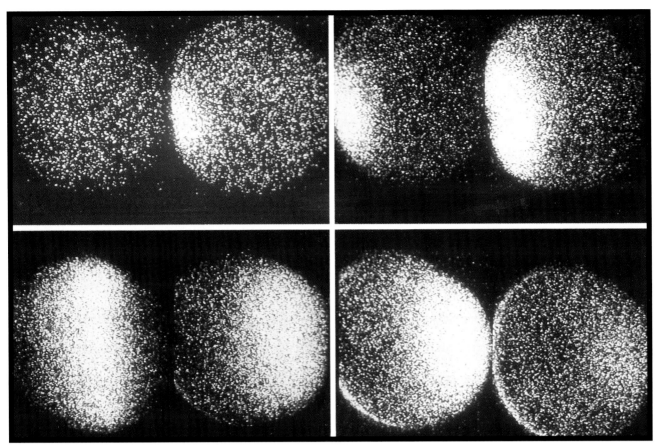

PLATE 12 (above)
Wave of calcium ions across sea urchin eggs during fertilization

When the sperm fuses with the egg, a wave of calcium begins at the site of sperm entry and propagates across the egg. This can be monitored by preloading the egg with a dye that fluoresces when it binds calcium. The wave takes 30 seconds to traverse the egg. Chapter 4. (Photographs courtesy of G. Schatten.)

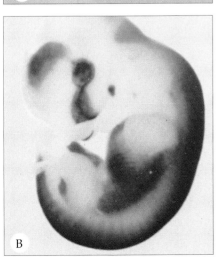

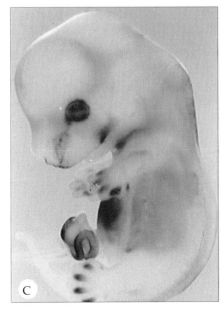

PLATE 13 (left)
Retinoic acid-responsive regions of the mouse embryo

A transgene consisting of a retinoic acid-responsive element fused to a β-galactosidase gene was inserted into a mouse embryo. Staining for β-galactosidase should reveal those cells that respond to endogenous concentrations of retinoic acid.
(A) 3-somite stage showing retinoic acid responsiveness in the medial region of the embryo; (B) 11.5-day embryos showing staining in the frontonasal region and forebrain; (C) 14.5-day embryo showing staining in jaw, optic region, whisker pads, and interdigital regions of the limb. Chapter 19. (Photographs courtesy of J. Rossant.)

PLATE 14 *(above)*
Pattern formation in *Drosophila*
(A) The anterior–posterior axis is specified by cytoplasmic mRNAs and proteins. The gradient of bicoid protein is especially important. High concentrations of this protein (yellow through red) cause head and thorax formation by activating the *hunchback* gene. (B) Proteins from "gap" genes such as *hunchback* interact to define the domains of the insect body. Here hunchback (orange) and Krüppel (green) proteins overlap to form a boundary (yellow). (C) These interactions activate the transcription of pair-rule genes (seen as dark bands) that divide the embryo into segments along the anterior–posterior axis. (D) At the extended germ band stage, the 14 bands of segment polarity gene *engrailed* can be seen. Chapter 15. (Photographs courtesy of (A) W. Driever and C. Nüsslein-Volhard; (B) C. Rushlow and M. Levine; (C) T. Karr; and (D) S. Carroll and S. Paddock.)

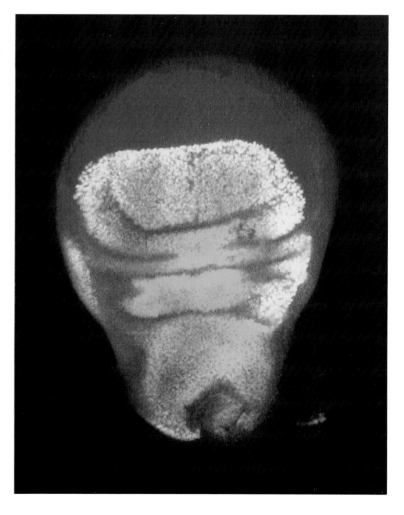

PLATE 15
Compartmentation
of *Drosophila* **wing imaginal disc**
The red immunofluorescent stain labels the cells where the vestigial protein is made (the future ventral wing); the green stain labels the cells expressing apterous protein (necessary for dorsal wing formation). The area of overlap is yellow. Chapter 20. (Photograph courtesy of S. Carroll.)

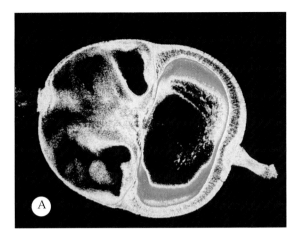

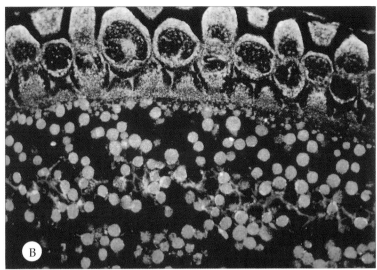

PLATE 16 (above and right)
**RNA polymerase II localization
in the oocytes of giant silk moths**
(A) Fluorescence photomicrograph (using confocal lens) of *Hyalophora cecropia* egg chamber. Orange fluorescence indicates the presence of RNA polymerase II (stained with labeled amanitin). Green background indicates the location of actin. (B) Higher magnification of an *Antherea polyphemus* oocyte cortical region and follicle cells. Orange indicates RNA polymerase II. The other colors are background staining of yolk granules and follicle cells. Chapter 22. (Photographs courtesy of S. Berry.)

PLATE 17 (above)
Gynandromorph moth
A sexual mosaic (gynandromorph) of an *Io* moth, divided bilaterally into a rose-brown female half and a smaller-winged, yellow male half. Such sex mosaics are caused when an X chromosome is lost from a nucleus during an early mitotic division. Chapter 21. (Photograph by T. R. Manley; courtesy of *The Journal of Heredity*.)

PLATE 18 (left)
Control of development by the environment
Caterpillars of *Nemoria arizonaria* that hatch in the spring eat oak flowers and develop a cuticle that mimics the flowers. Caterpillars of the same species that hatch in the summer (after the flowers are gone) eat oak leaves; these caterpillars develop cuticle that resembles the oak twigs. Chemicals in the leaves appear to modify cuticle development. Chapter 20. (Photographs courtesy of E. Greene.)

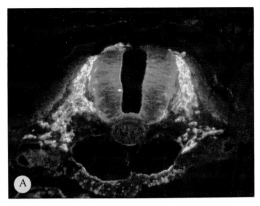

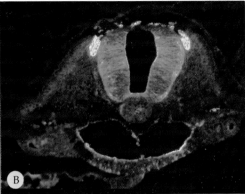

PLATE 19 (left)

Migration of chick neural crest cells

Chick neural crest cells can be followed in their migration by staining the cells with a fluorescently tagged monoclonal antibody. The neural crest cells (stained green) are found to migrate through the anterior (A) but not the posterior (B) regions of the somite tissue. This specific pattern of neural crest cell migration plays a role in determining the placement of peripheral neurons. Chapter 7. (Photographs courtesy of M. Bronner-Fraser.)

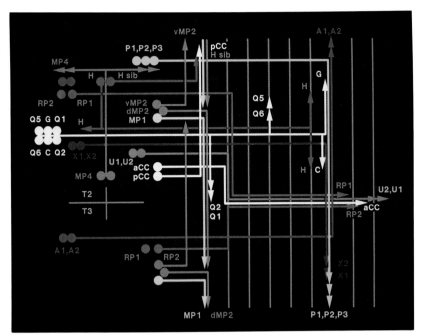

PLATE 20 (above)

Neural migration pathways in insects

Neural axons in insect embryos migrate in very specific patterns. Neurons derived from a common precursor (shown here in the same color) produce axons that selectively migrate along with other axons. The Q1 axon, for instance, travels until it meets the dMP2 axon and then travels with it, while the axon from the G neuron continues moving in a straight line until it meets the P1 axon. Chapter 8. (Photograph courtesy of C. Goodman.)

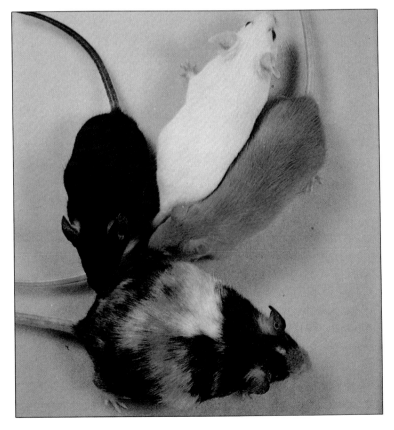

PLATE 21

A mouse with six parents

The multicolored mouse was formed by mixing together cells from three 4-cell stage embryos: an embryo from two black mice; an embryo from two white mice; and an embryo from two brown mice. Instead of forming a three-headed monster, the embryo regulated to form one normal-size mouse with contributions from each of the three embryos. That each of the three embryos also provided germ line cells was shown by mating this mouse with a recessive (white) mouse; this mating produced offspring of all three colors. Chapter 5. (Photograph courtesy of C. Markert and *The Journal of Heredity*.)

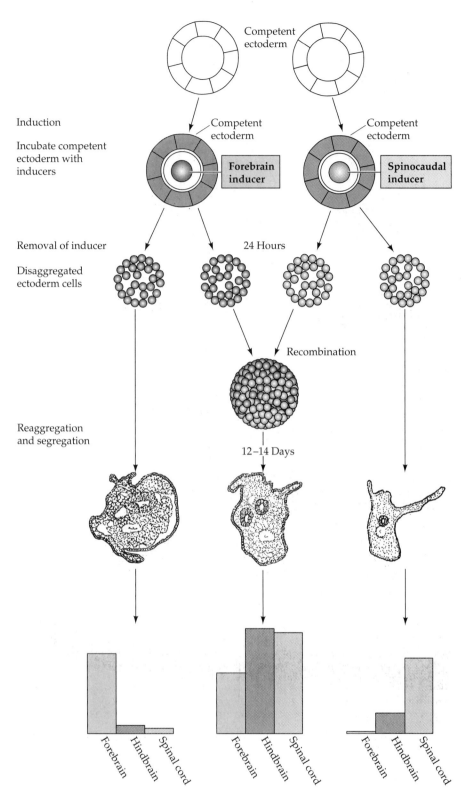

Competent ectoderm

Induction

Incubate competent ectoderm with inducers

Competent ectoderm

Competent ectoderm

Forebrain inducer

Spinocaudal inducer

Removal of inducer

24 Hours

Disaggregated ectoderm cells

Recombination

Reaggregation and segregation

12–14 Days

Forebrain Hindbrain Spinal cord

Forebrain Hindbrain Spinal cord

Forebrain Hindbrain Spinal cord

FIGURE 16.21
Evidence for a two-step process of induction. Competent ectoderm was incubated 24 hours with either a forebrain inducer or a spinocaudal inducer. The inducers were then removed and the ectoderm disaggregated. The ectodermal cells were reaggregated either separately or after complete mixing. When the ectoderm that had been induced by forebrain inducers was mixed with ectoderm that had been induced by spinocaudal inducers, the cells formed hindbrain as well as forebrain and spinocaudal structures. Thus, a neural determination step was followed by interactions between the induced cells.

gated either separately or as a mixed aggregate of the two induced types, and they were allowed to develop over a two-week period. The reaggregate of ectoderm induced by the archencephalic inducer produced forebrain structures, as expected. Similarly, the ectoderm induced by the spinocaudal inducer developed structures characteristic of the spinal cord.

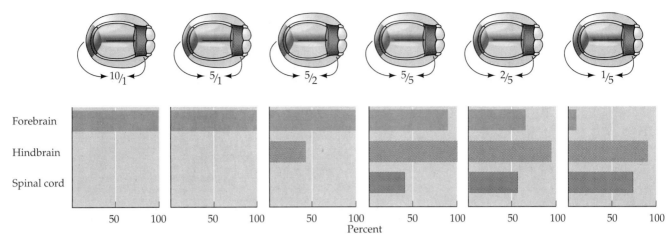

FIGURE 16.22

Evidence for the dual-gradient two-step induction within the amphibian embryo. The anterior region of the neural plate (i.e., cells that had already been naturally induced by a forebrain inducer; here seen as a gray crescent) and cells from the posterior notochord (color) were excised and mixed together in different ratios. The frequency of intermediate (hindbrain) structures increases as the ratio of anterior neural plate and mesodermal cells approaches one. This suggests that the regional specification occurs after the neural plate cells have been determined to be neural. (From Gilbert and Saxén, 1993.)

However, the reaggregate of ectodermal cells induced separately by the two inducers produced forebrain structures, spinocaudal structures, and intermediate structures characteristic of the hindbrain. These hindbrain structures (which were not present in either of the two separate inductions) were therefore determined during a second step, an interaction between the two types of induced cells. The double gradient model could then be rephrased to include the two-step temporal sequence of events: during the initial stages of induction, the competent ectoderm is determined to become either neural or mesodermal; in the second step, the regional properties are determined by the interactions between the neuralized and mesodermalized cells.

So far, however, this evidence came only from in vitro studies. Critical evidence that such gradients were found in the actual embryo came when Toivonen and Saxén (1968) reassociated cells from early neurula-stage salamander embryos. The first group of cells came from the anterior portion of the neural plate. The cells were already determined to become neural and would normally form forebrain tissue. The second set of cells were from the posterior chordamesoderm. These pieces were disaggregated, thoroughly mixed in different ratios, and allowed to reaggregate. The cultured reaggregates showed a spectrum of regional specificities that depended on how much presumptive mesodermal tissue had been added to the presumptive forebrain tissue (Figure 16.22). When sufficient mesodermal cells were added, forebrain structures became progressively rarer. The authors stated that "during the initial stage of induction the cells are determined to be neural, but they acquire no stable regional character. The regional specificity is subsequently controlled by the mesodermal cells and apparently in a quantitative way, since an increasing amount of mesoderm surrounding the neural cells shifts segregation in the caudal direction." Saha and Grainger (1992) have recently shown that when the neural plate is first formed, it does not express regionally specific markers. Rather, regional markers are expressed in low levels throughout the early neural plate, and the anterior markers appear to be expressed first (Sive et al., 1989). However, after a few hours, the regions of the neural ectoderm become demarcated. It is fair to say that most models of neural induction have converged on a scheme that includes (1) an initial "activation" step that determines that the cells are capable of binding forebrain neural cells, and (2) a "transformation" step in which a gradient of material from the posterior mesoderm causes the posteriorization of the neural specification.

The molecular mechanisms of primary embryonic induction

Despite the enormous amount of research that has been performed on amphibian embryos, we are just beginning to know the basic mechanisms of primary embryonic induction. In the past decade, numerous laboratories have focused their efforts on explaining embryonic induction in one amphibian—*Xenopus laevis*—and a there is a consensus concerning the general outline of primary embryonic induction in this organism.

The data point to an orchestration of induction that has at least four acts. The first act of induction takes place at fertilization. The unfertilized egg is radially symmetric around the animal–vegetal axis. The entry of the sperm breaks this symmetry by causing the internal cytoplasm of the egg to rotate relative to the cortex (Chapter 4). This asymmetry specifies the dorsal–ventral axis by mixing animal and vegetal cytoplasms in the vegetal cells that form opposite to the point of sperm entry. It appears that the cytoplasmic mixing activates dorsalizing determinants in these vegetal cells. In the second act, the descendants of these vegetal cells induce the cells above them to become the Spemann-Mangold organizer. The other vegetal cells induce the marginal cells above them to become the lateral and ventral mesoderm. Thus, there is an induction before "primary induction." In the third and fourth acts, the dorsal ectoderm is converted into neural tissue and the neural tissue is given its regional characteristics (forebrain, hindbrain, spinal cord, etc.).

The specification of dorsoventral polarity at fertilization

As we saw in Chapters 4 and 6, dorsoventral specification is accomplished by the rotation of the inner cytoplasm of the egg relative to the cortex. This rotation creates a region of vegetal cytoplasm where the dorsal information [what Wakahara (1989) calls the "anterodorsal structure-forming activity"] has been activated. If this rotation is inhibited by UV light, the embryo will not form dorsoanterior structures (Vincent and Gerhart, 1987). Render and Elinson (1986) and Wakahara (1989) cut eggs into fragments before and after this rotation. If the egg was cut before rotation, both sides developed dorsoanterior structures—head, notochord, neural tube. If the cut was made after the rotation had occurred, one fragment developed a head, heart, and some dorsal mesodermal structures, while the other half developed essentially into a *Bauchstück*, consisting almost solely of ventral cells, having little or no dorsal mesoderm and no nervous system. It appears, then, that this cytoplasmic rotation moves the dorsal-activating determinants to the future dorsal side of the egg.

It should be noted that in *Xenopus* (and other vertebrates), the formation of the anterior–posterior axis follows the formation of the dorsal–ventral axis. Once the dorsal portion of the embryo is established, the movement of the involuting mesoderm establishes the anterior–posterior axis. The mesoderm that migrates first through the dorsal blastopore lip gives rise to the anterior structures, the mesoderm in the ventral margin forms the posterior structures. This is in marked contrast to the situation in *Drosophila*, where the anterior–posterior and dorsal–ventral axes form independently of each other.

Localization of dorsoanterior determinants in the vegetal cells during cleavage

The cytoplasmic rotation that occurs during fertilization presumably creates an area of cytoplasm that is enriched with dorsoanterior-forming determinants (Figure 16.23). When the 4-cell embryo is separated into the two dorsal and the two ventral halves, each half behaves differently. If allowed to develop, the dorsal halves will, as expected from Spemann's experiment, form dorsoanterior structures. In fact, the dorsoanterior structures are larger than in control embryos. The ventral halves develop into embryos with marked dorsoanterior deficiencies (Kageura and Yamana, 1983; Cooke and Webber, 1985; Yamana and Kageura, 1987). These embryos are similar to those formed by cutting the eggs after rotation. By the 8-cell stage, the vegetal dorsal pair of blastomeres contains nearly all the dorsal-forming activity of the frog embryo. Removing this pair of blastomeres stops the formation of dorsoanterior structures, while the substitution of one of these vegetal dorsal blastomeres into the vegetal ventral region creates twins (Gimlich and Gerhart, 1984; Kageura and Yamana, 1986).

The location of the dorsoanterior structure-forming agents in the 32- and 64-cell embryos has been delineated by transplantation. Gimlich (1985, 1986) showed that the prospective chordamesoderm cells of the 32-cell embryo (i.e., the dorsal marginal zone cells; C1 and D1 in Figure 16.23), but not their ventral or lateral equatorial counterparts, were also able to induce the formation of the body axis. These dorsal marginal zone (DMZ) transplants also induced neighboring host cells to form the body axis, including the dorsoanterior somites and central neural cells (Figure 16.24). In Chapter 6, we discussed the dramatic transplantation experiments of Gimlich and Gerhart (1984). When the dorsalmost vegetal cell of the 64-cell *Xenopus* blastula (a descendant of the D1 cell in Figure 16.23) was transplanted into a 64-cell *Xenopus* embryo whose axis would not form because it had been irradiated at fertilization, the transplanted cell induced

FIGURE 16.23
Hypothetical boundaries for the location of "dorsal morphogenetic activities" in the 4- to 32-cell stage *Xenopus* embryo. The region is thought to remain stable throughout these cleavages. There is general agreement that dorsal information is contained in the C1 and D1 blastomeres of the 32-cell embryo. There is a possibility that it can extend into B1 at that time. (After Elinson and Kao, 1989.)

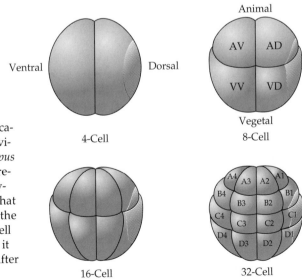

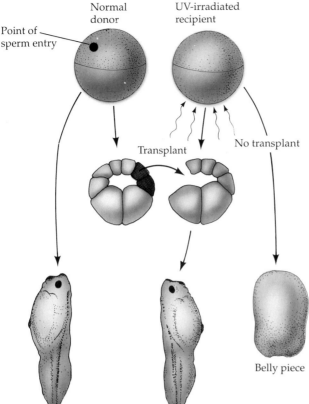

FIGURE 16.24
Rescue of irradiated embryos by the transplantation of dorsal marginal cells of normal 32-cell embryos. Without such a rescue, irradiated embryos fail to gastrulate, forming "belly pieces." If the dorsal marginal zone cells of a normal embryo are transplanted into the irradiated embryo, they will induce the formation of mesoderm and the body axis, thereby producing a normal tadpole. (After Gimlich, 1986.)

the formation of the dorsoanterior structures. The vegetal cell would go on to form endodermal structures, while the mesoderm it induced to form was derived from host tissue. The experiment diagrammed in Figure 18 of Chapter 6 can be seen as analogous to that of Spemann and Mangold. But, whereas Spemann and Mangold obtained two body axes by transplanting DMZ cells, Gimlich and Gerhart obtained the same phenomenon by transplanting the dorsalmost *vegetal* cells—the cells beneath the DMZ.

The conclusion that dorsoanterior information is contained within the D1 (dorsalmost vegetal) blastomere was further demonstrated by recombination experiments (Figure 16.25). Dale and Slack (1987) recombined single vegetal blastomeres from a 32-cell *Xenopus* embryo with the uppermost animal tier of a fluorescently labeled embryo of the same stage. The dorsalmost vegetal cell, as expected, induced the animal pole cells to become dorsal mesoderm. The remaining vegetal cells generally induced these animal cells to produce either intermediate or ventral mesodermal tissues. Thus, *dorsal vegetal cells* can induce animal cells to become dorsal mesodermal tissue.

Induction of mesodermal specificity by the endoderm

Nieuwkoop (1969, 1973, 1977) demonstrated the importance of the vegetal endoderm in inducing the mesoderm. He removed the equatorial cells

FIGURE 16.25
Regional specificity of mesoderm induction shown by recombining cells of 32-cell *Xenopus* embryos. The animal pole cells of 32-cell embryos were combined with individual vegetal blastomeres. The animal pole cells were labeled with fluorescent polymers to identify their descendants. The inductions resulting from these recombinations are summarized at right. (After Dale and Slack, 1987.)

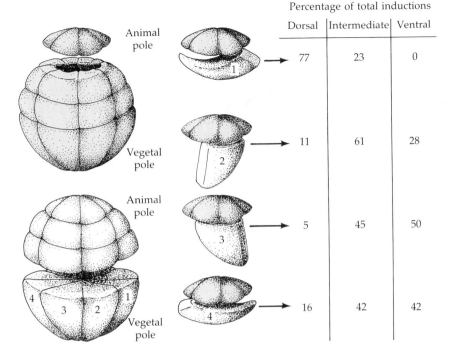

	Percentage of total inductions		
	Dorsal	Intermediate	Ventral
1	77	23	0
2	11	61	28
3	5	45	50
4	16	42	42

Dissected blastula fragments give rise to different tissues in culture:

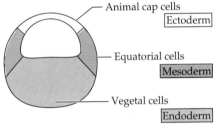

Animal + vegetal fragments give mesoderm:

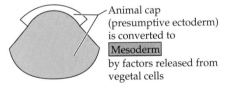

FIGURE 16.26

Summary of the experiments of Nieuwkoop and those of Nakamura and Takasaki, showing mesodermal induction by vegetal endoderm. Isolated animal cap cells become a mass of ciliated epidermis, isolated vegetal cells generate gutlike tissue, and isolated equatorial (marginal zone) cells become mesoderm. If the animal cap cells are combined with vegetal hemisphere cells, many of the cells of the animal cap generate mesodermal tissue.

from the blastula and showed that neither the animal nor the vegetal caps produce mesodermal tissue. However, when the two caps were recombined, the animal cap cells were induced to form mesodermal structures such as notochord, muscles, kidney cells, and blood cells (Figure 16.26). Moreover, the polarity of this induction (whether the region of animal cells formed notochord or muscles, and so on) depended on the dorsal–ventral polarity of the *endodermal* fragment. In a related series of experiments, Nakamura and Takasaki (1970) showed that explants from the equatorial region of the midblastula (the piece removed in Nieuwkoop's experiments) are able to form mesoderm in culture. Explants from the same regions of 32- and 64-cell embryos failed to form mesoderm, becoming ciliated ectodermal cells. These experiments suggested (1) that prospective mesoderm cells are not determined immediately but achieved their prospective fates by the progressive interaction of animal and vegetal cells, and (2) that the equatorial precursors already have their mesodermal fate determined before gastrulation.

These suggestions have been confirmed by transplantation experiments. At about the 32-cell stage, DMZ cells begin acquiring the ability to become dorsal mesoderm on their own (Gimlich, 1986; Jones and Woodland, 1987). Takasaki and Konishi (1989) demonstrated that DMZ blastomeres (B1 and C1) of the 32-cell stage from *Xenopus laevis* transplanted into the ventral marginal zone of *Xenopus borealis* produced descendants that differentiated into chordamesoderm and somite tissues and induced the host tissues to form a secondary neural tube with a head and tail. However, none of these embryos had complete anterior structures (full cement gland and nose), indicating that not all the information from the dorsalmost vegetal D1 blastomeres is present in these cells at this early stage. The frequency with which the DMZ transplants can produce the complete axis in irradiated host embryos starts off very low and increases with age. Conversely, the mesoderm-inducing activity of the dorsal *vegetal* cells declines over this period (Gimlich, 1985; Boterenbrood and Nieuwkoop, 1973). It appears, then, that the cells of the dorsal vegetal quadrant in some way activate (or produce) factors within the DMZ cells, and that the complete "inductive mesoderm" phenotype develops gradually as the vegetal cells influence the marginal cells above them. This set of factors capable of inducing the dorsal mesoderm has been called the **Nieuwkoop center** (Gerhart et al, 1989) and in *Xenopus laevis*, it resides in the dorsalmost vegetal cells of the blastula (Figure 16.27).

The ventral and lateral vegetal cells also have roles in specifying the mesoderm. Whereas the ventral and lateral vegetal cells specify the intermediate (muscle, mesenchyme) and ventral (mesenchyme, blood, pronephric kidney) types of mesoderm, the dorsalmost vegetal cells specify the axial mesoderm components (notochord and somites; Figure 16.25). Although this experiment demonstrated two distinct inductive processes (one from the dorsalmost vegetal cells to induce the dorsal marginal cells and one from the other vegetal cells to induce the intermediate and ventral mesoderm), it did not explain how ventral mesoderm is distinguished from intermediate mesoderm.

Dale and Slack (1987) provided evidence that can be explained by a third inductive signal, coming from the dorsal marginal cells, that "dorsalizes" the marginal cells adjacent to them. When the ventral marginal cells are isolated, they give rise primarily to ventral mesodermal tissues. However, if they are cultured adjacent to dorsal marginal cells, they generate intermediate mesodermal tissue. Thus, there is evidence for a three-step specification of the mesoderm (Figure 16.28): (1) the induction of the organizer activity by the dorsalmost vegetal cells, (2) the induction of the ventral mesoderm by the other vegetal cells, and (3) the dorsalization of the lateral marginal cells adjacent to the dorsal marginal cells to produce the intermediate mesoderm.

Vegetal cells induce mesodermal gene expression

Experiments by Smith and his colleagues (1991) have shown that at midblastula, the ventrolateral and the dorsal vegetal blastomeres of *Xenopus* induce the expression of the *Brachyury* gene in the marginal cells above them. The *Brachyury* mRNA encodes a transcription factor whose function is crucial to the *formation* of the mesoderm. It is expressed prior to α-actin and the other proteins that are the products of mesodermal cells, and if the *Brachyury* gene is expressed in cells where the transcription factor is normally inactive, those cells become mesodermal (Culiffe and Smith, 1992). If the animal hemisphere containing the marginal zone cells is removed from the vegetal hemisphere at midblastula, no mesoderm is formed in the animal hemisphere. However, if vegetal cells are added back to such animal hemispheres, the *Brachyury* gene becomes expressed, and those cells expressing the gene become mesodermal. Thus, the vegetal cells induce the expression of mesodermal genes in the marginal zone cells. Without this interaction, the marginal zone cells remain ectodermal.

Sargent and colleagues (1986) and Gurdon and his co-workers (1985) have also provided molecular confirmation of Nieuwkoop's observations. If Nieuwkoop's morphological studies were correct, α-actin gene activity should be induced in animal hemisphere cells when such cells (which would normally give rise to ectodermal tissues) are placed directly upon vegetal cells. Gurdon and his co-workers dissected away the marginal cells of midblastula embryos and recombined the animal pole cells with the vegetal cell mass (Figure 16.29). They found that the animal pole cells began transcribing mesoderm-specific α-actin message. The vegetal cells had induced a mesoderm-specific gene to be turned on in the presumptive ectodermal tissue. At the same time, these interactions appear to cause the loss of ectoderm-specific gene activity (Sargent et al., 1986).

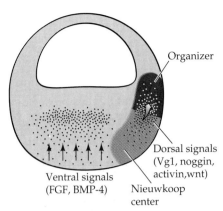

FIGURE 16.27
Model for mesoderm induction in *Xenopus*. A ventral signal (probably FGF-2) is released throughout the vegetal region of the embryo. This induces the marginal cells to become mesoderm. BMP-4 may specify the marginal cells to become posterior mesoderm. On the dorsal side (away from the point of sperm entry), a signal (probably initiated by Vg1 and propagated by activin, noggin, and wnt proteins) is released by the vegetal cells of the Nieuwkoop center. This dorsal signal induces the formation of the Spemann organizer in the overlying marginal zone cells. The molecules involved will be discussed presently. (After De Robertis et al., 1992.)

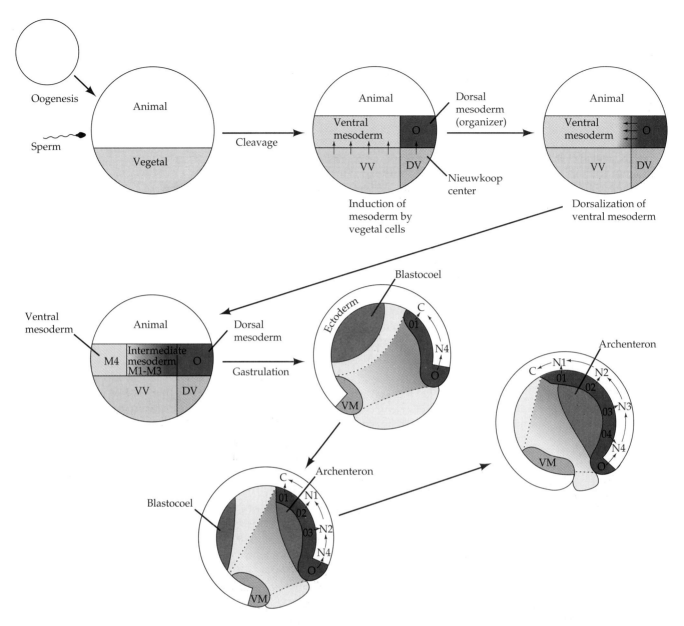

FIGURE 16.28

Inductive interactions during early *Xenopus* development. During oogenesis, the animal-vegetal axis arises. Fertilization causes cytoplasmic rearrangements that subdivide the vegetal region into dorsovegetal (DV) and ventrovegetal (VV) areas. During cleavage, mesodermal induction occurs such that the DV region induces the organizer activity (O) in the dorsal marginal cells above it, while the VV induces the cells above to become ventral mesoderm. A signal from the organizer converts the ventral mesoderm near it into lateral mesoderm (M1, M2, M3). During gastrulation, the ventral and lateral mesoderm go to the sides of the gastrula (not shown), while the dorsal mesoderm expands and induces the polarity in the ectodermal cells. This causes these ectodermal cells to become the different regions of the neural tube (N1, N2, N3, N4). C represents the cement gland, the most anterior structure of the tadpole. The polarity of the endoderm is thus transferred to the neural tissue. The uninduced ectoderm (A) becomes epidermis. (After Smith et al., 1985; Slack and Tannahill, 1992.)

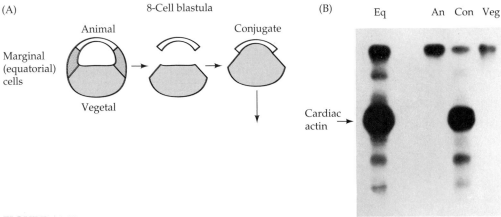

(A) 8-Cell blastula

Animal

Marginal (equatorial) cells

Conjugate

Vegetal

(B) Eq   An Con Veg

Cardiac actin →

FIGURE 16.29

Animal–vegetal recombination experiments. (A) The marginal cells are removed from an 8-cell blastula and the animal and vegetal cells allowed to come into contact. (B) Using an RNase protection assay for the detection of actin, muscle-specific actin was found in the marginal (equatorial) region and in the recombined animal and vegetal cells, but not in vegetal or animal cells alone. (After Gurdon et al., 1985.)

Peptide growth factors inducing dorsal mesoderm: The molecules of the Nieuwkoop center

Activin.   What are the identities of these molecules produced by the vegetal cells to induce the tissue above them to become mesoderm? The formation of dorsal mesoderm—the organizer mesoderm consisting of notochord and somites—appears to be controlled by the **activin, noggin,** and **Vg1** proteins.* These proteins are thought to constitute the activity of the Nieuwkoop center. When added to animal cap blastomeres (which would normally become ectoderm), activin induces the expression of *Brachyury* and causes these tissues to form dorsal mesodermal structures such as muscle and notochord (Figure 16.30A–D; Asashima et al., 1990; Sokol et al., 1990; Smith et al., 1990; Smith et al., 1991). Moreover, graded levels of activin cause animal cap cells to form different mesodermal structures that span the entire spectrum of mesodermal fates. Very low concentrations of activin cause the cells to differentiate into epidermis, but as the activin concentration increases, these cells form posterolateral mesoderm, muscle, notochord, and organizer tissue (Figure 16.30E; Green and Smith, 1990; Green et al., 1992). When activin mRNA was injected into ventral vegetal blastomeres of early *Xenopus* embryos, the resulting tadpoles had extra trunks, complete with notochord and often with part of the head (Plate 9; Thomsen et al., 1990). Activin can also induce axial structures in the chick epiblast, and *activin* mRNA has been identified in the hypoblast at the time when axial mesoderm is being induced (Mitrani et al., 1990). It seems, then, that activin (or an activin-like molecule) is an excellent candidate for a mesoderm inducer.

*These are probably not the only proteins involved in dorsal mesoderm induction. One of the ways these proteins might enable the dorsal marginal zone cells to become the organizer is to inhibit the expression of the *Xwnt-8* gene. Xwnt-8 protein is expressed in all the mesodermal cells that do not express goosecoid (a protein characteristic of anterior dorsal mesoderm); and if goosecoid is expressed ectopically in the ventral mesoderm, those ventral mesoderm cells lose their Xwnt-8 expression and become organizer tissue. Moreover, if plasmids encoding Xwnt-8 are caused to become expressed in the DMZ area, those cells become more ventral types of mesoderm, lacking organizer activity. Thus, Xwnt-8 may play a role in ventralizing the mesoderm, and the repression of this protein by goosecoid may be essential for organizer activity (Christian and Moon, 1993).

(A)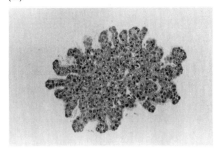

(B)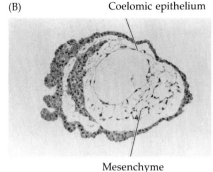

Coelomic epithelium

Mesenchyme

(C)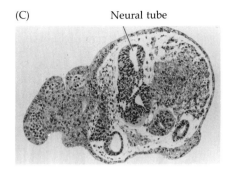

Neural tube

FIGURE 16.30

Activin is able to induce most of the dorsal structures of the *Xenopus* larvae. (A-D) Histological sections from intact animal caps of newt blastulae placed into control medium (A) or into medium containing activin (B–D). Those placed into activin-containing medium developed various mesodermal tissues, some of which (D) could induce the formation of neural tube-like structures. (E) Dissociated animal cap cells express different genes when exposed to different concentrations of activin. At low concentrations, they become epidermal cells (expressing keratin). However, after a sharp threshold concentration, the cells become posterior or lateral mesoderm (*XlHbox6, Xhox3, Brachyury*). If exposed to higher concentrations of activin, the cells become vacuolated and express genes characteristic of the notochord (MZ15 antigen and low levels of *goosecoid*) and at still higher concentrations, the cells acquire organizer activity and express high levels of *goosecoid*). The *Brachyury* gene is probably active in two separate places and times. *Brachyury* is first seen in the mesodermal precursors during gastrulation. Thereafter it is not seen in muscle cells, but is seen in notochord. (This would account for the break in the *Brachyury* expression band on the autoradiograph.) Cytoskeletal actin is the internal control found in all cells. (A–D from Moriya and Asashima, 1992; E from Green et al., 1992; photographs courtesy of the authors.)

However, while activin is clearly able to specify the dorsal characteristics of the mesoderm, it does not meet the complete requirements of the Nieuwkoop center mesoderm dorsalizer. First, the secondary axes induced by activin lack the most anterior portions of their heads. Second, injection of activin or its message into embryos that had been treated with ultraviolet irradiation does not rescue the dorsal axis. The UV-treated embryos still fail to gastrulate and lack their dorsal mesoderm and neural tube. Third, inhibitors of activin function do not inhibit the inducing activity of the dorsalmost vegetal blastomeres (Slack, 1991). Something else is needed.

noggin. One of those other agents appears to be the product of the *noggin* gene. Smith and Harland (1991, 1992) isolated this gene by constructing a cDNA library from dorsalized gastrulae. (When treated with lithium chloride during blastula stages, all the mesoderm acts as dorsal mesoderm and involutes around the entire embryo. Such gastrulae are "dorsalized".) RNAs synthesized from sets of these plasmids were injected into the ventralized embryos produced by UV irradiation. Those sets of plasmids whose RNAs rescued the dorsal axis were split into smaller sets, and so on, until single clones were isolated whose mRNAs were able to restore the dorsal axis in such embryos. One of these clones contained *noggin*. When injected into a ventral vegetal blastomere, *noggin* mRNA establishes a new Nieuwkoop center and induces the cells above them to become dorsal mesoderm (Smith and Harland, 1992). Moreover, injection of *noggin* mRNA into 1-cell, UV-irradiated embryos completely rescues the dorsal axis and allows the formation of a complete embryo (Plate 4). If too much noggin protein is synthesized at that time, the embryo becomes "hyperdorsal," forming *only* the head region (hence the name "noggin"). The mRNA for this noggin protein is already present in the fertilized egg, and the sequence of the protein (as deduced from the gene) suggests strongly that noggin is a secreted protein.

The Vg1 protein. Another critical Nieuwkoop center protein may be Vg1. This is the protein whose mRNA is tethered to the vegetal yolk mass during oogenesis and remains in the vegetal cytoplasm during cleavage (Chapters 4 and 13; Plate 8). After fertilization, Vg1 protein is made throughout the vegetal hemisphere of the blastula, but it is in an inactive precursor form that needs to be cleaved in order to become active. Thomsen and Melton (1993) speculate that the cortical rotation of the cytoplasm activates a protease that cleaves Vg1 precursor to its active form *solely* in the region that will produce the dorsalmost vegetal cells—that is, the cells of the Nieuwkoop center. The active Vg1 protein has a structure similar to that of activin, TGF-β1, and several other compounds capable of arti-

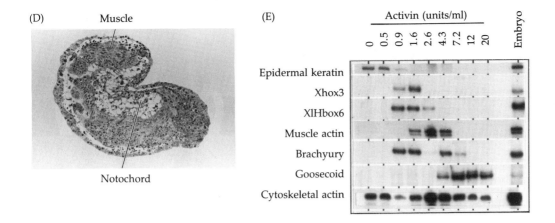

(D) Muscle / Notochord

(E)

Activin (units/ml)

	0	0.5	0.9	1.6	2.6	4.3	7.2	12	20	Embryo
Epidermal keratin										
Xhox3										
XlHbox6										
Muscle actin										
Brachyury										
Goosecoid										
Cytoskeletal actin										

ficially inducing dorsal mesoderm production. Activated Vg1 protein is able (1) to induce mesoderm in animal cap cells, (2) to induce an entire embryonic axis when transplanted into ventral vegetal cells, and (3) to rescue the dorsal axis of UV-irradiated eggs (Thomsen and Melton, 1983; Dale et al., 1983). While the other factors (activin, noggin) may be *products* of the Nieuwkoop center, it is probable that Vg1 protein is the *initiator* of the Nieuwkoop center activity.

Peptide growth factors inducing the ventrolateral mesoderm

Whereas Vg1, activin, and noggin proteins appear to be active in the dorsalmost vegetal blastomeres and induce their overlying marginal cells to become dorsal mesoderm, the ventral and lateral mesoderm appear to be induced by a combination of **fibroblast growth factor** (FGF) and **bone morphogenetic protein 4** (BMP-4). The importance of FGF was seen when isolated animal caps from *Xenopus* blastulae were incubated in FGF-2 (bFGF), one of the eight members of the FGF family. These bFGF-treated animal caps generated mesenchyme, muscle, and blood cells (Kimelman and Kirschner, 1987; Slack et al., 1987). These are the same types of ventral and lateral mesenchyme cells that are produced when ventral–vegetal and lateral–vegetal blastomeres are cultured beneath animal cap blastomeres (Dale and Slack, 1987). FGF-2 can also induce the expression of the *Brachyury* gene, which is thought to be responsible for forming mesoderm cells (Smith et al., 1991).

A close relative of FGF-2 has been found in cleavage-stage *Xenopus* embryos, and it has mesoderm-inducing activity very similar to that of FGF-2 (Isaacs et al., 1992). Moreover, it is a secreted molecule that can be produced in one cell and diffuse to its neighbors. The functional significance of such secreted FGF molecules was demonstrated by destroying the embryo's receptors for FGF. The FGF receptor is a dimer of two identical molecules that are embedded in the cell membrane. Amaya and co-workers (1991) injected the mRNA for a mutant form of this receptor protein into 2-cell *Xenopus* embryos. The mutated protein made by this message interferes with the activation of the normal FGF receptor proteins. This is what is called a **dominant negative receptor**, and if present in high enough concentrations, it can prevent normal FGF receptors from forming. Such blastulae are unable to respond to FGF (Figure 16.31). When this experiment was done, the embryos that lacked functional FGF receptors had dramatically reduced amounts of posterior and lateral mesoderm (Plate 7).

In addition to FGF, the induction of ventrolateral mesoderm also needs

FIGURE 16.31
The dominant negative receptor assay for the importance of a particular factor. The FGF receptor (FGFR) is a transmembrane protein with a cytoplasmic tyrosine protein kinase domain. (A) When FGF binds to its receptors, it causes them to dimerize, and the two protein kinase regions of the receptors phosphorylate one another. When phosphorylated, they can signal (through a pathway discussed in Chapter 18) the nucleus to express new genes. (B) The dominant negative receptor lacks the protein kinase domain. When it binds FGF, it creates an inactive dimer. Thus, the effect of FGF is not transmitted into the cell.

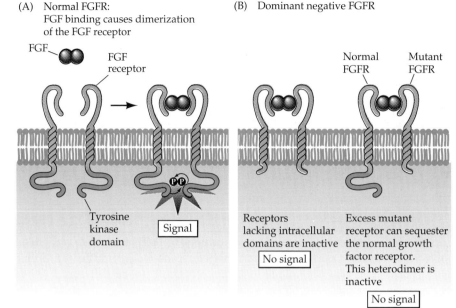

(A) Normal FGFR: FGF binding causes dimerization of the FGF receptor

FGF

FGF receptor

Tyrosine kinase domain

Signal

(B) Dominant negative FGFR

Normal FGFR   Mutant FGFR

Receptors lacking intracellular domains are inactive

No signal

Excess mutant receptor can sequester the normal growth factor receptor. This heterodimer is inactive

No signal

BMP-4. If the mRNA for BMP-4 is injected into 1-cell *Xenopus* eggs, all the mesoderm in the embryo becomes ventrolateral mesoderm, and no involution occurs at the blastopore lip (Dale et al., 1992; Jones et al., 1992). Further evidence for the role of BMP-4 in ventrolateral mesoderm induction came from implantation experiments. When animal caps from those embryos injected with BMP-4 message were isolated and implanted into the blastocoels of young *Xenopus* blastulae, they caused the formation of an extra tail (Figure 16.32). Thus the formation of the posterior (ventrolateral) mesoderm appears to be generated by the actions of FGF and BMP-4.

The molecular nature of the organizer I: The *goosecoid* gene product as the part of the Spemann organizer that controls mesodermal involution

There are many components to what we have been calling "the organizer". One function of the organizer is to initiate gastrulation by instructing the involution of cells to form the dorsal blastopore lip. A second function is to differentiate a dorsal notochord and to dorsalize the lateral mesoderm to the degree that it forms muscle, renal, and cardiac structures instead of mesenchyme and blood cells. A third function is to induce the overlying ectoderm to become neural tissue; and a fourth function might be necessary to cause the neural plate cells to form the neural tube. Within the past few years, several laboratories have begun identifying molecules that may be responsible for these organizer functions.

One possible mechanism for organizer activity is that the proteins of the Nieuwkoop center activate a transcription factor that might activate the genes encoding the agents of the organizer. By screening libraries of dorsal blastopore lip cDNA with probes to genes that are active in *Drosophila* axis formation, researchers have found several molecules that may function in the organizer. One of these genes actively expressed in the *Xenopus* organizer region is **goosecoid**, a gene that has a homeobox resembling *Drosophila*'s *bicoid* and *gooseberry* genes (Blumberg et al., 1991; Cho et al., 1991a). The *goosecoid* transcripts are first detected at the late blastula stage, indicating that this is a nuclear-controlled gene, and those tran-

FIGURE 16.32
The importance of BMP-4 in producing posterior structures can be seen when *BMP-4* mRNA was injected into embryos and the resulting animal cap cells were transplanted beneath the young gastrula ectoderm. The treated larvae usually developed an extra tail. (From Jones et al., 1992, courtesy of B. Hogan.)

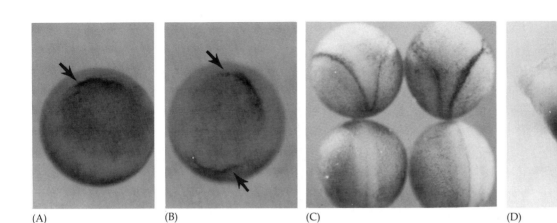

(A)   (B)   (C)   (D)

FIGURE 16.33

Ability of *goosecoid* mRNA to induce a new axis. (A) At gastrula, the control embryo (either uninjected or given an injection of *goosecoid*-like mRNA but lacking the homeobox) has one dorsal blastopore lip. (B) An embryo whose ventral vegetal blastomeres were injected at the 16-cell stage with *goosecoid* message. Note the secondary dorsal blastopore lip. (C) Top, two neurulae that had been injected with *goosecoid* mRNA, showing two axes; bottom, two control neurulae. (D) Twinned embryo produced by the *goosecoid* injection. Complete head structures have been induced. (After Cho et al., 1991a; Niehrs et al., 1993; courtesy of E. De Robertis.)

scripts accumulate in the area directly overlying the dorsal blastopore lip in the dorsal mesodermal precursor (C1 blastomere) cells. In animal cap cultures, either Vg1 protein or activin, but not FGF-2 or noggin, can induce the transcription of the *goosecoid* gene (Cho et al., 1991a; Thomsen and Melton, 1993). The expression of *goosecoid* mRNA also correlates with the organizer domain in experimentally treated animals. When LiCl is used to increase the dorsoanterior-inducing mesoderm in the marginal zone, the expression of *goosecoid* likewise is expanded. Conversely, when eggs are treated by UV light prior to first cleavage, both dorsoanterior induction and *goosecoid* expression are significantly inhibited. Injection of the full-length *goosecoid* message into the two ventral blastomeres of the 4-cell *Xenopus* embryo causes the progeny of these blastomeres to involute, undergo convergent extension, and form the dorsal mesoderm and head endoderm of the secondary axis (Figure 16.33). The *goosecoid* mRNA-injected cells do not act as a Nieuwkoop center, but as an organizer (Niehrs et al., 1993). While the Nieuwkoop center cells remain as endoderm, cells expressing goosecoid protein (and cells injected with the *goosecoid* message) involute inside the embryo and migrate toward the animal pole. Moreover, labeling experiments (Niehrs et al., 1993) show that the *goosecoid*-injected cells are able to recruit neighboring host cells into the dorsal axis, as well. Thus, the Nieuwkoop center activates the *goosecoid* gene encoding a DNA-binding protein that (1) activates the migration properties (involution and convergent extension) of the dorsal blastopore lip cells, (2) autonomously determines the head endodermal and dorsal mesodermal fates of those cells expressing it, and (3) enables the goosecoid-expressing cells to recruit neighboring cells into the dorsal axis.

The importance of goosecoid is also seen when other vertebrate embryos are studied. Blum and co-workers (1992) have shown that *goosecoid* mRNA in chick embryos is first seen in the posterior region of the epiblast, where the primitive streak is being formed. Shortly thereafter, the *goosecoid*-expressing cells are located only in the anterior-most tip of the primitive streak, that region that is analogous to the dorsal blastopore lip of amphibians. Moreover, transplantation of the *goosecoid*-expressing region of the 6-day mouse primitive streak into early *Xenopus* gastrulae causes the formation of a secondary axis. Other regions of the primitive streak either cause no inductions or induce tail-like structures. This suggests that the mouse embryo also has an organizer region, located at the Hensen's node region of the primitive streak.

The molecular nature of the organizer II: Possible soluble molecules from the dorsal blastopore lip and chordamesoderm

We now come to the stage during which the neural ectoderm is determined (i.e., what Spemann thought of as primary induction). What molecules (1) determine the ectoderm to become neural and (2) re-specify the lateral mesoderm from a ventral mesodermal fate (mesenchyme and blood) to an intermediate mesodermal fate (heart, muscle, and kidney)? Although Spemann thought that searching for an organizer molecule was folly, his students pressed forward. Indeed, "few compounds, other than the philosopher's stone, have been searched for more intensely than the presumed agent of primary embryonic induction in the amphibian embryo" (Løvtrup et al., 1978). Harrison (quoted by Twitty, 1966) referred to the amphibian gastrula as a "new Yukon to which eager miners were now rushing to dig for gold around the blastopore." Unfortunately, their picks and shovels proved too crude to uncover the molecules involved. The analysis of organizer molecules had to wait until recombinant DNA technologies enabled investigators to make cDNA clones from blastopore lip mRNA and see which of them might cause dorsalization of the embryo. We are now aware that there are at least two types of signals from the organizer. The first signals are soluble molecules that are synthesized by the chordamesoderm and interact vertically with the ectodermal cells above them. The second set of signals is produced by the dorsal blastopore lip and is sent horizontally through the plane of the ectoderm (Figure 16.34).

Evidence for diffusible organizer factors. Spemann considered two alternatives to explain the induction of a secondary neural tube by the dorsal blastopore lip. First, the organizer might be sending out its inducing signals horizontally through the plane of the ectoderm. Second, the dorsal blastopore lip cells might vertically induce the overlying ectoderm once they become chordamesoderm cells. Originally Spemann favored the first model, in which the dorsal lip of the blastopore directly induces the ectoderm (Spemann, 1918, 1927; Hamburger, 1988). However, the experiments by his students Mangold and Holtfreter supported the second model. Mangold, as discussed above, showed that pieces of the dorsal mesoderm could induce regionally specific neural tube structures, and Holtfreter (1933) prevented the interaction between the chordamesoderm and the dorsal ectoderm by removing the fertilization envelope and placing the embryos in a hypertonic solution. This caused the mesoderm to exogastrulate instead of involuting (Figure 16.35). In such embryos, where the dorsal mesoderm is separate from the dorsal ectoderm, Holtfreter saw no neural tube formation and concluded that the chordamesoderm is necessary to induce neural tube formation.

The ability of diffusible molecules from the chordamesoderm to induce neural induction was demonstrated in a series of transfilter experiments (Saxén, 1961; Toivonen et al., 1975; Toivonen and Wartiovaara, 1976). Newt dorsal lip was placed on one side of a filter thin enough so that no

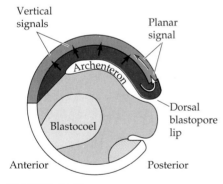

FIGURE 16.34

Two modes of inducing the dorsal axis. In the *planar* mechanism, molecules are transferred from the dorsal blastopore lip tissue through the plane of the ectoderm. In the *vertical* mechanism, soluble molecules from the dorsal blastopore lip-derived chordamesoderm induce the cells above them to become neural tissue. (After Doniach, 1993.)

(A) Normal gastrulation

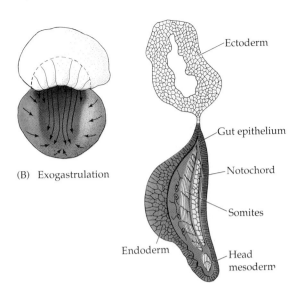
(B) Exogastrulation

Ectoderm

Gut epithelium

Notochord

Somites

Endoderm

Head mesoderm

(C) Differentiation in exogastrulation

FIGURE 16.35
Exogastrulation. (A) In normal gastrulation, the mesoderm involutes beneath the ectoderm. However, when the amphibian embryo is placed in a hypertonic salt solution, the mesoderm undergoes exogastrulation (B), turning outward from the ectoderm instead of involuting into the embryo. (C) The ectoderm in these exogastrulae does not form neural tissue. (After Holtfreter and Hamburger, 1955.)

processes could fit through the pores, and competent gastrula ectoderm was placed on the other side. After several hours, neural structures were observed in the ectodermal tissue. These molecules are being studied by recombinant DNA methods by which cDNA made from the dorsal blastopore lip and chordamesoderm mRNA are cloned and injected into ventralized embryos to see if they will rescue dorsal axis production.

The noggin protein as a diffusible organizer factor.　One of these molecules has been mentioned earlier: noggin. The sequence of this protein strongly suggests that it is a secreted peptide, and Smith and Harland (1992) have shown that newly transcribed (as opposed to maternal) *noggin* mRNA is first localized in the dorsal blastopore lip region and then becomes expressed in the notochord (see title page spread). Moreover, if the early embryo is treated with lithium chloride (LiCl) so that all the mesodermal mantle becomes notochord-like organizer tissue, then *noggin* mRNA is found throughout the mesodermal mantle. Treatment of the early embryo with ultraviolet light (which prevents dorsal blastopore lip formation) inhibits the synthesis of *noggin* mRNA. It seems, then, that noggin is an excellent candidate for mediating some of the functions of the organizer.

Recent evidence suggests that noggin protein can accomplish two major functions of the Spemann-Mangold organizer: it *induces neural tissue* from the dorsal ectoderm, and it *dorsalizes the mesoderm cells* that would otherwise contribute to the ventral mesoderm. Smith and co-workers (1993) have shown that noggin protein can dorsalize the ventral marginal zone cells at gastrulation and respecify their fate from ventral mesoderm (mesenchyme and blood cells) to more intermediate fates (muscle, heart, and pronephric kidney). Respecification at this relatively late stage cannot be accomplished by activin or fibroblast growth factor. When Smith and co-workers removed ventral marginal zones (the presumptive ventral mesoderm) from *Xenopus* gastrulae and placed them into medium containing soluble noggin protein, these explants produced muscle-specific mRNA, something that is normally reserved for the dorsal marginal explants. These explants also became elongated (another characteristic of dorsal development). However, the elongated explants did not stain for notochordal tissue. These experiments show that soluble noggin protein can induce gastrula ventral mesoderm cells to become muscle (but not notochord), and it therefore resembles the signal from the organizer that dorsalizes the lateral mesodermal tissue (see Figure 16.28).

The noggin protein is also able to induce neural tissue in gastrula ectoderm without the presence of any dorsal mesoderm (Lamb et al., 1993). When noggin is added to gastrula (or animal cap) ectoderm, the ectodermal cells are induced to express forebrain-specific neural markers. Moreover, the gene products for notochordal or muscle cells are not induced by the noggin protein. Since noggin is a secreted protein synthesized by the derivatives of the organizer (the head mesoderm and chordamesoderm) during gastrulation (when induction takes place), it is possible that noggin plays a major role in both the dorsalization of the mesoderm and the neuralization of the dorsal ectoderm (Plate 5).

Activin inhibitor as a soluble organizer factor. Another soluble protein involved in dorsal mesoderm formation may also play a role in neural tube formation. This protein is activin. Its role, however, may be a negative one. Hemmati-Brivanlou and Melton (1992) constructed genes for dominant negative activin receptors and injected them into *Xenopus* embryos. These mutant receptors were expressed, and it was found that they prevented activin from being received by the animal cap cells. Not only did these animal cap cells fail to form mesoderm—they became *neural* tissues. In the absence of activin, the ectoderm might be determined to become neural cells. This led them to suggest that neural induction may be caused by the notochord secreting an *inhibitor* of activin. Naturally occurring soluble inhibitors of activin (such as inhibin and follistatin) are found in the embryo, but their roles are not yet known. In addition to *noggin* and *goosecoid*, other homeobox-containing genes have been seen to be localized specifically in the dorsal blastopore lip and are being characterized.

The molecular nature of the organizer III: The involvement of protein kinase

What might the soluble molecules be doing? What might they be activating in the ectoderm? One of the problems that plagued Spemann's students as they tried to identify the organizer molecules was a lack of specificity; it seemed a huge variety of things could induce a neural plate. These compounds included turpentine, formaldehyde, and methylene blue dye, as well as dead archenteron and an assortment of adult tissues from several phyla. On the basis of such eclectic induction, Holtfreter (1948) suggested that the real inducer lies within the ectoderm itself and that it is released by mild cytolysis. Anything causing such sublethal damage would suffice, even a mildly alkaline or hypertonic medium. However, if there is a "masked inducer" that is released by these nonspecific means (as well as by natural induction), it has not yet been found.*

One possible reason nobody has been able to isolate *the* naturally occurring neural inducing factor is that there may be several simultaneous factors at work. There is now evidence (Davids et al., 1987; Davids, 1988; Otte et al., 1988, 1989) that whatever the inducing factors might be, they may act by initiating at least two reactions needed for neural induction. The first is the activation of protein kinase C (PKC) on the ectoderm cell surfaces, and the second is an increase in the concentration of cAMP within the ectodermal cells. These studies have shown that if one and not the other of these events occurs, no neural tissue is formed. However, if

*Spemann's laboratory, and those of his students, usually used salamander embryos for their experiments. It turns out that *frog* ectoderm is much more difficult to induce than that of these urodeles.

one artificially activates protein kinase C *and* adenyl cyclase on the ectodermal cell membranes, neural tissue is formed. In this model, neural induction is accomplished by two reactions, and each reaction might be initiated by a different molecule. The participation of PKC in natural neural induction was given further support when Otte and co-workers (1991; Otte and Moon, 1992) demonstrated that the PKC of the dorsal ectoderm differs from the PKC of the ventral ectoderm, both in its structure and in its ability to be activated by external compounds. Only the PKC found in the dorsal ectoderm correlates with the ability to respond to natural inducers.

The molecular nature of the organizer IV: Autonomous differences between dorsal and ventral ectoderm

Phillips and his co-workers (London et al., 1988; Savage and Phillips, 1989) have provided evidence that differences between prospective epidermis and prospective neural plate are present at the 8-cell stage, before the dorsal lip is formed. At that time, the animal pole blastomeres on the future dorsal side of the embryo lack a protein—Epi 1—that characterizes the animal pole cells on the future ventral side. During early gastrulation, the area of cells lacking Epi 1 is extended by signals transmitted from the dorsal blastopore lip through the ectoderm. This signal establishes a sharp boundary at what will become the lateral and anterior edges of the neural plate. Thus, they found that before the chordamesoderm involutes, the presumptive epidermal cells already express Epi 1 and the presumptive neural plate cells already lack it. Moreover, they also showed that the dorsal blastopore lip is capable of inducing these differences. When ventral ectoderm (prospective epidermis) was isolated from early gastrula cells, it expressed this Epi 1 protein (Figure 16.36). However, if this ventral ectoderm was placed in contact with dorsal blastopore lip from an early-to-mid gastrula, the prospective epidermis lost the protein. The loss of this protein is normally observed only on those ectoderm cells that will form the neural plate. Thus, certain cells appear to be biased to become neural cells even during early cleavage, and the early dorsal blastopore lip is able to extend this bias to the future neural plate cells even before gastrulation is underway.

The molecular nature of the organizer V: Dorsal lip factors transmitted through the ectoderm

During the past decade, studies on *Xenopus laevis* have shown that signals produced by the dorsal blastopore lip that travel *through the ectoderm* may be at least as important as the diffusible signal from the underlying chordamesoderm. Moreover, there may be several sets of signals traveling through the plane of the ectoderm. The first set of signals may amplify the differences between the presumptive dorsal ectoderm and the presumptive ventral ectoderm. It had long been assumed that all early-gas-

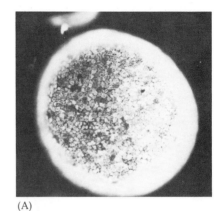

(A)

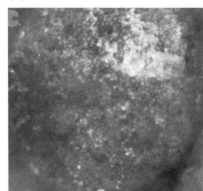

(B)

(C)

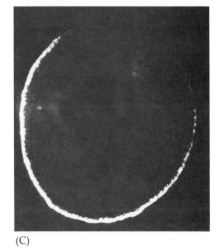

(D)

FIGURE 16.36
Expression of the protein Epi 1 in the *Xenopus* ectoderm. Expression of Epi 1 was monitored with antibodies to Epi 1. (A) Isolated ventral ectoderm expresses Epi 1. (B) Ventral ectoderm in contact with the dorsal blastopore lip loses Epi 1 expression. The cells of the neural plate are devoid of Epi 1 expression, as seen in a neurula observed in cross section (C) or looking down at the dorsal surface of the neurula (D). (A and B from Savage and Phillips, 1989; C and D from London et al., 1988; photographs courtesy of C. Phillips.)

FIGURE 16.37

Neural gene expression in induction experiments using dorsal or ventral ectoderm. Ectoderm from the ventral and dorsal third of the embryo was wrapped around a piece of chordamesoderm. The fragments were then separated and tested for the presence of neural mRNAs. The mesoderm had none, the ventral ectoderm had little, and the dorsal ectoderm had high concentrations of neural mRNAs. (After Sharpe et al., 1987.)

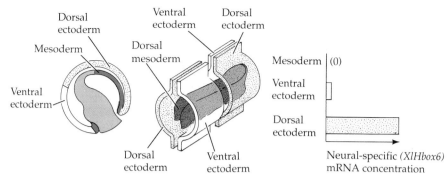

trula ectoderm had equal ability to be induced. However, we now understand that the dorsal ectoderm may already be biased to generate neural tissue in that it lacks Epi 1 and has a specific type of PKC. Sharpe and co-workers (1987) showed that the *dorsal* ectoderm of early gastrulae (before any contact with underlying mesoderm has occurred) is indeed much more readily induced than *ventral* ectoderm. As their markers for neural determination, they used two mRNAs for proteins found in neural, but not epidermal, tissues. Both these mRNAs are induced in the prospective neural region soon after neural plate formation begins. Sharpe and colleagues found that when *dorsal* ectoderm was combined with chordamesoderm, the neural mRNAs were induced in the ectoderm. However, when the same experiment was performed using *ventral* ectoderm, these mRNAs were not seen (Figure 16.37). This suggested that before the neurula stage, the dorsal ectoderm is already different from the ventral ectoderm.

Another set of dorsal lip signals may be responsible for actually inducing neural-specific genes through the plane of the ectoderm. Doniach and colleagues (1992) showed that instructive, positionally specific information is provided by planar signals passing through the ectoderm. When explants are taken from early *Xenopus* gastrulae such that the ectoderm retains contact with the dorsal blastopore lip but never contacts the mesoderm, not only are the pan-neural markers N-CAM and NF-3 induced in the ectoderm, but four position-specific neural markers—*engrailed-2, Krox-20, XlHbox1,* and *XlHbox6*—are expressed in the explant ectoderm in the appropriate anterior–posterior sequence (Figure 16.38). It appears, then, that the horizontally inductive signals from the dorsal blastopore lip are sufficient for inducing the anterior–posterior neural pattern. Ruiz i Altaba (1992) has also confirmed extensive neural patterning in these exogastrulae, showing that the pattern of neural markers in the exogastrulae reflects the normal pattern except in the forebrain and ventral regions. He also provides evidence that the transmission of these horizontal signals is through the notoplate, those ectodermal cells that also undergo convergent extension (in both normal embryos and exogastrulae) from the noninvoluting marginal zone. These notoplate cells form the midline of the neural plate and then the ventral floor plate of the neural tube. When they are removed from the neural plate, such plates lose their ability to induce neural differentiation in animal cap ectoderm.

If planar signals traveling from the dorsal blastopore lip through the ectoderm are responsible for neural induction, then the original source of such signals should be the *epithelium* of the dorsal marginal zone rather than the deep mesenchymal cells of that organizer zone. Shih and Keller (1992) have found this to be the case. They repeated the Spemann and Mangold experiment, but instead of using the entire dorsal marginal zone (DMZ), they transplanted either the epithelial cells or the deep cells of the DMZ (which they labeled with fluorescent dextran particles). The epithelial

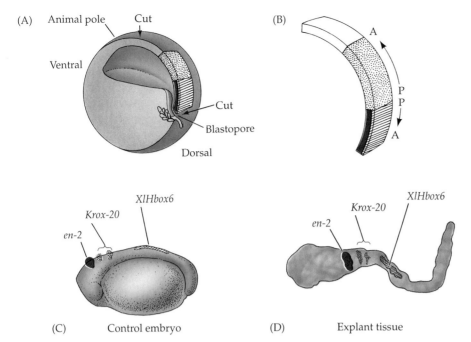

FIGURE 16.38

Expression pattern of neural markers induced by contact with the dorsal blastopore lip in plane of ectoderm. (A) Sagittal section of early *Xenopus* gastrula showing where cuts were made. (B) Explant depicting the anterior–posterior polarity expected from fate map: White region is epidermis; stippled region is presumptive mesenchyme; black region is dorsal mesoderm; striped region is archenteron roof. The explants were placed under coverslips to prevent migration of the mesoderm. (C) Expression of neural markers in the control embryo, stage 21. Homeobox genes *engrailed-2* and *XlHbox 6* are expressed at the hindbrain–midbrain border and in the spinal cord, respectively; zinc finger protein gene *Krox-20* is expressed in rhombomeres 3 and 5 of the hindbrain. (D) The same order of expression is seen in the ectoderm of those explants retaining a connection to the dorsal blastopore lip. (After Doniach et al., 1992.)

cells had all the inductive properties of the Spemann organizer and differentiated into mesodermal tissue. This epithelium was also able to rescue embryos that had been "ventralized" by UV irradiation. Neither the deep cells of the DMZ nor the ventral marginal cells could accomplish such organizer activities.

The molecular nature of the organizer VI: Vertical signals from the chordamesoderm

Recent evidence suggests that both vertical induction through the chordamesoderm and horizontal induction through the ectoderm are necessary for complete embryonic induction. Keller and co-workers (1992) have demonstrated that planar signals from the early-gastrula dorsal blastopore lip are both necessary and sufficient to induce the convergent extension of the presumptive hindbrain–spinal cord ectoderm directly adjacent to it. However, the induced ectodermal tissue does not roll into a tube or form the dorsal–ventral pattern typical of the normally induced neural tube. These latter functions have been ascribed to the notochord (Holtfreter, 1933; Smith and Schoenwolf, 1989; van Straaten, 1989; Yamada, et al. 1991) and probably represent actions of the underlying mesoderm upon the overlying ectoderm. (Thus, Holtfreter did not get a neural tube when he performed his famous exogastrulation experiment. However, recent research shows that the neural ectodermal genes are expressed.) According to Keller and co-workers, the horizontal and vertical induction mechanisms provide different functions during neural induction. Ruiz i Altaba (1992) also notes that while the horizontal induction is sufficient to generate the pattern of neural differentiation throughout most of the neural tube, the anterior region needs the influence of the notochord.

The molecular basis of regional specificity: Retinoic acid

Retinoic acid as possible posterior morphogen

As mentioned above, the major hypothesis for embryonic induction is a two-step model involving the activation of neural determination in the ectoderm, followed by regional specification. The evidence presented above also indicated that the signal for specification is a gradient of posteriorizing factor(s) emanating from the posterior of the embryo. Recent studies have suggested that retinoic acid (RA) may be that posteriorizing morphogen. Just as research (to be detailed in the next chapter) demonstrated that retinoic acid has a posteriorizing effect on mouse neural tube and paraxial mesoderm, similar work has shown that RA within the *Xenopus* gastrula may be able to specify the regional characteristics of its neural tube.

If early *Xenopus* gastrulae are treated with nanomolar to micromolar concentrations of RA, their forebrain and midbrain development is impaired in a concentration-dependent fashion (Figure 16.39A; Sharpe, 1991; Papalopulu et al., 1991). When lower concentrations are used, the actual induction of neural tissue does not appear to be inhibited, but fewer forebrain messages and structures are produced (Figure 16.39B; Durston et al., 1989, 1991; Sive et al, 1990). Retinoic acid appears to affect both the

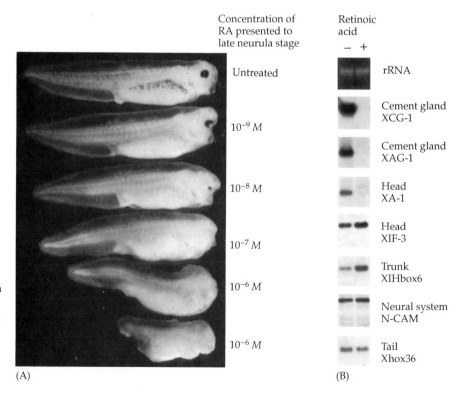

FIGURE 16.39
Retinoic acid causes posteriorization of neural structures. (A) Late neurula stage (16–18) embryos exposed continuously to different concentrations of RA and allowed to grow until controls reached tadpole stage. (B) Effect on neural marker mRNA expression when blastulae are treated with 10^{-6} M RA for 2 hours (sufficient to give acephalic tadpoles). Inhibitory effect can be seen on the genes expressed most anteriorly. (A after Ruiz i Altaba and Jessell, 1991; B after Sive et al., 1990.)

(A)

(B)

mesoderm and the ectoderm. Ruiz i Altaba and Jessell (1991) found that anterior dorsal mesoderm from RA-treated gastrulae was unable to induce head structures in host embryos, and Sive and Cheng (1991) found that RA-treated ectoderm was unable to respond to the anterior-inducing mesoderm of untreated gastrulae.

Retinoic acid acts on activin-treated cells to transform their specification from dorsal to lateral mesoderm (Cho et al., 1991b). Moreover, inductions of this RA-treated mesoderm resemble those of FGF-2, not activin (Figure 16.40). Cho and colleagues (1991b) and Sive and Cheng (1991) also showed that RA alters the expression of homeobox genes in a posterior direction in both ectoderm and mesoderm. In a variation of the two-step model proposed earlier, recent models (Figure 16.41; Otte et al., 1991; Sharpe, 1991) postulate that neural induction leads to the creation of an anterior (forebrain-type) neural determination that is acted upon by a posterior gradient of retinoic acid to create the regional specificities. RA would inhibit the expression of anterior-specific genes; such an RA gradient (tenfold higher in the posterior than in the anterior) has been detected in the dorsal mesoderm of early *Xenopus* neurulae (Chen et al., 1994).

Our discussion of retinoic acid, homeobox genes, and axis determination continues in the next chapter. Although our understanding of induction has increased enormously within the past five years, we still have much more work to do. Surveying the field in 1927, Spemann remarked,

What has been achieved is but the first step; we still stand in the presence of riddles, but not without hope of solving them. And riddles with the hope of solution—what more can a scientist desire?

The challenge still remains.

Competence and "secondary" induction

The primary inductive interactions, although complex, cannot construct the entire embryo. However, the formation of the neural tube, dorsal mesoderm, pharyngeal endoderm, and other tissues creates the conditions for a cascade of further inductive events. Interactions by which one tissue interacts with another to specifically direct its fate are called secondary inductions.

Any system of embryonic induction has at least two components: a tissue capable of producing the inducing stimulus, and a tissue capable of receiving and responding to it. So far we have been looking at the specificity of production; now we must look at the specificity of responding cells. This ability to respond in a specific manner to a given stimulus is called **competence**. We have already seen that in the early gastrula, an implanted blastopore lip can induce a new neural plate and embryonic axis just about anywhere in the embryo it can meet ectoderm. However, with increasing embryonic age, the ectoderm loses this ability to respond, and the implantation of a dorsal blastopore lip beneath the prospective epidermis of a neurula-stage embryo will not cause it to form a new neural plate. It has lost its competence to respond to the new blastopore lip.

Although the late-neurula ectoderm is no longer competent to respond to the blastopore lip, it has become competent to respond to new inducers. This competence can be localized to particular areas. During gastrulation and early neurulation, the *head* ectoderm (but not the *trunk* ectoderm)

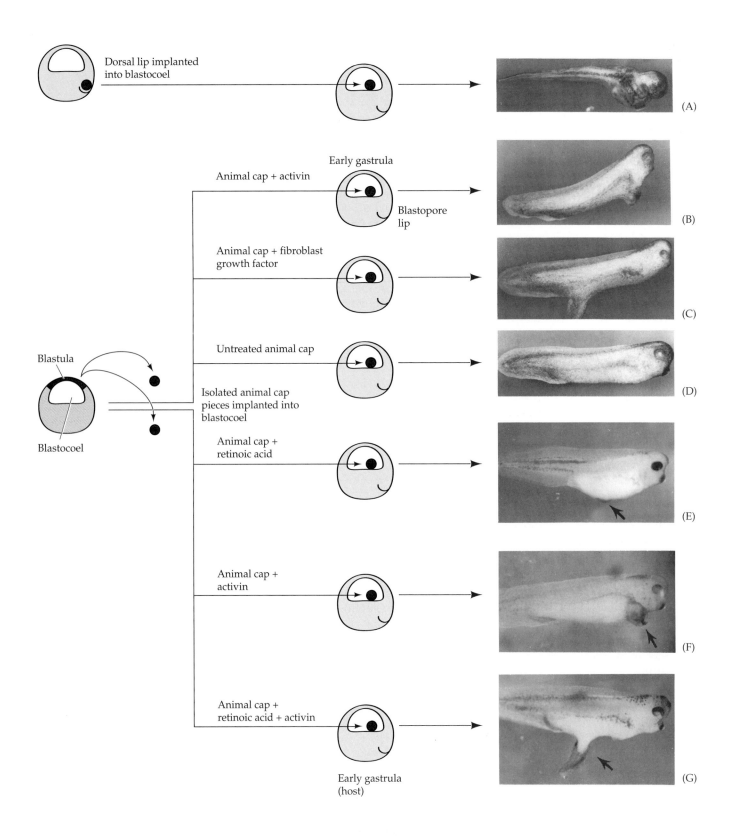

becomes competent to form the lens, nose, and ear placodes. This competence is acquired by its being acted upon by the neural plate region (Henry and Grainger, 1990). Thus, the head region of the neurula is now competent to respond to contact from the optic vesicle (derived from the forebrain) to become lens.

Moreover, once a tissue has been induced, it can induce other tissues.

FIGURE 16.40

Peptide growth factors induce mesoderm with different anterior–posterior specificity; retinoic acid can alter this specification. (A) Endogenous organizer tissue from untreated *Xenopus* gastrulae also induced head structures. (B–D) Animal caps were isolated from blastulae and treated in (B) activin, (C) bFGF-containing medium, or (D) a control saline solution. The caps were then implanted into the blastocoels of early gastrulae. A majority of the activin-treated caps induced heads, a majority of the bFGF-treated caps induced trunks and tails, and the control caps induced no extra structures. (E–G) In a separate series of experiments, animal caps treated with (E) retinoic acid also formed no separate structures, and (F) activin-treated animal cap cells formed mesoderm that could induce heads. (G) When the animal cap was treated with activin and RA, the animal cap became mesoderm that induced trunk structures. (A–D from Ruiz i Altaba and Melton, 1989; E–G from Cho et al., 1991b; photographs courtesy of D. A. Melton and E. M. De Robertis.)

This has been seen in the formation of the motor neurons of the spinal cord. Jessell and his colleagues (Placzek et al., 1990; Yamada et al., 1991) have shown that once the neural tube is induced, it can still respond to inductive signals from the notochord. Those neural tube cells adjacent to the notochord (the ventralmost midline cells) become the floor plate cells.* If notochord fragments are taken from one embryo and transplanted to the lateral sides of a host neural tube, the host neural tube will form another set of floor plate cells at the sides of the neural tube. If a piece of notochord is removed from an embryo, the neural tube adjacent to the deleted region has no floor plate cells (Figure 16.42). These floor plate cells, once induced, induce the formation of motor neurons on either side of them. (Later, as we saw in Chapter 8, they will direct the migration of the commisural axons to the ventral midline.)

We have seen, then, that the induction of the ectoderm to form neural cells also initiates a cascade wherein some of those induced cells begin to induce other cells. In one case, the induction of the lens is initiated by the anterior neural plate cells, and in the other case, the ventral floor plate cells induce the formation of motor neurons.

Although the inductions after "primary" embryonic induction have often been referred to as "secondary" inductions, there is no conceptual difference between them. We will return to secondary inductions in Chapter 18.

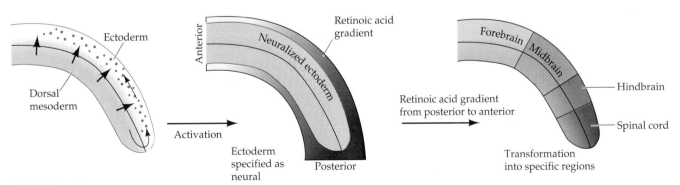

FIGURE 16.41

Recent two-step model based on those proposed by Sharpe (1991) and by Otte et al. (1991). An initial neuralizing set of signals is followed by a gradient of retinoic acid that causes the specification of the neural tube. The more RA, the more posterior the specification.

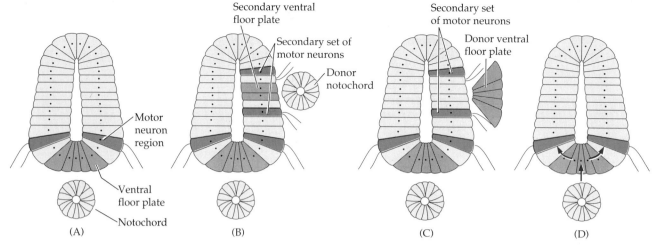

FIGURE 16.42

Cascade of inductions initiated by notochord on the newly formed neural tube. (A) Two cell types in the newly formed neural tube. Those cells closest to the notochord become the ventral floor plate cells. Motor neurons emerge on the ventrolateral sides. (B) If a second notochord is transplanted adjacent to the neural tube, it induces a new set of floor plate cells and two new sets of motor neurons. (C) If the ventral floor plate cells are transplanted adjacent to the neural tube, new sets of motor neurons differentiate. (D) Schematic diagram representing the inductive interactions between these cells. (After Placzek et al., 1990.)

Summary

In this and the preceding two chapters we have discussed the mechanisms of embryonic cell determination. We have seen that there are three general modes by which determination occurs. The first involves the partitioning of morphogenetic determinants into specific cells and the subsequent determination of those cells contingent upon which region of zygote cytoplasm they enclose. Each of these cells will differentiate autonomously, irrespective of its surrounding cells. Organisms in which this type of determination is prevalent tend toward a mosaic type of development. The second mode of determination involves interactions between parts of a syncytial embryo and is seen during early insect development.

The third mode of determination involves the interactions of cells somewhat later in development. Here, cells develop according to their positions within the embryo. Embryos whose cells are determined primarily in this interactive manner tend toward a regulative type of development. In this chapter, for instance, we considered a pathway in which endodermal cells induce the cells above them to become mesoderm. The dorsal mesoderm (notochord and dorsal blastopore lip) induce the neural determination and regional specification of the dorsalmost ectoderm cells. The notochord continues to induce the formation of the floor plate cells in the neural tube, and the floor plate cells induce the formation of motor neurons. By progressive cell–cell interactions, we have proceeded from the vegetal cells of the early frog blastula to the formation of specific types of neurons in the larval neural tube. Moreover, the pathways can diverge at each point. The notochord also induces cartilage formation in the somites, and the neural tube helps induce the formation of lenses.

It is important to realize that both autonomous and conditional mechanisms are used during the development of any particular organism and that there is a continuum from mosaic to regulative development. Mosaic animals such as tunicates and snails use regulative interactions to form certain tissues, and regulative animals such as frogs have regions of cytoplasm that contain morphogenic determinants such as polar granules. It is also important to note that both mosaic and regulative modes of determination are traceable to the heterogeneously distributed materials in the zygote cytoplasm.

LITERATURE CITED

Allen, G. E. 1985. Thomas Hunt Morgan: Materialism and reductionism in the development of modern genetics. *Trends Genet.* 3: 151–154; 186–190.

Asashima, M., Nikano, H., Shimada, K., Kinoshita, K., Ishii, K., Shibai, H. and Veno, N. 1990. Mesodermal induction in early amphibian embryos by activin A (erythroid differentiation factor). *Wilhelm Roux Arch. Dev. Biol.* 198: 330–335.

Ayama, E., Musci. T. J. and Kirschner, M. W. 1991. Expression of a dominant negative mutant of the FGF receptor disrupts mesoderm formation in *Xenopus* embryos. *Cell* 66: 257–270.

Berrill, N. J. and Liu, C. K. 1948. Germplasm, Weismann, and hydrozoa. *Q. Rev. Biol.* 23: 124–132.

Bijtel, J. H. 1931. Über die Entwicklung des Schwanzes bei Amphibien. *Wilhelm Roux Arch. Entwicklungsmech. Org.* 125: 448–486.

Blum, M., Gaunt, S. J., Cho, K. W. Y., Steinbeisser, H., Blumberg, B., Bittner, D. and De Robertis. 1992. Gastrulation in the mouse: The role of the homeobox gene *goosecoid*. *Cell* 69: 1097–1106.

Blumberg, B., Wright, C. V. E., De Robertis, E. M. and Cho, K. W. Y. 1991. Organizer-specific homeobox genes in *Xenopus laevis* embryos. *Nature* 253: 194–196.

Boterenbrood, E. C. and Nieuwkoop, P. D. 1973. The formation of the mesoderm in urodelean amphibians. V. Its regional induction by the endoderm. *Wilhelm Roux Arch. Entwicklungsmech. Org.* 173: 319–332.

Buss, L. 1987. *The Evolution of Individuality.* Princeton University Press, Princeton, NJ.

Chen, Y. P., Huang, L. and Solursh, M. 1994. A concentration gradient of retinoids in the early *Xenopus laevis* embryo. *Dev. Biol.* 161: 70–76.

Cho, K. W. Y., Blumberg, B., Steinbeisser, H. and De Robertis, E. 1991a. Molecular nature of Spemann's organizer: The role of the *Xenopus* homeobox gene *goosecoid*. *Cell* 67: 1111–1120.

Cho, K. W. Y., Morita, A. A., Wright, C. V. E. and De Robertis, E. M. 1991b. Overexpression of a homeodomain protein confers axis-forming activity to uncommitted *Xenopus* embryonic cells. *Cell* 65: 55–64.

Christian, J. L. and Moon, R. T. 1993. Interactions between *Xwnt-8* and Spemann organizer signalling pathways generate dorsoventral pattern in the embryonic mesoderm of *Xenopus*. *Genes Dev.* 7: 13–28.

Cooke, J. and Webber, J. A. 1985. Dynamics of the control of body pattern in the development of *Xenopus laevis*. *J. Embryol. Exp. Morphol.* 88: 85–99.

Culiffe, V. and Smith, J. C. 1992. Ectopic mesoderm formation in *Xenopus* embryos caused by widespread expression of a *Brachyury* homologue. *Nature* 358: 427–430.

Dale, L. and Slack, J. M. W. 1987. Regional specificity within the mesoderm of early embryos of *Xenopus laevis*. *Development* 100: 279–295.

Dale, L. Howes, G., Price, B. M. J. and Smith, J. C. 1992. Bone morphogenetic protein 4: a ventralizing factor in early *Xenopus* development. *Development* 115: 573–585.

Dale, L., Matthew, G. and Coleman, A. 1993. Secretion and mesoderm-inducing activity of the TGF-β-related domain of *Xenopus* Vg1. *EMBO J.* 12: 4471–4480.

Davids, M. 1988. Protein kinases in amphibian ectoderm induced for neural differentiation. *Wilhelm Roux Arch. Dev. Biol.* 197: 339–344.

Davids, M., Loppnow, B., Tiedemann, H. and Tiedemann, H. 1987. Neural differentiation of amphibian gastrula ectoderm exposed to phorbol ester. *Wilhelm Roux Arch. Dev. Biol.* 196: 137–140.

Davidson, E. H. 1989. Lineage-specific gene expression and the regulative capacities of the sea urchin embryo: a proposed mechanism. *Development* 105: 421–446.

De Robertis, E. M., Blum, M., Niehrs, C. and Steinbeisser, H. 1992. Goosecoid and the organizer. *Development 1992 Suppl.*: 167–171.

Doniach, T. 1993. Planar and vertical induction of anteroposterior pattern during the development of the amphibian central nervous system. *J. Neurobiol.* 24: 1256–1275.

Doniach, T., Phillips, C. R. and Gerhart, J. C. 1992. Planar induction of anteroposterior pattern in the developing central nervous system of *Xenopus laevis*. *Science* 257: 542–545.

Driesch, H. 1892. The potency of the first two cleavage cells in echinoderm development. Experimental production of partial and double formations. *In* B. H. Willier and J. M. Oppenheimer (eds.), *Foundations of Experimental Embryology*. Hafner, New York.

Driesch, H. 1893. Zur Verlagerung der Blastomeren des Echinideneies. *Anat. Anz.* 8: 348–357.

Driesch, H. 1894. *Analytische Theorie de organischen Entwicklung*. W. Engelmann, Leipzig.

Durston, A. and Otte, A. P. 1991. A hierarchy of signals mediates neural induction in *Xenopus laevis*. *In* J. Gerhart (ed.), *Cell–Cell Interactions in Early Development*. Wiley-Liss, New York, pp. 109–127.

Durston, A., Timmermans, A., Hage, W. J., Hendriks, H. F. J., de Vries, N. J., Heideveld, M. and Nieuwkoop, P. D. 1989. Retinoic acid causes an anteroposterior transformation in the developing central nervous system. *Nature* 330: 140–144.

Elinson, R. P. and Kao, K. R. 1989. The location of dorsal information in frog early development. *Dev. Growth Differ.* 31: 423–430.

Ettensohn, C. A. and McClay, D. R. 1988. Cell lineage conversion in the sea urchin embryo. *Dev. Biol.* 125: 396–409.

Gerhart, J. C., Danilchik, M., Doniach, T., Roberts, S., Browning, B. and Stewart, R. 1989. Cortical rotation of the *Xenopus* egg: Consequences for the anteroposterior pattern of embryonic dorsal development. *Development* [Suppl.] 107: 37–51.

Gilbert. S. F. In press. Looking at embryos: The visual and conceptual aesthetics of embryology. *In* A. I. Tauber (ed.), *Aesthetics and Science: The Elusive Synthesis*.

Gilbert, S. F. and Saxén, L. 1993. Spemann's Organizer: Models and molecules. *Mech. Dev.* 41: 73–89.

Gimlich, R. L. 1985. Cytoplasmic localization and chordamesoderm induction in the frog embryo. *J. Embryol. Exp. Morphol.* 89 89–111.

Gimlich, R. L. 1986. Acquisition of developmental autonomy in the equatorial region of the *Xenopus* embryo. *Dev. Biol.* 116: 340–352.

Gimlich, R. L. and Cook, J. 1983. Cell lineage and the induction of nervous systems in amphibian development. *Nature* 306: 471–473.

Gimlich, R. L. and Gerhart, J. C. 1984. Early cellular interactions promote embryonic axis formation in *Xenopus laevis*. *Dev. Biol.* 104: 117–130.

Green, J. B. A. and Smith, J. C. 1990. Graded changes in dose of a *Xenopus* activin A homologue elicit stepwise transitions in embryonic cell fate. *Nature* 347: 391–394.

Green, J. B. A., New, H. V. and Smith, J. C. 1992. Responses of embryonic Xenopus cells to activin and FGF are separated by multiple dose thresholds and correspond to distinct axes in the mesoderm. *Cell* 71: 731–739.

Gurdon, J. B., Fairman, S., Mohun, T. J. and Brennan, S. 1985. The activation of muscle-specific actin genes in *Xenopus* development by an induction between animal and vegetal cells of a blastula. *Cell* 41: 913–922.

Hamburger, V. 1984. Hilde Mangold, co-discoverer of the organizer. *J. Hist. Biol.* 17: 1–11.

Hamburger, V. 1988. *The Heritage of Experimental Embryology: Hans Spemann and the Organizer*. Oxford University Press, Oxford.

Haraway, D. J. 1976. *Crystals, Fabrics, and Fields: Metaphors of Organicism in Twentieth-Century Biology*. Yale University Press, New Haven.

Hemmati-Brivanlou, A. and Melton, D. A. 1992. A truncated activin receptor inhibits mesoderm induction and formation of axial structures in *Xenopus* embryos. *Nature* 359: 609–614.

Henry, J. J. and Grainger, R. M. 1990. Early tissue interactions leading to embryonic lens formation in *Xenopus laevis*. *Dev. Biol.* 141: 149–163.

Henry, J. J., Amemiya, S., Wray, G. A. and Raff, R. A. 1989. Early inductive interactions are involved in restricting cell fates of mesomeres in sea urchin embryos. *Dev. Biol.* 136: 140–153.

Hertwig, O. 1894. *Zeit- und Streitfragen der Biologie I. Präformation oder Epigenese? Grundzüge einer Entwicklungstheorie der Organismen.* Gustav Fischer, Jena.

Holtfreter, J. 1933. Die totale Exogastrulation, eine Selbstablösung des Ektoderms vom Entomesoderm. *Wilhelm Roux Arch. Entwicklungsmech. Org.* 129: 669–793.

Holtfreter, J. 1936. Regionale induktionen in xenoplastisch zusammengesetzten Explantaten. *Wilhelm Roux Arch. Entwicklungsmech. Org.* 134: 466–561.

Holtfreter, J. 1948. Concepts on the mechanism of embryonic induction and its relation to parthenogenesis and malignancy. *Symp. Soc. Exp. Biol.* 2: 17–48.

Holtfreter, J. and Hamburger, V. 1955. Amphibians. *In* B. H. Willier, P. A. Weiss and V. Hamburger (eds.), *Analysis of Development.* Saunders, Philadelphia, pp. 230–295.

Hörstadius, S. 1928. Über die Determination des Keimes bei Echinodermen. *Acta Zool.* 9: 1–191.

Hörstadius, S. 1935. Über die Determination im Verlaufe der Eiachse bei Seeigeln. *Publ. Staz. Zool. Napoli* 14: 251–479.

Hörstadius, S. 1939. The mechanics of sea urchin development studied by operative methods. *Biol. Rev.* 14: 132–179.

Hörstadius, S. and Wolsky, A. 1936. Studien Über die Determination der Bilateralsymmetrie des jungen Seeigelkeimes. *Wilhelm Roux Arch. Entwicklungsmech. Org.* 135: 69–113.

Hough-Evans, B. R., Franks, R. R., Cameron, R. A., Britten, R. J. and Davidson, E. H. 1987. Correct cell type-specific expression of a fusion gene injected into sea urchin eggs. *Dev. Biol.* 121: 576–579.

Huxley, J. S. and deBeer, G. R. 1934. *Elements of Experimental Embryology.* Cambridge University Press, Cambridge.

Isaacs, H. V., Tannahill, D. and Slack, J. M. W. 1992. Expression of a novel FGF in the *Xenopus* embryo. A new candidate inducing factor for mesoderm formation and anteroposterior specificiation. *Development* 114: 711–720.

Jacobson, M. 1984. Cell lineage analysis of neural induction: Origins of cells forming the induced nervous system. *Dev. Biol.* 102: 122–129.

Jones, C. M., Lyons, K. M., Lapan, P. M., Wright, C. V. E. and Hogan, B. L. M. 1992. DVR-4 (bone morphogenetic protein-4) as a posterior-ventralizing factor in *Xenopus* mesoderm induction. *Development* 115: 639–647.

Jones, E. A. and Woodland, H. R. 1987. The development of animal cap cells in *Xenopus*: a measure of the start of animal cap competence to form mesoderm. *Development* 101: 557–564.

Kageura, H. and Yamana, K. 1983. Pattern regulation in isolated halves and blastomeres of early *Xenopus laevis. J. Embryol. Exp. Morphol.* 74: 221–234.

Kageura, H. and Yamana, J. 1986. Pattern formation in 8–cell composite embryos of *Xenopus laevis. J. Embryol. Exp. Morphol.* 91: 79–100.

Keller, R., Shih, J., Sater, A. K. and Moreno, C. 1992. Planar induction of convergence and extension of the neural plate by the organizer of *Xenopus. Dev. Dynam.* 193: 218–234.

Khaner, O. and Wilt, F. 1990. The influence of cell interactions and tissue mass on differentiation of sea urchin mesomeres. *Development* 109: 625–634.

Khaner, O. and Wilt, F. 1991. Interactions of different vegetal cells with mesomeres during early vegetal stages of sea urchin development. *Development* 112: 881–890.

Kimelman, D. and Kirschner, M. 1987. Synergistic induction of mesoderm by FGF and TGF-β and the identification of an mRNA coding for FGF in the early *Xenopus* embryo. *Cell* 51: 869–877.

Lamb, T. M. and seven others. 1993. Neural induction by the secreted polypeptide noggin. *Science* 262: 713–718.

London, C., Akers, R. and Phillips, C. 1988. Expression of Epi 1, an epidermis-specific marker in *Xenopus laevis* embryos, is specified prior to gastrulation. *Dev. Biol.* 129: 380–389.

Løvtrup, S., Landström, U. and Løvtrup-Rein, H. 1978. Polarities, cell differentiation, and induction in the amphibian embryo. *Biol. Rev.* 53: 1–42.

Mangold, O. 1933. Über die Induktionsfahigkeit der verschiedenen Bezirke der Neurula von Urodelen. *Naturwissenschaften* 21: 761–766.

Marcus, N. H. 1979. Developmental aberrations associated with twinning in laboratory-reared sea urchins. *Dev. Biol.* 70: 274–277.

Maruyama, Y. K., Nakaseko, Y. and Yagi, S. 1985. Localization of the cytoplasmic determinants responsible for primary mesenchyme formation and gastrulation in the unfertilized eggs of the sea urchin *Hemicentrotus pulcherrimus. J. Exp. Zool.* 236: 155–163.

McClendon, J. F. 1910. The development of isolated blastomeres of the frog's egg. *Am. J. Anat.* 10: 425–430.

Mitrani, E., Ziv, P. T., Thomsen, G., Shimoni, Y., Melton, D. A. and Bril, A. 1990. Activin can induce the formation of axial structures and is expressed in the hypoblast of the chick. *Cell* 63: 495–501.

Moriya, N. and Asashima, M. 1992. Mesoderm and neural inductions on newt ectoderm by activin A. *Dev. Growth Differ.* 34: 589–594.

Nakamura, O. and Takasaki, H. 1970. Further studies on the differentiation capacity of the dorsal marginal zone in the morula of *Triturus pyrrhogaster. Proc. Japan Acad.* 46: 700–705.

Niehrs, C., Keller, R., Cho, K. W. Y. and De Robertis, E. M. 1993. The homeobox gene *goosecoid* controls cell migration in *Xenopus* embryos. *Cell* 72: 491–503.

Nieuwkoop, P. D. 1952. Activation and organization of the central nervous system in amphibians.III. Synthesis of a new working hypothesis. *J. Exp. Zool.* 120: 83–108.

Nieuwkoop, P. D. 1969. The formation of the mesoderm in urodele amphibians. I. Induction by the endoderm. *Wilhelm Roux Arch. Entwicklungsmech. Org.* 162: 341–373.

Nieuwkoop, P. D. 1973. The "organisation center" of the amphibian embryo: Its origin, spatial organisation and morphogenetic action. *Adv. Morphogen.* 10: 1–39.

Nieuwkoop, P. D. 1977. Origin and establishment of embryonic polar axes in amphibian development. *Curr. Top. Dev. Biol.* 11: 115–132.

Otte, A. P. and Moon, R. T. 1992. Protein kinase C isozymes have distinct roles in neural induction and competence in *Xenopus. Cell* 68: 1021–1029.

Otte, A. P., Koster, C. H., Snoek, G. T. and Durston, A. J. 1988. Protein kinase C mediates neural induction in *Xenopus laevis. Nature* 334: 818–620.

Otte, A. P., van Run, P., Heideveld, M., van Driel, R. and Durston, A. J. 1989. Neural induction is mediated by cross-talk between the protein kinase C and cyclic AMP pathways. *Cell* 58: 641–648.

Otte, A. P., Kramer, I. J. M. and Durston, A. J. 1991. Protein kinase C and regulation of the local competence of *Xenopus* ectoderm. *Science* 251: 570–573.

Papalopulu, N., Clarke, J. D. W., Bradley, L., Wilkinson, D., Krumlauf, R. and Holder, N. 1991. Retinoic acid causes abnormal development and segmental patterning of the anterior hindbrain in *Xenopus* embryos. *Development* 113: 1145–1158.

Placzek, M., Tessier-Lavigne, M., Yamada, T., Jessell, T. and Dodd, J. 1990. Mesodermal control of neural cell identity: Floor plate induction by the notochord. *Science* 250: 985–988.

Ransick, A. and Davidson, E. H. 1993. A complete second gut induced by transplanted micromeres in the sea urchin embryo. *Science* 259: 1134–1138.

Recanzone, G. and Harris, W. A. 1985. Demonstration of neural induction using nuclear markers in *Xenopus. Wilhelm Roux Arch. Dev. Biol.* 194: 344–354.

Render, J. and Elinson, R. P. 1986. Axis determination in polyspermic *Xenopus* eggs. *Dev. Biol.* 115: 425–433.

Roll-Hansen, N. 1978. *Drosophila* genetics: A reductionist research program. *J. Hist. Biol.* 11: 159–210.

Roux, W. 1888. Contributions to the developmental mechanics of the embryo. On the artificial production of half-embryos by destruction of one of the first two blastomeres and the later development (postgeneration) of the missing half of the body. *In* B. H. Willier and J. M. Oppenheimer (eds.), 1974, *Foundations of Experimental Embryology.* Hafner, New York, pp. 2–37.

Ruiz i Altaba, A. 1992. Planar and vertical signals in the induction and patterning of the *Xenopus* nervous system. *Development* 115: 67–80.

Ruiz i Altaba, A. and Jessell, T. 1991. Retinoic acid modifies mesodermal patterning in early Xenopus embryos. *Genes and Devel.* 5: 175–187.

Ruiz i Altaba, A. and Melton, D. A. 1989. Interaction between peptide growth factors and homeobox genes in the establishment of anteroposterior polarity in frog embryos. *Nature* 341: 33–38.

Saha, M. and Grainger, R. M. 1992. A labile period in the determination of the anterior–posterior axis during early neural development in *Xenopus*. *Neuron* 8: 1003–1014.

Sander, K. 1991a. Wilhelm Roux and his programme for developmental biology. *Wilhelm Roux Arch. Dev. Biol.* 200: 1–3.

Sander, K. 1991. "Mosaic work" *and* "assimilating effects" in embryogenesis: Wilhelm Roux's conclusions after disabling frog blastomeres. *Wilhelm Roux Arch. Dev. Biol.* 200: 237–239.

Sargent, T. D., Jamrich, M. and Dawid, I. 1986. Cell interactions and the control of gene activity during early development of *Xenopus laevis*. *Dev. Biol.* 114: 238–246.

Savage, R. and Phillips, C. 1989. Signals from the dorsal blastopore lip region during gastrulation bias the ectoderm toward a nonepidermal pathway of differentiation in *Xenopus laevis*. *Dev. Biol.* 133: 157–168.

Saxén, L. 1961. Transfilter neural induction of amphibian ectoderm. *Dev. Biol.* 3: 140–152.

Saxén, L. and Toivonen, S. 1962. *Embryonic Induction.* Prentice-Hall, Englewood Cliffs, NJ.

Saxén, L. Toivonen, S. and Vainio, T. 1964. Initial stimulus and subsequent interactions in embryonic induction. *J. Embryol. Exp. Morphol.* 12: 333–338.

Sharpe, C. R. 1991. Retinoic acid can mimic endogenous signals involved in transformation of the *Xenopus* nervous system. *Neuron* 7: 239–247.

Sharpe, C. R., Fritz, A., De Robertis, E. M. and Gurdon, J. B. 1987. A homeobox-containing marker of posterior neural differentiation shows the importance of predetermination in neural induction. *Cell* 50: 749–758.

Shih, J. and Keller, R. 1992. The epithelium of the dorsal marginal zone of *Xenopus* has organizer properties. *Development* 116: 887–899.

Sive, H. 1993. The frog prince-ss: A molecular formula for dorsoventral patterning in *Xenopus*. *Genes Dev.* 7: 1–12.

Sive, H. L. and Cheng, P. F. 1991. Retinoic acid perturbs the expression of *Xhox.lab* genes and alters mesodermal determination in *Xenopus laevis*. *Genes Dev.* 5: 1321–1332.

Sive, H., Hattori, K. and Weintraub, H. 1989. Progressive determination during formation of the anteroposterior axis in *Xenopus laevis*. *Cell* 58: 171–180.

Sive, H. L., Draper, B. W., Harland, R. M. and Weintraub, H. 1990. Identification of a retinoic acid sensitive period during primary axis formation in *Xenopus laevis*. *Genes and Devel.* 4: 932–942.

Slack, J. M. W. 1991. The nature of the mesoderm-inducing signal in *Xenopus*: A transfilter induction study. *Development* 113: 661–669.

Slack, J. M. W. and Tannahill, D. 1992. Mechanism of anteroposterior axis specification in vertebrates: Lessons from the amphibians. *Development* 114: 285–302.

Slack, J. M. W., Darlington, B. G., Heath, J. K. and Godsave, S. F. 1987. Mesoderm induction in early *Xenopus* embryos by heparin-binding growth factor. *Nature* 326: 197–200.

Smith, J. and Schoenwolf, G. 1989. Notochordal induction of cell wedging in the chick neural plate and its role in neural tube formation. *J. Exp. Zool.* 250: 49–62.

Smith, J. C. and Slack, J. M. W. 1983. Dorsalization and neural induction: Properties of the organizer in *Xenopus laevis*. *J. Embryol. Exp. Morphol.* 78: 299–317.

Smith, J. C., Dale, L. and Slack, J. M. W. 1985. Cell lineage labels and region-specific markers in the analysis of inductive interactions. *J. Embryol. Exp. Morphol.* [Suppl.] 89: 317–331.

Smith, J. C., Price, B. M. J., van Nimmen, K. and Huylebroeck, D. 1990. Identification of a potent *Xenopus* mesoderm-inducing factor as a homologue of activin A. *Nature* 345: 729–731.

Smith, J. C., Price, B. M. J., Green, J. B. A., Weigel, D. and Herrmann, B. G. 1991. Expression of a *Xenopus* homolog of *Brachyury* (*T*) is an immediate–early response to mesoderm induction. *Cell* 67: 79–87.

Smith, W. C. and Harland, R. M. 1991. Injected *wnt-8* RNA acts early in *Xenopus* embryos to promote formation of a vegetal dorsalizing center. *Cell* 67: 753–765.

Smith, W. C. and Harland, R. M. 1992. Expression cloning of *noggin*, a new dorsalizing factor localized to the Spemann organizer in *Xenopus* embryos. *Cell* 70: 829–840.

Smith, W. C. Knecht, A. K., Wu, M. and Harland, R. M. 1993. Secreted noggin mimics the Spemann organizer in dorsalizing *Xenopus* mesoderm. *Nature* 361: 547–549.

Sokol, S., Wong, G. and Melton, D. A. 1990. A mouse macrophage factor induces head structures and organizes a body axis in *Xenopus*. *Science* 249: 561–564.

Spemann, H. 1918. Über die Determination der ersten Organanlagen des Amphibienembryo. *Wilhelm Roux Arch. Entwicklungsmech. Org.* 43: 448–555.

Spemann, H. 1927. Neue Arbieten über Organisatoren in der tierischen Entwicklung. *Naturwissenschaften* 15: 946–951.

Spemann, H. 1938. *Embryonic Development and Induction.* Yale University Press, New Haven.

Spemann, H. and Mangold, H. 1924. Induction of embryonic primordia by implantation of organizers from a different species. *In* B. H. Willier and J. M. Oppenheimer (eds.), *Foundations of Experimental Embryology.* Hafner, New York, pp. 144–184.

Spofford, W. R. 1945. Observations on the posterior part of the neural plate in *Amblystoma*. *J. Exp. Zool.* 99: 35–52.

Storey, K., Crossley, J. M., De Robertis, E., Norris, W. E. and Stern, C. D. 1992. Neural induction and regionalisation in the chick embryo. *Development* 114: 729–741.

Sucov, H. M., Hough-Evans, B. R., Franks, R. R., Britten, R. J. and Davidson, E. H. 1988. A regulatory domain that directs lineage-specific expression of a skeletal matrix protein in the sea urchin embryo. *Genes Dev.* 2: 1238–1250.

Takasaki, H. and Konishi, H. 1989. Dorsal blastomeres in the equatorial region of the 32-cell *Xenopus* embryo autonomously produce progeny committed to the organizer. *Dev. Growth Differ.* 31: 147–156.

Tauber, A. I. and Sarkar, S. 1992. The human genome project: Has blind reductionism gone too far? *Perspec. Biol. Med.* 35: 220–235.

Thomsen, G. H. and Melton, D. A, 1993. Processed Vg1 protein is an axial mesoderm inducer in *Xenopus*. *Cell* 74: 433–441.

Thomsen, G., Woolf, T., Whitman, M., Sokol, S., Vaughan, J., Vale, W. and Melton, D. A. 1990. Activins are expressed early in *Xenopus* embryogenesis and can induce axial and mesoderm anterior structures. *Cell* 63: 485–493.

Toivonen, S. and Saxén, L. 1955. The simultaneous inducing action of liver and bone marrow of the guinea pig in implantation and explantation experiments with embryos of *Triturus*. *Exp. Cell Res.* [Suppl.] 3: 346–357.

Toivonen, S. and Saxén, L. 1968. Morphogenetic interaction of presumptive neural and mesodermal cells mixed in different ratios. *Science* 159: 539–540.

Toivonen, S. and Wartiovaara, J. 1976. Mechanism of cell interaction during primary induction studied in transfilter experiments. *Differentiation* 5: 61–66.

Toivonen, S., Tarin, D., Saxén, L., Tarin, P. J. and Wartiovaara, J. 1975. Transfilter studies on neural induction in the newt. *Differentiation* 4: 1–7.

Twitty, V. C. 1966. *Of Scientists and Salamanders.* Freeman, San Francisco. [p. 39]

van Straaten, H. W. M., Hekking, J. W. M., Beursgens, J. P. W. M., Terwindt-Rouwenhorst, E., and Drukker, J. 1989. Effect of the notochord on proliferation and differentiation in the neural tube of the chick embryo. *Development* 107: 793–803.

Vincent, J.-P. and Gerhart, J. C. 1987. Subcortical rotation in *Xenopus* eggs: An early step in embryonic axis specification. *Dev. Biol.* 123: 526–539.

Waddington, C. H. 1933. Induction by the primitive streak and its derivatives in the chick. *J. Exp. Biol.* 10: 38–46.

Waddington, C. H. 1940. *Organisers and Genes.* Cambridge University Press. Cambridge.

Wakahara, M. 1989. Specification and establishment of dorsal–ventral polarity in eggs and embryos of *Xenopus laevis*. *Dev. Growth Diff.* 31: 197–207.

Weismann, A. 1892. *Essays on Heredity and Kindred Biological Problems*. Translated by E. B. Poulton, S. Schoenland and A. E. Shipley. Clarendon, Oxford.

Weismann, A. 1893. *The Germ-Plasm: A Theory of Heredity*. Translated by W. Newton Parker and H. Ronnfeld. Walter Scott Ltd., London.

Wilkins, A. S. 1993. *Genetic Analysis of Animal Development*, 2nd ed. Wiley-Liss, New York.

Wilson, E. B. 1896. *The Cell in Development and Inheritance*. Macmillan, New York.

Yamada, T., Placzek, M., Tanaka, H., Dodd, J. and Jessell, T. M. 1991. Control of cell pattern in the developing nervous system: Polarizing activity of floor plate and notochord. *Cell* 64: 635–647.

Yamana, K. and Kageura, H. 1987. Reexamination of the "regulative development" of amphibian embryos. *Cell Differ.* 20: 3–10.

17

Constructing the mammalian embryo

Establishment of body axes and the mechanisms of teratogenesis

Between the fifth and tenth days the lump of stem cells differentiates into the overall building plan of the [mouse] embryo and its organs. It is a bit like a lump of iron turning into the space shuttle. In fact it is the profoundest wonder we can still imagine and accept, and at the same time so usual that we have to force ourselves to wonder about the wondrousness of this wonder.
MIROSLAV HOLUB (1990)

It is known that nature works constantly with the same materials. She is ingenious to vary only the forms. As if, in fact, she were restricted to the same primitive ideas, one sees her tend always to cause the same elements to reappear, in the same number, in the same circumstances, and with the same connections.
E. GEOFFROY SAINT-HILAIRE (1807)

This chapter concerns two related aspects of mammalian development. The first section details some of the research that has provided insights into how the body axes of mammals become established. Much of this work utilizes data provided by the analysis of embryos whose development has been malformed through mutations or disrupted by particular chemicals. The second section of the chapter looks at some of the ways that certain chemicals cause these disruptions of mammalian development.

THE ESTABLISHMENT OF THE MAMMALIAN BODY AXES

Despite the attempts of the quoted *savant* Geoffroy Saint-Hilaire (who believed that *all* the animals of the world shared a common body plan), most developmental biologists would not have predicted that the body plans of flies and mammals would be specified through the same set of genes. Having diverged over 500 million years ago, the fly's body and the vertebrate body seem exceptionally different. Most insects specify their axes in the common cytoplasm of the syncytial blastoderm, while vertebrate axes are specified by inductive interactions between groups of cells. In the *Drosophila* embryo, the general body plan is specified while the cells are a cylindrical monolayer surrounding the yolk; while in mammals, cells have undergone extensive movements by the time their body parts are specified. The formation of appendages in insects results from the extension of ectodermal imaginal discs, while the mammalian limb is generated by complex inductive interactions between the ectoderm and mesodermal cells that have migrated to these areas. However, recent studies have shown that the anterior–posterior axis of developing mammals is specified by the same homeotic genes that specify the *Drosophila* body axis. Indeed,

623

the homeobox has been called the Rosetta Stone of developmental biology (Riddihough, 1992; Slack and Tannahill, 1992), as it enables us to transfer our genetic knowledge of *Drosophila* embryos into the less well known realm of mammalian development.

Mammalian axis formation: The *Hox* code hypothesis

Initiating the anterior–posterior and left–right axes

The *goosecoid* gene of vertebrates resembles *Drosophila's* bicoid gene in several respects. First, both encode proteins with homeodomains; and second, these homeodomains bind to the same DNA sequences (Blumberg et al., 1991). It is possible that just as *bicoid* expression initiates anteroposterior axis formation in *Drosophila*, *goosecoid* expression may initiate anteroposterior axis formation in mammals.

In chick embryos, *goosecoid* mRNA is first seen in the cells in Koller's sickle, when the cells that form the primitive streak collect there. It is then detected in Hensen's node, as the streak moves forward to create the anteroposterior axis. When the node regresses, *goosecoid*-expressing cells remain in the head notochord and the pharyngeal endoderm (prechordal plate); as the head region forms, the only *goosecoid* expression occurs in the most anterior cells (Izpisúa-Belmonte et al., 1993). Thus, *goosecoid* expression appears in those cells that aggregate in the posterior and migrate anteriorly to form the anterior–posterior axis. Since the *goosecoid* gene is expressed in Hensen's node during mouse gastrulation (Blum et al., 1992), it is thought that the same sequence of events may define the mammalian embryonic axis.

While goosecoid appears to specify the cells of the anterior dorsal mesoderm, the middle and posterior dorsal axis appears to be specified by the *Brachyury* (*T*) protein (MacMurray and Shin, 1988; Yanagisawa, 1990; Stott et al., 1993). Notochord formation and differentiation require the *T* gene (Glucksohn-Schoenheimer, 1938; Herrmann, 1991; Rashbass et al., 1991), and mutations of the *Brachyury* gene cause posterior axial malformations.

Homology of the homeotic gene complexes between *Drosophila* and mammals

The *Drosophila* homeotic gene complex (HOM-C) on chromosome 3 contains the Antennapedia and Bithorax classes of homeotic genes and can be seen as a single functional unit. The genes are arranged in the same general order as their expression pattern along the anterior–posterior axis, the most 3' gene (*labial*) being required for producing the most anterior structures, the most 5' gene (*Abd-B*) specifying the development of the posterior abdomen. Mouse and human genomes contain four copies of HOM-C per haploid set (*Hox A* through *D* in the mouse, *HOX A* through *D* in humans; Boncinelli et al., 1988; McGinnis and Krumlauf, 1992; Scott, 1992). Not only are the same general *types* of homeotic genes found in both flies and mammals, but the *order* of these genes on the respective chromosomes is remarkably similar. And if this similarity isn't enough to argue for a common scheme of axis formation, it was soon discovered that the *expression pattern* of these genes follows the same pattern: those mammalian genes homologous to the *Drosophila labial*, *proboscipedia*, and *Deformed* genes are expressed anteriorly, while those genes homologous to

(A)

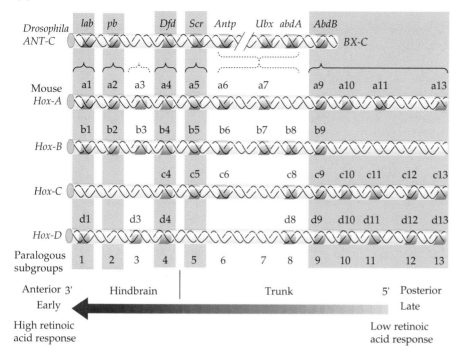

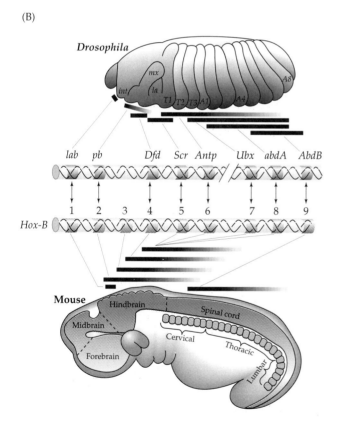

(B)

FIGURE 17.1
Evolutionary conservation of homeotic gene organization and transcriptional expression in flies and mice. (A) Conservation between the homeobox cluster on *Drosophila* chromosome 3 and the four *Hox* gene clusters in the mouse genome. The shaded regions show particularly strong structural similarities between species, and one can see that the order on the chromosomes has been conserved. Those genes at the 5' end (since all mouse homeobox genes are transcribed in the same direction) are those that are expressed more posteriorly, are expressed later, and can be induced only by high doses of retinoic acid. Genes having similar structures, the same relative positions on each of the four chromosomes, and similar expression patterns belong to the same paralogous group. (B) Comparison of the transcription patterns of the *HOM-C* and *Hox-B* genes of *Drosophila* (10 hours) and mice (12 days), respectively. Another set of genes that controls the formation of the fly head (*orthodonticle* and *empty spiracles*) have homologues in the mouse that show expression in midbrain and forebrain. (A after Krumlauf et al., 1993; B after McGinnis and Krumlauf, 1992.)

the *Drosophila Abdominal-B* gene are expressed posteriorly. The mammalian *Hox/HOX* genes are numbered from 1 to 13, starting from that end of the complex being expressed most anteriorly. Figure 17.1 shows the relationships between the *Drosophila* and mouse homeotic gene sets. The equivalent genes in each mouse complex (such as *Hoxa-1, Hoxb-1, Hoxc-1*, and

Hoxd-1) are called a **paralogous group**. It is thought that the four mammalian *Hox* complexes were formed from chromosome duplications. Since there is not a one-to-one correspondence between the *Drosophila* HOM-C genes and the mammalian *Hox* genes, it is probable that independent gene duplications have occurred since these two animal branches diverged (Hunt and Krumlauf, 1992).

Expression of *Hox* genes in the vertebrate nervous system and its derivatives

The branchial region of the head generates the hindbrain and its cranial ganglia; the cartilage of the ears, jaws, and neck; the aortic arches; and organs such as the thymus, thyroid, and parathyroid glands. As discussed in Chapter 7, the hindbrain neural tube becomes divided into segmental units called **rhombomeres** before and during the "birth" of the first neurons from this region. The migration of the cranial neural crest cells also appears to be organized on the rhombomeric pattern such that a specific cranial ganglion and the branchial arch it innervates originate from the crest above the same rhombomere (Lumsden et al., 1991). These neural crest cells appear to retain positional information from their original place on the anterior–posterior axis. When avian premigratory neural crest cells that would normally migrate to the *first* branchial arch (to form jaw cartilage) are placed into the region of the crest whose cells normally migrate to the *second* branchial arch (to form the hyoid cartilage), the grafted neural crest cells migrate into the *second* branchial arch, but they form the structures (jaw cartilage) characteristic of the *first* arch. Moreover, they will interact with the surface ectoderm and paraxial mesoderm to form the *first* arch structures (beak and jaw muscles). This strongly suggests that before they migrate, the neural crest cells are already committed to form structures appropriate to their level on the anterior–posterior axis (Noden, 1988).

This positional commitment may be the result of these cells' expressing particular combinations of *Hox* genes. For instance, the *Hox-B* genes are expressed in the presumptive mouse forebrain before neural crest formation, and when the neural crest cells migrate, they retain the *Hox-B* gene expression pattern characteristic of their place of origin (Hunt et al., 1991a). With only one known exception (*Hoxb-1*), the anterior boundary of each *Hox* gene stops at the rhombomere border two rhombomeres away from the anterior border of the next *Hox* gene (Wilkinson et al., 1989; Keynes and Lumsden, 1990). As is represented in Figure 17.2, the homeobox genes *Hoxb-2, 3,* and *4* are found throughout the spinal cord, but *Hoxb-2* stops at the border of rhombomeres 2 and 3; *Hoxb-3* stops at the 4/5 border, and *Hoxb-4* stops at the border between the sixth and seventh rhombomere. When the neural crest cells contact the surface ectoderm, they cause the ectodermal cells to express the same set of *Hox* genes (Figure 17.2A; Hunt et al., 1991b).

Some of the mammalian *Hox* genes are so similar to their *Drosophila* homologues that they can substitute for one another. The mouse *Hoxb-6* gene can perform some of the regulatory functions of the *Drosophila Antennapedia* gene when the murine gene is transfected into *Drosophila*. The human *HOXD-4* gene can also perform some of the regulatory functions of its *Drosophila* homologue, *Deformed*. (Malicki et al., 1990; McGinnis et al., 1990). Moreover, the enhancer region of the *Drosophila Deformed* gene (a gene specifying head-specific gene expression in *Drosophila*) can cause gene expression in the mouse hindbrain; and the regulatory sequences of human homologue of *Deformed* provide head-specific gene expression in *Drosophila* embryos (Awgulewitsch and Jacobs, 1992; Malicki et al., 1992).

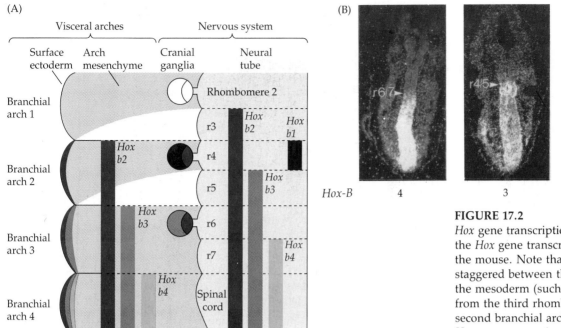

(A)

Visceral arches | Nervous system

Surface ectoderm | Arch mesenchyme | Cranial ganglia | Neural tube

Branchial arch 1 — Rhombomere 2

Hox b2 | *Hox b1* | r3

Branchial arch 2 — *Hox b2* — r4

Hox b3 | r5

Branchial arch 3 — *Hox b3* — r6

Hox b4 | r7

Branchial arch 4 — *Hox b4* | Spinal cord

Hox-B 4 3 2

FIGURE 17.2

Hox gene transcription. (A) Diagram of the *Hox* gene transcription pattern in the mouse. Note that the pattern is staggered between the neural tube and the mesoderm (such that crest cells from the third rhombomere enter the second branchial arch) and that the *Hox* gene expression borders coincide with rhombomere boundaries. (B) Transcription patterns of *Hox-B* homeotic genes in the 9.5-day mouse hindbrain. (A from McGinnis and Krumlauf, 1992; B from Hunt et al., 1991.)

A similar pattern of *Hox* gene expression seems to exist within the trunk as well. Here the *Hox* gene expression patterns correspond to somite (rather than rhombomere) boundaries (Kessel and Gruss, 1991), and some paralogous genes are expressed at slightly different somite boundaries (Figure 17.3).

Experimental analysis of a *Hox* code: Gene targeting

The expression patterns of the murine *Hox* genes suggest a combinatorial code whereby certain combination of *Hox* genes specify a particular region of the anterior–posterior axis (Hunt and Krumlauf, 1991). Particular sets of paralogous genes would provide segmental identity along the anterior–posterior axis of the body.* Evidence for such a code comes from gene targeting or "knockout" experiments (see Chapter 2) wherein mice are constructed that lack a normal pair of one of the *Hox* genes, and from **retinoic acid-induced homeosis** wherein mouse embryos given retinoic acid have a different pattern of homeotic gene expression along the anterior–posterior axis and abnormal differentiation of their axial structures.

When Chisaka and Capecchi (1991) knocked out the *Hoxa-3* gene from inbred mice, the homozygous mutant *Hoxa-3* mice died soon after they were born. Autopsies of these mice revealed that their neck cartilage was abnormally short and thick, and that they had severely deficient or absent thymuses, thyroids, and parathyroid glands. Their major heart and blood vessels were also malformed (Figure 17.4). This set of malformations is extremely similar to the human congenital disorder DiGeorge syndrome, in which these same deficiencies of neural crest-derived structures are found. It appears the *Hoxa-3* genes are responsible for specifying those

*This code is similar to the one that Lewis had proposed for *Drosophila* (Chapter 15), except that where the *Drosophila* code posits that the expression of the *Antennapedia* gene is located in segments different from those of the *Ultrabithorax* gene, and so forth, in vertebrates, the gene expression patterns are nested such that the expression of *Hoxb-2* is similar to that of *Hoxb-3* except for its anterior border.

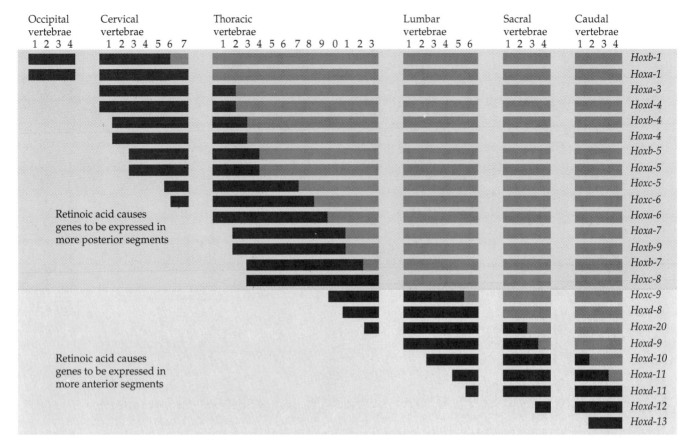

Occipital vertebrae 1 2 3 4	Cervical vertebrae 1 2 3 4 5 6 7	Thoracic vertebrae 1 2 3 4 5 6 7 8 9 0 1 2 3	Lumbar vertebrae 1 2 3 4 5 6	Sacral vertebrae 1 2 3 4	Caudal vertebrae 1 2 3 4	
						Hoxb-1
						Hoxa-1
						Hoxa-3
						Hoxd-4
						Hoxb-4
						Hoxa-4
						Hoxb-5
						Hoxa-5
						Hoxc-5
						Hoxc-6
						Hoxa-6
						Hoxa-7
						Hoxb-9
						Hoxb-7
						Hoxc-8
						Hoxc-9
						Hoxd-8
						Hoxa-20
						Hoxd-9
						Hoxd-10
						Hoxa-11
						Hoxd-11
						Hoxd-12
						Hoxd-13

Retinoic acid causes genes to be expressed in more posterior segments

Retinoic acid causes genes to be expressed in more anterior segments

FIGURE 17.3
The somite *Hox* code in the trunk and neck of the mouse embryo. Areas of major expression are indicated in darker color, while the posterior regions of expression are not as defined as might be suggested by the lighter color. The effect of retinoic acid is to push anterior gene expression more posteriorly and posterior gene expression more anteriorly. (After Kessel, 1992.)

cranial neural crest cells that give rise to the neck cartilage and fourth and sixth pharyngeal arch derivatives.

Another gene targeting experiment knocked out the *Hoxa-1* gene (Lufkin et al., 1991). The expression of *Hoxa-1* overlaps with the *Hoxa-3* gene, but it is also expressed more anteriorly than *Hoxa-3*. Those embryos without functional *Hoxa-1* genes show a constellation of abnormalities that

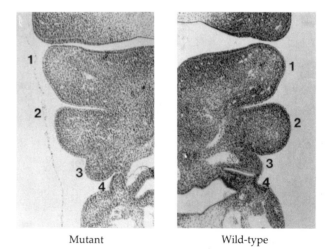

Mutant   Wild-type

FIGURE 17.4
Deficient development of neural crest-derived pharyngeal arch structures in *Hoxa-3*-deficient mice. Right, a 10.5-day embryo of heterozygous *Hoxa-3* mouse showing normal development of thymus (pouch 3), parathyroid (pouch 4), and other structures. Left, a homozygous mutant *Hoxa-3*-deficient mouse lacks the proper development of those structures. (From Chisaka and Capecchi, 1991.)

indicate deficient specification of rhombomeres 4 through 7. These mutants often fail to close their neural tube, lack inner ear structures, and are missing the hindbrain ganglia (which form the acoustic, glossopharyngeal, and vagus nerves) that are derived from these rhombomeres. However, no malformations of the pharyngeal arches, thymus, thyroid, parathyroid gland, or neck cartilage have been found. Thus, the the defects of *Hoxa-1* mutants are only seen in the anterior region of the *Hoxa-1* expression area. (It is possible that its functions are not needed or are redundant in the posterior portion of its range.) Unlike the defects of the *Hoxa-3*-deficient mice (which are confined to the neural crest), the *Hoxa-1* defects are seen in the central nervous system and placode-derived tissue, as well as in the paraxial mesoderm. Knockout of the *Hoxa-2* gene produces mice whose neural crest cells have been respecified. Cranial elements normally formed by the neural crest cells of the second branchial arch (stapes, styloid bones) are missing and are replaced by duplication of the structures of the first branchial arch (incus, malleus, etc.) (Gendron-Maguire et al., 1993; Rijli et al., 1993). Thus, without certain *Hox* genes, some regionally specific organs along the anterior–posterior axis fail to form or become respecified as other regions. The initial evidence supports the notion that (1) different sets of *Hox* genes are necessary for the complete specification of any region of the axis, and (2) a set of paralogous genes may be responsible for different subsets of organs within these regions.

Partial transformations of segments by knockout of *Hox* genes expressed in the trunk

If the *Hox* genes do indeed form a code that specifies the anterior–posterior axis, one would expect that changing the constellation of *Hox* genes expressed in any particular region of the embryo would change the structures formed in that region. This has been shown to be the case when the *Hoxc-8* gene is deleted from the embryo by gene targeting (Le Mouellic et al., 1992). In such mice, several axial skeletal segments resemble more anterior segments, much like what is seen in *Drosophila* loss-of-function homeotic mutations. As can be seen in Figure 17.5, the first lumbar vertebra of this mouse has formed a rib—something characteristic of the thoracic vertebrae anterior to it. The knockout of the *Hoxb-4* gene partially converts the

FIGURE 17.5
Homeotic transformations in the mouse induced by gene knockouts of homeobox genes expressed in the trunk. (A) Partial transformation of the first lumbar vertebra into a thoracic vertebra by the knockout of the *Hoxc-8* gene. Thoracic vertebrae, but not lumbar vertebrae, have ribs associated with them. (B,C) Partial transformation of the second cervical vertebra into a second copy of the first cervical vertebra by the knockout of the *Hoxb-4* gene. (B) The wild-type mouse has a first cervical vertebra characterized by a ventral tubercle. (C) In the mutant mouse, the second cervical vertebra also has this tubercle (arrow). (A from Le Mouellic et al., 1992; B from Ramirez-Solis et al., 1993; photographs courtesy of the authors.)

(A)

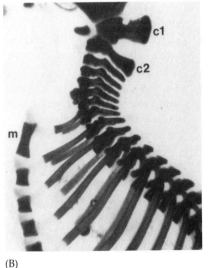

(B)

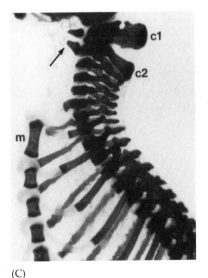

(C)

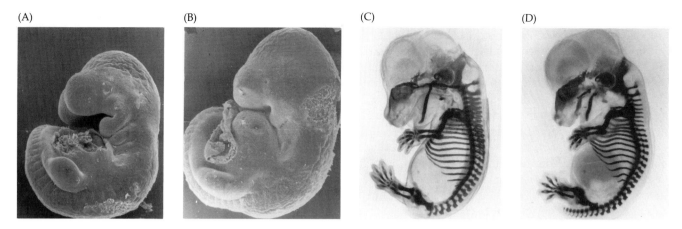

FIGURE 17.6

Mouse embryo cultured at day 8 in control medium (A and C) or in medium containing teratogenic retinoids (B and D). In two days (A,B), the first pharyngeal arch of the treated embryos has a shortened and flattened appearance and has apparently fused with the second pharyngeal arch. At day 17 (C,D), craniofacial malformations can be seen in the neural crest-derived cartilage of the treated embryos. Meckel's cartilage is completely displaced from the mandibular (lower jaw) to the maxillary (upper mouth) region The malleus and incus cartilages are also not formed. (A and B from Goulding and Pratt, 1986; C and D from Morriss-Kay, 1993; photographs courtesy of the authors.)

second cervical vertebra (the axis vertebra) into a copy of the first cervical vertebra (the atlas) and deletion of the *Hoxa-5* gene causes the posterior transformation of the seventh cervical (neck) vertebra into a rib-forming thoracic vertebra (Jeannotte et al., 1993; Ramirez-Solis et al., 1993).

Experimental analysis of the *Hox* code: Retinoic acid teratogenesis

Such homeotic changes are also seen when mouse embryos are given teratogenic doses of retinoic acid. Exogenous retinoic acid (RA) given to mouse embryos in utero can cause certain *Hox* genes to become expressed in groups of cells that usually do not express these genes (Conlon and Rossant, 1992; Kessel, 1992). Moreover, the craniofacial abnormalities of mouse embryos from mothers given teratogenic doses of RA (Figure 17.6) can be mimicked by causing the expression of *Hoxa-7* throughout the embryo (Balling et al., 1989). If high doses of RA can activate *Hox* genes in inappropriate cells along the anterior–posterior axis, and if the constellation of active *Hox* genes specifies the region of the anterior–posterior axis, then mice given RA in utero should show homeotic transformations wherein malformations occur along that axis. Kessel and Gruss (1991) found this to be the case. Wild-type mice have 7 cervical (neck) vertebrae, 13 thoracic vertebrae, and 6 lumbar vertebrae (in addition to the sacral and tail vertebrae). When exposed to RA on day 8 of gestation, the first one or two lumbar vertebrae were transformed into thoracic vertebrae, while the first sacral vertebra often became a lumbar vertebra (Figure 17.7). In some cases, the entire posterior region of the mouse embryo failed to form. These changes in structure were correlated with changes in the constellation of *Hox* genes expressed in these tissues. For example, when RA was given to embryos on day 8 (during gastrulation), *Hoxa-10* expression was shifted posteriorly, and an additional set of ribs formed on what had been the first lumbar vertebra. When posterior *Hox* genes did not become expressed at all, the caudal part of the embryo failed to form.

In the central nervous system, retinoic acid induces the anterior expression of *Hox* genes that are usually expressed only more posteriorly, and they cause rhombomeres 2 and 3 to assume the identity of rhombomeres 4 and 5 (Figure 17.8; Marshall et al., 1992; Kessel, 1993). In this instance, the trigeminal nerve (which usually comes from rhombomere 2) is transformed into another facial nerve (characteristic of rhombomere 4), and abnormalities of the first branchial arch indicate that the neural crest cells of the second and third rhombomeres have been transformed into more posterior phenotypes.

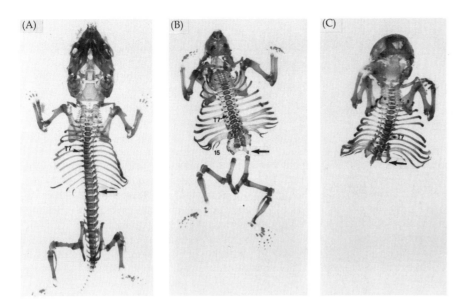

FIGURE 17.7

Retinoic acid administered to pregnant mice alters *Hox* gene expression and phenotype in fetuses. The figure shows changes to axial skeleton (vertebrae and ribs) caused by RA exposure in utero on day 8. (A) The wild-type has 7 cervical vertebrae, 13 thoracic vertebrae, 6 lumbar vertebrae, 4 fused sacral vertebrae, and tail vertebrae. (B,C) This arrangement is altered by RA given to the mothers. In some cases, RA caused the loss of the lumbar, sacral, and caudal vertebrae. (Photographs courtesy of M. Kessel and P. Gruss.)

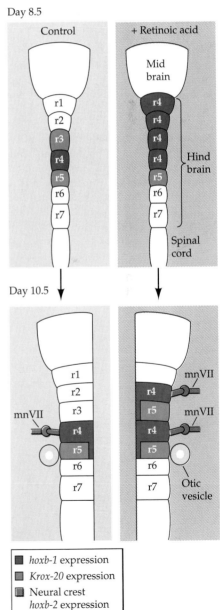

Retinoic acid probably plays a role in axis specification during normal development, and the source of retinoic acid is probably Hensen's node. Hogan and her colleagues (1992) hypothesize that the concentration of RA in Hensen's node increases as the embryo develops, so that the cells that migrate through the node early are exposed to less RA than those that migrate through at a later time. This would cause the specification of brain structures in the anterior portion of the neural tube, while the cells migrating through the node later would be specified to have posterior characteristics. Giving RA exogenously would mimic the situation normally encountered only by the posterior cells. The evidence points to a *Hox* code wherein different constellations of *Hox* genes specify the regional characteristics along the anterior–posterior axis. Moreover, as these expression patterns are similar for mammals and insects, it appears that there is a common developmental plan upon which the anterior–posterior axis of most all animals is constructed.

FIGURE 17.8

Retinoic acid mediates homeotic transformation of hindbrain regions. In untreated mouse embryos on day 8.5, *Hoxb-1* expression is limited to the r4 rhombomere. When exposed to retinoic acid at this time, *Hoxb-1* expression expands anteriorly toward the midbrain. After 2 days, *Hoxb-1* in normal embryos is expressed in the descendant cells of the r4 rhombomere and in the midlines cells of r5, which generates the facial motor nerve (mnVII). In the RA-treated embryos, the normal pattern of r4/5 has been duplicated in r2/3. The neural crest expression of *Hoxb-2* is also duplicated, and a second facial motor nerve is formed. This suggestions that retinoic acid mediates the homeotic transformation of r2/3 to r4/5. (After Krumlauf, 1993.)

Animals as variations on the same developmental theme

One of the most celebrated (and acrimonious) debates in biology was fought in Paris at the height of the French Revolution. There, in the Académie des Sciences, E. Geoffroy Saint-Hilaire contested with Georges Cuvier over the nature of the animal kingdom. Cuvier, the eminent comparative anatomist who had "made zoology a French science," emphasized the *differences* that separate the phyla from one another. There could be no "Chain of Being" linking all organisms, nor could there be any way that the parts of an insect could be seen as being homologous to those of a mollusc or a vertebrate. The only thing linking an insect leg, a mollusc foot, and a vertebrate leg was their locomotive function. Anatomically and embryologically, they were distinct and noncomparable entities.

Saint-Hilaire emphasized the *similarities* among all phyla. He claimed that all animals on the planet were organized according to the same basic principles, and an insect was but a vertebrate turned upside down. A head was formed at one end, a tail at the other, and all the animals had neural tubes, whether dorsal or ventral. Instead of nature composed of intrinsically different species, all animals were united in a kind of brotherhood, reminiscent of the Revolution's *egalité* and *fraternité* (Appel, 1987).

Since that time, different biological traditions have emphasized either the differences or the similarities among organisms. Comparative anatomy (following Cuvier) emphasizes the differences, while morphology (following Saint-Hilaire) celebrates the "underlying unities." Genetics and cell biology look at all animals (and plants) as composed in basically the same way and following the same laws, while embryology has traditionally seen each species as developing in a different manner.

Recently, though, embryology is being seen as providing evidence for the underlying unity of animal nature. Jonathan Slack and his colleagues (1993) have defined an animal as an organism that displays a particular spatial pattern of *Hox* gene expression. They propose that the body plan of each phylum is typified at a particular "phylotypic" stage during its development. For vertebrates, this would be the tailbud stage (where, despite their differences in cleavage and gastrulation, vertebrate embryos converge and have tailbuds and pharyngeal pouches); for insects, the fully segmented germband is where the embryos converge. At this stage, the homeotic gene expression pattern of the *Hox/HOM-C* genes is seen most clearly and is remarkably similar in all animals. Those genes resembling *Deformed* and *labial* are expressed in the anterior of the embryo; those resembling *Abdominal B* are expressed in the posterior. Even nematodes and hydra have homeotic gene clusters that appear to be expressed in the same anterior–posterior fashion (Schummer et al., 1992; Wang et al. 1993). Although fungi and plants have homeobox genes, they are not homologous with the ones in animals nor are they arranged in the same chromosomal order or expressed in the same anterior–posterior pattern. In this way, the spatial pattern of *Hox* gene expression is being used as the primary underlying characteristic defining animal existence. This observation has not been tested in several phyla, and it will be very interesting to see whether this general pattern is seen in throughout the animal kingdom.

Dorsal–ventral and left–right axes in mammals

Very little is known about how mammals form their dorsal–ventral axis. In chicks, the dorsal–ventral axis is determined by gravity placing the hypoblast on the ventral side (Chapter 5). In mice and humans, the hypoblast forms on the side of the inner cell mass that is exposed to the blastocyst fluid. As development proceeds, the notochord maintains dorsal–ventral polarity by inducing specific dorsal–ventral patterns of gene expression in the neural tube (Chapters 7 and 16; Goulding et al., 1993).

Similarly, we know very little about the formation of the left–right axis. The mammalian body is not symmetrical. The heart is assigned to the left side of the chest cavity, even though it forms in the center. The spleen is found solely on the left-hand side of the abdomen, while the major lobe of the liver is on the right-hand side of the abdomen. It is not known what regulates these asymmetries. However, recent findings suggest two levels in regulating the left–right axis: a global level and an organ-specific level. The global level of axis specification was discovered accidentally when Yokoyama and colleagues (1993) made transgenic mice wherein the transgene (for the tyrosinase enzyme) inserted randomly into the genome. In one instance, this gene inserted itself into a region of chromosome 4, knocking out the existing gene. Mice homozygous for this insertion were found to have *all* their asymmetrical organs on the wrong side of the body.

Evidence for the organ-specific regulation of polarity comes from another mouse mutant, *inversus viscerum* (Hummel and Chapman, 1959; Layton, 1976). In this mutant, each asymmetrical organ is given a "choice" of which side to be on, and that decision is made independently of the other organs. This mutation produces the same effect in humans, and such local asymmetry regulation is also seen in people with dynein-deficient cilia (see Chapter 4; Cockayne, 1938). Brown and Wolpert (1990) hypothesize that different organs respond locally to asymmetrical cues in their environment, but the molecular nature of the cues is still unknown. Yost (1992) has shown that the extracellular matrix is important in providing local left–right orientation in *Xenopus* embryos, while Fujinaga and colleagues (1992) have demonstrated that *situs inversus* can be induced in rats by adrenergic hormones given during certain critical periods in their development. After all these centuries, we still do not know how the heart comes to reside in the left side of our body; but we now have some clues to follow.

TERATOLOGY

Malformations and disruptions

The development of an organism is a complex orchestration of cell divisions, cell migrations, cell interactions, gene regulation, cell deaths, and differentiation. Any agent interfering with these processes can cause malformations in the embryo. If it seems amazing that any one of us survives to be born, you are correct; it is estimated that from one-half to two-thirds of the total number of human conceptions do not develop successfully to term (Figure 17.9). Many of these embryos express their abnormality so

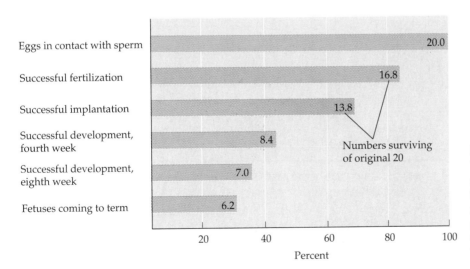

FIGURE 17.9
The hypothetical fates of 20 eggs that are naturally fertilized in the United States and Western Europe. Under normal conditions, only 6.2 eggs of the original 20 would be expected to develop successfully to term. (After Volpe, 1987.)

TABLE 17.1
Some agents thought to cause disruptions in human fetal development[a]

DRUGS AND CHEMICALS

Alcohol
Aminoglycosides (Gentamycin)
Aminopterin
Antithyroid agents (PTU)
Bromine
Cigarette smoke
Cocaine
Cortisone
Diethylstilbesterol (DES)
Diphenylhydantoin
Heroin
Lead
Methylmercury
Penicillamine
Retinoic acid (Isotretinoin, Accutane)
Streptomycin
Tetracycline
Thalidomide
Trimethadione
Valproic acid
Warfarin

IONIZING RADIATION (X-RAYS)

HYPOTHERMIA

INFECTIOUS MICROORGANISMS

Coxsackie virus
Cytomegalovirus
Herpes simplex
Parvovirus
Rubella (German measles)
Toxoplasma gondii (toxoplasmosis)
Treponema pallidum (syphilis)

METABOLIC CONDITIONS IN THE MOTHER

Autoimmune disease (including Rh incompatibility)
Diabetes
Dietary deficiencies, malnutrition
Phenylketonuria

Source: Adapted from Opitz, 1991.
[a]This list includes known and possible teratogenic agents and is not exhaustive.

early that they fail to implant in the uterus. Others implant but fail to establish a successful pregnancy. Thus, most abnormal embryos are spontaneously aborted before the woman even knows she is pregnant (Boué et al., 1985). Hertig and Rock (1949) showed that in 34 embryos between 1 and 17 days old, 13 showed some recognizable anomaly. Nearly 90 percent of a sample of fetuses naturally aborted before one month have developmental anomalies (Mikamo, 1970; Miller and Poland, 1970). Edmonds and co-workers (1982), using a sensitive immunological test that can detect the presence of human chorionic gonadotropin (hCG) 8 or 9 days after fertilization, monitored 112 pregnancies in normal women. Of these hCG-determined pregnancies, 67 failed to be maintained.

It appears, then, that many human embryos are impaired early in development and do not survive long in utero. Defects in the lungs, limbs, face, or mouth, however, would not be deleterious to the fetus (which does not depend on those organs while inside the mother), but can seriously threaten life once the baby is born. About 5 percent of all human births have a recognizable malformation, some of them mild, some very severe (McKeown, 1976).

Congenital ("at birth") abnormalities and the demise of embryos and fetuses prior to birth are caused both intrinsically and extrinsically. Those abnormalities caused by genetic events (mutations, aneuploidies, translocations) are called **malformations**. For instance, aniridia (absence of the iris) caused by the mutation of the *pax-3* gene is a malformation. Down syndrome, caused by a trisomy of chromosome 21, is likewise a malformation. Most of the early embryonic and fetal demise is probably due to chromosomal abnormalities that interfere with normal developmental processes.

Abnormalities due to exogenous agents (certain chemicals or viruses, radiation, or hyperthermia) are called **disruptions**. The agents responsible for these disruptions are called **teratogens** (Gk., "monster-formers"), and the study of how environmental agents disrupt normal development is called **teratology**. In this chapter, we have seen that abnormalities of the anterior–posterior axis can be caused either by mutation of certain *Hox* genes (a malformation) or by the application of a teratogen, retinoic acid, during gastrulation (a disruption). Teratogens work during certain critical periods of development. The most critical time for any organ is when it is growing and forming its particular structures. Different organs have different critical periods, although the time from day 15 to day 60 is critical for many organs. The heart forms primarily during weeks 3 and 4, while the external genitalia are most sensitive during weeks 8 and 9. The brain and skeleton are always sensitive, from the beginning of week 3 to the end of pregnancy and beyond.

The study of congenital abnormalities has provided insights into the possible relationships among organs as they develop in the mammal. As Meckel (1805) noted, "The constant involvement of certain organs together in congenital malformations allows the conclusion that their development is coordinated under normal conditions."*

Types of teratogenic agents

Different agents are teratogenic in different organisms. A partial list of agents that are teratogenic in humans is shown in Table 17.1.

*This does not translate well into English. Opitz (personal communication) points out that the metaphor is one of symphonic development, wherein the organs are *Zusammenstimmen*—voiced together in harmony.

Drugs and environmental chemicals

The major class of teratogens includes drugs and environmental chemicals. Some chemicals that are naturally found in the environment can cause birth defects. Even in the pristine alpine meadows of the Rocky Mountains, teratogens are found. Here grows the skunk cabbage *Veratrum californicum*, upon which sheep sometimes feed. If pregnant ewes eat this plant, their fetuses tend to develop severe neurological damage, including cyclopia, the fusion of two eyes in the center of the face (Figure 17.10). This condition also occurs in humans, pigs, and many other mammals; the affected organism dies shortly after birth (as a result of severe brain defects, including the lack of a pituitary gland).

Quinine and alcohol, two substances derived from plants, can also cause congenital malformations. Quinine can cause deafness, and alcohol (when more than 2–3 ounces per day are imbibed by the mother) can cause physical and mental retardation in the infant. Nicotine and caffeine have not been proved to cause congenital anomalies, but women who are heavy smokers (20 cigarettes a day or more) are likely to have infants that are smaller than those born to women who do not smoke. Smoking also significantly lowers the number and motility of sperm in the semen of males who smoke at least four cigarettes a day (Kulikauskas et al., 1985).

In addition, our industrial society produces hundreds of new artificial compounds that come into general use each year. Pesticides and organic mercury compounds have caused neurological and behavioral abnormalities in infants whose mothers have ingested them during pregnancy. A tragic demonstration of this occurred in 1965, when a Japanese firm dumped mercury into a lake, where it was ingested by the fish, which were eaten by pregnant women in the village of Minamata. The congenital brain damage and blindness in the resulting children became known as Minamata disease.

FIGURE 17.10
Head of a cyclopic lamb born of a ewe who had eaten *Veratrum californicum* early in pregnancy. The cerebral hemispheres fused, forming only one eye and no pituitary gland. (From Binns et al., 1964, courtesy of J. F. James and the USDA-ARS Poisonous Plant Laboratories.)

Retinoic acid as a teratogen

We have already discussed the effect of retinoic acid on mouse development. Retinoic acid is also a powerful teratogen in humans. Retinoic acids are analogues of vitamin A and can mimic the vitamin's effects on epithelial differentiation but are less toxic than high doses of the vitamin itself. These analogues, including 13-*cis*-retinoic acid, have been useful in treating severe cystic acne and have been available (under the name Accutane) since 1982. Because the deleterious effects resulting from the administration of large amounts of vitamin A or its analogues to various species of pregnant animals have been known since the 1950s (Cohlan, 1953; Giroud and Martinet, 1959; Kochhar et al., 1984), the drug contains a label warning that it should not be used by pregnant women. However, about 160,000 women of childbearing age (15–45 years) have taken this drug since it was introduced, and some of them have used it during pregnancy. Lammer and his co-workers (1985) studied a group of women who inadvertently exposed themselves to retinoic acid and who elected to remain pregnant. Of the 59 fetuses, 26 were born without any noticeable malformations, 12 aborted spontaneously, and 21 were born with obvious malformations. The malformed infants had a characteristic pattern of anomalies, including absent or defective ears, absent or small jaws, cleft palate, aortic arch abnormalities, thymic deficiencies, and abnormalities of the central nervous system.

This pattern of multiple congenital anomalies is similar to that seen in rat and mouse embryos whose pregnant mothers have been given these drugs. Goulding and Pratt (1986) have placed 8-day mouse embryos in a solution containing 13-cis-retinoic acid at very low concentrations (2 × 10⁻⁶ M). Even at this concentration, approximately one-third of the embryos developed a very specific pattern of anomalies, including dramatic reduction in the size of the first and second pharyngeal arches (see Figure 17.6). The first arch eventually forms the maxilla and mandible of the jaw and two ossicles of the middle ear, while the second arch forms the third ossicle of the middle ear as well as other facial bones.

The basis for this phenomenon appears to reside in the drug's ability to alter the expression of the *Hox* genes and thereby respecify portions of the anterior–posterior axis and inhibit neural crest cell migration from the cranial region of the neural tube. Radioactively labeled retinoic acid binds to the cranial neural crest cells and arrests both their proliferation and their migration (Johnston et al., 1985; Goulding and Pratt, 1986). The binding seems to be specific to the *cranial* neural crest-derived cells, and the teratogenic effect of the drug is confined to a specific developmental period (days 8–10 in mice; days 20–35 in humans). Animal models of retinoic acid teratogenesis have been extremely successful at elucidating the mechanisms of teratogenesis at the cellular level.

SIDELIGHTS & SPECULATIONS

Retinoic acid teratogenicity

Retinoids are carotene-like compounds related to vitamin A. Inside the body, vitamin A and 13-*cis* retinoic acid become isomerized to the developmentally active forms of retinoic acid, **all-*trans*-retinoic acid** and **9-*cis* retinoic acid** (Creech Kraft, 1992). Retinoic acid cannot bind directly to genes. In order to effect gene regulation, RA must bind to a group of transcription factors called the **retinoic acid receptors** (RARs). These proteins have the same general structure as the steroid and thyroid hormone receptors, and they are active only when they have bound retinoic acid (Linney, 1992). The retinoic acid receptors bind to specific enhancer elements in the DNA called **retinoic acid response elements**. Retinoic acid response elements contain at least two copies of the sequence GGTCA (Ruberte et al., 1990; 1991a). There are three major types of retinoic acid receptors: RAR-α, RAR-β, and RAR-γ. They each bind both forms of retinoic acid, and they both bind to the same retinoic acid response element.

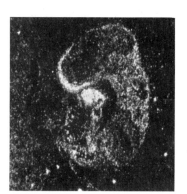

(A)

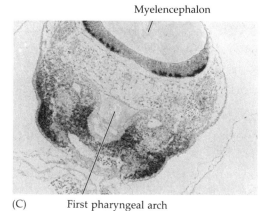

Frontonasal process

Myelencephalon

(B)

(C)   First pharyngeal arch

FIGURE 17.11

Accumulation of retinoic acid in specific places in the developing mouse embryo. (A) RA receptor-γ in the 8.5-day mouse embryo. Complementary DNA was cloned from the RAR-γ gene, and in situ hybridization and autoradiography were performed on sections of developing mouse embryos. (B) Binding of radioactive RA to the frontonasal process and pharyngeal arches of the 10-day mouse embryo. (C) In situ hybridization of the mRNA for CRABP showing expression in the neural crest-derived cells of the first pharyngeal arch and in the fifth cranial ganglion (of the myelencephalon). (A from Ruberte et al., 1990; B from Denker et al., 1990; C from Vaessen et al., 1990; photographs courtesy of the authors.)

In addition, there is a second family of retinoic acid receptor, the retinoid receptor, RXR, that only binds 9-*cis* retinoic acid. The activation of retinoic acid-sensitive promoters is thought to be accomplished by the binding of RAR-RXR heterodimers to the retinoic acid response elements. Some of the *Hox* genes have retinoic acid response elements in their promoters (Yu et al., 1991; Pöpperl and Featherstone, 1993).

The disruptions seen in mouse and human embryogenesis appear to correlate with the expression of retinoic acid receptor RAR-γ. (RAR-α is found throughout the body; RAR-β is found predominantly in epithelia and in the nervous system. In situ hybridization reveals RAR-γ mRNA in the facial (frontonasal) mesenchyme, pharyngeal arches, sclerotomes, and limb mesenchyme at 8.5 to 11.5 days after conception (Figure 17.11). When the RAR-γ genes are deleted from mouse embryos by gene targeting (Lohnes et al., 1993), one sees numerous abnormalities including homeotic transformations along the anterior–posterior axis (often involving the partial transformation of cervical vertebra 2 into cervical vertebra 1) and malformations of tracheal cartilage. It appears, though, that RAR-α is able to compensate for the loss of RAR-γ in the frontonasal mesenchyme.

In addition to the RAR proteins that act as transcription factors in the nucleus, there are also cytoplasmic proteins that bind retinoids. One of them, **cellular retinol-binding protein I** (CRBP I) is found in the yolk sac, where it binds retinol from the maternal circulation (Ruberte et al., 1991b). It brings the retinol to tissues where the retinol can be converted into retinoic acid. Once retinoic acid is in the cytoplasm, it can enter the nucleus and bind to an RAR; but it can also bind to **cellular retinoic acid-binding proteins** (CRABPs). At least one of these, CRABP1, binds to the retinoic acid and *prevents* its entering the nucleus. In cells where CRABP 1 is not present, retinoic acid can enter the nucleus and bind to an RAR (Figure 17.11C; Denker et al., 1990; Vaessen et al., 1990).

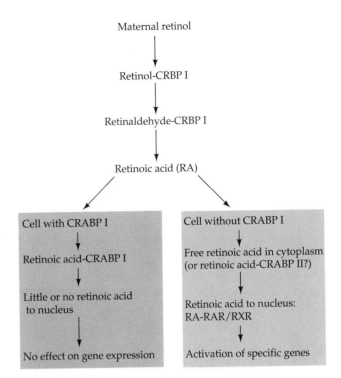

FIGURE 17.12

Hypothetical pathway summarizing the relationships between retinoic acid synthesis, retinoic acid-binding proteins, and receptors in the mammalian embryo. If CRABP were to be saturated by external supplies of retinoic acid, the RA would be able to enter the nucleus, bind to its receptors, and alter the normal pattern of gene expression. (After Morriss-Kay, 1993.)

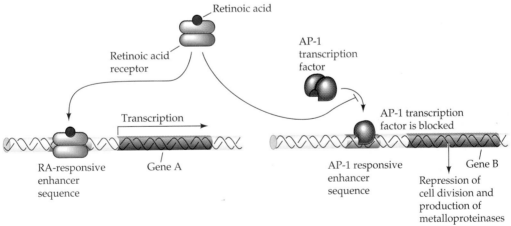

FIGURE 17.13

Model of the dual action of retinoic acid-bound RARs. (1) They activate the transcription of genes whose enhancers enable the RARs to bind. Some of these genes are the homeotic genes that specify the position of the cell along the anterior–posterior axis and therefore will cause the cell to display different characteris- tics. (2) They repress the activity of the AP-1 transcription factor. This transcription factor is important in activating those genes responsible for cell division and for the production of metalloproteinases that often permit cell migration. (After Desbois et al., 1991.)*

It is probable that exogenous retinoic acid works as a teratogen in those tissues that have both an RAR and the CRABP1 protein—cranial neural crest and hindbrain. In these tissues (and in the limb, which will be discussed in a later chapter), exogenous retinoic acid gets around the protective blockade of the CRABP and is able to bind to the nuclear RAR (Figure 17.12; Maden et al., 1991; Morriss-Kay, 1993). The retinoic acid-bound RARs have at least two modes of action. First, they can bind to their DNA enhancer sequences and activate particular genes that are not usually activated in these cells. These genes include certain homeotic genes that specify the anterior–posterior position along the body axis. In this way, they can cause homeotic transformations, generally converting anterior structures into more posterior structures. Second, they can inhibit those genes that are activated by another enhancer-binding transcription factor called AP-1. AP-1 is important in activating cell division, and retinoic acid may be able to inhibit normal cell division by preventing the activity of AP-1 (Figure 17.13; Desbois et al., 1991).

Thalidomide as a teratogen

Before 1961, there was very little evidence for drug-induced malformations in humans. But in that year Lenz and McBride independently accumulated evidence that a mild sedative, **thalidomide**, caused an enormous increase in a previously rare syndrome of congenital anomalies. The most noticeable of these anomalies was **phocomelia**, a condition in which the long bones of the limbs are absent (amelia) or severely deficient (peromelia), thus causing the resulting appendage to resemble a seal flipper (Figure 17.14). Over 7000 affected infants were born to women who had taken this drug, and a woman need only have taken one tablet to produce children with all four limbs deformed (Lenz, 1962;1966; Toms, 1962). Other abnormalities induced by the ingestion of thalidomide included heart defects, absence of the external ears, and malformed intestines. The drug was withdrawn from the market in November of 1961.

Nowack (1965) documented the **period of susceptibility** during which thalidomide caused these abnormalities. The drug was found to be teratogenic only during days 34 to 50 after the last menstruation (about 20 to 36 days postconception). The specificity of thalidomide action is shown in Figure 17.15. From day 34 to day 38, no limb abnormalities are seen. During this period, thalidomide can cause the absence or deficiency of ear components. Malformations of upper limbs are seen before those of the lower limbs since the arms form slightly before the legs during development.

The thalidomide tragedy showed the limits of animal models as tests of the potential teratogenic effects of drugs. Different species (and strains within species) metabolize thalidomide differently. Pregnant mice and rats—the animals usually used to test such compounds—do not generate malformed pups when given thalidomide. Rabbits produce some malformed offspring, but the defects are different from those seen in affected human infants. Primates such as the marmoset appear to have a susceptibility similar to that of humans, and affected marmoset fetuses have been studied in an attempt to discover how thalidomide causes these disruptions. McCredie (1976a,b) proposed that thalidomide might affect the differentiation of neural crest-derived cells, and McBride and Vardy (1983) reported that the most noticeable difference seen before limb malformations concerns the size of the dorsal root ganglia and their neurons. The number of neurons in these ganglia is markedly reduced (Figure 17.16). These authors speculate that the neurons from these ganglia are necessary for maintaining limb development and that thalidomide works by interfering with or destroying these neurons.

Another hypothesis concerning the mechanism of thalidomide teratogenicity has been advanced by Lash and Saxén (1972). Lash (1963) had

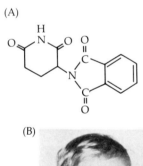

(A)

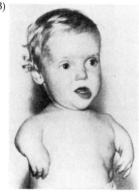

(B)

FIGURE 17.14
Thalidomide structure and effect. (A) Chemical structure of thalidomide. (B) Phocomelia in an infant whose mother took thalidomide during the first two months of pregnancy.

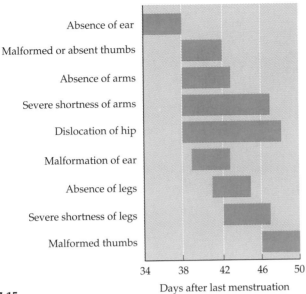

FIGURE 17.15
Timing of susceptibility to the teratogenic effects of thalidomide. (After Nowack, 1965.)

noticed that the primitive kidney, the mesonephros, induced cartilage growth in cultured limb tissue.* Lash and Saxén observed that thalidomide inhibited this mesonephros-induced cartilage growth in cultures of human organs obtained from electively aborted embryos. Moreover, radioactive thalidomide appeared to bind specifically to the human mesonephros. The molecular mechanism of this selective thalidomide teratogenicity is still not known, and this will remain a difficult problem to study so long as our only animal models are other primates.

The thalidomide tragedy also underscores another important principle: the metabolism of embryos is different from that of adults, and the *construction* of an organ can be affected by chemicals that have no deleterious effect on the normal *functioning* of that organ. Several medicines for adults are teratogenic to embryos. These include methotrexate (a drug used to stop tumor cell growth), anticonvulsants such as trimethadione and phenytoin, and anticoagulants such as warfarin. Smoking cigarettes during pregnancy has been associated with fetal growth retardation, but the drinking of coffee has not been seen to produce significant developmental abnormalities (see Friedman, 1992).

Alcohol as a teratogen

In terms of frequency and cost to society, the most devastating teratogen is undoubtedly ethanol. In 1968, Lemoine and colleagues noticed a syndrome of birth defects in the children of alcoholic mothers. Such a **fetal alcohol syndrome** (FAS) was also noted by Jones and Smith (1973). Babies with FAS were characterized as having small head size, an indistinct philtrum (the pair of ridges that run between the nose and mouth above the center of the upper lip), a narrow upper lip, and a low nose bridge. The brain of such a child may be dramatically smaller than normal and often shows defects in neuronal and glial migration (Figure 17.17; Clarren,

*This observation has been confirmed in vivo by Geduspan and Solursh (1992) who removed regions of the mesonephros that were adjacent to the limb bud and found poor limb development on the operated side of the chick; see Chapter 19.

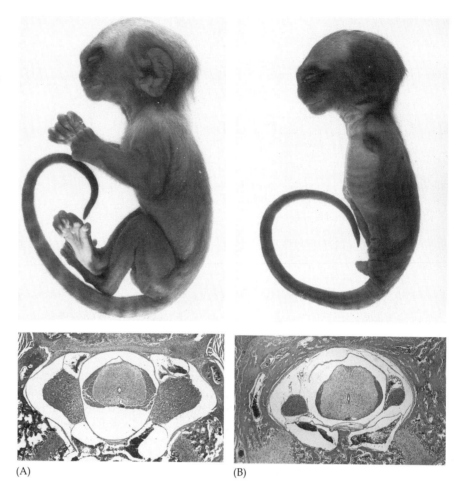

(A)   (B)

1986). Fetal alcohol syndrome is the third most prevalent type of mental retardation (behind fragile X syndrome and Down syndrome) and affects one out of every 500 to 750 children born in the United States (Abel and Sokol, 1987).

Children with fetal alcohol syndrome are developmentally and mentally retarded, with a mean IQ of around 68 (Streissguth and LaDue, 1987). Patients with a mean chronological age of 16.5 years were found to have the functional vocabulary of 6.5-year-olds and to have the mathematical abilities of fourth graders. Most of the adults and adolescents with FAS cannot handle money or their own lives, and they have difficulty learning from past experiences. Moreover, in many instances of FAS, the behavioral abnormalities exist without any gross physical changes in the head size or IQ (J. Opitz, personal communication). There is great variation in the ability of mothers and fetuses to metabolize ethanol, and it is thought that 30–40 percent of the children born to alcoholic mothers who drink during pregnancy will have FAS. It is also thought that lower amounts of ethanol ingestion by the mother can lead to *fetal alcohol effect*, a less severe form of FAS, but a condition that lowers the functional and intellectual abilities of the person.*

*For a remarkable account of raising a child with fetal alcohol syndrome, as well as an analysis of FAS in Native American culture in the United States, see Dorris (1989). The personal and sociological effects of FAS are well integrated with the scientific and economic data.

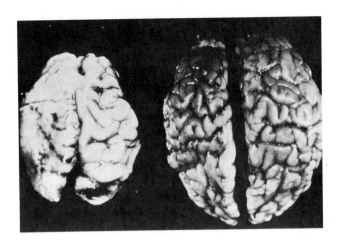

FIGURE 17.17
Comparison of a brain from an infant with fetal alcohol syndrome with that of a normal infant of the same age. The brain from the infant with FAS is significantly smaller, and the pattern of convolutions is obscured by glial cells that have migrated over the top of the brain. (Photograph courtesy of S. Clarren.)

Other teratogenic agents

Drugs and chemicals aren't the only extrinsic agents capable of causing disruptions in development. Another class of teratogens includes viruses. Gregg (1941) first documented the fact that women who had rubella (German measles) during the first third of their pregnancy had a 1 in 6 chance of giving birth to an infant with eye cataracts, heart malformations, or deafness. This was the first evidence that the mother could not fully protect the fetus from the outside environment. The earlier the rubella infection occurred during the pregnancy, the greater the risk that the embryo would be malformed. The first 5 weeks appear to be the most critical, because this is when the heart, eyes, and ears are being formed. The rubella epidemic of 1963–1965 probably resulted in about 20,000 fetal deaths and 30,000 infants with birth defects in the United States. Two other viruses, *Cytomegalovirus* and *Herpes simplex*, are also teratogenic. *Cytomegalovirus* infection of early embryos is nearly always fatal, but infection of later embryos can lead to blindness, deafness, cerebral palsy, and mental retardation.

Bacteria and protists are rarely teratogenic, but two of them can damage human embryos. *Toxoplasma gondii*, a protozoan carried by rabbits and cats (and their feces), can cross the placenta and cause brain and eye defects in the fetus. *Treponema pallidum*, the cause of syphilis, can kill early fetuses and produce congenital deafness in older ones.

Ionizing radiation can break chromosomes and alter DNA structure. For this reason, pregnant women are told to avoid unnecessary X-rays, even though there is no evidence for congenital anomalies resulting from diagnostic radiation (Holmes, 1979). Heat from high fevers is also a possible teratogen.

Unknown sources of congenital malformations

While we know the causes of certain malformations, most congenital abnormalities are not yet able to be explained. For instance, congenital cardiac anomalies occur in about 1 in every 200 live births. Genetic causes are responsible for about 8 percent of these heart abnormalities, and about 2 percent can be explained by known teratogens. That leaves 90 percent of the them unexplained (O'Rahilly and Müller, 1992). We still have a great deal of research to do. There are many reasons for our ignorance in this area. First, we have not screened most chemicals for their teratogenic effects. There are over 50,000 artificial chemicals presently used in our society and about 200 to 500 new materials being made each year (Johnson, 1980). The problem of screening these chemicals is of major importance, and standard protocols are expensive, long, and subject to interspecies differences in metabolism. There is still no consensus on how to test a substance's teratogenicity for human embryos.

Second, there are some drugs that appear to have teratogenic effects, but it is difficult to measure what these effects are. Cocaine is known to cause reduced blood flow, premature labor, and placental rips during pregnancy. It is also associated with brain lesions and impaired neuronal migration (Dominguez et al., 1991; Volpe, 1992). But it is not yet known whether these anatomical abnormalities are retained or if they are important for normal functioning, since the lesions would impair behaviors (such as attention span and emotional control) that are not observable in research animals and are not readily measured in children (Volpe, 1992). Moreover, it is difficult to separate the effects of a particular teratogen (such as cocaine) when a drug is often used with several others.

Third, most research pursued on mammalian teratogenesis has focused on the role of exogenous agents *during* development. Very little work has been done on those agents that may affect the gametes *prior* to fertilization and whose effects might be registered only during later embryonic stages. Davis and colleagues (1992) point out that epidemiological evidence suggests that many chemicals and processes effect changes in gamete function, and this is as true for the sperm as it is for the egg.

Although teratogenic compounds have always been with us, the fetus is placed at ever greater risk as more new and untested compounds enter our environment each year.

LITERATURE CITED

Abel, E. L. and Sokol, R. J. 1987. Incidence of fetal alcohol syndrome and economic impact of FAS-related anomalies. *Drug Alcohol Depend.* 19: 51–70.

Appel, T. A. 1987. *The Cuvier-Geoffroy Debate: French Biology in the Decades before Darwin.* Oxford University Press, New York.

Awgulewitsch, A. and Jacobs, D. 1992. *De*formed autoregulatory element from *Drosophila* functions in a conserved manner in transgenic mice. *Nature* 358: 341–344.

Balling, R., Mutter, G., Gruss, P. and Kessel, M. 1989. Craniofacial abnormalities induced by ectopic expression of the homeobox gene *Hox-1.1* in transgenic mice. *Cell* 58: 337–347.

Binns, W., James, L. F. and Shupe, J. L. 1964. Toxicosis of *Veratrum californicum* in ewes and its relationship to a congenital deformity in lambs. *Ann. N.Y. Acad. Sci.* 111: 571–576.

Blum, M., Gaunt, S. J., Cho, K. W. Y., Steinbeisser, H., Blumberg, B., Bittner, D. and De Robertis, E. M. 1992. Gastrulation in the mouse: The role of the homeobox gene *goosecoid*. *Cell* 69: 1097–1106.

Blumberg, B., Wrights, C. V. E., De Robertis, E. M. and Cho, K. W. Y. 1991. Organizer-specific homeobox genes in *Xenopus laevis* embryos. *Science* 253: 194–196.

Boncinelli, E., Somma, R., Acampora, D., Pannese, M., D'Esposito, M., Faiella, A. and Simeone, A. 1988. Organization of human homeobox genes. *Hum. Reprod.* 3: 880–886.

Boué, A., Boué, J. and Gropp, A. 1985. Cytogenetics of pregnancy wastage. *Adv. Hum. Genet.* 14: 1–57.

Brown, N. A. and Wolpert, L. 1990. The development of handedness in left/right asymmetry. *Development* 109: 1–9.

Chisaka, O. and Capecchi, M. 1991. Regionally restricted developmental defects resulting from targeted disruption of the mouse homeobox gene *Hox-1.5*. *Nature* 350: 473–479.

Clarren, S. K. 1986. Neuropathology in the fetal alcohol syndrome. *In* J. R. West (ed.), *Alcohol and Brain Development*. Oxford University Press, New York.

Cockayne, E. A. 1938. The genetics of transpositon of the viscera. *Q. J. Med.* 31: 479–493.

Cohlan, S. Q. 1953. Excessive intake of vitamin A as a cause of congenital anomalies in the rat. *Science* 117: 535–537.

Conlon, R. A. and Rossant, J. 1992. Exogenous retinoic acid rapidly induces anterior ectopic expression of murine *Hox-2* genes in vivo. *Development* 116: 357–368.

Creech Kraft, J. 1992. Pharmacokinetics, placental transfer, and teratogencity of 13-cis retinoic acid, its isomer and metabolites. *In* G. M. Morriss-Kay (ed.), *Retinoids in Normal Development and Teratogenesis*. Oxford University Press, Oxford, pp. 267–280.

Davis, D. L., Friedler, G., Mattison, D. and Morris, R. 1992. Male-mediated teratogenesis and other reproductive effects: Biologic and epidemiologic findings and a plea for clinical research. *Reprod. Toxic.* 6: 289–292.

Denker, L., Annerwall, E., Busch, C. and Eriksson, U. 1990. Localization of specific retinoid-binding sites and expression of cellular retinoic-acid-binding protein (CRABP) in the early mouse embryo. *Development* 110: 343–352.

Desbois, C., Aubert, D., Legrand, C., Pain, B. and Samarut, J. 1991. A novel mechanism of action for v-ErbA: Abrogation of the inactivation of transcription factor AP-1 by retinoic acid and thyroid hormone receptors. *Cell* 67: 731–740.

Dominguez, R., Vila-Coro, A. A., Slopsis, J. M. and Bohan, T. P. 1991. Brain and ocular abnormalities in infants with *in utero* exposure to cocaine and other street drugs. *Am. J. Dis. Child.* 145: 688–695.

Dorris, M. 1989. *The Broken Cord.* Harper and Row, New York.

Edmonds, D. K., Lindsay, K. S., Miller, J. F., Williamson, E. and Wood, P. J. 1982. Early embryonic mortality in women. *Fertil. Steril.* 38: 447–453.

Friedman, J. M. 1992. Effects of drugs and other chemicals on fetal growth. *Growth Genet. Horm.* 8(4): 1–5.

Fujinaga, M., Maze, M., Hoffman, B. B. and Baden, J. M. 1992. The activation of α-1 adrenergic receptors modulates the control of left/right sidedness in rat embryos. *Dev. Biol.* 150: 419–421.

Geduspan, J. S. and Solursh, M. 1992. A growth–promoting influence from the mesonephros during limb outgrowth. *Dev. Biol.* 151: 242–250.

Gendron-Maguire, M., Mallo, M., Zhang, M. and Gridley, T. 1993. *Hoxa-2* mutant mice exhibit homeotic transformation of skeletal elements derived from cranial neural crest. *Cell* 75: 1317–1331.

Giroud, A. and Martinet, M. 1959. Teratogenese pur hypervitaminose A chez le rat, la souris, le cobaye, et le lapin. *Arch. Fr. Pediatr.* 16: 971–980.

Gluecksohn-Schoenheimer, S. 1938. The development of two tailless mutants in the house mouse. *Genetics* 23: 573–584.

Goulding, E. H. and Pratt, R. M. 1986. Isotretinoin teratogenicity in mouse whole embryo culture. *J. Craniofac. Genet. Dev. Biol.* 6: 99–112.

Goulding, M. D., Lumsden, A. and Gruss, P. 1993. Signals from the notochord and floor plate regulate the region-specific expression of two *Pax* genes in the developing spinal cord. *Development* 117: 1001–1016.

Gregg, N. M. 1941. Congenital cataract following German measles in the mother. *Trans. Opthalmol. Soc. Aust.* 3: 35.

Herrmann, B. G. 1991. Expression pattern of the *Brachyury* gene in whole–mount T^{wis}/T^{wis} mutant embryos. *Development* 113: 913–917.

Hertig, A. T. and Rock, J. 1949. A series of potentially abortive ova recovered from fertile women prior to the first missed menstrual period. *Am. J. Obstet. Gynecol.* 58: 968–993.

Hogan, B. L. M., Thaller, C. and Eichele, G. 1992. Evidence that Hensen's node is a site of retinoic acid synthesis. *Nature* 359: 237–241.

Holmes, L. B. 1979. Radiation. *In* V. C. Vaughan, R. J. McKay and R. D. Behrman (eds.), *Nelson Textbook of Pediatrics*, 11th Ed. Saunders, Philadelphia.

Hummel, K. P. and Chapman, D. B. 1959. Visceral inversion and associated anomalies in the mouse. *J. Hered.* 50: 9–13.

Hunt, P. and Krumlauf, R., 1991. Deciphering the *Hox* code: Clues to patterning branchial regions of the head. *Cell* 66: 1075–1078.

Hunt, P. and Krumlauf, R. 1992. *Hox* codes and positional specification in vertebrate embryonic axes. *Annu. Rev. Cell Biol.* 8: 227–256.

Hunt, P. and seven others. 1991a. A distinct *Hox* code for the branchial region of the head. *Nature* 353: 861–864.

Hunt, P., Wilkinson, D. and Krumlauf, R. 1991b. Patterning the vertebrate head: Murine *Hox-2* genes mark distinct subpopulations of premigratory and migratory neural crest. *Development* 112: 43–50.

Izpisúa-Belmonte, J. C., De Robertis, E. M., Storey, K. G. and Stern, C. D. 1993. The homeobox gene *goosecoid* and the origin of organizer cells in the early chick blastoderm. *Cell* 74: 645–659.

Jeannotte, L., Lemieux, M., Cherron, J., Poirier, F. and Robertson, E. J. 1993. Specification of axial identity in the mouse: Role of the *Hoxa-5* (*Hox 1.3*) gene. *Genes Dev.* 7: 2085–2096.

Johnson, E. M. 1980. Screening for teratogenic potential: Are we asking the proper questions? *Teratology* 21: 259.

Johnston, M. C., Sulik, K. K., Webster, W. S. and Jarvis, B. L. 1985. Isotretinoin embryopathy in a mouse model: Cranial neural crest involvement. *Teratology* 31: 26A.

Jones, K. L. and Smith, D. W. 1973. Recognition of the fetal alcohol syndrome. *Lancet* 2: 999–1001.

Kessel, M. 1992. Respecification of vertebral identities by retinoic acid. *Development* 115: 487–501.

Kessel, M. 1993. Reversal of axonal pathways from rhombomere 3 correlates with extra *Hox* expression domains. *Neuron* 10: 379–393.

Kessel, M. and Gruss, P. 1991. Homeotic transformations of murine vertebrae and concomitant alteration of *Hox* codes induced by retinoic acid. *Cell* 67: 89–104.

Keynes, R. and Lumsden, A. 1990. Segmentation and the origin of regional diversity in the vertebrate central nervous system. *Neuron* 2: 1–9.

Kochhar, D. M., Penner, J. D. and Tellone, C. I. 1984. Comparative teratogenic activities of two retinoids: Effects on palate and limb development. *Teratogen. Carcinogen. Mutagen.* 4: 377–387.

Kulikauskas, V., Blaustein, A. B. and Ablin, R. J. 1985. Cigarette smoking and its possible effects on sperm. *Fertil. Steril.* 44: 526–528.

Krumlauf, R. 1993. *Hox* genes and pattern formation in the branchial region of the vertebrate head. *Trends. Genet.* 9: 106–112.

Lammer, E. J. and eleven others. 1985. Retinoic acid embryopathy. *N. Engl. J. Med.* 313: 837–841.

Lash, J. W. 1963. Studies on the ability of embryonic mesonephros explants to form cartilage. *Dev. Biol.* 6: 219–232.

Lash, J. W. and Saxén, L. 1972. Human teratogenesis: In vitro studies of thalidomide-inhibited chondrogenesis. *Dev. Biol.* 28: 61–70.

Layton, W. M. Jr. 1976. Random determination of a developmental process. *J. Hered.* 67: 336–338.

Lemoine, E. M., Harousseau, J. P., Borteyru, J. P. and Menuet, J. C. 1968. Les enfants de parents alcooliques: Anomalies observées *Oest. Med.* 21: 476–482.

Le Mouellic, H., Lallemand, Y. and Brûlet, P. 1992. Homeosis in the mouse induced by a null mutation in the *Hox-3.1* gene. *Cell* 69: 251–264.

Lenz, W. 1962. Thalidomide and congenital abnormalities. *Lancet* 1: 45. (First reported at a 1961 symposium.)

Lenz, W. 1966. Malformations caused by drugs in pregnancy. *Am. J. Dis. Child.* 112: 99–l06.

Linney, E. 1992. Retinoic acid receptors: transcription factors modulating gene expression, development, and differentiation. *Curr. Top. Dev. Biol.* 27: 309–350.

Lohnes, D., Kastner, P., Dierich, A., Mark, M., LeMur, M., Chambon, P. 1993. Function of retinoic acid receptor γ in the mouse. *Cell* 73: 643–658.

Lufkin, T., Dierich, A., LeMeur, M., Mark, M. and Chambon, P. 1991. Disruption of the *Hox-1.6* homeobox gene results in defects in a region corresponding to its rostral domain of expression. *Cell* 66: 1105–1119.

Lumsden, A., Sprawson, N. and Graham, A. 1991. Segmental origin and migration of neural crest cells in the hindbrain region of the chick embryo. *Development* 113: 1281–1291.

Maden, M., Hunt, P., Eriksson, U., Kuroiwa, A., Krumlauf, R. and Summerbell, D. 1991. Retinoic acid–binding protein, rhombomeres, and the neural crest. *Development* 111: 35–44.

MacMurray, A. and Shin, H.-S. 1988.The antimorphic nature of the T^c allele at the mouse *T* locus. *Genetics* 120: 545–550.

Malicki, J., Schugart, K. and McGinnis, W. 1990. Mouse *Hox-2.2* specifies thoracic segmental identity in *Drosophila* embryos and larvae. *Cell* 63: 961–967.

Malicki, J. Cianetti, L. C., Peschle, C. and McGinnis, W. 1992. Human HOX4B regulatory element provides head-specific expression in *Drosophila* embryos. *Nature* 358: 345–347.

Marshall, H., Nonchev, S., Sham, M. H., Muchamore, I., Lumsden, A. and Krumlauf, R. 1992. Retinoic acid alters hindbrain *Hox* code and induces transformation of rhombomeres 2/3 into a 4/5 identity. *Nature* 360: 737–741.

McBride, W. G. 1961. Thalidomide and congenital abnormalities. *Lancet* 2: 1358.

McBride, W. G. and Vardy, P. H. 1983. Pathogenesis of thalidomide teratogenesis in the marmot (*Callithrix jacchus*): Evidence suggesting a possible trophic influence of cholinergic nerves in limb morphogenesis. *Dev. Growth Differ.* 25: 361–374.

McCredie, J. 1976a. Neural crest defects: A neuroanatomic basis for classification of multiple malformations related to phocomelia. *J. Neurol. Sci.* 28: 373–387.

McCredie, J. 1976b. The pathogenesis of congenital malformations. *Australas. Radiol.* 19: 348–355.

McGinnis, W. and Krumlauf, R. 1992. Homeobox genes and axial patterning. *Cell* 68: 283–302.

McGinnis, N., Kuziora, M. A. and McGinnis, W. 1990. Human *Hox 4.2* and *Drosophila Deformed* encode similar regulatory specificities in *Drosophila* embryos and larvae. *Cell* 63: 969–976.

McKeown, T. 1976. Human malformations: An introduction. *Br. Med. Bull.* 32: 1–3.

Meckel, J. F. 1805. Über die Bildungsfehler des Herzen. *Reil's Arch. Physiol.* 6: 549–610.

Mikamo, K. 1970. Anatomical and chromosomal anomalies in spontaneous abortions. *Am. J. Obstet. Gynecol.* 103: 143–154.

Miller, J. R. and Poland, B. J. 1970. The value of human abortuses in the surveillance of developmental anomalies. *Can. Med. Assoc. J.* 103: 501–502.

Morriss-Kay, G. 1993. Retinoic acid and craniofacial development: Molecules and morphogenesis. *BioEssays* 15: 9–15.

Noden, D. 1988. Interactions and fates of avian craniofacial mesenchyme. *Development* [Suppl.] 103: 121–140.

Nowack, E. 1965. Die sensible Phase bei der Thalidomide-Embryopathie. *Humangenetik* 1: 516–536.

Opitz, J. 1991. Appendix I: Disruption. In *Second International Workshop on Fetal Genetic Pathology.*

O'Rahilly, R. and Müller, F. 1992. *Human Embryology and Teratology.* Wiley-Liss, New York.

Papalopulu, N., Clarke, J. D. W., Bradley, L., Wilkinson, D., Krumlauf, R. and Holder, N. 1991. Retinoic acid causes abnormal development and segmental patterning of the hindbrain of *Xenopus* embryos. *Development* 113: 1145–1158.

Pöpperl, H. and Featherstone, M. S. 1993. Identification of retinoic acid response element upstream from the mouse *Hox-4.2* gene. *Mol. Cell. Biol.* 13: 257–265.

Ramirez-Solis, R., Zheng, H., Whiting, J., Krumlauf, R. and Bradley, A. 1993. *Hoxb-4* (*Hox 2.6*) mutant mice show homeotic transformation of a cervical vertebra and defects in the closure of the sternal rudiments. *Cell* 73: 279–294.

Rashbass, P. R., Cook, L. A., Herrmann, B. G. and Beddington, R. S. P. 1991. A cell autonomous function of *Brachyury* in *T/T* embryonic stem cell chimeras. *Nature* 353: 348–351.

Riddihough, G. 1992. Homing in on the homeobox. *Nature* 357: 643–644.

Rijli, F. M., Mark, M., Lakkaraju, S., Dierich, A., Dollé, P. and Chambon, P. 1993. A homeotic transformation is generated in the rostral branchial region of the head by disruption on *Hoxa-2*, which acts as a selector gene. *Cell* 75: 1333–1349.

Ruberte, E., Dollé, P., Krust, A., Zalent, A., Morriss-Kay, G. and Chambon, P. 1990. Specific spatial and temporal distribution of retinoic acid receptor γ transcripts during mouse embryogenesis. *Development* 108: 213–222.

Ruberte, E. and seven others. 1991a. Retinoic acid receptors in the embryo. *Semin. Dev. Biol.* 2: 153–159.

Ruberte, E., Dollé, P., Chambon, P. and Morriss-Kay, G. M. 1991b. Retinoic acid receptors and cellular retinoid binding proteins. II. Their differential pattern of transcription during morphogenesis in mouse embryos. *Development* 115: 973–989.

Schummer, M., Scheurlen, I., Schaller, C. and Galliot, B. 1992. HOM/HOX homeobox genes are present in hydra (*Chlorohydra viridissima*) and are differentially expressed during regeneration. *EMBO J.* 11: 1815–1825.

Scott, M. 1992. Vertebrate homeobox gene nomenclature. *Cell* 71: 551–553.

Slack, J. M. W. 1983. *From Egg to Embryo: Determinative Events in Early Development.* Cambridge University Press, Cambridge.

Slack, J. M. W. and Tannahill, D. 1992. Mechanism of anteroposterior axis specification in vertebrates: Lessons from the amphibians. *Development* 114: 285–302.

Slack, J. M. W., Holland, P. W. H. and Graham, C. F. 1993. The zootype and the phylotypic stage. *Nature* 361: 490–492.

Stott, D., Kisbert, A. and Herrmann, B. G. 1993. Rescue of the tail defect of *Brachyury* mice. *Genes Dev.* 7: 197–203.

Streissguth, A. P. and LaDue, R. A. 1987. Fetal alcohol: Teratogenic causes of developmental disabilities. *In* S. R. Schroeder (ed.), *Toxic Substances and Mental Retardation.* American Association of Mental Deficiency, Washington, DC, pp. 1–32.

Toms, D. A. 1962. Thalidomide and congenital abnormalities. *Lancet* 2: 400.

Vaessen, M.-J., Meijers, J. H. C., Bootsma, D. and Kessel, G. van. 1990. The cellular retinoic acid-binding protein is expressed in tissues associated with retinoic-acid-induced malformations. *Development* 110: 371–378.

Volpe, E. P. 1987. Developmental biology and human concerns. *Am. Zool.* 27: 697–714.

Volpe, J. J. 1992. Effect of cocaine use on the fetus. *N. Engl. J. Med.* 327: 399–407.

Wang, B. B., Müller-Immergluck, M. M., Aystin, J., Robinson, N. T., Chisholm, A. and Kenyon, C. 1993. A homeotic gene cluster patterns the anteroposterior body axis of *C. elegans. Cell* 74: 29–42.

Wilkinson, D. G., Bhatt, S., Cook, M., Boncinelli, E. and Kruflauf, R. 1989. Segmental expression of *Hox-2* homeobox-containing genes in the developing mouse hindbrain. *Nature* 341: 405–409.

Yanagisawa, K. O. 1990. Does the *T* gene determine the anterior–posterior axis of the mouse embryo? *Japan. J. Genet.* 65: 287–297.

Yokoyama, T., Copeland, N . G., Jenkins, N. A., Montgomery, C. A., Elder, F. F. B. and Overbeek, P. A. 1993. Reversal of left–right symmetry: A *situs inversus* mutation. *Science* 260: 679–682.

Yost, H. J. 1992. Regulation of vertebrate left–right asymmetries by extracellular matrix. *Nature* 357: 158–161.

Yu, V. C. and nine others. 1991. RXRb: A coregulator that enhances binding of retinoic acid, thyroid hormone, and vitamin D receptors to their cognate response elements. *Cell* 67: 1251–1266.

V

CELLULAR INTERACTIONS DURING ORGAN FORMATION

18

Proximate tissue interactions
Secondary induction

Organs are complex structures composed of numerous types of tissues. If one considers an organ such as the vertebrate eye, for example, one finds that light is transmitted through the transparent corneal tissue; it is focused by the lens tissue, the diameter of which is controlled by muscle tissue; and it eventually impinges upon the tissue of the neural retina. The precise arrangement of tissues in this organ cannot be disturbed without damaging its function. Such coordination in the construction of organs is accomplished by one group of cells changing the behavior of an adjacent set of cells, thereby causing them to change their shape, mitotic rate, or differentiation. This action at close range, sometimes called **proximate interaction** or **secondary induction** enables one group of cells to respond to a second group of cells and, in changing, often to become able to alter a third set of cells.

In dealing with such a complex system as the developing embryo, it is futile to inquire whether a certain organ rudiment is "determined" and whether some feature of its surroundings, to the exclusion of others, "determines" it. A score of different factors may be involved and their effects most intricately interwoven. In order to resolve this tangle we have to inquire into the manner in which the system under consideration reacts with other parts of the embryo at successive stages of development and under as great a variety of experimental conditions as it is possible to impose.
R. G. HARRISON (1933)

The aspiration to truth is more precious than its assured possession.
G. E. LESSING (1778)

Instructive and permissive interactions

Howard Holtzer (1968) distinguished two major modes of proximate tissue interactions. One type is the **instructive interaction**. In this kind of interaction, a signal from the inducing cell is necessary for initiating new gene expression in the responding cell. Without the inducing cell, the responding cell would not be capable of differentiating in that particular way. For example, in Chapter 16 we discussed the ability of the notochord to induce the formation of floor plate cells in the neural tube. All neural tube cells are capable of responding to the notochord signal, but only those closest to the notochord are induced. The other cells became non-floor plate cells. Moreover, if one removed the notochord from the embryo, those cells that would normally have become floor plate cells would not differentiate into that type of cell, and if one added an additional notochord laterally to the neural plate, this new notochord would induce a secondary set of floor plate cells. The responding neural tube cells would somehow have been told to express a set of genes different from the set they would have expressed had they not been in contact with the notochord. The notochord is said to be an inducing tissue acting *instructively*. Wessells (1977) has proposed four general principles characteristic of most instructive interactions.

1. In the presence of tissue A, responding tissue B develops in a certain way.
2. In the absence of tissue A, responding tissue B does not develop in that way.
3. In the absence of tissue A, but in the presence of tissue C, tissue B does not develop in that way.
4. In the presence of tissue A, a tissue D, which would normally develop differently, is changed to develop like B.

The second type of proximate tissue interactions is **permissive interaction**. Here, the responding tissue contains all the potentials needed to be expressed, and it only needs an environment that allows the expression of these traits. For instance, many developing tissues need a solid substrate containing fibronectin or laminin in order to develop. The fibronectin or laminin does not alter the type of cell that is to be produced, but only enables its expression.*

Competence

It should be noted that in the fourth principle mentioned above, the responding tissue must be competent to respond. **Competence** is the ability to respond to an inductive signal (Waddington, 1940). It is not a passive state, but an actively acquired condition. When we detailed the induction of the neural tube, we discussed the observations that the gastrula ectoderm is capable of being induced by the dorsal blastopore lip or its mesodermal derivatives. Thus, the gastrula ectoderm is said to be *competent* to respond to the inductive stimuli. This competence for neural induction is acquired during late cleavage stages, and it is lost during the late gastrula stages. As this competence to respond to dorsal lip induction diminishes, the same ectoderm gains the competence to respond to lens inducers. Still later, the competence to lens inducers is lost, but the ectoderm can respond to ear placode inducers (Servetnick and Grainger, 1991). Therefore, competence is, itself, a differentiated phenotype that distinguishes cells spatially and temporally.

It is generally thought that competence can be attained in several ways. First, a cell can become competent by synthesizing a receptor for the inducer molecule. As we will see later in this chapter, a B cell is not competent to respond to induction by T cells until it has bound antigens. When the antigens are bound, they create a set of receptors that enable them to respond to the inducing molecules secreted by the T cells. This mechanism of competence is also seen in the induction of sympathetic neuron differentiation (Birren and Anderson, 1990; Cattanco and McKay, 1990). Since the early 1960s it has been known that the the differentiation of sympathetic neurons depends on nerve growth factor (NGF); but when the progenitor cells for these neurons were actually isolated, they did not respond to NGF. Moreover, they lacked the receptors capable of binding NGF. Rather, in order to differentiate, these cells first had to be exposed to fibroblast growth factor (FGF). The exposure to FGF resulted in these cells' expressing the NGF receptor on their cell membranes. These FGF-

*It is easy to distinguish permissive and instructive relationships by an analogy with a more familiar situation. This textbook is made possible by permissive and instructive interactions. The reviewers can convince me to change the material in the chapter. This is an *instructive interaction*, as the information is changed from what it would have been. However, the information in the book could not be expressed without *permissive interactions* by the publisher and printer.

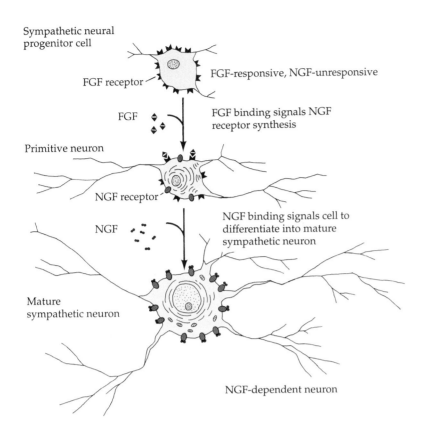

Sympathetic neural progenitor cell

FGF receptor

FGF-responsive, NGF-unresponsive

FGF

FGF binding signals NGF receptor synthesis

Primitive neuron

NGF receptor

NGF

NGF binding signals cell to differentiate into mature sympathetic neuron

Mature sympathetic neuron

NGF-dependent neuron

FIGURE 18.1
Induction and competence of a sympathetic neuron precursor lineage. The original stem cell is a mitotically active cell that has no NGF receptors but can respond to FGF. This gives rise to a primitive neural cell that has processes but still divides. This primitive neuron has receptors for NGF. The NGF-responsive cell can differentiate into the mature nondividing sympathetic neuron (characterized by its large soma, prominent nucleolus, extensive processes, and dependence on NGF for survival). (After Birren and Anderson, 1990.)

treated cells could respond to NGF (Figure 18.1). The original progenitor cell was not competent to be induced by NGF because it lacked the NGF receptor. When it was induced by FGF, it became competent to respond to NGF.

Second, a cell might achieve competence by synthesizing a molecule that allows the receptor to function. Receptors alone may bind the inducer, but that does not mean that the receptors are functional. Often, a receptor acts by sending a signal to the nucleus. As we have seen in Chapter 4, once a receptor has bound a ligand, it activates enzymes that manufacture the signal for division or differentiation. If one of these enzymes is not present, the signal is not transmitted. So a cell might achieve competence by synthesizing a missing link in the signaling pathway.

Third, competence may be acquired by the repression of an inhibitor. If the inhibitor is present, a cell might bind the inducer, send the signal to the nucleus, and still not be able to be become induced. For example, inducers often cause cell shape changes (as in the induction of the neural tube). If the cell were inhibited from changing its shape, it would not be able to respond.

Epithelio-mesenchymal interactions

Some of the best-studied cases of secondary induction are those involving the interactions of epithelial sheets with adjacent mesenchymal cells. These are called **epithelio-mesenchymal interactions**. The epithelium can come from any germ layer whereas the mesenchyme is usually derived from loose mesodermal or neural crest tissue. Examples of epithelio-mesenchymal interactions are listed in Table 18.1.

TABLE 18.1
Some epithelio-mesenchymal interactions

Organ	Epithelial component	Mesenchymal component
Cutaneous structures (hair, feathers, sweat glands, mammary glands)	Epidermis (ectoderm)	Dermis (mesoderm)
Limb	Epidermis (ectoderm)	Mesenchyme (mesoderm)
Gut organs (liver, pancreas, salivary glands)	Epithelium (endoderm)	Mesenchyme (mesoderm)
Pharyngeal and respiratory associated organs (lungs, thymus, thyroid)	Epithelium (endoderm)	Mesenchyme (mesoderm)
Kidney	Ureteric bud epithelium (mesoderm)	Mesenchyme (mesoderm)
Tooth	Jaw epithelium (ectoderm)	Mesenchyme (neural crest)

Regional specificity of induction

Using the induction of cutaneous structures as our examples, we will look at the properties of epithelio-mesenchymal interactions. The first phenomenon is the regional specificity of induction. Skin is composed of two main tissues: an outer epidermis derived from ectoderm, and a dermis derived from mesoderm. Chicken skin gives rise to three major cutaneous structures that are made almost entirely of ectodermal cells. These are the broad wing feathers, the narrow thigh feathers, and the scales and claws of the feet. After separating the embryonic epithelium and mesenchyme from each other, one can recombine them in different ways (Saunders et al., 1957). Some of the recombinations are illustrated in Figure 18.2. As you can see, the mesenchyme is responsible for the specificity of induction in the competent ectoderm. This same type of ectoderm develops according to the region from which the mesoderm was taken. Here, the mesenchyme has an instructive role, calling into play different sets of genes in the responding cells.

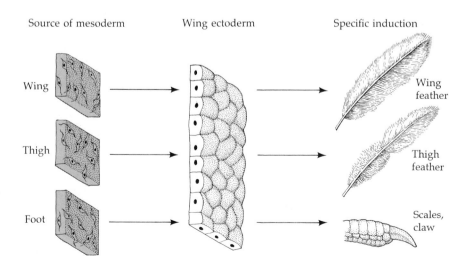

FIGURE 18.2
Regional specificity of induction. When cells of the dermis (mesoderm) are recombined with the epidermis (ectoderm) in the chick, the type of cutaneous structure made by the ectoderm is determined by the original location of the mesoderm. (Adapted from Saunders, 1980.)

Source of mesoderm   Wing ectoderm   Specific induction

Wing

Thigh

Foot

Wing feather

Thigh feather

Scales, claw

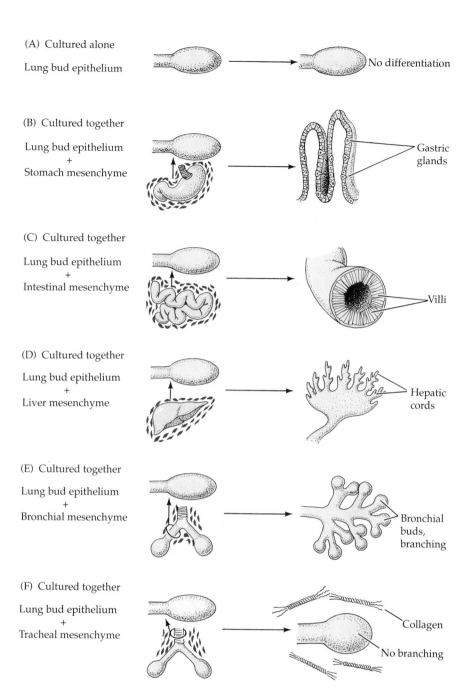

(A) Cultured alone

Lung bud epithelium

No differentiation

(B) Cultured together

Lung bud epithelium
+
Stomach mesenchyme

Gastric glands

(C) Cultured together

Lung bud epithelium
+
Intestinal mesenchyme

Villi

(D) Cultured together

Lung bud epithelium
+
Liver mesenchyme

Hepatic cords

(E) Cultured together

Lung bud epithelium
+
Bronchial mesenchyme

Bronchial buds, branching

(F) Cultured together

Lung bud epithelium
+
Tracheal mesenchyme

Collagen

No branching

FIGURE 18.3
Ability of presumptive lung epithelium to differentiate with respect to the source of the mesenchyme. (A) Lung epithelium does not differentiate when cultured in the absence of mesenchymal cells. (B–F) Mesenchyme-specific differentiation of epithelium. (Modified from Deuchar, 1975.)

This regional specificity of induction is critical during the development of the digestive system and the respiratory system. In the morphogenesis of the endodermal tubes, the endodermal epithelium is able to respond differently to different regionally specific mesenchymes. This enables the digestive tube and respiratory tube to develop different structures in different regions of the tube (Figure 18.3). Thus, as the digestive tube meets new mesenchymes, it differentiates into esophagus, stomach, small intestine, and colon (Gumpel-Pinot et al., 1978; Fukumachi and Takayama, 1980). This regional specificity of mesenchymal induction is dramatically apparent in the formation of the respiratory system. In the developing mammal, it responds in two distinct fashions. When in the region of the neck, it grows straight, forming the trachea. After entering the thorax, it branches, forming two bronchi and then the lungs. The respiratory epithelium can be isolated soon after it has split into two bronchi, and the

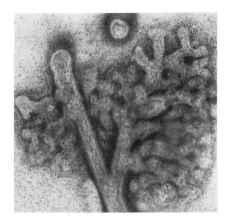

FIGURE 18.4
Ability of presumptive lung epithelium to differentiate with respect to the source of the inducing mesenchyme. After embryonic mouse lung epithelium has branched into two bronchi, the entire rudiment is excised and cultured. The right bronchus is left untouched, while the tip of the left bronchus is covered with tracheal mesenchyme. The tip of the right bronchus forms the branches characteristic of the lung, whereas no branching occurs in the tip of the left bronchus. (From Wessells, 1970, courtesy of N. Wessells.)

two sides can be treated differently. Figure 18.4 shows the result of such an experiment. The right bronchial epithelium retained its lung mesenchyme, whereas the left bronchus was surrounded by tracheal mesenchyme (Wessells, 1970). The right bronchus proliferated and branched under the influence of the lung mesenchyme, whereas the left side continued to grow in an unbranched manner. Thus, respiratory epithelium is extremely malleable and can differentiate according to its mesenchymal instructions.

Genetic specificity of induction

Whereas the mesenchyme may instruct the epithelium as to what sets of genes to activate, the responding epithelium can comply with this information only so far as its genome permits. In a classic experiment, Hans Spemann and Oscar Schotté (1932) transplanted flank ectoderm from an early *frog* gastrula to the region of a *newt* gastrula destined to become parts of the mouth. Similarly, the presumptive flank ectodermal tissue of *newt* gastrula was placed into the presumptive oral regions of *frog* embryos. The structures of the mouth region differ greatly between these salamander larvae and the frog larvae. The *Triturus* salamander larva has club-shaped balancers beneath its mouth, whereas the frog tadpoles produce mucus-secreting glands or suckers (Figure 18.5). The frog tadpoles also have a horny jaw without teeth, whereas the salamander has a set of calcareous teeth in its jaw. The larvae resulting from the transplants were chimeras. The salamander larvae had froglike mouths, and the frog tadpoles had salamander teeth and balancers. In other words, the mesodermal cells instructed the ectoderm to make a mouth, but the ectoderm

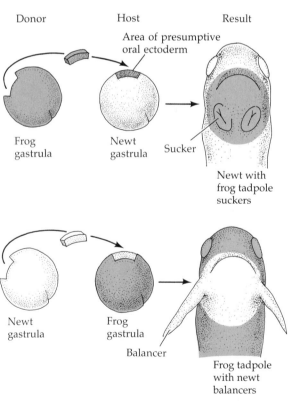

FIGURE 18.5
Genetic specificity of induction. Reciprocal transplantation between the presumptive oral ectoderm regions of newt and frog gastrulae leads to newt larvae with tadpole suckers and to frog tadpoles with newt balancers. (After Hamburgh, 1970.)

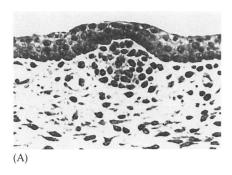
(A)

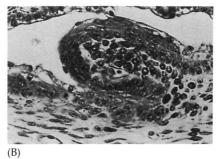
(B)

FIGURE 18.6

Genetic specificity of cutaneous induction. (A) Section of the corneal region of a 17-day chick embryo. At 5 days of incubation, the lens of this eye had been replaced by the flank dermis of an early mouse embryo. A condensation of the mouse embryo cells is located directly beneath the chick epithelium. (B) Feather forming from the corneal epithelium from such a specimen. Mouse cells are present in the feather rudiment. (From Coulombre and Coulombre, 1971, courtesy of A. J. Coulombre.)

responded by making the only mouth it "knew" how to make, no matter how inappropriate.[*]

The same genetic specificity is seen in combinations of chicken skin and mouse skin (Coulombre and Coulombre, 1971). When ectoderm normally destined to become cornea is isolated from chicken embryos and combined with chick skin mesoderm, the ectoderm produces feather buds typical of chick skin. Moreover, when the same tissue—presumptive cornea ectoderm—is combined with *mouse* skin mesoderm, feather buds also appear (Figure 18.6). The mouse mesoderm has instructed the chick cornea to make a cutaneous structure. This would normally be hair, for the mouse. The competent chick ectoderm, however, does the best it can, developing its cutaneous structures—namely, feathers.

Thus, the instructions sent by the mesenchymal tissue can cross species barriers. Salamanders respond to frog signals, and chick tissue responds to mammalian inducers. The response of the epithelium, however, is species-specific for that epithelium. So, whereas organ-type specificity (feather or claw) is usually controlled by the mesenchyme within a species, species specificity is usually controlled by the responding epithelium.

This observation has been used to generate a long-lost structure: hen's teeth (Kollar and Fisher, 1980). Epithelium from the jaw-forming region of 5-day-old chick embryos (the first and second pharyngeal arches) was isolated and combined with molar mesenchyme of 16- to 18-day-old mouse embryos. These recombined tissues were allowed to adhere to each other and were then cultured within the anterior chamber of a mouse eye. Several recombinations resulted in the formation of complete teeth that were unlike those of mammals (Figure 18.7). The cells of the chick pharyngeal arches, which have not made teeth for nearly 100 million years, still retain the genetic potential to do so in response to an appropriate inducer.[†]

[*]Spemann is reported to have put it as follows: "The ectoderm says to the inducer, 'you tell me to make a mouth; all right, I'll do so, but I can't make your kind of mouth; I can make my own and I'll do that.'" (Quoted in Harrison, 1933.)

[†]This may not be the case for all birds, however. When quail mandibular epithelium is recombined with mouse molar mesenchyme, teeth do not form. Such experiments must be done with remarkable care, because very small numbers of mesenchyme cells can instruct tooth morphogenesis in the epithelium (Arechaga et al., 1983). The importance of induction and competence during evolution will be detailed in Chapter 23.

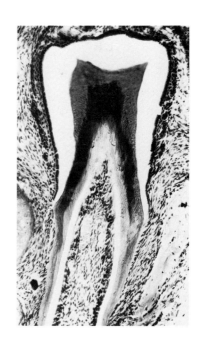

FIGURE 18.7

"Hen's tooth" formed after combination of chick pharyngeal (presumptive jaw) epithelium and mouse molar mesenchyme. (From Kollar and Fisher, 1980, courtesy of E. J. Kollar.)

Cascades of embryonic induction: Lens induction

The phenomena of lens induction

Proximate cell interactions provide a mechanism whereby coordinated organ development can occur, for a responding tissue can also become an inducing tissue. Recent studies have shown that secondary induction is a very complex process. Indeed, what we have traditionally been calling "secondary" inductions are usually only the last induction in a cascade that began much earlier in embryogenesis, and many tissues acquire their competence through a previous induction. Although these tissues may look unchanged through a microscope, they have been induced such that they can respond to a new inducer. This is probably true of the epidermal inductions mentioned above, and it is certainly true in the case of that most intensively studied secondary induction, the formation of the lens.

The optic cup model of lens induction. As discussed in Chapter 7, the cells that form the lens are derived from a region of head ectoderm that is contacted by the optic vesicle of the forebrain. This work was pioneered by Hans Spemann, and his review of these studies in 1938 has made lens induction the paradigm of secondary inductive events. The basic experiments were as follows. First, when Spemann (1901) destroyed the optic vesicle primordium of the frog *Rana temporaria*, lenses failed to develop. Thus, Spemann concluded that contact of the optic vesicle with the overlying ectoderm was essential for inducing the formation of the lens. Second, Warren Lewis (1904, 1907) confirmed and extended this conclusion. He removed optic vesicles from late-stage neurulae and transplanted them to the head ectoderm of regions that would not usually form lenses. He found that the head ectoderm of this region would then form lens-like structures, and he concluded that the optic vesicle was sufficient to induce the formation of lens tissue in ectoderm that would not otherwise have formed it. It appeared that contact with the optic vesicle was all that was needed to induce lenses in the overlying ectoderm.

Disagreements with the optic cup model. There were dissenters from this view, however. Mencl (1908) noticed that certain fish have congential defects wherein no eye forms. Nevertheless, these fish have lenses in their head ectoderm. More importantly, when King (1905) tried to repeat Spemann's experiments, she found, contrary to her expectations, that lenses still formed even when the optic vesicle rudiments had been obliterated. These and other investigators began to find that lenses could form without contact with optic vesicle.*

As more data began accumulating in various species, it seemed that there was a great deal of species diversity. Some species appeared to form lenses without the need for optic vesicles, while the lenses of other species appeared to be completely dependent on the optic vesicle contact. Spemann (1938) reconciled these results by arguing that an organism could evolve a margin of safety by developing two ways of forming a particular tissue. Thus, the lens would normally arise by contact with the optic vesicle, but, failing this, could arise separately if it had to. This concept

---

*The interpretation of these experiments has been extremely difficult because of species differences in the induction mechanisms, the temperatures at which maximal induction takes place, and the difficulty in getting uncontaminated pieces of tissue for transplantation. See Jacobson and Sater (1988) and Saha (1989, 1991) for reviews of these dissenters and their experiments.

was called the "double assurance" hypothesis. In 1966, Jacobson integrated more data into this model. He noted that the lens-forming ectoderm sequentially comes into contact with presumptive foregut endoderm, presumptive heart mesoderm, and the optic vesicle. Each of these tissues, he predicted, would act in an additive way to induce the formation of lenses in this tissue. In some species, the threshold for lens induction would be low, and contact with the endoderm would be sufficient. In other species, the threshold would be high, and all three inducers would have to be active. Here, lens formation would seem to be dependent upon the optic vesicle, but in actuality, the optic vesicle would be only the last of three inducers.

The cellular basis of lens induction

While not ruling out the role of the mesoderm and endoderm, recent studies in *Xenopus* stress the importance of the anterior neural plate as an early inducer of lens ectoderm. These experiments indicate that the presumptive lens ectoderm receives its ability to become lens very early in development (during late-gastrula to mid-neurula stages) and that the optic vesicle merely localizes the differentiation of this already autonomous tissue. In other words, the head ectoderm will form lenses without the contact of the optic cup, but the optic cup is needed for the full differentiation of the lens and its proper positioning with respect to the rest of the eye. This model (Figure 18.8; Saha et al., 1989; Grainger, 1992) divides the determination of lens ectoderm into three stages: competence, bias, and final determination. Competence to respond to the initial inducing signal is seen to be an autonomous process within the ectoderm, bias to produce lenses is provided by the anterior neural plate, and final determination is induced by the optic vesicle.

Ectodermal competence and bias. In 1987, Henry and Grainger demonstrated that the determination of lens-forming ability occurs very early in *Xenopus* development. They transplanted ectoderm from various regions of *Xenopus* gastrulae into the lens-forming region of the neurula. Would that ectoderm be able to form a lens when contacted hours later by the optic vesicle? Ectoderm from very young gastrulae did not form lenses. When ectoderm from the same prospective lens ectoderm in later embryos was grafted to the neurula, it was able to respond to the optic vesicle by forming a lens (Table 18.2). No other tissue responded in this way. Other ectodermal regions from gastrulae also had a limited ability to form lenses, but this ability was lost as development proceeded.

It appeared, then, that the lens-forming ectoderm achieved the ability to form lenses long before the optic vesicle contacted it. When was this lens-forming state achieved, and how was it effected? Experiments by Nieuwkoop (1952) had suggested that a signal from the neural plate could travel through the ectoderm. Could the neural plate induce the presumptive epidermis lateral to it to become lens-forming ectoderm? Henry and Grainger (1990) tested this hypothesis by combining the prospective anterior neural plate region of late-gastrula embryos with ectoderm from the region that would eventually become lens. While isolated ectoderm from the potential lens-forming region did not form lens proteins when cultured alone, the same region did form lens proteins when cultured next to the prospective anterior neural plate tissue. Although lens differentiation was often rudimentary, it was very specific. Lens proteins were not made when the gastrula ectoderm was combined with other tissues, including foregut endoderm or cardiac mesoderm (Figure 18.9A). These experiments show

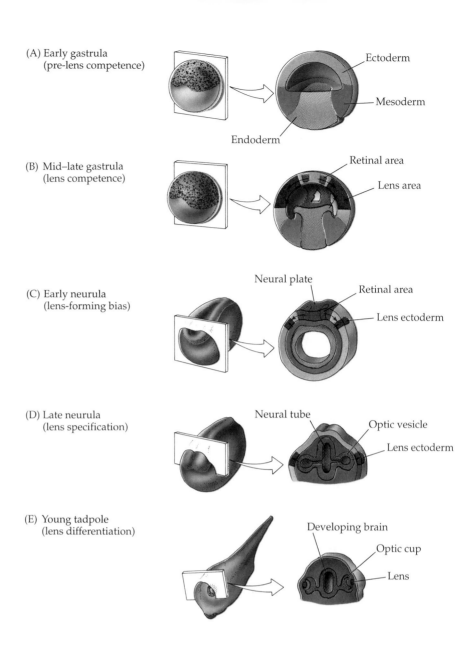

(A) Early gastrula
(pre-lens competence)

Ectoderm

Mesoderm

Endoderm

(B) Mid–late gastrula
(lens competence)

Retinal area

Lens area

(C) Early neurula
(lens-forming bias)

Neural plate

Retinal area

Lens ectoderm

(D) Late neurula
(lens specification)

Neural tube

Optic vesicle

Lens ectoderm

(E) Young tadpole
(lens differentiation)

Developing brain

Optic cup

Lens

that the anterior portion of the prospective neural plate (which contains the future retinal regions) provides a signal that biases this tissue to become lens.

But are all tissues able to respond to the signal coming from the anterior neural plate? Servetnick and Grainger (1991) showed that only mid- to late-gastrula ectoderm is competent to respond to these signals. They removed animal cap ectoderm from various gastrula stages and then transplanted them into the presumptive lens region of neural plate-stage embryos (Figure 18.9B). Ectoderm from early gastrulae showed little or no competence to form lenses (as assayed by the production of crystallin proteins), but ectoderm from slightly later stages was able to form lenses. By the end of gastrulation, this ability to respond to the neural plate signal had been lost. This competence was seen to be inherent within the ectoderm itself and was not induced by other surrounding tissues. The animal cap ectoderm from various embryonic stages could be removed, cultured in glass for a certain period of time, and then placed back into neural plate-stage embryos. Such ectoderm showed the same pattern of compe-

FIGURE 18.8
A current model for lens induction. The inductive signals are indicated by arrows. (A) At early gastrula, the ectoderm has not achieved the competence to become lens (although it does have competence to become neural tissue). (B) During midgastrula, the lens-forming ectoderm becomes competent to respond to lens-inducing signal from the presumptive neural plate (possibly the presumptive retinal cells). During late gastrula, this signal from neuralized ectoderm (most likely the presumptive eye region) induces the presumptive lens-forming ectoderm. An additional inductive signal may be coming from the prospective foregut endoderm. (C) At early neurula, the signals from the anterior neural region and endoderm have caused a lens-forming bias in the head ectoderm. This signal may be reinforced by induction from the anterior lateral mesoderm. (D) At the late neurula stage, the optic vesicle contacts the lens-forming ectoderm, signaling the final determination of this tissue into lens. (E) At tadpole stage, the presumptive lens ectoderm differentiates into lens tissue. (After Saha et al., 1989; Grainger, 1992.)

tence even though it had spent part of its development inside a petri dish. It appears, then, that the ectoderm acquires the competence to respond to inducing signals from the anterior neural plate at the early midgastrula stages, and during late gastrula, the anterior neural plate induces a lens-forming bias in this tissue. Similar experiments have shown that dorso-lateral mesoderm at this stage, while it will not induce the lens-forming bias in the ectoderm, will enhance the lens-forming bias imparted by the neural plate.

Final determination of the lens. Although the optic vesicle does not appear to play a major role in the initial induction of lens formation in *Xenopus*, it does play a role in enabling the complete lens phenotype to become expressed. The lenses that form in the absence of the optic vesicle are usually very rudimentary. It is not known whether the influence of the optic vesicle is a directly positive one, promoting the differentiation of the lens placode into a fully differentiated lens, or if the influence is to remove an inhibitor of lens differentiation. It has been proposed (von Woellwarth, 1961; Henry and Grainger, 1987) that the cranial neural crest

TABLE 18.2
Increasing responsiveness of prospective lens ectoderm with age

Donor stage	Operation Donor	Host neurula	Number examined	Induced lens (%)	Lens-like body (%)	Ectodermal thickening (%)	Non-lens body (%)	No response (%)	Total positive
Midgastrula	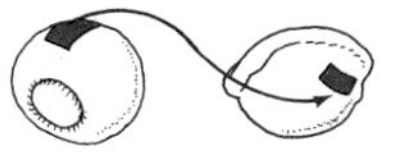		24	0	4	38	8	50	1 (4%)
Late gastrula			21	10	14	42	10	24	5 (24%)
Early neurula	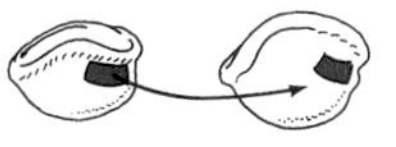		24	75	8	0	4	13	20 (83%)
Late neurula	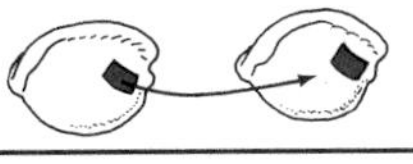		20	95	5	0	0	0	20 (100%)

Source: After Henry and Grainger, 1987.

FIGURE 18.9

Early determination of lens-forming ability in *Xenopus* ectoderm. (A) The source of lens-determining signal was found to be the anterior neural plate. Presumptive lens ectoderm was cultured with either lateral endoderm/ mesoderm or with anterior neural plate (the two major tissues adjacent to it). The ectoderm formed lens proteins only when cultured with the neural plate. (B) The time at which the anterior neural plate cells could induce competence in the ectoderm was determined by transplanting presumptive ectoderm from different stage donor gastrulae into the lens forming region of the neurula. Only the ectoderm from midgastrula embryos was competent to respond to the signals. (After Grainger, 1992.)

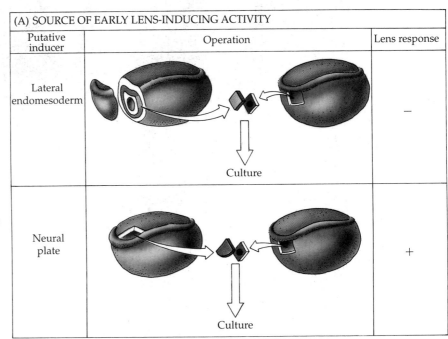

(A) SOURCE OF EARLY LENS-INDUCING ACTIVITY

Putative inducer	Operation	Lens response
Lateral endomesoderm	Culture	−
Neural plate	Culture	+

(B) DETERMINATION OF LENS-COMPETENT PERIOD

Stage	Operation	Lens response
Early gastrula		−
Mid gastrula		+
Late gastrula		−

cells prevent lens differentiation, and that contact with the optic vesicle shields the lens placode from these inhibiting signals.

The lens is placed between the anterior and vitrous chambers of the eye, and it is thought that the differentiation of the lens (discussed in Chapter 7) is mediated by growth factors emanating from these two chambers. The anterior chamber appears to concentrate a mitogenic protein (whose identity remains unknown) that is specific for causing mitosis and inhibiting differentiation in the lens-forming epithelium. This protein is thought to come into the anterior chamber from the blood capillaries. In the vitreous chamber, the acidic and basic forms of fibroblast growth factor stimulate the elongation and differentiation of the lens cells and block the

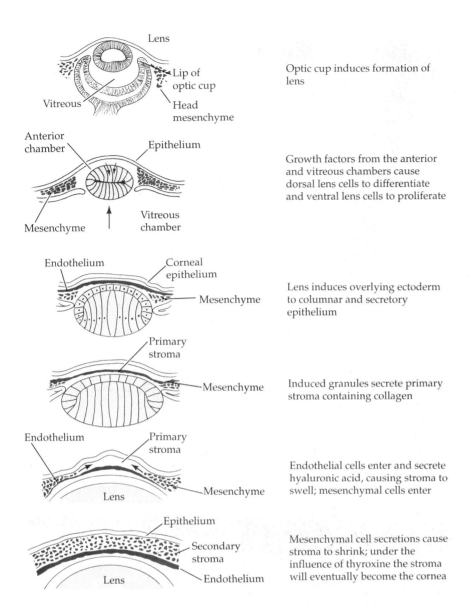

Optic cup induces formation of lens

Growth factors from the anterior and vitreous chambers cause dorsal lens cells to differentiate and ventral lens cells to proliferate

Lens induces overlying ectoderm to columnar and secretory epithelium

Induced granules secrete primary stroma containing collagen

Endothelial cells enter and secrete hyaluronic acid, causing stroma to swell; mesenchymal cells enter

Mesenchymal cell secretions cause stroma to shrink; under the influence of thyroxine the stroma will eventually become the cornea

FIGURE 18.10
Lens and corneal development. The optic cup induces the final determination of the lens. Mitogenic proteins (black arrow) in the anterior chamber maintain a line of proliferating cells in the ventral surface of the lens, while fibroblast growth factors (colored arrows) stimulate the differentiation of the dorsal lens epithelium. Under the inductive influence of the lens, the corneal epithelium differentiates and secretes a primary stroma consisting of collagen layers. Endothelial cells then secrete hyaluronic acid into this region, enabling mesenchymal cells from the neural crest to enter. Afterward, hyaluronidase (secreted by either the mesenchyme or endothelium) digests the hyaluronic acid, causing the primary stroma to shrink. (After Hay and Revel, 1969; Hyatt and Beebe, 1993.)

mitogenic activity of the anterior chamber growth factor (Hyatt and Beebe, 1993; Schultz et al., 1993). The result is the elongation of those lens cells on the dorsal surface of the lens placode, and the continued proliferation of those cells on the ventral side of the lens placode (Figure 18.10).

Cornea formation

Once the lens placode has invaginated, it becomes covered by two cell layers from the adjacent ectoderm. Now the developing lens can act as an inducer. The ectoderm that is to become the cornea has probably also been determined during an earlier stage of development (Meier, 1977). Now, under the influence of the lens, corneal differentiation takes place. The overlying ectoderm becomes columnar and fills with secretory granules. These granules migrate to the bases of the cells and secrete a primary stroma containing about 20 layers of types I and II collagen (Figure 18.10). Neighboring capillary endothelial cells migrate into this region (upon the primary stroma) and secrete hyaluronic acid into this matrix. The hyaluronic acid causes the matrix to swell and to become a good substrate for the migration of two waves of mesenchymal cells derived from the neural crest. Upon entering the matrix, the second wave of mesenchymal cells remains there, secreting type I collagen and hyaluronidase. The hyaluronidase causes the stroma to shrink. Under the influence of thyroxine from the developing thyroid gland, this secondary stroma is dehydrated and the collagen-rich matrix of epithelial and mesenchymal tissues becomes the transparent cornea (see Hay, 1980; Bard, 1990).

We can see, then, that "simple" inductive interactions are actually well-coordinated dramas in which the actors must come on stage and speak their lines at the correct times and positions. In acquiring new information, they can also impart information for others to use. With this in mind, we can now study some principles about secondary induction obtained from other developing organs.

Formation of parenchymal organs

Epithelio-mesenchymal interactions are also seen in the formation of duct-forming organs such as the kidney, liver, lung, mammary gland, and pancreas. In the formation of these organs we see the reciprocal induction of the mesenchyme upon the epithelium and the epithelium upon the mesenchyme.

Morphogenesis of the mammalian kidney

The progression of renal tubules. Like the eye, the mammalian kidney is an exceedingly intricate structure. Its functional unit, the nephron, contains over 10,000 cells and at least 12 different cell types, each cell type located in a particular place in relation to the other cell types along the length of the nephron. The development of the mammalian kidney progresses through three major stages. Early in development (day 22 in humans; day 8 in mice), the **pronephric duct** arises in the mesoderm just ventral to the anterior somites. The cells in this duct migrate caudally (as discussed for amphibian embryos in Chapter 3), and the anterior region of the duct induces the adjacent mesenchyme to form the **pronephric kidney tubules** (Figure 18.11A). While these pronephric tubules form functioning kidneys in fish and in amphibian larvae, they are not thought to be active in mammalian amniotes. In mammals, the pronephric tubules and the anterior portion of the pronephric duct degenerate, but the more caudal portions of the pronephric duct persist and become the central component of the excretory system throughout its development (Saxén, 1987). This remaining duct is often referred to as the **nephric** or **Wolffian duct**.

As the pronephric tubules degenerate, the midportion of the nephric duct initiates a new set of kidney tubules in the adjacent mesenchyme. This set of tubules constitutes the **mesonephros**, or **mesonephric kidney**. In humans, about 30 mesonephric tubules form, beginning around day 25. However, as more tubules are induced caudally, the anterior mesonephric tubules begin to regress (Figure 18.11B). In female mammals, this regression is complete. However, as we will discuss in Chapter 21, some of these mesonephric tubules persist in male mammals to become the sperm-carrying tubes (the vas deferens and efferent ducts) of the testes.

The permanent kidney of amniotes, the **metanephros**, is generated by some of the same components as the earlier, transient, kidney types, and it is thought to originate by a five-step process. In the first two steps, the **metanephrogenic mesenchyme** forms in posteriorly located regions of the

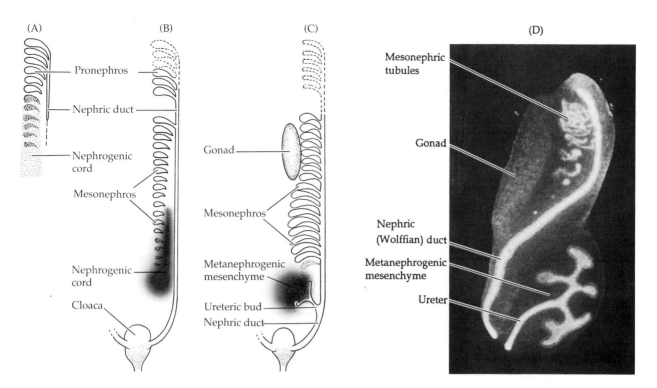

(A) Pronephros
Nephric duct
Nephrogenic cord
Mesonephros
Nephrogenic cord
Cloaca

(B)

(C) Gonad
Mesonephros
Metanephrogenic mesenchyme
Ureteric bud
Nephric duct

(D) Mesonephric tubules
Gonad
Nephric (Wolffian) duct
Metanephrogenic mesenchyme
Ureter

FIGURE 18.11
General scheme of development in the vertebrate kidney. (A) The original tubules, constituting the pronephric kidney, are induced from the nephrogenic mesenchyme by the pronephric duct as it migrates caudally. (B) As the pronephros degenerates, the mesonephric tubules form. (C) The final mammalian kidney, the metanephros, is induced by the ureteric bud. (D) Section of a mouse kidney showing the initiation of the metanephric kidney (bottom) while the mesonephros is still apparent. The duct tissue is stained with a fluorescent antibody to a cytokeratin found in the pronephric duct and its derivatives. (A–C after Saxén, 1987; D courtesy of S. Vainio.)

intermediate mesoderm, and it induces the formation of a branch from each of the paired nephric ducts. These epithelial tubes are called the **ureteric buds**. These buds eventually separate from the nephric duct to become the ureters that take the urine to the bladder. When the ureteric buds emerge from the nephric duct, they enter the metanephrogenic mesenchyme. In the third and fourth steps, the ureteric buds induce this mesenchymal tissue to condense around the buds and to differentiate into the nephrons of the mammalian kidney. The fifth step of kidney initation occurs when this nephron-forming tissue induces the further branching of the ureter bud (Figure 18.11C,D).

Reciprocal induction during kidney development. These two mesodermal tissues, the ureteric bud and the metanephrogenic mesenchyme, interact and reciprocally induce each other (Figure 18.12). The metanephrogenic mesenchyme causes the ureteric bud to elongate and branch. At the tip of these branches, the ureteric bud induces the loose mesenchymal cells to form an epithelial aggregate. Each aggregate of about 20 cells will divide and differentiate into the intricate structure of the renal nephron. First, each node elongates into a "comma" shape and then forms a characteristic S-shaped tube. Soon after the S-shaped tube is formed, the cells of this epithelium begin to differentiate into the regionally specific cell types such as the capsule cells, the podocytes, and the distal and proximal tubule cells. At this time. a connection develops between the ureteric bud and the newly formed tube, thereby enabling material to pass from one to the other. The newly formed tubes derived from the mesenchyme form the secretory nephrons of the functioning kidney, and the branched ureteric bud gives rise to the renal collecting ducts and to the ureter, which drains the urine from the kidney.

Clifford Grobstein (1955, 1956) documented this **reciprocal induction** in vitro. He separated the ureteric bud from the mesenchyme and cultured them individually or together. In the absence of mesenchyme, the ureteric

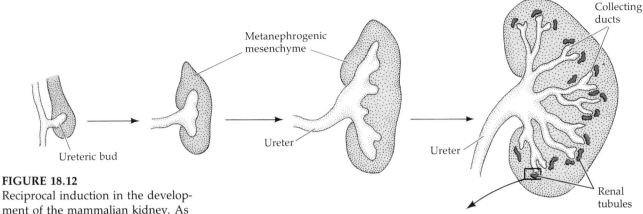

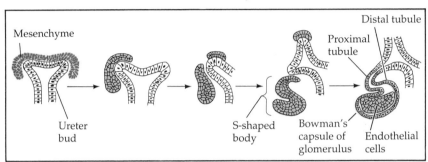

FIGURE 18.12
Reciprocal induction in the development of the mammalian kidney. As the ureteric bud enters the metanephrogenic mesenchyme, the mesenchyme induces the bud to branch. At the tips of the branches, the epithelium induces the mesenchyme to aggregate and cavitate to form the renal tubules. The formation of the nephron from the mesenchymal cells is shown in the insert. After aggregating at the branches, the mesenchymal cells form an epithelial node that extends into an S-shaped tube, which fuses with the ureter bud epithelium. (Insert after Romanoff, 1960.)

bud does not branch. In the absence of the ureteric bud, the mesenchyme soon dies. When they are placed together, however, the ureteric bud grows and branches, and tubules form throughout the mesenchyme (Figure 18.13). Although certain other tissues (notably neural tube) will enable the metanephrogenic mesenchyme to form kidney tubules, the ureteric bud branches only under instructions from the metanephrogenic mesenchyme. Mesenchymes that induce branching in other epithelia (such as salivary

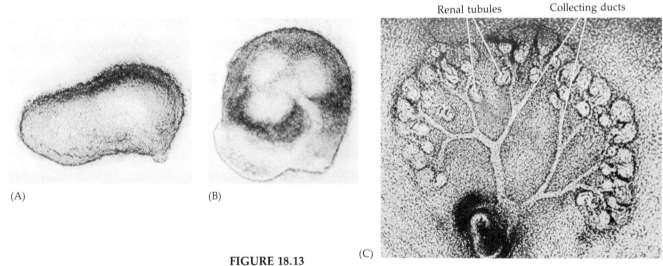

(A)   (B)   (C)

FIGURE 18.13
Kidney induction observed in vitro. (A) An 11-day mouse metanephric rudiment includes both ureteric bud and metanephrogenic mesenchyme. (B) After the first day of culture, tubules can be seen at the tips of the branching ureters. (C) The branching collecting ducts formed by the ureteric bud and the kidney tubules formed by the mesenchymal condensations at the tips of these buds can be clearly seen after eight days of culture. (A and B from Saxén and Sariola, 1987; C from Grobstein, 1955; all photographs courtesy of the authors.)

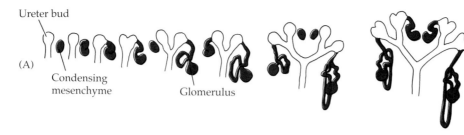

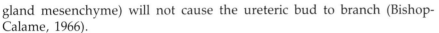

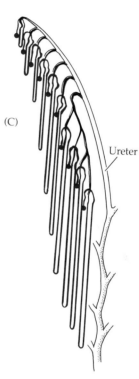

gland mesenchyme) will not cause the ureteric bud to branch (Bishop-Calame, 1966).

The first metanephric nephrons, then, are immediately joined to the collecting ducts. These nephrons are carried outward into the metanephrogenic mesenchyme as the ureter continues to grow (Figure 18.14A). The ends of the ureteric buds, however, retain their ability to induce tubule formation in this mesenchyme, and the result is the formation of tubular arcades (Figure 18.14B). As the ureteric branch migrates further through the mesenchyme, new nephrons are formed and become joined to the same collecting duct (Figure 18.14C; Osathanondh and Potter, 1963).

The mechanisms of kidney organogenesis

The metanephrogenic mesenchyme. It is one thing to say that the ureteric bud induces the metanephrogenic mesenchyme to become the epithelium of the nephrons. It is another thing to understand how this process occurs. Like the development of the lens by the optic vesicle, it is thought that induction by the ureteric bud is only the final step that triggers a cascade of events in the competent mesenchyme. Only the metanephrogenic mesenchyme has the ability to respond to the ureteric bud to form kidney tubules, and if induced by other epithelia (such as embryonic salivary gland or neural tube), the metanephrogenic mesenchyme will respond by forming kidney tubules and no other structures (Saxén, 1970; Sariola et al., 1982). Thus, the metanephrogenic mesenchyme cannot become any other tissue than kidney tubules (if induced) or renal stromal cells (if uninduced). Based on his experiments on the induction of the chick mesonephros, Etheridge (1968) has suggested that this determination is the result of interactions with the endoderm earlier in development.

Although this mesenchyme appears homogeneous, recent evidence (Sariola, 1989; Sainio et al., in press) suggest a dual origin for this tissue. Part of it would indeed be mesoderm-derived mesenchyme, and these cells would form the nephron. The other cells have neural characteristics and may be derived from the neural crest. These latter cells may be important in inducing the nephron to form.

FIGURE 18.14
Schematic representation of human nephron development. (A) Formation of early nephrons directly attached to the ureteric bud epithelium. (B) Formation of nephron arcades where several nephrons are joined to the same collecting duct. (C) General arrangement of human nephrons at birth. The deeper nephrons constitute an arcade, while the nephrons closer to the surface are directly connected to the collecting duct of the ureter. (After Osathanondh and Potter, 1963.)

Formation of the ureteric bud.   The second step in kidney development is the formation of the two ureter buds from the nephric ducts. While the substance causing this branching has not yet been identified, the inducer appears to be made by the mesenchyme cells and is regulated by the *WT-1* gene. This gene, discussed in Chapter 10, encodes a zinc-finger protein that is thought to be a transcription factor. In situ hybridization shows that this gene is normally first expressed in the intermediate mesoderm prior to kidney formation, and is then expressed in the developing kidney, gonad, and mesothelium (Pritchard-Jones et al., 1990; van Heyningen et al., 1990; Armstrong et al., 1992). The importance of this gene in kidney development has been shown by gene knockout experiments that delete *WT-1* from the mouse genome (Kreidberg et al., 1993). These *WT-1*-deficient mice lack kidneys and gonads, and they die before birth. In those cells that would form the kidney, the lack of *WT-1* appears to destroy the ability of the metanephrogenic mesenchyme to induce the formation of the ureter bud. This bud does not grow out from the ureter, and hence the kidney is not induced. In another mouse mutation, the *Danforth short-tail* mutant, the ureter bud is initated, but doesn't enter into the metanephrogenic mesenchyme (Gluecksohn-Schoenheimer, 1943). Here, too, the kidney does not form.

Conversion of mesenchyme cells into an epithelium.   The third stage in renal development consists of the interactions between ureter bud and metanephrogenic mesenchyme whereby some of the mesenchymal cells organize themselves into an epithelium. As discussed in Chapter 3, such changes would involve substantial remodeling of the extracellular matrix, and indeed, the ureteric bud causes dramatic changes in the extracellular matrix of the metanephrogenic mesenchyme cells. The uninduced mesenchyme secretes an extracellular matrix consisting largely of fibronectin and collagen types I and III. Upon induction, these proteins disappear and are replaced by an epithelial basal lamina made of laminin and type IV collagen. The cytoskeleton also changes from one typical of mesenchymal cells to one typical of epithelia (Ekblom et al., 1983; Lehtonen et al., 1985). In this manner, the loose mesenchymal cells are linked together as a polarized epithelium on a basal lamina.

Prior to these changes, the newly induced metanephrogenic mesenchyme synthesizes two critical proteins. The first is the membrane proteoglycan **syndecan** (Vainio et al., 1989a). This adhesive proteoglycan is first seen around the mesenchymal cells surrounding the ureteric bud as the bud first enters the region of mesenchyme. As the ureteric bud initiates its first branch, the entire mesenchymal region around the branch stains positively for syndecan (Figure 18.15). By using inducers from different species and species-specific antibodies for syndecan, Vainio and co-workers demonstrated that the expression of syndecan is under translational regulation. Syndecan *mRNA* is present in the uninduced kidney mesenchyme, but it is not translated into protein unless the mesenchyme is induced (Figure 18.16; Vainio et al., 1992).

Syndecan might not only regulate the *condensation* of the mesenchyme into an epithelium, but it also might promote the *proliferation* of these cells. By labeling proliferating cells with bromodeoxyuridine (which is incorporated into DNA only if the cells are dividing) and labeling the syndecan-expressing cells with fluorescent antibodies to syndecan, Vainio and colleagues (1992) demonstrated a close correlation between the dividing cells and those expressing syndecan. Thus, syndecan may be playing two roles in kidney morphogenesis. First, it may promote the aggregation of the induced mesenchymal cells through its adhesive properties. Second, it

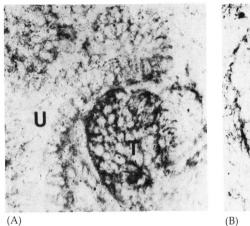

(A)

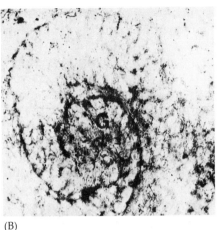

(B)

FIGURE 18.15
The extracellular matrix proteoglycan syndecan is not synthesized or secreted by mesenchymal cells until after induction. This molecule is probably involved in structuring the new tubule epithelium, and it distinguishes the cells of the tubule from the remaining mesenchyme. (A) Immunological staining of syndecan shows its presence in the newly induced mesenchyme cells (T) that are becoming epithelial. Some staining (U) is also seen on the ureter bud epithelium. (B) Intense syndecan staining is seen in the region of the developing tubule that is to become the renal glomerulus (G). (From Vainio et al., 1989a, courtesy of L. Saxén.)

may promote proliferation by binding growth factors. If this is correct, one of the major functions of induction is to enable the translation of syndecan mRNA into protein within the metanephrogenic mesenchyme. The second protein synthesized in the induced mesenchyme is Pax-2. When antisense RNA to Pax-2 prevents the translation of the *Pax-2* mRNA that is transcribed as a response to induction, the mesenchyme cells of cultured kidney rudiments fail to condense (Rothenpieler and Dressler, 1993).

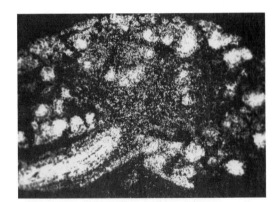

(A)

(B)

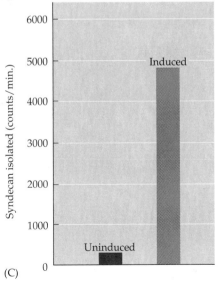

(C)

FIGURE 18.16
Syndecan expression in induced and uninduced kidney mesenchymes. (A) In situ hybridization localizing syndecan mRNA in the mesenchymal aggregates of a 15-day embryonic mouse kidney. Visualization of the autoradiograph is by dark-field illuminescence. (B) Isolated kidney mesenchyme (M) induced by spinal cord (SPC) shows intense syndecan expression when stained with fluorescent antibodies to syndecan. Uninduced mesenchyme does not. (C) The amount of syndecan (labeled with radioactive sulfur) isolated from an induced kidney mesenchyme is tenfold greater than that isolated from a similar amount of uninduced mesenchyme. (After Vainio et al., 1992, courtesy of S. Vainio.)

Conversion of the aggregated cells into a nephron. In the fourth stage of kidney formation, the condensed epithelium is specified into the different cell types of the nephron. While the epithelial cells of the nephron are probably derived from the mesodermal cells of the metanephrogenic mesenchyme, their specification may involve their interacting with other mesenchymal cells (the ones displaying neural proteins). In addition to causing the aggregation of the first group of mesenchyme cells, the ureter buds may prevent the apoptosis (programmed cell death) of the second (neurally marked) mesenchymal cells (Koseki et al., 1992; Coles et al., 1993; Sainio et al., in press). Once these cells begin their interaction, the ureteric bud is no longer essential for further nephron development.

In this stage, the genes responsible for cell specification are activated. In recent years, three genes have been found whose products may be important for this specification. The first is the gene for gap junction protein, connexin 43. This protein is seen in the condensed mesenchyme and connects the cells of the S-shaped body (Sainio et al., 1992). The second gene is the *Pax-2* gene. *Pax-2* is active in the condensed mesenchyme and is turned off as the cells differentiate. If it remains active, the podocytes, glomerulus, and proximal tubule cells form abnormally (Dressler et al., 1992).

The third of these genes encodes the low-affinity receptor for nerve growth factors, **NGFR**. This NGFR is probably not the receptor for the first signal, because it is not found on the uncondensed mesenchyme. Thus, it probably does not receive the signal to make syndecan and condense. However, it is abundantly present on the condensed cells that later form the nephrons. When antisense oligonucleotides to NGFR were added to rat kidney rudiments in culture, they were taken up by the mesenchymal cells and prevented the translation of NGFR mRNA. These kidneys had gross defects (Figure 18.17; Sariola et al., 1991). First, the condensed mesenchyme cells failed to form epithelial tubules. Thus, no secretory nephrons were formed. Second, without the reciprocating signal from the epithelializing tubule, the ureter buds were not instructed to branch. Therefore, NGFR appears to play a critical role in turning the condensing mesenchyme into the nephron.

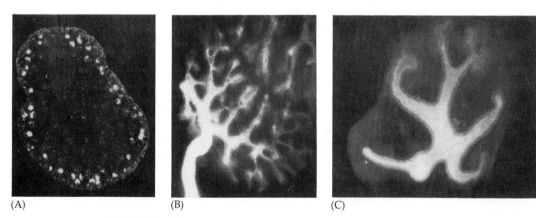

(A)   (B)   (C)

FIGURE 18.17
The role of the low affinity NGF receptor in kidney morphogenesis. (A) In situ hybridization shows the localization of NGFR mRNA in the condensed mesenchymes of an 18-day rat embryonic kidney. (B) Higher magnification of the branching pattern of the ureter bud (stained with antibodies to an epithelial specific cytokeratin) in a 13-day kidney cultured 5 days in vitro. (C) Ureter bud from a kidney similar to that in (B) but cultured in antisense oligonucleotides to NGFR mRNA. (From Sariola et al., 1991, courtesy of H. Sariola.)

Coordinated differentiation and morphogenesis

During the morphogenesis of any organ, numerous dialogues are occurring between the interacting tissues. In epithelio-mesenchymal interactions, the mesenchyme influences the epithelium; the epithelial tissue, once changed by the mesenchyme, can secrete factors that change the mesenchyme. Such interactions continue until an organ is formed with organ-specific mesenchymal cells and organ-specific epithelia. The identification of the chemicals involved in these intertissue conversations is underway in several laboratories. Some of the most extensively studied interactions are those that form the mammalian tooth. Here, the jaw epithelium differentiates into the enamel-secreting **ameloblasts**, while the neural crest-derived mesenchymal cells become the dentin-secreting **odontoblasts**.

First, the epithelium causes the mesenchyme to aggregate at specific sites. At this time, the *epithelium* possesses the potential to generate tooth structures out of several types of mesenchymal cells (Mina and Kollar, 1987; Lumsden, 1988). However, this tooth-forming potential soon becomes transferred to the mesenchyme that aggregated beneath it. These *mesenchymal* cells form the dental papilla and are now able to induce tooth morphogenesis in other epithelia (Kollar and Baird, 1970). At this stage, the jaw epithelium has lost its ability to instruct tooth formation in other mesenchymes. Thus, the "odontogenic potential" has shifted from the epithelium to the mesenchyme. At the basement membrane that separates the epithelium from the mesenchyme, the epithelium induces the mesenchyme to become the odontoblasts, while the mesenchyme induces the epithelium to become the ameloblast cells (Figure 18.18; Thesleff et al., 1989).

This shift in the odontogenic potential coincides with a shift in the synthesis of bone morphogenetic protein 4 (**BMP-4**, which was discussed in Chapters 9 and 16). During the earliest phases of tooth development, BMP-4 is synthesized in the epithelium. This epithelial BMP-4 induces the differentiation of the mesenchyme and stimulates the mesenchyme to express three transcription factors, including homeodomain-containing proteins msx-1 and msx-2 The induction of mesenchymal differentiation can be mimicked by placing BMP-4 on agarose beads and applying them to the mesenchymal mass (Vainio et al., 1993). Thus, BMP-4 appears to be a critical morphogenetic signal from the epithelium to the mesenchyme.

A summary of recent research correlating mesenchyme induction and differentiation is shown in Figure 18.18. As can be seen, the mesenchyme at one stage is different from the mesenchyme at other stages. The mesenchymal cells are first induced (by BMP-4) to express a set of transcription

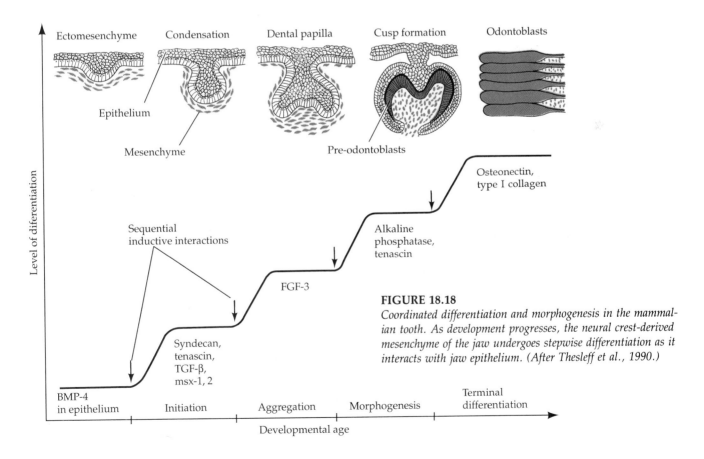

FIGURE 18.18

Coordinated differentiation and morphogenesis in the mammalian tooth. As development progresses, the neural crest-derived mesenchyme of the jaw undergoes stepwise differentiation as it interacts with jaw epithelium. (After Thesleff et al., 1990.)

factors, including msx-1 and msx-2. As these mesenchymal cells condense, they are induced (probably by some other signal, not BMP-4) to synthesize the membrane protein syndecan and the extracellular matrix protein tenascin. These proteins (which can bind each other) appear at the time the epithelium induces mesenchymal aggregation, and Thesleff and colleagues (1990) have proposed that these two molecules may interact to bring about this condensation. As in the kidney, syndecan expression also correlates with the proliferation of the clustered mesenchyme cells, suggesting that it is regulating cell division as well as aggregation (Vainio et al., 1991).

After the mesenchymal cells have aggregated, they begin to secrete a mitosis-promoting protein, fibroblast growth factor-3 (Wilkinson et al., 1989). It is not certain whether this protein is stimulating the growth of the mes-

enchymal cells, the epithelial cells, or both. As the mesenchymal cells begin to differentiate into odontoblasts, tenascin is induced to be expressed at much higher levels and at the same sites as alkaline phosphatase. Both these proteins have been associated with bone and cartilage differentiation, and they may promote the mineralization of the extracellular matrix (Mackie et al., 1987).

Finally, as the odontoblast phenotype emerges, osteonectin and type I collagen are secreted as components of the extracellular matrix. By this steplike process, the cranial neural crest cells of the jaw can be transformed into the dentin-secreting odontoblasts. These interactions occur at specific times of development and are correlated to the maturation of the epithelium. Under normal circumstances, two independent phenomena—morphogenesis and cell differentiation—are coordinated in organ formation.

The nature of proximity in epithelio-mesenchymal inductions

Proximate induction occurs when an inducing tissue is brought near a competent responding tissue. Do these cells have to make physical contact or do they interact over a small distance? Three types of interactions can be postulated: cell–cell contact, cell–matrix contact, and diffusion of soluble signals (Figure 18.19; Grobstein, 1955; Saxén et al., 1976). In some tissues, cell-cell contact appears to be required. For example, the induction of kidney tubules by the ureteric bud appears to depend upon their intimate contact. When metanephrogenic mesenchyme is separated from its inducer by a porous filter, induction only takes place when small projections from the inducing tissue are able to cross the pores to contact the mesenchyme (Figure 18.20; Lehtonen, 1975; Lehtonen et al., 1975).

In other organs, the extracellular matrix of one cell type is seen to be necessary and permissive for differentiation of another set of cells. When rat Sertoli cells—the major structural cells of the testis—are cultured on plastic petri dishes, they lose their differentiated morphology, forming a flat monolayer of cells. However, when they are cultured atop a basement membrane (containing laminin, type IV collagen, heparan sulfate, nidogen, and entactin), they form columnar monolayers identical to normal Sertoli cells in the testis (Figure 18.21; Hadley et al., 1985). Matrix-induced differentiation has also been seen in endothelial cells (Grant et al., 1989) and in the cornea. In this last case, the differentiation of the corneal epithelium depends upon an inductive influence from the collagen in the lens capsule (Hay and Revel, 1969; Hay and Dodson, 1973; Meier and Hay, 1974, 1975).

The extracellular matrix can also provide positional information for secondary inductions. In 5- to 6-day chick embryos, the dermal cells of the skin begin to condense at particular sites. These places are the sites of feather development. The dermal condensations are not randomly arranged but follow a precise pattern (Figure 18.22), and the interactions that produce feathers occur only at these condensations. Stuart and coworkers (1972) suggested that the dermal condensations arise from the migration of the mesenchymal cells along a preformed collagen matrix (Figure 18.22C). By treating back skin ectoderm with collagenase, they

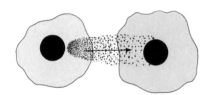

Diffusion of inducers from one cell to another

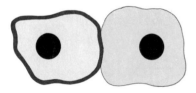

Contact of matrix of one cell with another

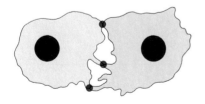

Contact between the inducing and responding cells

FIGURE 18.19

Three possible ways that inductive interactions might occur. (After Grobstein, 1956.)

(A)

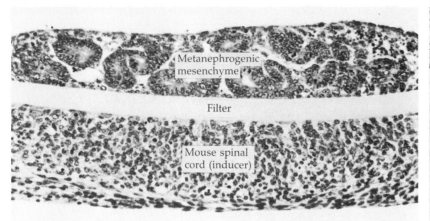

(B)

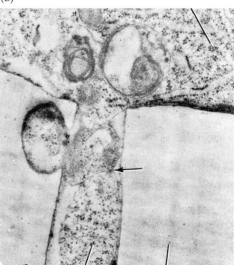

Metanephrogenic
mesenchymal cell

Neural tube cell   Filter

FIGURE 18.20

Transfilter induction of kidney tubules. (A) Metanephrogenic mesenchyme is above the filter and the inducer (in this case, mouse spinal cord, which mimics the effect of the ureteric bud) is placed below. Tubules have been induced in the mesenchyme. (B) Electron micrograph showing cell contact (arrow) through the pore of a filter "separating" the metanephrogenic mesenchyme from the inducing spinal cord. (Courtesy of L. Saxén.)

were able to destroy the hexagonal pattern of collagen deposition and inhibit the condensation of dermal cells. This hexagonal pattern is lacking in the skin of chick embryos carrying the *scaleless* mutation; and such embryos are unable to form these dermal condensations (Goetinck and Sekellick, 1972). The ectodermal component of the skin appears to be the

FIGURE 18.21

Extracellular matrix and cell differentiation. Light micrographs of rat Sertoli cells grown for two weeks (A) on tissue culture plastic dishes, and (B) on tissue culture plastic dishes coated with basal lamina. The photographs are taken at the same magnification (×1200). Some Sertoli cell nuclei are labeled S. (From Hadley et al. 1985, courtesy of M. Dym.)

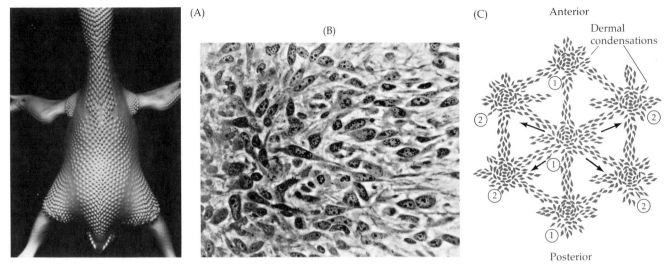

(A)

(B)

(C) Anterior

Dermal condensations

Posterior

FIGURE 18.22

Positioning of feather tracts in chick embryo. (A) Feather tracts on the dorsum of a 9-day chick embryo. Note that each feather primordium is located between the primordia of adjacent rows. (B,C) Pattern of dermal condensations giving rise to feather rudiments on the dorsum. (B) Condensation of dermal mesenchyme cells on the left, with nonaggregated cells oriented along the axis connecting the condensation to a neighboring one. (C) Hexagonal pattern of dermal cell aggregation and alignment: (1) primary row of dermal condensations (papillae) and cells aligned along the axis connecting them; (2) secondary rows of papillae and their connections. (A courtesy of P. Sengal; B from Wessells and Evans, 1968, courtesy of N. Wessells; C after Saunders, 1980.)

source of the mutation, as both lattice and feather germs formed when normal epidermis was combined with *scaleless* dermis, but no lattice or feather germ appeared when mutant epidermis was combined with normal dermis.

There are some inductive systems in which contact is not needed. One of these is induction of the neural tube by the action of chordamesoderm cells upon the overlying ectoderm. No contact is seen between inducing and responding cell sets, and this induction can occur through filters separating the two components (Toivonen, 1979). The secretion of BMP-4 in tooth development, fibroblast growth factors in limb development (to be discussed in the next chapter), and paracrine factors in the differentiation of *Xenopus* mesoderm (Chapter 16) are further examples of soluble signals between cells. In some cases, such as the induction of immunocompetence in B cells, both contact and paracrine* signals are used. This will be discussed later.

Mechanisms of branching in the formation of parenchymal organs

The generation of organ-specific epithelial branching patterns has remained a largely unexplored area. Earlier studies (for reviews, see Bard, 1990; Mizuno and Yasugi, 1990) revealed three major patterns whereby the mesenchyme regulates branching specificity. In organs such as the kidney, only one type of mesenchyme can cause this branching to occur (Saxén, 1987). In organs such as the salivary and mammary glands, the mesenchyme specifies the branching pattern, but the differentiation of the epithelium is determined autonomously by the epithelium (Lawson, 1974;

*Physiologists have described three major modes by which soluble molecules effect changes in cells. *Paracrine factors* are those soluble molecules that effect changes in cells adjacent to or nearby the secreting cell. *Endocrine factors* (hormones) are those soluble molecules that travel via the blood to effect changes in cells at a distance from the secreting cell. *Autocrine factors* are those molecules that effect changes in the cell that secreted them. For autocrine effects to happen, the cell synthesizes a molecule for which it has its own receptor. Such autocrine stimulation is not common, but it is seen in the case of placental cytotrophoblast cells that sythesize and secrete the platelet-derived growth whose receptor is on the cytotrophoblast cell membrane (Goustin et al., 1985). The result is the explosive proliferation of that tissue.

FIGURE 18.23
Scanning electron micrograph of collagen fibril accumulation within an early cleft of a 12 day mouse salivary gland. (From Nakanishi et al., 1986b, courtesy of Y. Nakanishi.)

Mesenchymal cell

Epithelial cell

Collagen in cleft between epithelial cells

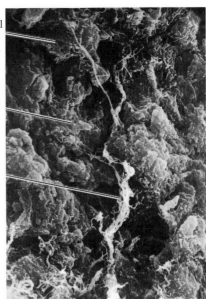

Sakakura et al., 1976). In those epithelial tubes that form tracts containing several differently branching regions (such as respiratory tract, digestive tract, and reproductive tracts), the regional mesenchymes specify both the branching patterns *and* the types of proteins in each region (Wessells, 1970; Cunha et al., 1976a,b; Hilfer et al., 1985; Haffen et al., 1987). For instance, in the region of the endodermal tube that is going to become liver, the mRNA for albumin (a liver-specific protein) is synthesized in the hepatic region of the epithelium even before the cells aggregate to form the liver rudiment. All that is needed for albumin mRNA synthesis is that the epithelium be in close contact with the mesenchyme cells of that area. In both liver and pancreas, these early interactions with the region-specific mesenchyme produce a low level of specific gene expression in that region of the proliferating endodermal tube (Rutter et al., 1964; Cascio and Zaret, 1991). This initial pattern will be amplified when the organs form their morphological structures.

The extracellular matrix as a critical element in branch formation

The mechanisms for such branching may have both general and specific components (Grobstein, 1967) and may depend on the interaction between those forces promoting cell growth and those forces promoting intercellular cohesion. The general components are thought to involve the selective degradation of the epithelial basement membrane at the branching sites (Bernfield et al., 1984; Mizuno and Yasugi, 1990).

As seen in the kidney and many other organs, the mesenchyme can interact with an epithelial tube by causing it to branch. Branching occurs when the epithelial outgrowths are divided by clefts, yielding lobules on either side of the cleft. These lobules grow to create branches. The branching of epithelial buds depends upon the presence of the mesenchyme. In some cases, such as the interaction of respiratory epithelium with several different mesenchymes, the interaction is instructive. In most cases, however, these interactions are merely permissive. The buds are prepared to branch and form acini, but they need support from the mesenchyme. It is now thought that the mesenchyme causes cleft formation and branching by splitting the lobule and selectively digesting away part of the epithelial tissue's basal lamina.

The control of cleft formation appears, in part, to be a function of collagen molecules. Collagen III fibrils are produced by the mesenchymal cells but accumulate only within the clefts of lobules (Figure 18.23; Grobstein and Cohen, 1965; Nakanishi et al., 1988). Moreover, the extent of branching can be artificially regulated by preserving or removing collagen

FIGURE 18.24

Control of epithelial cleft formation by mesenchymal collagen. The 12-day mouse salivary gland rudiments were cultured and observed at 1, 18, and 25 hours. (Row A) Normal development, showing three major lobes. (Row B) Growth of lobule but no branching when exogenous collagenase (5 μg/ml) was added to the medium. (Row C) Supernumerary branches when collagenase inhibitor (5 μg/ml) was added to the medium to suppress endogenous collagenase activity. (From Nakanishi et al., 1986a, courtesy of Y. Nakanishi.)

1 hr 18 hr 25 hr

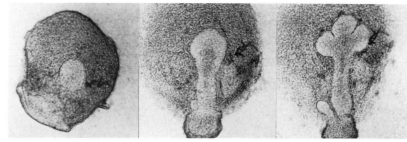

(A) Control

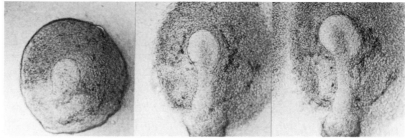

(B) Collagenase added

(C) Collagenase inhibitor added

molecules (Nakanishi et al., 1986a). Figure 18.24 shows the branching of a 12-day rudiment of a submandibular salivary gland under conditions that prevent the degradation of collagen fibrils (collagenase inhibitor was added to the medium) or accelerate its degradation (exogenous collagenase was added to the medium). Without the collagen, no clefts are seen, but when the endogenous collagenase is unable to remove excess collagen, supernumerary clefts appear.

The mechanism by which collagen initiates this branching is still unclear. Nakanishi and co-workers (1986b) have proposed that the mesenchymal cells align collagen fibrils by their traction to form ridges that cut into the lobular epithelium to form clefts. These clefts then become clearly defined as more migrating mesenchymal cells further deform the lobule by their traction. This initial cleft formation does not depend on epithelial cell proliferation (Nakanishi et al., 1987). The collagen fibrils may also be responsible for the development of the cleft into distinct branches. Bernfield and Banerjee (1982) have proposed that collagen can protect the basal lamina of the epithelial cells against hyaluronidase secreted by the mesenchymal cells. They showed that the mesenchymal cells do indeed digest glycosaminoglycan (GAG) from the lobule (Banerjee and Bernfield, 1979) and that the GAGs at the tips are more susceptible than those in the clefts. When heparan sulfate GAGs are removed from cultured salivary gland rudiments, branching ceases (Nakanishi et al., 1993). The breakdown of the basal lamina would enable the expansion of the branch by the in-

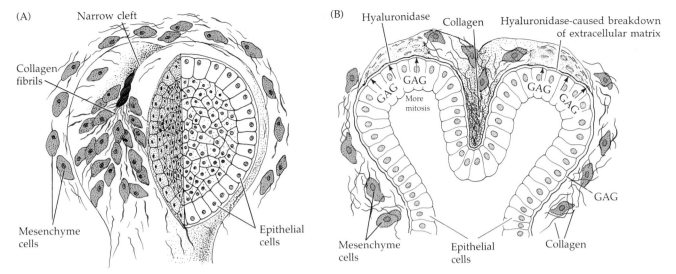

FIGURE 18.25

Possible model for cleft formation and branching in the mouse salivary gland rudiment. (A) A furrow is made in the lobule by contracting a bundle of collagen fibrils (shown here as a twisted, ropelike structure) by the traction of the mesenchymal cells. As shown in Figure 18.23, the fibrils extend between two groups of mesenchymal cells. (B) The elongation of the two separated lobules into branches may ensue, as the GAGs at the tips of the lobules are more sensitive to hyaluronidase since they do not have the protection of the collagen fibrils. The stalk of the lobule is stabile, while the increased division (stimulated by the mesenchyme) at the tips pushes the lobule forward. (A from Nakanishi et al., 1986b; B after Wessells, 1977.)

creased mitoses stimulated in that area. In this model, shown in Figure 18.25, the mesenchyme promotes epithelial growth, degrades the GAG, and deposits collagen fibers in the cleft. The epithelium synthesizes the basal lamina materials and stimulates mesenchymal collagen synthesis. The result is a differential breakdown of the basal lamina at the tips of the lobes, thus enabling the dividing cells of the lobe to form branches. Here, the interaction of mesenchymal cells with the extracellular matrix of the epithelium would determine the branching pattern of the organ.

Collagen is also important in stablilizing the branches once they have formed. If one adds collagenase to salivary gland rudiments after branching has occurred, the collagen is removed and the branches coalesce into a globe (Grobstein and Cohen, 1965; Wessels and Cohen, 1968).

Paracrine factors effecting branching patterns

We still do not know for certain the identities of those molecules secreted by the mesenchyme that are responsible for inducing these epithelial branching patterns. Recent evidence has implicated two proteins in these events. The first candidate is **transforming growth factor β1** (TGF-β1). This molecule is abundant in embryonic organs. When exogenous TGF-β1 is added to cultured mammary glands or embryonic kidney rudiments, it prevents the epithelium from branching (Figure 18.26; Silberstein and Daniel,1987; Silberstein et al., 1990; Ritvos et al., in preparation). TGF-β1 is known to promote the synthesis of extracellular matrix proteins and to inhibit the metalloproteinases that can digest these matrices (Penttinen et al., 1988; Nakamura et al., 1990). It is possible that TGF-β1 plays a role in stabilizing the branches after they have forked.

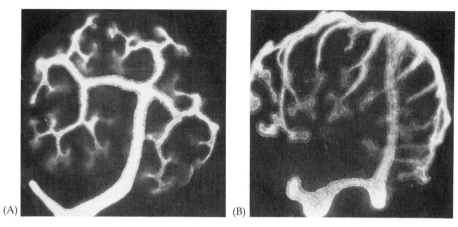

(A)   (B)

FIGURE 18.26
The effect of TGF-β1 on the morphogenesis of kidney epithelium. (A) An 11-day mouse kidney cultured 4 days in control medium has normal branching pattern. (B) An 11-day mouse kidney cultured in TGF-β1 shows no branching until reaching the periphery of the mesenchyme, and the branches formed are elongated. (After Ritvos et al., in preparation.)

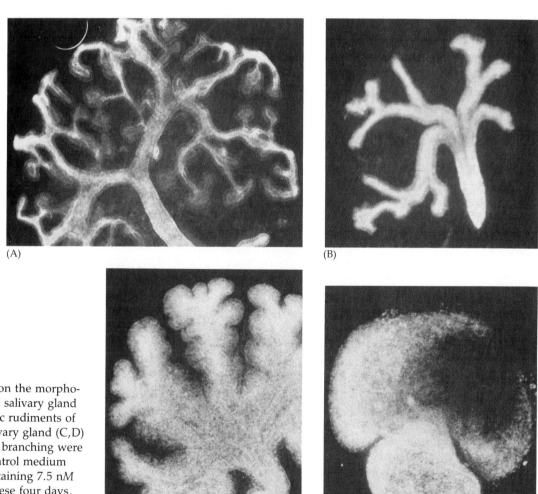

(A)   (B)

FIGURE 18.27
The effects of activin on the morphogenesis of kidney and salivary gland epithelium. Embryonic rudiments of kidney (A,B) and salivary gland (C,D) that had just initiated branching were cultured 4 days in control medium (A,C) or medium containing 7.5 n*M* activin (B,D). After these four days, the organs were fixed and stained for epithelial cytokeratin. (From Ritvos et al., in preparation.)

(C)   (D)

A second molecule that may be important in epithelial branching is **activin**. Activin is known to be important in specifying amphibian mesoderm (see Chapter 16), and it has been detected in salivary glands, pancreas, and kidneys of embryonic mice. When activin is added exogenously to embryonic mouse kidney, salivary, or pancreatic rudiments, it severely distorts the normal branching pattern (Figure 18.27; Gilbert et al., 1991; Ritvos et al., in preparation). The epithelial cells are not dead and are still capable of inducing the mesenchymal cells to form nephrons, but the branches are grossly disordered. The similarities between the salivary gland rudiments treated with collagenase and those treated with activin suggests that activin may be used to digest the extracellular matrix at the site of a new branch and that adding it exogenously causes the breakdown of the extracellular matrix throughout the epithelium.

SIDELIGHTS & SPECULATIONS

Induction of plasma cells by helper T cells

It is extremely difficult to study the processes of embryonic induction. The amounts of tissue that one isolates from embryos are painfully small, and the extraction of defined inducing agents from them is accomplished only with great difficulty. However, animals never stop developing, and certain adult cells are constantly differentiating from stem cells. In the adult vertebrate, the differentiation of the antibody-secreting plasma cell has been the subject of intense scrutiny for several decades. As mentioned earlier, the plasma cell is derived from a B cell (B lymphocyte). Each B cell can make a specific type of antibody and place it on its cell membrane as an antigen receptor. When an antigen binds to it, the B cell divides several times, develops the characteristic structures of a secretory cell (differentiates), and secretes its specific antibody (Chapter 10).

But antigen is not enough to stimulate the differentiation of the B cell to a plasma cell. The second type of cell involved in B-cell differentiation is the T cell (T lymphocyte). When T cells are absent, B cells do not proliferate or differentiate, even if antigens are present. Patients lacking T cells (as happens to people whose thymuses fail to develop) lack the ability to make antibodies to most substances, even though they have a population of B cells. Thus, the induction of the B cell to become an antigen-secreting plasma cell involves antigen and **helper T cells**.

It is possible to test theories of induction with these cells because of tumors that represent a state of arrested development. Some of the helper T-cell tumors are arrested in the state of development wherein they are making large quantities of inducer substances. While the inducer substances of embryonic cells might be made in small quantities for a brief duration, the T-cell tumor lines enable us to obtain large quantities of such a substance. Similarly, the development of certain B-cell tumors is arrested in a state wherein the B cell (or its precursors) can respond to these

substances. In these studies tumors have become for developmental biologists what mutations have been for geneticists (Auerbach, 1972; Pierce, 1985) and have given us our first insights into the molecular nature of vertebrate induction.

The interactions of helper T cells with B cells is a multistep process that will only be outlined here. First, the immunoglobulin molecules on the B cell act as membrane-bound receptors to bind the antigen (such as a virus or bacterium). This cross-links the immunoglobulins together (Figure 18.28). The cross-linking of immunoglobulins activates phospholipase C, which splits inositol phospholipids (PIP_2) into inositol phosphate and diacylglycerol. Transcription of new mRNAs is seen shortly thereafter, followed by the ability of the B cell to re-enter the cell cycle. Within a day of antigen stimulation, the B cell express a new set of receptors. These receptors enable the B cell to bind and respond to a set of proteins called **B cell growth and differentiation cytokines**. These cytokine proteins include **interleukin-4, interleukin-5, and interleukin-6** (IL-4, IL-5, IL-6).

When the B cell binds the antigen onto its cell surface, the antigen is endocytosed into the cell and is incorporated into a digestive vacuole. Here the antigen is partially digested, and the partially digested fragments are placed back on the cell membrane, where they are bound by certain membrane glycoproteins called the class II major histocompatibility complex glycoproteins. The helper T cell has a membrane receptor that recognizes certain antigen fragments on the B cell surface when the antigen fragments are bound to these major histocompatibility complex glycoproteins. Thus, the antigen fragments form a bridge linking the T cell and B cell together (Figure 18.28; Lanzavecchia, 1985; Janeway et al., 1987).

The binding of the helper T cell to the antigen fragments crosslinks the antigen receptors in the T cell's membrane and initiates an analogous cascade of events leading to the T cell's activation. The T cell secretes its cytokines,

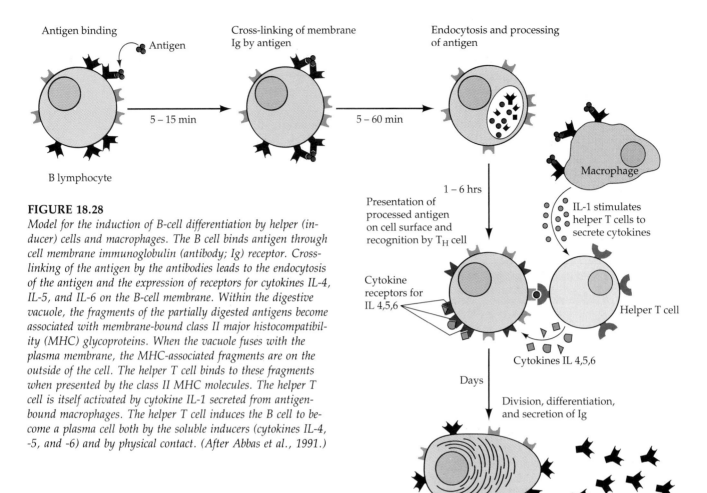

FIGURE 18.28

Model for the induction of B-cell differentiation by helper (inducer) cells and macrophages. The B cell binds antigen through cell membrane immunoglobulin (antibody; Ig) receptor. Cross-linking of the antigen by the antibodies leads to the endocytosis of the antigen and the expression of receptors for cytokines IL-4, IL-5, and IL-6 on the B-cell membrane. Within the digestive vacuole, the fragments of the partially digested antigens become associated with membrane-bound class II major histocompatibility (MHC) glycoproteins. When the vacuole fuses with the plasma membrane, the MHC-associated fragments are on the outside of the cell. The helper T cell binds to these fragments when presented by the class II MHC molecules. The helper T cell is itself activated by cytokine IL-1 secreted from antigen-bound macrophages. The helper T cell induces the B cell to become a plasma cell both by the soluble inducers (cytokines IL-4, -5, and -6) and by physical contact. (After Abbas et al., 1991.)

including IL-4 which causes the resting B cell to enter the cell cycle (from G_0 to G_1); IL-5, which stimulates the B cell to proliferate (taking it from G_1 to S phase of the cell cycle); and IL-6, which causes the B cells to differentiate and secrete its antibodies (Killar et al., 1987; Abbas et al., 1991).

In addition to cytokine-mediated effects, the physical interaction of the T cell and B cell also promotes the proliferation and differentiation of the B cell. Supernatants containing cytokines secreted by helper T cells are not nearly as effective in triggering antibody production by antigen-bound B cells as when the T cells are able to physically associate with the B cells through their antigen bridges. Thus, both soluble paracrine factors (the cytokines) and contact-mediated factors (as yet unknown but possibly the major histocompatibility complex glycoproteins) are important in inducing the differentiation of the plasma cell from the B cell.

The helper T cell is the critical regulatory cell of the immune system. Not only is it responsible for inducing B cells to become antibody-secreting plasma cells, but it is also responsible for inducing the activation of other types of T cells (that destroy tumor cells and virally infected cells) and macrophages (which digest bacteria). If the helper T cells are destroyed or inactivated, these inductions cannot take place and the immune system is severely impaired. This appears to be the case when human helper T lymphocytes are specifically destroyed by the **human immunodeficiency virus** (HIV), which attaches to an adhesion molecule (the CD4 glycoprotein) on their cell surfaces (Dal-

gleish et al., 1984; Klatzmann et al., 1984).* The loss of these inducer T cells causes the acquired immune deficiency syndrome (**AIDS**), which is characterized by the body's inability to combat infectious microorganisms. During pregnancy, however, it is important to *maintain* a local immune deficiency within the uterus to prevent the fetus from being destroyed (as a graft would be). The placenta contains several proteins whose function appears to involve blocking the ability of T cells to induce the immune response in this area (Brown et al., 1986; Daya et al., 1987).

We have a system, then, in which one set of cells—the helper T lymphocytes—can induce the differentiation of another group of cells—the B lymphocytes. The B lymphocytes, although capable of synthesizing a small amount of differentiated protein, do not secrete it. Upon inductive signals—both paracrine and contact-mediated—from another cell type, differentiation takes place. Also, as with embryonic induction, these final signals are only the last events of a complex process by which the cells became competent to respond to particular signals in their immediate environment.

*In humans, these T cells are called *helper/inducer T cells*, a name that recognizes their developmental role. The CD4 glycoprotein is normally involved in mediating nonspecific cell adhesion between the helper/inducer T cells and B lymphocytes (Doyle and Strominger, 1987).

Induction at the single-cell level

Sevenless and bride of sevenless

Embryonic induction occurs when interactions between inducing and responding cells bring about changes in the developmental pathway of the responding cell (Jacobson and Sater, 1988). Without the induction, the responding cell would become one cell type; with the induction, it becomes another. Our discussions of induction, however, have usually concerned tissues, not cells. Recent research into the developmental genetics of *Drosophila* and *Caenorhabditis* have shown that induction does indeed occur on the cell-to-cell level. Some of the best-studied examples involve the formation of the retinal photoreceptors in the *Drosophila* eye. The retina consists of about 800 units called **ommatidia** (Figure 18.29). Each ommatidium is composed of 20 cells arranged in a precise pattern. The eye develops in the flat epithelial layer of the eye imaginal disc of the larva. There are no cells directly above or below this layer, so the interactions are confined to neighboring cells. The differentiation of the randomly arranged epithelial cells into retinal photoreceptors and their surrounding lens tissue occurs during the last (third) larval stage. An indentation forms at the posterior margin of the imaginal disc, and this furrow begins to travel forward toward the anterior of the epithelium (Figure 18.30). As the furrow passes through a region of cells, those cells begin to differentiate in a specific order. The first cell to develop is the central (R8) photoreceptor.

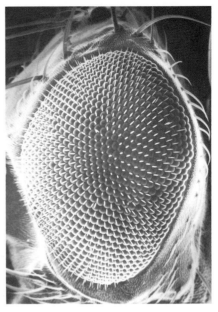

FIGURE 18.29
Scanning electron micrograph of a compound eye in *Drosophila*. Each facet is a single ommatidium. A sensory bristle projects from each ommatidium. (Courtesy of T. Venkatesh.)

FIGURE 18.30
Differentiation of photoreceptors in the late larval eye imaginal disc. The morphogenetic furrow (arrow) crosses the disc from posterior (left) to anterior (right). Behind the furrow, the photoreceptor cells differentiate in a defined sequence (shown below). The first photoreceptor cell to differentiate is R8. R8 appears to induce the differentiation of R2 and R5, and the cascade of induction continues until the R7 photoreceptor is differentiated. (After Tomlinson, 1988, photograph courtesy of T. Venkatesh.)

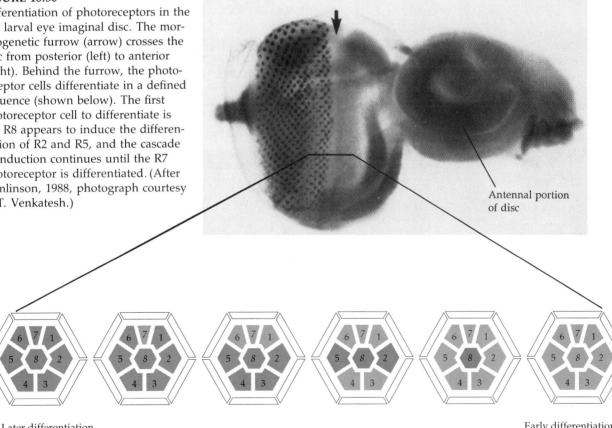

Antennal portion of disc

Later differentiation (posterior to morphogenetic furrow)

Early differentiation (entering morphogenetic furrow)

(It is not yet known how the furrow instructs certain cells to become R8 photoreceptors.*) The R8 cell is thought to induce the cell above it and the cell below it (with respect to the furrow) to become the R2 and R5 photoreceptors, respectively. The R2 and R5 photoreceptors are functionally equivalent, so the signal from R8 is probably the same to both cells (Tomlinson and Ready, 1987). Signals from these cells induce four more adjacent cells to become the R3, R4, and then the R1 and R6 photoreceptors. Last, the R7 photoreceptor appears. The other cells around these photoreceptors become the lens cells. Lens determination is the "default" condition if the cells are not induced.

A series of mutations has been found that blocks some of the steps of this induction cascade. The *rough* (*ro*) mutation, for instance, blocks the induction of the R3 and R4 photoreceptors. The *sevenless* (*sev*) mutation and the *bride of sevenless* (*boss*) mutations can each prevent the R7 cell from differentiating. (This cell becomes a lens cell instead.) Analysis of these mutations has shown that they are involved in the inductive process. The *sevenless* gene is required in the R7 cell itself. If mosaic embryos are made such that some of the cells of the eye disc are heterozygous (normal) and some are homozygous for the *sevenless* mutation, the R7 photoreceptor is seen to develop only if the R7 precursor cell has the wild-type *sevenless* allele (Basler and Hafen, 1989; Bowtell et al., 1989). Antibodies to this protein find it in the cell membrane, and the sequence of the *sevenless* gene suggests that it is a transmembrane protein with a tyrosine kinase site in its cytoplasmic domain (Banerjee et al., 1987; Hafen et al., 1987). This is consistent with the protein's being a receptor for some signal.*

This signal for the R7 precursor to differentiate into the R7 photoreceptor probably comes directly from a protein encoded by the wild-type allele of *bride of sevenless* (*boss*). Flies homozygous for the *boss* mutation also do not have the R7 photoreceptors. Gene mosaic studies wherein some of the cells of the imaginal disc are normal and some of the cells are homozygous for the *boss* mutation show that the wild-type *boss* gene is not needed in the R7 precursor cell. Rather, the R7 photoreceptor only differentiates if the wild-type *boss* gene is expressed in the *R8* cell. Thus, the *bride of sevenless* gene is encoding some protein whose existence in the R8 cell is necessary for the differentiation of the R7 cell.[†] The signal produced by the boss protein probably works by cell contact. Wild-type *boss* genes in an R8 cell in one ommatidium will not correct the deficiency of a mutant *boss* allele in the adjacent ommatidia, and the extracellular domain of the boss protein is sufficient to activate the *sevenless* tyrosine kinase in a neighboring cell (Reinke and Zipursky, 1988; Hart et al., 1993). A summary of the known cell-to-cell inductions in the *Drosophila* retina (Figure 18.31) shows that individual cells are able to induce other individual cells to create the precise arrangement of cells in particular tissues.

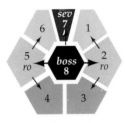

FIGURE 18.31
Summary of genes known to be involved in the induction of *Drosophila* photoreceptors. For development to continue beyond the differentiation of the R8, R2, and R5 photoreceptors, the *rough* gene (*ro*) must be present in both the R2 and the R5 cells. For the differentiation of the R7 photoreceptor, the *sevenless* gene (*sev*) has to be active in the R7 precursor cell, while the *bride of sevenless* gene (*boss*) must be active in the R8 photoreceptor. (After Rubin, 1989.)

*The morphogenetic furrow is likely to induce the differentiation of the R8 photoreceptor through a complex series of inductions involving the *decapentaplegic* (*dpp*) gene, a *Drosophila* homologue of TGF-β. Differentiating ommatidia (posterior to the furrow) secrete hedgehog protein that diffuses anteriorly. This protein mediates the induction of dpp protein expression in the morphogenetic furrow. As the new ommatidia form, the morphogenetic wave is moved anteriorly by new secretion of hedgehog and dpp proteins. The dpp protein is known to induce gene activity (see Chapter 15), and may be important in initiating the differentiation of the R8 cells (Heberlein et al., 1993).

†All the photoreceptor precursors synthesize the sev protein, and the boss signal given by the R8 photoreceptor is probably given to and received by all the surrounding cells. What, then, prevents R1–R6 cells from also becoming R7 cells? The restrictive agent is probably the product of the *seven-up* (*sup*) gene. In *sup*-deficient mutants, the R1, R3, R4, and R6 precursors all develop the R7 phenotype. The *sup* gene encodes a transcription factor of the steroid-receptor family (Mlodzik et al., 1990).

FIGURE 18.32
Model for the generation of two cell types (anchor cell and ventral uterine precursor) from two equivalent cells (Z1.ppp and Z4.aaa). (A) The cells start off as equivalent, with fluctuating amounts of signal (arrow) and receptor (inverted arrow). The *lin-12* gene is thought to encode the receptor. Reception of the signal turns down signal production and upregulates lin-12. (B) A stochastic (chance) event causes one cell to be producing more signal substance than the other cell at some particular critical time. This stimulates more lin-12 activity in the neighboring cell. (C) This difference is amplified, since the cell with the more lin-12 doesn't produce as much signal. (D) Eventually, one cell delivers the signal, and the other cell receives it. The signaling cell becomes the anchor cell; the receiving cell becomes the ventral uterine precursor. (After Greenwald and Rubin, 1992.)

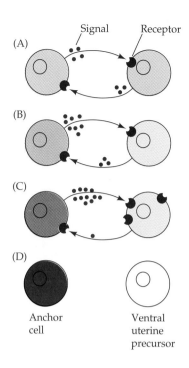

Vulval induction in *Caenorhabditis elegans*

The determination of the anchor cell. The development of the vulva in *C. elegans* shows several instances of induction on the cellular level. The first instance concerns induction of the gonadal anchor cell. The formation of the anchor cell is mediated by the *lin-12* gene, which encodes a cell surface receptor protein. In wild-type hermaphrodites, two adjacent cells, Z1.ppp and Z4.aaa, have the potential to become the gonadal anchor cell. They interact in a manner that causes one of them to be the anchor cell while the other one becomes the precursor of the uterine tissue. In recessive *lin-12* mutants, both cells become anchor cells, while in dominant mutations, both cells become uterine precursors (Greenwald et al., 1983). Studies using genetic mosaics and cell ablations have shown that this decision is made in the second larval stage and that the *lin-12* gene only needs to function in that cell destined to become the uterine precursor cell. The presumptive anchor cell does not need it. Seydoux and Greenwald (1989) speculate that these two cells originally synthesize both the signal for uterine differentiation (which is still unidentified) and the receptor for this molecule, the lin-12 protein (Figure 18.32). During a particular time in larval development, the cell that, by chance, is secreting more of this differentiation signal causes its neighbor to cease its production of the signal molecule and to increase its production of lin-12 protein. The cell secreting the signal becomes the gonadal anchor cell, while the cell receiving the signal through its lin-12 protein becomes the ventral uterine precursor cell. Thus, the two cells are thought to determine each other prior to their respective differentiation events.

SIDELIGHTS & SPECULATIONS

The role of cell–cell interactions in determining the identities of equivalent cells

The anchor cell/ventral uterine precursor decision illustrates two important aspects of determination between two originally equivalent cells. First, the initial difference between them is created by chance. Second, these initial differences are reinforced by feedbacks. Such determination is also seen in the determination of which of the originally equivalent epidermal cells of the insect embryo generate the neurons of the peripheral nervous system. Here, the choice is between becoming a skin (hypodermal) cell or a neuroblast. The *Notch* mutation of *Drosophila* also channels a bipotential cell into one of two alternative paths. Soon after gastrulation, a region of about 1800 ectodermal cells lies along the ventral midline of the *Drosophila* embryo. These cells have the potential to form the ventral nerve

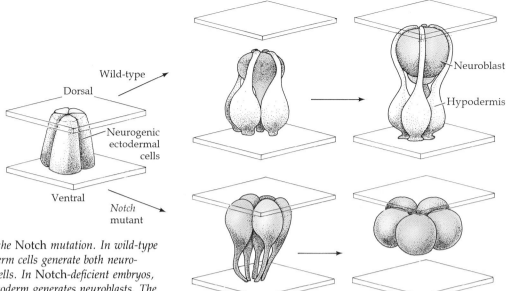

FIGURE 18.33

Representation of the effect of the Notch *mutation. In wild-type embryos, the neurogenic ectoderm cells generate both neuroblasts and skin (hypodermal) cells. In* Notch*-deficient embryos, however, all the neurogenic ectoderm generates neuroblasts. The proportion of neuroblasts to hypodermal cells differs between regions of the embryo.*

cord of the insect, and about one-quarter of these cells become neuroblasts while the rest become the precursors of the hypodermis. The cells that give rise to neuroblasts are intermingled with those cells destined to give rise to hypodermal precursors. Thus, each ectodermal cell in the nerve-forming regions of the fly embryo can give rise to either hypodermal or neural precursor cells (Hartenstein and Campos-Ortega, 1984). In the absence of *Notch* gene transcription in the embryo, the cells develop into neural precursors rather than into a mixture of hypodermal and neural precursor cells (Figure 18.33; Artavanis-Tsakonis et al., 1983; Lehmann et al., 1983). These embryos die, having a gross excess of neural cells at the expense of the ventral and head hypodermis (Poulson, 1937; Hoppe and Greenspan, 1986). The *Notch* gene has been cloned (Kidd et al.,

1983; Yedvobnick et al., 1985) and found to be transcribed during the early half of embryogenesis (and later in the early pupal stage). Both Notch and lin-12 proteins share remarkable sequence homologies to one another. They are both transmembrane proteins that may act as receptors for signals from adjacent cells (Yochem et al., 1988).

Heitzler and Simpson (1991) have proposed that the Notch protein, like lin-12, serves as a receptor for intercellular signals involved in distinguishing equivalent cells. Moreover, they provide evidence that another transmembrane protein, the product of the *delta* gene (whose absence creates a phenotype very similar to that of *Notch*) is the ligand for Notch. Genetic mosaics show that whereas *Notch* is needed in the cells that are to become epidermis, the *delta* gene is needed in the cells that induce the epidermal phenotype.

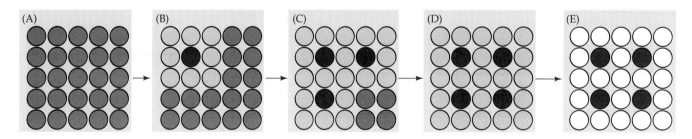

FIGURE 18.34

Model to explain the spacing patterns of neuroblasts among the initially equivalent neurogenic ectoderm cells. Based on the model for two cells shown in Figure 18.32, each cell both gives and receives the same signal. (A) A field of equivalent cells, all of which signal and receive equally. (B) A chance event causes one of the cells (darker shading) to produce more signal. Its surrounding cells receive this higher amount of signal and reduce their own signal level (lighter shading). (C) The rest of the pattern is now constrained. Those cells that have down-regu- *lated their own signaling (in response to the events in B) are less likely to express more signal than their neighboring cells. The cells surrounded by more down-regulated signalers are more likely to become signalers. (D,E) The fates of the cells throughout the field become specified as the amplification of the signals creates populations of signalers surrounded by populations of receivers. In the case of the neurogenic genes, the signal is thought to be from the delta protein, the receiver being the Notch protein. (After Greenwald and Rubin, 1992.)*

Greenwald and Rubin (1992) have proposed a model based on the lin-12 hypothesis to explain the spacing of neuroblasts in the proneural clusters of epidermal and neural precursors (Figure 18.34). Initially, all the cells have equal potentials and signaling. However, when one of the cells, by chance, produces more signal (say, delta product), it activates the receptors on adjacent cells and reduces their signaling level. Since the signaling levels on adjacent cells are low, the neighbors of those low-signaling cells will tend to be high. In this way, a spacing of neuroblasts is produced.

The role of chance in cell determination is not as uncommon as may be supposed. As we will discuss in Chapter 22, the maturation of only one ovum per month in humans is determined largely on the chance number of hormone receptors on the follicle cells. Similarly, the decision as to whether or not a cell becomes part of the embryo or part of the trophoblast—certainly a major decision in mammalian development—is also determined by the chance position of the cell during compaction. Such chance factors can cause interactions that are amplified and eventually distinguish two cell types from what had been a homogeneous cell population.

The determination of the vulval precursor cells. Once it is determined, the anchor cell next determines the fates of six **vulval precursor cells**. Each vulval precursor cell is equipotent and can acquire any one of three possible fates. These fates are determined largely by their proximity to the anchor cell. The anchor cell secretes a protein, the **lin-3 protein**, which resembles the mammalian epidermal growth factor and TGF-α in its composition (Hill and Sternberg, 1992). This signal acts in a position-dependent manner (Figure 18.35). The vulval precursor cell closest to the anchor cell (usually P6.p) is given a "primary" fate and is instructed to divide symmetrically three times to form vulval cells. The two cells to the side of this primary cell are given "secondary" fates, instructed by the anchor cell to divide in an asymmetric manner to generate further vulval cells. The other three potential vulval precursors are not instructed to make vulval cells. Their "tertiary" fate is to divide once to provide cells for the hypodermis of the nematode (Ferguson et al., 1987). If the anchor cell is ablated by focused laser beams, all six vulval precursor cells remain uninduced and form hypodermis (Kimble, 1981). The same phenotype is seen in those mutants that lack the lin-3 protein (Hill and Sternberg, 1992).

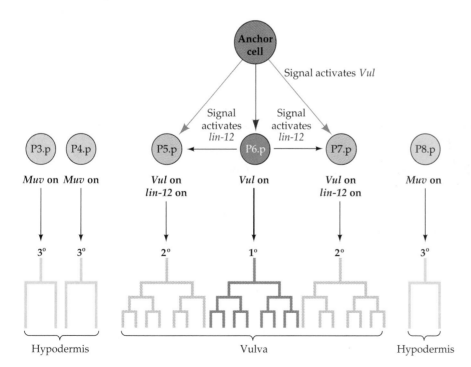

FIGURE 18.35
Model for the determination of vulval cell lineages in *C. elegans*. Signal from the anchor cell causes the *Vulvaless* genes to be activated in three of the six potential precursor cells. The cell closest to the anchor cell becomes the primary vulval precursor. In the three cells that do not receive the anchor cell stimulus, the *Multivulva* genes repress the *Vulvaless* genes, and these cells generate hypodermis rather than vulval tissue. The precursor cell closest to the anchor cell secretes a short-range signal that induces the neighboring cells to activate the *lin-12* gene. This gene causes them to become secondary vulval precursors.

In addition to the signal provided by the anchor cell, a second signal from the primary vulval precursor appears to stabilize the cell fates. This signal acts at very close range to inhibit the secondary vulval precursors from assuming a primary fate. Thus, there is **lateral inhibition** of the secondary vulval precursor cells by the primary vulval precursor cell (Sternberg, 1988).

SIDELIGHTS & SPECULATIONS

Molecular regulators of development: The ras pathway

How does a signal get from the cell membrane to the nucleus? How does the binding of a ligand on the plasma membrane cause alterations in gene transcription? Studies on the development of human tumors, nematode vulva, and fly ommatidia have recently teased out a pathway that connects events in the cell membrane with altered patterns of gene transcription. This is sometimes called the **receptor kinase/ras pathway**, and it is a major route for signal transduction in our cells (Figure 18.36).

Receptor and ligand

We have just seen that the lin-3 protein of *C. elegans* is similar in structure to mammalian epidermal growth factor and TGF-α. Whereas EGF and TGF-α are used in mammals to differentiate skin and other epithelia (Savage and Cohen, 1972), the lin-3 protein is used to determine the set of cells who progenitors construct the vulva. Moreover, the *Drosophila* bride-of-sevenless protein is homologous in structure to these other proteins, as it contains a series of EGF-

like domains in its extracellular region. These three proteins bind to cell membrane receptors that are also similar in structure. In mammals, EGF binds to the EGF/TGF-α receptors. In *C. elegans*, the homologue of the EGF receptor is the let-23 protein, a protein that binds the lin-3 EGF-like ligand (Aroian et al., 1990). In *Drosophila*, the homologue of the EGF receptor turns out to be the sevenless protein—a protein that binds to the EGF-like domains of the bride-of-sevenless ligand (Greenwald and Rubin, 1992). In all three systems, the pathway is activated by the binding of the EGF-like ligand by the cell membrane receptor.

Receptor tyrosine kinases

The EGF receptor and its nematode and fly homologues are called **receptor tyrosine kinases**. These proteins have

FIGURE 18.36

Summary of the tyrosine kinase–ras activation pathway for three separate developing systems: the determination of vulva cells in C. elegans, *the proliferation of epidermal cells in mammals, and the determination of the seventh photoreceptor in* Drosophila *eyes. The pathway in each appears to be formally identical.*

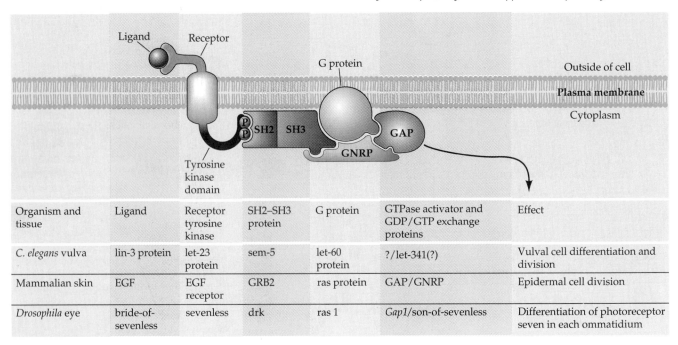

Organism and tissue	Ligand	Receptor tyrosine kinase	SH2–SH3 protein	G protein	GTPase activator and GDP/GTP exchange proteins	Effect
C. elegans vulva	lin-3 protein	let-23 protein	sem-5	let-60 protein	?/let-341(?)	Vulval cell differentiation and division
Mammalian skin	EGF	EGF receptor	GRB2	ras protein	GAP/GNRP	Epidermal cell division
Drosophila eye	bride-of-sevenless	sevenless	drk	ras 1	*Gap1*/son-of-sevenless	Differentiation of photoreceptor seven in each ommatidium

dormant ATP–protein tyrosine kinase activity in their cytoplasmic domains, which is activated when the ligand binds to the receptor. Thus, when the EGF-like ligand is bound by the receptor, the tyrosine kinase in the cytoplasmic region of the receptor is able to phosphorylate specific tyrosine residues of target proteins. (The FGF receptor discussed in Chapter 16 is another receptor tyrosine kinase.) The first target of the tyrosine kinase is the cytoplasmic region of a neighboring EGF receptor. The binding of the ligand to the receptor causes the **autophosphorylation** of the cytoplasmic domain of the receptor.

The importance of autophosphorylation was shown by viral versions of the EGF receptor, which are truncated. The viral oncogene v-*erb*-B* of the avian erythroblastosis virus encodes a protein that is almost identical to residues 551 to 1154 of the human EGF receptor (Figure 18.37; Yamamoto et al., 1983; Downward et al., 1984). This viral protein represents the cytoplasmic and transmembrane domains, but lacks the region of the protein that actually binds EGF. Usually the EGF receptor has a small amount of tyrosine kinase activity, which is greatly stimulated when it binds EGF. The truncated EGF receptor produced by v-*erb*-B appears *always* to have high tyrosine kinase activity, which continuously stimulates the cell to divide, even in the absence of EGF (Sefton et al., 1980; Hayman et al., 1983; Klarlund, 1985; Martin-Zanca et al., 1986). The EGF receptor, sevenless protein, and let-23 protein are all receptor tyrosine kinases whose enzymatic activity is stimulated by binding the ligand.

*The tumor-causing genes (oncogenes) of the tumor-causing viruses are prefaced with a "v-" (for viral) designation. The wild-type gene (or homologue) found in normal cells is designated with the "c-" (cellular) sign.

FIGURE 18.37

*The v-*erb*-B protein as a truncated EGF receptor. The v-*erb*-B protein has six amino acids from a virally encoded gene (gag) joined to 609 amino acids encoded by what appears to be a gene for the carboxyl half of an EGF receptor. This part includes the protein kinase domain, the transmembrane domain, and a small portion of the extracellular domain. (After Hunter, 1985.)*

SH2–SH3 domain proteins

The phosphorylation of the receptor tyrosine kinase is the first step in the pathway to the nucleus. The phosphorylated tyrosine on the receptor protein is then recognized by a series of proteins called **SH2–SH3 proteins**. In the pathway from epidermal growth factor receptor, the SH2–SH3 protein is called GRB2 (Lowenstein et al., 1992); the SH2 and SH3 domains found in this protein stand for "src homology domains 2 and 3," and give the protein class its name. The SH2 domain in GRB2 binds to the region of the receptor containing the autophosphorylated tyrosines. (Another of these SH2-domain proteins is a phospholipase C that can trigger the inositol phosphate pathway to cell division.) In *C. elegans*, the SH2–SH3 protein is the product of the *sem-5* gene, and in *Drosophila* the homologous SH2–SH3 protein is called drk (*downstream of receptor kinase*) (Clark et al., 1992; Olivier et al., 1993; Simon et al., 1993). The structural similarities between the sem-5, GRB2, and drk proteins are striking. Moreover, they are functionally interchangeable: the genes for human GRB2 and *Drosophila* drk can correct the phenotypic defects of *sem-5*-deficient nematodes, and the sem-5 and drk proteins can bind through their SH2 regions to the phosphorylated form of the human EGF receptor (Stern et al., 1993).

The ras protein and its regulators

The SH2–SH3 proteins serve as a bridge that links the phosphorylated receptor kinase to a powerful intracellular signaling system. While an SH2–SH3 protein binds to the phosphorylated receptor kinase through its SH2 domain, it uses its SH3 domain to regulate the activator of a **ras G protein**. The ras protein was first found to be encoded by the v-*ras* oncogene of the Harvey and Kirsten sarcoma viruses. Normally, the wild-type ras protein is in an inactive, GDP-binding state. When activated by the ligand-bound receptor, it exchanges a phosphate from another GTP to transform the bound GDP into GTP. This catalysis is aided by the **guanine nucleotide exchange protein**. The GTP-bound ras is the active form of the protein that transmits the signal. After delivering the signal, the GTP on the ras protein is hydrolyzed into GDP. This catalysis is greatly stimulated by the normal complexing of the ras protein to

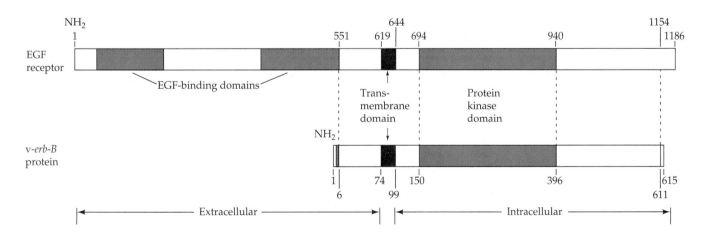

(A) Wild-type ras protein

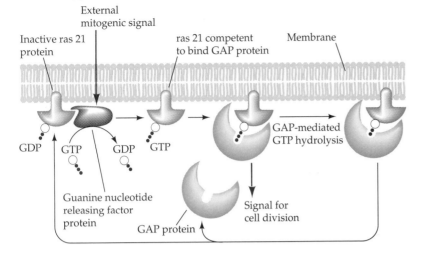

(B) Mutant ras protein

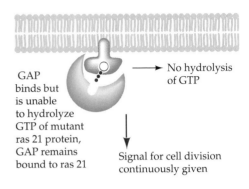

(C) ras phosphorylation cascade

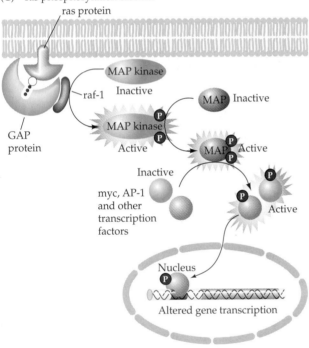

FIGURE 18.38

Mechanism for ras protein signaling. (A) In cells with wild-type ras proteins, an external mitogenic signal causes the phosphorylation of GDP bound to inactive ras protein. This reaction is aided by a guanine nucleotide exchange protein bound to ras. The GTP-bound ras protein is able to complex the GAP protein and transduce its signal for cell division. The GAP protein also hydrolyzes the GTP back to GDP and leaves the ras protein, which is now inactive. (B) In mutant ras proteins, the GAP protein binds the GTP-bound form of ras and allows the mitogenic signal to be given. However, it cannot hydrolyze the GTP to GDP, so it remains bound to the ras protein, a situation causing the continuous emission of its mitogenic signal. (C) A possible pathway (the ras phosphorylation cascade) linking the mitogenic signal at the cell surface to changes in transcription. The ras/GAP complex can bind the protein Raf-1. When Raf-1 is so bound, it catalyzes the phosphorylation of MAP kinase kinase, which then phosphorylates MAP kinase. This activates MAP kinase to phophorylate MAP (the so-called mitogen-activated protein. MAP can phosphorylate transcription factors such as AP-1 and myc, causing them to become active and thereby altering gene expression. (A and B after McCormick, 1989; C after Marx, 1993.)

the **GTPase activating protein** (GAP). This 120-kDa protein increases its GTP hydrolyzing activity over 100 fold (Trahey and McCormick, 1987; Gibbs et al., 1988), which returns the ras to its inactive state (Figure 18.38A). Indeed, mutations in the *ras* gene account for a large proportion of human tumors (Shih and Weinberg, 1982), and the mutations of the *ras* gene that make it oncogenic all inhibit the binding of the GAP protein. Without GAP protein, ras protein cannot catalyze GTP well and so remains in its active configuration (Figure 18.38B; Cales et al., 1988; McCormick, 1989).

The SH2–SH3 proteins link the phosphorylated receptor kinase (bound by the SH2 domain) to the guanosine nucleotide exchange protein (bound by the SH3 domain). This activates the exchange protein, allowing more of the

ras protein to be in its GTP-bound (active) state. Thus, the binding of the EGF-like ligand to its receptor tyrosine kinase results in the activation of the ras G protein (Beitel et al., 1990).

The downstream kinase pathway to the nucleus

Once the ras G protein is activated, it can stimulate the Raf-1 protein (Figure 18.38C). Raf-1 protein is a kinase that can phosphorylate a series of other kinases (the MAP kinases). The MAP kinases are able to phosphorylate various transcription factors that enter the nucleus. The kinase cascade in mammalian cells can phosphorylate the c-jun, c-fos, and SRF transcription factors, enabling them to become active in the nucleus. In *Drosophila* the cascade is thought to activate the sina (sevenless-in-absentia) transcription factor, whose presence is necessary in order for the *sevenless* gene to be transcribed (Carthew and Rubin, 1990; Dickson et al., 1992).

It seems, then, that mammals, nematodes, and flies all use the same general pathway to induce cell differentiation or proliferation. Other organisms and organs may also be using this route. The torso protein of *Drosophila* is a tyrosine kinase that (when bound by its ligand) activates ras and Raf-1 proteins and directs the differentiation of the anterior and posterior termini (Lu et al., 1993; Perrimon,

1993; Sprenger et al., 1993). Similarly, c-ros—the mammalian homologue of the *Drosophila* sevenless tyrosine kinase is found at the tips of the ureteric bud epithelium during the period of induction. The expression of this gene is transient; it is turned on during induction and turned off when the mesenchymal condensate fuses to the distal end of the bud (Tessarollo et al., 1992). Such examples show that the receptor kinase–ras activation pathway may be one of the major mechanisms for transducing a signal from the cell surface to the nucleus during secondary induction.

Paracrine factors are extremely important in coordinating the events of organ formation. Moreover, the pathways by which a ligand/receptor complex on the cell membrane transmits information to the nucleus are now being elucidated. In the next chapter, we will focus our attention on one particular type of organ and the theories concerning its formation.

LITERATURE CITED

Abbas, A. K., Lichtman, A. H. and Pober, J. S. 1991. *Cellular and Molecular Immunology*. W. B. Saunders, Philadelphia.

Arechaga, J., Karcher-Djvricic, V. and Ruch, J. V. 1983. Neutral morphogenetic activity of epithelium in heterologous tissue recombinations. *Differentiation* 25: 142–147.

Armstrong, J. F., Pritchard-Jones, K., Bickmore, W. A., Hastie, N. D. and Bard, J. B. L. 1992. The expression of the Wilms' tumor gene, WT-1, in the developing mammalian embryo. *Mech. Dev.* 40: 85–97.

Aroian, R. V., Koga, M., Mendel, J. E., Oshima, Y. and Sternberg, P. W. 1990. The *let-23* gene necessary for *Caenorhabditis elegans* vulval induction encodes a tyrosine kinase of the EGF receptor subfamily. *Nature* 348: 693–699.

Artavanis-Tsakonis, S., Muskavitch, M. A. T. and Yedvobnick, Y. 1983. Molecular cloning of *Notch*, a locus affecting neurogenesis in *Drosophila melanogaster*. *Proc. Natl. Acad. Sci. USA* 80: 1977–1981.

Auerbach, R. 1972. The use of tumors in the analysis of inductive tissue interaction. *Dev. Biol.* 28: 304–309.

Banerjee, S. and Bernfield, M. 1979. Developmentally regulated neutral hyaluronidase activity during epithelial mesenchymal interaction. *J. Cell Biol.* 83: 469a.

Banerjee, U., Renfranz, P. J., Pollock, J. A. and Benzer, S. 1987. Molecular characterization and expression of *sevenless*, a gene involved in neuronal pattern formation in the *Drosophila* eye. *Cell* 49: 281–291.

Bard, J. 1990. *Morphogenesis: The Cellular and Molecular Processes of Developmental Anatomy*. Cambridge University Press, Cambridge.

Basler, K. and Hafen, E. 1989. Ubiquitous expression of *sevenless*: Position-dependent specification of cell fate. *Science* 243: 931–934.

Beitel, G. J., Clark, S. G. and Horvitz, H. R. 1990. *Caenorhabditis elegans ras* gene *let-60* acts as a switch in the pathway of vulval induction. *Nature* 348: 503–509.

Bernfield, M. and Banerjee, S. D. 1982. The turnover of basal lamina glycosaminoglycan correlates with epithelia morphogenesis. *Dev. Biol.* 90: 291–305.

Bernfield, M., Banarjee, S. D., Koda, J. E. and Rapraeger, A. C. 1984. Remodelling of basement membrane as a mechanism of morphogenetic tissue interaction. *In* R. L. Trelstad (ed.), *The Role of Extracellular Matrix in Development*. Alan R. Liss, New York, pp. 545–547.

Birren, S. J. and Anderson, D. J. 1990. A v-myc-immortalized sympathoadrenal progenitor cell line in which neuronal differentiation is initiated by FGF but not NGF. *Neuron* 4: 189–201.

Bishop-Calame, S. 1966. Étude experimentale de l'organogenese du systéme urogénital de l'embryon de poulet. *Arch. Anat. Microsc. Morphol. Exp.* 55: 215–309.

Bowtell, D. D. L., Simon, M. A. and Rubin, G. M. 1989. Ommatidia in the developing *Drosophila* eye require and can respond to sevenless for only a restricted period. *Cell* 56: 931–936.

Brown, K. M., Muchmore, A. V. and Rosenstreich, D. L. 1986. Uromodulin, an immunosuppressive protein derived from pregnancy urine, is an inhibitor of interleukin 1. *Proc. Natl. Acad. Sci. USA* 83: 9119–9123.

Cales, C., Hancock, J. F., Marshall, C. J. and Hall, A. 1988. The cytoplasmic protein GAP is implicated as a target for regulation by the *ras* gene product. *Nature* 332: 548–551.

Carthew, R. W. and Rubin, G. M. 1990. *seven-in-absentia*, a gene required for specification of R7 cell fate in the *Drosophila* eye. *Cell* 63: 561–577.

Cascio, S. and Zaret, K. S. 1991. Hepatocyte differentiation initiates during endodermal-mesenchymal interactions prior to liver formation. *Development* 113: 217–225.

Cattanco, E. and McKay, R. D. G. 1990. Proliferation and differentiation of neuronal stem cells regulated by nerve growth factor. *Nature* 347: 762–765.

Clark, S. G., Stern, M. J. and Horvitz, H. R. 1992. *C. elegans* signalling gene *sem-5* encodes a protein with SH2 and SH3 domains. *Nature* 356: 340–344.

Coles, H. S. R., Burne, J. F. and Raff, M. C. 1993. Large-scale normal cell death in the developing rat kidney and its reduction by epidermal growth factor. *Development* 118: 777–784.

Coulombre, J. L. and Coulombre, A. J. 1971. Metaplastic induction of scales and feathers in the corneal anterior epithelium of the chick embryo. *Dev. Biol.* 25: 464–478.

Cunha, G. R. 1976a. Epithelial–stromal interactions in the development of the urogenital tract. *Int. Rev. Cytol.* 47, 137–194.

Cunha, G. R. 1976b. Stromal induction and specification of morphogenesis and cytodifferentiation of the epithelia of the Mullerian ducts and urogenital sinus during development of the uterus and vagina in mice. *J. Exp. Zool.* 196, 361–370.

Dalgleish, A. G., Beverley, P. C. L., Clapham, P. R., Crawford, D. H., Greaves, M. F. and Weiss, R. A. 1984. The CD4 (T4) antigen is an essential component of the receptor for the AIDS retrovirus. *Nature* 312: 763–767.

Daya, S., Rosenthal, K. L. and Clark, D. A. 1987. Immunosuppressor factor(s) produced by decidua-associated suppressor cells: A proposed mechanism for fetal allograft survival. *Am. J. Obstet. Gynecol.* 156: 344–350.

Deuchar, E. M. 1975. *Cellular Interactions in Animal Development*. Chapman and Hall, London.

Dickson, B., Sprenger, F., Morrison, D. and Hafen, E. 1992. Raf functions downstream in the *sevenless* signal transduction pathway. *Nature* 360: 600–603.

Downward, J. and 8 others. 1984. Close similarity of epidermal growth factor receptor and v-*erb*-B oncogene protein sequences. *Nature* 307: 521–527.

Doyle, C. and Strominger, J. L. 1987. Interaction between CD4 and class II MHC molecules mediates cell adhesion. *Nature* 330: 256–259.

Dressler, G. R., Wilkinson, J. E., Rothenpieler, V. W., Patterson, L. T., Williams-Simons, L., and Westphal, H. 1993. Deregulation of *Pax-2* expression in transgenic mice generates severe kidney abnormalities. *Nature* 362: 65–67.

Ekblom, P., Thesleff, I., Saxén, L., Miettinen, A. and Timpl, R. 1983. Transferrin as a fetal growth factor: Acquisition of responsiveness related to embryonic induction. *Proc. Natl. Acad. Sci. USA* 80: 2651–2655.

Etheridge, A. L. 1968. Determination of the mesonephric kidney. *J. Exp. Zool.* 169: 357–370.

Ferguson, E. C., Sternberg, P. W. and Horvitz, R. 1987. A genetic pathway for the specification of vulval cell lineages of *Caenorhabditis elegans. Nature* 326: 259–267.

Fukumachi, H. and Takayama, S. 1980. Epithelial-mesenchymal interaction in differentiation of duodenal epithelium of fetal rats in organ culture. *Experientia* 36: 335–336.

Gibbs, J. B., Scaber, M. D., Allard, W. J., Sigal, I. S. and Scolnick, E. M. 1988. Purification of *ras* GTPase activating protein from bovine brain. *Proc. Natl. Acad. Sci. USA* 85: 5026–5030.

Gilbert, S. F., Ritvos, O., Vaahtokari, A., Erämaa, M. and Saxén, L. 1991. Effects of activin and TGF-β1 on epithelial branching in developing murine organs. *J. Cell Biol.* 115: 145a.

Gluecksohn-Schoenheimer, S. 1943. The morphological manifestations of a dominant mutation in mice affecting tail and urogenital system. *Genetics* 28: 341–348.

Goetinck, P. F. and Sekellick, M. J. 1972. Observations on collagen synthesis, lattice formation, and the morphology of scaleless and normal embryonic skin. *Dev. Biol.* 28: 636–648.

Goustin, A. S. and nine others. 1985. Coexpression of the *sis* and *myc* proto-oncogenes in developing human placenta suggests autocrine control of trophoblast growth. *Cell* 41: 301–312.

Grainger, R. M. 1992. Embryonic lens induction: Shedding light on vertebrate tissue determination. *Trends Genet.* 8: 349–355.

Grant, D. S., Tashiro, K.-I., Segui-Real, B., Yamada, Y., Martin, G. R. and Kleinman, H. K. 1989. Two different laminin domains mediate the differentiation of human endothelial cells into capillary-like structures in vitro. *Cell* 933–943.

Greenwald, I. and Rubin, G. M. 1992. Making a difference: The role of cell–cell interactions in establishing separate identities for equivalent cells. *Cell* 68: 271–281.

Greenwald, I., Sternberg, P. W. and Horvitz, H. R. 1983. The *lin-12* locus specifies cell fates in *Caenorhabditis elegans. Cell* 34: 435–444.

Grobstein, C. 1955. Induction interaction in the development of the mouse metanephros. *J. Exp. Zool.* 130: 319–340.

Grobstein, C. 1956. Trans-filter induction of tubules in mouse metanephrogenic mesenchyme. *Exp. Cell Res.* 10: 424–440.

Grobstein, C. 1967. Mechanisms of organogenetic tissue interaction. *Natl. Cancer Inst. Monogr.* 26: 279–299.

Grobstein, C. and Cohen, J. 1965. Collagenase: Effect on the morphogenesis of embryonic salivary epithelium in vitro. *Science* 150: 626–628.

Gumpel-Pinot, M., Yasugi, S. and Mizuno, T. 1978. Différenciation d'épithéliums endodermiques associaés au mésoderme splanchnique. *Comp. Rend. Acad. Sci.* (Paris) 286: 117–120.

Hadley, M. A., Byers, S. W., Suárez-Quian, C. A., Kleinman, H. K. and Dym, M. 1985. Extracellular matrix regulates Sertoli cell differentiation, testicular formation, and germ cell development in vitro. *J. Cell Biol.* 101: 1511–1522.

Hafen, E., Basler, K., Edstrom, J. E. and Rubin, G. M. 1987. *sevenless*, a cell-specific homeotic gene of *Drosophila*, encodes a putative transmembrane receptor with a tyrosine kinase domain. *Science* 236: 55–63.

Haffen, K., Kedinger, M. and Simonassmann, P. 1987. Mesenchyme-dependent differentiation of epithelial progenitor cells in the gut. *J. Pediat. Gastroent.* 6: 15–23.

Hamburgh, M. 1970. *Theories of Differentiation*. Elsevier, New York.

Harrison, R. G. 1920. Experiments on the lens in *Amblystoma. Proc. Soc. Exp. Biol. Med.* 17: 199–200.

Harrison, R.G. 1933. Some difficulties of the determination problem. *Am. Nat.* 67: 306–321.

Hart, A. C., Krämer, H. and Zipursky, S. L. 1993. Extracellular domain of the boss transmembrane ligand acts as an antagonist of the the sev receptor. *Nature* 361: 732–736.

Hartenstein, V. and Campos-Ortega, J. A. 1984. Early neurogenesis in wild-type *Drosophila melanogaster. Wilhelm Roux Arch. Dev. Biol.* 193: 308–325.

Hay, E. D. 1980. Development of the vertebrate cornea. *Int. Rev. Cytol.* 63: 263–322.

Hay, E. D. and Dodson, J. W. 1973. Secretion of collagen by corneal epithelium. I. Morphology of the collagenous products produced by isolated epithelia grown on frozen-killed lens. *J. Cell Biol.* 57: 190–213.

Hay, E. D. and Revel, J.-P. 1969. Fine structure of the developing avian cornea. In A. Wolsky and P. S. Chen (eds.), *Monographs in Developmental Biology*. Karger, Basel.

Hayman, M. J., Ramsay, G. M., Savin, K., Kitchener, G., Graf, T. and Beug, H. 1983. Identification and characterization of the avian erythroblastosis virus *erb*-B gene product as a membrane glycoprotein. *Cell* 32: 579–588.

Heberlein, U., Wolff, T. and Rubin, G. M. 1993. The TGF-β homolog *dpp* and the segment polarity gene *hedgehog* are required for propagation of a morphogenetic wave in the *Drosophila* retina. *Cell* 75: 913–926.

Heitzler, P. and Simpson, P. 1991. The choice of cell fate in the epidermis of *Drosophila. Cell* 64: 1083–1092.

Henry, J. J. and Grainger, R. M. 1987. Inductive interactions in the spatial and temporal restriction of lens-forming potential in embryonic ectoderm. *Dev. Biol.* 124: 200–214.

Henry, J. J. and Grainger, R. M. 1990. Early tissue interactions leading to embryonic lens formation in *Xenopus laevis. Dev. Biol.* 141: 149–163.

Hilfer, S. R., Rayner, R. M. and Brown, J. W. 1985. Mesenchymal control of branching pattern in the fetal mouse lung. *Tissue Cell* 17: 523–538.

Hill, R. J. and Sternberg, P. W. 1992. The gene *lin-3* encodes an inductive signal for vulval development in *C. elegans. Nature* 358: 470–476.

Holtzer, H. 1968. Induction of chondrogenesis: A concept in terms of mechanisms. In R. Gleischmajer and R. E. Billingham (eds.), *Epithelial-Mesenchymal Interactions*. Williams & Wilkins, Baltimore, pp. 152–164.

Hoppe, P. E. and Greenspan, R. J. 1986. Local function of the *Notch* gene for embryonic ectodermal pathway choice in *Drosophila. Cell* 46: 773–783.

Hunter, T. 1985. Oncogenes and growth control. *Trends Biochem. Sci.* 10: 275–280.

Hyatt, G. A. and Beebe, D. C. 1993. Regulation of lens cell growth and polarity by embryonic-specific growth factor and by inhibitors of lens cell proliferation and differentiation. *Development* 117: 701–709.

Jacobson, A. G. 1966. Inductive processes in embryonic development. *Science* 152: 25–34.

Jacobson, A. G. and Sater, A. K. 1988. Features of embryonic induction. *Development* 104: 341–359.

Janeway, C. A., Ron, J. and Katz, M. E. 1987. The B cell is the initiating antigen-presenting cell in peripheral lymph nodes. *J. Immunol.* 138: 1051–1055.

Kidd, S., Lockett, T. J. and Young, M. W. 1983. The *Notch* locus of *Drosophila melanogaster. Cell* 34: 421–433.

Killar, L., MacDonald, G., West, J., Woods, A. and Bottomly, K. 1987. Cloned, Ia-restricted T cells that do not produce IL4 (BSF-1) fail to help antigen-specific B cells. *J. Immunol.* 138: 1674–1679.

Kimble, J. 1981. Alterations in cell lineage following laser ablation of cells in the somatic gonad of *Caenorhabditis elegans*. *Dev. Biol.* 87: 286–300.

King, H. D. 1905. Experimental studies on the eye of the frog embryo. *Wilhelm Roux Arch. Entwicklungsmech. Org.* 19: 85–107.

Klarlund, J. D. 1985. Transformation of cells by an inhibitor of phosphatases acting on phosphotyrosine in proteins. *Cell* 41: 707–717.

Klatzmann, D. and ten others. 1984. Selective tropism of lymphadenopathy associated virus (LAV) for helper-inducer T lymphocytes. *Science* 225: 59–63.

Kollar, E. J. and Baird, G. 1970. Tissue interaction in developing mouse tooth germs. II. The inductive role of the dental papilla. *J. Embryol. Exp. Morphol.* 24: 173–186.

Kollar, E. J. and Fisher, C. 1980. Tooth induction in chick epithelium: Expression of quiescent genes for enamel synthesis. *Science* 207: 993–995.

Koscki, C., Herzlinger, D. and Al-Auqati, Q. 1992. Apoptosis in metanephric development. *J. Cell Biol.* 119: 1327–1333.

Kreidberg, J. A., Sariola, H., Loring, J. M., Maeda, M., Pelletier, J., Housman, D. and Jaenisch, R. 1993. WT–1 is required for early kidney development. *Cell* 74: 679–691.

Lanzavecchia, A. 1985. Antigen-specific interaction between T and B cells. *Nature* 314: 537–539.

Lawson, K. A. 1974. Mesenchyme specificity in rodent salivary gland development: the response of salivary epithelium to lung mesenchyme in vitro. *J. Embryol. Exp. Morphol.* 32, 469–493.

Lehmann, R., Jimenez, F., Dietrich, U. and Campos–Ortega, J. A. 1983. On the phenotype and development of mutants of early neurogenesis in *Drosophila melanogaster*. *Wilhelm Roux Arch. Dev. Biol.* 192: 62–74.

Lehtonen, E. 1975. Epithelio-mesenchymal interface during mouse kidney tubule induction in vivo. *J. Embryol. Exp. Morphol.* 34: 695–705.

Lehtonen, E., Wartiovaara, J., Nordling, S. and Saxén, L. 1975. Demonstration of cytoplasmic processes in Millipore filters permitting kidney tubule induction. *J. Embryol. Exp. Morphol.* 33: 187–203.

Lehtonen, E., Virtanen, I. and Saxén, L. 1985. Reorganization of the intermediate cytoskeleton in induced mesenchyme cells is independent of tubule morphogenesis. *Dev. Biol.* 108: 481–490.

Lewis, W. 1904. Experimental studies on the development of the eye in amphibia. I. On the origin of the lens, *Rana palustris*. *Am. J. Anat.* 3: 505–536.

Lewis, W. 1907. Experimental studies on the development of the eye in Amphibia. III. On the origin and differentiation of the lens. *Am. J. Anat.* 6: 473–509.

Lowenstein, E. J. and eight others. 1992. The SH2 and SH3 domain–containing protein GRB2 links receptor tyrosine kinases to ras signalling. *Cell* 70: 431–442.

Lu, X.-Y., Chou, T. B., Williams, N. G., Roberts, T. and Perrimon, N. 1993. Control of cell fate determination by p21^{Ras}/Ras-1, and essential component of torso signaling in *Drosophila*. *Genes Devel.* 7: 621–632.

Lumsden, A. G. S. 1988. Spatial organization of the epithelium and the role of neural crest cells in the initiation of the mammalian tooth germ. *Development* 103 [Suppl.]: 155–169.

Mackie, E. J., Thesleff, I. and Chiquet-Ehrismann, R. 1987. Tenascin is associated with chondrogenic and osteogenic differentiation *in vivo* and promotes chondrogenesis in vivo. *J. Cell Biol.* 105: 2569–2579.

Martin-Zanca, D., Hughes, S. H. and Barbacid, M. 1986. A human oncogene formed by the fusion of truncated tropomyosin and protein tyrosine kinase sequences. *Nature* 347: 667–669.

Marx, J. 1993. Forging a path to the nucleus. *Science* 260: 1588–1590.

McCormick, F. 1989. *ras* GTPase activating protein: Signal transmitter and signal terminator. *Cell* 56: 5–8.

Meisenhelder, J., Suh, P.-G., Rhee, S. G. and Hunter, T. 1989. Phospholipase Cγ is a substrate for the PDGF and EGF receptor protein–tyrosine kinase in vivo and in vitro. *Cell* 57: 1109–1122.

Meier, S. 1977. Initiation of corneal differentiation prior to cornea–lens association. *Cell Tissue Res.* 184: 255–267.

Meier, S. and Hay, E. D. 1974. Control of corneal differentiation by extracellular materials. Collagen as promoter and stabilizer of epithelial stroma production. *Dev. Biol.* 38: 249–270.

Meier, S. and Hay, E. D. 1975. Stimulation of corneal differentiation by interaction between the cell surface and extracellular matrix. I. Morphometric analysis of transfilter induction. *J. Cell Biol.* 66: 275–291.

Mencl, E. 1908. Neue Tatsachen zur Selbstdifferenzierung der Augenlinse. *Wilhelm Roux Arch. Entwicklungsmech. Org.* 25: 431–450.

Mina, M. and Kollar, E. J. 1987. The induction of odontogenesis in non–dental mesenchyme combined with early murine mandibular arch epithelium. *Arch. Oral Biol.* 32: 123–127.

Mizuno, T. and Yasugi, S. 1990. Susceptibility of epithelia to directive influences of mesenchymes during organogenesis: Uncoupling of morphogenesis and cytodifferentiation. *Cell Differ. Devel.* 31, 151–159.

Mlodzik, M., Hiromi, Y., Weber, U., Goodman, C. S. and Rubin, G. M. 1990. The *Drosophila seven-up* gene, a member of the steroid receptor gene superfamily, controls photoreceptor cell fates. *Cell* 60: 211–224.

Nakamura, T. Okuda, S., Miller, D., Ruoslahti, E. and Border, W. 1990. Transforming growth factor-β (TGF-β) regulates production of extracellular matrix (ECM) components by glomerular epithelial cells. *Kidney Int.* 37, 221.

Nakanishi, Y., Sugiura, F., Kishi, J.-I. and Hayakawa, T. 1986a. Collagenase inhibitor stimulates cleft formation during early morphogenesis of mouse salivary gland. *Dev. Biol.* 113: 201–206.

Nakanishi, Y., Sugiura, F., Kishi, J.-I. and Hayakawa, T. 1986b. Scanning electron microscopic observations of mouse embryonic submandibular glands during initial branching: Preferential localization of fibrillar structures at the mesenchymal ridges participating in cleft formation. *J. Embryol. Exp. Morphol.* 96: 65–77.

Nakanishi, Y., Morita, T. and Nogawa, H. 1987. Cell proliferation is not required for the initiation of early cleft formation in mouse embryonic submandibular epithelium in vitro. *Development* 99: 429–437.

Nakanishi, Y., Nogawa, H., Hashimoto, Y., Kishi, J.-I. and Hayakawa, T. 1988. Accumulation of collagen III at the cleft points of developing mouse submandibular gland. *Development* 104: 51–59.

Nakanishi, Y., Uematsu, J., Takamatsu, H., Fukuda, Y. and Yoshida, K. 1993. Removal of heperan sulfate chains halted epithelial branching morphogenesis of developing mouse submandibular gland in vitro. *Dev. Growth Differ.* 35: 371–384.

Nieuwkoop. P. 1952. Activation and organization of the central nervous system in amphibians. *J. Exp. Zool.* 120: 1–108.

Olivier, J. P. and eight others. 1993. A *Drosophila* SH2-SH3 adaptor protein implicated in a coupling of sevenless tyrosine kinase to an activator of ras guanosine nucleotide exchange, Sos. *Cell* 73: 179–191.

Osathanondh, V. and Potter, E. 1963. Development of human kidney as shown by microdissection. III. Formation and interrelationships of collecting tubules and nephrons. *Arch. Pathol.* 76: 290–302.

Penttinen, R. P., Kobayashi, S. and Bornstein, P. 1988. Transforming growth factor-β increases mRNA for matrix proteins in the presence and in the absence in the changes in mRNA stability. *Proc. Natl. Acad. Sci. USA* 85: 1105–1108.

Perrimon, N. 1993. The torso receptor protein–tyrosine kinase signaling pathway: An endless story. *Cell* 74: 219–222.

Pierce, B. G. 1985. Carcinoma is to embryology as mutation is to genetics. *Am. Zool.* 25: 707–712.

Poulson, D. F. 1937. Chromosomal deficiencies and the embryonic development of *Drosophila melanogaster*. *Proc. Natl. Acad. Sci. USA* 23: 133–137.

Pritchard-Jones, K. and eleven others. 1990. The candidate Wilms' tumour gene is involved in genitourinary development. *Nature* 346: 194–197.

Reinke, R. and Zipursky, A. L. 1988. Cell–cell interaction in the *Drosophila* retina: The *bride of sevenless* gene is required in photoreceptor cell R8 for R7 cell development. *Cell* 55: 321–330.

Ritvos, O. Gilbert, S. F., Erämaa, M. and Saxén, L. In preparation. Activin A disrupts epithelial branching in developing murine organs. Manuscript in preparation.

Romanoff, A. L. 1960. *The Avian Embryo.* Macmillan, New York.

Rothenpieler, U. W. and Dressler, G. R. 1993. Pax-2 is required for mesenchyme-to-epithelium conversion during kidney development. *Development* 119: 711–720.

Roux, W. 1894. The problems, methods, and scope of developmental mechanics. In *Lectures of the Marine Biological Laboratory.* Ginn & Co., Boston, pp. 149–190.

Rubin, G. M. 1989. Development of the *Drosophila* retina: Inductive events studied at single cell resolution. *Cell* 57: 519–520.

Rutter, W. J., Wessells, N. K. and Grobstein, C. 1964. Controls of specific synthesis in the developing pancreas. *Natl. Cancer Inst. Monogr.* 13: 51–65.

Saha, M. S. 1991. Spemann seen through a lens. *In* S. F. Gilbert (ed.), *A Conceptual History of Modern Embryology.* Plenum, New York, pp.91–108.

Saha, M. S., Spann, C. L. and Grainger, R. M. 1989. Embryonic lens induction: More than meets the optic vesicle. *Cell Differ. Dev.* 28: 153–172.

Sainio, K. and seven others. 1992. Differential expression of gap junction mRNAs and proteins in the developing murine kidney and in experimentally induced nephric mesenchymes. *Development* 115: 827–837.

Sainio, K., Saarma, M., Palgi, J., Saxén, L. and Sariola, H. In press. Neuronal characteristics in embryonic renal stroma.

Sakakura, T., Nishizuka, Y. and Dawe, C. J. 1976. Mesenchyme-dependent morphogenesis and epithelium-specific cytodifferentiation in mouse mammary gland. *Science* 194, 1439–1441.

Sariola, H., Ekblom, P. and Saxén, L. 1982. Restricted developmental options of the metanephric mesenchyme. *In* M. Burger and R. Weber (eds.), *Embryonic Development, Part B: Cellular Aspects.* Alan R. Liss, New York, pp. 425–431.

Sariola, H., Holm-Sainio, K. and Henke-Fahle, S. 1989. The effect of neuronal cells on kidney differentiation. *Int. J. Dev. Biol.* 33: 149–155.

Sariola, H. and seven others. 1991. Dependence of kidney morphogenesis on the expression of nerve growth factor receptor. *Science* 254: 571–573.

Saunders, J. W., Jr. 1980. *Developmental Biology.* Macmillan, New York.

Saunders, J. W., Jr., Cairns, J. M. and Gasseling, M. T. 1957. The role of the apical ectodermal ridge of ectoderm in the differentiation of the morphological structure of and inductive specificity of limb parts of the chick. *J. Morphol.* 101: 57–88.

Savage, C. R. and Cohen, S. 1972. Epidermal growth factor and a new derivative. *J. Biol. Chem.* 247: 7609–7611.

Saxén, L. 1970. Failure to demonstrate tubule induction in heterologous mesenchyme. *Dev. Biol.* 23: 511–523.

Saxén, L. 1987. *Organogenesis of the Kidney.* Cambridge University Press, Cambridge.

Saxén, L. and Sariola, H. 1987. Early organogenesis of the kidney. *Pediat. Nephrol.* 1: 385–392.

Saxén, L., Lehtonen, E., Karkinen-Jääskeläinen, M., Nordling, S. and Wartiovaara, J. 1976. Are morphogenetic tissue interactions mediated by transmissible signal substances or through cell contacts? *Nature* 259: 662–663.

Schulz, M. W., Chamberlain, C. G., de longh, R. U. and McAvoy, J. W. 1993. Acidic and basic FGF in ocular media and lens: Implications for lens polarity and growth patterns. *Development* 118: 117–126.

Sefton, B. M., Hunter, T. Beemon, K. and Eckhart, W. 1980. Evidence that the phosphorylation of tyrosine is essential for cellular transformation by Rous sarcoma virus. *Cell* 20: 807–816.

Servetnick, M. and Grainger, R. M. 1991. Changes in neural and lens in *Xenopus* ectoderm: Evidenc efor an autonomous developmental timer. *Development* 112: 177–188.

Seydoux, G. and Greenwald, I. 1989. Cell autonomy of *lin-12* function in a cell fate decision in *C. elegans. Cell* 57: 1237–1245.

Shih, C. and Weinberg, R. A. 1982. Isolation of a transforming sequence from a human bladder carcinoma cell line. *Cell* 29: 161–169.

Silberstein, G. B. and Daniel, C. W. 1987. Reversible inhibition of mammary gland growth by transforming growth factor beta. *Science* 237, 291–293.

Silberstein, G. B., Strickland, P., Coleman, S. and Daniel, C. W. 1990. Epithelium-dependent extracellular matrix synthesis in transforming growth factor-β1-growth-inhibited mouse mammary gland. *J. Cell Biol.* 110, 2209–2219.

Simon, M. A., Dodson, G. S. and Rubin, G. M. 1993. An SH3–SH2–SH3 protein is required for p21^{ras1} activation and binds to sevenless and Sos proteins in vitro. *Cell* 73: 169–177.

Spemann, H. 1901. Über Correlationen in der Entwicklung des Auges. *Verh. Anat. Ges. 15 Vers. Bonn.* 61–79.

Spemann, H. 1938. *Embryonic Development and Induction.* Yale University Press, New Haven.

Spemann, H. and Schotté, O. 1932. Über xenoplatische Transplantation als Mittel zur Analyse der embryonalen Induction. *Naturwissenschaften* 20: 463–467.

Sprenger, F., Trosclair, M. M. and Morrison, D. K. 1993. Biochemical analysis of torso and D-Raf during *Drosophila* embryogenesis: Implications for terminal signal transduction. *Mol. Cell. Biol.* 13: 1163–1172.

Stern, M. J. and eight others. 1993. The human *GRB2* and *Drosophila drk* genes can functionally replace the *Caenorhabditis elegans* cell signalling gene *sem-5. Mol. Biol. Cell* 4: 1175–1188.

Sternberg, P. W. 1988. Lateral inhibition during vulval induction in *Caenorhabditis elegans. Nature* 335: 551–554.

Stuart, E. S., Garber, B. and Moscona, A. 1972. An analysis of feather germ formation in normal development and in skin treated with hydrocortisone. *J. Exp. Zool.* 179: 97–110.

Tessarollo, L., Nagarnjan, L. and Prada, L. 1992. c-ros, the vertebrate homolog of the sevenless tyrosine kinase receptor is tightly regulated during organogenesis in mouse embryonic development. *Development* 115: 11–20.

Thesleff, I., Vainio, S. and Jalkanen, M. 1989. Cell–matrix interaction in tooth development. *Int J. Dev. Biol.* 33: 91–97.

Thesleff, I., Vaahtokari, A. and Vainio, S. 1990. Molecular changes during determination and differentiation of the dental mesenchyme cell lineage. *J. Biol. Bucalle* 18: 179–188.

Toivonen, S. 1979. Transmission problem in primary induction. *Differentiation* 15: 177–181.

Tomlinson, A. 1988. Cellular interactions in the developing *Drosophila* eye. *Development* 104: 183–193.

Tomlinson, A. and Ready, D. F. 1987. Cell fate in the *Drosophila* ommatidium. *Dev. Biol.* 123: 264–275.

Trahey, M. and McCormick, F. 1987. A cytoplasmic protein stimulates normal N-ras p21 GTPase, but does not affect oncogenic mutants. *Science* 238: 542–544.

Vainio, S., Lehtonen, E., Jalkanen, M., Bernfield, M. and Saxén, L. 1989a. Epithelial-mesenchymal interactions regulate the stage-specific expression of a cell surface proteoglycan, syndecan, in the developing kidney. *Dev. Biol.* 134: 382–391.

Vainio, S., Jalkanen, M., Vaahtokaari, A., Sahlberg, C., Mali, M., Bernfield, M. and Thesleff, I. 1991. Expression of syndecan gene is induced early, is transient, and correlates with changes in mesenchymal proliferation during tooth organogenesis. *Dev. Biol.* 147: 322–333.

Vainio, S., Jalkanen, M., Bernfield, M. and Saxén, L. 1992. Transient expression of syndecan in mesenchymal cell aggregates of the embryonic kidney. *Dev. Biol.* 152: 221–232.

Vainio, S. Karavanova, I., Jowett, A. and Thesleff, I. 1993. Identification of BMP-4 as a signal mediating secondary induction between epithelial and mesenchymal tissues during early tooth development. *Cell* 75: 45–58.

van Heyningen, V. and eleven others. 1990. Role for Wilms tumor gene in genital development? *Proc. Natl. Acad. Sci. USA* 87: 5383–5386.

von Woellwarth, V. 1961. Die Rolle des neuralleistenmaterials und der Temperatur bei der Determination der Augenlinse. *Embriologia* 6: 219–242.

Waddington, C. H. 1940. *Organisers and Genes.* Cambridge University Press, Cambridge.

Wessells, N. K. 1970. Mammalian lung development: Interactions in formulation and morphogenesis of tracheal buds. *J. Exp. Zool.* 175: 455–466.

Wessells, N. K. 1977. *Tissue Interaction and Development*. Benjamin, Menlo Park, CA.

Wessells, N. K. and Cohen, J. H. 1968. Effects of collagenase on developing epithelia in vitro: lung, ureteric bud and pancreas. *Dev. Biol.* 18: 294–309.

Wessells, N. K. and Evans, J. 1968. The ultrastructure of oriented cells and extracellular materials between developing feathers. *Dev. Biol.* 18: 42–61.

Wilkinson, D. G., Bhatt, S. and McMahon, A. P. 1989. Expression of the FGF-related proto-oncogene *int-2* suggests multiple roles in fetal development. *Development* 105: 131–136.

Yamamoto, T. and seven others. 1983. Similarity of protein encoded by the c-*erb*B2 gene to epidermal growth factor receptor. *Nature* 319: 230–234.

Yedvobnick, B., Muskavitch, M. A. T., Wharton, K. A., Halpern, M. E., Paul, E., Grimwade, B. G. and Artavanis-Tsakonas, S. 1985. Molecular genetics of *Drosophila* neurogenesis. *Cold Spring Harbor Symp. Quant. Biol.* 50: 841–854.

Yochem, J., Weston, K. and Greenwald, I. 1988. The *Caenorhabditis elegans lin-12* gene encodes a transmembrane protein with overall similarity to *Drosophila Notch*. *Nature* 335: 547–550.

19

Development of the tetrapod limb

Pattern formation in the limb

Pattern formation is the activity by which embryonic cells form ordered spatial arrangements of differentiated tissues. The ability to carry out this process is one of the most dramatic properties of developing organisms, and one that has provoked a sense of awe in scientists and laypeople alike. How is it that the embryo is able not only to generate the different cell types of the body, but also to produce them in a way that forms functional tissues and organs? It is one thing to differentiate the chondrocytes and osteocytes that synthesize the cartilage and bone matrices, respectively; it is another thing to produce these cells in a temporal–spatial orientation that generates a functional bone. It is still another thing to make that bone a humerus and not a pelvis or a femur. The ability of limb cells to sense their relative positions and to differentiate with regard to their position has been the subject of intense debate and experimentation. How are the cells that differentiate into the cartilage of the embryonic bone specified so as to form digits at one end and a shoulder at the other? (It would be quite a useless appendage if the order were reversed.) Here the cell types are the same, but the patterns they form are different.

The vertebrate limb is an extremely complex organ with an asymmetric pattern of parts. The bones of the forelimb, be it wing, hand, flipper, or fin, consist of a proximal humerus (adjacent to the body wall), a radius and an ulna in the middle region, and the distal bones of the wrist and the digits (Figure 19.1). Originally, these structures are cartilaginous, but eventually most of the cartilage is replaced by bone. The position of each of the bones and muscles in the limb is precisely organized. Polarity exists in other dimensions as well. In humans, it is obvious that each hand develops as a mirror image of the other. It is possible for other arrangements to exist—such as the thumb developing on the left side of both hands—but this is not generally seen. In some manner, the three-dimensional pattern of forelimb is routinely produced.

The basic "morphogenetic rules" for forming a limb appear to be the same in all tetrapods (see Hinchliffe, 1991). Fallon and Crosby (1977) showed that grafted pieces of reptile or mammal limb bud can direct the formation of chick limb development, and Sessions and co-workers (1989) have found that frog and salamander limb buds can direct the formation of the other class's limbs. Moreover, the regeneration of salamander limbs appears to follow the same rules as those of developing limbs (Muneoka and Bryant, 1982). But what are these "morphogenetic rules"?

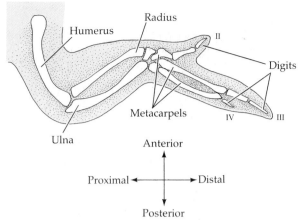

FIGURE 19.1
Skeletal pattern of the chick wing. According to convention, digits are numbered
II, III, IV. Digits I and V are not found in chick wings. (After Saunders, 1982).

Formation of the limb bud

The limb field

A **morphogenetic field** can be described as a group of cells whose position
and fate are specified with respect to the same set of boundaries (Weiss,
1939; Wolpert, 1977). A particular field of cells will give rise to its particular
organ (forelimb, eye, tail, etc.) when transplanted to a different part of
the embryo, and the cells of the field can regulate their fates to make up
for missing cells in the field (Huxley and De Beer, 1934; Opitz, 1985; De
Robertis et al., 1991). One of the first fields identified was the limb field.

The mesodermal cells that give rise to the vertebrate limb can be
identified by (1) removing certain groups of cells and observing whether
a limb develops in their absence (Detwiler, 1918; Harrison, 1918), (2)
transplanting certain groups of cells to new locations and observing
whether they form a limb (Hertwig, 1925), and (3) marking groups of cells
with dyes or radioactive precursors and observing which descendants of
marked cells partake in limb development (Rosenquist, 1971). By these
procedures, the prospective limb area has been precisely localized in many
vertebrate embryos. Figure 19.2 shows the prospective forelimb area in
the tailbud stage of the salamander *Ambystoma maculatum*. The center of

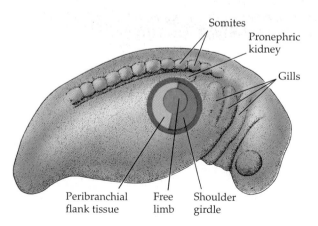

FIGURE 19.2
Prospective forelimb field of the salamander *Ambystoma maculatum*. The
central area contains those cells destined to form the free limb; the cells
surrounding the free limb are those
that give rise to the peribrachial flank
tissue and the shoulder girdle. The
cells outside these regions usually are
not included in limbs but can form a
limb if the more central tissues are extirpated. (After Stocum and Fallon,
1982.)

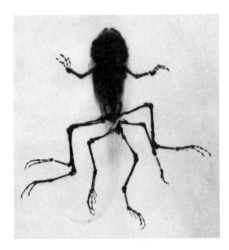

FIGURE 19.3
Regulative ability of the limb field, seen when the early hindlimb fields of a *Hyla regila* tadpole were split by numerous trematode eggs. (Courtesy of S. Sessions.)

this disc marks the lateral plate mesoderm cells normally destined to give rise to the limb itself. Adjacent to it are the cells that will form the peribrachial flank tissue and the shoulder girdle. These two regions encompass the classic "limb disc" that will be used in the experiments to be mentioned in this chapter. However, if all these cells are extirpated from the embryo, a limb will still form, albeit somewhat later, from an additional ring of cells that surrounds this area. If this last ring of cells is included in the extirpated tissue, no limb will develop. This larger region, representing all the cells in the area capable of forming a limb, is called the **limb field**.

The limb field originally has the ability to regulate for lost or added parts. In the tailbud-stage *Ambystoma*, any half of the limb disc is able to regenerate the entire limb when grafted to a new site (Harrison, 1918). This potential can also be shown by splitting the limb disc vertically into two or more segments and placing thin barriers between these segments to prevent their reunion. When this is done, each part develops into a full limb. The regulative ability of the limb bud was recently highlighted by a remarkable experiment of nature. In a pond in Santa Cruz, California, numerous multilegged frogs and salamanders have been found (Figure 19.3). The presence of these extra appendages has been linked to the infestation of the larval abdomen with parasitic trematode worms. The eggs of these worms apparently split the limb bud in several places while the tadpole was first forming these structures (Sessions and Ruth, 1990). Thus, like an early sea urchin embryo, the limb field represents a "harmonious equipotential system" wherein a cell can be instructed to form any part of the limb.

Growth of the early limb bud: Interactions with the mesonephros

Limb development begins when mesenchymal cells are released from the somatic layer of the limb field lateral plate mesoderm (limb skeletal precursors) and from the somites (limb muscle precursors) (Figure 19.4). The cells migrate laterally and accumulate under the epidermal tissue of the neurula. The circular bulge on the surface of the embryo is called the **limb bud**. The mesenchymal cells in the limb bud multiply to create a bulge that will proliferate to form a limb. The initial stages of this proliferation

FIGURE 19.4
Limb bud formation. Migration of mesodermal cells from the somatic region of the lateral plate mesoderm causes the limb bud in the amphibian embryo to bulge out. These cells generate the skeletal elements of the limb. (Migration of somitic cells into the limb bud generates limb musculature.)

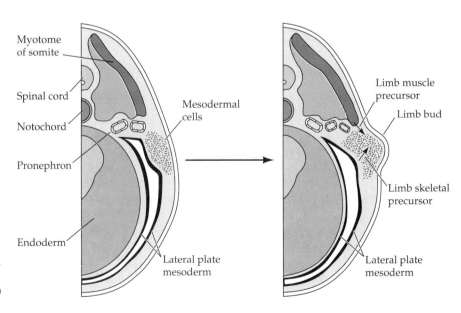

appear to be under the regulation of the nearby mesonephros. If the mesonephros on one side of the embryo is removed at this stage or if a thin impermeable membrane is imposed between the mesonephros and a limb bud, the limb mesenchyme cells of that particular limb bud stop dividing (Figure 19.5; Stephens et al., 1991; Geduspan and Solursh, 1992a). The identity of the growth factor(s) produced by the mesonephros is still unknown, but insulin-like growth factor I (ILGF-I) is a good candidate for at least some of the properties of the mesonephros mitogen. If limb buds are cultured with mesonephros tissue, their mesenchymal cells are able to proliferate and condense into cartilage that resembles the shape of the normal skeleton. Without this mesonephros tissue, only a small amount of cartilage is produced, and it is amorphous. When ILGF-I is added to limb buds cultured without mesonephros tissue, the skeletal growth and differentiation approximate those cultures wherein the limb bud was cultured with mesonephros tissue. Moreover, if antibodies to ILGF-I are added to limb buds cultured in the presence of the mesonephros, the medium is ineffective at promoting cartilage growth, although cartilage differentiation proceeds normally (Geduspan and Solursh, 1993). This link between kidney development and limb development may explain the *acrorenal syndromes*, conditions characterized by deformities of both limb skeleton and kidney formation (Dieker and Opitz, 1969; Evans et al., 1992).

Induction in the chick limb bud: Mesenchyme-AER interactions

The growth and differentiation of the limb bud is made possible by a series of interactions between the limb bud mesoderm and its overlying ectoderm (Harrison, 1918). In birds and mammals, the mesoderm is seen to induce the ectodermal cells to elongate and form a special structure: the **apical ectodermal ridge** (AER) (Figure 19.6). Saunders and his co-workers (1976) have also shown that the AER is a self-sustaining population of cells that does not exchange cells with its surroundings as the limb develops. Chick and quail cells can readily be distinguished by their nuclear heterochromatin (Chapter 9), and when a quail AER is placed upon the stump of a chick limb bud, the limb continues to grow. However, the AER of that limb remains composed of quail cells, and the surrounding chick ectoderm is not seen to become part of the AER. Once induced, this AER becomes essential to limb growth and interacts with the mesenchyme (Figure 19.7).

1. When the AER is removed at any time during limb development, further distal limb development ceases.
2. When an extra AER is grafted onto an existing limb bud, supernumerary structures are formed, usually toward the distal end of the limb.
3. When leg mesenchyme is placed directly beneath the wing AER, distal hindlimb structures (toes) develop at the end of the limb. (However, if this mesenchyme is placed further from the AER, the hindlimb mesenchyme becomes integrated into wing structures.)
4. When nonlimb mesoderm is grafted beneath the AER, the AER regresses and limb development ceases.

Thus, although the mesenchymal cells induce and sustain the AER and instruct the AER to produce a certain type of limb, the AER is responsible

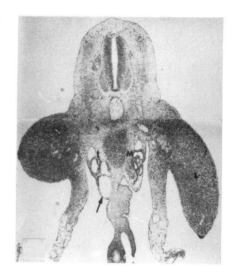

FIGURE 19.5
Dependence of limb mesodermal cell proliferation on the presence of the mesonephric kidney. The kidney was removed from the left side of the embryo, and after two days, the limb bud on the left side has not grown to the extent as the control limb bud on the right side of the embryo. (From Geduspan and Solursh, 1992a, courtesy of M. Solursh.)

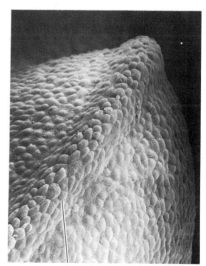

Apical
ectodermal ridge

FIGURE 19.6
Scanning electron micrograph of an early chick forelimb bud, with its apical ectodermal ridge in the foreground. (Courtesy of K. W. Tosney.)

FIGURE 19.7

Summary of the inductive role of the apical ectodermal ridge (AER) upon the underlying mesenchyme. (Modified from Wessells, 1977.)

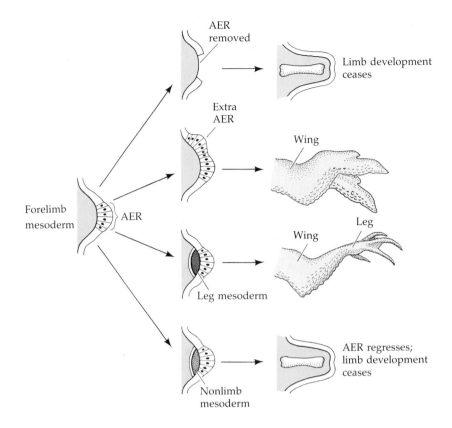

for the sustained outgrowth and development of the limb (Zwilling, 1955; Saunders et al., 1957; Saunders, 1972; Krabbenhoft and Fallon, 1989). The AER keeps the mesenchyme directly beneath it in a state of mitotic proliferation and prevents the mesenchyme cells from forming cartilage. Hurle and co-workers (1989) found that if they cut away a small portion of the AER in a region that would normally fall between the digits of the chick leg, an extra digit emerged at that place (Figure 19.8). It thus appears that the AER also has the effect of inhibiting induction of cartilage in the mesenchyme cells directly beneath it.

The relationships between the AER and the limb bud mesenchyme can best be seen by mutations of chick limb development. The *polydactylous* mutation, as the name implies, confers extra digits on each limb. By recombining mutant and wild-type tissues (Table 19.1), the defect can be traced to mesodermal cells that induce too broad an AER. In the mutant *eudiplopodia* (Greek, "two good feet"), there are not only extra digits, but two complete rows of toes on each hindlimb (Figure 19.9). Similar recon-

FIGURE 19.8

Cross section through the distal region of a chick limb 3 days after a wedge of the AER was removed from the area that would form interdigital tissue. Instead of degenerating, the remaining interdigital tissue formed an extra digit. (From Hurle et al., 1989, courtesy of the authors.)

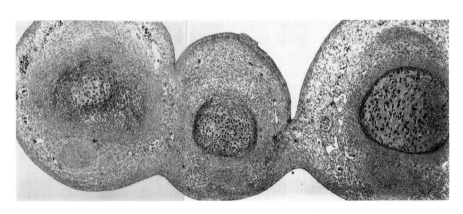

TABLE 19.1
Mutations affecting the reciprocal interactions between AER and its underlying mesenchyme[a]

Mesoderm	Epidermis	Result	Conclusion
POLYDACTYLOUS			
Polydactylous	Wild-type	Polydactylous	Mesoderm is affected
Wild-type	Polydactylous	Wild-type	by the mutation
EUDIPLOPODIA			
Eudiplopodia	Wild-type	Wild-type	Ectoderm is affected
Wild-type	Eudiplopodia	Eudiplopodia	by the mutation
LIMBLESS			
Limbless	Wild-type	Wild-type	Ectoderm is affected
Wild-type	Limbless	Limbless	by the mutation

[a] By reciprocal transplantation between wild-type and mutant AER and mesenchyme, the aberrant compartment of the induction can be identified.

stitution experiments show that here the defect is in the ectodermal tissue. Chick embryos homozygous for the *limbless* mutation initiate the limb bud formation, but the AER fails to form. Recombination experiments show that the *limbless* ectoderm is unable to form an AER even when placed on wild-type limb mesoderm (Carrington and Fallon, 1988). In mice homozygous for mutant alleles of the *limb deformity* gene, the AER is malformed, and there is a marked reduction of the most distal structures of the limb (and, interestingly, a suppression of kidney development). In these mutants, there is a deficiency in a nuclear protein encoded in both the AER and the underlying mesenchyme (Jackson-Grusby et al., 1992).

In addition, there are some naturally limbless vertebrates whose lack of limb formation can be traced to deficiencies in the AER–mesenchyme interactions. The curse against snakes in the Book of Genesis seems to have been directed at the distal end of the limb bud, since the AER of these reptiles degenerates prematurely and at the same time as cell death in the underlying mesenchyme (Lande, 1978). It is not known whether the initial defect is in the mesenchyme or the AER.

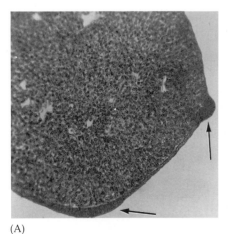

(A)

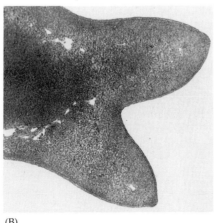

(B)

FIGURE 19.9
Cross sections of hindlimb buds from *eudiplopodia* chick embryos. (A) Two AERs on hindlimb bud; extra outgrowth on the dorsal side will form an extra set of toes. (B) Both outgrowth regions are covered by an AER. (From Goetinck, 1964, courtesy of P. Goetinck.)

The regeneration of salamander limbs

As we observed in the previous chapter, it is often useful to find adult models of embryonic development. For two centuries, the regeneration of the amphibian limb has been not only one of the most awesome instances of regulation, but also a model for vertebrate limb development. When a salamander limb is amputated, the remaining cells are able to reconstruct a complete limb with all its differentiated cells arranged in the proper order. It is remarkable that not only has the limb regenerated, but the remaining cells retain the information specifying their position and the position of the cells that have been removed. In other words, the new cells construct only the missing structures and no more; for example, when a wrist is amputated, the salamander forms a new wrist and not a new elbow (Figure 19.10). Oscar Schotté declared that he would give his right arm to know the secret of limb regeneration (in Goss, 1991).

Upon amputation, epidermal cells from the remaining stump migrate to cover the wound surface. This single layer of cells then proliferates to form the **apical ectodermal cap**. The cells beneath this cap undergo a dramatic dedifferentiation: bone cells, cartilage cells, fibroblasts, myocytes, and neural cells lose their differentiated characteristics and become detached from each other. The well-structured limb region at the cut edge of the stump thus forms a proliferating mass of indistinguishable, dedifferentiated cells just beneath the apical ectodermal cap. This dedifferentiated cell mass is called the **regeneration blastema**, and these cells will continue to proliferate and differentiate to form the new structures of the limb. If the blastema cells are destroyed, no regeneration takes place (Butler, 1935). Moreover, once the cells have dedifferentiated to form a blastema, they have regained their embryonic plasticity. Carlson (1972) has shown that when at least 99 percent of the muscle cells are removed from the stump of a newly amputated salamander limb, the regenerated limb contains a normal supply of muscles in their appropriate positions. Thus, other cells in the blastema—cells derived from nonmuscle tissue—must be able to form the muscles of the regenerated limb.

In most instances, however, neural tissue is essential for the formation of the new limb by other cells. Singer (1954) demonstrated that a minimum number of nerve fibers must be present for regeneration to take place. It is thought that the neurons release a mitosis-stimulating factor that increases the proliferation of the blastema cells (Singer and Caston, 1972; Mescher and Tassava, 1975). One candidate for this crucial neural substance is **glial growth factor** (GGF). This peptide is known to be produced by newt neural cells, is present in the blastema, and is lost upon denervation. When GGF is added to a denervated blastema, the mitotically arrested cells are able to divide

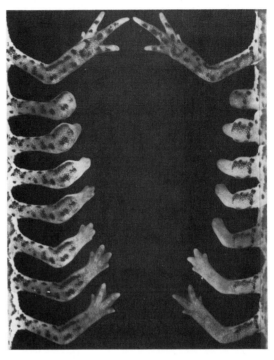

FIGURE 19.10
Regeneration of a salamander forelimb. On the left, the amputation was made below the elbow; the amputation shown on the right cut through the humerus. In both cases the correct positional information is respecified. (From Goss, 1969, courtesy of R. J. Goss.)

again (Brockes and Kinter, 1986). Another candidate is **transferrin**, an iron-transport protein that is necessary for mitosis in all dividing cells (since ribonucleotide reductase, the rate-limiting enzyme of DNA synthesis, requires a ferric ion in its active site). When hindlimbs are severed, the sciatic nerve transports transferrin along the axon and releases large quantities of this protein into the blastema (Munaim and Mescher, 1986; Mescher, 1992).

So we are faced with a situation wherein the adult cells of an organism can return to an "embryonic" condition and begin the formation of the limb anew. Just as in embryonic development, the blastema forms successively more distal structures (Rose, 1962). Thus, the blastema must contain some positional information that directs a blastema on a stump containing a humerus neither to make another humerus nor to start immediately producing digits. Not only does the blastema regenerate those structures beginning at the appropriate proximal–distal level in the limb, but the polarity of the anterior–posterior ("thumb–pinky") and dorsal–ventral ("wrist–palm") axes also correspond to those of the stump.

Development of the proximal–distal limb axis

The positional information needed to construct a limb has to function in a three-dimensional coordinate system.* The axes of polarity appear to be determined in the following sequence: anterior–posterior (as in the line between the thumb and the little finger); dorsal–ventral (as in the line between the upper and lower surfaces of the hand); and proximal–distal (as in a line connecting the shoulder and the fingertip).

The proximal–distal axis is defined only after the induction of the apical ectodermal ridge by the underlying mesoderm. The limb bud elongates by the proliferation of the mesenchymal cells underneath the AER. This region of cell division is called the **progress zone**, and it extends about 200 μm from the AER. Molecules from the AER are thought to keep these cells dividing, and when the mesenchymal cells leave the progress zone, they differentiate in a regionally specific manner. The first cells leaving the progress zone form proximal structures; those cells that have undergone numerous divisions in the progress zone become the more distal structures (Saunders, 1948; Summerbell, 1974). Therefore, when the AER is removed from an early-stage wing bud, the cells of the progress zone stop differentiating and only a humerus forms. When the AER is removed slightly later, humerus and radius and ulna form (Figure 19.11; Rowe et al., 1982).

*Actually, it is a four-dimensional system in which time is the fourth axis. Developmental biologists get used to seeing nature in four dimensions.

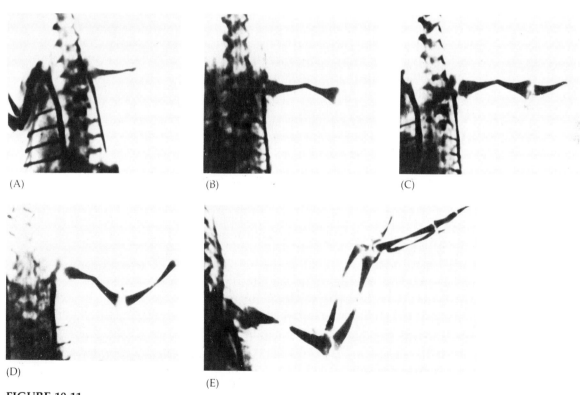

(A) (B) (C)

(D) (E)

FIGURE 19.11
Dorsal view of chick skeletal pattern after removal of the entire AER from the right wing bud of embryos at various stages. The last picture is of a normal wing skeleton. (From Iten, 1982, courtesy of L. Iten.)

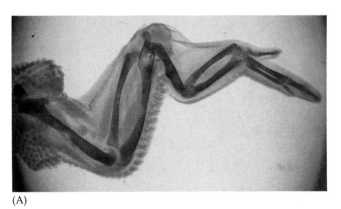

(A)

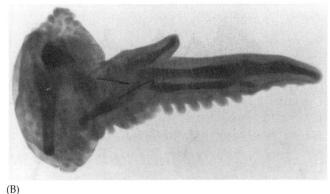

(B)

FIGURE 19.12
Control of proximal–distal specification by the cells of the progress zone (PZ).
(A) Extra set of ulna and radius formed when early-bud PZ is transplanted to
late wing bud that has already formed ulna and radius. (B) Lack of intermediate
structures seen when late-limb bud PZ is transplanted to early limb bud. The
hinges indicate the location of the grafts. (From Summerbell and Lewis, 1975,
courtesy of D. Summerbell.)

Proximal–distal polarity resides in the mesodermal compartment of
the limb. If the AER provides the positional information—somehow instructing the undifferentiated mesoderm beneath it as to what structures
to make—then older AERs should produce more distal structures when
placed on young mesoderm. This was not found to be the case, however
(Rubin and Saunders, 1972), as the normal complete sequence of limb
development occurred when young mesoderm was combined with any
stage AER. But when the entire progress zone, including the mesoderm,
from an early embryo was placed on the limb bud of a later-stage embryo,
new proximal structures were produced beyond those already present.
Conversely, when old progress zones were added to young limb buds,
distal structures immediately developed so that digits were seen to emerge
from the humerus without the intervening ulna and radius (Figure 19.12;
Summerbell and Lewis, 1975).

The molecular signals by which the AER and the progress zone mesenchyme interact are starting to be identified. Fallon and colleagues (in
press) have shown that fibroblast growth factor 2 (FGF-2) is synthesized
by the AER and is secreted into the mesenchyme under it. Moreover, if
the AER is removed, the implantation of FGF-2-releasing beads will replace
it (Figure 19.13). When the AER was removed from 2-day chick embryos,
growth ceased and only a humerus was formed. When beads soaked in
saline solution were placed into the progress zone mesenchyme of such
AER-deficient limb buds, they did not further limb growth or development. However, when the implanted beads had been soaked in FGF-2, a
single bead enabled the wing bud to grow a bit further and to form the
humerus, radius, and ulna. Fallon and colleagues found that the FGF-2
on the beads became depleted at this time, and that the addition of a
second FGF-2-containing bead to the progress zone mesenchyme 24 hours
later enabled the full growth of the limb bud, complete with humerus,
radius, ulna, carpels, and digits. Thus, FGF-2 may be the molecule that
keeps the progress zone mesenchyme proliferating in chick embryos. FGF-
4 is also able to substitute for the AER under experimental conditions, but
it is not known whether it is actually synthesized throughout the chick
AER (Niswander and Martin, 1993; Niswander et al., 1993).

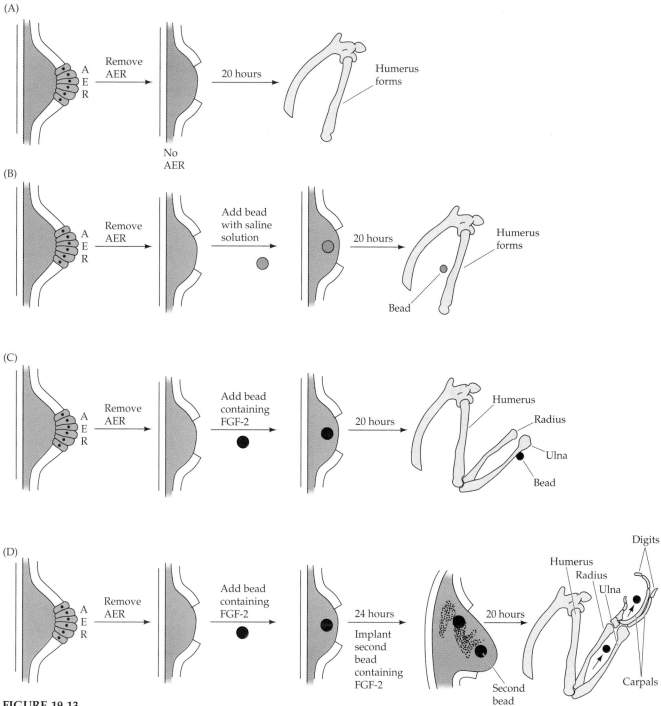

FIGURE 19.13
Ability of FGF-2 to replace apical ectodermal ridge in developing chick forelimb bud. (A) When the AER is removed from stage 20 chick wing buds, only the humerus is able to form. (B) When a slow-release gel bead soaked in saline solution is placed into the progress zone mesenchyme, the limb is still truncated and only forms a humerus. (C) When a bud soaked in FGF-2 is placed in the progress zone, growth of the limb bud continues and the ulna and radius are formed. (D) If a second FGF-2-containing bead is placed into the progress zone after most of the FGF-2 of the first bead has dissipated, the limb bud continues to grow and to make the metacarpals and digits. (After Fallon et al., in press.)

Specification of the limb skeleton by reaction–diffusion processes

The limb skeleton is a highly polar structure. In our own case, the limb starts at the shoulder/pelvis to form a single humerus/femur. After a break, two further bones, the ulna/tibia and radius/fibula, extend outward to the bones of the wrist/ankle and fingers/toes. The mesenchymal cells of the limb bud are derived from two distinct sources. Quail-chick chimeras (Chevallier et al., 1977; Christ et al., 1977; Geduspan and Solursh, 1992b) and lineage-specific monoclonal antibodies (George-Weinstein et al., 1988) have shown that the precursor cells for cartilage and fibroblasts arise from the limb field somatopleure (of the lateral plate mesoderm), while the muscle cell precursors come from the somites. Thus, the muscle cells are already determined before limb development. However, which cells become the cartilage-forming chondrocytes and which form the fibroblasts has not yet been determined. Cartilage formation begins in the proximal region and ends with the digits (Figure 19.14). At each stage, there is a condensation of mesenchymal cells to form tight clusters of cells. This condensation is probably mediated by fibronectin, whose presence is seen at various foci concomitant with such condensation. Fibronectin may promote the clustering of these cells by linking together the proteoglycans on adjacent cell surfaces (Frenz et al., 1989). Cyclic AMP levels increase in these condensing mesenchymal cells, followed by the modification of chromatin proteins and the transcription of new cartilage-specific mRNAs. Shortly thereafter, the cartilage-specific extracellular matrix molecules are secreted (Solursh, 1984; Newman, 1988).

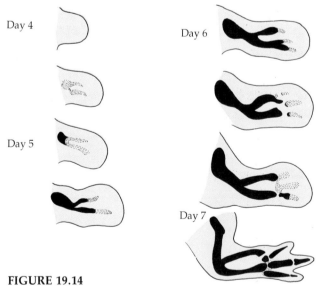

FIGURE 19.14

Cartilaginous condensations in the chick wing bud between days 4 and 7. Stippled regions indicate new areas of mesenchymal cell condensation; solid black regions indicate definitive cartilage. Areas of cartilage are seen to be foreshadowed by mesenchymal condensations about 12 hours earlier (After Newman, 1988.)

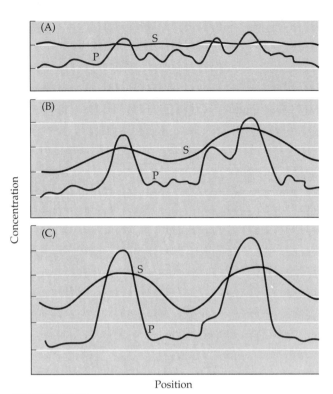

FIGURE 19.15

Generation of periodic spatial heterogeneity when two reactants, S and P, are mixed together under the conditions that S inhibits P, P catalyzes the formation of both S and P, and S diffuses faster than P. (A) The distributions are initially random, but quickly begin to fluctuate. (B,C) As P increases locally, it produces more S, which inhibits other peaks of P in the near vicinity. The result is a series of P peaks ("standing waves") at regular intervals.

But which cells are to form these precartilaginous clusters and which cells are to be excluded? Simple gradient models do not suffice. Why should one cartilage condensation (say, the humerus) be formed in the most proximal region, followed by two (say, ulna and radius) in the midlimb, followed by a whole array of cartilaginous condensations to form the digits? Alan Turing (1952), one of the founders of computer science (and the mathematician who cracked the German "Enigma" code during World War II), proposed a model wherein two homogeneously distributed solutions would interact to produce stable patterns during morphogenesis. These patterns would represent regional differences in the concentrations of the two substances. These interactions would produce an ordered structure out of random chaos.

This **reaction–diffusion model** involves two substances. One of them, substance S, inhibits the production of the other, substance P. Substance P promotes the production of more substance P as well as more substance S. If S diffuses more readily than P, Turing's mathematics show that sharp waves of concentration differences will be generated for substance P (Figure 19.15). These waves have been observed in certain chemical reactions (Prigogine and

Nicolis, 1967; Winfree, 1974) and can be observed in three dimensions (Figure 19.16; Welsh et al., 1983).

The reaction–diffusion model predicts alternating areas of high and low concentrations of some substance. When the concentration of such a substance is above a threshold level, a cell (or group of cells) may be instructed to differentiate in a certain way. An important part of Turing's model is that particular chemical wavelengths will be amplified while all others will be suppressed. As local concentrations of P increase, the values of S form a peak centering on the P peak but becoming broader and shallower because of its more rapid diffusion. These S peaks inhibit other P peaks from forming. But which of the many P peaks will survive? This depends on the size and shape of the tissues in which the oscillating reaction is occurring. (This is analogous to the harmonics of vibrating strings, as in a guitar. Only certain resonance vibrations are permitted, based on the boundaries of the string.)

The mathematics describing which wavelengths are selected at a particular length consist of complex polynomial equations called Bessel functions. Such functions have been used to model the spiral patterning of *Dictyostelium* (Chapter 1), which depends upon the spatial concentration of cAMP in the environment. Using these equations, we can predict that the size of the original limb bud would allow only one such peak—and hence, only one cartilaginous condensation (Figure 19.17; Newman and Frisch, 1979; Oster et al., 1988). As the progress zone mesenchyme becomes confined into progressively smaller regions behind the AER, the number of parallel condensations that

FIGURE 19.16

Photograph of a test tube 10-mm in diameter wherein standing waves have been generated from a homogeneous solution. In this reaction, malonic acid is oxidized by bromate in the presence of cerium ions. A dye (ferroin) reflects the periodic changes in cerium^{3+}:cerium^{4+} levels. (From Welsh et al., 1983.)

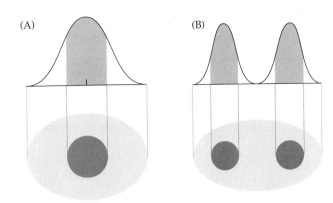

FIGURE 19.17

The number of standing waves fitting into a region depends upon the size of the region. As the region grows, it becomes possible to fit more complete wavelengths into the region. (A) When the limb cross section is only able to accommodate a single wavelength, a single condensation emerges. (B) If the domain changes shape so that two complete wavelengths can fit into the cross section, then two condensations arise. (After Oster et al., 1988.)

can form along that proximal–distal axis increases. First a single condensation can fit ("humerus"), then two ("ulna and radius"), then several ("wrist, digits").

In this reaction–diffusion hypothesis, the aggregations of cartilage actively recruit more cells from the surrounding area and laterally inhibit the formation of other foci of condensation. The number of foci, then, depends on the geometry of the tissue and the strength of the lateral inhibition. If the inhibition remains the same, tissue volume must increase in order to get two foci forming where one had been allowed before. This model predicts that slight size changes of the distal limb bud alter the number of digits. This is indeed found to be the case, and it may be a very simple way of gaining or losing digits during evolution (see Chapter 23). These standing waves constitute the prepattern of the limb. This reaction–diffusion mechanism does not rule out prelocalized morphogens or gradients (for which there is good evidence), for it can exist as a separate morphogenetic mechanism to structure the embryo.

The molecule forming these standing waves (and therefore responsible for organizing these precartilage condensations) has not yet been identified. Newman and co-workers (1988) suggest that transforming growth factor β (TGF-β) may be involved in creating a prepattern of fibronectin in the limb buds. These nodes of fibronectin would link cells together into precartilagenous nodules. TGF-β is known to stimulate both its own synthesis and the production of fibronectin. In cultured limb mesenchyme, TGF-β1 induces fibronectin synthesis, followed by the condensation of these cells and their subsequent differentiation into cartilage (Leonard et al., 1991). Other pattern-generating mechanisms (such as the polar coordinate model to be discussed later in this chapter) may also be active in the creation or maintenance of limb polarity. (For a discussion of the various mathematical models for embryonic polarity and periodicity, see Held, 1992.)

Specification of limb anterior–posterior axis

The zone of polarizing activity

The self-differentiation of the anterior–posterior axis is the first change from the pluripotent condition. In chicks, this axis is specified long before a limb bud is recognizable. Hamburger (1938) showed that as early as the 16-somite stage, prospective wing mesoderm transplanted to the flank area develops into a limb with the anterior–posterior and dorsal–ventral polarities of the donor graft and not those of the host tissue (Figure 19.18).

Although the differentiation of the proximal–distal structures is thought to depend upon how many divisions a cell undergoes while in the progress zone, positional information instructing a cell as to its position on the anterior–posterior and dorsal–ventral axes must come from other sources. There are two major models to account for the specificity of the anterior–posterior axis. Several experiments (Saunders and Gasseling, 1968; Tickle et al., 1975; Summerbell, 1979) suggest that the anterior–posterior axis is specified by a small block of mesodermal tissue near the posterior junction of the young limb bud and the body wall. When this tissue from a young limb bud is transplanted into a position on the anterior

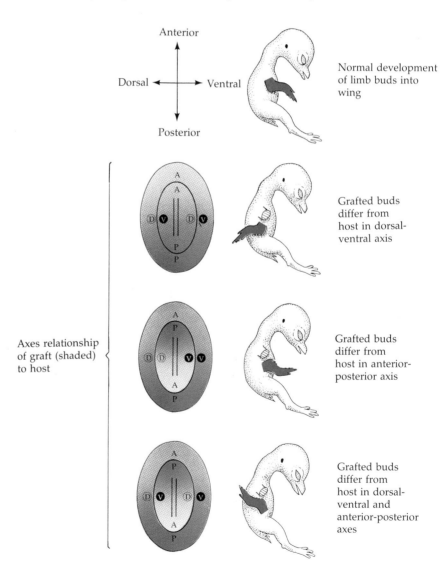

FIGURE 19.18
Specification of the anterior–posterior and dorsal–ventral axes in the chick wing. The grafted limb bud develops in accordance with its own polarity and does not adopt the polarity of its host. Wings that develop from grafted limb buds are shown in color. For the sake of clarity, the host's normally developed wing is not shown. (After Hamburger, 1938.)

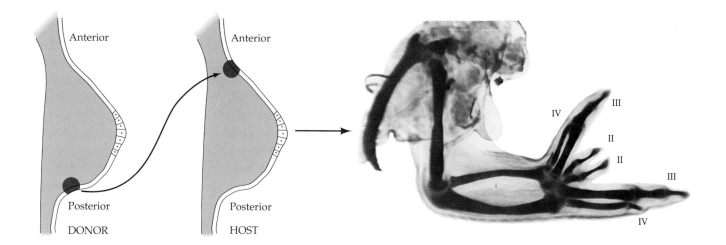

FIGURE 19.19
Duplicated digits emerge as mirror image of normal digits when ZPA is grafted to anterior limb bud mesoderm. (From Honig and Summerbell, 1985, photograph courtesy of D. Summerbell.)

side of another limb bud (Figure 19.19), the number of digits of the resulting wing is doubled. Moreover, the structures of the extra set of digits are mirror images of the normally produced structures. The polarity has been maintained, but the information is now coming from both an anterior and a posterior direction. This region of the mesoderm has been called the **zone of polarizing activity** (ZPA).

The distribution and strength of the ZPA's positional signaling activity in the chick wing and leg buds have been mapped (Hinchliffe and Sansom, 1985; Honig and Summerbell, 1985). As shown in Figure 19.20, the polarizing activity (measured after grafting the posterior marginal cells into the anterior margin of the limb bud) is highest in a particular region of the posterior margin and tapers off from there. It weakens as development progresses.

The mediator of ZPA activity

One of the classic quests in developmental biology has been the search for the molecular agent(s) of the ZPA. Seekers of the "ZPA factor" usually went on one of two paths.

The morphogen model. The first path, cleared by Lewis Wolpert (1969), is known as the **ZPA diffusible morphogen model**. Here, the anterior–posterior pattern-forming mechanism is a *gradient*: a homogeneous population of cells is instructed to act in certain ways by the concentration of a soluble molecule. In the ZPA morphogen hypothesis, the ZPA tissue is thought to operate by secreting a soluble morphogen that diffuses from its source to form a concentration gradient from the posterior end to the anterior end of the limb bud. Those cells nearest the ZPA would be exposed to the highest amount of this compound, whereas those farthest

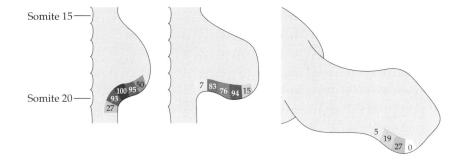

FIGURE 19.20
Map of positional signaling activity as the limb develops. Different regions of the limb bud were excised and grafted into the anterior margin. 100 percent indicates that all grafts showed complete duplication as shown in Figure 19.19. Drawing to scale. (After Honig and Summerbell, 1985.)

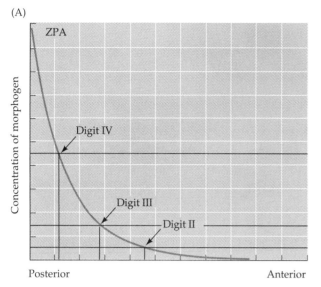 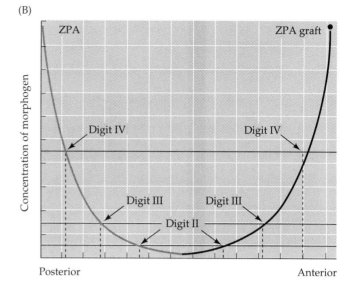

FIGURE 19.21
Model for a diffusible morphogen secreted by ZPA tissue. (A) Normal limb, wherein the ZPA produces a diffusible compound whose gradient declines anteriorly. As it falls through certain threshold levels, it instructs the cells to make specific digits. The posterior digit is 4; the anterior is 2. (B) When an extra ZPA is placed in the anterior of the limb bud, the concentration of the postulated morphogen changes, establishing a symmetrical distribution of the morphogen. The threshold conditions still apply, so the result is a symmetrical duplication.

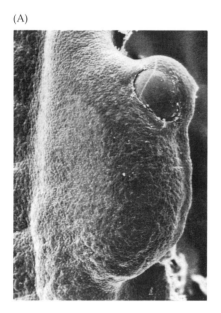 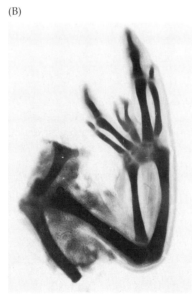 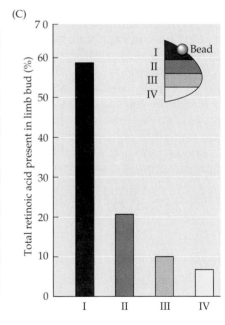

FIGURE 19.22
Retinoic acid effects in chick limb development. (A) Scanning electron micrograph of RA-soaked resin bead placed into early limb bud of chick embryo. (B) Duplicated wing resulting when bead soaked with 10 μg/ml RA is placed into limb bud. (C) Gradient of retinoic acid seen in limb bud 6 hours after placing bead into the limb bud. (From Tickle et al., 1985, courtesy of C. Tickle.)

from the ZPA would be exposed to a relatively low concentration. The original support for this hypothesis came from a series of experiments (Summerbell, 1979) in which the anterior limb bud was separated from the posterior limb bud by a nonpermeable barrier. In such cases, the anterior structures were no longer formed. These results were interpreted to mean that the ZPA was secreting a morphogen that organized the anterior–posterior gradient in the limb bud and that the anterior limb structures were unable to form because they could not receive this morphogen (Figure 19.21). The polarizing activity of the ZPA could be attenuated by reducing the number of cells grafted from the ZPA to the anterior margin (Tickle, 1981).

For several years retinoic acid was thought to be the endogenous morphogen that was secreted from the ZPA. Beads soaked in retinoic acid could substitute for ZPA tissue and induce a mirror-image reversal of anterior–posterior polarity, just like the ZPA does (Figure 19.22; Tickle et al., 1982, 1985), and a single retinoic acid-soaked bead could replace a ZPA when the normal ZPA tissue was removed (Eichele, 1989). However, the retinoic acid content of the ZPA does not seem high enough to activate retinoic acid-responsive genes (Plate 13; Noji et al., 1991; Rossant et al., 1991), and theoretical considerations (see Wanek et al., 1991) make it unlikely that retinoic acid is the active agent of the ZPA.

The polar coordinate model. The second path followed by seekers of the ZPA factor is the **polar coordinate model**. In this scenario, pattern is generated not by diffusible molecules, but by fixed molecules on cell membranes that can alter the specification of the partners.

Research on developing and regenerating limb buds indicates that adjacent cell surfaces may be critical for anterior–posterior pattern formation in the limb. When anterior and posterior cells from regeneration blastema or limb buds are juxtaposed, the results are dramatic and strange: the resulting limb contains supernumerary structures, often with two or three sets of digits (Figure 19.23; Bryant and Iten, 1976; Javois and Iten, 1986). This patterning mechanism appears to be the same for both regenerating and normally developing limbs. When axolotl limb buds were transferred to regenerating axolotl blastema stumps in a way that maintained the original polarity with respect to the stump, normal limbs developed. However, when the polarity of the anterior–posterior axis was reversed with respect to the stump, mirror-image supernumerary digits emerged (Figure 19.24; Muneoka and Bryant, 1982). These results strongly suggest that the patterning rules are the same for developing and regenerating limbs and that regenerating limbs can be used as models for developing limbs.

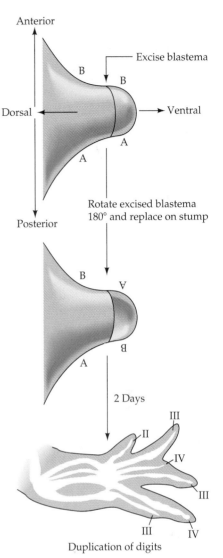

Duplication of digits

FIGURE 19.23
Limb bud rotation experiment. When a newly emerged limb bud (or regeneration blastema) is severed at its base and rotated 180 degrees on its stump, the result is a limb with three areas of outgrowth. In this case, the limb bud proceeded to form a radius and ulna with two new ulnas between them. At their tips were digits representing three partial wings. (After Javois and Iten, 1986.)

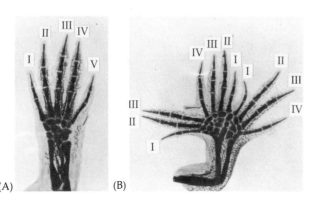

(A)   (B)

FIGURE 19.24
Ability of regenerating salamander limb blastema to be controlled by progress zone of developing limb bud. (A) Control showing normal five-digit right hindlimb where right hindlimb bud was placed on right regenerating hindlimb stump without rotation. (B) Limb resulting when a graft from the left hindlimb bud was placed onto a right regenerating hindlimb stump. Here the anterior–posterior axes are reversed between the host and the graft. (From Muneoka and Bryant, 1982, courtesy of K. Muneoka.)

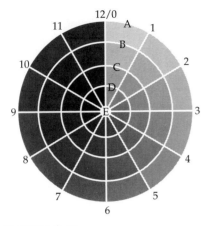

FIGURE 19.25
Polar coordinate model for the specification of positional information. Each cell has a circumferential value (0–12) specifying the anterior–posterior axis and a radial value (A–E) specifying the proximal (A) to distal (E) axis. (From French et al., 1976.)

By comparing such results in regenerating vertebrate limbs, insect limbs, and insect imaginal discs, French and his colleagues (1976; Bryant et al., 1981) have proposed a series of empirical rules that predict the outcome of a wide variety of experimental perturbations with regenerating appendages. Starting from Wolpert's (1969) premise that pattern arises from a cell's recognition of its relative position in a developing population, they speculate that a cell assesses its physical location in a system of polar (clocklike) coordinates (Figure 19.25). In this system, each cell has a circumferential value (from 0 to 12) as well as a radial value (from A to E). In regenerating limbs, the outer circle represents the proximal (shoulder) boundary of the limb field; the innermost circle represents the most distal regions.

As we have seen, when normally nonadjacent tissues of a field are juxtaposed, duplications often arise. Yet other transplanted tissue (not of the field) will not cause these duplications. The polar coordinate model has been extremely useful in predicting the extent of these duplicated structures. The **shortest intercalation rule** states that when two normally nonadjacent cells are juxtaposed, growth occurs at the junction until the cells between these two points have all the positional values between the original points (Figure 19.26). The circular sequence, like a clock, is continuous, 0 being equal to 12 and having no intrinsic value in itself. Being circular, however, means that there are two paths by which intercalation can occur between any two points. For example, when cells having the values 4 and 7 are placed next to each other, there are two possible routes between them: 4, 5, 6, 7 and 4, 3, 2, 1, 12, 11, 10, 9, 8, 7. According to this model, the shortest route is taken. The exception, of course, is when the cells have values that fall exactly opposite each other in the coordinate system, so that there is no one shortest route. In this case, all values are formed between the two opposites.

The second rule is the **complete circle rule** for distal transformation. Once the complete circle of positional values has been established on the wound surface, the cells proliferate and produce the more distal structures. The mechanism by which this is thought to occur is outlined in Figure 19.26 and again involves intercalation of structures between cells having different positional information (Bryant et al., 1981). The predictive value of these rules can be seen when a transplant is made between regeneration blastemas, a transplant in which the anterior and posterior axes are reversed (Figure 19.27). The result is a limb with three distal portions (Iten and Bryant, 1975). This outcome can be explained by viewing the anterior–posterior axis on the grid as having two opposite numbers—say, 3 and 9. In juxtaposing the values of 3 and 9, one generates a complete circle of values at each of the extreme sites, and a smaller intercalating series at all other sites. The result is three complete circles, which, by the law of distal transformation, will generate three complete limbs from that point on.

The polar coordinate model also predicts the effects of regulation when a portion of the tissue is lost. The newly formed cells would have positional values intermediate between those of the remaining cells and would reconstruct the appropriate part of the tissue. Because the basis of regeneration appears to be the recognition of differences between adjacent tissues, it is probable that epimorphic pattern formation during regeneration (and normal amphibian pattern formation during embryonic limb development) are the result of the proximate interactions between adjacent cells rather than the result of long-range gradients (Bryant et al., 1981).

Sonic hedgehog as the ZPA factor. Recently some researchers traveled a new pathway in the search for the elusive ZPA factor. Their approach was to take the genes that encode morphogens in *Drosophila* and see if their vertebrate homologues encode any important molecule in vertebrates. As you will recall from Chapter 17, the homeotic genes of *Drosophila* have vertebrate counterparts that play critical developmental functions. And, as was mentioned in Chapter 18, the segment polarity gene *hedgehog* is thought to encode a diffusible protein that interacts with neighboring cells. Would it be too much to ask that there be a vertebrate homologue that performs a similar function?

Using the known sequence of the *Drosophila hedgehog* gene, Riddle and his co-workers (1993) used the polymerase chain reaction to identify a *hedgehog*-like message in chick limb buds. They dubbed the gene *sonic hedgehog*. In situ hybridization showed that *sonic hedgehog* expression was not found throughout the entire limb bud, but was localized exactly to the region previously shown by Honig and Summerbell to contain the highest ZPA activity (Figure 19.20). Moreover, *sonic hedgehog* expression could be induced by retinoic acid-soaked beads. This would explain how such beads can produce the ZPA effect—they could do so by inducing the cells next to them to synthesize sonic hedgehog. Thus, it appears that in the normal course of development, retinoic acid is not the ZPA morphogen, but could work earlier in development to direct the placement of the ZPA.

Riddle and co-workers showed that, in all probability, the secretion of sonic hedgehog protein would be sufficient for ZPA activity. They transfected embryonic chick fibroblasts (which normally would never synthesize this protein) with a viral vector containing the *sonic hedgehog* gene. The gene became expressed and translated in these fibroblasts, which were then inserted into the anterior ridge of an early chick limb bud. The ZPA-like digit polarity reversals were seen. Thus, sonic hedgehog looks like the active agent of the ZPA.

It is not yet known how sonic hedgehog protein functions. It could work in a gradient manner, as envisioned by proponents of the morphogen gradient model. It could work locally, by changing the fate of its neighboring cells and initiating a cascade of such changes—a view similar to that envisioned by champions of the polar coordinate model. It could also work in other ways, perhaps through the AER.

The homeotic gene explanation of limb development

The fundamental problem of morphogenesis—how specific structures arise in particular places—is exemplified in limb development. How is it that one part of the lateral plate mesoderm develops limb-forming capacities? How is it that the fingers form at one end of the limb and nowhere else? How is it that the little finger develops at one edge of the limb and the big finger at the other? As we have seen in earlier chapters, homeotic genes have been used to establish a coordinate system throughout the

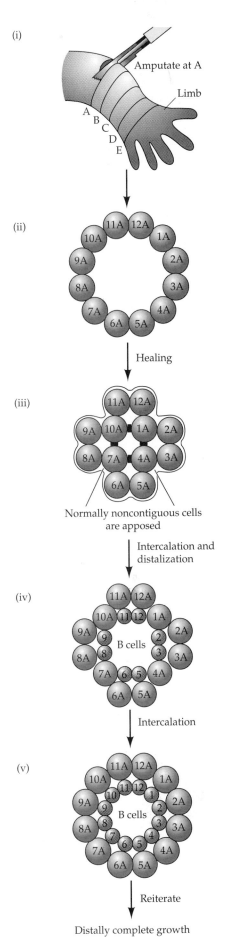

FIGURE 19.26

Model for distal outgrowth. (i,ii) Limb is cut at position A, proximal to positions B–E. This exposes circumferential positions at A. (iii) Healing leads to the apposing of normally noncontiguous cells (such as 10A and 1A) near the blastema tip. Cells between the newly apposed tissues proliferate and acquire positional specification between the two sites. However, these cells are adjacent to pre-existing cells sharing the same circumferential values. By the complete circle rule, such cells acquire a more distal positional value. (iv,v) Intercalation then occurs between these newly specified cells, creating a new surface that contains all the circumferential values. This scheme is repeated until the limb is complete. (After Bryant et al., 1981.)

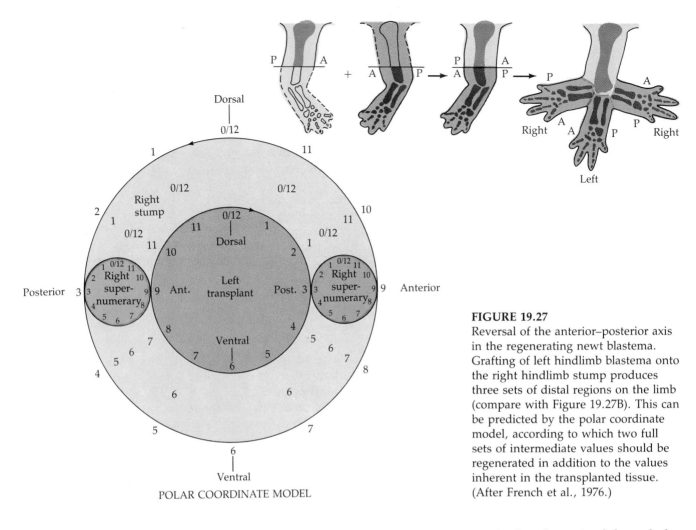

FIGURE 19.27
Reversal of the anterior–posterior axis in the regenerating newt blastema. Grafting of left hindlimb blastema onto the right hindlimb stump produces three sets of distal regions on the limb (compare with Figure 19.27B). This can be predicted by the polar coordinate model, according to which two full sets of intermediate values should be regenerated in addition to the values inherent in the transplanted tissue. (After French et al., 1976.)

POLAR COORDINATE MODEL

embryo. Recent research suggests that the limb is determined through the expression of particular homeotic genes in particular cells. Indeed, homeotic genes have been invoked to explain all the major phenomena of limb development.

Determination of the limb field by homeotic genes

The concept of morphogenetic fields was very popular in embryology until the 1940s (reviewed in Huxley and De Beer, 1934; Weiss, 1939; Opitz, 1985). Thereafter, this concept declined, probably because these "field properties" could not be correlated with any particular molecules. However, it is now probable that the boundaries of these fields, and the inductive abilities associated with them, are created by the expression patterns of homeotic genes (De Robertis et al., 1991). The forelimb field of *Xenopus* mesoderm lies completely within a thin band of cells whose nuclei express the XlHbox 1 (Hoxc-6) protein (Oliver et al., 1988). However, there is a three-week delay between the determination of the limb field and limb formation in *Xenopus*. In zebra fish, limb formation occurs early in embryogenesis, and the lateral plate mesoderm extends as a thin layer around the yolk, making it easy to study. Staining a 19-hour zebra fish embryo with antibodies to the XlHbox 1 protein allows the visualization of a circular region in the lateral plate mesoderm (Molven et al., 1990). This circular patch can be followed throughout its development, and it corresponds to the forelimb (pectoral fin) field.

Bryant and Gardiner (1992) suggest that the gradient of retinoic acid along the anterior–posterior axis might activate certain homeotic genes in particular cells and thereby specify them to become included in the limb field. This might explain a bizarre observation made by Mohanty-Hejmadi and colleagues (1992) and repeated by Maden (1993). When *tails* of tadpoles were amputated and the stump exposed to retinoic acid during the first days of regeneration, these tadpoles regenerated several *legs* from their tail stump (Figure 19.28). It appeared that the retinoic acid may have respecified the tail cells into limb fields.

FIGURE 19.28
Legs regenerating from the tail blastema of a marbled balloon frog tadpole. The tail blastema had been treated with retinoic acid after amputation. (From Mohanty-Hejmadi et al., 1992, courtesy of P. Mohanty-Hejmadi.)

Progress zone interactions as being mediated by *msx-1* and *msx-2*

The interactions between the apical ectodermal ridge and the underlying mesenchyme seem to be mediated, in part, by the products of the *msx-1* and *msx-2* (*Msx1* and *Msx2*) genes. These homeobox-containing genes are not related to the HOM/Hox gene complex, but are related to another homeotic *Drosophila* gene, *muscle segment homeobox (msh)*, which is not in the cluster. The *msx-1* gene is initially expressed throughout the early limb bud, but its expression becomes limited to the progress zone mesenchyme directly beneath the AER. In *limbless* mutants that lack AERs, no *msx-1* message is induced in this mesenchyme, and in polydactylous limb buds with an overly broad AER, the mesenchyme transcribes the *msx-1* gene over an extended area beneath the AER (Figure 19.29). In mutants such as *eudiplopodia* wherein two AERs form, each giving a limb axis, *msx-1* gene expression is seen under both AERs. In experimental manipulations, removal of the AER causes the cessation of *msx-1* expression, while the addition of an AER (even a mouse AER to a chick limb bud) causes the expression of *msx-1* in the mesenchyme beneath it and directs the formation of a secondary limb axis (Robert et al., 1991; Coelho et al., 1991; 1993). Therefore, it appears that *msx-1* is induced in the mesenchyme as a direct response to signals from the AER. Since *msx-1* probably encodes a transcription factor, this protein may be critical for keeping this population of cells proliferating. When myogenic cells are transfected with actively transcribing *msx-1* genes, they lose their differentiated phenotype and divide without the need of further growth factors (Song et al., 1992). The ability of the AER to maintain a zone of proliferating cells immediately beneath it is critical to limb development, and *msx-1* may be providing this key function in limb development.

Although *msx-1* is expressed in the progress zone mesenchyme, the *msx-2* gene is primarily expressed in the AER and anterior mesenchyme. In mutations giving rise to polydactylous limbs, the expression of *msx-2* expands throughout the broad AER and leaves the anterior mesenchyme (Coelho et al., 1993). As shown in Figure 19.29, there are other nonchondrogenic regions of limb mesenchyme that express these two genes but are probably not involved in the proximal–distal axis.

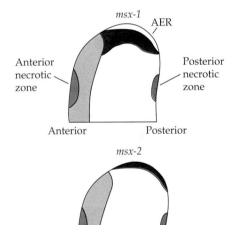

Determination of anterior–posterior axis and finger identity by homeotic genes

All modern tetrapods (four-legged vertebrates) as well as most fossil tetrapods have limbs characterized by five or fewer digits. Moreover, in individuals with an extra digit, this supernumerary digit is a replica of an existing one and not a new type of digit; and when fossil species are characterized by limbs supporting eight digits, there still appear to be only five different types (Tabin, 1992).

In the 1970s, experiments by Finch and Zwilling (1971) and by Patou (1973) suggested that the positional information in the limb that is capable

FIGURE 19.29
Diagrammatic representation of the *msx-1* and *msx-2* expression patterns in normal chick wing buds. The *msx-1* mRNA is seen primarily in the distal mesenchyme below the AER, but also in the anterior region of the limb bud and in the anterior and posterior necrotic zones where cells will die. The mRNA from *msx-2* is expressed in a similar fashion to that for *msx-1*, except that the main region of expression is in the AER. (After Coelho et al., 1993.)

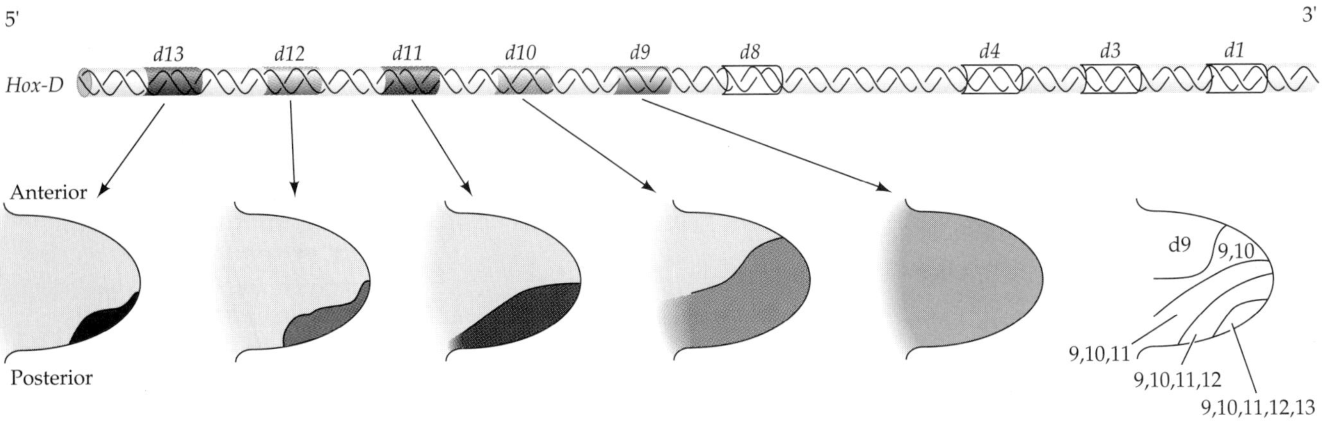

FIGURE 19.30

Expression of the *Hox-D* gene series in the developing chick limb bud. The most posterior portion of the limb bud expresses *Hoxd*-9 through *Hoxd*-13. As the tissue becomes more anterior, the *Hox* gene expression declines in order of appearance on the chromosome, such that the anterior structures are expressing only *Hoxd*-9 of this series. (After Tabin, 1992.)

of specifying the *number* of digits is independent of the information that specifies the *type* of digits. These researchers dissociated the mesenchyme of the developing limb and then repacked it back into its ectodermal shell, thereby scrambling any anterior–posterior information in the mesenchyme. The reassociated limb buds were able to form limblike structures, but their digits lacked any distinctive identities.

In 1989, Dollé and co-workers showed that the expression pattern of the 5′ end of the *Hox-D* gene family (*Hoxd*-9 through *Hoxd*-13) overlap at the posterior edge of the mouse limb bud. This meant that the cells generating the little finger at the posterior margin express all these genes, while the thumb-forming cells at the anterior edge express only *Hoxd*-9 (Figure 19.30). The same distribution of *Hox-D* messages was seen in the chick limb bud. Moreover, the transplantation of either the ZPA or retinoic acid-containing beads to the anterior margin leads to the formation of mirror-image digit patterns and mirror-image patterns of *Hox-D* gene expression (Izpisúa-Belmonte et al., 1991; Nohno et al., 1991). It appears, then, that there is a code wherein the expression of different *Hox-D* genes specifies the anterior–posterior digit pattern.

Such experiments, however, cannot rule out the possibility that the new *Hox-D* expression pattern was a consequence of the cell fate change and not the cause of it. Like the experiments demonstrating a *Hox* code for anterior–posterior specification in the vertebrate body axis, one would have to "alter the code" and see if a transformation in digit identity took place. This was accomplished by Morgan and colleagues (1992), who transfected the embryo with a retrovirus containing an active *Hoxd*-11 gene. In this way, *Hoxd*-11 was expressed throughout the body, including the anterior limb bud region (of the "big toe") where this gene is not usually expressed (Figure 19.31). In these *Hoxd*-11-infected chickens, the most anterior toe took on the properties of the second, "index," toe. These chickens had two index toes per foot instead of a big toe and an index toe.* Changing the combination of *Hox-D* genes in the anterior region caused a posteriorization of the phenotype. As this investigation was underway, another part of that same laboratory was investigating a natural mutation of chick limbs, the *talpid* (*ta³*) mutant. Chicks homozygous for this mutant allele have many anteroposterior axial defects (such as vertebral fusions) as well as polydactylous wings whose digits lack clear anter-

---

*When referring to the hand, one has an orderly set of names to specify each digit (*digitus pollicis, d. indicis, d. medius, d. annularis,* and *d. minimus,* respectively, from thumb to little finger). No such nomenclature exists for the pedal digits, but the plan proposed by Phillips (1991) has much merit. The pedal digits, from hallux to small toe would be named *porcellus fori, p. domi, p. carnivorus, p. non voratus,* and *p. plorans domi,* respectively.

(A) Wild-type leg

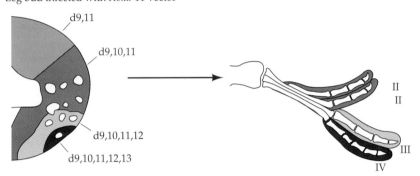

d9
d9,10
d9,10,11
d9,10,11,12
d9,10,11,12,13

I
II
III
IV

(B) Leg bud infected with *Hoxd-11* vector

d9,11
d9,10,11
d9,10,11,12
d9,10,11,12,13

II
II
III
IV

FIGURE 19.31
The expression pattern of the *Hox-D* genes coincides with the specification of individual digits. When the *Hox-D* expression pattern is altered such that all cells express *Hoxd-11*, the anterior toe (which had been in a region that only expressed *Hoxd-9* but is now in a region expressing *Hoxd-9* and *d-11*) is respecified as the second toe. (After Morgan et al., 1992.)

oposterior specification. The *Hox-D* genes in these limbs are expressed throughout the limb bud, and this absence of *Hox-D* polarity correlates with the lack of digit specification (Izpisúa-Belmonte et al., 1992). Moreover, Riddle and colleagues (1993) have shown that Hoxd-13 is able to be induced by sonic hedgehog, the putative ZPA factor. These experiments provide evidence for the existence of a *Hox* code that specifies the anterior–posterior axis of the vertebrate limb and gives each digit its unique properties.

Expression of the *Hox-D* genes, however, can only provide a portion of the needed specificity. A tibia and a big toe are both anterior elements, and they have to be distinguished from each other. So it is not surprising that other *Hox* genes are also active in the limb bud and have their expression pattern at an angle to the *Hox-D* gene set. Like the *Hox-D* genes, the most 5' *Hox-A* gene is active only at the distal edge of the limb bud. Other *Hox-A* genes are expressed more proximally (Figure 19.32; Yokouchi et al., 1991). Whereas the *Hox-D* expression pattern distinguishes the digits from each other, the expression pattern of the *Hox-A* genes distinguishes regions of the wrist from the long bones of the limb.

Dorsal–ventral polarity

Very little is known about the generation of dorsal–ventral (back-of-hand to palm) polarity in the limb. MacCabe and co-workers (1974) demonstrated that the dorsal–ventral polarity of the limb bud is determined by the ectoderm encasing it. If the ectoderm were rotated 180° with respect to the limb bud mesenchyme, the dorsal–ventral axis is reversed. The homeobox-containing gene *Engrailed-1* (discussed in Chapter 7 with respect to brain development) is expressed specifically in the ventral limb ectoderm (prior to AER formation), and it is possible that this specifies the future ventral region (Davis and Joyner, 1988; Davis et al., 1991).

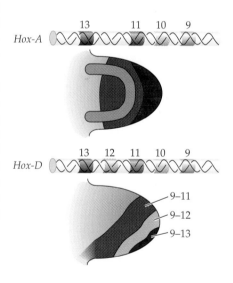

Hox-A   13   11   10   9

Hox-D   13   12   11   10   9

9–11
9–12
9–13

FIGURE 19.32
Nested expression patterns of the *Hox-D* genes are set at different angles to the nested expression patterns of the *Hox-A* genes. In the relatively late limb bud (Stage 25), the paralogous genes are expressed in slightly different sets of cells, enabling the distinction between cell types. (After Izpisúa-Belmonte and Duboule, 1992.)

Cell death and homeotic genes

Cell death also plays a role in sculpting the limb. Indeed, it is essential if joints are to form and if our fingers are to become separate (Zaleske, 1985). The death (or lack of death) in specific cells in the vertebrate limb is genetically programmed and has been selected for during evolution. One such case involves the webbing or nonwebbing of feet. The difference between a duck's foot and that of a chicken is the presence or absence of cell death between the digits (Figure 19.33A,B). Saunders and his coworkers (1962; Saunders and Fallon, 1966) have shown that after a certain stage, chick cells between the digit cartilage are destined to die and will do so even if transplanted to another region of the embryo or placed into culture. Before that time, however, transplantation to a duck limb will save them. Between the time when the cell's death is determined and when death actually takes place, levels of DNA, RNA, and protein synthesis decrease dramatically (Pollak and Fallon, 1976).

In addition to the **interdigital necrotic zone**, there are three other regions that are "sculpted" by cell death. The ulna and radius are separated from each other by an **interior necrotic zone**, and two other regions, the

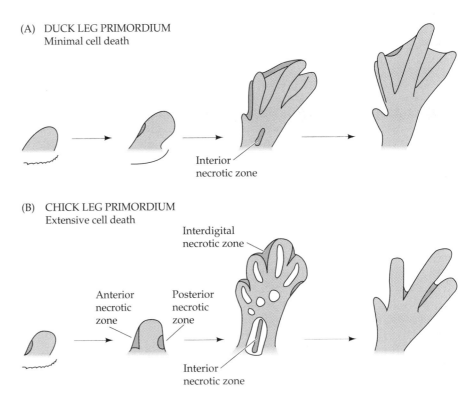

(A) DUCK LEG PRIMORDIUM
Minimal cell death

Interior necrotic zone

(B) CHICK LEG PRIMORDIUM
Extensive cell death

Interdigital necrotic zone

Anterior necrotic zone

Posterior necrotic zone

Interior necrotic zone

(C) Gene expression in region of necrosis prior to cell deaths

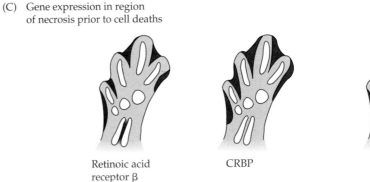

Retinoic acid receptor β

CRBP

msx-1

FIGURE 19.33
Patterns of cell death in leg primordia of (A) duck and (B,C) chick embryos. Shading indicates areas of cell death. In the duck, the cell death is very slight; whereas there are regions of extensive cell death in the interdigital tissue of the chicken leg. (C) The expression of certain genes presages cell death in certain regions of the chick hindlimb bud. (After Saunders and Fallon, 1966; Tabin, 1991.)

anterior and **posterior necrotic zones** further shape the end of the limb (Figure 19.33B; Saunders and Fallon, 1966). There are several genes whose expression prefigures those regions where cells are destined to die. One of them is *msx-1*, which at the end of limb development is expressed solely in the interdigital necrotic zones (Figure 19.33C; Hill et al., 1989; Robert et al, 1989). This pattern of expression is also seen by the gene for cellular retinol binding protein I (CRBP1). Retinoic acid receptor β is found in the interdigital and interior necrotic zones (Dollé et al., 1989). Studies using transgenic animals or organ rudiments cultured with antisense oligonucleotides to these messages should enable us to determine if they are responsible for the cell death in these areas of the limb. Moreover, such studies may provide insights into the mechanisms of evolution that enable the forelimbs of bats and the hindlimbs of sea fowl to retain and expand their embryonic webbing.

The limb has been a cornerstone in the study of pattern generation in vertebrates. It arises from three processes that control limb outgrowth and skeletal patterning: (1) the induction of mesenchymal outgrowth by mesonephric kidney and by the AER, (2) the generation of spaced precartilagenous condensations of mesenchyme that will form the cartilage and bone tissue, and (3) the imposition of asymmetry on the limb bud by the ZPA. There are still lively debates concerning the mechanisms by which these processes occur, and there is considerable controversy over the molecules that might regulate such phenomena.

LITERATURE CITED

Brockes, J. P. and Kinter, C. R. 1986. Glial growth factor and nerve-dependent proliferation in the regeneration blastema of urodele amphibians. *Cell* 45: 301–306.

Bryant, S. V. and Gardiner, D. M. 1992. Retinoic acid, local cell–cell interactions, and pattern formation in vertebrate limbs. *Dev. Biol.* 152: 1–25.

Bryant, S. V. and Iten, L. E. 1976. Supernumerary limbs in amphibians: Experimental production in *Notophthalamus viridescens* and a new interpretation of their formation. *Dev. Biol.* 50: 212–234.

Bryant, S. V, French, V. and Bryant, P. J. 1981. Distal regeneration and symmetry. *Science* 212: 993–1002.

Butler, E. G. 1935. Studies on limb regeneration in X-rayed *Ambystoma* larvae. *Anat. Rec.* 62: 295–307.

Carlson, B. M. 1972. Muscle morphogenesis in axolotl limb regenerates after removal of stump musculature. *Dev. Biol.* 28: 487–497.

Carrington, J. L. and Fallon, J. F. 1988. Initial limb budding is independent of apical ectodermal ridge activity: Evidence from a limbless mutant. *Development* 104: 361–367.

Chevallier, A., Kieny, M., Mauger, M. and Sengel, P. 1977. Limb–somite relationships: Origin of limb musculature. *J. Embryol. Exp. Morphol.* 41: 245–258.

Christ, B., Jacob, H. J. and Jacob, M. 1977. Experimental analysis of the origin of wing musculature in avian embryos. *Anat. Embryol.* 150: 171–186.

Coelho, C. N. D., Krabbenhoft, K. M., Upholt, W. B., Fallon, J. F. and Kosher, R. A. 1991. Altered expression of the chick homeobox-containing genes *GHox-7* and *GHox-8* in the limb buds of *limbless* mutant chick embryos. *Development* 113: 1487–1493.

Coelho, C. N. D., Upholt, W. B. and Kosher, R. A. 1993. The expression pattern of the chicken homeobox-containing gene *GHox-7* in developing polydactylous limb buds suggests its involvement in apical ectodermal ridge–directed outgrowth of limb mesoderm and in programmed cell death. *Differentiation* 52: 129–137.

Davis, C. A. and Joyner, A. L. 1988. Expression patterns of the homeobox-containing genes *En-1* and *En-2* and the protooncogene *int-1* diverge during mouse development. *Genes Dev.* 2: 1736–1744.

Davis, C. A. Holmyard, D. P., Millen, K. J. and Joyner, A. L. 1991. Examining pattern formation in mouse, chicken, and frog embryos with an En-specific antiserum. *Development* 111: 287–298.

De Robertis, E. M., Morita, E. A. and Cho, K. W. Y. 1991. Gradient fields and homeobox genes. *Development* 112: 669–678.

Detwiler, S. R. 1918. Experiments on the development of the shoulder girdle and the anterior limb of *Amblystoma punctatum. J. Exp. Zool.* 25: 499–538.

Dieker, H. and Opitz, J. M. 1969. Associated acral and renal malformations. *Birth Defects Orig. Art. Ser.* 5: 68–77.

Dollé, P., Izpisúa-Belmonte, J.-C., Falkenstein, H., Renucci, A. and Duboule, D. 1989. Coordinate expression of the murine *Hox-5* complex homeobox-containing genes during limb pattern formation. *Nature* 342: 767–772.

Eichele, G. 1989. Retinoic acid induces a pattern of digits in anterior half wing buds that lack the zone of polarizing activity. *Development* 107: 863–867.

Evans, J. A., Vitez, M. and Czeizel, A. 1992. Patterns of acrorenal malformation associations. *Am. J. Med. Genet.* 44: 413–419.

Fallon, J. F. and Crosby, G. M. 1977. Polarizing zone activity in limb buds of amniotes. *In* D. A. Ede, J. R. Hinchliffe and M. Balls (eds.), *Vertebrate Limb and Somite Morphogenesis.* Cambridge University Press, Cambridge.

Fallon, J. F., Olwin, B. B. and Simandl, B. K. In press. FGF-2 replaces the apical ectodermal ridge during chick limb development in vivo.

Finch, R. and Zwilling, E. 1971. Culture stability of morphogenetic properties of chick limb bud mesenchyme. *J. Exp. Zool.* 176: 397–408.

French, V., Bryant, P. J. and Bryant, S. V. 1976. Pattern regulation in epimorphic fields. *Science* 193: 969–981.

Frenz, D. A., Jaikaria, N. S. and Newman, S. A. 1989. The mechanism of precartilage mesenchymal condensation: A major role for interaction of the cell surface with the amino-terminal heparin-binding domain of fibronectin. *Dev. Biol.* 136: 97–103.

Geduspan, J. S. and Solursh, M. 1992a. A growth-promoting influence from the mesonephros during limb outgrowth. *Dev. Biol.* 151: 242–250.

Geduspan, J. S. and Solursh, M. 1992b. Cellular contribution of the different regions of the somatopleure to the developing limb. *Dev. Dynam.* 195: 177–187.

Geduspan, J. S. and Solursh, M. 1993. Effects of the mesonephros and insulin-like growth factor I on chondrogenesis of limb explants. *Dev. Biol.* 156: 500–518.

George-Weinstein, M., Decker, C. and Horwitz, A. 1988. Combinations of monoclonal antibodies distinguish mesenchyme, myogenic, and chondrogenic precursors of the developing chick embryo. *Dev. Biol.* 125: 34–50.

Goetinck, P. F. 1964. Studies on limb morphogenesis. II. Experiments with the polydactylous mutant *eudiplopodia*. *Dev. Biol.* 10: 71–91.

Goss, R. J. 1969. *Principles of Regeneration*. Academic Press, New York.

Goss, R. J. 1991. The natural history (and mystery) of regeneration. *In* C. E. Dinsmore (ed.), *A History of Regeneration Research*. Cambridge University Press, New York, pp. 7–23.

Hamburger, V. 1938. Morphogenetic and axial self–differentiation of transplanted limb primordia of 2-day chick embryos. *J. Exp. Zool.* 77: 379–400.

Hamburger, V. 1939. The development and innervation of transplanted limb primordia of chick embryos. *J. Exp. Zool.* 80: 347–389.

Harrison, R. G. 1918. Experiments on the development of the forelimb of *Amblystoma*, a self–differentiating equipotential system. *J. Exp. Zool.* 25: 413–461.

Held, L. I., Jr. 1992. *Models for Embryonic Periodicity*. Karger, New York.

Hertwig, O. 1925. Haploidkernige Transplante als Organisatoran diploidkeniger Extremitaten be *Triton*. *Anat. Anz.* [Suppl.] 60: 112–118.

Hill, R. E. and eight others. 1989. A new family of mouse homeobox-containing genes: Molecular structure, chromosomal location, and developmental expression of *Hox-7.1*. *Genes Dev.* 3: 26–37.

Hinchliffe, J. R. 1991. Developmental approaches to the problem of transformation of limb structure in evolution. *In* J. R. Hinchliffe (ed.), *Developmental Patterning of the Vertebrate Limb*. Plenum, New York, pp. 313–323.

Hinchliffe, J. R. and Sansom, A. 1985. The distribution of the polarizing zone (ZPA) in the legbud of the chick embryo. *J. Embryol. Exp. Morphol.* 86: 169–175.

Honig, L. S. and Summerbell, D. 1985. Maps of strength of positional signaling activity in the developing chick wing bud. *J. Embryol. Exp. Morphol.* 87: 163–174.

Hurle, J. M., Gañan, Y. and Macias, D. 1989. Experimental analysis of the in vivo chondrogenic potential of the interdigital mesenchyme of the chick limb bud subjected to local ectodermal removal. *Dev. Biol.* 132: 368–374.

Huxley, J. S. and De Beer, G. R. 1934. *The Elements of Experimental Embryology*. Cambridge University Press, Cambridge.

Iten, L. E. 1982. Pattern specification and pattern regulation in the embryonic chick limb bud. *Am. Zool.* 22: 117–129.

Iten, L. E. and Bryant, S. V. 1975. The interaction between blastema and stump in the establishment of the anterior–posterior and proximal–distal organization of the limb regenerate. *Dev. Biol.* 44: 119–147.

Izpisúa-Belmonte, J.-C. and Duboule, D. 1992. Homeobox genes and pattern formation in the vertebrate limb. *Dev. Biol.* 152: 26–36.

Izpisúa-Belmonte, J.-C., Tickle, C., Dollé, P., Wolpert, L. and Duboule, D. 1991. Expression of the homeobox *Hox-4* genes and the specification of position in chick wing development. *Nature* 350: 585–589.

Izpisúa-Belmonte, J.-C., Ede, D. A., Tickle, C. and Duboule, D. 1992. The mis-expression of posterior *Hox-4* genes in *talpid* (*ta³*) mutant wings correlates with the absence of anteroposterior polarity. *Development* 114: 959–963.

Jackson-Grusby, L., Kuo, A. and Leder, P. 1992. A variant *limb deformity* transcript expressed in the embryonic mouse limb defines a novel formin. *Genes Dev.* 6: 29–37.

Javois, L. C. and Iten, L. E. 1986. The handedness and origin of supernumerary limb structures following 180° rotation of the chick limb bud on its stump. *J. Embryol. Exp. Morphol.* 91: 135–152.

Krabbenhoft, K. M. and Fallon, J. F. 1989. The formation of leg or wing specific structures by leg bud cells grafted to the wing bud is influenced by proximity to the apical ridge. *Dev. Biol.* 131: 373–382.

Lande, R. 1978. Evolutionary mechanisms of limb loss in tetrapods. *Evolution* 32: 73–92.

Leonard, C. M., Fuld, H. M., Frenz, D. A., Downie, S. A., Massagué, J. and Newman, S. A. 1991. Role of transforming growth factor-β in chondrogenic pattern formation in the embryonic limb: Stimulation of mesenchymal condensation and fibronectin gene expression by exogenous TGF-β and evidence for endogenous TGF-β-like activity. *Dev. Biol.* 145: 99–109.

MacCabe, J. A., Errick, J. and Saunders, J. W. Jr. 1974. Ectodermal control of dorsoventral axis in leg bud of chick embryo. *Dev. Biol.* 39: 69–82.

Maden, M. 1993. The homeotic transformation of tails into limbs in *Rana temporaria* by retinoids. *Dev. Biol.* 159: 379–391.

Mescher, A. L. 1992. Trophic activity of regenerating peripheral nerves. *Comments Dev. Neurobiol.* 1: 373–390.

Mescher, A. L. and Tassava, R. A. 1975. Denervation effects on DNA replication and mitosis during the initiation of limb regeneration in adult newts. *Dev. Biol.* 44: 187–197.

Mohanty-Hejmadi, P., Dutta, S. K. and Mahapatra, P. 1992. Limbs generated at the site of tail amputation in marbled balloon frog after vitamin A treatment. *Nature* 355: 352–353.

Molven, A., Wright, C. V. E., Bremiller, R., De Robertis, E. M. and Kimmel, C. B. 1990. Expression of a homeobox gene product in normal and mutant zebrafish embryos: Evolution of the tetrapod body plan. *Development* 109: 279–288.

Morgan, B. M., Izpisúa-Belmonte, J.-C., Duboule, D. and Tabin, C. J. 1992. Targeted misexpression of *Hox-4.6* in the avian limb bud causes apparent homeotic transformations. *Nature* 358: 236–239.

Munaim, S. I. and Mescher, A. L. 1986. Transferrin and the trophic effect of neural tissue on amphibian limb reneration blastemas. *Dev . Biol.* 116: 138–142.

Muneoka, K. and Bryant, S. V. 1982. Evidence that patterning mechanisms in developing and regenerating limbs are the same. *Nature* 298: 369–371.

Newman, S. A. 1988. Lineage and pattern in the developing vertebrate limb. *Trends Genet.* 4: 329–332.

Newman, S. A. and Frisch, H. L. 1979. Dynamics of skeletal pattern formation in developing chick limb. *Science* 205: 662–668.

Newman, S. A., Frisch, H. L. and Percus, J. K. 1988. The stationary state analysis of reaction–diffusion mechanisms for biological pattern formation. *J. Theoret. Biol.* 134: 183–197.

Niswander, L. and Martin, G. M. 1993. FGF-4 and BMP-2 have opposite effects on limb growth. *Nature* 361: 68–71.

Niswander, L., Tickle, C., Vogel, A., Booth, I. and Martin, G. R. 1993. FGF-4 replaces the apical ectodermal ridge and directs outgrowth and patterning of the limb. *Cell* 75: 579–587.

Nohno, T. and seven others. 1991. Involvement of the *Chox-4* chicken homeobox genes in determination of anteroposterior axial polarity during limb development. *Cell* 64: 1197–1205.

Noji, S. and ten others. 1991. Retinoic acid induces polarizing activity but is unlikely to be a morphogen in the chick limb bud. *Nature* 350: 83–86.

Oliver, G., Wright, C. V. E., Hardwicke, J. and De Robertis, E. M. 1988. A gradient of homeodomain protein in developing forelimbs of *Xenopus* and mouse embryos. *Cell* 55: 1017–1024.

Opitz, J. M. 1985. The developmental field concept. *Am. J. Med. Genet.* 21: 1–11.

Oster, G. F., Shubin, N., Murray, J. D. and Alberch, P. 1988. Evolution and morphogenetic rules: The shape of the vertebrate limb in ontogeny and phylogeny. *Evolution* 42: 862–884.

Patou, M.-P. 1973. Analyse de la morphogenèse du pied des Oiseaux à l'aide de mélange cellulaires interspecifiques. I. Etude morphologie. *J. Embryol. Exp. Morphol.* 29: 175–196.

Phillips, J. 1991. Higgledy, piggledy. *N.Engl. J. Med.* 324: 497.

Pollak, R. D. and Fallon, J. F. 1976. Autoradiographic analysis of macromolecular synthesis in prospectively necrotic cells of the chick limb bud. II. Nucleic acids. *Exp. Cell Res.* 100: 15–22.

Prigogine, I. and Nicolis, G. 1967. On symmetry-breaking instabilities in dissipative systems. *J. Chem. Phys.* 46: 3542–3550.

Riddle, R. D., Johnson, R. L., Laufer, E. and Tabin, C. 1993. *Sonic hedgehog* mediates the polarizing activity of the ZPA. *Cell* 75: 1401–1416.

Robert, B., Lyons, G., Simandl, B. K., Kuroiwa, A. and Buckingham, M. 1991. The apical ectodermal ridge regulates *Hox-7* and *Hox-8* gene expression in developing chick limb buds. *Genes Dev.* 5: 2363–2374.

Rose, S. M. 1962. Tissue-arc control of regeneration in the amphibian limb. *In* D. Rudnick (ed.), *Regeneration.* Ronald, New York, pp. 153–176.

Rosenquist, G. C. 1971. The origin and movement of the limb-bud epithelium and mesenchyme in the chick embryo as determined by radioautographic mapping. *J. Embryol. Exp. Morphol.* 25: 85–96.

Rossant, J., Zirngibl, R., Cado, D., Shago, M. and Giguère, V. 1991. Expression of a retinoic acid response element-*hsplacZ* transgene defines specific domains of transcriptional activity during mouse embryogenesis. *Genes Dev.* 5: 1333–1344.

Rowe, D. A., Cairnes, J. M. and Fallon, J. F. 1982. Spatial and temporal patterns of cell death in limb bud mesoderm after apical ectodermal ridge removal. *Dev. Biol.* 93: 83–91.

Rubin, L. and Saunders, J. W., Jr. 1972. Ectodermal–mesodermal interactions in the growth of limbs in the chick embryo: Constancy and temporal limits of the ectodermal induction. *Dev. Biol.* 28: 94–112.

Saunders, J. W., Jr. 1948. The proximal–distal sequence of origin of the parts of the chick wing and the role of the ectoderm. *J. Exp. Zool.* 108: 363–404.

Saunders, J. W., Jr. 1972. Developmental control of three-dimensional polarity in the avian limb. *Ann. N.Y. Acad. Sci.* 193: 29–42.

Saunders, J. W., Jr. 1982. *Developmental Biology.* Macmillan, New York.

Saunders, J. W., Jr. and Fallon, J. F. 1966. Cell death in morphogenesis. *In* M. Locke (ed.), *Major Problems of Developmental Biology.* Academic Press, New York, pp. 289–314.

Saunders, J. W., Jr. and Gasseling, M. T. 1968. Ectodermal–mesodermal interactions in the origin of limb symmetry. *In* R. Fleischmajer and R. E. Billingham (eds.), *Epithelial–Mesenchymal Interactions.* Williams & Wilkins, Baltimore, pp. 78–97.

Saunders, J. W., Jr., Cairns, J. M. and Gasseling, M. T. 1957. The role of the apical ridge of ectoderm in the differentiation of the morphological structure and inductive specificity of limb parts of the chick. *J. Morphol.* 101: 57–88.

Saunders, J. W., Jr., Gasseling, M. T. and Saunders, L.C. 1962. Cellular death in morphogenesis of the avian wing. *Dev. Biol.* 5: 147–178.

Saunders, J. W., Jr., Gasseling, M. T. and Errick, J. E. 1976. Inductive activity and enduring cellular constitution of a supernumerary apical ectodermal ridge grafted to the limb bud of the chick embryo. *Dev. Biol.* 50: 16–25.

Sessions, S. and Ruth, S. B. 1990. Explanation for naturally occurring supernumerary limbs in amphibians. *J. Exp. Zool.* 254: 38–47.

Sessions, S. K., Gardiner, D. M. and Bryant, S. V. 1989. Compatible limb patterning mechanisms in urodeles and anurans. *Dev. Biol.* 131: 294–301.

Singer, M. 1954. Induction of regeneration of the forelimb of the postmetamorphic frog by augmentation of the nerve supply. *J. Exp. Zool.* 126: 419–472.

Singer, M. and Caston, J. D. 1972. Neurotrophic dependence of macromolecular synthesis in the early limb regenerate of the newt, *Triturus. J. Embryol. Exp. Morphol.* 28: 1–11.

Solursh, M. 1984. Ectoderm as a determinant of early tissue pattern in the limb bud. *Cell Differ.* 15: 17–24.

Song, K., Wang, Y. and Sassoon, D. 1992. Expression of *Hox-7.1* in myoblasts inhibits terminal differentiation and induces cell transformation. *Nature* 360: 477–481.

Stephens, T. D. and seven others. 1991. Axial and paraxial influences on limb morphogenesis. *J. Morphol.* 208: 367–379.

Stocum, D. L. and Fallon, J. F. 1982. Control of pattern formation in urodele limb ontogeny: A review and a hypothesis. *J. Embryol. Exp. Morphol.* 69: 7–36.

Summerbell, D. 1974. A quantitative analysis of the effect of excision of the AER from the chick limb bud. *J. Embryol. Exp. Morphol.* 32: 651–660.

Summerbell, D. 1979. The zone of polarizing activity: Evidence for a role in abnormal chick limb morphogenesis. *J. Embryol. Exp. Morphol.* 50: 217–233.

Summerbell, D. and Lewis, J. H. 1975. Time, place and positional value in the chick limb bud. *J. Embryol. Exp. Morphol.* 33: 621–643.

Tabin, C. J. 1991. Retinoids, homeoboxes, and growth factors: Toward molecular models for limb development. *Cell* 66: 199–217.

Tabin, C. J. 1992. Why we have (only) five fingers per hand: *Hox* genes and the evolution of paired limbs. *Development* 116: 289–296.

Thaller, C., Hofmann, C. and Eichele, G. 1993. 9-*cis* retinoic acid, a potent inducer of digit formation in the chick wing bud. *Development* 118: 957–965.

Tickle, C. 1981. The number of polarizing region cells required to specify additional digits in the developing chick wing. *Nature* 289: 295–298.

Tickle, C., Summerbell, D. and Wolpert, L. 1975. Positional signaling and specification of digits in chick limb morphogenesis. *Nature* 254: 199–202.

Tickle, C., Alberts, B., Wolpert, L. and Lee, J. 1982. Local application of retinoic acid to the limb bud mimics the action of the polarizing region. *Nature* 296: 564–566.

Tickle, C., Lee, J. and Eichele, G. 1985. A quantitative analysis of the effect of all-*trans*-retinoic acid on the pattern of chick wing development. *Dev. Biol.* 109: 82–95.

Tosney, K. W. and Landmesser, L. T. 1985. Development of the major pathways for neurite outgrowth in the chick hindlimb. *Dev. Biol.* 109: 193–214.

Turing, A. M. 1952. The chemical basis of morphogenesis. *Philos. Trans. R. Soc. Lond.* [B] 237: 37–72.

Wanek, N., Gardiner, D. M., Mueoka, K. and Bryant, S. V. 1991. Conversion by retinoic acid of anterior cells into ZPA cells in the chick wing bud. *Nature* 350: 81–83.

Welsh, B. J., Gomatam, J. and Burgess, A. E. 1983. Three-dimensional chemical waves in the Belousov-Zhabotinski reaction. *Nature* 304: 611–614.

Weiss, P. 1939. *Principles of Development.* Holt, Rinehart & Winston, New York.

Wessells, N. K. 1977. *Tissue Interaction and Development.* Benjamin, Menlo Park, CA.

Winfree, A. T. 1974. Rotating chemical reactions. *Sci. Am.* 230(6): 82–95.

Wolpert, L. 1969. Positional information and the spatial pattern of cellular formation. *J. Theoret. Biol.* 25: 1–47.

Wolpert, L. 1977. *The Development of Pattern and Form in Animals.* Carolina Biological, Burlington, NC.

Yokouchi, Y., Sasaki, H. and Kuroiwa, A. 1991. Homeobox gene expression correlated with bifurcation process of limb cartilage development. *Nature* 353: 443–445.

Zaleske, D. J. 1985. Development of the upper limb. *Hand Clin.* 1985(3): 383–390.

Zwilling, E. 1955. Ectoderm–mesoderm relationship in the development of the chick embryo limb bud. *J. Exp. Zool.* 128: 423–441.

Cell interactions at a distance

Hormones as mediators of development

Organ formation in animals is accomplished by the interactions of numerous cell types. In the preceding two chapters, we have seen how developmental interactions can be mediated by adjacent cell populations. In this chapter, we will discuss the regulation of development by diffusible molecules that travel long distances from one cell type to another. Diffusible regulators of development that travel through the blood to cause changes in the differentiation or morphogenesis of other tissues are called **hormones**.

Metamorphosis: The hormonal redirecting of development

Because minute quantities of hormones are enough to accomplish their actions, it is exceedingly difficult to isolate them from embryos. Some of the most thorough analysis of the hormonal control of development has therefore centered upon the dramatic "reprogramming" of development known as **metamorphosis**.

In many species of animals, embryonic development leads to a larval stage with characteristics very different from those of the adult organism. Very often the larval forms are specialized for some function, such as growth or dispersal. The pluteus larva of the sea urchin, for instance, can travel on ocean currents, whereas the adult urchin leads a sedentary existence. The caterpillar larvae of butterflies and moths are specialized for feeding, whereas their adult forms are specialized for flight and reproduction and often lack the mouth parts necessary for eating. The division of functions between larva and adult is often remarkably distinct (Wald, 1981). Mayflies hatch from eggs and develop for several months. All this development enables them to spend one day as fully developed winged insects, mating quickly before they die. As might be expected from the above discussion, the larval form and the adult form often live in different environments. The adult viceroy butterfly mimics the unpalatable monarch butterfly, but the viceroy caterpillar does not resemble the beautiful larva of the monarch. Rather, the viceroy larva escapes detection by resembling bird droppings (Begon et al., 1986).

During metamorphosis, developmental processes are reactivated by specific hormones, and the entire organism changes to prepare itself for its new mode of existence. These changes are not solely ones of form. In amphibian tadpoles, metamorphosis causes the developmental maturation of liver enzymes, hemoglobin, and eye pigments, as well as the remodeling of the nervous, digestive, and reproductive systems. Thus, metamorphosis is a time of dramatic developmental change affecting the entire organism.

This chapter focuses on three cases in which hormones reactivate the developmental processes after birth: amphibian metamorphosis, insect metamorphosis, and mouse breast development.

Amphibian metamorphosis

In amphibians, metamorphosis is generally associated with the changes that prepare an aquatic organism for a terrestrial existence. In urodeles (salamanders), the changes include the resorption of the tail fin, the destruction of the external gills, and the change of skin structure. In anurans (frogs and toads), the metamorphic changes are most striking, and almost every organ is subject to modification (Table 20.1). The changes in form

TABLE 20.1
Summary of some metamorphic changes in anurans

System	Larva	Adult
Locomotory	Aquatic; tail fins	Terrestrial; tailless tetrapod
Respiratory	Gills, skin, lungs; larval hemoglobins	Skin, lungs; adult hemoglobins
Circulatory	Aortic arches; aorta; anterior, posterior, and common cardinal veins	Carotid arch; systemic arch; jugular veins
Nutritional	*Herbivorous*: Long spiral gut—intestinal symbionts; small mouth—horny jaws, labial teeth	*Carnivorous*: Short gut—proteases; large mouth—long tongue
Nervous	Lack of nictitating membrane, porphyropsin, lateral line system—Mauthner's neurons	Development of ocular muscles, nictitating membrane, rhodopsin, loss of lateral line system—degeneration of Mauthner's neurons; tympanic membrane
Excretory	Largely ammonia, some urea (ammonotelic)	Largely urea, high activity of enzymes of ornithine–urea cycle (ureotelic)
Integumental	Thin, bilayered epidermis with thin dermis; no mucus glands or granular glands	Stratified squamous epidermis with adult keratins; well-developed dermis contains mucus glands and granular glands secreting antimicrobial peptides

Source: Data from Turner and Bagnara, 1976; Silver et al., personal communication.

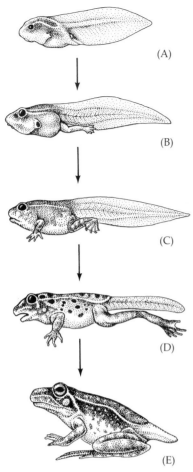

FIGURE 20.1
Sequence of metamorphosis in the frog *Rana pipiens*. (A) Premetamorphic tadpole. (B) Prometamorphic tadpole showing hindlimb growth. (C) Onset of metamorphic climax as forelimbs emerge. (D,E) Climax stages.

are very obvious (Figure 20.1). Regressive changes include the loss of the tadpole's horny teeth and internal gills, as well as the destruction of the tadpole's tail. At the same time, constructive processes such as limb development and dermoid gland construction are also evident. For locomotion, the paddle tail recedes while the hindlimbs and forelimbs differentiate. The horny teeth constructed for tearing pond plants disappear as the mouth and jaw take a new shape and the tongue muscle develops. Meanwhile the large intestine characteristic of herbivores shortens to suit the more carnivorous diet of the adult frog. The gills regress and the gill arches degenerate. The lungs enlarge, and muscles and cartilage develop for pumping air in and out of the lungs. The sensory apparatus changes, too, as the lateral line system of the tadpole degenerates and the eye and ear undergo further differentiation. In the ear, the middle ear develops, as does the tympanic membrane so characteristic of frogs and toads. In the eye, both nictitating membrane and eyelids emerge. Moreover, the eye pigment changes. In tadpoles, as in freshwater fishes, the major retinal photopigment is porphyropsin, a complex between the protein opsin and the aldehyde of vitamin A_2. In adult frogs, the pigment changes to rhodopsin, the characteristic photopigment of terrestrial and marine vertebrates. Rhodopsin consists of opsin conjugated to the aldehyde of vitamin A1 (Wald, 1945, 1981; Smith-Gill and Carver, 1981).

Other biochemical events are also associated with metamorphosis. Tadpole hemoglobin binds oxygen faster and releases it more slowly than does adult hemoglobin (McCutcheon, 1936). Moreover, Riggs (1951) showed that the binding of oxygen by tadpole hemoglobin is independent of pH, whereas frog hemoglobin (like most other vertebrate hemoglobins) shows increased oxygen binding as the pH rises (Bohr effect). Another biochemical change in the metamorphosis of some frogs is the induction of those enzymes necessary for the production of urea. Tadpoles, like most freshwater fishes, are **ammonotelic**; that is, they excrete ammonia. Many adult frogs (such as the genus *Rana*, but not *Xenopus*) are **ureotelic**, excreting urea, like most terrestrial vertebrates. During metamorphosis, the liver develops those enzymes necessary to create urea from carbon dioxide and ammonia. These enzymes constitute the urea cycle, and each of them is seen to arise during metamorphosis (Figure 20.2).

Hormonal control of amphibian metamorphosis

All these diverse changes are brought about by the secretion of the hormones **thyroxine** (T_4) and **triiodothyronine** (T_3) from the thyroid during metamorphosis (Figure 20.3). It is now believed that T_3 is the active hormone, as it will cause the metamorphic changes in thyroidectomized tadpoles in much lower concentrations than will T_4 (Kistler et al., 1977; Robinson et al., 1977). The control of metamorphosis by thyroid hormones was shown by Gudernatsch (1912), who found that tadpoles metamorphosed prematurely when fed powdered sheep thyroid gland. Allen (1916) and Hoskins and Hoskins (1917) found that when they removed the thyroid rudiment from early tadpoles, the larvae never metamorphosed, becoming giant tadpoles instead.

Regionally specific changes. The various organs of the body respond differently to hormonal stimulation. The same stimulus causes certain tissues to degenerate while causing others to develop and differentiate. For instance, tail degeneration is clearly associated with the increasing levels of thyroid hormones. The degeneration of tail structures is relatively

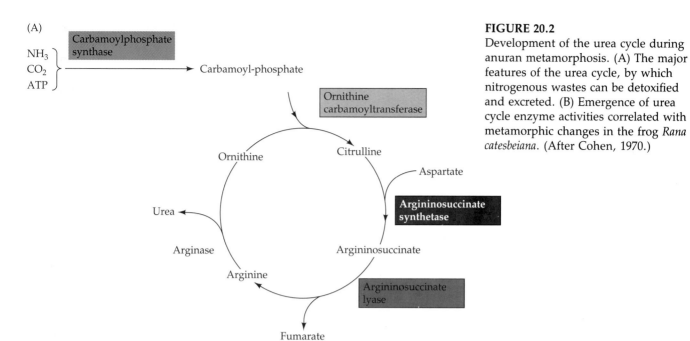

FIGURE 20.2

Development of the urea cycle during anuran metamorphosis. (A) The major features of the urea cycle, by which nitrogenous wastes can be detoxified and excreted. (B) Emergence of urea cycle enzyme activities correlated with metamorphic changes in the frog *Rana catesbeiana*. (After Cohen, 1970.)

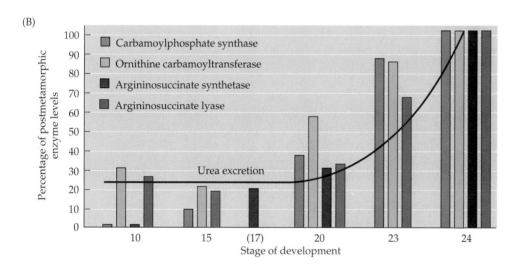

FIGURE 20.3

Formulas of thyroxine (T_4) and triiodothyronine (T_3).

FIGURE 20.4

Regression of isolated tail ends under the influence of thyroxine. (A) Control tips from *Xenopus* tadpoles cultured in Holtfreter's salt solution for 6, 8, 10, and 12 days. (B) Treated tail tips at the same ages as controls; thyroxine was added to their salt solutions. The bar represents 1 mm. (From Weber, 1965, courtesy of R. Weber.)

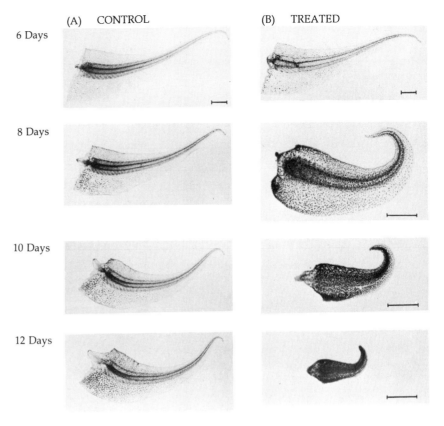

(A) CONTROL   (B) TREATED

6 Days

8 Days

10 Days

12 Days

rapid as the bony skeleton does not extend to the tail, which is supported only by the notochord (Wassersug, 1989). This degeneration can be shown in vitro (Weber, 1967) when isolated tail pieces are placed in agar dishes and subjected to chemical treatments (Figure 20.4). Those tails grown in untreated medium remain healthy, whereas those placed into medium containing thyroid hormones undergo characteristic regression. Moreover, prolactin inhibits the degeneration of the tail induced by thyroid hormones (Brown and Frye, 1969). The regression of the tail is thought to occur in four stages. First, protein synthesis decreases in the striated muscle cells of the tail (Little et al., 1973). Next, there is an increase in the lysosomal enzymes. Concentrations of proteases, RNase, DNase, collagenase, phosphatase, and glycosidases all rise in the epidermis, notochord, and nerve cord cells (Fox, 1973). Cell death is probably caused by the release of these enzymes into the cytoplasm. The epidermis helps to digest the muscle tissue, probably by releasing these digestive enzymes. If the epidermis is surgically removed from the tail tips, the tips will not regress when cultured in thyroxine (Eisen and Gross, 1965; Niki et al., 1982). After this death, macrophages collect in the tail region, digesting the debris with their own proteolytic enzymes (Kaltenbach et al., 1979). The result is that the tail becomes a large sac of proteolytic enzymes (Figure 20.5). The major proteolytic enzyme appears to be a collagenase, and if a collagenase inhibitor (TIMP) is added to the tails, it prevents the thyroid hormone-induced tail regression (Oofusa and Yoshizato, 1991).

The response to thyroid hormones is intrinsic to the organ itself and is not dependent on surrounding tissues. In the epidermis, the response to thyroid hormones depends on which part of the body the epidermis covers. Tadpole head and body epidermal cells usually undergo a small turnover (as expected in skin), and T_3 does not change this rate. In the

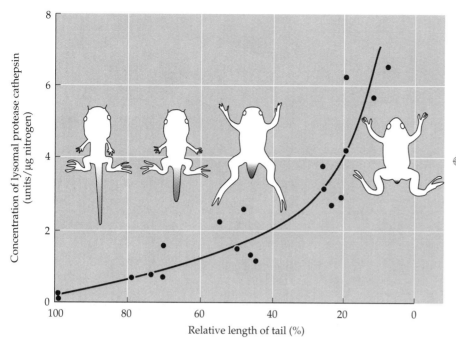

FIGURE 20.5
Increase in lysosomal protease activity during tail regression in *Xenopus laevis*. The lysosomal enzymes are thought to be responsible for digesting the tail cells. (After Karp and Berrill, 1981.)

tail, however, T$_3$ causes a rapid increase in the keratinization and death of these cells. It also causes a tail-specific suppression of stem cell divisions that could give rise to more epidermal cells. The result is the death of the tail epidermal cells while the head and body epidermis continues to function (Nishikawa et al., 1989). These local epidermal responses appear to be controlled by the regional specificity of the dermal mesoderm. If tail dermatome cells (that generate the tail dermis) are transplanted into the trunk, the epidermis they contact will degenerate upon metamorphosis. Conversely, when trunk dermatome is transplanted into the tail, those regions of skin persist. Changing the ectoderm does not alter the regional response to thyroid hormones (Kinoshita et al., 1989).

This organ-specific response is dramatically demonstrated when tail tips are transplanted to the trunk region or when eye cups are placed in the tail (Schwind, 1933; Geigy, 1941). The extra tail tip placed into the trunk is not protected from degeneration, but the eye retains its integrity despite the fact that it lies within the degenerating tail (Figure 20.6). Thus, the degeneration of the tail represents a cell-autonomous programmed cell death. Only specific tissues die when a signal is given. Such programmed cell deaths are important in molding the body. In humans, such programmed degeneration occurs in the tissues between our fingers and toes, and the degeneration of the human tail during week 4 of development resembles the regression of the tadpole tail (Fallon and Simandl, 1978).

Coordination of developmental changes.   One of the major problems of metamorphosis is the coordination of developmental events. The tail should not degenerate until some other means of locomotion—the limbs—has developed, and the gills should not regress until the animal can utilize its newly developed lung muscles. The means of coordinating the metamorphic events appears to be that different amounts of hormone are needed to produce different specific effects (Kollros, 1961). This model is called the **threshold concept**. As the concentration of thyroid hormones gradually builds up, different events occur at different concentrations of the hormone. If tadpoles are deprived of their thyroids and are placed in

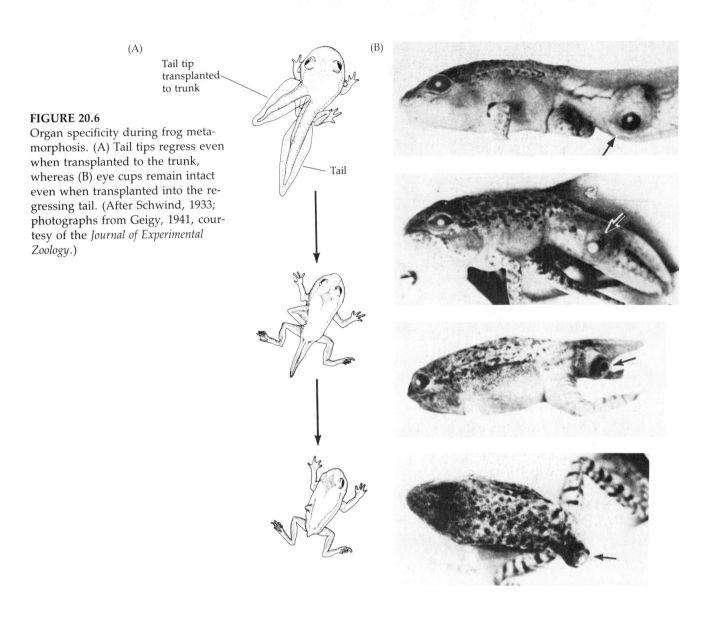

FIGURE 20.6
Organ specificity during frog metamorphosis. (A) Tail tips regress even when transplanted to the trunk, whereas (B) eye cups remain intact even when transplanted into the regressing tail. (After Schwind, 1933; photographs from Geigy, 1941, courtesy of the *Journal of Experimental Zoology*.)

(A)

Tail tip transplanted to trunk

Tail

(B)

a dilute solution of thyroid hormones, the only morphological effects are the shortening of the intestines and accelerated hindlimb growth. However, at higher concentrations of thyroid hormones, tail regression is seen before the hindlimbs are formed. These experiments suggest that as thyroid hormone levels rise, the hindlimbs develop first and then the tail regresses. Thus, the timing of metamorphosis is regulated by the competency of different tissues to respond to thyroid hormones.

Neuronal changes. But what happens to the nervous system when the animal is constructing a new organism from the old? The adaptational anatomy of a frog certainly differs from that of its tadpole. One readily observed consequence of anuran metamorphosis is the movement of the eyes forward from their originally lateral position* (Figure 20.7). The lateral eyes of the tadpole are typical of preyed-upon herbivores, whereas the frontally located eyes of the frog befit its more predatory lifestyle. In order

*One of the most spectacular movements of the eyes during metamorphosis occurs in flatfish such as flounder. Originally the eyes are on opposite sides of the face. However, during metamorphosis one of the eyes migrates dorsally to meet the other at the top of the head, allowing the fish to dwell on the bottom, looking upward (Martin and Drewry, 1978).

FIGURE 20.7
Eye migration and associated neuronal changes during metamorphosis of the *Xenopus laevis* tadpole. The eyes of the tadpole are laterally placed, so there is relatively little binocular field. Eyes migrate dorsally and rostrally during metamorphosis, creating a large binocular field for the adult frog. Below the metamorphosing tadpole is a representation of the optic region of its brain. When horseradish peroxidase is injected into the retina, the optic neurons translocate horseradish peroxidase to the contralateral (opposite) side of the brain (small arrow), but not to the ipsilateral side. As metamorphosis continues, the ipsilateral projections (involved in binocular vision) begin to be seen (large arrow). (From Hoskins and Grobstein, 1984, courtesy of P. Grobstein.)

to catch its prey, the frog needs to see in three dimensions. That is, it has to acquire a binocular field of vision wherein input from both eyes converges in the brain. In the tadpole, the right eye innervates the left side of the brain and vice versa. There are no ipsilateral (same-side) projections of the retinal neurons. During metamorphosis, however, these additional ipsilateral pathways emerge, enabling input from both eyes to reach the same area of the brain (Currie and Cowan, 1974; Hoskins and Grobstein, 1985a). In *Xenopus*, these new neuronal pathways result not from the remodeling of existing neurons, but from the formation of new neurons that differentiate in response to thyroid hormones (Hoskins and Grobstein, 1985a,b). Both the movement of the eyes to their new positions and the differentiation of new neurons that extend processes ipsilaterally to the brain are thyroid hormone-dependent changes.

Other neurons also undergo profound changes. Some nerve cells, such as those that innervate the tadpole tail muscles, die (Forehand and Farel, 1982). This neuronal death does not seem to be caused by the death of the target tissue but appears to be a separate response to thyroid hormones. Other neurons, such as certain motor neurons in the tadpole jaw, switch their allegiances from larval muscle to the newly formed adult muscle (Alley and Barnes, 1983). Still other neurons, such as those innervating the tongue (a newly formed muscle not present in the larva), have lain dormant during the tadpole stage and first form connections during metamorphosis (Grobstein, 1987). The brain also undergoes changes in its structure during metamorphosis. Thus, the anuran nervous system undergoes enormous restructuring during metamorphosis. Some neurons die, others are born, and others change their specificity.

Behavioral changes. Metamorphosis also brings behavioral changes; obviously, the behavior of a frog differs from that of its tadpole. The study of tropical frogs has recently demonstrated surprising behaviors involving tadpole-frog interrelationships. The poison arrow frog, *Dendrobates*, is found in the rain forests of Central America. Most of the time these highly toxic frogs live in the leaf litter of the forest floor. After laying eggs in a damp leaf, the parent (sometimes male, sometimes female) stands guard over the eggs. When the eggs mature into tadpoles, the guarding frog allows them to wriggle onto its back (see the title page of this book). The frog then climbs into the canopy until it finds a bromeliad with a small pool of water in its leaf base. Here it deposits one of its tadpoles, then goes back for another, and so on until the brood has been placed in numerous small pools. Then each day the female returns to these pools and deposits a small number of unfertilized eggs into them, replenishing the dwindling food supply for the tadpoles until they finish metamorphosis (Mitchell, 1986; vanWijngaarden and Bolanos, 1992; Brust, 1993). It is not known how the female frog remembers—or is informed about—where the tadpoles have been deposited.

Molecular responses to thyroid hormones during metamorphosis

Evidence from inhibitor experiments suggested that thyroid hormones controlled metamorphosis at the level of transcription. Weber (1967) demonstrated that the injection of actinomycin D into normal prometamorphic tadpoles inhibited tail regression and head remodeling. In the liver (which is remodeled during metamorphosis rather than being destroyed or replaced), the metamorphic changes are accompanied by dramatic increases in ribosomal and messenger RNA synthesis, the rate of protein synthesis increasing nearly 100-fold within 4 hours of thyroid hormone stimulation (Cohen et al., 1978). Many of these new mRNAs are those coding for the new enzymes of the adult liver. Mori and co-workers (1979) have shown that much of the increase in carbamoyl-phosphate synthase can be attributed to the increased transcription from that gene.

Experiments hybridizing radioactive mRNA from premetamorphic and metamorphosing bullfrog tadpoles to cloned genes (as dot blots) have demonstrated three types of responses to thyroid hormones. The transcription of one set of genes increases in response to either natural or experimentally induced metamorphosis; the transcription of another set of genes is dramatically decreased; while a third set of genes remains unaffected by thyroid hormones (Lyman and White, 1987; Mathison and Miller, 1987). The transcription of mRNAs for albumin, adult globin, and adult skin keratin are all seen to be controlled by T_3.

But these are relatively late gene responses to T_3. The earliest reponse to T_3 is the transcriptional activation of the **thyroid hormone receptor** (TR) genes (Yaoita and Brown, 1990; Kawahara et al., 1991). Thyroid hormone receptors are members of the steroid hormone receptor superfamily of transcription factors. There are two major types of TRs, TRα and TRβ, and the mRNAs of both are present at relatively low levels before metamorphosis begins (Table 20.2; Kawahara et al., 1991; Baker and Tata, 1992). However, the synthesis of these mRNAs accelerates dramatically as metamorphosis begins. The injection of exogenous T_3 causes a two- to fivefold increase in TRα message and a 20 to 50-fold increase in the mRNA for TRβ. This "autoinduction" of T_3 receptor message by T_3 may play a significant role in the acceleration of metamorphosis (Figure 20.8). The more T_3 receptors a tissue has, the more competent it should be to respond to small amounts of T_3. Thus, **metamorphic climax**, that time when the visible changes of metamorphosis occur rapidly, may be brought about by the enhanced production and induction of more T_3 receptors. The mechanism for this induction is not known, but Kanamori and Brown (1992)

TABLE 20.2
Relative accumulation of TRα and β mRNA in *Xenopus* tadpoles following treatment with T_3 and prolactin

Treatment	Relative units	
	TRα	TRβ
None	505	24
T_3	1290	368
Prolactin + T_3	799	<10
Prolactin	405	43

Source: After Baker and Tata, 1992.

(A) PREMETAMORPHOSIS

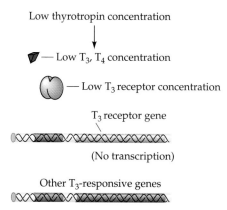

Low thyrotropin concentration

— Low T$_3$, T$_4$ concentration

— Low T$_3$ receptor concentration

T$_3$ receptor gene

(No transcription)

Other T$_3$-responsive genes

(B) EARLY METAMORPHOSIS

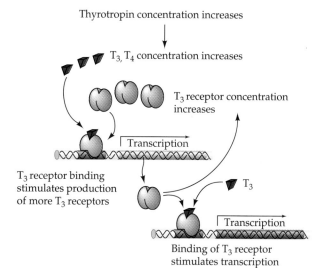

Thyrotropin concentration increases

T$_3$, T$_4$ concentration increases

T$_3$ receptor concentration increases

Transcription

T$_3$ receptor binding stimulates production of more T$_3$ receptors

T$_3$

Transcription

Binding of T$_3$ receptor stimulates transcription of other genes

(C) METAMORPHIC CLIMAX

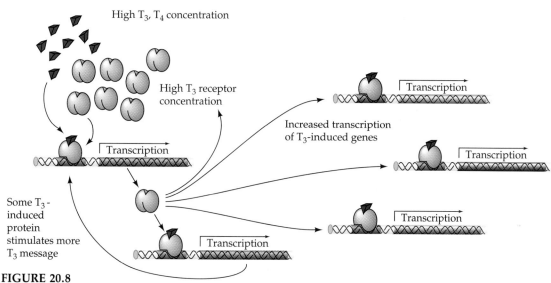

High thyrotropin concentration

High T$_3$, T$_4$ concentration

High T$_3$ receptor concentration

Transcription

Increased transcription of T$_3$-induced genes

Transcription

Transcription

Some T$_3$-induced protein stimulates more T$_3$ message

Transcription

Transcription

FIGURE 20.8
Hypothetical model for the acceleration of metamorphosis in *Xenopus* by the autoinduction of T$_3$ receptors by T$_3$. (A) In the tadpole, premetamorphosis is characterized by low levels of thyrotropin (thyroid hormone-releasing factor), thyroid hormones, and T$_3$ receptors. (B) At the onset of metamorphosis, the levels of thyrotropin are increased (probably as a part of the developmental maturation of the pituitary gland). This increases the amount of T$_3$. The T$_3$ binds to the small amount of T$_3$ receptor to stimulate the transcription of more T$_3$ receptor mRNA. Some other T$_3$-induced proteins are also needed for the transcription of more T$_3$ message. (C) At metamorphic climax, the large concentrations of T$_3$ induce the synthesis of still more T$_3$ receptors, which cause a more rapid response to T$_3$.

have shown that the upregulation of TRβ mRNA is significantly blocked by inhibitors of protein synthesis. Thus, other proteins are probably involved in the T$_3$-responsiveness of the TR genes.

Prolactin has also been found to inhibit the upregulation of TRα and TRβ mRNAs. Moreover, if the upregulation of the TRs is blocked by prolactin, the tail is not resorbed and the adult-specific keratin gene is not activated (Tata et al., 1991; Baker and Tata, 1992). Injections of prolactin stimulate larval growth and inhibit metamorphosis (Bern et al., 1967; Etkin and Gona, 1967), but there is dispute as to whether this reflects the natural role of prolactin (Takahashi et al., 1990; Buckbinder and Brown, 1993). We still do not know the mechanisms by which levels of thyroid hormone are regulated in the tadpole, nor do we know how the reception of thyroid hormone elicits different responses (proliferation, differentiation, cell death) in different tissues.

SIDELIGHTS & SPECULATIONS

Heterochrony

Most species of animals develop through a larval phase. These are often highly specialized stages that allow the organism to find and digest specific types of food and to survive in a particular environment. However, some species have modified their life cycles by either greatly extending or shortening their larval period. The phenomenon wherein animals change the relative time of appearance and rate of development of characters already present in their ancestors is called **heterochrony**. Here we will discuss three of the extreme types of heterochrony. **Neoteny** refers to the retention of the juvenile form produced by the retardation of body development relative to the germ cells and gonads, which achieve maturity at the normal time. **Progenesis** also involves the retention of the juvenile form, but in this case the gonads and germ line develop at a faster rate than normal, and they become sexually mature while the rest of the body is still in a juvenile phase. In **direct development**, the embryos abandon the stages of larval development entirely and proceed to construct a small adult.

Neoteny

In certain salamanders, sexual maturity occurs in what is usually considered a larval state. The reproductive system (and germ cells) mature, while the rest of the body retains its juvenile form throughout its life. In most instances, metamorphosis fails to occur and sexual maturity takes place in a "larval" body.

The Mexican axolotl, *Ambystoma mexicanum*, does not undergo metamorphosis in nature because its pituitary gland does not release an active thyroid-stimulating hormone (TSH) to activate T_3 synthesis in its thyroid glands (Prahlad and DeLanney, 1965; Norris et al., 1973; Taurog et al., 1974). Thus, when investigators gave *A. mexicanum* either thyroid hormones or TSH, they found that the salamander metamorphosed into an adult not seen in nature (Huxley, 1920). Other species, such as *A. tigrinum*, metamorphose only if given cues from the environment. Otherwise, they become neotenic, successfully mating as larvae. In part of its range, *A. tigrinum* is a neotenic salamander, paddling its way through the cold ponds of the Rocky Mountains. However, in the warmer region of its range, the larval form of *A. tigrinum* is transitory, leading to the land-dwelling tiger salamander. Neotenic populations from the Rockies can be induced to undergo meta-

(A)   (B)

FIGURE 20.9
Metamorphosis induced in the axolotl. (A) Normal condition of the axolotl. (B) Specimen treated with thyroxine to induce metamorphosis. (Courtesy of G. Malacinski.)

morphosis simply by placing them in water at higher temperatures. It appears that the hypothalamus of this species cannot produce TSH-releasing factor at the low temperatures.

Some salamanders, however, are permanently neotenic, even in the laboratory. Whereas thyroxine is able to produce the long-lost adult form of *A. mexicanum* (Figure 20.9), the neotenic species of *Necturus* and *Siren* remain unresponsive to thyroid hormones (Frieden, 1981); their neoteny is permanent. Yaoita and Brown (1990) noted that the mRNA for thyroid hormone receptor β is absent in *Necturus* and thus cannot be induced by T_3. The genetic lesions thought to be responsible for neoteny in several species are shown in Figure 20.10.

De Beer (1940) and Gould (1977) have speculated that neoteny is a major factor in the evolution of more complex taxa. By retarding the development of somatic tissues, natural selection is given a flexible substrate. According to Gould, neoteny would "provide an escape from specialization. Animals can relinquish their highly specialized adult forms, return to the lability of youth, and prepare themselves for new evolutionary directions."

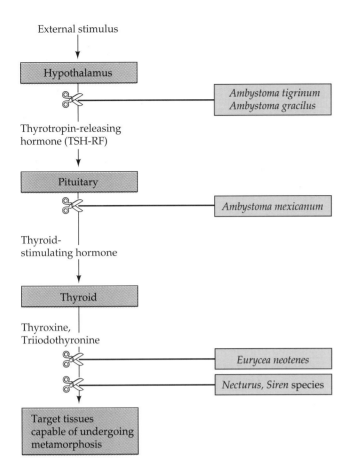

External stimulus

↓

Hypothalamus ✂ ——— *Ambystoma tigrinum*
Ambystoma gracilus

↓

Thyrotropin-releasing
hormone (TSH-RF)

↓

Pituitary ✂ ——— *Ambystoma mexicanum*

↓

Thyroid-
stimulating hormone

↓

Thyroid ✂ ——— *Eurycea neotenes*
✂ ——— *Necturus, Siren* species

↓

Target tissues
capable of undergoing
metamorphosis

FIGURE 20.10

*Stages along the hypothalamus–pituitary–thyroid axis of sala-
manders where various species are thought to have blocked
metamorphosis.* Eurycea, Necturus *and* Siren *appear to have
a receptor defect in the responsive tissues.* Eurycea *will metamorphose when exposed to extremely high concentrations of thyroxine, while* Necturus *and* Siren *do not respond to any dose.*
(After Frieden, 1981.)

the eggs of *E. coqui* are fertilized while they are still within
the female's body. Each egg is about 3.5 mm in diameter
(roughly 20 times the volume of *Xenopus* eggs). After the
eggs are laid, the male gently sits upon the developing
embryos, protecting them from predators and desiccation
(Taigen et al., 1984). Early development is like that of most
frogs. Cleavage is holoblastic, gastrulation is initiated at a
subequatorial position (Figure 20.11A), and the neural folds
become elevated from the surface (Figure 20.11B). However, shortly after the neural tube closes, limb buds appear
on the surface (Figure 20.11C). This early emergence of
limb buds is the first indication that development is direct
and will not pass through a limbless tadpole stage. Moreover, the emergence of the limbs is not dependent upon
thyroid hormones (Lynn and Peadon, 1955). What emerges
from the egg jelly three weeks after fertilization is not a
tadpole but a little frog (Figure 20.11D). The froglet has a
tail for the first part of its life, but it is used for respiration
rather than locomotion. Such direct-developing frogs do
not need ponds for their larval stages and can therefore
colonize new regions inaccessible to other frogs.

Progenesis

In *progenesis*, gonadal maturation is accelerated while the
rest of the body develops normally to a certain stage. Progenesis has enabled some salamander species to find new
ecological niches. *Bolitoglossa occidentalis* is a tropical salamander that, unlike other members of its genus, lives in
trees. This salamander has webbed feet and a small body
size that suit it for arboreal existence, the webbed feet
producing suction for climbing and the small body size
making such traction efficient. Alberch and Alberch (1981)
have shown that *B. occidentalis* resembles juveniles of the
related species *B. subpalmata* and *B. rostrata* (whose young
are small, with digits that have not yet grown past their
webbing). It is thought that *B. occidentalis* became a sexually
mature adult at a much smaller size than its predecessors.
This gave it a phenotype that made tree-dwelling a possibility.

Direct development

While some animals have extended their larval period of
life, others have "accelerated" their development by abandoning their "normal" larval forms. This latter phenomenon is called *direct development* and is typified by frog species that lack tadpoles and by sea urchins that lack pluteus
larvae. Elinson and his colleagues (Elinson, 1987; del Pino
and Elinson, 1983) have studied *Eleutherodactylus coqui*, a
small frog that is one of the most populous animals on the
island of Puerto Rico. Unlike eggs of *Rana* and *Xenopus*,

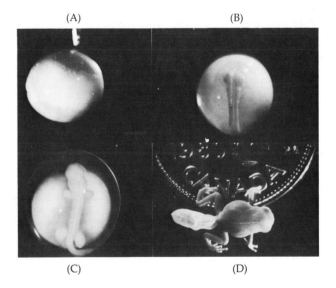

(A)   (B)

(C)   (D)

FIGURE 20.11

Direct development of the frog Eleuthorodactylus coqui. *(A)
Early gastrula showing blastopore lip. (B) Dorsal view of neurula showing raised neural folds. (C) A day after the closure of
the neural folds, limb buds are seen. (D) Three weeks after fertilization, a tiny froglet hatches, seen here beside a Canadian
penny (the inflation of the tail is an artifact caused by the chemical fixatives used to prepare the specimen). (From Elinson,
1987, courtesy of R. P. Elinson.)*

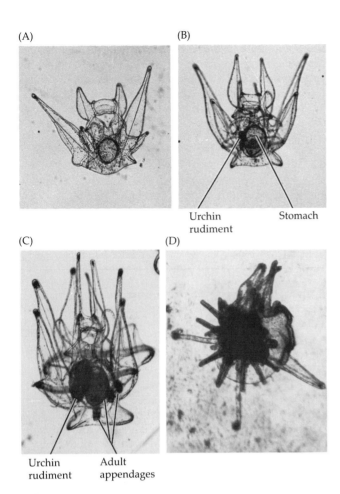

(A)

(B)

Urchin rudiment   Stomach

(C)

(D)

Urchin   Adult
rudiment   appendages

FIGURE 20.12
Normal metamorphosis of pluteus larva into adult in the sea urchin Lytechinus pictus. (A) Pluteus larva 8 days after fertilization. (B) An 11-day pluteus larva with urchin rudiment on left coelomic sac. (C) A 19-day pluteus with developing sea urchin rudiment. (D) About 11 minutes after attaching to substrate, the larval arms are being resorbed. (From Hinegardner, 1969, courtesy of R. T. Hinegardner.)

Raff (1987) has studied direct development in sea urchins. In the "typical" sea urchin, the primary mesenchyme cells invaginate and secrete the calcium carbonate skeleton of pluteus larvae. These larvae feed and grow until coelomic vesicles (also derived from the micromeres) form at the sides of the gut (Pehrson and Cohen, 1986). The left coelom grows further to produce a hydrocoel, which induces the overlying ectoderm to invaginate to form a vestibule. The hydrocoel and vestibule form a rudiment that grows within the larva until it is released at metamorphosis to become a juvenile sea urchin (Figure 20.12).

Several different sea urchin species have suppressed stages of the pluteus larvae while accelerating the development of the adult rudiment. Like the above-mentioned case of direct development in frogs, direct development in sea urchins also depends upon a large, yolky egg. In fact, Raff has found a correlation between egg volume and the extent of direct development (Table 20.3). North American

and European sea urchins have eggs whose diameters range from 60 to 200 μm. These species undergo indirect development through the pluteus larvae. Eggs in the 300 to 350 μm range produce partial pluteus larvae that have larval skeletons but no gut (and therefore cannot feed). These species show an accelerated growth of the adult rudiment so a feeding, juvenile urchin is quickly formed. There are some yolky eggs that reach 2 mm in diameter (about the same volume as *Xenopus* eggs). These embryos develop directly without any pluteus stage. The feeding stage is not needed because nutrition can be provided by the yolk.

Nature has provided an excellent comparison in two Australian species of the sea urchin genus *Heliocidaris*. *Heliocidaris erythrogramma* and *H. tuberculata* are common species that, according to morphological and DNA-sequencing data, are very closely related. They live side by side and spawn during the same time in the summer. However, *H. erythrogramma* has an egg with a diameter of 425 μm and is a direct developer; *H. tuberculata* produces an egg with a diameter of 95 μm and develops through a typical pluteus larva. The comparison between these species (Figure 20.13) reveals that the direct developer has eliminated the larval stages and proceeds directly to coelom formation and the construction of the juvenile sea urchin. The pluteus larva is designed for swimming and feeding (Strathmann, 1971, 1975), using its arms as supports for bands of cilia that sweep food particles into the mouth. The cells of the direct developer have changed their fates such that no larval skeleton or mouth forms. In the gastrulation of direct-developing sea urchins, one does not see the descendants of the micromere cells invaginating to form the larval skeleton. Rather, these cells are immediately involved in forming the calcareous spine of the young adult. Also, the tip of the archenteron of the direct-developing urchin forms an extensive hydrocoel that interacts with the ectodermal vestibule to form the sea urchin rudiment at gastrulation. In indirect developers, only two cells initiate the formation of the vestibule, and these interact with the hydrocoel after the pluteus structure has been established (Wray and Raff,

TABLE 20.3
Relationship of developmental mode to egg size in sea urchins

Number of species	Egg size range (μm)	Developmental mode
83	60–345	Feeding pluteus larva
1	280	Facultative feeding pluteus
2	300–350	Abbreviated pluteus, nonfeeding
19	400–2000	Pluteus lost; direct development

Source: After Raff, 1987.

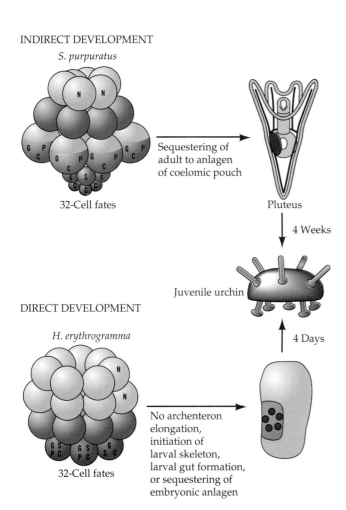

INDIRECT DEVELOPMENT

S. purpuratus

32-Cell fates

Sequestering of adult to anlagen of coelomic pouch

Pluteus

4 Weeks

Juvenile urchin

4 Days

DIRECT DEVELOPMENT

H. erythrogramma

32-Cell fates

No archenteron elongation, initiation of larval skeleton, larval gut formation, or sequestering of embryonic anlagen

1990, 1991). Thus, we have an interesting paradox. In one sense, the development of larval stages seems highly constrained. Larvae from different classes of echinoderms are very similar, and the tadpoles of different groups of frogs are also very much the same. However, these constraints can be eliminated by abandoning the need for a feeding larval stage. Increasing the amount of yolk available to the embryo seems to make this possible.

FIGURE 20.13

Changes in cell fate and gastrulation in an indirect-developing sea urchin and a direct-developing sea urchin. The fate maps at the 32-cell stage show the differences in cells fate. The vegetal fates (indicated by shading) include coelom (C), gut (G), pigment cells (P), and skeletogenic mesenchyme (S). Those cells giving rise to neural tissues are denoted as N. Note that the direct developer has not produced separate micromeres and macromeres. The indirect developer forms a pluteus, and within this larval structure, interactions form the rudiment of the juvenile sea urchin (color). In the direct developer, such interactions between coelom and vestibule cells occur immediately at gastrulation, and the juvenile rudiment (color) is formed without any feeding larval stage. Both types of development generate the same adult structures. (After Raff, in press.)

Metamorphosis in insects

Eversion and differentiation of the imaginal discs

Whereas amphibian metamorphosis is characterized by the remodeling of existing tissues, insect metamorphosis often involves the destruction of larval tissues and their replacement by an entirely different population of cells.

There are three major patterns of insect development. A few insects, such as springtails, have no larval stage and undergo direct development. Other insects, notably grasshoppers and bugs, undergo a gradual, **hemimetabolous** metamorphosis (Figure 20.14A). Adult organs are formed without any profound discontinuity. The rudiments of the wing, genital organs, and other adult structures are present at hatching, and they become more mature with each molt. At the last molt, the emerging insect is a winged and sexually mature adult. The larval form of a hemimetabolous insect is called a **nymph**.

In the **holometabolous** insects (flies, beetles, moths, and butterflies), there is a dramatic and sudden transformation between the larval and

FIGURE 20.14
(A) Hemimetabolous (incomplete) metamorphosis. (B) Holometabolous (complete) metamorphosis.

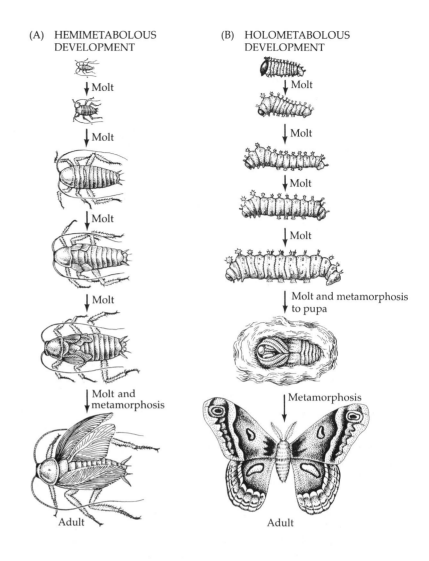

(A) HEMIMETABOLOUS DEVELOPMENT

↓ Molt

↓ Molt

↓ Molt

↓ Molt

↓ Molt

↓ Molt and metamorphosis

Adult

(B) HOLOMETABOLOUS DEVELOPMENT

↓ Molt

↓ Molt

↓ Molt

↓ Molt

↓ Molt and metamorphosis to pupa

↓ Metamorphosis

Adult

adult stages (Figure 20.14B). The juvenile larva (caterpillar, grub, maggot) undergoes a series of molts as it becomes larger. The newly hatched insect larva is covered by a hard **cuticle**. To grow, the insect must produce a new, larger, cuticle and shed the old one. Thus, the postembryonic development of these insects consists of a succession of molts. The number of molts before becoming an adult is characteristic for the species, although environmental factors can increase or decrease the number. The stages between these molts are called **instars**. Here, growth of the larva is the major characteristic. After the last instar stage, the larva undergoes a metamorphic molt to become a **pupa**. The pupa does not feed, and its energy must come from those foods ingested while a larva.

Drosophila undergoes four molts in its life cycle. The embryo develops into the first-instar larva and then molts to become the second-instar larva. Subsequent molts separate the second instar from the third instar, the third instar from the pupa, and the pupa from the adult. At each molt, the epidermal cells separate from the cuticle and secrete a "moting fluid" into the intervening space. When the epidermal cells have secreted a new cuticle, they degrade the old one by activating enzymes in the molting fluid (Hepburn, 1985).

It is within the pupal cuticle that the transformation of juvenile into adult occurs. Most of the old body of the larva is systematically destroyed as new adult organs develop from undifferentiated nests of cells, the

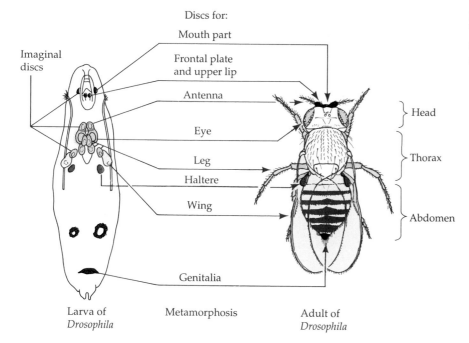

Discs for:

Mouth part

Frontal plate
and upper lip

Antenna

Eye

Leg

Haltere

Wing

Genitalia

Imaginal
discs

Head

Thorax

Abdomen

Larva of
Drosophila

Metamorphosis

Adult of
Drosophila

FIGURE 20.15
The locations and developmental fates
of the imaginal discs in *Drosophila melanogaster*. (After Fristrom et al., 1969.)

imaginal discs (and, in some insects, **histoblasts**). When the adult organism (**imago**) is developed, the **imaginal molt** results in the shedding of the pupal cuticle, and the mature insect emerges. In holometabolous larvae, then, there are two cell populations: the larval cells, which are used for the functions of the juvenile, and the thousand or so imaginal cells, which lie in clusters awaiting the signal to differentiate. Figure 20.15 shows the location of the *Drosophila* imaginal discs and the structures into which they develop.

In *Drosophila*, there are 10 major pairs of imaginal discs, which reconstruct the entire adult (except for the abdomen), and a genital disc, which forms the reproductive structures. The abdominal epidermis forms from a small group of imaginal cells called histoblasts, which lie in the region of the larval gut, and other nests of histoblasts located throughout the larva form the internal organs of the adult. The imaginal discs can be seen in the newly hatched larva as local thickenings of the epidermis. In *Drosophila*, these newly hatched eye-antenna, wing, haltere, leg, and genital discs contain 70, 38, 20, 36–45, and 64 cells, respectively (Madhavan and Schneiderman, 1977). Whereas most of the larval cells have a very limited mitotic capacity, the imaginal discs divide rapidly at specific characteristic times. As the cells proliferate, they form a tubular epithelium that folds in upon itself in a compact spiral (Figure 20.16). The largest disc, that of the wing, contains some 60,000 cells, whereas the leg and haltere discs

FIGURE 20.16
Imaginal disc elongation. Scanning electron micrograph of *Drosophila* third-instar leg disc (A) before and (B) after elongation. (From Fristrom et al., 1977, courtesy of D. Fristrom.)

(A)

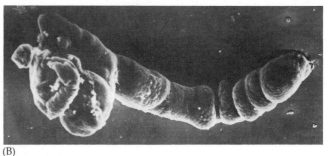

(B)

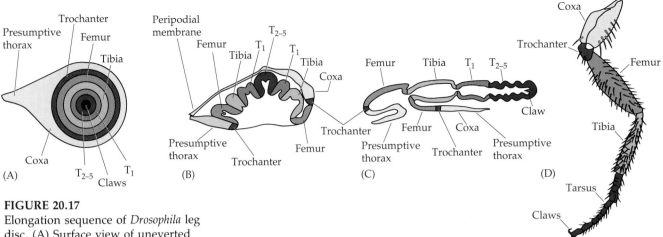

FIGURE 20.17

Elongation sequence of *Drosophila* leg disc. (A) Surface view of uneverted disc. (B,C) Longitudinal section through elongating and fully everted leg disc. t_1, basitarsus; t_{2-5}, tarsal segments 2–5. (D) Adult leg. (From Fristrom and Fristrom, 1975, courtesy of D. Fristrom.)

contain around 10,000 (Fristrom, 1972). At metamorphosis, these cells differentiate and elongate. The fate map and elongation sequence of the leg disc is shown in Figure 20.17. The cells at the center of the disc telescope out to become the most distal portions of the leg—the claws and the tarsus—and the outside cells become the proximal structures—the coxa and the adjoining epidermis. After differentiating, the cells of the appendages and epidermis secrete a cuticle appropriate for the specific region. Although the discs are composed primarily of epidermal cells, a small number of **adepithelial** cells migrate into the disc early in development. During the pupal period, these cells give rise to the muscles and nerves that serve that structure.

The process of elongation can be initiated in culture by placing imaginal discs in a solution containing the molting hormone, 20-hydroxyecdysone. Moreover, such eversion can be inhibited by adding any of three sets of drugs. (1) Inhibitors of RNA synthesis and protein synthesis inhibit eversion when added to cultured imaginal discs at the same time as 20-hydroxyecdysone. It is known that RNA and protein syntheses occur prior to elongation and that some of these proteins are needed for this elongation to occur. (2) Cytochalasin B, an inhibitor of microfilament function,

FIGURE 20.18

Changes in cell shape during the elongation of a *Drosophila* leg imaginal disc. Top: Optical sections through the elongating second leg disc. The arrows mark the basitarsal segments, and the calibration bar is 100 μm. Bottom: Higher magnification (calibration bar is 10 μm) of cell apices through the basitarsal area. The cell boundaries are stained by fluorescently labeled phalloidin. (A) Beginning prepupa stage. (B) Six-hour prepupa. (C) Leg disc from a beginning prepupa treated with trypsin. The basitarsal cells are initially compressed along the proximal–distal axis. Upon hydroxyecdysone treatment or trypsinization, the compression is released and the cells expand to elongate the tissue. (From Condic et al., 1990, courtesy of the authors.)

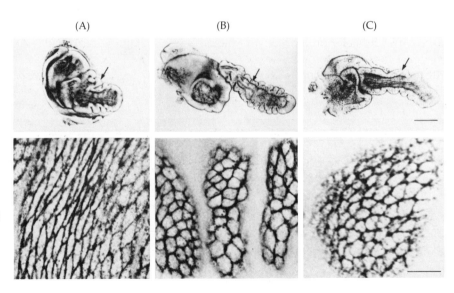

(A) (B) (C)

also inhibits elongation, thereby indicating a need for actin microfilaments. (3) Protease inhibitors also inhibit elongation (Pino-Heiss and Schubiger, 1989), as cell surface proteases are required for the releasing of constraints upon cell shape. Taken together, these data suggest that the eversion of the imaginal discs requires new protein synthesis, a well-developed system of actin microfilaments, and some cellular communication by the cell surface (Fristrom et al., 1977).

Recent studies by Condic and her colleagues (1990) have demonstrated that the elongation of the imaginal disc is due primarily to cell shape change within the disc epithelium. Using fluorescently labeled phalloidin to stain the peripheral microfilaments of the leg disc cells, they showed that the cells of the early third-instar discs are tightly compressed along the proximal–distal axis. This compression is maintained through several rounds of cell division. Then, when the tissue begins elongating, the compression is removed and the cells "spring" into their longer state (Figure 20.18). This conversion of an epithelium of compressed cells into a longer epithelium of noncompressed cells represents a novel mechanism for the extension of an organ during development.

SIDELIGHTS & SPECULATIONS

The determination of the leg and wing imaginal discs

Determination of discs from ectoderm

The molecular biology of insect metamorphosis begins with the specification of certain epidermal cells to become imaginal disc precursors. As we discussed in Chapter 15, the organ rudiments in *Drosophila* may be specified on an orthagonal grid by intersecting anterior–posterior and dorsal–ventral signals. Cohen and colleagues (1993) have demonstrated that the leg and wing imaginal precursors are specified at the intersection between the anteroposterior stripes of wingless (wg) protein expression and the horizontal band of cells expressing the decapentaplegic (dpp) protein. Both proteins are soluble and have a limited range. In the early *Drosophila* embryo (at germ band extension some 4.5 hours after fertilization), a single group of cells at the intersection of these domains forms the imaginal disc precursors in the abdomen. These cells (and only these cells) express the distal-less protein. As the cells expressing dpp are moved dorsally, these distal-less-expressing cells move to establish a secondary cluster of imaginal cells (derived from the original ventral cluster). The initial clusters form the leg imaginal discs, while the secondary cluster forms the wing and haltere discs. Thus, the leg and wing discs have a common origin (Figure 20.19).

Determination of disc identity

Despite their common origin, it is obvious that the leg and wing discs are determined to become different structures. As we detailed earlier, the specification of these discs into their particular fates is probably accomplished by the in-

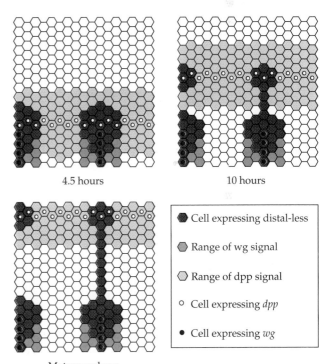

4.5 hours   10 hours

Mature embryo

● Cell expressing distal-less

⬡ Range of wg signal

⬡ Range of dpp signal

○ Cell expressing *dpp*

● Cell expressing *wg*

FIGURE 20.19
Schematic model for the allocation and separation of the leg-wing disc in the Drosophila *thorax. The embryo is divided into an orthogonal grid with vertical stripes of wingless (wg) and a horizontal band of decapentaplegic (dpp) synthesis and secretion. The initial disc forms at the intersection of these secretory domains. The dpp-secreting cells migrate dorsally, bringing with them some of the imaginal disc cells. These dorsal disc cells generate the wing disc, while the cells remaining form the leg disc. (After Cohen et al., 1993.)*

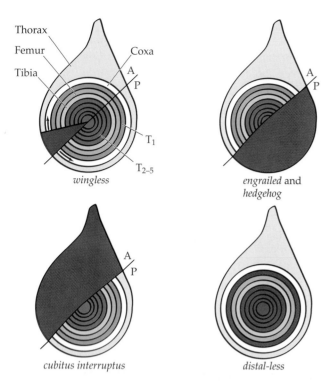

FIGURE 20.20

Patterns of gene expression in the leg imaginal disc. The wingless *gene is expressed in the anterior cells of the A/P border. The* engrailed *and* hedgehog *genes are expressed in the future posterior cells, while* cubitus interruptus *is seen only in the anterior cells. In some instances, such as* distal-less, *the gene is expressed in the precursors of particular cell types. (After Bryant, 1993.)*

teractions of homeotic genes. Even so, we still do not know the molecules that specify the leg discs to be different from the wing discs, or the eye discs to be different from the antenna discs. We do know that if certain homeotic genes are expressed in the wrong places (such as the expression of *Antennapedia* in the eye-antennal disc), the discs become respecified (such that legs grow from the antennal disc).

Determination of the disc polarity

Recent evidence suggests that the leg and wing discs become specified by polar coordinates similar to those discussed in the previous chapter for vertebrate limb development. Indeed, some of the best evidence for the polar coordinate model comes from the regeneration and duplications of experimentally manipulated insect leg discs (French et al., 1976). The genes that specify the polar coordinates appear to be the segment polarity genes already known to regulate segmentation (Figure 20.20; Bryant, 1993; Causo et al., 1993). These genes are expressed in specific concentric regions of the leg disc, and the loss of these genes in late development (by temperature-sensitive genes that lose their function when the larvae are placed at high temperatures) causes the loss of these specific structures. The exception seems to be the *wingless* gene. The loss of *wingless* function late in development leads not to the loss of the ventral structures where it is expressed. Rather, the loss of *wingless* function leads to the mirror-image duplication of dorsal structures in this region.

The *wingless–engrailed* system (discussed in Chapter 15) is crucial for leg disc growth and determination. First, the activity of the *wingless* gene appears to organize the dorsal–ventral patterning of the limb. The cells that express this protein become the most ventral cells of the leg, and the diffusion (or transport) of the wingless protein to neighboring cells causes them to become ventralized (Struhl and

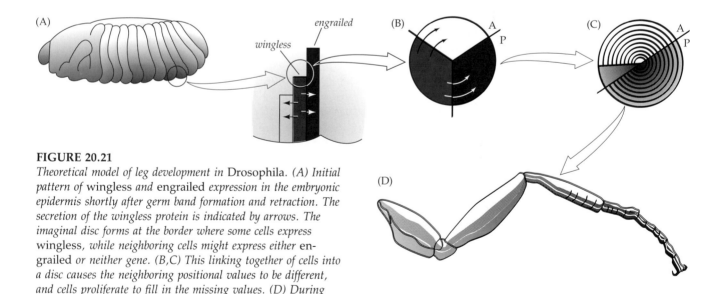

FIGURE 20.21

Theoretical model of leg development in Drosophila. *(A) Initial pattern of* wingless *and* engrailed *expression in the embryonic epidermis shortly after germ band formation and retraction. The secretion of the wingless protein is indicated by arrows. The imaginal disc forms at the border where some cells express* wingless, *while neighboring cells might express either* engrailed *or neither gene. (B,C) This linking together of cells into a disc causes the neighboring positional values to be different, and cells proliferate to fill in the missing values. (D) During metamorphosis, the disc differentiates and unfolds. (After Causo et al., 1993.)*

Basler, 1993). Second, the original parallel arrangement of the *wingless* and *engrailed* transcription patterns becomes integrated into a circular coordinate system when imaginal discs are formed (Figure 20.21). According to the polar coordinate model, the positional values of the cells expressing the *wingless* gene, the *engrailed* gene, or neither gene differ, and this causes the cells to proliferate and produce the cells of intermediate positional value. The more distal values arise from the center of the disc, and the proximal values from the periphery. The expression of engrailed protein is found only in the posterior of the limb, while wingless expression is found in the medial portion. Thus, the orthogonal (rectilinear) coordinate system that specified the position and identity of the imaginal discs within the embryonic ectoderm appears to have become a polar coordinate system within the larval imaginal disc.

In the wing and haltare imaginal discs, a similar phenomenon is thought to be taking place (Williams et al., 1993). Again, the *wingless* gene appears to be the earliest acting locus for specifying regional fate. The expression of *wingless* in the notum (shoulder) distinguishes that subfield from that of the wing (which does not express *wingless*),

and its expression in a band of cells along the presumptive margin of the wing separates those cells destined to form the dorsal and ventral wing surfaces. In addition, the diffusion of the wingless protein also is needed for the restriction of the *apterous* gene expression to dorsal cells and to promote the expression of other genes (such as *vestigial* and *scalloped*) that act to shape the wing (see Plate 15). The *apterous* gene encodes a transcription factor whose presence is necessary for establishing the dorsal cell fate in the wing and haltare, and its presence is required for normal wing development. The boundary cells expressing both *wingless* and *apterous* may be necessary for producing the signal that causes the growth of the wing disc inside the larva (Diaz-Benjumea and Cohen, 1993). In one of the first papers in developmental genetics, C. H. Waddington (1939) described the development of the normal wing and the effects of various mutations (such as *vestigial* and *apterous*) on wing development. "The wing," he remarked, "appears favorable for investigations on the developmental action of genes." We are finally able to use his mutants for the analysis of gene action during metamorphosis.

Remodeling of the nervous system

As in anuran metamorphosis, insect metamorphosis causes a major restructuring of the organism's nervous system. Some nerves die, while other nerves take over new functions. In Chapter 18, we saw the development of photoreceptors from the epithelial cells of the eye disc. Here, a new set of neurons is generated to take on a new function. The neurons that have been connected to dying tissues either die with the tissue or are respecified for new functions. The nerve that innervates the proleg muscle of the caterpillar of the *Manduca* moth is independently sensitive to ecdysone and perishes simultaneously with its larval target tissue. However, the motor neuron innervating the second oblique muscle of the larva survives the death of its target to innervate a newly formed adult muscle (the fourth dorsal external muscle) that differentiates during metamorphosis (Truman et al., 1985).

In some instances, larval functions are taken over by different regions in the adult. The larval firefly has its paired lanterns in the eighth (last) abdominal segment. The neurons from the eighth abdominal segment control its luminescence. During pupation, the sixth and seventh segments also develop the light-producing photocytes and the nerves to control the timing of the flash. By the end of pupation, only the sixth and seventh segments have functional lanterns. Moreover, if the larval lanterns are removed, the adult lanterns will still form (Strause et al., 1979). Thus, what had been a neural function of the eighth segment ganglia has become a function of the ganglia of the sixth and seventh segments.

Hormonal control of insect metamorphosis

The hormonal control of insect metamorphosis was shown by the dramatic experiments of Wigglesworth (1934), who studied *Rhodnius prolixus*, a blood-sucking bug that has five instars before undergoing a striking meta-

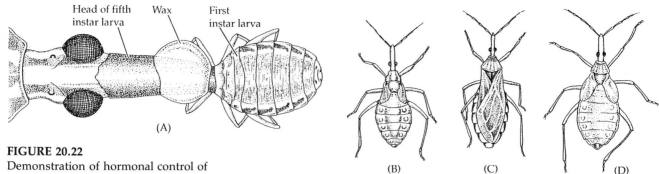

FIGURE 20.22
Demonstration of hormonal control of insect metamorphosis. (A) Precocious "adult" produced from the first-instar larva of *Rhodnius* by fusing it to the head of a molting fifth-instar larva. (B) Normal fifth instar larva of *Rhodnius*. (C) Normal adult *Rhodnius*, (D) "Sixth instar larva" produced when corpora allata from a fourth-instar larva were implanted into the abdomen of a fifth-instar larva. (After Wigglesworth, 1939.)

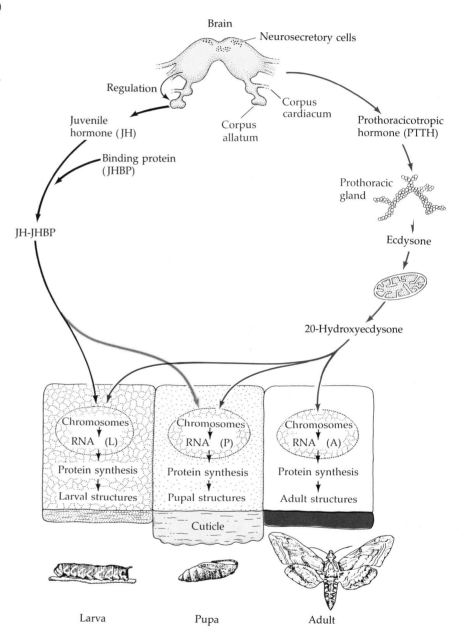

FIGURE 20.23
Schematic diagram illustrating the control of molting and metamorphosis in the tobacco hornworm moth. (After Gilbert and Goodman, 1981.)

morphosis. When a first-instar larva of *Rhodnius* was decapitated and fused to a molting fifth-instar larva, the minute first instar developed the cuticle, body structure, and genitalia of the adult (Figure 20.22A). This showed that blood-borne hormones are responsible for the induction of metamorphosis. Wigglesworth also showed that the **corpora allata**, near the insect brain, produces a hormone that counteracts this tendency to undergo metamorphosis. If the corpora allata was removed from a third-instar larva, the next molt turned the larva into a precocious adult. Conversely, if the corpora allata from fourth-instar larvae were implanted into fifth-instar larvae, these larvae would molt into extremely large "sixth-instar" larvae rather than into adults (Figure 20.22B-D).

Transplantation of insect tissues carried out in several laboratories eventually generated an integrated account of how metamorphosis takes place. Although the detailed mechanisms of metamorphosis differ between species, the general pattern of hormone action is usually very similar (Figure 20.23). Like amphibian metamorphosis, the metamorphosis of insects appears to be regulated by effector hormones controlled by neurosecretory peptide hormones in the brain (for reviews, see Granger and Bollenbacher, 1981; Gilbert and Goodman, 1981). The molting process is initiated in the brain, where neurosecretory cells release **prothoracicotropic hormone** (PTTH) in response to neural, hormonal, or environmental factors. PTTH is a family of peptide hormones with a molecular weight of approximately 40,000, and it stimulates the production of ecdysone by the prothoracic gland (Figure 20.24). Ecdysone, however, is not an active hormone, but a prohormone that must be converted into an active form. This conversion is accomplished by a heme-containing oxidase in the mitochondria and microsomes of peripheral tissues such as the fat body. Here the ecdysone is changed to the active hormone **20-hydroxyecdysone** (Figure 20.25).*

Each molt is occasioned by one or more pulses of 20-hydroxyecdysone. For a molt from a larva, the first pulse produces a small rise in the hydroxyecdysone concentration in the larval hemolymph (blood) and elicits a change in cellular commitment. The second, large pulse of hydroxyecdysone initiates the differentiation events associated with molting. The hydroxyecdysone produced by these pulses commits and stimulates the epidermal cells to synthesize enzymes that digest and recycle the components of the cuticle. In some cases, environmental conditions can control molting, as in the case of the silkworm moth *Hyalophora cecropia*. Here, PTTH secretion ceases after the pupa has formed. The pupa remains in this suspended state, called **diapause**, throughout the winter. If not exposed to cold weather, diapause lasts indefinitely. Once exposed to two weeks of cold, however, the pupa can molt when returned to a warmer temperature (Williams, 1952, 1956).

The second major effector hormone in insect development is **juvenile hormone** (JH). The structure of a common juvenile hormone active in butterfly and moth caterpillars is shown in Figure 20.25A. JH is secreted by the corpora allata. The secretory cells of the corpora allata are active during larval molts but are inactive during the metamorphic molt. This hormone is responsible for preventing metamorphosis. As long as JH is present, the hydroxyecdysone-stimulated molts result in a new larval instar. In the last larval instar, the medial nerve from the brain to the corpora allata inhibits the gland from producing juvenile hormone, and

*Since its discovery in 1954, when Butenandt and Karlson isolated 25 milligrams of ecdysone from 500 kilograms of silkworm moth pupae, 20-hydroxyecdysone has gone under several names, including β-ecdysone, ecdysterone, and crustecdysone.

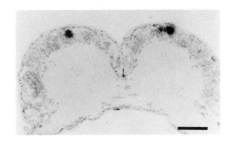

FIGURE 20.24
Cellular localization of PTTH mRNA in the *Bombyx mori* (silkworm moth) larva. In situ hybridization of a radioactive cloned gene for the 224-amino acid peptide localizes the PTTH mRNA to two neurosecretory cells in the left hemisphere of the brain and two neurosecretory cells in the right hemisphere. In this section, one PTTH-secreting cell can be seen on each side. The bar equals 100 μm. (From Kawakami et al., 1990, courtesy of H. Ishizaki and A. Kawakami.)

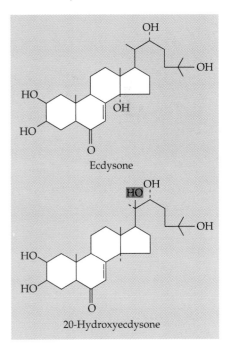

Juvenile hormone

Ecdysone

20-Hydroxyecdysone

FIGURE 20.25
Structures of a commonly occurring juvenile hormone, ecdysone, and of the active molting hormone 20-hydroxyecdysone.

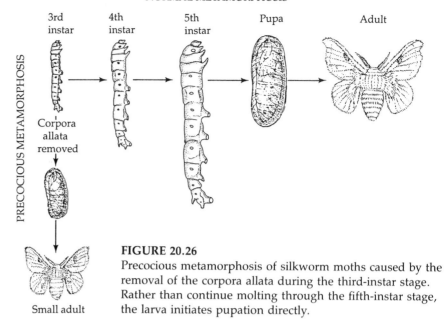

FIGURE 20.26
Precocious metamorphosis of silkworm moths caused by the removal of the corpora allata during the third-instar stage. Rather than continue molting through the fifth-instar stage, the larva initiates pupation directly.

there is a simultaneous increase in the body's ability to degrade existing JH (Safranek and Williams, 1989). Both these mechanisms cause JH levels to drop below a critical threshold value. This triggers the release of PTTH from the brain (Nijhout and Williams, 1974; Rountree and Bollenbacher, 1986). PTTH, in turn, stimulates the prothoracic glands to secrete a small amount of ecdysone. The resulting hydroxyecdysone, in the relative scarcity of JH, commits the cells to pupal development. The subsequent molt, occurring in the relative scarcity of JH, shifts the organism from larva to pupa. During pupation, the corpora allata do not release any JH, and the hydroxyecdysone-stimulated pupa will eventually metamorphose into the adult insect. Removal of the corpora allata from larvae causes the premature metamorphosis of the larvae into small pupae and small adults (Figure 20.26), while the absorption of exogenously supplied juvenile hormone into last-instar larvae can elicit an extra larval molt and thereby delay the pupal molt. Metamorphosis to the adult, then, occurs when the insect is exposed to hydroxyecdysone in the absence of JH.

The molecular biology of hydroxyecdysone activity

Binding of hydroxyecdysone to DNA. During molting and metamorphosis, certain regions of the polytene chromosomes of *Drosophila* become puffed out in certain cells (Figure 2.14; Clever, 1966; Ashburner, 1972; Ashburner and Berondes, 1978). These chromosome puffs represent areas where the DNA is being actively transcribed. Moreover, the organ-specific pattern of puffing can be reproduced by culturing the larval tissue and adding hormones to the medium or by adding hydroxyecdysone to an earlier stage larva. When hydroxyecdysone is added to larval salivary glands, certain puffs are produced and others regress (Figure 20.27). This puffing is mediated by the binding of hydroxyecdysone to specific places on the chromosomes. This can be shown by cross-linking the hydroxyecdysone to the chromatin so that it stays where it has bound. Then rabbit antibodies recognizing hydroxyecdysone are added and the unbound antibodies washed away. Fluorescein-labeled goat antibody against rabbit immunoglobulins is added last, and the unbound goat antibodies are

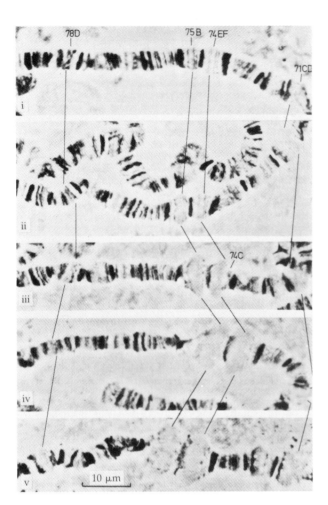

FIGURE 20.27
FIGURE 20.27
Ecdysone-induced puffs in cultured salivary gland cells of *D. melanogaster*. The chromosome region here is the same as in Figure 2.14. Puffing is induced by hydroxyecdysone. (i) Uninduced control. (ii–v) Hydroxyecdysone-stimulated chromosomes at 25 minutes, 1 hour, 2 hours, and 4 hours, respectively. (Courtesy of M. Ashburner.)

washed away. Ultimately, the fluorescent tag should be located wherever a goat antibody has bound to a rabbit antibody, and rabbit antibodies should bind only to hydroxyecdysone. In this manner, a fluorescent label should appear at every site on the chromosome where hydroxyecdysone has been bound (Gronemeyer and Pongs, 1980). The result (Figure 20.28) shows that almost all the hydroxyecdysone-sensitive puff sites bind hydroxyecdysone.

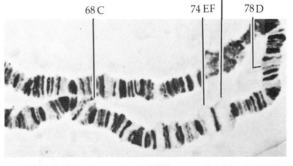

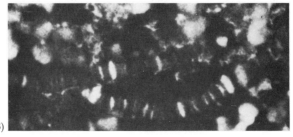

FIGURE 20.28
Localization of ecdysone on the polytene chromosomes of *Drosophila*. (A) Phase-contrast photomicrograph of a segment of salivary gland chromosomes. (B) The same field seen under ultraviolet light, showing immunofluorescence at ecdysone-sensitive areas. (Courtesy of O. Pongs.)

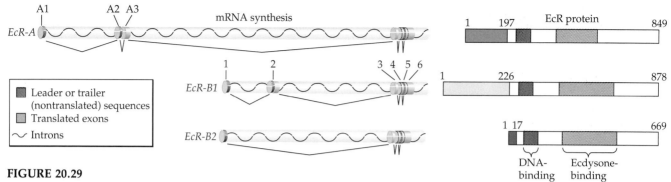

FIGURE 20.29
Formation of the ecdysone receptors. Alternative mRNA splicing of the ecdysone receptor (EcR) transcript creates three types of EcR mRNAs. These generate proteins having the same DNA-binding site and hydroxyecdysone-binding site, but with very different amino termini. (After Talbot et al., 1993.)

Different hydroxyecdysone receptors in different tissues. The tissues of a late-instar larvae can be grossly divided into two types based on their responses to hydroxyecdysone: the strictly larval tissues that degenerate in response to hydroxyecdysone, and the imaginal tissues that generate the adult structures when exposed to hydroxyecdysone. It is not known how one group of cells proliferates while another group of cells degenerates when given the same signal, but recent studies (Talbot et al., 1993; Truman et al., 1994) suggest that not all ecdysone receptors are the same in every tissue. The gene for the **ecdysone receptor** (EcR) can be alternatively spliced into three mRNAs that will yield three different, but related, proteins, EcR-A, EcR-B1, and EcR-B2 (Figure 20.29). All cells appear to have some of each, but the strictly larval tissues and regressing neurons are characterized by their great abundance of EcR-B1 compared with EcR-A. Imaginal discs and differentiating neurons, on the other hand, show a preponderance of the EcRA isoform over EcR-B1. It is therefore possible that the different receptors activate different sets of genes when they bind hydroxyecdysone.

Different ecdysone receptors within a single cell. In addition to the heterogeneity of responses to hydroxyecdysone between tissues, there is also a heterogeneity of responses to hydroxyecdysone within an individual cell. Hydroxyecdysone-sensitive puffs occurring during the late stages of the third-instar larva (as it prepares to form the pupa) can be grossly divided into three categories. There are those puffs that hydroxyecdysone causes to regress; there are those puffs that it induces rapidly; and there are those puffs that are first seen several hours after stimulation. For example, in the larval salivary gland, about six puffs emerge within a few minutes of hydroxyecdysone treatment. These genes do not need protein synthesis to be active. A much larger set of genes was found to be induced later in development, and these genes do need protein synthesis in order to become transcribed. Ashburner (1974, 1990) predicted that the "early" genes make a protein product that is essential for the activation of the "late" genes. Moreover, this protein itself would turn off the transcription of the early gene (Figure 20.30).

Recent investigations support this view and suggest that the early genes represent transcription factors that can mediate the ecdysone effect. The ecdysone receptor is a transcription factor that binds this steroid hormone and bring it to the specific region of DNA. Like the vertebrate steroid-binding receptors, it forms a heterodimer. The EcR does not bind either its hormone or its DNA sequence without first forming a heterodimer with the product of the *ultraspiracle* (*USP*) gene (the *Drosophila* analogue of the retinoid receptor; Yao et al., 1992; Thomas et al., 1993). When

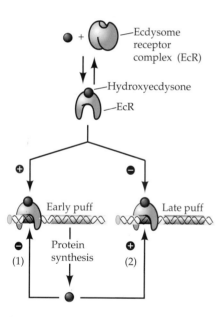

FIGURE 20.30
The Ashburner model of hydroxyecdysone regulation of transcription. Hydroxyecdysone binds to its receptor, and this compound binds to a early puff gene and a late puff gene. The early puff gene is activated, and its protein product (1) represses the transcription of its own gene and (2) activates the late puff gene, perhaps by displacing the ecdysone receptor. (After Richards, 1992.)

the EcR/USP heterodimer has been formed, it binds hydroxyecdysone and activates the earliest ecdysone-responsive genes. These early puffs have been found to encode a family of transcription factors that may also form heterodimers with the ecdysone receptor. One of the early ecdysone-sensitive genes, *E75*, encodes three steroid receptor proteins, the different forms being created by alternative RNA splicing. *E78*, one of the early-to-late genes turned on by an E75 protein, encodes another steroid receptor (Koelle et al., 1991; Stone and Thummel, 1993). Richards (1978) showed that in the absence of JH, hydroxyecdysone induces the synthesis of the proteins from the late puffs. When JH is present during the initial hydroxyecdysone pulse, the early puffs are produced as usual, but the later puffs are repressed. Thus, the specificity of the response to hydroxyecdysone might be mediated through cascades of different transcription factors, each of which could bind with the original hydroxyecdysone-bearing receptor (Figure 20.31).

It is worth noting that hydroxyecdysone not only stimulates the puffing of certain regions but also causes the regression of certain existing larval puffs. One of these latter puffs is present at position 68C of the left arm of chromosome 3. This puff site contains the gene for Sgs3, a "glue protein" responsible for attaching the pupal case to a solid surface (Korge, 1975). Only in the salivary gland does this region of the chromosome puff out at the last instar. The addition of hydroxyecdysone to cultured salivary glands causes the rapid regression of this puff and the cessation of transcription from this gene (Crowley and Meyerowitz, 1984). Thus, the glue can be synthesized when it is needed to adhere the larva to a substrate; but afterward the gene is shut off while metamorphosis occurs.

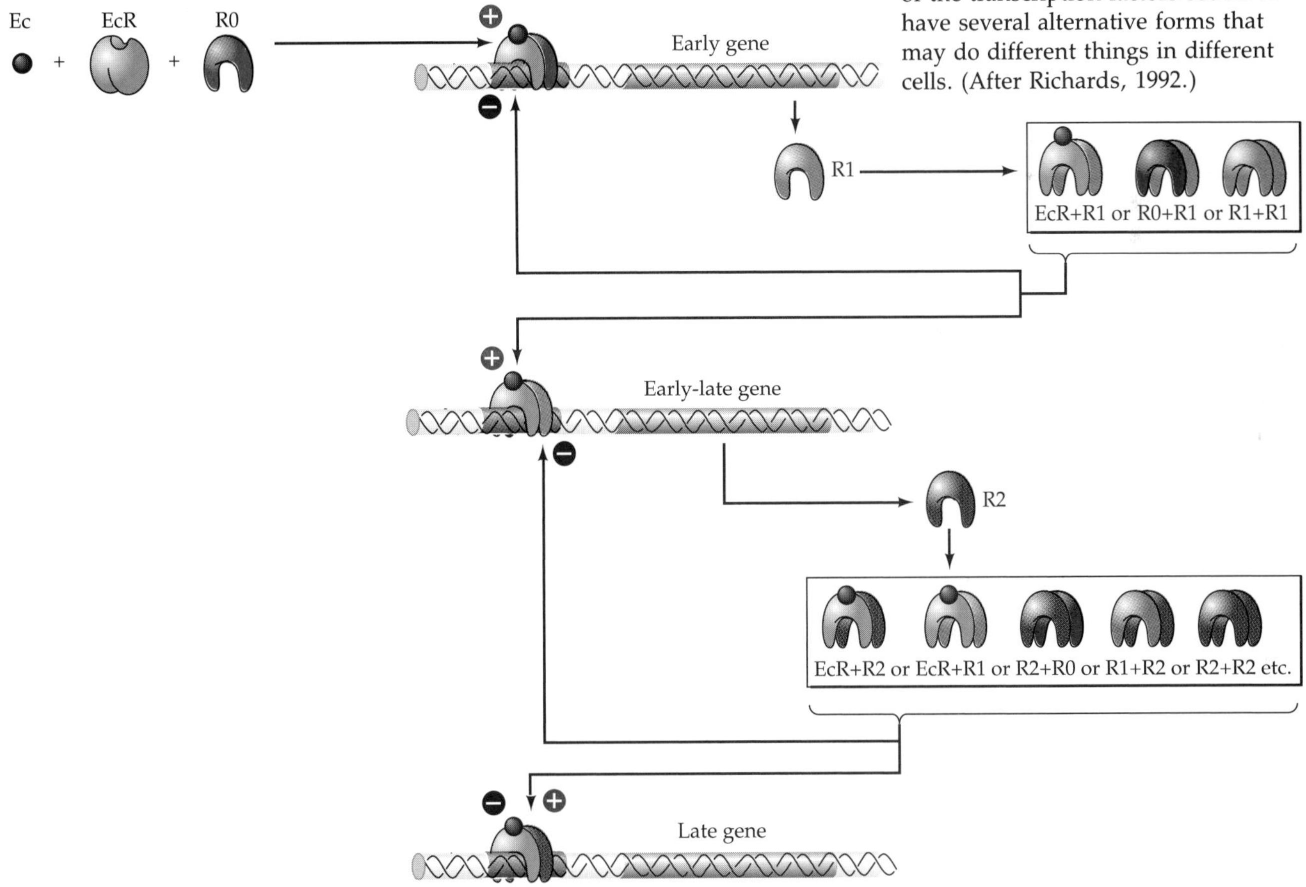

FIGURE 20.31
A more complex circuit diagram for the sequential effects of hydroxyecdysone. (1) The ecdysone receptor (EcR) dimerizes with the ultraspiracle protein (R0) to form the active receptor and binds hydroxyecdysone. (2) This complex activates the early puff genes. One of the products of these genes is the R1 protein, a steroid-binding transcription factor similar to R0, but different in its binding specificity to DNA. The EcR protein switches from R0 to R1. The EcR-R1 (or the R1-R1 or R0-R1) complex inhibits the early puff gene, while the EcR-R1 activates a new set of ecdysone-responsive genes (the early late genes). The product of one of these genes is R2, which can likewise bind the EcR to activate the late genes, and so forth. We can be certain that this scheme is too simple, as each of the transcription factors seems to have several alternative forms that may do different things in different cells. (After Richards, 1992.)

Environmental control over larval form and functions

Most discussions of development are limited to within the developing organism's body. However, an organism's development can sometimes be regulated significantly by environmental factors outside the body. There are several types of developmental phenomena where chemicals produced by one organism (often of another species) induce changes in the development of another organism.

Childhood's end: Signals to initiate metamorphosis

In 1880, William Keith Brooks, an embryologist at Johns Hopkins University (and thesis adviser to T. H. Morgan, E. B. Wilson, R. G. Harrison, and E. G. Conklin) was asked to help the ailing oyster industry of Chesapeake Bay. For decades, oysters had been dredged from the bay and there had always been a new crop to take their place. But recently, each year brought fewer oysters. What was responsible for the decline? Experimenting with larval oysters, Brooks discovered that the American oyster (unlike its better-studied European cousin) needs a hard substrate on which to metamorphose. For years, oystermen had thrown the shells back into the sea, but with the advent of suburban sidewalks, the oystermen were selling the shells to the cement factories. Brooks's solution: throw the shells back into the bay. The oyster population responded, and the Baltimore wharves still sell their descendants.

Brooks's study raises an important point. Larvae will not metamorphose under just any set of conditions. In almost all cases, the larvae need a substrate upon which to settle. Cameron and Hinegardner (1974) found that the plutei of two sea urchins, *Arbacia punctulata* and *Lytechinus pictus*, settle only on containers with a bacterial film on them. The bacteria produce a factor that induces the sea urchins to settle. This factor can pass through a dialysis membrane that holds back any molecule larger than 5 kDa. If the plutei are kept in clean containers, they float for about two months and then deteriorate (Hinegardner, 1969). Other echinoderm plutei settle for different conditions.

The adult sand dollar *Dendraster excentricus* helps its larvae to metamorphose by secreting a chemical (probably a peptide less than 10 kDa) into the sand. This chemical attracts *Dendraster* larvae and induces their metamorphosis. The result is that larvae tend to associate within or near existing sand dollar beds, causing some areas of a beach to have several hundred adults per square meter while nearby areas have none. Highsmith (1982) has speculated that the adults in such clusters aid individual survival by protecting the newly metamorphosed *Dendraster* from a particular crustacean predator.

As one might predict, those individuals survive best that can be induced to metamorphose close to protection or potential food sources. It would not be advantageous for a larva to settle where there was no food for it, espe-

cially if the adult form were relatively sedentary. In molluscs, there are often very specific cues for settlement (Table 20.4). Most nudibranch (sea slug) larvae undergo metamorphosis only if triggered by living adult prey (which differs from species to species). In some cases, the soluble product of the prey that triggers metamorphosis has been identified (Hadfield, 1977). The larva of the shipworm *Teredo navalis* is induced to settle by compounds released by wood, and soluble material eluted from oyster shells induces the settlement of oyster larvae.

TABLE 20.4
Specific settlement substrates of molluscan larvae

Molluscan species	Substrate
GASTROPODA (SNAILS, NUDIBRANCHS)	
Nassarius obsoletus	Mud from adult habitat
Philippia radiata	*Porites lobata* (a cnidarian)
Adalaria proxima	*Electra pilosa* (a bryozoan)
Doridella obscura	*Electra crustulenta* (a bryozoan)
Phestilla sibogae	*Porites compressa* (a cnidarian)
Rostanga pulchra	*Ophlitaspongia pennata* (a sponge)
Trinchesia aurantia	*Tubularia indivisa* (a cnidarian)
Elysia chlorotica	Primary film of microorganisms from adult habitat
Haminoea solitaria	Primary film of microorganisms from adult habitat
Aplysia californica	*Laurencia pacifica* (a red alga)
Aplysia juliana	*Ulva* spp. (green algae)
Aplysia parvula	*Chondrococcus hornemanni* (a red alga)
Stylocheilus longicauda	*Lyngbya majuscula* (a cyanobacterium)
Onchidoris bilamellata	Living barnacles
AMPHINEURA (CHITONS)	
Tonicella lineata	*Lithophyllum* sp. and *Lithothamnion* sp. (red algae)
LAMELLIBRANCHIA (BIVALVES)	
Teredo sp.	Wood
Bankia gouldi	Wood
Mercenaria mercenaria	Clam liquor; sand
Placopecten magellanicus	Adult shell; sand; etc.
Mytilus edulis	Filamentous algae; other nonbiological silk material
Crassostrea virginica	Shell liquor; body extract; "shellfish glycogen"
Ostrea edulis	Muscle extract and extrapallial fluid of adult

Source: After Hadfield, 1977.

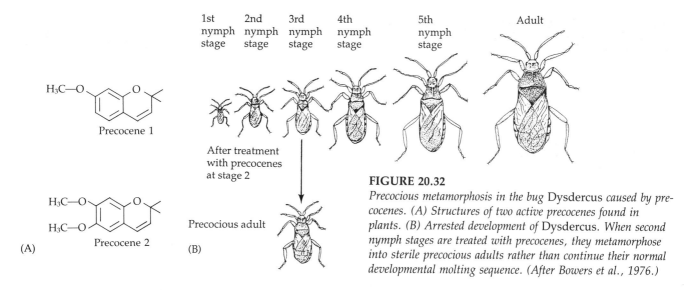

FIGURE 20.32

Precocious metamorphosis in the bug Dysdercus *caused by precocenes. (A) Structures of two active precocenes found in plants. (B) Arrested development of* Dysdercus. *When second nymph stages are treated with precocenes, they metamorphose into sterile precocious adults rather than continue their normal developmental molting sequence. (After Bowers et al., 1976.)*

The red abalone *Haliotis rufescens* has larvae that only settle when they physically contact coralline red algae. A brief amount of contact is all that is required for the competent larvae to stop swimming and begin metamorphosis. The chemical agent responsible for this change has not yet been isolated, but recognition of an algal peptide induces metamorphosis in competent larvae. Larvae that are not competent to induce metamorphosis do not appear to have this receptor. The receptor is thought to be linked to a G protein similar to those found in vertebrates, and the activation of this G protein may be necessary for inducing larval settlement and metamorphosis (Morse et al., 1984; Morse, 1991; Baxter and Morse, 1992).

In insects, the *size* of the larva plays a major role in initiating metamorphosis. The tobacco hornworm caterpillar will not metamorphose until it weighs 3 grams, and in some cases it needs an extra larval molt to get there (Safranek and Williams, 1984). It appears that upon reaching this size, the brain releases a substance that inhibits juvenile hormone release (Bhaskaran et al., 1980). In bees, the size of the female larva at its pupal molt determines whether the individual is to be a worker or a queen. A larva fed nutrient-rich "royal jelly" retains the activity of its corpora allata during its last instar stage. The juvenile hormone secreted by this organ delays pupation, thus allowing the resulting bee to emerge larger and (in some species) more specialized in her anatomy (Plowright and Pendrel, 1977; Brian, 1980).

Derangement of metamorphosis by plant compounds: Precocenes and juvenile hormones

When Karel Sláma came from Czechoslovakia to work in Carroll Williams's laboratory at Harvard, he brought with him his chief experimental animal, the European plant bug *Pyrrhocoris apterus*. To the consternation of the entire laboratory, these bugs failed to undergo metamorphosis at the end of the fifth instar. Rather, they became large sixth-instar larvae—something never before observed in nature or in the laboratory—and ultimately died before becoming adults. After many variables were tested, the paper towels lining the dishes were tested for their effect on the larvae.

The results were as conclusive as they were surprising: larvae reared on European paper (including pages of the journal *Nature*) underwent metamorphosis as usual, while larvae reared on American paper (such as shredded copies of the journal *Science*) did not undergo metamorphosis. It was eventually determined that the source of the American paper was the balsam fir, a tree indigenous to the northern United States and Canada. This tree synthesizes a compound that closely resembles juvenile hormone (Sláma and Williams, 1966; Bowers et al., 1966; Williams, 1970), and it probably employs this juvenile hormone analogue to get rid of certain insect predators.

Some other plants have compounds that produce the same effect—the death of insect predators—but do so by eliciting metamorphosis too early. Two compounds that have been isolated from composite herbs have been found to cause the premature metamorphosis of certain insect larvae into sterile adults (Bowers et al., 1976). These compounds are called **precocenes**; their chemical structures are shown in Figure 20.32A. When the larvae or nymphs of these insects are dusted with either of these compounds, they undergo one more molt and then metamorphose into the adult form (Figure 20.32B). Precocenes accomplish this by causing the selective death of the corpora allata cells in the immature insect (Schooneveld, 1979; Pratt et al., 1980). These cells are responsible for synthesizing juvenile hormone. Thus, without juvenile hormone, the larva commences its metamorphic and imaginal molts. Moreover, juvenile hormone is also responsible for the maturation of the insect egg (Chapter 22). Without this hormone, females are sterile. So the precocenes are able to protect the plant by causing the premature metamorphosis of certain insect larvae into sterile adults.

Plant-induced polyphenism

Plants and animals adapt their life cycles to each other in many ways, and they produce numerous chemicals to induce changes in the other species. In some cases, the morphology of the larva depends on cues from the plant. This ability of an individual to express one phenotype under one set of circumstances and another phenotype under another set of environmental conditions is called **polyphenism** or **phenotypic plasticity**. One dramatic example of such polyphenism occurs in the moth *Nemoria arizonaria*. This moth has a fairly typical life cycle. Eggs hatch in the spring, and the caterpillars feed on young oak flowers (catkins). These larvae metamorphose in the late spring, mate in the summer, and produce another brood of caterpillars on the oak trees. These caterpillars eat the oak leaves, metamorphose, and mate. Their eggs overwinter to start the cycle over again next spring. What is remarkable is that the caterpillars that hatch in the spring look nothing like their progeny that hatch in the summer (Plate 18). The caterpillars that hatch in the spring and eat oak catkins are yellow-brown, rugose, and beaded, resembling nothing else but an oak catkin. They are magnificently camouflaged against predation. But what of the caterpillars that hatch in the summer, after all the catkins are gone? They, too, are well camouflaged, appearing like year-old oak twigs. What controls this? By doing reciprocal feeding experiments, Greene (1989) was able to convert the spring morphs into the summer forms by feeding them oak leaves. The reciprocal experiment did not turn the later morphs into catkin-like caterpillars. Thus, it appears that the catkin form is the "default state" and that something induces the twig-like morphology. This substance is probably a tannin that is concentrated in the oak leaves as they mature.

Predator-induced polyphenism

In *Nemoria*, the diet alters the phenotype and protects the individual from predation. Some animals have taken this a step further: The development of a juvenile is changed by chemicals released by a predator such that the juveniles can better escape those same predators. This is sometimes called **predator-induced cyclomorphosis** (or **predator-induced polyphenism**). Several *Daphnia* and rotifer species will alter their morphology when they develop in pond water in which their predators were cultured (Dodson, 1989; Adler and Harvell, 1990). The predatory rotifer *Asplanchna* releases into its water a soluble compound that induces the eggs of a prey species, *Keratella slacki*, to develop into individuals with slightly larger bodies, but with anterior spines 130 percent longer than they would otherwise develop. These changes make them more difficult to eat (Gilbert and Stemberger, 1984). Thus, these prey organisms can alter their development in response to a chemical produced by their predators.*

The development of an organism proceeds through several stages, each of which must be integrated into its environmental context. Each phase becomes part of what Darwin called the "tangled web of interrelationships" that link one species to the other.

---

*A recent report indicates that symbionts may alter the morphogenesis of their hosts (Montgomery and McFall-Ngai, 1993). The luminous bacteria *Vibrio fischeri* usually infects the light organ of the juvenile form of the squid *Euprymia scolopes*. When the bacteria enter this organ, the cells of the region undergo differentiation and regional cell death to form the mature light organ. Light organs of uninfected juveniles remain in a state of "arrested development."

Multiple hormonal interactions in mammary gland development

Hormones have both constructive and destructive effects in development. In metamorphosis, hormones instruct some cells to die while instructing other cells to form new organs. In breast development, different hormones provide different information to the rudimentary tissue. Mammary development can be divided into four stages: the embryonic stage, the adolescent stage, pregnancy, and lactation. The differentiated products of the mammary glands—casein and other milk proteins—are made only during the final stage (Topper and Freeman, 1980).

Embryonic stage

In the normal development of the female mouse, two bands of raised epidermal tissue appear on both sides of the ventral midline on day 11 of gestation. This tissue is called the **mammary ridge**. Within each ridge, cells collect at centers of concentration and remain there, forming the **mammary buds** (Figure 20.33). In the mouse, there are five of these buds on each side; in humans, only one per side. In the days immediately prior to birth, the epithelial cells at these places proliferate rapidly, giving rise

(A)

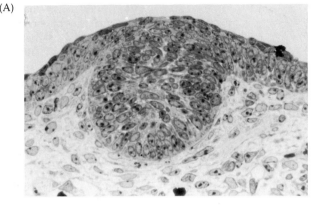

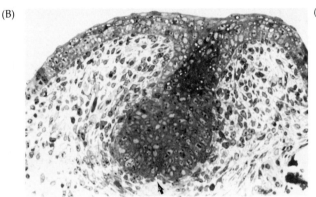

(B)

(C)

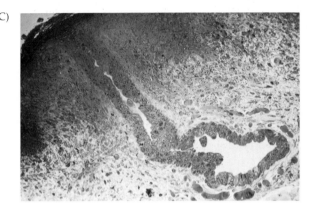

FIGURE 20.33
Sequence of early mammary gland development in the female mouse. (A) Mammary bud of 12-day fetus. Epithelial ectoderm cells protrude into mesenchyme. (B) Mammary cord of 15-day fetus. A small cleft at the bottom signals the initiation of branching. (C) Cord cavity extending to form a hollow lumen in the 20-day fetus. (From Hogg et al., 1983, courtesy of C. Tickle.)

to the **mammary cord**. This cord opens at the skin at one end, forming the nipple, while its other end begins branching into ducts. Here development ceases until puberty.

The development of mammary tissue in male mice is identical to that of females until days 13–15 of gestation. At this time, the mesenchyme condenses around the center of the mammary bud, and the cells of the cord die. Thus, a small cord of epithelial cells is detached from the skin (Figure 20.34), and the mammary gland does not extend to the surface. No further development occurs.

This cell death in the mammary cord of males has been studied by culturing the mouse mammary buds in vitro. Such buds from female mice normally develop lobes connected to the surface (Figure 20.35). However, when testosterone is added to the culture medium, the buds degenerate. Mammary buds from male mice also produce lobes if they are cultured in the absence of testosterone; thus, the hormone testosterone prevents mammary development in the male. Testosterone causes this specific cell death by instructing the mesenchymal cells to destroy the epithelial cord. This was shown by a series of recombination experiments. There exists in mice (and in humans as well) a mutation called **androgen insensitivity syndrome**, in which chromosomally male (XY) individuals do not make a functional testosterone receptor. Thus, even though these individuals have testes that are actively secreting testosterone, they are unable to respond to it. One of the results is that these individuals have female breast development (see Figure 21.11). Kratochwil and Schwartz (1976) isolated mesenchyme and epithelial cells from normal and mutant mammary buds and cultured them in various combinations. Some cultures were given testosterone and some were not. The results are shown in Figure 20.36. When both mesenchyme and epithelium were wild-type, the rudiment

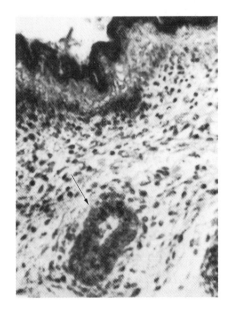

FIGURE 20.34
Mammary rudiment in a male mouse fetus. The rudiment (arrow) has separated from the epidermis. (From Raynaud, 1961.)

FIGURE 20.35
Role of testosterone in mediating the detachment of the mammary cord. (A) Female mouse mammary tissue, either in vivo or in culture, will grow downward from the epidermis and branch. (B) When female mouse mammary tissue is cultured in the presence of testosterone, the bud elongates, but mesemchymal cells aggregate around the stalk and the lower portion is cut off, just as in normal male development. (C) When male mouse mammary tissue is cultured in the absence of testosterone, it develops as it would in the female mouse. (After Kratochwil, 1971.)

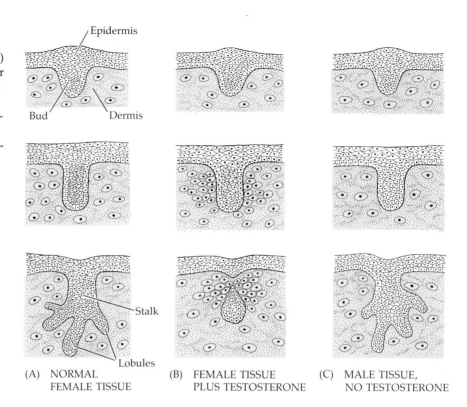

(A) NORMAL FEMALE TISSUE

(B) FEMALE TISSUE PLUS TESTOSTERONE

(C) MALE TISSUE, NO TESTOSTERONE

developed into breast tissue. When testosterone was added, the mesenchyme condensed around the bud and the cord was severed. When normal epithelium was cultured with mutant mesenchyme (which could not respond to testosterone), normal breast development occurred in the presence of testosterone. However, when the mesenchyme was normal and the epithelium was mutant, testosterone was able to cause the degenera-

FIGURE 20.36
Evidence that the mesenchymal cell is the target of testosterone in the arrest of mammary development. (A) Cultured mammary rudiment from 14-day female embryo. (B) Mammary rudiment from 14-day male embryo beginning its response to testosterone. (C) Recombined mammary bud containing wild-type epithelial cells and androgen-insensitive mesenchyme, cultured with testosterone. No androgen response is seen. (D) Recombined mammary bud containing androgen-insensitive epithelial cells and wild-type mesenchyme, cultured with testosterone. Mesenchyme cells are condensing at the neck of the bud. (From Kratochwil and Schwartz, 1976, courtesy of K. Kratochwil.)

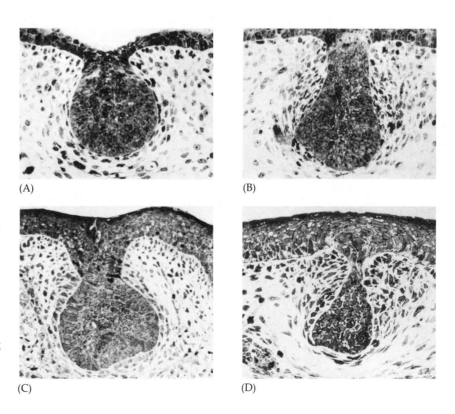

(A)

(B)

(C)

(D)

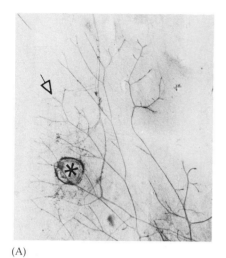 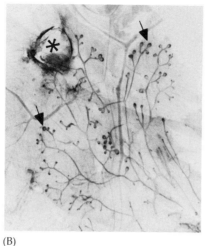 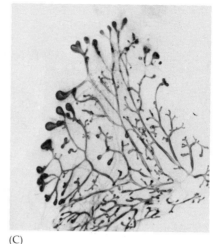

(A) (B) (C)

FIGURE 20.37
EGF-dependent mammary gland growth in the absence of estrogen. (A) No duct growth or differentiation is seen in estrogen-deficient mice when a pellet of bovine serum albumin (✲) is implanted into the mammary gland. (B) When a pellet containing EGF is implanted into the estrogen-deficient mammary gland, nearby ducts increase their size and develop lobular tissue at their tips (arrows). (C) Normal mammary duct development in a control 5-week virgin mouse. (From Coleman et al., 1988, courtesy of S. Coleman.)

tion of the mammary cord. Thus, the target of testosterone is the mesenchyme, not the epithelium. The mesenchyme must be responsive to testosterone for its action to occur. In males, the testosterone induces the mammary mesenchyme to destroy its adjacent epithelium. The effect is specific for the organ in that no other mesenchyme will kill the mammary epithelium and no other epithelium can be destroyed by mammary mesenchyme* (Dürnberger and Kratochwil, 1980).

Adolescence

During adolescence (which in the mouse occurs from week 4 to week 6), the duct system of the mammary gland proliferates extensively. The milk-secreting alveolar cells at the tips of the ducts have not differentiated yet, and no milk is produced. The extensive cell division is under the control of estrogen and growth hormones and appears to be concentrated at the ductal tips. Recent studies by Coleman and her colleagues (1988) implicate epidermal growth factor (EGF) as being responsible for controlling ductal growth during this period. They implanted EGF in slow-release plastic pellets into the mammary glands of 5-week mice. These mice had had their ovaries removed, and their mammary development was thus halted. The ducts adjacent to the EGF implant reinitiated their growth and morphological development, while the ducts further away from the implant did not (Figure 20.37). Moreover, when sections of the mammary gland were incubated with radioactive EGF, the EGF was seen to bind to the tip of the ducts and to be associated with the cells that were undergoing mitosis. It is probable that EGF may act directly to cause the growth of the mammary glands during adolescence.

*This phenomenon seems to resemble that of the mesenchymal specificity in anuran metamorphosis that we discussed earlier in this chapter.

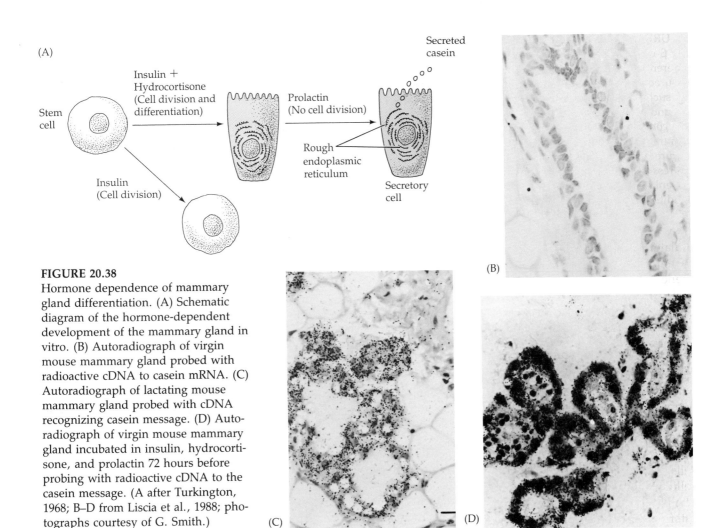

FIGURE 20.38
Hormone dependence of mammary gland differentiation. (A) Schematic diagram of the hormone-dependent development of the mammary gland in vitro. (B) Autoradiograph of virgin mouse mammary gland probed with radioactive cDNA to casein mRNA. (C) Autoradiograph of lactating mouse mammary gland probed with cDNA recognizing casein message. (D) Autoradiograph of virgin mouse mammary gland incubated in insulin, hydrocortisone, and prolactin 72 hours before probing with radioactive cDNA to the casein message. (A after Turkington, 1968; B–D from Liscia et al., 1988; photographs courtesy of G. Smith.)

Pregnancy and lactation

Between adolescence and pregnancy, mouse breast cells are mitotically dormant and undifferentiated. This state is changed during the second half of pregnancy. Under the influence of the hormones estrogen and progesterone (the latter from the placenta), new ducts are formed and the distal cells of the ducts begin to develop the characteristics of secretory tissue.

When midpregnancy mammary glands are cultured in vitro, most of the cells have little rough endoplasmic reticulum and Golgi apparatus, and no casein granules. When insulin or another promoter of DNA synthesis is added to these cultures, the cells become responsive to other hormones (Turkington et al., 1965). Glucocorticoids then induce the formation of the rough endoplasmic reticulum, where casein and the other proteins are synthesized. When the mouse gives birth, prolactin is secreted. Prolactin causes the casein gene to be transcribed and stabilizes the casein message once it is made (Figure 20.38).

During the period of lactation (when the young are suckling), a female mouse can produce about 10 percent of her body weight in milk per day. About 80 percent of the proteins in that milk are caseins, and of these, β-casein is the most abundant. The promoter of the mouse β-casein gene is located immediately upstream from the β-casein gene, and it is bound by a transcription factor, MGF. High levels of this transcription factor accu-

FIGURE 20.39

The β-casein mRNA levels in cultured mouse mammary gland cells placed in different culture conditions. (A) Endogenous β-casein mRNA when cells were cultured for six days on either extracellular matrix or plastic in medium containing such hormones as insulin, hydrocortisone, or prolactin. The extracellular matrix and prolactin were essential. (B) Reporter gene CAT expression when fused to a construct containing the β-casein enhancer and promoter. The fused gene was transfected into cultured mouse mammary cells and maintained six days in culture under various substrate and hormonal conditions. The fused gene was expressed only in the presence of prolactin and the extracellular matrix. Without the enhancer (having only the promoter), no transcription was seen under any conditions. (After Schmidhauser et al., 1992.)

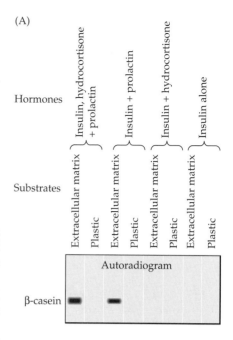

mulate during late pregnancy and lactation, and the activity of MGF can be eliminated by phosphorylation. MGF appears to be phosphorylated (hence, inactivated) whenever suckling stops. Withdrawal of pups from their mothers during lactation results in a rapid decrease in MGF activity. Restoration of the suckling pups restores activity to its maximum level within 4 hours. The effect may be mediated by pituitary or hypothalamic hormones that are responsive to suckling (Schmitt-Ney et al., 1992).

Casein is synthesized in competent mammary cells only in response to prolactin when the cells are anchored to an extracellular matrix (Figure 20.39). The enhancer of β-casein is responsive to both prolactin and the extracellular matrix. Using a reporter gene (CAT) linked to different regions of the 5′ flanking sequence, Schmidhauser and colleagues (1992) found a 160-base pair sequence 1517 base pairs from the transcription initiation site (Figure 20.40). This enhancer site only functions in mammary cells, and it is responsive to both prolactin and the presence of the extracellular matrix (Figure 20.40B).

The development of the mammary gland, then, involves a complex interplay of several hormones, paracrine proteins, and environmental factors at four different stages of life: embryonic, adolescent, pregnant, and lactating. The mammary gland never develops in normal males and does not become a fully differentiated organ in females until the middle of pregnancy in the adult organism. Studies of this organ have given us insights into the complexity of local and hormonal control in mammalian development.

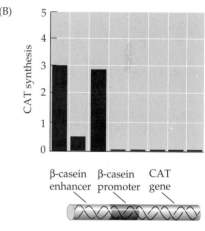

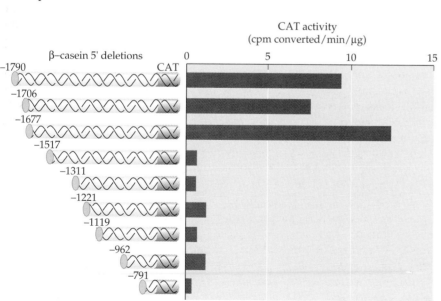

FIGURE 20.40

Constructs important for identifying the enhancer of the mouse β-casein gene. The CAT gene was used as a reporter and was fused to the 5′ end of the mouse β-casein gene. Exonuclease chopped off successively larger pieces from the 5′ flanking region of the gene. While the fused gene containing 1677 base pairs of the 5′ flanking sequence gave full activity, the sequence containing only 1517 base pairs gave very little activity. Thus, the enhancer was postulated to exist within these 160 base pairs. (After Schmidhauser et al., 1992.)

We have seen that diffusible regulation of cell–cell interactions is also important in regulating development. In studying the reactivation of development that occurs during metamorphosis and breast development, we can identify the roles of hormones in eliciting new patterns of differentiation and morphogenesis. We can also see the interactions between the development of the organism and the ecosystem of which it is part. In the next chapter, we will study the roles of cell-autonomous and diffusible factors in the processes responsible for gonad development and sex determination.

LITERATURE CITED

Adler, F. R. and Harvell, C. D. 1990. Inducible defenses, phenotypic variability and biotic environments. *Trends Ecol. Evol.* 5: 407–410.

Alberch, P. and Alberch, J. 1981. Heterochronic mechanisms of morphological diversification and evolutionary change in the neotropical salamander *Bolitoglossa occidentalis* (Amphibia: Plethodontidae). *J. Morphol.* 167: 249–264.

Allen, B. M. 1916. Extirpation experiments in *Rana pipiens* larva. *Science* 44: 755–757.

Alley, K. E. and Barnes, M. D. 1983. Birthdates of trigeminal motor neurons and metamorphic reorganization of the jaw myoneural system in frogs. *J. Comp. Neurol.* 218: 395–405.

Ashburner, M. 1972. Patterns of puffing activity in the salivary glands of *Drosophila*. VI. Induction by ecdysone in salivary glands of *D. melanogaster* cultured in vitro. *Chromosoma* 38: 255–281.

Ashburner, M. 1974. Sequential gene activation by ecdysone in polytene chromosomes of *Drosophila melanogaster*. II. Effects of inhibitors of protein synthesis. *Dev. Biol.* 39: 141–157.

Ashburner, M. 1990. Puffs, genes, and hormones revisited. *Cell* 61: 1–3.

Ashburner, M. and Berondes, H. D. 1978. Puffing of polytene chromosomes. In *The Genetics and Biology of Drosophila*, Vol. 2B. Academic Press, New York, pp. 316–395.

Baker, B. S. and Tata, J. R. 1992. Prolactin prevents the autoinduction of thyroid hormone receptor mRNAs during amphibian metamorphosis. *Dev. Biol.* 149: 463–467.

Baxter, G. T. and Morse, D. E. 1992. Cilia from abalone larvae contain a receptor-dependent G-protein transduction system similar to that in mammals. *Biol. Bull.* 183: 147–154.

Begon, M., Harper, J. L. and Townsend, C. R. 1986. *Ecology: Individuals, Populations, and Communities*. Blackwell Scientific, Oxford.

Bern, H. A., Nicoll, C. S. and Strohman, R. C. 1967. Prolactin and tadpole growth. *Proc. Soc. Exp. Biol. Med.* 126: 518–521.

Bhaskaran, G., Jones, G. and Jones, D. 1980. Neuroendocrine regulation of corpus allatum activity in *Manduca sexta*: Sequential neurohormonal and nervous inhibition in the last instar larva. *Proc. Natl. Acad. Sci. USA* 77: 4407–4411.

Bowers, W. S., Fales, H. M., Thompson, M. J. and Uebel, E. C. 1966. Identification of an active compound from balsam fir. *Science* 154: 1020–1021.

Bowers, W. S., Ohta, T., Cleere, J. S. and Marsella, P. A. 1976. Discovery of insect anti-juvenile hormones in plants. *Science* 193: 542–547.

Brian, M. V. 1980. Social control over sex and caste in bees, wasps and ants. *Biol. Rev.* 55: 379–415.

Brooks, W. K. 1880. The development of the American oyster. *Studies Biol. Lab. Johns Hopkins Univ.* 1: 1–81.

Brown, P. S. and Frye, B. E. 1969. Effect of prolactin and growth hormone on growth and metamorphosis of tadpoles of the frog *Rana pipiens*. *Gen. Comp. Endocrinol.* 13: 139–145.

Brust, D. G. 1993. Maternal brood care by *Dendrobates pumilio*: A frog that feeds its young. *J. Herpatol.* 26: 102–105.

Bryant, P. J. 1993. The polar coordinate model goes molecular. *Science* 259: 471–472.

Buckbinder, L. and Brown, D. D. 1993. Expression of the *Xenopus* prolactin and thyrotropin genes during metamorphosis. *Proc. Natl. Acad. Sci. USA* 90: 3820–3824.

Butenandt, A. and Karlson, P. 1954. Über die Isolierung eines Metamorphosen-Hormons der Insekten in kristallisierter Form. *Z. Naturforsch., Teil B* 9: 389–391.

Cameron, R. A. and Hinegardner, R. T. 1974. Initiation of metamorphosis in laboratory-cultured sea urchins. *Biol. Bull.* 146: 335–342.

Causo, J. P., Bate, M. and Martánez-Arias, A. 1993. A *wingless*-dependent polar coordinate system in *Drosophila* imaginal discs. *Science* 259: 484–489.

Clever, U. 1966. Induction and repression of a puff in *Chironomus tentans*. *Dev. Biol.* 14: 421–438.

Cohen, B., Simcox, A. A. and Cohen, S. M. 1993. Allocation of the thoracic imaginal primordia in the *Drosophila* embryo. *Development* 117: 597–608.

Cohen, P. P. 1970. Biochemical differentiation during amphibian metamorphosis. *Science* 168: 533–543.

Cohen, P. P., Brucker, R. F. and Morris, S. M. 1978. Cellular and molecular aspects of

thyroid–hormone action during amphibian metamorphosis. *Horm. Prot. Peptides* 6: 273–381.

Coleman, S., Silberstein, G. B. and Daniel, C. W. 1988. Ductal morphogenesis in the mouse mammary gland: Evidence supporting a role for epidermal growth factor. *Dev. Biol.* 127: 304–315.

Condic, M. L., Fristrom, D. and Fristrom, J. W. 1990. Apical cell shape changes during *Drosophila* imaginal leg disc elongation: A novel morphogenetic mechanism. *Development* 111: 23–33.

Crowley, T. E. and Meyerowitz, E. M. 1984. Steroid regulation of RNAs transcribed from the *Drosophila* 68C polytene chromosome puff. *Dev. Biol.* 102: 110–124.

Currie, J. and Cowan, W. M. 1974. Evidence for the late development of the uncrossed retinothalamic projections in the frog *Rana pipiens*. *Brain Res.* 71: 133–139.

De Beer, G. 1940. *Embryos and Ancestors*. Clarendon Press, Oxford.

del Pino, E. M. and Elinson, R. P. 1983. A novel development pattern for frogs: Gastrulation produces an embryonic disk. *Nature* 306: 589–591.

Diaz-Benjumea, F. J. and Cohen, S. M. 1993. Interaction between dorsal and ventral cells in the imaginal disc directs wing development in *Drosophila*. *Cell* 75: 741–752.

Dodson, S. 1989. Predator-induced reaction norms. *BioSci.* 39: 447–452.

Dürnberger, H. and Kratochwil, K. 1980. Specificity of tissue interaction and origin of mesenchymal cells in the androgen response of the embryonic mammary gland. *Cell* 19: 465–471.

Eisen, A. Z. and Gross, J. 1965. The role of epithelium and mesenchyme in the production of a collagenolytic enzyme and a hyaluronidase in the anuran tadpole. *Dev. Biol.* 12: 408–418.

Elinson, R. P. 1987. Change in developmental patterns: Embryos of amphibians with large eggs. *In* R. A. Raff and E. C. Raff (eds.), *Development as an Evolutionary Process*. Alan R. Liss, New York, pp. 1–21.

Etkin, W. and Gona, A. G. 1967. Antagonism between prolactin and thyroid hormone in amphibian development. *J. Exp. Zool.* 165: 249–258.

Fallon, J. F. and Simandl, B. K. 1978. Evidence of a role for cell death in the disappearance of the embryonic human tail. *Am. J. Anat.* 152: 111–130.

Forehand, C. J. and Farel, P. B. 1982. Spinal cord development in anuran larvae. I. Primary and secondary neurons. *J. Comp. Neurol.* 209: 386–394.

Fox, H. 1973. Ultrastructure of tail degeneration in *Rana temporaria* larva. *Folia Morphol.* 21: 103–112.

French, V., Bryant, P. J. and Bryant, S. V. 1976. Pattern regulation in epimorphic fields. *Science* 193: 969–981.

Frieden, E. 1981. The dual role of thyroid hormones in vertebrate development and calorigenesis. *In* L. I. Gilbert and E. Frieden (eds.), *Metamorphosis: A Problem in Developmental Biology.* Plenum, New York, pp. 545–564.

Fristrom, D. and Fristrom, J. W. 1975. The mechanisms of evagination of imaginal disks of *Drosophila melanogaster.* I. General considerations. *Dev. Biol.* 43: 1–23.

Fristrom, J. W. 1972. The biochemistry of imaginal disc development. *In* H. Ursprung and R. Nothiger (eds.), *The Biology of Imaginal Discs.* Springer-Verlag, Berlin, pp. 109–154.

Fristrom, J. W., Raikow, R., Petri, W. and Stewart, D. 1969. In vitro evagination and RNA synthesis in imaginal discs of *Drosophila melanogaster. In* E. W. Hanley, *Problems in Biology: RNA in Development.* University of Utah Press, Salt Lake City.

Fristrom, J. W., Fristrom, D., Fekete, E. and Kuniyuki, A. H. 1977. The mechanism of evagination of imaginal discs of *Drosophila melanogaster. Am. Zool.* 17: 671–684.

Geigy, R. 1941. Die metamorphose als Folge gewebsspezifischer determination. *Rev. Suisse Zool.* 48: 483–494.

Gilbert, J. J. and Stemberger, R. S. 1984. *Asplanchna*-induced polymorphism in the rotifer *Keratella slacki. Limnol. Oceanogr.* 29: 1309–1316.

Gilbert, L. I. and Goodman, W. 1981. Chemistry, metabolism, and transport of hormones controlling insect metamorphosis. *In* L. I. Gilbert and E. Frieden (eds.), *Metamorphosis: A Problem in Developmental Biology.* Plenum, New York, pp. 139–176.

Gould, S. J. 1977. *Ontogeny and Phylogeny.* Harvard University Press, Cambridge, MA, p. 283.

Granger, N. A. and Bollenbacher, W. E. 1981. Hormonal control of insect metamorphosis. *In* L. I. Gilbert and E. Frieden (eds.), *Metamorphosis: A Problem in Developmental Biology.* Plenum, New York, pp. 105–138.

Greene, E. 1989. A diet-induced developmental polymorphism in a caterpillar. *Science* 243: 643–646.

Grobstein, P. 1987. On beyond neuronal specificity: Problems in going from cells to networks and from networks to behavior. *In* P. Shinkman (ed.), *Advances in Neural and Behavioral Development,* Vol. 3, Ablex, pp. 1–58.

Gronemeyer, H. and Pongs, O. 1980. Localization of ecdysterone on polytene chromosomes of *Drosophila melanogaster. Proc. Natl. Acad. Sci. USA* 77: 2108–2112.

Gudernatsch, J. F. 1912. Feeding experiments on tadpoles. I. The influence of specific organs given as food on growth and differentiation. A contribution to the knowledge of organs with internal secretion. *Wilhelm Roux Arch. Entwicklungsmech. Org.* 35: 457–483.

Hadfield, M. G. 1977. Metamorphosis in marine molluscan larvae: An analysis of stimulus and response. *In* R.-S. Chia and M. E. Rice (eds.), *Settlement and Metamorphosis of Marine Invertebrate Larvae.* Elsevier, New York, pp. 165–175.

Hepburn, H. R. 1985. Structure of the integument. *In* G. A. Kerkut and L. I. Gilbert (eds.), *Comprehensive Insect Physiology, Biochemistry, and Pharmacology,* Vol 3. Pergamon Press, Oxford, pp. 1–58.

Highsmith, R. C. 1982. Induced settlement and metamorphosis of sand dollar (*Dendraster excentricus*) larvae in predator-free sites: Adult sand dollar beds. *Ecology* 63: 329–337.

Hinegardner, R. T. 1969. Growth and development of the laboratory cultured sea urchin. *Biol. Bull.* 137: 465–475.

Hogg, N. A. S., Harrison, D. J. and Tickle, C. 1983. Lumen formation in the mammary gland. *J. Embryol. Exp. Morphol.* 73: 39–57.

Hoskins, E. R. and Hoskins, M. M. 1917. On thyroidectomy in amphibia. *Proc. Soc. Exp. Biol. Med.* 14: 74–75.

Hoskins, S. G. and Grobstein, P. 1984. Thyroxine induces the ipsilateral retinothalamic projection in *Xenopus laevis. Nature* 307: 730–733.

Hoskins, S. G. and Grobstein, P. 1985a. Development of the ipsilateral retinothalamic projection in the frog *Xenopus laevis.* II. Ingrowth of optic nerve fibers and production of ipsilaterally projecting cells. *J. Neurosci.* 5: 920–929.

Hoskins, S. G. and Grobstein, P. 1985b. Development of the ipsilateral retinothalamic projection in the frog *Xenopus laevis.* III. The role of thyroxine. *J. Neurosci.* 5: 930–940.

Huxley, J. 1920. Metamorphosis of axolotl caused by thyroid feeding. *Nature* 104: 436.

Kaltenbach, J. C., Fry, A. E. and Leius, V. K. 1979. Histochemical patterns in the tadpole tail during normal and thyroxine-induced metamorphosis. II. Succinic dehydrogenase, Mg- and Ca-adenosine triphosphatases, thiamine pyrophosphatase, and 5' nucleotidase. *Gen. Comp. Endocrinol.* 38: 111–121.

Kanamori, A. and Brown, D. D. 1992. The regulation of thyroid hormone receptor β genes by thyroid hormone in *Xenopus laevis. J. Biol. Chem.* 267: 739–745.

Karp, G. and Berrill, N. J. 1981. *Development.* McGraw-Hill, New York.

Kawahara, A., Baker, B. S. and Tata, J. R. 1991. Developmental and regional expression of thyroid hormone receptor genes during *Xenopus* metamorphosis. *Development* 112: 933–943.

Kawakami, A. and 9 others 1990. Molecular cloning of the *Bombyx mori* prothoracicotropic hormone. *Science* 247: 1333–1335.

Kinoshita, T., Takahama, H., Sasaki, F. and Watanabe, K. 1989. Determination of cell death in the developmental process of anuran larval skin. *J. Exp. Zool.* 251: 37–46.

Kistler, A., Yoshizato, K. and Frieden, E. 1977. Preferential binding of tri-substituted thyronine analogs by bullfrog tadpole tail fin cytosol. *Endocrinology* 100: 134–137.

Koelle, M. R., Talbot, W. S., Segraves, W. A., Bender, M. T., Cherbas, P. and Hogness, D. S. 1991. The *Drosophila* EcR gene encodes an ecdysone receptor, a new member of the steroid receptor superfamily. *Cell* 67: 59–77.

Kollros, J. J. 1961. Mechanisms of amphibian metamorphosis: Hormones. *Am. Zool.* 1: 107–114.

Korge, G. 1975. Chromosome puff activity and protein synthesis in larval salivary glands of *Drosophila melanogaster. Proc. Natl. Acad. Sci. USA* 72: 4550–4554.

Kratochwil, K. 1971. In vitro analysis of the hormonal basis for sexual dimorphism in the embryonic development of the mouse mammary gland. *J. Embryol. Exp. Morphol.* 25: 141–153.

Kratochwil, K. and Schwartz, P. 1976. Tissue interaction in androgen response of embryonic mammary rudiment of mouse: Identification of target tissue for testosterone. *Proc. Natl. Acad. Sci. USA* 73: 4041–4044.

Liscia, D. S., Doherty, P. J. and Smith, G. H. 1988. Localization of α-casein gene transcription in sections of epoxy resin-embedded mouse mammary tissues by *in situ* hybridization. *J. Histochem. Cytochem.* 36: 1503–1510.

Little, G., Atkinson, B. G. and Frieden, E. 1973. Changes in the rates of protein synthesis and degradation in the tail of *Rana catesbeiana* tadpoles during normal metamorphosis. *Dev. Biol.* 30: 366–373.

Lyman, D. F. and White, B. A. 1987. Molecular cloning of hepatic mRNAs in *Rana catesbeiana* response to thyroid hormone during induced and spontaneous metamorphosis. *J. Biol. Chem.* 262: 5233–5237.

Lynn, W. G. and Peadon, A. M. 1955. The role of the thyroid gland in direct development of the anuran *Eleutherodactylus martinicenis. Growth* 19: 263–286.

Madhavan, M. M. and Schneiderman, H. A. 1977. Histological analysis of the dynamics of growth of imaginal disc and histoblast nests during the larval development of *Drosophila melanogaster. Wilhelm Roux Arch. Dev. Biol.* 183: 269–305.

Martin, F. D. and Drewry, G. E. 1978. *Development of Fishes of the Mid- Atlantic Bight,* Vol. 6. U. S. Department of the Interior, Washington, D.C.

Mathison, P. M. and Miller, L. 1987. Thyroid hormone induction of keratin genes: A two-step activation of gene expression during development. *Genes Dev.* 1: 1107–1117.

McCutcheon, F. H. 1936. Hemoglobin function during the life history of the bullfrog. *J. Cell. Comp. Physiol.* 8: 63–81.

Mitchell, A. W. 1988. *The Enchanted Canopy.* Macmillan, New York.

Montgomery, M. K. and McFall-Ngai, M. J. 1993. Inductive role of bacterial symbionts in morphogenesis of squid light organ. *Am. Zool.* 33: 100A.

Mori, M., Morris, S. M., Jr. and Cohen, P. P. 1979. Cell-free translation and thyroxine induction of carbamylphosphate synthetase I messenger RNA in tadpole liver. *Proc. Natl. Acad. Sci. USA* 76: 3179–3183.

Morse, A. N. C. 1991. How do planktonic larvae know where to settle? *Amer. Sci.* 79: 154–167.

Morse, A. N. C., Froyd, C. A. and Morse, D. E. 1984. Molecules from cyanobacteria and red algae that induce larval settlement and metamorphosis in the mollusc *Haliotis rufescens*. *Marine Biol.* 81: 293–298.

Nijhout, H. F. and Williams, C. M. 1974. Control of moulting and metamorphosis in the tobacco hornworm, *Manduca sexta*: Cessation of juvenile hormone secretion as a trigger for pupation. *J. Exp. Biol.* 61: 493–501.

Niki, K., Namiki, H., Kikuyama, S. and Yoshizato, K. 1982. Epidermal tissue requirement for tadpole tail regression induced by thyroid hormone. *Dev. Biol.* 94: 116–120.

Nishikawa, A., Kaiho, M. and Yoshizato, K. 1989. Cell death in the anuran tadpole tail: Thyroid hormone induces keratinization and tail-specific growth inhibition of epidermal cells. *Dev. Biol.* 131: 337–344.

Norris, D. O., Jones, R. E. and Criley, B. B. 1973. Pituitary prolactin levels in larval, neotenic, and metamorphosed salamanders (*Ambystoma tigrinum*). *Gen. Comp. Endocrinol.* 20: 437–442.

Oofusa, K. and Yoshizato, K. 1991. Biochemical and immunological characterization of collagenase in tissues of metamorphosing bullfrog tadpoles. *Dev. Growth Differ.* 33: 329–339.

Pehrson, J. R. and Cohen, L. H. 1986. The fate of the small micromeres in sea urchin development. *Dev. Biol.* 113: 522–526.

Pino-Heiss, S. and Schubiger, G. 1989. Extracellular protease production by *Drosophila* imaginal discs. *Dev. Biol.* 132: 282–291.

Plowright, R. C. and Pendrel, B. A. 1977. Larval growth in bumble-bees. *Can. Entomol.* 109: 967–973.

Prahlad, K. V. and DeLanney, L. E. 1965. A study of induced metamorphosis in the axolotl. *J. Exp. Zool.* 160: 137–146.

Pratt, G. E., Jennings, R. C., Hammett, A. F. and Brooks, G. T. 1980. Lethal metabolism of precocene-I to a reactive epoxide by locust corpora allata. *Nature* 284: 320–323.

Raff, R. A. 1987. Constraint, flexibility, and phylogenetic history in the evolution of direct development in sea urchins. *Dev. Biol.* 119: 6–19.

Raff, R. A. In press. Developmental mechanisms in the evolution of animal form: Origins and evolvability of body plans. In *Early Life on Earth*. Columbia University Press, New York.

Raynaud, A. 1961. Morphogenesis of the mammary gland. *In* S. K. Kon and A. T. Cowrie (eds.), *Milk: The Mammary Gland and Its Secretion*, Vol. 1. Academic Press, New York, pp. 3–46.

Richards, G. 1978. Sequential gene activity in polytene chromosomes of *Drosophila melanogaster*. VI. Inhibition by juvenile hormones. *Dev. Biol.* 66: 32–42.

Richards, G. 1992. Switching partners? *Curr. Biol.* 2: 657–658.

Riggs, A. F. 1951. The metamorphosis of hemoglobin in the bullfrog. *J. Gen. Physiol.* 35: 23–40.

Robinson, H., Chaffee, S. and Galton, V. A. 1977. Sensitivity of *Xenopus laevis* tadpole tail tissue to the action of thyroid hormones. *Gen. Comp. Endocrinol.* 32: 179–186.

Rountree, D. B. and Bollenbacher, W. E. 1986. The release of the prothoracicotropic hormone in the tobacco hornworm, *Manduca sexta*, is controlled intrinsically by juvenile hormone. *J. Exp. Biol.* 120: 41–58.

Safranek, L. and Williams, C. M. 1984. Critical weights for metamorphosis in the tobacco hornworm, *Manduca sexta*. *Biol. Bull.* 167: 555–567.

Safranek, L. and Williams, C. M. 1989. Inactivation of the corpora allata in the final instar of the tobacco hornworm, *Manduca sexta*, requires integrity of certain neural pathways from the brain. *Biol. Bull.* 177: 396–400.

Schmidhauser, C., Casperson, G. F., Myers, C. A., Sanzo, K. T., Bolten, S. and Bissell, M. J. 1992. A novel transcriptional enhancer is involved in the prolactin- and extracellular matrix-dependent regulation of β-casein gene expression. *Mol. Biol. Cell* 3: 699–709.

Schmitt-Ney, M., Happ, B., Ball, R. K. and Groner, B. 1992. Developmental and environmental regulation of a mammary gland-specific nuclear factor essential for the transcription of the gene encoding β–casein. *Proc. Natl. Acad. Sci. USA* 89: 3130–3134.

Schooneveld, H. 1979. Precocene-induced collapse and resorption of corpora allata in nymphs of *Locusta migratoria*. *Experientia* 35: 363–364.

Schwind, J. L. 1933. Tissue specificity at the time of metamorphosis in frog larvae. *J. Exp. Zool.* 66: 1–14.

Sláma, K. and Williams, C. M. 1966. The juvenile hormone. V. The sensitivity of the bug, *Pyrrhocoris apterus*, to a hormonally active factor in American paper-pulp. *Biol. Bull.* 130: 235–246.

Smith-Gill, S. J. and Carver, V. 1981. Biochemical characterization of organ differentiation and maturation. *In* L. I. Gilbert and E. Frieden (eds.), *Metamorphosis: A Problem in Developmental Biology*. Plenum, New York, pp. 491–544.

Stone, B. L. and Thummel, C. S. 1993. *Drosophila* 78C early-late puff contains E78, an ecdysone-inducible gene that encodes a novel member of the nuclear hormone superfamily. *Cell* 75: 1–20.

Strathmann, R. R. 1971. The feeding behavior of planktotrophic echinoderm larvae: Mechanisms, regulation and rates of suspension feeding. *Exp. Mar. Biol. Ecol.* 6: 109–160.

Strathmann, R. R. 1975. Larval feeding in echinoderms. *Am. Zool.* 15: 717–730.

Strause, L. G., DeLuca, M. and Case, J. F. 1979. Biochemical and morphological change accompanying light organ development in the firefly, *Photuris pennsylvanica*. *J. Insect Physiol.* 125: 339–347.

Struhl, G. and Basler, K. 1993. Organizing activity of wingless protein in *Drosophila*. *Cell* 72: 527–540.

Taigen, T. L., Plough, F. H. and Stewart, M. M. 1984. Water balance of terrestrial anuran (*Eleutherodactylus coqui*) eggs: Importance of paternal care. *Ecology* 65: 248–255.

Takahashi. N., Yoshihama, K., Kikuyama, S., Yamamoto, K., Wakabayashi, K. and Kato, Y. 1990. Molecular cloning and nucleotide sequence of complementary DNA for bullfrog prolactin. *J. Mol. Endocrinol.* 5: 281–287.

Talbot, W. S., Swyryd, E. A. and Hogness, D. S. 1993. *Drosophila* tissues with different metamorphic responses to ecdysone express different ecdysone receptor isoforms. *Cell* 73: 1323–1337.

Tata, J. R., Kawahara, A. and Baker, B. S. 1991. Prolactin inhibits both thyroid hormone-induced morphogenesis and cell death in cultured amphibian larval tissues. *Dev. Biol.* 146: 72–80.

Taurog, A., Oliver, C., Porter, R. L., McKenzie, J. C. and McKenzie, J. M. 1974. The role of TRH in the neoteny of the Mexican axolotl (*Ambystoma mexicanum*). *Gen. Comp. Endocrinol.* 24: 267–279.

Thomas, H. E., Stunnenberg, H. G. and Stewart, A. F. 1993. Heterodimerization of the *Drosophila* ecdysone receptor with retinoid X receptor and *ultraspiracle*. *Nature* 362: 471–475.

Topper, Y. J. and Freeman, C. S. 1980. Multiple hormone interactions in the developmental biology of the mammary gland. *Physiol. Rev.* 60: 1049–1106.

Truman, J. W., Weeks, J. and Levine, R. B. 1985. Developmental plasticity during the metamorphosis of an insect nervous system. *In* M. J. Cohen and F. Strumwasser (eds.), *Comparative Neurobiology*. Wiley, New York, pp. 25–44.

Truman, J. W., Talbot, W. S., Fahrbach, S. E. and Hogness, D. S. 1994. Ecdysone receptor expression in the CNS correlates with stage-specific responses to ecdysteroids during *Drosophila* and *Manduca* development. *Development* 120: 219–234.

Turkington, R. W. 1968. Hormone-dependent differentiation of mammary gland in vitro. *Curr. Top. Dev. Biol.* 3: 199–218.

Turkington, R. W., Juergens, W. G. and Topper, Y. J. 1965. Hormone-dependent synthesis of casein *in vitro*. *Biochem. Biophys. Acta* 111: 573–576.

Turner, C. D. and Bagnara, J. T. 1976. *General Endocrinology*, 6th ed. Saunders, Philadelphia.

vanWijngaarden, R. and Bolanos, F. 1992. Parental care in *Dendrobates granuliferus* (Anura, Dendrobatidae) with a description of the tadpole. *J. Herpetol.* 26: 102–105.

Waddington, C. H. 1939. Preliminary notes on the development of wings in normal and mutant strains of Drosophila. *Proc. Natl. Acad. Sci. USA* 25: 299–307.

Wald, G. 1945. The chemical evolution of vision. *Harvey Lect.* 41: 117–160.

Wald, G. 1981. Metamorphosis: An overview. *In* L. I. Gilbert and E. Frieden (eds.), *Metamorphosis: A Problem in Developmental Biology*. Plenum, New York, pp. 1–39.

Wassersug, R. J. 1989. Locomotion in amphibian larvae (or why aren't tadpoles built like fish). *Am. Zool.* 29: 65–84.

Weber, R. 1965. Inhibitory effect of actinomycin D on tail atrophy in *Xenopus laevis* larvae at metamorphosis. *Experientia* 21: 665–666.

Weber, R. 1967. Biochemistry of amphibian metamorphosis. *In* R. Weber (ed.), *The Biochemistry of Animal Development*, Vol. 3. Academic Press, New York, pp. 227–301.

Wigglesworth, V. B. 1934. The physiology of ecdysis in *Rhodnius prolixus* (Hemiptera). II. Factors controlling moulting and metamorphosis. *Q. J. Microsc. Sci.* 77: 121–222.

Wigglesworth, V. B. 1939. *Principles of Insect Physiology*. Chapman and Hall, London.

Williams, C. M. 1952. Physiology of insect diapause. IV. The brain and prothoracic glands as an endocrine system in the *Cecropia* silkworm. *Biol. Bull.* 103: 120–138.

Williams, C. M. 1956. The juvenile hormone of insects. *Nature* 178: 212–213.

Williams, C. M. 1970. Hormonal interactions between plants and insects. *In* E. Sondheimer and J. B. Simeone (eds.), *Chemical Ecology*. Academic Press, New York, pp. 103–132.

Williams, J. A., Paddock, S. W. and Carroll, S. B. 1993. Pattern formation in a secondary field: A hierarchy of regulatory genes subdivides the developing *Drosophila* wing disc into discrete subregions. *Development* 117: 571–584.

Wray, G. A. and Raff, R. A. 1990. Novel origins of lineage founder cells in the direct-developing sea urchin, *Heliocidaris erythrogramma. Dev. Biol.* 141: 41–54.

Wray, G. A. and Raff, R. A. 1991.Rapid evolution of gastrulation mechanisms in a sea urchin with lecithotrophic larvae. *Evolution* 45: 1741–1750.

Yao, T.-P., Segraves, W. A., Oro, A. E., McKeown, M. and Evans, R. M. 1992. *Drosophila ultraspiracles* modulates ecdysone receptor function via heterodimer formation. *Cell* 71: 63–72.

Yaoita, Y. and Brown, D. D. 1990. A correlation of thyroid hormone receptor gene expression with amphibian metamorphosis. *Genes Dev.* 4: 1917–1924.

21

Sex determination

The mechanisms by which an individual's sex is determined has been one of the great questions of embryology since antiquity. Aristotle, who collected and dissected embryos, claimed that sex was determined by the heat of the male partner during intercourse. The more heated the passion, the greater the probability of male offspring. (Aristotle counseled elderly men to conceive in the summer if they wished to have male heirs.) Aristotle (ca. 335 B.C.E.) promulgated a very straightforward hypothesis of sex determination: women were men whose development arrested too early. The female was "a mutilated male" whose development stopped because the coldness of the mother's womb overcame the heat of the man's semen. Women were colder and more passive than men, and female sexual organs had not matured to the point where they could provide active seeds. This view was accepted by the Christian church and by Galen (whose anatomy texts were to be the standard for over 1000 years). Around the year 200, Galen wrote:

> just as mankind is the most perfect of all animals, so within mankind, the man is more perfect than the woman, and the reason for this perfection is his excess heat, for heat is Nature's primary instrument . . . the woman is less perfect than the man in respect to the generative parts. For the parts were formed within her when she was still a fetus, but could not because of the defect in heat emerge and project on the outside.

The view that women were but poorly developed men and that their genitalia were like men's, only turned inside out, was a very popular view for over a thousand years. Even in 1543, Andreas Vesalius, the Paduan anatomist who overturned much of Galen's anatomy (and who risked censure by the church for arguing that men and women have the same number of ribs) held this view. The illustrations from his two major works, *De Humani Corporis Fabrica* and *Tabulae Sex*, show that he saw the female genitalia as internal representations of the male genitals (Figure 21.1). Vesalius' books, though, sparked a revolution in anatomy, and by the end of the 1500s, anatomists dismissed Galen's representation of female anatomy. During the 1600s and 1700s, females were seen as producing eggs that could transmit parental traits, and the physiology of the sex organs began to be studied. Still, there was no consensus about how the sexes became determined (see Horowitz, 1976; Tuana, 1988; Schiebinger, 1989).

During that time, the environment—heat and nutrition, in particular—was believed to be important in determining sex. In 1890, Geddes and Thomson summarized all available data on sex determination and came to the conclusion that the "constitution, age, nutrition, and environment

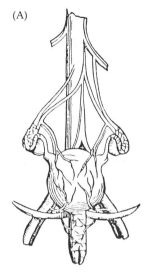
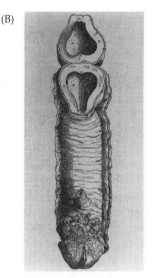

FIGURE 21.1
Vesalius' representations (1538, 1543) of the female reproductive organs. (A) Vesalius' rendering of Galen's conception of the female tract from the vagina to the uterus. (B) The female reproductive system. (Reprinted in Schiebinger, 1989.)

of the parents must be especially considered" in any such analysis. They argued that those factors favoring the storage of energy and nutrients predisposed one to have female offspring, whereas those factors favoring the utilization of energy and nutrients influenced one to have male offspring.

This environmental view of the determination of sex remained the only major scientific theory until the rediscovery of Mendel's work in 1900 and the rediscovery of the sex chromosome by McClung in 1902. Based on his knowledge of Mendelism, Correns speculated that the 1:1 sex ratio of most species could be achieved if the male was heterozygous and the female homozygous for some sex-determining factor. It was not until 1905, however, that the correlation (in insects) of the female sex with XX sex chromosomes and the male sex with XY or XO chromosomes was established (Stevens, 1905; Wilson, 1905). This suggested strongly that a specific nuclear component was responsible for directing the development of the sexual phenotype. Thus, evidence accumulated that sex determination occurs by nuclear inheritance rather than by environmental happenstance.

Today we find that both environmental and internal mechanisms of sex determination can operate in different species. We will first discuss the chromosomal mechanisms of sex determination and then consider the ways by which the environment regulates the sexual phenotype.

Chromosomal sex determination in mammals

Primary sex determination

Primary sex determination concerns the determination of the gonads. In mammals, the determination of sex is strictly chromosomal and is not usually influenced by the environment. In most cases, the female is XX and the male is XY. Every individual must have at least one X chromosome. Since the female is XX, each of her eggs has a single X chromosome. The male, being XY, can generate two types of sperm: half bear the X chromosome, half the Y. If the zygote receives another X chromosome from the sperm, the resulting individual forms ovaries and is female; if the zygote receives a Y chromosome from the sperm, the individual forms testes and is male. The Y chromosome carries a gene that encodes a testis-determining factor. This factor (probably the product of the *SRY* gene) organizes the gonad into a testis rather than an ovary. If the sperm contains neither an X nor Y, the resulting individual (XO) forms fibrous gonads that do not form either egg or sperm. Thus, the second X or the Y is needed to form and maintain the functional gonad. Unlike the case in *Drosophila* (discussed later), the mammalian Y chromosome is a crucial factor for determining sex in mammals. A person with five X chromosomes and one Y chromosome (XXXXXY) would be male. Furthermore, an individual with only a single X chromosome and no second X or Y develops as a female and begins making ovaries, but is unable to maintain the ovarian follicles.

Secondary sex determination

Secondary sex determination concerns the bodily phenotype outside the gonads. A male mammal has a penis, seminal vesicles, prostate gland, and often sex-specific size, vocal cartilage, and musculature. A female mammal has a vagina, cervix, uterus, oviducts, mammary glands, and often sex-specific size, vocal cartilage, and musculature. The secondary sexual characteristics are usually determined by hormones secreted from the gonads. However, in the absence of gonads, the female phenotype is generated. When Jost (1953) removed fetal rabbit gonads before they had differentiated, he found that the resulting rabbits were female, regardless of whether they were XX or XY. They all had oviducts, uteruses, and vaginas, and they lacked penises and male accessory structures.

The scheme of mammalian sex determination is shown in Figure 21.2.

FIGURE 21.2
Sequence of events leading to the formation of the sexual phenotypes in mammals. In the presence of the Y chromosome, the indifferent gonad is converted into a testis. The testis cells secrete the hormones that cause differentiation of the body in the male direction. In the absence of the Y chromosome, ovarian genes act to form the ovary and develop the female phenotype.

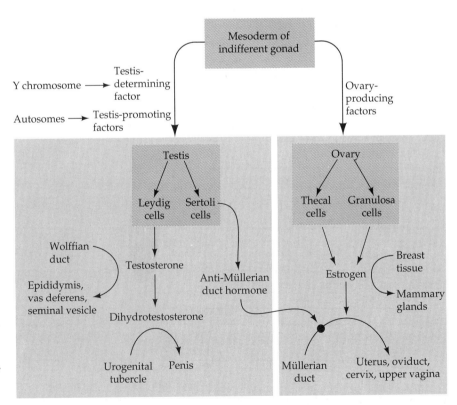

If no Y chromosome is present, the gonadal primordia develop into ovaries. The estrogenic hormones produced by the ovary enable the development of the **Müllerian duct** into the vagina, cervix, uterus, and oviducts. If the Y chromosome is present and the testis forms, it secretes two major hormones. The first hormone—**anti-Müllerian duct hormone** (often referred to as anti-Müllerian hormone, AMH, or Müllerian-inhibiting substance, MIS)—destroys the tissue that would otherwise become the uterus, oviduct, cervix, and upper portion of the vagina. The second hormone—**testosterone**—masculinizes the fetus, stimulating the formation of the penis, scrotum, and other portions of the male anatomy, as well as inhibiting the development of the breast primordia. Thus, the body has the female phenotype unless it is changed by the two hormones elaborated from the fetal testes. We will now take a more detailed look at these events.

The developing gonads

The development of gonads is a unique embryological situation. All other organ rudiments can normally differentiate into only one type of organ. A lung rudiment can become only a lung, and a liver rudiment can develop only into a liver. The gonadal rudiment, however, has two normal options. When it differentiates, it can develop into either an ovary or a testis. The type of differentiation taken by this rudiment determines the future sexual development of the organism. Before this decision is made, the mammalian gonad first develops through an **indifferent stage**, during which time it has neither female nor male characteristics. In humans, the gonadal rudiment appears in the intermediate mesoderm during week 4 and remains sexually indifferent until week 7. During this indifferent stage, the epithelium of the genital ridge proliferates into the loose connective mesenchymal tissue above it (Figure 21.3A,B). These epithelial layers form the sex cords, which will surround the germ cells that migrate into the human gonad during week 6. In both XY and XX gonads, the sex cords remain connected to the surface epithelium.

If the fetus is XY, then the sex cords continue to proliferate through the eighth week, extending deeply into the connective tissue. These cords fuse with each other, forming a network of internal (medullary) sex cords and, at its most distal end, the thinner **rete testis** (Figure 21.3C,D). Eventually, the testis cords lose contact with the surface epithelium and become separated from it by a thick extracellular matrix, the **tunica albuginea**. Thus, the germ cells are found in the cords *within* the testes. During fetal life and childhood, these cords remain solid. At puberty, however, the cords hollow out to form the seminiferous tubules, and the germ cells begin sperm production. The sperm are transported from the inside of the testis through the rete testis, which joins the **efferent ducts**. These efferent tubules are the remnants of the mesonephric kidney, and they link the testis to the Wolffian duct. This duct used to be the collecting tube of the mesonephric kidney. In males, the Wolffian duct differentiates to become the **vas deferens**, the tube through which the sperm pass into the urethra and out of the body. Meanwhile, during fetal development the interstitial mesenchymal cells of the testes have differentiated into **Leydig cells**, which make testosterone. The cells of the testis cords differentiate into **Sertoli cells**, which nurture the sperm and secrete the anti-Müllerian duct hormone.

In females, the germ cells will reside near the outer surface of the gonad. Unlike the male sex cords, which continue their proliferation, the initial sex cords of XX gonads degenerate. However, the epithelium soon

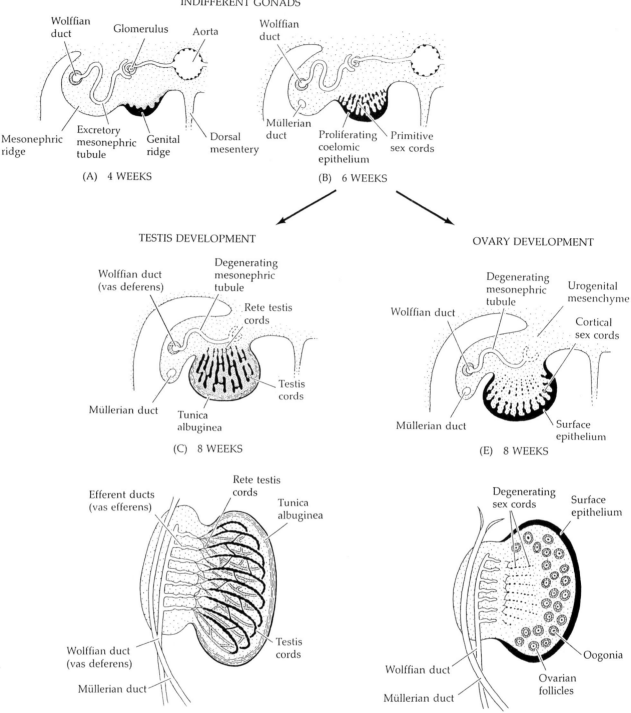

INDIFFERENT GONADS

(A) 4 WEEKS

Wolffian duct
Glomerulus
Aorta
Mesonephric ridge
Excretory mesonephric tubule
Genital ridge
Dorsal mesentery

(B) 6 WEEKS

Wolffian duct
Müllerian duct
Proliferating coelomic epithelium
Primitive sex cords

TESTIS DEVELOPMENT

(C) 8 WEEKS

Wolffian duct (vas deferens)
Degenerating mesonephric tubule
Rete testis cords
Müllerian duct
Tunica albuginea
Testis cords

(D) 16 WEEKS

Efferent ducts (vas efferens)
Rete testis cords
Tunica albuginea
Wolffian duct (vas deferens)
Müllerian duct
Testis cords

OVARY DEVELOPMENT

(E) 8 WEEKS

Degenerating mesonephric tubule
Urogenital mesenchyme
Cortical sex cords
Wolffian duct
Müllerian duct
Surface epithelium

(F) 20 WEEKS

Degenerating sex cords
Surface epithelium
Wolffian duct
Müllerian duct
Ovarian follicles
Oogonia

FIGURE 21.3
Differentiation of human gonads shown in transverse section. (A) Genital ridge of a 4-week embryo. (B) Genital ridge of a 6-week indifferent gonad showing primitive sex cords. (C) Testis development in the eighth week. The sex cords lose contact with the cortical epithelium and develop the rete testis. (D) By the sixteenth week of development, the testis cords are continuous with the rete testis and connect with the Wolffian duct. (E) Ovary development in an 8-week human embryo, as primitive sex cords degenerate. (F) The 20-week human ovary does not connect to the Wolffian duct, and new cortical sex cords surround the germ cells that have migrated into the genital ridge. (After Langman, 1981.)

produces a new set of sex cords, which do not penetrate deeply into the mesenchyme but stay near the outer surface (cortex) of the organ. Thus, they are called cortical sex cords. These cords are split into clusters, each cluster surrounding a germ cell (Figure 21.3E,F). The germ cell will become the ova and the surrounding epithelial sex cords will differentiate into the **granulosa cells**. The mesenchymal cells of the ovary differentiate into the **thecal cells**. Together, the thecal and granulosa cells form the **follicles** that envelop the germ cells and secrete steroid hormones. Each follicle will contain a single germ cell. In females, the Müllerian duct remains intact, and it will differentiate into the oviducts, uterus, cervix, and upper vagina; the Wolffian duct, deprived of testosterone, degenerates. A summary of the development of mammalian reproductive systems is shown in Figure 21.4.

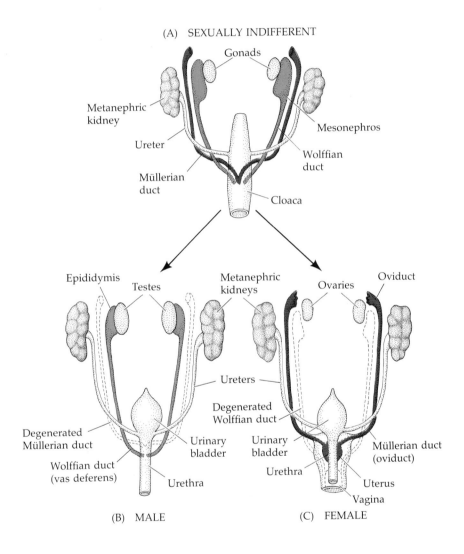

(A) SEXUALLY INDIFFERENT

Gonads

Metanephric kidney

Mesonephros

Ureter

Wolffian duct

Müllerian duct

Cloaca

Epididymis

Testes

Metanephric kidneys

Ovaries

Oviduct

Ureters

Degenerated Wolffian duct

Degenerated Müllerian duct

Urinary bladder

Urinary bladder

Müllerian duct (oviduct)

Wolffian duct (vas deferens)

Urethra

Urethra

Uterus

Vagina

(B) MALE

(C) FEMALE

GONADS		
Gonadal type	Testis	Ovary
Sex cords	Medullary (internal)	Cortical (external)
DUCTS		
Remaining duct for germ cells	Wolffian	Müllerian
Duct differentiation	Vas deferens, epididymis, seminal vesicle	Oviduct, uterus, cervix, upper portion of vagina

FIGURE 21.4

Summary of the development of gonads and their ducts in mammals. Note that both the Wolffian and Müllerian ducts are present at the indifferent-gonad stage. The regional development of the Wolffian ducts depends upon the mesenchyme that they encounter. Lower portions of the Wolffian duct that would normally form the epididymis will form seminal vesicle tissue if cultured with mesenchyme associated with the upper (seminal vesicle) portions of the duct (Higgins et al., 1989.)

Mammalian primary sex determination: Y-chromosomal genes for testis determination

Several genes have been found whose function is necessary for normal sexual differentiation. In *Drosophila* and *Caenorhabditis elegans*, genetic studies have identified numerous genes in the pathways leading to the sexual phenotypes. In mammals, where genetic methods are more difficult to employ, clinical studies have also identified several genes active in determining whether the organism is male or female.

In mice and humans, the presence of a portion of the Y chromosome is essential for initiating testis development, and the differentiation of the indifferent gonad into the testicular pathway appears to depend on the expression of Y chromosomal genes in the epithelial sex cord cells. This was shown by making chimeric mice from XX and XY blastomeres. When such XX/XY chimeras are made, they are usually fertile males (McLaren et al., 1984). The Leydig cells of these testes are composed of both XX and XY cells (showing that XX cells can form testicular structures), but the Sertoli cells are almost always derived from the XY embryo (Palmer and Burgoyne, 1991a; Patek et al., 1991). Thus, the critical event in the differentiation of testes appears to be the expression of the Y chromosome sex-determining genes in the Sertoli cell lineage. Since the Sertoli cells are the first type of testis cell to differentiate (Magre and Jost, 1980), it is possible that the Y chromosome sex-determining gene functions solely in these cells and all other events of testis formation follow as a consequence.

In humans, the gene for the testis-determining factor (*TDF*) resides on the short arm of the Y chromosome. Individuals who are born with the short arm but not the long arm of the Y chromosome are male, while individuals born with the long arm of the Y chromosome but not the short arm are female. Since 1985, the region of the short arm containing the *TDF* gene has been identified by combining clinical studies with DNA hybridization. DNA specific for the Y chromosome can be isolated from recombinant DNA libraries made from male cells by screening DNA prepared from XX and XY cells. Y-specific sequences will bind only to DNA prepared from XY cells but not from XX cells (Figure 21.5). By screening such Y-specific sequences on DNA from individuals missing parts of the Y chromosome, one can determine the portion of the Y chromosome from which that particular sequence is derived (Page, 1986).

In the human population, there exist both XX males (about 1 out of every 20,000 males) and XY females. The DNA hybridization results (Page et al., 1985; Vergnaud et al., 1986) demonstrated that XX males had Y-specific DNA from Y chromosome region 1 translocated onto one of their chromosomes. Moreover, all the XY females tested were missing this region of the Y chromosome (Figure 21.6). It seems most likely, then, that the gene for the testis-determining factor maps to region 1 of the Y chromosome, near the tip of the short arm. Similar evidence from mice also suggested that this region of DNA contains the testis-determining gene of the Y chromosome (Singh and Jones, 1982; Eicher et al., 1983).

By analyzing the DNA of XX men and XY women, the position of this testis-determining gene has been further narrowed down to a 35,000-base pair region of the Y chromosome located just before the pseudoautosomal end (Figure 21.7). In this region, Sinclair and colleagues (1990) found a sequence of male-specific DNA that could encode a peptide of 223 amino acids. Such a peptide would probably be a transcription factor, since it would contain a DNA binding domain called the **HMG box**. This HMG (high-mobility group) box is found in several transcription factors and

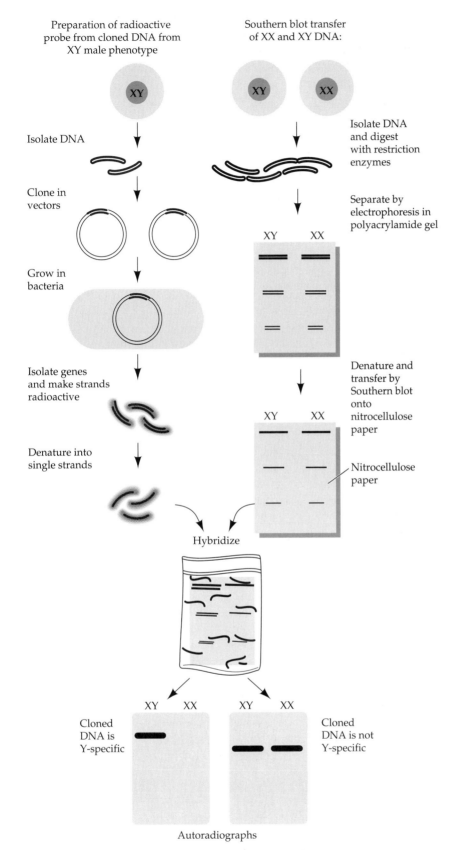

Preparation of radioactive
probe from cloned DNA from
XY male phenotype

Southern blot transfer
of XX and XY DNA:

Isolate DNA

Clone in
vectors

Grow in
bacteria

Isolate genes
and make strands
radioactive

Denature into
single strands

Hybridize

Cloned
DNA is
Y-specific

Cloned
DNA is not
Y-specific

Autoradiographs

Isolate DNA
and digest
with restriction
enzymes

Separate by
electrophoresis in
polyacrylamide gel

Denature and
transfer by
Southern blot
onto
nitrocellulose
paper

Nitrocellulose
paper

FIGURE 21.5
Protocol for screening Y-specific DNA
clones. DNA from male cells is cloned
into viruses and grown in *E. coli*. Each
clone is separately cultured, its human
DNA is isolated, and its strands are
made radioactive. Each DNA fragment
is denatured into single strands and
hybridized with DNA from male or
female cells that has been electrophoresed and Southern blotted. The Y-
specific DNA fragments should bind
specifically to the DNA from XY cells
but not to that from XX cells.

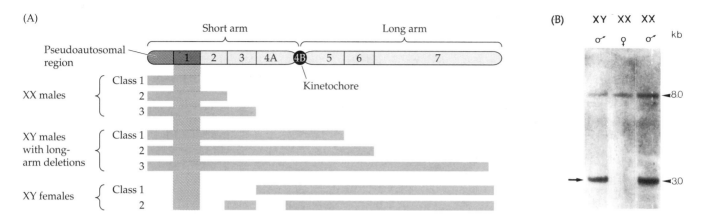

FIGURE 21.6

Deletion mapping of the gene for the testis-determining factor(s) in humans. (A) The Y chromosome has been divided into nine intervals. Interval 4B is the kinetochore, and the stippled region is the "pseudoautosomal" portion, which contains DNA sequences in common with the X chromosome (and allows it to pair with the X during meiosis). XX males have DNA from region 1 whereas XY females have a deletion of this material from their Y chromosomes. As these regions are often too small to be seen microscopically, the presence or absence of the region was determined by DNA hybridization (as in Figure 21.5). Results of such a hybridization are shown in B, where the radioactive DNA was from region 1 of the Y chromosome. This Y-specific DNA is thought to contain the TDF, since the autoradiograph of the Southern blot demonstrates its presence in XY males and XX males, but not in XX females. (A after Page, 1985; B from Vergnaud et al., 1986, courtesy of D. Page.)

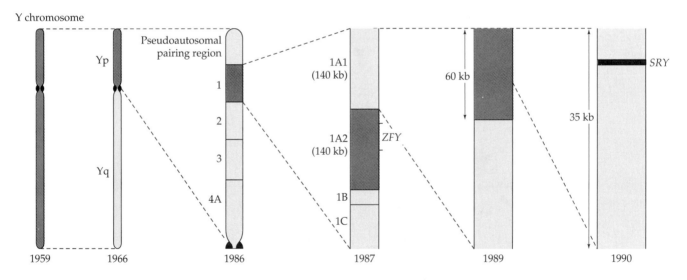

FIGURE 21.7

In 1959, the Y chromosome was shown to contain the testis determining factor in both humans and mice. By 1966, translocations and deletions of human chromosomes indicated that the TDF resides in the short arm of the Y chromosome (Yp). In the late 1980s, deletion mapping with male-specific DNA probes showed that the gene(s) resided in region 1 of the small arm of the Y chromosome and it was thought that *ZFY* was the testis-determining factor. In 1990, studies of XY females, XX males, and transgenic mice localized the TDF to a single region on the Y chromosome containing the *SRY* gene. (After McLaren, 1990.)

nonhistone chromatin proteins, and it induces bending in the region of DNA to which it binds (Giese et al., 1992). This DNA was called the **SRY** (sex-determining region of the Y) gene, and there is evidence that it is indeed the human testis determining factor gene. It is found in XY males and in the rare XX males, and it is absent from normal XX females and from many of the XY females. Another group of XY females were found to have point or frameshift mutations in the *SRY* gene (Berta et al., 1990; Jäger et al., 1990). Mobility shift assays (Harley et al., 1992) show that the DNA binding activity is important for sex determination, and XY females with point mutations in the *SRY* gene form an SRY protein that does not bind well to DNA. At least two genes involved in secondary sex determination (the genes for AMF and the P450 aromatase involved in steroid synthesis) contain SRY binding sites in their promoters (Haqq et al., 1993).

If *SRY* were actually to encode the major testis-determining factor, one would expect that it would act in the genital ridge immediately before or during testis differentiation. This prediction has been met in studies of the homologous gene found in mice. The mouse gene (called *Sry*) also correlates with the presence of testes; it is present in XX males and absent in XY females. The *Sry* gene is expressed in the somatic cells of the indifferent mouse gonad immediately before or during its differentiating into a testis. This is the only time and place this gene is seen to be expressed in the mouse embryo (Gubbay et al., 1990; Koopman et al., 1990). The most impressive evidence for *Sry* being the gene for testis-determining factor comes from transgenic mice. If *Sry* were acting to induce testes, then inserting *Sry* DNA into the genome of a normal XX mouse zygote should cause that XX mouse to form testes. Koopman and colleagues (1991) took the 14-kilobase region of DNA that included the *Sry* gene (and presumably its regulatory elements) and microinjected this sequence into the pronuclei of newly fertilized mouse zygotes. In several instances, the XX embryos injected with this *Sry*-containing sequence developed testes, male accessory organs, and penises (Figure 21.8). (Functional sperm were not formed; but they were not expected because the presence of two X chromosomes prevents sperm formation in XXY mice and men). Therefore, there are good reasons to think that *Sry/SRY* is the major gene on the Y chromosome for testis determination in mammals.*

*It is useful to remember that we are discussing the determination of the testis, not of males *per se*. Although the formation of the testis leads to the formation of the male phenotype, other genes are also involved in "male sex determination," Other genes on the Y chromosome, for instance, are critical for sperm formation. As Ursula Mittwoch (1992) has noted, "Spermiogenesis is an extremely complex process and preparing for it may be particularly arduous for eutherian males, who undergo their early development in a female hormonal environment. The timely interaction of multiple genes leading to the early differentiation of a competent testis appears to be one of the requirements for success."

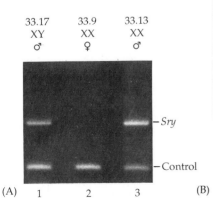

(A)   1   2   3

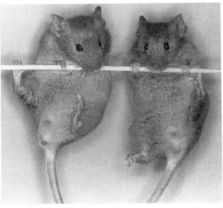

(B)

FIGURE 21.8

XX mouse transgenic for *Sry* is male. (A) Polymerase chain reaction showing the presence of *Sry* gene in normal XY males and a transgenic XX *Sry* mouse. The gene is absent in the female XX littermate. (B) The external genitalia of that transgenic mouse was male (right) and essentially the same as an XY male (left). (From Koopman et al., 1991, photograph courtesy of the authors.)

Mammalian primary sex determination: Autosomal genes in testis determination

The testis-determining gene of the Y chromosome is necessary but not sufficient for the development of the mammalian testis. While it is not yet known what genes are being regulated by the testis-determining gene on the Y chromosome, studies on mice have shown that the Y chromosome gene has to be coordinated with certain autosomal genes. Eicher and Washburn (1983) discovered that when the Y chromosome of *Mus domesticus* was bred into a particular inbred laboratory mouse strain, ovarian tissue developed in the XY individuals (Figure 21.9). Half of these XY mice were completely sex-reversed (having only ovarian tissue as gonads), and the other half of this group contained both testicular and ovarian tissue (Eicher and Washburn, 1983, 1986). The testis-determining gene of the Y chromosome of each strain of mouse can cooperate with its own autosomal testis-determining gene (*Tda-1*) but not necessarily with that of another. It is possible that the *time* the gene products are expressed differs between the two strains, which would mean they would not be present at the same time if the chromosome were from the other strain (Palmer and Burgoyne, 1991b).

Another autosomal sex-determining gene is the *T-associated sex reversal* gene (*Tas*). This gene segregates with the *T/t* gene complex on chromosome 17 of *Mus musculus*, and different alleles of this gene have been found in different inbred strains of mice (Washburn and Eicher, 1983, 1989). The C57 and AKR strains of *M. musculus* both form males about 50 percent of the time. However, the AKR Y chromosome is unable to direct testis morphogenesis if chromosome 17 comes from the C57 strain. In such cases, normal testicular determination is disrupted and each gonad contains a mixture of ovarian and testicular tissue. It is not known how these genes interact with the *Sry* gene product.

In humans, the need for testis-determining genes on chromosomes other than the Y has been shown by the occurrence of extremely rare XY,SRY$^+$ females and XX,SRY$^-$ males. McElreavey and colleagues (1993) postulate the existence of a gene, Z, whose activity inhibits testis differ-

FIGURE 21.9
Ovarian tissue in XY gonads when a *Mus domesticus* Y chromosome is in the genome of the C57BL strain of laboratory mouse. Gonads and attached mesonephros were dissected from 14.5- to 16-day-old fetuses. (A) Normal ovary from XX fetus. (B) Testis from normal C57BL male fetus. (C) Ovary from XY female. (D) "Ovotestis" of XY fetus containing areas of ovarian and testicular tissue. (From Eicher et al., 1982, courtesy of E. Eicher and L. Washburn.)

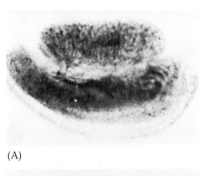
(A)

(B)

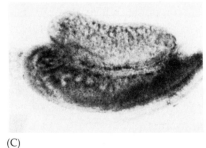
(C)

(D)

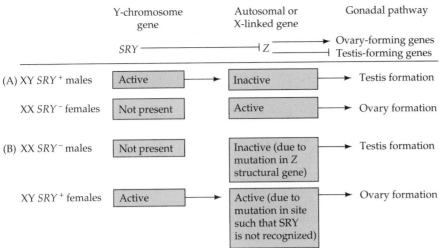

FIGURE 21.10

Hypothetical model of human primary sex determination. The *SRY* gene inactivates the expression of the Z gene product, whose activity suppresses testicular development and stimulates ovarian development. (A) Wild-type males and females. (B) Postulated mutations in the Z gene would account for XX,SRY⁻ males and XY,SRY⁺ females. (After McElreavey et al., 1993.)

entiation and stimulates the formation of ovaries. The SRY protein would inhibit the Z gene (or its product), thereby allowing testicular development (Figure 21.10A). In XX males lacking SRY, the Z gene is postulated to be mutated in such a way that no functional product is formed. Thus, the testicular phenotype would occur. XY females with SRY might be explained if the Z gene (or its product) were insensitive to the inhibitory effects of the SRY protein. The Z gene would then be active and the ovarian phenotype would be expressed (Figure 21.10B). It should be realized that both testis and ovary development are active processes. In mammalian primary sex determination, neither is a "default state" (Eicher and Washburn, 1986). Although remarkable progress has been made in recent years, we still do not know what the testis- or ovary-determining genes are doing, and the problem of primary sex determination remains (as it has since prehistory) one of the great unsolved problems of biology.

Secondary sex determination in mammals

Hormonal regulation of the sexual phenotype

Primary sex determination involves the formation of either an ovary or a testis from the indifferent gonad. This, however, does not give the complete sexual phenotype. Secondary sex determination concerns the development of the female and male phenotypes from the hormones elaborated by the ovaries and testes. Both female and male secondary sex determination have two major temporal components. The first occurs within the embryo during organogenesis; the second occurs during adolescence.

As mentioned earlier, if the indifferent gonads are removed from an embryonic animal, the female phenotype is realized. The Müllerian ducts develop while the Wolffian duct decays. This is also seen in certain humans who are born without functional gonads. Individuals whose cells have only one X chromosome (and no Y chromosome) originally develop ovaries, but these ovaries atrophy before birth and the germ cells die before

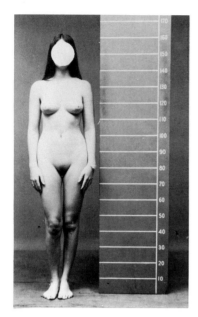

FIGURE 21.11
An XY individual with androgen insensitivity syndrome. Despite the XY karyotype and the presence of testes, the individual develops female secondary sex characteristics. Internally, however, the woman lacks the Müllerian duct derivatives. (Courtesy of C. B. Hammond.)

puberty. Under the influence of estrogen derived from the mother and from the placenta, these infants are born with a female genital tract (Langman and Wilson, 1982).*

The formation of the male phenotype involves the secretion of testicular hormones that promote Wolffian duct development and cause the Müllerian duct to atrophy. The first of these hormones is the anti-Müllerian duct hormone—the Sertoli cell hormone that causes the degeneration of the Müllerian duct. The second of these hormones is the steroid testosterone, which is secreted from the testicular Leydig cells. This hormone causes the Wolffian duct to differentiate into the epididymis, vas deferens, and seminal vesicles, and it causes the urogenital swellings and sinus to develop into the scrotum and penis. The existence of these two independent systems of masculinization is demonstrated by people having **androgen insensitivity syndrome**. These XY individuals have the testis-determining factor gene and so have testes that make testosterone and AMH. However, these people lack the testosterone receptor protein and therefore cannot respond to the testosterone made in their testes (Meyer et al., 1975). They are able to respond to estrogen made in their adrenal glands, so they are distinctly female in appearance (Figure 21.11). However, despite their female appearance, these individuals do have testes, and even though they cannot respond to testosterone, they do respond to AMH. Thus, their Müllerian ducts degenerate. These people develop as normal but sterile women, lacking uteruses and oviducts.[†]

So there are two distinct masculinizing hormones, testosterone and AMH. There is evidence, though, that testosterone might not be the active hormone in certain tissues. In 1974, Siiteri and Wilson showed that testosterone is converted to **5α-dihydrotestosterone** in the urogenital sinus and swellings, but not in the Wolffian duct. In the Dominican Republic, Imperato-McGinley and her colleagues (1974) found a small community in which several inhabitants had a genetic deficiency of the enzyme 5α-ketosteroid reductase, the enzyme that converts testosterone to dihydrotestosterone. Affected individuals lack a functional gene for this enzyme (Andersson et al., 1991; Thigpen et al., 1992). Although these XY individuals have functioning testes, they have a blind vaginal pouch and an enlarged clitoris. They appear to be girls and are raised as such. Their internal anatomy, however, is male: testes, Wolffian duct development, and Müllerian duct degeneration. Thus, it appears that the formation of external genitalia is under the control of dihydrotestosterone, whereas the Wolffian duct differentiation is controlled by testosterone itself (Figure 21.12). Interestingly enough, the external genitalia become responsive to testosterone at puberty, causing obvious masculinization in a person originally thought to be a girl.

*The mechanisms by which estrogen promotes the differentiation of the Müllerian duct are not well understood. During embryonic development, the duct is extremely sensitive to estrogenic compounds, as is known from the teratogenic effects of diethylstilbesterol (DES). This compound is a synthetic estrogen that was given to women from the 1940s through the 1960s to help maintain pregnancies. Daughters born to women who used this drug have a high incidence of Müllerian duct anomalies, including malformation of the vaginal and cervical epithelia, structural abnormalities of the oviducts and uterus, and a higher than normal incidence of vaginal cancer (Robboy et al., 1982; Bell, 1986).

[†]Androgen insensitivity syndrome is one of several conditions called *pseudohermaphroditism*. True hermaphrodites (rare in humans and most mammals, but the norm in some invertebrates) contain both male and female gonads. True hermaphrodites can occur when the Y chromosome is translocated to the X chromosome. If the translocated X is inactivated, the Y will be shut off. Thus, some of the gonadal cells will be XX and others will be XY (Berkovitz et al., 1992). Thus, mammalian hermaphroditism involves primary sex determination. In a pseudohermaphrodite condition, there is only one type of gonad, but the secondary sexual characteristics differ from what is expected from the gonadal sex.

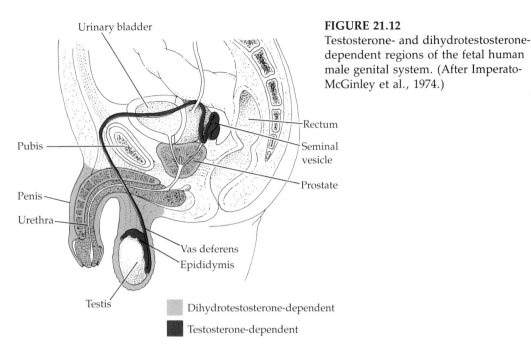

FIGURE 21.12
Testosterone- and dihydrotestosterone-dependent regions of the fetal human male genital system. (After Imperato-McGinley et al., 1974.)

Urinary bladder

Rectum

Seminal vesicle

Pubis

Prostate

Penis

Urethra

Vas deferens

Epididymis

Testis

Dihydrotestosterone-dependent

Testosterone-dependent

The anti-Müllerian duct hormone (AMH) is a 560-amino acid glycoprotein (Cate et al., 1986) made in the Sertoli cells (Tran et al., 1977). When fragments of fetal testes or isolated Sertoli cells are placed adjacent to cultured segments containing portions of the Wolffian and Müllerian ducts, the Müllerian duct atrophies even though no change occurs in the Wolffian duct (Figure 21.13). This atrophy is caused both by cell death and by the epithelial cells of the duct becoming mesenchymal and migrating away (Trelstad et al., 1982). The Müllerian duct is sensitive to the action of this hormone for only a brief period of time (Figure 21.14), and the testes stop making this factor at or near the time of birth.* The mouse *AMH* gene has a promoter sequence that binds Sry protein, so it is likely that Sry directly activates transcription of the *AMH* gene (Zwingman et al., 1993).

We see, then, that once the testes are formed, they elaborate two

*In birds, AMH is able to alter *primary* sex determination by changing the steroid production of the ovary. Chick sex is generally determined by the chromosomes, but it remains labile to hormonal influences while it is in the indifferent stage. If an inhibitor of the enzyme P_{450} aromatase is added to the chick embryo prior to day 7, testosterone is made instead of estrogen, and 100 percent of the chickens hatched become roosters with functional, sperm-producing testes (Elbrecht and Smith, 1992). Although there is evidence that AMH promotes testis determination in some mammals (Vigier et al., 1987, 1989; McLaren, 1988), it probably does not do so in humans. Some XY men exist who have normal testicular development, but no functional AMH (Knebelmann et al., 1991). Such men retain the Müllerian duct system.

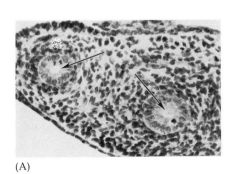

(A)

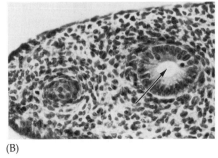

(B)

FIGURE 21.13
Assay for anti-Müllerian duct hormone activity in the anterior segment of a 14.5-day fetal rat reproductive tract, after 3 days in culture. (A) Both the Müllerian duct (arrow at left) and Wolffian duct (arrow at right) are open. (B) The Wolffian duct (arrow) is open, but the Müllerian duct has degenerated and closed. (Courtesy of N. Josso.)

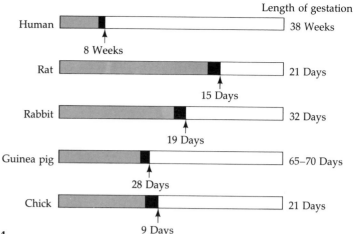

FIGURE 21.14

Sensitivity of Müllerian ducts to AMH. The shaded area represents the duration of the indifferent gonad. The black zone represents the period during which the Müllerian ducts are responsive to AMH. (After Josso et al., 1977.)

hormones that cause the masculinization of the fetus. One of these hormones—testosterone—may be converted into a more active form by the tissues that create the external genitalia. In females, estrogen secreted from the fetal ovaries appears sufficient to induce the differentiation of the Müllerian duct into the uterus, oviducts, and cervix. In such a manner, the sex chromosomes control the sexual phenotype of an individual.

The central nervous system

One of the most controversial areas of secondary sex determination involves the development of sex-specific behaviors. In songbirds, testosterone is seen to regulate the growth of male-specific neuronal clusters in the brain. Male canaries and zebra finches sing eloquently, whereas females sing little, if ever. These songs are used to mark territories and to attract mates. The ability to sing is controlled by six different clusters of neurons (nuclei) in the avian brain (Figure 21.15). Neurons connect each of these regions to one another. In male canaries, these nuclei are several times larger than the corresponding cluster of neurons in female canaries; and in zebra finches, the females may lack one of these regions entirely (Arnold, 1980; Konishi and Akutagawa, 1985).

Testosterone plays a major role in song production. In adult male zebra finches, Pröve (1978) demonstrated a linear correlation between the amount of song and the concentration of serum testosterone. It has been

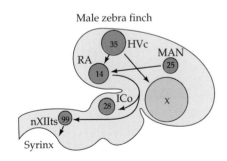

Male zebra finch

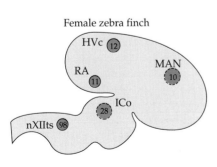

Female zebra finch

FIGURE 21.15

Sexual dimorphism in the avian brain. Schematic diagram indicates the major neural areas thought to be involved in the production of bird song in the zebra finch. Circles represent specific brain areas; the size of each circle is proportional to the volume occupied by that region. Circles with dashed lines are estimated volumes. The numbers within each circle represent the percentage of cells therein that incorporate radiolabeled testosterone. The volume differences in three of these regions (HVc, RA, and nXIIts) are significant between the sexes, and area X has not been observed in the brains of female finches. The differences in testosterone binding in regions HVc and MAN are significant, and no sex differences in steroid hormone binding have been observed in other regions of the brain. The arrows indicate the axonal paths connecting the regions in the male finch. (After Arnold, 1980.)

shown that the seasonal changes in testosterone levels are correlated with the seasonal singing patterns of these birds. When testosterone levels are low, there is not only a decrease in bird song but also a decrease in the size of the male-specific brain nuclei (Nottebohm, 1981). In adult chaffinches, castration eliminates song, but injection of testosterone induces such birds to sing even in November, when they are normally silent (Thorpe, 1958). In several species of birds, the females can be induced to sing by injecting them with testosterone (Nottebohm, 1980). The four song-controlling regions of the brain grow 50–69 percent in such birds, whereas other brain regions show no such growth. Autoradiographic studies (Arnold et al., 1976) have shown that the neurons of the song-controlling nuclei incorporate radioactive testosterone, whereas other regions of the brain do not. It is apparent, then, that gonadal hormones can play a major role in the development of the regions of the nervous system that generate sex-specific behaviors.

The situation is not as clear in mammals, for there are fewer behaviors that exclusively characterize one sex. Penile thrusting in rats is one such behavior, and it is controlled by motor neurons to the levator ani and bulbocavernosus muscles. Both these neurons originate from a spinal nucleus that can specifically concentrate testosterone. In female rats, these muscles are vestigial, and the volume of the controlling neurons is greatly reduced (Breedlove and Arnold, 1980; Tobin and Joubert, 1992). Testosterone appears to cause two types of changes in these responsive neurons. In fetal and newborn rats, testosterone prevents a "normally" occurring death of neurons in this region. Female rats lose up to 70 percent of the neurons of this spinal nucleus, while male rats lose only 25 percent. In adult rats, testosterone acts upon this nucleus to maintain the size of the nerve cells and their dendrites. The soma area and dendrite length of this spinal nucleus are reduced by half when an adult rat is castrated. This reduction is reversed by the injection of testosterone (Nordeen et al., 1985; Kurz et al., 1986).

Testosterone is not the only steroid capable of mediating behavior. In the mammalian brain, estrogen-sensitive neurons are also seen. These neurons are located (Figure 21.16) at positions in neural circuits that are known to mediate reproductive behavior: the hypothalamus, the pituitary, and the amygdala (McEwen, 1981). Pfaff and McEwen (1983) demonstrated that estrogen alters the electrical and chemical features of those hypotha-

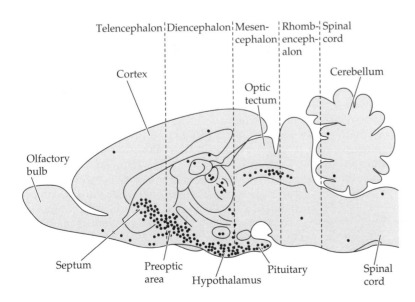

FIGURE 21.16
Diagrammatic representation of the estrogen-binding regions in a female rat brain. (After Kandel and Schwartz, 1985.)

lamic neurons capable of binding estrogen in their chromatin. Terasawa and Sawyer (1969) previously had found that the electrical activity of these neurons varies during the seasonal estrogen cycle of the rat, becoming elevated at the time of ovulation. Moreover, estrogen appears to stimulate those neurons in the regions that induce female reproductive behavior. Ovariectomized rats given estrogen injections directly to the hypothalamus displayed **lordosis**, a position that stimulates the male to mount them, whereas control ovariectomized rats did not show that behavior (Barfield and Chen, 1977; Rubin and Barfield, 1980). The mechanism by which estrogen causes the specific neuronal activity at these times is thought to involve increasing the permeability of these neurons to potassium (Nabekura et al., 1986).

SIDELIGHTS & SPECULATIONS

The development of sexual behaviors

The organization/activation hypothesis

Does prenatal (or neonatal) exposure to particular hormones impose permanent sex-specific changes in the central nervous system? Such sex-specific neural changes have been shown in those regions of the brain that regulate "involuntary" sexual physiology. The cyclical secretion of luteinizing hormone by the adult female rat pituitary is dependent on the lack of testosterone during the first week of its life. The luteinizing hormone secretion of female rats can be made noncyclical by giving them testosterone at four days after birth; conversely, the luteinizing hormone secretions of males can be made cyclical by removing their testes within a day of their birth (Barraclough and Gorski, 1962). It is thought that sex hormones can act during the fetal or neonatal stage of a mammal's life to organize the nervous system, and that during adult life the same hormones may have transitory, activational, effects. This is called the **organization/activation hypothesis**.

Interestingly, the hormone chiefly responsible for determining the male brain pattern is estradiol.* Testosterone in fetal or neonatal blood can be converted into estradiol by P450 aromatase, and this conversion occurs in the hypothalamus and limbic system—two areas of the brain known to regulate reproductive and hormonal behaviors (Reddy et al. 1974; McEwen et al., 1977). Thus, testosterone is able to cause its effects by being converted into estradiol. But the fetal environment is rich in estrogens from the gonads and placenta. What stops the estrogens from masculinizing the nervous system of a female fetus? Fetal estrogen (both male and female) is bound by α-fetoprotein; this protein is made in the fetal liver and becomes a major component of the fetal blood and cerebrospinal fluid. It will bind estrogen but not testosterone.

*The terms *estrogen* and *estradiol* are often used interchangeably. However, estrogen refers to a class of steroid hormones responsible for establishing and maintaining specific female characteristics. Estradiol is one of these hormones, and in most mammals (including humans) it is the most potent of the estrogens.

Attempts to extend the organization/activation hypothesis to "voluntary" sexual behaviors are more controversial. This is due to there being no truly sex-specific behavior that distinguishes the two sexes of many mammals, and to the multiple effects of hormone treatment upon the developing mammal. For instance, injecting testosterone into a week-old female rat will increase her pelvic thrusts and diminish her amount of lordosis (Phoenix et al., 1959; Schwartz et al. 1993). These changes can be ascribed to testosterone-mediated changes in the central nervous system, but can also be due to hormonal effects on other tissues. Testosterone enables the growth of the muscles that allow pelvic thrusting. And, since testosterone causes females to grow larger and to close their vaginal orifices, one cannot conclude that the lack of lordosis is due solely to testosterone-mediated changes in the neural circuitry (Harris and Levine, 1965; De Jonge, 1988; Moore, 1990; Moore et al., 1992).

Extrapolating from rats to humans is a very risky business, as no sex-specific behavior has yet been identified in humans, and what is "masculine" in one culture may be considered "feminine" in another (see Jacklin, 1981; Bleier, 1984; Fausto-Sterling, 1992). As one review (Schwartz et al., 1993) concludes, "there is ample evidence that the neural organization of reproductive behaviors, while biased by hormonal events during a critical prenatal period, does not exert an immutable influence over adult sexual behavior or even over an individual's sexual orientation. Within the life of an individual, religious, social, or economic motives can prompt biologically similar persons to diverge widely in their sexual habits."

Male homosexuality

Certain behaviors are often said to be part of the "complete" male or female phenotypes. That is, the brain of a mature man is said to be formed such that it causes the man to desire mating with a mature woman, and the brain of a mature woman causes her to desire to mate with a man. However, as important as these desires are in our lives, they cannot be detected by in situ hybridization or isolated by monoclonal antibodies. We do not yet know if these desires are instilled into us by our social education or are "hardwired" in our brains by genes or hormones during our intrauterine development.

In 1991, Simon LeVay proposed that part of the anterior hypothalamus of homosexual men has the anatomical form typical of women and not of heterosexual men. The hypothalamus is thought to be the source of our sexual urges, and rats have a sexually dimorphic area in their anterior hypothalamus that appears to regulate sexual behavior. Thus, this study generated a great deal of publicity and discussion. The major results are shown in Figure 21.17. The interstitial nuclei of the anterior hypothalamus (INAH) were divided into four regions. Three of them showed no signs of sexual dimorphism. One of them, INAH3, showed a significant statistical difference between males and females; it was claimed that the male INAH3 is, on average, more than twice as large as the female INAH3. Moreover, LeVay's data suggested that the INAH3 of homosexual males was similar to that of the females and is less than half the size of the heterosexual mens' INAH3. This finding, he claimed, "suggests that sexual orientation has a biological substrate."

While this study can be read in that manner, there have been several criticisms of this interpretation. First, the data are from populations, not individuals. One can also say that there is a statistical range, and that men and women have the same general range. Indeed, one of the INAHs from a "homosexual male" was greater than all but one of the 16 "heterosexual males." Second, the "heterosexual men" were not necessarily heterosexual, nor were the "homosexual men" necessarily homosexual; the brains came from corpses whose sexual preferences were not known. This brings up another issue: homosexuality has many forms, and is thus not a phenotype in the usual sense. Third, the brains of the "homosexual men" were taken from patients who had died of AIDS. AIDS affects the brain, and its effect on the hypothalamic neurons is not known. Fourth, as the study was done on the brains of dead subjects, one cannot infer cause and effect. As mentioned in Chapter 1, such data show only correlations, not causation. It is as likely that behaviors can affect regional neuronal density size as the regional neuronal density can affect behaviors. If one interprets the data as indicating that the INAH3 of male homosexuals is smaller than that of male heterosexuals, one still does not know whether that is a result of homosexuality or a cause. Fifth, given the interpretation of the data that says the difference exists, it does not say whether the difference has anything to do with sexuality. Sixth, these studies do not indicate when such differences (if they exist) emerge. The question of whether differences in male, female, and homosexual male INAH3 occur during embryonic development or whether they occur shortly after birth, during the first few years of life, during adolescence, or at some other time was not addressed.

In 1993, a correlation was made between a particular DNA sequence on the X chromosome and a particular subgroup of male homosexuals (those homosexual men

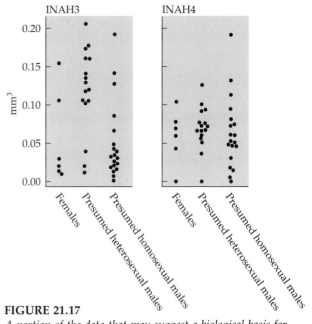

FIGURE 21.17
A portion of the data that may suggest a biological basis for homosexuality. INAH4 and INAH3 are two groups of hypothalamic neurons. INAH4 shows no sexual dimorphism in volume, while INAH3 shows a statistically significant clustering, although the range is similar. The INAH3 from autopsies of "homosexual" male brains clusters toward the female distribution. However, no cause and effect relationship can be posited. (After LeVay, 1991.)

who had a homosexual brother). Out of 40 pairs of homosexual brothers who inherited a particular region of the X chromosome from their mothers, 33 of them had brothers who also inherited this region (Hamer et al., 1993). One would have expected only 20 of them to have done so, on average. Again, this is only a statistical concordance, and one that could be coincidental. Moreover, the control (seeing if the same marker exists in the "non-homosexual" males of these families) was not reported, and the statistical bias of the observations has been called into question (Risch et al., 1993). No gene has been cloned from this region (which is large enough to carry hundreds of genes), and no mechanism has been proposed as to how it might lead to homosexuality. Despite the reports of this paper in the public media, no "gay gene" has been found.

It is worth remembering that genes encode RNAs and proteins, not behaviors. While genes may bias behavioral outcomes, we have no evidence for their "controlling" them. The existence of people with "multiple personality syndrome" indicates that one genotype can support a wide range of personalities. This is certainly a problem with any definition of a "homosexual phenotype," since many people can alternate between homosexual and heterosexual behavior. Thus, while these results are provocative, the question as to whether homosexual desires are formed by genes within the nucleus, by sex hormones during fetal development, or by experiences after birth is still an open question.

Chromosomal sex determination in *Drosophila*

The sexual development pathway

Both mammals and insects such as *Drosophila* use an XX/XY system of sex determination. However, the sex-determining mechanisms in these two groups are very different. In mammals, the Y chromosome plays a pivotal role in determining the male sex. Thus, XO mammals are *females*, with ovaries, uteruses, and oviducts (but usually very few, if any, ova). In *Drosophila*, sex determination is achieved by a balance of female determinants on the X chromosome and male determinants on the autosomes (non-sex chromosomes). If there is but one X chromosome in a diploid cell (1X:2A), the organism is male. If there are two X chromosomes in a diploid cell (2X:2A), the organism is female (Bridges, 1921, 1925). Thus, XO *Drosophila* are sterile *males*. Table 21.1 shows the different X to autosome ratios and the resulting sex.

In *Drosophila*, and in insects in general, one can observe **gynandromorphs**—animals in which certain regions are male and other regions are female (Figure 21.18; Plate 17). This can happen when an X chromosome is lost from one embryonic nucleus. The progeny of that nucleus, instead of being XX (female), are XO (male). Because there are no sex hormones in insects to modulate such events, each cell makes its own sexual "decision." The XO cells display male characteristics, whereas the XX cells display female traits. This situation provides a beautiful example of the association between X chromosomes and sex. As can be seen from this example, the Y chromosome plays no role whatever in *Drosophila* sex determination. Rather, it is necessary only to ensure fertility in males. The Y chromosome is active only late in development, during sperm formation.

Any theory of *Drosophila* sex determination must explain how the X:autosome ratio is read and how this information is transmitted to the genes controlling the male or female phenotypes. Although we do not yet know the intimate mechanisms by which the X:A ratio is made known to the cells, research in the past five years has revolutionized our view of *Drosophila* sex determination. Much of this research has concerned the identification and analysis of genes that are necessary for sexual differentiation and the placing of those genes in a developmental sequence. Loss-of-function mutations in most of these genes—*Sex-lethal* (*Sxl*), *transformer* (*tra*), and *transformer-2* (*tra-2*)—transform XX individuals into males. Such mutations have no effect on male sex determination. Homozygosity

TABLE 21.1
Ratios of X chromosomes to autosomes in different sexual phenotypes in *Drosophila melanogaster*

X chromosomes	Autosome sets (A)	X:A ratio	Sex
3	2	1.50	Metafemale
4	3	1.33	Metafemale
4	4	1.00	Normal female
3	3	1.00	Normal female
2	2	1.00	Normal female
2	3	0.66	Intersex
1	2	0.50	Normal male
1	3	0.33	Metamale

Source: After Strickberger, 1968.

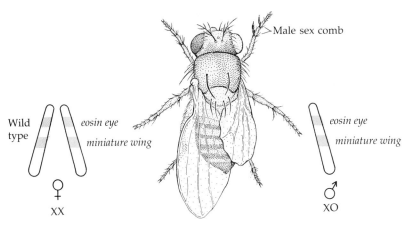

FIGURE 21.18
Gynandromorph of *D. melanogaster* in which the left side is female (XX) and the right side is male (XO). The male side has lost an X chromosome bearing the wild-type alleles of eye color and wing shape, thereby allowing the expression of the recessive alleles *eosin eye* and *miniature wing* on the remaining X chromosome. (After Morgan, 1919.)

of the *intersex* (*ix*) gene causes XX flies to develop an intersex phenotype having portions of male and female tissue in the same organ. The *doublesex* (*dsx*) gene is important for the sexual differentiation of both sexes. Homozygous recessive *dsx* alleles will turn both XX and XY flies into intersexes (Baker and Ridge, 1980; Belote et al., 1985a).

The position of these genes in a developmental pathway is based on (1) the interpretation of genetic crosses resulting in flies bearing two or more of these mutations, and (2) the determination of what happens when there is a complete absence of the products of one of these genes. Such studies have generated the model of this regulatory cascade seen in Figure 21.19.

FIGURE 21.19
Proposed regulation cascade for *Drosophila* somatic sex determination. Arrows represent activation while a block at the end of a line indicates suppression. The *msl* loci, under the control of the *Sxl* gene, regulate the dosage compensatory transcription of the male X chromosome. (After Baker et al., 1987.)

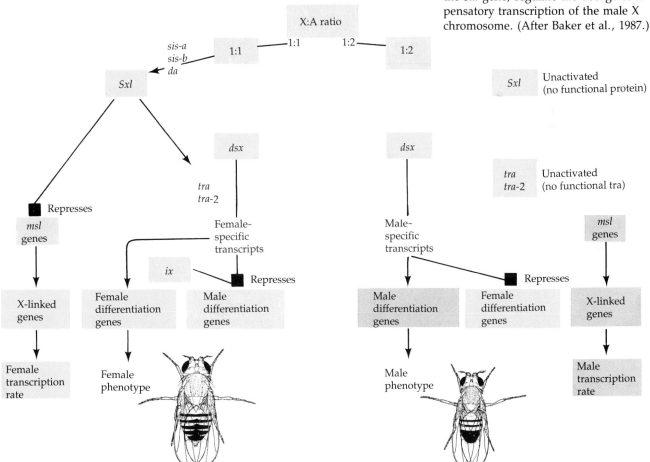

The *Sex-lethal* gene as pivot for sex determination

Interpreting the X:A ratio. The first phase of *Drosophila* sex determination involves reading the X:A ratio. What elements on the X chromosome are "counted" and how is this information used? It appears that high values of the X:A ratio are responsible for activating the feminizing switch gene *Sex-lethal* (*Sxl*). At low values (males), *Sxl* remains inactive during the early stages of development (Cline, 1983; Salz et al., 1987). In XX *Drosophila*, *Sxl* is activated during the first 2 hours after fertilization, and this gene transcribes a particular embryonic type of *Sxl* mRNA that is only found for about 2 hours more (Salz et al., 1989). Once it is activated, the *Sxl* gene remains on despite any further changes in the X:A ratio (Sánchez and Nöthiger, 1983). Early *Sxl* function is necessary for XX embryos to initiate the female developmental pathway and to maintain an appropriate level of transcription from the two X chromosomes. In those XX individuals where *Sxl* is not functional, the male phenotype emerges and the X chromosomes transcribe more rapidly (as would a single X in a normal male embryo, which has to produce as much gene product as the two X chromosomes of a normal female). In XX embryos this overstimulation of X chromosome transcription is lethal.

This female-specific transcription of *Sxl* is thought to be stimulated by "numerator elements" on the X chromosome that constitute the "X" part of the X:A ratio. Cline (1988) has demonstrated that two of these numerator elements are the X chromosome genes *sisterless-a* and *sisterless-b*. In this study, he discovered a variant *Drosophila* autosome with two regions of the X chromosome inserted into it. This chromosome was lethal if it was present in males, because males who received this chromosome activated their *Sxl* gene just as if they had inherited two X chromosomes. The activated *Sxl* gene caused the dosage compensation of the X chromosome to be female (that is, to have less transcription), but as the male had only one X chromosome, this proved to be lethal. Using this assay for the numerator genes, Cline found the *sisterless-a* and *sisterless-b* genes. Having two copies of these genes proved lethal for XY embryos, and the simultaneous deletion of one copy of each *sis* locus proved lethal to females (whose embryos did not activate the *Sxl* gene even though they were XX). Therefore, the presence of two copies of each of the *sis* genes is equivalent to having two copies of the X chromosome. The *sis* genes are probably not the only numerator elements, as Cline has documented cases in which the genetic background of the wild-type male fly is not sensitive to double doses of these genes.

The *Sxl* gene does not seem to sense these numerator elements without the presence of products of the *runt* and *daughterless* (*da*) genes. The lack of daughterless protein prevents the activation of *Sxl*. This does not affect XY embryos (since they do not activate *Sxl* anyway), but it is lethal in female embryos, since the mechanism for dosage compensation causes the two X chromosomes to be transcribed at the faster (male) rate (Cline, 1986; Cronmiller and Cline, 1987; Duffy and Gergen, 1991)—hence the name of the mutation. Thus, shortly after fertilization, the *sis-a*, *sis-b*, *runt*, and *da* genes enable *Sxl* to be transcriptionally active only in female embryos.

The "denominator elements" are those genes that get counted from the autosomes. One of the major denominator elements appears to be the *deadpan* gene (Younger-Shepherd, 1992). Males with too high a ratio of *sis-b* to *deadpan* activate *Sxl* and die, while females with too low a ratio of *sis-b* to *deadpan* fail to activate *Sxl* and die. The *daughterless*, *sis-a*, *sis-b*, and *deadpan* genes are all helix-loop-helix (HLH) transcription factors, and it is

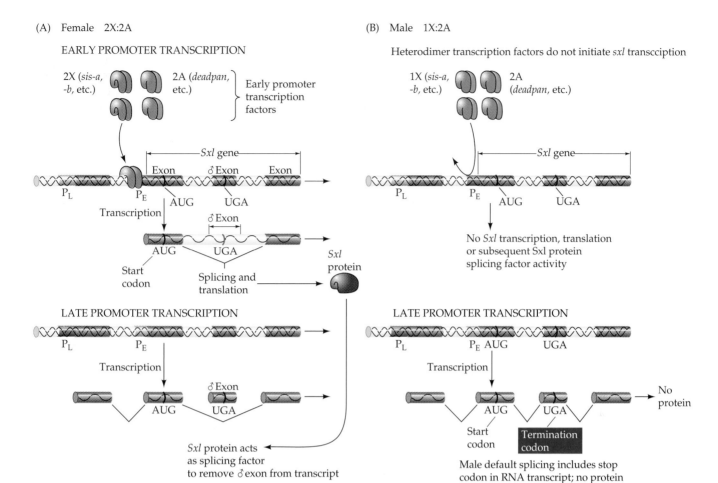

(A) Female 2X:2A

EARLY PROMOTER TRANSCRIPTION

2X (sis-a, -b, etc.) 2A (deadpan, etc.) Early promoter transcription factors

Sxl gene

Exon ♂Exon Exon

P_L P_E AUG UGA

Transcription

♂Exon

AUG UGA

Start codon

Splicing and translation

Sxl protein

LATE PROMOTER TRANSCRIPTION

P_L P_E

Transcription

♂Exon

AUG UGA

Sxl protein acts as splicing factor to remove ♂exon from transcript

(B) Male 1X:2A

Heterodimer transcription factors do not initiate sxl transccription

1X (sis-a, -b, etc.) 2A (deadpan, etc.)

Sxl gene

P_L P_E AUG UGA

No Sxl transcription, translation or subsequent Sxl protein splicing factor activity

LATE PROMOTER TRANSCRIPTION

P_L P_E AUG UGA

Transcription

AUG UGA

Start codon

Termination codon

No protein

Male default splicing includes stop codon in RNA transcript; no protein translated

FIGURE 21.20

Schematic representation of the differential activation of the Sxl gene in females and males. (A) In wild-type Drosophila with two X chromosomes and two sets of autosomes (XX; AA) the numerator transcription factor subunits (sis-a, sis-b, etc.) are not fully complexed by inhibitory subunits derived from genes (such as deadpan) on the autosomes. These numerator factors activate the early promoter of the Sxl gene, which produces a transcript that is automatically spliced to a female-specific mRNA that encodes functional Sxl protein. Eventually, constitutive transcription of Sxl starts from the late promoter. If Sxl is already available (i.e., from early transcription), the Sxl pre-mRNA is spliced to form the functional female-specific message. (B) In wild-type Drosophila with one X chromosome and two sets of autosomes (XY; AA), the numerator transcription factors are bound by the denominator subunits and cannot activate the early promoter. When the Sxl gene is transcribed from the late promoter, RNA splicing does not exclude the male-specific exon in the mRNA. The resulting message encodes a truncated and nonfunctional peptide, since the male-specific exon contains a translation termination codon. (After Keyes et al., 1992.)

possible that the denominator and numerator elements can form heterodimers with one another. Presumably the denominator proteins are able to form heterodimers that block those of the activator (sis and daughterless) proteins. (Figure 21.20A).

Maintenance of Sxl function. Shortly after this transcription has taken place, a second promoter on the Sex-lethal gene is activated, and this gene is transcribed in both males and females. However, analysis of the cDNA from Sxl mRNA shows that the Sxl mRNA of males differs from the Sxl mRNA of females (Bell et al., 1988). This is the result of differential RNA processing. Moreover, the *Sxl* protein appears to bind to its own mRNA precursor to splice it in the female manner. Since males do not have any available *Sxl* protein, their new *Sxl* transcripts are processed in the male manner (Keyes et al., 1992). The male *Sxl* mRNA is nonfunctional. While the female-specific *Sxl* message encodes a protein of 354 amino acids, the male-specific *Sxl* transcript contains a translation termination codon (UGA) after amino acid 48. The differential RNA processing that puts this termination codon into the male-specific mRNA is shown in Figures 21.20B and 21. In males, the nuclear transcript is spliced in a manner that yields three exons, and the termination codon is within the central exon. In females, RNA processing yields only two exons, and the male-specific central exon is now spliced out as a large intron. Thus, the female-specific mRNA lacks the termination codon.

The protein made by the female-specific *Sxl* transcript can be predicted from its nucleotide sequence. This protein would contain two regions that are important for binding to RNA. These regions are shared with nuclear RNA-binding proteins such as those in snRNPs. Bell and colleagues (1988) have proposed that there are two targets for the RNA-binding protein encoded by *Sxl*. One of these targets is the pre-mRNA of *Sxl* itself. This would be the mechanism that would maintain the female state of the pathway after the initial activating event has passed. The second target of the female-specific *Sxl* protein would be the pre-mRNA of the next gene on the pathway, *transformer*.

The *transformer* genes

The *Sxl* gene regulates somatic sex determination by controlling the processing of the *transformer* gene transcript. As we saw in Chapter 12, the *transformer* gene (*tra*) is alternatively spliced in males and females. There is a female-specific mRNA and also a nonspecific mRNA that is found in both females and males. Like the male *Sxl* message, the nonspecific *tra* mRNA contains a termination codon early in the message, making the protein nonfunctional (Boggs et al., 1987). In *tra*, the second exon of the nonspecific mRNA has the termination codon. This exon is not utilized in the female-specific message (Figure 21.21). How is it that the females should make a different transcript than the males? It is thought that the female-specific protein from the *Sxl* gene activates a female-specific 3' splice site in the *transformer* pre-mRNA, causing them to be processed in a way that splices out the second exon. To do this, the Sxl protein blocks the binding of U2AF to the nonspecific splice site by specifically binding

FIGURE 21.21
The pattern of sex-specific RNA splicing in three major *Drosophila* sex-determining genes. The pre-mRNAs are located in the center of the diagram and are identical in both the male and the female nuclei. In each case, the female-specific transcript is shown at the left, while the default transcript (whether male or nonspecific) is shown to the right. Exons are numbered, and the positions of the termination codons and poly(A) sites are marked. (After Baker, 1989.)

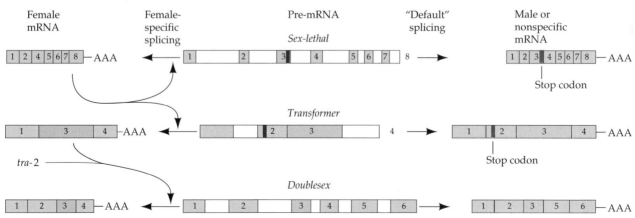

to the polypyrimidine tract adjacent to it. This causes U2AF to bind to the lower-affinity (female-specific) 3' splice site and generate a female-specific mRNA (Valcárcel et al., 1993). The protein encoded by this message is critical in female sex determination. If the female-specific transcript is artificially produced in XY flies, those flies become female. The nonspecific transcript has no effect on either males or females (McKeown et al., 1988).

The female-specific *tra* product acts in concert with the *transformer-2* (*tra-2*) gene to help generate the female phenotype. (The *tra-2* gene is not needed for male sex determination, although it is needed for male spermatogenesis later in development.) The *tra-2* gene is constitutively active and makes the same protein product in both males and females. This tra-2 protein, like that of the female-specific Sxl protein, contains an RNA-binding domain (Amrein et al., 1988; Goralski et al., 1988). It is proposed that the *tra-2* gene can bind to the transcript of the *doublesex* gene, but only in the presence of the female-specific tra protein (Baker, 1989).

doublesex: The switch gene of sex determination

The *doublesex* gene is active in both males and females, but its primary transcript is processed in a sex-specific manner (see Figure 12.13; Baker et al., 1987). Male and female transcripts are identical through the first three exons. The 3' exons differ in a way reminiscent of the two types of IgM molecules discussed in Chapter 12. What is an exon for the female-specific transcripts is part of the untranslated 3' end of the male-specific message. Moreover, molecular analyses of the dominant *dsx* mutations reveal them to be insertions in the female-specific exon. If a dominant *dsx* allele exists in an XX individual, the fly becomes male.

The alternative RNA processing appears to be the result of the *transformer* genes (Figure 21.21). The tra-2 and female-specific tra-1 proteins bind specifically to a DNA sequence adjacent to the female-specific 3' splice site of the *dsx* pre-mRNA, and they recruit nonspecific splicing factors to this site (Tian and Maniatis, 1993). If *tra* is not produced, the *doublesex* transcript is spliced in the male-specific manner. The downstream 3' splice site is used, and a male-specific transcript is made. This encodes an active protein that inhibits female traits and promotes male traits. On the other hand, if the *transformer* gene is making its active, female-specific protein, a different type of processing is done (Ryner and Bruce, 1991). The transformer proteins bind to sequences within the female-specific exon and activate the female-specific 3' splice site. (The alternative would have been their blocking the male-specific 3' splice site). This activation of an otherwise unused female-specific 3' splice site produces an mRNA encoding a female-specific transcript that inhibits male development.

The functions of the doublesex proteins can be seen in the formation of the *Drosophila* genitalia. Male and female genitalia in *Drosophila* are derived from separate cell populations. In male (XY) flies, the female primordium is repressed and the male primordium differentiates into the adult genital structures. In female (XX) flies, the male primordium is repressed and the female primordium differentiates. If the *doublesex* gene is absent (and thus neither transcript is made) *both* the male and female primordia develop and intersexual genitalia are produced. Thus, one of the roles of the sex-specific doublesex transcripts is to actively inhibit the development of the inappropriate genitalia. Male *dsx* transcripts inhibit female development; female-specific *dsx* transcripts inhibit male development (Nöthiger et al., 1977; Schüpbach et al., 1978). According to this model (Baker, 1989), the sexual determination cascade comes down to what type of mRNA is going to be processed from the *doublesex* transcript.

If the X:A ratio is 1, then *Sxl* makes a female-specific splicing factor that causes the *tra* gene transcript to be spliced in a female-specific manner. This female-specific protein interacts with the *tra-2* splicing factor to cause the *doublesex* pre-mRNA to be spliced in a female-specific manner. If the *doublesex* transcript is not acted upon in this way, it will be processed in a "default" manner to make the male-specific message.

Target genes for the sex determination cascade

Numerous proteins in *Drosophila* are present in one sex and not the other. In females, these include yolk proteins and eggshell (chorion) proteins. In males, the sex combs of the legs are sex-specific structures. Coschigano and Wensink (1993) have shown that the male and female *doublesex* transcripts both bind to three sites within the 127-base pair enhancer of the *yolk protein* genes. Their binding and mutagenesis studies demonstrate that the male-specific doublesex product inhibits transcription by its binding to these sites, whereas the female-specific doublesex protein activates gene transcription from the same sites. In addition to inhibiting female-specific gene products and organs, the male doublesex protein may also have a positive role in promoting the differentiation of the male sex combs (Jursnich and Burtis, 1993).

Temperature-sensitive mutations of sex determining genes can enable researchers to determine the critical times at which certain target genes are sensitive to a sex-determining switch. When *temperature-sensitive (ts)* alleles of the *tra-2* gene were used, the sexual development pathways in *Drosophila* were shown to be active from the late larval stages through the adult period. The *tra-2^{ts}* gene is a temperature-sensitive allele in which the female phenotype is expressed at permissive (colder) temperatures and the male phenotype is expressed at nonpermissive (warmer) temperatures. During late larval and pupal stages, raising the temperature from permissive to nonpermissive causes an XX larva or pupa to develop into a male. Moreover, when adult mutants are kept at low temperatures, the adult fat body makes yolk proteins that will enter the oocyte. When shifted to a higher, nonpermissive temperature, transcription of the yolk protein genes ceases (Belote et al., 1985b). One remarkable finding has been that if adult XX *tra-2^{ts}* flies are kept at the nonpermissive temperature for several days, they begin to exhibit male courtship behaviors (Belote and Baker, 1987).

Hermaphroditism

Hermaphroditism in the nematode *C. elegans*

The nematode *Caenorhabditis elegans* usually has two sexual types: hermaphrodite and male. Most individuals of this species are **hermaphroditic**,* having both testes and ovaries. As larvae, these hermaphrodites make sperm, which is stored in the nematode's genital tract (Figure 21.22). The adult ovary produces eggs, and these eggs become fertilized as they migrate into the uterus. (Therefore, the sperm is already present in the hermaphroditic adult.) Self-fertilization almost always produces more hermaphrodites. Only 0.2 percent of the progeny are males. These males,

*Hermaphrodites are named after the son of Hermes (Mercury) and Aphrodite (Venus). Having inherited the beauty of both parents, he excited the love of the nymph of the Salmacis fountain. As he bathed in this fountain, she embraced him, praying to the gods that they might forever be united. She got her wish in the most literal of fashions.

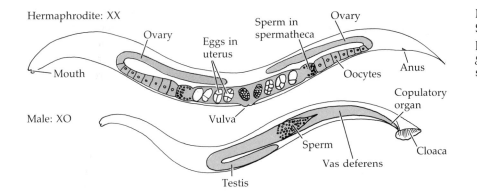

Hermaphrodite: XX

Ovary

Mouth

Eggs in uterus

Sperm in spermatheca

Ovary

Anus

Oocytes

Copulatory organ

Vulva

Male: XO

Sperm

Vas deferens

Cloaca

Testis

FIGURE 21.22
Schematic diagrams of the herma-phrodite and male *Caenorhabditis ele-gans* emphasizing their reproductive systems. (From Hodgkin, 1985.)

however, can mate with hermaphrodites; and because their sperm has a competitive advantage over endogenous hermaphroditic sperm, the sex ratio resulting from such matings is about 50 percent hermaphrodites and 50 percent males (Hodgkin, 1985).

In *C. elegans*, the hermaphrodite is XX and the male is XO. As in *Drosophila*, sex is determined by the ratio of X chromosomes to autosomes. In closely related species of nematodes, XX females are found, suggesting that the hermaphrodites evolved from females. Somatically, the females and hermaphrodites are identical, the only difference being that the her-maphrodites make sperm during their early development before switching over to egg production. In *C. elegans*, there even exists a dominant mu-tation ($tra-1^D$) that transforms XX or XO individuals into fertile females. In colonies with such an allele, three sexes are possible and functioning (Hodgkin, 1980).

As in *Drosophila*, sex determination in *C. elegans* involves several au-tosomal genes that read and respond to the X:A ratio. Some of these are essential for hermaphrodite development (*tra* genes), whereas others are necessary for the expression of the male phenotype (*her* and *fem* genes). Homozygous mutant *tra* genes will transform XX individuals into males, while the mutant *her* and *fem* alleles will transform males into hermaphro-dites. By creating genotypes carrying different combinations of these mu-tations, Hodgkin (1980) was able to construct a model for this develop-mental pathway (Figure 21.23). He found, for example, that the *tra* mutations all suppressed the *her-1* mutation, indicating that *her-1* is earlier in the pathway.

The crucial gene in the pathway for sex determination appears to be *tra-1*. If the wild-type *tra-1* gene is active, the individual is a hermaphro-dite. If *tra-1* is not functional, the individual is a male. The other genes appear to regulate this single switch gene. The first known gene in this pathway is *xol-1* (*XO-lethal*), which can turn off the pathway for herma-phroditic development, thereby turning the animal into a male. It appears to accomplish this by repressing the *sdc* genes, whose activities are needed for activating *tra-1* and making the animal a hermaphrodite (Miller et al., 1988). The *sdc-1* and *sdc-2* genes may be analogous to the *Sxl* gene of

FIGURE 21.23
Schematic model of somatic sex deter-mination in *C. elegans*. The *sdc-1* gene is postulated to be involved in trans-mitting the X/A ratio. It controls X-chromosome dosage compensation as well as suppressing the *her-1* gene if the ratio is 1. The high/low designa-tion reflects functional gene activity. The activity of the *sdc* genes eventually leads to the activity of the *tra-1* gene whose activity promotes the herma-phroditic phenotype. The *scd* genes can be inhibited by the *xol* gene, which is only active in XO (males). (After Hodgkin, 1985; Miller et al., 1988.)

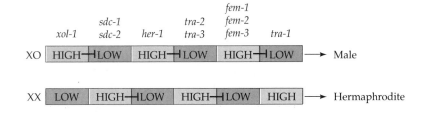

Drosophila in that they are X-linked and affect both sex determination and X chromosome dosage compensation. It is proposed that the *sdc* products repress *her-1* activity in XX animals (Villeneuve and Meyer, 1987; Nusbaum and Meyer, 1989). When *her-1* activity is high, *tra-2* and *tra-3* activities are low. This combination of activities is thought to allow the expression of the *fem* genes, which turn off *tra-1*. At low levels of *tra-1* product, the somatic cells of *C. elegans* become male. Conversely, if *her-1* activity is low, *tra-1* activity is high. This leads to the hermaphroditic developmental pattern.

But what does this linear genetic pathway have to do with the actual cellular events leading to sex determination? Recent studies have indicated that some of these genes may encode proteins of a signaling pathway between cells. Analysis of genetic mosaics suggests that *sdc-1* and *her-1* are not necessarily acting in the cells that make them. Rather, these genes appear to make secreted products. In contrast, tra-1 acts in a cell-autonomous fashion and is therefore likely to be part of a signal-receiving apparatus. The sequence of the *tra-1* gene suggests that it encodes a zinc finger transcription factor (Hunter and Wood, 1990; Zarkower and Hodgkin, 1992; Perry et al., 1993). Kuwabara and Kimble (1992) have recently proposed a model that integrates this genetic pathway with the cellular biology of sex determination. The her-1 protein is thought to promote male development in XO nematodes by inhibiting tra-2. The *tra-2* gene product, however, is not a transcription factor or a splicing factor, but an integral membrane protein with multiple transmembrane domains. Moreover, its mRNA is found (in different amounts) in both males and females. According to this speculative model (Figure 21.24), the fem proteins combine to create one large fem protein complex, and this complex is bound by the tra-2 membrane protein. In XX individuals, the fem protein complex is kept bound to the membrane, and the tra-1 protein can enter the nucleus. In XO nematodes, however, the her-1 protein binds to the extracellular region of the tra-2 protein, causing the tra-2 protein to release the

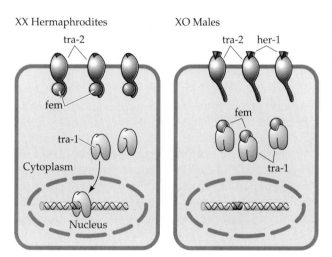

FIGURE 21.24
Hypothetical scheme for the actions of sex-determining genes in *C. elegans*. In XX individuals, the fem proteins are sequestered near the cell membrane by the products of the *tra-2* genes. In the absence of the fem proteins, tra-1 protein enters the nucleus to transcribe genes needed for hermaphroditic development. In XO individuals, the her-1 protein binds to the tra-2 product, causing the tra-2 product to release the fem proteins. Once free in the cytoplasm, the fem proteins can bind the tra-1 product, preventing it from entering the nucleus. (After Kuwabara and Kimble, 1992.)

fem complex. This fem complex, once free in the cytoplasm, can bind the tra-1 protein and prevent its entering the nucleus. Since the tra-1 protein (a putative transcription factor) cannot enter the nucleus, it cannot activate the hermaphrodite-specific genes. More studies need to be done to confirm or disprove this model, but it is useful both for suggesting new research and for visualizing how the genes might generate the pathway for sex determination in *C. elegans*.

Hermaphroditism in fish

One of the most interesting problems of *C. elegans* is its hermaphroditism. How did such a condition arise in an organism that probably had a male/female sex system? What gene changes arose, and were there other solutions that could have been used? The sex-determining genes of the closely related species *C. ramanei* (with male and female individuals) are now being identified so that such questions may be answered.

While hermaphroditism is not uncommon in worms and insects, it is rarely seen in vertebrates. In birds and mammals, hermaphroditism is a pathological condition causing infertility. The most common vertebrate hermaphrodites are fishes, which display several types of hermaphroditism (Yamamoto, 1969). Some fishes are **gonochoristic**; that is, they have a chromosomally determined sex that is either male or female. The hermaphroditic fish species can be divided into three groups. The first are the **synchronous** hermaphrodites, in which ovaries and testicular tissues exist at the same time and in which both sperm and eggs are produced. One such species is *Servanus scriba*. In nature and in aquaria, these fish form spawning pairs. As soon as one of the fish spawns its eggs, the other fish fertilizes them. Then the fish reverse their roles and the fish that was formerly male spawns its eggs so that they can be fertilized by the sperm of its partner (Clark, 1959).

In other hermaphroditic species, an animal undergoes a genetically programmed sex change during its development. In these cases, the gonads are dimorphic, having both male and female areas. One or the other is predominant during a certain phase of life. In protogynous ("female-first") hermaphrodites, an animal begins its life as a female, but later becomes male. The reverse is the case in protandrous ("male-first") species. Figure 21.25 shows the gonadal changes of the protandrous hermaphroditic fish *Sparus auratus*. At first, testicular tissue predominates, and later, after a transition period during which both testicular and ovarian tissues are seen, the ovarian cells take over.

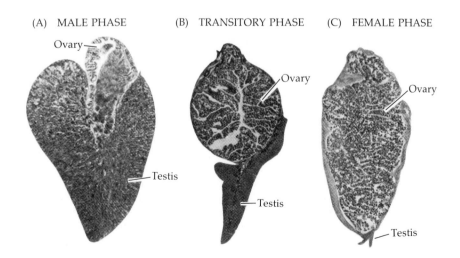

(A) MALE PHASE (B) TRANSITORY PHASE (C) FEMALE PHASE

Ovary — Testis

Ovary — Testis

Ovary — Testis

FIGURE 21.25
Gonadal changes in the hermaphroditic fish *Sparus auratus*, shown in section through (A) the gonad of the male phase, (B) the transitory phase, and (C) the final, female phase. (Courtesy of the family of T. Yamamoto.)

Why are there males?

Hermaphrodites appear to have many of the advantages that population biologists look for in fit individuals or successful species. First, nearly all the individuals of a population are capable of bearing young. (Males are considered an extravagance by population biologists—and by many animal breeders.) Second, it only takes one individual to colonize a new area. A single nematode or aphid can give rise to a large population very rapidly. Third, the hermaphroditic organism spends hardly any time or energy on courtship and mating (Maynard Smith, 1979).

What, then, keeps males around? Males are thought to be needed for supplying new combinations of genes. Without males, the gene pool of the species would be exceedingly limited. Van Valen (1973) has proposed that species cannot remain stable. They must change or they become extinct. This follows from the idea that one's predators and parasites are evolving to better eat their prey and that if the prey does not evolve, the predators could wipe out the species. This proposal is called the Red Queen hypothesis (after the character in *Through the Looking Glass* who claimed that you have to run as fast as you can just to remain in the same place). Moreover, clonal animals would not be able to purge their lineage of harmful mutations (Muller, 1964). Computer simulations (Howard and Lively, 1994) suggest that normal mutation rates in the presence of moderate predation would greatly favor retaining two sexes.

Environmental sex determination

Temperature-dependent sex determination in reptiles

While the sex of most snakes and most lizards is determined by sex chromosomes at the time of fertilization, the sex of most turtles and all species of crocodilians is determined by the environment after fertilization. In these reptiles, the temperature of the eggs during a certain period of development is the deciding factor in determining sex (Bull, 1980), and small changes of temperature can cause dramatic changes in the sex ratio. Generally, eggs incubated at low temperatures (22–27°C) produce one sex, whereas eggs incubated at higher temperatures (30°C and above) produce the other. There is only a small range of temperatures that permits both males and females to hatch from the same brood of eggs. Figure 21.26 shows the abrupt temperature-induced change in sex ratios for certain species of turtles. If eggs are incubated below 28°C, all the turtles hatching from them will be male. Above 32°C, every egg gives rise to a female. Thus, a brood of eggs usually produces individuals of the same sex. Variations on this theme also exist. Snapping turtle eggs, for instance, become female at either cold (≤20°C) or hot (≥30°C) temperatures. In between these extremes, males predominate.

The developmental period during which sex determination occurs can be studied by incubating eggs at the male-producing temperature for a certain amount of time and then shifting the eggs to an incubator of the female-producing temperature. In map turtles and snapping turtles, the middle third of development appears to be the time when the temperature exerts its effect. Ferguson and Joanen (1982) have studied sex determination in the Mississippi alligator, both in the laboratory and in the field; they have concluded that sex is determined between 7 and 21 days of incubation. Eggs raised at 30°C or below produce female alligators, whereas those incubated at 34°C or above produce all males. Moreover, nests constructed on levees (close to 34°C) give rise to males, whereas

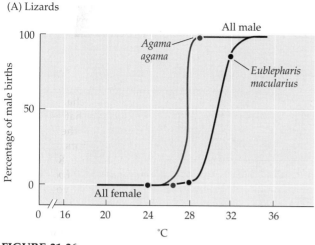

(A) Lizards

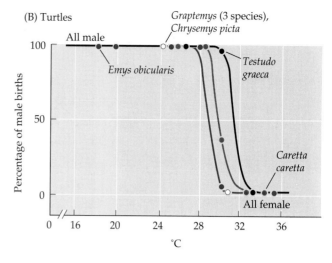

(B) Turtles

FIGURE 21.26
Relationship between sex ratio and incubation temperature in reptiles. (A) Two species of lizards in which higher temperatures result in the generation of male offspring. (B) Seven species of turtles in which higher temperatures result in female offspring. (After Bull, 1980.)

those built in wet marshes (close to 30°C) produce females. Thus, the sex of many turtle and alligator species is based on the temperature of the egg's environment.

Temperature-dependent sex determination has its advantages and its disadvantages. One advantage is that it probably gives the species the benefits of sexual reproduction without tying the species to a 1:1 sex ratio. The sex ratio in crocodiles, for example, may be as great as 10 females to each male (Woodward and Murray, 1993). The major disadvantage of temperature-dependent sex determination may be in its narrowing the temperature limits within which a species can exist. This would mean that thermal pollution (either locally or by "global warming") could conceivably eliminate a species in a given area (Janzen and Paukstis, 1991). Ferguson and Joanen (1982) speculate that dinosaurs may have had temperature-dependent sex determination and that their sudden demise may have been caused by a slight change in temperature that created conditions wherein only males or females hatched from their eggs.

Location-dependent sex determination in *Bonellia viridis* and *Crepidula fornicata*

The sex of the echiuroid worm *Bonellia* depends upon where a larva settles. The female *Bonellia* is a marine, rock-dwelling animal, with a body about 10 cm long (Figure 21.27). It has a proboscis, however, that can extend to over a meter in length. This proboscis serves two functions. First, it sweeps food from the rocks into the digestive tract of the female *Bonellia*. Second, should a larva land on the proboscis, it enters the female's mouth, migrates to the uterus, and differentiates into a 1- to 3-mm long symbiotic male. Thus, when a larva settles on a rocky surface, it becomes a female, but had that same larva settled on the proboscis of a female, it would have become a male. The male *Bonnellia* spends its life within the body of the female, fertilizing her eggs.

Baltzer (1914) demonstrated that when larvae were cultured in the absence of adult females, about 90 percent of them became females. However, when these larvae were cultured in the presence of an adult female or its isolated proboscis, 70 percent of them adhered to the proboscis and

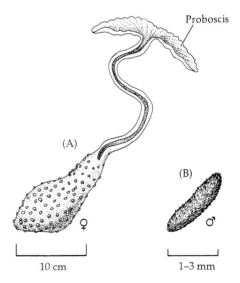

FIGURE 21.27
Extreme sexual dimorphism in *Bonellia viridis*. (A) Female, around 10 cm, with a proboscis capable of extending over a meter. (B) Symbiotic male (highly magnified compared with the female), 1–3 mm long. (After Barnes, 1968.)

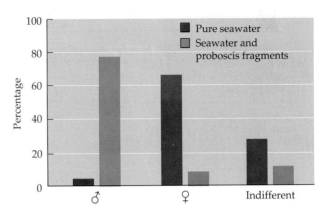

FIGURE 21.28

In vitro analysis of *Bonellia* differentiation. Larval *Bonellia* were placed either in normal seawater or in seawater containing fragments of the female proboscis. A majority of the animals cultured in the presence of the proboscis fragments became males, whereas normally they would have become females.

developed the male structures. These results have been more recently confirmed by Leutert (1974) (Figure 21.28).

The molecule(s) responsible for masculinizing the larvae can be extracted from the proboscis of adult females. When larvae are cultured in normal seawater in the absence of adult females, most become females. When cultured in seawater containing aqueous extracts of proboscis tissue, most become either male or an intermediate form, neither completely male nor completely female (Nowinski, 1934; Agius, 1979). The compound or compounds that attract the larvae to the proboscis and cause its masculinization are presently being purified.

Another example in which sex determination is affected by the position of the organism is that of the slipper snail *Crepidula fornicata*. Here, individuals pile up on top of each other to form a mound (Figure 21.29). Young individuals are always male. This phase is followed by the degeneration of the male reproductive system and a period of lability. The next phase can be either male or female, depending on the animal's position in the mound. If the snail is attached to a female, it will become male. If such a snail is removed from its attachment, it will become female. Similarly, the presence of large numbers of males will cause some of the males to become females. However, once the individual becomes female, it will not revert to being male (Coe, 1936).

Nature has provided many variations on her masterpiece. In some species, sex is determined solely by chromosomes, whereas in other species, sex is a matter of environmental conditions. Within these two large categories, numerous variations also exist. A complete catalog of known sex-determining mechanisms would take a separate (and very interesting) volume. But the sex-determining mechanism functions as the part of the somatic physiology necessary for the maintenance and propagation of the germ cells. The gonads will die with the body, but the germ cells that resided within them have the potential to renew life. It is to these germ cells that we now turn our attention.

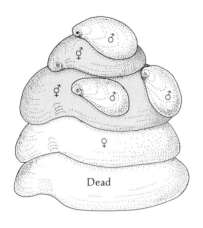

FIGURE 21.29

Cluster of *Crepidula* snails. Two individuals are changing from male to female. After these molluscs become female, they will be fertilized by the male above them. (After Coe, 1936.)

Agius, L. 1979. Larval settlement in the echiuran worm *Bonellia vividis*: Settlement on both the adult proboscis and body trunk. *Mar. Biol.* 53: 125–129.

Amrein, H., Gorman, M. and Nöthiger, R. 1988. The sex-determining gene *tra-2* of *Drosophila* encodes a putative RNA binding protein. *Cell* 55: 1025–1035.

Andersson, S., Berman, D. M., Jenkins, E. P. and Russell, D. W. 1991. Deletion of steroid 5α-reductase 2 gene in male pseudohermaphroditism. *Nature* 354: 159–161.

Aristotle. The generation of animals. Translated by A. Platt. *In* J. Barnes (ed.), 1984, *The Complete Works of Aristotle*, Vol. 8. Princeton University Press, Princeton, NJ.

Arnold, A. P. 1980. Sexual differences in the brain. *Am. Sci.* 68: 165–173.

Arnold, A. P., Nottebohm, F. and Pfaff, D. W. 1976. Hormone concentrating cells in vocal control and other brain regions of the zebra finch (*Poephila guttata*). *J. Comp. Neurol.* 165: 487–512.

Baker, B. S. 1989. Sex in flies: The splice of life. *Nature* 340: 521–524.

Baker, B. S. and Ridge, K. A. 1980. Sex and the single cell. I. On the action of major loci affecting sex determination in *Drosophila melanogaster*. *Genetics* 94: 383–423.

Baker, B. S., Nagoshi, R. N. and Burtis, K. C. 1987. Molecular genetic aspects of sex determination in *Drosophila*. *BioEssays* 6: 66–70.

Baltzer, F. 1914. Die Bestimmung und der Dimorphismus des Geschlechtes bei *Bonellia*. *Sber. Phys.-Med. Ges. Würzb.* 43: 1–4.

Barfield, R. J. and Chen, J. J. 1977. Activation of estrous behavior in ovariectomized rats by intracerebral implants of estradiol benzoate. *Endocrinology* 101: 1716–1725.

Barnes, R. D. 1968. *Invertebrate Zoology*. Saunders, Philadelphia.

Bell, L. R., Maine, E. M., Schedl, P. and Cline, T. W. 1988. *Sex-lethal*, a *Drosophila* sex determination switch gene, exhibits sex-specific RNA splicing and sequence similarity to RNA binding proteins. *Cell* 55: 1037–1046.

Barraclough, C. A. and Gorski, R. A. 1962. Studies on mating behavior in the androgen-sterilized female rat in relation to the hypothalamic regulation of sexual behavior. *J. Endocrinol.* 25: 175–182.

Bell, S. E. 1986. A new model of medical technology development: A case study of DES. *Sociol. Health Care* 4: 1–32.

Belote, J. M. and Baker, B. S. 1987. Sexual behavior: Its genetic control during development and adulthood in *Drosophila melanogaster*. *Proc. Natl. Acad. Sci. USA* 84: 8026–8030.

Belote, J. M., McKeown, M. B., Andrew, D. J., Scott, T. N., Wolfner, M. F. and Baker, B. S. 1985a. Control of sexual differentiation in *Drosophila melanogaster*. *Cold Spring Harbor Symp. Quant. Biol.* 50: 605–614.

Belote, J. M., Handler, A. M., Wolfner, M. F., Livak, K. L. and Baker, B. S. 1985b. Sex-specific regulation of yolk protein gene expression in *Drosophila*. *Cell* 40: 339–348.

Berkovitz, G. D. and seven others. 1992. The role of the sex-determining region of the Y chromosome (SRY) in the etiology of 46,XX true hermaphrodites. *Hum. Genet.* 88: 411–416.

Berta, P., Hawkins, J. R., Sinclair, A. H., Taylor, A., Griffiths, B. L., Goodfellow, P. N. and Fellous, M. 1990. Genetic evidence equating SRY and the testis-determining factor. *Nature* 348: 448–450.

Bleier, R. 1984. *Science and Gender*. Pergamon, New York, pp. 80–114.

Boggs, R. T., Gregor, P., Idriss, S., Belote, J. M. and McKeown, M. 1987. Regulation of sexual differentiation in *D. melanogaster* via alternative splicing of RNA from the *transformer* gene. *Cell* 50: 739–747.

Breedlove, S. M. and Arnold, A. P. 1980. Hormone accumulation in a sexually dimorphic motor nucleus of the rat spinal cord. *Science* 210: 565–566.

Bridges, C. B. 1921. Triploid intersexes in *Drosophila melanogaster*. *Science* 54: 252–254.

Bridges, C. B. 1925. Sex in relation to chromosomes and genes. *Am. Nat.* 59: 127–137.

Bull, J. J. 1980. Sex determination in reptiles. *Q. Rev. Biol.* 55: 3–21.

Burgoyne, P. S., Buehr, M., Koopman, P., Rossant, J. and McLaren, A. 1988. Cell-autonomous action of the testis-determining gene: Sertoli cells are exclusively XY in XX/XY chimaeric mouse testes. *Development* 102: 443–450.

Cate, R. L. and eighteen others. 1986. Isolation of the bovine and human genes for Müllerian inhibiting substance and expression of the gene in animal cells. *Cell* 45: 685–698.

Clark, E. 1959. Functional hermaphroditism and self-fertilization in a serranid fish. *Science* 129: 215–216.

Cline, T. W. 1983. The interaction between *daughterless* and *Sex-lethal* in triploids: A novel sex-transforming maternal effect linking sex determination and dosage compensation in *Drosophila melanogaster*. *Dev. Biol.* 95: 260–274.

Cline, T. W. 1986. A female-specific lethal lesion in an X–linked positive regulator of the *Drosophila* sex determination gene, *Sex-lethal*. *Genetics* 113: 641–663.

Cline, T. W. 1988. Evidence that *sisterless-a* and *sisterless-b* are two of several discrete "numerator elements" of the X/A sex determination signal in *Drosophila* that switch Sxl between two alternative stable expression states. *Genetics* 119: 829–862.

Coe, W. R. 1936. Sexual phases in *Crepidula*. *J. Exp. Zool.* 72: 455–477.

Coschigano, K. T. and Wensink, P. C. 1993. Sex-specific transcriptional regulation of the male and female doublesex proteins of *Drosophila*. *Genes Dev.* 7: 42–54.

Cronmiller, C. and Cline, T. W. 1987. The *Drosophila* sex determination gene *daughterless* has different functions in the germ line versus the soma. *Cell* 48: 479–487.

De Jonge, F. H., Muntjewerff, J.-W., Louwerse, A. L. and Van de Poll, N. E. 1988. Sexual behavior and sexual orientation of the female rat after hormonal treatment during various stages of development. *Hormones Behav.* 22: 100–115.

Duffy, J. B. and Gergen, J. P. 1991. The *Drosophila* segmentation gene *runt* acts as a position-specific numerator element necessary for the uniform expression of the sex determining gene *sex-lethal*. *Genes Dev.* 5: 2176–2187.

Eicher, E. M. and Washburn, L. L. 1983. Inherited sex reversal in mice: Identification of a new primary sex-determining gene. *J. Exp. Zool.* 228: 297–304.

Eicher, E. M. and Washburn, L. L. 1986. Genetic control of primary sex determination in mice. *Annu. Rev. Genet.* 20: 327–360.

Eicher, E. M., Washburn, L. L., Whitney, J. B. III and Morrow, K. E. 1982. *Mus poschiavinus* Y chromosome in the C57Bl/6J murine genome causes sex reversal. *Science* 217: 535–537.

Eicher, E. M., Phillips, S. J. and Washburn, L. L. 1983. The use of molecular probes and chromosomal rearrangements to partition the mouse Y chromosome into functional regions. *In* A. Messer and I. H. Porter (eds.), *Recombinant DNA and Medical Genetics*. Academic Press, New York, pp. 57–71.

Elbrecht, A. and Smith, R. G. 1992. Aromatase enzyme activity and sex determination in chickens. *Science* 255: 467–469.

Fausto-Sterling, A. 1992. *Myths of Gender*. Basic Books, New York.

Ferguson, M. W. J. and Joanen, T. 1982. Temperature of egg incubation determines sex in *Alligator mississippiensis*. *Nature* 296: 850–853.

Galen, C. *On the Usefulness of the Parts of the Body*. Translated by M. May, 1968. Cornell University Press, Ithaca, N.Y.

Geddes, P. and Thomson, J. A. 1890. *The Evolution of Sex*. Walter Scott, London.

Giese, K., Cox, J. and Grosschedl, R. 1992. The HMG domain of lymphoid enhancer factor 1 bends DNA and facilitates the assembly of functional nucleoprotein structures. *Cell* 69: 185–195.

Goralski, T. J., Edström, J.-E. and Baker, B. S. 1988. The sex determination locus *transformer-2* of *Drosophila* encodes a polypeptide with similarity to RNA binding proteins. *Cell* 56: 1011–1018.

Gubbay, J. and eight others. 1990. A gene mapping to the sex-determining region of the mouse Y chromosome is a member of a novel family of embryonically expressed genes. *Nature* 346: 245–250.

Hamer, D. H., Hu, S., Magnuson, V. L., Hu, N. and Pattatucci, A. M. L. 1993. A linkage between DNA markers on the X chromosome and male sexual orientation. *Science* 261: 321–327.

Haqq, C. M., King, C. Y., Donahoe, P. K. and Weiss, M. A. 1993. Sry recognizes conserved DNA sites in sex-specific promoters. *Proc. Natl. Acad. Sci. USA* 90: 1097–1101.

Harley, V. R. and seven others. 1992. DNA binding activity of recombinant *SRY* from normal males and XY females. *Science* 255: 453–456.

Harris, G. W. and Levine, S. 1965. Sexual physiology of the brain and its experimental control. *J. Physiol.* 181: 379–400.

Higgins, S. J., Young, P. and Cunha, G. R. 1989. Induction of functional cytodifferentiation in the epithelium of tissue recombinants II. Instructive induction of Wolffian duct epithelia by neonatal seminal vesicle mesenchyme. *Development* 106: 235–250.

Hodgkin, J. 1980. More sex-determination mutants of *Caenorhabditis elegans*. *Genetics* 96: 649–664.

Hodgkin, J. 1985. Males, hermaphrodites, and females: Sex determination in *Caenorhabditis elegans*. *Trends Genet.* 1: 85–88.

Horowitz, M. C. 1976. Aristotle and women. *J. Hist. Biol.* 9: 183–213.

Howard, R. S. and Lively, C. M. 1994. Parasitism, mutation accumulation, and the maintenance of sex. *Nature* 367: 554–557.

Hunter, C. P. and Wood, W. B. 1990. The *tra-1* gene determines sexual phenotype cell-autonomously in *C. elegans*. *Cell* 63: 1193–1204.

Imperato-McGinley, J., Guerrero, L., Gautier, T. and Peterson, R. E. 1974. Steroid 5α-reductase deficiency in man: An inherited form of male pseudohermaphroditism. *Science* 186: 1213–1215.

Jacklin, D. 1981. Methodological issues in the study of sex-related differences. *Dev. Rev.* 1: 266–273.

Jäger, R. J., Anvret, M., Hall, K. and Scherer, G. 1990. A human XY female with a frame shift mutation in the candidate testis-determining gene *SRY*. *Nature* 348:452–454.

Janzen, F. J. and Paukstis, G. L. 1991. Environmental sex determination in reptiles: Ecology, evolution, and experimental design. *Q. Rev. Biol.* 66: 149–179.

Josso, N., Picard, J.-Y. and Tran, D. 1977. The anti-Müllerian hormone. *Recent Prog. Horm. Res.* 33: 117–167.

Jost, A. 1953. Problems of fetal endocrinology: The gonadal and hypophyseal hormones. *Recent Prog. Horm. Res.* 8: 379–418.

Jursnich, V. A. and Burtis, K. C. 1993. A positive role in differentiation for the male doublesex protein of *Drosophila*. *Dev. Biol.* 155: 235–249.

Kandel, E. R., Schwartz, J. H. and Jessell, T. M. 1991. *Principles of Neural Science*, 3rd Ed. Elsevier, New York.

Keyes, L. N., Cline, T. W. and Schedl, P. 1992. The primary sex determination signal of *Drosophila* acts at the level of transcription. *Cell* 68: 933–943.

Knebelmann, B., Boussin, L., Guerrier, D., Legeai, L., Kahn, A., Josso, N. and Picard, J.-Y. 1991. Anti-Müllerian hormone Bruxelles: A nonsense mutation associated with the persistent Müllerian duct syndrome. *Proc. Natl. Acad. Sci. USA* 88: 3767–3771.

Konishi, M. and Akutagawa, E. 1985. Neuronal growth, atrophy, and death in a sexually dimorphic song nucleus in the zebra finch brain. *Nature* 315: 145–147.

Koopman, P., Münsterberg, A., Capel, B., Vivian, N. and Lovell-Badge, A. 1990. Expression of a candidate sex-determining gene during mouse testis differentiation. *Nature* 348: 450–452.

Koopman, P., Gubbay, J., Vivian, N., Goodfellow, P. and Lovell-Badge, R. 1991. Male development of chromosomally female mice transgenic for *Sry*. *Nature* 351: 117–121.

Kurz, E. M., Sengelaub, D. R. and Arnold, A. P. 1986. Androgens regulate the dendritic length of mammalian motoneurons in adulthood. *Science* 232: 395–397.

Kuwabara, P. E. and Kimble, J. 1992. Molecular genetics of sex determination in *C. elegans*. *Trends. Genet.* 8: 164–168.

Langman, J. 1981. *Medical Embryology*, 4th Ed. Williams & Wilkins, Baltimore.

Langman, J. and Wilson, D. B. 1982. Embryology and congenital malformations of the female genital tract. *In* A. Blaustein (ed.), *Pathology of the Female Genital Tract*, 2nd Ed. Springer-Verlag, New York, pp. 1–20.

Leutert, T. R. 1974. Zur Geschlechtsbestimmung und Gametogenese von *Bonellia vividis* Rolando. *J. Embryol. Exp. Morphol.* 32: 169–193.

LeVay, S. 1991. A difference in hypothalamic structure between heterosexual and homosexual men. *Science* 253: 1034–1037.

Magre, S. and Jost, A. 1980. Initial phases of testicular organogenesis in the rat. An electron microscope study. *Arch. Anat. Micr. Morph.* 69: 297–318.

Maynard Smith, J. 1979. *The Evolution of Sex*. Cambridge University Press, London.

McClung, C. E. 1902. The accessory chromosome—sex determinant? *Biol. Bull.* 3: 72–77.

McElreavey, K., Vilain, E., Abbas, N., Herskowitz, I. and Fellows, M. 1993. A regulatory cascade hypothesis for mammalian sex determination: SRY represses a negative regulator of male development. *Proc. Natl. Acad. Sci. USA* 90: 3368–3372.

McEwen, B. S. 1981. Neural gonadal steroid actions. *Science* 211: 1303–1311.

McEwen, B. S., Leiberburg, I., Chaptal, C. and Krey, L. C. 1977. Aromatization: Important for sexual differentiation of the neonatal rat brain. *Horm. Behav.* 9: 249–263.

McKeown, M., Belote, J. M. and Boggs, R. T. 1988. Ectopic expression of the female *transformer* gene product leads to female differentiation of chromosomally male *Drosophila*. *Cell* 53: 887–895.

McLaren, A. 1988. Sex determination in mammals. *Trends Genet.* 4: 153–157.

McLaren, A. 1990. The making of male mice. *Nature* 351: 96.

McLaren, A., Simpson, E., Tomonari, K., Chandler, P. and Hogg, H. 1984. Male sexual differentiation in mice lacking the H-Y antigen. *Nature* 312: 552–555.

Meyer, W. J., Migeon, B. R. and Migeon, C. J. 1975. Locus on human X chromosome for dihydrotestosterone receptor and androgen insensitivity. *Proc. Natl. Acad. Sci. USA* 72: 1469–1472.

Miller, L. M., Plenefisch, J. D., Casson, L. P. and Meyer, B. 1988. xol-1: A gene that controls the male mode of both sex determination and X chromosome dosage compensation in *C. elegans*. *Cell* 55: 167–183.

Mittwoch, U. 1992. Sex determination and sex reversal: Genotype, phenotype, dogma, and semantics. *Hum. Genet.* 89: 467–479.

Moore, C. L. 1990. Comparative development of vertebrate sexual behavior: Levels, cascades, and webs. *In* D. A. Dewsbury (ed.), *Issues in Comparative Psychology*. Sinauer Associates, Sunderland, MA., pp. 278–299.

Moore, C. L., Dou, H. and Juraska, J. M. 1992. Maternal stimulation affects the number of motor neurons in a sexually dimorphic nucleus of the lumbar spinal cord. *Brain Res.* 572: 52–56.

Morgan, T. H. 1919. *The Physical Basis of Heredity*. Lippincott, Philadelphia.

Muller, H. J. 1964. The relationship of recombination to mutational advance. *Mutat. Res.* 1: 2–7.

Nabekura, J., Oomura, Y., Minami, T., Mizuno, Y. and Fukuda, A. 1986. Mechanism of the rapid effect of 17β-estradiol on medial amygdala neurons. *Science* 233: 226–228.

Nordeen, E. J., Nordeen, K. W., Sengelaub, D. R. and Arnold, A. P. 1985. Androgens prevent normally occurring cell death in a sexually dimorphic spinal nucleus. *Science* 229: 671–673.

Nöthiger, R., Dübendorfer, A. and Epper, F. 1977. Gynandromorphs reveal two separate primordia for male and female genitalia in *Drosophila melanogaster*. *Wilhelm Roux Arch.* 181: 367–373.

Nottebohm, F. 1980. Testosterone triggers growth of brain vocal control nuclei in adult female canaries. *Brain Res.* 189: 429–436.

Nottebohm, F. 1981. A brain for all seasons: Cyclical anatomical changes in song control nuclei of the canary brain. *Science* 214: 1368–1370.

Nowinski, W. 1934. Die vermännlichende Wirkung fraktionierter Darmextrakte des Weibchens auf die Larven der *Bonellia viridis*. *Pubbl. Staz. Zool. Napoli* 14: 110–145.

Nusbaum, C. and Meyer, B. J. 1989. The *Caenorhabditis elegans* gene sdc-2 controls sex determination and dosage compensation in XX animals. *Genetics* 122: 579–593.

Page, D. C. 1986. Sex-reversal: Deletion mapping of the male-determining function of the human Y chromosome. *Cold Spring Harbor Symp. Quant. Biol.* 51: 229–235.

Page, D. C., de la Chappelle, A. and Weissenbach, J. 1985. Chromosome Y-specific DNA in related human XX males. *Nature* 315: 224–226.

Palmer, S. J. and Burgoyne, P. S. 1991a. In situ analysis of fetal, prepubertal and adult XX ↔ XY chimaeric mouse testes: Sertoli cells are predominantly, but not exclusively, XY. *Development* 112: 265–268.

Palmer, S. J. and Burgoyne, P. S. 1991b. The *mus musculus domesticus Tdy* allele acts later than the *Mus musculus musculus Tdy* allele: A basis for XY sex-reversal in C57BL/6-Y^{POS} mice. *Development* 113: 709–714.

Patek, C. E. and eight others. 1991. Sex chimaerism, fertility, and sex determination in the mouse. *Development* 113: 311–325.

Perry, M. D., Li, W., Trent, C., Robertson, B., Fire, A., Hageman, J. M. and Wood, W. B. 1993. Molecular characterization of the *her-1* gene suggests a direct role in cell signaling during *Caenorhabditis elegans* sex determination. *Genes Dev.* 7: 216–228.

Pfaff, D. W. and McEwen, B. S. 1983. The actions of estrogens and progestins on nerve cells. *Science* 219: 808–814.

Phoenix, C. H., Goy, R. W., Gerall, A. A. and Young, W. C. 1959. Organizing action of prenatally administered testosterone propionate on the tissues mediating mating behavior in the female guinea pig. *Endocrinology* 65: 369–382.

Pröve, E. 1978. Courtship and testosterone in male zebra finches. *Z. Tierpsychol.* 48: 47–67.

Reddy, V. R., Naftolin, F. and Ryan, K. J. 1974. Conversion of androstenedione to estrone by neural tissues from fetal and neonatal rats. *Endocrinology* 94: 117–121.

Risch, N., Squires-Wheeler, E. and Keats, B. J. B. 1993. Male sexual orientation and genetic evidence. *Science* 262: 2063–2065.

Robboy, S. J., Young, R. H. and Herbst, A. L. 1982. Female genital tract changes related to prenatal diethylstilbesterol exposure. *In* A. Blaustein (ed.), *Pathology of the Female Genital Tract*, 2nd Ed. Springer-Verlag, New York, pp. 99–118.

Rubin, B. S. and Barfield, R. J. 1980. Priming of estrus responsiveness by implants of 17β-estradiol in the ventromedial hypothalamic nuclei of female rats. *Endocrinology* 106: 504–509.

Ryner, L. C. and Bruce, B. S. 1991. Regulation of *doublesex* pre-mRNA processing occurs by 3′-splice site activation. *Genes Dev.* 5: 2071–2085.

Salz, H. K., Cline, T. W. and Schedl, P. 1987. Functional changes associated with structural alterations induced by mobilization of a P element inserted into the *Sex-lethal* gene of Drosophila. *Genetics* 117: 221–231.

Salz, H. K., Maine, E. M., Keyes, L. N., Samuels, M. E., Cline, T. W. and Schedl, P. 1989. The *Drosophila* female-specific sex-determination gene, *Sex-lethal*, has stage-, tissue-, and sex-specific RNAs suggesting multiple modes of regulation. *Genes Dev.* 3: 708–719.

Sánchez, L. and Nöthiger, R. 1983. Sex determination and dosage compensation in *Drosophila melanogaster*: Production of male clones in XX females. *EMBO J.* 1: 485–491.

Schiebinger, L. 1989. *The Mind Has No Sex?* Harvard University Press, Cambridge.

Schbach, T., Wieschaus, E. and Nöthiger, R. 1978. The embryonic organization of the genital disc studied in genetic mosaics of *Drosophila melanogaster*. *Wilhelm Roux Arch.* 185: 249–270.

Siiteri, P. K. and Wilson, J. D. 1974. Testosterone formation and metabolism during male sexual differentiation in the human embryo. *J. Clin. Endocrinol. Metab.* 38: 113–125.

Sinclair, A. H. and nine others. 1990. A gene from the human sex-determining region encodes a protein with homology to a conserved DNA-binding motif. *Nature* 346: 240–244.

Singh, L. and Jones, K. W. 1982. Sex reversal in the mouse (*Mus musculus*) is caused by a recurrent nonreciprocal crossover involving the X and an aberrant Y chromosome. *Cell* 28: 205–216.

Stevens, N. M. 1905. Studies in spermatogenesis with especial reference to the "accessory chromosome." *Carnegie Inst. Washington Rep.* 36.

Strickberger, M. W. 1968. *Genetics*. Macmillan, New York, p. 463.

Terasawa, E. and Sawyer, C. H. 1969. Changes in electrical activity in rat hypothalamus related to electrochemical stimulation of adenohypophyseal function. *Endocrinology* 85: 143–149.

Thigpen, A. E., Davis, D. L., Imperato-McGinley, J. and Russell, D. W 1992. The molecular basis of steroid 5α-reductase deficiency in a large Dominican kindred. *New Engl. J. Med.* 327: 1216–1219.

Thorpe, W. H. 1958. The learning of song patterns by birds with especial reference to the song of the chaffinch, *Fringilla coelebs*. *Ibis* 100: 535–570.

Tian, M. and Maniatis, T. 1993. A splicing enhancer complex controls alternative splicing of *doublesex* pre-mRNA. *Cell* 74: 105–114.

Tobin, C. and Joubert, Y. 1992. Testosterone-induced development of the rat levator ani muscle. *Dev. Biol.* 146: 131–138.

Tran, D., Meusy-Dessolle, N. and Josso, N. 1977. Anti-Müllerian hormone is a functional marker of foetal Sertoli cells. *Nature* 269: 411–412.

Trelstad, R. L., Hayashi, A., Hayashi, K. and Donahoe, P. K. 1982. The epithelial–mesenchymal interface of the male rate Müllerian duct: Loss of basement membrane integrity and ductal regression. *Dev. Biol.* 92: 27–40.

Tuana, N. 1988. The weaker seed. *Hypatia* 3: 35–59.

Valcárcel, J., Singh, R., Zamore, P. and Greene, M. R. 1993. The protein Sex-lethal antagonizes the splicing factor U2AF to regulate alternative splicing of *transformer* pre-mRNA. *Nature* 362: 171–175.

Van Valen, L. 1973. A new evolutionary law. *Evol. Theor.* 1: 1–30.

Vergnaud, G. and eight others. 1986. A deletion map of the human Y chromosome based on DNA hybridization. *Am. J. Hum. Genet.* 38: 109–124.

Vigier, B., Watrin, F. Magre, S., Tran, D. and Josso, N. 1987. Purified bovine AMH induces a characteristic freemartin effect in fetal rat prospective ovaries exposed to it in vitro. *Development* 100: 43–55.

Vigier, B., Forest, M. G., Eychenne, B., Bezard, J., Garrigou, O., Robel, P. and Josso, M. 1989. Anti-Müllerian hormone produces endocrine sex reversal in fetal ovaries. *Proc. Natl. Acad. Sci. USA* 86: 3684–3688.

Villeneuve, A. M. and Meyer, B. J. 1987. *sdc-1*: A link between sex determination and dosage compensation in *C. elegans*. *Cell* 48: 25–37.

Wachtel, S. S., Hall, J. L., Müller, U. and Chaganti, R. S. K. 1980. Serum-borne H–Y antigen in the fetal bovine freemartin. *Cell* 21: 917–926.

Washburn, L. L. and Eicher, E. M. 1983. Sex reversal in XY mice caused by a dominant mutation on chromosome 17. *Nature* 303: 338–340.

Washburn, L. L. and Eicher, E. M. 1989. Normal testis determination in the mouse depends on genetic interaction of a locus on chromosome 17 and the Y chromosome. *Genetics* 123: 173–179.

Wilson, E. B. 1905. The chromosomes in relation to the determination of sex in insects. *Science* 22: 500–502.

Woodward, D. E. and Murray, J. D. 1993. On the effect of temperature-dependent sex determination on sex ratio and survivorship in crocodilians. *Proc. R. Soc. Lond.* [B] 252: 149–155.

Yamamoto, T.-O. 1969. Sex differentiation. *In* W. S. Hoar and D. J. Randall (eds.), *Fish Physiology*, Vol. 3. Academic Press, New York, pp. 117–175.

Younger-Shepherd, S., Vaessin, H., Bier, E., Jan, L. Y. and Jan, Y. N. 1992. *deadpan*, an essential pan-neural gene encoding an HLH protein, acts as a denominator in *Drosophila* sex determination. *Cell* 70: 911–922.

Zarkower, D. and Hodgkin, J. 1992. Molecular analysis of the *C. elegans* sex-determining gene *tra-1*: A gene encoding two zinc finger proteins. *Cell* 70: 237–249.

Zwingman, T., Erickson, R. P., Boyer, T. and Ao, A. 1993. Transcription of the sex-determining region genes *Sry* and *Zfy* in the mouse preimplantation embryo. *Proc. Natl. Acad. Sci. USA* 90: 814–817.

22

The saga of the germ line

*And the end of all our exploring
Will be to arrive where we started
And know the place for the first
time.*

T. S. ELIOT (1942)

We began our analysis of animal development by discussing fertilization, and we will finish by investigating **gametogenesis**, the collection of processes by which the sperm and the egg are formed. Germ cells provide the continuity of life between generations, and the mitotic ancestors of our own germ cells once resided in the gonads of reptiles, amphibians, fishes, and invertebrates. In many animals, such as insects, roundworms, and vertebrates, there is a clear and early separation of germ cells from somatic cell types. In several animal phyla (and in the entire plant kingdom), this division is not as well established. In these species (which include cnidarians, flatworms, and tunicates) somatic cells can readily become germ cells even in adult organisms. The zooids, buds, and polyps of many invertebrate phyla testify to the ability of somatic cells to give rise to new individuals.

In those organisms where there is an established germ line that separates early in development, the germ cells do not arise from within the gonad itself. Rather, their precursors—the **primordial germ cell**s (PGCs)—migrate into the developing gonads. The first step in gametogenesis, then, involves forming the PGCs and getting them into the genital ridge as the gonad is forming. The initiation of the germ cell lineage (the germ line) in amphibians, insects, and roundworms was discussed in Chapter 14. We resume our story of the germ line with the migration of the PGCs from their place of origin into the gonads.

Germ cell migration

Germ cell migration in amphibians

As discussed in Chapter 14, the germ plasm of anuran amphibians—frogs and toads—collects around the vegetal pole in the one-cell embryo. During cleavage, this material is brought upward through the yolky cytoplasm, and the RNA-rich granules become associated with the endodermal cells lining the floor of the blastocoel (Figure 22.1; Bounoure, 1934; Ressom and Dixon, 1988). Although we don't know the identities of the amphibian germ cell determinants, recent studies (Kloc et al., 1993) show that a particular type of RNA migrates to the same places as the germ plasm in the *Xenopus* egg. The PGCs become concentrated in the posterior region of the larval gut, and as the abdominal cavity forms, the anuran PGCs emigrate along the dorsal side of the gut, first along the dorsal mesentery (which connects the gut to the region where the mesodermal organs are

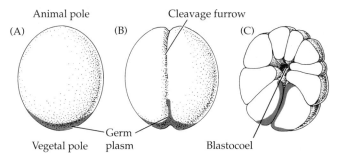

Animal pole

(A)   (B)   Cleavage furrow   (C)

Vegetal pole   Germ plasm   Blastocoel

FIGURE 22.1
Changes in the position of germ plasm (color) in an early frog embryo. Originally located near the vegetal pole of the uncleaved egg (A), the germ plasm advances along the cleavage furrows (B) until it becomes localized at the floor of the blastocoel (C). (After Bounoure, 1934.)

forming) and then along the abdominal wall and into the genital ridges. They migrate up this tissue until they reach the developing gonads (Figure 22.2). *Xenopus* PGCs move by extruding a single filopodium and then streaming their yolky cytoplasm into the filopodium while retracting their tail. Contact guidance in this migration seems likely as both the cells and the extracellular matrix over which they migrate are oriented in the direction of that migration (Wylie et al., 1979). Furthermore, PGC adhesion and migration can be inhibited if the mesentery is treated with antibodies against *Xenopus* fibronectin (Heasman et al., 1981). Thus, the pathway for germ cell migration in these frogs appears to be composed of an oriented fibronectin-containing extracellular matrix. The fibrils over which the PGCs travel lose this polarity soon after migration has ended.* As they migrate, *Xenopus* PGCs divide about three times, and approximately 30 PGCs colonize the gonads (Whitington and Dixon, 1975; Wylie and Heasman, 1993). These will divide to form the germ cells.

The primordial germ cells of urodele amphibians (salamanders) have an apparently different origin, which has been traced by reciprocal transplantation experiments to the regions of the mesoderm that involute through the ventrolateral lips of the blastopore. Moreover, there does not seem to be any particular localized "germ plasm" in salamander eggs. Rather, the interaction of the dorsal endoderm cells and animal hemisphere cells creates the conditions needed to form germ cells in the particular areas that involute through the ventrolateral lips (Sutasurya and Nieuwkoop,1974). So in salamanders, the PGCs are formed by induction within the mesodermal region and presumably follow a different path into the gonad.

Germ cell migration in mammals

Mammalian germ cells are not morphologically distinct during early development. However, by using monoclonal antibodies that recognize cell surface differences between the PGCs and their surrounding cells, Hahnel and Eddy (1986) showed that mouse PGCs originally reside in the epiblast of the gastrulating embryo. Ginsburg and her colleagues (1990) localized this region to the extraembryonic mesoderm just posterior to the primitive streak of the 7-day mouse embryo. Here about eight large, alkaline phosphatase-staining cells are seen. If this area is removed, the remaining

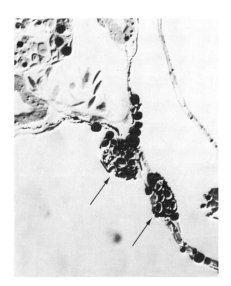

FIGURE 22.2
Migration of primordial germ cells in a frog. This phase-contrast photomicrograph of a section through the body wall and dorsal mesentery of a *Xenopus* embryo shows the migration of two large primordial germ cells (arrows) along the dorsal mesentery. (From Heasman et al., 1977, courtesy of the authors.)

*This does not necessarily hold true for all anurans. In the frog *Rana pipiens*, the germ cells follow a similar route but may be passive travelers along the mesentery rather than actively motile cells (Subtelny and Penkala, 1984).

FIGURE 22.3

Pathway for the migration of mammalian primordial germ cells. (A) Primordial germ cells seen in the yolk sac near the junction of the hindgut and allantois. (B) Migration through gut and, dorsally, up the dorsal mesentery into the genital ridge. (After Langman, 1981.)

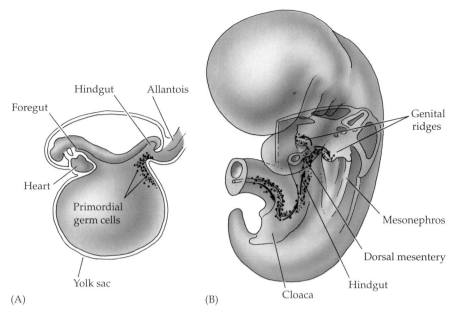

(A) (B)

embryo becomes devoid of germ cells, while the isolated segment develops a large number of primordial germ cells. The germ cell precursors in the extraembryonic mesoderm then migrate back into the embryo, first into the mesoderm of the primitive streak, and then to the endoderm by way of the allantois. The route of the mammalian PGC migration from the allantois (Figure 22.3) resembles that of the anuran PGC migration. After collecting at the allantois by day 7.5 (Chiquoine, 1954; Mintz, 1957), the mammalian PGCs migrate to the adjacent yolk sac (Figure 22.4). By this time, they have already split into two populations that will migrate to either the right or the left genital ridge. The PGCs then move caudally from the yolk sac through the newly formed hindgut and up the dorsal mesentery into the genital ridge (Figure 22.4B). Most of the PGCs have reached the developing gonad by the eleventh day after fertilization. During this trek, they have proliferated from an initial population of 10–100 cells to the 2500–5000 PGCs present in the gonads by day 12. Like the

FIGURE 22.4

Mouse primordial germ cells at different stages of their migration. (A) PGCs in hindgut of mouse embryo (near the allantois and yolk sac). Four large PGCs (dark circles) stain positively for high levels of alkaline phosphatase. (B) PGCs, stained for alkaline phosphatase, can be seen migrating up the dorsal mesentery and entering the genital ridges. (A from Heath, 1978; B from Mintz, 1957; photographs courtesy of the authors.)

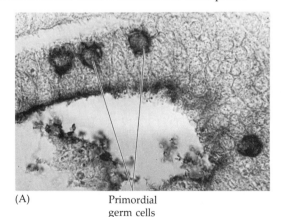

(A) Primordial germ cells

(B) Dorsal mesentery Genital ridges

PGCs of *Xenopus*, mammalian PGCs appear to be closely associated with the cells over which they migrate, and they move by extending filopodia over the underlying cell surfaces. These cells are also capable of penetrating cell monolayers and migrating through the cell sheets (Stott and Wylie, 1986). The mechanism by which the primordial germ cells know the route of this journey is still unknown. Fibronectin is likely to be an important substrate on which the PGCs migrate (ffrench-Constant et al., 1991), and in vitro evidence suggests that the genital ridges of 10.5-day mouse embryos secrete a diffusible TGF-β1-like protein that is capable of attracting mouse primordial germ cells (Godin et al, 1990; Godin and Wylie, 1991). Whether the genital ridge is able to provide such cues in vivo has still to be tested.

The proliferation of the PGCs appears to be promoted by the **stem cell factor**, the same growth factor needed for the proliferation of neural crest-derived melanoblasts and hematopoietic stem cells (see Chapter 7). The stem cell factor is produced by the cells along the migration pathway and remains bound to their cell membranes. It appears that the presentation of this protein on membranes is important for its activity. Mice homozygous for the *White Spotting* mutation (*W*) are deficient in germ cells (and melanocytes and blood cells), as their stem cells lack the receptor for stem cell growth factor. Mice homozygous for the *Steel* mutation have a similar phenotype, as they lack the ability to make this growth factor. Mice homozygous for the *Steel-Dickie* (*Sl^d*) allele have reduced numbers of germ cells, as these mice can make the stem cell growth factor, but it does not remain bound to their membranes (Dolci et al., 1991; Matsui et al., 1991). The addition of stem cell factor to PGCs taken from 11-day mice will stimulate their proliferation for about 24 hours, and appears to prevent programmed cell death that would otherwise occur (Pesce et al., 1993).

SIDELIGHTS & SPECULATIONS

Teratocarcinomas and embryonic stem cells

Stem cell factor increases the proliferation of migrating mouse primordial germ cells in culture, and this proliferation can be further increased by adding another growth factor, leukemia inhibition factor (LIF). However, the life span of these cells is short, and the cells soon die. But if an additional mitotic regulator—basic fibroblast growth factor—is added, a remarkable change takes place. The cells continue to proliferate, producing a pluripotent embryonal stem cell with characteristics resembling the cells of the inner cell mass (Matsui et al., 1992). We have discussed such embryonic stem cells earlier, as these are the cells that can be transfected with recombinant genes and inserted into the blastocyst to create transgenic mice.

Such a mammalian germ cell or stem cell contains within it all the information needed for subsequent development. What would happen if such a cell became malignant? In one type of tumor, the germ cells become embryonic stem cells as in the experiment mentioned above. This type of tumor is called a **teratocarcinoma**. Whether spon-

taneous or experimentally produced, a teratocarcinoma contains an undifferentiated stem cell population that has biochemical and developmental properties remarkably similar to those of cells of the inner cell mass (Graham, 1977). Moreover, these stem cells not only divide but can also differentiate into a wide variety of tissues including gut and respiratory epithelia, muscle, nerve, cartilage, and bone (Figure 22.5). Once differentiated, these cells no longer divide and are therefore not malignant. Such tumors can give rise to most of the tissue types in the body. Thus, the teratocarcinoma stem cells mimic early mammalian development, but the tumor they form is characterized by random, haphazard development.

In 1981, Stewart and Mintz formed a mouse from cells derived in part from a teratocarcinoma stem cell! Stem cells that had arisen in a teratocarcinoma of an agouti (yellow-tipped) strain of mice were cultured for several cell generations and were seen to maintain the characteristic chromosome complement of the parental mouse. Individual stem cells of this type were injected into the blastocysts of black mice. The blastocysts were then transferred to the uterus of a foster mother, and live mice were born. Some

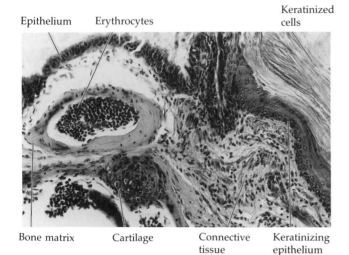

Epithelium   Erythrocytes   Keratinized
                                              cells

Bone matrix   Cartilage   Connective   Keratinizing
                                 tissue   epithelium

FIGURE 22.5
Photomicrograph of a section through a teratocarcinoma, show-ing numerous differentiated cell types. (From Gardner, 1982; photograph from C. Graham, courtesy of R. L. Gardner.)

of these mice had coats of two colors, indicating that the tumor cell had integrated itself into the embryo. Moreover, when mated to a mouse carrying an appropriate marker, the chimeric mouse was able to generate mice having some of the phenotypes of the tumor "parent." The malignant embryonal carcinoma cell had produced many, if not all, types of normal somatic cells and even had produced nor-mal, functional germ cells! When mice having a tumor cell for one parent were mated together, the resultant litter contained mice that were homozygous for a large number of genes from the tumor cell (Figure 22.6).

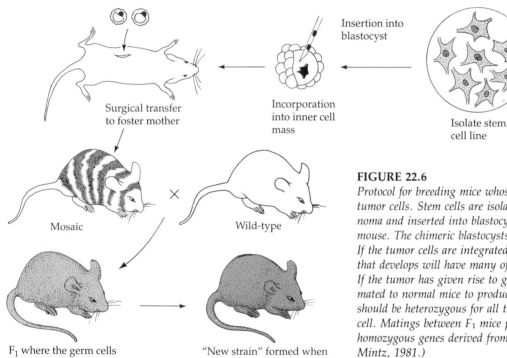

Insertion into blastocyst

Incorporation into inner cell mass

Isolate stem cell line

Malignant teratocarcinoma

Surgical transfer to foster mother

Mosaic   ×   Wild-type

F_1 where the germ cells were tumor-derived

"New strain" formed when two F_1 mice are mated

FIGURE 22.6
Protocol for breeding mice whose genes are derived largely from tumor cells. Stem cells are isolated from a mouse teratocarci-noma and inserted into blastocysts from a different strain of mouse. The chimeric blastocysts are placed into a foster mother. If the tumor cells are integrated into the blastocyst, the mouse that develops will have many of its cells derived from the tumor. If the tumor has given rise to germ cells, the mosaic mice can be mated to normal mice to produce an F_1 generation. The F_1 mice should be heterozygous for all the chromosomes of the tumor cell. Matings between F_1 mice produce F_2 mice having some homozygous genes derived from tumor cells. (After Stewart and Mintz, 1981.)

Germ cell migration in birds and reptiles

In birds and reptiles, primordial germ cells are derived from epiblastic cells that migrate from the central region of the area pellucida to a crescent-shaped zone in the hypoblast at the anterior border of the area pellucida (Figure 22.7; Eyal-Giladi et al., 1981; Ginsburg and Eyal-Giladi, 1987). This extraembryonic region is called the **germinal crescent**, and the primordial germ cells multiply in this region. Unlike the PGCs in amphibians and mammals, germ cells in birds and reptiles migrate primarily by means of the bloodstream (Figure 22.8). When the blood vessels form in the ger-minal crescent, the PGCs enter the vessels and are carried by the circula-

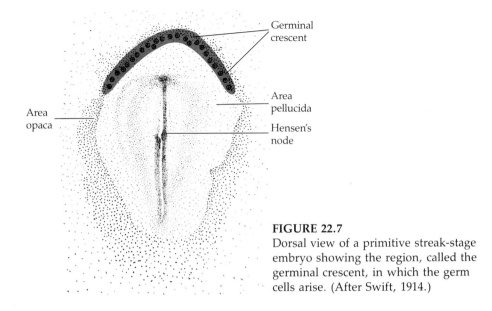

FIGURE 22.7
Dorsal view of a primitive streak-stage embryo showing the region, called the germinal crescent, in which the germ cells arise. (After Swift, 1914.)

tion to the region where the hindgut is forming. Here, they exit from the circulation, become associated with the mesentery, and migrate into the genital ridges (Swift, 1914; Kuwana, 1993). The PGCs of the germinal crescent appear to enter the blood vessels by **diapedesis**, a type of movement common to lymphocytes and macrophages that enables cells to squeeze between the endothelial cells of small blood vessels.

The PGCs thus enter the embryo by being transported in the blood (Pasteels, 1953; Dubois, 1969). The PGCs must also "know" to get out of the blood when they reach the developing gonad (Figure 22.8B). When the germinal crescent of a chick embryo is removed, and the circulation of that embryo is joined with that of a normal chick embryo, the primordial germ cells from the normal embryo will migrate into both sets of gonads (Simon, 1960). It is not known what causes this attraction for the genital ridges. One possibility is that the developing gonad produces a chemotactic substance that attracts PGCs and retains them in the capillaries bordering the gonad (Rogulska, 1969). (Such substances are known to be secreted by lymphocytes at the sites of infection in order to attract macrophages to that area and to permit them to pass through the capillary wall by diapedesis.) Evidence for such chemotaxis came from studies (Kuwana et al., 1986) in which circulating chick PGCs were isolated from

FIGURE 22.8
Primordial germ cells in the chick embryo. (A) Scanning electron micrograph of chick PGC in a capillary of a gastrulating embryo. The PGC can be identified by its large size and the microvilli on its surface. (B) Transverse section near the prospective gonadal region of a chick embryo. Several PGCs within the blood vessel cluster next to the epithelium. One PGC is crossing through the blood vessel endothelium and another PGC is already located within the epithelium. (A from Kuwana, 1993, courtesy of T. Kuwana; B after Romanoff, 1960.)

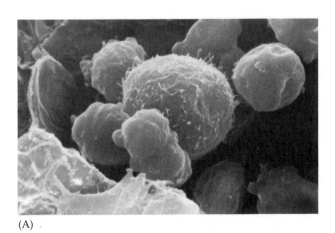

(A)

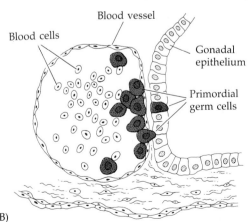

(B)

the blood and cultured between gonadal rudiments and other embryonic tissues. The PGCs migrated into the gonadal rudiments during a 3-hour incubation.

Another possibility is that the endothelial cells of the gonadal capillaries have a cell surface compound that causes the PGCs to adhere there specifically. Using monoclonal antibodies that recognize different cell surface molecules, Auerbach and Joseph (1984) have shown that the endothelial cells of several capillary networks have different cell membrane components and that the endothelial cells from ovarian capillaries differ from all others tested.* Both chemotaxis and differential cell adhesion mechanisms may be working. Whatever these factors may be, they are not species-specific. The chicken gonad attracts circulating PGCs from the turkey and even the mouse (Reynaud, 1969; Rogulska et al., 1971).

Meiosis

Once in the gonad, primordial germ cells continue to divide mitotically, producing millions of potential gametes. The PGCs of both male and female gonads are then faced with the necessity of reducing their chromosome number from the diploid to the haploid condition. In the haploid condition, each chromosome is represented only once, whereas diploid cells have two copies of each chromosome. To accomplish this reduction, the male and female germ cells undergo meiosis.

After the last mitotic division, a period of DNA synthesis occurs, so that the cells initiating meiosis have twice the normal amount of DNA in their nuclei. In this state, each chromosome consists of two sister chromatids attached at a common centromere. Meiosis (shown in Figure 1.13) entails two cell divisions. (In other words, the diploid nucleus contains four copies of each chromosome, but the chromosomes are seen as two chromatids bound together.) In the first division, homologous chromosomes (for example, the chromosome 3 pair in the diploid cell) come together and are then separated into different cells. Hence, the first meiotic division separates homologous chromosomes into the two daughter cells such that each cell has only one copy of each chromosome. But each of the chromosomes has already replicated. The second meiotic division then separates the two sister chromatids from each other. Consequently, each of the four cells produced by meiosis has a single (haploid) copy of each chromosome.

The first meiotic division begins with a long prophase, which is subdivided into five parts. During the **leptotene** (Gk., "thin thread") stage, the chromatin of the chromatids is stretched out very thinly so that it is not possible to identify individual chromosomes. DNA replication has already occurred, however, and each chromosome consists of two parallel chromatids. At the **zygotene** (Gk., "yoked threads") stage, homologous chromosomes pair side by side. This pairing is called **synapsis**, and it is characteristic of meiosis. Such pairing does not occur during mitotic divi-

*A similar situation is thought to occur when lymphocytes migrate through the bloodstream and leave the circulation when they enter the capillary bed of a particular lymphoid organ. The mechanism for this "homing" and organ specificity involves the lymphocytes' ability to specifically adhere to the blood vessel endothelial cells in these organs. Peripheral lymph node endothelial cells contain a glycoprotein, a **selectin**, in their cell membranes that is essential for the binding and exit of those lymphocytes that can recognize it. For each selectin on these endothelial cells, there is a complementary molecule on the lymphocytes that can recognize it (Gallatin et al., 1983, 1986).

(A)

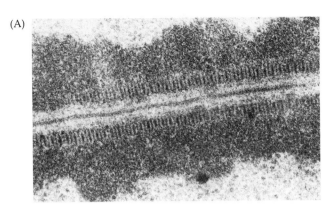

(B)

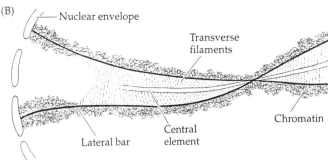

- Nuclear envelope
- Transverse filaments
- Chromatin
- Central element
- Lateral bar

FIGURE 22.9

The synaptonemal complex. (A) Homologous chromosomes held together at the pachytene stage during meiosis in the *Neottiella* oocyte. (B) Interpretive diagram of the synaptonemal complex structure. (A from von Wettstein, 1971, courtesy of D. von Wettstein; B after Moens, 1974.)

sions. Although the mechanism whereby each chromosome recognizes its homologue is not known, pairing seems to require the presence of the nuclear membrane and the formation of a proteinaceous ribbon called the **synaptonemal complex**. This complex is a ladderlike structure with a central element and two lateral bars (von Wettstein, 1984). The chromatin is associated with the two lateral bars and the chromatids are thus joined together (Figure 22.9). Examinations of the meiotic cell nuclei with the electron microscope (Moses, 1968; Moens, 1969) suggest that paired chromosomes are bound to the nuclear membrane, and Comings (1968) has suggested that the nuclear envelope aids in bringing together the homologous chromosomes. The configuration formed by the four chromatids and the synaptonemal complex is referred to as a **tetrad** or a **bivalent**.

During the next stage of meiotic prophase, the chromatids thicken and shorten. This stage has therefore been called the **pachytene** (Gk., "thick thread") stage. Individual chromatids can now be distinguished under the light microscope, and crossing over may occur. Crossing over represents exchanges of genetic material whereby genes from one chromatid are exchanged with homologous genes from another chromatid. This crossing over continues into the next stage, the **diplotene** (Gk., "double threads"). Here, the synaptonemal complex breaks down and the two homologous chromosomes start to separate. Usually, however, they are seen to remain attached at various places called **chiasmata**, which are thought to represent regions where crossing over is occurring (Figure 22.10). The diplotene stage is characterized by a high level of gene transcription. In some species, the chromosomes of both male and female germ cells take on the "lampbrush" appearance characteristic of chromosomes that are actively making RNA. During the next stage, **diakinesis** (Gk., "moving apart"), the centromeres move away from each other and the chromosomes remain joined only at the tips of the chromatids. This last stage of meiotic prophase ends with the breakdown of the nuclear membrane and the migration of the chromosomes to the metaphase plate.

During anaphase I, homologous chromosomes are separated from each other in an independent fashion. This stage leads to telophase I, during which two daughter cells are formed, each cell containing one partner of the homologous chromosome pair. After a brief **interkinesis**, the second division of meiosis takes place. During this division, the centromere of each chromosome divides during anaphase so that each of the new cells gets one of the two chromatids; the final result being the creation of four haploid cells. Note that meiosis has also reassorted the chromosomes into new groupings. First, each of the four haploid cells has a different assortment of chromosomes. In humans, where there are 23 different chromosome pairs, there can be 2^{23} (nearly 10 million) different

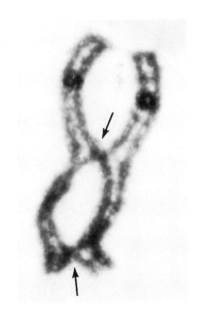

FIGURE 22.10

Chiasmata in diplotene bivalent chromosomes of salamander oocytes. Centromeres are visible as darkly staining circles; arrows point to the two chiasmata. (Courtesy of J. Kezer.)

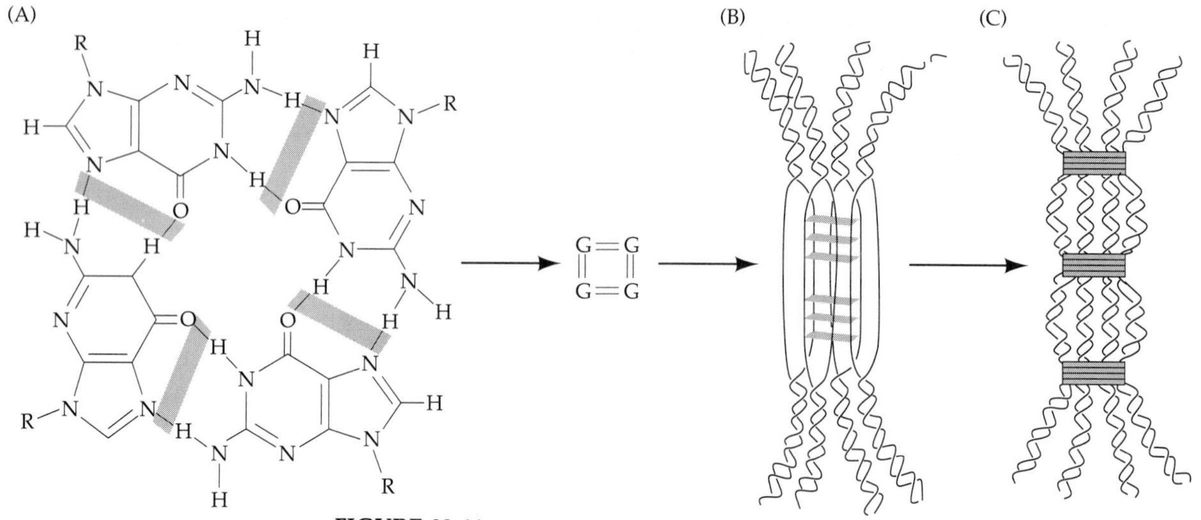

FIGURE 22.11

Sen and Gilbert hypothesis of homologue pairing during meiosis. (A) Four DNA strands may form stable tetrads through hydrogen bonding among four guanosine residues. (B) Diagram of such a stabilized complex showing regions of guanosine tetrads (colored squares). In the region of these tetrads, normal hydrogen bonding between complementary strands is broken. (C) Diagram of the four associated chromatids on the nuclear envelope. (After Sen and Gilbert, 1988.)

types of haploid cells formed from the genome of a single person. In addition, the crossing over that occurs during the pachytene and diplotene stages of prophase I further increases genetic diversity and makes the number of different gametes incalculable.

The mechanism for synapsis is unknown. One hypothesis is that certain regions of DNA may not replicate during the premeiotic S phase. The chromosomes would be coupled through the base pairs at these sites. Hotta and co-workers (1984) found just such sequences in the lily, wherein certain single-stranded regions remain in the DNA throughout the formation of the synaptonemal complex and disappear as the complex disassembles. This differentially replicating DNA is rich in guanosine and cytosine, and this pattern of late replication is not seen in the germ cells before or after this period, nor is it seen in any nongerm cell. Sen and Gilbert (1988) have extended these observations into a model in which the meiotic chromatids are held together by these late-replicating regions of the DNA helices. Following up an observation that certain DNA fragments would not migrate properly in a gel, Sen and Gilbert found that these guanosine-rich sequences self-associated into four-stranded structures at physiological conditions. Chemical analysis showed that these sequences were running parallel to each other, not antiparallel as in the normal double helices. Such four-stranded parallel helices would be possible because of strong hydrogen bonding between the guanosines (Figure 22.11). According to this model, similar G-rich regions at the telomeres would bring homologous chromatids together at the nuclear envelope. (This is where synapsis is often seen to be initiated.) Once joined at the telomeres, the "zipping up" of the homologous chromatids would be accomplished by matching the internal G-rich regions together. The G-rich regions may be accessible due to local unwinding that happens when large stretches of G are on one strand of a helix. The G-rich regions would act as "velcro" for the zipper. In this way, the interaction of the four double helices at meiosis would be caused by the interaction of G-rich regions on four of the strands.

Big decisions: Mitosis or meiosis? Sperm or egg?

In many species, the germ cells migrating into the gonad are bipotential and can differentiate into sperm or ova, depending on their gonadal environment. When the ovaries of salamanders are transformed into testes (by grafting a mature testis into the female), the resident germ cells cease their oogenic differentiation and begin developing as sperm (Burns, 1930; Humphrey, 1931). Similarly, in the housefly and mouse, the gonad is able to direct the differentiation of the germ cell (McLaren, 1983; Inoue and Hiroyoshi, 1986). Thus, in most organisms, the sex of the gonads and its germ cells are the same.

But what about hermaphroditic animals, in which the change from sperm production to egg production is a naturally occurring physiological event? How is the same animal capable of producing sperm during one part of its life and oocytes during another part? Using *Caenorhabditis elegans*, Kimble and her colleagues have identified two "decisions" that presumptive germ cells have to make. The first involves whether to enter meiosis or to remain a mitotically dividing stem cell. The second decision concerns whether the meiotic cell is to become an egg or a sperm. The mitotic/meiotic decision is controlled by a single nondividing cell at the end of each gonad, the distal tip cell. The germ cell precursors near this cell divide mitotically, forming the pool of germ cells; but as these cells get further away from the distal tip cell, they enter meiosis. If the distal tip cells are destroyed by a focused laser beam, *all* the germ cells enter meiosis, and if the distal tip cell is placed into a different location in the gonad, germ line stem cells are generated near this new position (Figure 22.12; Kimble, 1981; Kimble and White, 1981). It appears that the distal tip cells secrete some substance that maintains these cells in mitosis and inhibits their meiotic differentiation.

Although we do not yet know how the distal tip cell interacts with the presumptive germ cells, Austin and Kimble (1987) have isolated a mutation that mimics the phenotype obtained when the distal tip cells are removed. All the germ cell precursors of nematodes homozygous for the recessive mutation *glp-1* initiate meiosis, leaving no mitotic population. Instead of the 1500 germ cells usually found in the fourth larval stage of hermaphroditic development, these mutants produce 5–8 sperm cells. When genetic chimeras are made in which wild-type germ cell precursors are found within a mutant larva, the chimeras are able to respond to the distal tip cells and undergo mitosis. However, when mutant germ cell precursors are found within wild-type larvae, they each enter meiosis. Thus, the *glp-1* gene appears to be responsible for enabling the germ cells to respond to the distal tip cell's signal.*

*The *glp-1* gene appears to be involved in a number of inductive interactions in *C. elegans*. You will no doubt recall that *glp-1* is also needed by the AB blastomere to receive inductive signals from the EMS blastomere to form pharyngeal muscles (Chapter 14).

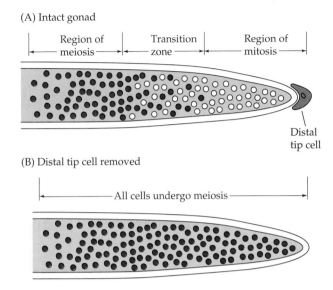

(A) Intact gonad

Region of meiosis — Transition zone — Region of mitosis

Distal tip cell

(B) Distal tip cell removed

All cells undergo meiosis

FIGURE 22.12

Regulation of the mitosis-or-meiosis decision by the distal tip cell of the C. elegans *ovotestis. (A) Intact gonad early in development with regions of mitosis (shaded cells) and meiosis. (B) Gonad after laser ablation of the distal tip cell. All germ cells enter meiosis.*

After the cells begin their meiotic divisions, they still must become either sperm or ova. Generally, in each ovotestis, the most proximal germ cells produce sperm while the most distal (near the tip) become eggs (Hirsh et al., 1976). The genetics of this switch are currently being analyzed. As discussed in the previous chapter, the genes for sex determination generate either a female body that is functionally hermaphroditic or a male body. Certain mutations can alter these phenotypes. For example, *mog* (masculinization of germ line) mutant homozygotes develop as sperm-producing males, and *fem-1* homozygous mutants develop as egg-producing females (Figure 22.13). The double mutants homozygous for both *tra-1* and *fem-1* have a unique phenotype. They are somatically male but are female in the germ line (Doniach and Hodgkin, 1984). This suggests that *tra-1* is the primary sex-determining gene of the somatic tissues, but that the *fem* genes are responsible for the sperm/oocyte decision.

The laboratories of Hodgkin (1985) and Kimble (1986) have recently isolated several genes needed for germ cell pathway selection. Figure 22.14 presents a scheme for how these genes might function in the switch from sperm formation to oocyte formation. During early development, the *fem* genes, especially *fem-3*, are critical for the specification of sperm cells. Loss-of-function mutations of these genes convert XX nematodes into females (i.e., spermless hermaphrodites). As long as the fem proteins are made in the germ cells, sperm are produced. The active *fem* genes are thought to activate the *fog* genes (whose loss of function mutations cause the feminization of the germ line and eliminate spermatogenesis). The *fog* gene products activate the

(A) Wild-type: Sperm and oocytes

Spermatheca (sperm storage region) | Mature sperm | First oocyte

Early stages of spermatogenesis

(B) Feminized: Ooctyes only

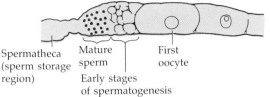

Spermatheca (empty) | First oocyte

(C) Masculinized: Sperm only

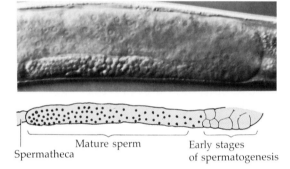

Mature sperm | Early stages of spermatogenesis

Spermatheca

FIGURE 22.13

Gonads of wild-type and mutant C. elegans. *(A) Wild-type hermaphrodite producing first sperm and then eggs. (B) Female animal produced by the* fem-1 *mutation. Only eggs are produced. (C) Masculinized hermaphrodite produced by loss-of-function mutations of* mog *genes (or mutations of the 3' UTR of* fem-3*) produce only sperm, (Photographs courtesy of J. Kimble.)*

genes involved in transforming the germ cell into sperm and also inhibit those genes that would otherwise direct the germ cells to initiate oogenesis. Oogenesis can begin only when the fem activity is suppressed. This suppression appears to act at the level of RNA translation. The 3' untranslated region (3'UTR) of the *fem-3* mRNA contains a sequence that binds a repressor during normal development. If this region is mutated such that the repressor protein cannot bind, the *fem-3* mRNA remains translatable, and oogenesis never occurs. The result is a hermaphrodite body that only produces sperm (Ahringer and Kimble, 1991; Ahringer et al., 1992). The *trans*-acting repressor factor has not yet been identified, but it is likely to be the product of one of the *mog* genes (Graham and Kimble, 1993). It is thought that proteins or messages stored in the oocyte may control the timing of this process, such that spermatogenesis occurs as long as there are repressors of *mog* expression. When these maternal inhibitors of *mog* expression decay, the mog proteins are able to inhibit the synthesis of fem proteins, thereby switching the gametogenesis from sperm to eggs (Ellis and Kimble, in press).

In *Drosophila*, the germ cells are instructed to differentiate into either sperm or eggs by the gonadal cells. Female gonadal cells make a product that is received by the germ cell and which activates a series of proteins whose activity is critical for the early transcription of the germ cell *Sxl* gene. The proper X:autosome ratio is also needed. By this mechanism, the XX flies get to make eggs while the XY flies make sperm (Burtis, 1993; Oliver et al., 1993).

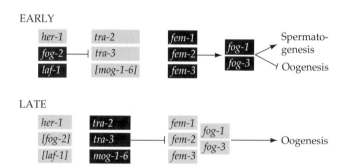

FIGURE 22.14

Model of sex determination in the germ line of C. elegans *hermaphrodites, based on analysis of mutations. Actively expressed genes are shown in color boxes; inactive genes are in gray boxes. Arrows indicate positive regulation, while blunted lines indicate negative regulation. The expression of those genes in brackets has not been ascertained by genetic means. (After Ellis and Kimble, in press.)*

Spermatogenesis

Once the vertebrate primordial germ cells arrive at the genital ridge of male embryos, they become incorporated into the sex cords. They remain there until maturity, at which time the sex cords hollow out to form the

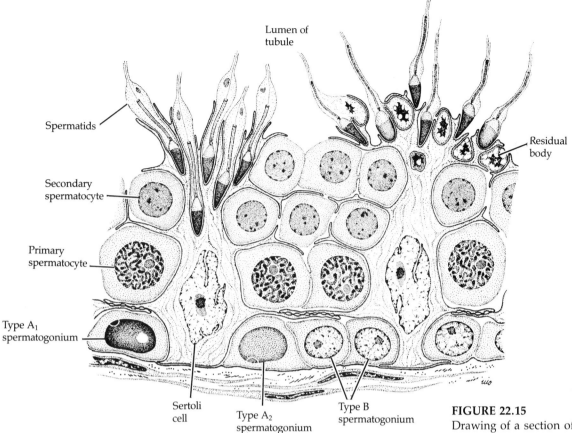

Lumen of
tubule

Spermatids

Secondary
spermatocyte

Primary
spermatocyte

Type A₁
spermatogonium

Residual
body

Sertoli
cell

Type A₂
spermatogonium

Type B
spermatogonium

FIGURE 22.15
Drawing of a section of the seminiferous tubule, showing the relationship between Sertoli cells and the developing sperm. As cells mature, they progress toward the lumen of the seminiferous tubule. (After Dym, 1977.)

seminiferous tubules and the epithelium of the tubules differentiates into the Sertoli cells. The spermatogenic cells are bound to the Sertoli cells by N-cadherin molecules on their respective cell surfaces, and by galactosyltransferase molecules on the spermatogenic cells that bind a receptor on the Sertoli cells (Newton et al., 1993; Pratt et al., 1993). The Sertoli cells nourish and protect the developing sperm cells, and **spermatogenesis**—the meiotic divisions giving rise to the sperm—occurs in the recesses of the Sertoli cells (Figure 22.15). The processes by which the PGCs generate sperm have been studied in detail in several organisms, but we shall focus here on spermatogenesis in mammals. After reaching the gonad, the PGCs divide to form **type A1 spermatogonia**. These cells are smaller than the PGCs and are characterized by an ovoid nucleus that contains chromatin associated with the nuclear membrane. The A1 spermatogonia are found adjacent to the outer basement membrane of the sex cords. At maturity, these spermatogonia are thought to divide so as to make another type A1 spermatogonium as well as a second, paler type of cell, the type **A2 spermatogonium**. Thus, each type A1 spermatogonium is a stem cell capable of regenerating itself as well as producing a new cell type. The A2 spermatogonia divide to produce the A3 spermatogonia, which then beget the type A4 spermatogonia, which beget the intermediate spermatogonia. These **intermediate spermatogonia** are cells that are committed to become spermatozoa. They mitotically divide once to form the **type B spermatogonia**. These cells are the precursors to the spermatocyte and are the last cell that undergoes mitosis. These cells divide once to generate the **primary spermatocytes**—the cells that enter meiosis.

Looking at Figure 22.16, we find that during the spermatogonial divisions, cytokinesis is not complete. Rather, the cells form a syncytium whereby each cell communicates with the other via cytoplasmic bridges

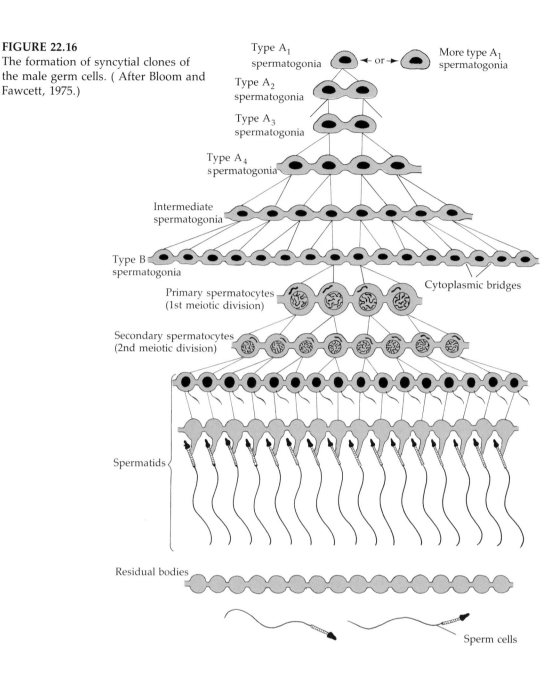

FIGURE 22.16
The formation of syncytial clones of the male germ cells. (After Bloom and Fawcett, 1975.)

Type A_1 spermatogonia ← or → More type A_1 spermatogonia

Type A_2 spermatogonia

Type A_3 spermatogonia

Type A_4 spermatogonia

Intermediate spermatogonia

Type B spermatogonia

Cytoplasmic bridges

Primary spermatocytes (1st meiotic division)

Secondary spermatocytes (2nd meiotic division)

Spermatids

Residual bodies

Sperm cells

about 1 μm in diameter (Dym and Fawcett, 1971). The successive divisions produce clones of interconnected cells, and because ions and molecules readily pass through these intercellular bridges, each cohort matures synchronously.

Each primary spermatocyte undergoes the first meiotic division to yield a pair of **secondary spermatocytes**, which complete the second division of meiosis. The haploid cells formed are called **spermatids**, and they are still connected to each other through their cytoplasmic bridges. The spermatids that are connected in this manner have haploid nuclei but are functionally diploid, since the gene product made in one cell can readily diffuse into the cytoplasm of its neighbors (Braun et al., 1989a). During the divisions from type A1 spermatogonium to spermatid, the cells move farther and farther away from the basement membrane of the seminiferous tubule and closer to its lumen (Figure 22.15). Thus, each type of cell can be found in a particular layer of the tubule. The spermatids are located at the border of the lumen, and here they lose their cytoplasmic connections and differentiate into sperm cells.

Spermiogenesis

The haploid spermatid is a round, unflagellated cell that looks nothing like the mature vertebrate sperm. The next step in sperm maturation, then, is **spermiogenesis** (or **spermateliosis**), the differentiation of the sperm cell. In order for fertilization to occur, the sperm has to meet and bind with the egg, and spermiogenesis differentiates the sperm for these functions of motility and interaction. The processes of mammalian sperm differentiation can be seen in Figure 4.2. The first steps involve the construction of the acrosomal vesicle from the Golgi apparatus. The acrosome forms a cap that covers the sperm nucleus. As the cap is formed, the nucleus rotates so that the acrosomal cap will then be facing the basal membrane of the seminiferous tubule. This rotation is necessary because the flagellum is beginning to form from the centriole on the other side of the nucleus, and this flagellum will extend into the lumen. During the last stage of spermiogenesis, the nucleus flattens and condenses, the remaining cytoplasm (the "cytoplasmic droplet") is jettisoned, and the mitochondria form a ring around the base of the flagellum. The resulting sperm then enter the lumen of the tubule.

In the mouse, the entire development from stem cell to spermatozoan takes 34.5 days. The spermatogonial stages last 8 days, meiosis lasts 13 days, and spermiogenesis takes up another 13.5 days. In humans, spermatic development takes 74 days to complete. Because the type A1 spermatogonia are stem cells, spermatogenesis can occur continuously. Each day, some 100 million sperm are made in each human testicle, and each ejaculation releases 200 million sperm. Unused sperm are either resorbed or passed out of the body in urine.

Gene expression during sperm development

Gene transcription during spermatogenesis takes place predominantly during the diplotene stage of meiotic prophase. This transcription has been observed in many organisms, but the best-documented case is probably that of Y chromosome transcription in *Drosophila hydei*. Here, RNA transcripts originating from the Y chromosome are seen to be essential for controlling spermiogenesis. When we recall the function of the Y chromosome in *Drosophila*, this is not surprising, for the Y chromosome is not involved in sex determination here. Rather, it is needed for the formation of viable sperm. The difference between XY *Drosophila* and XO *Drosophila* is that the latter are sterile. Both are male. In *D. hydei*, the Y chromosome extends five prominent loops of DNA. If any one of these loops is deleted, the organization of the sperm tail is abnormal. All the component parts of the sperm are present, but they are not properly organized (Hess, 1973). It appears, then, that Y-specific RNA made during meiotic prophase is utilized later during spermiogenesis.

The genes that are transcribed specifically during spermatogenesis are often those whose products are necessary to sperm motility or binding to the egg. In *Drosophila melanogaster*, one of the sperm-specific genes transcribed is for β2-tubulin. This isoform of β-tubulin is only seen during spermatogenesis, and it is responsible for forming the meiotic spindles, the axoneme, and the microtubules associated with the lengthening mitochondria.* Hoyle and Raff (1990) have shown that another β-tubulin isoform, β3-tubulin (which is normally expressed in mesodermal cells and epidermis) cannot substitute for the β2-tubulin. When they fused the 5′

*Making the sperm axoneme in *Drosophila* is a large undertaking. The sperm tail is 2 mm long—as long as the entire male fly! Even more remarkable, the egg incorporates the entire sperm (Karr, 1991).

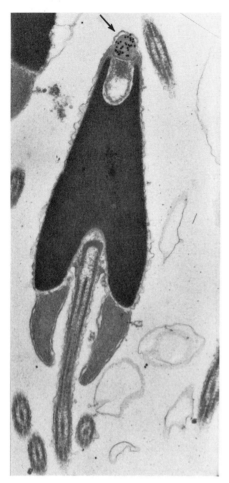

FIGURE 22.17
Localization of bindin in sperm acrosome by gold-labeled antibindin antibodies. The gold atoms enable the antibodies to show up as black dots in the electron micrograph. These sperm are still inside the sea urchin testis. (Courtesy of D. Nishioka.)

regulatory region from the β2-tubulin gene to the coding sequences of the β3-tubulin gene, the β3-tubulin gene was able to be expressed in the developing sperm. When this gene was expressed in the absence of the β2-tubulin gene, the resulting germ cells failed to undergo meiosis, axoneme assembly, or nuclear shaping. Only the mitochondrial elongation occurred. This indicates that the formation of the meiotic spindles and axoneme of sperm cells cannot be accomplished by just any β-tubulin and that the transcription of sperm-specific isoforms is important.

Those genes whose products are necessary for the binding of the sperm and the extracellular matrices of the egg are also transcribed during spermatogenesis. The gene for sea urchin bindin is transcribed relatively late in spermatogenesis, and its mRNA is translated into bindin shortly after being made (Nishioka et al., 1990). The bindin accumulates in vesicles that fuse together to form the single acrosomal vesicle of the mature sea urchin sperm. Figure 22.17 shows the localization of the bindin protein in the acrosomal vesicle of the sperm while it is still in the testis.

In addition to gene transcription during meiotic prophase, there is also evidence that certain genes are transcribed in the spermatids (reviewed in Palmiter et al., 1984). This evidence for **haploid gene expression** comes from studies involving heterozygous mice in which two different populations of sperm are seen to exist—one population expressing the mutant phenotype and one population expressing the wild-type trait. If the synthesis of the RNA or protein were to occur while the cells were still diploid, all the sperm would show the same phenotype. In mice, the sperm protamine proteins that compact the sperm during spermiogenesis appear to be transcribed during the spermatid stage (Peschon et al., 1987). Similarly, the gene for the β1,4 galactosyltransferase that binds the sperm to the zona pellucida is transcribed only during the haploid phase of mouse sperm maturation (Hardvin-Lepers et al., 1993).

In some species, sperm provide important developmental information that cannot be compensated for by the egg. We have discussed the imprinting of mammalian chromosomes wherein the sperm and egg DNA differ in their methylation patterns (Chapters 4 and 11). There are also cases of **paternal effect genes**. Here, homozygous recessive alleles in the male cause abnormal development in the embryo even if the female is homozygous for the wild-type allele, while the reciprocal cross, where the father is wild-type and the mother is homozygous for the mutant allele, leads to normal embryos. One such paternal effect gene is *spe-11* in *C. elegans*. The sperm containing mutant alleles at this locus are unable to direct chromosomal movements that orient the mitotic spindle of the embryo, suggesting that the mutation affects the microtubule-organizing regions such as centrioles (Figure 22.18; Hill et al., 1989).

Eventually, the haploid genome is condensed as the histones are re-

FIGURE 22.18
Immunofluorescence photomicrographs of mitotic spindles in a first cleavage *C. elegans* embryo when sperm is (A) from a wild-type male, or (B) from a male homozygous for the paternal effect gene *spe-11*. In (B), three microtubule organizing centrioles can be seen instead of the usual two poles of mitosis. (From Hill et al., 1989, courtesy of S. Strome.)

(A)

(B)

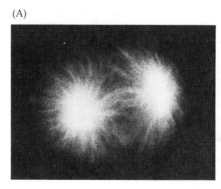
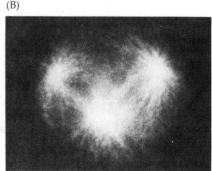

placed by protamines or by specifically modified histones. Many of the sperm histones become modified in the late spermatid stage during spermiogenesis. These modifications (such as dephosphorylating the *N*-terminal regions of certain histones) causes the chromatin to condense. Condensation results in severely reduced transcription. Thus, transcription from the male genome is not detected again until it is reactivated sometime during development (Poccia, 1986; Green and Poccia, 1988).

Oogenesis

Oogenic meiosis

Oogenesis—the differentiation of the ovum—differs from spermatogenesis in several ways. Whereas the gamete formed by spermatogenesis is essentially a motile nucleus, the gamete formed by oogenesis contains all the factors needed to initiate and maintain metabolism and development. Therefore, in addition to forming a haploid nucleus, oogenesis also builds up a store of cytoplasmic enzymes, mRNAs, organelles, and metabolic substrates. So while the sperm becomes differentiated for motility, the oocyte develops a remarkably complex cytoplasm.

The mechanisms of oogenesis vary more than those of spermatogenesis. This difference should not be surprising, since the patterns of reproduction vary so greatly among species. In some species, such as sea urchins and frogs, the female routinely produces hundreds or thousands of eggs at a time, whereas in other species, such as humans and most mammals, only a few eggs are produced during the lifetime of an individual. In those species that produce thousands of ova, the oogonia are self-renewing stem cells that endure for the lifetime of the organism. In those species that produce fewer eggs, the oogonia divide to form a limited number of egg precursor cells. In humans, the thousand or so oogonia divide rapidly from the second to the seventh month of gestation to form roughly 7 million germ cells (Figure 22.19). After the seventh month of embryonic development, however, the number of germ cells drops precipitously. Most oogonia die during this period, while the remaining oo-

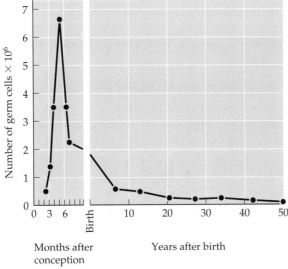

FIGURE 22.19
Changes in the number of germ cells in the human ovary. (After Baker, 1970.)

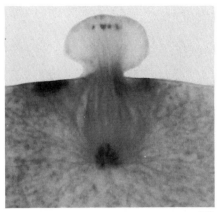

(A)

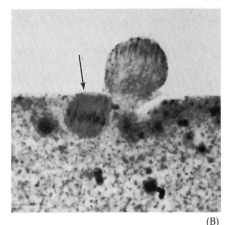

(B)

FIGURE 22.20
Polar body formation in the oocyte of the whitefish *Coregonus*. (A) Anaphase of first meiotic division, showing the first polar body pinching off with its chromosomes. (B) Metaphase (within the oocyte, arrow) of the second meiotic division, with the first polar body still in place. The first polar body may or may not divide again. (From Swanson et al., 1981, courtesy of C. P. Swanson.)

gonia enter prophase of the first meiotic division (Pinkerton et al., 1961). These latter cells, called the **primary oocytes**, progress through the first meiotic prophase until the diplotene stage, at which point they are maintained until puberty in the first meiotic prophase. With the onset of adolescence, groups of oocytes periodically resume meiosis. Thus, in the human female, the first part of meiosis is begun in the embryo, and the signal to resume meiosis is not given until roughly 12 years later. In fact, some oocytes are maintained in meiotic prophase for nearly 50 years. As Figure 22.19 indicates, primary oocytes continue to die even after birth. Of the millions of primary oocytes present at birth, only about 400 mature during a woman's lifetime.

Oogenic meiosis also differs from spermatogenic meiosis in its placement of the metaphase plate. When the primary oocyte divides, the nucleus of the oocyte, called the **germinal vesicle**, breaks down and the metaphase spindle migrates to the periphery of the cell. At telophase, one of the two daughter cells contains hardly any cytoplasm, whereas the other cell has nearly the entire volume of cellular constituents (Figure 22.20). The smaller cell is called the **first polar body**, and the larger cell is referred to as the **secondary oocyte**. During the second division of meiosis, a similar unequal cytokinesis takes place. Most of the cytoplasm is retained by the mature egg (ovum), and a second polar body receives little more than a haploid nucleus. Thus, oogenic meiosis serves to conserve the volume of oocyte cytoplasm in a single cell rather than splitting it equally among four progeny.

In a few species of animals, meiosis is severely modified such that the resulting gamete is diploid and need not be fertilized to develop. Such animals are said to be **parthenogenetic**. In the fly *Drosophila mangabeirai*, one of the polar bodies acts as a sperm and "fertilizes" the oocyte after the second meiotic division. In other insects (such as *Moraba virgo*) and the lizard *Cnemidophorus uniparens*, the oogonia double their chromosome number before meiosis, so that the halving of chromosomes restores the diploid number. The germ cells of the grasshopper *Pycnoscelus surinamensis* dispense with meiosis altogether, forming diploid ova by two mitotic divisions (Swanson et al., 1981).

In the above examples, the species consists entirely of females. In other species, haploid parthenogenesis is widely used, not only as a means of reproduction, but also as a mechanism of sex determination. In the Hymenoptera (bees, wasps, and ants), unfertilized eggs develop into males whereas fertilized eggs, being diploid, develop into females. The haploid males are able to produce sperm by abandoning the first meiotic division, thereby forming two sperm cells through second meiosis.

Maturation of the oocyte in amphibians

The egg is responsible for initiating and directing development, and in some species (as seen earlier) fertilization is not even necessary. The accumulated material in the oocyte cytoplasm includes energy sources and organelles (the yolk and mitochondria); the enzymes and precursors for DNA, RNA, and protein syntheses; stored messenger RNAs; structural proteins; and morphogenetic regulatory factors that control early embryogenesis. A partial catalogue of the materials stored in the oocyte cytoplasm is shown in Table 22.1. Most of this accumulation takes place during meiotic prophase I, and this stage is often subdivided into **previtellogenic** (Gk., "before yolk formation") and **vitellogenic** (yolk-forming) phases.

Eggs of fishes and amphibians are derived from a stem cell oogonial population, which can generate a new cohort of oocytes each year. In the frog *Rana pipiens*, oogenesis lasts three years. During the first two years, the oocyte increases its size very gradually. During the third year, however, the rapid accumulation of yolk in the oocytes causes them to swell to their characteristic large size (Figure 22.21). Eggs mature in yearly batches, the first cohort maturing shortly after metamorphosis; the next group matures a year later.

Vitellogenesis occurs when the oocyte reaches the diplotene stage of meiotic prophase. Yolk is not a single substance, but a mixture of materials used for embryonic nutrition. The major yolk component is a 470-kDa protein called **vitellogenin**. It is not made in the frog oocyte (as are the major yolk proteins of organisms such as annelids and crayfish), but it is synthesized in the liver and carried by the bloodstream to the ovary (Flickinger and Rounds, 1956). This large protein passes between the follicle cells of the ovary and is incorporated into the oocyte by **micropinocytosis**, the pinching off of membrane-bounded vesicles at the base of microvilli (Dumont, 1978). In the mature oocyte, vitellogenin is split into two smaller proteins: the heavily phosphorylated **phosvitin** and the lipoprotein **lipovitellin**. These two proteins are packaged together into membrane-bounded **yolk platelets** (Figure 22.22A). Glycogen granules and lipochondrial inclusions store the carbohydrate and lipid components of the yolk, respectively.

Most eggs are highly asymmetrical, and it is during oogenesis that the animal–vegetal axis of the egg is specified. Danilchik and Gerhart (1987) have shown that although the concentration of yolk in *Xenopus* oocytes increases nearly tenfold as one goes from the animal to the vegetal poles of the mature egg, vitellogenin uptake is uniform around the surface of the oocyte. What differs is its movement *within* the oocyte, and this depends upon where the yolk proteins enter. When yolk platelets are formed in the future animal hemisphere, they move inward toward the center of

TABLE 22.1
Cellular components stored in the mature oocyte of *Xenopus laevis*

Component	Approximate excess over amount in larval cells
Mitochondria	100,000
RNA polymerases	60,000–100,000
DNA polymerases	100,000
Ribosomes	200,000
tRNA	10,000
Histones	15,000
Deoxyribonucleoside triphosphates	2,500

Source: After Laskey, 1974.

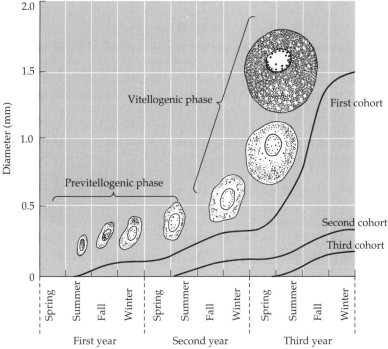

FIGURE 22.21
Growth of oocytes in the frog. During the first three years of life, three cohorts of oocytes are produced. The drawings follow the growth of the first-generation oocytes. (After Grant, 1953.)

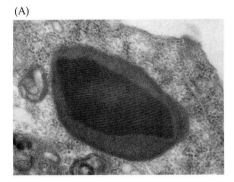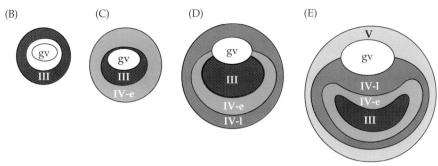

FIGURE 22.22
Yolk distribution in *Xenopus*. (A) An amphibian yolk platelet. (B–E) Establishment of animal–vegetal polarity of yolk platelets in the *Xenopus* oocyte. (B) In the late stage III (600-μm) oocyte, yolk platelets enter the cell equally at all points of the surface. (C,D) As the oocyte grows, the platelets at the future animal pole are displaced toward the vegetal pole, while those at the vegetal pole remain there. More yolk still enters the egg on all sides. (E) By the end of vitellogenesis, the earliest platelets (III) are all in the vegetal hemisphere, which has now concentrated roughly 75 percent of the oocyte yolk. Time of entry of yolk into oocyte platelets is indicated by shading and Roman numerals: III, stage III platelets; IV-e, early stage IV platelets; IV-l, late stage IV platelets; V, stage V platelets; gv, germinal vesicle. (After Danilchik and Gerhart, 1987; photograph courtesy of L. K. Opresko.)

the cell. The vegetal yolk platelets, however, do not actively move but remain at the periphery for long periods of time, enlarging as they stay there. They are slowly displaced from the cortex as new yolk platelets come in from the surface. As a result of this differential intracellular transport, the amount of yolk steadily increases in the vegetal hemisphere, until the vegetal half of a mature *Xenopus* oocyte contains nearly 75 percent of the yolk (Figure 22.22B–E). The mechanism of this translocation remains unknown.

As the yolk is being deposited, the organelles also become arranged asymmetrically. The cortical granules begin to form from the Golgi apparatus and are originally scattered randomly through the oocyte cytoplasm. They later migrate to the periphery of the cell. The mitochondria replicate at this time, dividing to form millions of organelles that will be apportioned to the different cells during cleavage. (In *Xenopus*, new mitochondria are not formed until after gastrulation is initiated.) As vitellogenesis nears an end, the oocyte cytoplasm becomes stratified. The cortical granules, mitochondria, and pigment granules are found at the periphery of the cell, within the actin-rich oocyte cortex. Within the inner cytoplasm, distinct gradients emerge. While the yolk platelets become more heavily concentrated at the vegetal pole of the oocyte, the glycogen granules, ribosomes, lipid vesicles, and endoplasmic reticulum are found more toward the animal pole. Even specific mRNAs stored in the cytoplasm become localized to certain regions of the oocyte.

While the precise mechanisms for establishing these gradients remain unknown, studies using inhibitors have shown that the cytoskeleton is critically important in localizing specific RNAs and morphogenetic factors. Yisraeli and his co-workers (1990) have shown that the Vg1 mRNA is translocated into the vegetal cortex in a two-step process. In the first phase, microtubules are needed to bring the Vg1 mRNA into the vegetal hemisphere. In the second phase, microfilaments are responsible for anchoring the Vg1 message to the cortex. The portion of the Vg1 mRNA that binds to these cytoskeletal elemnts appears to reside in the 3' untranslated region. When a specific 340-base sequence is placed onto a β-globin message, that β-globin mRNA is similarly localized to the vegetal cortex (Chapter 13; Mowry and Melton, 1992).

In *Xenopus*, the leptotene stage of meiosis lasts only 3 to 7 days, zygotene takes from 5 to 9 days, and pachytene persists for roughly 3 weeks. The diplotene stage, however, can last for years. Even so, vitellogenesis occurs in only part of the diplotene, and the signal for the breakdown of the nucleus (the germinal vesicle) occurs after vitellogenesis is completed. The regulation of these events is controlled by the hormonal interactions between the hypothalamus, pituitary gland, and follicle cells of the ovary (Figure 22.23). When the hypothalamus receives the cues that the mating season has arrived, it secretes gonadotropin-releasing hor-

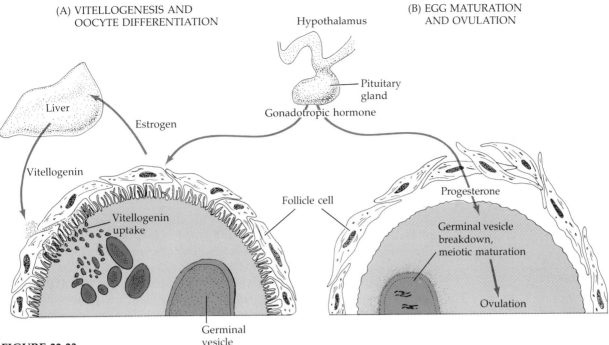

(A) VITELLOGENESIS AND OOCYTE DIFFERENTIATION

Hypothalamus

(B) EGG MATURATION AND OVULATION

Pituitary gland

Gonadotropic hormone

Liver

Estrogen

Vitellogenin

Vitellogenin uptake

Follicle cell

Germinal vesicle

Progesterone

Germinal vesicle breakdown, meiotic maturation

Ovulation

FIGURE 22.23

Control of amphibian oocyte growth and egg maturation by estrogen and progesterone. (A) Gonadotropic hormone stimulates the follicle cells to produce estrogen, which instructs the liver to secrete vitellogenin. This protein is absorbed by the oocyte. (B) After vitellogenesis, again under the influence of gonadotropic hormone, the follicle cells secrete progesterone. Within 6 hours of progesterone stimulation, the germinal vesicle breaks down, initiating the reactions leading to ovulation. (After Browder, 1980.)

mone, which is received by the pituitary. The pituitary responds by secreting the gonadotropic hormones into the blood. These hormones stimulate the follicle cells to secrete estrogen, and the estrogen instructs the liver to synthesize and secrete vitellogenin. Upon estrogen stimulation, the liver cells change drastically (Figure 22.24). These changes are usually brought about during the annual mating season, but when estrogen is injected into adult frogs (either male or female) at any time, these changes can be induced and vitellogenin secreted (Skipper and Hamilton, 1977).

FIGURE 22.24

Effects of estrogen (estradiol) on the ultrastructure of *Xenopus laevis* liver cells. (A) Hepatocytes prior to estrogen injection. (B) After estrogen injection, hepatocytes show enormous expansion of the rough endoplasmic reticulum and Golgi apparatus needed for production and secretion of vitellogenin. (From Skipper and Hamilton, 1977, courtesy of the authors.)

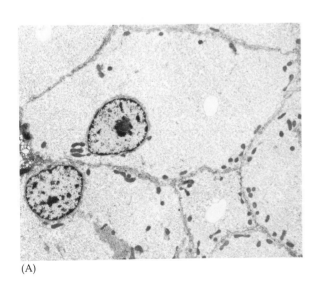

(A)

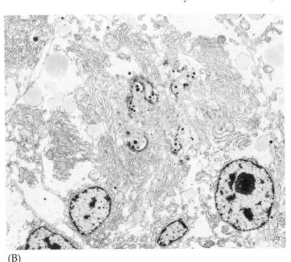

(B)

Estrogen induces vitellogenin at both the transcriptional and translational levels. Before estrogen release, there are no detectable vitellogenin mRNAs in the liver. Following hormone administration, each cell contains some 50,000 vitellogenin mRNA molecules, constituting roughly half the total cellular mRNA. The transcriptional activation of the vitellogenin genes would normally generate only about 1500 of these mRNA molecules per cell. However, estrogen also specifically stabilizes that message, somehow increasing its half-life from 16 hours to 3 weeks. Because these effects occur in the absence of new protein synthesis, they appear to be direct consequences of estrogen (Brock and Shapiro, 1983).

Completion of meiosis: Progesterone and fertilization

Amphibian oocytes can stay years in the diplotene stage of meiotic prophase. Resumption of meiosis in the amphibian primary oocyte requires progesterone. This hormone is secreted by follicle cells in response to the gonadotropic hormones secreted by the pituitary gland. Within 6 hours of progesterone stimulation, **germinal vesicle breakdown (GVBD)** occurs, the microvilli retract, the nucleoli disintegrate, and the lampbrush chromosomes contract and migrate to the animal pole to begin division. Soon afterward, the first meiotic division occurs, and the mature ovum is released from the ovary by a process called **ovulation**. The ovulated egg is in second meiotic metaphase when it is released.

How does progesterone enable the egg to break its dormancy and resume meiosis? To understand the mechanisms by which this activation is accomplished, it is necessary to briefly review the model for early blastomere division that was presented in Chapter 5. Maturation-promoting factor (MPF) is responsible for the resumption of meiosis. Its activity is cyclic, being high during cell division and undetectable during interphase. MPF is a protein kinase that contains an enzymatic subunit (cyclin). Since all the components of MPF are present in the oocyte, it is generally thought that progesterone somehow converts a pre-MPF complex into active MPF, perhaps by the activation of the cdc25 phosphatase (see Chapter 5; Minishull, 1993).

The mediator of the progesterone signal is probably the **c-mos** protein. Studies by Sagata and colleagues (1988, 1989) show that progesterone reinitiates meiosis by causing the egg to translate the maternal c-mos mRNA that has been stored in its cytoplasm. This message translates into a 39-kDa phosphoprotein, pp39^{mos}. This protein is detectable only during oocyte maturation and is destroyed quickly upon fertilization. Yet during its brief lifetime, it plays a major role in releasing the egg from its dormancy. If the translation of pp39^{mos} is inhibited (by injecting *mos*-antisense mRNA into the oocyte), pp39^{mos} does not appear, and germinal vesicle breakdown and the renewal of oocyte maturation do not occur. Having stimulated the reinitiation of meiosis, pp39^{mos} enables the oocyte to undergo a meiotic division, but freezes the second meiotic cycle in metaphase. This metaphase block is caused by the combined actions of pp39^{mos} and another protein, cyclin-dependent kinase 2 (cdk2; Gabrielli et al., 1993). These two proteins are now thought to constitute the **cytostatic factor** that is found in mature frog eggs, which can block cell cycles in metaphase (Masui, 1974). The cdk2 protein, in cooperation with pp39^{mos}, stimulates the second meiotic division and "arrests" this division in metaphase. It is possible that these proteins act to prevent the degradation of cyclin.

The next question concerns the mechanisms by which fertilization enables the oocyte that is at second metaphase to complete that division

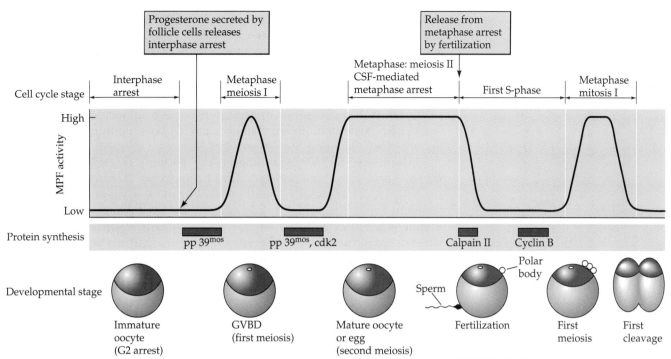

FIGURE 22.25
Schematic representation of *Xenopus* oocyte maturation, showing the regulation of meiotic cell division by pp39^mos, cdk2, and calpain II. The solid line on the graph represents the relative levels of active MPF. The bars below the graph show the periods when the synthesis of particular proteins are needed for entry into the next M phase. Oocyte morphology is diagrammed below. (After Minishull, 1993.)

to form a haploid gamete. It had been thought that events at fertilization inactivated pp39^mos, and in the absence of this mos protein, the cyclin component of MPF would be degraded. However, new evidence indicates that the inactivation of pp39^mos and the inactivation of MPF are independent events (Watanabe et al., 1991). Cyclin is probably degraded by a pathway activated by the calcium ion flux accompanying fertilization. Shortly thereafter, pp39^mos is degraded, most likely by the activation of **calpain II**, a calcium-dependent protease that specifically cleaves pp39^mos (Watanabe et al., 1989). The pathways by which these events occur has not been mapped in detail. However, it appears that progesterone acts to reinitiate meiosis through the synthesis of pp39^mos, which can take the chromosomes to the interphase of the second meiotic division. Here a combination of pp39^mos and cdk2 are needed to take the cell into the second meiotic division and suspend the division in metaphase. The calcium ion flux released at fertilization is then able to complete meiosis by inactivating the pp39^mos (Figure 22.25).

Gene transcription in oocytes

In most animals (insects being a major exception), the growing oocyte is active in transcribing genes where products are either (1) necessary for cell metabolism, (2) necessary for oocyte-specific processes, or (3) needed for early development before the nuclei begin to function. In mice, for instance, the growing dictyate oocyte is actively transcribing the genes for zona pellucida proteins ZP1, ZP2, and ZP3. Moreover, these genes are *only* transcribed in the oocyte and not in any other cell (Chapter 2).

The amphibian oocyte has certain periods of very active RNA synthesis. During the diplotene stage, the compacted chromosome streches out large loops of DNA, causing the chromosome to resemble a lampbrush (a handy instrument for cleaning test tubes in the days before microfuges). These **lampbrush chromosomes** (Plate 2) can be seen as the sites of mRNA synthesis by in situ hybridization. Oocyte chromosomes can be prepared,

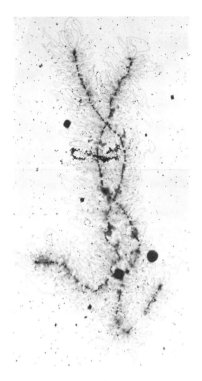

FIGURE 22.26
Localization of histone genes on a lampbrush chromosome. Genes have been visualized by in situ hybridization and autoradiography. (From Old et al., 1977, courtesy of H. G. Callan.)

denatured, and incubated with radioactive RNA that encodes a specific protein. After the unbound RNA is washed away, autoradiography visualizes the precise location of the gene. Figure 22.26 shows diplotene chromosome I of the newt *Triturus cristatus* after incubation with radioactive histone mRNA. It is obvious that a histone gene (or set of histone genes) is located on one of these loops of the lampbrush chromosome (Old et al., 1977). Electron micrographs of gene transcripts from lampbrush chromosomes also enable one to see chains of mRNA coming off each gene as it is transcribed (Hill and MacGregor, 1980).

In addition to mRNA synthesis, the patterns of rRNA and tRNA transcription are also regulated during oogenesis. Figure 22.27 shows ribosomal and transfer RNA synthesis during the *Xenopus* oogenesis. Transcription appears to begin in early (stage I, 25–40 μm) oocytes, during the diplotene stage of meiosis. At this time, all the ribosomal and transfer RNAs needed for protein synthesis until the midblastula stage are made, and all the maternal mRNAs for early development are transcribed. This stage lasts months in *Xenopus*. After reaching a certain size, the chromosomes of the mature (stage VI) oocyte condense and the genes are not actively transcribing. This "mature oocyte" condition can also be months long. Upon hormonal stimulation, the oocyte completes its first meiotic division and is ovulated. The mRNAs stored by the oocyte now join with the ribosomes to initiate protein synthesis. Within hours, the second meiotic division has begun and the secondary oocyte has been fertilized. The genes do not begin active transcription until the midblastula transition (Davidson, 1986).

As we have seen in Chapter 13, the oocytes of several species make two classes of mRNAs—those for immediate use in the oocyte and those that are stored for use during early development. In sea urchins, the translation of stored maternal messages is initiated by fertilization, while in frogs the signal for such translation is initiated by progesterone as the egg is about to be ovulated. One of the actions of the MPF kinase activity induced by progesterone may be the phosphorylation of CPE-binding proteins on the stored oocyte mRNAs. The phosphorylation of these factors is associated with the lengthening of the poly(A) tails in the stored messages and with translation of the stored mRNAs (Paris et al., 1991).

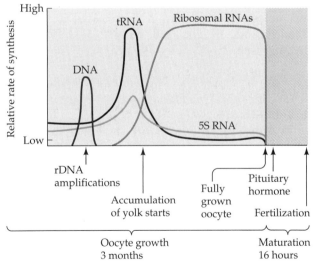

FIGURE 22.27
Relative rates of DNA, tRNA, and rRNA synthesis in amphibian oogenesis during the last three months before ovulation. (From Gurdon, 1976.)

Synthesizing ribosomes

In most species, transcription does not occur immediately upon fertilization. Rather, the initial protein synthesis is based on the translation of mRNAs that are already present within the oocyte. This means that a large number of ribosomes must be present to translate these messages. In many respects, the ribosome can be considered as a differentiated cell product of the oocyte. This is especially so in amphibians, where the oocyte contains around 200,000 times the number of ribosomes found in most somatic cells. For one cell, that number of ribosomes is truly spectacular. In fact, in those zygotes lacking the nucleolar organizing regions that make ribosomal RNA, development can still continue on the stored ribosomes up until the feeding tadpole stage (at which time the organism finally dies from the lack of ribosomes needed for growth).

Synthesis of ribosomal RNA in amphibian oocytes occurs during the prolonged prophase of the first meiotic division. Specifically, rRNA synthesis takes place during the months-long diplotene stage, after pairing and replication of the homologous chromosomes. The "large" rRNA genes of the frog *Xenopus laevis* are diagrammed in Figure 22.28. There is a 5′ leader sequence followed by the coding sequence for 18S rRNA. This is followed by a transcribed spacer, the gene for the 5.8S rRNA gene, another transcribed spacer, and then the gene for the 28S rRNA. There are no introns within these genes, and the entire unit is transcribed into a 40S rRNA precursor, which is then processed to form the three rRNAs.

Ribosomal gene amplification

One of the mechanisms for synthesizing so many ribosomes is the repetition and amplification of the rRNA genes.* There are two homologous regions of large-rRNA genes per diploid frog cell. Each region contains roughly 450 copies of the unit described in the preceding section, and the copies are separated within the region by nontranscribed spacers. All the rRNA gene units on a chromosome are identical, but the nontranscribed spacers vary in length from unit to unit, even on the same chromosome. We would expect, then, that the diplotene oocyte (in which the DNA has already replicated) would contain 4×450, or roughly 1800, genes coding for rRNA. But apparently this is not enough! Rather than a few thousand copies, one finds nearly a million rRNA genes in the maturing oocyte. One sees not four nucleoli, but thousands (Figure 22.29). These extra nucleoli reside within the nucleus but are not attached to any chromosome, as nucleoli usually are. When the hybridization of radioactive 28S rRNA to oocyte DNA was compared with hybridization to somatic cell DNA, it was determined that the rRNA genes were amplified 1500 times over their normal amount. If each nucleolus contains

Repeated genes denotes inherited multiplicity wherein a given sequence is present numerous times in the genome of every cell. The 5S rRNA genes, represented thousands of times per genome, constitutes a set of repeated genes. *Amplified* genes are those DNA sequences that are reiterated in specific cells only. Their multiplicity is not inherited. The 28S, 18S, and 5.8S rRNA genes are repeated (about 900 times per cell) *and* amplified (in the oocyte).

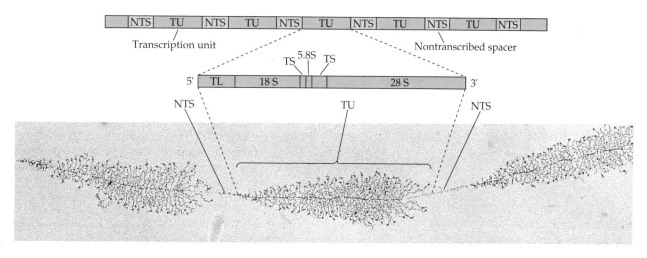

FIGURE 22.28
Organization of the genes for 18S, 5.8S, and 28S ribosomal RNA in amphibians. Each transcription unit (TU) is flanked by nontranscribed spacer regions (NTS). The transcription unit contains a transcribed leader sequence (TL), the 18S rRNA gene, a transcribed spacer (TS), the 5.8S rRNA gene, another

transcribed spacer, and the gene for 28S rRNA. Beneath the diagram is an electron micrograph of tandemly arranged rRNA genes actively transcribing in the newt oocyte. The RNA gets progressively larger as it is transcribed, and dozens of transcripts are made simultaneously. (Photograph courtesy of O. L. Miller, Jr.)

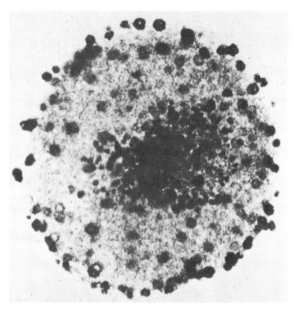

FIGURE 22.29
Nucleus isolated from an oocyte of X. laevis. *The deeply stained spots represent the extrachromosomal nucleoli. (From Brown and Dawid, 1968, courtesy of D. D. Brown.)*

450 gene copies, we would estimate that there are 6.8×10^5 genes for the large-rRNA precursors in the amphibian oocyte (Brown and Dawid, 1968). These extra rDNA genes are made during the pachytene stage of meiosis and are transcribed during the diplotene stage. When the oocyte nucleus disintegrates during the first meiotic division, the extra nucleoli are thrown into the cytoplasm and are destroyed.

Visualization of transcription from amplified genes

The amplified ribosomal RNA genes of amphibian oocytes were the first eukaryotic genes to be isolated and purified. They were also the first genes whose transcription could actually be seen by electron microscopy. Knowing that these genes would be intensely active in transcribing rRNA during the diplotene stage of meiosis, Miller and Beatty (1969) used low-ionic-strength buffers to disperse the chromatic strands for electron microscopy. The circles of DNA unwound to reveal the configuration seen in Figure 22.28. There is an axial core of DNA upon which rRNA is being synthesized. Numerous rRNA chains are being transcribed from any one unit, and several such units are seen in this picture. In each unit, transcription begins at the small end of the "Christmas tree" and continues until the large-rRNA precursor is formed. At each junction of the DNA with an RNA transcript, there is a molecule of RNA polymerase. The length of the RNA in these large transcripts corresponds to about 7200 bases, which is roughly the size estimated for the ribosomal precursor. Between the transcribing units of DNA are the "nontranscribed spacers." These electron micrographs beautifully reveal eukaryotic DNA in the act of transcription.

Regulation of the 5S rRNA gene

In addition to the large-rRNA transcription unit, there is the transcription unit of the 5S ribosomal RNA. The 5S rRNA is made from another region of the genome, and these 5S genes are present in thousands of copies in the genome (and are not amplified.) In each haploid genome, there are 20,000 "oocyte" 5S genes (that are only active in the oocyte) and 400 "somatic" 5S genes (that are functional in every cell). They are *all* transcribed during early oogenesis. During late oogenesis, all the 5S genes are turned off until the midblastula transition takes place. At midblastula, both oocyte and somatic 5S genes are expressed, but by the end of the blastula stage, only the somatic 5S genes are active (Korn, 1982). The RNA polymerase controlling the production of 5S rRNA genes is RNA polymerase III, and it has a unique promoter. Whereas the promoters for RNA polymerases I and II are in the 5′ flanking regions of the gene (as they are in bacteria), the promoters for RNA polymerase III are in the center of the gene (Figure 22.30; Bogenhagen et al., 1980; Sakonju et al., 1980).

As in the case of RNA polymerase I and II, the binding of eukaryotic RNA polymerases to their promoters is mediated by transcription factors. A 38.5-kDa protein binds to the promoter region of the 5S rRNA gene and directs the RNA polymerase III to bind (Ng et al., 1979; Engelke et al., 1980). This protein, **TFIIIA** (i.e., the first transcription factor for RNA polymerase **III**) is specific for the 5S rRNA gene (it does not bind to the promoters of other RNA polymerase III-dependent genes) and is itself developmentally regulated. There are 10^{12} TFIIIA molecules per cell in the oocyte, but only 10^3 TFIIIA molecules per somatic cell (Ginsburg et al., 1984; Scotto et al., 1989). The binding of TFIIIA to the internal promoter of the 5S rRNA genes is very weak. It is greatly stabilized by the nonspecific factor TFIIIC (Lasser et al., 1983; Pieler et al., 1987). This tightly bound complex binds another factor, TFIIIB, which can bind RNA polymerase III (Wolffe and Brown, 1988). Once RNA polymerase III binds to this complex, transcription begins. Like the transcription assemblies discussed in Chapter 10, TFIIIA and TFIIIC are assembly factors, and TFIIIB contains the TATA-binding subunit (even though it does not bind to TATA or any other DNA).

It appears, then, that the type and amount of 5S rRNA depends on the level of TFIIIA in the nucleus. But how is that control effected? Brown and Schlissel (1985) injected cloned oocyte and somatic 5S rRNA genes into oocytes and blastula nuclei and showed that TFIIIA has a much higher (over 200-fold) affinity for the control region of the somatic 5S rRNA genes than it does for the oocyte 5S rRNA genes. Moreover, when purified TFIIIA was injected into late-blastula-stage embryos (in which only the somatic 5S rRNA genes are active), the oocyte 5S genes become activated as well. These observations indicated that at low concentrations of TFIIIA, the only genes binding TFIIIA will be of the somatic type; but at higher levels, the oocyte 5S rRNA genes will also become activated. The basis for this difference in the ability of the 5S genes to be activated by TFIIIA was found to reside in three nucleotides that differed between the oocyte and somatic 5S genes. While these three

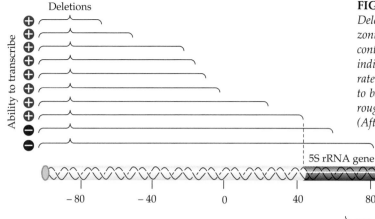

Deletions

Ability to transcribe

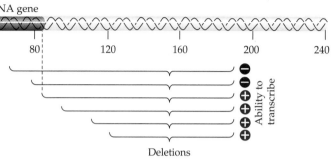

5S rRNA gene

−80   −40   0   40   80   120   160   200   240

Deletions

Ability to transcribe

FIGURE 22.30

Deletion map of a 5S rRNA gene of X. laevis. *The heavy horizontal lines represent the deleted portions of DNA in plasmids containing the 5S rRNA gene. A plus mark adjacent to a line indicates that the remaining plasmid DNA still supports accurate transcription of 5S rRNA. A negative sign indicates failure to be transcribed accurately. A promoter region within the gene, roughly 30 base pairs in length, is defined by these deletions. (After Brown, 1981.)*

base substitutions were responsible for changing the stability of the 5S gene–TFIIIA complex, they did not directly alter TFIIIA–promoter binding. Rather, the changes in nucleotides caused the TFIIIA protein to become displaced from the promoter by TFIIIC (Wolffe and Brown, 1988).

DNase I "footprinting" has shown that TFIIIA binds to the 5S rRNA gene in three places. TFIIIA is incubated with a 5S rRNA gene that has been radiolabeled at one end (as is done for sequencing). The DNA is then digested with DNase I, a relatively nonspecific DNase. Each strand should be broken (on average) only once. The resulting DNA is placed in a gel and separated by electrophoresis as if it were being sequenced. One obtains the same ladders expected of sequencing gels, where each oligonucleotide is one nucleotide shorter than the one above it. But in those regions where the protein (in this case, TFIIIA) has protected the DNA from DNase I treatment, the corresponding oligonucleotides are absent. Such a pattern for the binding of TFIIIA to the *Xenopus* oocyte 5S rRNA gene is shown in Figure 22.31, and it shows the three sites where TFIIIA binds to the internal promoter region (Sakonju and Brown, 1982).

rRNA binding by TFIIIA

TFIIIA is a zinc finger transcription factor with nine zinc fingers. Miller and co-workers (1985) suggest that this arrangement enables the TFIIIA to "hold on" to the gene as the RNA polymerase III moves along. Some of the binding regions maintain the hold on the DNA as others relinquish it. TFIIIA is also an *RNA*-binding protein, and complexes with the transcribed 5S rRNA in the oocyte cytoplasm. The same zinc fingers that bind to the central promoter region of the DNA enable it to bind to the RNA (Theunissen et al., 1992). In *Xenopus*, the 5S rRNA appears before the large-rRNA transcript. When it enters the cytoplasm, it becomes complexed with TFIIIA. However, when the large transcript is being synthesized and the large ribosomal subunit is being made in the germinal vesicle, the TFIIIA appears to be exchanged for ribosomal protein L5, and the 5S rRNA–L5 complex shuttles back into the germinal vesicle, where it will become incorporated into the 60S ribosomal subunit (Allison et al., 1991).

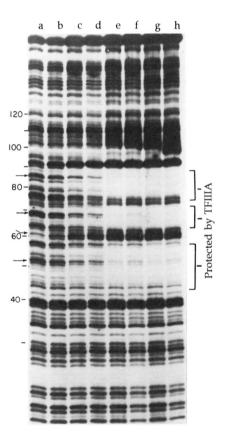

Protected by TFIIIA

a b c d e f g h

120 −

100 −

80 −

60 −

40 −

FIGURE 22.31

Autoradiographs of radioactive 5S DNA incubated with increasing amounts of TFIIIA and subsequently cleaved with DNase I. The fragments generated by this technique were separated by electrophoresis, blotted, and autoradiographed. The regions where DNA was protected by TFIIIA show up as blank areas (From Sakonju and Brown, 1982, courtesy of D. D. Brown.)

Meroistic oogenesis in insects

There are several types of oogenesis in insects, but most studies have focused on those insects, such as *Drosophila* and moths, that undergo **meroistic** oogenesis. In meroistic oogenesis, cytoplasmic connections remain between the cells produced by the oogonium. In *Drosophila*, each oogonium divides four times to produce a clone of 16 cells connected to each other through **ring canals**. The production of these interconnected cells (called **cystocytes**) involves a highly ordered array of cell divisions (Figure 22.32). Only those two cells having four interconnections are capable of developing into oocytes, and of these two, only one prevails. The other begins meiosis but does not complete it. Thus, only 1 of the 16 cystocytes can become an ovum. All the other cells become **nurse cells.** As it turns out, the cell destined to become the oocyte is that cell residing at the most posterior tip of the egg chamber that encloses the 16-cell clone.

Transport of RNA from nurse cells to oocyte. The oocytes of meroistic insects do not pass through a transcriptionally active stage, nor do they have lampbrush chromosomes. Rather, autoradiographic evidence shows that RNA synthesis is largely confined to the nurse cells and that the RNA made by these cells is actively transported into the oocyte cytoplasm. This can be seen in Figure 22.33. When the egg chambers of a housefly are incubated in radioactive cytidine, the nuclei of the nurse cells show intense labeling. When the labeling is stopped and the cells are incubated for 5 more hours in nonradioactive media, the labeled RNA is seen to enter the oocyte from the nurse cells (Bier, 1963). Oogenesis takes place in only 12 days, so the nurse cells are very metabolically active during this time. They are aided in their transcriptional efficiency by becoming polytene.

(A)

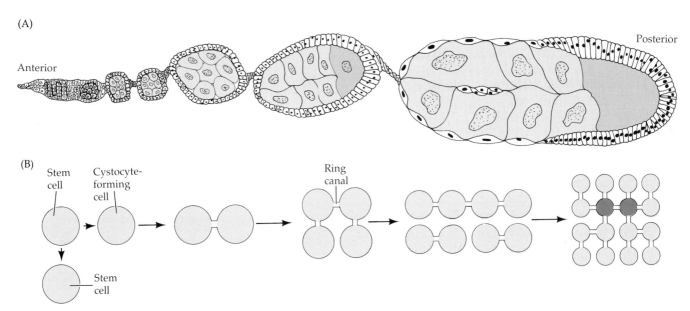

Posterior

Anterior

(B)

Stem cell

Cystocyte-forming cell

Ring canal

Stem cell

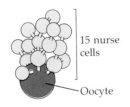

15 nurse cells

Oocyte

FIGURE 22.32
The formation of 16 interconnected cystocytes in *Drosophila*. (A) Diagram of adult ovariole showing sequence of oogenesis as younger germinal cysts mature within the ovariole. (B) Division of the cystocyte-forming cells. The cells are represented schematically as dividing in a single plane. The stem cell divides to produce another stem cell plus a cell that is committed to form the cystocytes. Only 1 of the 16 cystocytes becomes an oocyte; the others become nurse cells, connected to the oocyte by ring canals (cytoplasmic bridges). (A after Ruohola et al., 1991; B after Koch et al., 1967.)

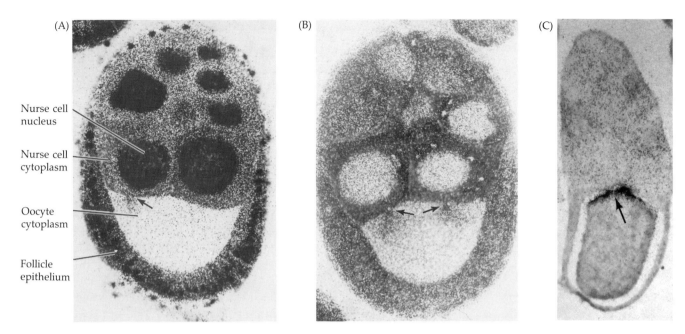

Nurse cell
nucleus

Nurse cell
cytoplasm

Oocyte
cytoplasm

Follicle
epithelium

Instead of having two copies of each chromosome, they replicate their chromosomes until they have produced 512 copies. The 15 nurse cells are known to pass ribosomal and messenger RNAs as well as proteins into the oocyte cytoplasm, and entire ribosomes may be transported as well (Plate 16). The mRNAs do not associate with polysomes, which suggests that they are not immediately active in protein synthesis (Paglia et al., 1976; Telfer et al., 1981).

One of the earliest mRNAs to enter the oocyte is the *bicoid* transcript. As discussed earlier (Chapter 15), this gene is critical for the construction of the anterior–posterior axis of the embryo, and it is a maternal product placed into the oocyte cytoplasm. The *bicoid* gene is transcribed during oogenesis by the nurse cells and its message accumulates at the anterior end of the oocyte (Figure 22.33C; Frigerio et al., 1986; Stephanson et al., 1988). Leakage of anterior cytoplasm from wild-type eggs leads to embryos with reduced head structures, and if the mother lacks wild type bicoid genes, her embryos lack heads and have tails at both ends. Since there is such polarity in *Drosophila* eggs, the next question concerns how this polarity is established and maintained. What keeps the *bicoid* message from diffusing throughout the oocyte? What localizes polar granule components to the posterior of the egg?

The *cis* component of bicoid localization is present in the *bicoid* message. The *bicoid* mRNA appears to stick to the anterior components of the egg via its 3′ untranslated region. If other mRNAs are given this sequence, they, too, localize to the anterior pole of the *Drosophila* egg (MacDonald et al., 1993). There seem to be at least three *trans* components that bind the bicoid message. The major component is probably the microtubules in the anterior cortex of the egg. If the eggs chambers are incubated in drugs such as nocodazone that disrupt microtubule assembly, the *bicoid* message fails to be transported to the oocyte. If the microtubule inhibitors are added after transport, the *bicoid* mRNA fails to be localized in the anterior cortex (Pokrywka and Stephenson, 1991). The other components are the products of the *swallow* and *exuperantia* genes (Frohnhöfer and Nüsslein-Volhard, 1987; Manseau and Schübach, 1989). In eggs produced by mothers with either of these mutations, *bicoid* message extends into more posterior positions in the egg. The product of the *exuperantia* gene seems to be

FIGURE 22.33

Transport of mRNA from nurse cells into fly oocytes. (A,B) Autoradiographs of the follicle cell of the housefly, *Musca domestica*, after incubation with [³H]cytidine. (A) Egg chamber fixed immediately after label was introduced. The nuclei of the nurse cells are heavily labeled, indicating that they are synthesizing new RNA. The oocyte remains unlabeled except where some RNA is escaping into the oocyte through the cytoplasmic connection between it and a nurse cell (arrow). (B) A similar egg chamber fixed 5 hours later. Label is gone from the nurse cell nuclei but has moved into the cytoplasm. Moreover, radioactive RNA can be seen passing into the oocyte cytoplasm through the two channels between the nurse cells and the oocyte. (C) Autoradiograph of *Drosophila* egg chamber stained by a radioactive probe to the *bicoid* mRNA. This message is transported from the nurse cells and remains in the anteriormost portion of the oocyte. (A and B from Bier, 1963, courtesy of D. Ribbert; C from Stephanson et al., 1988, courtesy of E. C. Stephanson.)

needed for the initial localization of the *bicoid* message to the anterior, but it does not keep it there, while the product of the *swallow* gene appears to be necessary for the stabilization of the message on the microtubular scaffold (Macdonald et al., 1991; Marcey et al., 1991; Pokrywka and Stephenson, 1991). Therefore, the *Drosophila* ovary must not only make a cytoplasmically polarized product but must also synthesize some of the oocyte components that provide this polarity.

The meroistic ovary confronts us with some interesting problems. If all the cells are connected so that proteins and RNAs shuttle freely between them, why should they have different developmental fates? Why should one cell become the oocyte while the others become "RNA-synthesizing factories"? Why is the flow of protein and RNA in one direction only? It is now thought that the cytoskeleton is actively involved in transporting mRNAs from the nurse cells into the cytoplasm. Before most of the mRNAs go into the oocyte, the centrioles from the 15 cells destined to become nurse cells dissociate from their nuclei and migrate through the ring canals into the cell destined to become the oocyte (Mahowald and Strassheim, 1970). The centrioles form a microtubular array that connects the 16 cells. If this microtubular lattice is disrupted (either chemically or by mutations such as *Bicaudal-D* or *egalitarian*), gene products are transmitted in all directions and all 16 cells differentiate into nurse cells (Gutzeit, 1986; Spradling, 1993; Theurkauf et al., 1992, 1993). It is possible that some compounds transported from the nurse cells into the oocyte become associated with transport proteins such as kinesin that would enable them to travel along the tracks of microtubules extending through the ring canal (Theurkauf et al., 1992; Sun and Wyman, 1993). Actin may become important for maintaining this distinction during later stages of oogenesis. Mutations that prevent actin microfilaments from lining the ring canals prevent the transport mRNAs from nurse cell to oocyte, and the disruption of the actin microfilaments randomizes the distribution of mRNA (Cooley et al., 1992; Watson et al., 1993). Thus, the microtubule and microfilamentous cytoskeleton appears to control the movement of organelles and RNAs between nurse cells and oocyte such that developmental cues are exchanged only in the appropriate direction.

Transport of yolk proteins into the egg. The three major yolk proteins in *Drosophila* are made in the fat body and ovary, but not in the oocyte itself (Bownes, 1982; Brennen et al., 1982). Yolk synthesis is controlled by several interacting agents, including sex, juvenile hormone levels, ecdysone, and nutrition. These physiological agents are "integrated" by the enhancer region between the two yolk protein genes of *Drosophila* (Figure 22.34A; Bownes et al., 1988). These genes are only active in female flies, and this is regulated by the binding of the female-specific doublesex protein to this enhancer region. It is thought that a brain hormone, responding to environmental cues,* stimulates the corpora allata to secrete juvenile hormone (Figure 22.34B). Juvenile hormone (1) regulates the uptake of the yolk peptides at the surface of the oocyte, (2) stimulates the synthesis of ovarian yolk proteins (which are each identical to those made by the fat body), and (3) causes the ovarian follicles and other abdominal cells to secrete

---

*In *Drosophila*, the environmental cue appears to be the photoperiod. In the common mosquito, the cue is the blood meal. Only female mosquitoes bite, and they make no vitellogenin before this meal. Some factor in the blood stimulates the mosquito's brain to release juvenile hormone and the corpus cardiacum-stimulating factor. The latter factor causes the release of the egg development neurosecretory hormone (EDNH). EDNH stimulates the ovary to secrete ecdysone, which, in concert with JH, stimulates the fat body to synthesize vitellogenin (Hagedorn, 1983; Borovsk et al., 1990).

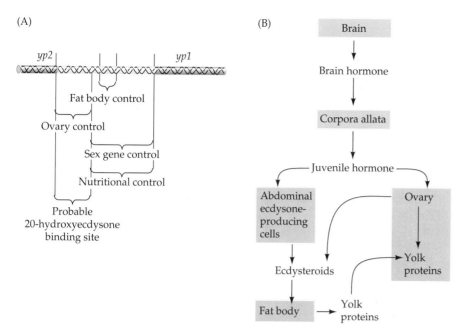

(A)

yp2   *yp1*

Fat body control

Ovary control

Sex gene control

Nutritional control

Probable
20-hydroxyecdysone
binding site

(B)

Brain

Brain hormone

Corpora allata

Juvenile hormone

Abdominal
ecdysone-
producing
cells

Ovary

Ecdysteroids

Yolk
proteins

Fat body ⟶ Yolk
proteins

FIGURE 22.34
Yolk protein synthesis in *Drosophila melanogaster*. (A) Enhancer sites between the two yolk protein genes (*yp1* and *yp2*). These genes are positively regulated by ovarian *trans* factors, fat body *trans* factors, and (since they only function in females), the female form of the doublesex protein. These genes are negatively regulated by the *male*-specific form of the doublesex protein. A factor regulated by the nutritional state of the adult fly binds to the same 890-base pair region as the sex-determining proteins, but the regulation by hormones occurs on a different region. (B) Model for the hormonal regulation of yolk peptide synthesis in *D. melanogaster*. In response to a brain hormone, the corpora allata produce juvenile hormone, which causes the ovary to make yolk proteins and ecdysteroids. JH also induces ecdysteroid synthesis in abdominal cells. These ecdysteroids cause the fat body to produce yolk proteins that are transported to the ovary. (A after Bownes et al., 1988; B after Bownes, 1982.)

ecdysone. Ecdysone is metabolized into its active form—20-hydroxyecdysone—and stimulates the fat body to produce yolk proteins just as estradiol stimulates the amphibian liver to do so. Similarly, the administration of ecdysone to adult males causes their fat bodies to secrete yolk proteins as well (Postlethwait et al., 1980), and yolk protein gets taken into insect oocytes through receptor-mediated endocytosis (Raikhel and Dhadialla, 1992). The receptors for vitellogenin are located in the regions of oocyte membrane at the base of and between the microvilli. The receptor–vitellogenin complexes are internalized and the vitellogenin freed from the receptor inside the endocytic vacuole. This vacuole fuses with other such endosomes to form the yolk storage granule full of vitellogenin.

Oogenesis in mammals

The maturation and ovulation of the mammalian egg follows one of two basic patterns, depending upon the species. One type of ovulation is stimulated by the physical act of intercourse itself. Physical stimulation of the cervix triggers the release of gonadotropins from the pituitary. These gonadotropins signal the egg to resume meiosis and initiate the events expelling the ovum from the ovary. This method ensures that most copulations lead to fertilized ova, and animals that utilize this method of ovulation—rabbits and minks—have a reputation for procreative success.

Most mammals, however, have a periodic type of ovulation. The female ovulates only at specific times of the year, called **estrus** (or its English equivalent, "heat"). In these cases, environmental cues, most notably the amount and type of light during the day, stimulate the hypothalamus to release gonadotropin-releasing factor. This factor stimulates the pituitary to release its gonadotropins—follicle-stimulating hormone (FSH) and luteinizing hormone (LH)—which cause the follicle cells to proliferate and to secrete estrogen. The estrogen subsequently binds to certain neurons and evokes the pattern of mating behavior characteristic of the species. Gonadotropins also stimulate follicular growth and the initiation of ovulation. Thus, estrus and ovulation occur close together.

Humans have a variation on the theme of periodic ovulation. Although human females have a cyclic ovulation (averaging about 29.5 days) and

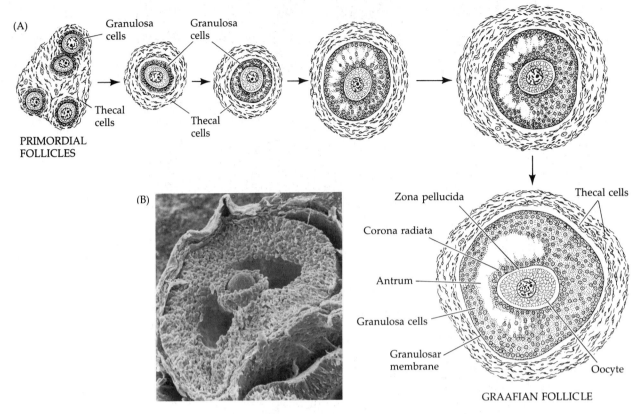

(A)

Granulosa cells

Granulosa cells

Thecal cells

Thecal cells

PRIMORDIAL FOLLICLES

(B)

Zona pellucida

Thecal cells

Corona radiata

Antrum

Granulosa cells

Granulosar membrane

Oocyte

GRAAFIAN FOLLICLE

FIGURE 22.35
The ovarian follicle of mammals. (A) Maturation of the ovarian follicle. When mature, it is often called a Graafian follicle. (B) Scanning electron micrograph of a mature follicle in the rat. The oocyte (center) is surrounded by the smaller granulosa cells that will make the corona. (A after Carlson, 1981; B courtesy of P. Bagavandoss.)

no definitive yearly estrus, most of human reproductive physiology is shared with other primates. The characteristic primate periodicity in maturing and releasing ova is called the **menstrual cycle** because it entails the periodic shedding of blood and cellular debris from the uterus at monthly intervals.* The menstrual cycle represents the integration of three very different activities: (1) the ovarian cycle, the function of which is to mature and release an oocyte, (2) the uterine cycle, the function of which is to provide the appropriate environment for the developing blastocyst to implant, and (3) the cervical cycle, the function of which is to allow sperm to enter the female reproductive tract only at the appropriate time. These three functions are integrated through the hormones of the pituitary, hypothalamus, and ovary.

The majority of the oocytes within the adult human ovary are maintained in the prolonged diplotene stage of the first meiotic prophase (often referred to as the **dictyate** state). Each oocyte is enveloped by a primordial follicle consisting of a single layer of epithelial granulosa cells and a less-organized layer of mesenchymal thecal cells (Figure 22.35). Periodically a group of primordial follicles enter a stage of follicular growth. During this time, the oocyte undergoes a 500-fold increase in volume (corresponding to an increase in oocyte diameter from 10 μm in a primordial follicle to 80

*The periodic shedding of the uterine lining has recently been proposed as a mechanism by which the uterus is protected against infections from semen or other environmental agents (Profet, 1993).

μm in a fully developed follicle). Concomitant with oocyte growth is an increase in the numbers of follicular granulosa cells, which form concentric layers about the oocyte. Throughout this growth period, the oocyte remains in the dictyate stage. The fully grown follicle thus contains a large oocyte surrounded by several layers of granulosa cells. Many of these cells will stay with the ovulated egg, forming the **cumulus**, which surrounds the egg in the oviduct. In addition, during the growth of the follicle, an **antrum** (cavity) forms, which becomes filled with a complex mixture of proteins, hormones, cAMP, and other molecules. At any given time, a small group of follicles is maturing. However, after progressing to a more mature stage, most oocytes and their follicles die. In order to survive, the follicle must find a source of gonadotropic hormones and, "catching the wave" at the right time, must ride it until it peaks. Thus, for oocyte maturation to occur, the follicle needs to be at a certain stage of development when the waves of gonadotropin arise.

Day 1 of the menstrual cycle is considered to be the first day of "bleeding" (Figure 22.36). This bleeding from the vagina represents the sloughing off of extrauterine tissue and blood vessels that would have aided the implantation of the blastocyst. In the first part of the cycle (called the **proliferative** or **follicular phase**), the pituitary gland starts secreting increasingly large amounts of FSH. The group of maturing follicles, which have already undergone some development, respond to this hormone by

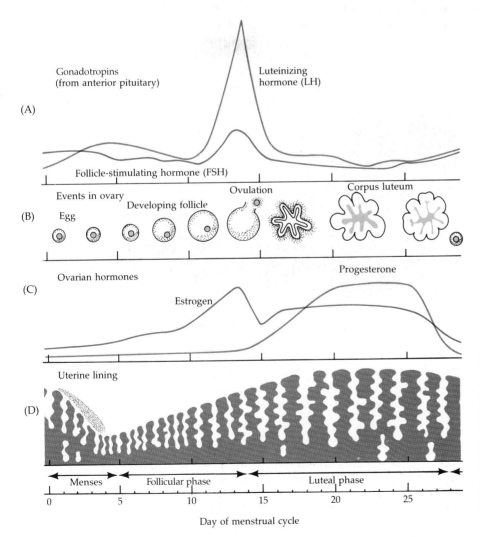

FIGURE 22.36
The human menstrual cycle. The coordination of (B) ovarian and (D) uterine cycles is controlled by (A) the pituitary and (C) the ovarian hormones. During the follicular phase, the egg matures within the follicle and the uterine lining is prepared to receive the embryo. The mature egg is released around day 14. If an embryo does not implant into the uterus, the uterine wall begins to break down, leading to menses.

further growth and cellular proliferation. FSH also induces the formation of LH receptors on the granulosa cells. Shortly after this period of initial follicle growth, the pituitary begins secreting LH. In response to LH, the meiotic block is broken. The nuclear membranes of competent oocytes break down, and the chromosomes assemble to undergo the first meiotic division. One set of chromosomes is kept inside the oocyte and the other is given to the small polar body. Both are encased by the zona pellucida, which has been synthesized by the growing oocyte. It is in this stage that the egg will be ovulated.

The two gonadotropins, acting together, cause the follicle cells to produce increasing amounts of estrogen, which has at least five major activities in regulating the further progression of the menstrual cycle.

1. It causes the uterine mucosa to begin its proliferation and to become enriched with blood vessels.
2. It causes the cervical mucus to thin, thereby permitting sperm to enter the inner portions of the reproductive tract.
3. It causes an increase in the number of FSH receptors on the follicular granulosa cells (Kammerman and Ross, 1975) while causing the pituitary to lower its FSH production. It also stimulates the granulosa cells to secrete the peptide hormone inhibin, which also suppresses pituitary FSH secretion (Rivier et al., 1986; Woodruff et al., 1988).
4. At low concentrations, it inhibits LH production, but at high concentrations, it stimulates it.
5. At very high concentrations and over long durations, estrogen interacts with the hypothalamus, causing it to secrete gonadotropin-releasing factor.

As estrogen levels increase as a result of follicular production, FSH levels decline. LH levels, however, continue to rise as more estrogen is secreted. As estrogens continue to be made (days 7 to 10), the granulosa cells continue to grow. Starting at day 10, estrogen secretion rises sharply. This rise is followed at midcycle by an enormous surge of LH and a smaller burst of FSH. Experiments with female monkeys have shown that exposure of the hypothalamus to greater than 200 pg of estrogen per milliliter of blood for more than 50 hours results in the hypothalamic secretion of gonadotropin-releasing factor. This factor causes the subsequent release of FSH and LH from the pituitary. Within 10 to 12 hours after the gonadotropin peak, the egg is ovulated (Figure 22.37; Garcia et al., 1981). Although the detailed mechanism of ovulation is not yet known, the physical expulsion of the mature oocyte from the follicle appears to be due to an LH-induced increase in collagenase, plasminogen activator, and prostaglandin within the follicle (Lemaire et al., 1973). The mRNA for plasminogen activator has been dormant in the oocyte cytoplasm. LH causes this message to be polyadenylated and translated into this powerful protease (Huarte et al., 1987). Prostaglandins may cause localized contractions in the smooth muscles in the ovary and may also increase the flow of water from the ovarian capillaries (Diaz-Infante et al., 1974; Koos and Clark, 1982). If ovarian prostaglandin synthesis is inhibited, ovulation does not take place. In addition to the prostaglandin-induced pressure, collagenases and the plasminogen activator protease loosen and digest the extracellular matrix of the follicle (Beers et al., 1975; Downs and Longo, 1983). The result of LH, then, would be increased follicular pressure coupled with the degradation of the follicle wall. A hole would be digested through which the ovum could burst.

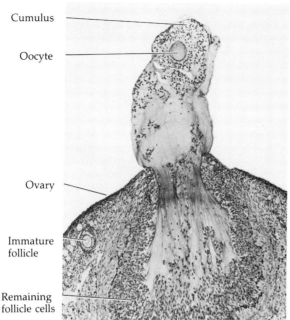

Cumulus

Oocyte

Ovary

Immature
follicle

Remaining
follicle cells

FIGURE 22.37
Ovulation in the rabbit. The ovary of a living, anesthetized rabbit was exposed
and observed. When the follicle started to ovulate, the ovary was removed,
fixed, and stained. (Courtesy of R. J. Blandau.)

Following ovulation, the **luteal phase** of the menstrual cycle begins.
The remaining cells of the ruptured follicle under the continued influence
of LH become the **corpus luteum**. (They are able to respond to this LH
because the surge in FSH stimulates them to develop even more LH
receptors.) The corpus luteum secretes some estrogen, but its predominant
secretion is **progesterone**. This steroid hormone circulates to the uterus,
where it completes the job of preparing the uterine tissue for blastocyst
implantation, stimulating the growth of the uterine wall and its blood
vessels. Blocking the progesterone receptor with the synthetic steroid
mifepristone (RU486) stops the uterine wall from thickening and prevents
the implantation of the blastocyst into the uterus* (Couzinet et al., 1986).
Progesterone also inhibits the production of FSH, thereby preventing the
maturation of any more follicles and ova. (For this reason such a combination of estrogen and progesterone has been used in birth control pills.
The growth and maturation of new ova are prevented so long as FSH is
inhibited.)

If the ovum is not fertilized, the corpus luteum degenerates, progesterone secretion ceases, and the uterine wall is sloughed off. With the
decline in serum progesterone levels, the pituitary secretes FSH again,
and the cycle is renewed. However, if fertilization occurs, the trophoblast
secretes a new hormone, **luteotropin**, which causes the corpus luteum to
remain active and serum progesterone levels to remain high. Thus, the
menstrual cycle enables the periodic maturation and ovulation of human
ova and allows the uterus to periodically develop into an organ capable
of nurturing a developing organism for nine months.

*RU486 is thought to compete for the progesterone receptor inside the nucleus. RU486 can
bind to the progesterone site in the receptor, and the receptor–RU486 complex appears to
form heterodimers with the normal progesterone-carrying progesterone receptor. When this
RU486–progesterone complex binds to the progesterone-responsive enhancer elements on
the DNA, transcription from this site is inhibited (Vegeto et al., 1992; Spitz and Bardin,
1993). In Europe, RU486 has become a widely used alternative to surgical abortion (Palka,
1989; Maurice, 1991).

The reinitiation of meiosis in mammalian oocytes

If numerous follicles are capable of maturing when follicle-stimulating hormone is secreted, how is it that usually only one follicle and its oocyte prevail? It appears that the follicle capable of producing the most estrogen in response to FSH is the one that matures, while all the others die. Those sets of follicles initially receiving FSH not only begin to proliferate, but also produce new luteinizing hormone receptors on their thecal cells (Figure 22.38). The reception of LH causes these thecal cells to initiate estrogen production. As we have seen, estrogen has two disparate effects involving the future reception of FSH. At one level, it turns down the pituitary secretion of FSH; while at another level, it increases the FSH receptors on the follicle cells. Thus, the more estrogen a follicle produces, the more FSH receptors it has while less FSH remains in circulation. As FSH concentrations get progressively lower, only one follicle can bind the available FSH. Only this follicle can still grow, and the other follicles die.

What does LH do that causes the reinitiation of meiosis? To address this question, the nature of the meiotic block has been intensely studied. As with amphibian oocytes, the dictyate stage is extremely important, because that is the time during which oocytes grow, differentiate the structures specific to oocytes, and acquire the ability to resume meiosis (Sorensen and Wassarman, 1976). Early experiments demonstrated that follicle-enclosed oocytes, in

(A)

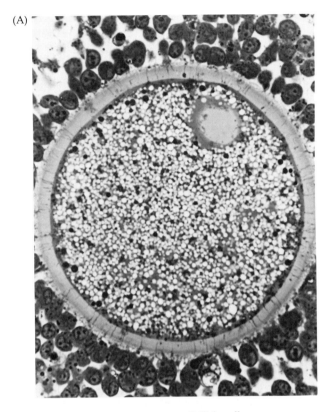

Follicle cell
process

(B)

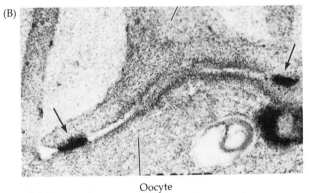

Oocyte

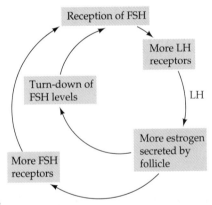

FIGURE 22.38
Positive feedback cycle in mammalian follicle cells. Reception of follicle-stimulating hormone (FSH) leads to the production of more luteinizing hormone (LH) receptors. The follicle cells secrete estrogen when stimulated by LH; the estrogen causes both an increase in the number of FSH receptors and a decrease in the amount of pituitary FSH production. Eventually, very few follicles are able to receive the small amounts of FSH produced, thereby amplifying their ability to receive LH. These few follicles are able to mature.

FIGURE 22.39
Communication between oocyte and granulosa cells. (A) Sheep oocyte surrounded by the zona pellucida and follicle cells. The granulosa cells of the follicle are extending processes through the zona pellucida and touching the oocyte. (B) Electron micrograph of follicle cell processes establishing gap junction connections with a rhesus monkey oocyte. Gap junctions (arrows) are stained with ionic lanthanum. (A from Moor and Cran, 1980, courtesy of the authors; B from Anderson and Albertini, 1976, courtesy of D. Albertini.)

vivo or in vitro, do not undergo maturation unless exposed to gonadotropins, whereas oocytes removed from the follicle spontaneously resume meiosis even without the hormonal stimulus (Pincus and Enzmann, 1935).

It appears, then, that meiosis is normally inhibited by the follicle cells and that it can be reinitiated by gonadotropins. This hypothesis—that the follicle cells are important regulators of meiosis—is strengthened by observations that granulosa cells communicate with the oocyte through processes extending to the oocyte through the zona. These processes have gap junctions that enable small molecules to pass between the oocyte and the granulosa cells of the follicle (Figure 22.39; Anderson and Albertini, 1976; Gilula et al., 1978).

Because the elevation of cAMP levels inhibits oocyte maturation (Cho et al., 1974), it has been proposed that the meiotic arrest is maintained by the transfer of cAMP through the gap junctions from the follicular granulosa cell to the oocyte (Dekel and Beers, 1978, 1980). The luteinizing hormone surge could trigger maturation by terminating the gap junction communication, thereby inhibiting the transfer of cAMP into the oocyte. Several lines of evidence now support this hypothesis. First, the decline of cAMP appears critical for the resumption of meiosis. Germinal vesicle breakdown can be prevented by inhibiting the degradation of cAMP in follicle-free eggs or by providing such eggs directly with cAMP (Bornslaeger et al., 1986). The decline in oocyte cAMP concentration occurs immediately prior to the resumption of meiosis (Figure 22.40; Schultz et al., 1983).

Second, the gonadotropins can cause the loss of communication between the follicle cells and the oocyte. The follicle cells appear to be important sources of oocyte cAMP, and changes of cAMP concentration in the follicle cells are reflected in oocyte cAMP levels (Bornslaeger and Schultz, 1985; Racowsky, 1985). This observation explains why oocytes remain in meiotic arrest when they are surrounded by the follicle cells but resume meiosis when they are removed from them.

The gonadotropin surge, however, can elevate the *follicle cell* cAMP concentrations to new levels. In response to this elevation, the mature follicle cells synthesize hyaluronic acid, which causes a physical disruption of the contact between the follicle cell processes and the oocyte (Eppig, 1979; Larsen et al., 1986). The bridges through which cAMP flows from follicular granulosa cell to oocyte have thereby

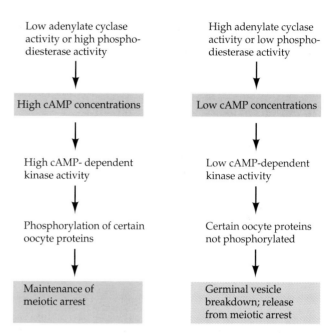

FIGURE 22.40
Summary of proposed mechanism whereby the oocyte cAMP level regulates resumption of meiosis by the oocyte. The cAMP levels within the oocyte are provided, at least in part, by cAMP from the follicle cells. Cyclic AMP cannot cross cell membranes, but it can enter the oocyte through gap junctions connecting the oocyte with its follicle cells. When the connections are released, oocyte cAMP levels decline, leading to the release of meiotic arrest.

been removed, allowing the mammalian oocyte to resume meiosis (Dekel and Sherizly, 1985; Racowsky and Satterlie, 1985).

Like amphibian oocytes, the ovulated mouse oocyte is suspended in the second meiotic metaphase and is fertilized in that state. Paules and his co-workers (1989) have shown that maturing mouse oocytes also contain the pp39^{mos} cytostatic factor. If the translation of the *c-mos* gene that makes this factor is specifically interrupted, the oocyte does not release its first polar body and fails to undergo normal maturation. Evidently, similar events must take place for the maturation of the amphibian and mammalian oocytes.

The egg and the sperm will both die if they do not meet. We are now back where we began—the stage is set for fertilization to take place. As F. R. Lillie recognized in 1919, "The elements that unite are single cells, each on the point of death; but by their union a rejuvenated individual is formed, which constitutes a link in the eternal process of Life."

Adamson, E., Evans, M. J. and Magrane, G. G. 1977. Biochemical markers of the progress of differentiation in cloned teratocarcinoma cell lines. *Eur. J. Biochem.* 79: 607–615.

Ahringer, J. and Kimble, J. 1991. Control of the sperm–oocyte switch in *Caenorhabditis elegans* hermaphrodites by the fem-3 3′untranslated region. *Nature* 349: 346–348.

Ahringer, J., Rosequist, T. A., Lawson, D. N. and Kimble, J. 1992. The *C. elegans* sex determining gene, *fem-3*, is regulated posttranscriptionally. *EMBO J.* 11: 2303–2310.

Allison, L. A., Romaniuk, P. J. and Bakken, A. H. 1991. RNA–protein interactions of stored 5S RNA with TFIIIA and ribosomal protein L5 during *Xenopus* oogenesis. *Dev. Biol.* 144: 129–144.

Ambrosio, L. and Schedl, P. 1985. Two discrete modes of histone gene expression during oogenesis in *Drosophila melanogaster*. *Dev. Biol.* 111: 220–231.

Anderson, E. and Albertini, D. F. 1976. Gap junctions between the oocyte and companion follicle cells in the mammalian ovary. *J. Cell Biol.* 71: 680–686.

Auerbach, R. and Joseph, J. 1984. Cell surface markers on endothelial cells: A developmental perspective. *In* E. A. Jaffe (ed.), *The Biology of Endothelial Cells.* Nijhoff, The Hague, pp. 393–400.

Austin, J. and Kimble, J. 1987. glp-1 is required in the germ line for regulation of the decision between mitosis and meiosis in *C. elegans*. *Cell* 51: 589–599.

Baker, T. G. 1970. Primordial germ cells. *In* C. R. Austin and R. V. Short (eds.), *Reproduction in Mammals,* Vol. 1: *Germ Cells and Fertilization.* Cambridge University Press, Cambridge, pp. 1–13.

Becker, S. and Grabel, L. 1992. Localization of endoderm-specific messenger RNAs in differentiating F9 erythroid bodies. *Mech. Devel.* 37: 3–12.

Beers, W. H., Strickland, S. and Reich, E. 1975. Ovarian plasminogen activator: Relationship to ovulation and hormonal regulation. *Cell* 6: 387–394.

Bier, K. 1963. Autoradiographische Untersuchungen über die Leistungen des Follikelepithels und der Nahrzellen bei der Dottbildung und Eiweisssynthese im Fliegenova. *Wilhelm Roux Arch. Entwicklungsmech. Org.* 154: 552–575.

Blanco, A. 1980. On the functional significance of LDH-X. *Johns Hopkins Med. J.* 146: 231–235.

Bloom, W. and Fawcett, D. W. 1975. *Textbook of Histology,* 10th Ed. Saunders, Philadelphia.

Bogenhagen, D. F., Sakonju, S. and Brown, D. D. 1980. A control region in the center of the 5S RNA gene directs specific initiation of transcription. II. The 3′ border of the region. *Cell* 19: 27–35.

Bornslaeger, E. A. and Schultz, R. M. 1985. Regulation of mouse oocyte maturation: Effect of elevating cumulus cell cAMP on oocyte cAMP levels. *Biol. Reprod.* 33: 698–704.

Bornslaeger, E. A., Mattei, P. and Schultz, R. M. 1986. Involvement of cAMP–dependent protein kinase and protein phosphorylation in regulation of mouse oocyte maturation. *Dev. Biol.* 114: 453–462.

Borovsk, D., Carlson, D. A., Griffin, P. R., Shabanowitz, J. and Hunt, D. F. 1990. Mosquito oostatic factor: A novel decapeptide modulating trypsin-like enzyme biosynthesis in the midgut. *FASEB J.* 4: 3015–3020.

Bounoure, L. 1934. Recherches sur lignée germinale chez la grenouille rousse aux premiers stades au développement. *Ann. Sci. Zool. Ser.* 17, 10: 67–248.

Bownes, M. 1982. Hormonal and genetic regulation of vitellogenesis in *Drosophila*. *Q. Rev. Biol.* 57: 247–274.

Bownes, M., Scott, A. and Shirras, A. 1988. Dietary components modulate yolk protein transcription in *Drosophila melanogaster*. *Development* 103: 119–128.

Braun, R. E., Behringer, R. R., Peschon, J. J., Brinster, R. L. and Palmiter, R. D. 1989a. Genetically haploid spermatids are phenotypically diploid. *Nature* 337: 373–376.

Braun, R. E., Peschon, J. J., Behringer, R. R., Brinster, R. L. and Palmiter, R. D. 1989b. Protamine 3′ untranslated sequences regulate temporal translational control and subcellular localization of growth hormone in spermatids of transgenic mice. *Genes Dev.* 3: 793–802.

Brennen, M. D., Weiner, A. J., Goralski, T. J. and Mahowald, A. P. 1982. The follicle cells are a major site of vitellogenin synthesis in *Drosophila melanogaster*. *Dev. Biol.* 89: 225–236.

Brock, M. L. and Shapiro, D. J. 1983. Estrogen stabilizes vitellogenin mRNA against cytoplasmic degradation. *Cell* 34: 207–214.

Browder, L. 1980. *Developmental Biology.* Saunders, Philadelphia.

Brown, D. D. 1981. Gene expression in eukaryotes. *Science* 211: 667–674.

Brown, D. D. and Dawid, I. B. 1968. Specific gene amplification in oocytes. *Science* 160: 272–280.

Brown, D. D. and Schlissel, M. S. 1985. A positive transcription factor controls the differential expression of two 5S RNA genes. *Cell* 42: 759–767.

Burns, R. K., Jr. 1930. The process of sex-transformation parabiotic *Amblystoma.* I. Transformation from female to male. *J. Exp. Zool.* 55: 123–169.

Burtis, K. C. 1993. The regulation of sex determination and sexually dimorphic differentiation in *Drosophila*. *Curr. Opin. Cell Biol.* 5: 1006–1014.

Carlson, B. M. 1981. *Patten's Foundations of Embryology.* McGraw-Hill, New York.

Chiquoine, A. D. 1954. The identification, origin, and migration of the primordial germ cells in the mouse embryo. *Anat. Rec.* 118: 135–146.

Cho, W. K., Stern, S. and Biggers, J. D. 1974. Inhibitory effect of dibutyryl cAMP on mouse oocyte maturation in vitro. *J. Exp. Zool.* 187: 383–386.

Comings, D. E. 1968. The rationale for an ordered arrangement of chromatin in the interphase nucleus. *Am. J. Hum. Genet.* 20: 440–460.

Cooley, L., Verheyen, E. and Ayers, K. 1992. *chickadee* encodes a profilin required for intercellular cytoplasm transport during *Drosophila* oogenesis. *Cell* 69: 173–184.

Couzinet, B., Le Strat, N. Ulmann, A., Baulieu, E. E. and Schaison, G. 1986. Termination of early pregnancy by the progesterone antagonist RU486 (Mifepristone). *N. Engl. J. Med.* 315: 1565–1570.

Danilchik, M. V. and Gerhart, J. C. 1987. Differentiation of the animal–vegetal axis in *Xenopus laevis* oocytes: Polarized intracellular translocation of platelets establishes the yolk gradient. *Dev. Biol.* 122: 101–112.

Davidson, E. 1986. *Gene Activity in Early Development,* 3rd Ed. Academic Press, Orlando, FL.

Dekel, N. and Beers, W. H. 1978. Rat oocyte maturation in vitro: Relief of cyclic cAMP inhibition by gonadotropins. *Proc. Natl. Acad. Sci. USA* 75: 4369–4373.

Dekel, N. and Beers, W. H. 1980. Development of the rat oocyte in vitro: Inhibition and induction of maturation in the presence or absence of the cumulus oophorus. *Dev. Biol.* 75: 247–254.

Dekel, N. and Sherizly, I. 1985. Epidermal growth factor induces maturation of rat follicle-enclosed oocytes. *Endocrinology* 116: 406–409.

Diaz-Infante, A., Wright, K. H. and Wallach, E. E. 1974. Effects of indomethacin and PGF2a on ovulation and ovarian contraction in the rabbit. *Prostaglandins* 5: 567–581.

Dolci, S. and eight others. 1991. Requirement for mast cell growth factor for primordial germ cell survival in culture. *Nature* 352: 809–811.

Doniach, T. and Hodgkin, J. 1984. A sex-determining gene, *fem-1*, required for both male and hermaphroditic development in *C. elegans*. *Dev. Biol.* 106: 223–235.

Downs, S. and Longo, F. J. 1983. Prostaglandins and preovulatory follicular maturation in mice. *J. Exp. Zool.* 228: 99–108.

Dubois, R. 1969. Le mécanisme d'entrée des cellules germinales primordiales dans le réseau vasculaire, chez l' embryon de poulet. *J. Embryol. Exp. Morphol.* 21: 255–270.

Dumont, J. N. 1978. Oogenesis in *Xenopus laevis.* VI. Route of injected tracer transport in follicle and developing oocyte. *J. Exp. Zool.* 204: 193–200.

Dym, M. 1977. The male reproductive system. *In* L. Weiss and R. O. Greep (eds.), *Histology,* 4th Ed. McGraw–Hill, New York, pp. 979–1038.

Dym, M. and Fawcett, D. W. 1971. Further observations on the number of spermatogonia, spermatocytes, and spermatids connected by intercellular bridges in the mammalian testis. *Biol. Reprod.* 4: 195–215.

Ellis, R. E. and Kimble, J. In press. Control of germ cel differentiation in *Caenorhabditis elegans*. CIBA Foundation.

Engelke, D. R., Ng, S.-Y., Shastry, B. S. and Roeder, R. G. 1980. Specific interaction of a purified transcription factor with an internal control region of 5 S RNA genes. *Cell* 19: 717–728.

Eppig, J. J. 1979. FSH stimulates hyaluronic acid synthesis by oocyte–cumulus cell complexes from mouse preovulatory follicles. *Nature* 281: 483–484.

Eyal-Giladi, H., Ginsburg, M. and Farbarou, A. 1981. Avian primordial germ cells are of epiblastic origin. *J. Embryol. Exp. Morphol.* 65: 139–147.

ffrench–Constant, C., Hollingsworth, A., Heasman, J. and Wylie, C. 1991. Response to fibronectin of mouse primordial germ cells before, during, and after migration. *Development* 113: 1365–1373.

Flickinger, R. A. and Rounds, D. E. 1956. The maternal synthesis of egg yolk proteins as demonstrated by isotopic and serological means. *Biochim. Biophys. Acta* 22: 38–72.

Frigerio, D., Burri, M., Bopp, D., Baumgartner, S. and Noll, M. 1986. Structure of the segmentation gene *paired* and the *Drosophila* PRD gene set as part of a gene network. *Cell* 47: 735–746.

Frohnhöfer, H. G. and Nüsslein-Volhard, C. 1987. Maternal genes required for the anterior localization of *bicoid* activity in the embryo of *Drosophila*. *Genes Dev.* 1: 880–890.

Gabrielli, B., Roy, L. M. and Maller, J. L. 1993. Requirement for cdk2 in cytostatic factor-mediated metaphase II arrest. *Science* 259: 1766–1769.

Gallatin, W. M., Weissman, I. L. and Butcher, E. C. 1983. A cell surface molecule involved in organ-specific homing of lymphocytes. *Nature* 304: 30–35.

Gallatin, W. M., St. John, T. P., Siegelman, M., Reichert, R., Butcher, E. C. and Weissman, I. L. 1986. Lymphocyte homing receptors. *Cell* 44: 673–680.

Garcia, J. E., Jones, G. S. and Wright, G. L. 1981. Prediction of the time of ovulation. *Fert. Steril.* 36: 308–315.

Gardner, R. L. 1982. Manipulation of development. *In* C. R. Austin and R. V. Short (eds.), *Embryonic and Fetal Development*, Cambridge University Press, Cambridge, pp. 159–180.

Gilula, N. B., Epstein, M. L. and Beers, W. H. 1978. Cell-to-cell communication and ovulation. A study of the cumulus–oocyte complex. *J. Cell Biol.* 78: 58–75.

Ginsberg, A. M., King, D. O. and Roeder, R. G. 1984. *Xenopus* 5S gene transcription factor, TFIIIA: Characterization of a cDNA clone and measurement of RNA levels throughout development. *Cell* 39: 479–489.

Ginsburg, M. and Eyal-Giladi, H. 1987. Primordial germ cells of the young chick blastoderm originate from the central zone of the area pellucida irrespective of the embryo-forming process. *Development* 101: 209–219.

Ginsburg, M., Snow, M. H. L. and McLaren, A. 1990. Primordial germ cells in the mouse embryo during gastrulation. *Development* 110: 521–528.

Godin, I. and Wylie, C. C. 1991. TGFβ1 inhibits proliferation and has a chemotactic effect on mouse primordial germ cells in culture. *Development* 113: 1451–1457.

Godin, I., Wylie, C. and Heasman, J. 1990. Genital ridges exert long–range effects on primordial germ cell numbers and direction of migration in culture. *Development* 108: 357–363.

Graham, C. E. 1977. Teratocarcinoma cells and normal mouse embryogenesis. In M. I. Sherman (ed.), *Concepts of Mammalian Embryogenesis*. M.I.T. Press, Cambridge, MA, pp. 315–394.

Graham, P. L. and Kimble, J. 1993. The *mog-1* gene is required for the switch from spermatogenesis to oogenesis in *C. elegans*. *Genetics* 133: 919–931.

Grant, P. 1953. Phosphate metabolism during oogenesis in *Rana temporaria*. *J. Exp. Zool.* 124: 513–543.

Green, G. R. and Poccia, D. L. 1988. Interaction of sperm histone variants and linker DNA during spermiogenesis in the sea urchin. *Biochemistry* 27: 619–625.

Gurdon, J. B. 1976. *The Control of Gene Expression in Animal Development*. Harvard University Press, Cambridge, MA.

Gutzeit, H. O. 1986. The role of microfilaments in cytoplasmic streaming in *Drosophila* follicles. *J. Cell Sci.* 80: 159–169.

Hagedorn, H. H. 1983. The role of ecdysteroids in the adult insect. *In* G. Downer and H. Laufer (eds.), *Endocrinology of Insects*. Alan R. Liss, New York, pp. 241–304.

Hahnel, A. C. and Eddy, E. M. 1986. Cell surface markers of mouse primordial germ cells defined by two monoclonal antibodies. *Gamete Res.* 15: 25–34.

Hardvin-Lepers, A., Shaper, J. and Shaper, N. L. 1993. Characterization of two *cis*-regulatory regions in the murine β1,4-galactosyltransferase gene. *J. Biol. Chem.* 268: 14348–14359.

Heasman, J., Mohun, T. and Wylie, C. C. 1977. Studies on the locomotion of primordial germ cells from *Xenopus laevis* in vitro. *J. Embryol. Exp. Morphol.* 42: 149–162.

Heasman, J., Hynes, R. D., Swan, A. P., Thomas, V. and Wyle, C. C. 1981. Primordial germ cells of *Xenopus* embryos: The role of fibronectin in their adhesion during migration. *Cell* 27: 437–447.

Heath, J. K. 1978. Mammalian primordial germ cells. *Dev. Mammals* 3: 272–298.

Hess, D. 1973. Local structural variations in the Y chromosome of *Drosophila hydei* and their correlation to genetic activity. *Cold Spring Harbor Symp. Quant. Biol.* 38: 663–672.

Hill, D. P., Shakes, D. C., Wards, S. and Strome, S. 1989. A sperm-supplied product essential for initiation of normal embryogenesis in *Caenorhabditis elegans* is encoded by the paternal effect embryonic–lethal gene, *spe–11*. *Dev. Biol.* 136: 154–166.

Hill, R. S. and MacGregor, H. C. 1980. The development of lampbrush chromosome-type transcription in the early diplotene oocytes of *Xenopus laevis*: An electron microscope analysis. *J. Cell Sci.* 44: 87–101.

Hirsh, D., Oppenheim, D. and Klass, M. 1976. Development of the reproductive system of *Caenorhabditis elegans*. *Dev. Biol.* 49: 200–219.

Hodgkin, J., Doniach, T. and Shen, M. 1985. The sex determination pathway in the nematode *Caenorhabditis elegans*: Variations on a theme. *Cold Spring Harbor Symp. Quant. Biol.* 50: 585–593.

Hotta, Y., Tabata, S. and Stern, H. 1984. Replicating and nicking of zygotene DNA sequences: control by a meiosis-specific protein. *Chromosoma* 90: 243–253.

Hoyle, H. D. and Raff, E. C. 1990. Two *Drosophila* β-tubulin isoforms are not functionally equivalent. *J. Cell Biol.* 111: 1009–1026.

Humphrey, R. R. 1931. Studies of sex reversal in *Amblystoma*. III. Transformation of the ovary of *A. tigrinum* into a functional testis through the influence of a testis resident in the same animal. *J. Exp. Zool.* 58: 333–365.

Iatrou, K., Spira, A. and Dixon, G. H. 1978. Protamine messenger RNA: Evidence for early synthesis and accumulation during spermatogenesis in rainbow trout. *Dev. Biol.* 64: 82–98.

Inoue, H. and Hiroyoshi, T. 1986. A maternal-effect sex-transformation mutant of the housefly, *Musca domestica* L. *Genetics* 112: 469–481.

Kammerman, S. and Ross, J. 1975. Increase in numbers of gonadotropin receptors on granulosa cells during follicle maturation. *J. Clin. Endocrinol.* 41: 546–550.

Karr, T. L. 1991. Intracellular sperm–egg interaction in *Drosophila*: A three-dimensional structural analysis of a paternal product in the developing egg. *Mech. Devel.* 34: 101–111.

Kimble, J. E. 1981. Strategies for control of pattern formation in *Caenorhabditis elegans*. *Philos. Trans. R. Soc. Lond.* [B] 295: 539–551.

Kimble, J. E. and White, J. G. 1981. Control of germ cell development in *Caenorhabditis elegans*. *Dev. Biol.* 81: 208–219.

Kimble, J., Barton, M. K., Schedl, T. B., Rosenquist, T. A. and Austin, J. 1986. Controls of postembryonic germ line development in *Caenorhabditis elegans*. *In* J. Gall (ed.), *Gametogenesis and the Early Embryo*. Alan R. Liss, New York, pp. 97–110.

Kloc, M., Spohr, G. and Etkin, L. 1993. Translocation of repetitive RNA sequences with the germ plasm in *Xenopus* oocytes. *Science* 262: 1712–1714.

Koch, E. A., Smith, P. A. and King, R. C. 1967. The division and differentiation of *Drosophila* cystocytes. *J. Morphol.* 121: 55–70.

Koos, R. D. and Clark, M. R. 1982. Production of 6–keto–prostaglandin $F_{1\alpha}$ by rat granulosa cells in vitro. *Endocrinology* 111: 1513–1518.

Korn, L. 1982. Transcription of *Xenopus* 5S ribosomal RNA genes. *Nature* 295: 101–105.

Kuwana, T. 1993. Migration of avian primordial germ cells toward the gonadal anlage. *Dev. Growth Differ.* 35: 237–243.

Kuwana, T., Maeda-Suga, H. and Fujimoto, T. 1986. Attraction of chick primordial germ cells by gonadal anlage in vitro. *Anat. Rec.* 215: 403–406.

Langman, J. 1981. *Medical Embryology*, 4th Ed. Williams & Wilkins, Baltimore.

Lantz, V., Ambrosio, L. and Schedl, P. 1992. The *Drosophila orb* gene is predicted to encode sex-specific germline RNA-binding proteins and has localized transcripts in ovaries and embryos. *Development* 115: 75–88.

Larsen, W. J., Wert, S. E. and Brunner, G. D. 1986. A dramatic loss of cumulus cell gap junctions is correlated with germinal vesicle breakdown in rat oocytes. *Dev. Biol.* 113: 517–521.

Laskey, R. A. 1974. Biochemical processes in early development. *In* A. T. Bull, J. R. Lagnado, J. O. Thomas and K. F. Tipton (eds.), *Companion to Biochemistry*, Vol. 2. Longman, London, pp. 137–160.

Lasser, A. B., Martin, P. L. and Roeder, R. G. 1983. Transcription of class III genes: Formation of preinitiation complexes. *Science* 222: 740–748.

Lemaire, W. J., Yang, N. S. T., Behram, H. H. and Marsh, J. M. 1973. Preovulatory changes in concentration of prostaglandin in rabbit graafian follicles. *Prostaglandins* 3: 367–376.

Lillie, F. R. 1919. *Problems of Fertilization.* University of Chicago Press, Chicago.

Macdonald, P. M., Luk, S. K.-S. and Kilpatrick, M. 1991. Protein encoded by the *exuperantia* gene is concentrated at sites of *bicoid* mRNA accumulation in *Drosophila* nurse cells but not in oocytes or embryos. *Genes Dev.* 5: 2455–2466.

Macdonald, P. M., Kerr, K., Smith, J. L. and Leask, A. 1993. RNA regulatory element BLE1 directs the early steps of *bicoid* mRNA localization. *Development* 118: 1233–1243.

Mahowald, A. P. and Strassheim, J. M. 1970. Intercellular migration of centrioles in the germinarium of *Drosophila melanogaster*: An electron microscopic study. *J. Cell Biol.* 45: 306–320.

Manseau, L. J. and Schüpbach, T. 1989. *cappuccino* and *spire*: two unique maternal-effect loci required for both the antero-posterior and dorsoventral patterns of the *Drosophila* embryo. *Genes Dev.* 3: 1437–1452.

Marcey, D., Watkins, W. S. and Hazelrigg, T. 1991. The temporal and spatial distribution pattern of maternal *exuparentia* protein: Evidence for a role in establishment but not maintainance of *bicoid* mRNSA localization. *EMBO J.* 10: 4259–4266.

Marushige K. and Dixon, G. H. 1969. Developmental changes in chromosome composition and template activity during spermatogenesis in trout testis. *Dev. Biol.* 19: 397–414.

Masui, Y. 1974. A cytostatic factor in amphibian: Its extraction and partial characterization. *J. Exp. Zool.* 187: 141–147.

Matsui, Y., Toksoz, D., Nishikawa, S., Nishikawa, S.-I., Williams, D., Zsebo, K. and Hogan, B. L. M. 1991. Effect of Steel factor and leukemia inhibitory factor on murine primordial germ cells in culture. *Nature* 353: 750–752.

Matsui, Y., Zsebo, K. and Hogan, B. L. M. 1992. Derivation of pluripotential embryonic stem cells from murine primordial germ cells in culture. *Cell* 70: 841–847.

Maurice, J. 1991. Improvements seen for RU-486 abortions. *Science* 254: 198–200.

McLaren, A. 1983. Does the chromosomal sex of a mouse cell affect its development? *Symp. Br. Soc. Dev. Biol.* 7: 225–227.

Miller, D. L., Jr. and Beatty, B. R. 1969. Visualization of nucleolar genes. *Science* 164: 955–957.

Miller, J., McLachlan, A. D. and Klug, A. 1985. Repetitive zinc-binding domains in the protein transcription factor IIIA from *Xenopus* oocytes. *EMBO J.* 4: 1609–1614.

Minishull, J. 1993. Cyclin synthesis: Who needs it? *BioEssays* 15: 149–155.

Mintz, B. 1957. Embryological development of primordial germ cells in the mouse: Influence of a new mutation. *J. Embryol. Exp. Morphol.* 5: 396–403.

Moens, P. B. 1969. The fine structure of meiotic chromosome polarization and pairing in *Locusta migratoria. Chromosoma* 28: 1–25.

Moens, P. B. 1974. Coincidence of modified crossover distribution with modified synaptonemal complexes. *In* R. F. Grell (ed.), *Mechanisms of Recombination.* Plenum, New York, pp. 377–383.

Moor, R. M. and Cran, D. G. 1980. Intercellular coupling of mammalian oocytes. *Dev. Mammals* 4: 3–38.

Moses, M. J. 1968. Synaptonemal complex. *Annu. Rev. Genet.* 2: 363–412.

Mowry, K. L. and Melton, D. A. 1992. Vegetal messenger RNA localization directed by a 340-nt RNA sequence element in *Xenopus* oocytes. *Science* 255: 991–994.

Newton, S. C., Blaschuk, O. W. and Millette, C. F. 1993. N-cadherin mediates Sertoli cell–spermatogenic cell adhesion. *Dev. Dynam.* 197: 1–13.

Ng, S.-Y., Parker, C. S. and Roeder, R. G. 1979. Transcription of cloned *Xenopus laevis* RNA polymerase III in reconstituted systems. *Proc. Natl. Acad. Sci. USA* 76: 136–140.

Nishioka, D., Ward, D., Poccia, D., Costacos, C. and Minor, J. E. 1990. Localization of bindin expression during sea urchin spermatogenesis. *Mol. Reprod. Dev.* 27: 181–190.

Old, R. W., Callan, H. G. and Gross, K. W. 1977. Localization of histone gene transcripts in newt lampbrush chromosomes by in situ hybridization. *J. Cell Sci.* 27: 57–80.

Oliver, B., Kim, Y.-J. and Baker, B. S. 1993. *Sex-lethal*, master and slave: A hierarchy of germ-line sex determination in *Drosophila. Development* 119: 897–908.

Paglia, L. M., Berry, J. and Kastern, W. H. 1976. Messenger RNA synthesis, transport, and storage in silkmoth ovarian follicles. *Dev. Biol.* 51: 173–181.

Palka, J. 1989. The pill of choice? *Science* 245: 1319–1323.

Palmiter, R. D., Wilkie, T. M., Chen, H. Y. and Brinster, R. L. 1984. Transmission distortion and mosaicism in an unusual transgenic mouse pedigree. *Cell* 36: 869–877.

Paris, J., Swenson, K., Piwnice-Worms, H. and Richter, J. D. 1991. Maturation-specific polyadenylation: In vitro activation by $p34^{cdc2}$ and phosphorylation of a 58-kD CPE-binding protein. *Genes Dev.* 5: 1697–1708.

Pasteels, J. 1953. Contributions à l'étude du developpement des reptiles. I. Origine et migration des gonocytes chez deux Lacertiens. *Arch. Biol.* 64: 227–245.

Paules, R. S., Buccione, R., Moscel, R. C., Vande Woude, G. F. and Eppig, J. J. 1989. Mouse *mos* protoncogene product is present and functions during oogenesis. *Proc. Natl. Acad. Sci. USA* 86: 5395–5399.

Pesce, M., Farrace, M. G., Piacentini, M., Dolci, S. and De Felici, M. 1993. Stem cell factor and leukemia inhitory factor promote primordial germ cell survival by suppressing programmed cell death (apoptosis). *Development* 118: 1089–1094.

Peschon, J. J., Behringer, R. R., Brinster, R. L. and Palmiter, R. D. 1987. Spermatid-specific expression of protamine-1 in transgenic mice. *Proc. Natl. Acad. Sci. USA* 84: 5316–5319.

Pieler, T., Hamm, J. and Roeder, R. G. 1987. The 5 S gene internal control region is composed of three distinct sequence elements, organized as two functional domains with variable spacing. *Cell* 48: 91–100.

Pincus, G. and Enzmann, E.V. 1935. The comparative behavior of mammalian eggs in vivo and in vitro. I. The activation of ovarian eggs. *J. Exp. Med.* 62: 665–675.

Pinkerton, J. H. M., McKay, D. G., Adams, E. C. and Hertig, A. T. 1961. Development of the human ovary: A study using histochemical techniques. *Obstet. Gynecol.* 18: 152–181.

Poccia, D. 1986. Remodeling of nucleoproteins during gametogenesis, fertilization, and early development. *Int. Rev. Cytol.* 105: 1–65.

Pokrywka, N. J and Stephenson, E. C. 1991. Microtubules mediate the localization of *bicoid* RNA during *Drosophila* oogenesis. *Development* 113: 55–66.

Postlethwait, J. H., Brownes, M. and Jowett, T. 1980. Sexual phenotype and vitellogenin synthesis in *Drosophila melanogaster. Dev. Biol.* 79: 379–387.

Pratt, S. A., Scully, N. F. and Shur, B. D. In press. Cell surface β-1,4-galactosyltransferase on primary spermatocytes facilitates their initial adhesion to Sertoli cells in vitro. *Biol. Reprod.*

Profet, M. 1993. Menstruation as a defense against pathogens transported by sperm. *Q. Rev. Biol.* 68: 335–385.

Racowsky, C. 1985. Effect of forskolin on the spontaneous maturation and cyclic AMP content of hamster and oocyte–cumulus complexes. *J. Exp. Zool.* 234: 87–96.

Racowsky, C. and Satterlie, R. A. 1985. Metabolic, fluorescent dye and electrical coupling between hamster oocytes and cumulus cells during meiotic maturation in vivo and in vitro. *Dev. Biol.* 108: 191–202.

Raikhel, A. S. and Dhadialla, T. S. 1992. Accumulation of yolk proteins in insect oocytes. *Annu. Rev. Entomol.* 37: 217–251.

Reynaud, G. 1969. Transfert de cellules germinales primordiales de dindon à l'embryon de poulet par injection intravasculaire. *J. Embryol. Exp. Morphol.* 21: 485–507.

Ressom, R. E. and Dixon, K. E. 1988. Relocation and reorganization of germ plasm in *Xenopus* embryos after fertilization. *Development* 103: 507–518.

Rivier, C., Rivier, J. and Vale, W. 1986. Inhibin-mediated feedback control of follicle-stimulating hormone secetion in the female rat. *Science* 234: 205–208.

Rogulska, T. 1969. Migration of chick primordial germ cells from the intracoelomically transplanted germinal crescent into the genital ridge. *Experientia* 25: 631–632.

Rogulska, T. Ozdzenski, W. and Komer, A. 1971. Behavior of mouse primordial germ cells in chick embryo. *J. Embryol. Exp. Morphol.* 25: 155–164.

Romanoff, A. L. 1960. *The Avian Embryo.* Macmillan, New York.

Ruohola, H., Bremer, K. A., Baker, D., Swedlow, J. R., Jan, L. Y. and Jan, Y. N. 1991. Role of neurogenic genes in establishment of follicle cell fate and oocyte polarity during oogenesis in *Drosophila. Cell* 66: 433–449.

Sagata, N., Oskarsson, M., Copeland, T., Brumbaugh, J. and Vande Woude, G. F. 1988. Function of c-*mos* proto-oncogene product in meiotic maturation in *Xenopus* oocytes. *Nature* 335: 519–525.

Sagata, N., Watanabe, N., Vande Woude, G. F. and Ikawa, Y. 1989. The c-*mos* proto-oncogene product is a cytostatic factor responsible for meiotic arrest in vertebrate eggs. *Nature* 342: 512–518.

Sakonju, S. and Brown, D. D. 1982. Contact points between a positive transcription factor and the *Xenopus* 5S RNA gene. *Cell* 31: 595–405.

Sakonju, S., Bogenhagen, D. F. and Brown, D. D. 1980. A control region in the center of the 5S RNA gene directs specific initiation of transcription. I. The 5' border of the region. *Cell* 19: 13–25.

Schultz, R. M., Montgomery, R. R. and Belanoff, J. R. 1983. Regulation of mouse oocyte maturation: Implications of a decrease in oocyte cAMP and protein dephosphorylation in commitment to resume meiosis. *Dev. Biol.* 97: 264–273.

Scotto, K. W., Kaulen, H. and Roeder, R. G. 1989. Positive and negative regulation of the gene for transcription factor TFIIIA in *Xenopus laevis* oocytes. *Genes Dev.* 3: 651–662.

Sen, D. and Gilbert, W. 1988. Formation of parallel four-stranded complexes by guanine-rich motifs in DNA and its implications for meiosis. *Nature* 334: 364–366.

Shaper, J. 1993. Can an analysis of galactosyltransferase gene expression provide insights into murine sperm–egg recognition? Ninth Annual Mid-Atlantic Regional Developmental Biology Conference.

Simon, D. 1960. Contribution à l'étude de la circulation et du transport des gonocytes primaires dans les blastodermes d'oiseau cultivé in vitro. *Arch. Anat. Microsc. Morphol. Exp.* 49: 93–176.

Skipper, J. K. and Hamilton, T. H. 1977. Regulation by estrogen of the vitellogenin gene. *Proc. Natl. Acad. Sci. USA* 74: 2384–2388.

Sorensen, R. and Wassarman, P. M. 1976. Relationship between growth and meiotic maturation of the mouse oocyte. *Dev. Biol.* 50: 531–536.

Spitz, I. M. and Bardin, C. W. 1993. Mifeprisone (RU486): A modulator of progestin and glucocorticoid action. *N. Engl. J. Med.* 329: 404–412.

Spradling, A. C. 1993. Germine cysts: Communes that work. *Cell* 72: 649–651.

Stephanson, E. C., Chao, Y.-C. and Fackenthal, J. D. 1988. Molecular analysis of the *swallow* gene of *Drosophila melanogaster. Genes Dev.* 2: 1655–1665.

Stewart, T. A. and Mintz, B. 1981. Successful generations of mice produced from an established culture line of euploid teratocarcinoma cells. *Proc. Natl. Acad. Sci. USA* 78: 6314–6318.

Stott, D. and Wylie, C. C. 1986. Invasive behaviour of mouse primordial germ cells in vitro. *J. Cell Sci.* 86: 133–144.

Subtelny, S. and Penkala, J. E. 1984. Experimental evidence for a morphogenetic role in the emergence of primordial germ cells from the endoderm of *Rana pipiens. Differentiation* 26: 211–219.

Sun, Y.-A. and Wyman, R. J. 1993. Reevaluation of electrophoresis in the *Drosophila* egg chamber. *Dev. Biol.* 155: 206–215.

Sutasurya, L. A. and Nieuwkoop, P. D. 1974. The induction of primordial germ cells in the urodeles. *Wilhelm Roux Arch. Entwicklungsmech. Org.* 175: 199–220.

Swanson, C. P., Merz, T. and Young, W. J. 1981. *Cytogenetics: The Chromosome in Division, Inheritance and Evolution.* Prentice-Hall, Englewood Cliffs, NJ.

Swift, C. H. 1914. Origin and early history of the primordial germ-cells in the chick. *Am. J. Anat.* 15: 483–516.

Telfer, W. H., Woodruff, R. I. and Huebner, E. 1981. Electrical polarity and cellular differentiation in meroistic ovaries. *Am. Zool.* 21: 675–686.

Theunissen O., Rudt, F., Guddat, U., Mentzel, H. and Pieler, T. 1992. RNA and DNA binding zinc fingers in *Xenopus* TFIIIA. *Cell* 71: 679–690.

Theurkauf, W. E., Smiley, S., Wong, M. L. and Alberts, B. M. 1992. Reorganization of the cytoskeleton during *Drosophila* oogenesis: Implications for axis specification and intercellular transport. *Development* 115: 923–936.

Theurkauf, W. E., Alberts, B. M., Jan, Y. N. and Jongens, T. A. 1993. A control code for microtubules in the differentiation of *Drosophila* oocytes. *Development* 118: 1169–1180.

Vegeto, E., Allan, G. F., Schrader, W. T., Tsai, M.-J., McDonnell, D. P. and O'Malley, B. W. 1992. The mechanism of RU486 antagonism is dependent on the conformation of the carboxy-terminal tail of the human progesterone receptor. *Cell* 69: 703–713.

von Wettstein, D. 1971. The synaptonemal complex and four-strand crossing over. *Proc. Natl. Acad. Sci. USA* 68: 851–855.

von Wettstein, D. 1984. The synaptonemal complex and genetic segregation. *In* C. W. Evans and H. G. Dickinson (eds.), *Controlling Events in Meiosis.* Cambridge University Press, Cambridge, pp. 195–231.

Watanabe, N., Vande Woude, G. F., Ikawa, Y. and Sagata, N. 1989. Specific proteolysis of the c-*mos* proto-oncogene product by calpain on fertilization of *Xenopus* eggs. *Nature* 342: 505–517.

Watanabe, N., Hunt, T., Ikawa, Y. and Sagata, N. 1991. Independent inactivation of MPF and cytostatic factor (Mos) upon fertilization of *Xenopus* eggs. *Nature* 352: 247–248.

Watson, C. A., Sauman, I. and Berry, S. J. 1993. Actin is a major structural and functional element of the egg cortex of giant silkmoths during oogenesis. *Dev. Biol.* 155: 315–323.

Whitington, P. M. and Dixon, K. E. 1975. Quantitative studies of germ plasm and germ cells during early embryogenesis of *Xenopus laevis. J. Embryol. Exp. Morphol.* 33: 57–74.

Woodruff, T. K., D'Agostino, J., Schwartz, N. B. and Mayo, K. E. 1988. Dynamic changes in inhibin messenger RNAs in rat ovarian follicles during the reproductive cycle. *Science* 239: 1296–1299.

Wolffe, A. P. and Brown, D. D. 1988. Developmental regulation of two 5S ribosomal RNA genes. *Science* 241: 1626–1632.

Wylie, C. C. and Heasman, J. 1993. Migration, proliferation, and potency of primordial germ cells. *Semin. Dev. Biol.* 4: 161–170.

Wylie, C. C., Heasman, J., Swan, A. P. and Anderton, B. H. 1979. Evidence for substrate guidance of primordial germ cells. *Exp. Cell Res.* 121: 315–324.

Yisraeli, J. K., Sokol, S. and Melton, D. A. 1990. A two-step model for the localization of a maternal mRNA in *Xenopus* oocytes: Involvement of microtubules and microfilaments in translocation and anchoring of Vg1 mRNA. *Development* 108: 289–298.

23

How does newness come into the world? How is it born? Of what fusions, translations, conjoinings is it made? How does it survive, extreme and dangerous as it is? What compromises, what deals, what betrayals of its secret nature must it make to stave off the wrecking crew, the exterminating angel, the guillotine?
SALMAN RUSHDIE (1988)

The first Bird was hatched from a Reptile's egg.
WALTER GARSTANG (1922)

Developmental mechanisms of evolutionary change

Charles Darwin was heir to centuries of speculation concerning the origins of the diversity of animal life. Darwin's own education was steeped in the British tradition of natural theology that held that God's wisdom, omnipotence, and benevolence could be seen in the works of His creation. The dominant part of this tradition was an account of Creation proclaiming that the species were intricately designed works of the Creator. The fingers of the human hand were seen as exquisitely (some said perfectly) designed contrivances that allowed humans mastery of their environment. The shovel-like claw of the mole was, again, perfectly adapted for its "office of existence," as were the wings of a bird and the fins of a fish. A more sophisticated form of natural theology, championed in Britain by anatomist and embryologist Richard Owen, held that adaptations were merely of secondary importance. What was really important was the idea that the human hand, the mole's claw, the bird's wing, and the fish's fin were based on the same plan. In abstracting the plan of the limb, we could determine the grand design upon which God had constructed all vertebrate appendages. To Owen (1848), the homologies underlying animal diversity were what counted, not the secondary adaptations of these basic unities.

"Unity of Type" and "Conditions of Existence"

Charles Darwin's synthesis

Darwin acknowledged his debt to these earlier debates when he wrote (1859), "It is generally acknowledged that all organic beings have been formed on two great laws—Unity of Type, and Conditions of Existence." Darwin went on to explain that his theory would explain unity of type by descent. The changes creating these types and causing the marvelous adaptations to the conditions of existence, moreover, were explained by natural selection. Darwin called this "descent with modification." After reading Johannes Müller's summary of von Baer's laws in 1842, Darwin saw that embryonic resemblances would be a very strong argument in favor of the genetic connectedness of different animal groups. "Community of embryonic structure reveals community of descent," he would conclude in *Origin of Species*.

Larval forms had been used for taxonomic classification even before Darwin. J. V. Thompson, for instance, had demonstrated that larval barnacles were almost identical to larval crabs, and he therefore counted barnacles as arthropods, not molluscs (Figure 23.1; Winsor, 1969). Darwin, an expert on barnacle taxonomy, celebrated this finding: "Even the illustrious Cuvier did not perceive that a barnacle is a crustacean, but a glance at the larva shows this in an unmistakable manner." Darwin's evolutionary interpretation of von Baer's laws set a paradigm that was to be followed for many decades, namely, that relationships between groups can be discovered by finding common larval forms. Kowalevsky (1871) would soon make a similar discovery (publicized in Darwin's *Descent of Man*) that tunicate larvae have notochords and form their neural tube and other organs in a manner very similar to that of the primitive chordate amphioxus. The tunicates, another enigma of classification schemes (usually placed, along with barnacles, as a mollusc), thereby found a home with the chordates. Darwin also noted that embryonic organisms sometimes make structures that are inappropriate for their adult form but that show their relatedness to other animals. He pointed out the existence of eyes in embryonic moles, pelvic rudiments in embryonic snakes, and teeth in embryonic baleen whales. In this book, we noted that mammalian embryos form a rudimentary yolk sac, send blood vessels to this yolk sac, and undergo gastrulation in a manner resembling that of birds and reptiles, whose development is constrained by the yolk.

Darwin also argued that adaptations that depart from the "type" and allow an organism to survive in its particular environment develop late in the embryo. He noted that the differences between species and genera are, as predicted by von Baer's laws, produced only late in development; he even chloroformed pigeons (with great reluctance) to prove to himself that this was indeed the case. Thus, Darwin recognized two ways of looking at "descent with modification." One could emphasize the common *descent* by pointing out embryonic homologies between two or more groups of animals, or one could emphasize the *modifications* by showing how development was altered to produce structures that enabled animals to adapt to particular conditions.

Darwin did not attempt to construct complete phylogenies from embryological data, but his work influenced many of his contemporaries to do so. One of the first scientists to realize the evolutionary importance of von Baer's studies was Elie Metchnikoff. Metchnikoff appreciated that evolution consists of modifying embryonic organisms, not adult ones. He wrote (1891):

> Man appeared as a result of a one-sided, but not total, improvement of organism, by joining not so much adult apes, but rather their unevenly developed fetuses. From the purely natural historical point of view, it would be possible to recognize man as an ape's 'monster,' with an enormously developed brain, face and hands.

Thus, organisms were seen to evolve through changes in their embryonic development.

E. B. Wilson and F. R. Lillie

If changes in embryonic development effected evolutionary changes, how did these developmental changes take place? During the late 1800s, many investigators attempted to link development to phylogeny through the analysis of cell lineages. They meticulously observed each cell in devel-

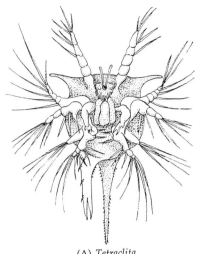

(A) *Tetraclita*

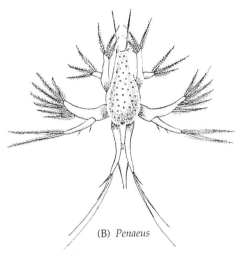

(B) *Penaeus*

FIGURE 23.1
Nauplius larvae of (A) a barnacle (*Tetraclita*, seen in ventral view) and (B) a shrimp (*Penaeus*, seen in dorsal view). The shrimp and barnacle share a similar larval stage despite their radical divergence in later development. (After F. Müller, 1864.)

oping embryos and compared the ways that different organisms formed their tissues. In 1898, two eminent embryologists gave cell lineage lectures at the Marine Biology Laboratories at Woods Hole, Massachusetts, and their lectures served to emphasize the two ways that embryology was being used to support evolutionary biology. The first lecture, presented by E. B. Wilson, was a landmark in the use of embryonic homologies to establish phylogenetic relationships. Wilson had observed the spiral cleavage patterns of flatworms, molluscs, and annelids, and he had discovered that, in each case, the same organs came from the same groups of cells. For him this meant that these phyla all had a common ancestor. The various groups of cleavage-stage cells in flatworms, molluscs, and annelids

> . . . show so close a correspondence both in origin and in fate that it seems impossible to explain the likeness save as a result of community of descent. The very differences, as we shall see, give some of the most interesting and convincing evidence of genetic affinity; for processes which in the lower forms play a leading role in the development are in the higher forms so reduced as to be no more than vestiges or reminiscences of what they were, and in some cases seem to have disappeared as completely as the teeth of birds or the limbs of snakes.

The next lecturer was F. R. Lillie, who had also done his research on the development of mollusc embryos and on modifications of cell lineage. He stressed the modifications, not the similarities, of cleavage. His research on *Unio*, a mussel whose cleavage is altered to produce the "bear-trap" larva that enables it to survive in flowing streams, was highlighted in Chapter 5. Lillie argued that "modern" evolutionary studies would do better to concentrate on changes in embryonic development that allowed for survival in particular environments rather than to focus on ancestral homologies that united animals into lines of descent.

In 1898, then, the two main avenues of approach to evolution and development were clearly defined: finding underlying unities that unite disparate groups of animals, and detecting the differences in development that enable species to adapt to particular environments. (These same lines of argument characterized the two types of natural theologies before Darwin.) Darwin had thought these to be temporally distinguished—that is, one would find underlying unities in the earliest stages while the later stages would diverge to allow specific adaptations (see Ospovat, 1981). However, Wilson and Lillie were both discussing the cleavage stage of embryogenesis. These two ways of characterizing evolution and development are still the major approaches today.

The evolution of early development: *E pluribus unum*

The emergence of embryos

In the evolution and the development of living organisms, one sees the emergence of multicellularity from single-celled organisms. A new whole is formed from component cellular parts. This is a fundamental step in the emergence of a new level of complexity. The volvocaceans and the dictyostelids mentioned in Chapter 1 represent but two of the 17 types of protists in which multicellularity was attained (Buss, 1987). However, only three groups (those that generated fungi, plants, and animals) evolved the ability to form multicellular aggregates that could differentiate into particular cell types—in other words, an embryo.

The first embryos had to solve a fundamental problem. Since each of the component cells had the genetic apparatus and the cytoplasmic architecture needed to divide, why shouldn't each cell continue its own proliferation? What would cause them to sacrifice their proliferative capacity to form a collective individual? There may have been more than one solution. Buss suggests that in these early embryos, there was a sharp dichotomy between proliferation and differentiation and that our protist ancestors never learned the trick of dividing once they had differentiated cilia. While some other protist groups (especially ciliates) could make more microtubule organizing centers, our ancestors could not. To this day, no ciliated metazoan cell divides (although ciliated metazoan cells can lose their cilia and then divide). Buss speculates that the ancestors of today's metazoans stopped their cellular proliferation by differentiating into a blastula of ciliated cells. (The early embryos of the first metazoan phyla—sponges and cnidarians—are characterized as balls of ciliated cells, just like the sea urchin embryos discussed in Chapter 5.) These ciliated blastulae could move, but it would appear that all their development had ceased, for ciliated cells do not divide, nor do they become any other differentiated cell type. To develop into an organism, this dilemma had to be solved.

This problem was solved by retaining or producing a population of nonciliated cells. These nonciliated cells could proliferate new cells, while the ciliated cells allowed the embryo to move. But these dividing cells could not go just anywhere. They could not grow on top of the ciliated cells or movement would cease. They could not grow into the water or they would be a drag on the embryo's movement. Rather, they would have to migrate *inside* the blastocoel (Figure 23.2). This movement and proliferation of cells is thought to be the origin of gastrulation. Thus, the blastula arose as a means of joining autonomous cells into a federation. The gastrula arose as a compromise within this federation that allowed the embryo to develop while moving* (Buss, 1987).

The earliest embryos probably developed in this mosaic fashion. However, induction provided a second mechanism to ensure that totipotent blastomeres remained together to form a single individual. Here, each cell sacrificed its autonomy to create a coherent community. Henry and co-workers (1989) have found that whereas individual sea urchin blastomeres can be totipotent, aggregates made of these same cells are not. Rather, each cell restricts the potency of its neighbor (Chapter 16). This restrictive regulation is also seen in allophenic mice (Chapter 5), wherein mammalian blastomeres combine to form a single chimeric mouse rather than two individual mice. There appear to be very important restrictions on cell potency once cells are brought together. Moreover, once the inside population can interact with the outside population of cells and with other parts of the inside population, inductive events can occur to give rise to new organs.

Whichever way this community of cells has been formed, their integration into a unified embryo is accomplished by maternal input into the

---

*This is a modification of the theory originally proposed by Metchnikoff (1886) to account for the origin of multicellular organisms. Using hydroid and sponge embryos, Metchnikoff pointed out that certain cells from the wall of the blastula "draw in their flagellum, become amoeboid and mobile, multiply by division, fill the cavity of the blastula, and become capable of digesting." This embryonic state, he felt, "is therefore entitled to be considered the prototype of multicellular beings." Metchnikoff attempted to make a phylogeny of all organisms on the basis of their germ layers, and he believed that all mesodermal cells could be characterized by their ability to phagocytize foreign substances. His discoveries in comparative embryology eventually allowed him to formulate the conceptual foundations of a new science, immunology. (For details of Metchnikoff's theory of multicellular origins, see Chernyak and Tauber, 1988, 1991.)

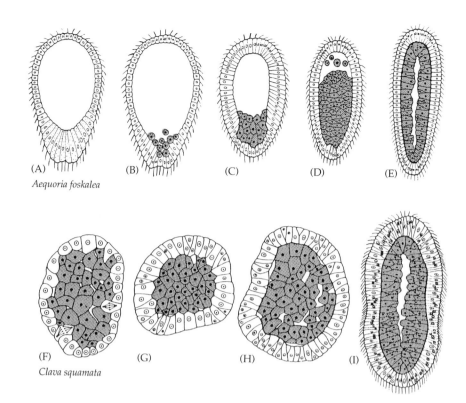

FIGURE 23.2

Gastrulation in two hydroid cnidarians. (A–E) Gastrulation in *Aequoria foskalea*, wherein a ciliated blastula is formed. Cells at the vegetal pole lose their cilia and migrate into the blastocoel to form a mitotically dividing population. (F–I) Gastrulation in *Clava squamata*, wherein a cell-filled stereoblastula is formed and the outer layer then becomes ciliated. Both plans converge on the ciliated planula larva characteristic of cnidarians. (Epiboly of a nonciliated ectoderm is not seen in embryos that are free-swimming.) (After Buss, 1987.)

Aequoria foskalea

Clava squamata

egg cytoplasm. It is this set of instructions that causes cells to cleave in certain arrangements, to stick to one another, and to differentiate at particular times. As we saw in Chapter 13, the sea urchin embryo becomes a ciliated blastula even in the absence of nuclear transcription. Only at gastrulation does the nucleus begin to regulate development. Thus, selection at the level of cell propagation (which had been the rule of survival among the protists) has been superceded by selection at the level of the individual multicellular organism.

Formation of the phyla: Modifying developmental pathways

Only about 33 animal body plans are presently being used on this planet (Margulis and Schwartz, 1988). These constitute the animal phyla. This is not to say that these body plans are the only possible ones. The Burgess Shale, a repository of early Cambrian soft-body fossils, is interpreted as containing representatives of 20 more phyla that never evolved descendants in the upper strata (Figure 23.3). In addition, this small band of sediment, about the size of a city block, contains about a dozen previously unknown classes of arthropods. These animals are not "primitive" members of existing phyla or classes, but are specialized examples of their own groups (Whittington, 1985; Gould, 1989). There are also two specimens in the Burgess Shale that may be related to ancestral forms of existing phyla. One is a peripatus-like animal that may be close to the ancestral form of insects; the other appears to be a well-preserved chordate called *Pikaia gracilens* that may be related to the ancestral chordates (Figure 23.3B). This latter fossil has several features that recommend its being classified in our phylum: it appears to have a notochord, and the zigzag bands along its sides look very much like the somite-derived musculature found in *Amphioxus* (Conway Morris and Whittington, 1979). Thus, all the known metazoan phyla (and many heretofore unknown ones) appear to have been formed in the Cambrian radiation that began 544 million years ago and lasted some 10 million years (Bowring et al., 1993).

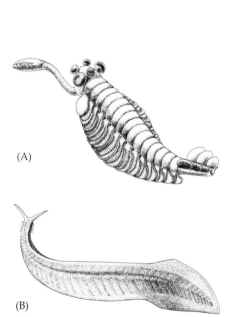

(A)

(B)

FIGURE 23.3

Two fossil organisms from the mid-Cambrian Burgess Shale. (A) *Opabina*, an organism having five eyes on its head, a frontal appendage with a terminal claw, body segments with dorsal gills, and a three-segment tail piece. (B) *Pikaia gracilens*, a possible chordate. (From Gould, 1989.)

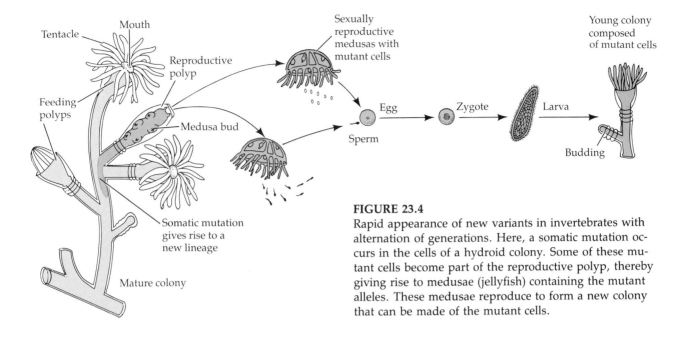

FIGURE 23.4
Rapid appearance of new variants in invertebrates with alternation of generations. Here, a somatic mutation occurs in the cells of a hydroid colony. Some of these mutant cells become part of the reproductive polyp, thereby giving rise to medusae (jellyfish) containing the mutant alleles. These medusae reproduce to form a new colony that can be made of the mutant cells.

If the interpretations of the fossil record are correct, eukaryotic cells emerged some 1.4 billion years ago, the phyla diverged during the brief Cambrian radiation, and no new morphological pattern (*Bauplan*) has been added in the past half-billion years. How did these basic body plans come into existence in such a relatively short time, and why has no other phylum emerged since then? Kauffman (1993) has a mathematical model that predicts that any evolving system (whether it be a phylum, species, automobile, or religion) displays this pattern of divergence followed by the locking in on a particular subset of the original diversity. Kauffman uses the metaphor of a rugged fitness landscape wherein there are peaks and valleys of fitness, and all organisms start out with the same average fitness value (in the middle of a peak). If they take large jumps, they have a 50 percent chance of becoming fitter organisms. Eventually, the chance of finding a fitter body plan decreases if an organism takes a jump far from where it is situated. Long jumps become risky, and the chance that these higher peaks are already occupied increases. Instead, small jumps (on the same peak) may make the organism somewhat fitter than the surrounding population. Then what we see is a diversification around a few successful models. In general, the duration between successful long jumps doubles with each attempt. During the early Cambrian, it is possible that the genome had not become stabilized into the sets of interactions that we see today. Moreover, in many invertebrate groups, there is an alternation of generations, wherein a sexual form generates an asexual form (zooid, polyp, bud) that then gives rise to the sexual form again. In such cases, somatic mutations in the asexual form can enter into the body of the sexual form and be propagated very rapidly (Figure 23.4; Buss, 1987).

How, then, can one modify one *Bauplan* to create another *Bauplan*? The first way would be to modify the earliest stages of development. According to von Baer (Chapter 7), animals of different species but of the same genus diverge very late in development. The more divergent the species are from one another, the earlier one can distinguish their embryos. Thus, embryos of the snow goose are indistinguishable from those of the blue goose until the very last stages. However, snow goose development diverges from chick development a bit earlier, and goose embryos can be distinguished from lizard embryos at even earlier stages. It appears, then,

(A) *Podarke*

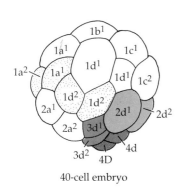

40-cell embryo

Fate map

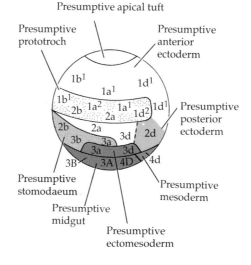

Presumptive apical tuft

Presumptive prototroch

Presumptive anterior ectoderm

Presumptive posterior ectoderm

Presumptive mesoderm

Presumptive ectomesoderm

Presumptive midgut

Presumptive stomodaeum

Trochophore larva

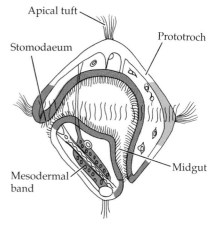

Apical tuft

Stomodaeum

Prototroch

Midgut

Mesodermal band

(B) *Tubifex*

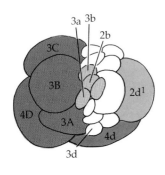

Cleaving embryo

Fate map

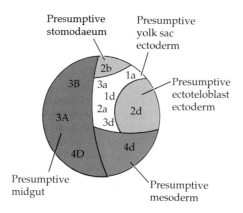

Presumptive stomodaeum

Presumptive yolk sac ectoderm

Presumptive ectoteloblast ectoderm

Presumptive mesoderm

Presumptive midgut

Gastrulation

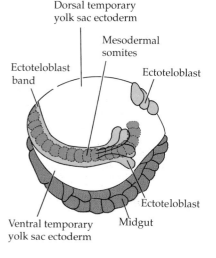

Dorsal temporary yolk sac ectoderm

Mesodermal somites

Ectoteloblast band

Ectoteloblast

Ectoteloblast

Midgut

Ventral temporary yolk sac ectoderm

FIGURE 23.5
Comparison of the development of two classes of annelid worms. (A) The polychaete *Podarke*, and (B) the oligo-chaete *Tubifex*. Their cleaving embryos, blastula fate maps, and products of gastrulation are seen. In *Podarke*, gastrulation leads to the formation of a trochophore larva. In *Tubifex*, there is no larval stage, and the embryo develops directly into a segmented body. (After Anderson, 1973.)

that mutations that create new *Baupläne* could do so by altering the earliest stages of development.

These early developmental changes can be effected by changing the localization of cytoplasmic determinants, changing the rate of cell division of one cell or group of cells relative to the others, or changing the positions of the cells as they divide. In Chapter 5, we saw that a modification of molluscan cleavage can give the bulk of cytoplasm to the ectodermal cells that form the larval shell. This is due to changing the manner in which the blastomeres divide and apportion cytoplasm. In annelid worms, the differences between polychaetes and oligochaetes stem from differences in the cytoplasmic localization of morphogens within the egg (Figure 23.5). While they both undergo spiral cleavage, they apportion their morphogens into different cells. Polychaetes undergo a relatively standard spiral cleavage to give rise to the trochophore larva. Oligochaetes, however, put most of their cytoplasm into those cells destined to form adult, rather than larval, structures. This group then skips the larval stage. If a mutation were to place a certain cytoplasmic morphogen into one region of the egg instead of another, or if a mutation caused a change in the axis of cell division so that different sets of cells acquired these determinants, then a radically different phenotype could be produced. As E. G. Conklin wrote

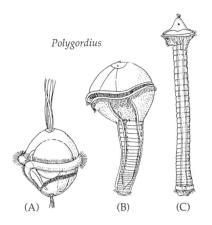

Polygordius

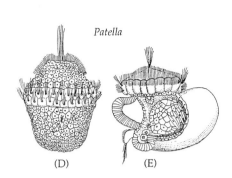

Patella

Vestimentiferan

(A)   (B)   (C)   (D)   (E)   (F)

in 1915, "We are vertebrates because our mothers were vertebrates and produced eggs of the vertebrate pattern."

Another way of evolving new phyla may involve modifying the larva. Darwin and others thought that similarities of larval form signified common descent. However, this can be reinterpreted to mean that changes that give rise to different phyla may occur in larvae. Snails, echiuroids, and polychaetes have very similar patterns of division and form trochophore larvae (Figure 23.6). In fact, the placement of the newly discovered phylum Vestimentifera (the bright red, gutless invertebrates found in the deep ocean trenches) near the annelids was made in part on the basis of vestimentiferans' having trochophore larvae (Jones and Gardiner, 1989). Thus, one of the principal mechanisms for establishing new phyla and classes may be the rearrangement of development during the larval stage so that metamorphosis brings about new types of organization. Garstang (1928) showed how the veliger larva of certain snails could have arisen by mutation and then been selected because the new arrangement of head and shell allowed the head to retract beneath the shell for safety. He also framed the hypothesis that chordates arose from ancestral tunicate larvae that had become neotenic. Unfortunately, soft-bodied larvae rarely fossilize, so we know very little about the mechanisms by which chordates and other phyla may have arisen from early Cambrian larvae.*

FIGURE 23.6
Divergence of development after the trochophore larval stage. (A–C) Metamorphosis of the polychaete annelid *Polygordius* from its free-swimming trochophore larva shows the formation of a segmented trunk. Eventually, the larval structures shrink at the anterior end as the head forms. (D–F) Metamorphosis of the prosobranch (clam) mollusc *Patella*. After the trochophore stage, it develops a molluscan foot, shell gland, and visceral hump. (F) Scanning electron micrograph of the trochophore larva of a vestimentiferan. (A–E after Grant, 1978; F from Jones and Gardiner, 1989, courtesy of the authors.)

The developmental-genetic mechanisms of evolutionary change

Isolation

Recent research is attempting to circumvent this gap in our understanding by studying the genes that control embryonic and larval development.

---

*Larval forms often bridge the gap between the different adult forms. The larval form is seen either as being ancestral to two groups or as "breaking away" by neoteny and forming a different type of organism. This has often been hypothesized as the mechanism by which chordates emerged from invertebrates and vertebrates arose from chordates. The tornaria larva of hemichordates is formed in a deuterostome manner similar to that of echinoderm larvae and looks enough like echinoderm larvae to have been originally mistaken for them. This would link echinoderms and chordates. Garstang (1928) and Berrill (1955) hypothesized that the larvae of certain tunicates could have evolved into chordates such as amphioxus by neotenic development. In this way, the tunicates would keep the notochord, larval musculature, and feeding apparatus of the larval tunicate while becoming sexually mature. There are, in fact, neotenic free-swimming tunicates (such as *Larvacea*). Modifications of this view (using a different protochordate stock) have recently been suggested by Jefferies (1986). The origin of chordates remains a difficult problem.

One way developmental changes can bring about evolutionary change is to cause the reproductive isolation of a group of organisms within a population. Such isolation is thought to be essential for the formation of new species. The snail shell coiling mutations discussed in Chapter 5 are mutations that act during early development to change the position of the mesodermal organs. Mating between left-coiling and right-coiling snails is mechanically very difficult, if not impossible, in some species (Clark and Murray, 1969). As this mutation is inherited as a maternal effect gene, a group of related snails would emerge that could mate with one another but not with other members of the original population. These reproductively isolated snails could expand their range and, by the accumulation of new mutations, form a new species (Alexandrov and Sergievsky, 1984).

Homeosis

Major developmental changes can occur when the interactions between genes are changed. This can give rise to new cellular phenotypes, and, in some instances, to new organismal phenotypes as well. We have seen in Chapter 11 that when a structural gene is set next to a new enhancer, it becomes expressed at the time or place specified by the enhancer. A globin gene will be expressed at midblastula transition if placed next to certain *Xenopus* enhancers, and a proto-oncogene leads to uncontrolled lymphocyte growth when translocated near an immunoglobulin gene enhancer in a B-cell precursor. One possible way of changing a cell's differentiated phenotype would be to alter a gene's enhancer, either by mutation or by translocation.

Rabinow and Dickinson (1986) are testing this hypothesis on Hawaiian *Drosophila*. These flies have undergone an extremely rapid rate of speciation, and several species have changes in the tissue specificity of gene expression. The larva of *D. hawaiiensis*, for instance, expresses its alcohol dehydrogenase (ADH) in its fat bodies and carcass, whereas the larva of *D. formella* expresses its alcohol dehydrogenase only in its fat bodies. The two ADHs can be distinguished electrophoretically. These species can mate with one another to generate hybrid larvae. When the fat bodies of these larvae were examined, both *D. hawaiiensis* and *D. formella* ADHs were seen. However, when the carcasses were analyzed, only the *D. hawaiiensis* protein was observed. This is the pattern one would expect if the ADH genes were controlled by *cis*-regulatory elements and if they differed between the two closely related species. (Were expression based on *trans*-regulatory elements, both genes would either be on or off.) It appears, then, that enhancers may change during evolution, thereby causing different products to be synthesized in different tissues.

If cellular phenotypes can change when the *cis*-regulatory elements of structural genes are altered, the alterations can be even more far-reaching when the *trans*-regulatory factors are altered. In Chapter 15, we discussed homeotic transformations in *Drosophila*. These mutations are caused when certain *trans*-regulatory enhancer-binding factors are altered. In 1940, Richard Goldschmidt and Conrad H. Waddington pointed out that these homeotic mutations provided the means to link together development, genetics, and evolution. Waddington (1940) stressed that homeotic genes were embryonic "switches," analogous to switches of railroad yards that directed trains into one path rather than another. Goldschmidt (1940) saw homeotic genes as generating several large changes in the embryo simultaneously, thereby causing the formation of new species. In focusing attention on homeotic transformations, they were far ahead of the science of their day.

In many ways, the phenotypes of the homeotic deletions resemble the

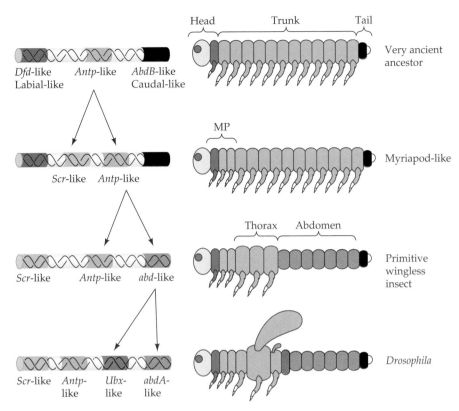

Head Trunk Tail

Dfd-like *Antp*-like *AbdB*-like
Labial-like *Caudal*-like

Very ancient
ancestor

MP

Scr-like *Antp*-like

Myriapod-like

Thorax Abdomen

Scr-like *Antp*-like *abd*-like

Primitive
wingless
insect

Scr-like *Antp*- *Ubx*- *abdA*-
like like like

Drosophila

FIGURE 23.7
A scheme for the origin of differentially specified segments during the evolution of insects from a primitive myriapod-like ancestor. Originally only the head, trunk, and tail were specified. The trunk then became specified into abdomen, thorax, and palp-bearing segments. Later, each of these regions became divided into different structures. It is thought that this cascade of specifications was accomplished through the duplication and divergence of homeotic genes of the bithorax-Antennapedia complex. (After Akam et al., 1988.)

postulated earlier stages of *Drosophila* evolution. It is hypothesized (Raff and Kaufman, 1983; Akam et al., 1988) that the differentiation of insect segments evolved by the duplication of homeotic genes that eventually acquired new roles (Figure 23.7). Such evolutionary trends are suggested by the phenotypes created when the genes are deleted.

As we saw in Chapter 17, vertebrate segments are also specified by homeotic genes. Moreover, when Rijli and co-workers (1993) deleted the *Hoxa-2* gene in mice, one of the skull bones was respecified to form the pterygoquadrate bone—a structure found in reptiles but not in mammals. Homeotic genes have been invoked as a means of getting new types of structures and also as evidence of the continuity of all animal life on this planet. As we discussed in Chapter 17, all animal species studied have certain homeotic genes, have them in the same order on the chromosome, and have similar expression patterns along the anterior–posterior axis. In the homeotic gene complex, we see descent with modification on the molecular level.

New combinations of genes could also create new cell types. One of the major changes in phyla has been the occurrence of cell types with new characteristics. Kauffman (1993) has mathematically modeled the generation of new cell types from a random genome consisting of 10,000 genes, each regulated by 2 other genes. In such cases, he finds only 100 stable states (out of 210,000 possible states). Each of these stable states represents a differentiated cell type. In some cases, the mutation of a regulatory gene is sufficient for the restructuring of the interactions, and a new cell type is created. Most genes, however, remain unaffected by this new arrangement. The creation of a new cell type is a rare event in nature and can often change the nature of the beast. Neural crest cells, for instance, distinguish vertebrates from the protochordates and invertebrates (Figure 23.8). The protochordates have a dorsal neural tube and notochord, but

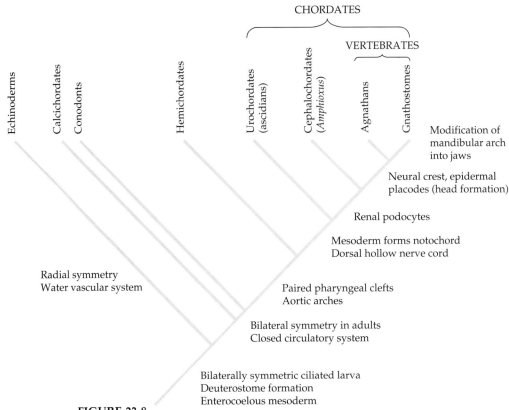

FIGURE 23.8

Developmental changes in the evolution of invertebrate to vertebrate. The original deuterostome invertebrates were able to form the echinoderms and those other organisms that eventually gave rise to the vertebrate lineage. The ability of the mesoderm to form a notochord and its overlying ectoderm to become a neural tube separated the chordates from the remaining invertebrates. The development of neural crest cells and the epidermal placodes that give rise to the sensory nerves of the face distinguish the vertebrates from the protochordates. (After Gans, 1989; Langille and Hall, 1989.)

no real "head." The cranial neural crest cells are largely responsible for the creation of the face, skull, and the branchial arches. It is thought that the development of the head originally allowed for more efficient predation, placing the sensory structures adjacent to the prey-capturing jaws (Gans and Northcutt, 1983; Langille and Hall, 1989; Hall, 1992). As Figure 23.8 shows, the vertebrates are thought to have arisen from invertebrates in several steps that involved the formation and modification of new cell types.

Development and evolution within established *Baupläne*

Developmental constraints

We have alluded to the fact that there are relatively few *Baupläne*, and that one can easily imagine types of animals that do not exist within existing phyla. Why aren't there more major body types among the animals? To answer this, we have to consider the constraints imposed upon evolution. There are three major classes of constraints on morphogenetic evolution.

First, there are **physical constraints** on the construction of the organism. These constraints of diffusion, hydraulics, and physical support allow only certain mechanisms of development to occur. One cannot have a vertebrate on wheeled appendages (of the sort that Dorothy saw in Oz) because blood cannot circulate to a rotating organ; this entire possibility of evolution has been closed off. Similarly, structural parameters and fluid dynamics forbid the existence of 5-foot-tall mosquitos.

Second, there are constraints involving **morphogenetic construction rules** (Oster et al., 1988). Bateson (1894) noted that when organisms depart from their normal development, they do so in only a limited number of ways. Research in this area attempts to find the architectural parameters upon which organisms are constructed and seeks to show how these parameters can be modified during evolution. Some of the best examples of these types of constraints come from the analysis of limb formation in vertebrates. Holder (1983) pointed out that although there have been many modifications of the vertebrate limb over 300 million years, some modifications are not found (Figure 23.9). Moreover, analyses of natural populations suggest that there are a relatively small number of ways that limb changes can occur (Wake and Larson, 1987). If a longer limb is favorable in a given environment, the humerus may become elongated. One never sees two smaller humeri joined together in tandem, although one could imagine the selective advantages that such an arrangement might have. This indicates a construction scheme that has certain rules.

The principal rules of vertebrate limb formation have been summarized by Oster and his co-workers (1988). They find that a reaction–diffusion mechanism can explain the known morphologies of the limb and can explain why other morphologies are forbidden. As discussed in Chapter 19, this model posits that the aggregations of cartilage actively recruit more cells from the surrounding area and laterally inhibit the formation of other foci of condensation. The number of foci depends on the geometry of the tissue and the strength of the lateral inhibition. If the inhibition remains the same, the size of the tissue volume must increase in order to get two foci forming where one had been allowed before. At a certain threshold (called a bifurcation threshold), this size is reached, and the limb can branch into two foci.

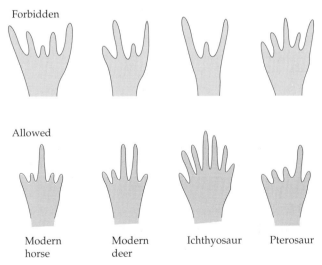

FIGURE 23.9
Some "forbidden" and some permissible tetrapod limb structures. (From Thomson, 1988.)

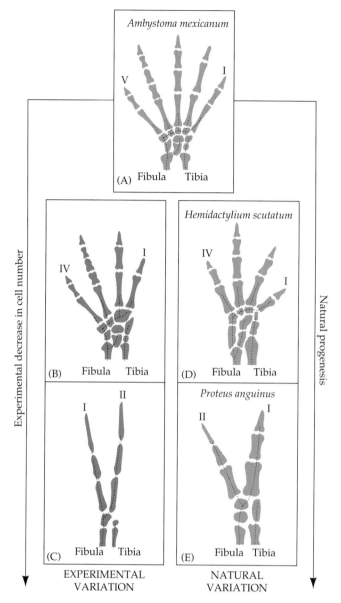

FIGURE 23.10
Relationship between cell number and number of digits in salamanders. (A) The hindlimb of an axolotl (*Ambystoma mexicanum*) with its five symmetrical digits. (B,C) Digits on axolotl hindlimb after hindlimb bud is incubated in colchicine to reduce cell number. (D,E) Two wild salamanders formed by progenesis, each having a smaller limb bud. (D) *Hemidactylium scutatum* and (E) *Proteus anguinus*. The parallels between the experimental variation and the natural variation can be seen, and the common denominator is reduced cell numbers in the limb buds. (After Oster et al., 1988.)

Evidence for this mathematical model comes from experimental manipulations and comparative anatomy. When an axolotl limb bud is treated with the antimitotic drug colchicine, the dimensions of the limb are reduced. In such limbs, there is not only a reduction of digits, but a reduction of certain digits in a certain order, as expected by the mathematical model and from the "forbidden" morphologies. Moreover, these reductions of specific digits are very similar to those limbs of progenetic salamanders, those species that reach maturity at a smaller stage than did their ancestors and whose limbs develop from smaller limb buds (Figure 23.10; Alberch and Gale, 1983, 1985). Thus, the use of reaction–diffusion mechanisms to construct limbs may constrain the possibilities that can be generated during development, because only certain types of limbs are possible using these rules.

Phyletic constraints comprise the third set of constraints on the evolution of new types of structures (Gould and Lewontin, 1979). These are the historical restrictions based on the genetics of an organism's development. For instance, once a structure is generated by inductive interactions, it is difficult to start over again. The notochord, which is still functional in adult protochordates (Berrill, 1987), is considered vestigial in adult birds and mammals. Yet it may be transiently necessary in the embryo to specify the neural tube. Similarly, Waddington (1938) noted that although the pronephric kidney of the chick embryo is considered vestigial (since it has no ability to concentrate urine), it is the source of the ureteric bud that induces the formation of a functional kidney during chick development.

This type of phyletic constraint has recently been reviewed by Raff and colleagues (1991). Until recently it was thought that the earliest stages of development would be the hardest to change, because altering them would either destroy the embryo or generate a radically new phenotype. But recent work (and the reappraisal of older work) has shown that alterations can be made to early cleavage without upsetting the final form. Modifications of morphogens in mollusc embryos can give rise to new types of larvae that still metamorphose into molluscs, and changes in sea urchin cytoplasmic morphogens can generate sea urchins that develop without larvae but still become sea urchins. In fact, looking at vertebrates, one can see there is an entire history leading up to the famous diagram of von Baer's law shown in Chapter 7. All the vertebrates arrive at this particular stage of development (called the *pharyngula*), but they can do so by different means (Figure 23.11). Birds, reptiles, and fish arrive there after meroblastic cleavages of different sorts; amphibians get to the same stage by way of radial holoblastic cleavage; and mammals reach the same place after constructing a blastocyst, chorion, and amnion. The earliest stages of development, then, appear to be extremely plastic. Similarly, the later stages are very different, as the different phenotypes of mice, sunfish, snakes, and newts amply demonstrate. There is something in the *middle* of development that appears to be invariant.

Raff argues that the formation of new *Baupläne* is inhibited by the need for global sequences of induction during the neurula stage (Figure 23.12). Before that stage, there are few inductive events. After that period, there are a great many inductive events, but almost all of them are local. During early organogenesis, however, there are several inductive events occurring simultaneously that are global in nature. In vertebrates, to use von Baer's example, the earliest stages involve specifying axes and undergoing gastrulation. Induction has not happened on a large scale. Moreover, as Raff and colleagues have shown (Henry et al., 1989), there is a great deal of regulative ability here, so small changes in morphogen distribution or the position of cleavage planes can be accommodated. After the major body plan is fixed, inductions occur all over the body but are compartmentalized into discrete organ-forming systems. The lens induces the cornea, but if it fails to do so, only the eye is affected. Similarly, there are inductions in the skin that form feathers, scales, or fur. If they do not occur, the skin or patch of skin may lack these structures. But during early organogenesis, the interactions are more global (Slack, 1983). Failure to have the heart at a certain place can affect the induction of eyes (Chapter 18). Failure to induce the mesoderm in a certain region leads to malformations of the kidneys, limbs, and tail. It is this stage that constrains evolution and that typifies the vertebrate phylum. Thus, once a vertebrate, it is difficult to evolve into anything else.

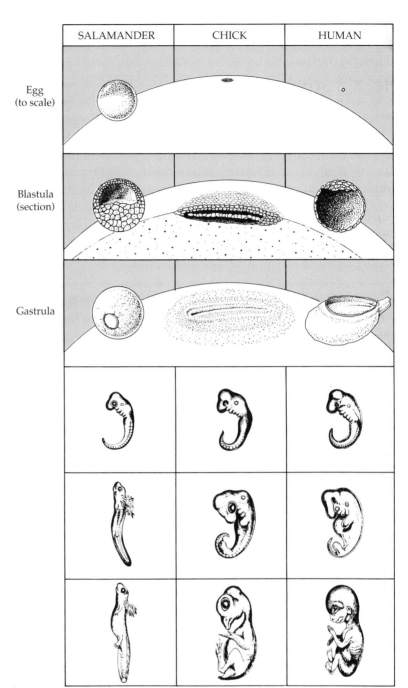

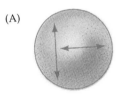

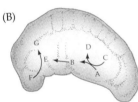

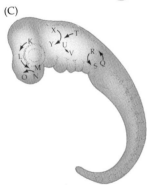

FIGURE 23.11
The bottleneck at the "pharyngula" stage of vertebrate development. The bottom of this chart is the standard depiction of von Baer's law (as shown in Chapter 7), demonstrating the divergence of vertebrate classes after a common embryonic stage. The top of this chart represents the divergent beginnings of development. Von Baer himself (1886) was aware of this bottleneck. (After Elinson, 1987.)

FIGURE 23.12
Mechanism for the bottleneck at the pharyngula stage of vertebrate development. (A) In the cleaving embryo, global interactions exist, but there are very few of them (mainly to specify the axes of the organism). (B) At the neurula to pharyngula stages, there are many global interactions. (C) After the pharyngula stage, there are even more inductive interactions, but these are primarily local in effect, confined to their own fields. (After Raff, in press.)

Ernst Haeckel and the "Biogenetic Law"

A disastrous union of embryology and evolutionary biology was forged in the last half of the nineteenth century by the German embryologist and philosopher Ernst Haeckel. Based on the assumption that the laws by which species arose on this planet (phylogeny) were identical to the laws by which the individuals of the species developed (ontogeny), he viewed adult organisms as the embryonic stages of more advanced organisms. This view was summarized by his "Biogenetic Law": ontogeny recapitulates phylogeny. In other words, development of advanced species was seen to pass through stages represented by adult organisms of more primitive species.

In this view, the creation of new phyla is a step toward the completion of human development. In earlier epochs, only the initial stages of this development occurred, producing protists and cnidarians. Later, more stages were added sequentially, until finally a human being evolved. According to Haeckel, three laws sufficed to explain how this advancing ontogeny could generate new species. First, there was the law of correspondence. The human zygote, for instance, was represented by the "adult" stage of the protists; the colonial protists represented the advancement of development to the blastula stage; the gill slit stage of human embryos was represented by adult fishes. Haeckel even postulated an extinct organism, *Gastraea*, a two-layered sac corresponding to the gastrula, which he considered the ancestor of all metazoan species. Second, there was the law of terminal addition. The embryo evolved new species by adding a step at the end of the previous ones. In such a view, humans evolved when the embryo of the next highest ape added on a new stage. This provided a linear, not a branching, phylogeny. There was also the law of truncation, which held that preceding development could be foreshortened. This law was needed to prevent gestation time from being enormous. It also was needed since embryologists did not observe all these stages in all animals.

This notion of ontogeny recapitulating phylogeny was not Darwinism. In fact, Haeckel's synthesis was an attempt to fuse the works of Darwin, Lamarck, and Goethe. In Haeckel's scheme, animals advanced to new, "higher" levels by adding stages to existing embryonic development. In Darwinism, contemporary species are seen as having a common ancestor. The result is a multibranched "bush." (A tree metaphor has also been used, but trees have a central axis on which scientists have frequently placed the lineage leading up to *Homo sapiens*.) Humans are not "higher" than chimps, but humans and chimps have one ancestor from which both groups diverged.

Interestingly, von Baer (1828) had disproven the "biogenetic law" before Haeckel ever invented it. In ridiculing the pre-evolutionary forms of this law, von Baer fantasized what would happen if birds were writing the embryology textbooks.

> Let us imagine that birds had studied their own development and that it was they who investigated the structure of the adult mammal and of man. Wouldn't their physiological textbooks teach the following? 'Those four- and two-legged animals bear many resemblances to our own embryos, for their cranial bones are separated, and they have no beak, just as we do in the first five or six days of our incubation; their extremities are all very much alike, as ours are for about the same period; there is not a true feather on their body, rather only thin feather-shafts, so that we, as fledglings in the nest, are more advanced than they will ever be. . . . And these mammals that cannot find their own food for such a long time after birth, that can never rise freely from the earth, want to consider themselves more highly organized than we?'

By observing development, von Baer (Chapter 7) noted that embryos never pass through the adult stages of other animals. However, there are stages that related embryos do share. All vertebrate embryos pass through a stage in which there are embryonic gill slits. Fish elaborate them into true gills, while the slits become part of the jaw or ear apparatus in other vertebrates. But a frog or human embryo never passes through a stage in which it has the structures of an adult fish. However, even though von Baer and others had discredited the recapitulation notion, it became one of the most popular notions in biology.*

*Even more than in biology, Haeckel's "biogenetic law" was adapted uncritically by many of the newly forming social sciences. Early anthropologists espoused the view that other cultures were "primitive" in the embryological sense in that their development had stopped short of our own. Indeed, the word "underdeveloped" is still used to define such a culture. Haeckel also felt that within the human species, races could be ranked in a linear fashion and thereby gave scientific "validity" to the racial prejudices that culminated in the genocidal biopolicy of the Third Reich. (For details, see Gasman, 1971; Gould, 1977a,b; Stein, 1988.) Gould shows that recapitulation has a limited value in looking at the formation of related species but that it is not a universal phenomenon.

Inductive interactions and the generation of novel structures

In addition to constraining development, inductive interactions can be altered to produce evolutionary novelties. This was first suggested by Walter Garstang. During the first decades of the twentieth century, the dominant way of relating embryology to evolution was through Ernst Haeckel's "Biogenetic Law" (see Sidelights & Speculations). Garstang delivered a stunning critique of this "law" in 1922, showing that it was incompatible with Mendelian genetics and with the observations of embryology. Garstang stood the biogenetic law on its head: ontogeny does not recapitulate phylogeny, it creates phylogeny. The animals that arose late in evolutionary history did not arise through a terminal addition onto an existing phylogeny. Rather, they arose by mutations that affected the interactions of modules already existing in the *Bauplan* of the organism: "A house is not a cottage with an extra story on the top. A house represents a higher grade in the evolution of a residence, but the whole building is altered—foundation, timbers, and roof—even if the bricks are the same."

Thus, when we say that the contemporary one-toed horse evolved from a five-toed ancestor, we are saying that hereditable changes occurred in the differentiation of limb mesoderm into chondrocytes during embryogenesis in the horse lineage. In this perspective, evolution is the result of hereditary changes affecting development. This is the case whether the mutation is one that changes the reptilian embryo into a bird or one that changes the color of *Drosophila* eyes.

Therefore, even after the 30 or so contemporary *Baupläne* had been established, changes during development could still effect drastic changes in evolution. There are many ways by which changes in induction can lead to diverse morphogenic phenomena. Changes in induction can change scales into feathers (as in the case of bantam chicks) and are responsible for such remarkable adaptations as the avian lung, the rumi-

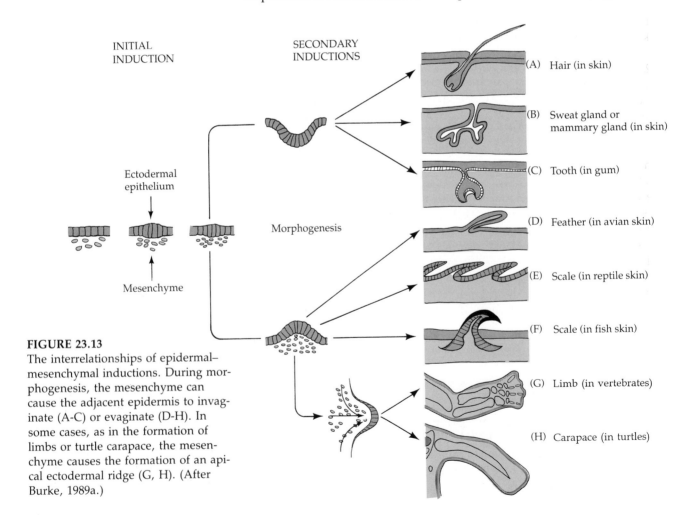

INITIAL INDUCTION

SECONDARY INDUCTIONS

Ectodermal epithelium

Mesenchyme

Morphogenesis

(A) Hair (in skin)

(B) Sweat gland or mammary gland (in skin)

(C) Tooth (in gum)

(D) Feather (in avian skin)

(E) Scale (in reptile skin)

(F) Scale (in fish skin)

(G) Limb (in vertebrates)

(H) Carapace (in turtles)

FIGURE 23.13
The interrelationships of epidermal–mesenchymal inductions. During morphogenesis, the mesenchyme can cause the adjacent epidermis to invaginate (A-C) or evaginate (D-H). In some cases, as in the formation of limbs or turtle carapace, the mesenchyme causes the formation of an apical ectodermal ridge (G, H). (After Burke, 1989a.)

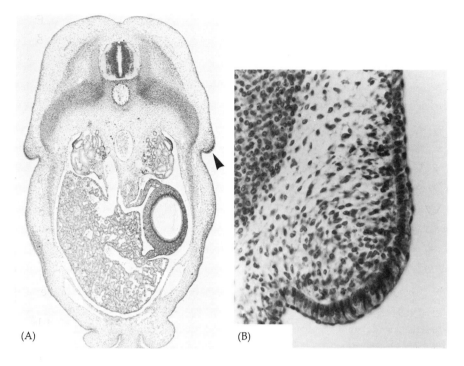

FIGURE 23.14
Midtrunk cross section through the embryo of the turtle *Chelydra serpentina*. (A) The carapacial ridge (arrow) is formed at the boundary of the somitic and lateral plate mesoderm and now represents the dorsal-ventral boundary. The thickened mesodermal bands extending from the center into the carapace area are the rib condensations. (B) Higher magnification of the carapacial ridge. (From Burke, 1989b, courtesy of the author.)

nant stomach, and the tusks of elephants (modified incisors) and walruses (modified upper canine teeth). The secretory glands of the epidermis are each modifications of the same type of induction—mammary glands are embryologically modified sweat glands. Likewise, the fearsome rows of sharks' teeth are modifications of its body scales. Indeed, Burke (1989a) has proposed that all secondary inductions have a common ancestry (Figure 23.13).

One interesting point in Burke's diagram concerns the mechanism used to create limbs (an apical ectodermal ridge accompanied by a malleable cartilage-forming mesoderm). She finds that one family of vertebrates uses this same mechanism in another region of the body. The carapace shell of the turtle results from such a placement of an epithelial–mesenchymal interaction in a new area. As can be seen in Figure 23.14, the carapacial ridge is formed in a manner analogous to the apical ectodermal ridge of limbs. The patterns of cell proliferation and the distribution of adhesive glycoproteins such as N-CAM and fibronectin also show similarity to those in limb formation (Burke, 1989b). Induction is also responsible for the creation of "extra" arms in echinoderms. The formation of the five arms of a starfish are caused by the induction of the epidermis by the five protrusions of the hydrocoel. However, in the six-armed starfish, *Leptasterias hexactis*, a sixth protrusion of the hydrocoel occurs, and nine protrusions cause the formation of the nine-armed *Solaster* sea stars (Gemmill, 1912).

Correlated progression

When a drastic remodeling of morphology takes place during evolution, one has to account for the developmental changes that have caused it. As early as 1871, St. George Mivart pointed out that large evolutionary changes were not due to the simple alteration of one structure of an organism. Rather, an entire group of structures changed. Raissa Berg (1960) has called these co-varying constellations of characters "correlation pleiades." The mechanism that enables such changes to occur has been called correlated development and was a major feature of embryology at the turn of the century. Spemann (1901) wrote a major work on this topic that formed the context for his work on lens formation. Lenses, he said, form only when induced by other tissues (such as optic vesicle tissue) that form eye structures. Similarly, skeletal cartilage informs the placement of muscles, and muscles induce the placement of nerve axons. Thomson (1988) has used this concept to explain how one part of an organism can change without causing the entire structure to become uncoordinated. If one structure changes, it will induce other structures to change with it. This is correlated progression.

One of the best examples of such correlated progression involves the formation of the vertebrate jaw and ear ossicles (see Gould, 1990, for a review). Two remarkable transitions have occurred in the evolution of the vertebrate jaw. The first is the creation of jaws from the gill arches of unjawed fish. The second is the use of the bones that had articulated the upper and lower jaws in reptiles to become the malleus and incus (hammer and stirrup) bones of the middle ear. In the first vertebrates, a series of gills opened behind the jawless mouth. When gill slits became supported by cartilaginous elements, the first set of these gill supports appears to have surrounded the mouth to form the jaw. There is ample evidence that jaws are modified gill supports. In both cases, these bones are made from neural crest cells rather than from mesodermal tissue as are most bones. Second, both structures form from upper and lower bars that bend forward and are hinged in the middle. Third, the jaw musculature seems to be homologous to the original gill support musculature. Thus, the first transformation of the first branchial arch cartilage was from gill apparatus to jaw apparatus. But the story does not end here.

The upper portion of the second branchial arch supporting the gill becomes the hyomandibular bone of jawed fishes. This element supports the skull and links the jaw to the cranium (Figure 23.15). As we saw in Chapter 7, this hyomandibular bone functions in mammals as the *stapes*, one of the middle ear bones. But fish do not use this bone for hearing, so how did a bone used for gill support and then for cranial support become part of the mammalian auditory apparatus? As fish came up onto land, they had a new problem: how to hear in a medium as thin as air. The hyomandibular bone happens to be near the otic capsule, and bony material is excellent for transmitting sound. Thus, while still functioning as a cranial brace, the hyomandibular bone of the first amphibians also began functioning as a sound transducer (Clack, 1989). As the terrestrial vertebrates altered their locomotion, jaw structure, and posture, the cranium became firmly attached to the rest of the skull and did not need the hyomandibular brace. It then seems to have become specialized into the stapes bone of the middle ear. What had been this bone's secondary function became its primary function.

The original jaw bones changed also. The first branchial arch generates the jaw apparatus. In amphibians, reptiles, and birds, the posterior portion of this cartilage forms the quadrate bone of the upper jaw and the articular bone of the lower jaw. These bones connect to one another and are responsible for articulating the upper and lower jaws. However, in mammals, this articulation occurs at another region (the dentary and squamosal bones), thereby "freeing" these bony elements to acquire new functions. The quadrate bone of the reptilian upper jaw evolved into the mammalian incus bone, and the articular bone of the reptile's lower jaw has become our malleus. This latter process was first described by Reichert in 1837, when he observed in the pig embryo that the mandible (jawbone) ossifies on the side of Meckel's cartilage, while the posterior region of Meckel's

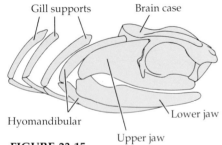

FIGURE 23.15
Homologies of the jaws and the gill arches as seen in the skull of the paleozoic shark *Cobeledus aculentes*. (After Zangerl and Williams, 1975.)

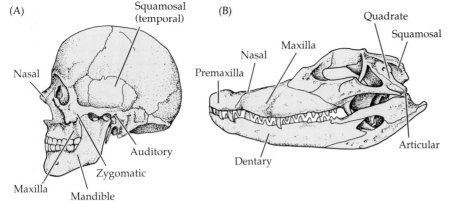

(A) Squamosal (temporal)

Nasal

Maxilla

Mandible

Zygomatic

Auditory

(B) Quadrate

Squamosal

Nasal

Maxilla

Premaxilla

Dentary

Articular

FIGURE 23.16
(A) Lateral view of the human skull showing the junction of the lower jaw with the squamosal (temporal) region of the skull. (B) Lateral view of an alligator skull. The articular portion of the lower jaw articulates with the quadrate bone of the skull. In mammals, the quadrate becomes internalized to form the incus of the middle ear. The articular bone retains its contact with the quadrate, becoming the malleus of the middle ear.

cartilage ossifies, detaches from the rest of the cartilage, and enters into the region of the middle ear to become the malleus* (Figure 23.16).

These dramatic changes in bone arrangement from agnathans to jawed fish, from jawed fish to amphibians, and from reptiles to mammals were coordinated with changes in jaw structure, jaw musculature, tooth deposition and shape, and modifications of the cranial vault and ear (Kemp, 1982; Thomson, 1988). It therefore seems that one particular change may co-vary with several others.

Evidence for correlated progression

These changes in jaw and ear formation are the products of development and natural selection that occurred hundreds of millions of years ago. Can one breed for such changes in existing organisms and see if correlated progression occurs? Fortunately for us, humans have a great talent for selecting hereditary variants involving those neural crest cells that form the frontonasal and mandibular processes. Figure 23.17 shows the skull profiles from different breeds of dogs. Each variation is genetically determined, and it is important to note that each represents a harmonious rearrangement of the different bones with each other and with their muscular attachments. In some cases, such as that of bull dogs, the breed is selected for a wide face with very little angle between head and jaw. Other breeds, such as the collie, are selected for narrow snouts with a long jaw protruding away from the head. All breeds can move their jaws, shake their heads, and bark, despite the differences in the ways their bones are shaped or positioned. As the skeletal elements were selected, so were the muscles that moved them, the nerves that controlled these movements, and the blood vessels that fed them.[†]

*The lack of transition forms is often cited by Creationists as a criticism of evolution. For instance, in the transition from reptiles to mammals, discussed above, three of the bones of the reptilian jaw became the incus and malleus, leaving only one bone (the dentary) in the lower jaw. Gish (1973), a Creationist, says that this is an impossible situation, since no fossil has been discovered showing two or three jaw bones and two or three ear ossicles. Such an animal, he claims, would have dragged its jaw on the ground. However, such a specific transition form (and there are over a dozen documented transition forms between reptile and mammalian skulls) need never have existed. Hopson (1966) has shown on embryological grounds how the bones of the jaw could have divided and been used for different functions, and Romer (1970) has found reptilian fossils wherein the new jaw articulation was already functional while the older bones were becoming useless. There are several species of therapsid reptiles that had two jaw articulations, with the stapes brought into close proximity with the upper portion of the quadrate bone (which would become the incus).

[†]This coordination is not quite universal, however. In dogs with greatly shortened faces (such as bulldogs), the skin has not coordinated its development with the bones and therefore hangs in folds from the head (Stockard, 1941).

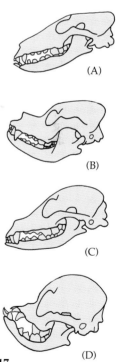

FIGURE 23.17
Skulls of various breeds of dogs. (A) Borzoi. (B) St. Bernard. (C) Windspiel. (D) Japanese spaniel. The jaw musculature and facial skeleton develop in a correlated fashion. (After Waddington, 1956.)

	(A) Embryonic skeletal patterns	(B) Final skeletal patterns	(C) Final muscle patterns
Archaeopteryx			
Modern bird			Popliteal muscle
Experimental bird			
Reptile (*Crocodylus*)			

FIGURE 23.18

Experimental "atavisms" produced by altering embryonic fields in the limb. (A–C) Results of Müller's experiments wherein gold foil split the chick hindlimb field. (A) and (B) show the embryonic and final bone pattern, indicating that the fibulare structure was retained by the experimental chick limb, as it is in extant reptiles and as it is thought to be in *Archaeopteryx*. (C) shows some of the muscle changes in the experimental chick embryos. The popliteal muscle is present in the chick, but is absent from reptile limbs and from the experimental limb. The fibularis brevis muscle, which normally originates from both the tibia and fibula in chicks, takes on the reptilian pattern of originating solely from the fibula in the operated limbs. (D) Fossil *Archaeopteryx* in limestone. Imprints of feathers can be clearly seen. Were it not for the feathers, this toothed organism would probably have been classified as a reptile. (A–C after Müller, 1989; photograph courtesy of B. A. Miller/Biological Photo Service.)

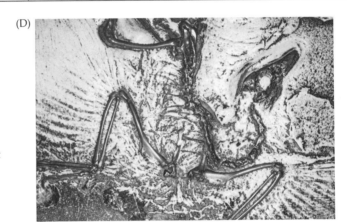

(D)

Correlated progression has also been shown experimentally. Repeating earlier experiments of Hampé (1959), Gerd Müller (1989) inserted barriers of gold foil into the prechondrogenic hindlimb buds of a 3.5-day chick embryo. This barrier separated the regions of tibia formation and fibula formation. The results of these experiments are twofold. First, the tibia is shortened and the fibula bows and retains its connection to the fibulare. Such relationships between the tibia and fibula are not usually seen in birds, but they are characteristic of reptiles (Figure 23.18). Second, the *musculature* of the hindlimb undergoes parallel changes with the bones. Three of the muscles that attach to these bones now show characteristic reptilian patterns of insertion. It seems, therefore, that experimental manipulations that alter the development of one part of the mesodermal limb-forming field also alter the development of other mesodermal components as well. As with the correlated progression seen in face development, these changes all appear to be due to interactions within a field, in this case, the chick hindlimb field. These are not global effects and can occur independently of the other portions of the body.

In some cases, correlated progression may be a pleiotropic effect having a single origin. This could be a "key adaptation" that alters other tissues as a consequence (Larson et al., 1981). For instance, if the bones of the limb or mouth are changed, the mesodermal regions that become bone-associated muscle are simultaneously altered. This situation could be caused either by a single inducer that affects both cell types or by a

(A)

(B)

FIGURE 23.19
Mimicry in *P. memnon*. (A) Unpalatable model, *A. coon*, and its mimic, *P. memnon*. (B) Unpalatable model, *A. aristolochiae*, with its mimic, another genetic form of *P. memnon*. The ability of *P. memnon* to mimic either species depends on alleles at a gene complex containing several linked loci. (After Clarke et al., 1968.)

Unpalatable	Mimic	Unpalatable	Mimic
A. coon	*P. memnon*	*A. aristolochia*	*P. memnon*

cascade of inductions such that the muscles that insert in the upper portion of the bone are induced to insert there by the bone itself. In other cases, correlated progression might be selected for at the level of gene translocations. One such case involves Batesian mimicry in the swallowtail butterfly *Papilio memnon* (Clarke et al., 1968). This butterfly is a tasty morsel for several bird species, but it survives by mimicking two relatively unpalatable butterflies, *Atrophaneura coon* and *A. aristolochiae* (Figure 23.19). *P. memnon* carries a "supergene" complex containing at least five closely linked genes. One locus controls the presence (T) or absence (t) of tails. A second gene controls whether the abdomen color is yellow (B^Y), yellow-tipped (b^y), or black (b). A third gene controls the hindwing color pattern mimicking the wings of *A. coon* (W^{al}) or *A. aristolochiae* (W^d). A fourth and fifth gene in this cluster control forewing color pattern and the color of the spot at the base of the forewings. In each population of *P. memnon* studied, there are different alleles at this locus that allow the butterfly to mimic the distasteful model. For example, in Hong Kong, *P. memnon* has the gene constellation $TW^{al}B^Y$, which forms a tailed butterfly with abdomen and hindwing colors similar to the *A. aristolochiae* populations found there. This constellation of characters, however, is inherited as a unit because its genes have been selected to be so closely linked together that recombinations are relatively rare events. This enables the butterfly to mimic the model completely. Mimicking only part of the animal is not as good. Thus, correlated variation in some species (such as dogs) probably comes from interactions between correlated components during development, while in other species (such as the swallowtail butterflies), correlated variation can be at the genetic level.

Heterochrony and allometry

Heterochrony is a shift in the relative timing of two developmental processes during embryogenesis from one generation to the next. We have come across this concept in our discussions of neoteny and progenesis in salamanders (Chapter 20). Heterochrony can be caused in different ways. In salamander heterochronies, where the larval stage is retained, heterochrony is caused by gene mutations in the induction-competence system. Other heterochronic phenotypes, however, are caused by the heterochronic expression of certain genes. The direct development of the adult sea urchin rudiment (Chapter 20) involves the early activation of adult genes and the suppression of larval gene expression (Raff and Wray, 1989). As discussed earlier in the text, heterochrony can affect evolution in several ways. It can "return" an organism to a larval state, free from the specialized adaptations of the adult. Heterochrony can also give larval characteristics to an adult organism, as in the small size and webbed feet of arboreal salamanders or the fetal growth rate of human newborn brain tissue.

Allometry occurs when different parts of the organism grow at different rates. Allometry can be very important in forming variant body plans within a *Bauplan*. Such differential growth changes can involve altering a target cell's sensitivity to growth factors or altering the amounts of growth factors produced. Again, the vertebrate limb can provide a useful illustration. Local differences in chondrocytes cause the central toe of the horse to grow at a rate 1.4 times that of the lateral toes (Wolpert, 1983). This means that as the horse grew larger during evolution, this regional difference caused the five-toed horse to become a one-toed horse. A particularly dramatic example of allometry in evolution comes from skull development. In the very young (4- to 5-mm) whale embryo, the nose is in the usual mammalian position. However, the enormous growth of the maxilla and premaxilla (upper jaw) pushes over the frontal bone and forces the nose to the top of the skull (Figure 23.20). This new position of the nose (blowhole) allows the whale to have a large and highly specialized jaw apparatus and to breathe while parallel to the water's surface (Slijper, 1962).

Allometry can also generate evolutionary novelty by small, incremental changes that eventually cross some developmental threshold (sometimes called a bifurcation point). Eventually, a change in quantity becomes a change in quality when such a threshold is crossed. This type of mechanism has been postulated to have produced the external fur-lined "neck" pouches of pocket gophers and kangaroo rats that live in deserts. External pouches differ from internal ones in that (1) they are fur-lined, and (2) they have no internal connection to the mouth. They are very useful in that they allow these animals to store seeds without running the risk of desiccation. Brylski and Hall (1988) have dissected the heads of pocket gopher and kangaroo rat embryos and have looked at the way that the external cheek pouch is constructed. When data from these animals were compared with data from animals that form internal cheek pouches (such

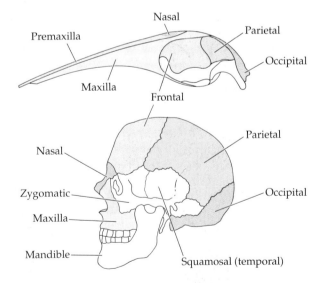

FIGURE 23.20
Allometric growth in the whale head. The jaw has pushed forward, causing the nose to move to the top of the skull. (The premaxilla is present in the early human fetus, but it fuses with the maxilla by the end of the third month of gestation. The human premaxilla was discovered by Wolfgang Goethe, among others, in 1786.) (After Slijper, 1962.)

as hamsters), the investigators found that the pouches are formed in very similar manners. In both cases, the pouches are formed within the embryonic cheeks by outpocketings of the cheek (buccal) epithelium into the facial mesenchyme (Figure 23.21). In animals with internal cheek pouches, these evaginations stay within the cheek. However, in animals that form external pouches, the elongation of the snout draws up the outpocketings into the region of the lip. As the lip epithelium rolls out of the oral cavity, so do the outpockets. What had been internal becomes external. The fur lining is probably derived from the external pouches' coming in contact with dermal mesenchyme, which can induce hair to form in epithelia (see Chapter 18). Such a pouch has no internal opening to the mouth. Indeed, the transition from internal to external pouch is one of threshold. The placement of the evaginations anteriorly or posteriorly determines whether the pouch is internal or not. There is no "transition stage" having two openings, one internal and one external. One could envision this externalization occurring by a chance mutation that shifted the outpocketing to a slightly more anterior location. Such a trait would be selected for in the desert. As Van Valen reflected in 1976, evolution can be defined as "the control of development by ecology."

Transfer of competence

Another way that induction can be used to create developmental changes that may be selected for during evolution is by **transfer of competence**. This idea, popularized by C. H. Waddington and I. I. Schmalhausen, enables the selection of new phenotypes that are favored for survival. In 1936, Waddington, Needham, and Brachet had made an unexpected discovery. They had been hoping to find a specific factor from the notochord that would induce neural plates but instead found that a large variety of natural and artificial compounds were able to cause this induction. This led them to reinterpret their data and to suggest that the actual neuralizing factor lies dormant in the competent ectoderm and that a variety of factors are able to release the neuralizing factor from its inhibitor. Waddington then began to focus on the competent cells rather than on the inducing cells. He thought that many things could act as inducers, but the competent tissue had to have something within it that allowed it to respond to these chemicals. The inducer, he wrote in 1940, is only the push. It was the competence that is genetically controlled and that is responsible for the details of the development.

Since competence could be achieved independently from an inducer, and since different compounds could induce the same developmental process in this competent tissue, Waddington proposed that a given competent tissue could transfer its ability to respond from one inducing stimulus to another. Moreover, these inducers could be either internal or external. External inducers were already known, and Waddington used the case of sex determination in *Bonnelia* (Chapter 21) as such an example. He also noted that the ability to form calluses on those areas of skin that abrade the ground is such an example. Here, the skin cells have the ability to form a callus if induced by friction. The genes can respond by causing the proliferation of cells to form the callus structure. While such examples of environmentally induced callus formation are widespread, the ostrich is *born* with them. Waddington hypothesized that since the skin cells are already competent to be induced by friction, they could be induced by other things as well. As ostriches evolved, a mutation appeared that enabled the skin cells to respond to a substance within the embryo. In this way, a trait that had been induced by the environment became part

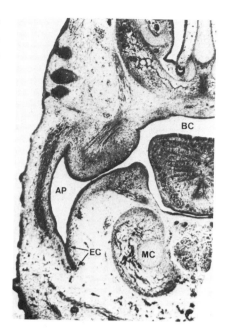

FIGURE 23.21
Transverse section through the anterior region of the pocket gopher (*Thomomys*) embryo showing the anterior opening of the pouch (AP) and the continuity between the pouch at this stage and the buccal cavity (BC) across the developing lip area. EC, epithelial cells; MC, Meckel's cartilage; T, tongue. (From Brylski and Hall, 1988, courtesy of the authors.)

of the genome of the organism and could be selected. He called this phenomenon "genetic assimilation."

Proving transfer of competence is extremely difficult since it means that one has to first establish that a tissue can respond in a particular way to environmental inducers and then that it evolved a receptor for an internal stimulus that would move the cell into the same pathway of differentiation. As in much of evolutionary theory, one has to use circumstantial evidence for such evolution. There are, however, some candidates for evolution by transfer of competence.

Possible transfer of competence in ant populations. One place where competence may be transferred from an environmental stimulus to a genomic stimulus is in the evolution of caste separation in ant species. Ant colonies are predominantly female, and the females can be extremely polymorphic. The two major types of females are the worker and the gyne. The gyne is a potential queen. In more specialized species, a larger worker, the soldier, is also seen. These castes are determined by the levels of juvenile hormone seen by the developing larva. If the larva has more juvenile hormone, the larva grows, and this allometric growth forms larger jaws or allows the development of reproductive organs. The determination of the amount of juvenile hormone in the larva can be nutritionally controlled or regulated internally through maternal hormones that act during embryogenesis.

The developmental mechanisms of caste determination have been analyzed by Diana Wheeler (1986) and are summarized in Figure 23.22. In most species, ant larvae are bipotential until near pupation. In *Myrmica rubra*, only larvae that overwinter remain bipotential. After winter, the queen stimulates workers to *under*feed the last-instar larvae. This means that as long as there is a queen, no new queens can result. If the larvae are fed, they can becomes gynes. Thus, larvae remain bipotential until late in their last instar.

In *Pheidole pallidula*, the queens appear to control caste determination during embryogenesis. There are no bipotential larvae in this species, caste determination being decided completely within the embryo. Gynes are produced from those eggs laid just after hibernation has ended, so it is assumed that eggs laid at different times are biochemically different. Although there is controversy over how the gyne-eggs and the worker-eggs differ, we have here an example of caste determination occurring within the embryo.

These ants show that what had been an environmental type of determination can be internalized. Notice, too, that in *Pheidole* there is polymorphism in the worker caste. Soldiers—large workers—have been formed. Wilson (1971) has pointed out that of the 260 known genera of ants, only seven have complete dimorphism of the workers. If having polymorphic workers is advantageous (Oster and Wilson, 1978), why don't more genera have them? Wheeler answers that worker polymorphism is only seen when the gyne determination has occurred very early. Once that has been determined, the nutritional variations can be used to create different worker castes.

Possible transfer of competence in fish sex determination. If competent tissues can change their developmental triggers from environmental stimuli to internal inducers encoded by the genome, then it would stand to reason that the reverse could also occur. Under some selective conditions, one would expect genetic induction to be replaced by environmental induction. This phenomenon might be the case in certain types of environmental sex determination. Environmental sex determination is found

(A)

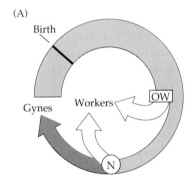

(B)

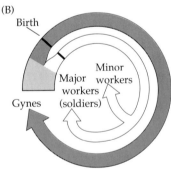

FIGURE 23.22
Possible transfer of competence in gyne (queen) formation in ants. Lightly colored areas represent bipotentiality to become either workers or gynes. The circled N represents a nutritional switch controlled by the larva's environment. (A) *Myrmica rubra*, wherein only the larvae that overwinter (OW) remain bipotential. In the last instar, the nutritional switch determines caste. (B) *Pheidole pallidula*, wherein the queen controls gyne determination through hormones that act during embryogenesis. (After Wheeler, 1986.)

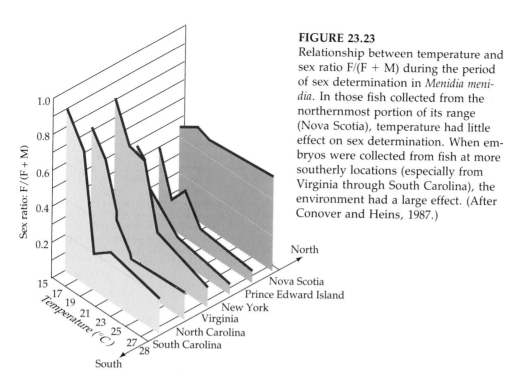

FIGURE 23.23
Relationship between temperature and sex ratio F/(F + M) during the period of sex determination in *Menidia menidia*. In those fish collected from the northernmost portion of its range (Nova Scotia), temperature had little effect on sex determination. When embryos were collected from fish at more southerly locations (especially from Virginia through South Carolina), the environment had a large effect. (After Conover and Heins, 1987.)

among certain invertebrates and among vertebrates such as fish, amphibians, and reptiles (Chapter 21). Charnov and Bull (1977) have argued that environmental sex determination would be adaptive in certain habitats characterized by patchiness and having certain regions in which it is more advantageous to be male and other regions in which it is more advantageous to be female. Conover and Heins (1987) provide evidence that in certain fish, females benefit from being larger, since size translates into higher fecundity. It is an advantage to be born early in the breeding season if you are a female *Menidia*, since you would have a longer feeding season and would grow larger. In the males, size is of no importance. Conover and Heins showed that in the southern range of *Menidia*, females are indeed born early in the breeding season. Temperature appears to play a major role. However, in the northern reaches of its range, the same species shows no environmental sex determination. Rather, a 1:1 ratio is generated at all temperatures (Figure 23.23). The authors speculate that this is because the more northern populations have a very short feeding season so that there is no advantage for a female to be born earlier. Thus, this species of fish has environmental sex determination in those regions where it is adaptive and genotypic sex determination in those regions where it is not. Here again, one sees that the environment can induce sexual phenotype, or sexual phenotype can be a property of the genome, as it is with most mammals.

Delayed implantation and diapause. One probable example of competence being transferred from one inducer to another involves a species of which some subspecies have internal inductive stimuli and other subspecies have the same phenomenon under the control of another inducer. This phenomenon can be seen in the implantation of blastocysts into the uterus of the spotted skunk. The eastern subspecies of the spotted skunk *Spirogale putorius* has normal implantation. It mates in April and has a gestation period of about two months; the young are born in June or July, when food is abundant. The western subspecies, however, lives at higher altitudes and is not able to mate in April. Rather, the western skunks mate in September. If the embryos developed at the same rate and had a two-month gestation, the young would be born in November and December and would starve. Instead, the western subspecies has evolved **delayed implantation**. Although fertilization and early cleavage start in September, implantation of the blastocyst is delayed until April (the same time as implantation occurs in the eastern subspecies). The timing of the delay is probably regulated by day length, since blinded skunks have prolonged periods of delayed implantation. (In other animals with delayed implantation, the photoperiod is sometimes perceived by the pineal gland.) At the end of the period of delay, progesterone is made, which enables the uterus to receive a blastocyst that has been waiting some 210–230 days (Renfree, 1982). Delayed implantation is part of the life cycle of numerous mammals.

Many species of insects have evolved a similar strategy, called **diapause**. Diapause is a suspension of development that can occur at the embryonic, larval, pupal, or adult stage, depending on the species. In some species, diapause is facultative and occurs only when induced by harsh environmental conditions; in other species the diapause period has become an obligatory part of the life cycle. The latter is often seen in temperate-zone insects, where diapause is induced by shortened day length. When the hours of sunlight get below a certain threshold, the larvae molt into pupae that can withstand the winter's cold and lack of food. In some species, the larvae enter diapause after pupation independent of the environmental conditions (Doane, 1973).

SIDELIGHTS & SPECULATIONS

Transfer of competence by transfer of receptors

According to Waddington, the competent cell evolved a genetically controlled pathway by which it could respond to an inducer. When the competency of the cell was transferred from one inducer to another, the receptor for the signal changed, but the pathway distal to the signal remained the same. We can see this in the competencies that different cells have to mitogenic agents. Let us recall (from Chapter 18) the division of a competent B lymphocyte. Upon binding antigen through its surface immunoglobulins, the B cell acquires the cell surface receptors for growth factors. In other words, specific competence is provided by the antigen receptor. If a different antigen is presented, a different set of B cells becomes competent to divide and differentiate into antibody-secreting plasma cells. Thus, the binding of antigen to a B cell antigen receptor triggers into action a *preset program* for cell division and differentiation.

We also see transfer of competence when we look at the cell division pathway of the B cell. A G protein that is bound to the cell surface immunoglobulin activates phospholipase C. Phospholipase C splits phosphatidylinositol bisphosphate (PIP_2) into inositol trisphosphate, which releases calcium from the endoplasmic reticulum, and into diacylglycerol, which activates protein kinase C with a resultant elevation in intracellular pH. Together, these signals activate the nuclear proto-oncogenes that initiate cell division. Interestingly enough, these are the same reactions that happen when antigens bind to the T cell receptor. So it appears that lymphocytes, both B and T, use the same pathway for cell division. Similarly, several neurohormones, such as serotonin and acetylcholine, function by activating G proteins and initiating the same pathway. The reactions are the same. All that changes is the cell surface receptor.

But this should be an already-familiar story. Indeed, lymphocytes and neurons are latecomers to this pathway, for this pathway is the series of reactions by which every cell in the body divides. A more general scheme is known in which a growth factor binds to the growth factor receptor. This, in turn, activates the G protein, which activates phospholipase C, which starts the process in motion. But neither is this the entire story. The same pathway of division is seen in the reactions that activate the fertilized egg (Chapter 4). Here, the activator is not an antigen or a peptide growth factor; it is a sperm. One can see that the pathway is independent of the stimulus that initiates the cell division. Different receptors can be inserted into the membrane, causing the pathway to become activated by new compounds. Indeed, one can experimentally add the mRNA for acetylcholine and serotonin receptors and activate *Xenopus* eggs when one adds these neurotransmitters (Kline et al., 1988). Here an internal inducer (neurotransmitter) has substituted for an external inducer (sperm) merely by changing the receptor. Therefore, by changing the cell surface receptor, the competence of the cell is changed, and the cell can respond to a different inducer.

A new evolutionary synthesis

One of the major events in evolutionary theory has been the "modern synthesis" of evolutionary biology and Mendelian genetics (Mayr and Provine, 1980). One outcome of this hard-won merger is that evolution has been redefined to mean changes in gene frequencies in a population over time. "Since evolution is a change in the genetic composition of populations," wrote Dobzhansky (1937), "the mechanisms of evolution constitute problems of population genetics." This definition became largely agreed upon in the United States, England, and the Soviet Union, and thus morphology and development were seen to play little role in modern evolutionary theory (Adams, 1991). The large morphological changes seen during evolutionary history could be explained by the accumulation of small genetic changes. In other words, macroevolution (the large morphological changes seen between species, classes, and phyla) could be explained by the mechanisms of microevolution, the "differential adaptive values of genotypes or deviations from random mating or both these factors acting together" (Torrey and Feduccia, 1979).

However, this view has had its critics (its heretics, some would say). Perhaps the foremost of these was Richard Goldschmidt. Goldschmidt began his book *The Material Basis of Evolution* (1940), with a challenge to the modern synthesis. It may be able to explain the *survival* of the fittest, but not the *arrival* of the fittest:

> I may challenge the adherents of the strictly Darwinian view, which we are discussing here, to try to explain the evolution of the following features by accumulation and selection of small mutants: hair in mammals, feathers in birds, segmentation in arthropods and vertebrates, the transformation of the gill arches in phylogeny including the aortic arches, muscles, nerves, etc.; further, teeth, shells of molluscs, ectoskeletons, compound eyes, blood circulation, alternation of generations, statocysts, ambulacral systems of echinoderms, pedicellaria of the same, cnidocysts, poison apparatus of snakes, whalebone, and finally chemical differences like hemoglobin vs. hemocyanin . . .

Goldschmidt claimed that new species did not arise from the mechanisms of microevolution, and that population genetics was unable to explain new types of structures that involve several components changing simultaneously. Such macroevolutionary change "requires another evolutionary method than that of sheer accumulation of micromutations." Goldschmidt saw homeotic mutants as "macromutations" that could change one structure into another and possibly create new structures or new combinations of structures. These mutations would not be in the structural genes but in the regulatory genes. A new species, he asserted, would start as a "hopeful monster" (a rather unfortunate phrase having its antecedent in Metchnikoff's prose).

At the same time, Waddington was attempting to find developmental mechanisms for producing such new species. He, too, looked at homeotic mutations in flies as models for drastically new phenotypes, and he formulated the notion of competence transfer ("genetic assimilation") to explain certain aspects of morphological evolution. Few scientists paid attention to Goldschmidt or Waddington because they were not writing in the population genetics paradigm of the modern synthesis and their scientific programs were suspect. (Goldschmidt did not believe in Morgan's notion of the gene as a particulate entity, and Waddington's work was misinterpreted as supporting the inheritance of acquired traits.) However, in the 1970s, events in paleontology (the punctuated equilibrium theory), events in society (the Creationists giving the microevolutionary contest to the biologists but contesting macroevolution), and events in molecular biology (notably King and Wilson's 1975 paper showing that chimps and humans have DNA that is greater than 99 percent identical) prompted scientists to consider seriously the view that mutations in regulatory genes can create large changes in morphology.

A new developmental synthesis is emerging that retains the best of the microevolution-yields-macroevolution model and the macroevolution-as-separate-phenomenon model. From the latter it derives the concept that mutations in regulatory genes can create "jumps" from one phenotype to another without necessary intermediate steps. From the former, it derives the notion that genetic mutations can account for such variants and that selection acts upon them to delete them or retain them in populations. It also retains a multiplicity of paradigms. In some instances (such as the creation of neural crest cells) a qualitative change occurs, whereas in other

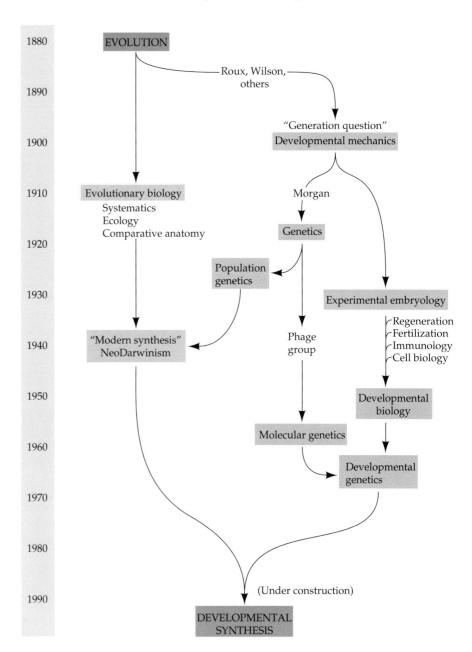

FIGURE 23.24
Disciplinary road map of the evolutionary side of biology, 1880 to the present. For sake of clarity, other paths (such as those from genetics to human genetics or from evolution to immunology) have not been shown.

cases (such as the formation of the pocket gopher pouch) quantity becomes quality when a threshold is passed.

We are at a remarkable point in our understanding of nature, for a synthesis of developmental genetics with evolutionary biology may transform our appreciation of the mechanisms underlying evolutionary change and animal diversity. Such a synthesis is actually a return to a broader-based evolutionary theory that fragmented at the turn of the past century (Figure 23.24). In the late 1800s, evolutionary biology contained the sciences that we now call evolutionary biology, systematics, ecology, genetics, and development. By the beginning of the twentieth century, the "question of heredity"—genetics and embryology—separated from the rest of evolutionary biology. Genetics eventually split into (among other rubrics) population genetics and molecular genetics, while embryology became developmental biology (Gilbert, 1978, 1988). During the mid-twentieth century, population genetics merged with evolutionary biology to produce the evolutionary genetics of the modern synthesis, while molecular genetics merged with developmental biology to produce developmental genetics. These two vast areas, developmental genetics and evolutionary genetics, are on the verge of a merger that may unite these long-separated strands of biology and that may produce a developmental genetic theory capable of explaining macroevolution.

When Wilhelm Roux (1894) announced the creation of "developmental mechanics," he did not fully break with evolutionary biology. Rather, he stated that "an *ontogenetic* and a *phylogenetic developmental mechanics* are to be perfected." He noted further that the developmental mechanics of embryos (the ontogenetic branch) would proceed faster than the phylogenetic studies, but posited that "in consequence of the intimate causal connections between the two, many of the conclusions drawn from the investigation of ontogeny [would] throw light on the phylogenetic processes." One hundred years later, we are now at the point where we can attend to the second of Roux's developmental mechanics and create a unified theory of evolution.

LITERATURE CITED

Adams, M. 1991. Soviet perspectives on evolutionary theory. *In* L. Warren and M. Meselson (eds.), *New Perspectives in Evolution.* Liss/Wiley, New York.

Akam, M., Dawson, I. and Tear, G. 1988. Homeotic genes and the control of segment diversity. *Development* [Suppl.] 104: 123–133.

Alberch, P. and Gale, E. 1983. Size dependency during the development of the amphibian foot. Colchicine induced digital loss and reduction. *J. Embyol. Exp. Morphol.* 76: 177–197.

Alberch, P. and Gale, E. 1985. A developmental analysis of an evolutionary trend. Digit reduction in amphibians. *Evolution* 39: 8–23.

Alexandrov, D. A. and Sergievsky, S. O. 1984. A variant of sympatric speciation in snails. *Malacol. Rev.* 17: 147.

Anderson, D. T. 1973. *Embryology and Phylogeny of Annelids and Arthropods.* Pergamon Press, Oxford.

Bateson, W. 1894. *Materials for the Study of Variation.* Cambridge University Press, Cambridge.

Berg, R. L. 1960. Evolutionary significance of correlation pleiades. *Evolution* 14: 171–180.

Berrill, N. J. 1955. *The Origins of the Vertebrates.* Oxford University Press, New York.

Berrill, N. J. 1987. Early chordate evolution. I. Amphioxus, the riddle of the sands. *Int. J. Invert. Repr. Dev.* 11: 1–27.

Bowring, S. A., Grotzinger, J. P. Isachsen, C. E., Knoll, A. H., Pelechaty, S. M. and Kolosov, P. 1993. Calibrating rates of early Cambrian evolution. *Science* 261: 1293–1298.

Brylski, P. and Hall, B. K. 1988. Ontogeny of a macroevolutionary phenotype: The external cheek pouches of geomyoid rodents. *Evolution* 42: 391–395.

Burke, A. C. 1989a. Epithelial–mesenchymal interactions in the development of the chelonian Bauplan. *Fortschr. Zool.* 35: 206–209.

Burke, A. C. 1989b. Development of the turtle carapace: Implications for the evolution of a novel bauplan. *J. Morphol.* 199: 363–378.

Buss, L. W. 1987. *The Evolution of Individuality.* Princeton University Press, Princeton, NJ.

Charnov, E. L. and Bull, J. J. 1977. When is sex environmentally determined? *Nature* 266: 828–830.

Chernyak, L. and Tauber, A. I. 1988. The birth of immunology: Metchnikoff, the embryologist. *Cell. Immunol.* 117: 218–233.

Chernyak, L. and Tauber, A. I. 1991. *From Metaphor to Theory: Metchnikoff and the Origin of Immunology.* Oxford University Press, New York.

Clack, J. A. 1989. Discovery of the earliest known tetrapod stapes. *Nature* 342: 425–427.

Clark, B. and Murray, J. 1969. Ecological genetics and speciation in land snails of the genus *Partula.* *Biol. J. Linn. Soc.* 1: 31–42.

Clarke, C. A., Sheppard, P. M. and Thornton, I. W. B. 1968. The genetics of the mimetic butterfly *Papilio memnon.* *Philos. Trans. R. Soc. Lond.* [B] 254: 37–89.

Conklin, E. G. 1915. *Heredity and Environment in the Development of Men.* Princeton University Press, Princeton, NJ.

Conover, D. O. and Heins, S. W. 1987. Adaptive variation in environmental and genetic sex determination in a fish. *Nature* 326: 496–498.

Conway Morris, S. and Whittington, H. B. 1979. The animals of the Burgess Shale. *Sci. Am.* 240(1): 122–133.

Darwin, C. 1859. *The Origin of Species.* John Murray, London.

Doane, W. W. 1973. Role of hormones in insect development. *In* S. J. Counce and C. H. Waddington (eds.), *Developmental Systems: Insects.* Academic Press, New York, pp. 291–498.

Dobzhansky, T. G. 1937. *Genetics and the Origin of Species.* Columbia University Press, New York.

Elinson, R. P. 1987. Changes in developmental patterns: Embryos of amphibians with large eggs. *In* R. A. Raff and E. C. Raff (eds.), *Development as an Evolutionary Process.* Alan R. Liss, New York; pp. 1–21.

Gans, C. 1989. Stages in the origin of vertebrates: Analysis by means of scenarios. *Biol. Rev.* 64: 221–268.

Gans, C. and Northcutt, R. G. 1983. Neural crest and the origin of vertebrates: A new head. *Science* 220: 268–274.

Garstang, W. 1922. The theory of recapitulation: a critical restatement of the biogenetic law. *J. Linn. Soc. Zool.* 35: 81–101.

Garstang, W. 1928. Presidential address to the British Association for the Advancement of Science, Section D. Republished in *Larval Forms and Other Zoological Verses,* 1985. University of Chicago Press, Chicago, pp. 77–98.

Gasman, D. 1971. *The Scientific Origins of National Socialism: Social Darwinism in Ernst Haeckel and the German Monist League.* Mac-Donald, London.

Gemmill, J. F. 1912. The development of the starfish *Solaster endica* (Forbes). *Trans. Zool. Soc. Lond.* 20: 1–72.

Gilbert, S. F. 1978. The embryological origins of the gene theory. *J. Hist. Biol.* 11: 307–351.

Gilbert, S. F. 1988. Cellular politics: Ernest Everett Just, Richard B. Goldschmidt, and the attempt to reconcile embryology and genetics. *In* R. Ranger, K. R. Benson and J. Maienschein (eds.), *The American Development of Biology.* University of Pennsylvania Press, Philadelphia, pp. 311–346.

Gish, D. T. 1973. *Evolution? The Fossils Say No!* Creation-Life Publishers, San Diego.

Goldschmidt, R. B. 1940. *The Material Basis of Evolution.* Yale University Press, New Haven.

Gould, S. J. 1977a. *Ever Since Darwin.* Norton, New York.

Gould, S. J. 1977b. *Ontogeny and Phylogeny.* Harvard University Press, Cambridge, MA.

Gould, S. J. 1989. *Wonderful Life.* Norton, New York.

Gould, S. J. 1990. An earful of jaw. *Natural History* 1990(3): 12–23.

Gould, S. J. and Lewontin, R. C. 1979. The spandrels of San Marcos and the Panglossian paradigm: A critique of the adaptationist program. *Proc. R. Soc. London* [B] 205: 581–598.

Grant, P. 1978. *Biology of Developing Systems.* Holt, Rinehart & Winston, New York.

Haeckel, E. 1900. *The Riddle of the Universe.* Harper & Bros., New York.

Hall, B. K. 1992. *Evolutionary Developmental Biology.* Chapman and Hall, London.

Hampé, A. 1959. Contribution à l'étude du développement et la régulation des déficiences et excédents dans la patte de l'embryon de poulet. *Arch. Anat. Microsc. Morph. Exp.* 48: 347–479.

Henry, J. J., Amemiya, S., Wray, G. A. and Raff, R. A. 1989. Early inductive interactions are involved in restricting cell fates of mesomeres in sea urchin embryos. *Dev. Biol.* 136: 140–153.

Holder, N. 1983. Developmental constraints and the evolution of vertebrate limb patterns. *J. Theor. Biol.* 104: 451–471.

Hopson, J. A. 1966. The origin of the mammalian middle ear. *Am. Zool.* 6: 437–450.

Jefferies, R. P. S. 1986. *The Ancestry of the Vertebrates.* British Museum of Natural History, London.

Jones, M. L. and Gardiner, S. L. 1989. On the early development of the vestimentiferan tube worm *Ridgeia* sp. and observations on the nervous system and trophosome of *Ridgeia* sp. and *Riftia pachyptila.* *Biol. Bull.* 177: 254–276.

Kauffman, S. A. 1993 *Origins of Order.* Oxford University Press, New York

Kemp, T. S. 1982. *Mammal-Like Reptiles and the Origin of Mammals.* Academic Press, New York.

Kline, D., Simoncini, L., Mandel, G., Maue, R. A., Kado, R. T. and Jaffe, L. A. 1988. Fertilization events induced by neurotransmitters after injection of mRNA in *Xenopus* eggs. *Science* 241: 464–467.

Kowalevsky, A. 1871. Weitere Studien II. Die Entwicklung der einfachen Ascidien. *Arch. Micr. Anat.* 7: 101–130.

Langille, R. M. and Hall, B. K. 1989. Developmental processes, developmental sequences and early vertebrate phylogeny. *Biol. Rev.* 64: 73–91.

Langman, J. 1981. *Medical Embryology.* William and Wilkins, Baltimore.

Larson, A., Wake, D. B., Maxson, L. R. and Highton, R. 1981. A molecular phylogenetic perspective on the origins of morphological novelties in the salamander of the tribe plethodontini (Amphibia, Plethodontidae). *Evolution* 35: 405–422.

Lillie, F. R. 1898. Adaptation in cleavage. *Biological Lectures from the Marine Biological Laboratories, Woods Hole, Massachusetts.* Ginn, Boston, pp. 43–67.

Margulis, L. and Schwartz, K. V. 1988. *The Five Kingdoms,* 2nd Ed. W. H. Freeman, San Francisco.

Mayr, E. and Provine, W. 1980. *The Evolutionary Synthesis: Perspectives on the Unification of Biology.* Harvard University Press, Cambridge, MA.

Metchnikoff, E. 1886. *Embryologische Studien an Medusen. Ein Beitrag zur Genealogie der Primitivorgane.* Vienna.

Metchnikoff, E. 1891. Zakon zhizni. Po povodu nektotorykh proizvedenii gr. L. Tolstogo. *Vest. Evropy* 9: 228–260. Quoted and translated in Chernyak and Tauber, 1990, *From Metaphor to Theory: Metchnikoff and the Origin of Immunology.* Oxford University Press, New York.

Mivart, St. G. 1871. *On the Genesis of Species.* Macmillan, London.

Müller, F. 1864. *Für Darwin.* Engelmann, Leipzig.

Müller, G. B. 1989. Ancestral patterns in bird limb development: A new look at Hampé's experiment. *J. Evol. Biol.* 1: 31–47.

Ospovat, D. 1981. *The Development of Darwin's Theory.* Cambridge University Press, Cambridge.

Oster, G. F. and Wilson, E. O. 1978. *Caste and Ecology in the Social Insects.* Princeton University Press, Princeton, NJ.

Oster, G. F., Shubin, N., Murray, J. D. and Alberch, P. 1988. Evolution and morphogenetic rules: The shape of the vertebrate limb in ontogeny and phylogeny. *Evolution* 42: 862–884.

Owen, R. 1848. *On the Archetype and Homologies of the Vertebrate Skeleton.* London.

Rabinow, L. and Dickinson, W. J. 1986. Complex *cis*–acting regulators and locus structure of *Drosophila* tissue-specific ADH variants. *Genetics* 112: 523–537.

Raff, R. A. In press. Developmental mechanisms in the evolution of animal form: Origins and evolvability of body plans. In *Early Life on Earth* (Nobel Symposium 84). Columbia University Press.

Raff, R. A. and Kaufman, T. C. 1983. *Embryos, Genes, and Evolution.* Macmillan, New York.

Raff, R. A. and Wray, G. A. 1989. Heterochrony: Developmental mechanisms and evolutionary results. *J. Evol. Biol.* 2: 409–434.

Raff, R. A., Wray, G. A. and Henry, J. J. 1991. Implications of radical evolutionary changes in early development for concepts of developmental constraint. *In* L. Warren and M. Meselson (eds.), *New Perspectives in Evolution.* Liss/Wiley, New York.

Reichert, C. B. 1837. Entwicklungsgeschichte der Gehörknöchelchen der sogenannte Meckelsche Forsatz des Hammers. *Müller's Arch. Anat. Phys. wissensch. Med.* 177–188.

Renfree, M. B. 1982. Implantation and placentation. *In* C. R. Austin and R. V. Short (eds.), *Reproduction in Mammals 2: Embryonic and Fetal Development,* Cambridge University Press, Cambridge.

Rijli, F. M., Mark, M., Lakkaraju, S., Dierich, A., Dollé, P., and Chambon, P. 1993. A homeotic transformation is generated in the rostral branchial region of the head by the disruption of *Hoxa-2,* which acts as a selector gene. *Cell* 75: 1333–1349.

Romer, A. S. 1970. The Chanares (Argentina) Triassic reptile fauna VI. A chiniquodontid cynodont with an incipient squamosal-dentary jaw articulation. *Breviora* 344: 1–18.

Roux, W. 1894. The problems, methods, and scope of developmental mechanics. *Biological lectures of the Marine Biology Laboratory, Woods Hole*. Ginn, Boston., pp. 149–190.

Schmalhausen, I. I. 1949. *Factors of Evolution: The Theory of Stabilizing Selection*. Blakiston, Philadelphia.

Slack, J. M. W. 1983. *From Egg to Embryo: Determinative Events in Early Development*. Cambridge University Press.

Slijper, E. J. 1962. *Whales*. (Translated by A. J. Pomerans.) Basic Books, New York.

Spemann, H. 1901. Über Correlationen in der Entwicklung des Auges. *Verh. Anat. Ges.* [Vers. Bonn] 15: 61–79.

Stein, G. J. 1988. Biological science and the roots of Nazism. *Am. Sci.* 76: 50–58.

Stockard, C. R. 1941. The genetic and endocrine basis for differences in form and behaviour as elucidated by studies of contrasted pure–line dog breeds and their hybrids. *Am. Anat. Memoirs* 19.

Thomson, K. S. 1988. *Morphogenesis and Evolution*. Oxford University Press, New York.

Torrey, T. W. and Feduccia, A. 1979. *Morphogenesis of the Vertebrates*. Wiley, New York.

Van Valen, L. M. 1976. Energy and evolution. *Evol. Theor.* 1: 179–229.

von Baer, K. E. 1828. *Entwicklungsgeschichte der Thiere: Beobachtung und Reflexion*. Bornträger, Königsberg.

von Baer, K. E. 1886. *Autobiography of Dr. Karl Ernst von Baer*. (Translated by H. Schneider.) Science History Publications, Canton, MA, pp. 261–262.

Waddington, C. H. 1938. The morphogenetic function of a vestigial organ in the chick. *J. Exp. Biol.* 15: 371–376.

Waddington, C. H. 1940. *Organisers and Genes*. Cambridge University Press, Cambridge.

Waddington, C. H. 1956. *Principles of Embryology*. Allen and Unwin, London.

Waddington, C. H., Needham, J. and Brachet, J. 1936. Studies on the nature of the amphibian organization centre. III. The activation of the evocator. *Proc. R. Soc. London* [B] 120: 173–198.

Wake, D. B. and Larson, A. 1987. A multidimensional analysis of an evolving lineage. *Science* 238: 42–48.

Wheeler, D. 1986. Developmental and physiological determinants of caste in social hymenoptera: Evolutionary implications. *Am. Nat.* 128: 13–34.

Whittington, H. B. 1985. *The Burgess Shale*. Yale University Press, New Haven.

Wilson, E. B. 1898. Cell lineage and ancestral reminiscence. *Biological Lectures from the Marine Biological Laboratories, Woods Hole, Massachusetts*. Ginn, Boston, pp. 21–42.

Wilson, E. O. 1971. *The Insect Societies*. Harvard University Press, Cambridge, MA.

Winsor, M. P. 1969. Barnacle larvae in the nineteenth century: A case study in taxonomic theory. *J. Hist. Med. Allied Sci.* 24: 294–309.

Wolpert, L. 1983. Constancy and change in the development and evolution of pattern. *In* B. C. Goodwin, N. Holder and C. C. Wylie (eds.), *Development and Evolution*. Cambridge University Press, Cambridge.

Zangerl, R. and Williams, M. E. 1975. New evidence on the nature of the jaw suspension in Paleozoic anacanthus sharks. *Paleontology* 18: 333–341.

Sources for chapter-opening quotations

Bard, J. 1990. *Morphogenesis: The Cellular and Molecular Processes of Developmental Anatomy.* Cambridge University Press, New York, p. 9.

Borges, J. L. 1962. "The Library of Babel" in *Ficciones.* Grove Press, New York, pp. 87–88.

Claude, A. 1974. The coming of age of the cell. Nobel lecture, reprinted in *Science* 189 (1975): 433–435.

Conrad, J. 1920. *The Rescue: A Romance of the Shallows.* Doubleday, Page and Co., Garden City, NJ 1924, p. 447.

Darwin, C. 1871. *The Descent of Man.* Murray, London, p. 893.

Darwin, E. 1791. Quoted in M. T. Ghiselin, *The Economy of Nature and the Evolution of Sex.* University of California Press 1974, Berkeley, p. 49.

Doyle, A. C. 1891. "A Case of Identity" in *The Adventures of Sherlock Holmes.* Reprinted in *The Complete Sherlock Holmes Treasury,* 1976. Crown, New York, p. 31.

Eliot, T. S. 1936. "The Hollow Men," Part V in *Collected Poems 1909–1962.* Harcourt, Brace and World, New York, pp. 81–82. Copyright T. S. Eliot.

Eliot, T. S. 1942. "Little Gidding" in *Four Quartets.* Harcourt, Brace and Company, New York, 1943, p. 39. Copyright T. S. Eliot.

Fuentes, C. 1989. *Christopher Unborn.* Trans. A. MacAdam. Farrar, Straus, and Giroux, New York, p. 281.

Garstang, W. 1922. The theory of recapitulation: A critical restatement of the biogenetic law. *J. Linn. Soc. Zool.* 35: 81–101.

Geoffroy Saint-Hilaire, E. 1807. Considérations sur les pièces de la tête osseuse de animaux vertebrés, et particulièrment sur celles du crâne des oiseaux. *Ann. Mus. Hist. Nat.* 10: 342–343.

Hardin, G. 1968. *Exploring New Ethics for Survival: The Voyage of the Spaceship Beagle.* Viking Press, New York, p. 45.

Harrison, R. G. 1933. Some difficulties of the determination problem. *Am. Nat.* 67: 306–321.

Holub, M. 1990. "From the Intimate Life of Nude Mice" in *The Dimension of the Present Moment.* Trans. D. Habova and D. Young, Faber and Faber, London, p. 38.

Huxley, T. 1882. On science and art in relation to education. In T. H. Huxley, *Science and Education,* Appleton, New York, p. 178.

Just, E. E. 1939. *The Biology of the Cell Surface.* Blakiston, Philadelphia, p. 288.

Lessing, G. E. 1778. "Eine Duplik." Reprinted in F. Muncker (ed.), *Sämtliche,* Schriften 13, Göschen, Leipzig, 1897, p. 23.

Levi-Montalcini, R. 1988. *In Praise of Imperfection.* Basic Books, New York, p. 90.

Monod, J. and Jacob, F. 1961. Teleonomic mechanism in cellular metabolism, growth, and differentiation. *Cold Spring Harbor Symp. Quant. Biol.* 26: 389–401.

Muller, H. J. 1922. Variation due to change in the individual gene. *Am. Nat.* 56: 32–50.

Needham, J. 1967. *Order and Life.* M.I.T. Press, Cambridge, MA, p. xv.

Ozick, C. 1989. *Metaphor and Memory.* Alfred A. Knopf, New York, p. 111.

Ramón y Cajal, S. 1937. *Recollections of My Life.* Trans. E. H. Craigie and J. Cano. MIT Press, Cambridge, MA. pp. 36–37.

Rostand, J. 1960. *Carnets d'un Biologiste.* Librairie Stock, Paris.

Roux, W. 1894. The problems, methods, and scope of developmental mechanics. *Biol. Lect. Woods Hole* 3: 149–190.

Rushdie, S. 1989. *The Satanic Verses.* Viking, New York, p. 8.

Schultz, J. 1935. Aspects of the relation between genes and development in *Drosophila. Am. Nat.* 69: 30–54.

Shelley, M. W. 1817. *Frankenstein, or The Modern Prometheus.* Oxford University Press 1969, p. 36.

Spemann, H. 1943. *Forschung und Leben.* Quoted in T. J. Horder, J. A. Witkowski and C. C. Wylie, *A History of Embryology,* Cambridge University Press 1986, Cambridge, p. 219.

Stern, C. 1936. Genetics and ontogeny. *Am. Nat.* 70: 29–35.

Tennyson, A. 1886. *Idylls of the King.* Macmillan, London 1958, p. 292.

Thomas, L. 1979. "On Embryology" in *The Medusa and the Snail,* Viking Press, New York, p. 157.

Thomson, J. A. 1926. *Heredity.* Putnam, New York, p. 477.

Virchow, R. 1858. *Die Cellularpathologie in ihre Begrundung auf physiologische und pathologische Gewebelehre.* Berlin, p. 493.

Virgil. 37 B.C.E. *Georgics II,* p. 490.

Waddington, C. H. 1956. *Principles of Embryology.* Macmillan, New York, p. 5.

Weiss, P. 1960. Ross Granville Harrison, 1870–1959. Memorial minute. Rockefeller Inst. Quarterly, p. 6.

Whitehead, A. N. 1934. *Nature and Life.* Cambridge University Press, Cambridge, p. 41.

Whitman, W. 1855. "Song of Myself" in *Leaves of Grass and Selected Prose.* S. Bradley (ed.), Holt, Rinehart & Winston, New York 1949, p. 25.

Whitman, W. 1867. "Inscriptions" in *Leaves of Grass and Selected Prose.* S. Bradley (ed.), Holt, Rinehart & Winston, New York 1949, p. 1.

Williams. C. M. 1959. Hormonal regulation of insect metamorphosis. In W. D. McElroy and B. Glass (eds.), *The Chemical Basis of Development.* Johns Hopkins University Press, Baltimore, p. 794.

Wilson, E. B. 1923. *The Physical Basis of Life.* Yale University Press, New Haven, p. 10.

Wolpert, L. 1983. Quoted in J. M. W. Slack, *From Egg to Embryo: Determinative Events in Early Development.* Cambridge University Press 1983, Cambridge, p. 1.

Author index

Subject index

A23187, calcium ion storage
   within eggs, 145
ABa cell, 514, 515–516
abdominal A gene, 553
Abdominal B gene, 553
ABp cell, 515–516
ACE, *see* Adenylation control element
Acetabularia, developmental morphogenesis, 8–10
Acetylation, transcription, 430
Acetylcholinesterase, cytoplasmic segregation, 500–503
achaete gene, 569
achaete-scute gene, 395
Acquired immune deficiency syndrome (AIDS), *see* HIV
Acrosomal process, 123
   bindin localization, 132, 133
   extension in acrosome reaction, 130
Acrosomal vesicle, 123
   exocytosis in acrosome reaction, 130
Acrosome reaction
   G proteins, 153
   sperm activation, 129–131, 153
   ZP3 induction, 136–137
Actin
   axon growth cones, 266
   par-3 mutant, 512
α-Actinin, 110
Action-at-a-distance mechanism
   fertilization, 128–131, 134
   mammalian gametes, 134
Activin
   branching pattern effects, 674, 675
   dorsal mesoderm induction, 603–604
   inhibitor as soluble organizer factor, 610
   synthesis, 67
Activity-dependent development, axons, 313–314
Adaptation, embryonic cleavage modification, 175
Adaptive enzymes, 46
Adenovirus, RNA transport, 457
Adenylation control element (ACE), 475
Adepithelial cells, 732
Adherens junction, 94
Adhesion, *see* Cell adhesion; Substrate adhesion
Adhesion proteins, sperm-zona, 135–136
Adhesive gradients, axonal guidance, 298
Adhesive specificities, *see also* Specific adhesion

differential, 299–300
   optic tectum, 312–313
Adipogenesis, 398
Adolescence, mammary gland development, 747
Adrenal medullary cells, 282, 283
Adrenergic neurons, 280
Adult hemoglobin, 418
Aequorin, 144
AER, *see* Apical ectodermal ridge
Affinity
   cell, 79–86
   substrate, 105–113
Agametic mutation, *Drosophila*, 523
Agar, DNA cloning, 56
Aggregation
   amphibians, 80
   catenins, 94, 95
   conversion of cells into nephron, 666
   germ cells, 526–527
   histotypic, 83
   thermodynamic model, 83–85
AIDS, *see* HIV
Albumin, hormonal mRNA stabilization, 466
Alcohol, teratogenesis, 639–641
Allantois, 31, 342
Alleles
   gene targeting, 70
   pronuclear transplantation, 149
   snail shell coiling direction, 174
Allometry, 850–851
Allophenic humans, 185
Allophenic mice, 182–183
   skeletal muscle formation, 329
all-*trans* retinoic acid, 636
AluI, 54
Ambystoma, *see also* Salamander
   fertilization, 150n
Ameloblasts, 667
AMH, *see* Anti-Müllerian duct hormone
Amino acids, extracellular cadherin binding specificity, 94
Aminopterin, monoclonal antibody synthesis, 89
Ammonia, *Dictyostelium* differentiation, 27–28
Ammonotelic, defined, 718
Amnion, 31, 341
   structure, 235, 236
Amniotes, 341
   egg, 31
   neurulation, 246
Amniotic fluid, 235
Amoeboflagellate, differentiation, 10–12

Amphibian(s), *see also specific type*
   behavioral changes in metamorphosis, 723–724
   cell fate, 586–591
   cell membrane formation, 197–198
   cleavage, 170–172
   cleavage preparation, 158
   coordination of developmental changes, 721–722
   embryonic genome activation, 479
   gap junctions, 100
   gastrulation, 211–222, 588–589
   germ cells, 79–81, 526–527, 788–789
   heterochrony, 726–729
   metamorphosis, 717–729
   neurulation, 246, 247–248, 588–589, 722–723
   oocyte maturation, 804–809
   regulative development, 586–591
Amphibian cloning
   nuclear potency restriction, 41–42
   somatic cell pluripotency, 42–44
Ampicillin, DNA cloning, 56
Amplification, ribosomal, 811–812
Amplified genes, 811n
   transcription visualization from, 812
Anchor cell, vulval induction, 679, 681–682
Androgen insensitivity syndrome, 745, 766
Androgenones, mammalian gastrulation, 240
Anencephaly, 252
Angelman syndrome, 426
Angioblasts, vasculogenesis, 348
Angiogenesis, 348–349
   tumor-induced, 349–351
Angiogenesis factors, 348
Angiogenic clusters, 347–348
Angiogenin, 349
Animal cap, 220
   goosecoid gene, 607
   mesodermal specificity induction, 600
Animal pole, 165
   snail, 174
Animal–vegetal axis, cell fate, 584–586
Antennapedia complex, 552
Antennapedia gene, 57–58, 552–557, 734
   homeodomain proteins and,

557–559
   homologous specification, 569
Anterior necrotic zone, 713
Anterior neuropore, 252
Anterior organizing center, gradient of bicoid protein, 537–541
Anterior–posterior axis
   initiation, 624
   limb, 702–707, 708, 709–711
Anterior–posterior polarity
   anterior organizing center, 537–541
   cis-regulatory elements, 559–562
   Drosophila, 533–562
   gap genes, 546–548
   gradient models of positional information, 535–536
   homeotic selector genes, 552–562
   maternal effect genes, 534, 535–544
   origins, 533–562
   posterior organizing center, 542–543
   regulation by oocyte cytoplasm, 535
   segmentation genes, 544–552
   terminal gene group, 543–544
anterobithorax gene, 560
Antibodies, *see also* Immunoglobulin *entries*
   cell adhesion, 25
   monoclonal, 88–91
   N-CAM, 99
   production, 389–393
   specificity, 389
Antigens
   differentiation, 88
   histocompatibility, 239–240
Anti-Müllerian duct hormone (AMH), 757, 766, 767, 768
Antisense RNA, 70–71, 72
Antiserum, compaction prevention, 181
Antrum, 819
Ants, *see also* Insect
   transfer of competence, 852
Anurans, *see* Amphibian(s)
Aorta, 343
AP-1 protein, 397
   muscle cells, 396
Apical ectodermal ridge (AER)
   mesenchyme interactions, 693–695
   proximal–distal limb axis, 698–699
Apoptosis, 518
Apo VLDL II, hormonal mRNA stabilization, 466

tor (BDNF), 315–316
trunk neural crest cell potency, 282
Branching mechanisms, parenchymal organ formation, 670–675
Branch sites, 445–446
Breast development, *see* Mammary gland development
bride of sevenless gene, 677–679
5-Bromo-4-chloroindole, 56n
Buffers, DNA hybridization, 57–58
Burkitt lymphoma, 399
Butterfly wings, gradient model of spot development, 536
buttonhead gene, 541
bZip proteins, *see* Basic leucine zipper transcription factors

C_0t value, hnRNA, 441–442
CAAT box, 376
CACCC sequence, 376, 377
transcription factor interactions, 403–404
Cactus protein, separation from dorsal protein, 566–568
Cadherins, 91–95, *see also* Cell adhesion molecules (CAMs); *specific type*
brain, 94
extracellular region, 94
intracellular region, 94, 95
retinal, 94, 310
Caenorhabditis
cell specification, 510–518
gamete determination, 476
hermaphroditism, 778–781, 798
pharyngeal progenitor cells, 513–518
regulatory interactions of cells, 515–518
vulval induction, 679, 681–682
Calcium, endochondral ossification, 337
Calcium-dependent adhesion molecules, *see* Cadherins
Calcium ions
cell division, 158
cortical granule reaction, 143–144
egg, 145, 153–155
fertilization role, 143–144, 151, 153–155
G protein–phosphoinositide cascade, 153–155
release from storage, 143–144
sperm activation, 153
Callus, plant cloning, 46
Calpain II, 809
cAMP
Dictyostelium cell aggregation, 23
Dictyostelium differentiation, 26, 28
meiosis reinitiation, 823
CAMs, *see* Cell adhesion molecules
Cancers, *see also specific type*

enhancers, 398–399
Capacitation, mammalian fertilization, 134
Cap-binding protein, 461–462
cappucino gene, 525
Cap sequence, 372
Cardia bifida, 343
Cardiac anomalies, 641–642
"Cardiac jelly," 345
Cardiac neural crest, 273, 284, 286
Carrot, cloning, 45–46
Cartesian coordinate model, 568–569
Cartilage
chondrocytes, 326, *see also* Chondrocytes
chondrogenesis, 335
ossification front, 335–336
proteoglycans, 108
reaction–diffusion model of limb specification, 701
Casein, hormonal mRNA stabilization, 466
β-casein, lactation, 748–749
Catenins, 94, 95
CAT gene, *see* Chloramphenicol acetyltransferase gene
caudal gene, 553
Cavitation, 180
CCAAT enhancer-binding protein (C/EBP), 397–398
CD4 glycoprotein, helper/inducer T cells, 676
cdc2 kinase, 192
cdc25 phosphatase, mitosis, 193–194
cDNA
dot blotting, 64, 65
preparation, 54
RNA in situ hybridization, 63
subtraction cloning, 64, 65
cDNA libraries, mRNA analysis, 61–62
C/EBP, *see* CCAAT enhancer-binding protein
ced-9 gene, 518
Cell(s), *see also specific type*
accumulation in primitive streak, 229–232
commitment and differentiation, 493–495, *see also* Differentiation
commitment in Hensen's node, 232
cytoplasmic determinants of types, 493–528
determination, 493
specification, 493–528
Cell adhesion
differential, 82, 85–86, 112–113
glycosyltransferases, 112
inhibition, 110
molecular basis, 86–88
multiple adhesion systems, 112–113
specificity, 91
Cell adhesion molecules (CAMs), 91–101, *see also specific type*

alternative RNA splicing, 448–449
cadherins, 91–95
classes, 86–88
classification, 91, 92
combinations, 112–113
Dictyostelium, 24–25
differential expression, 98
distribution at tissue boundaries, 99
gap junctions and, 100–101
immunoglobulin superfamily, 91, 92, 95–99
retinal axons, 310
roles in development, 91
Cell affinity, differential, 79–86
Cell–cell adhesion, *see* Cell adhesion
Cell–cell interactions, 716–750
identities of equivalent cells, 679–681
metamorphosis, *see* Metamorphosis
progressive, 575–618, *see also* Cell fate
thermodynamic model, 83–86
Cell cycle
regulation, 191–195
transcription, 190
Cell death
differentiation and, 19–20
homeotic genes, 712–713
Cell division, *see also* Cleavage
cyclin, 158
cytoplasmic segregation, 501–503
eukaryotic, 6–7
retinal development, 269
Cell fate, 518, 575–618, *see also* Fate maps; Induction; Regulative development
amphibian, 586–591
germ plasm, 522–523
late-cleavage embryos, 581–582, 583
progressive determination, 586–589
prospective potency and, 577–580
sea urchin, 577–586
specification, 544
Cell interactions, *see* Cell–cell interactions
Cell junctional molecules, 88
Cell lineage, 500–501, *see also* Autonomous cell specification
Cell lines, clonal, 389–394
Cell membrane, *see* Plasma membrane
Cell migration, *see also* Cell movement; Gastrulation
cranial neural crest cells, 283–285
fibronectin, 109, 220–221
germ cells, 788–794
glycosyltransferases, 112
integrins, 110–111
lymphocytes, 794n
melanocytes, 279–280

molecular bases of specificity, 101–105
neural crest cells, 273–279, 283–285, 326
neuronal, 260, 261, 263
primitive streak, 226–228
trunk neural crest cells, 273–279
unit of activity, 202–203
Cell movement, *see also* Cell migration
amphibian gastrulation, 211–214, 216–219, 222
chick gastrulation, 224–228
human gastrulation, 235–236
primitive streak, 224–225
Cell potency, *see* Potency
Cell receptors
glycosyltransferases, 111–112
integrins, 110–111
Cell sheet
folding, 203
spreading, 203
Cell surface
compaction mechanism, 180–181
egg, 126, 127
gastrulation, 203
structure, 86–87
Cell surface cytoplasm, subcortical cytoplasm rotation relative to, 157–158
Cellular blastoderm, 187, 188
Cellular retinoic acid-binding proteins (CRABPs), 637
Cellular retinol-binding protein I (CRBP-I), 637, 713
Central nervous system, *see also* Brain; Neural *entries*; Neurulation; *specific parts*
myelination, 267
N-cadherin, 92, 93
secondary sex determination, 768–770
tissue architecture, 256–258, 259
Centrifugation, DNA cloning, 56
Centrolecithal eggs, cleavage, 165
Cephalic furrow, 531
Cephalic neural crest, 272–273, 283–286
cell developmental potency, 285–286
Cerebellum, 253
organization, 258, 260–262
wnt-1 gene, 255, 256
Cerebral hemispheres, 253
Cerebral organization, 262–264
c-fos, *proliferin* transcription and, 404–405
CFU-M,L, hematopoiesis and, 356
CFU-S, hematopoiesis and, 355–356, 360
Charge-to-mass ratio, DNA fragments, 57n
Chemicals, teratogenic, 635
Chemoaffinity hypothesis, 312

DNA fragments, Southern blots, 57–58
DNA hybridization, 53–54
    within and across species, 57–58
DNA ligase, 55
DNA polymerase, PCR, 64, 66–67
DNase(s), chromatin attachment to nuclear matrix, 431–432
DNase-hypersensitive sites, 415–416, 432
DNase I, 414
    5S rRNA gene, 813
DNA sequencing, 58–60
    hnRNA processing, 442–443
DNA synthesis, 191
    central nervous system, 256–257
    egg responses to fertilization, 152, 153
    myoblast, 328
    nucleosomes, 412–413
Dominant control region, *see* Locus control region (LCR)
Dominant negative receptor, 605
Dorsal ectoderm, autonomous differences from ventral ectoderm, 611
dorsal gene, 562, 568
Dorsal lip
    blastopore, 213, 217–218, 590–591, 608–610, 611–613
    factors transmitted through ectoderm, 611–613
Dorsal marginal zone (DMZ), 220–221
    primary embryonic induction, 612–613
Dorsal mesoderm, 323, 324, 325–338
    paraxial, 325
    peptide growth factors inducing, 603–605
Dorsal protein, 562–563
    asymmetric signal for translocation, 563–565
    separation from cactus protein, 566–568
    targets for binding, 566–568
Dorsal root ganglia, 273
    growth cone repulsion, 300–301
    neural crest populations, 276
Dorsal skin, 327
Dorsal–ventral axis, 632–633
    cytoplasmic rearrangements, 158
Dorsal–ventral polarity
    asymmetric signal, 563–565
    Drosophila, 562–568
    homeotic genes, 711
    morphogen, 562–563
    polar lobe, 508–509
    specification at fertilization, 597–598
    spinal cord, 257–258, 259
Dorsoanterior determinants, localization in vegetal cells, 598–599

Dorsolateral hinge points (DLHPs), neural plate bending, 250
Dorsolateral pathway, neural crest cell migration, 273, 275
Dosage compensation, 427–431
Dot blots, 63–64, 65
Double gradient model, 593–594
    neural induction, 594–597
doublesex gene, 773, 777–778
Drosophila, see also Insect
    alternative RNA processing, 450–452
    anterior–posterior polarity, 533–562
    antisense RNA, 70–71, 72
    axis specification, 531–570
    bHLH proteins, 395
    cell cycle, 191, 193–194
    cleavage, 187–190
    development summary, 531–533
    differential RNA synthesis, 49–50, 51
    DNA hybridization, 57–58
    DNA insertion into cells, 68
    dorsal–ventral polarity, 562–568
    embryonic genome activation, 480
    gastrulation, 531, 532, 567
    germ cell determination, 520–526
    gradient models of positional information, 536
    homeodomain proteins, 386, 557–559
    homeotic selector genes, 552–562, 624–626
    homologous specification, 569
    integrins, 111
    maternal effect genes, 534, 535–544
    mitosis, 193–194
    monoclonal antibodies, 89, 91
    oogenesis, 816–817
    organ primorida, 568–569
    pair-rule genes, 548–550
    Pax proteins, 255
    pole plasm components, 523–526
    reverse genetics, 91
    segmentation genes, 544–552
    segment polarity genes, 550–552
    sex determination, 450–452, 772–778
    transcription initiation, 190
    transcription regulation, 405
Drugs, teratogenic, 635
D segment, antibody heavy chain genes, 391–393
Dual-gradient two-step induction, 593–597
Ductus arteriosus, 354
Dwarf mouse, 387
Dyes
    DNA cloning, 56
    egg cytoplasm rearrangement,

157
    monitoring intracellular calcium release, 144
    vital, 211–213
Dynein, 124

E. coli
    DNA cloning, 56
    operon model, 46–48
easter gene, 565
Ebony mutation, *Drosophila*, 522–523
E-cadherin, 92
    compaction, 180, 181
    epithelial integrity, 94
    extracellular region, 94
    intracellular region, 94
Ecdysone, 737, 739, 740
    meroistic oogenesis, 817
Ecdysone receptor (EcR), 740–741
Echinoderm
    acrosome reaction in sperm, 129
    cleavage, 166–167
    embryonic genome activation, 479
EcoRI, 54
EcR, *see* Ecdysone receptor
Ectoderm, 4, 244–290, 245, 295–320, 324, *see also* Neurulation
    apical ridge, 693–695
    autonomous differences of dorsal and ventral, 611
    determination of imaginal discs from, 733
    dorsal lip factor transmission, 611–613
    epiboly, *see* Epiboly
    induction, *see* Induction
Ectodermal bias, lens induction, 655–657
Ectodermal competence, lens induction, 655–657
Ectodermal mRNA, cDNA libraries, 62
Ectodermal placodes, cranial, 268
Ectopic pregnancy, 183
EDNH, *see* Egg development neurosecretory hormone
Efferent ducts, gonadal development, 757
EGF, *see* Epidermal growth factor
Egg, *see also* Oocyte; Oogenesis
    amniote, 31
    bindin receptors, 133
    calcium ions, 145
    cell membrane fusion with sperm cell membrane, 138–140, 155
    centrolecithal, 165
    cleavage, *see* Cleavage
    early responses, 150–151
    endoplasmic reticulum, 145
    G protein–phosphoinositide pathway activation, 153–155
    isolecithal, 165
    late responses, 151–153
    mammalian, 44, 45, 125–128

    meiosis, 797–798
    mesolecithal, 165
    metabolism activation, 149–158
    nuclei transplantation into enucleated oocytes, 41–42
    parthenogenesis, 148–149
    polyspermy prevention, 141–146
    rotation and gastrulation, 215–216
    sperm entry, 138–140
    sperm recognition, 128–137
    stages of maturation, 126, 127
    structure, 125–128
    telolecithal, 165
    tunicate, 177
    yolk protein transport into, 816–817
Egg cytoplasm, 125–127, *see also* Cytoplasm
    cleavage plane orientation, 173
    cleavage preparation, 158
    dorsalizing signal from follicle cell, 565
    polarity regulation, 535
    rearrangement, 155–158, 216
    tunicate, 177
Egg development neurosecretory hormone (EDNH), 816n
Egg jelly, 128
    sperm activation, 129–131
Egr-1 protein, 399
eIF2, heme and, 482–483
eIF2-GTP, 461
eIF4E, 461–462
Eighth cranial nerve, *see* Vestibuloacoustic nerve
Electrical current, polyspermy, 142
Electrophoresis, *see* Gel electrophoresis
Electroporation, 68
Elongation, translation phase, 462
Embryo
    autonomous specification, *see* Autonomous cell specification
    defined, 3
    differential cell affinity, 80–81
    gene targeting, 69–70
    late-cleavage, 581–582, 583
    mammalian, 623–642
    preimplantation, 67
Embryogenesis, 4
Embryology
    defined, 3
    gene theory and, 34–37
    genetics split from, 37–38
Embryonic axes, formation in avian gastrulation, 228–229
Embryonic circulation, 351–352
Embryonic genome, activation, 478–481
Embryonic hemoglobin, 417
Embryonic induction, *see also* Induction
    primary, 589–591, 597–614
    "secondary," 615–618

Embryonic mammary tissue, 744–747

Embryonic red blood cells
globin gene switching, 421
promoter methylation and gene inactivity, 424

Embryonic stem (ES) cells, *see also* Stem cell(s)
teratocarcinomas, 791–792
transgenic mice, 68
twins, 184–185

empty spiracles gene, 541

EMS blastomere, maternal control of identity, 513–516

Endocardial cushion, 346

Endocardium, 343

Endochondral ossification, 333, 335–338

Endocrine factors, 670n

Endocrine system, neural crest derivatives, 274

Endoderm, 4, 245, 361–364
formation during avian gastrulation, 226–228
mesodermal specificity induction, 599–601
vegetal, 599–601
yolk sac, 235

Endodermal mRNA
cDNA libraries, 62
in situ hybridization, 63

Endonucleases, restriction, *see* Restriction enzymes

Endoplasmic reticulum, calcium ion storage in eggs, 145

Endothelial cells, heart, 348

Energids, 189

engrailed gene, 255, 256, 545, 550–552, 612, 613, 734–735

Enhancer(s), *see also* Promoter(s); Transcription
cancers, 398–399
eukaryotic genes, 72–73
β-globin gene, 419–421
glucocorticoid, 400–405
immunoglobulin, 392–393
interactions, 403–405
master, *see* Locus control region (LCR)
NF-μNR protection of heavy chain gene enhancer, 433–434
positional information, 536
prolactin, 387–389
structure and function, 382–384
transcription pattern regulation, 383–384

Enhancer element effect, 382

Entelechy, 579

Enteric ganglia, *see* Parasympathetic ganglia

Enucleated oocytes
maternal control of development, 467–468
nuclei transplantation into, 41–44

Environmental chemicals, teratogenic, 635

Environmental factors
larval form and functions, 742–744
meroistic oogenesis, 816
sex determination, 782–784

Enzymes, *see* Restriction enzymes; *specific enzymes*
adaptive, 46
hatching, 170

Ependyma, 257

Ependymal cells, 264

Epi1 protein, 611, 612

Epiblast, 186
avian gastrulation, 223–224
mammalian gastrulation, 235

Epiboly, 202
avian ectoderm, 232–233
Xenopus ectoderm, 214, 222

Epidermal cells, origin, 286–288

Epidermal growth factor (EGF), mammary gland development, 747

Epidermis, 286–289

Epigenesis, 495–498

Epimyocardium, 343

Epiphyseal plates, 336, 337

Epithelial branching patterns, parenchymal organ formation, 670–675

Epithelial cadherin, *see* E-cadherin

Epithelial cells
basal laminae, 106
cadherins in interactions, 94
dorsal marginal zone, 612–613
laminin, 109–110
morphogenesis, 78, 79
somite formation, 326
uterine, 183

Epithelio-mesenchymal interactions, 649–653, *see also* Proximate tissue interactions
genetic specificity, 652–653
nature of proximity, 668–670
regional specificity, 650–652

Epithelium, mesenchyme cell conversion into, 664

Equatorial cleavage, 167

Equipotentiality, 584–585

Equivalence, nuclear, 587

Equivalence group, 517

Equivalent cells, cell–cell interactions in determining identities, 679–681

Erythroblast, 356, 358

Erythrocyte(s), 358, 494
globin gene switching, 421
promoter methylation and gene inactivity, 424

Erythropoietin, 356, 358

ES cells, *see* Embryonic stem cells

Escherichia coli
DNA cloning, 56
operon model, 46–48

Esophagus, 364

Estradiol, sexual behavior, 770

Estrogen
menstrual cycle, 820
mRNA stabilization, 466

oogenesis, 806–807

osteoclast development, 359

prolactin gene transcription regulation, 387–389

secondary sex determination, 766

sex-specific behavior, 769–770

Estrus, 817

Euchromatin, 427

Eudorina, 17

Eukaryotes, 6–8
colonial, 16–18
nucleic acid hybridization, 53–54
translation mechanisms, 461–463
unicellular, 8–16

Eukaryotic genes
"anatomy," 72–73
transcription, *see* Transcription

Eukaryotic initiation factor 2 (eIF2-GTP), 461

Eukaryotic mRNA, 5′ cap, 374, 375

even–skipped gene, 549

Eversion, imaginal discs, 729–733

Evidence, types, 25

Evolution
allometry, 850–851
approaches to study, 828–830
"biogenetic law," 843, 844
correlated progression, 845–849
developmental constraints, 838–842
developmental–genetic mechanisms, 835–838
developmental mechanisms, 828–857
differentiation, 16–18
emergence of embryos, 830–832
emerging theory, 855–857
established *Baupläne*, 838–854
heterochrony, 849–850
inductive interactions, 843–845
phyla formation, 832–835
transfer of competence, 851–854

Evolutionary divergences, 28, 29

Exocytosis
acrosomal vesicle, 129–131
cortical granule, 143, 144, 155
synaptic vesicle, 130n

Exons, 371–374
splice site, 445–448
transcription regulation, 406

Experience, changes in inherent mammalian visual pathways, 317–320

Expression cloning, 57n

Expression vectors, 57n

External germinal layer, cerebellum, 258, 260

Extracellular matrix
axonal guidance, 297–302
branch formation, 671–673
cell differentiation, 668, 669
cell receptors, 110–112

collagen, 105–106
constituents, 105, 106
differential substrate affinity, 105–110
fibronectin and mesodermal migration, 220–221
glycoproteins, 108–110
metalloproteinases degrading, 113
mineralization, 337–338
sea urchin blastocoel, 205–208
trunk neural crest cell migration, 276–279

Extraembryonic membranes
formation, 237–240, 340–342
monozygotic twins, 184

extramacrochaetae gene, 395

exuperantia gene, 537, 539, 541

Eye, 317–320, *see also* Lens *entries*; Optic *entries*; Retina
vertebrate, 267–272

Eyeless mouse, 268

"Eyespot" patterns, butterfly wings, 536

Fab fragments, N-CAM, 96

Facial nerve, 254

FAS, *see* Fetal alcohol syndrome

Fascicles, 303

Fasciclin I, 303

Fasciclin II, *see* Neural cell adhesion molecule (N-CAM)

Fasciclins, 99

Fasciculation, 99

Fat cell formation, 398

Fate maps, *see also* Cell fate
dorsal protein in *Drosophila*, 567
Hensen's node, 232
Xenopus blastula, 212, 213

Feedback regulation, heme synthesis, 482

Female pronucleus, 146, *see also* Pronucleus

Ferret, cerebrum, 264

Ferritin, translational regulation, 485, 486

Fertilization, 4, 121–158
amphibian, 808–809
cytoplasmic determinant segregation, 498, 499
cytoplasmic rearrangement, 155–158
dorsal–ventral polarity specification, 597–598
egg metabolism activation, 149–155
fusion of genetic material, 146–148
gamete fusion, 138–140
gamete structure, 121–128
G protein–phosphoinositide pathway, 153–155
human, 177
nuclear events, 147–148
polyspermy prevention, 140–146
preparation for cleavage, 158
recognition of egg and sperm, 128–137

sea urchin, 150
Fertilization cone, 138
Fertilization envelope, 142, 143, 144
Fertilization tube, 15
Fetal alcohol syndrome (FAS), 639–641, *see also* Teratogenesis
Fetal hemoglobin, 354, 417
  hereditary persistence, 421–422
Fetal neurons, transplants to adult hosts, 305
Fibroblast growth factor (FGF), *see also* Basic fibroblast growth factor (FGF-2)
  myoblasts, 328
  ventrolateral mesoderm induction, 605
Fibroblast growth factor-2 (FGF-2), proximal–distal limb axis, 698, 699
Fibroblast growth factor-5 (FGF-5), 315–316
Fibronectin, 108–109
  mesodermal migration pathways, 220–221
  myoblast fusion, 328
  sea urchin gastrulation, 207–208
  somite formation, 326
Fibronectin receptors, 110
Fifth cranial nerve, *see* Trigeminal nerve
Filopodia, 204
  sea urchin gastrulation, 204, 208, 209–211
Finger identity, determination by homeotic genes, 709–711
First polar body, 804
Fish
  cleavage, 186–187
  hermaphroditism, 781
  lamellipodium, 104
  neurulation, 247n
  Pax proteins, 256
  sex determination, 852–853
  transfer of competence, 852–853
5′ ends, hemoglobin production, 484
5′ splice site, 445–448, *see also* RNA splicing
5S rRNA gene, 812–813
Flagellum, sperm, 123–125
Floor plate cells, neural tube, 258
Fluid mosaic model, 86–87
Fluorescein, egg cytoplasm rearrangement, 157
Follicle cells
  dorsalizing signal, 563–564
  insect oviduct, 494
  signal to oocyte cytoplasm, 565
Follicles
  gonadal development, 759
  ovarian, 818–819
Follicle–stimulating hormone (FSH), 817
  meiosis, 822
  menstrual cycle, 819–821
Follicualr phase of menstrual

cycle, 819–820
Foramen ovale, 354
Forebrain, *see* Prosencephalon
Foregut, partitioning, 364
Founder cells, 511
Fourth germinal layer, *see* Neural crest
Fovea, 317
F protein, 140
Fraternal twins, *see* Dizygotic twins
French teratological experiments, epigenesis, 497–498
Frog, *see also* Amphibian(s); *specific type*
  cleavage, 164–165, 170–172
  cloning, 41–44
  developmental history, 5
  dorsal–ventral axis, 158
  dot blots, 63–64, 65
  egg cytoplasm rearrangement, 156–158
  extracellular cadherin region, 94
  neurulation, 247
FSH, *see* Follicle–stimulating hormone
fs(1)*Nasrat*[fs(1)*N*] gene, 537
fs(1)*pole hole*[fs(1)*ph*] gene, 537
F-spondin, axon chemotaxis, 306
Fura-2, 144
fushi tarazu gene, 545, 550, 555
Fusion
  gamete, 138–140, 155
  genetic material, 146–148
  heart rudiments, 342–344
  myoblast, 327–330

GAGs, *see* Glycosaminoglycans
Gain–of–function evidence, 25
β-Galactosidase
  DNA cloning, 56
  E. coli synthesis, 46
Galactosyltransferases, 112
Gallbladder, 362
Galvanotaxis, 103
Gametes, 4
  binding, 135–137
  contact, 131–133, 135–137
  determination in *C. elegans*, 476
  fusion, 138–146, 155
  G protein–phosphoinositide model of activation, 153–155
  mammalian, 134, 135–137
  recognition in mammals, 135–137
  structure, 121–128
Gametogenesis, 5
Ganglia
  dorsal root, *see* Dorsal root ganglia
  parasympathetic, 273
  retinal, 269, 270, 309–310
  sympathetic, 273
Gap genes, 534, 544, 545, 546–548
Gap junctions, 100–101
Gastrulation, 4, 202–241
  amphibian, 211–222, 588–589
  avian, 223–234

blastopore positioning, 214–216, 217
  cadherins, 92
  cell movements, 211–213, 216–219
  cell potency, 588–589
  cleavage and, 164
  Drosophila, 531, 532, 567
  endoderm-specific mRNA, 63
  epiboly, 222
  extraembryonic membrane formation, 237–240
  fibronectin, 109
  gray crescent, 157, *see also* Gray crescent
  initiation, 216–219
  mammalian, 234–240
  maternal control, 467
  mesoderm formation, 219
  paraxial mesoderm, 325
  primitive streak, 224–228
  sea urchin, 203–211, 467, 585
  subtraction cloning, 64, 65
gastrulation-defective gene, 565
GATA–1, globin gene switching, 421
gcl gene, 523–526
GCN4 transcription factor, 397
G-CSF, *see* Granulocyte colony–stimulating factor
Gel electrophoresis, DNA blots, 57–58
Gel mobility shift assay, 393
Gene, imprinting, 425–427
Gene(s), *see also specific gene or type of gene*
  alternative proteins created from one, 448–452
  constancy, 46n
  DNA methylation and activity, 423–427
  mutations, 19, 20, *see also specific type*
Gene amplification, ribosomal, 811–812
Gene cloning, 55
Gene expression, *see also* Differential gene expression
  Cartesian coordinate model, 568–569
  homeotic genes, 552–557
  Hox genes, 626–627
  larval, 476
  locus control region, 419
  mesodermal, 601, 603
  neural, 612
  RNA processing, 443–445
  sperm development, 801–803
  transcriptional regulation, 371–407, 411–434, *see also* Transcription
Gene function, determination, 67–70, 71
Gene "knockout," *see* Gene targeting
Gene regulation, eukaryotic versus prokaryotic, 7–8
Gene targeting, 69–70, 71
  Hox genes, 627–630

Gene theory, embryological origins, 34–37
Geneticist, developmental, 39
Genetic material, fusion in fertilization, 146–148
Genetics, split from embryology, 37–38
Genetic specificity, induction, 652–653
Gene transcription, *see* Gene expression; Transcription
Genomes
  embryonic activation, 478–481
  hybrid, 68n
  maternal versus paternal in gastrulation, 240
  sperm-derived versus egg-derived, 148–149
Genomic DNA, cloning from, 54–57
Genomic equivalence, evidence, 39–44
Genomic identity, polytene chromosomes, 50
Genomic imprinting, 425–427
Germ band, 533
Germ cell(s), 5
  cytoplasmic localization of determinants, 518–527
  differential affinity, 79–81
  meiosis, 794–798
  migration, 788–794
  oogenesis, 803–823
  primordial, 526, 527, 788
  sperm formation, 123, 798–803, *see also* Sperm
Germ cell determination
  amphibians, 526–527
  insects, 520–526
  nematodes, 519–520
germ cell-less gene, 523–526
Germinal crescent, 226, 792
Germinal layer
  external, 258, 260
  fourth, *see* Neural crest
  retinal development, 269
Germinal vesicle, 804
Germinal vesicle breakdown (GVBD), 808
Germ layers, 4, 324, *see also* Ectoderm; Endoderm; Mesoderm
Germ line, 788–823, *see also* Germ cell(s)
Germ line chimeras, 565
Germ line granules, 511
Germ plasm, 521–526
  theory, 575–577
GGF, *see* Glial growth factor
GHF-1, *see* Pit-1
giant gene, 546, 547
Glial cell processes, neuronal migration upon, 260, 261
Glial cells
  dorsal protein, 567
  retinal axons, 310
Glial growth factor (GGF), salamander limb regeneration, 696

β-Globin, production, 374
Globin gene, *see also* Hemoglobin
  methylation, 423–424
  structure, 372–374
  switching, 419–423
  transcription, 417–423
Glochidia, cleavage modification,
  175
Glossopharyngeal nerve, 254
glp-1 gene, 516, 517, 797
Glucocorticoid enhancer, 400–403
  interactions with other tran-
    scription factors, 403–405
Glucose, mRNA stabilization,
  466
Glycoproteins, *see specific type*
Glycosaminoglycans (GAGs),
  107–108
  somites, 326
Glycosyltransferases, 111–112
  sperm cell membrane, 136
GM-CSF, *see* Granulocyte-
  macrophage colony-stimulating factor
GnRH, *see* Gonadotropin-releas-
  ing hormone
Goldberg-Hogness box, *see* TATA
  box
Golden hamster, sperm entry
  into egg, 139
Gonadotropin-releasing hormone (GnRH), *see also specific*
gonadotropins
  Kallmann syndrome, 301–302
  oogenesis, 806–807
Gonads
  development, 757–759
  WT-1 protein, 400
  Y chromosome, 760–763
Gonidia, 19
Gonium, 16–17
Gonochoristic hermaphroditism,
  781
goosecoid gene, 606–608
G protein, fertilization, 153–155
G protein–phosphoinositide
  pathway, 153–155, *see also*
  Phosphatidylinositol path-
  way
Gradient models
  bicoid protein, 537–541
  dorsal protein, 567–568
  double, 593–597
  hunchback protein,
    537–542542–543
  positional information,
    535–536
  potency and oocyte gradients,
    580–582, 583
Grandchildless mutation,
  Drosophila, 523
Granular cell, 260
Granular layer, internal, 260
Granulocyte colony-stimulating
  factor (G-CSF), 358
Granulocyte-macrophage
  colony-stimulating factor
  (GM-CSF), 358

Granulosa cells, 759, 820
Grasshopper
  fasciclins, 99
  pattern of neuroblasts, 302, 303
Gray crescent
  amphibian gastrulation, 216,
    588
  cleavage preparation, 158
  egg cytoplasm rearrangement,
    157
  progressive determination of
    embryonic cells, 588
Gray matter, 257
Grex, *Dictyostelium*, 27–28
Growth, 4
  bone, 335
Growth cone, 265–266, *see also*
  Axon(s)
  adhesive gradients, 298
  axon-specific migratory cues,
    302–305
  dorsal root ganglia, 300–301
  specific repulsion, 300–301
Growth factors, *see also specific
  type*
  dorsal mesoderm induction,
    603–605
  hematopoietic, 358
  proteogylcans, 108
  retinoic acid and regional
    specificity, 615, 616
  spotted mice, 279–280
  ventrolateral mesoderm induc-
    tion, 605–606
Growth hormone
  endochondral ossification, 336,
    337
  hormonal mRNA stabilization,
    466
GTPase activating protein, 684
Guanine nucleotide exchange
  protein, 683–684
Guanosine nucleotide-binding
  protein, *see* G protein
Guinea pigs, sperm secondary
  binding to zona pellucida,
    137
gurken gene, 564
GVBD, *see* Germinal vesicle
  breakdown
Gynandromorphs, 772, 773
Gynogenones, 240

H1, *see* Histones
*Hae*III, 54
Hair, 288–289
hairy gene, 549
Hamster
  acrosome reaction in sperm,
    130
  egg structure, 127
  nuclear fusion, 147–148
  polyspermy prevention,
    145–146
  sperm entry into egg, 139
Haploid gene expression, sperm
  development, 802
Haploid nucleus
  polyspermy prevention,

  140–146
  sperm, 122–123
HA protein, 140
Haptotaxis, 102–103, 298
Hatched blastula, 170, 183
Hatching enzyme, 170
Head mesenchyme, 323
Head process, 226
Heart, 342–346
  cardiac neural crest, 273, 284,
    286
  chamber formation, 345–346
  congenital anomalies, 641–642
  fusion of rudiments, 342–344
Heavy chain genes, antibody,
  391–392, 434
Hebbian synapse, 319n
Heliocidaris, 728
Helper/inducer T cells, 676n
Helper T cells, plasma cell induc-
  tion, 675–676
Hematopoiesis, sites, 359–361
Hematopoietic growth factors,
  358
Hematopoietic inductive mi-
  croenvironments (HIMs),
    358–359
Hematopoietic stem cells,
  355–359
Heme, hemoglobin production,
  482–483
Hemimetabolous metamorphosis, 729
Hemoglobin
  adult, 418
  coordinated protein synthesis,
    481–484
  embryonic, 417
  fetal, 354, 417, 421–422
  production, 374, 481–484
Hensen's node, 225
  cell commitment, 232
  retinoic acid, 631
"Hen's tooth," 653
Heparan sulfate proteoglycans,
  108
Heparin, angiogenesis inhibition,
  350
Hepatic diverticulum, 362
Hepatocyte, 494
Hereditary persistence of fetal
  hemoglobin (HPFH),
    421–422
Heredity, control, 34–36
Hermaphroditism, 766n, 782, 798
  fish, 781
  nematode, 778–781
Herpes simplex, teratogenesis,
  641
Heterochromatin, X-chromosome
  inactivation, 427
Heterochrony, 726–729, 849–850
Heterogamy, 18
Heterogeneous nuclear RNA,
  438–439, 440–445
*Hha*I, 54
HIMs, *see* Hematopoietic induc-
  tive microenvironments
Hindbrain, *see* Rhomben-

cephalon
*Hind*III, 54
Hindlimb buds, 695
Hippocampal neuron, axonal
  connections, 296
Histoblasts, 731
Histocompatibility antigens,
  239–240
Histones, 6, 411, *see also* Chromatin
  acetylation, 430
  cdc2 kinase, 192
  chromatin compaction, 413
  fusion of genetic material, 146
  transcription factors and,
    412–413
Histotypic aggregation, 83
HIV
  helper T cells, 676
  NF-κB protein, 394n
  RNA transport, 457
HMG box, sex determination,
  760
HNK-1, avian gastrulation,
  229–230, 231, 233–234
hnRNA, 438–439
  processing in control of early
    development, 440–445
Holist organicism, 580n
Holoblastic cleavage, 165
  bilateral, 176–177
  parasitic wasp, 190
  radial, 166–172
  rotational, 177–185
  spiral, 172–176
Holometabolous metamorphosis,
  729–730
Holtfreter's recombination experiments, 217
Homeobox, 557, *see also Hox*
  genes
Homeodomain proteins,
  384–385, 386, 557–559
  multiple regulatory modes,
    558–559
Homeosis
  evolutionary change, 836–838
  retinoic acid-induced, 627
Homeotic genes, 534, 546,
  552–562
  cell death, 712–713
  cis-regulatory elements,
    559–562
  determination of anterior–pos-
    terior axis and finger iden-
    tity, 709–711
  expression patterns, 552–554
  homology between *Drosophila*
    and mammals, 624–626
  initiation of expression,
    555–556
  limb development, 707–713
  limb field determination,
    708–709
  maintenance of expression,
    556–557
Homeotic transformations
  gene knockouts of *Hox* genes,
    629–630

pelle gene, 566
Pentalitomastix, 190
Peptide growth factors, *see also* Growth factors
dorsal mesoderm induction, 603–605
retinoic acid and regional specificity, 615, 616
ventrolateral mesoderm induction, 605–606
Pericardial cavity, 342
Periderm, 286
Periosteum, 334
Peripheral nervous system
myelination, 267
neural crest derivatives, 274
Permissive interaction, 648, *see also* Proximate tissue interactions
PGCs, *see* Primordial germ cells
P granules, 511
pH, egg responses to fertilization, 151–152
PH-20 protein, 137
Phages, cDNA libraries, 62
Pharyngeal arches, 273, 283, 285, 361
Hoxa-3 gene, 628
Pharyngeal pouches, 361
Pharyngeal progenitor cells, genetic control, 513–518
Pharyngula stage of vertebrate development, 841, 842
Pharynx, 361
Phenotype
cell fate, 518
sexual, 765–768
Phenotypic plasticity, 744
Phloem cells, totipotency, 45
Phocomelia, 638
Phosphatidylinositol 4,5-bisphosphate, *see* PIP$_2$
Phosphatidylinositol pathway, *see also* G protein–phosphoinositide
pathway
compaction, 181
Phosphoinositides, fertilization, 153–155
Phospholipase C
G protein activation, 158, 159
myoblast fusion, 328, 329
Phospholipids, surfactant, 364
Phosphorylation
eIF2, 482–483
transcription factor regulation, 405–406
Phosvitin, 805
Photoperiod, meroistic oogenesis, 816n
Photoreceptors, 269, *see also* Retinal cells
induction at single-cell level, 677–678
vision in infants, 270–271
Phyla
formation, 832–835
underlying unity, 632
Phyletic constraints, evolution,

841
Physical constraints, evolution, 839
Pig, *see also* Mammals
placenta, 238
Pigment cells, neural crest derivatives, 274
Pigmented retina, 269
Pigments, photoreceptive, 270
PIP$_2$, gamete activation, 153
pipe gene, 565
Pit-1 protein, 385–389, 405
prolactin gene transcription regulation, 387–389
Pituitary, oogenesis, 807
PKC, *see* Protein kinase C
Placenta, 166, 234
components, 238
types, 238n
X-chromosome inactivation, 431n
Placental cadherin, 92
Placental lactogen, *see* Chorionic somatomammotropin
Placodes
cranial ectodermal, 268
lens, 271
Plants
cloning, 45–46
metamorphosis derangement, 743
polyphenism induced by, 743–744
Plasma cells, induction by helper T cells, 675–676
Plasma membrane (cell membrane), 86
compaction, 181
egg, 138–140, 141, 155
formation, 197–198
fusion of egg and sperm cell membranes, 138–140, 155
monitoring changes through monoclonal antibodies, 88–89
photoreceptor, 270
polyspermy prevention, 141–142
Plasmids, cloning vectors, 55–57
Plasmin, 113
Plasminogen, activation, 113
Plasminogen activator, 183
metastasis, 350
Plasticity, phenotypic, 744
Platelets, yolk, 805–806, *see* Yolk platelets
Pleodorina, 17
Plexus, axon pathways, 308
Pluripotency
somatic cells, 42–44
trunk neural crest cells, 280–281
Pluripotential hematopoietic stem cells, 355–359
Pluteus hnRNA, 440–442
Pluteus larvae, metamorphosis, 728
Poiseuille's law, 347
Polar granules, 521

Polarity, *see* Anterior–posterior polarity; Dorsal–ventral polarity
imaginal disc, 734–735
proximal–distal, 697–699
Polarization, 180, 181
ZPA, *see* Zone of polarizing activity (ZPA)
Polar lobe, 505–509
Pole cells, 521–522
oskar gene, 525
Pole plasm, 521–523
components, 523–526
Poly(A)-binding protein II, 453
Poly(A) polymerase, 453
Poly(A) tail hypothesis, 473–475
Polygerm, parasitic wasp, 190
Polyinvagination islands, 223
Polylinker, 55–56
Polymerase chain reaction (PCR), 64, 66–67
Polyphenism, 743–744
Polyphosphoinositide cascade, 153–155
Polyspermy
fast block, 141–142
prevention, 140–146
slow block, 142–146
Polysplenia, heart chambers, 346
Polytene chromosomes, *see also* Chromosome puffs
RNA synthesis, 49–53
Porifera, developmental patterns, 28–29
Positional information, gradient models, 535–536
"Positional memory," regeneration blastema, 85–86
Posterior morphogen, 614–615, 616
Posterior necrotic zone, 713
Posterior neuropore, 252
Posterior organizing center, nanos protein, 537, 542–543
Posterior pole plasm, 521–523
posterobithorax gene, 560
Posttranslational control, 487
Naeglaria, 11–12
Potassium ions, egg resting membrane potential, 141
Potency, *see also* Pluripotency
cephalic neural crest cells, 285–286
changes during gastrulation, 588–589
oocyte gradients, 580–582
prospective fate and, 577–580
restriction in neighboring cells, 584–586
trunk neural crest cells, 280–283
POU domain, 385
POU transcription factors, 385–389
Prader-Willi syndrome, 426
Pre-B cells, 394
Precocenes, 743
Precocious larvae, parasitic wasp, 190–191
Precursor cells

germ cell, 790, *see also* Germ cell(s)
muscle, *see* Myoblasts
retinal, 269, 270
vulval, 681–682
Predators, polyphenism induced by, 744
Preformationism, 495–498
Pregnancy
ectopic, 183
mammary glands, 748–749
teratogenesis, *see* Teratogenesis
Preimplantation, PCR, 67
Premitotic gap, 191
Pre-mRNAs, spliceosomes, 447–448
Prereplication gap, 191
Prespore cells, 22
Pressure-plate experiment, regulative development, 578, 579
Prestalk cells, 21–22
Previtellogenic phase of oogenesis, 804
Primary embryonic induction, 589–591, *see also* Induction
dorsoanterior determinant localization, 598–599
dorsoventral polarity specification, 597–598
mesodermal gene expression, 601, 603
mesodermal specificity, 599–601, 602
molecular mechanisms, 597–614
organizer, 606–614
peptide growth factors, 603–606
retinoic acid, 614–615, 616
Primary hypoblast, 223
Primary mesenchyme, ingression, 204–208
Primary villi, 238
Primers
DNA sequencing, 56
hybridization, 54
PCR, 67
Primitive groove, 225
Primitive knot, *see* Hensen's node
Primitive streak
avian gastrulation, 224–228, 229–232
mammalian gastrulation, 236–237
regression, 227, 228
Primordia, axis specification, 568–569
Primordial germ cells (PGCs), 788, *see also* Germ cell(s)
amphibian, 526, 527
Proacrosin, 137
Probes
hybridization, 54, *see also* Hybridization
PCR, 67
proboscipedia gene, 552–553
Proerythroblast, 356
Progenesis, 726, 727

Progenitor cells, pharyngeal, 513–518
Progesterone, 239
  amphibian oogenesis, 808–809
  menstrual cycle, 821
  mRNA stabilization, 466
Progressive cell–cell interactions, cell fate specification, 575–618, *see also* Cell fate
Progressive determination, 586–589
Progress zone
  interactions mediated by *msx-1* and *msx-2*, 709
  proximal–distal limb axis, 697
Prokaryotes, 6
Prolactin
  lactation, 748, 749
  mRNA stabilization, 466
  thyroid hormones in metamorphosis, 725–726
Prolactin gene, transcription regulation, 387–389
Proliferative phase of menstrual cycle, 819–820
proliferin gene, 404–405
Promoter(s), 375–381, *see also* Enhancer(s); Transcription
  eukaryotic genes, 72–73
  function, 377–379
  β-globin gene, 419–421
  interactions, 403–405
  lacking TATA elements, 381
  methylation and gene inactivity, 423–424
  positional information, 536
  RNA polymerase and *trans*-regulatory factors, 379–381
  structure, 376–377
Promoter region, 372
Pronephric duct, 660
Pronephric duct cells, haptotaxis, 102–103
Pronephric kidney tubules, 660
Pronucleus
  female, 146
  male, 146
  mammalian movement, 147–148
  mammalian nonequivalence, 148–149
  transplantation, 149
Prosencephalon, 253
Prospective potency, prospective fate and, 577–580
Proteases, trophoblast secretion, 183
Protein(s), *see also specific type*
  differential modification, 371
  differential RNA processing, 448–452
  egg, 125
  "fusogenic," 140
  gap junction, 100–101
  homeodomain, 384–385, 386, 557–559
  phosphorylation, 405–406
  receptor, 87
  reverse genetics, 90–91

sexual inducer, 20
transcription factors, *see* Transcription factors
transport, 87
Protein kinase C (PKC)
  dorsal ectoderm, 612
  gamete activation, 155
  organizer III, 610–611
Protein synthesis, 6
  coordinated, 481–484
  differential longevity of mRNA, 463–466
  egg responses to fertilization, 152
  translational control, 481–484, *see also* Translation
Proteoglycans, 106, 107–108
  structure, 107
Prothoracicotropic hormone (PTTH), insect metamorphosis, 737
Protist "differentiation," *see also* Differentiation
  Naegleria, 10–12
Protostomes
  developmental patterns, 30–31
  fertilization, 150
Proximal–distal limb axis, 697–699
Proximate tissue interactions, 647–684, *see also* Secondary induction
  cell–cell interactions and equivalent cell identities, 679–681
  competence, 648–649
  coordinated differentiation and morphogenesis, 667–668
  epithelio–mesenchymal, 649–653, 668–670
  instructive, 647–648
  lens induction, 654–660
  parenchymal organ formation, 660–666, 670–675
  permissive, 648
  plasma cell induction, 675–676
  single-cell level, 677–679
Proximity, nature in epithelio–mesenchymal inductions, 668–670
Pseudohermaphroditism, 766n
PTTH, *see* Prothoracicotropic hormone
pUC18, 55–56
Puffs, chromosomal, *see* Chromosome puffs
pumilio gene, 537, 543
Punctuated equilibrium theory, 856
Pupa, 730
Purkinje neurons, 260

Quail
  chemotaxis, 101–102
  myoblasts, 328

Rabbit
  ovulation, 821
  polyspermy prevention, 146
  reticulocyte lysate, 482

Radial cleavage, 31, 166–172
  versus rotational cleavage, 178
Radiata, 30
Radiation, teratogenesis, 641
Radioactivity, 56, *see also* Autoradiography
Radiolabeled probes, 54, *see also* Hybridization; Probes
Rana, *see also* Frog
  cloning, 41–42
ras G protein, 683–684
ras pathway, 682–685
Rat
  fetal neurons in adult hosts, 305
  retinal development, 269, 270
Rathke's pouch, 361, 387
RB protein, muscle cells, 396
R-cadherin, 94
Reaction–diffusion model, limb skeleton specification, 700–701
Realisator genes, 556–557
Receptor kinase/ras pathway, 682–685
Receptor proteins, 87
  glycosyltransferases, 111–112
  integrins, 110–111
Receptors, *see also specific type*
  ras pathway, 682
  transfer of competence, 854
Receptor tyrosine kinases, 682–683
Reciprocal induction, kidney development, 661–663
Recognition sites, restriction enzymes, 54
Recombinant DNA, 55
Recombinant phages, cDNA libraries, 62
Recombinant plasmids, 55, 56
Recombinases, 392
Recombination
  Holtfreter's experiments, 217
  homologous, 70
Recombination experiments, regulative development, 580
Reductionism, holist organicism versus, 580n
Regeneration
  Acetabularia, 9–10
  blastema, 85–86
  salamander limb, 85–86, 696
  Wolffian, 40
Regional specificity
  amphibian metamorphosis, 718–721
  epithelio–mesenchymal interactions, 650–652
  induction, 592–597, 650–652, *see also* Primary embryonic induction
  molecular basis, 614–615, 616
Regulative development, 495, 575–577
  amphibian, 586–591
  Davidson's theory, 582–584, 585
  Driesch's theory, 577–580

Hörstadius's theory, 580–582, 583
  late-cleavage embryos, 581–582, 583
  sea urchin, 577–586
  Spemann and Mangold's theory, 589–591
  Spemann's theory, 586–589
Renal tubules
  progression, 660–661
  transfilter induction, 668, 669
Renaturation, 53
Repeated genes, 811n
Reporter gene, 73
Reproduction, 4
  asexual, 12, 17
  sexual, 12–21
Reptiles
  germ cell migration, 792–794
  temperature-dependent sex determination, 782–783
Resact, sperm attraction, 128–129
Respiratory system, regional specificity of induction, 651–652
Respiratory tube, 364
Resting membrane potential, polyspermy prevention, 141–142
Restriction endonucleases, *see* Restriction enzymes
Restriction enzymes, 54, 55
  methylation analysis, 426
Restriction fragments, 55
Rete testis, 757, 758
Reticular lamina, 105
Reticulocyte, 358
Reticulocyte lysate, translation system, 482
Retina
  axons, 309–311
  induction at single-cell level, 677
  neural, 269–270
  pigmented, 269
  precursor cell, 269, 270
Retinal cadherin, 94
Retinal cells, 269–270
  cell surface specificity, 89
  differential affinity, 83–84
  immunoglobulin superfamily CAMs, 95–96
  vision in infants, 270–271
Retinoblastoma protein, *see* RB protein
Retinoic acid
  chick limb development, 704, 705
  homeosis induced by, 627
  regional specificity, 614–615, 616
  teratogenesis, 630–631, 635–638
Retinoic acid receptors, 636
Retinoic acid response elements, 636
Retroviral vector, 68
Reverse genetics, 88–91
Reverse transcriptase hybridization, 54

PCR, 67
Rhesus monkey
  implantation, 180
  "inside–out" gradient of cere-
    bral cortex formation, 263
  tissue derivations, 234, 235
Rhombencephalon, 253
rhomboid gene, 567, 568
Rhombomeres, 253–254
  Hox gene expression, 626
Ribonuclease protection assay,
  443, 444, 445
Ribonucleic acid, *see* RNA
Ribonucleotide reductase, translation, 474
Ribosomal gene amplification,
  811–812
Ribosomal RNA, *see* rRNA
Ribosomes
  egg, 125–126
  synthesis, 811–813
Ring canals, 814
RNA
  antisense, 70–71, 72
  heterogeneous nuclear, *see*
    hnRNA
  localization techniques, 63–64,
    65
  messenger, *see* mRNA
  nuclear, 438–445
  radioactive, 56
  ribosomal, *see* rRNA
  selective processing, 371
  small nuclear, 446–448
  transfer, *see* tRNA
  translation, *see* Translation
  transport from nurse cells to
    oocyte, 814–816
RNA blots, 63, 73
RNA editing, 485–487
RNA hybridization, 53–54
  in situ, 63
RNA polymerase
  cdc2 kinase, 192
  nuclear matrix, 432
  operon model, 47
  promoters, 72, 379–381
RNA polymerase I, 379
RNA polymerase II, 379
  TATA element binding,
    379–381
  TFIIB and, 380
RNA polymerase III, 379
RNA processing, 438–458
  determination of 3′ end,
    452–455, 456
  differential, 448–452
  spliceosomes, 445–448
  transport out of nucleus,
    455–457
RNA splicing, 445–448
  alternative, 449–452
  transcription regulation, 406
RNA synthesis, differential,
  48–53
Rods and cone, *see* Retinal cells
Rotation
  amphibian gastrulation,
    215–216

avian gastrulation, 228–229
Rotational holoblastic cleavage,
  177–185
  compaction, 178–181
  escape from zona pellucida, 183
  inner cell mass formation,
    181–183
  radial cleavage versus, 178
rough gene, 678
rRNA
  oogenesis, 810
  synthesis, 811–813
RU486, 821
Rubella, teratogenesis, 641
runt gene, 549

Salamander, *see also*
    Amphibian(s)
  cleavage, 170
  embryonic genome activation,
    479–480
  fertilization, 150n
  gastrulation, 220–221
  haptotaxis, 102–103
  limb regeneration, 85–86, 696
  neurulation, 248
  regeneration blastema, 85–86
  regulative development, 588
*Sal*I, 54
salm gene, 557
Sanger "dideoxy" sequencing
    technique, 56–58
scaleless mutation, 669–670
Scatter factor, 230–231
SCF, *see* Stem cell factor
Schwann cell, 267
Sclerotomes, 273, 326, *see also*
    Neural crest
Screening
  cDNA libraries, 62
  clones, 56–57
scute gene, 569
Sea cucumber, cleavage, 166–167
Sea urchin
  acrosome reaction in sperm,
    130–131, 153
  cell cycle, 191
  cell fate, 577–586
  cleavage, 167–170, 581–582,
    583
  direct development, 728–729
  egg, 125, 126, 138–140, 145
  embryonic genome activation,
    480
  endoderm-specific mRNA, 63
  endoplasmic reticulum sur-
    rounding cortical granule,
    145
  fertilization events, 150
  fertilization process interrela-
    tionships, 151
  fusion of genetic material, 146,
    147
  gamete contact, 131–133
  gastrulation, 203–211, 467, 585
  hnRNA, 440–445
  hybrid, 467
  maternal control of mes-
    enchyme cell number, 467

Northern blots, 64
polyspermy prevention, 140,
    142, 143, 144
protein synthesis in fertilized
    eggs, 152
regulative development,
    577–586
species-specific recognition,
    131–133
sperm attraction, 128–129
sperm entry into egg, 138–140
Sebaceous glands, 289
Sebum, 289
Secondary hypoblast, 223
Secondary induction, 615–618,
    647–684, *see also* Proximate
    tissue interactions
Secondary villi, 238
Second messengers, G
    protein–phosphoinositide
    model of gamete activation,
    153–155
Segmental plate, 325
  somitomeres, 326
Segmentation genes, 534,
    544–552
Segment polarity genes, 534, 545,
    550–552
Selectin, 794n
Self-differentiation, 590–591
Semen, 121–122, *see also* Sperm
Sendai virus, F protein, 140
Sequence complexity, RNA,
    439–440, 442–443
Sequencing, DNA, 58–60,
    442–443
Sequestered mRNA, 477–478
Sertoli cells, 757, 766, 767
sevenless gene, 677–679
Seventh cranial nerve, *see* Facial
    nerve
seven-up gene, 678n
Sex
  without reproduction, 12–13
  smell and, 301–302
Sex comb reduced gene, 552, 568
Sex determination, 754–784
  alternative RNA processing,
    450–452
  autosomal genes, 764–765
  central nervous system,
    768–770
  Drosophila, 450–452
  environmental, 782–784
  fish, 852–853
  gonad development, 757–759
  hermaphroditism, 766n,
    778–782
  mammalian, 755–771
  meiosis, 797–798
  primary, 755–756, 760–765
  secondary, 756–757, 765–770
  target genes for cascade, 778
  transfer of competence,
    852–853
  Y chromosome, 760–763
Sex-lethal gene, 451, 772
  maintenance of function,
    775–776

sex determination, 774–776
Sexual behaviors, 770–771
Sexual inducer protein, 20
Sexual phenotype, hormonal reg-
    ulation, 765–768
Sexual reproduction, *see also* Fer-
    tilization
  origins, 12–16
  volvocaceans, 17–21
SH2–SH3 domain proteins, 683
Sheep, transgenic, 378, 379
Shell gland, 505n
Shortest intercalation rule, 706
Sialic acid, N-CAM, 97–98
sina transcription factor, 685
Singed mutation, *Drosophila*, 523
Single-copy DNA, 439–440
  hnRNA processing and,
    440–441
Sinus venosus, 343
Skeletal muscle, *see also*
    Muscle(s)
  differentiation, 327–332
Skeletal muscle cells, 327–328
Skeleton
  axial, 326
  limb, 700–701
  osteogenesis, 333–338
Skin, *see also* Cutaneous struc-
    tures
  dorsal, 327
Skin cells, differential affinity,
    82–83
skn-1 protein, 513–514
*Sma*I, 54
Small nuclear ribonucleoprotein
    particles (snRNPs), 446–448
Small nuclear RNA (snRNA),
    446–448
Smell, sex and, 301–302
Smittia, polarity regulation by
    oocyte cytoplasm, 535
Snail
  cleavage, 173–174
  location-dependent sex deter-
    mination, 784
  morphogenetic determinants,
    506–509
  shell coiling direction, 174
snail gene, 563, 567, 568
snake gene, 565
snRNA, 446–448
snRNPs ("snurps"), 446–448
Sodium ions, egg resting mem-
    brane potential, 141, 142
Soma, 265
Somatic cells, 5
  cell cycle, 191
  nuclear potency restriction,
    41–42
  pluripotency, 42–44
  X-chromosome inactivation,
    429
Somatic mesoderm, 339
Somatic sex determination, *see*
    Sex determination
Somatopleure, 340
Somatotrophs, transcription fac-
    tors, 387

Threonine-161, cdc2 kinase, 192, 193

Threshold concept, amphibian metamorphosis, 721–722

Thyroid hormone receptor (TR) genes, 724–725

Thyroid hormones, *see also* Thyroxine (T₄); Triiodothyronine (T₃)

Thyrotrophs, transcription factors, 387

Thyroxine (T₄)
  amphibian metamorphosis, 718, 720, 724–726
  mRNA stabilization, 466
  structure, 719

Tight junctions
  mammalian cleavage, 179
  somite formation, 326

Tissue determinants, cytoplasmic segregation, 500–503

T lymphocytes, 494, *see also* Lymphocyte(s)
  helper, 675–676
  NF-κB protein, 394n

Toad, *see* Amphibian(s)

Toes, nomenclature, 710n

Toll gene, 565

Topoisomerases, gene transcription, 433–434

torpedo gene, 563–564

torso gene, 537, 543, 544

torsolike gene, 543–544

Totipotency, 42–43
  blastomere, 44
  plant cells, 45
  pluripotency versus, 44

Toxoplasma gondii, teratogenesis, 641

tra-1 gene, hermaphroditism, 779–781

trans-acting factors, RNA, 453–454

trans-activating domain, transcription factors, 384, 400

Transcription, 411–434, *see also* Transcription factors
  Acetabularia, 10
  activation, 479
  basal, 381
  bithorax complex, 561–562
  chromatin activation, 412–419, *see also* Chromatin
  differential, 371
  DNA methylation, 423–427
  DNA supercoiling, 433
  Drosophila, 190
  exons and introns, 371–374
  gap genes, 546, 547
  globin gene, 417–423
  homeotic genes, 625
  immunoglobulin light chain genes, 389–394
  methylation in pattern maintenance, 424–425, 430–431
  mRNA, 61–62
  MyoD family of regulators, 330–332
  nuclear matrix, 431–434

oocyte, 809–810

prokaryotes, 8

promoters and enhancers, 375–405, *see also* Enhancer(s); Promoter(s)
  regulation, 405
  segment polarity genes, 550–552
  temporal and spatial pattern regulation, 383–384
  topoisomerases, 433–434
  unacetylated nucleosome, 430
  visualization from amplified genes, 812
  X-chromosome dosage compensation, 427–431
  Xenopus, 190

Transcription factors, 376, 379–381, 384–407, *see also* Transcription; *specific factors*
  activity regulation, 405–406
  bHLH, 395–396
  binding to promoters and enhancers, 71
  bZip, 397–398
  cell fate, 582–584
  DNase-hypersensitive sites, 415–416
  5S rRNA gene, 812–813
  homeodomain proteins, 384–385, 386
  hormone receptors, 400–403
  interactions, 403–405
  nucleosome removal, 412–414
  POU, 385–389
  zinc finger, 399–400

Transdifferentiation, 40

Transfection, 68

Transfer of competence, 851–854

Transferrin receptor, translational regulation, 485, 486

Transfer RNA, *see* tRNA

Transfilter induction, 668, 669

transformer gene, 450–452, 772, 776–777

Transforming growth factor-α (TGF-α), 287–288

Transforming growth factor-β1 (TGF-β1)
  branching pattern effects, 673–674
  reaction–diffusion model, 701

Transgenic cells and organisms, 67–70, 71

Transgenic mouse, 69
  luciferinase mRNA and ZP3, 73
  promoter function, 377–378

Transgenic sheep, promoter function, 378, 379

Translation, 461–486
  defined, 461
  differential longevity of mRNA, 463–466
  embryonic genome activation, 478–481
  eukaryotic, 461–463
  hunchback gene, 542
  leader sequence, 372
  masked maternal message hy-

pothesis, 472–473
  mRNA, 371
  poly(A) tail hypothesis, 473–475
  posttranslational regulation, 11–12, 487
  prokaryotes, 8
  RNA editing, 485–487
  sequestered messages, 477–478
  stored mRNAs, 468–471
  uncapped messages, 477

Translational control
  Acetabularia, 10
  coordinated protein synthesis, 481–484
  Naegleria, 11
  oocyte messages, 466–481
  widespread use, 485–487

Translational efficiency hypothesis, 477

Translation termination codon, 373, 462–463

Translocation
  c-myc gene, 399
  dorsal protein, 562–565

Transplantation
  mesodermal specificity induction, 600–601
  nuclear, 41–44
  pronuclear, 149

Transport proteins, 87

Transposable element, 68

trans regulators, 375, *see also* Transcription factors
  chick globin gene switching, 419
  chromatin activation, 412–416
  interactions, 403–405
  locus control region, 419
  RNA polymerase and, 379–381

Transverse folds, 531

Trembler mouse, 267

Treponema pallidum, teratogenesis, 641

TR genes, *see* Thyroid hormone receptor genes

Trigeminal nerve, 254

Triiodothyronine (T₃)
  amphibian metamorphosis, 718, 720, 724–726
  structure, 719

Triturus, primary embryonic induction, 590

tRNA
  egg, 126
  oogenesis, 810

Trochus, cleavage, 173

Trophectoderm cells, 179

Trophoblast cells
  compaction, 179
  P-cadherin, 92
  protease secretion, 183

β-Tropomyosin, alternative RNA splicing, 449

Truncus arteriosus, 343

Trunk, knockouts *Hox* gene expressed in, 629–630

trunk gene, 537

Trunk neural crest, 273–283

cell developmental potency, 280–283
  cell migration pathways, 273–276
  extracellular matrix and cell migration, 276–279

Tubal pregnancy, 183

tube gene, 566

Tubule cell, hen oviduct, 494

Tubules, renal, 660–661, 668, 669

Tubulin, flagella, 11

tudor gene, 525, 537, 542

Tumor angiogenesis factors, 348

Tumor-induced angiogenesis, 349–351

Tunica albuginea, 757

Tunicate
  autonomous specification, 499–503
  cleavage, 176–177
  cleavage preparation, 158
  egg cytoplasm rearrangement, 156
  epigenesis, 497–498
  morphogenetic determinants, 499, 503–505

Twins
  amphibian gastrulation, 215–216
  embryonic stem cells, 184–185
  formation by polar lobe suppression, 509

twisted gene, 563

twist gene, 567, 568

Tyrosine-15, cdc2 kinase, 192, 193

Tyrosine kinases, c-kit, 279–280

UDP-galactose, sperm–zona binding, 136

Ultrabithorax gene, 553–556, 561–562

Ultraviolet light, germ cell determination, 527

Umbilical artery, 352

Umbilical vein, 352

Uncapped mRNA, 477

Unicellular eukaryotes, *see also* *specific organisms*
  development, 8–16
  sexual reproduction, 13–16

Unio, cleavage, 175

Unsegmented mesoderm, 325

Untranslated region (UTR), 373, 476
  mRNA, 464–465, 798

Urea cycle, amphibian metamorphosis, 718, 719

Urechis
  egg activation, 155
  sperm effects on egg membrane potential, 142

Ureteric buds, 661
  formation, 664
  interaction with mesenchyme, 661–663

Urokinase, 113

Urotelic, defined, 718

Uterine epithelium, trophoblast, 183

ABOUT THE BOOK

Editor: Andrew D. Sinauer
Project Editor: Carol J. Wigg
Production Manager: Christopher Small
Book Production: Janice Holabird
Illustration Program: J/B Woolsey Associates
Book and Cover Design: Rodelinde Graphic Design
Copy Editor: Gretchen Becker
Composition: DEKR Corporation
Prepress: Jay's Publishers Services, Inc.
Cover Manufacture: John P. Pow Company
Book Manufacture: Courier Westford, Inc.